BEFORE CALCULUS

Third Edition

BEFORE CALCULUS

Functions, Graphs, and Analytic Geometry

Louis Leithold

HarperCollinsCollegePublishers

Sponsoring Editor: George Duda
Development Editor: Richard Wallis
Art Administrator: Jess Schaal
Cover Art: Dan Douke
Production Administrator: Randee S. Wire
Project Coordination and Text Design: Elm Street Publishing Services, Inc.
Compositor: Interactive Composition Corporation
Printer and Binder: R. R. Donnelley & Sons Company

Before Calculus, Third Edition

Library of Congress Cataloging-in-Publication Data
Leithold, Louis.
    Before calculus : functions, graphs, and analytic geometry / Louis Leithold. — 3rd ed.
        p.   cm.
    Includes index.
    ISBN 0-673-46911-5 (Student Edition)
    ISBN 0-673-99096-6 (Annotated Instructor's Edition)
    1. Mathematics.   I. Title.
QA39.2.L44   1993
512'.13—dc20   93–34084

To
FARLEY, JENNY, ORSON, AGNES, and SYDNEY
for their patience, understanding, and love while the three
editions of this text were being written.

1994

Cover Artist

Dan Douke, a painter working in Southern California and currently professor of art at California State University at Los Angeles, exhibits his work regularly at Tortue Gallery in Santa Monica and O.K. Harris Works of Art in New York. Professor Douke prepared the following statement regarding the body of work which began with the "monolithic-like" image he created for the front-cover art:

"Looking at the real world and the things in it, I have always been interested in the idea of observable reality and the dilemmas of abstract thinking. My current work is a response to the electronic age of information that seems to touch everyone. A dichotomous distrust and a respect of this technology directs my work. The pieces, positioned in the ambiguous zone between painting and sculpture, are invented objects, recognizable and fantastic that are both calculated and random. After constructing the desired shape, the pieces are fitted with found objects, such as cigar packs, water containers, food holders, packing materials, wire, cable, plastic pieces, and radio parts. The pieces are then painted in various ways by airbrushing, splattering, staining, and grinding to achieve the desired effect. I want the work to look oddly familiar, perhaps like a part of something bigger or more powerful, somehow old, yet futuristic. I refer to the works as *techno junk.*"

Table of Contents

Preface

My paramount goal with this third edition of *Before Calculus (BC3)* is for students to appreciate the beauty of mathematics as a logical science while they gain a solid foundation for the study of calculus. To achieve this goal, I have incorporated the following features:

GRAPHICS CALCULATOR "ACTIVE"

Unlike the first two editions, *BC3* uses throughout the presentation, beginning with Section 1.5, modern technology in the form of the hand-held graphics calculator—not only powerful and fascinating as a learning device but vital as a problem-solving tool. As guidelines for using graphics calculator numerical and visual methods to enhance the teaching and learning of precalculus, I have followed the philosophy endorsed by leading mathematics educators and organizations and summarized by TICAP (Technology Intensive Calculus for Advanced Placement) by the following three strategies:

1. Do *analytically* (with paper and pencil); then **support** *numerically and graphically* (with a graphics calculator).
2. Do *numerically and graphically*; then **confirm** analytically.
3. Do *numerically and graphically* because other methods are *impractical or impossible.*

MATHEMATICAL MODELING AND WORD PROBLEMS

A significant feature of this edition is mathematical modeling of practical situations, stated as word problems, in such diverse fields as business, economics, psychology, sociology, physics, chemistry, engineering, biology, and medicine. Section 2.3 contains step-by-step suggestions for obtaining an equation as a mathematical model (see page 84). Functions as mathematical models are first introduced in Sections 4.3 and 4.4 and appear prominently throughout the rest of the text.

WRITING IN MATHEMATICS

To complete the solution of each word-problem example, a *conclusion* that answers the questions of the problem is stated. The student is required to write a similar conclusion consisting of one or more complete sentences for each word-problem exercise. Included at the end of nearly every exercise set is a writing exercise that may ask a question pertaining to *how* or *why* a specific procedure works, or that may require the student to *describe, explain*, or *justify* a particular process.

EXERCISES

More than 4,000 exercises, revised from previous editions and graded in difficulty, provide a wide variety of problem types, ranging from computational to applied and theoretical problems incuding calculator-active and writing exercises as indicated above. They occur at the end of sections and as review exercises following the last section of each chapter.

EXAMPLES AND ILLUSTRATIONS

Examples and illustrations—650 in all—appear in every section. The examples, carefully chosen to prepare students for the exercises, should be used as models for their solutions. Each example has a title, related to the section's goals and stating the purpose of the example. An illustration serves to demonstrate a particular concept, definition, or theorem; it is a prototype of the idea being presented.

VISUAL ART PROGRAM (FIGURES)

All the figures, twice as many as in the previous edition, are original for *BC3*. Graphs plotted on a graphics calculator are shown on a graphics calculator screen surrounded by a color border. Graphs sketched by hand are purposely shown in one color to avoid the possibility of reproduction errors caused by improper registration during the printing process beyond the author's control. When more than one curve appears in the same figure, the curves are distinguished by using different shades.

PEDAGOGICAL DEVICES

Each chapter begins with an introduction entitled "Looking Ahead," that provides students with an incentive for studying the chapter, and ends with a summary called "Looking Back," that provides students with a reinforcement of the chapter's ideas. Furthermore, each section of a chapter starts with a list of "Goals" to be covered in that section. Together, these three features serve as an overall review of the chapter when a student studies for a test.

CALCULUS-ORIENTED COVERAGE

Topics that preview calculus are delineated with the symbol .

Chapters 1 and 2

Intermediate-algebra topics necessary to develop skills needed in calculus are reviewed. Examples and exercises in these chapters include algebraic expressions similar to those occurring in calculus (see Examples 1, 4, and 5 in Section 1.2; Examples 7 and 10 in Section 2.2; and Example 5 in Section 2.7).

Chapters 3 and 4

Functions and graphs are the heart of *BC3*. Prior to the discussion of functions in Chapter 4, however, thorough coverage of lines and parabolas are presented in Chapter 3 so that these curves are then available in Chapter 4 for the treatment of graphs of linear and quadratic functions, respectively. With this order of presentation, the discussion of functions does not have to be interrupted with the analytic geometry necessary for their graphs. Circles are also included in Chapter 3 because they are used in Chapter 4 as illustrations of curves that are not graphs of functions. Chapter 4 contains examples and exercises pertaining to concepts arising in calculus: difference quotients (see Example 3 in Section 4.1); piecewise-defined functions (see Example 1 in Section 4.2); and composite functions (see Example 3 in Section 4.5). Extreme-value problems involving quadratic functions are also included to give the student a preview of this important application of calculus (see Examples 3–6 in Section 4.3).

Chapter 5

The treatment of graphs of polynomial and rational functions is as complete as possible without using calculus. For instance, the theorem that gives the number of possible relative extrema for a polynomial function is stated but not proved (see Example 8 in Section 5.1), and vertical and horizontal asymptotes of graphs of rational functions are defined after an intuitive introduction to limits involving infinity (see Examples 3–5 in Section 5.5).

Chapter 6

Inverse functions appear here in the same chapter where they are first applied to define a logarithmic function as the inverse of an exponential function. The number *e* is introduced in Section 6.2 by considering interest on an investment at a rate compounded continuously. This discussion is followed by an intuitive demonstration of why *e* can be defined in calculus

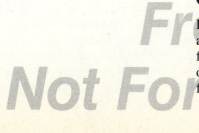

as the number approached by the expression $\left(1 + \dfrac{1}{x}\right)^x$ as x increases without bound. Applications of exponential and logarithmic functions in this chapter involve exponential growth and decay, bounded growth, and logistic growth, all of which are studied in calculus (see Examples 2–5 in Section 6.3).

Chapters 7 and 8

The study of trigonometry is begun in Chapter 7 with trigonometric functions of real numbers, essential for calculus, and then they are applied to periodically repetitive phenomena (see Examples 3 and 4 in Section 7.2 and Example 3 in Section 7.4). Simple harmonic motion, an important application in calculus, is included (see Examples 1 and 2 in Section 7.4). The Taylor polynomials for sine and cosine that the student will encounter in calculus appear in Section 7.2 (see Example 5 in that section). The graph of the function defined by $\dfrac{\sin t}{t}$ is sketched, which suggests that the quotient approaches 1 as t approaches zero, a fact with significant consequences in calculus (see Example 3 in Section 7.5). Chapter 8 gives a full treatment of trigonometric functions of angles with applications involving solutions of triangles.

Chapter 9

Trigonometric identities and equations, as well as inverse trigonometric functions, are analytic trigonometry topics presented in this chapter. In Section 9.1, a step-by-step summary of suggestions for proving trigonometric identities is given (see page 508). The computational skills acquired here will be beneficial in calculus when it is necessary to convert a trigonometric expression from one form to another. Inverse trigonometric functions are applied to solve trigonometric equations. I stress that the choice of the range of the inverse secant function is made so that certain computations in calculus are simplified. I show how the inverse tangent function can be used in calculus to determine an observer's "best view" of an object (see Example 7 in Section 9.5).

Chapter 10

Vectors, vector-valued functions, and parametric equations are calculus subjects introduced in Sections 10.1 and 10.2. An introduction to motion of a projectile, a vector application of calculus, is also presented (see Example 4 in Section 10.2). The treatment of polar coordinates and graphs of polar equations includes a discussion of when the limaçon has a dent and when it does not have a dent (see Example 3 in Section 10.4). The polar form of a

complex number is used to obtain the product and quotient of complex numbers and to determine powers and roots of complex numbers by De Moivre's theorem.

Chapter 11

An unusually thorough coverage of conic sections, a major topic in analytic geometry, serves as preparation for their use in calculus. Cartesian equations of ellipses and hyperbolas are derived from their definitions as sets of points in a plane in a manner similar to the way cartesian equations of parabolas are derived in Chapter 3.

Chapter 12

Partial fractions, the topic of Section 12.3, are used for computational work in calculus, and here provide an application of systems of linear equations, which are discussed in the first two sections of this chapter along with matrices. Sequences and series make up an important part of a calculus course, and in Section 12.4, a brief introduction to these topics, based on the function concept, is given. This section also includes a discussion of sigma notation, which is applied in calculus to write a summation. A section is also devoted to mathematical induction, a device for proving certain theorems in calculus. Two types of infinite series important in calculus are the geometric series and the binomial series, both presented in the final section of the text.

FLEXIBILITY

Diversity of opinion regarding the content of a precalculus text exists among mathematics educators. Consequently, this book contains more topics than may be taught in a course of three or four semester-hours, so that instructors may choose the appropriate material for their classes. Because Chapters 1 and 2 consist of mostly review material, some sections may be treated lightly or not at all. Chapters 3 through 7 and Chapter 9 should be included in most courses. Some sections in Chapter 8, involving trigonometric functions of angles, may be omitted if the student has seen this material elsewhere. Each of Chapters 10 through 12 is self-contained. Coverage of these chapters depends on the needs of the mathematics department using this text. For instance, sections involving complex numbers and matrices, not normally part of a calculus course, may be omitted. Furthermore, the amount of analytic geometry taught here will depend on how much of this material will be covered in the student's calculus course.

Louis Leithold
Mendocino, CA

Acknowledgments

Reviewers of BC3

Robert Barefoot, *Chaparral High School*
L. Rena Brakebill, *Georgia Institute of Technology*
Philip S. Clarke, *Los Angeles Valley College*
Charles S. Johnson, *Los Angeles Valley College*
Clay Kaufman, *The Field School*
Jack Leddon, *Dundalk Community College*
Pamela M. Levy, *Penn State University—Ogontz*
Charlotte McGreaham, *Westminster Schools*
Latchmy Narine, *Dawson College*
Charles Stone, *DeKalb College*
Charles E. Sweatt, *Odessa College*
Fancher E. Wolfe, *Metropolitan State University*

Preparation of Solutions and Answers for Exercises in BC3

Leon Gerber, *St. John's University*
    assisted by Shmuel Gerber

Checkers of Answers for Exercises in BC3

Katherine B. Braungart, *University of Georgia*
Carl Cuneo, *Essex Community College*
Kathryn Gillespie, *Heartland Community College*
Patrick Hughes, *Cuesta College*
Kumara Jayasuriya, *University of Wisconsin—Milwaukee*
Steven S. Terry, *Ricks College*

Cover Artist

Don Douke
Courtesy of Tortue Gallery, Santa Monica

To these people, to the staff of HarperCollins *College Publishers*, and to all the users of the first two editions of *Before Calculus*, I express my deep appreciation.

L.L.

Numbers, Algebraic Expressions, and Graphs of Equations

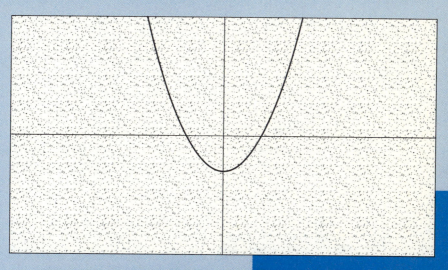

Even though you have had a course in intermediate algebra or the equivalent, you may need a review of some of that material but not a complete redevelopment. Consequently the pace is quick for the algebra topics appearing in this chapter. The sections may be covered in detail, treated as a review, or omitted.

If Section 1.1 is omitted in class, I recommend that you read the section on your own because in this text and in your calculus course you will need to know facts about the set of real numbers and its subsets.

Because graphs and graphics calculators play prominent roles throughout this text, you should be thoroughly familiar with the content of Section 1.5.

1.1 THE SET OF REAL NUMBERS

GOALS

1. Learn about the set of real numbers and its subsets.
2. Learn properties of the real-number system.
3. Establish a one-to-one correspondence between the real numbers and the points on the real-number line.
4. Learn interval notation to define a set of numbers.
5. Determine if one set of numbers is a subset of another.
6. Determine the union or intersection of sets of numbers.
7. Use set notation and inequality symbols to denote a set of numbers.
8. Illustrate a set of numbers on the real-number line.
9. Define the absolute value of a real number.
10. Represent numbers by points on the real-number line and find distances between points.

Algebra, like arithmetic, involves numbers on which operations such as addition, multiplication, subtraction, and division are performed. While arithmetic is concerned with operations on specific numbers such as $2 + 5$ or $8 \cdot 9$, in algebra we deal with operations on unspecified or unknown numbers. These are designated by symbols, letters such as x, y, z, a, b, or c. The word *algebra* originated from the Arabic word *al-jabr* that is in the title *ilm al-jabr w'al muqâbalah* (translated as "the science of reduction and cancellation"), an early ninth-century work. The algebraic symbolism used to generalize the operations of arithmetic were formulated in the sixteenth and seventeenth centuries.

From time to time we will use some **set** notation and terminology. We can say that a set is a collection of objects, and the objects in a set are called the **elements** of the set. Each particular object must be either in the set or not in the set.

A pair of braces, { }, used with words or symbols can describe a set. If S is the set of natural numbers less than 6, we can write set S as

{1, 2, 3, 4, 5}

or as

{x, such that x is a natural number less than 6}

where x is called a **variable,** a symbol used to represent any element of the given set. The given set is called the **domain** of the variable. The set S can be written as follows, with **set-builder notation,** where a vertical bar is used in place of the words *such that:*

$\{x \mid x$ is a natural number less than 6$\}$

which is read "the set of all x such that x is a natural number less than 6."

Two sets A and B are said to be **equal,** written $A = B$, if and only if A and B have identical elements. For example,

$\{1, 2, 3\} = \{3, 1, 2\}$

The **union** of two sets A and B, denoted by $A \cup B$ and read "A union B," is the set of all elements that are in A or in B or in both A and B. The **intersection** of A and B, denoted by $A \cap B$ and read "A intersection B," is the set of all elements that are in both A and B. The set that contains no elements is called the **empty set** and is denoted by $\varnothing$.

▷ **ILLUSTRATION 1**

Suppose $A = \{2, 4, 6, 8, 10, 12\}$, $B = \{1, 4, 9, 16\}$, and $C = \{2, 10\}$. Then

$A \cup B = \{1, 2, 4, 6, 8, 9, 10, 12, 16\}$ $\qquad A \cap B = \{4\}$

$B \cup C = \{1, 2, 4, 9, 10, 16\}$ $\qquad B \cap C = \varnothing$ ◀

Observe in Illustration 1 that the intersection of sets B and C is the empty set. These two sets have no elements in common, and they are called **disjoint sets.**

The symbol $\in$ indicates that a specific element belongs to a set. Hence for the set C of Illustration 1 we may write $2 \in C$, which is read "2 is an element of C." The notation $a, b \in S$ indicates that both a and b are elements of S. The symbol $\notin$ is read "is not an element of." Therefore we read $5 \notin A$ as "5 is not an element of A."

If every element of a set S is also an element of a set T, then S is a **subset** of T, written $S \subseteq T$. In Illustration 1 every element of C is also an element of A; thus C is a subset of A, and we write $C \subseteq A$. The symbol $\nsubseteq$ is read "is not a subset of." Thus we may write $\{1, 2, 3, 4\} \nsubseteq \{1, 2, 3\}$.

We have referred to the set of **natural numbers,** which we denote by N, so that

$N = \{1, 2, 3, \ldots\}$

where the three dots are used to indicate that the list goes on and on with no last number.

The number *zero*, denoted by the symbol 0, is the number having the property that if it is added to any number, the result is that number. The set of numbers whose elements are the natural numbers and zero is called the set of **whole numbers.** The set of natural numbers is also called the set of **positive integers.**

Corresponding to each positive integer n there is a negative integer such that if the negative integer is added to n, the result is 0. For example, the negative integer -5, read "negative five," is the number that when added to 5 gives a result of 0. The set of **negative integers** can be denoted by $\{-1, -2, -3, \ldots\}$. The set of numbers whose elements are the positive integers, the negative integers, and zero is called the set of **integers,** denoted by J; thus

$$J = \{\ldots, -3, -2, -1, 0, 1, 2, 3, \ldots\}$$

The set of integers then is the union of three disjoint subsets: the set of positive integers, the set of negative integers, and the set consisting of the single number 0. Note that the number 0 is an integer, but it is neither positive nor negative. Sometimes we refer to the set of **nonnegative integers,** which is the set consisting of the positive integers and 0 or, equivalently, the set of whole numbers. Similarly, the set of **nonpositive integers** is the set consisting of the negative integers and 0.

Consider now the set whose elements are those numbers that can be represented by the quotient of two integers p and q, where q is not 0, that is, the numbers that can be represented symbolically as

$$\frac{p}{q} \qquad \text{where } q \text{ is not } 0$$

This set of numbers is called the set of **rational numbers,** which is denoted by Q. Thus

$$Q = \left\{ x \,\middle|\, x \text{ can be represented by } \frac{p}{q}, p \in J, q \in J, q \text{ is not } 0 \right\}$$

Some numbers in the set Q are $\frac{1}{2}$, $\frac{3}{4}$, $\frac{11}{5}$, $\frac{-2}{3}$, and $\frac{-31}{12}$. Every integer is a rational number because every integer can be represented as the quotient of itself and 1; that is, 8 can be represented by $\frac{8}{1}$, 0 can be represented by $\frac{0}{1}$, and -15 can be represented by $\frac{-15}{1}$. Hence $J \subseteq Q$.

Any rational number can be written as a decimal. You are familiar with the process of using long division to do this. For example, $\frac{3}{10}$ can be written 0.3, $\frac{9}{4}$ can be written 2.25, and $\frac{83}{16}$ can be written 5.1875. These decimals are called **terminating decimals.** There are rational numbers whose decimal representation is nonterminating and repeating; for example, $\frac{1}{3}$ has the decimal representation 0.333 . . . , where the digit 3 is repeated, and $\frac{47}{11}$ can be represented as 4.272727 . . . , where the digits 2 and 7 are repeated in that order. It can be proved that the decimal representation of every rational number is either a terminating decimal or a nonterminating repeating decimal. We shall show in Section 12.8 that every nonterminating repeating decimal is a representation of a rational number.

The following question now arises: Are there numbers whose decimal representations are nonterminating and nonrepeating? The answer is yes, and an example of such a number is the principal square root of 2, denoted symbolically by $\sqrt{2}$ and indicated by a nonterminating nonrepeating decimal as 1.41421 . . . . Another such number is π (pi), which is the ratio of the circumference of a circle to its diameter and indicated by a nonterminating nonrepeating decimal as 3.14159 . . . . The numbers whose decimal representations are nonterminating and nonrepeating cannot be expressed as the quotient of two integers and hence are not rational numbers. This set of numbers is called the set of **irrational numbers,** which we denote by H.

The union of the set of rational numbers and the set of irrational numbers is the set of **real numbers.** Denoting the set of real numbers by R, we define R symbolically by

$$R = Q \cup H$$

Figure 1 shows the relationships among the sets of numbers discussed above. Examples of each classification of numbers appear in the corresponding rectangle.

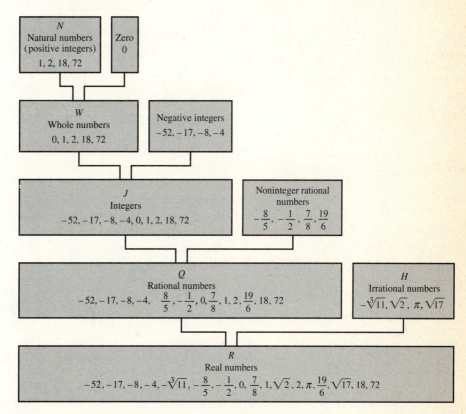

FIGURE 1

▶   **EXAMPLE 1**   *Determining If One Set of Numbers Is a Subset of Another*

The sets N, J, Q, H, and R are the sets of numbers defined in this section. Insert either $\subseteq$ or $\not\subseteq$ to make the statement correct. **(a)** N _____ J; **(b)** Q _____ J; **(c)** $\{\sqrt{2}, \pi, 3.5\}$ _____ H; **(d)** $\{0\}$ _____ Q; **(e)** N _____ R.

Solution
(a) Because every natural number (or positive integer) is an integer, $N \subseteq J$.
(b) Because there are rational numbers that are not integers, $Q \not\subseteq J$.
(c) $\sqrt{2}$ and π are irrational numbers, but 3.5 is a rational number; therefore, $\{\sqrt{2}, \pi, 3.5\} \not\subseteq H$.
(d) Zero is a rational number, and thus $\{0\} \subseteq Q$.
(e) Every positive integer is a real number, hence $N \subseteq R$.   ◀

▶   **EXAMPLE 2**   *Determining the Union or Intersection of Two Sets of Numbers*

Which one of the sets N, J, Q, H, R, and $\varnothing$ is equal to the following sets: **(a)** $J \cup Q$; **(b)** $J \cap Q$; **(c)** $N \cap H$; **(d)** $H \cup R$?

Solution
(a) The union of J and Q is the set of numbers that are either integers or rational. Because the set of integers is a subset of the set of rational numbers, this union is the set of rational numbers. Therefore $J \cup Q = Q$.
(b) The intersection of J and Q is the set of numbers that are both integers and rational. This intersection is the set of integers, and thus $J \cap Q = J$.
(c) Because the set of positive integers and the set of irrational numbers have no elements in common, $N \cap H = \varnothing$.
(d) The union of H and R is the set of numbers that are either irrational or real. Because the set of irrational numbers is a subset of the set of real numbers, $H \cup R = R$.   ◀

The **real-number system** consists of the set of real numbers and two operations called **addition** and **multiplication**. Addition is denoted by the symbol $+$, and multiplication is denoted by the symbol $\cdot$ (or $\times$). If a and b are real numbers, $a + b$ denotes the **sum** of a and b, and $a \cdot b$ (or ab) denotes their **product.** A discussion of the real-number system appears in Section A.1 of the appendix.

Subtraction and division of real numbers are defined in terms of addition and multiplication, respectively. The definition of subtraction is as follows: If a and b are real numbers, the operation of **subtraction** assigns to a and b a real number, denoted by $a - b$, called the **difference** of a and b,

and

$$a - b = d \quad \text{if and only if} \quad a = b + d$$

▷ **ILLUSTRATION 2**

$$7 - 4 = 3 \quad \text{because} \quad 7 = 4 + 3 \qquad ◀$$

In the preceding definition the "if and only if" qualification is used to combine two statements:

1. $a - b = d$ if $a = b + d$.
2. $a - b = d$ only if $a = b + d$, which is equivalent to the statement $a = b + d$ if $a - b = d$.

We now define division.

If a and b are real numbers and $b \neq 0$, the operation of **division** assigns to a and b a real number, denoted by $a \div b$, called the **quotient** of a and b, and

$$a \div b = q \quad \text{if and only if} \quad a = bq. \tag{1}$$

▷ **ILLUSTRATION 3**

$$24 \div 6 = 4 \quad \text{because} \quad 24 = 6 \cdot 4 \qquad ◀$$

Observe in the definition of division that $b \neq 0$. The reason for this restriction can be seen by allowing b to be 0 in statement (1). For instance, if in (1) $b = 0$ and $a = 3$ (any other nonzero value of a can be used instead of 3), the statement becomes

$$3 \div 0 = q \quad \text{if and only if} \quad 3 = 0 \cdot q$$

Of course, there is no value of q satisfying this statement because $0 \cdot q = 0$ and $3 \neq 0$. Furthermore, if in (1) $b = 0$ and $a = 0$, the statement becomes

$$0 \div 0 = q \quad \text{if and only if} \quad 0 = 0 \cdot q$$

Because $0 \cdot q = 0$ for any value of q, $0 \div 0$ could equal any real number; that is, $0 \div 0$ is indeterminate. Therefore, for every real number a, no meaning can be attached to $a \div 0$. Hence

division by zero is undefined

An ordering of the set of real numbers can be accomplished by means of a relation denoted by the symbols $<$ (read "is less than") and $>$ (read "is greater than"). In the following definition of these symbols we use the concept of a positive number given by Axiom 8 in Section A.1.

DEFINITION **The Symbols $<$ and $>$**

If a and b are real numbers,

(i) $a < b$ if and only if $b - a$ is positive
(ii) $a > b$ if and only if $a - b$ is positive

▷ **ILLUSTRATION 4**

$3 < 5$ because $5 - 3 = 2$, and 2 is positive

$-10 < -6$ because $-6 - (-10) = 4$, and 4 is positive

$7 > 2$ because $7 - 2 = 5$, and 5 is positive

$-2 > -7$ because $-2 - (-7) = 5$, and 5 is positive

$\frac{3}{4} > \frac{2}{3}$ because $\frac{3}{4} - \frac{2}{3} = \frac{1}{12}$, and $\frac{1}{12}$ is positive   ◀

Observe that

$3 > 0$ because $3 - 0 = 3$, and 3 is positive

$-4 < 0$ because $0 - (-4) = 4$, and 4 is positive

These statements are special cases of the following properties that are obtained from the definitions of $>$ and $<$:

$a > 0$ if and only if a is positive

$a < 0$ if and only if a is negative

If we write $a \leq b$ (read "a is less than or equal to b") we mean that either $a < b$ or $a = b$. Similarly, $a \geq b$ (read "a is greater than or equal to b") indicates that either $a > b$ or $a = b$.

The statements $a < b$, $a > b$, $a \leq b$, and $a \geq b$ are called **inequalities**. In particular, $a < b$ and $a > b$ are **strict** inequalities, whereas $a \leq b$ and $a \geq b$ are **nonstrict** inequalities.

A number x is **between** a and b if $a < x$ and $x < b$. We can write this as a **continued inequality** as follows:

$$a < x < b$$

Therefore

$$2 < 3 < 4 \qquad -5 < 1 < \frac{4}{3} \qquad \frac{1}{2} < \frac{2}{3} < \frac{3}{4}$$

Another continued inequality is

$$a \leq x \leq b$$

which means that both $a \leq x$ and $x \leq b$. Other continued inequalities are $a \leq x < b$ and $a < x \leq b$.

▶ **EXAMPLE 3** *Using Set Notation and Inequality Symbols to Denote a Set of Numbers*

With set notation and one or more of the symbols $<$, $>$, $\leq$, and $\geq$, denote the set: **(a)** the set of all x such that x is between -2 and 2; **(b)** the set of all t such that $4t - 1$ is nonnegative; **(c)** the set of all y such that $y + 3$ is positive and less than or equal to 15; **(d)** the set of all z such that $2z$ is greater than or equal to -5 and less than -1.

Solution

(a) $\{x \mid -2 < x < 2\}$ **(b)** $\{t \mid 4t - 1 \geq 0\}$

(c) $\{y \mid 0 < y + 3 \leq 15\}$ **(d)** $\{z \mid -5 \leq 2z < -1\}$ ◀

We now give a geometric interpretation to the set R of real numbers by associating them with points on a horizontal line called an **axis.** A point, called the **origin,** is chosen to represent the number 0. A unit of distance is selected arbitrarily. Then each positive integer n is represented by the point at a distance of n units to the right of the origin, and each negative integer $-n$ is represented by the point at a distance of n units to the left of the origin. We call these points **unit points.** They are labeled with the numbers with which they are associated. For example, 4 is represented by the point 4 units to the right of the origin and -4 is represented by the point 4 units to the left of the origin. Figure 2 shows the unit points representing 0 and the first 12 positive integers and their corresponding negative integers.

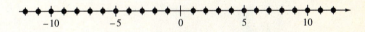

FIGURE 2

The rational numbers are associated with points on the axis of Figure 2 by dividing the segments between points representing successive integers. For instance, if the segment from 0 to 1 is divided into seven equal parts, the endpoint of the first such subdivision is associated with $\frac{1}{7}$, the endpoint of the second is associated with $\frac{2}{7}$, and so on. The point associated with the number $\frac{24}{7}$ is three-sevenths of the distance from the unit point 3 to the unit point 4. A negative rational number, in a similar manner, is associated with a point to the left of the origin. Figure 3 shows some of the points associated with rational numbers.

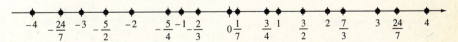

FIGURE 3

Geometric constructions can be used to find points corresponding to certain irrational numbers, such as $\sqrt{2}$, $\sqrt{3}$, $\sqrt{5}$, and so on. (See Exercises 35 and 36.) Points corresponding to other irrational numbers can be found by using decimal approximations. For example, a point corresponding to the number π can be approximated using some of the digits in the decimal representation 3.14159 . . . .

Every irrational number can be associated with a unique point on the axis, and every point that does not correspond to a rational number can be associated with an irrational number. This indicates that a one-to-one correspondence between the set of real numbers and the points on the horizontal axis can be established. For this reason the horizontal axis is referred to as the **real-number line.** Because the points on this line are identified with the numbers they represent, the same symbol is used for that number and the point.

▷ **ILLUSTRATION 5**

Consider the set $\{x \mid -6 < x \leq 4\}$. This set is represented on the real-number line in Figure 4. The bracket at 4 indicates that 4 is in the set, and the parenthesis at -6 indicates that -6 is not in the set.

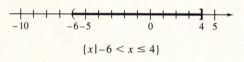

$$\{x \mid -6 < x \leq 4\}$$

FIGURE 4   ◀

The set of all numbers x satisfying the continued inequality $a < x < b$ is called an **open interval** and is denoted by (a, b). Therefore,

$$(a, b) = \{x \mid a < x < b\}$$

The **closed interval** from a to b is the open interval (a, b) together with the two endpoints a and b and is denoted by $[a, b]$. Thus,

$$[a, b] = \{x \mid a \leq x \leq b\}$$

Figure 5 illustrates the open interval (a, b), and Figure 6 shows the closed interval $[a, b]$.

The **interval half-open on the left** is the open interval (a, b) together with the right endpoint b. It is denoted by $(a, b]$; so

$$(a, b] = \{x \mid a < x \leq b\}$$

We define an **interval half-open on the right** in a similar way and denote it by $[a, b)$. Thus

$$[a, b) = \{x \mid a \leq x < b\}$$

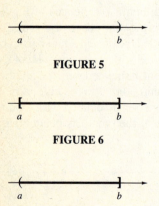

FIGURE 5

FIGURE 6

FIGURE 7

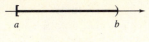

FIGURE 8

FIGURE 9

FIGURE 10

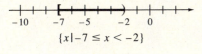

$\{x\,|\,-7 \le x < -2\}$

(a)

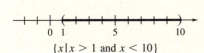

$\{x\,|\,x > 1 \text{ and } x < 10\}$

(b)

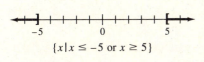

$\{x\,|\,x \le -5 \text{ or } x \ge 5\}$

(c)

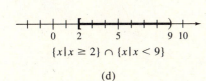

$\{x\,|\,x \ge 2\} \cap \{x\,|\,x < 9\}$

(d)

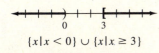

$\{x\,|\,x < 0\} \cup \{x\,|\,x \ge 3\}$

(e)

FIGURE 11

The interval $(a, b]$ appears in Figure 7, and the interval $[a, b)$ is shown in Figure 8.

We shall use the symbol $+\infty$ (positive infinity) and the symbol $-\infty$ (negative infinity); however, take care not to confuse these symbols with real numbers, for they do not obey the properties of the real numbers. We have the following intervals:

$$(a, +\infty) = \{x \mid x > a\}$$
$$(-\infty, b) = \{x \mid x < b\}$$
$$[a, +\infty) = \{x \mid x \ge a\}$$
$$(-\infty, b] = \{x \mid x \le b\}$$
$$(-\infty, +\infty) = R$$

Figure 9 shows the interval $(a, +\infty)$, and Figure 10 illustrates the interval $(-\infty, b)$. Note that $(-\infty, +\infty)$ denotes the set of all real numbers.

▶ **EXAMPLE 4** *Illustrating a Set of Numbers on the Real-Number Line and Defining the Set by Interval Notation*

Show the set on the real-number line and represent the set by interval notation.

(a) $\{x \mid -7 \le x < -2\}$ (b) $\{x \mid x > 1 \text{ and } x < 10\}$
(c) $\{x \mid x \le -5 \text{ or } x \ge 5\}$ (d) $\{x \mid x \ge 2\} \cap \{x \mid x < 9\}$
(e) $\{x \mid x < 0\} \cup \{x \mid x \ge 3\}$

Solution   The sets are shown on the real-number line in Figure 11(a), (b), (c), (d), and (e), respectively. With interval notation, we have

(a) $\{x \mid -7 \le x < -2\} = [-7, -2)$
(b) $\{x \mid x > 1 \text{ and } x < 10\} = (1, 10)$
(c) $\{x \mid x \le -5 \text{ or } x \ge 5\} = (-\infty, -5] \cup [5, +\infty)$
(d) $\{x \mid x \ge 2\} \cap \{x \mid x < 9\} = [2, 9)$
(e) $\{x \mid x < 0\} \cup \{x \mid x \ge 3\} = (-\infty, 0) \cup [3, +\infty)$ ◀

▶ **EXAMPLE 5** *Illustrating a Set of Numbers on the Real-Number Line and Defining the Set by Set Notation and Inequality Symbols*

Show the interval on the real-number line and use set notation and inequality symbols to denote the interval: (a) $(-2, 4)$; (b) $[3, 7]$; (c) $[1, 6)$; (d) $(-4, 0]$; (e) $[0, +\infty)$; (f) $(-\infty, 5)$

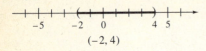

(-2, 4)

(a)

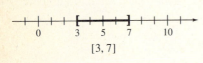

[3, 7]

(b)

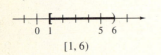

[1, 6)

(c)

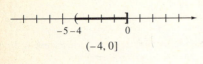

(-4, 0]

(d)

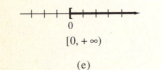

[0, +∞)

(e)

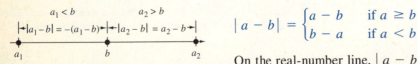

(-∞, 5)

(f)

FIGURE 12

Solution The intervals are shown on the real-number line in Figure 12(a), (b), (c), (d), (e), and (f), respectively. With set notation we have

(a) $(-2, 4) = \{x \mid -2 < x < 4\}$ (b) $[3, 7] = \{x \mid 3 \le x \le 7\}$

(c) $[1, 6) = \{x \mid 1 \le x < 6\}$ (d) $(-4, 0] = \{x \mid -4 < x \le 0\}$

(e) $[0, +\infty) = \{x \mid x \ge 0\}$ (f) $(-\infty, 5) = \{x \mid x < 5\}$ ◀

Associated with each real number is a nonnegative number called its *absolute value*.

DEFINITION   **Absolute Value**

If a is a real number, the **absolute value** of a, denoted by $|a|$, is a if a is nonnegative and is $-a$ if a is negative. With symbols we write

$$|a| = \begin{cases} a & \text{if } a \ge 0 \\ -a & \text{if } a < 0 \end{cases}$$

▷ **ILLUSTRATION 6**

If in the preceding definition we take a as 6, 0, and -6, we have, respectively,

$$|6| = 6 \quad |0| = 0 \quad |-6| = -(-6)$$
$$= 6$$

◀

The absolute value of a number can be considered as its distance (without regard to direction, left or right) from the origin. In particular, the points 6 and -6 are each six units from the origin.

From the definition of absolute value

$$|a - b| = \begin{cases} a - b & \text{if } a - b \ge 0 \\ -(a - b) & \text{if } a - b < 0 \end{cases}$$

or, equivalently

$$|a - b| = \begin{cases} a - b & \text{if } a \ge b \\ b - a & \text{if } a < b \end{cases}$$

On the real-number line, $|a - b|$ units can be interpreted as the distance between a and b without regard to direction. See Figure 13.

$a_1 < b$ $a_2 > b$

$|a_1 - b| = -(a_1 - b)$ $|a_2 - b| = a_2 - b$

a_1 b a_2

FIGURE 13

▶ **EXAMPLE 6** *Representing Numbers by Points on the Real-Number Line and Finding the Distance Between the Points*

Show the points corresponding to the numbers $-10, -7, -5, -3, 0, 3, 5, 7$, and 10 on the real-number line. Find the distance between u and v in the following cases:

(a) $u = 10, v = 3$ (b) $u = 3, v = 7$ (c) $u = 5, v = -3$
(d) $u = -7, v = 0$ (e) $u = -3, v = -5$ (f) $u = -10, v = -7$

Solution   Figure 14 shows the points corresponding to the given numbers on the real-number line. In each part the distance between u and v is $|u - v|$ units:

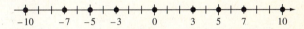

FIGURE 14

(a) $\begin{aligned}|u - v| &= |10 - 3| \\ &= |7| \\ &= 7\end{aligned}$ (b) $\begin{aligned}|u - v| &= |3 - 7| \\ &= |-4| \\ &= 4\end{aligned}$

(c) $\begin{aligned}|u - v| &= |5 - (-3)| \\ &= |8| \\ &= 8\end{aligned}$ (d) $\begin{aligned}|u - v| &= |-7 - 0| \\ &= |-7| \\ &= 7\end{aligned}$

(e) $\begin{aligned}|u - v| &= |-3 - (-5)| \\ &= |2| \\ &= 2\end{aligned}$ (f) $\begin{aligned}|u - v| &= |-10 - (-7)| \\ &= |-3| \\ &= 3\end{aligned}$ ◀

EXERCISES 1.1

In Exercises 1 through 10, N is the set of natural numbers, J is the set of integers, Q is the set of rational numbers, H is the set of irrational numbers, and R is the set of real numbers.

In Exercises 1 and 2, insert $\in$ or $\notin$ in the blank to make the statement correct.

1. (a) 15 _____ N (b) 1.41421 _____ H
   (c) -3 _____ J (d) π _____ Q

2. (a) 0 _____ Q (b) 2007 _____ J
   (c) $\frac{3}{7}$ _____ R (d) -5 _____ H

In Exercises 3 and 4, use the symbol $\subseteq$ to give a correct statement involving the two sets.

3. (a) N and Q (b) R and Q (c) J and N
   (d) J and R

4. (a) R and N (b) J and Q (c) H and R
   (d) $\{0\}$ and J

In Exercises 5 and 6, insert either $\subseteq$ *or* $\nsubseteq$ *to make the statement correct.*

5. (a) J _____ R **(b)** $\{0\}$ _____ N **(c)** H _____ R
   (d) $\{0, \frac{1}{3}, 1.732\}$ _____ Q

6. (a) N _____ Q **(b)** Q _____ H **(c)** Q _____ R
   (d) $\{-\sqrt{3}, 0, \sqrt{3}\}$ _____ H

In Exercises 7 and 8, determine which of the sets N, J, Q, H, R, and $\varnothing$ *is equal to the set.*

7. (a) $Q \cap R$ **(b)** $Q \cup H$ **(c)** $J \cup N$
   (d) $H \cap J$

8. (a) $H \cap R$ **(b)** $Q \cup R$ **(c)** $H \cap Q$
   (d) $J \cap N$

In Exercises 9 and 10, for the set S define each of the following sets: (a) S $\cap$ N; (b) S $\cap$ Q; (c) S $\cap$ H; (d) S $\cap$ J.

9. $S = \{12, \frac{5}{3}, \sqrt{7}, 0, -38, -\sqrt{2}, 571, \pi, -\frac{1}{10},$
$$0.666 \ldots, 16.34\}$$

10. $S = \{-\frac{1}{4}, 26, \sqrt{3}, 1.23, -\sqrt{9},$
$$-0.333 \ldots, -6214, \frac{1}{2}\pi, \frac{4}{7}, 1\}$$

In Exercises 11 and 12, arrange the elements of the given subset of R in the same order as their corresponding points from left to right on the real-number line.

11. $\{-2, 3, 21, 5, -7, \frac{2}{3}, \sqrt{2}, -\frac{7}{4}, -\sqrt{5}, -10,$
$$0, \frac{3}{4}, -\frac{5}{3}, -1\}$$

12. $\{\frac{11}{3}, \pi, -8, -\sqrt{2}, 3, -\sqrt{3}, 4, \frac{21}{4}, -\frac{3}{2}, 1.26, \frac{1}{2}\pi\}$

In Exercises 13 through 16, use set notation and one or more of the symbols $<, >, \leq,$ *and* $\geq$ *to denote the set.*

13. (a) The set of all x such that x is greater than -9 and less than 8; **(b)** the set of all y between -12 and -3; **(c)** the set of all z such that $4z - 5$ is negative.

14. (a) The set of all x between -5 and 3; **(b)** the set of all y such that y is greater than or equal to -26 and less than -16; **(c)** the set of all t such that $8t - 4$ is positive.

15. (a) The set of all x such that $2x + 4$ is nonnegative; **(b)** the set of all r such that r is greater than or equal to 2 and less than 8; **(c)** the set of all a such that $a - 2$ is greater than -5 and less than or equal to 7.

16. (a) The set of all s such that $2s + 3$ is nonpositive; **(b)** the set of all x such that $3x$ is greater than 10 and less than or equal to 20; **(c)** the set of all z such that $2z + 5$ is between and including -1 and 15.

In Exercises 17 through 24 do the following: (i) show the set on the real-number line; (ii) represent the set by interval notation; (iii) describe the set in words similar to the descriptions in Exercises 13 through 16.

17. (a) $\{x \mid x > 2\}$ **(b)** $\{x \mid -4 < x \leq 4\}$
18. (a) $\{x \mid x \leq 8\}$ **(b)** $\{x \mid 3 < x < 9\}$
19. (a) $\{x \mid x > 2 \text{ and } x < 12\}$
   (b) $\{x \mid x \leq -4 \text{ or } x > 4\}$
20. (a) $\{x \mid x \geq -5 \text{ and } x \leq 5\}$
   (b) $\{x \mid x < 3 \text{ or } x > 6\}$
21. (a) $\{x \mid x > 2\} \cap \{x \mid x < 12\}$
   (b) $\{x \mid x \leq -4\} \cup \{x \mid x > 4\}$
22. (a) $\{x \mid x \geq -5\} \cap \{x \mid x \leq 5\}$
   (b) $\{x \mid x < 3\} \cup \{x \mid x > 6\}$
23. (a) $\{x \mid x > -4\} \cap \{x \mid x \leq 0\}$
   (b) $\{x \mid x \leq 0\} \cup \{x \mid x \leq 7\}$
24. (a) $\{x \mid x > -8\} \cap \{x \mid x \leq 0\}$
   (b) $\{x \mid x > 2\} \cup \{x \mid x > 10\}$

In Exercises 25 through 28, show the interval on the real-number line and use set notation and inequality symbols to denote the interval.

25. (a) $(2, 7)$ **(b)** $[-3, 6]$
   (c) $(-5, 4]$ **(d)** $[-10, -2)$
26. (a) $(-5, 5)$ **(b)** $[1, 9]$
   (c) $[-8, 3)$ **(d)** $(-7, 0]$
27. (a) $[3, +\infty)$ **(b)** $(-\infty, 0)$
   (c) $(-4, +\infty)$ **(d)** $(-\infty, +\infty)$
28. (a) $(-\infty, -2]$ **(b)** $(-1, +\infty)$
   (c) $(-\infty, 10)$ **(d)** $[0, +\infty)$

In Exercises 29 and 30, write the number without absolute-value bars.

29. (a) $|7|$ **(b)** $\left|-\frac{3}{4}\right|$
   (c) $|3 - \sqrt{3}|$ **(d)** $|\sqrt{3} - 3|$
30. (a) $\left|\frac{1}{3}\right|$ **(b)** $|-8|$
   (c) $|\pi - 2|$ **(d)** $|3 - \pi|$

In Exercises 31 through 34, show the points corresponding to u and v on the real-number line and then find the distance between u and v.

31. (a) $u = 8, v = 2$ **(b)** $u = -8, v = 2$
   (c) $u = 8, v = -2$ **(d)** $u = -8, v = -2$

32. **(a)** $u = 6, v = 4$ **(b)** $u = -6, v = 4$
    (c) $u = 6, v = -4$ **(d)** $u = -6, v = -4$

33. **(a)** $u = t, v = 2t$, and $t > 0$
    (b) $u = t, v = 2t$, and $t < 0$

34. **(a)** $u = t, v = \frac{1}{2}t$, and $t > 0$
    (b) $u = t, v = \frac{1}{2}t$, and $t < 0$

35. To determine the point on the real-number line corresponding to the irrational number $\sqrt{2}$, use the construction indicated in the figure below: From the point 1, a line segment of length one unit is drawn perpendicular to the axis. Then a right triangle is formed by connecting the endpoint of this segment with the origin. The length of the hypotenuse of this right triangle is $\sqrt{2}$ units. (This fact follows from the Pythagorean theorem, which states that c^2 has the same value as $a^2 + b^2$, where c units is the length of the hypotenuse, and a units and b units are the lengths of the other two sides.) An arc of a circle having its center at the origin and a radius of $\sqrt{2}$ is then drawn; the point where this arc intersects the axis is $\sqrt{2}$. Use this method to determine the point corresponding to $\sqrt{5}$.

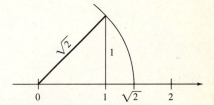

36. Determine the point on the real-number line corresponding to the irrational number $\sqrt{10}$. (*Hint:* See Exercise 35.)

37. Write a description of the set of real numbers by specifying the relationships among the sets of numbers shown in Figure 1.

1.2 ALGEBRAIC EXPRESSIONS

GOALS

1. **Learn notation and terminology associated with algebraic expressions.**
2. **Simplify complex rational expressions similar to types occurring in calculus.**
3. **Learn laws of rational exponents.**
4. **Apply laws of exponents to simplify algebraic expressions.**
5. **Learn properties of radicals.**
6. **Simplify algebraic expressions that occur in calculus.**
7. **Rationalize the numerators of fractions that occur in calculus.**

Operations with polynomial and rational expressions, including laws of exponents and radicals, are discussed in this section. First we review some notation and terminology.

To indicate a product, we use the centered dot, ·, or parentheses around one or more symbols. Sometimes we omit the symbol for multiplication. For example, the product of a and b can be written in the following ways:

$$a \cdot b \quad (a)(b) \quad a(b) \quad (a)b \quad ab$$

The numbers a and b are called **factors** of the product ab.

Suppose we have the product of two factors of x. We can use the notation x^2 to indicate the product, where the numeral 2 written to the upper right of the symbol x is called an *exponent;* thus

$$x^2 = x \cdot x$$

In particular, $3^2 = 3 \cdot 3$; that is, $3^2 = 9$.

In general, if a is a real number and n is a positive integer,

$$a^n = a \cdot a \cdot a \cdot \ldots \cdot a \quad (n \text{ factors of } a)$$

where n is called the **exponent,** a is called the **base,** and a^n called the **nth power of a.** For example, x^2 is the second power of x, and y^5 is the fifth power of y, where

$$y^5 = y \cdot y \cdot y \cdot y \cdot y$$

The fourth power of -2 is denoted by $(-2)^4$ and

$$
\begin{aligned}
(-2)^4 &= (-2)(-2)(-2)(-2) \\
&= 16
\end{aligned}
$$

When a symbol is written without an exponent, the exponent is understood to be 1. Hence $x = x^1$. It is customary to read x^2, the second power of x, as "x squared" and x^3, the third power of x, as "x cubed."

The representation of positive-integer powers by exponents was introduced by René Descartes (1596–1650) in 1637.

In Section 1.1 we stated that a **variable** is a symbol used to represent any element of a given domain. If the domain is R, then the variable represents a real number. A **constant** is a symbol whose domain contains only one element. For example, in the sum

$$6x^2 + 2x + 5$$

x is a symbol for a variable, and 6, 2, and 5 are constants. Sometimes letters are used as symbols for constants to designate fixed but unspecified numbers. For instance, we can write the sum

$$ax + b$$

where a and b are symbols for constants and x is a symbol for a variable. The sum $6x^2 + 2x + 5$ is a particular *algebraic expression*. The terminology **algebraic expression** is used to mean a constant, a variable, or a combination of variables and constants involving a finite number of indicated operations (addition, subtraction, multiplication, division, raising to a power, and extraction of a root) on them. Examples of algebraic expressions are

$$3x^2y^5 \qquad 5x^2 - 8x + 2 \qquad \frac{3x^2 - 6xy + y^2}{x + 2y} \qquad \frac{\sqrt{x + y} - 4}{(z + 2)^3 - \sqrt[3]{x}}$$

An algebraic expression involving only nonnegative-integer powers of one or more variables and containing no variable in a denominator is called a **polynomial.** For example,

$$2x \qquad 5x^2 - 8x + 2 \qquad 4x^5 - 6x^3 + 3x^2 - 2x + 1$$

are polynomials in the variable x. Examples of polynomials in the variables x and y are

$$3x^2y^5 \qquad 6x^2 + 8y^2 \qquad 8xy - 7x + y - 3$$

A **term** of a polynomial is a constant or a constant multiplied by nonnegative-integer powers of variables. A polynomial can be considered as the sum of a finite number of terms. For example, $5x^2 - 8x + 2$ can be written as $5x^2 + (-8x) + 2$, and the terms are $5x^2$, $-8x$, and 2. The polynomial $8xy - 7x + y - 3$ can be written as $8xy + (-7x) + y + (-3)$, and the terms are $8xy$, $-7x$, y, and -3.

Any factor of a product is said to be the **coefficient** of the other factors. For instance, in the product $5xyz$ the coefficient of yz is $5x$, the coefficient of x is $5yz$, the coefficient of $5z$ is xy, and so on. If a coefficient is a constant, then it is called a **constant coefficient.** Hence in the product $5xyz$, 5 is the constant coefficient of xyz. Terms that may differ only in their constant coefficients are called **like terms.** For example, $6x^2$ and $3x^2$ are like terms as are x and $-4x$. Like terms of a polynomial are combined algebraically by using the distributive law. In particular,

$$
\begin{aligned}
6x^2 + x + 7 + 3x^2 - 4x &= (6x^2 + 3x^2) + (x - 4x) + 7 \\
&= (6 + 3)x^2 + (1 - 4)x + 7 \\
&= 9x^2 - 3x + 7
\end{aligned}
$$

If after combining like terms a polynomial has one term, it is called a **monomial;** if it has two terms, it is call a **binomial;** and if it has three terms, it is called a **trinomial.** Polynomials $2x$ and $3x^2y^5$ are monomials, $6x^2 + 8y^2$ is a binomial, and $5x^2 - 8x + 2$ is a trinomial.

By the **degree** of a monomial in one variable, we mean the exponent of that variable. In particular, the monomial $5x^3$ has degree 3. If a monomial has more than one variable, the **degree** of the monomial is the sum of the exponents of all the variables that appear. For example, the degree of $3x^2y^5$ is 7. The degree of $-2xyz$ is 3. The degree of a nonzero-constant monomial, such as the constant 4, is zero. The constant 0 has no degree.

The **degree** of a polynomial is the same as the degree of the term with the highest degree in the polynomial. Therefore, $7x^2 - 4x + 2$ is a second-degree polynomial, and $3x + 6$ is a first-degree polynomial. The degree of $6x^2y^2 - 4x^3 + 2y$ is 4 because $6x^2y^2$ is the term with the highest degree, 4.

If a and b are real numbers and $b \neq 0$, the operation of division of a by b can be denoted by

$$\frac{a}{b}$$

Other notations for the division of a by b are

$$a \div b \quad \text{and} \quad a/b$$

The numerals $\frac{a}{b}$ and a/b are called **fractions,** the number a is called the **numerator,** and b is called the **denominator.** If the numerator and denominator of a fraction are polynomials, then the fraction is called a **rational expression.** Examples of rational expressions are

$$\frac{5}{y - 7} \qquad \frac{3x + 2}{x^2 - 4} \qquad \frac{2}{5rs} \qquad \frac{3t^2 + t + 5}{t^4 + 1}$$

Because the denominator cannot be zero, it is understood that in the preceding rational expressions, $y \neq 7$, $x \neq \pm 2$, $r \neq 0$, and $s \neq 0$. Because a rational expression denotes a quotient of real numbers, properties of fractions also hold for rational expressions.

If a fraction contains a fraction in either the numerator or denominator or both, it is called a **complex fraction.** In contrast, a fraction that is not complex is referred to as a **simple fraction.**

▷ ILLUSTRATION 1

The fraction

$$\frac{\dfrac{3}{5} - \dfrac{4}{7}}{\dfrac{2}{5} + \dfrac{1}{3}}$$

is complex. We find an equivalent simple fraction by multiplying the numerator and denominator by the LCD (lowest common denominator) of each of the fractions appearing.

$$\frac{\dfrac{3}{5} - \dfrac{4}{7}}{\dfrac{2}{5} + \dfrac{1}{3}} = \frac{105 \cdot \dfrac{3}{5} - 105 \cdot \dfrac{4}{7}}{105 \cdot \dfrac{2}{5} + 105 \cdot \dfrac{1}{3}}$$

$$= \frac{63 - 60}{42 + 35}$$

$$= \frac{3}{77} \qquad \qquad ◀$$

EXAMPLE 1 *Simplifying a Complex Rational Expression*

Simplify

$$\frac{\dfrac{1}{(x+h)^2} - \dfrac{1}{x^2}}{h} \qquad h \neq 0$$

which is an expression occurring in calculus.

Solution   The LCD is $x^2(x+h)^2$.

$$\frac{\dfrac{1}{(x+h)^2} - \dfrac{1}{x^2}}{h} = \frac{x^2(x+h)^2\dfrac{1}{(x+h)^2} - x^2(x+h)^2\dfrac{1}{x^2}}{x^2(x+h)^2 h}$$

$$= \frac{x^2 - (x+h)^2}{hx^2(x+h)^2}$$

$$= \frac{x^2 - (x^2 + 2hx + h^2)}{hx^2(x+h)^2}$$

$$= \frac{x^2 - x^2 - 2hx - h^2}{hx^2(x+h)^2}$$

$$= \frac{-h(2x+h)}{hx^2(x+h)^2}$$

$$= -\frac{2x+h}{x^2(x+h)^2} \qquad \blacktriangleleft$$

We now state five laws of positive-integer exponents you learned in a previous algebra course and follow each law by an illustration demonstrating its application.

If n and m are positive integers and a is a real number,

$$a^n \cdot a^m = a^{n+m}$$

▷ **ILLUSTRATION 2**

(a) $2^3 \cdot 2^2 = 2^{3+2}$ (b) $x^4 \cdot x = x^{4+1}$ (c) $y^3 \cdot y^7 = y^{3+7}$

         $= 2^5$ $= x^5$ $= y^{10}$

         $= 32$ $\blacktriangleleft$

If n and m are positive integers and a is a real number,

$$(a^n)^m = a^{nm}$$

▷ **ILLUSTRATION 3**

(a) $(2^3)^2 = 2^{3 \cdot 2}$

$\qquad = 2^6$

$\qquad = 64$

(b) $(y^4)^6 = y^{4 \cdot 6}$

$\qquad = y^{24}$

◀

If n and m are positive integers and a is a real number where $a \neq 0$,

$$\frac{a^n}{a^m} = \begin{cases} a^{n-m} & \text{if } n > m \\[2mm] \dfrac{1}{a^{m-n}} & \text{if } n < m \\[2mm] 1 & \text{if } n = m \end{cases}$$

▷ **ILLUSTRATION 4**

(a) $\dfrac{x^6}{x^2} = x^{6-2}$

$\qquad = x^4$

(b) $\dfrac{x^3}{x^7} = \dfrac{1}{x^{7-3}}$

$\qquad = \dfrac{1}{x^4}$

(c) $\dfrac{x^5}{x^5} = 1$

◀

If n is a positive integer and a and b are real numbers,

$$(ab)^n = a^n b^n$$

▷ **ILLUSTRATION 5**

(a) $(2 \cdot 5)^3 = 2^3 \cdot 5^3$

$\qquad = 8 \cdot 125$

$\qquad = 1000$

(b) $(x^3 y^5)^4 = (x^3)^4 (y^5)^4$

$\qquad = x^{12} y^{20}$

◀

If n is a positive integer and a and b are real numbers where $b \neq 0$,

$$\left(\frac{a}{b}\right)^n = \frac{a^n}{b^n}$$

▷ **ILLUSTRATION 6**

(a) $\left(\dfrac{2}{3}\right)^5 = \dfrac{2^5}{3^5}$

$\qquad = \dfrac{32}{243}$

(b) $\left(\dfrac{x^4 z^2}{y^3}\right)^3 = \dfrac{(x^4 z^2)^3}{(y^3)^3}$

$\qquad = \dfrac{(x^4)^3 (z^2)^3}{y^9}$

$\qquad = \dfrac{x^{12} z^6}{y^9}$

◀

The definition of a^n as the product of n factors of a has meaning only when the exponent n is a positive integer. When the exponent is zero or a negative integer, we must, therefore, define a^n another way. These definitions should be made so that the same laws that apply for positive-integer exponents also hold for zero and negative-integer exponents. In particular, if the formula $a^n \cdot a^m = a^{n+m}$ is to hold for a zero exponent, then if $a \neq 0$,

$$a^n \cdot a^0 = a^{n+0}$$
$$a^n \cdot a^0 = a^n$$

Because 1 is the number having the property that $a^n \cdot 1 = a^n$, we must define a^0 as 1.

Now suppose that n is a positive integer, and thus $-n$ is a negative integer. If the formula $a^n \cdot a^m = a^{n+m}$ is to hold for a negative-integer exponent, then

$$a^n \cdot a^{-n} = a^0$$
$$a^n \cdot a^{-n} = 1$$

Hence we must define a^{-n} as $\dfrac{1}{a^n}$. We state the definitions formally.

DEFINITION   **Zero and Negative-Integer Exponents**

If n is a positive integer and a is a real number where $a \neq 0$,

$$a^0 = 1$$

$$a^{-n} = \frac{1}{a^n}$$

▷ **ILLUSTRATION 7**

(a) $6^0 = 1$ **(b)** $4^{-2} = \dfrac{1}{4^2}$ **(c)** $(-5)^0(-2)^{-3} = 1 \cdot \dfrac{1}{(-2)^3}$

$$= \frac{1}{16}$$

$$= \frac{1}{-8}$$

$$= -\frac{1}{8} \quad ◄$$

The laws for positive-integer exponents also hold for zero and negative-integer exponents.

▶ **EXAMPLE 2** *Applying Laws of Exponents to Negative-Integer Exponents*

Write the quantity as a rational number with an exponent of 1:

(a) $(2^{-3} \cdot 3^{-2})^{-1}$

(b) $(2^{-3} + 3^{-2})^{-1}$

Solution

(a)
$$(2^{-3} \cdot 3^{-2})^{-1} = 2^{-3(-1)} \cdot 3^{-2(-1)}$$
$$= 2^3 \cdot 3^2$$
$$= 72$$

(b)
$$(2^{-3} + 3^{-2})^{-1} = \frac{1}{(2^{-3} + 3^{-2})^1}$$
$$= \frac{1}{\dfrac{1}{2^3} + \dfrac{1}{3^2}}$$
$$= \frac{1}{\dfrac{1}{8} + \dfrac{1}{9}}$$
$$= \frac{1}{\dfrac{17}{72}}$$
$$= \frac{72}{17}$$ ◀

Before discussing fractional exponents, we define an *nth root* of a real number.

DEFINITION **An *n*th Root of a Real Number**

If n is a positive integer greater than 1 and a and b are real numbers such that

$$b^n = a$$

then b is an **nth root** of a.

▷ **ILLUSTRATION 8**

(a) Because $2^2 = 4$, 2 is a square root of 4; furthermore, because $(-2)^2 = 4$, -2 is also a square root of 4.
(b) Because $3^4 = 81$, 3 is a fourth root of 81. Also, because $(-3)^4 = 81$, -3 is a fourth root of 81.
(c) Because $4^3 = 64$, 4 is a cube root of 64.
(d) Because $(-4)^3 = -64$, -4 is a cube root of -64. ◀

Observe in part (a) of Illustration 8 that there are two square roots of 4 and in part (b) there are two fourth roots of 81. To distinguish between the two roots in such cases, we introduce the concept of *principal nth root*.

DEFINITION   **The Principal *n*th Root of a Real Number**

If n is a positive integer greater than 1, a is a real number, and $\sqrt[n]{a}$ denotes the principal nth root of a, then

  (i) if $a > 0$, $\sqrt[n]{a}$ is the positive nth root of a,
  (ii) if $a < 0$ and n is odd, $\sqrt[n]{a}$ is the negative nth root of a, and
(iii) $\sqrt[n]{0} = 0$.

In the preceding definition, the symbol $\sqrt{\ }$ is called a **radical sign.** The entire expression $\sqrt[n]{a}$ is called a **radical,** where the number a is the **radicand.** The number n is the **index** and indicates the **order** of the radical. If no n appears, the order is understood to be 2.

▷ **ILLUSTRATION 9**

In parts (a), (b), and (c) we use (i) of the preceding definition, and in part (d) we use (ii).

(a) $\sqrt{4} = 2$ (read "the principal square root of 4 equals 2")
Note that -2 is also a square root of 4, but it is not the principal square root. However, we can write $-\sqrt{4} = -2$.
(b) $\sqrt[4]{81} = 3$ (read "the principal fourth root of 81 equals 3")
The number -3 is also a fourth root of 81, and we can write $-\sqrt[4]{81} = -3$.
(c) $\sqrt[3]{64} = 4$ (read "the principal cube root of 64 equals 4")
(d) $\sqrt[3]{-64} = -4$ (read "the principal cube root of -64 equals -4") ◄

Observe that if $a < 0$, $\sqrt[n]{a}$ is defined only if n is odd. For instance, $\sqrt{-16}$ is not defined as a real number because there is no real number whose square is -16. Complex numbers, discussed in Section 1.3, are needed to define an even-order root of a negative number.

The principal nth root of a real number b is a rational number if and only if b is the nth power of a rational number. For instance, $\sqrt[4]{625} = 5$ because $5^4 = 625$, and $\sqrt[3]{-\frac{1}{27}} = -\frac{1}{3}$ because $(-\frac{1}{3})^3 = -\frac{1}{27}$.

Recall from Section 1.1 that a real number that is not rational is called an irrational number and cannot be represented by a terminating decimal or a nonterminating repeating decimal. Because 3 is not the square of a rational number, $\sqrt{3}$ is an irrational number. Other examples of irrational numbers are

$$\sqrt{2} \quad \sqrt[3]{4} \quad \sqrt[3]{-5} \quad \sqrt[4]{15}$$

We are now ready to define a rational exponent of the form $1/n$ where n is a positive integer. If the formula $(a^m)^n = a^{mn}$ is to hold when m is $1/n$, then we must have

$$(a^{1/n})^n = a^{n/n}$$
$$(a^{1/n})^n = a$$

This equality states that the nth power of $a^{1/n}$ equals a. Thus we define $a^{1/n}$ as the principal nth root of a.

DEFINITION $a^{1/n}$

If n is a positive integer greater than 1 and a is a real number, then if $\sqrt[n]{a}$ is a real number,

$$a^{1/n} = \sqrt[n]{a}$$

▷ **ILLUSTRATION 10**

(a) $25^{1/2} = \sqrt{25}$ **(b)** $(-8)^{1/3} = \sqrt[3]{-8}$ **(c)** $\left(\dfrac{1}{81}\right)^{1/4} = \sqrt[4]{\dfrac{1}{81}}$

$$= 5 \qquad\qquad\qquad\qquad = -2 \qquad\qquad\qquad = \dfrac{1}{3} \quad ◀$$

Consider now how we should define expressions such as

$$9^{3/2} \quad 8^{2/3} \quad (-27)^{4/3} \quad 7^{3/4}$$

If the formula $a^{pq} = (a^p)^q$ is to hold for rational exponents as well as for integer exponents, then $a^{m/n}$ must be defined in such a way that

$$a^{m/n} = (a^{1/n})^m$$

In the definition we place the restriction that m and n are relatively prime, which means that they contain no common positive-integer factors other than 1.

DEFINITION $a^{m/n}$

If m and n are positive integers that are relatively prime and a is a real number, then if $\sqrt[n]{a}$ is a real number

$$a^{m/n} = (\sqrt[n]{a})^m \quad \Leftrightarrow \quad a^{m/n} = (a^{1/n})^m$$

The double arrow ⇔ is used to mean that the statement preceding it and the statement following it are equivalent.

▷ ILLUSTRATION 11

In the following we apply the definition of $a^{m/n}$.

(a) $9^{3/2} = (\sqrt{9})^3$ (b) $8^{2/3} = (\sqrt[3]{8})^2$ (c) $(-27)^{4/3} = (\sqrt[3]{-27})^4$
$\quad\quad = 3^3$ $\quad\quad\quad\quad = 2^2$ $\quad\quad\quad\quad\quad = (-3)^4$
$\quad\quad = 27$ $\quad\quad\quad\quad = 4$ $\quad\quad\quad\quad\quad = 81$ ◄

Because the commutative law holds for rational exponents, $(a^m)^{1/n} = (a^{1/n})^m$. Therefore, $a^{m/n}$ can be evaluated by finding either $(\sqrt[n]{a})^m$ or $\sqrt[n]{a^m}$. The computation of $(\sqrt[n]{a})^m$, however, is simpler than that for $\sqrt[n]{a^m}$. See Exercises 17 and 18.

The laws of positive-integer exponents are satisfied by positive-rational exponents with one exception. If $a < 0$, for certain values of p and q, $(a^p)^q \neq a^{pq}$.

▷ ILLUSTRATION 12

(a) $[(-9)^2]^{1/2} = 81^{1/2}$ and $(-9)^{2(1/2)} = (-9)^1$
$\quad\quad\quad\quad\quad\quad = 9$ $\quad\quad\quad\quad\quad\quad\quad = -9$

Therefore, $[(-9)^2]^{1/2} \neq (-9)^{2(1/2)}$.

(b) $[(-9)^2]^{1/4} = 81^{1/4}$ and $(-9)^{2(1/4)} = (-9)^{1/2}$ (not a real number)
$\quad\quad\quad\quad\quad\quad = 3$

Therefore, $[(-9)^2]^{1/4} \neq (-9)^{2(1/4)}$. ◄

The problems that arise in Illustration 12 are avoided by adopting the following rule: If m and n are positive even integers and a is a real number,

$$(a^m)^{1/n} = |a|^{m/n}$$

A particular case of this equality occurs when $m = n$. We then have

$$(a^n)^{1/n} = |a| \quad \text{(if } n \text{ is a positive even integer)}$$
$$\Leftrightarrow \quad \sqrt[n]{a^n} = |a| \quad \text{(if } n \text{ is even)}$$

If n is 2, we have

$$\sqrt{a^2} = |a|$$

▷ ILLUSTRATION 13

(a) $[(-9)^2]^{1/2} = |-9|$ (b) $[(-9)^2]^{1/4} = |-9|^{2/4}$
$\quad\quad\quad\quad\quad\quad = 9$ $\quad\quad\quad\quad\quad\quad\quad = 9^{1/2}$
$\quad\quad\quad\quad\quad\quad\quad\quad\quad\quad\quad\quad\quad = 3$ ◄

For the formula $a^{pq} = (a^p)^q$ to hold for negative rational exponents, we must have

$$a^{-m/n} = (a^{1/n})^{-m}$$

By the definition of a negative-integer exponent, if $a \neq 0$,

$$(a^{1/n})^{-m} = \frac{1}{(a^{1/n})^m}$$

Therefore, we give the following definition.

> **DEFINITION   A Negative Rational Exponent**
>
> If m and n are positive integers that are relatively prime, a is a real number, and $a \neq 0$, then if $\sqrt[n]{a}$ is a real number,
>
> $$a^{-m/n} = \frac{1}{a^{m/n}}$$

The first complete explanation of fractional and negative exponents was given by John Wallis (1616–1703) in 1655. Sir Isaac Newton (1642–1727) also used such exponents in his work.

Rational exponents (positive, negative, and zero) satisfy the laws of positive-integer exponents, with the understanding that $(a^m)^{1/n} = |a|^{m/n}$ when m and n are even integers.

▶ **EXAMPLE 3** *Simplifying an Algebraic Expression by Applying Laws of Exponents*

Simplify the expression. Each variable can be any real number.

(a) $(u^2v^4)^{1/4}$ **(b)** $[(-x)^2(y-3)^2]^{1/2}$

Solution

(a)
$$(u^2v^4)^{1/4} = (u^2)^{1/4}(v^4)^{1/4}$$
$$= |u|^{2/4}|v|^{4/4}$$
$$= |u|^{1/2}|v|$$

(b)
$$[(-x)^2(y-3)^2]^{1/2} = [(-x)^2]^{1/2}[(y-3)^2]^{1/2}$$
$$= |-x||y-3|$$
$$= |x||y-3|$$ ◀

▶ **EXAMPLE 4** *Simplifying an Algebraic Expression that Occurs in Calculus*

Simplify:

$$\frac{4x^3(x^2+1)^{1/2} - x^4[\frac{1}{2}(x^2+1)^{-1/2}(2x)]}{[(x^2+1)^{1/2}]^2}$$

Solution The two terms in the numerator contain the common factor $(x^2 + 1)^{-1/2}$.

$$\frac{4x^3(x^2 + 1)^{1/2} - x^4[\frac{1}{2}(x^2 + 1)^{-1/2}(2x)]}{[(x^2 + 1)^{1/2}]^2} = \frac{(x^2 + 1)^{-1/2}[4x^3(x^2 + 1) - x^5]}{x^2 + 1}$$

$$= \frac{4x^5 + 4x^3 - x^5}{(x^2 + 1)^{1/2}(x^2 + 1)}$$

$$= \frac{3x^5 + 4x^3}{(x^2 + 1)^{3/2}} \qquad \blacktriangleleft$$

  In calculus, you sometimes need to *rationalize the numerator* of a fraction; that is, you wish to obtain an equivalent fraction containing no radical in the numerator.

Recall the product

$$(a + b)(a - b) = a^2 - b^2$$

Each factor is called the *conjugate* of the other. The concept of conjugate is used to rationalize the numerator of a fraction when the numerator is a binomial containing a radical of order 2. The following example demonstrates the procedure.

 ▶ **EXAMPLE 5** *Rationalizing the Numerator of a Fraction That Occurs in Calculus*

Rationalize the numerator:

$$\frac{\sqrt{x + h} - \sqrt{x}}{h} \qquad \text{where } x > 0, x + h > 0, \text{ and } h \neq 0$$

Solution We multiply the numerator and denominator by the conjugate of the numerator.

$$\frac{\sqrt{x + h} - \sqrt{x}}{h} = \frac{(\sqrt{x + h} - \sqrt{x})(\sqrt{x + h} + \sqrt{x})}{h(\sqrt{x + h} + \sqrt{x})}$$

$$= \frac{(\sqrt{x + h})^2 - \sqrt{x}\sqrt{x + h} + \sqrt{x}\sqrt{x + h} - (\sqrt{x})^2}{h(\sqrt{x + h} + \sqrt{x})}$$

$$= \frac{(x + h) - x}{h(\sqrt{x + h} + \sqrt{x})}$$

$$= \frac{h}{h(\sqrt{x + h} + \sqrt{x})}$$

$$= \frac{1}{\sqrt{x + h} + \sqrt{x}}$$

EXERCISES 1.2

 In Exercises 1 through 8, the complex rational expression is similar to a type occurring in calculus. Find a simple rational expression equivalent to it. In each exercise, $h \neq 0$.

1. $\dfrac{\dfrac{1}{x + h} - \dfrac{1}{x}}{h}$

2. $\dfrac{\dfrac{3 + h}{4 + h} - \dfrac{3}{4}}{h}$

3. $\dfrac{\dfrac{1}{3x + 3h + 2} - \dfrac{1}{3x + 2}}{h}$

4. $\dfrac{\dfrac{1}{2x + 2h - 5} - \dfrac{1}{2x - 5}}{h}$

5. $\dfrac{\dfrac{1}{(4 + h)^2} - \dfrac{1}{16}}{h}$

6. $\dfrac{\dfrac{x + h}{(x + h)^2 + 1} - \dfrac{x}{x^2 + 1}}{h}$

7. $\dfrac{\dfrac{1}{(x + h)^3} - \dfrac{1}{x^3}}{h}$

8. $\dfrac{\dfrac{1}{(1 + h)^3 + 1} - \dfrac{1}{2}}{h}$

In Exercises 9 through 16, write the quantity as a rational number with an exponent of 1.

9. (a) $(3^2)^4$ **(b)** $(2 \cdot 3)^5$

   (c) $(\frac{7}{13})^2$ **(d)** $\left(\dfrac{2^3 \cdot 3^2}{5}\right)^3$

10. (a) $(2^2)^5$ **(b)** $(3 \cdot 5)^4$

   (c) $(\frac{3}{11})^3$ **(d)** $\left(\dfrac{2^4 \cdot 5^2}{7}\right)^2$

11. (a) $(-5)^{-3}$ **(b)** $(-6)^{-2}$ **(c)** $36^{1/2}$ **(d)** $27^{2/3}$

12. (a) $\dfrac{2}{4^{-3}}$ **(b)** $9 \cdot 3^0$ **(c)** $(-8)^{1/3}$ **(d)** $(\frac{2}{5})^{-4}$

13. (a) $(-\frac{1}{8})^{-2/3}$ **(b)** $-0.16^{3/2}$

   (c) $2^{-3} \cdot 7^{-1}$ **(d)** $2^{-4} - 2^4$

14. (a) $(-\frac{1}{125})^{-4/3}$ **(b)** $-0.0016^{-3/4}$

     $(-3)^{-4}$ **(d)** $(4^{-1} \cdot 2^{-3})^{-1}$

   (b) $[(-9)^2]^{1/2}$

   (b) $[(-36)^2]^{1/4}$

17. In Illustration 11, we applied the definition: $a^{m/n} = (\sqrt[n]{a})^m$. Perform the computations in parts (a)–(c) of Illustration 11 by letting $a^{m/n} = \sqrt[n]{a^m}$.

18. Simplify in two ways: (i) let $a^{m/n} = (\sqrt[n]{a})^m$; (ii) let $a^{m/n} = \sqrt[n]{a^m}$.
   (a) $4^{3/2}$ **(b)** $16^{3/4}$ **(c)** $(-125)^{2/3}$

In Exercises 19 through 24, simplify the expression where each variable can be any real number.

19. (a) $(x^6 y^8)^{1/4}$ **(b)** $(4s^4 t^{10})^{1/2}$

20. (a) $(9x^2 y^4)^{1/2}$ **(b)** $(a^4 b^{12})^{1/4}$

21. (a) $[(-3y)^4(y - 2)^2]^{1/2}$
   (b) $[(-2)^8(x - 2)^8(2 - y)^4]^{1/4}$

22. (a) $[(-2a)^4(b + 2)^8]^{1/4}$ **(b)** $[(-3)^6 x^2(x^2 + 9)^2]^{1/2}$

23. $[(-4)^4(u + 1)^8(u - 4)^4]^{1/4}$

24. $\left[\dfrac{(-5)^6 x^2}{(x^2 + 4)^2}\right]^{1/2}$

 In Exercises 25 through 30, the expression occurs in calculus. Simplify in a manner similar to Example 4.

25. $\dfrac{(x + 1)^{1/2} - x[\frac{1}{2}(x + 1)^{-1/2}]}{[(x + 1)^{1/2}]^2}$

26. $\dfrac{(x - 1)^{2/3} - x[\frac{2}{3}(x - 1)^{-1/3}]}{[(x - 1)^{2/3}]^2}$

27. $\dfrac{3x^2(2x + 5)^{1/2} - x^3[\frac{1}{2}(2x + 5)^{-1/2}(2)]}{[(2x + 5)^{1/2}]^2}$

28. $\dfrac{5(x^2 - 1)^{1/2} - 5x[\frac{1}{2}(x^2 - 1)^{-1/2}(2x)]}{[(x^2 - 1)^{1/2}]^2}$

29. $\dfrac{4x^3(3x^2 + 1)^{1/2} - x^4[\frac{1}{2}(3x^2 + 1)^{-1/2}(6x)]}{[(3x^2 + 1)^{1/2}]^2}$

30. $\dfrac{\frac{1}{3}(3x + 1)^{-2/3}(3)(2x - 3)^{1/2} - (3x + 1)^{1/3}[\frac{1}{2}(2x - 3)^{-1/2}(2)]}{[(2x - 3)^{1/2}]^2}$

 In Exercises 31 through 38, the expression occurs in calculus. Rationalize the numerator. All the radicands and all the variables represent positive numbers; none of the denominators is zero.

31. $\dfrac{\sqrt{4 + h} - 2}{h}$ **32.** $\dfrac{\sqrt{3 + h} - \sqrt{3}}{h}$

33. $\dfrac{\sqrt{2(x + h) + 1} - \sqrt{2x + 1}}{h}$

34. $\dfrac{\sqrt{3(x + h) - 2} - \sqrt{3x - 2}}{h}$

35. $\dfrac{\dfrac{1}{\sqrt{x + h}} - \dfrac{1}{\sqrt{x}}}{h}$

36. $\dfrac{\dfrac{2}{\sqrt{2h + 9}} - \dfrac{2}{3}}{h}$

37. $\dfrac{\dfrac{h + 2}{\sqrt{h + 1}} - 2}{h}$

38. $\dfrac{\dfrac{\sqrt{h + 2}}{h + 1} - \sqrt{2}}{h}$

In Exercises 39 and 40, use

$$a^3 - b^3 = (a - b)(a^2 + ab + b^2)$$

 39. Rationalize the numerator of $\dfrac{\sqrt[3]{x + h} - \sqrt[3]{x}}{h}$.

 40. Rationalize the numerator of $\dfrac{(h + 1)^{2/3} - 1}{h}$.

41. (a) Simplify the expression

$$(x^2 + 6x + 9)^{1/2} - (x^2 - 6x + 9)^{1/2}$$

(b) For what values of x is the expression in part (a) equivalent to 6?

42. (a) Simplify the expression

$$(x^2 - 8x + 16)^{1/2} + (x^2 + 8x + 16)^{1/2}$$

(b) For what values of x is the expression in part (a) equivalent to $2x$?

43. If $b^n = a$, and b and $\sqrt[n]{a}$ are real numbers, explain why we cannot conclude that $b = \sqrt[n]{a}$. In your explanation include a particular numerical example.

44. A positive-integer exponent of a real number a can be defined in the following way: $a^1 = a$, and if k is any positive integer such that a^k is defined, then $a^{k+1} = a^k \cdot a$. Explain why this definition is equivalent to

$$a^n = a \cdot a \cdot a \cdot \ldots \cdot a. \qquad (n \text{ factors of } a)$$

where n is any positive integer.

1.3 THE SET OF COMPLEX NUMBERS

GOALS

1. **Learn about the set of complex numbers.**
2. **Define a square root of any real number.**
3. **Define the principal square root of a negative number.**
4. **Learn properties of complex numbers.**
5. **Perform computations with complex numbers.**
6. **Compute powers of i.**

Even though in calculus you will work with real numbers only, it is important that you be familiar with the *set of complex numbers*. This fact will be apparent in the next chapter when complex numbers appear as solutions to some equations. Furthermore, complex roots of equations are used to solve certain differential equations in calculus. Consequently, we devote this section to an informal treatment of complex numbers.

In Section 1.2, we defined an *n*th root of a real number. This definition stated that if n is a positive integer greater than 1 and a and b are real numbers such that

$$b^n = a$$

then b is an *n*th root of a. We indicated that if $a < 0$ and n is an even pos' integer, there is no real *n*th root of a because an even power of a real ʳ is a nonnegative number. For instance, suppose we have the equation

$$b^2 = -25$$

There is no real number that can be substituted for b in this equation. Therefore, there is no real square root of -25. It follows in a similar manner that there is no real square root of any negative number.

To consider square roots of negative numbers, we must deal with numbers other than real numbers. We must develop a set of numbers that contains the set R of real numbers as a subset and also contains square roots of negative numbers. We denote such a set of numbers by C and refer to it as the set of *complex numbers*. We first require that the set C is such that the real number -1 has a square root. Let i be a symbol for a number in C whose square is -1; that is,

$$i^2 = -1$$

Because every real number is to be an element of C, it follows that if $b \in R$, then $b \in C$. We would like the set C to be closed under the operations of addition and multiplication. For the closure law of multiplication to hold, the number $b \cdot i$, abbreviated bi, must be an element of C. Furthermore, if $a \in R$, then $a \in C$, and if the closure law of addition is to hold, the number $a + bi$ must be an element of C. We now have a set C, which we call the set of **complex numbers.** With symbols, we write

$$C = \{a + bi \mid a, b \in R, i^2 = -1\}$$

For the complex number $a + bi$ the number a is called the **real part,** and the number b is called the **imaginary part.**

▷ ILLUSTRATION 1

(a) The number $-3 + 6i$ is a complex number whose real part is -3 and whose imaginary part is 6.
(b) The number $7 + (-4)i$ is a complex number whose real part is 7 and whose imaginary part is -4.   ◀

If $-p$ is a negative number, then the complex number $a + (-p)i$ can be written as $a - pi$. Hence

$$7 + (-4)i = 7 - 4i$$

A real number is a complex number whose imaginary part is 0; that is, if a is a real number,

$$a = a + 0i$$

Therefore R is a subset of C. Another subset of C is the set I of **imaginary numbers,** defined by

$$I = \{a + bi \mid a, b \in R, i^2 = -1, b \neq 0\}$$

The number $0 + bi$ can be written more simply as bi; that is,

$$bi = 0 + bi$$

This number is called a **pure imaginary number.**

▷ **ILLUSTRATION 2**

(a) The complex number $-5 + 2i$ is an imaginary number.
(b) The complex number $8i$ is a pure imaginary number.
(c) The real number -3 is a complex number, and it can be written as $-3 + 0i$.
(d) The real number 0 is a complex number, and it can be written as $0 + 0i$.

◀

The terminology *imaginary number* is an unfortunate but historical choice and arose from the fact that an equation such as $x^2 = -1$ has no solution in the set of real numbers. As far back as the fifteenth century, mathematicians found it desirable to treat an equation such as $x^2 = -1$ in the same manner as an equation such as $x^2 = 4$, which has real solutions. From their viewpoint a number whose square is -1 was not real but something imaginary. René Descartes in 1637 introduced the words *real* (*vraye*) and *imaginary* (*imaginaire*) in connection with sets of numbers, and Leonhard Euler (1707–1783) in 1748 used the symbol i to represent a number whose square is -1. The terminology and symbolism have become standard even though imaginary numbers are just as "real things" as the numbers 7, -4, and $\sqrt{3}$.

We now state a definition that allows every real number (positive, negative, or zero) to have a square root.

DEFINITION **A Square Root of Any Real Number**

A number s is said to be a **square root** of a real number r if and only if

$$s^2 = r$$

You have learned that any positive number has two square roots, one positive and one negative, and the number 0 has only one square root, 0. What about a square root of a negative number? In particular, consider the number -2. Because

$$(i\sqrt{2})^2 = i^2(\sqrt{2})^2 \quad \text{and} \quad (-i\sqrt{2})^2 = (-1)^2 i^2(\sqrt{2})^2$$
$$= (-1)(2) \qquad\qquad\qquad = (1)(-1)(2)$$
$$= -2 \qquad\qquad\qquad\qquad = -2$$

it follows from the definition that both $i\sqrt{2}$ and $-i\sqrt{2}$ are square roots of -2. More generally, if $-p$ is any negative number, then p is a positive number, and both $i\sqrt{p}$ and $-i\sqrt{p}$ are square roots of $-p$. As we did with square roots of positive numbers, we distinguish between the two square roots by using the concept of *principal square root*.

DEFINITION   **Principal Square Root of a Negative Number**

If p is a positive number, then the **principal square root** of $-p$, denoted by $\sqrt{-p}$, is defined by

$$\sqrt{-p} = i\sqrt{p}$$

The two square roots of $-p$ are written as $\sqrt{-p}$ and $-\sqrt{-p}$, or as $i\sqrt{p}$ and $-i\sqrt{p}$.

▷ ILLUSTRATION 3

(a) $\sqrt{-5} = i\sqrt{5}$
    The two square roots of -5 are $i\sqrt{5}$ and $-i\sqrt{5}$.
(b) $\sqrt{-16} = i\sqrt{16}$
    $\qquad\quad = 4i$
    The two square roots of -16 are $4i$ and $-4i$.
(c) $\sqrt{-1} = i\sqrt{1}$
    $\qquad\; = i$
    The two square roots of -1 are i and $-i$.   ◀

A complex number is said to be in **standard form** when it is written as $a + bi$ where a and b are real numbers.

▶ EXAMPLE 1   *Writing a Complex Number in Standard Form*

Write in the standard form $a + bi$. (a) $\sqrt{-9}$; (b) $5 - 6\sqrt{-4}$;
(c) $-\sqrt{\frac{16}{49}} + \sqrt{-25}$; (d) $\sqrt{24} + 5\sqrt{-27}$.

Solution

(a) $\sqrt{-9} = i\sqrt{9}$ (b) $5 - 6\sqrt{-4} = 5 - 6(i\sqrt{4})$
    $\qquad\quad = 3i$ $\qquad\qquad\quad\; = 5 - 6(2i)$
    $\qquad\quad = 0 + 3i$ $\qquad\qquad\quad\; = 5 + (-12i)$
(c) $-\sqrt{\frac{16}{49}} + \sqrt{-25} = -\frac{4}{7} + i\sqrt{25}$
    $\qquad\qquad\qquad = -\frac{4}{7} + 5i$

(d) $\sqrt{24} + 5\sqrt{-27} = \sqrt{4}\sqrt{6} + 5(i\sqrt{9}\sqrt{3})$

$$= 2\sqrt{6} + 5(i \cdot 3\sqrt{3})$$

$$= 2\sqrt{6} + 15\sqrt{3}i \qquad \blacktriangleleft$$

DEFINITION   Equality of Two Complex Numbers

Two complex numbers $a + bi$ and $c + di$ are said to be **equal** if and only if $a = c$ and $b = d$.

▷ **ILLUSTRATION 4**

If

$$x + 4i = -6 + yi$$

then $x = -6$ and $y = 4$. $\qquad \blacktriangleleft$

We wish to define addition and multiplication of complex numbers so that the axioms for these operations on the set of real numbers are valid. To arrive at such definitions, we consider two complex numbers $a + bi$ and $c + di$ as if they were polynomials in i and then simplify the result by letting $i^2 = -1$. Thus

$$(a + bi) + (c + di) = a + c + bi + di$$

$$= (a + c) + (b + d)i$$

and

$$(a + bi)(c + di) = ac + adi + bci + bdi^2$$

$$= ac + (ad + bc)i + bd(-1)$$

$$= (ac - bd) + (ad + bc)i$$

We have then the following definition.

DEFINITION   Sum and Product of Two Complex Numbers

If $a + bi$ and $c + di$ are complex numbers, then

$$(a + bi) + (c + di) = (a + c) + (b + d)i$$

$$(a + bi)(c + di) = (ac - bd) + (ad + bc)i$$

By using the preceding definition, it can be shown that the set C is closed under the operations of addition and multiplication. It also can ⸺ proved that addition and multiplication on C are commutative and ⸺ ciative and that multiplication is distributive over addition; th⸺

$u, v, w \in C$, then

$$u + v = v + u \qquad\qquad uv = vu$$
$$(u + v) + w = u + (v + w) \qquad (uv)w = u(vw)$$
$$u(v + w) = uv + uw$$

You are advised to compute with complex numbers, as in the following example, rather than memorize the definitions.

▶ **EXAMPLE 2** *Adding and Multiplying Complex Numbers*

Find the sum and product of the complex numbers.

$$5 - 4i \quad \text{and} \quad -2 + 6i$$

Solution

$$(5 - 4i) + (-2 + 6i) = 5 - 2 - 4i + 6i$$
$$= 3 + 2i$$
$$(5 - 4i)(-2 + 6i) = -10 + 30i + 8i - 24i^2$$
$$= -10 + 38i - 24(-1)$$
$$= -10 + 38i + 24$$
$$= 14 + 38i \qquad\qquad ◀$$

The additive identity element in the set of complex numbers is 0, which can be written as $0 + 0i$. The additive inverse of the complex number $a + bi$ is $-a - bi$ because

$$(a + bi) + (-a - bi) = [a + (-a)] + [b + (-b)]i$$
$$= 0 + 0i$$

Therefore

$$-(a + bi) = -a - bi$$

As with real numbers, subtraction of complex numbers is defined in terms of addition; that is,

$$(a + bi) - (c + di) = (a + bi) + [-(c + di)]$$
$$= (a + bi) + (-c - di)$$
$$= (a - c) + (b - d)i$$

▶ **EXAMPLE 3** *Subtracting Complex Numbers*

Find the difference of the complex numbers of Example 2.

Solution

$$(5 - 4i) - (-2 + 6i) = 5 - 4i + 2 - 6i$$
$$= 7 - 10i$$ ◀

Because

$$(a + bi)(1 + 0i) = a \cdot 1 + a \cdot 0i + 1 \cdot bi + bi \cdot 0i$$
$$= a + 0i + bi + 0i^2$$
$$= a + bi$$

it follows that the multiplicative identity in C is $1 + 0i$, which is the real number 1.

The **conjugate** of the complex number $a + bi$ is $a - bi$.

▷ **ILLUSTRATION 5**

(a) The conjugate of $3 + 2i$ is $3 - 2i$.
(b) The conjugate of $-4 - 5i$ is $-4 - (-5i)$, or, equivalently, $-4 + 5i$. ◀

Let us compute the sum and product of a complex number and its conjugate:

$$(a + bi) + (a - bi) = 2a \qquad (a + bi)(a - bi) = a^2 - b^2i^2$$
$$= a^2 - b^2(-1)$$
$$= a^2 + b^2$$

Observe that in each case we obtain a real number. The concept of the conjugate is useful in certain computations with complex numbers. For instance, to write the quotient

$$\frac{a + bi}{c + di}$$

in standard form $u + vi$, we multiply the numerator and denominator by the conjugate of the denominator. We do this in the following example.

▶ **EXAMPLE 4** *Dividing Complex Numbers*

Find the quotient of the complex numbers $5 - 4i$ and $-2 + 6i$.

Solution

$$\frac{5 - 4i}{-2 + 6i} = \frac{(5 - 4i)(-2 - 6i)}{(-2 + 6i)(-2 - 6i)}$$

$$= \frac{-10 - 22i + 24i^2}{4 - 36i^2}$$

$$= \frac{-10 - 22i + 24(-1)}{4 - 36(-1)}$$

$$= \frac{-34 - 22i}{40}$$

$$= -\frac{17}{20} - \frac{11}{20}i$$ ◀

The **multiplicative inverse** (or **reciprocal**) of the complex number $a + bi$ is defined as $\dfrac{1}{a + bi}$. We use the method of Example 4 to write the multiplicative inverse in standard form, as shown in the following illustration.

▷ **ILLUSTRATION 6**

The multiplicative inverse of $4 - 3i$ is $\dfrac{1}{4 - 3i}$. To write this complex number in the standard form $a + bi$, we multiply the numerator and denominator by the conjugate of $4 - 3i$. We have then

$$\frac{1}{4 - 3i} = \frac{1 \cdot (4 + 3i)}{(4 - 3i)(4 + 3i)}$$

$$= \frac{4 + 3i}{16 - 9i^2}$$

$$= \frac{4 + 3i}{16 - 9(-1)}$$

$$= \tfrac{4}{25} + \tfrac{3}{25}i$$

Thus the multiplicative inverse of $4 - 3i$ is $\tfrac{4}{25} + \tfrac{3}{25}i$. That is,

$$(4 - 3i)(\tfrac{4}{25} + \tfrac{3}{25}i) = \tfrac{16}{25} + \tfrac{12}{25}i - \tfrac{12}{25}i - \tfrac{9}{25}i^2$$

$$= \tfrac{16}{25} - \tfrac{9}{25}(-1)$$

$$= \frac{16 + 9}{25}$$

$$= 1$$ ◀

In summary, we have the following facts about the set C.

The Set of Complex Numbers

1. The set C is closed under the operations of addition and multiplication.
2. Addition and multiplication on C are commutative and associative; multiplication is distributive over addition.
3. There is an identity element for addition and an identity element for multiplication.
4. Every element in C has an additive inverse and every element in C, except $0 + 0i$, has a multiplicative inverse.

These facts are the field axioms discussed in Section A.1 of the appendix. Therefore, the set C is a field under the operations of addition and multiplication. Consequently, the laws of exponents apply to positive-integer powers of i.

▷ **ILLUSTRATION 7**

$$
\begin{aligned}
i^3 &= i^2 i & i^4 &= i^2 i^2 & i^5 &= i^4 i & i^6 &= i^4 i^2 \\
&= (-1)i & &= (-1)(-1) & &= (1)i & &= (1)(-1) \\
&= -i & &= 1 & &= i & &= -1 \quad ◀
\end{aligned}
$$

In Illustration 7 we see that we obtain the results i, $-i$, 1, and -1. By noting that $i^4 = 1$, we can find any positive-integer power of i, and it will be one of these four numbers obtained in Illustration 7.

▶ **EXAMPLE 5** *Computing Powers of i*

Find: **(a)** i^9; **(b)** i^{23}; **(c)** $\dfrac{1}{i^3}$.

Solution

(a)
$$
\begin{aligned}
i^9 &= i^8 i \\
&= (i^4)^2 i \\
&= (1)^2 i \\
&= i
\end{aligned}
$$

(b)
$$
\begin{aligned}
i^{23} &= i^{20} i^2 i \\
&= (i^4)^5 (-1)i \\
&= (1)^5 (-1)i \\
&= -i
\end{aligned}
$$

(c)
$$
\begin{aligned}
\frac{1}{i^3} &= \frac{1}{-i} \\
&= \frac{1 \cdot i}{-i^2} \\
&= \frac{i}{-(-1)} \\
&= i \quad ◀
\end{aligned}
$$

▷ **ILLUSTRATION 8**

$$\sqrt{-4}\sqrt{-25} = (i\sqrt{4})(i\sqrt{25})$$
$$= (2i)(5i)$$
$$= 10i^2$$
$$= -10 \qquad \blacktriangleleft$$

Observe in Illustration 8 that before multiplying we expressed $\sqrt{-4}$ and $\sqrt{-25}$ as $i\sqrt{4}$ and $i\sqrt{25}$, respectively. To avoid making an error, you should replace the symbol $\sqrt{-p}$ when $p > 0$ by $i\sqrt{p}$ before performing any multiplication or division (see Exercises 49 and 50).

▶ **EXAMPLE 6** *Multiplying and Dividing Complex Numbers Involving Radicals*

Perform the indicated operations and express the result in the form $a + bi$.

(a) $\sqrt{-5}(\sqrt{15} - \sqrt{-5})$ **(b)** $(2 - \sqrt{-9}) \div (2 + \sqrt{-9})$

Solution

(a) $\sqrt{-5}(\sqrt{15} - \sqrt{-5}) = i\sqrt{5}(\sqrt{3 \cdot 5} - i\sqrt{5})$
$$= i\sqrt{3 \cdot 5^2} - i^2\sqrt{5^2}$$
$$= i\sqrt{5^2}\sqrt{3} - (-1)\sqrt{5^2}$$
$$= 5 + 5i\sqrt{3}$$

(b) $\dfrac{2 - \sqrt{-9}}{2 + \sqrt{-9}} = \dfrac{(2 - 3i)(2 - 3i)}{(2 + 3i)(2 - 3i)}$

$$= \frac{4 - 6i - 6i + 9i^2}{4 - 9i^2}$$

$$= \frac{4 - 12i + 9(-1)}{4 - 9(-1)}$$

$$= \frac{-5 - 12i}{13}$$

$$= -\frac{5}{13} - \frac{12}{13}i \qquad \blacktriangleleft$$

EXERCISES 1.3

In Exercises 1 through 4, write the complex number in the form $a + bi$.

1. (a) 5   **(b)** $\sqrt{-49}$
   (c) $3 + \sqrt{-25}$ **(d)** $3 - \sqrt{-25}$

2. (a) -4 **(b)** $\sqrt{-36}$
   (c) $-2 + \sqrt{-16}$ **(d)** $-2 - \sqrt{-16}$

3. (a) $8 - 5\sqrt{-1}$ **(b)** $-8 + 5\sqrt{-1}$

    (c) $-\sqrt{36} + \sqrt{-36}$ **(d)** $\dfrac{1}{3} - \dfrac{1}{5}\sqrt{-45}$

4. (a) $-4 + \sqrt{-4}$ **(b)** $48 - \sqrt{-48}$

    (c) $2 - \sqrt{-\dfrac{25}{16}}$ **(d)** $54 + \sqrt{-162}$

In Exercises 5 through 36, perform the indicated operations and express the result in the form a + bi.

5. (a) $(5 + 2i) + (7 + i)$
    (b) $(3 - 6i) - (2 - 4i)$

6. (a) $(4 - 3i) + (-6 + 8i)$
    (b) $(7 + 10i) - (1 - 5i)$

7. (a) $(-9 - 4i) + (3 + 4i)$ **(b)** $7i - (5 - i)$

8. (a) $(3 + 8i) + (-3 - 6i)$ **(b)** $9 - (2 - 4i)$

9. $(5 + 2\sqrt{-9}) + (3 + 4\sqrt{-25})$

10. $(4 - 3\sqrt{-16}) + (-1 - \sqrt{-4})$

11. $(-3 - \sqrt{-20}) - (6 - \sqrt{-45})$

12. $(4 - \sqrt{-18}) - (2 - \sqrt{-2})$

13. (a) $\sqrt{-9}\sqrt{-25}$ **(b)** $\sqrt{-2}\sqrt{-8}$

14. (a) $\sqrt{-4}\sqrt{-16}$ **(b)** $\sqrt{-5}\sqrt{-75}$

15. $\sqrt{-12}\sqrt{-16}\sqrt{-27}$

16. $\sqrt{-27}\sqrt{-54}\sqrt{-162}$

17. $\sqrt{-8}(3\sqrt{-9} - \sqrt{-8})$

18. $\sqrt{-18}(\sqrt{-2} - 9\sqrt{-18})$

19. $(2 - 7i)(2 + 7i)$

20. $(4 - 3i)(-1 + 2i)$

21. $(3 + 2\sqrt{-3})(-2 + 3\sqrt{-3})$

22. $(2 - \sqrt{-2})(2 - 3\sqrt{-2})$

23. $(-3 - 3\sqrt{-3})^2$

24. $(\sqrt{-3} - \sqrt{-2})^2$ **25.** $-5 \div i$

26. $7 \div 3i$ **27.** $1 \div (2i - 3)$

28. $-4 \div (6 + i)$ **29.** $(3 + 2i) \div (2 - i)$

30. $(2 - 5i) \div 3i$ **31.** $(3 + 2i) \div 4i$

32. $(2 - 6i) \div (2i - 3)$ **33.** $1 \div (3 + \sqrt{-2})$

34. $(\sqrt{-5} - 3)(2\sqrt{-5} + 4)$

35. $1 \div (3 + 2i)^2$ **36.** $1 \div (4 - 2i)^2$

In Exercises 37 through 44, simplify the expression.

37. (a) i^{11} **(b)** i^{33} **(c)** i^{26}

38. (a) i^{22} **(b)** i^{37} **(c)** i^{47}

39. (a) $\dfrac{1}{i^5}$ **(b)** $\dfrac{1}{i^{15}}$ **(c)** $\dfrac{1}{i^6}$

40. (a) $\dfrac{1}{i^7}$ **(b)** $\dfrac{1}{i^9}$ **(c)** $\dfrac{1}{i^{14}}$

41. $(i^4 + i^3 - i^2 + 1)^2$

42. $(2i + 3i^3 - 4i^5 - i^7)^2$

43. $(2i + 3i^2 + 4i^3 - i^6)^3$

44. $(i - 1)^2 - (-i - 1)^2 + i^4$

In Exercises 45 through 48, find the value of the expression for the indicated value of x.

45. $x^2 - 2x + 3$; $x = 1 - i\sqrt{2}$

46. $x^2 - 2x + 4$; $x = 1 - \sqrt{-3}$

47. $4x^2 + 4x + 3$; $x = \frac{1}{2}(-1 + \sqrt{-2})$

48. $3x^2 - 2x + 2$; $x = \frac{1}{3}(1 - \sqrt{-5})$

49. Compute:

    (a) $\sqrt{4}\sqrt{25}$ and $\sqrt{4 \cdot 25}$

    (b) $\sqrt{-4}\sqrt{-25}$ and $\sqrt{(-4)(-25)}$

    (c) $\sqrt{4}\sqrt{-25}$ and $\sqrt{4(-25)}$

    (d) Under what conditions are $\sqrt{a} \cdot \sqrt{b}$ and $\sqrt{ab}$ equal?

50. Compute:

    (a) $\dfrac{\sqrt{16}}{\sqrt{4}}$ and $\sqrt{\dfrac{16}{4}}$

    (b) $\dfrac{\sqrt{-16}}{\sqrt{-4}}$ and $\sqrt{\dfrac{-16}{-4}}$

    (c) $\dfrac{\sqrt{-16}}{\sqrt{4}}$ and $\sqrt{\dfrac{-16}{4}}$

    (d) $\dfrac{\sqrt{16}}{\sqrt{-4}}$ and $\sqrt{\dfrac{16}{-4}}$

    (e) Under what conditions are $\dfrac{\sqrt{a}}{\sqrt{b}}$ and $\sqrt{\dfrac{a}{b}}$ equal?

1.4 THE NUMBER PLANE

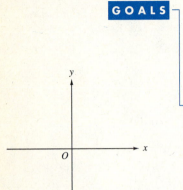

FIGURE 1

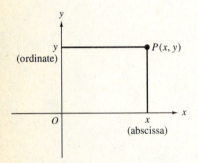

FIGURE 2

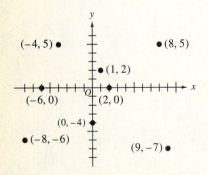

FIGURE 3

The origination of analytic geometry is credited to René Descartes (1596–1650), a French mathematician and philosopher. In his book *Geometry,* published in 1635, Descartes established the union of algebra and geometry by a *rectangular cartesian* (named for Descartes) *coordinate system.* This coordinate system utilizes *ordered pairs* of real numbers.

Any two real numbers form a pair, and when the order of appearance of the numbers is significant, we call it an **ordered pair.** If x is the first real number and y is the second, this ordered pair is denoted by (x, y). Observe that the ordered pair $(5, 2)$ is different from the ordered pair $(2, 5)$.

The set of all ordered pairs of real numbers is called the **number plane,** denoted by R^2, and each ordered pair (x, y) is a **point** in the number plane. Just as R, the set of real numbers, can be identified with points on an axis (a one-dimensional space), we can identify R^2 with points in a geometric plane (a two-dimensional space). A horizontal line, called the ***x* axis,** is chosen in the geometric plane. A vertical line, called the ***y* axis,** is selected, and the point of intersection of the x axis and the y axis is called the **origin,** denoted by the letter O. The units of measurement along the two axes are usually the same. We establish the positive direction on the x axis to the right of the origin, and the positive direction on the y axis above the origin. See Figure 1.

We now associate an ordered pair of real numbers (x, y) with a point in the geometric plane. At the point x on the horizontal axis and the point y on the vertical axis, line segments are drawn perpendicular to the respective axes. The intersection of these two perpendicular line segments is the point P associated with the ordered pair (x, y). Refer to Figure 2. The first number x of the pair is called the **abscissa** (or ***x* coordinate**) of P, and the second number y is called the **ordinate** (or ***y* coordinate**) of P. If the abscissa is positive, P is to the right of the y axis; and if it is negative, P is to the left of the y axis. If the ordinate is positive, P is above the x axis; and if it is negative, P is below the x axis. The abscissa and ordinate of a point are called the **rectangular cartesian coordinates** of the point. There is a one-to-one correspondence between the points in a geometric plane and R^2; that is, with each point there corresponds a unique ordered pair (x, y), and with each ordered pair (x, y) there is associated only one point. This one-to-one correspondence is called a **rectangular cartesian coordinate system.** Fig-

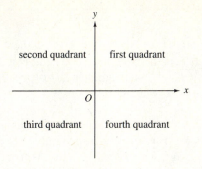

second quadrant | first quadrant

third quadrant | fourth quadrant

FIGURE 4

ure 3 illustrates a rectangular cartesian coordinate system with some points located.

The x and y axes are called **coordinate axes.** They divide the plane into four parts, called **quadrants.** The first quadrant is the one in which the abscissa and the ordinate are both positive, that is, the upper right quadrant. The other quadrants are numbered in the counterclockwise direction, with the fourth being the lower right quadrant. See Figure 4.

Because of the one-to-one correspondence, we identify R^2 with the geometric plane. For this reason we call an ordered pair (x, y) a point.

We now discuss the problem of finding the distance between two points in R^2. If A is the point (x_1, y_1) and B is the point (x_2, y_1) (that is, A and B have the same ordinate but different abscissas), then the **directed distance** from A to B is denoted by $\overline{AB}$, and we define

$$\overline{AB} = x_2 - x_1$$

▷ ILLUSTRATION 1

Refer to Figure 5(a), (b), and (c). If A is the point $(3, 4)$ and B is the point $(9, 4)$, then $\overline{AB} = 9 - 3$; that is, $\overline{AB} = 6$. If A is the point $(-8, 0)$ and B is the point $(6, 0)$, then $\overline{AB} = 6 - (-8)$; that is, $\overline{AB} = 14$. If A is the point $(4, 2)$ and B is $(1, 2)$, then $\overline{AB} = 1 - 4$; that is, $\overline{AB} = -3$. We see that $\overline{AB}$ is positive if B is to the right of A, and $\overline{AB}$ is negative if B is to the left of A.

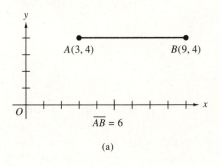

$\overline{AB} = 6$

(a)

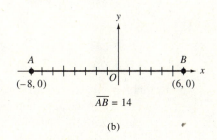

$\overline{AB} = 14$

(b)

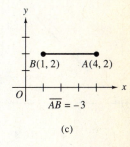

$\overline{AB} = -3$

(c)

FIGURE 5

If C is the point (x_1, y_1) and D is the point (x_1, y_2), then the **directed distance** from C to D, denoted by $\overline{CD}$, is defined by

$$\overline{CD} = y_2 - y_1$$

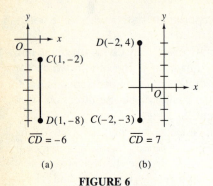

$\overline{CD} = -6$ $\overline{CD} = 7$

(a)   (b)

FIGURE 6

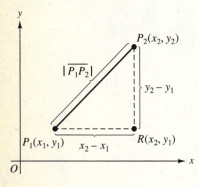

FIGURE 7

FIGURE 8

▷ **ILLUSTRATION 2**

Refer to Figure 6(a) and (b). If C is the point $(1, -2)$ and D is $(1, -8)$, then $\overline{CD} = -8 - (-2)$; that is, $\overline{CD} = -6$. If C is the point $(-2, -3)$ and D is $(-2, 4)$, then $\overline{CD} = 4 - (-3)$; that is, $\overline{CD} = 7$. The number $\overline{CD}$ is positive if D is above C, and $\overline{CD}$ is negative if D is below C.   ◀

Observe that the terminology *directed distance* indicates both a distance and a direction (positive or negative). If we are concerned only with the length of the line segment between two points P_1 and P_2 (that is, the distance between the points P_1 and P_2 without regard to direction), then we use the terminology *undirected distance*. We denote the **undirected distance** from P_1 to P_2 by $|\overline{P_1 P_2}|$, which is a nonnegative number. If we use the word *distance* without an adjective *directed* or *undirected,* it is understood that we mean an undirected distance.

We now wish to obtain a formula for computing the distance $|\overline{P_1 P_2}|$ if $P_1(x_1, y_1)$ and $P_2(x_2, y_2)$ are any two points in the plane. We use the Pythagorean theorem from plane geometry, which we now state. Refer to Figure 7.

THEOREM   *Pythagorean Theorem*
In a right triangle, if a and b are the lengths of the perpendicular sides and c is the length of the hypotenuse, then $$a^2 + b^2 = c^2$$

Figure 8 shows P_1 and P_2 in the first quadrant and the point $R(x_2, y_1)$. Note that $|\overline{P_1 P_2}|$ is the length of the hypotenuse of the right triangle $P_1 R P_2$. From the Pythagorean theorem, we have

$$|\overline{P_1 P_2}|^2 = |\overline{P_1 R}|^2 + |\overline{R P_2}|^2$$
$$|\overline{P_1 P_2}| = \sqrt{|\overline{P_1 R}|^2 + |\overline{R P_2}|^2}$$
$$|\overline{P_1 P_2}| = \sqrt{(x_2 - x_1)^2 + (y_2 - y_1)^2}$$

In this formula we do not have a $\pm$ symbol in front of the radical because $|\overline{P_1 P_2}|$ is a nonnegative number. The formula holds for all possible positions of P_1 and P_2 in all four quadrants. The length of the hypotenuse is always $|\overline{P_1 P_2}|$, and the lengths of the legs are always $|\overline{P_1 R}|$ and $|\overline{R P_2}|$. We have then the following formula.

Distance Formula

The distance between two points $P_1(x_1, y_1)$ and $P_2(x_2, y_2)$ is given by
$$|\overline{P_1 P_2}| = \sqrt{(x_2 - x_1)^2 + (y_2 - y_1)^2}$$

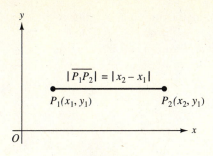

FIGURE 9

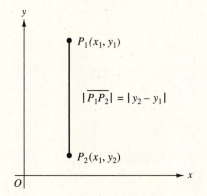

FIGURE 10

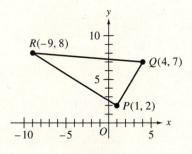

FIGURE 11

If P_1 and P_2 are on the same horizontal line, as in Figure 9, then $y_2 = y_1$ and

$$|\overline{P_1P_2}| = \sqrt{(x_2 - x_1)^2 + 0^2}$$
$$\Leftrightarrow \quad |\overline{P_1P_2}| = |x_2 - x_1| \quad \text{(because } \sqrt{a^2} = |a|\text{)}$$

Furthermore, if P_1 and P_2 are on the same vertical line, as in Figure 10, then $x_2 = x_1$, and

$$|\overline{P_1P_2}| = \sqrt{0^2 + (y_2 - y_1)^2}$$
$$\Leftrightarrow \quad |\overline{P_1P_2}| = |y_2 - y_1|$$

In the following example, we use the converse of the Pythagorean theorem.

THEOREM *Converse of the Pythagorean Theorem*

If a, b, and c are the lengths of the sides of a triangle and $a^2 + b^2 = c^2$, then the triangle is a right triangle, and c is the length of the hypotenuse.

▶ **EXAMPLE 1** *Applying the Distance Formula*

Prove that the points $P(1, 2)$, $Q(4, 7)$, and $R(-9, 8)$ are vertices of a right triangle.

Solution Refer to Figure 11 showing the triangle. We compute the lengths of the sides by applying the distance formula.

$$|\overline{PQ}| = \sqrt{(4 - 1)^2 + (7 - 2)^2} \qquad |\overline{PR}| = \sqrt{(-9 - 1)^2 + (8 - 2)^2}$$
$$= \sqrt{9 + 25} \qquad\qquad\qquad = \sqrt{100 + 36}$$
$$= \sqrt{34} \qquad\qquad\qquad\quad = \sqrt{136}$$
$$|\overline{QR}| = \sqrt{(-9 - 4)^2 + (8 - 7)^2}$$
$$= \sqrt{169 + 1}$$
$$= \sqrt{170}$$

Therefore

$$|\overline{PQ}|^2 + |\overline{PR}|^2 = 34 + 136$$
$$= 170$$
$$= |\overline{QR}|^2$$

Thus the triangle is a right triangle, and the hypotenuse is the side connecting the points Q and R. ◀

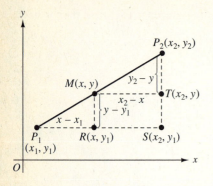

FIGURE 12

We now obtain the formulas for finding the midpoint of a line segment. Let $M(x, y)$ be the midpoint of the line segment from $P_1(x_1, y_1)$ to $P_2(x_2, y_2)$. Refer to Figure 12. Because triangles P_1RM and MTP_2 are congruent,

$$|\overline{P_1R}| = |\overline{MT}| \quad \text{and} \quad |\overline{RM}| = |\overline{TP_2}|$$

Thus

$$x - x_1 = x_2 - x \qquad y - y_1 = y_2 - y$$
$$2x = x_1 + x_2 \qquad 2y = y_1 + y_2$$
$$x = \frac{x_1 + x_2}{2} \qquad y = \frac{y_1 + y_2}{2}$$

Midpoint Formulas

If $M(x, y)$ is the midpoint of the line segment from $P_1(x_1, y_1)$ to $P_2(x_2, y_2)$, then

$$x = \frac{x_1 + x_2}{2} \qquad y = \frac{y_1 + y_2}{2}$$

In the derivation of the formulas, we assumed that $x_2 > x_1$ and $y_2 > y_1$. The same formulas are obtained by using any orderings of these numbers.

▶ **EXAMPLE 2** *Applying the Midpoint Formulas and the Distance Formula*

(a) Determine the coordinates of the midpoint M of the line segment from $A(5, -3)$ to $B(-1, 6)$.
(b) Locate the points A, M, and B, and show that $|\overline{AM}| = |\overline{MB}|$.

Solution
(a) From the midpoint formulas, if M is the point (x, y)

$$x = \frac{5 - 1}{2} \qquad y = \frac{-3 + 6}{2}$$

$$= 2 \qquad\qquad = \frac{3}{2}$$

Thus M is the point $\left(2, \frac{3}{2}\right)$.

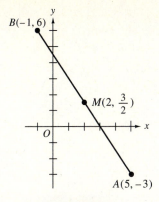

FIGURE 13

(b) Figure 13 shows the points A, M, and B. From the distance formula

$$|\overline{AM}| = \sqrt{(2-5)^2 + \left(\frac{3}{2}+3\right)^2} \qquad |\overline{MB}| = \sqrt{(-1-2)^2 + \left(6-\frac{3}{2}\right)^2}$$

$$= \sqrt{9 + \frac{81}{4}} \qquad\qquad = \sqrt{9 + \frac{81}{4}}$$

$$= \frac{3}{2}\sqrt{13} \qquad\qquad = \frac{3}{2}\sqrt{13}$$

Therefore $|\overline{AM}| = |\overline{MB}|$. ◀

Theorems from plane geometry can be proved by analytic geometry by using coordinates and techniques of algebra. The following example demonstrates the procedure.

▶ **EXAMPLE 3** *Proving a Theorem from Plane Geometry by Analytic Geometry*

Use analytic geometry to prove that the line segments joining the midpoints of the opposite sides of any quadrilateral bisect each other.

Solution   We draw a general quadrilateral. Because the coordinate axes can be chosen anywhere in the plane, and because the choice of the position of the axes does not affect the truth of the theorem, we take the origin at one vertex and the x axis along one side. These selections simplify the coordinates of the two vertices on the x axis. See Figure 14.

We now state the hypothesis and conclusion of the theorem.

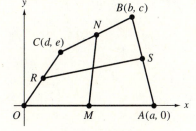

FIGURE 14

Hypothesis: $OABC$ is a quadrilateral. M is the midpoint of OA, N is the midpoint of CB, R is the midpoint of OC, and S is the midpoint of AB.

Conclusion: MN and RS bisect each other. ◀

Proof   To prove that two line segments bisect each other, we show that they have the same midpoint. From the midpoint formulas we obtain the coordinates of M, N, R, and S. The point M is $(\frac{1}{2}a, 0)$, N is the point $(\frac{1}{2}(b+d), \frac{1}{2}(c+e))$, R is the point $(\frac{1}{2}d, \frac{1}{2}e)$, and S is the point $(\frac{1}{2}(a+b), \frac{1}{2}c)$.

The abscissa of the midpoint of MN is $\frac{1}{2}[\frac{1}{2}a + \frac{1}{2}(b + d)] = \frac{1}{4}(a + b + d)$.

The ordinate of the midpoint of MN is $\frac{1}{2}[0 + \frac{1}{2}(c + e)] = \frac{1}{4}(c + e)$.

Therefore the midpoint of MN is the point $(\frac{1}{4}(a + b + d), \frac{1}{4}(c + e))$.

The abscissa of the midpoint of RS is $\frac{1}{2}[\frac{1}{2}d + \frac{1}{2}(a + b)] = \frac{1}{4}(a + b + d)$.

The ordinate of the midpoint of RS is $\frac{1}{2}[\frac{1}{2}e + \frac{1}{2}c] = \frac{1}{4}(c + e)$.

Therefore the midpoint of RS is the point $(\frac{1}{4}(a + b + d), \frac{1}{4}(c + e))$.

Thus the midpoint of MN is the same point as the midpoint of RS.

Therefore MN and RS bisect each other.   ■

EXERCISES 1.4

In Exercises 1 and 2, locate the point P on a rectangular cartesian coordinate system and state the quadrant in which it lies.

1. (a) $P(3, 7)$ **(b)** $P(-4, -6)$
    (c) $P(2, -5)$ **(d)** $P(-1, 4)$

2. (a) $P(5, 6)$ **(b)** $P(8, -1)$
    (c) $P(-7, -2)$ **(d)** $P(-9, 3)$

In Exercises 3 through 8, locate the point P and each of the following points as may apply: (a) The point Q such that the line through Q and P is perpendicular to the x axis and bisected by it. Give the coordinates of Q. (b) The point R such that the line through P and R is perpendicular to and bisected by the y axis. Give the coordinates of R. (c) The point S such that the line through P and S is bisected by the origin. Give the coordinates of S. (d) The point T such that the line through P and T is perpendicular to and bisected by the 45° line through the origin bisecting the first and third quadrants. Give the coordinates of T.

3. $P(1, -2)$ **4.** $P(-2, 2)$ **5.** $P(2, 2)$
6. $P(-2, -2)$ **7.** $P(-1, -3)$ **8.** $P(0, -3)$

In Exercises 9 through 12, describe the set of points P(x, y) on a rectangular cartesian coordinate system that satisfies the given condition.

9. (a) $y = 4$ **(b)** $x = -2$
    (c) $x > 0$ **(d)** $y \le 0$

10. (a) $x = 7$ **(b)** $y = -3$
    (c) $x \le 0$ **(d)** $y > 0$

11. (a) $x = 0$ **(b)** $x < 4$
    (c) $y \ge -2$ **(d)** $xy > 0$

12. (a) $y = 0$ **(b)** $x \ge 3$
    (c) $y < -1$ **(d)** $xy < 0$

In Exercises 13 through 16, for the points A and B, find the directed distances: (a) $\overline{AB}$; (b) $\overline{BA}$.

13. $A(-1, 7)$ and $B(6, 7)$ **14.** $A(-2, 3)$ and $B(-4, 3)$

15. $A(3, -4)$ and $B(3, -8)$ **16.** $A(-4, -5)$ and $B(-4, 6)$

17. If A is the point $(-2, 3)$ and B is the point $(x, 3)$, find x such that **(a)** $\overline{AB} = -8$; **(b)** $\overline{BA} = -8$.

18. If A is the point $(-4, y)$ and B is the point $(-4, 3)$, find y such that **(a)** $\overline{AB} = -3$; **(b)** $\overline{BA} = -3$.

In Exercises 19 through 22, do the following: (a) locate the points A and B and draw the line segment between them; (b) find the distance between A and B; (c) find the midpoint of the line segment from A to B.

19. $A(1, 3)$ and $B(-2, 7)$ **20.** $A(-4, -1)$ and $B(4, 5)$
21. $A(8, 5)$ and $B(3, -7)$ **22.** $A(6, -5)$ and $B(2, -2)$

In Exercises 23 through 26, do the following: (a) determine the coordinates of the midpoint M of the line segment from A to B; (b) locate the points A, M, and B and show that $|\overline{AM}| = |\overline{MB}|$.

23. $A(-4, 7)$ and $B(1, -3)$ **24.** $A(3, 4)$ and $B(4, -3)$
25. $A(1, 3)$ and $B(4, 0)$ **26.** $A(0, -2)$ and $B(2, 0)$

In Exercises 27 and 28, draw the triangle having vertices at A, B, and C and find the lengths of the sides.

27. $A(4, -5)$, $B(-2, 3)$, $C(-1, 7)$
28. $A(2, 3)$, $B(3, -3)$, $C(-1, -1)$

29. A median of a triangle is a line segment from a vertex to the midpoint of the opposite side. Find the length of the medians of the triangle having vertices $A(2, 3)$, $B(3, -3)$, and $C(-1, -1)$.

30. Find the length of the medians of the triangle having vertices $A(-3, 5)$, $B(2, 4)$, and $C(-1, -4)$.

31. Prove that the triangle with vertices $A(3, -6)$, $B(8, -2)$, and $C(-1, -1)$ is a right triangle. (*Hint:* Use the converse of the Pythagorean theorem.)

32. Find the midpoints of the diagonals of the quadrilateral whose vertices are $(0, 0)$, $(0, 4)$, $(3, 5)$, and $(3, 1)$.

33. Prove that the points $A(-7, 2)$, $B(3, -4)$, and $C(1, 4)$ are the vertices of an isosceles triangle.

34. Prove that the points $A(-4, -1)$, $B(-2, -3)$, $C(4, 3)$, and $D(2, 5)$ are the vertices of a rectangle.

35. By using the distance formula, prove that the points $(-3, 2)$, $(1, -2)$, and $(9, -10)$ lie on a line.

36. Determine whether the points $(14, 7)$, $(2, 2)$, and $(-4, -1)$ lie on a line by using the distance formula.

37. Prove that the points $A(6, -13)$, $B(-2, 2)$, $C(13, 10)$, and $D(21, -5)$ are the vertices of a square. Find the length of a diagonal.

38. If one end of a line segment is the point $(-4, 2)$ and the midpoint is $(3, -1)$, find the coordinates of the other end of the line segment.

39. If one end of a line segment is the point $(6, -2)$ and the midpoint is $(-1, 5)$, find the coordinates of the other end of the line segment.

40. By showing that the three sides have the same length, prove that the three points $A(-5, 0)$, $B(3, 0)$, and $C(-1, 4\sqrt{3})$ are the vertices of an equilateral triangle. Draw the triangle.

41. The abscissa of a point is -6, and its distance from the point $(1, 3)$ is $\sqrt{74}$. Find the ordinate of the point.

42. Given the two points $A(-3, 4)$ and $B(2, 5)$, find the coordinates of a point P on the line through A and B and not between A and B such that P is **(a)** twice as far from A as from B and **(b)** twice as far from B as from A.

In Exercises 43 through 46, use analytic geometry to prove the given theorem from plane geometry.

43. The lengths of the diagonals of a rectangle are equal.

44. The midpoint of the hypotenuse of any right triangle is equidistant from each of the vertices.

45. The line segment joining the midpoints of two opposite sides of any quadrilateral and the line segment joining the midpoints of the diagonals of the quadrilateral bisect each other.

46. For a parallelogram the sum of the squares of the lengths of the diagonals is equal to the sum of the squares of the lengths of the sides.

1.5 GRAPHS OF EQUATIONS

GOALS
1. Learn the definition of the graph of an equation.
2. Plot graphs of equations on a graphics calculator.
3. Choose the appropriate viewing rectangle on a graphics calculator.
4. Learn the definitions of symmetry.
5. Learn and apply the symmetry tests.

In Section 1.4 we demonstrated how a rectangular cartesian coordinate system can be used to obtain geometric facts by algebra. We now show how such a coordinate system enables us to associate a *graph* (a geometric concept) with an *equation* (an algebraic concept).

An **algebraic equation** in two variables x and y is a statement that two algebraic expressions in x and y are equal. When x and y are replaced by specific numbers, say a and b, the resulting statement may be either true or false. If it is true, the ordered pair (a, b) is called a solution of the equation.

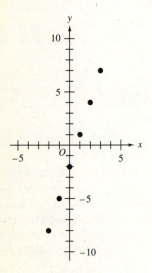

FIGURE 1

▷ **ILLUSTRATION 1**

Consider the equation

$$y = 3x - 2 \qquad\qquad (1)$$

where (x, y) is a point in R^2. If x is replaced by 2 in the equation, we see that $y = 4$; thus the ordered pair $(2, 4)$ is a solution. If any number is substituted for x in the right side of (1), a corresponding value for y is obtained. Therefore, (1) has an unlimited number of solutions. The solutions obtained from Table 1 are $(-2, -8)$, $(-1, -5)$, $(0, -2)$, $(1, 1)$, $(2, 4)$, and $(3, 7)$.

Table 1

x	-2	-1	0	1	2	3
$y = 3x - 2$	-8	-5	-2	1	4	7

◄

DEFINITION   **The Graph of an Equation**

The **graph of an equation** in R^2 is the set of all points in R^2 whose coordinates are solutions of the equation.

Because Equation (1) has an unlimited number of solutions, its graph consists of an unlimited number of points. The six points, given by Table 1 and shown in Figure 1, appear to lie on a line. In fact, you will learn in Section 3.1 that every solution of Equation (1) corresponds to a point on the line, and conversely, the coordinates of each point on the line satisfy (1). The line is, therefore, the graph of the equation. This graph is sketched in Figure 2, where the arrowheads indicate that the line continues in both directions. The coordinates of any point (x, y) on the line satisfy (1), and the coordinates of any point not on the line do not satisfy the equation.

High-speed automatic graphing devices, such as graphics calculators and computers with appropriate software, permit us to display graphs of equations in an instant. We shall take advantage of these devices and apply them through the text. They operate in a similar manner, but for student use, graphics calculators are obviously more practical than desktop computers. We shall, therefore, refer to *graphics calculators,* with the understanding that personal computers with graphing software can be used just as well.

Graphics calculators are not strictly automatic because they require a human operator to press specific keys, but because those keys depend on the manufacturer and model of the calculator, you should consult your calculator owner's manual for information on how to perform specific operations.

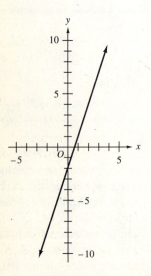

FIGURE 2

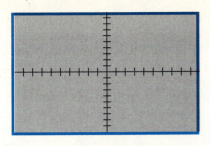

[−10, 10] by [−10, 10]
FIGURE 3

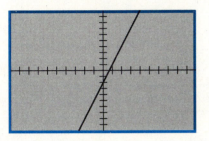

[−10, 10] by [−10, 10]
$y = 3x - 2$
FIGURE 4

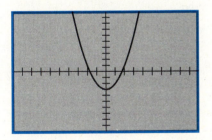

[−10, 10] by [−10, 10]
$y = x^2 - 3$
FIGURE 5

[−10, 10] by [−10, 10]
$y = x^2 + 14$
FIGURE 6

The display screen of your calculator shows a portion of the number plane R^2 called a **viewing rectangle**. The viewing rectangle, denoted $[X_{min}, X_{max}]$ by $[Y_{min}, Y_{max}]$ is the set of points in R^2 for which $X_{min} \leq x \leq X_{max}$ and $Y_{min} \leq y \leq Y_{max}$. Figure 3 shows the viewing rectangle $[-10, 10]$ by $[-10, 10]$, the standard default values, those used on most calculators if you do not specify otherwise. We shall call this rectangle the **standard viewing rectangle.** In Figure 3 the scale on each axis is 1 unit, the distance between each tick mark on the coordinate axes. Changing the scale affects the distance between the tick marks but does not alter the size of the viewing rectangle.

▷ **ILLUSTRATION 2**

The graph of the equation $y = 3x - 2$, plotted in the standard viewing rectangle on a graphics calculator, appears in Figure 4. Compare Figures 4 and 2, showing the same line.   ◄

As we proceed, we will obtain graphs in two ways: by hand and by a graphics calculator. When obtaining a graph by hand, as we did for Figure 2, we will use the terminology: *sketch the graph*. When obtaining a graph on a graphics calculator, as we did for Figure 4, we will state: *plot the graph*.

▷ **ILLUSTRATION 3**

The graph of the equation

$$y = x^2 - 3$$

plotted in the standard viewing rectangle appears in Figure 5.   ◄

▷ **ILLUSTRATION 4**

Consider the equation

$$y = x^2 + 14$$

If we attempt to plot the graph of this equation in the standard viewing rectangle we obtain Figure 6, which of course contains no points on the graph. This fact should come as no surprise because from the equation, the

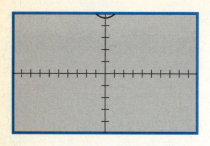

[−10, 10] by [−15, 15]
$y = x^2 + 14$
FIGURE 7

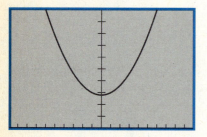

[−10, 10] by [0, 50]
$y = x^2 + 14$
FIGURE 8

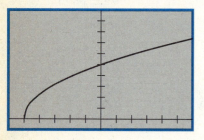

[−30, 30] by [−1, 10]
$y = \sqrt{x + 25}$
FIGURE 9

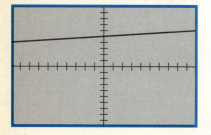

[−10, 10] by [−10, 10]
$y = \sqrt{x + 25}$
FIGURE 10

smallest value for y is 14, which occurs when $x = 0$. Changing the viewing rectangle to [−10, 10] by [−15, 15] gives Figure 7, which shows only a small portion of the graph. Figure 8 shows the graph plotted in a more convenient viewing rectangle, [−10, 10] by [0, 50]. Notice the scale on the y axis is 5 units, whereas the scale on the x axis is 1 unit. Very often it is helpful to change the scales on the axes when changing the viewing rectangle. ◄

The graphs of the equations in Illustrations 3 and 4, shown in Figures 5 and 8, respectively, are *parabolas*. We discuss *parabolas* in detail in Section 3.3.

▶ **EXAMPLE 1**   *Choosing the Appropriate Viewing Rectangle to Plot the Graph of an Equation*

Plot the graph of the equation

$$y = \sqrt{x + 25}$$

by choosing the appropriate viewing rectangle.

Solution   Because y is the principal square root of a number, y must be nonnegative. Furthermore, because the radicand is $x + 25$, which cannot be a negative number, x can be any number greater than or equal to −25. Thus an appropriate viewing rectangle would be [−30, 30] by [−1, 10]. Figure 9 shows the graph of the equation plotted in this viewing rectangle, where the scale on the x axis is 5. ◄

▷ **ILLUSTRATION 5**

If we attempt to plot the graph of the equation of Example 1 in the standard viewing rectangle we obtain Figure 10, which is a distorted view of the graph, resembling a line. Figure 11 shows another distorted view of the graph, plotted in the viewing rectangle [−10, 10] by [−30, 30], where the scale on the y axis is 5. ◄

Illustrations 4 and 5 and Example 1 demonstrate the importance of choosing the relevant viewing rectangle when plotting graphs on a graphics calculator.

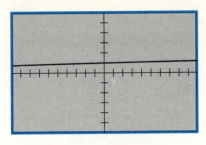

[−10, 10] by [−30, 30]
$y = \sqrt{x} + 25$
FIGURE 11

Most graphics calculators can plot graphs for more than one equation in the same viewing rectangle. We do this in the following example.

▶ **EXAMPLE 2**   *Obtaining the Same Curve by Hand and on a Graphics Calculator*

(a) Sketch the graph of the equation

$$y^2 = 4x$$

by locating the points for which x is 0, 1, 2, 3, and 4 and connecting these points by a curve.

(b) Plot the graphs of the two equations

$$y = 2\sqrt{x} \qquad \text{and} \qquad y = -2\sqrt{x}$$

in the same viewing rectangle.

(c) Why are the curves obtained in parts (a) and (b) identical?

Solution

(a) Because y^2 is nonnegative, values of x are restricted to nonnegative numbers. For each positive value of x there are two values of y. Table 2 gives the values of y when x is 0, 1, 2, 3, and 4. By locating the points whose coordinates are the x and y values in the table and connecting these points we obtain the graph sketched in Figure 12.

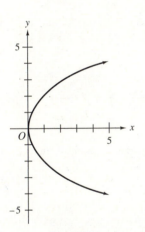

FIGURE 12

Table 2

x	0	1	1	2	2	3	3	4	4
$y^2 = 4x$	0	2	−2	$2\sqrt{2}$	$-2\sqrt{2}$	$2\sqrt{3}$	$-2\sqrt{3}$	4	−4

(b) On a graphics calculator we let

$$y_1 = 2\sqrt{x} \qquad \text{and} \qquad y_2 = -2\sqrt{x}$$

and plot the graphs of these two equations in the same viewing rectangle $[-1, 5]$ by $[-5, 5]$ as shown in Figure 13.

(c) The equation $y^2 = 4x$ is equivalent to the two equations $y = 2\sqrt{x}$ and $y = -2\sqrt{x}$. Therefore the union of the graphs of y_1 and y_2 plotted in part (b) is the same as the graph sketched in part (a). ◀

The curve in Example 2 is also a *parabola*.

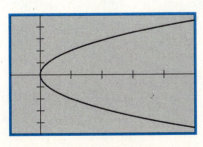

[−1, 5] by [−5, 5]
$y_1 = 2\sqrt{x}$ and $y_2 = -2\sqrt{x}$
FIGURE 13

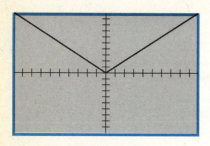

[−10, 10] by [−10, 10]
y = ABS (x)
FIGURE 14

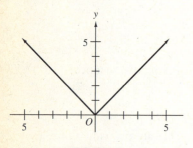

FIGURE 15

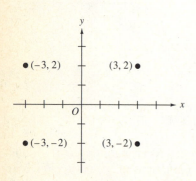

FIGURE 16

▶ **EXAMPLE 3** *Obtaining the Same Graph on a Graphics Calculator and by Hand*

(a) Plot the graph of the equation $y = |x|$.

(b) Sketch the graph of the equation in part (a).

Solution

(a) Computing the absolute value of a number is *built-in* on a graphics calculator and on most calculators is denoted by ABS. We let

$$y = \text{ABS}(x)$$

and plot the graph in the standard viewing rectangle shown in Figure 14.

(b) From the definition of the absolute value of a number,

$$y = \begin{cases} x & \text{if } x \geq 0 \\ -x & \text{if } x < 0 \end{cases}$$

Table 3 gives some values of x and y satisfying the equation, and the graph is sketched in Figure 15, which agrees with Figure 14.

Table 3

x	0	1	2	3	4	−1	−2	−3	−4
$y = |x|$	0	1	2	3	4	1	2	3	4

◀

Symmetry is an important property of graphs, especially helpful when sketching graphs.

DEFINITION   Symmetry of Two Points

Two points P and Q are said to be **symmetric with respect to a line** if and only if the line is the perpendicular bisector of the line segment PQ. Two points P and Q are said to be **symmetric with respect to a third point** if and only if the third point is the midpoint of the line segment PQ.

▷ **ILLUSTRATION 6**

The points $(3, 2)$ and $(3, -2)$ are symmetric with respect to the x axis, the points $(3, 2)$ and $(-3, 2)$ are symmetric with respect to the y axis, and the points $(3, 2)$ and $(-3, -2)$ are symmetric with respect to the origin. See Figure 16.

◀

In general, the points (x, y) and $(x, -y)$ are symmetric with respect to the x axis, (x, y) and $(-x, y)$ are symmetric with respect to the y axis, and (x, y) and $(-x, -y)$ are symmetric with respect to the origin.

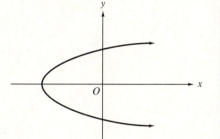

FIGURE 17

DEFINITION   Symmetry of a Graph

The graph of an equation is **symmetric with respect to a line** l if and only if for every point P on the graph there is a point Q also on the graph such that P and Q are symmetric with respect to l. The graph of an equation is **symmetric with respect to a point** R if and only if for every point P on the graph there is a point S also on the graph such that P and S are symmetric with respect to R.

Figure 17 shows a graph symmetric with respect to the x axis, Figure 18 shows one symmetric with respect to the y axis, and Figure 19 shows one symmetric with respect to the origin.

From the definition of symmetry of a graph, it follows that if a point (x, y) is on a graph symmetric with respect to the x axis, then the point $(x, -y)$ also must be on the graph. And if both the points (x, y) and $(x, -y)$ are on the graph, then the graph is symmetric with respect to the x axis. Therefore the coordinates of the point $(x, -y)$ as well as (x, y) must satisfy an equation of the graph. Hence the graph of an equation in x and y is symmetric with respect to the x axis if and only if an equivalent equation is obtained when y is replaced by $-y$ in the equation. We have thus proved part (i) in the following symmetry tests. The proofs of parts (ii) and (iii) are similar.

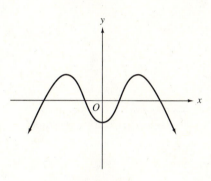

FIGURE 18

Symmetry Tests

The graph of an equation in x and y is

  (i) symmetric with respect to the x axis if and only if an equivalent equation is obtained when y is replaced by $-y$ in the equation;
 (ii) symmetric with respect to the y axis if and only if an equivalent equation is obtained when x is replaced by $-x$ in the equation;
(iii) symmetric with respect to the origin if and only if an equivalent equation is obtained when x is replaced by $-x$ and y is replaced by $-y$ in the equation.

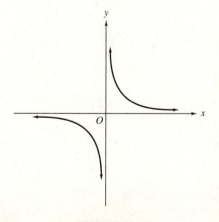

FIGURE 19

▷ **ILLUSTRATION 7**

Refer to the graph in Figure 5, symmetric with respect to the y axis and having the equation

$$y = x^2 - 3$$

If x is replaced by $-x$, we obtain the equation

$$y = (-x)^2 - 3$$
$$\Leftrightarrow \quad y = x^2 - 3$$ ◀

▷ **ILLUSTRATION 8**

The graph sketched in Figure 12 is symmetric with respect to the x axis, and its equation is

$$y^2 = 4x$$

Replacing y by $-y$ in this equation, we obtain

$$(-y)^2 = 4x$$
$$\Leftrightarrow \quad y^2 = 4x$$ ◀

▶ **EXAMPLE 4** *Applying the Symmetry Tests and Plotting a Graph*

Test for symmetry the graph of the equation

$$y = \frac{1}{2}x^3$$

Then plot the graph.

Solution We test for symmetry. If x is replaced by $-x$ and y is replaced by $-y$ in the given equation, we have

$$-y = \frac{1}{2}(-x)^3$$

$$-y = -\frac{1}{2}x^3$$

$$\Leftrightarrow \quad y = \frac{1}{2}x^3$$

Therefore, by symmetry test (iii), the graph is symmetric with respect to the origin. The graph is neither symmetric with respect to the x axis nor symmetric with respect to the y axis, as symmetry tests (i) and (ii) will verify.

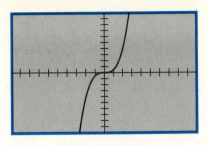

$[-10, 10]$ by $[-10, 10]$

$$y = \frac{1}{2}x^3$$

FIGURE 20

The graph of the given equation, plotted in the standard viewing rectangle, appears in Figure 20. Observe the symmetry with respect to the origin. ◀

EXERCISES 1.5

In Exercises 1 through 8, plot the graph of the equation in the standard viewing rectangle.

1. (a) $y = x + 2$ **(b)** $y = 5 - 2x$ **(c)** $y = 5$

2. (a) $y = 5 - x$ **(b)** $y = 4x - 3$ **(c)** $y = -4$

3. (a) $y = -\frac{1}{3}x - 6$ **(b)** $y = 4 + 7x$
(c) $y = -\frac{1}{2}x$

4. (a) $y = \frac{1}{2}x - 3$ **(b)** $y = 2x + 8$
(c) $y = \frac{1}{4}x$

5. (a) $y = x^2$ **(b)** $y = 2x^2$
(c) $y = x^2 + 2$

6. (a) $y = -x^2$ **(b)** $y = -2x^2$
(c) $y = -x^2 - 2$

7. (a) $y = 4 - x^2$ **(b)** $y = x^2 - 4$
(c) $y = (x - 4)^2$

8. (a) $y = x^2 + 2$ **(b)** $y = -x^2 - 2$
(c) $y = (x + 2)^2$

In Exercises 9 through 16, choose an appropriate viewing rectangle and plot the graph of the equation.

9. (a) $y = x + 15$ **(b)** $y = x - 15$
(c) $y = 15 - x$

10. (a) $y = x + 30$ **(b)** $y = x - 30$
(c) $y = 30 - x$

11. (a) $y = 7x + 20$ **(b)** $y = \frac{1}{9}x - 2$
(c) $y = -\frac{1}{5}x + 18$

12. (a) $y = 9x + 24$ **(b)** $y = \frac{1}{10}x - 5$
(c) $y = -\frac{1}{8}x + 12$

13. (a) $y = \sqrt{x + 16}$ **(b)** $y = \sqrt{x - 16}$
(c) $y = \sqrt{16 - x}$

14. (a) $y = \sqrt{x + 27}$ **(b)** $y = \sqrt{x - 27}$
(c) $y = \sqrt{27 - x}$

15. (a) $y = \sqrt{x} + 16$ **(b)** $y = \sqrt{x} - 16$
(c) $y = 16 - \sqrt{x}$

16. (a) $y = \sqrt{x} + 27$ **(b)** $y = \sqrt{x} - 27$
(c) $y = 27 - \sqrt{x}$

In Exercises 17 through 24, do the following: (a) test the graph of equation (i) for symmetry; (b) sketch the graph of equation (i); (c) plot the graphs of equations (ii) and (iii) in the same viewing rectangle; (d) compare the curves obtained in parts (b) and (c).

17. (i) $y^2 = 9x$ **(ii)** $y = 3\sqrt{x}$ **(iii)** $y = -3\sqrt{x}$

18. (i) $y^2 = \frac{1}{4}x$ **(ii)** $y = \frac{1}{2}\sqrt{x}$ **(iii)** $y = -\frac{1}{2}\sqrt{x}$

19. (i) $y^2 = -\frac{1}{4}x$ **(ii)** $y = \frac{1}{2}\sqrt{-x}$
(iii) $y = -\frac{1}{2}\sqrt{-x}$

20. (i) $y^2 = -4x$ **(ii)** $y = 2\sqrt{-x}$
(iii) $y = -2\sqrt{-x}$

21. (i) $y^2 = 1 - x^2$ **(ii)** $y = \sqrt{1 - x^2}$
(iii) $y = -\sqrt{1 - x^2}$

22. (i) $y^2 = 9 - x^2$ **(ii)** $y = \sqrt{9 - x^2}$
(iii) $y = -\sqrt{9 - x^2}$

23. (i) $y^2 = x^2 - 4$ **(ii)** $y = \sqrt{x^2 - 4}$
(iii) $y = -\sqrt{x^2 - 4}$

24. (i) $y^2 = x^2 - 16$ (ii) $y = \sqrt{x^2 - 16}$
(iii) $y = -\sqrt{x^2 - 16}$

In Exercises 25 and 26, sketch the graph of the equation.

25. (a) $y = |x - 2|$ (b) $y = |x + 2|$
(c) $y = |x| - 2$ (d) $y = |x| + 2$

26. (a) $y = 2|x - 3|$ (b) $y = 2|x + 3|$
(c) $y = 2|x| - 6$ (d) $y = 2|x| + 6$

In Exercises 27 through 30, test the graph of the equation for symmetry and then plot the graph.

27. (a) $y = x^3$ (b) $y = -x^3$
(c) $y = \frac{1}{8}x^3$ (d) $y = -\frac{1}{8}x^3$

28. (a) $y = 2x^3$ (b) $y = -2x^3$
(c) $y = \frac{1}{6}x^3$ (d) $y = -\frac{1}{6}x^3$

29. (a) $y = 2x^4$ (b) $y = -2x^4$
(c) $y = \frac{1}{4}x^4$ (d) $y = -\frac{1}{4}x^4$

30. (a) $y = 4x^4$ (b) $y = -4x^4$
(c) $y = \frac{1}{2}x^4$ (d) $y = -\frac{1}{2}x^4$

In Exercises 31 through 38, plot the graph of the equation. Experiment until you have an appropriate viewing rectangle giving the important characteristics of the graph.

31. $y = 4x^4 - 8x - 1$ **32.** $y = 2x^2 + 8x + 4$

33. $y = x^3 - 3x$ **34.** $y = 3x^3 + 2x - 5$

35. $y = 2x^4 - 50$ **36.** $y = 100 - x^4$

37. $y = x^4 - 8x^2 + 17$ **38.** $y = 4x^4 - 16x^2 - 20$

In Exercises 39 through 44, plot the graph of each equa-

tion in the same viewing rectangle.

39. $y = x^2$, $y = (x + 2)^2$, $y = (x - 2)^2$, $y = (x - 4)^2$

40. $y = x^2$, $y = x^2 + 2$, $y = x^2 - 2$, $y = x^2 - 4$

41. $y = x^2$, $y = 2x^2$, $y = -2x^2$, $y = 4x^2$

42. $y = x^2$, $y = \frac{1}{2}x^2$, $y = -\frac{1}{2}x^2$, $y = \frac{1}{4}x^2$

43. $y = |x|$, $y = |x| + 3$, $y = |x| - 3$, $y = |x| - 4$

44. $y = |x|$, $y = |x + 3|$, $y = |x - 3|$, $y = |x - 4|$

45. Describe how the graphs in Exercise 39 are similar and how they differ. Do the same for the graphs in Exercise 41.

46. Follow the instructions of Exercise 45 for the graphs in Exercises 40 and 42.

47. Follow the instructions of Exercise 45 for the graphs in Exercises 43 and 44.

48. Use the definitions of symmetry to explain why a graph symmetric with respect to both coordinate axes is also symmetric with respect to the origin.

49. Explain why in Illustration 5 we get distorted views of the graph of the equation $y = \sqrt{x + 25}$ in the standard viewing rectangle and in the viewing rectangle $[-10, 10]$ by $[-30, 30]$.

50. Plot the graph of $y = \sqrt{x + 50}$ in each of the following viewing rectangles: $[-10, 10]$ by $[-10, 10]$; $[-100, 100]$ by $[-10, 10]$; $[-10, 10]$ by $[-100, 100]$. Which one of these viewing rectangles is appropriate for showing the important characteristics of the graph? Explain why the other two rectangles give a distorted view of the graph.

CHAPTER 1   REVIEW

▶ LOOKING BACK

1.1   We discussed the set of real numbers and its subsets, the rational and irrational numbers. We associated the real numbers with points on an axis called the real-number line, and we used this geometric representation to define open and closed intervals. The absolute value of a real number was considered as the undirected distance of the number from the origin.

1.2   The algebraic expressions treated in this section were either polynomial or rational. We defined terms associated with these expressions and simplified various types of rational expressions that frequently occur in calculus.

1.3 The introduction to complex numbers was motivated by our need to consider the square root of a negative number. By showing that the set *C* of complex numbers satisfies the axioms of a field, we were then able to perform algebraic operations on these numbers.

1.4 After introducing the rectangular cartesian coordinate system in the plane, we used it to obtain the distance between two points and the midpoint of a line segment.

We then showed how to use the coordinate system to prove by analytic geometry some theorems from plane geometry.

1.5 The use of the graphics calculator to plot graphs of equations was emphasized in this section. We also discussed symmetry of graphs and tests for symmetry, which are important when sketching graphs by hand.

▶ REVIEW EXERCISES

In Exercises 1 through 4, N is the set of natural numbers, J is the set of integers, Q is the set of rational numbers, H is the set of irrational numbers, and R is the set of real numbers. In Exercises 1 and 2, determine which of these sets and Ø is equal to the given set. In Exercises 3 and 4, list the elements of the set if

$$S = \left\{-4, \sqrt{2}, 15, \frac{3}{4}, 7\pi, 0.5, 0\right\}$$

and

$$T = \left\{-\frac{1}{3}, 2, -\sqrt{3}, -5, 0.75, \frac{1}{2}, \frac{2}{3}\pi\right\}$$

1. (a) $N \cup Q$ (b) $Q \cap H$
     (c) $J \cap N$ (d) $Q \cup J$

2. (a) $Q \cap J$ (b) $J \cup N$
     (c) $J \cap H$ (d) $Q \cup H$

3. (a) $S \cap N$ (b) $S \cap J$
     (c) $T \cap Q$ (d) $T \cap H$

4. (a) $S \cap Q$ (b) $S \cap H$
     (c) $T \cap N$ (d) $T \cap J$

In Exercises 5 and 6, show the set on the real-number line and represent the set by interval notation.

5. (a) $\{x \mid x \le 4\}$ (b) $\{x \mid 2 < x < 8\}$
     (c) $\{x \mid x \le 0\} \cup \{x \mid x > 2\}$
     (d) $\{x \mid x > 1\} \cap \{x \mid x \le 5\}$

6. (a) $\{x \mid x > 5\}$ (b) $\{x \mid -2 < x \le 3\}$
     (c) $\{x \mid x \le -4\} \cup \{x \mid x \ge 4\}$
     (d) $\{x \mid x > -1\} \cap \{x \mid x < 6\}$

In Exercises 7 and 8, write the number without absolute-value bars.

7. (a) $|7|$ (b) $|-\sqrt{5}|$
     (c) $|2 - \sqrt{3}|$ (d) $|\sqrt{3} - 2|$

8. (a) $\left|\frac{1}{2}\right|$ (b) $|-\pi|$
     (c) $|\sqrt{10} - 3|$ (d) $|2\sqrt{5} - 5|$

In Exercises 9 and 10, show the interval on the real-number line and use set notation and inequality symbols to denote the interval.

9. (a) $(-4, 6)$ (b) $[3, 11]$
     (c) $[-10, 10)$ (d) $(-\infty, 0)$

10. (a) $[-8, 12]$ (b) $(-4, 4)$
     (c) $(5, 6]$ (d) $[0, +\infty)$

In Exercises 11 through 22, find the numerical value written without exponents.

11. (a) $(2^3)^2$ (b) $\left(\frac{1}{5}\right)^{-3}$

12. (a) $(-64)^{2/3}$ (b) $\left(\frac{1}{16}\right)^{-3/4}$

13. (a) $\left(\frac{5}{3^2}\right)^3$ (b) $(27)^{4/3}$

14. (a) $\left(\frac{4}{25}\right)^{-3/2}$ (b) $\frac{(-2)^{-3}}{(-7)^{-2}}$

15. $(1^0 + 2^0 + 3^0 + 4^0)^{-5/2}$ **16.** $2^0 + 2^{-1} + 2^{-2} + 2^{-3}$

17. $\dfrac{2^{-5} + 4^{-2}}{2^{-3} + 4^0}$ **18.** $(3^{-3} + 9^{-2})^{-1/2}$

19. $\dfrac{2^{-1} + 3^{-2}}{2^{-2}}$ **20.** $(2^2 + 3^2 \cdot 2^{-2})^{1/2}$

21. $\left(\dfrac{5^{-1}}{5^{-2} + 4 \cdot 5^{-3}}\right)^{-1/2}$ **22.** $\left(\dfrac{3^{-1} + 2^{-2}}{3^{-2} + 2^{-1}}\right)^{-1}$

In Exercises 23 and 24, use laws of exponents to simplify the expression where each variable can be any real number.

23. (a) $(4x^2y^4z^6)^{1/2}$ (b) $[(y + 1)^4(x^2 + 4)^4]^{1/4}$

24. (a) $(81r^4s^8t^{12})^{1/4}$ (b) $[(u^4 + 81)^2(v - 1)^2]^{1/2}$

In Exercises 25 through 30, perform the indicated operations and express the result in the form a + bi.

25. (a) $(8 + 3i) + (10 - 2i)$ **(b)** $(\frac{1}{2} - i) - (\frac{1}{4} - \frac{1}{3}i)$

26. (a) $(11 - 2i) - (-5 + 6i)$ **(b)** $(3 + \frac{2}{3}i) + (-1 - i)$

27. (a) $\sqrt{-9}\sqrt{-49}$ **(b)** $(-4 + 2i)(-3 + i)$

28. (a) $\sqrt{-8}\sqrt{-24}\sqrt{-48}$ **(b)** $(-5 - i)^2$

29. $(5 - 2i) \div (-4 - 3i)$

30. $i \div (-6 - i)$

In Exercises 31 and 32, simplify the expression.

31. (a) i^9 **(b)** i^{23} **(c)** i^{-10}

32. (a) i^{19} **(b)** i^{66} **(c)** i^{-7}

In Exercises 33 and 34, do the following: (a) locate the points A and B and draw the line segment between them; (b) find the distance between A and B; (c) find the midpoint of the line segment between A and B.

33. $A(7, -1)$ and $B(3, 2)$ **34.** $A(-5, 2)$ and $B(3, -4)$

In Exercises 35 and 36, do the following: (a) determine the coordinates of the midpoint M of the line segment between A and B; (b) locate the points A, M, and B and show that $|\overline{AM}| = |\overline{MB}|$.

35. $A(6, 5)$ and $B(-4, -9)$

36. $A(-1, -8)$ and $B(7, 6)$

37. Find the lengths of the sides of the triangle having vertices at $A(4, 1)$, $B(5, -5)$, and $C(1, -3)$.

38. Find the lengths of the medians of the triangle of Exercise 37.

39. Use the converse of the Pythagorean theorem to prove that the points $(2, 4)$, $(1, -4)$, and $(5, -2)$ are vertices of a right triangle and find the area of the triangle.

40. Prove that the triangle with vertices at $(-8, 1)$, $(-1, -6)$, and $(2, 4)$ is isosceles, and find its area.

In Exercises 41 and 42, use the distance formula to determine whether the three points lie on a line.

41. $(-2, -3)$, $(-1, 4)$, $(1, 17)$

42. $(0, -3)$, $(1, 4)$, $(2, 11)$

43. Use the distance formula to prove that the quadrilateral having vertices at $(1, 2)$, $(5, -1)$, $(11, 7)$, and $(7, 10)$ is a rectangle.

44. Use analytic geometry to prove that the length of the line segment joining the midpoints of any two sides of a triangle is one-half the length of the third side.

45. Use analytic geometry to prove that the lengths of two medians of an isosceles triangle are equal.

46. Use analytic geometry to prove that if the lengths of two medians of a triangle are equal, the triangle is isosceles.

In Exercises 47 through 50, the complex rational expression is similar to a type occurring in calculus. Find a simple rational expression equivalent to it. In each exercise, $h \neq 0$.

47. $\dfrac{\dfrac{5 + h}{7 + h} - \dfrac{5}{7}}{h}$ **48.** $\dfrac{\dfrac{1}{(3 + h)^2} - \dfrac{1}{9}}{h}$

49. $\dfrac{\dfrac{1}{2x + 2h + 1} - \dfrac{1}{2x + 1}}{h}$

50. $\dfrac{\dfrac{x + h + 1}{x + h - 2} - \dfrac{x + 1}{x - 2}}{h}$

In Exercises 51 and 52, the expression occurs in calculus. Simplify in a manner similar to Example 4 in Section 1.2.

51. $\dfrac{6x(2x + 1)^{1/2} - 3x^2[\frac{1}{2}(2x + 1)^{-1/2}(2)]}{[(2x + 1)^{1/2}]^2}$

52. $\dfrac{3x^2(x^2 - 1)^{1/3} - x^3[\frac{1}{3}(x^2 - 1)^{-2/3}(2x)]}{[(x^2 - 1)^{1/3}]^2}$

In Exercises 53 through 58, the rational expression occurs in calculus. Rationalize the numerator. All the radicands and all the variables represent positive numbers; none of the denominators is zero.

53. $\dfrac{\sqrt{9 + 2h} - 3}{h}$ **54.** $\dfrac{\sqrt{16 + h} - 4}{h}$

55. $\dfrac{\sqrt{x + h - 2} - \sqrt{x - 2}}{h}$

56. $\dfrac{\sqrt{3(x + h) + 1} - \sqrt{3x + 1}}{h}$

57. $\dfrac{\dfrac{h + 3}{\sqrt{3h + 4}} - \dfrac{3}{2}}{h}$ **58.** $\dfrac{\dfrac{\sqrt{h + 5}}{h + 3} - \dfrac{\sqrt{5}}{3}}{h}$

In Exercises 59 through 70, plot the graph of the equation in an appropriate viewing rectangle.

59. (a) $y = 2x - 7$ **(b)** $y = 7x + 2$

60. (a) $y = 9 - 4x$ **(b)** $y = 9x - 4$

61. (a) $y = \frac{1}{4}x^2$ **(b)** $y = -4x^2$

62. (a) $y = 8x^2$ **(b)** $y = -\frac{1}{8}x^2$

63. (a) $y = \frac{1}{4}x^3 - 2$ **(b)** $y = \frac{1}{4}(x - 2)^3$

64. (a) $y = \frac{1}{9}x^3 + 1$ **(b)** $y = \frac{1}{9}(x + 1)^3$

65. $y = 2x^2 - 7x - 15$ **66.** $y = 9x^2 - 12x + 4$

67. $y = 16 - x^4$ **68.** $y = 4x^3 - 6x + 1$

69. $y = 3x^3 - 4x^2 + 1$ **70.** $y = 2x^4 - 6x^2 + 7$

In Exercises 71 through 76, do the following: (a) test the graph of equation (i) for symmetry; (b) sketch the graph of equation (i); (c) plot the graphs of equations (ii) and (iii) in the same viewing rectangle; (d) compare the curves obtained in parts (b) and (c).

71. (i) $y^2 = \frac{1}{4}x$ **(ii)** $y = \frac{1}{2}\sqrt{x}$
    (iii) $y = -\frac{1}{2}\sqrt{x}$

72. (i) $y^2 = -9x$ **(ii)** $y = 3\sqrt{-x}$
    (iii) $y = -3\sqrt{-x}$

73. (i) $y^2 = -4(x - 3)$ **(ii)** $y = 2\sqrt{3 - x}$
    (iii) $y = -2\sqrt{3 - x}$

74. (i) $y^2 = \frac{1}{9}(x + 4)$ **(ii)** $y = \frac{1}{3}\sqrt{x + 4}$

    (iii) $y = -\frac{1}{3}\sqrt{x + 4}$

75. (i) $y^2 = x^2 - 25$ **(ii)** $y = \sqrt{x^2 - 25}$
    (iii) $y = -\sqrt{x^2 - 25}$

76. (i) $y^2 = 25 - x^2$ **(ii)** $y = \sqrt{25 - x^2}$
    (iii) $y = -\sqrt{25 - x^2}$

In Exercises 77 through 80, sketch the graph of the equation.

77. (a) $y = 4|x - 1|$ **(b)** $y = 4|x + 1|$
    (c) $y = 4|x| - 1$ **(d)** $y = 4|x| + 1$

78. (a) $y = 3|x - 4|$ **(b)** $y = 3|x + 4|$
    (c) $y = 3|x| - 4$ **(d)** $y = 3|x| + 4$

79. (a) $y = x|x|$ **(b)** $y = \dfrac{|x|}{x}$

    (c) $y = \dfrac{x}{|x|}$

80. (a) $y = |x| + x$ **(b)** $y = |x| - x$

    (c) $y = \dfrac{|x - 2|}{x - 2}$

81. Consider the graph of the equation $|x| + |y| = 1$. **(a)** Prove that the graph is symmetric with respect to both axes and the origin. **(b)** Sketch the portion of the graph in the first quadrant, and use the symmetry properties to sketch the complete graph. **(c)** The graph represents what geometrical figure?

82. Plot the graph of the equation in Exercise 81 on your graphics calculator, and write a description of how you did it.

83. Do parts (a) and (b) of Exercise 81 for the equation $|x| - |y| = 1$. Then plot the graph on your graphics calculator, and write a description of how you did it.

Equations and Inequalities

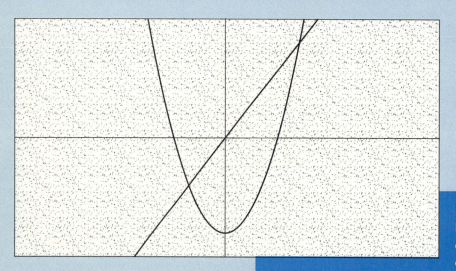

Facility in solving equations and in-equalities is crucial in calculus. In this chapter, we are concerned with solving equations and inequalities involving one unknown.

The first three sections constitute subject matter from intermediate algebra; you should review them at this time. Study carefully the procedures suggested in Section 2.3 for obtaining an equation as a mathematical model to solve a word problem.

Section 2.4 very likely contains material you will encounter for the first time. Sections 2.5–2.7 should be covered in detail. The procedures for obtaining solution sets of inequalities and those required to replace an inequality by an equivalent simpler one are needed in calculus.

2.1 LINEAR EQUATIONS IN ONE UNKNOWN

1. Solve linear equations algebraically.
2. Show the solution of a linear equation on a graph.
3. Solve linear equations on a graphics calculator.
4. Solve rational equations leading to a linear equation.
5. Obtain and apply an equation as a mathematical model.
6. Solve literal equations leading to a linear equation.

We introduced algebraic equations in two variables in Section 1.5. In this chapter we deal with algebraic equations in one variable, sometimes called an **unknown.** The **domain of the unknown** in an equation is the set of numbers for which the algebraic expressions in the equation are defined.

▷ ILLUSTRATION 1

In the following algebraic equations, let x be a real number:

$$x - 5 = 0 \tag{1}$$
$$x^2 + 12 = 7x \tag{2}$$
$$x + 5 = 5 + x \tag{3}$$
$$x + 2 = x + 3 \tag{4}$$
$$\frac{x}{x} = 1 \tag{5}$$
$$\frac{3}{x + 4} = \frac{2}{3x - 2} \tag{6}$$

For Equations (1) through (4), the domain is R. Because the left side of (5) is not defined if x is 0, the domain is the set of all real numbers except zero. The left side of (6) is not defined if x is -4, and the right side is not defined if x is $\frac{2}{3}$; therefore, the domain is the set of all real numbers except -4 and $\frac{2}{3}$. ◀

When the variable in an equation is replaced by a specific number, the resulting statement may be either true or false. If it is true, then that number is called a **solution** (or **root**) of the equation. The set of all solutions is called the **solution set** of the equation. A number that is a solution of an equation is said to **satisfy the equation.**

▷ **ILLUSTRATION 2**

(a) If in Equation (1), x is replaced by 5, the resulting statement is true, but if x is replaced by a number other than 5, the resulting statement is false. Thus the solution set is $\{5\}$.

(b) In Equation (2), if x is replaced by either 3 or 4, a true statement is obtained, and if x is replaced by a number other than 3 or 4, we get a false statement. Therefore, the equation has two solutions, 3 and 4, and the solution set is $\{3, 4\}$.

(c) If in Equation (3), any real number is substituted for x, a true statement is obtained. Hence the solution set is R.

(d) When any real number is substituted for x in Equation (4), we get a false statement. Thus the solution set is $\varnothing$.

(e) In Equation (5), if any number other than 0 is substituted for x, a true statement is obtained. Therefore, the solution set is $\{x \mid x \in R, x \neq 0\}$.

(f) The solution set of Equation (6) is not apparent by inspection; however, we show how to find it in Example 3.   ◄

If the solution set of any equation in one unknown is the same as the domain of the unknown, the equation is called an **identity.** Equations (3) and (5) are identities. If there is at least one number in the domain of the unknown that is not in the solution set of the equation, we have a **conditional equation.** Equations (1), (2), (4), and (6) are conditional equations.

An important type of equation is the polynomial equation in one unknown, which can be written in the form $P = 0$, where P is a polynomial in one variable. The degree of the polynomial is the degree of the equation. Particular examples of polynomial equations in one unknown are

$$7x - 21 = 0 \qquad \text{(first degree)}$$
$$2y^2 - 3y - 5 = 0 \qquad \text{(second degree)}$$
$$4z^3 - 8z^2 - z + 2 = 0 \qquad \text{(third degree)}$$
$$9w^4 - 13w^2 + 4 = 0 \qquad \text{(fourth degree)}$$

Determining the solution set of a conditional equation is called **solving an equation.** As shown by Equation (1) in Illustration 2, it is sometimes possible to solve an equation by inspection. In general, however, we use the concept of **equivalent equations,** which are equations having the same solution set. For example, the following equations are equivalent:

$$7x - 21 = 0 \qquad 7x = 21 \qquad x = 3$$

We can often solve an equation by replacing it by a succession of equivalent equations, each in some way simpler than the preceding one, so that we eventually obtain an equation for which the solution set is apparent. We can apply properties of real numbers to replace an equation by an equivalent one. For instance, the solution set of an equation is not changed if the same algebraic expression is added to or subtracted from both sides of an equation. Furthermore, the solution set is not changed if both sides of an equation are multiplied or divided by the same algebraic expression, provided that the algebraic expression is not zero. It is advisable to check a solution of an equation in case you have made an error in algebra or arithmetic. A solution is checked by substituting it into the original equation.

A first-degree polynomial equation is called a *linear equation* because, as you will learn in Section 3.1, its graph is a line.

DEFINITION   **A Linear Equation**

A linear equation in the variable x is an equation of the form

$$ax + b = 0$$

where a and b are real numbers and $a \neq 0$.

To solve the equation $ax + b = 0$ for x, we subtract b from both sides and then divide both sides by a, which we can do because $a \neq 0$. We have then the following equivalent equations:

$$ax + b = 0$$
$$ax + b - b = 0 - b$$
$$ax = -b$$
$$\frac{ax}{a} = \frac{-b}{a}$$
$$x = -\frac{b}{a}$$

We have proved the following theorem.

THEOREM 1

The linear equation $ax + b = 0$ (where $a \neq 0$) has exactly one solution, $-\dfrac{b}{a}$.

To show the solution of the linear equation

$$ax + b = 0$$

on a graph, we set y equal to the left-hand side and sketch the graph of that equation. The graph is a line that intersects the x axis at the point where

$x = -\dfrac{b}{a}$, the solution of the equation. The number $-\dfrac{b}{a}$ is called the **x intercept** of the line.

▶ **EXAMPLE 1** *Solving a Linear Equation Algebraically and Showing the Solution on a Graph*

Solve the equation

$$5x - 2x - 11 = 2 + x - 8$$

and show the solution on a graph.

Solution   We combine terms on each side of the equation and then add $-x$ and 11 to both sides. We have the following equivalent equations:

$$3x - 11 = x - 6$$
$$3x - 11 - x + 11 = x - 6 - x + 11$$
$$3x - x = 11 - 6$$
$$2x = 5$$
$$x = \frac{5}{2}$$

Therefore the solution set is $\{\frac{5}{2}\}$. We check the solution. Does

$$5\left(\frac{5}{2}\right) - 2\left(\frac{5}{2}\right) - 11 = 2 + \frac{5}{2} - 8?$$

$$5\left(\frac{5}{2}\right) - 2\left(\frac{5}{2}\right) - 11 = \frac{25}{2} - 5 - 11 \qquad\qquad 2 + \frac{5}{2} - 8 = \frac{5}{2} - 6$$

$$= -\frac{7}{2} \qquad\qquad\qquad\qquad\qquad = -\frac{7}{2}$$

Thus the solution checks.

Because the given equation is equivalent to $2x - 5 = 0$, we plot the graph of

$$y = 2x - 5$$

on a graphics calculator and obtain the line shown in Figure 1. The x intercept of this line is 2.5, the solution of the equation.   ◀

A strictly graphical method of solving the equation of Example 1 appears in the next example.

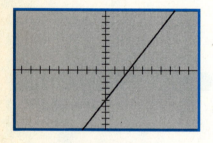

[−10, 10] by [−10, 10]
$y = 2x - 5$
FIGURE 1

▶ **EXAMPLE 2** *Solving a Linear Equation on a Graphics Calculator*

Use a graphics calculator to solve the equation of Example 1.

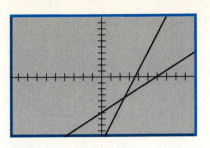

$[-10, 10]$ by $[-10, 10]$
$y_1 = 5x - 2x - 11$ and $y_2 = 2 + x - 8$

FIGURE 2

Solution We let

$$y_1 = 5x - 2x - 11 \quad \text{and} \quad y_2 = 2 + x - 8$$

and plot the graph of these two equations in the same viewing rectangle. We obtain the lines shown in Figure 2. The x coordinate of the point of intersection of the lines is 2.5, which agrees with our solution in Example 1. ◀

To solve an equation involving rational expressions, we multiply both sides of the equation by the LCD (lowest common denominator) of the fractions as shown in the following example.

▶ **EXAMPLE 3** *Solving a Rational Equation*

Find the solution set of the equation

$$\frac{3}{x + 4} = \frac{2}{3x - 2}$$

Solution Observe that when x is -4 or $\frac{2}{3}$, we obtain 0 in the denominator of one of the fractions in the given equation. Therefore, -4 and $\frac{2}{3}$ are not in the domain of the unknown x.

The LCD is $(x + 4)(3x - 2)$. Multiplying both sides of the equation by the LCD, we get

$$(x + 4)(3x - 2)\frac{3}{x + 4} = (x + 4)(3x - 2)\frac{2}{3x - 2}$$

$$3(3x - 2) = 2(x + 4)$$

$$9x - 6 = 2x + 8$$

$$9x - 6 + 6 - 2x = 2x + 8 + 6 - 2x$$

$$9x - 2x = 8 + 6$$

$$7x = 14$$

$$x = 2$$

Therefore the solution set is $\{2\}$. We check the solution. Does $\frac{3}{2 + 4} = \frac{2}{3(2) - 2}$?

$$\frac{3}{2 + 4} = \frac{3}{6} \qquad\qquad \frac{2}{3(2) - 2} = \frac{2}{4}$$

$$= \frac{1}{2} \qquad\qquad\qquad = \frac{1}{2}$$

Thus the solution checks. ◀

When multiplying both sides of an equation by the LCD, the resulting equation may not be equivalent to the given one. When this happens, you can obtain a number as a possible solution that does not satisfy the original equation because it is not in the domain of the unknown. This situation occurs in Exercises 25 and 26.

In the following example, we obtain an equation as an algebraic representation of a practical situation. The equation obtained is called a *mathematical model* of the practical situation. Equations as mathematical models are discussed in detail in Section 2.3.

▶ **EXAMPLE 4** *Obtaining and Applying an Equation as a Mathematical Model*

The cost of hiring a one-passenger plane from a private airline is $310 for the plane and pilot plus 75 cents per mile flown. What is the cost of a trip of **(a)** 100 miles; **(b)** 200 miles; and **(c)** 300 miles? **(d)** If y dollars is the cost of a trip of x miles, write an equation involving y and x. Use the equation in part (d) to determine **(e)** the cost for a trip of 270 miles; and **(f)** the length of a trip costing $720.

Solution

(a) The cost of a trip of 100 miles is $310 + $0.75(100) or $385.

(b) The cost of a trip of 200 miles is $310 + $0.75(200) or $460.

(c) The cost of a trip of 300 miles is $310 + $0.75(300) or $535.

(d) To obtain an equation we observe in the answers to parts (a), (b), and (c) that the number of dollars in the total cost of a trip is obtained by adding 310 and the product of 0.75 and the number of miles in the length of the trip. Therefore if y dollars is the cost of a trip of x miles,

$$y = 310 + 0.75x$$

(e) For a trip of 270 miles, $x = 270$ and

$$y = 310 + 0.75(270)$$
$$= 512.50$$

The cost of the trip is, therefore, $512.50.

(f) For a trip costing $720, $y = 720$ and

$$720 = 310 + 0.75x$$
$$0.75x = 720 - 310$$
$$0.75x = 410$$
$$x = \frac{410}{0.75}$$
$$x = 546.67$$

The length of the trip is, therefore, 546.67 miles.   ◀

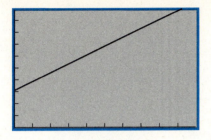

[0, 1000] by [0, 1000]
$y = 310 + 0.75x$
FIGURE 3

Figure 3 shows the graph of the equation in Example 4 plotted in the viewing rectangle [0, 1000] by [0, 1000] with a scale of 100 on both axes. In Exercise 40 you are asked to zoom in on this graph on your graphics calculator to estimate the answers in parts (e) and (f) of Example 4.

An equation may contain more than one variable or it may contain symbols such as a and b, representing constants. An equation of this type is sometimes called a **literal equation**. To solve for one of the variables in terms of the other variables or symbols, we treat the variable for which we are solving as the unknown and the other variables and symbols as known.

▶ **EXAMPLE 5** *Solving a Literal Equation*

If F degrees is the Fahrenheit temperature reading and C degrees is the Celsius temperature reading, then

$$F = \frac{9}{5}C + 32$$

Solve this equation for C.

Solution   We first multiply each side by 5 and obtain

$$5F = 9C + 160$$
$$5F - 160 = 9C$$
$$9C = 5(F - 32)$$
$$C = \frac{5}{9}(F - 32)$$

◀

▶ **EXAMPLE 6** *Solving a Literal Equation*

Solve the equation

$$y = xtz + xyz$$

(a) for x and **(b)** for y.

Solution
(a) To solve for x, we first factor the right side.

$$y = x(tz + yz)$$

We now divide both sides by $tz + yz$ and obtain

$$x = \frac{y}{tz + yz} \qquad (tz + yz \neq 0)$$

(b) To solve for y, we add $-xyz$ to each side and then factor the left side, and we get

$$y - xyz = xtz$$
$$y(1 - xz) = xtz$$

Dividing both sides by $1 - xz$, we have

$$y = \frac{xtz}{1 - xz} \qquad (xz \neq 1)$$

◄

EXERCISES 2.1

In Exercises 1 through 4, find the solution set of the equation and show the solution on a graph on your graphics calculator.

1. $2x + 7 = 0$ **2.** $3x - 2 = 0$

3. $5x - 8 = 0$ **4.** $4x + 9 = 0$

In Exercises 5 through 14, solve the equation two ways: (i) on your graphics calculator as in Example 2; (ii) algebraically.

5. $7x + 4 = 25$ **6.** $7 = y + 10$

7. $4w - 3 = 11 - 3w$ **8.** $x - 9 = 3x + 3$

9. $2(t - 5) = 3 - (4 + t)$

10. $1 - 3(2x - 4) = 4(6 - x) - 8$

11. $x = x + 1$ **12.** $x + 3 = 1 + x + 2$

13. $3(4y + 9) = 7(2 - 5y) - 2y$

14. $-2[s - (5 - 4s)] + 4 = -3s$

In Exercises 15 through 24, find the solution set of the equation.

15. $\dfrac{3x - 2}{3} + \dfrac{x - 3}{2} = \dfrac{5}{6}$ **16.** $\dfrac{3}{8} + \dfrac{1}{2x} = \dfrac{2}{x}$

17. $\dfrac{5}{2y} - \dfrac{1}{y} = \dfrac{3}{4}$ **18.** $\dfrac{1}{4 - t} + \dfrac{3}{6 + t} = 0$

19. $\dfrac{2}{3x - 4} = \dfrac{5}{6x - 7}$

20. $\dfrac{3}{x^2 - 9} - \dfrac{7}{x - 3} = -\dfrac{4}{x + 3}$

21. $\dfrac{4}{25w^2 - 1} + \dfrac{3}{5w - 1} = \dfrac{2}{5w - 1}$

22. $\dfrac{3}{x^2 - x - 6} = \dfrac{4}{2x^2 + x - 6}$

23. $\dfrac{5}{x^2 + 6x - 7} = \dfrac{2}{x^2 - 1}$ **24.** $\dfrac{2}{y + 1} - \dfrac{3}{1 - y} = \dfrac{5}{y}$

In Exercises 25 and 26, show that the solution set of the equation is $\varnothing$.

25. $\dfrac{2x + 16}{x^2 + 16x + 55} + \dfrac{4}{x + 5} = \dfrac{1}{x + 11}$

26. $\dfrac{1}{2x + 5} - \dfrac{4}{2x - 1} = \dfrac{4x + 4}{4x^2 + 8x - 5}$

In Exercises 27 through 32, solve for x or y in terms of the other symbols.

27. $3ax + 6ab = 7ax + 3ab$

28. $\dfrac{a + 3x}{b} = \dfrac{c}{2}$

29. $a(y - a) - 2b(y - 3b) = ab$

30. $5a(5a + x) = 2a(2a - x)$

31. $\dfrac{x + b}{3a - 4b} = \dfrac{x - a}{2a - 5b}$ **32.** $\dfrac{1}{c - y} + \dfrac{2}{c + y} = \dfrac{1}{y}$

In Exercises 33 through 39, solve for the indicated quantity in the given formula.

33. $A = \frac{1}{2}(a + b)h$; for h **34.** $A = \frac{1}{2}(a + b)h$; for b

35. $E = I(R + r)$; for r **36.** $A = P\left(1 + \frac{r}{n}\right)$; for r

37. $\frac{1}{f} = \frac{1}{p} + \frac{1}{q}$; for p **38.** $E = I\left(R + \frac{r}{n}\right)$; for n

39. $S = \frac{a - rl}{1 - r}$; for r

40. Zoom in on the graph of Figure 3 on your graphics calculator to estimate the answers in parts (e) and (f) of Example 4.

41. Suppose that the gasoline tank of your automobile has a capacity of 16 gallons and you can drive 20 miles on one gallon of gas. If you start on a trip with a full tank, how many gallons are left in the tank after you have driven **(a)** 100 miles; **(b)** 200 miles; and **(c)** 300 miles? **(d)** If y gallons of gasoline are left in the tank after you have driven x miles, write an equation relating y and x. Use your answer in part (d) to determine **(e)** how much gasoline is left in the tank after you have driven 240 miles, **(f)** how far you have driven when your fuel gauge indicates you have one-fourth of a tank of gasoline left, and **(g)** how far you can drive on a full tank before you run out of gas.

42. The cost of long distance telephone service from Mendocino to San Francisco at full weekday rates is 40 cents for the first minute and 31 cents for each additional minute. What is the cost of a call lasting **(a)** 4 minutes; **(b)** 7 minutes; and **(c)** 10 minutes? **(d)** If y dollars is the cost of a call lasting x minutes, write an equation relating y and x. Use your equation in part (d) to determine **(e)** the cost of a call lasting 16 minutes, and **(f)** the length of a call costing $4.74.

43. Plot the graph of your equation in part (d) of Exercise 41 on your graphics calculator. Zoom in on this graph to estimate the answers in parts (e), (f), and (g) of Exercise 41.

44. Plot the graph of your equation in part (d) of Exercise 42 on your graphics calculator. Zoom in on this graph to estimate the answers in parts (e) and (f) of Exercise 42.

45. Find the solution set of each of the following equations, if x is a real number:

(a) $\frac{x}{x} = 1$ **(b)** $\frac{x}{x} = 0$ **(c)** $\frac{x}{x} = 2$

46. Find the solution set of each of the following equations, if x is a real number:

(a) $\frac{x - 2}{x - 2} = 1$ **(b)** $\frac{x - 2}{x - 2} = 0$

(c) $\frac{x - 2}{x - 2} = 2$

2.2 QUADRATIC EQUATIONS IN ONE UNKNOWN

GOALS

1. Learn the zero-factor theorem.
2. Solve quadratic equations by factoring.
3. Show the real-number solutions of a quadratic equation on a graph.
4. Solve quadratic equations by square root.
5. Complete the square of $x^2 + kx$.
6. Learn the quadratic formula.
7. Solve quadratic equations by the quadratic formula.
8. Use the discriminant to determine the character of the roots of a quadratic equation.

9. Learn the relationship between the character of the roots of a quadratic equation and the x intercepts of the corresponding parabola.
10. Solve rectilinear motion problems graphically and algebraically.
11. Solve literal equations leading to a quadratic equation.
12. Apply the concept of completing the square to equations occurring in analytic geometry and algebraic expressions occurring in calculus.

An equation that can be written as

$$ax^2 + bx + c = 0$$

where a, b, and c are real-number constants and $a \neq 0$ is called a second-degree polynomial equation, or **quadratic equation,** in the variable x. The word *quadratic* comes from the French word *quadrate,* meaning square or rectangular. When a quadratic equation is written in the preceding manner (where all the nonzero terms are on the left side and 0 is on the right side), it is said to be in **standard form.** It is convenient to have a standard form for a particular type of equation so that we have a way of referring to its properties when stating formulas and theorems.

Following are examples of quadratic equations in x, written in standard form, with the indicated values of a, b, and c.

$$6x^2 + 7x - 3 = 0 \quad (a = 6, b = 7, c = -3)$$
$$x^2 - 7 = 0 \quad (a = 1, b = 0, c = -7)$$
$$3x^2 - 4x = 0 \quad (a = 3, b = -4, c = 0)$$

Note that it is possible for either b or c to be 0, as in the second and third equations, respectively. However, the restriction that $a \neq 0$ is necessary to have a second-degree equation.

One method of finding the solution set of a quadratic equation involves the following theorem. Its proof is based on properties of the set of complex numbers given in Section 1.3.

THEOREM *Zero-Factor Theorem*

If r and s are complex numbers, then

$$rs = 0 \quad \text{if and only if} \quad r = 0 \text{ or } s = 0$$

This theorem can be extended to a product of more than two factors. For instance, if $r, s, t, u \in C$, then $rstu = 0$ if and only if at least one of the numbers r, s, t, or u is 0.

Using the Zero-Factor Theorem

The zero-factor theorem can be applied to any quadratic equation in standard form for which the left side can be factored.

1. Factor the left side.
2. Set each factor equal to zero.
3. Find the solutions of these linear equations.
4. The solution set of the given quadratic equation is the union of the solution sets of the two linear equations.
5. Check by substituting in the original equation.

The following example demonstrates the procedure.

▶ **EXAMPLE 1** *Solving a Quadratic Equation by Factoring*

Find the solution set of the equation

$$x^2 + 3x - 10 = 0$$

Solution   We factor the left side and obtain

$$(x + 5)(x - 2) = 0$$

By applying the zero-factor theorem, it follows that the equation gives a true statement if and only if

$$x + 5 = 0 \quad \text{or} \quad x - 2 = 0$$

The solution of the first of these equations is -5, and the solution of the second is 2. Therefore the solution set of the given quadratic equation is $\{-5, 2\}$.

We check the solutions by substituting -5 and 2 into the original equation as follows:

Does $(-5)^2 + 3(-5) - 10 = 0$?   Does $2^2 + 3(2) - 10 = 0$?

$(-5)^2 + 3(-5) - 10 = 25 - 15 - 10$ $2^2 + 3(2) - 10 = 4 + 6 - 10$
$$= 0 \qquad\qquad\qquad\qquad\qquad = 0$$

Therefore both solutions check.   ◀

The graph of the equation obtained by setting y equal to the left side of a quadratic equation in standard form is a parabola. The x intercepts of the parabola are the real-number solutions of the quadratic equation.

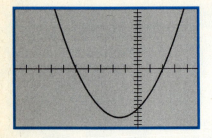

[−10, 5] by [−15, 15]
$y = x^2 + 3x - 10$
FIGURE 1

▶ **EXAMPLE 2** *Showing the Real-number Solutions of a Quadratic Equation on a Graph*

For the quadratic equation in Example 1, show the solutions graphically.

Solution We set y equal to the left side of the equation:

$$y = x^2 + 3x - 10$$

We plot the graph of this equation and obtain the parabola appearing in Figure 1. The x intercepts of the parabola are -5 and 2, which are the solutions obtained in Example 1. ◀

Suppose we have a quadratic equation of the form

$$x^2 = d$$

that is, there is no first-degree term. Then an equivalent equation is

$$x^2 - d = 0$$

and factoring the left member, we obtain

$$(x - \sqrt{d})(x + \sqrt{d}) = 0$$

We set each factor equal to zero and solve the equations.

$$x - \sqrt{d} = 0 \qquad x + \sqrt{d} = 0$$
$$x = \sqrt{d} \qquad x = -\sqrt{d}$$

Therefore the solution set of the equation $x^2 = d$ is $\{\sqrt{d}, -\sqrt{d}\}$. We can abbreviate this solution set as $\{\pm\sqrt{d}\}$. Thus

$$x^2 = d \qquad \text{if and only if} \qquad x = \pm\sqrt{d}$$

where $x = \pm\sqrt{d}$ refers to the two equations $x = \sqrt{d}$ and $x = -\sqrt{d}$. You can think of solving the equation $x^2 = d$ by taking the square root of both sides of the equation and writing the symbol $\pm$ to indicate both square roots of d.

▶ **EXAMPLE 3** *Solving a Quadratic Equation by Square Root*

Find the solution set of each of the following equations:

(a) $x^2 = 25$ **(b)** $x^2 = 11$ **(c)** $x^2 = -9$

Solution We take the square root of both sides of the equation.

(a) $x^2 = 25$ **(b)** $x^2 = 11$ **(c)** $x^2 = -9$
$\quad x = \pm\sqrt{25}$ $x = \pm\sqrt{11}$ $x = \pm\sqrt{-9}$
$\quad x = \pm 5$ $x = \pm 3i$ ◀

▷ **ILLUSTRATION 1**

The solution set of the equation

$$(x - 4)^2 = 5$$

can be found by taking the square root of both sides of the equation. We then have

$$x - 4 = \pm\sqrt{5}$$
$$x = 4 \pm \sqrt{5}$$

Thus the two solutions are $4 + \sqrt{5}$ and $4 - \sqrt{5}$, and the solution set can be written as $\{4 \pm \sqrt{5}\}$.   ◄

We can apply the method of Illustration 1 to find the solution set of any quadratic equation. The first step is to write the equation in a form similar to the given equation in the illustration. That is, the left side will be the square of an algebraic expression containing the variable, and the right side will be a constant. We use this procedure in the following illustration.

▷ **ILLUSTRATION 2**

To find the solution set of the equation

$$x^2 + 6x - 1 = 0$$

we first add 1 to each side and obtain

$$x^2 + 6x = 1 \qquad\qquad (1)$$

We now add to each side the square of one-half of the coefficient of x, or 3^2. We obtain

$$x^2 + 6x + 9 = 1 + 9 \qquad\qquad (2)$$

The left side is now the square of $x + 3$. Thus we have

$$(x + 3)^2 = 10 \qquad\qquad (3)$$

Taking the square root of both sides of the equation, we have

$$x + 3 = \pm\sqrt{10}$$
$$x = -3 \pm \sqrt{10}$$

Therefore the solution set of the given equation is $\{-3 \pm \sqrt{10}\}$.   ◄

In Illustration 2, the solutions $-3 + \sqrt{10}$ and $-3 - \sqrt{10}$ are the **exact** solutions of the quadratic equation. From a calculator, $\sqrt{10} \approx 3.16227766$, where the symbol $\approx$ is read "approximately equals." Therefore, **decimal approximations** of the solutions to seven decimal

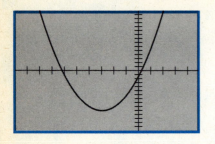

[−10, 5] by [−15, 15]
$y = x^2 + 6x - 1$
FIGURE 2

places are −6.1622777 and 0.1622777. The graph of $y = x^2 + 6x - 1$, shown in Figure 2, indicates that our solutions are reasonable approximations to the x intercepts of the parabola.

The method used to get Equation (3) in Illustration 2 is called **completing the square.** The important step is obtaining Equation (3) equivalent to Equation (1). Note that in (1) the left side is $x^2 + 6x$ (the coefficient of x^2 is 1, and the coefficient of x is 6). We added the square of one-half of 6 to each side of (1), which gives (2), in which the left side is $(x + 3)^2$. More generally:

To complete the square of $x^2 + kx$, add the square of one-half the coefficient of x; that is, add $\left(\dfrac{k}{2}\right)^2$.

Observe that this rule for completing the square applies only to a quadratic expression of the form $x^2 + kx$, where the coefficient of the second-degree term is 1. Also note that when $\left(\dfrac{k}{2}\right)^2$ is added to $x^2 + kx$, we have the square of a binomial:

$$x^2 + kx + \left(\frac{k}{2}\right)^2 = \left(x + \frac{k}{2}\right)^2$$

If k is $2a$, this formula becomes

$$x^2 + 2ax + a^2 = (x + a)^2$$

▶ **EXAMPLE 4** *Completing the Square of $x^2 + kx$*

Add a term to each of the following algebraic expressions to make it the square of a binomial. Then write the resulting expression as the square of a binomial.

(a) $x^2 + 12x$ **(b)** $x^2 - 5x$ **(c)** $x^2 + \dfrac{3}{4}x$

Solution The coefficient of x^2 in each of the given expressions is 1. Therefore, to complete the square, we add the square of one-half of the coefficient of x.

(a) $x^2 + 12x + 6^2 = x^2 + 12x + 36$
$$= (x + 6)^2$$

(b) $x^2 - 5x + \left(-\dfrac{5}{2}\right)^2 = x^2 - 5x + \dfrac{25}{4}$
$$= \left(x - \frac{5}{2}\right)^2$$

(c) $x^2 + \dfrac{3}{4}x + \left(\dfrac{3}{8}\right)^2 = x^2 + \dfrac{3}{4}x + \dfrac{9}{64}$

$$= \left(x + \dfrac{3}{8}\right)^2 \qquad \blacktriangleleft$$

Consider now the general quadratic equation in standard form:

$$ax^2 + bx + c = 0$$

We solve this equation for x in terms of a, b, and c by completing the square. We first divide both sides of the equation by a (remember $a \neq 0$). Then we add $-\dfrac{c}{a}$ to both sides to obtain

$$x^2 + \dfrac{b}{a}x = -\dfrac{c}{a}$$

We now add the square of one-half the coefficient of x to both sides:

$$x^2 + \dfrac{b}{a}x + \left(\dfrac{b}{2a}\right)^2 = -\dfrac{c}{a} + \left(\dfrac{b}{2a}\right)^2$$

$$\left(x + \dfrac{b}{2a}\right)^2 = \dfrac{b^2}{4a^2} - \dfrac{c}{a}$$

$$\left(x + \dfrac{b}{2a}\right)^2 = \dfrac{b^2 - 4ac}{4a^2}$$

$$x + \dfrac{b}{2a} = \pm\dfrac{\sqrt{b^2 - 4ac}}{2a}$$

$$x = -\dfrac{b}{2a} \pm \dfrac{\sqrt{b^2 - 4ac}}{2a}$$

$$x = \dfrac{-b \pm \sqrt{b^2 - 4ac}}{2a}$$

These two values of x are the solutions of the equation $ax^2 + bx + c = 0$. We have obtained the quadratic formula, which we now state formally.

Quadratic Formula

If $a \neq 0$, the solutions of the equation $ax^2 + bx + c = 0$ are given by

$$x = \dfrac{-b \pm \sqrt{b^2 - 4ac}}{2a}$$

Simply substituting the values of a, b, and c into the quadratic formula allows us to solve any quadratic equation in standard form.

▶ **EXAMPLE 5** *Solving a Quadratic Equation by the Quadratic Formula*

Use the quadratic formula to find the solution set of the equation

$$2x^2 = 5(2x - 3)$$

Solution We write the given equation in standard form as

$$2x^2 - 10x + 15 = 0$$

From the quadratic formula where a is 2, b is -10, and c is 15,

$$x = \frac{-b \pm \sqrt{b^2 - 4ac}}{2a}$$

$$= \frac{10 \pm \sqrt{100 - 120}}{4}$$

$$= \frac{10 \pm \sqrt{-20}}{4}$$

$$= \frac{10 \pm 2i\sqrt{5}}{4}$$

$$= \frac{5 \pm i\sqrt{5}}{2}$$

The solution set is $\left\{ \dfrac{5 \pm i\sqrt{5}}{2} \right\}$. ◄

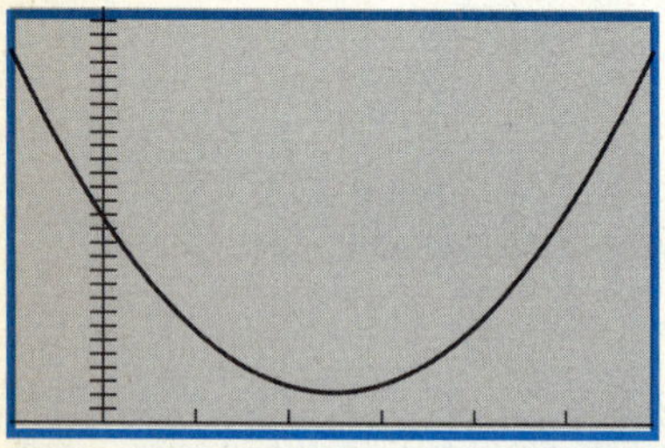

The parabola $y = 2x^2 - 10x + 15$, associated with the equation in Example 5 and shown in Figure 3, does not intersect the x axis because the equation has no real roots.

[−1, 6] by [0, 30]
$y = 2x^2 - 10x + 15$
FIGURE 3

Summary of the Methods for Solving Quadratic Equations

1. Factoring (Example 1)
2. Square root (Example 3 and Illustration 1)
3. Completing the square (Illustration 2)
4. Quadratic formula (Example 5)

Any quadratic equation can be solved by the quadratic formula; so method 4 is used most frequently. If methods 1 or 2 seem appropriate, try them first because they are usually easier to apply. Completing the square, seldom used to solve a quadratic equation, is important because of its prominence in the derivation of the quadratic formula.

We now show how to obtain information about the character of the roots of a quadratic equation without actually solving the equation. Let r and s denote the roots of

$$ax^2 + bx + c = 0$$

where a, b, and c are real numbers, with

$$r = \frac{-b + \sqrt{b^2 - 4ac}}{2a} \quad \text{and} \quad s = \frac{-b - \sqrt{b^2 - 4ac}}{2a}$$

The number represented by $b^2 - 4ac$ is called the **discriminant** of the quadratic equation. The character of the roots can be determined by finding the value of the discriminant.

1. If $b^2 - 4ac = 0$, then

$$r = \frac{-b}{2a} \quad \text{and} \quad s = \frac{-b}{2a}$$

and therefore r and s are equal real numbers. In such a case the number $-\dfrac{b}{2a}$ is called a **double root** (or a **root of multiplicity 2**). The graph of the equation $y = ax^2 + bx + c$ is a parabola tangent to the x axis at the point where $x = -\dfrac{b}{2a}$. See Figure 4.

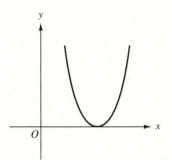

FIGURE 4

2. If $b^2 - 4ac > 0$, then r and s are unequal real numbers. This situation occurred in Example 1. The graph of the equation $y = ax^2 + bx + c$ is a parabola having two x intercepts. See Figure 1.

3. If $b^2 - 4ac < 0$, then r and s are unequal imaginary numbers, each one being the complex conjugate of the other. This happened in Example 5. The graph of the equation $y = ax^2 + bx + c$ is a parabola that does not intersect the x axis. See Figure 3.

We summarize these results.

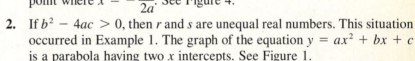

Character of the Roots of the Equation $ax^2 + bx + c = 0$

1. $b^2 - 4ac = 0$: roots are real and equal; a double root
2. $b^2 - 4ac > 0$: roots are real and unequal
3. $b^2 - 4ac < 0$: roots are imaginary and unequal; they are complex conjugates of each other

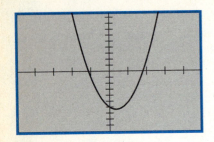

$[-5, 5]$ by $[-10, 10]$
$y = 3x^2 - 2x - 6$
FIGURE 5

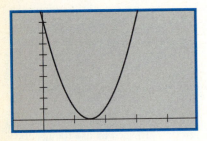

$[-1, 5]$ by $[-1, 10]$
$y = 4x^2 - 12x + 9$
FIGURE 6

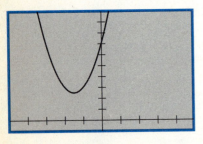

$[-5, 5]$ by $[-1, 10]$
$y = 2x^2 + 6x + 7$
FIGURE 7

▶ **EXAMPLE 6** *Using the Discriminant to Determine the Character of the Roots of a Quadratic Equation and Verifying the Conclusion by a Graph*

Determine the character of the roots of the given equation. Then plot the graph of the corresponding parabola and verify your conclusion.

(a) $3x^2 - 2x - 6 = 0$ **(b)** $4x^2 - 12x + 9 = 0$

(c) $2x^2 + 6x + 7 = 0$

Solution

(a) In the given equation a is 3, b is -2, and c is -6. Thus

$$b^2 - 4ac = (-2)^2 - 4(3)(-6)$$
$$= 76$$

Because the discriminant is positive, the roots are real and unequal. The graph having two x intercepts appears in Figure 5.

(b) Because a is 4, b is -12, and c is 9,

$$b^2 - 4ac = (-12)^2 - 4(4)(9)$$
$$= 0$$

The discriminant is zero; therefore the roots are equal real numbers. Observe that because $4x^2 - 12x + 9 = (2x - 3)^2$, the equation can be written as

$$(2x - 3)^2 = 0$$
$$(2x - 3)(2x - 3) = 0$$

and each factor in the equation gives the number $\frac{3}{2}$ as a solution. The graph, shown in Figure 6, is a parabola tangent to the x axis at the point $(1.5, 0)$.

(c) Because a is 2, b is 6, and c is 7,

$$b^2 - 4ac = 6^2 - 4(2)(7)$$
$$= -20$$

The discriminant is negative; the roots are, therefore, imaginary and unequal, and complex conjugates of each other. Figure 7 shows the graph, a parabola that does not intersect the x axis.   ◀

An application of calculus in physics involves *rectilinear motion,* the motion of a particle on a line. An object thrown vertically upward is a particular example. In such a situation, if s feet is the distance of the object from the starting point, then t seconds after it is thrown

$$s = -16t^2 + v_0 t \tag{4}$$

where v_0 feet per second is the initial velocity of the object. Equation (4) is called an *equation of motion*. When you study calculus, you will be able to determine from the equation of motion the velocity of the object at any particular time.

▶ **EXAMPLE 7** *Solving Graphically a Rectilinear Motion Problem*

A ball is thrown vertically upward from the ground with an initial velocity of 72 ft/sec.

(a) Plot the graph of the equation of motion on a graphics calculator.
(b) From the graph in part (a) determine the distance of the ball from the ground at the end of 1 sec.
(c) From the graph, at the end of how many seconds will the ball on its way down be the same distance from the ground as in part (b)?
(d) From the graph, how many seconds does it take the ball to reach the ground?

Solution

(a) From Equation (4) with $v_0 = 72$, the equation of motion is

$$s = -16t^2 + 72t$$

Note that both s and t must be nonnegative. We plot the graph of the equation in the $[0, 10]$ by $[0, 100]$ viewing rectangle and obtain Figure 8.

(b) By tracing and zooming in, we obtain $s = 56$ when $t = 1$. Therefore, the ball is 56 ft from the ground at the end of 1 sec.

(c) In the same viewing rectangle we draw the line $s = 56$. See Figure 9. By zooming in, we determine that the line intersects the graph of the equation of motion where $t = 3.5$. Therefore, the ball on its way down will be 56 ft above the ground at the end of 3.5 sec.

(d) The ball is on the ground when $s = 0$, which is when the graph intersects the horizontal axis. This occurs, of course, when $t = 0$ and, by zooming in, when $t = 4.5$. Therefore it takes 4.5 sec for the ball to reach the ground. ◀

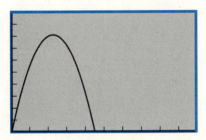

$[0, 10]$ by $[0, 100]$
$s = -16t^2 + 72t$
FIGURE 8

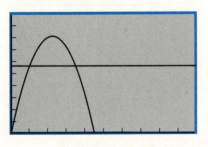

$[0, 10]$ by $[0, 100]$
$s = -16t^2 + 72t$ and $s = 56$
FIGURE 9

▶ **EXAMPLE 8** *Solving Algebraically a Rectilinear Motion Problem*

Find the answers in parts (b)–(d) in Example 7 by algebra.

Solution

(b) In the equation of motion

$$s = -16t^2 + 72t$$

we let $t = 1$, and we get

$$s = -16(1)^2 + 72(1)$$
$$= 56$$

(c) Replacing s by 56 in the equation of motion, we have

$$56 = -16t^2 + 72t$$
$$16t^2 - 72t + 56 = 0$$
$$2t^2 - 9t + 7 = 0$$
$$(2t - 7)(t - 1) = 0$$
$$2t - 7 = 0 \qquad t - 1 = 0$$
$$t = 3.5 \qquad t = 1$$

(d) Replacing s by 0 in the equation of motion, we have

$$0 = -16t^2 + 72t$$
$$-8t(2t - 9) = 0$$
$$t = 0 \qquad 2t - 9 = 0$$
$$t = 4.5$$

These answers agree with those found in Example 7.   ◀

▶ **EXAMPLE 9** *Solving a Literal Equation Leading to a Quadratic Equation*

Solve Equation (4) for t.

Solution   Writing the equation as a standard form of a quadratic equation in t, we have

$$16t^2 - v_0t + s = 0$$

We use the quadratic formula where a is 16, b is $-v_0$, and c is s, and we have

$$t = \frac{-b \pm \sqrt{b^2 - 4ac}}{2a}$$
$$= \frac{-(-v_0) \pm \sqrt{(-v_0)^2 - 4(16)(s)}}{2(16)}$$
$$= \frac{v_0 \pm \sqrt{v_0{}^2 - 64s}}{32} \qquad ◀$$

The concept of completing the square has applications other than its importance in deriving the quadratic formula. In analytic geometry, it is used to write equations of conics in standard form; see Exercises 35 through 42. In calculus, it is applied in techniques of integration to write certain algebraic expressions in recognizable forms; see the following example and Exercises 43 through 46.

▶ **EXAMPLE 10** *Applying the Concept of Completing the Square to an Algebraic Expression Occurring in Calculus*

 Write $\sqrt{27 + 6x - x^2}$ in the form $\sqrt{a^2 - (x - h)^2}$ where a and h are constants.

Solution

$$\sqrt{27 + 6x - x^2} = \sqrt{27 - (x^2 - 6x)}$$

To complete the square of the expression $x^2 - 6x$, we must add $(-\frac{6}{2})^2$, which is 9. Because a minus sign precedes the parentheses we actually need -9 in the radicand. We therefore add $9 - 9$ to the radicand, which does not affect the value of the radicand. We have

$$\sqrt{27 - (x^2 - 6x)} = \sqrt{27 + 9 - 9 - (x^2 - 6x)}$$
$$= \sqrt{36 - (x^2 - 6x + 9)}$$
$$= \sqrt{6^2 - (x - 3)^2}$$

which is of the required form. ◀

EXERCISES 2.2

In Exercises 1 through 12, find the solution set of the equation. Plot the graph of the corresponding parabola and verify that the x intercepts of the parabola are the roots of the equation.

1. $x^2 = 49$ **2.** $25x^2 - 16 = 0$

3. $5t^2 - 12 = 0$ **4.** $3y^2 - 5 = 0$

5. $4x^2 = x$ **6.** $\frac{1}{6}x^2 + x = 0$

7. $x^2 = 8x - 15$ **8.** $y^2 - 11y + 28 = 0$

9. $8w^2 + 10w - 3 = 0$

10. $14x^2 - x - 3 = 0$

11. $49x^2 + 84x + 36 = 0$

12. $64y^2 - 80y + 25 = 0$

In Exercises 13 through 22, find the solution set of the equation by using the quadratic formula. Plot the corresponding parabola. If the equation has real roots, verify that the x intercepts of the parabola are the roots of the equation. If the equation has no real roots, verify that the parabola does not intersect the x axis.

13. $x^2 - 3x - 4 = 0$ **14.** $x^2 + 2x - 3 = 0$

15. $2x + 2 = x^2$ **16.** $x^2 + 1 = 6x$

17. $5y^2 - 4y - 2 = 0$ **18.** $4s^2 - 10s + 5 = 0$

19. $x^2 + \frac{1}{2} = x$ **20.** $\frac{x^2}{4} + 2 = x$

21. $t^2 - 4t + 7 = 0$ **22.** $25y^2 - 20y + 7 = 0$

In Exercises 23 through 26, find the discriminant and determine the character of the roots of the quadratic equation; do not solve the equation. Then plot the graph of the corresponding parabola and verify your conclusion.

23. (a) $6x^2 - 11x - 10 = 0$ **(b)** $3x^2 - 4x = 3$

24. (a) $4x^2 + 12x + 9 = 0$ **(b)** $4t^2 + 2t = 1$

25. (a) $25x^2 - 40x + 16 = 0$ **(b)** $3y = 2y^2 + 5$

26. (a) $14y^2 = 11y - 15$
(b) $14y^2 + 11y - 15 = 0$

In Exercises 27 through 30, find the solution set of the equation.

27. $\dfrac{x + 4}{x} = \dfrac{3x}{x + 2}$

28. $\dfrac{3t}{3t + 4} + \dfrac{2}{5} = \dfrac{t}{3t - 4}$

29. $\dfrac{70}{y^2 - 4y + 3} = \dfrac{23}{1 - y} - 3$

30. $\dfrac{32}{x^2 + 3x + 2} - 3 = \dfrac{x - 3}{x + 1}$

In Exercises 31 through 34, complete the square by adding a term to the algebraic expression; then write the resulting expression as the square of a binomial.

31. (a) $x^2 + 6x$ **(b)** $x^2 - 5x$

32. (a) $x^2 - 4x$ **(b)** $x^2 + 7x$

33. (a) $x^2 - \frac{2}{3}x$ **(b)** $x^2 + \frac{3}{5}x$

34. (a) $x^2 + \frac{4}{3}x$ **(b)** $x^2 - \frac{5}{6}x$

In Exercises 35 and 36, the graph of the equation is a circle. Use the concept of completing the square to write the equation in standard form: $(x - h)^2 + (y - k)^2 = r^2$; h, k, and r are constants.

35. $x^2 + y^2 - 4x + 6y = 13$

36. $4x^2 + 4y^2 + 8x - 32y + 59 = 0$

*In Exercises 37 and 38, the graph of the equation is a parabola. Use the concept of completing the square to write the equation in standard form:
$(x - h)^2 = 4a(y - k)$; h, k, and a are constants.*

37. $3x^2 - 6x - 5y + 13 = 0$

38. $x^2 - 4x - 16y + 52 = 0$

In Exercises 39 and 40, the graph of the equation is an ellipse. Use the concept of completing the square to write the equation in standard form: $\dfrac{(x - h)^2}{a^2} + \dfrac{(y - k)^2}{b^2} = 1$; *h, k, a, and b are constants.*

39. $4x^2 + 9y^2 - 16x - 18y - 11 = 0$

40. $16x^2 + 25y^2 + 96x - 100y + 228 = 0$

In Exercises 41 and 42, the graph of the equation is a hyperbola. Use the concept of completing the square to write the equation in standard form: $\dfrac{(x - h)^2}{a^2} - \dfrac{(y - k)^2}{b^2} = 1$; *h, k, a, and b are constants.*

41. $25x^2 - 4y^2 + 100x - 40y - 25 = 0$

42. $x^2 - 4y^2 - 2x + 32y - 99 = 0$

$\boxed{\int\!\!\int \frac{dy}{dx}}$ *In Exercises 43 through 46, use the concept of completing the square to write the indicated expression in the required form; a and h are constants, and p is a positive integer. This procedure is often necessary when studying techniques of integration in calculus.*

43. Write $\sqrt{x^2 + 2x - 3}$ in the form $\sqrt{(x - h)^2 - a^2}$

44. Write $\sqrt{65x - x^2}$ in the form $\sqrt{a^2 - (x - h)^2}$

45. Write $\sqrt{-4x^2 + 24x - 35}$ in the form
$p\sqrt{a^2 - (x - h)^2}$

46. Write $\sqrt{36x^2 + 72x + 32}$ in the form
$p\sqrt{(x - h)^2 - a^2}$

In Exercises 47 through 50, solve the formula for the indicated variable. All the variables represent positive numbers.

47. $V = \frac{1}{3}\pi r^2 h$; for r **48.** $A = 2\pi r(r + h)$; for r

49. $F = \dfrac{kMv^2}{r}$; for v **50.** $g = \dfrac{4\pi^2}{t^2}$; for t

In Exercises 51 through 54, solve for x in terms of the other symbols.

51. $5ax^2 - 3x - 2a = 0$ **52.** $6dx^2 - 3dx + 5 = 0$

53. $x^2 - 2xy - 4x - 3y^2 = 0$

54. $9x^2 - 6xy + y^2 - 3y = 0$

55. The standard form of an equation of a parabola having a vertical axis is $y = ax^2 + bx + c$. Solve for x in terms of y, a, b, and c.

56. If a regular polygon of 10 sides is inscribed in a circle of radius r units, then if s units is the length of a side,

$$\frac{r}{s} = \frac{s}{r - s}$$

Solve this formula for s in terms of r.

 57. An object is shot vertically upward from the ground with an initial velocity of 128 ft/sec. **(a)** From Equation (4), write an equation of motion of the object. Find algebraically: **(b)** the distance of the ball from the ground at the end of 2 sec; **(c)** at the end of how many seconds the distance of the object from the ground is again the same distance as in part (b); **(d)** at the end of how many seconds the object reaches the ground.

 58. Plot on your graphics calculator the graph of the equation of motion in part (a) of Exercise 57. Solve parts (b)–(d) in Exercise 57 from your graph in part (a).

 59. If an object is thrown straight up from a point s_0 feet above ground level with an initial velocity of v_0 feet per second, then if s feet is the distance of the object from the ground t seconds after it is thrown

$$s = -16t^2 + v_0 t + s_0$$

(a) Use this equation to write an equation of motion for a ball thrown upward from the edge of a rooftop 68 ft above the ground with an initial velocity of 76 ft/sec. **(b)** Plot the graph of your equation in part (a). **(c)** From your graph in part (b), determine the height of the ball above the ground at the end of 2 sec.

(d) From the graph, at the end of how many seconds will the ball on its way down be the same distance from the ground as in part (c)? **(e)** From the graph, at the end of how many seconds will the ball on its way down pass the edge of the rooftop? **(f)** From the graph, at the end of how many seconds will the ball reach the ground?

 60. Determine algebraically the answers to parts (c)–(f) in Exercise 59.

 61. In calculus you will learn that if an object is thrown vertically upward at an initial velocity of v_0 feet per second, then t seconds after it is thrown the velocity of the object is v feet per second where

$$v = -32t + v_0$$

Discuss how you would use this equation and Equation (4) to determine how many seconds it will take for the object to reach its greatest height and how high the object will go.

 62. Use the method you described in Exercise 61 and the equation for v in that exercise to determine how long it will take the ball of Example 7 to reach its greatest height and how high the ball will go.

63. In his ninth-century work, *ilm al-jabr w'al muqâbalah*, Al–Khowarizmi described entirely in Arabic words the solution of quadratic equations by completing the square. You do the same entirely in English words for the quadratic equation

$$3x^2 + 6x - 7 = 0$$

2.3 EQUATIONS AS MATHEMATICAL MODELS

GOALS
1. **Find an equation as a mathematical model of a word problem.**
2. **Estimate the solution of a mathematical model on a graphics calculator.**
3. **Solve a mathematical model algebraically.**
4. **Write a conclusion for a word problem.**

In many applications of algebra, *word problems* give relationships between known numbers and unknown numbers to be determined. Solving word problems strengthens not only your mathematical skills but your reading and writing skills as well. In this section we solve word problems by using an equation as a mathematical model of the problem. You were introduced to this process in Section 2.1.

Although no one specific method is always used to obtain a mathematical model, here are some steps that give a possible procedure for you to follow. As you read through the examples, refer to these steps to see how they are applied.

Suggestions for Obtaining an Equation As a Mathematical Model

1. Read the problem carefully so that you understand it. To gain understanding, it is often helpful to make up a specific example that involves a similar situation in which all the quantities are known. Another aid is to draw a diagram if feasible, as shown in Examples 1, 3, and 4.
2. Determine the known and unknown quantities. Use a variable to represent one of the unknown quantities in the equation you will obtain. When using only one equation, any other unknown quantities should be expressed in terms of this one variable. Because the variable represents a number, the variable's definition should indicate this fact. For example, if the unknown quantity is time and time is measured in seconds, then if t is the variable, t should be defined as the number of seconds in the time or, equivalently, t seconds is the time.
3. Write down any numerical facts known about the variable. In many word problems, these facts can be incorporated in a table as indicated in Examples 2 and 3.
4. From the information in step 3, determine two algebraic expressions for the same number and form an equation from them. The use of a table as suggested in step 3 will help you to discover equal algebraic expressions.

You now have a mathematical model of the problem.

The following steps should be followed to complete the problem once you have a mathematical model:

5. Find the solution set of the equation you obtained as a mathematical model.

6. Check your solutions by determining whether the conditions of the problem are satisfied. This check verifies the accuracy of the equation obtained in step 4 as well as the accuracy of its solution set.
7. Write a conclusion, consisting of one or more complete sentences, that answers the questions of the problem. For instance, suppose the problem requires you to find the length of a rectangle and the variable x is defined so that x feet is the length. Then if the solution set of the equation is {5}, your conclusion could be: The length of the rectangle is 5 ft.

Bear in mind that the equation you obtain as a mathematical model is an equality of numbers. Thus the units of measurement do not appear in the equation. Furthermore, the units of measurement do not appear in the solution set of the equation. These units, however, are included in the definition of the variable, as indicated in step 2. The units also appear in your conclusion as indicated in step 7.

▶ **EXAMPLE 1** *Solving a Word Problem with a Linear Equation as a Mathematical Model*

If a rectangle has a length 3 cm less than four times its width, and its perimeter is 19 cm, what are the dimensions?

Solution   We wish to determine the number of centimeters in each dimension of the rectangle. We now use a variable to represent these numbers.

w: the number of centimeters in the width of the rectangle
$4w - 3$: the number of centimeters in the length of the rectangle

Refer to Figure 1. The perimeter of the rectangle is the total distance around it. The number of centimeters in the perimeter, therefore, can be represented by either

$$w + (4w - 3) + w + (4w - 3)$$

or 19; thus we have the equation

$$w + (4w - 3) + w + (4w - 3) = 19$$
$$10w - 6 = 19$$
$$10w = 25$$
$$w = \frac{5}{2} \qquad 4w - 3 = 4\left(\frac{5}{2}\right) - 3$$
$$= 7$$

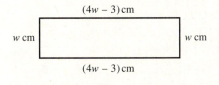

(4w − 3) cm

w cm $\qquad\qquad$ w cm

(4w − 3) cm

FIGURE 1

Check: If the width of the rectangle is 2.5 cm and the length is 7 cm, the perimeter is $(2.5 + 7 + 2.5 + 7)$ cm, which equals 19 cm.

Conclusion: The width of the rectangle is 2.5 cm and the length is 7 cm.   ◀

The word problem in the next example can be classified as an **investment problem** because it is one involving income from an investment. The income can be in the form of interest, and in that case we use the formula

$$I = P \cdot R$$

where I dollars is the annual interest earned when P dollars is invested at a rate R per year. The rate is usually given as a percent; thus if the rate is 8 percent, then $R = 0.08$.

▶ **EXAMPLE 2** *Finding a Linear Equation as a Mathematical Model of a Word Problem, Estimating the Solution of the Equation on a Graphics Calculator, and Solving the Equation Algebraically*

A woman has $15,000 invested in two accounts from which she receives an annual income of $1456. If the riskier investment pays 12 percent per year and the other pays 8 percent, determine how much she has invested at each rate by doing the following: **(a)** Find a linear equation as a mathematical model of the situation; **(b)** estimate the solution of the equation on a graphics calculator; and **(c)** solve the equation algebraically. **(d)** Write a conclusion.

Solution

(a) We wish to determine the number of dollars she has invested at each rate. We define these numbers.

x: the number of dollars invested at 12 percent
$15,000 - x$: the number of dollars invested at 8 percent
We use the formula $I = P \cdot R$ to get Table 1.

Table 1

	Number of Dollars Invested	× Rate =	Number of Dollars in Interest
12 percent investment	x	0.12	$0.12x$
8 percent investment	$15,000 - x$	0.08	$0.08(15,000 - x)$

Because the annual income from the two investments is $1456, the sum of the entries in the last column of the table is 1456; we have,

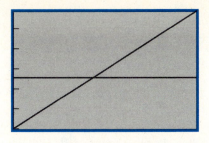

[0, 15,000] by [1200, 1800]
$y_1 = 0.12x + 0.08(15,000 - x)$ and $y_2 = 1456$
FIGURE 2

therefore, the equation

$$0.12x + 0.08(15,000 - x) = 1456$$

(b) On our graphics calculator, we let

$$y_1 = 0.12x + 0.08(15,000 - x) \quad \text{and} \quad y_2 = 1456$$

We need to determine the permissible values for x and y_1. When all of the money is invested at 8 percent, x is 0 and y_1 is $0.08(15,000)$, which is 1200. When all of the money is invested at 12 percent, x is 15,000 and y_1 is $0.12(15,000)$, which is 1800. Because $0 \le x \le 15,000$ and $1200 \le y_1 \le 1800$, we select [0, 15,000] by [1200, 1800] as our viewing rectangle. See Figure 2. Using zoom-in, we estimate the x coordinate of the point of intersection of the two graphs as 6400.

(c) We solve the equation algebraically.

$$0.12x + 0.08(15,000 - x) = 1456$$
$$0.12x + 1200 - 0.08x = 1456$$
$$0.04x = 256$$
$$x = 6400$$

This value of x agrees with our estimate in part (b). When $x = 6400$, $15,000 - x = 8600$.

Check: With $x = 6400$, the number of dollars in the annual interest from the 12 percent investment is $0.12(6400) = 768$. With $15,000 - x = 8600$, the number of dollars in the annual interest from the 8 percent investment is $0.08(8600) = 688$. Because $768 + 688 = 1456$, the solution checks.

(d) **Conclusion:** The woman has $6400 invested at 12 percent and $8600 at 8 percent. ◄

A **mixture problem** can involve mixing solutions containing different percents of a substance to obtain a solution containing a certain percent of the substance. For instance, in the next example a chemist wants to obtain 6 liters of a 10 percent acid solution by mixing a 7 percent acid solution with a 12 percent acid solution. Another kind of mixture problem for which the method of solving is similar involves mixing commodities of different values to obtain a combination worth a specific sum of money. A problem of this type appears in Exercise 12.

EXAMPLE 3 *Finding a Linear Equation as a Mathematical Model of a Word Problem, Estimating the Solution of the Equation on a Graphics Calculator, and Solving the Equation Algebraically*

A chemist wants to obtain 6 liters of a 10 percent acid solution by mixing a 7 percent acid solution and a 12 percent acid solution. Determine how much of each solution the chemist should use by doing the following: **(a)** Find an equation as a mathematical model of the situation; **(b)** estimate the solution of the equation on a graphics calculator; and **(c)** solve the equation algebraically. **(d)** Write a conclusion.

Solution

(a) We want to determine the number of liters of each solution the chemist should use. We define these numbers.

x: the number of liters of the 7 percent acid solution

$6 - x$: the number of liters of the 12 percent acid solution

The diagram in Figure 3 gives a visual interpretation of the problem. The information appearing there is incorporated in Table 2.

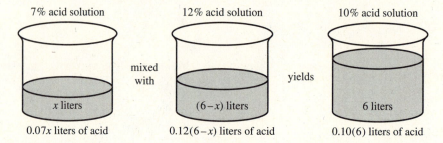

7% acid solution   mixed with   12% acid solution   yields   10% acid solution

x liters $(6-x)$ liters   6 liters

$0.07x$ liters of acid $0.12(6-x)$ liters of acid $0.10(6)$ liters of acid

FIGURE 3

Table 2

	Percent of Acid	$\times$	Number of Liters of Solution	$=$	Number of Liters of Acid
7% acid solution	7%		x		$0.07x$
12% acid solution	12%		$6 - x$		$0.12(6 - x)$
Mixture	10%		6		0.60

From the last column in the table, we see that the total number of liters of acid in the mixture can be represented by either 0.60 or $0.07x + 0.12(6 - x)$. Thus we have the equation

$$0.07x + 0.12(6 - x) = 0.60$$

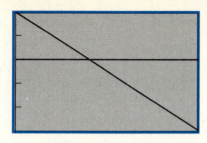

[0, 6] by [0.42, 0.72]
$y_1 = 0.07x + 0.12(6 - x)$ and $y_2 = 0.60$
FIGURE 4

(b) On our graphics calculator, we let

$$y_1 = 0.07x + 0.12(6 - x) \quad \text{and} \quad y_2 = 0.60$$

We need to determine the permissible values for x and y_1. If the chemist uses only the 12 percent acid solution, x is 0 and y_1 is 0.12(6), which is 0.72. If the chemist uses only the 7 percent acid solution, x is 6 and y_1 is 0.07(6), which is 0.42. Because $0 \leq x \leq 6$ and $0.42 \leq y_1 \leq 0.72$, we select [0, 6] by [0.42, 0.72] as our viewing rectangle. See Figure 4. By zooming in on the calculator, we estimate the x coordinate of the point of intersection of the two graphs as 2.4.

(c) We solve the equation algebraically.

$$0.07x + 0.12(6 - x) = 0.60$$
$$0.07x + 0.72 - 0.12x = 0.60$$
$$-0.05x = 0.60 - 0.72$$
$$-0.05x = -0.12$$
$$x = \frac{-0.12}{-0.05}$$
$$x = 2.4$$

This value of x agrees with our estimate in part (b). When $x = 2.4$, $6 - x = 3.6$.

Check: With $x = 2.4$, the number of liters of acid from the 7 percent acid solution is 0.07(2.4) = 0.168. With $6 - x = 3.6$, the number of liters of acid from the 12 percent acid solution is 0.12(3.6) = 0.432. Because 0.168 + 0.432 = 0.60, our answers check.

(d) Conclusion: The chemist should use 2.4 liters of the 7 percent acid solution and 3.6 liters of the 12 percent acid solution. ◀

▶ **EXAMPLE 4** *Finding a Quadratic Equation as a Mathematical Model of a Word Problem, Estimating the Solution of the Equation on a Graphics Calculator, and Solving the Equation Algebraically*

A park contains a flower garden 50 m long and 30 m wide and a path of uniform width around it having an area of 600 m². Determine the width of the path by doing the following: **(a)** Find an equation as a mathematical model of the situation; **(b)** estimate the solution of the equation on a graphics calculator; and **(c)** solve the equation algebraically. **(d)** Write a conclusion.

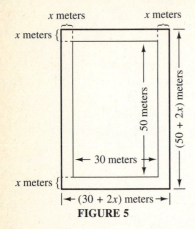

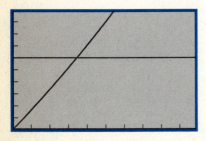

FIGURE 5

Solution

(a) We begin by defining the number we wish to find.

x: the number of meters in the width of the path.

Refer to Figure 5. The area of the park minus the area of the garden equals the area of the path; thus we have the equation

$$(50 + 2x)(30 + 2x) - 1500 = 600$$

(b) On our graphics calculator, we let

$$y_1 = (50 + 2x)(30 + 2x) - 1500 \quad \text{and} \quad y_2 = 600$$

It is reasonable to assume $0 \leq x \leq 10$. Because $y_2 = 600$, a suitable interval for y is $[0, 1000]$. We therefore select $[0, 10]$ by $[0, 1000]$ as our viewing rectangle. Figure 6 shows the graphs of y_1 and y_2 plotted in this viewing rectangle. By zooming in, we estimate the x coordinate of the point of intersection of the two graphs as 3.45.

(c) We solve the equation algebraically.

$$(50 + 2x)(30 + 2x) - 1500 = 600$$
$$1500 + 160x + 4x^2 - 1500 = 600$$
$$4x^2 + 160x - 600 = 0$$
$$x^2 + 40x - 150 = 0$$

We solve this equation by the quadratic formula where a is 1, b is 40, and c is -150.

$$x = \frac{-b \pm \sqrt{b^2 - 4ac}}{2a}$$
$$= \frac{-40 \pm \sqrt{(40)^2 - 4(1)(-150)}}{2(1)}$$
$$= \frac{-40 \pm \sqrt{2200}}{2}$$
$$= \frac{-40 \pm 10\sqrt{22}}{2}$$
$$= -20 \pm 5\sqrt{22}$$

Because x must be a positive number, we reject the negative root. To two decimal places, $\sqrt{22} \approx 4.69$. Therefore

$$x \approx -20 + 5(4.69)$$
$$\approx 3.45$$

This answer agrees with our estimate in part (b).

$[0, 10]$ by $[0, 1000]$
$y_1 = (50 + 2x)(30 + 2x) - 1500$ and $y_2 = 600$
FIGURE 6

Check: With $x = 3.45$, the number of meters in the length of the park is $50 + 2(3.45) = 56.90$, and the number of meters in the width is $30 + 2(3.45) = 36.90$. The number of square meters in the area of the park is, therefore, $(56.90)(36.90) = 2100$. The area of the garden is 1500 m^2, and the area of the path is 600m^2; and $2100 - 1500 = 600$.

(d) **Conclusion:** The width of the path is 3.45 m. ◀

EXERCISES 2.3

In each exercise solve the word problem by using an equation as a mathematical model of the situation. Complete the exercise by writing a conclusion.

1. The sum of two numbers is 9 and their difference is 6. What are the numbers?

2. Find two numbers whose sum is 7 and one of which is three times the smaller.

3. The width of a rectangle is 2 cm more than one-half its length and its perimeter is 40 cm. What are the dimensions?

4. The longest side of a triangle is twice as long as the shortest side and 2 cm longer than the third side. The perimeter of the triangle is 33 cm. What is the length of each side?

5. Admission tickets to a motion picture theater cost $6 for adults and $4.50 for students. If 810 tickets were sold and the total receipts were $4279.50, how many of each type of ticket were sold?

6. An investor wants to realize a return of 12 percent on a total of two investments. If he invests $10,000 at 10 percent, how much additional money should he invest at 16 percent?

7. A woman invested $25,000 in two business ventures. Last year she made a profit of 15 percent from the first venture but lost 5 percent from the second venture. Her income last year from the two investments was equivalent to a return of 8 percent on the entire amount invested. How much has she

invested in each venture? Estimate the solution of your equation on your graphics calculator. Then solve the equation algebraically.

8. A retail merchant invested $6500 in three kinds of cameras. The profit on sales of camera A was 25 percent; on the sales of camera B, the profit was 12 percent; and there was a loss of 1 percent on the sales of camera C. The merchant invested an equal amount on cameras A and B and the overall profit on the total investment was 14 percent. How much was invested in each kind of camera? Estimate the solution of your equation on your graphics calculator. Then solve the equation algebraically.

9. A goldsmith obtained 40 grams of an alloy containing 70 percent gold by combining one alloy containing 80 percent gold and another alloy containing 55 percent gold. How many grams of each alloy did the goldsmith combine? Estimate the solution of your equation on your graphics calculator. Then solve the equation algebraically.

10. Determine how much water is required to dilute 15 liters of a solution that is 12 percent dye so that a 5 percent dye solution is obtained. Estimate the solution of your equation on your graphics calculator. Then solve the equation algebraically.

11. How much water must be evaporated from the 15 liters of the 12 percent dye solution in Exercise 10 to obtain a solution that is 20 percent dye? Assume that the total amount of dye is not affected by the process of evaporation. Estimate the solution of your equation on your graphics calculator. Then solve the equation algebraically.

12. Perfume to sell for $80 an ounce is to be a blend of perfume selling for $104 an ounce and perfume selling for $50 an ounce. If 270 ounces of the blend are desired, how much of each kind of perfume should be used? Estimate the solution of your equation on your graphics calculator. Then solve the equation algebraically.

13. An open bin with a square bottom, rectangular sides, and a height of 3 m is to be constructed from $63 worth of material. If the material for the bottom costs $5.40 per square meter and the material for the sides costs $2.40 per square meter, what will be the volume of the bin? Estimate the solution of your equation on your graphics calculator. Then solve the equation algebraically.

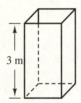

14. A page of print containing 144 cm² of printed region has a margin of 4.5 cm at the top and bottom and a margin of 2 cm at the sides. What are the dimensions of the page if the width across the page is four-ninths of the length? Estimate the solution of your equation on your graphics calculator. Then solve the equation algebraically.

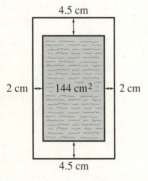

15. A park in the shape of a rectangle, having dimensions 60 m by 100 m, contains a rectangular garden enclosed by a concrete terrace of uniform width.

If the area of the garden is one-half the area of the park, how wide is the terrace? Estimate the solution of your equation on your graphics calculator. Then solve the equation algebraically.

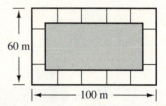

16. What is the width of a strip that must be plowed around a rectangular field 100 m long by 60 m wide so that two-thirds of the field will be plowed? Estimate the solution of your equation on your graphics calculator. Then solve the equation algebraically.

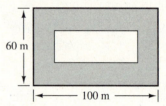

17. Every freshman student at a particular college is required to take an English aptitude test. A student who passes the examination enrolls in English Composition, and a student who fails the test must enroll in English Fundamentals. In a freshman class of 1240 students there are more students enrolled in English Fundamentals than in English Composition. However, if 30 more students had passed the test, each course would have the same enrollment. How many students are taking each course?

18. The annual sophomore class picnic is planned by a committee consisting of 17 members. A vote to determine if the picnic should be held at the beach or in the mountains resulted in victory for the beach location. However, if two committee members had changed their vote from favoring the beach to favoring the mountains, the mountain site would have won by one vote. How many votes did each picnic location receive?

19. The sum of the reciprocals of two consecutive even integers is $\frac{9}{40}$. What are the integers?

20. Are there two consecutive even integers the sum of whose reciprocals is $\frac{8}{45}$? If your answer is yes, find them. If your answer is no, show how you arrived at your answer.

2.4 OTHER EQUATIONS IN ONE UNKNOWN

GOALS

1. **Solve algebraically equations involving radicals.**
2. **Solve graphically equations involving radicals.**
3. **Solve equations quadratic in form.**
4. **Find exact solutions of some particular equations of degree greater than 2.**
5. **Use a graphics calculator to find approximate values of irrational roots of cubic equations.**
6. **Solve graphically word problems having a cubic equation as a mathematical model.**

If an algebraic equation contains radicals or rational exponents, we can solve it by raising both sides of the equation to the same integer power. When we do this, we must apply the following theorem.

THEOREM 1

If E and F are algebraic expressions, then

$$E = F$$

is an algebraic equation, and its solution set is a subset of the solution set of the equation

$$E^n = F^n$$

where n is any positive integer.

The theorem follows immediately from the fact that if a and b are complex numbers and $a = b$, then $a^n = b^n$, where n is any positive integer.

▷ **ILLUSTRATION 1**

The equation

$$x = 5$$

has the solution set {5}. If we square each side, we obtain the equation

$$x^2 = 25$$

having the solution set {±5}. The solution set of the first equation is a subset of the solution set of the second. This result agrees with Theorem 1. ◄

Theorem 1 is used to solve an equation having terms involving one or more radicals. The first step is to obtain an equivalent equation having the term involving the most complicated radical on one side and all the other terms on the other side. Then applying Theorem 1, we raise both sides of the equation to the power corresponding to the index of the isolated radical.

▷ **ILLUSTRATION 2**

To solve the equation

$$\sqrt{2x + 5} + x = 5$$

we first add $-x$ to both sides and get the equivalent equation

$$\sqrt{2x + 5} = 5 - x \tag{1}$$

Now with the radical alone on one side of the equation we apply Theorem 1 with $n = 2$ and square both sides to obtain

$$(\sqrt{2x + 5})^2 = (5 - x)^2$$
$$2x + 5 = 25 - 10x + x^2 \tag{2}$$
$$x^2 - 12x + 20 = 0$$
$$(x - 10)(x - 2) = 0$$
$$x - 10 = 0 \qquad x - 2 = 0$$
$$x = 10 \qquad x = 2$$

Thus the solution set of Equation (2) is {2, 10}. According to Theorem 1, the solution set of Equation (1) is a subset of {2, 10}; that is, both of the numbers 2 and 10 may be solutions, only one may be a solution, or neither may be a solution. To determine which case applies, we substitute each number into (1) to see if the equation is satisfied.

$$\sqrt{2(2) + 5} \stackrel{?}{=} 5 - 2 \qquad \sqrt{2(10) + 5} \stackrel{?}{=} 5 - 10$$
$$\sqrt{9} \stackrel{?}{=} 3 \qquad \sqrt{25} \stackrel{?}{=} -5$$
$$3 = 3 \qquad 5 \neq -5$$

Hence 2 is a solution and 10 is not. Therefore, the solution set of Equation (1) is {2}.

If you plot the graph of $y = \sqrt{2x + 5} + x - 5$, you will observe that 2 is the only x intercept. ◄

In Illustration 2, the number 10 is called an **extraneous solution** of Equation (1); it was introduced when both sides were squared. The reason that this extraneous solution was introduced should be apparent after you read Illustration 3.

▷ ILLUSTRATION 3

If both sides of the equation

$$-\sqrt{2x + 5} = 5 - x \tag{3}$$

are squared, we obtain

$$(-\sqrt{2x + 5})^2 = (5 - x)^2$$
$$2x + 5 = 25 - 10x + x^2$$

which is Equation (2). In Illustration 2 we showed that the solution set of Equation (2) is {2, 10}. Hence, by Theorem 1, the solution set of Equation (3) is a subset of {2, 10}. Substituting each of these numbers into Equation (3), we have

$$
\begin{array}{ll}
-\sqrt{2(2) + 5} \overset{?}{=} 5 - 2 & \qquad -\sqrt{2(10) + 5} \overset{?}{=} 5 - 10 \\
-\sqrt{9} \overset{?}{=} 3 & \qquad -\sqrt{25} \overset{?}{=} -5 \\
-3 \neq 3 & \qquad -5 = -5
\end{array}
$$

Hence 10 is a solution and 2 is an extraneous solution. The solution set of Equation (3) is, therefore, {10}.

Plot the graph of $y = -\sqrt{2x + 5} + x - 5$ and note that 10 is the only x intercept. ◄

Observe in Illustration 2 that Equation (1) states that the principal square root of $2x + 5$ is $5 - x$, and in Illustration 3 that Equation (3) states that the negative square root of $2x + 5$ is $5 - x$. Thus, when squaring both sides of either equation, we obtain Equation (2); so in each case an extraneous solution is introduced. This discussion should convince you that when Theorem 1 is applied, all solutions obtained must be checked in the original equation. The check is for possible extraneous solutions, not just for computational errors.

If an equation contains more than one radical, it is sometimes necessary to apply Theorem 1 more than once before obtaining an equation free of radicals.

▶ **EXAMPLE 1** *Solving an Equation Involving Radicals*

Find the solution set of the equation

$$\sqrt{2x + 3} - \sqrt{x - 2} - 2 = 0$$

Solution We first write an equivalent equation in which one radical is isolated on one side. Then after applying Theorem 1, we obtain another equation involving a radical. So we use the same procedure for the new equation.

$$\sqrt{2x + 3} = \sqrt{x - 2} + 2$$
$$(\sqrt{2x + 3})^2 = (\sqrt{x - 2} + 2)^2$$
$$2x + 3 = x - 2 + 4\sqrt{x - 2} + 4$$
$$x + 1 = 4\sqrt{x - 2}$$
$$(x + 1)^2 = 16(\sqrt{x - 2})^2$$
$$x^2 + 2x + 1 = 16(x - 2)$$
$$x^2 + 2x + 1 = 16x - 32$$
$$x^2 - 14x + 33 = 0$$
$$(x - 3)(x - 11) = 0$$
$$x - 3 = 0 \qquad x - 11 = 0$$
$$x = 3 \qquad\quad x = 11$$

Therefore, the solution set of the given equation is a subset of {3, 11}. We substitute these numbers into the given equation.

$$\sqrt{2(3) + 3} - \sqrt{3 - 2} - 2 \stackrel{?}{=} 0 \qquad \sqrt{2(11) + 3} - \sqrt{11 - 2} - 2 \stackrel{?}{=} 0$$
$$\sqrt{9} - \sqrt{1} - 2 \stackrel{?}{=} 0 \qquad\qquad \sqrt{25} - \sqrt{9} - 2 \stackrel{?}{=} 0$$
$$3 - 1 - 2 \stackrel{?}{=} 0 \qquad\qquad 5 - 3 - 2 \stackrel{?}{=} 0$$
$$0 = 0 \qquad\qquad 0 = 0$$

Each of the numbers 3 and 11 is a solution of the given equation; so its solution set is {3, 11}. ◀

You can check the solutions in Example 1 by plotting the graph of $y = \sqrt{2x + 3} - \sqrt{x - 2} - 2$ in the $[0, 15]$ by $[-1, 1]$ viewing rectangle with an x scale of 1 and a y scale of 0.1. The x intercepts of the graph are 3 and 11.

The equation

$$\sqrt{3 - 3x} - \sqrt{3x + 2} - 3 = 0$$

is similar to the one in Example 1 because it also contains two radicals. The similarity ends here, however, because this equation has no solutions; that is, its solution set is ∅. You are asked to show this in Exercise 37. Observe on your graphics calculator that the graph of $y = \sqrt{3 - 3x} - \sqrt{3x + 2} - 3$ does not intersect the x axis.

An equation in a single variable x is said to be **quadratic in form** if it can be written as

$$au^2 + bu + c = 0$$

where $a \neq 0$ and u is an algebraic expression in x.

In the next two examples we solve equations quadratic in form.

▶ **EXAMPLE 2** *Solving an Equation Quadratic in Form*

Find the solution set of the equation

$$x^4 - 2x^2 - 15 = 0$$

Solution   This equation is quadratic in x^2 because if $u = x^2$, the equation becomes

$$u^2 - 2u - 15 = 0$$

We solve this equation for u.

$$(u - 5)(u + 3) = 0$$
$$u - 5 = 0 \qquad u + 3 = 0$$
$$u = 5 \qquad\quad u = -3$$

Now we replace u with x^2 and solve the resulting equations.

$$x^2 = 5 \qquad\quad x^2 = -3$$
$$x = \pm\sqrt{5} \qquad x = \pm i\sqrt{3}$$

The solution set of the original equation is therefore $\{\pm\sqrt{5}, \pm i\sqrt{3}\}$. ◀

▶ **EXAMPLE 3** *Solving an Equation Quadratic in Form*

Find the solution set of the equation

$$2x^{2/3} - 5x^{1/3} - 3 = 0$$

Solution This equation is quadratic in $x^{1/3}$. If $u = x^{1/3}$, the equation becomes

$$2u^2 - 5u - 3 = 0$$
$$(2u + 1)(u - 3) = 0$$

$$2u + 1 = 0 \qquad u - 3 = 0$$

$$u = -\frac{1}{2} \qquad u = 3$$

Replacing u by $x^{1/3}$, we get

$$x^{1/3} = -\frac{1}{2} \qquad x^{1/3} = 3$$

$$(x^{1/3})^3 = \left(-\frac{1}{2}\right)^3 \qquad (x^{1/3})^3 = 3^3$$

$$x = -\frac{1}{8} \qquad x = 27$$

The solution set of the original equation is $\{-\frac{1}{8}, 27\}$. ◀

We have solved quadratic equations in standard form by factoring the left side and setting each factor equal to zero. This procedure can also be used to solve some higher-degree equations. For instance in Example 2, we solved the equation

$$x^4 - 2x^2 - 15 = 0$$

by letting $u = x^2$. An alternate method is to factor the left side and get

$$(x^2 - 5)(x^2 + 3) = 0$$

and then set each factor equal to zero to obtain

$$x^2 - 5 = 0 \qquad x^2 + 3 = 0$$

The solutions of these equations give the solution set of the original equation. In the following example we apply the same method to a third-degree equation.

▶ **EXAMPLE 4** *Finding Exact Solutions of a Particular Equation of Degree Greater than 2*

Find the solution set of the equation

$$x^3 = 8$$

Solution   This equation is equivalent to

$$x^3 - 8 = 0$$

Recall that $a^3 - b^3 = (a - b)(a^2 + ab + b^2)$. Using this formula where a is x and b is 2, we can factor the left side to obtain

$$(x - 2)(x^2 + 2x + 4) = 0$$

Setting each factor equal to zero and solving the resulting equations, we have

$$x - 2 = 0 \qquad x^2 + 2x + 4 = 0$$
$$x = 2$$
$$x = \frac{-2 \pm \sqrt{4 - 16}}{2}$$
$$= \frac{-2 \pm \sqrt{-12}}{2}$$
$$= \frac{-2 \pm 2i\sqrt{3}}{2}$$
$$= -1 \pm i\sqrt{3}$$

Therefore the solution set of the original equation is $\{2, -1 \pm i\sqrt{3}\}$.

Observe on your graphics calculator that the graph of $y = x^3 - 8$ intersects the x axis only at the point $(2, 0)$. ◀

In Example 4, because $x^3 = 8$, x is a cube root of 8. The solutions are, therefore, cube roots of 8; one root is real and two are imaginary. This equation is an example of a polynomial equation of the third degree, or *cubic equation*.

The general **cubic equation** in the variable x is

$$a_3x^3 + a_2x^2 + a_1x + a_0 = 0$$

where a_3, a_2, a_1, and a_0 are real number constants and $a_3 \neq 0$. Every cubic equation has exactly three roots because Theorem 2 in Section 5.4 guarantees that a polynomial equation of degree n has exactly n roots. Even though there are formulas for the exact roots of a cubic equation, as well as for a fourth degree equation, these formulas are complicated and we will not introduce them in this text. However, for any cubic equation having at least one rational root, the exact roots can be found fairly easily by methods discussed in Section 5.3. You saw a particular case of such an equation in Example 4, where we found exact values of the roots because we could factor $x^3 - 8$ as the difference of two cubes.

If a cubic equation has no rational roots, it has at least one irrational root. This fact follows from Theorem 4 in Section 5.4, which guarantees that any polynomial equation, whose degree is an odd number, has at least one real root. In the next example we use a zoom-in process on a graphics calculator to approximate an irrational root of a cubic equation.

▶ **EXAMPLE 5** *Approximating the Value of an Irrational Root of a Cubic Equation on a Graphics Calculator*

On a graphics calculator find the approximate value, to the nearest one-hundredth, of the positive root of the equation

$$x^3 = 4x + 8$$

Solution   We write the equation in the form $x^3 - 4x - 8 = 0$ and set y equal to the left side:

$$y = x^3 - 4x - 8$$

The graph of this equation in the $[-10, 10]$ by $[-10, 10]$ viewing rectangle, with an axis scale of 1, appears in Figure 1. Because the graph intersects the x axis at only one point, there is exactly one real root between 2 and 3. To get our next approximation, we select the $[2, 3]$ by $[-0.1, 0.1]$ viewing rectangle with an axis scale of 0.1. Figure 2, showing this graph, indicates that the root is between 2.6 and 2.7. We now choose the $[2.6, 2.7]$ by $[-0.01, 0.01]$ viewing rectangle with an axis scale of 0.01 and plot the graph appearing in Figure 3. Because the x intercept of this graph is between 2.64 and 2.65 and much closer to 2.65, we conclude that, to the nearest one-hundredth, the root is 2.65.

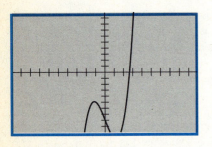

$[-10, 10]$ by $[-10, 10]$
$y = x^3 - 4x - 8$
FIGURE 1

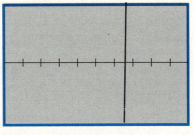

$[2, 3]$ by $[-0.1, 0.1]$
$y = x^3 - 4x - 8$
FIGURE 2

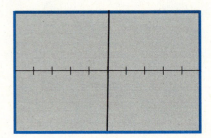

$[2.6, 2.7]$ by $[-0.01, 0.01]$
$y = x^3 - 4x - 8$
FIGURE 3   ◀

▶ **EXAMPLE 6** *Solving Graphically a Word Problem Having a Cubic Equation as a Mathematical Model*

A cardboard box manufacturer makes open boxes of volume 140 in.^{3} from rectangular pieces of cardboard with dimensions 10 in. by 17 in. by cutting equal squares from the four corners and turning up the sides. On a graphics calculator find, to the nearest one-hundredth of an inch, the length of a side of the square cut out.

Solution Figure 4 represents a piece of cardboard and Figure 5 represents the box obtained from the cardboard. We define the unknown numbers.

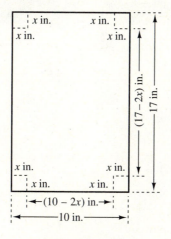

FIGURE 4

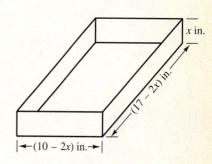

FIGURE 5

x: the number of inches in the length of the side of the square cut out
$17 - 2x$: the number of inches in the length of the base of the box
$10 - 2x$: the number of inches in the width of the base of the box

Of course, the height of the box is also x inches. The volume of the box is the product of the three dimensions. Setting this product equal to 140, we have the equation

$$x(17 - 2x)(10 - 2x) = 140$$
$$x(170 - 54x + 4x^2) = 140$$
$$4x^3 - 54x^2 + 170x - 140 = 0$$

We set y equal to the left side:

$$y = 4x^3 - 54x^2 + 170x - 140$$

The graph of this equation in the $[-5, 10]$ by $[-250, 100]$ viewing rectangle, with an x axis scale of 1, appears in Figure 6. The graph intersects the x axis at three points: one between 1 and 2; one between 2 and 3; and one between 9 and 10. Because the width of the cardboard is 10 in., the length of the side of the square cut out cannot be greater than 5 in. Therefore we need to consider only the first two points for values of x. Using zoom-in for each point, we obtain 1.33 and 2.81.

Conclusion: The length of the side of the square cut out is either 1.33 in. or 2.81 in. ◀

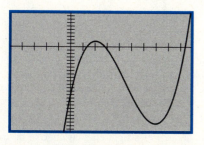

$[-5, 10]$ by $[-250, 100]$
$y = 4x^3 - 54x^2 + 170x - 140$
FIGURE 6

We will return to the situation in the previous example in Section 4.4 and determine how to cut a square so that the resulting box has maximum volume.

EXERCISES 2.4

In Exercises 1 through 12, find the solution set of the equation algebraically. Then check your solutions on your graphics calculator.

1. $\sqrt{x} - 5 = 3$

2. $\sqrt{x} + 4 = 7$

3. $\sqrt{x} + 5 = 3$

4. $\sqrt{x} - 4 = 7$

5. $\sqrt{2x - 3} = 2$

6. $\sqrt{2x + 5} = 3$

7. $\sqrt{y} + y = 6$

8. $\sqrt{t} + 6 = t$

9. $\sqrt{3x + 7} = x + 1$

10. $2\sqrt{x + 4} - 1 = x$

11. $\sqrt{x + 5} - \sqrt{x} = 1$

12. $\sqrt{2t + 1} - 2\sqrt{t - 1} = 0$

In Exercises 13 through 24, estimate the solutions of the equation on your graphics calculator. Then find the solution set algebraically.

13. $\sqrt{3x - 4} + 8 = 0$

14. $\sqrt{5x + 1} + \sqrt{3x + 1} = 0$

15. $\sqrt{5w + 1} - \sqrt{3w - 1} = 0$

16. $\sqrt{2 + 4y} + \sqrt{3 - 4y} = 3$

17. $\sqrt{2x + 11} + 1 = \sqrt{5x + 1}$

18. $7 - \sqrt{8 - x} = \sqrt{2x + 22}$

19. $\sqrt{y - 2\sqrt{y} + 3} + 6 = 0$

20. $\sqrt{w + 2\sqrt{w} - 1} + 2 = 0$

21. $\sqrt{4 - 3x} + \sqrt{3x - 9} = \sqrt{3x - 14}$

22. $\sqrt{2x - 1} + \sqrt{x + 4} = 3\sqrt{x - 1}$

23. $\sqrt[3]{3x + 1} = x + 1$

24. $\sqrt[3]{x^2 - 1} = x - 1$

In Exercises 25 through 36, find the solution set of the equation.

25. $x^4 - 5x^2 + 4 = 0$

26. $9x^4 - 8x^2 - 1 = 0$

27. $t^4 - 5t^2 + 6 = 0$

28. $6w^4 - 17w^2 + 12 = 0$

29. $8x^4 - 6x^2 - 5 = 0$

30. $8x^4 + 6x^2 - 9 = 0$

31. $5t^{-2} - 9t^{-1} - 2 = 0$

32. $y^{-4} - 37y^{-2} + 9 = 0$

33. $y^6 - 35y^3 + 216 = 0$

34. $27z^6 - 35z^3 + 8 = 0$

35. $\sqrt[4]{x} + 2\sqrt{x} = 3$

36. $\sqrt{x} - 5\sqrt[4]{x} + 4 = 0$

In Exercises 37 and 38, show that the solution set of the equation is $\varnothing$.

37. $\sqrt{3 - 3x} - \sqrt{3x + 2} = 3$

38. $\sqrt[4]{3x^2 - 3x + 1} = \sqrt{x - 2}$

39. Find three cube roots of 1 by solving the equation $x^3 = 1$.

40. Find three cube roots of 27 by solving the equation $x^3 = 27$.

41. Find three cube roots of -8 by solving the equation $x^3 = -8$.

42. Find four fourth roots of 1 by solving the equation $x^4 = 1$.

43. Find four fourth roots of 81 by solving the equation $x^4 = 81$.

44. Find six sixth roots of 64 by solving the equation $x^6 = 64$.

In Exercises 45 through 48, find on your graphics calculator the approximate value, to the nearest one-hundredth, of the indicated root.

45. $2x^3 - 2x^2 - 4x + 1 = 0$: **(a)** the negative root; **(b)** the smallest positive root.

46. $2x^3 - 8x^2 + x + 16 = 0$: **(a)** the negative root; **(b)** the largest positive root.

47. $x^4 - 10x^2 + 5 = 0$: **(a)** the smallest positive root; **(b)** the largest positive root.

48. $2x^4 - 2x^3 + x^2 + 3x - 4 = 0$: the negative root.

49. A spherical balloon is being inflated so that its volume is increasing at the rate of $\frac{148}{3}\pi$ ft^3/min. If the radius is now 3 ft, what will the radius be in 1 min? (The formula for the volume of a sphere is $V = \frac{4}{3}\pi r^3$.)

50. The lateral surface area of a right-circular cone of base radius r units and height h units is S square units where $S = \pi r\sqrt{r^2 + h^2}$. Solve this formula for **(a)** h and **(b)** r.

51. A spherical solid having radius r units and specific gravity k will sink in water to a depth of x units, where x is a positive root of the equation

$$x^3 - 3rx^2 + 4kr^3 = 0$$

On your graphics calculator find, to the nearest one-hundredth of an inch, the depth to which a spherical buoy of radius 10 in. and specific gravity of 0.1 will sink.

52. Use the equation of Exercise 51 and your graphics calculator to find, to the nearest one-hundredth of an inch, the depth to which a spherical ball will sink if the radius is 3 in. and the specific gravity is 0.5.

In Exercises 53 through 57, solve the word problem by finding an equation as a mathematical model of the situation. Complete the exercise by writing a conclusion.

53. The length of the diagonal of a rectangle is 50 in. and the area is 1200 in.^{2} What are the dimensions?

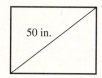

54. Solve Exercise 53 if the length of the diagonal is 13 in. and the area is 60 in.^{2}

55. A manufacturer of open tin boxes of volume 85 in.^{3} makes use of pieces of tin with dimensions 8 in. by 15 in. by cutting equal squares from the four corners and turning up the sides. On your graphics calculator find, to the nearest one-hundredth of an inch, the length of a side of the square to be cut out.

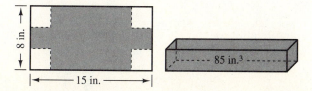

56. A cardboard box manufacturer makes open boxes of volume 120 cm^3 from square pieces of cardboard of side 12 cm by cutting equal squares from the four corners and turning up the sides. On your graphics calculator find, to the nearest one-hundredth of a centimeter, the length of a side of the square to be cut out.

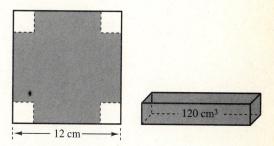

57. An open rectangular box is to be made of wood of uniform thickness on the sides and bottom. The inside dimensions are to be 5 ft in length, 3 ft in width, and 2 ft in height. If the wood weighs 40 lb/ft^3, use your graphics calculator to determine, to the nearest one-hundredth of an inch, what the thickness of the wood should be if the weight of the empty box is to be 160 lb.

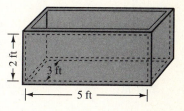

58. How do you determine on a graphics calculator whether a polynomial equation has real-number solutions and how many there are?

59. Why can extraneous solutions of an equation be introduced when both sides of the equation are squared? Are extraneous solutions always introduced when both sides of an equation are squared? Give examples of specific equations to substantiate your answers.

2.5 LINEAR INEQUALITIES

GOALS

1. **Learn properties of order.**
2. **Solve linear inequalities algebraically.**
3. **Solve linear inequalities graphically.**
4. **Solve word problems having a linear inequality as a mathematical model.**

In Section 1.1, a relation denoted by the symbols $<$ and $>$ gave an ordering for the set R of real numbers. In that section, we also introduced inequalities and used them to define intervals on the real-number line. You may wish to review these concepts at this time.

In calculus you need to be as proficient working with inequalities as you are with equations because inequalities occur in calculus almost as frequently as equations. In this section and the next two, we are concerned with inequalities involving one variable. Preliminary to our discussion we present the *trichotomy* and *transitive properties of order*.

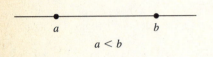

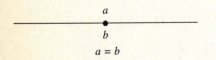

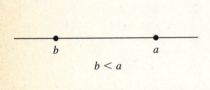

FIGURE 1

Trichotomy Property of Order

If a and b are real numbers, exactly one of the following three statements is true:
$$a < b \qquad b < a \qquad a = b$$

The geometric interpretation of the trichotomy property is that either point a or b on the real-number line lies to the left of the other or else they are the same point. Figure 1 shows the three possibilities.

Transitive Property of Order

If a, b, and c are real numbers, and if $a < b$ and $b < c$, then $a < c$.

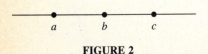

FIGURE 2

The geometric interpretation of the transitive property is shown in Figure 2; if point a is to the left of point b and b is to the left of point c, then a is to the left of c.

To prove the transitive property, we use the fact that the sum of two positive numbers is positive. We also apply the definition of $<$, as given in Section 1.1: If a and b are real numbers,

$a < b$ if and only if $b - a$ is positive

Proof of the Transitive Property   Because $a < b$, and $b < c$, it follows from the definition of $<$ that

$b - a$ is positive   and $c - b$ is positive

Because the sum of two positive numbers is positive,

$(b - a) + (c - b)$ is positive

By applying the commutative and associative properties of real numbers to this statement, we conclude that

$c - a$ is positive

and therefore

$a < c$ ∎

▷ **ILLUSTRATION 1**

If $x < 5$ and $5 < y$, then by the transitive property, it follows that $x < y$.
◀

The **domain** of a variable in an inequality is the set of real numbers for which the members of the inequality are defined. Examples of *linear inequalities* having the set R of real numbers as domain are

$$3x - 8 < 7 \qquad \frac{x - 7}{4} \le x \qquad 2 < 4x + 6 \le 14$$

An example of a *quadratic inequality* having domain R is

$$x^2 + 2x > 15$$

The inequality

$$\frac{3x}{x + 2} < 5$$

is rational. Because the left side is not defined when x is -2, the domain of x is the set of all real numbers except -2.

Any number in the domain for which the inequality is true is a **solution** of the inequality, and the set of all solutions is called the **solution set.** An

absolute inequality is one that is true for every number in the domain. For instance, if x is a real number,

$$x + 1 < x + 2 \quad \text{and} \quad x^2 \geq 0$$

are absolute inequalities. A **conditional inequality** is one for which there is at least one number in the domain that is not in the solution set. To find the solution set of a conditional inequality, we proceed in a manner similar to that used to solve an equation; that is, we obtain **equivalent inequalities** (those having the same solution set) until we have one whose solution set is apparent. The following properties are used to get equivalent inequalities.

Properties of $<$

If a, b, and c are real numbers and

(i) if $a < b$, then $a + c < b + c$ (addition property)
(ii) if $a < b$, then $a - c < b - c$ (subtraction property)
(iii) if $a < b$ and $c > 0$, then $ac < bc$ (multiplication property)
(iv) if $a < b$ and $c < 0$, then $ac > bc$ (multiplication property)

These properties can be proved by using the definition of $<$ and the fact that the product of two positive numbers is positive. We first prove the addition property (i). Because $a < b$, it follows from the definition of $<$ that

$$b - a > 0$$

From the commutative and associative properties of real numbers

$$b - a = (b + c) - (a + c)$$

Thus

$$(b + c) - (a + c) > 0$$

from which we conclude that

$$a + c < b + c$$

which proves property (i). Property (ii) follows from property (i) and the fact that subtracting c is the same as adding $-c$.

We now prove the multiplication property (iii). Because $a < b$, $b - a > 0$. Then because $c > 0$ and the product of two positive numbers is positive, we have

$$(b - a)c > 0$$
$$bc - ac > 0$$
$$ac < bc$$

which proves property (iii). You are asked to prove property (iv) in Exercise 49.

In this section, we concentrate on linear inequalities, that is, those that can be written in the form

$$ax + b < 0 \qquad \text{(the symbol } < \text{ can be replaced by } >, \leq, \text{ or } \geq\text{)}$$

where a and b are real numbers and $a \neq 0$. The following three examples pertain to finding algebraically the solution set of a linear inequality and representing the solution set on the real-number line.

▶ **EXAMPLE 1** *Solving a Linear Inequality Algebraically*

Find and show on the real-number line the solution set of the inequality

$$3x - 8 < 7$$

Solution   The following inequalities are equivalent:

$$3x - 8 < 7$$
$$3x - 8 + 8 < 7 + 8$$
$$3x < 15$$
$$\frac{1}{3}(3x) < \frac{1}{3}(15)$$
$$x < 5$$

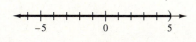

FIGURE 3

Therefore the solution set of the given inequality is $\{x \mid x < 5\}$, which is the interval $(-\infty, 5)$. This interval appears in Figure 3.   ◀

▶ **EXAMPLE 2** *Solving a Linear Inequality Algebraically*

Find and show on the real-number line the solution set of the inequality

$$\frac{x - 7}{4} \leq x$$

Solution　The following inequalities are equivalent:

$$\frac{x-7}{4} \le x$$

$$(4)\frac{x-7}{4} \le (4)x$$

$$x - 7 \le 4x$$

$$x - 7 - 4x + 7 \le 4x - 4x + 7$$

$$-3x \le 7$$

$$-\frac{1}{3}(-3x) \ge \left(-\frac{1}{3}\right)7$$

$$x \ge -\frac{7}{3}$$

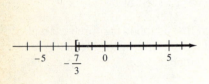

FIGURE 4

Thus the solution set of the given inequality is $\{x \mid x \ge -\frac{7}{3}\}$, which is the interval $[-\frac{7}{3}, +\infty)$ shown in Figure 4.　◀

▶　**EXAMPLE 3**　*Solving a Linear Inequality Algebraically*

Find and show on the real-number line the solution set of the inequality

$$3 < 4x + 7 \le 15$$

Solution　A solution of the given inequality must be a solution of both of the inequalities

$$3 < 4x + 7 \quad \text{and} \quad 4x + 7 \le 15$$

We solve each of these inequalities separately.

$$
\begin{array}{ll}
3 < 4x + 7 & 4x + 7 \le 15 \\
3 - 7 < 4x + 7 - 7 & 4x + 7 - 7 \le 15 - 7 \\
-4 < 4x & 4x \le 8 \\
\frac{1}{4}(-4) < \frac{1}{4}(4x) & \frac{1}{4}(4x) \le \frac{1}{4}(8) \\
-1 < x & x \le 2
\end{array}
$$

A value of x will be a solution of the given inequality if and only if

$$-1 < x \quad \text{and} \quad x \le 2$$
$$\Leftrightarrow \quad -1 < x \le 2$$

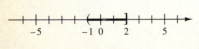

FIGURE 5

Therefore the solution set of the given inequality is the interval $(-1, 2]$ shown in Figure 5.

The work can be shortened by performing the same computations with the given continued inequality as follows:

$$3 < 4x + 7 \le 15$$

$$3 - 7 < 4x + 7 - 7 \le 15 - 7$$

$$-4 < 4x \le 8$$

$$\frac{-4}{4} < \frac{4x}{4} \le \frac{8}{4}$$

$$-1 < x \le 2 \qquad \blacktriangleleft$$

It is usually easier to solve linear inequalities algebraically than graphically. We demonstrate, however, how to use your graphics calculator to solve linear inequalities.

▶ **EXAMPLE 4** *Solving a Linear Inequality Graphically*

Use a graphics calculator to solve the linear inequality of Example 1.

Solution   We first rewrite the inequality so that all the nonzero terms are on the left side:

$$3x - 15 < 0$$

We set y equal to the left side:

$$y = 3x - 15$$

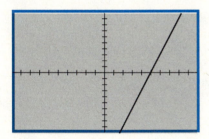

[−10, 10] by [−10, 10]
$y = 3x - 15$
FIGURE 6

The graph of this equation on our graphics calculator appears in Figure 6. The graph is below the x axis when $y < 0$. This occurs when $x < 5$, which agrees with the solution of Example 1.   ◀

The next example shows another graphical method of solving a linear inequality.

▶ **EXAMPLE 5** *Solving a Linear Inequality Graphically*

Use a graphics calculator to solve the linear inequality of Example 2.

Solution   The inequality is

$$\frac{x - 7}{4} \le x$$

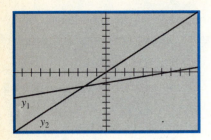

$[-10, 10]$ by $[-10, 10]$

$y_1 = \dfrac{x-7}{4}$ and $y_2 = x$

FIGURE 7

We set y_1 equal to the left side and y_2 equal to the right side:

$$y_1 = \frac{x-7}{4} \qquad y_2 = x$$

The solution set of the given inequality consists of those values of x for which $y_1 \leq y_2$. The graphs of the two equations appear in Figure 7. They intersect at the point $(-\frac{7}{3}, -\frac{7}{3})$, which can be verified by substituting the coordinates into the equation for y_1. The graph for y_1 intersects or lies below the graph for y_2 when $x \geq -\frac{7}{3}$, which agrees with the solution set of Example 2. ◀

▶ **EXAMPLE 6** *Solving a Linear Inequality Graphically*

Use a graphics calculator to solve the linear inequality of Example 3.

Solution   The inequality is

$$3 < 4x + 7 \leq 15$$

We let

$$y_1 = 3 \qquad y_2 = 4x + 7 \qquad y_3 = 15$$

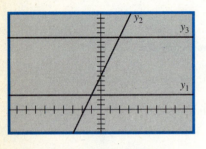

$[-10, 10]$ by $[-5, 20]$

$y_1 = 3, y_2 = 4x + 7$ and $y_3 = 15$

FIGURE 8

We wish to find the values of x for which $y_1 < y_2 \leq y_3$. In Figure 8 we have plotted the graphs of the three equations. The line for y_2 is above the line for y_1 and is either below or intersects the line for y_3 when $-1 < x \leq 2$, which agrees with the solution of Example 3. ◀

▶ **EXAMPLE 7** *Solving a Word Problem Having a Linear Inequality as a Mathematical Model*

A company that builds and sells cabinets has a weekly overhead, including salaries and plant cost, of \$3400. The cost of materials for each cabinet is \$40 and it is sold for \$200. How many cabinets must be built and sold each week so that the company is guaranteed a profit?

Solution   We begin by defining the unknown numbers.

x: the number of cabinets built and sold each week
$200x$: the number of dollars in the total weekly revenue
$3400 + 40x$: the number of dollars in the total weekly cost
P: the number of dollars in the weekly profit
Because profit equals revenue minus cost, we have

$$P = 200x - (3400 + 40x)$$
$$P = 160x - 3400$$

For a profit, P must be positive; that is,

$$160x - 3400 > 0$$
$$160x > 3400$$
$$x > 21.25$$

Because x must be a positive integer, we conclude that $x \geq 22$.

Conclusion: The company must build and sell at least 22 cabinets each week to be guaranteed a profit. ◀

EXERCISES 2.5

In Exercises 1 through 28, find algebraically the solution set of the inequality and write it with interval notation. Show the solution set on the real-number line.

1. $20 \leq 4x$

2. $3x - 5 < 7$

3. $2x - 1 < 6$

4. $3x + 1 \geq 4x - 3$

5. $2x + 1 > x - 4$

6. $5x + 6 \leq x - 2$

7. $\dfrac{2x - 5}{4} \leq 3$

8. $\dfrac{2x - 9}{7} > 0$

9. $-3 > \dfrac{3x + 5}{4}$

10. $\dfrac{x - 5}{4} \leq x$

11. $\dfrac{2x - 5}{3} < x + 1$

12. $6x - 1 < \dfrac{5x - 1}{3}$

13. $\frac{1}{2}x + 2 \leq \frac{1}{4}x + 3$

14. $5 - \frac{1}{3}x \geq \frac{1}{4}x - 2$

15. $10 - 3x > \dfrac{4x - 5}{-3}$

16. $\dfrac{3x + 8}{-5} < 4 - 2x$

17. $5 \leq 2x - 3 < 13$

18. $11 \geq 3x - 5 > 2$

19. $-7 < 2x + 1 < 3$

20. $1 \leq 3x - 2 \leq 16$

21. $1 < \dfrac{4x - 1}{3} < 5$

22. $10 < 4 - 3x < 19$

23. $6 \leq 2 - x \leq 8$

24. $-1 < \dfrac{7 - 2x}{5} \leq 5$

25. $2 > -3 - 3x \geq -7$

26. $-1 < 2 - 2x \leq 3$

27. $-4 < \dfrac{5x + 3}{-6} \leq 2$

28. $-4 \leq \dfrac{5 + 3x}{-2} \leq 5$

In Exercises 29 through 40, find the solution set of the inequality of the indicated exercise on your graphics calculator.

29. Exercise 1

30. Exercise 2

31. Exercise 7

32. Exercise 8

33. Exercise 9

34. Exercise 10

35. Exercise 15

36. Exercise 16

37. Exercise 17

38. Exercise 18

39. Exercise 23

40. Exercise 24

41. If the temperature using the Fahrenheit scale is F degrees and using the Celsius scale is C degrees, then

$$C = \frac{5}{9}(F - 32)$$

What is the set of values of F if C is between 10 and 20?

42. When the temperature of water is greater than or equal to 100 degrees Celsius, the water boils. Use the formula in Exercise 41 to determine the temperature Fahrenheit when water boils.

In Exercises 43 through 48, solve the word problem by finding an inequality as a mathematical model of the situation. Complete the exercise by writing a conclusion.

43. An investor has $8000 invested at 9 percent and wishes to invest some additional money at 16 percent in order to realize a return of at least 12 percent on the total investment. What amount of money should be invested?

44. Part of $20,000 is to be invested at 9 percent and the remainder is to be invested at 12 percent. What is the least amount of money that can be invested at 12 percent to have a yearly income of at least $2250 from the two investments?

45. A lamp manufacturer sells only to wholesalers through its showroom. The weekly overhead, including salaries, plant cost, and showroom rental, is $6000. If each lamp sells for $168 and the material used in its production costs $44, how many lamps must be made and sold each week so that the manufacturer realizes a profit?

46. If in a particular course a student has an average score of less than 90 and not below 80 on four examinations, the student will receive a grade of B in the course. If the student's scores on the first three examinations are 87, 94, and 73, what score on the fourth examination will result in a B grade?

47. A silversmith wishes to obtain an alloy containing at least 72 percent silver and at most 75 percent silver. Determine the greatest and least amounts of an 80 percent silver alloy that should be combined with a 65 percent silver alloy to have 30 grams of the required alloy.

48. What amount of pure alcohol must be added to 24 liters of a 20 percent alcohol solution to obtain a mixture that is at least 30 percent alcohol?

49. Prove the multiplication property: If $a < b$ and $c < 0$, then $ac > bc$. *Hint:* If $c < 0$, then $-c > 0$.

50. Prove that if $x < y$, then $x < \frac{1}{2}(x + y) < y$.

51. Prove that if $a, b, c,$ and d are real numbers, and

$$\text{if } a < b \text{ and } c < d, \text{ then } a + c < b + d$$

Hint: Use the definition of $<$ and the fact that the sum of two positive numbers is positive.

52. If $a > b$ and $c > d$, we cannot conclude that ac is necessarily greater than bd. Why? Give an example.

53. Examples 4 and 5 give two methods for solving a linear inequality graphically. Explain in words the two methods applied to the inequality of Exercise 9.

54. In Example 6, we solve a continued linear inequality graphically. Explain in words the method used in this example to solve the inequality of Exercise 25.

2.6 POLYNOMIAL INEQUALITIES

GOALS
1. **Solve quadratic inequalities algebraically.**
2. **Solve quadratic inequalities graphically.**
3. **Solve some particular polynomial inequalities of degree greater than 2.**
4. **Solve graphically polynomial inequalities of degree greater than 2 where the solution set has irrational endpoints.**
5. **Solve word problems having a polynomial inequality as a mathematical model.**

We now discuss algebraic and graphical methods for solving inequalities containing polynomials of degree greater than 1.

A **quadratic inequality** is one of the form

$$ax^2 + bx + c < 0 \quad \text{(the symbol } < \text{ can be replaced by } >, \leq, \text{ or } \geq)$$

where $a, b,$ and c are real numbers and $a \neq 0$. To solve a quadratic inequality, we use the concepts of *critical number* and *test number*.

A **critical number** of the preceding inequality is a real root of the quadratic equation

$$ax^2 + bx + c = 0$$

Suppose r_1 and r_2 are critical numbers and $r_1 < r_2$. Then the polynomial $ax^2 + bx + c$ can change algebraic sign only at r_1 and r_2. Thus the sign (plus or minus) of $ax^2 + bx + c$ will be constant on each of the intervals

$$(-\infty, r_1) \quad (r_1, r_2) \quad (r_2, +\infty)$$

To determine the sign on a particular one of these intervals, we compute the value of $ax^2 + bx + c$ at an arbitrary **test number** in the interval. From the results we can obtain the solution set of the inequality. The procedure is shown in the following illustration and example.

▷ ILLUSTRATION 1

To solve the inequality

$$x^2 - 8 < 2x$$

we first write an equivalent inequality having all the nonzero terms on one side of the inequality sign. Thus we have

$$x^2 - 2x - 8 < 0$$
$$(x + 2)(x - 4) < 0$$

We observe from the factored form of the inequality that the equation $x^2 - 2x - 8 = 0$ has the roots -2 and 4, which are the critical numbers of the inequality. We locate on the real-number line the points corresponding to these numbers. See Figure 1. These points separate the line into the following three intervals:

$$(-\infty, -2) \quad (-2, 4) \quad (4, +\infty)$$

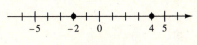

FIGURE 1

On each of these intervals the sign of $(x + 2)(x - 4)$ is constant. To determine the sign on an interval, we choose an arbitrary test number in the interval and compute the sign of each of the factors $x + 2$ and $x - 4$ at this test number. We select -3 in $(-\infty, -2)$, 0 in $(-2, 4)$, and 5 in $(4, +\infty)$. The results are summarized in Table 1.

Table 1

Interval	Test Number k	Sign of $x + 2$ at k	Sign of $x - 4$ at k	Sign of $(x + 2)(x - 4)$ on Interval
$(-\infty, -2)$	-3	$-$	$-$	$+$
$(-2, 4)$	0	$+$	$-$	$-$
$(4, +\infty)$	5	$+$	$+$	$+$

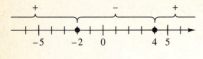

FIGURE 2

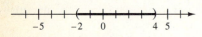

FIGURE 3

From the table we obtain Figure 2, which indicates on the real-number line the points -2 and 4 and the intervals on which $(x + 2)(x - 4)$ is positive or negative. We conclude that the solution set of the inequality is the interval $(-2, 4)$ shown in Figure 3.

The inequality can be solved graphically in two ways. For one method we let

$$y_1 = x^2 - 8 \qquad y_2 = 2x$$

The graphs of these two equations, plotted on a graphics calculator, appear in Figure 4. Observe that the graph for y_1 is below the graph for y_2 (that is, $y_1 < y_2$) when x is in the open interval $(-2, 4)$.

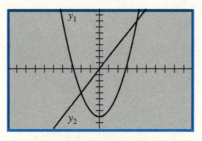

[−10, 10] by [−10, 10]
$y_1 = x^2 - 8$ and $y_2 = 2x$
FIGURE 4

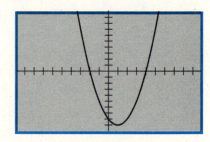

[−10, 10] by [−10, 10]
$y = x^2 - 2x - 8$
FIGURE 5

For another graphical method we let

$$y = x^2 - 2x - 8$$

The graph of this equation, plotted in Figure 5, is below the x axis (that is, $y < 0$) when x is in the open interval $(-2, 4)$.   ◄

▶ **EXAMPLE 1** *Solving a Quadratic Inequality Algebraically and Graphically*

Find algebraically the solution set of the inequality

$$x^2 + 2x \geq 15$$

and show the solution set on the real-number line. Verify the result graphically.

Solution   The given inequality is equivalent to

$$x^2 + 2x - 15 \geq 0$$
$$(x + 5)(x - 3) \geq 0$$

The critical numbers are -5 and 3. The points corresponding to these numbers are located in Figure 6, and the following intervals are determined:

$$(-\infty, -5) \quad (-5, 3) \quad (3, +\infty)$$

Table 2 summarizes the results obtained by choosing a test number in each of these intervals and determining the sign of $(x + 5)(x - 3)$ on the intervals.

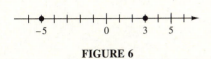

FIGURE 6

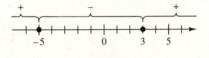

FIGURE 7

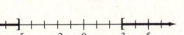

FIGURE 8

Table 2

Interval	Test Number k	Sign of $x + 5$ at k	Sign of $x - 3$ at k	Sign of $(x + 5)(x - 3)$ on Interval
$(-\infty, -5)$	-6	$-$	$-$	$+$
$(-5, 3)$	0	$+$	$-$	$-$
$(3, +\infty)$	4	$+$	$+$	$+$

Figure 7 shows on the real-number line the points -5 and 3 as well as the sign of $(x + 5)(x - 3)$ on the intervals $(-\infty, -5)$, $(-5, 3)$, and $(3, +\infty)$. Therefore $(x + 5)(x - 3) > 0$ if x is in either $(-\infty, -5)$ or $(3, +\infty)$. Furthermore, -5 and 3 are in the solution set because $(x + 5)(x - 3) = 0$ if x is either of these numbers. Thus the solution set of the given inequality is $(-\infty, -5] \cup [3, +\infty)$, appearing in Figure 8.

We plot the graph of

$$y = x^2 + 2x - 15$$

and obtain Figure 9. This graph intersects or is above the x axis (that is, $y \geq 0$) when x is in $(-\infty, -5] \cup [3, +\infty)$, which agrees with the result found algebraically.   ◀

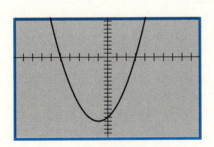

$[-10, 10]$ by $[-20, 10]$
$y = x^2 + 2x - 15$
FIGURE 9

▶ **EXAMPLE 2** *Solving Algebraically a Quadratic Inequality Whose Solution Set Is ∅*

Find the solution set of the inequality

$$5x^2 - 2x + 1 < x^2 + 2x$$

Solution   The given inequality is equivalent to

$$4x^2 - 4x + 1 < 0$$
$$(2x - 1)^2 < 0$$

Because there is no value of x for which $(2x - 1)^2$ is negative, there is no solution. Therefore, the solution set is $\varnothing$.   ◀

▶ **EXAMPLE 3** *Solving Algebraically an Absolute Quadratic Inequality*

Find the solution set of the inequality

$$-6x^2 - 8x + 1 \leq 3x^2 + 4x + 5$$

Solution   The given inequality is equivalent to

$$-9x^2 - 12x - 4 \leq 0$$
$$9x^2 + 12x + 4 \geq 0$$
$$(3x + 2)^2 \geq 0$$

Because $(3x + 2)^2$ is nonnegative for all values of x, the solution set is the set R of all real numbers. Therefore the inequality is absolute.   ◀

▶ **EXAMPLE 4** *Solving Graphically a Rectilinear Motion Problem Having a Quadratic Inequality as a Mathematical Model*

In Example 7 of Section 2.2, a ball is thrown vertically upward from the ground with an initial velocity of 72 ft/sec. The equation of motion is

$$s = -16t^2 + 72t$$

where s feet is the distance of the ball from the ground t seconds after it is thrown. Find graphically when the ball will be more than 25 ft above the ground.

Solution   In the viewing rectangle $[0, 10]$ by $[0, 100]$, we plot the graph of the equation of motion and the line $s = 25$. See Figure 10. We wish to determine the values of t for which

$$-16t^2 + 72t > 25$$

that is, the values of t for which the parabola is above the line. Using trace and zoom-in, we find that the parabola and the line intersect at the points where t is 0.38 and 4.12. Therefore from Figure 10, the parabola is above the line when $0.38 < t < 4.12$.

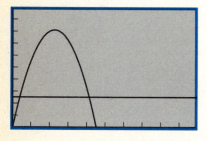

[0, 10] by [0, 100]
$s = -16t^2 + 72t$ and $s = 25$
FIGURE 10

Conclusion: The ball will be more than 25 ft above the ground when the time since the ball was thrown is between 0.38 sec and 4.12 sec. ◄

In the next example, we solve a *cubic* inequality by the same method we used in Illustration 1 and Example 1 to solve a quadratic inequality.

► **EXAMPLE 5** *Solving a Cubic Inequality Algebraically and Graphically*

Find and show on the real-number line the solution set of the inequality

$$(x + 1)(2x^2 - 5x + 2) > 0$$

Verify the result graphically.

Solution The given inequality is equivalent to

$$(x + 1)(2x - 1)(x - 2) > 0$$

The critical numbers are $-1, \frac{1}{2}$, and 2, and the points corresponding to these numbers are located on the real-number line in Figure 11. These points determine the following intervals:

$$(-\infty, -1) \quad (-1, \tfrac{1}{2}) \quad (\tfrac{1}{2}, 2) \quad (2, +\infty)$$

We compute the sign of $(x + 1)(2x - 1)(x - 2)$ in each interval by selecting a test number there. The results are summarized in Table 3.

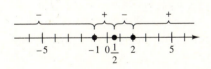

FIGURE 11

Table 3

Interval	Test Number k	Sign of $x + 1$ at k	Sign of $2x - 1$ at k	Sign of $x - 2$ at k	Sign of $(x + 1)(2x - 1)(x - 2)$ on Interval
$(-\infty, -1)$	-2	$-$	$-$	$-$	$-$
$(-1, \frac{1}{2})$	0	$+$	$-$	$-$	$+$
$(\frac{1}{2}, 2)$	1	$+$	$+$	$-$	$-$
$(2, +\infty)$	3	$+$	$+$	$+$	$+$

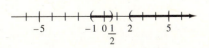

FIGURE 12

FIGURE 13

From the table we obtain Figure 12, showing on the real-number line the points $-1, \frac{1}{2}$, and 2 as well as the sign of $(x + 1)(2x - 1)(x - 2)$ on the intervals $(-\infty, -1)$, $(-1, \frac{1}{2})$, $(\frac{1}{2}, 2)$, and $(2, +\infty)$. Therefore $(x + 1)(2x - 1)(x - 2) > 0$ if x is in either $(-1, \frac{1}{2})$ or $(2, +\infty)$. Hence the solution set of the given inequality is $(-1, \frac{1}{2}) \cup (2, +\infty)$, which appears in Figure 13.

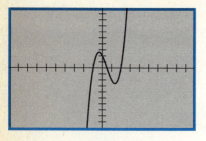

$[-10, 10]$ by $[-10, 10]$
$y = (x + 1)(2x^2 - 5x + 2)$
FIGURE 14

Figure 14 shows the graph of $y = (x + 1)(2x^2 - 5x + 2)$ plotted on our graphics calculator. The graph is above the x axis (that is, $y > 0$) if x is in $(-1, \frac{1}{2}) \cup (2, +\infty)$, which agrees with our algebraic solution. ◀

The cubic inequality in Example 5 was given in factored form. Often a mathematical model obtained in a practical situation is an inequality involving a polynomial of degree greater than 2 that cannot be factored. We solve such inequalities graphically by zooming in as shown in the following example.

▶ **EXAMPLE 6** *Solving Graphically a Polynomial Inequality of Degree Greater than 2 Where the Solution Set has Irrational-number Endpoints*

Find the solution set of the inequality

$$x^3 < 2x + 7$$

Express any irrational-number endpoints to the nearest one-hundredth.

Solution The given inequality is equivalent to

$$x^3 - 2x - 7 < 0$$

We let

$$y = x^3 - 2x - 7$$

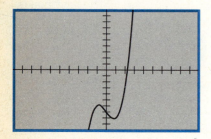

$[-10, 10]$ by $[-10, 10]$
$y = x^3 - 2x - 7$
FIGURE 15

Figure 15 shows the graph of this equation plotted on our graphics calculator. We observe that the graph intersects the x axis between 2 and 3. We zoom in and determine that this x intercept is 2.26 to the nearest one-hundredth. Because the graph is below the x axis (that is, $y < 0$ when $x < 2.26$, the solution set of the given inequality is the interval $(-\infty, 2.26)$. ◀

Note in Example 6 that even though 2.26, the right endpoint of the solution set, is an approximation and not exact, we are treating it as exact in interval notation. We shall use this convention throughout the text.

Inequalities involving polynomials of degree greater than 3 are solved graphically by the same method as in Example 6.

▶ **EXAMPLE 7** *Solving a Word Problem Having a Quadratic Inequality as a Mathematical Model*

A decorator designs and sells wall fixtures and can sell at a price of $75 each all the fixtures produced. If x fixtures are manufactured each day, then the number of dollars in the daily total cost of production is $x^2 + 25x + 96$. How many fixtures should be produced each day so that the decorator is guaranteed a profit?

Solution   We define the unknown numbers.

x: the number of fixtures manufactured each day
$75x$: the number of dollars in the total revenue from the sale of x fixtures
P: the number of dollars in the daily total profit

Because profit equals revenue minus cost,

$$P = 75x - (x^2 + 25x + 96)$$
$$= -x^2 + 50x - 96$$

For the decorator to be guaranteed a profit, we must have $P > 0$; that is,

$$-x^2 + 50x - 96 > 0$$
$$x^2 - 50x + 96 < 0$$
$$(x - 2)(x - 48) < 0$$

We wish to solve this inequality. Because x is the number of fixtures, the solution set is restricted to nonnegative values of x. The critical numbers are 2 and 48. The points corresponding to these two numbers separate the nonnegative side of the real-number line into the following three intervals:

$$[0, 2) \quad (2, 48) \quad (48, +\infty)$$

Table 4 summarizes the results obtained by choosing a test number in each of these intervals to determine the sign of $(x - 2)(x - 48)$ on the interval.

Table 4

Interval	Test Number k	Sign of $x - 2$ at k	Sign of $x - 48$ at k	Sign of $(x - 2)(x - 48)$ on Interval
$[0, 2)$	1	$-$	$-$	$+$
$(2, 48)$	3	$+$	$-$	$-$
$(48, +\infty)$	49	$+$	$+$	$+$

From the table, the solution set of the inequality is the open interval $(2, 48)$.

Conclusion: For the decorator to be guaranteed a profit, the number of fixtures produced and sold each day must be more than 2 and less than 48.

◀

In Section 4.3 we return to the situation in Example 7 and learn how to determine the number of fixtures that should be produced and sold each day for the decorator to have the greatest profit.

EXERCISES 2.6

In Exercises 1 through 30, find algebraically the solution set of the inequality, write the solution set with interval notation, and show the solution set on the real-number line. Then verify your result graphically.

1. $x^2 > 9$
2. $x^2 < 4$
3. $(x + 3)(x - 4) < 0$
4. $(x - 1)(x - 5) > 0$
5. $(2x + 1)(2x - 7) > 0$
6. $(3x + 5)(2x - 3) < 0$
7. $x^2 - 4x + 3 \le 0$
8. $x^2 + 6x + 8 \ge 0$
9. $4 - 3x - x^2 \ge 0$
10. $2x^2 + x - 1 \le 0$
11. $x^2 > 8 - 2x$
12. $x^2 < 15 + 2x$
13. $x \le 6 - 2x^2$
14. $4x^2 \ge 9 - 9x$
15. $16t^2 + 1 \ge 8t$
16. $9y^2 < 30y - 25$
17. $x(11 - 3x) < 10$
18. $x(6x + 1) \ge 15$
19. $(y + 3)(y - 1)(y - 4) > 0$
20. $(x + 4)(x^2 - 4) < 0$
21. $x^3 \le 16x$
22. $6t^3 > 7t^2 + 3t$
23. $x^2 - x - 1 < 0$
24. $x^2 + 2x - 2 \ge 0$
25. $x^2 + x + 3 > 0$
26. $x^2 - 2x + 2 < 0$
27. $x^4 - 13x^2 + 36 < 0$
28. $(x^2 + 3x - 4)(x^2 - x - 2) \le 0$
29. $(x + 3)(x^2 - x - 2)(x^2 - 8x + 15) \ge 0$
30. $x^5 - 5x^3 + 4x > 0$

In Exercises 31 through 38, find graphically the solution set of the inequality, and express the endpoints of intervals to the nearest one-hundredth.

31. $x^3 + x^2 + x - 4 > 0$
32. $x^3 + 6x < 13$
33. $x^3 + 3x^2 < x + 2$
34. $x^3 + 6x > 5x^2 + 1$
35. $2x^4 - 14x^3 + 24x^2 + x - 4 > 0$
36. $3x^4 + 36x^2 - 8 < 21x^3 - 2x$
37. $9x^2 + 8x - 14 < x^4 + 2x^3$
38. $x^5 + 2x^4 - 8x^3 - 9x^2 + 8x + 6 > 0$

In Exercises 39 through 50, solve the word problem by finding an inequality as a mathematical model of the situation. Complete the exercise by solving the inequality and writing a conclusion.

39. In Exercise 57 of Exercises 2.2, an object is thrown vertically upward from the ground with an initial velocity of 128 ft/sec. Find graphically when the object will be more than 50 ft above the ground.

40. Find graphically when the object of Exercise 39 will be less than 30 ft above the ground.

41. In Exercise 59 of Exercises 2.2, a ball is thrown upward from the edge of a rooftop 68 ft above the ground with an initial velocity of 76 ft/sec. Find graphically when the ball will be **(a)** less than 100 ft above the ground and **(b)** at most 25 ft above the ground.

42. Find graphically when the ball of Exercise 41 will be **(a)** more than 80 ft above the ground and **(b)** at least 50 ft above the ground.

43. A firm can sell at a price of $100 per unit all of a particular commodity it produces. If x units are produced each day, the number of dollars in the total cost of each day's production is $x^2 + 20x + 700$. How many units should be produced each day so that the firm is guaranteed a profit?

44. A company that builds and sells desks can sell at a price of $400 per desk all the desks it produces. If x desks are built and sold each week, then the number of dollars in the total cost of the week's production is $2x^2 + 80x + 3000$. How many desks should be built each week so that the manufacturer is guaranteed a profit?

45. A rectangular field is to be fenced off along the bank of a river; no fence is required along the river. The material for the fence costs $8 per running foot for the two ends and $16 per running foot for the side parallel to the river. If the area of the field is to be 12,000 ft² and the cost of the fence is not to exceed $3520, what are the restrictions on the dimensions of the field?

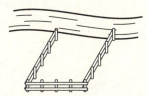

46. A rectangular plot of ground is to be enclosed by a fence and then divided down the middle by another fence. The fence down the middle costs $3 per

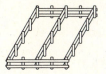

running foot and the other fence costs $6 per running foot. If the area of the plot is to be 1800 ft² and the cost of the fence is not to exceed $2310, what are the restrictions on the dimensions of the plot?

47. For the park containing a garden in Exercise 15 of Exercises 2.3, determine graphically the possible widths of the concrete terrace if the total area of the terrace is to be at least 2500 m² and at most 3600 m².

48. For the field in Exercise 16 of Exercises 2.3, determine graphically the possible widths of the plowed strip if at least two-thirds of the field is to be plowed and if at most three-fourths is to be plowed.

49. For the manufacturer in Exercise 55 of Exercises 2.4, determine graphically the possible lengths of the sides of the cutout squares if the volume of the tin boxes must be at least 75 in.³ and at most 86 in.³

50. For the manufacturer in Exercise 56 of Exercises 2.4, determine graphically the possible lengths of the sides of the cutout squares if the volume of the cardboard boxes must be more than 110 cm³ and less than 122 cm³.

2.7 EQUATIONS AND INEQUALITIES INVOLVING ABSOLUTE VALUE

GOALS
1. **Solve algebraically equations involving absolute value.**
2. **Solve graphically equations involving absolute value.**
3. **Solve algebraically inequalities involving absolute value.**
4. **Solve graphically inequalities involving absolute value.**
5. **Learn theorems about absolute value.**
6. **Show that one inequality involving absolute value is equivalent to a simpler one.**
7. **Learn the triangle inequality.**

Both equations and inequalities involving absolute value occur in calculus, and learning to solve them will help you understand certain concepts in a calculus course. To solve an equation involving absolute value, we apply the

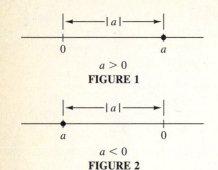

$a > 0$
FIGURE 1

$a < 0$
FIGURE 2

definition in Section 1.1, which states that the absolute value of a real number a, denoted by $|a|$, is given by

$$|a| = \begin{cases} a & \text{if } a \geq 0 \\ -a & \text{if } a < 0 \end{cases}$$

Recall from Section 1.1 that on the real-number line, $|a|$ is the distance (without regard to direction, left or right) from the origin to point a. See Figure 1, where a is positive, and Figure 2, where a is negative.

▶ **EXAMPLE 1** *Solving Algebraically and Graphically an Equation Involving Absolute Value*

Find algebraically the solution set of the equation

$$|3x + 5| = 9$$

Check the solutions on a graphics calculator.

Solution   The given equation is satisfied if either

$$\begin{array}{lll} 3x + 5 = 9 & \text{or} & -(3x + 5) = 9 \\ 3x = 4 & & -3x - 5 = 9 \\ x = \dfrac{4}{3} & & -3x = 14 \\ & & x = -\dfrac{14}{3} \end{array}$$

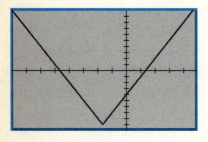

$[-8, 5]$ by $[-10, 10]$
$y = |3x + 5| - 9$
FIGURE 3

The solution set is $\{\frac{4}{3}, -\frac{14}{3}\}$.

To check the solutions on our graphics calculator, we let

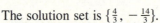
$$y = |3x + 5| - 9$$

and plot the graph. The x intercepts of the graph, shown in Figure 3, appear to agree with our solutions $-\frac{14}{3}$ and $\frac{4}{3}$.   ◀

▶ **EXAMPLE 2** *Solving Algebraically and Graphically an Equation Involving Absolute Value*

Find algebraically the solution set of the equation

$$|2x - 3| = |7 - 3x|$$

Check the solutions graphically.

Solution   The given equation is satisfied if either

$$2x - 3 = 7 - 3x \qquad \text{or} \qquad 2x - 3 = -(7 - 3x)$$

$$2x + 3x = 7 + 3 \qquad\qquad\qquad 2x - 3 = -7 + 3x$$

$$5x = 10 \qquad\qquad\qquad\qquad 2x - 3x = -7 + 3$$

$$x = 2 \qquad\qquad\qquad\qquad\qquad -x = -4$$

$$x = 4$$

The solution set is $\{2, 4\}$.

To check the solutions, we let

$$y_1 = |2x - 3| \qquad \text{and} \qquad y_2 = |7 - 3x|$$

The graphs of these two equations are plotted in Figure 4. They intersect at the points where $x = 2$ and $x = 4$; these solutions agree with the solutions found algebraically. ◄

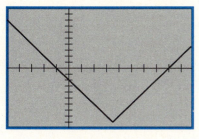

$[-5, 5]$ by $[-1, 10]$

$y_1 = |2x - 3|$ and $y_2 = |7 - 3x|$

FIGURE 4

We begin our discussion of inequalities involving absolute value by solving such an inequality graphically.

▷ **ILLUSTRATION 1**

To solve the inequality

$$|2x - 7| - 9 < 0$$

graphically, we let

$$y = |2x - 7| - 9$$

and plot the graph shown in Figure 5. We want to determine when $y < 0$, that is, when the graph of the equation is below the x axis. From Figure 5, we observe that this occurs when $-1 < x < 8$. The solution set is, therefore, the open interval $(-1, 8)$. ◄

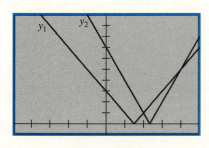

$[-5, 10]$ by $[-10, 10]$

$y = |2x - 7| - 9$

FIGURE 5

▷ **ILLUSTRATION 2**

The inequality of Illustration 1 is equivalent to

$$|2x - 7| < 9$$

This form gives us another graphical method of solving the inequality. Let

$$y_1 = |2x - 7| \qquad \text{and} \qquad y_2 = 9$$

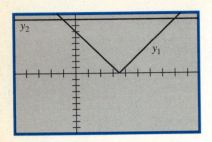

$[-5, 10]$ by $[-10, 10]$
$y_1 = |2x - 7|$ and $y_2 = 9$

FIGURE 6

The graphs of y_1 and y_2 plotted in the same viewing rectangle appear in Figure 6. Observe that the graph of y_1 is below the graph of y_2 when $-1 < x < 8$. Thus as in Illustration 1, the solution set is the open interval $(-1, 8)$. ◀

Before solving algebraically an inequality involving absolute value, we consider an additional property of absolute value.

$-3 < x < 3$

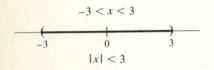

$|x| < 3$

FIGURE 7

▷ **ILLUSTRATION 3**

Suppose we have the inequality

$$|x| < 3$$

This inequality states that on the real-number line the distance from the origin to the point x is less than 3 units; that is, $-3 < x < 3$. Therefore, x is in the open interval $(-3, 3)$. See Figure 7. It appears then that the solution set of the inequality $|x| < 3$ is $\{x \mid -3 < x < 3\}$. We now show that this is the case.

Because $|x| = x$ if $x \geq 0$ and $|x| = -x$ if $x < 0$, it follows that the solution set of the inequality $|x| < 3$ is the union of the sets

$$\{x \mid x < 3 \text{ and } x \geq 0\} \quad \text{and} \quad \{x \mid -x < 3 \text{ and } x < 0\}$$

Observe that the first of these sets is equivalent to $\{x \mid 0 \leq x < 3\}$, and the second is equivalent to $\{x \mid -3 < x < 0\}$ because $-x < 3$ is equivalent to $x > -3$. Thus the solution set of $|x| < 3$ is

$$\{x \mid 0 \leq x < 3\} \cup \{x \mid -3 < x < 0\}$$
$$\Leftrightarrow \{x \mid -3 < x < 3\}$$ ◀

In the preceding illustration, by comparing the given inequality and its solution set, we conclude that the inequality

$$|x| < 3 \quad \text{is equivalent to} \quad -3 < x < 3$$

More generally, if $b > 0$,

$$|x| < b \quad \text{is equivalent to} \quad -b < x < b$$

The proof of this statement is exactly the same as the argument given in Illustration 3 if we replace 3 by b where $b > 0$. Furthermore, if instead of

x we have an algebraic expression E and $b > 0$, then the inequality

$$|E| < b \qquad \text{is equivalent to} \qquad -b < E < b \tag{1}$$

This statement is valid if the symbol $<$ is replaced by $\leq$.

▶ **EXAMPLE 3** *Solving Algebraically an Inequality Involving Absolute Value*

Find algebraically and show on the real-number line the solution set of the inequality of Illustration 1.

Solution The inequality

$$|2x - 7| < 9$$

is equivalent to

$$-9 < 2x - 7 < 9$$
$$-9 + 7 < 2x < 9 + 7$$
$$-2 < 2x < 16$$
$$-1 < x < 8$$

FIGURE 8

The solution set is, therefore, the open interval $(-1, 8)$ shown in Figure 8. ◀

▷ **ILLUSTRATION 4**

Consider the inequality

$$|x| > 2$$

This inequality states that on the real-number line the distance from the origin to the point x is greater than 2 units; that is, either $x > 2$ or $x < -2$. Therefore x is in $(-\infty, -2) \cup (2, +\infty)$. See Figure 9. Thus it appears that the solution set of $|x| > 2$ is $\{x \mid x > 2\} \cup \{x \mid x < -2\}$. We now use properties of absolute value to show that this is the situation.

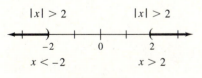

FIGURE 9

From the definition of $|x|$, the solution set of the inequality $|x| > 2$ is the union of the sets

$$\{x \mid x > 2 \text{ and } x \geq 0\} \qquad \text{and} \qquad \{x \mid -x > 2 \text{ and } x < 0\}$$

The first of these sets is equivalent to $\{x \mid x > 2\}$, and the second is equivalent to $\{x \mid x < -2\}$ because $-x > 2$ is equivalent to $x < -2$. Thus the solution set of $|x| > 2$ is

$$\{x \mid x > 2\} \cup \{x \mid x < -2\} \qquad ◀$$

By comparing the given inequality and its solution set, we observe from the preceding illustration that the inequality

$$|x| > 2 \qquad \text{is equivalent to} \qquad x > 2 \text{ or } x < -2$$

More generally, if $b > 0$,

$$|x| > b \qquad \text{is equivalent to} \qquad x > b \text{ or } x < -b$$

The proof of this statement is identical to the discussion in Illustration 4 if the number 2 is replaced by b, where $b > 0$. If instead of x we have an algebraic expression E and $b > 0$, then the inequality

$$|E| > b \qquad \text{is equivalent to} \qquad E > b \text{ or } E < -b \tag{2}$$

That is, the solution set of the inequality $|E| > b$ is the union of the solution sets of the inequalities $E > b$ and $E < -b$.

From (1), $|E| < b$ is equivalent to the continued inequality $-b < E < b$; however, $|E| > b$ is not equivalent to any continued inequality.

Statement (2) is valid if the symbol $>$ is replaced by $\geq$ and the symbol $<$ is replaced by $\leq$.

▶ **EXAMPLE 4** *Solving Algebraically and Graphically an Inequality Involving Absolute Value*

Find algebraically and show on the real-number line the solution set of the inequality

$$\left| \frac{2}{3}x - 5 \right| \geq 3$$

Verify the solutions on a graphics calculator.

Solution   The solution set of the given inequality is the union of the solution sets of the inequalities

$$\frac{2}{3}x - 5 \geq 3 \qquad \frac{2}{3}x - 5 \leq -3$$

$$2x - 15 \geq 9 \qquad 2x - 15 \leq -9$$

$$2x \geq 24 \qquad 2x \leq 6$$

$$x \geq 12 \qquad x \leq 3$$

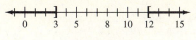

FIGURE 10

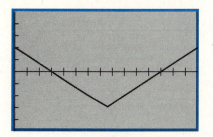

$[0, 15]$ by $[-5, 5]$

$y = \left| \dfrac{2}{3}x - 5 \right| - 3$

FIGURE 11

Thus the solution set is $\{x \mid x \leq 3\} \cup \{x \mid x \geq 12\}$ or, with interval notation, $(-\infty, 3] \cup [12, +\infty)$. The solution set is shown on the real-number line in Figure 10.

To verify the solutions, we let

$$y = \left| \frac{2}{3}x - 5 \right| - 3$$

Figure 11 shows the graph of this equation plotted on our graphics calculator. We observe that the graph intersects or is above the x axis, that is, $y \geq 0$, when x is in $(-\infty, 3] \cup [12, +\infty)$, which agrees with the result found algebraically.   ◀

The following theorems about absolute value are useful in calculus.

THEOREM 1

If a and b are real numbers, then

$$|ab| = |a| \cdot |b|$$

Proof   Recall from Section 1.2 that $|x| = \sqrt{x^2}$. Therefore

$$\begin{aligned} |ab| &= \sqrt{(ab)^2} \\ &= \sqrt{a^2 b^2} \\ &= \sqrt{a^2} \cdot \sqrt{b^2} \\ &= |a| \cdot |b| \end{aligned}$$
■

THEOREM 2

If a and b are real numbers and $b \neq 0$, then

$$\left| \frac{a}{b} \right| = \frac{|a|}{|b|}$$

The proof is similar to that of Theorem 1 and is left as an exercise (see Exercise 49).

In calculus we often want to replace one inequality with an equivalent but simpler one, as in the following example.

▶ **EXAMPLE 5**   *Showing that One Inequality Involving Absolute Value Is Equivalent to a Simpler One*

Show that the inequality

$$|(3x + 2) - 8| < 1 \quad \text{is equivalent to} \quad |x - 2| < \frac{1}{3}$$

Solution   The following inequalities are equivalent:

$$|(3x + 2) - 8| < 1$$
$$|3x - 6| < 1$$
$$|3(x - 2)| < 1$$
$$|3||x - 2| < 1$$
$$3|x - 2| < 1$$
$$|x - 2| < \frac{1}{3}$$

◀

 The following theorem, called the *triangle inequality,* is used frequently in proving theorems in calculus.

THEOREM 3   *The Triangle Inequality*
If a and b are real numbers, then $|a + b| \le |a| + |b|$

In Exercise 50 you are asked to prove the triangle inequality. We demonstrate the content of the triangle inequality in the following illustration with four particular cases.

▷ **ILLUSTRATION 5**

If $a = 3$ and $b = 4$, then

$$|a + b| = |3 + 4| \qquad |a| + |b| = |3| + |4|$$
$$= |7| \qquad\qquad\qquad = 3 + 4$$
$$= 7 \qquad\qquad\qquad\quad = 7$$

If $a = -3$ and $b = 4$, then

$$|a + b| = |-3 + 4| \qquad |a| + |b| = |-3| + |4|$$
$$= |1| \qquad\qquad\qquad\quad = 3 + 4$$
$$= 1 \qquad\qquad\qquad\quad\, = 7$$

If $a = 3$ and $b = -4$, then

$$|a + b| = |3 + (-4)| \qquad |a| + |b| = |3| + |-4|$$
$$= |-1| \qquad\qquad\qquad = 3 + 4$$
$$= 1 \qquad\qquad\qquad\qquad = 7$$

If $a = -3$ and $b = -4$, then

$$|a + b| = |-3 + (-4)| \qquad |a| + |b| = |-3| + |-4|$$
$$= |-7| \qquad\qquad\qquad = 3 + 4$$
$$= 7 \qquad\qquad\qquad\qquad = 7$$

In each case $|a + b| \le |a| + |b|$. ◄

EXERCISES 2.7

In Exercises 1 through 14, find algebraically the solution set of the equation and check the solutions graphically.

1. $|x - 5| = 4$
2. $|2x + 3| = 7$

3. $|3y - 8| = 4$
4. $|4t - 9| = 11$

5. $|4x + 5| = 15$
6. $|8 - x| = 4$

7. $|7 - 2w| = 9$
8. $|3x - 2| = |2x + 3|$

9. $|x - 4| = |5 - 2x|$
10. $|5y| = |6 - y|$

11. $\left|\dfrac{x + 3}{x - 3}\right| = 7$
12. $\left|\dfrac{2x + 1}{x - 1}\right| = 3$

13. $|3t - 2| = t^2$
14. $|x - 1| = \dfrac{x^2}{4}$

In Exercises 15 through 22, solve the inequality graphically and write it with interval notation. Verify your solutions algebraically and show the solution set on the real-number line.

15 $|x| \le 5$
16. $|x| > 6$

17. $|x - 1| > 7$
18. $|x + 1| < 5$

19. $|x - 5| \le 3$
20. $|3x - 4| \le 2$

21. $|2x - 7| < 9$
22. $|3x - 4| \ge 2$

In Exercises 23 through 30, find algebraically the solution set of the inequality and write it with interval notation. Show the solution set on the real-number line. Verify your solutions graphically.

23. $|7 - 2x| > 9$
24. $6 < |4x + 7|$

25. $4 < |3x + 12|$
26. $|5 - 3x| \ge 10$

27. $|5x - 7| + 4 \le 6$
28. $|4x - 3| - 3 > 6$

29. $\left|\frac{3}{2} - 2x\right| > \frac{1}{2}$
30. $\left|2 - \frac{3}{4}x\right| \le \frac{1}{2}$

$\iint\frac{dy}{dx}$ *In Exercises 31 through 36, show that the two inequalities are equivalent.*

31. $|(2x - 3) - 9| < 1;\ |x - 6| < \frac{1}{2}$

32. $|(2x + 3) - 1| < 1;\ |x + 1| < \frac{1}{2}$

33. $|(3x - 5) - 1| < \frac{1}{2};\ |x - 2| < \frac{1}{6}$

34. $|(5x - 2) - 3| < \frac{1}{2};\ |x - 1| < \frac{1}{10}$

35. $|(\frac{1}{2}x - 5) + 7| < \frac{1}{8};\ |x + 4| < \frac{1}{4}$

36. $|(\frac{1}{4}x - 1) + 2| < \frac{1}{6};\ |x + 4| < \frac{2}{3}$

In Exercises 37 through 42, find algebraically the solution set of the inequality and write it with interval notation. Verify your solutions graphically.

37. $|x^2 - 5| < 4$ **38.** $|y^2 - 10| \le 6$

39. $|t^2 - 5t| \le 6$ **40.** $|x^2 - 3x - 1| < 3$

41. $|x^2 - 17| \ge 8$ **42.** $|x^2 + x - 4| > 2$

In Exercises 43 through 46, solve the equation graphically. Verify your solutions algebraically. Hint: Apply the equality $|a| = \sqrt{a^2}$ and use the method of Section 2.4 for solving equations involving radicals.

43. $|x - 1| + |x| = 3$ **44.** $|x + 3| + |x| = 6$

45. $|2x + 1| = 3 - |2x|$

46. $|x - 1| + |x + 1| = 8$

In Exercises 47 and 48, verify the triangle inequality for the values of a and b.

47. (a) $a = 5$ and $b = 7$ **(b)** $a = 5$ and $b = -7$
   (c) $a = -5$ and $b = 7$ **(d)** $a = -5$ and $b = -7$

48. (a) $a = \frac{1}{2}$ and $b = \frac{1}{3}$ **(b)** $a = -\frac{1}{2}$ and $b = \frac{1}{3}$
   (c) $a = \frac{1}{2}$ and $b = -\frac{1}{3}$ **(d)** $a = -\frac{1}{2}$ and $b = -\frac{1}{3}$

49. Prove Theorem 2.

50. Prove the triangle inequality. *Hint:* First show that

$$-|a| \le a \le |a| \quad \text{and} \quad -|b| \le b \le |b|$$

Then apply the result of Exercise 51 of Exercises 2.5 and statement (1) of this section where $<$ is replaced by $\le$.

$\iint \dfrac{dy}{dx}$ *In Exercises 51 and 52, use the triangle inequality to prove the statement.*

51. If $|x - 1| < \frac{1}{3}$ and $|y + 1| < \frac{1}{4}$, then $|x + y| < \frac{7}{12}$

52. If $|x - 1| < \frac{1}{3}$ and $|y - 1| < \frac{1}{4}$, then $|x - y| < \frac{7}{12}$

53. Prove: If a and b are real numbers, then $|a - b| \le |a| + |b|$. *Hint:* Write $a - b$ as $a + (-b)$ and use the triangle inequality.

54. Prove: If a and b are real numbers, then $|a| - |b| \le |a - b|$. *Hint:* Let $|a| = |(a - b) + b|$ and use the triangle inequality.

55. Explain why the inequalities $|a| > |b|$ and $a^2 > b^2$ are equivalent. Use this fact to explain how the inequality $|2x - 1| > |x - 3|$ can be solved algebraically the same way we solved the quadratic inequality in Example 1 of Section 2.6.

56. Refer to Exercise 55 and explain how the inequality of that exercise can be solved graphically the same way we solved the quadratic inequality in Example 1 of Section 2.6.

CHAPTER 2 REVIEW

▶ LOOKING BACK

2.1 We showed how to find the exact real-number solution of a linear equation. We also solved linear equations on a graphics calculator. The use of an equation as a mathematical model of a practical situation was first introduced in this section.

2.2 The exact solutions, either real or imaginary, of quadratic equations were found by factoring, square root, and the quadratic formula. From the discriminant of a quadratic equation, we determined the character of its roots. We applied the graphics calculator to illustrate that the real-number solutions of quadratic equations are the x intercepts of the corresponding parabola. The process of completing the square was first

applied to derive the quadratic formula, and then we showed how this idea is used in calculus to write certain algebraic expressions in recognizable forms.

2.3 We discussed in detail how to obtain an equation as a mathematical model of a word problem. We then solved the equation either algebraically or graphically and completed the problem by writing a conclusion that answered the questions of the problem.

2.4 Equations involving radicals, equations quadratic in form, and particular polynomial equations of degree greater than 2 were solved algebraically and graphically.

2.5 We began the discussion of inequalities by stating the trichotomy and transitive properties of order, and we then listed the properties of $<$ used to obtain equivalent inequalities to solve inequalities. The emphasis in this section was on solving linear inequalities.

2.6 The algebraic method of solving polynomial inequalities involved the use of tables to summarize the sign of a polynomial expression on a particular interval. We discussed two methods for solving polynomial inequalities graphically.

2.7 Equations and inequalities involving absolute value were solved both algebraically and graphically. Three theorems, including the triangle inequality, important in calculus, were also included in this section.

▶ REVIEW EXERCISES

In Exercises 1 and 2, find the solution set of the equation and show the solution on a graph on your graphics calculator.

1. $4x - 11 = 0$

2. $2x + 9 = 0$

In Exercises 3 through 6, solve the equation in two ways: (i) on your graphics calculator; (ii) algebraically.

3. $6x + 5 = 23$

4. $5x - 6 = 2 + 9x$

5. $2(5x - 4) = 11 - (3 + 2x)$

6. $5(2t - 4) = 11 - (3 + 2t)$

In Exercises 7 through 16, find the solution set of the equation. Plot the graph of the corresponding parabola on your graphics calculator. If the equation has real roots, verify that the x intercepts of the parabola are the roots of the equation. If the equation has no real roots, verify that the parabola does not intersect the x axis.

7. $49x^2 - 64 = 0$

8. $5x^2 = x$

9. $x^2 - 3x - 10 = 0$

10. $5x^2 + 17x - 12 = 0$

11. $10x^2 + 7x - 6 = 0$

12. $x^2 + 2x - 1 = 0$

13. $2x^2 - 4x - 5 = 0$

14. $4x^2 + 2x + 1 = 0$

15. $3x^2 - 2x + 2 = 0$

16. $(3x + 10)(x - 3) = 2x + 14$

In Exercises 17 through 42, find the solution set of the equation.

17. $\dfrac{2}{5 - y} - \dfrac{1}{y - 2} = 0$

18. $\dfrac{x - 1}{x - 1} = 1$

19. $\dfrac{x - 1}{x - 1} = 0$

20. $\dfrac{x}{2x - 2} = \dfrac{x - 4}{2x - 4}$

21. $\dfrac{7}{x - 4} = x + 2$

22. $\dfrac{x + 11}{x + 8} - \dfrac{3x - 2}{x - 2} = 0$

23. $\dfrac{6}{t^2 - 1} + \dfrac{1}{t - 1} = \dfrac{1}{2}$

24. $\dfrac{w^2 - 3}{w^2 - 6w + 5} = \dfrac{2w + 3}{w - 5} - \dfrac{w + 3}{w - 1}$

25. $\sqrt{2x + 5} + x = 5$

26. $\sqrt{x + 2} + 2 + x = 0$

27. $\sqrt{3t + 7} + \sqrt{t + 6} = 3$

28. $\sqrt{y + 2} + \sqrt{y + 5} - \sqrt{8 - y} = 0$

29. $x^3 - 8 = 0$

30. $x^4 - 8x^2 = 9$

31. $36x^4 - 13x^2 + 1 = 0$

32. $8z^3 + 27 = 0$

33. $(x^2 + 3x)^2 - 3(x^2 + 3x) - 10 = 0$

34. $6\left(y + \dfrac{1}{y}\right)^2 - 35\left(y + \dfrac{1}{y}\right) + 50 = 0$

35. $2\sqrt{t - 5} + \sqrt[4]{t - 5} = 3$

36. $15t^{-2} - 14t^{-1} - 8 = 0$

37. $y^{-2/3} + y^{-1/3} - 6 = 0$

38. $3x^{1/3} + 5 = 2x^{-1/3}$

39. $|2x + 5| = 7$

40. $|3y - 4| = 8$

41. $\left|\dfrac{w + 1}{w - 1}\right| = 3$

42. $|4x - 1| = |x + 5|$

In Exercises 43 through 48, solve for x in terms of the other symbols.

43. $Ax + By + C = 0$

44. $\dfrac{d}{10x} - \dfrac{d}{5} = \dfrac{1}{x}$

45. $x^2 + b^2 = 2bx + a^2$

46. $x^2 + xy + 2x - 1 = 0$

47. $6x^2 - 2xy - x + y - 1 = 0$

48. $rsx^2 + s^2x + rtx + st = 0$

In Exercises 49 and 50, add a term to the algebraic expression in order to make it a square of a binomial; also write the resulting expression as a square of a binomial.

49. (a) $x^2 - 8x$ **(b)** $y^2 + 3y$ **(c)** $x^2 + \frac{5}{3}x$

50. (a) $w^2 + 10w$ **(b)** $x^2 - \frac{6}{5}x$ **(c)** $x^2 + 7bx$

In Exercises 51 and 52, find the discriminant and determine the character of the roots of the quadratic equation; do not solve the equation. Then plot the graph of the corresponding parabola on your graphics calculator and verify your conclusion.

51. (a) $4x^2 + 20x + 25 = 0$ **(b)** $8x^2 - 10x = 3$
(c) $5t - 3 = 4t^2$

52. (a) $15x^2 - 19x = 10$ **(b)** $2y^2 = 9 - 4y$
(c) $9x^2 - 42x + 49 = 0$

In Exercises 53 through 56, find on your graphics calculator the approximate value, to the nearest one-hundredth, of the indicated root.

53. $x^3 + x = 2x^2 + 1$: the positive root

54. $x^3 - 5x^2 + 7x + 8 = 0$: the negative root

55. $x^4 + x^3 - 3x^2 - x - 4 = 0$: the positive root

56. $x^4 - 4x^3 - 4x^2 + 17x - 4 = 0$: **(a)** the largest positive root; **(b)** the smallest positive root.

In Exercises 57 through 78, find algebraically the solution set of the inequality; write the solution set with interval notation, and show the solution set on the real-number line. Then verify your result graphically.

57. $3x - 1 \le 11$ **58.** $5x + 7 \ge 2x - 2$

59. $3x - 2 > 7x + 3$ **60.** $4x < 8x - 7$

61. $\dfrac{x+1}{2} - 3 \ge \dfrac{x+2}{4}$ **62.** $\dfrac{2x+1}{3} \le \dfrac{9-2x}{6}$

63. $-4 < 1 - 5x < 11$ **64.** $7 < 2x + 3 \le 15$

65. $-5 \le \dfrac{4x-5}{-3} < 7$ **66.** $-7 \le \dfrac{7-3x}{-4} \le 8$

67. $|x + 1| \ge 2$ **68.** $|x - 4| \le 6$

69. $|2x - 5| < 7$ **70.** $|3x + 7| > 11$

71. $\dfrac{4}{3} < \left| \dfrac{3}{2} - x \right|$ **72.** $\left| \dfrac{3-2x}{5} \right| < \dfrac{1}{3}$

73. $x^2 < 25$ **74.** $x^2 + 3x - 10 \ge 0$

75. $2x^2 - 3x \ge 5$ **76.** $3t^2 < 4(t + 1)$
77. $4y^3 < 5y^2 - 6y$ **78.** $x^3 > x$

In Exercises 79 and 80, find algebraically the solution set of the inequality and write it with interval notation. Verify your solutions graphically.

79. $|x^2 - 3x - 7| < 3$ **80.** $|y^2 - 26| \ge 10$

In Exercises 81 through 86, use the concept of completing the square to write the equation in one of the following forms where x and y represent variables and the other letters represent constants:

$$(x - h)^2 + (y - k)^2 = r^2 \qquad (x - h)^2 = 4p(y - k)$$

$$\frac{(x-h)^2}{a^2} + \frac{(y-k)^2}{b^2} = 1 \qquad \frac{(x-h)^2}{a^2} - \frac{(y-k)^2}{b^2} = 1$$

81. $y = 4(x^2 - 6x - 8)$

82. $x^2 + y^2 + 8x - 2y - 8 = 0$

83. $x^2 + 4y^2 = 8(x - 2y - 2)$

84. $4x^2 - 9y^2 = 4(17 - 2x - 9y)$

85. $2x^2 - 5y^2 + 4x - 20y - 68 = 0$

86. $x^2 = 8(4 - 2y - x)$

 In Exercises 87 and 88, use the concept of completing the square to write the indicated expression in the required form where a and h are constants and p is a positive integer.

87. $\sqrt{16x^2 + 32x + 12}$ in the form $p\sqrt{(x - h)^2 - a^2}$
88. $\sqrt{-9x^2 + 18x - 8}$ in the form $p\sqrt{a^2 - (x - h)^2}$

 In Exercises 89 and 90, show that the two inequalities are equivalent.

89. $|(4x + 3) - 11| < \frac{1}{2}; |x - 2| < \frac{1}{8}$

90. $|(2x - 5) - 7| < \frac{4}{5}; |x - 6| < \frac{2}{5}$

91. The cost of hiring a chauffeur-driven limousine from a car service is $180 for the car and driver plus 52 cents per mile. What is the cost of a trip of **(a)** 30 miles; **(b)** 40 miles; and **(c)** 100 miles? **(d)** If y dollars is the cost of a trip of x miles, write an equation involving y and x. **(e)** Plot the graph of your equation in part (d) and from the graph estimate the cost of a trip of 83 miles. **(f)** Check your estimate in part (e) algebraically. **(g)** From the graph in part (e), estimate the length of a trip costing $205. **(h)** Check your estimate in part (g) algebraically.

92. If S is the sum of the first n consecutive natural numbers, then $S = \frac{1}{2}n(n + 1)$. **(a)** Solve for n in terms of S, and **(b)** find n when S is 435.

In Exercises 93 through 102, solve the word problem by finding an equation as a mathematical model of the situation. Complete the exercise by writing a conclusion.

93. A group of four students decides to share the cost of hiring a tutor for a review session before an examination. If two more students join the group, the cost to each student is reduced by $6. What is the cost per student if four are in the group?

94. A company obtained two loans totaling $30,000 and having interest rates of 16 percent and 12 percent. The total interest for the two loans equaled the interest the company would have paid if the entire amount had been borrowed at 15 percent. What was the amount of the two loans? Estimate the solution of your equation on your graphics calculator. Then solve the equation algebraically.

95. Your automobile radiator contains 8 liters of a solution that is 10 percent antifreeze and 90 percent water. You wish to replace the solution with one containing 25 percent antifreeze by draining part of that solution and replacing it with pure antifreeze. How much pure antifreeze should be added? Estimate the solution of your equation on your graphics calculator. Then solve the equation algebraically.

96. To get a solution containing 35 percent glycerine a solution containing 55 percent glycerine was added to 25 liters of a solution containing 28 percent glycerine. How much of the 55 percent glycerine solution was added? Estimate the solution of your equation on your graphics calculator. Then solve the equation algebraically.

97. To form an open box having a volume of 400 cm³, squares of side 4 cm are cut from each corner of a

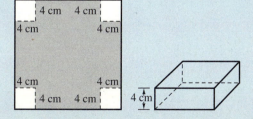

square piece of tin and then the sides of the tin are turned up. Determine the length of the side of the original piece of tin. Estimate the solution of your equation on your graphics calculator. Then solve the equation algebraically.

98. A farmer plowed a strip around a rectangular field that is 400 rods long and 240 rods wide. One-half of the field was plowed when the farmer finished. Determine the width of the strip. Estimate the solution of your equation on your graphics calculator. Then solve the equation algebraically.

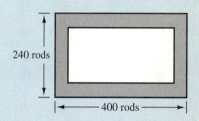

240 rods

400 rods

99. Suppose the square piece of tin in Exercise 97 has an area of 400 cm². On your graphics calculator determine to the nearest one-hundredth of a centimeter the length of the side of the square to be cut out for the box to have a volume of 400 cm³.

100. An open metal pan having a volume of 1 gal (231 in.³) is to be made by cutting out squares of the same size from the corners of a rectangular piece of metal 14 in. by 18 in. and turning up the sides. On your graphics calculator determine, to the nearest one-hundredth of an inch, the length of a side of the squares to be cut from each corner.

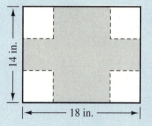

14 in.

18 in.

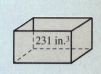

231 in.³

101. A closed rectangular safe is to be made of lead of uniform thickness on the top and bottom and sides. The inside dimensions are to be 4 ft by 4 ft by 6 ft, and 450 ft³ of lead is to be used in the construction. On your graphics calculator find, to the nearest one-hundredth of a foot, the thickness of the sides.

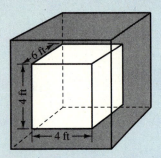

102. A hollow spherical container, whose capacity is 800 cm³, has an outer radius of 6 cm. On your graphics calculator find, to the nearest one-hundredth of a centimeter, the thickness of the wall of the container. The formula for the volume of a sphere is $V = \frac{4}{3}\pi r^3$.

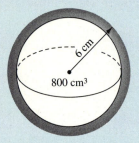

In Exercises 103 and 104, solve the word problem by finding an inequality as a mathematical model of the situation. Complete the exercise by writing a conclusion.

103. The perimeter of a rectangle must not be greater than 30 cm and the length must be 8 cm. What is the range of values for the width?

104. A student must receive an average score of at least 90 on five examinations to earn a grade of A in a particular course. If the student's scores on the first four examinations are 93, 95, 79, and 88, what must the score on the fifth examination be for the course grade to be A?

Lines, Parabolas, Circles, and Translation of Axes

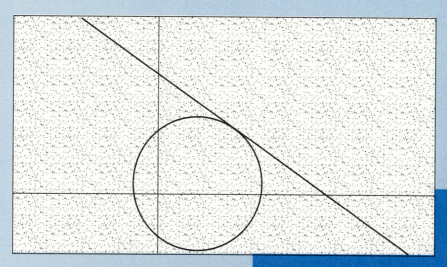

Analytic geometry forms a foundation for the study of calculus. The subject matter of analytic geometry, begun in Section 1.4, continues in this chapter with the traditional material on lines, parabolas, and circles. Graphs of these curves are presented early in the text so that they are available prior to our study of functions and graphs in Chapter 4. The discussion of systems of linear equations in two unknowns is a natural extension of our study of lines. Translation of axes is used to obtain more general equations of some curves in the plane and paves the way for the vertical and horizontal translation rules applied later to graphs of functions.

3.1 LINES

Recall from Example 4 of Section 2.1 that if the cost of hiring a one-passenger plane from a private airline is \$310 for the plane and pilot plus 75 cents per mile flown, then if y dollars is the cost of a trip of x miles,

$$y = 0.75x + 310$$

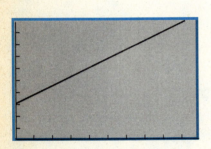

[0, 1000] by [0, 1000]
$y = 0.75x + 310$
FIGURE 1

Figure 1 shows the graph of this equation plotted in the viewing rectangle [0,1000] by [0,1000], with scales of 100 on the axes. The graph appears to be a line. That the graph is indeed a line is established in Theorem 2 of this section. Observe that for each 100-unit increase in x, y increases by 75 units, or, equivalently, for each 1-unit increase in x, y increases by 0.75 unit. Thus the ratio of the change in y to the change in x is a constant 0.75. This constant ratio is called the *slope* of the line. We proceed now to arrive at a formal definition of slope.

Let l be a nonvertical line and $P_1(x_1, y_1)$ and $P_2(x_2, y_2)$ be any two distinct points on l. Figure 2 shows such a line. In the figure R is the point (x_2, y_1), and the points P_1, P_2, and R are vertices of a right triangle; furthermore, $\overline{P_1 R} = x_2 - x_1$ and $\overline{RP_2} = y_2 - y_1$. The number $y_2 - y_1$ gives the measure of the change in the ordinate from P_1 to P_2, and it may be positive, negative, or zero. The number $x_2 - x_1$ gives the measure of the change in the abscissa from P_1 to P_2, and it may be positive or negative. Because the line l is not vertical, $x_2 \neq x_1$, and therefore $x_2 - x_1$ may not be zero. Let

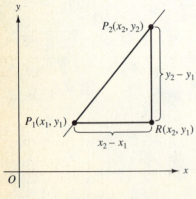

FIGURE 2

$$m = \frac{y_2 - y_1}{x_2 - x_1} \tag{1}$$

The value of m computed from this equation is independent of the choice of the two points P_1 and P_2 on l. To show this, suppose we choose two different points $\overline{P}_1(\overline{x}_1, \overline{y}_1)$ and $\overline{P}_2(\overline{x}_2, \overline{y}_2)$ on line l, and compute a number $\overline{m}$ from (1).

$$\overline{m} = \frac{\overline{y}_2 - \overline{y}_1}{\overline{x}_2 - \overline{x}_1}$$

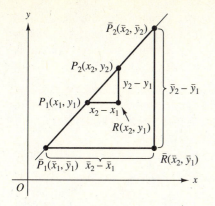

FIGURE 3

We shall show that $\overline{m} = m$. Refer to Figure 3. Triangles $\overline{P}_1 \overline{R} \overline{P}_2$ and $P_1 R P_2$ are similar; so the lengths of corresponding sides are proportional. Therefore

$$\frac{\overline{y}_2 - \overline{y}_1}{\overline{x}_2 - \overline{x}_1} = \frac{y_2 - y_1}{x_2 - x_1}$$

or

$$\overline{m} = m$$

Thus the value of m computed from (1) is the same number no matter what two points on l are selected. This number m is called the *slope* of the line.

DEFINITION   Slope

If $P_1(x_1, y_1)$ and $P_2(x_2, y_2)$ are any two distinct points on line l, which is not parallel to the y axis, then the **slope** of l, denoted by m, is given by

$$m = \frac{y_2 - y_1}{x_2 - x_1}$$

Multiplying on both sides of the preceding equation by $x_2 - x_1$, we obtain

$$y_2 - y_1 = m(x_2 - x_1)$$

It follows from this equation that if we consider a particle moving along a line, the change in the ordinate of the particle is equal to the product of the slope and the change in the abscissa.

▷ **ILLUSTRATION 1**

If l is the line through the points $P_1(2, 1)$ and $P_2(4, 7)$ and m is the slope of l, then

$$m = \frac{7 - 1}{4 - 2}$$
$$= 3$$

Refer to Figure 4. If a particle is moving along line l, the change in the ordinate is 3 times the change in the abscissa. That is, if the particle is at

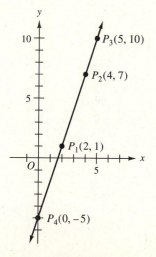

FIGURE 4

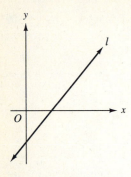

FIGURE 5

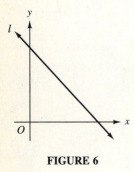

FIGURE 6

$P_2(4, 7)$ and the abscissa is increased by 1 unit, then the ordinate is increased by 3 units, and the particle is at the point $P_3(5, 10)$. Similarly, if the particle is at $P_1(2, 1)$ and the abscissa is decreased by 2 units, then the ordinate is decreased by 6 units, and the particle is at $P_4(0, -5)$.   ◀

If the slope of a line is positive, then as the abscissa of a point on the line increases, the ordinate increases. Such a line is shown in Figure 5. A line whose slope is negative appears in Figure 6. For this line, as the abscissa of a point on the line increases, the ordinate decreases.

If a line is parallel to the x axis, then $y_2 = y_1$; so the slope of the line is zero. If a line is parallel to the y axis, $x_2 = x_1$; thus the fraction $\dfrac{y_2 - y_1}{x_2 - x_1}$ is meaningless because we cannot divide by zero. For this reason lines parallel to the y axis are excluded in the definition of slope. Thus the slope of a vertical line is not defined.

▶ **EXAMPLE 1** *Sketching a Line Through Two Points and Determining its Slope*

For each pair of points, sketch the line through them and find its slope: **(a)** $A(3, 7)$ and $B(-2, -4)$; **(b)** $A(-2, 5)$ and $B(2, -3)$; **(c)** $A(-3, 4)$ and $B(5, 4)$; **(d)** $A(5, 3)$ and $B(5, -1)$.

Solution   The lines appear in Figure 7(a)–(d). We compute the slope from the definition.

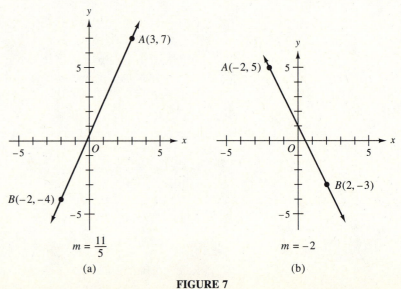

FIGURE 7

(Figure 7 continued on the next page)

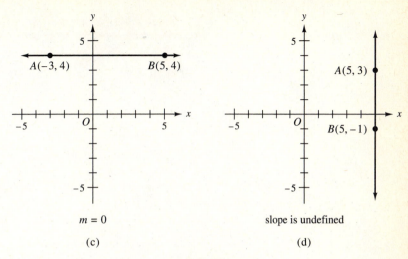

$$m = 0$$

(c)

slope is undefined

(d)

FIGURE 7

(Figure 7 continued from the previous page)

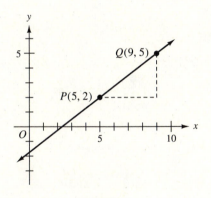

FIGURE 8

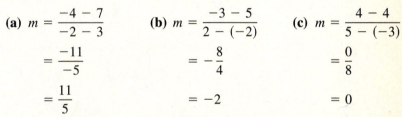

(a) $m = \dfrac{-4 - 7}{-2 - 3}$ **(b)** $m = \dfrac{-3 - 5}{2 - (-2)}$ **(c)** $m = \dfrac{4 - 4}{5 - (-3)}$

$$= \dfrac{-11}{-5} \qquad\qquad\quad = -\dfrac{8}{4} \qquad\qquad\quad = \dfrac{0}{8}$$

$$= \dfrac{11}{5} \qquad\qquad\quad = -2 \qquad\qquad\quad = 0$$

(d) Because the line is vertical, the slope is not defined. If you attempt to use the definition to compute the slope, you obtain zero in the denominator. ◄

▷ **ILLUSTRATION 2**

(a) Suppose a line contains the point $P(5, 2)$ and its slope is $\frac{3}{4}$. To determine another point on the line, we start at P, and because the slope is $\frac{3}{4}$, we go 4 units to the right and then 3 units upward. We have then the point $Q(9, 5)$ also on the line. The line is sketched through points P and Q as in Figure 8.

(b) If a line through $P(5, 2)$ has the negative slope $-\frac{3}{4}$, we obtain another point on the line by starting at P and then going 4 units to the right and 3 units downward. This gives the point $Q(9, -1)$. Figure 9 shows the line through these points P and Q. ◄

In Section 1.5 we defined the graph of an equation. We now define what we mean by an *equation of a graph.*

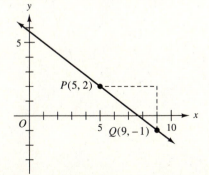

FIGURE 9

> **DEFINITION**   **An Equation of a Graph**
>
> An **equation of a graph** is an equation that is satisfied by the coordinates of those, and only those, points on the graph.

From this definition, it follows that an equation of a graph has the following properties:

1. If a point $P(x, y)$ is on the graph, then its coordinates satisfy the equation.
2. If a point $P(x, y)$ is not on the graph, then its coordinates do not satisfy the equation.

To obtain an equation of a line, we use the fact that a point $P_1(x_1, y_1)$ and a slope m determine a unique line. Let $P(x, y)$ be any point on the line except P_1. Then because the slope of the line through P_1 and P is m, we have from the definition of slope

$$\frac{y - y_1}{x - x_1} = m$$

$$y - y_1 = m(x - x_1)$$

This equation is called the **point-slope form** of an equation of the line. It gives an equation of the line if a point $P_1(x_1, y_1)$ on the line and the slope of the line are known.

▷ ILLUSTRATION 3

To find an equation of the line through the points $A(6, -3)$ and $B(-2, 3)$, we first compute m.

$$m = \frac{3 - (-3)}{-2 - 6}$$

$$= \frac{6}{-8}$$

$$= -\frac{3}{4}$$

Using the point-slope form of an equation of the line with A as P_1, we have

$$y - (-3) = -\frac{3}{4}(x - 6)$$

$$4y + 12 = -3x + 18$$

$$3x + 4y - 6 = 0$$

If B is taken as P_1 in the point-slope form, we have

$$y - 3 = -\frac{3}{4}[x - (-2)]$$

$$4y - 12 = -3x - 6$$

$$3x + 4y - 6 = 0$$

which of course is the same equation.   ◄

If in the point-slope form we choose the particular point $(0, b)$ (that is, the point where the line intersects the y axis) for the point (x_1, y_1), we have

$$y - b = m(x - 0)$$

$$\Leftrightarrow \qquad y = mx + b$$

The number b, the ordinate of the point where the line intersects the y axis, is the **y intercept** of the line. Consequently, the preceding equation is called the **slope-intercept form** of an equation of the line. This form is especially useful because it enables us to find the slope of a line from its equation. It is also important because it expresses the y coordinate of a point on the line explicitly in terms of its x coordinate.

► **EXAMPLE 2** *Determining the Slope of a Line from its Equation*

Find the slope of the line having the equation

$$6x + 5y - 7 = 0$$

Solution   We solve the equation for y.

$$5y = -6x + 7$$

$$y = -\frac{6}{5}x + \frac{7}{5}$$

This equation is in the slope-intercept form where $m = -\frac{6}{5}$ and $b = \frac{7}{5}$. Therefore the slope is $-\frac{6}{5}$.   ◄

Because the slope of a vertical line is undefined, we cannot apply the point-slope form to obtain its equation. We use instead the following theorem, which also gives an equation of a horizontal line.

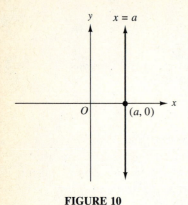

FIGURE 10

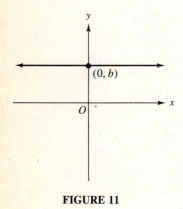

FIGURE 11

> ### THEOREM 1
>
> **(i)** An equation of the vertical line having x intercept a is
>
> $$x = a$$
>
> **(ii)** An equation of the horizontal line having y intercept b is
>
> $$y = b$$

Proof

(i) Figure 10 shows the vertical line that intersects the x axis at the point $(a, 0)$. This line contains those and only those points on the line having the same abscissa. So $P(x, y)$ is any point on the line if and only if

$$x = a$$

(ii) The horizontal line that intersects the y axis at the point $(0, b)$ appears in Figure 11. For this line, $m = 0$. Therefore, from the slope-intercept form an equation of this line is

$$y = b \qquad \blacksquare$$

We have shown that an equation of a nonvertical line is of the form $y = mx + b$, and an equation of a vertical line is of the form $x = a$. Because each of these equations is a special case of an equation of the form

$$Ax + By + C = 0 \qquad (2)$$

where A, B, and C are constants and A and B are not both zero, it follows that every line has an equation of the form (2). The converse of this fact is given by the next theorem.

> ### THEOREM 2
>
> The graph of the equation
>
> $$Ax + By + C = 0$$
>
> where A, B, and C are constants and where not both A and B are zero, is a line.

The proof of this theorem is left as an exercise. See Exercise 55.

Because the graph of (2) is a line, it is called a **linear equation;** it is the general equation of the first degree in x and y.

To plot a line on a graphics calculator, we first write its equation in slope-intercept form and proceed as we did in Section 2.1. To sketch a line by hand, we need only determine the coordinates of two points on the line, locate the points, and then draw the line. Any two points will suffice, but for convenience we usually choose the two points where the line intersects the axes.

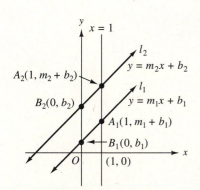

FIGURE 12

▷ **ILLUSTRATION 4**

To sketch the line having the equation

$$4x - 3y = 6$$

we first find the x intercept a and the y intercept b. In the equation, we substitute 0 for y and get $a = \frac{3}{2}$. Substituting 0 for x, we obtain $b = -2$. Thus we have the line appearing in Figure 12.   ◀

An application of slopes is given by the following theorem.

THEOREM 3

If l_1 and l_2 are two distinct nonvertical lines having slopes m_1 and m_2, respectively, then l_1 and l_2 are parallel if and only if $m_1 = m_2$.

Proof   Let equations of l_1 and l_2 be, respectively,

$$y = m_1 x + b_1 \qquad \text{and} \qquad y = m_2 x + b_2$$

See Figure 13, showing the two lines intersecting the y axis at the points $B_1(0, b_1)$ and $B_2(0, b_2)$. Let the vertical line $x = 1$ intersect l_1 at the point $A_1(1, m_1 + b_1)$ and l_2 at the point $A_2(1, m_2 + b_2)$. Then

$$\left| \overline{B_1 B_2} \right| = b_2 - b_1 \qquad \text{and} \qquad \left| \overline{A_1 A_2} \right| = (m_2 + b_2) - (m_1 + b_1)$$

The two lines are parallel if and only if the vertical distances $\left| \overline{B_1 B_2} \right|$ and $\left| \overline{A_1 A_2} \right|$ are equal; that is, l_1 and l_2 are parallel if and only if

$$b_2 - b_1 = (m_2 + b_2) - (m_1 + b_1)$$
$$b_2 - b_1 = m_2 + b_2 - m_1 - b_1$$
$$m_1 = m_2$$

Thus l_1 and l_2 are parallel if and only if $m_1 = m_2$.   ■

FIGURE 13

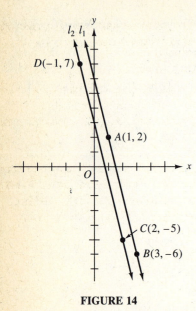

FIGURE 14

▷ **ILLUSTRATION 5**

Let l_1 be the line through the points $A(1, 2)$ and $B(3, -6)$ and m_1 be the slope of l_1; and let l_2 be the line through the points $C(2, -5)$ and $D(-1, 7)$ and m_2 be the slope of l_2. See Figure 14. Then

$$m_1 = \frac{-6 - 2}{3 - 1} \qquad m_2 = \frac{7 - (-5)}{-1 - 2}$$

$$= \frac{-8}{2} \qquad\qquad = \frac{12}{-3}$$

$$= -4 \qquad\qquad = -4$$

Because $m_1 = m_2$, it follows from Theorem 3 that l_1 and l_2 are parallel. ◀

Any two distinct points determine a line. Three distinct points may or may not lie on the same line. If three or more points lie on the same line, they are said to be **collinear**. Hence three points A, B, and C are collinear if and only if the line through the points A and B is the same as the line through the points B and C. Because the line through A and B and the line through B and C both contain the point B, they are the same line if and only if their slopes are equal.

▶ **EXAMPLE 3   Using Slopes to Determine if Three Given Points are Collinear**

Determine by means of slopes if the points $A(-3, -4)$, $B(2, -1)$, and $C(7, 2)$ are collinear.

Solution   If m_1 is the slope of the line through A and B, and m_2 is the slope of the line through B and C, then

$$m_1 = \frac{-1 - (-4)}{2 - (-3)} \qquad m_2 = \frac{2 - (-1)}{7 - 2}$$

$$= \frac{3}{5} \qquad\qquad = \frac{3}{5}$$

Hence $m_1 = m_2$. Therefore the line through A and B and the line through B and C have the same slope and contain the common point B. Thus they are the same line, and therefore A, B, and C are collinear.   ◀

We now state and prove a theorem regarding the slopes of two perpendicular lines.

THEOREM 4

Two nonvertical lines l_1 and l_2 having slopes m_1 and m_2, respectively, are perpendicular if and only if $m_1 m_2 = -1$.

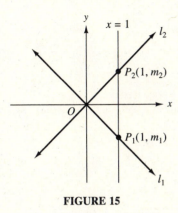

FIGURE 15

Proof   Let us choose the coordinate axes so that the origin is at the point of intersection of l_1 and l_2. See Figure 15. Because neither l_1 nor l_2 is vertical, these two lines intersect the line $x = 1$ at points P_1 and P_2, respectively. The abscissa of both P_1 and P_2 is 1. Let $\bar{y}$ be the ordinate of P_1. Because l_1 contains the points $(0, 0)$ and $(1, \bar{y})$ and its slope is m_1, then

$$m_1 = \frac{\bar{y} - 0}{1 - 0}$$

Thus $\bar{y} = m_1$. Similarly, the ordinate of P_2 is shown to be m_2. From the Pythagorean theorem and its converse, triangle $P_1 O P_2$ is a right triangle if and only if

$$|\overline{OP_1}|^2 + |\overline{OP_2}|^2 = |\overline{P_1 P_2}|^2 \tag{3}$$

Applying the distance formula, we obtain

$$|\overline{OP_1}|^2 = (1 - 0)^2 + (m_1 - 0)^2 \qquad |\overline{OP_2}|^2 = (1 - 0)^2 + (m_2 - 0)^2$$
$$= 1 + m_1{}^2 \qquad\qquad\qquad = 1 + m_2{}^2$$
$$|\overline{P_1 P_2}|^2 = (1 - 1)^2 + (m_2 - m_1)^2$$
$$= m_2{}^2 - 2m_1 m_2 + m_1{}^2$$

Substituting into (3), we can conclude that $P_1 O P_2$ is a right triangle if and only if

$$1 + m_1{}^2 + 1 + m_2{}^2 = m_2{}^2 - 2m_1 m_2 + m_1{}^2$$
$$2 = -2m_1 m_2$$
$$m_1 m_2 = -1 \qquad\blacksquare$$

Because $m_1 m_2 = -1$ is equivalent to

$$m_1 = -\frac{1}{m_2} \quad \text{and} \quad m_2 = -\frac{1}{m_1}$$

Theorem 4 states that two nonvertical lines are perpendicular if and only if the slope of one of them is the negative reciprocal of the slope of the other.

▶ **EXAMPLE 4**   *Finding an Equation of a Line Through a Given Point and Either Parallel or Perpendicular to a Given Line*

Given the line l having the equation

$$5x + 4y - 20 = 0$$

find an equation of the line through the point $(2, -3)$ and **(a)** parallel to l and **(b)** perpendicular to l.

Solution   We first determine the slope of l by writing its equation in the slope-intercept form. Solving the equation for y, we have

$$4y = -5x + 20$$

$$y = -\frac{5}{4}x + 5$$

The slope of l is the coefficient of x, which is $-\frac{5}{4}$.

(a) The slope of a line parallel to l is also $-\frac{5}{4}$. Because the required line contains the point $(2, -3)$, we use the point-slope form, which gives

$$y - (-3) = -\frac{5}{4}(x - 2)$$

$$4y + 12 = -5x + 10$$

$$5x + 4y + 2 = 0$$

(b) The slope of a line perpendicular to l is the negative reciprocal of $-\frac{5}{4}$, which is $\frac{4}{5}$. From the point-slope form, an equation of the line through $(2, -3)$ and having slope $\frac{4}{5}$ is

$$y - (-3) = \frac{4}{5}(x - 2)$$

$$5y + 15 = 4x - 8$$

$$4x - 5y - 23 = 0$$ ◀

▶ **EXAMPLE 5** *Proving that Four Given Points are Vertices of a Rectangle*

Prove by means of slopes that the four points $A(6, 2)$, $B(8, 6)$, $C(4, 8)$, and $D(2, 4)$ are vertices of a rectangle.

Solution   See Figure 16, where l_1 is the line through A and B, l_2 is the line through B and C, l_3 is the line through D and C, and l_4 is the line through A and D; m_1, m_2, m_3, and m_4 are their respective slopes.

$$m_1 = \frac{6 - 2}{8 - 6} \qquad m_2 = \frac{8 - 6}{4 - 8} \qquad m_3 = \frac{8 - 4}{4 - 2} \qquad m_4 = \frac{4 - 2}{2 - 6}$$

$$= 2 \qquad\qquad = -\frac{1}{2} \qquad\qquad = 2 \qquad\qquad = -\frac{1}{2}$$

Because $m_1 = m_3$, l_1 is parallel to l_3; and because $m_2 = m_4$, l_2 is parallel to l_4. Because $m_1 m_2 = -1$, l_1 and l_2 are perpendicular. Therefore the quadrilateral has its opposite sides parallel, and a pair of adjacent sides are perpendicular. Thus the quadrilateral is a rectangle.   ◀

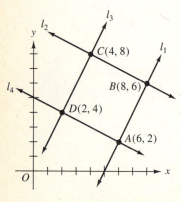

FIGURE 16

EXERCISES 3.1

In Exercises 1 through 6, sketch the line through points A and B and determine the slope of the line.

1. (a) $A(1, 4)$, $B(6, 5)$ **(b)** $A(2, -3)$, $B(-4, 3)$

2. (a) $A(5, 2)$, $B(-2, -3)$ **(b)** $A(-4, 2)$, $B(8, 5)$

3. (a) $A(-4, 3)$, $B(0, 0)$ **(b)** $A(\frac{1}{3}, \frac{1}{2})$, $B(-\frac{5}{6}, \frac{2}{3})$

4. (a) $A(7, 0)$, $B(0, -6)$ **(b)** $A(-\frac{3}{4}, \frac{1}{8})$, $B(\frac{5}{4}, -\frac{1}{2})$

5. (a) $A(1, 5)$, $B(-2, 5)$
  (b) $A(-2.1, 0.3)$, $B(2.3, 1.4)$

6. (a) $A(3, -5)$, $B(3, 4)$
  (b) $A(5.2, -3.5)$, $B(-6.3, -1.4)$

In Exercises 7 and 8, sketch the line through point P and having slope m.

7. (a) $P(3, 4)$, $m = \frac{2}{5}$ **(b)** $P(-1, 6)$, $m = -3$

8. (a) $P(4, 3)$, $m = 2$ **(b)** $P(2, -5)$, $m = -\frac{4}{3}$

In Exercises 9 through 20, find an equation of the line satisfying the conditions.

9. (a) The slope is 4 and through the point $(2, -3)$;
  (b) through the two points $(-1, -5)$ and $(3, 6)$.

10. (a) The slope is -2 and through the point $(-4, 3)$;
  (b) through the two points $(3, 1)$ and $(-5, 4)$.

11. (a) The slope is $-\frac{2}{3}$ and the y intercept is 1;
  (b) the slope is 2 and the x intercept is $-\frac{4}{3}$.

12. (a) The slope is $\frac{3}{5}$ and the y intercept is -4;
  (b) the slope is -2 and the x intercept is 4.

13. (a) Through the point $(1, -7)$ and parallel to the x axis; **(b)** through the point $(2, 6)$ and parallel to the y axis.

14. (a) Through the point $(-5, 2)$ and parallel to the x axis; **(b)** through the point $(-3, -4)$ and parallel to the y axis.

15. (a) The x intercept is -3 and the y intercept is 4;
  (b) through the origin and bisecting the angle between the axes in the first and third quadrants.

16. (a) The x intercept is 5 and the y intercept is -6;
  (b) through the origin and bisecting the angle between the axes in the second and fourth quadrants.

17. Through the point $(-2, 3)$ and parallel to the line whose equation is $2x - y - 2 = 0$.

18. Through the point $(1, 4)$ and parallel to the line whose equation is $2x - 5y + 7 = 0$.

19. Through the point $(2, 4)$ and perpendicular to the line whose equation is $x - 5y + 10 = 0$.

20. Through the origin and perpendicular to the line whose equation is $2x - 5y + 6 = 0$.

In Exercises 21 through 24, find the slope and y intercept of the line having the given equation, and sketch the line.

21. (a) $x + 3y - 6 = 0$ **(b)** $4y - 9 = 0$

22. (a) $8x - 4y = 5$ **(b)** $3y - 5 = 0$

23. (a) $7x - 8y = 0$ **(b)** $x = 6 - 2y$

24. (a) $x = 4y - 2$ **(b)** $4x = 3y$

In Exercises 25 and 26, find an equation of the line through the two points, and write the equation in slope-intercept form; sketch the line.

25. $(1, 3)$ and $(2, -2)$ **26.** $(3, -5)$ and $(1, -2)$

27. Show that the lines having the equations $3x + 5y + 7 = 0$ and $6x + 10y - 5 = 0$ are parallel, and sketch their graphs.

28. Show that the lines having the equations $4x - 3y + 12 = 0$ and $8x - 6y + 15 = 0$ are parallel, and sketch their graphs.

29. Show that the lines having the equations $2x - 3y + 6 = 0$ and $3x + 2y - 12 = 0$ are perpendicular, and sketch their graphs.

30. Show that the lines having the equations $2y = 10 - 5x$ and $5y = 2x + 20$ are perpendicular, and sketch their graphs.

31. Find the value of k such that the lines whose equations are $3x + 6ky = 7$ and $9kx + 8y = 15$ are parallel.

32. Find the value of k such that the lines whose equations are $3kx + 8y = 5$ and $6y - 4kx = -1$ are perpendicular.

In Exercises 33 through 36, determine by means of slopes if the points are collinear.

33. (a) $(2, 3)$, $(-4, -7)$, $(5, 8)$
    (b) $(2, -1)$, $(1, 1)$, $(3, 4)$

34. (a) $(4, 6)$, $(1, 2)$, $(-5, -4)$
    (b) $(-3, 6)$, $(3, 2)$, $(9, -2)$

35. (a) $(2, 5)$, $(-1, 4)$, $(3, -2)$
    (b) $(0, 2)$, $(-3, -1)$, $(4, 6)$

36. (a) $(-1, 2)$, $(7, 4)$, $(2, -1)$
    (b) $(4, -9)$, $(4, 1)$, $(4, 8)$

37. Show by means of slopes that the four points $(0, 0)$, $(-2, 1)$, $(3, 4)$, and $(5, 3)$ are vertices of a parallelogram (a quadrilateral with opposite sides parallel).

38. Show by means of slopes that the four points $(-4, -1)$, $(3, \frac{8}{3})$, $(8, -4)$, and $(2, -9)$ are vertices of a trapezoid (a quadrilateral with one pair of opposite sides parallel).

39. Show by means of slopes that the three points $(3, 1)$, $(6, 0)$, and $(4, 4)$ are the vertices of a right triangle, and find the area of the triangle.

40. Show by means of slopes that the points $(-6, 1)$, $(-4, 6)$, $(4, -3)$, and $(6, 2)$ are the vertices of a rectangle.

41. The producer of a particular commodity has a total cost consisting of a weekly overhead of $3000 and a manufacturing cost of $25 per unit. (a) If x units are produced per week and y dollars is the total weekly cost, write an equation involving x and y. (b) Sketch the graph of the equation in part (a).

42. A producer's total cost consists of a manufacturing cost of $20 per unit and a fixed daily overhead. (a) If the total cost of producing 200 units in 1 day is $4500, determine the fixed daily overhead. (b) If x units are produced per day and y dollars is the total daily cost, write an equation involving x and y. (c) Sketch the graph of the equation in part (b).

43. Do Exercise 42 if the producer's cost is $30 per unit, and the total cost of producing 200 units in 1 day is $6600.

44. The graph of an equation relating the temperature reading in Celsius degrees and the temperature reading in Fahrenheit degrees is a line. Water freezes at 0° Celsius and 32° Fahrenheit, and water boils at 100° Celsius and 212° Fahrenheit. (a) If y degrees Fahrenheit corresponds to x degrees Celsius, write an equation involving x and y. (b) Sketch the graph of the equation in part (a). (c) What is the Fahrenheit temperature corresponding to 20° Celsius? (d) What is the Celsius temperature corresponding to 86° Fahrenheit?

45. Find the ordinate of the point whose abscissa is -3 and that is collinear with the points $(3, 2)$ and $(0, 5)$.

46. The equation

$$\frac{x}{a} + \frac{y}{b} = 1$$

where a and b are the x and y intercepts, respectively, is the *intercept form* of an equation of a line. Explain how you can obtain this form from the slope-intercept form and the relationship between the slope and the intercepts.

47. If you know the coordinates of the three vertices A, B, and C of a triangle, explain how you would find an equation of the median from A to the side through B and C.

48. For the triangle of Exercise 47, explain how you would find an equation of the altitude from A to the side through B and C.

49. Apply your explanation in Exercise 47 to find equations of the three medians of the triangle having vertices at $(3, -2)$, $(2, 4)$, and $(-1, 1)$.

50. Apply your explanation in Exercise 48 to find equations of the three altitudes of the triangle of Exercise 49.

In Exercises 51 through 54, use analytic geometry to prove the given theorem from plane geometry.

51. The diagonals of a parallelogram bisect each other.

52. The line segments joining consecutive midpoints of the sides of any quadrilateral form a parallelogram.

53. If the diagonals of a quadrilateral bisect each other, then the quadrilateral is a parallelogram.

54. The diagonals of a rhombus (an equilateral parallelogram) are perpendicular.

55. Prove Theorem 2: The graph of the equation $Ax + By + C = 0$, where A, B, and C are constants and where not both A and B are zero, is a line. *Hint:* Consider two cases $B \neq 0$ and $B = 0$. If $B \neq 0$, show that the equation is that of a line having slope $-A/B$ and y intercept $-C/B$. If $B = 0$, show that the equation is that of a vertical line.

3.2 SYSTEMS OF LINEAR EQUATIONS IN TWO UNKNOWNS

GOALS

1. Sketch and plot the graphs of two linear equations in two unknowns on the same coordinate axes.
2. Determine both graphically and algebraically if a system's equations are (i) consistent and independent, (ii) inconsistent, or (iii) dependent.
3. Solve a system of linear equations by the substitution method.
4. Solve a system of linear equations by the elimination method.
5. Solve a system of equations linear in the reciprocals of the unknowns.
6. Solve word problems having a system of linear equations as a mathematical model.

Many applications of mathematics lead to more than one equation in several unknowns. The resulting equations are called a **system of equations,** and the solution set consists of all solutions common to the equations in the system.

In Section 3.1 we proved that the graph of an equation of the form

$$ax + by = c$$

is a line and that all ordered pairs (x, y) satisfying the equation are coordinates of points on the line. A system of two linear equations in two unknowns x and y can be written as

$$\begin{cases} a_1x + b_1y = c_1 \\ a_2x + b_2y = c_2 \end{cases}$$

where a_1, b_1, c_1, a_2, b_2, and c_2 are real numbers. The left brace is used to indicate that the two equations form a system. If an ordered pair (x, y) is to satisfy a system of two linear equations, the corresponding point (x, y) must lie on the two lines that are the graphs of the equations.

To solve a system of two linear equations in two unknowns on your graphics calculator, plot the graphs of the two lines in the same viewing rectangle. If the lines intersect at exactly one point, the coordinates of that point give the ordered-pair solution of the system.

▷ ILLUSTRATION 1

A particular system of two linear equations is

$$\begin{cases} 2x + y = 3 \\ 5x + 3y = 10 \end{cases}$$

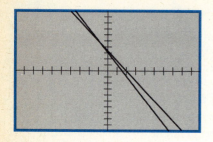

[−10, 10] by [−10, 10]
$2x + y = 3$ and $5x + 3y = 10$
FIGURE 1

Figure 1 shows graphs of the two lines in the system. From the figure it is apparent that the lines intersect at exactly one point. This point $(-1, 5)$ can be verified by substituting into the equations as follows:

$$2(-1) + 5 = 3$$
$$5(-1) + 3(5) = 10$$

The only ordered pair that is common to the solution sets of the two equations is $(-1, 5)$. Hence the solution set of the system is $\{(-1, 5)\}$. ◀

▷ ILLUSTRATION 2

Consider the system

$$\begin{cases} 6x - 3y = 5 \\ 2x - y = 4 \end{cases}$$

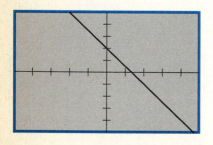

[−5, 5] by [−5, 5]
$6x - 3y = 5$ and $2x - y = 4$
FIGURE 2

As may be seen in Figure 2, the lines having these equations appear to be parallel. It can easily be proved that the lines are indeed parallel by writing each of the equations in the slope-intercept form $y = mx + b$. Solving for y in each equation, we have

$$\begin{aligned} 6x - 3y &= 5 \\ -3y &= -6x + 5 \\ y &= 2x - \frac{5}{3} \end{aligned} \qquad \begin{aligned} 2x - y &= 4 \\ -y &= -2x + 4 \\ y &= 2x - 4 \end{aligned}$$

For each equation the slope $m = 2$. The two lines are not the same because the y intercepts, $-\frac{5}{3}$ and -4, are not equal. Therefore the lines are parallel and have no point in common. The solution set of the system is $\varnothing$. ◀

▷ ILLUSTRATION 3

For the system

$$\begin{cases} 3x + 2y = 4 \\ 6x + 4y = 8 \end{cases}$$

[−5, 5] by [−5, 5]
$3x + 2y = 4$ and $6x + 4y = 8$
FIGURE 3

the graphs of the two equations are the same line. See Figure 3. This fact is evident when the equations are written in the slope-intercept form. Solving each of the equations for y, we have

$$\begin{aligned} 3x + 2y &= 4 \\ 2y &= -3x + 4 \\ y &= -\frac{3}{2}x + 2 \end{aligned} \qquad \begin{aligned} 6x + 4y &= 8 \\ 4y &= -6x + 8 \\ y &= -\frac{3}{2}x + 2 \end{aligned}$$

◀

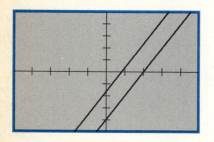

> ### Three Possibilities Regarding the Solution Sets of Two Linear Equations in Two Unknowns
>
> *Possibility 1.* The intersection of the two solution sets contains exactly one ordered pair, as in Illustration 1. The graphs intersect in exactly one point. The equations are said to be **consistent** and **independent.**
>
> *Possibility 2.* The intersection of the two solution sets is the empty set, as in Illustration 2. The graphs are distinct parallel lines. The equations are said to be **inconsistent.**
>
> *Possibility 3.* The solution sets of the two equations are equal, as in Illustration 3. The graphs are the same line. The equations are said to be **dependent.**

When two linear equations in two unknowns are consistent and independent, as in Illustration 1, the solution obtained from the graphs is generally only an approximation because reading numbers from the graphs depends upon measurement. To obtain exact solutions of systems of linear equations, we must use algebraic methods. These methods consist of replacing the given system by an **equivalent system,** one that has exactly the same solution set.

If any equation in a given system is replaced by an equivalent equation, the resulting system is equivalent to the given system. Furthermore, if any two equations of a given system are interchanged, the resulting system is equivalent to the given system.

One method for finding the solution set of a system of two linear equations in two unknowns is called the **substitution method.** For any ordered pair (x, y) in the solution set of a system of equations, the variables x and y in one equation represent the same numbers as the variables x and y in the other equation. Therefore, if we replace one of the variables in one of the equations by its equal from the other equation, we have an equivalent system. The following example shows the procedure.

▶ **EXAMPLE 1** *Solving a System of Linear Equations by the Substitution Method*

Use the substitution method to find the solution set of the system in Illustration 1:

$$\begin{cases} 2x + y = 3 \\ 5x + 3y = 10 \end{cases}$$

Solution   We solve the first equation for y and obtain the equivalent system

$$\begin{cases} y = 3 - 2x \\ 5x + 3y = 10 \end{cases}$$

We replace y in the second equation by its equal, $3 - 2x$, from the first equation. We then have the equivalent system

$$\begin{cases} y = 3 - 2x \\ 5x + 3(3 - 2x) = 10 \end{cases}$$

Simplifying the second equation, we have

$$\begin{cases} y = 3 - 2x \\ -x + 9 = 10 \end{cases}$$

Solving the second equation for x, we get

$$\begin{cases} y = 3 - 2x \\ x = -1 \end{cases}$$

Finally we substitute the value of x from the second equation into the first equation, and we have

$$\begin{cases} y = 5 \\ x = -1 \end{cases}$$

This system is equivalent to the given one. Hence the solution set is $\{(-1, 5)\}$. ◀

Another approach to solving a system of two linear equations is called the **elimination method.** In this method we replace one of the equations in the system by an equation obtained in the following way: Multiply each equation by a nonzero real number, and add the resulting equations. An equivalent system is obtained. (It is understood that "to multiply an equation by a number" means to multiply each side of the equation by that number and "to add two equations" means to add the corresponding sides of the equations.) We choose the multipliers in such a way that adding the resulting equations eliminates one of the unknowns. We demonstrate the elimination method in the following example.

▶ **EXAMPLE 2** *Solving a System of Linear Equations by the Elimination Method*

Use the elimination method to find the solution set of the system of equations in Example 1.

Solution   Our goal is to eliminate one of the unknowns of the system

$$\begin{cases} 2x + y = 3 \\ 5x + 3y = 10 \end{cases}$$

Observe that the coefficient of y is 1 in the first equation and 3 in the second equation. To obtain an equation not involving y, we therefore replace the second equation by the sum of the second equation and -3 times the first. We begin by multiplying the first equation by -3 and writing the equivalent system:

$$\begin{cases} -6x - 3y = -9 \\ 5x + 3y = 10 \end{cases}$$

Adding the two equations gives the following computation:

$$\begin{array}{rcl} -6x - 3y &=& -9 \\ \underline{5x + 3y} &=& \underline{10} \\ -x &=& 1 \end{array}$$

With this equation and the first equation in the given system, we can write the following equivalent system:

$$\begin{cases} 2x + y = 3 \\ -x = 1 \end{cases}$$

If we now multiply both sides of the second equation by -1, we have the equivalent system

$$\begin{cases} 2x + y = 3 \\ x = -1 \end{cases}$$

We next substitute -1 for x in the first equation to obtain

$$\begin{cases} 2(-1) + y = 3 \\ x = -1 \end{cases}$$

$$\Leftrightarrow \qquad \begin{cases} y = 5 \\ x = -1 \end{cases}$$

Therefore the solution set is $\{(-1, 5)\}$, which agrees with the result of Example 1.   ◀

The next two illustrations show what happens when the elimination method is used if the two equations in the given system are inconsistent or dependent.

▷ **ILLUSTRATION 4**

The system of Illustration 2 is

$$\begin{cases} 6x - 3y = 5 \\ 2x - y = 4 \end{cases}$$

We now write the equivalent system obtained by multiplying the second equation by -3. We have

$$\begin{cases} 6x - 3y = 5 \\ -6x + 3y = -12 \end{cases}$$

This system is equivalent to the following one, obtained by replacing the second equation by the sum of the two equations:

$$\begin{cases} 6x - 3y = 5 \\ 0 = -7 \end{cases}$$

The solution set of this latter system is the empty set $\emptyset$ because there is no ordered pair (x, y) for which the second equation is a true statement. Hence the solution set of the given system is $\emptyset$. The two equations are inconsistent.

◀

▷ **ILLUSTRATION 5**

The system of Illustration 3 is

$$\begin{cases} 3x + 2y = 4 \\ 6x + 4y = 8 \end{cases}$$

Multiplying the second equation by $-\frac{1}{2}$, we have the equivalent system

$$\begin{cases} 3x + 2y = 4 \\ -3x - 2y = -4 \end{cases}$$

Replacing the second equation by the sum of these two equations, we have the equivalent system

$$\begin{cases} 3x + 2y = 4 \\ 0 = 0 \end{cases}$$

The second equation of this latter system is an identity, that is, it is a true statement for any ordered pair (x, y). Therefore the solution set of the system is the same as the solution set of the first equation. The solution set can be written as $\{(x, y) \mid 3x + 2y = 4\}$. The equations are dependent. Another way of indicating the solution set arises by solving the equation $3x + 2y = 4$ for x in terms of y and getting

$$x = \frac{4}{3} - \frac{2}{3}y$$

Then the solution set is the set of ordered pairs $\{(\frac{4}{3} - \frac{2}{3}y, y)\}$. Alternatively, by letting y be any real number t, x is $\frac{4}{3} - \frac{2}{3}t$, and the solution set can be written as $\{(\frac{4}{3} - \frac{2}{3}t, t)\}$. ◄

The system in the following example is said to be linear in the reciprocals of x and y.

► **EXAMPLE 3** *Solving a System of Equations Linear in the Reciprocals of x and y*

Find the solution set of the system

$$\begin{cases} \dfrac{4}{x} + \dfrac{3}{y} = 4 \\[2mm] \dfrac{2}{x} - \dfrac{6}{y} = -3 \end{cases} \qquad \text{(I)}$$

Solution   The system can be written as

$$\begin{cases} 4\left(\dfrac{1}{x}\right) + 3\left(\dfrac{1}{y}\right) = 4 \\[2mm] 2\left(\dfrac{1}{x}\right) - 6\left(\dfrac{1}{y}\right) = -3 \end{cases}$$

Thus the system is linear in $\dfrac{1}{x}$ and $\dfrac{1}{y}$. If we make the substitutions $u = \dfrac{1}{x}$ and $v = \dfrac{1}{y}$, the system is equivalent to

$$\begin{cases} 4u + 3v = 4 \\ 2u - 6v = -3 \end{cases} \qquad \text{(II)}$$

To solve this system, we first multiply the second equation by -2 and obtain

$$\begin{cases} 4u + 3v = 4 \\ -4u + 12v = 6 \end{cases}$$

We now replace the second equation by the sum of the two equations. We get

$$\begin{cases} 4u + 3v = 4 \\ \phantom{4u + {}} 15v = 10 \end{cases}$$

Solving the second equation for v, we have

$$\begin{cases} 4u + 3v = 4 \\ \qquad v = \dfrac{2}{3} \end{cases}$$

We substitute $\frac{2}{3}$ for v in the first equation and obtain

$$\begin{cases} 4u + 3\left(\dfrac{2}{3}\right) = 4 \\ \qquad v = \dfrac{2}{3} \end{cases}$$

$$\Leftrightarrow \quad \begin{cases} 4u + 2 = 4 \\ \quad\; v = \dfrac{2}{3} \end{cases}$$

$$\Leftrightarrow \quad \begin{cases} 4u = 2 \\ \; v = \dfrac{2}{3} \end{cases}$$

$$\Leftrightarrow \quad \begin{cases} u = \dfrac{1}{2} \\ v = \dfrac{2}{3} \end{cases}$$

Therefore the solution of system (II) is $u = \dfrac{1}{2}$ and $v = \dfrac{2}{3}$. To obtain the solution of system (I), we replace u and v by $\dfrac{1}{x}$ and $\dfrac{1}{y}$, respectively. Thus

$$\dfrac{1}{x} = \dfrac{1}{2} \qquad \dfrac{1}{y} = \dfrac{2}{3}$$

$$x = 2 \qquad y = \dfrac{3}{2}$$

Hence the solution set of system (I) is $\{(2, \frac{3}{2})\}$.   ◀

In Example 3 the variables are treated as $\dfrac{1}{x}$ and $\dfrac{1}{y}$. If each of the equations in system (I) is multiplied by xy (the LCD), we obtain the system

$$\begin{cases} 4y + 3x = 4xy \\ 2y - 6x = -3xy \end{cases} \tag{III}$$

Because $4xy$ and $-3xy$ are second-degree terms, the equations in system (III) are of the second degree. To solve this system for x and y requires a more complicated procedure than that used in Example 3. Observe that

(0, 0) is a solution of system (III), whereas the equations of system (I) are not defined when $x = 0$ or $y = 0$.

In our previous discussions of word problems, we obtained a single equation as a mathematical model by representing each of the unknown numbers by using only one variable. In this section we obtain a system of equations as a mathematical model by using a different variable to represent each of the unknown numbers. You will see that some problems having a system of two linear equations as a mathematical model can also be solved by using a single linear equation in one variable as a mathematical model.

▶ **EXAMPLE 4** *Solving a Word Problem Having a System*
of Linear Equations as a Mathematical Model

Two pounds of Indian tea and 5 lb of China tea can be purchased for $50.16, and 3 lb of Indian tea and 4 lb of China tea cost $50.88. What is the price per pound of each kind of tea?

Solution   Because we wish to obtain the price of each of the two kinds of tea, we use two variables.

   x: the number of cents in the cost per pound of Indian tea

   y: the number of cents in the cost per pound of China tea

Because the cost of 2 lb of Indian tea plus the cost of 5 lb of China tea is $50.16, we have the equation

$$2x + 5y = 5016$$

Because the cost of 3 lb of Indian tea plus the cost of 4 lb of China tea is $50.88, we have the equation

$$3x + 4y = 5088$$

Thus we have the system

$$\begin{cases} 2x + 5y = 5016 \\ 3x + 4y = 5088 \end{cases}$$

Multiplying the first equation by 3 and the second equation by -2, we have the equivalent system

$$\begin{cases} 6x + 15y = 15{,}048 \\ -6x - 8y = -10{,}176 \end{cases} \tag{IV}$$

In the following equivalent system, the first equation is the first equation of the original system, and the second equation is the sum of the two equations in system (IV).

$$\begin{cases} 2x + 5y = 5016 \\ 7y = 4872 \end{cases}$$

$$\Leftrightarrow \quad \begin{cases} 2x + 5y = 5016 \\ y = 696 \end{cases}$$

$$\Leftrightarrow \quad \begin{cases} 2x + 5(696) = 5016 \\ y = 696 \end{cases}$$

$$\Leftrightarrow \quad \begin{cases} 2x + 3480 = 5016 \\ y = 696 \end{cases}$$

$$\Leftrightarrow \quad \begin{cases} x = 768 \\ y = 696 \end{cases}$$

Check: With $x = 768$ and $y = 696$, the number of dollars in the value of 2 lb of Indian tea is $2(7.68) = 15.36$, and the number of dollars in the value of 5 lb of China tea is $5(6.96) = 34.80$; $15.36 + 34.80 = 50.16$. The number of dollars in the value of 3 lb of Indian tea is $3(7.68) = 23.04$, and the number of dollars in the value of 4 lb of China tea is $4(6.96) = 27.84$; $23.04 + 27.84 = 50.88$.

Conclusion: The price per pound of Indian tea is $7.68 and the price per pound of China tea is $6.96.   ◀

EXERCISES 3.2

In Exercises 1 through 10, plot the graph of the system of equations. Classify the equations as (i) consistent and independent, (ii) inconsistent, or (iii) dependent. If the equations are consistent and independent, determine the solution set of the system from the graphs, and verify your solution by substituting into the two equations.

1. $\begin{cases} x - y = 8 \\ 2x + y = 1 \end{cases}$

2. $\begin{cases} y = 8 + 2x \\ 6x + 3y = 0 \end{cases}$

3. $\begin{cases} 2x + y = 6 \\ 8x = 6y + 9 \end{cases}$

4. $\begin{cases} 9x - 3y = 7 \\ y = 3x - \dfrac{5}{2} \end{cases}$

5. $\begin{cases} y = 2x - 4 \\ 6x - 3y - 12 = 0 \end{cases}$

6. $\begin{cases} 2x - 3y = -1 \\ 5x - 4y = 8 \end{cases}$

7. $\begin{cases} 4x - 2y - 7 = 0 \\ x = \dfrac{1}{2}y + 5 \end{cases}$

8. $\begin{cases} 3x - y = 1 \\ 6x + 5y = 2 \end{cases}$

9. $\begin{cases} 2x + 6y = -11 \\ 4x - 3y = -2 \end{cases}$

10. $\begin{cases} y = 3x - 5 \\ 6x - 2y = 10 \end{cases}$

In Exercises 11 through 22, find the solution set of the system algebraically by using either the substitution or elimination method. Verify your solution by plotting the graph of the system of equations.

11. $\begin{cases} 5x - 2y - 5 = 0 \\ 3x + y - 3 = 0 \end{cases}$

12. $\begin{cases} 3x + 4y - 4 = 0 \\ 5x + 2y - 8 = 0 \end{cases}$

13. $\begin{cases} 4x + 3y + 6 = 0 \\ 3x - 2y - 4 = 0 \end{cases}$

14. $\begin{cases} 8x - 3y = 5 \\ 5x - 2y = 4 \end{cases}$

15. $\begin{cases} 5x + 3y = 3 \\ x + 9y = 2 \end{cases}$

16. $\begin{cases} 5x + 6y = -5 \\ 15x - 3y = 13 \end{cases}$

17. $\begin{cases} 3x + 4y - 4 = 0 \\ 6x - 2y - 3 = 0 \end{cases}$

18. $\begin{cases} 18x + 3y - 10 = 0 \\ 2x - 2y - 5 = 0 \end{cases}$

19. $\begin{cases} 8x + 5y = 3 \\ 7x + 3y = -7 \end{cases}$ **20.** $\begin{cases} 2x - 5y = -21 \\ 5x + 3y = -6 \end{cases}$

21. $\begin{cases} \dfrac{x}{3} + \dfrac{y}{2} = 1 \\[2mm] \dfrac{x}{4} - \dfrac{y}{3} = -1 \end{cases}$ **22.** $\begin{cases} \dfrac{x}{2} - \dfrac{y}{6} = 1 \\[2mm] \dfrac{x}{3} + \dfrac{y}{2} = -1 \end{cases}$

In Exercises 23 through 26, find the solution set of the system algebraically.

23. $\begin{cases} \dfrac{6}{x} + \dfrac{3}{y} = -2 \\[2mm] \dfrac{4}{x} + \dfrac{7}{y} = -2 \end{cases}$ **24.** $\begin{cases} \dfrac{2}{x} + \dfrac{3}{y} = 2 \\[2mm] \dfrac{4}{x} - \dfrac{3}{y} = 1 \end{cases}$

25. $\begin{cases} \dfrac{3}{x} - \dfrac{2}{y} = 14 \\[2mm] \dfrac{6}{x} + \dfrac{3}{y} = -7 \end{cases}$ **26.** $\begin{cases} \dfrac{1}{x} - \dfrac{10}{y} = 6 \\[2mm] \dfrac{2}{x} + \dfrac{5}{y} = 2 \end{cases}$

In Exercises 27 through 36, solve the word problem by finding a system of equations as a mathematical model of the situation. Complete the exercise by writing a conclusion.

27. Three pounds of tea and 8 lb of coffee cost $39.70, and 5 lb of tea and 6 lb of coffee cost $47.10. What is the cost per pound of tea, and what is the cost per pound of coffee?

28. The cost of sending a telegram is based on a flat rate for the first 10 words and a fixed charge for each additional word. If a telegram of 15 words costs $11.65 and a telegram of 19 words costs $14.57, what is the flat rate, and what is the fixed charge for each additional word?

29. A group of women decided to contribute equal amounts toward obtaining a speaker for a book review. If there were 10 more women, each would have paid $2 less. However, if there were 5 less women, each would have paid $2 more. How many women were in the group and how much was the speaker paid?

30. A woman has a certain amount of money invested. If she had $6000 more invested at a rate 1 percent lower, she would have the same yearly income from the in-

vestment. Furthermore, if she had $4500 less invested at a rate 1 percent higher, her yearly income from the investment would also be the same. How much does she have invested, and at what rate is it invested?

31. A chemist has two acid solutions. One contains 15 percent acid and the other contains 6 percent acid. How many cubic centimeters of each solution should be used to obtain 400 cm^3 of a solution that is 9 percent acid?

32. A tank contains a mixture of insect spray and water in which there are 5 gal of insect spray and 25 gal of water. A second tank also contains 5 gal of spray but only 15 gal of water. If it is desired to have 7.5 gal of a mixture of which 20 percent is spray, how many gallons should be taken from each tank?

33. If a girl works for 8 min and her brother works for 15 min, they can wash the front windows of their house. Also, if the girl works for 12 min and her brother works for 10 min, they can wash the same windows. How long will it take each person alone to wash the windows? *Hint:* If it takes t minutes for a person to do a job alone, then in 1 min, the person can do $\dfrac{1}{t}$ of the job.

34. A painter and his son can paint a room together in 8 hr. If the father works alone for 3 hr and then is joined by his son, the two together can complete the job in 6 hr more. How long will it take each person alone to paint the room? See the hint for Exercise 33.

35. If either 4 is added to the denominator of a fraction or 2 is subtracted from the numerator of the fraction, the resulting fraction is equivalent to $\frac{1}{2}$. What is the fraction?

36. If the numerator and denominator of a fraction are both increased by 5, the resulting fraction is equivalent to $\frac{2}{3}$. However, if the numerator and denominator are both decreased by 5, the resulting fraction is equivalent to $\frac{3}{7}$. What is the fraction?

37. Discuss how you would determine algebraically and graphically if two systems of two linear equations in two unknowns are equivalent.

38. Discuss how you would determine graphically whether a system of three linear equations in two unknowns is consistent or inconsistent.

Apply your answer in Exercise 38 to the systems of Exercises 39 through 42 by plotting the graphs of the three equations in the same viewing rectangle. Are the equations consistent or inconsistent? Verify your answer algebraically: To show they are inconsistent, solve a system of two of the equations and show that no member of the solution set satisfies the third equation; to show they are consistent, find the solution set.

39. $\begin{cases} 4x - y = 1 \\ 2x + y = 5 \\ 5x - 2y = -3 \end{cases}$ **40.** $\begin{cases} 3x + 4y = 4 \\ 2x - y = 10 \\ x + 3y = -2 \end{cases}$

41. $\begin{cases} 2x - 5y = 4 \\ 3x - 2y = -5 \\ -3x + 4y = 1 \end{cases}$ **42.** $\begin{cases} 2x + y = 10 \\ 3x - 4y = -5 \\ 4x - 3y = 0 \end{cases}$

3.3 PARABOLAS

GOALS

1. **Define a parabola.**
2. **Sketch parabolas from their equations.**
3. **Plot parabolas from their equations.**
4. **Learn properties of parabolas.**
5. **Find properties of parabolas from their equations.**
6. **Find equations of parabolas from their properties.**
7. **Solve word problems having an equation of a parabola as a mathematical model.**

We introduced parabolas in Section 1.5. The graphs in Figures 5 and 8 of that section are parabolas. These curves have many important applications. They are used in the design of parabolic mirrors, searchlights, and automobile headlights. The path of a projectile is a parabola if motion is considered to be in a plane and air resistance is neglected. Arches are sometimes parabolic in appearance; and the cable of a suspension bridge could hang in the form of a parabola. Dish antennas for receiving satellite television signals are also parabolic in shape.

In the definition of a parabola we refer to the distance from a point to a line. By such a distance, we mean the length of the perpendicular line segment from the point to the line. See Figure 1, where $|\overline{PQ}|$ is the distance from point P to line l.

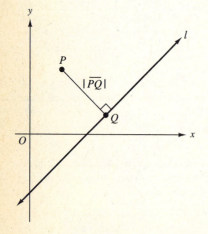

FIGURE 1

DEFINITION   A Parabola

A **parabola** is the set of all points in a plane equidistant from a fixed point and a fixed line. The fixed point is called the **focus,** and the fixed line is called the **directrix.**

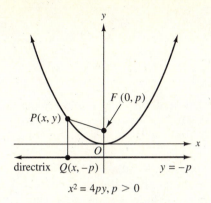

$x^2 = 4py, p > 0$

FIGURE 2

We now derive an equation of a parabola from the definition. For this equation to be as simple as possible, we choose the y axis as perpendicular to the directrix and containing the focus. The origin is taken as the point on the y axis midway between the focus and the directrix. Observe that we are choosing the axes (*not* the parabola) in a special way. See Figure 2.

Let p be the directed distance $\overline{OF}$. The focus is the point $F(0, p)$, and the directrix is the line having the equation $y = -p$. A point $P(x, y)$ is on the parabola if and only if P is equidistant from F and the directrix. That is, if $Q(x, -p)$ is the foot of the perpendicular line from P to the directrix, then P is on the parabola if and only if

$$|\overline{FP}| = |\overline{QP}|$$

Because

$$|\overline{FP}| = \sqrt{x^2 + (y - p)^2}$$

and

$$|\overline{QP}| = \sqrt{(x - x)^2 + (y + p)^2}$$

the point P is on the parabola if and only if

$$\sqrt{x^2 + (y - p)^2} = \sqrt{(y + p)^2}$$

By squaring on both sides of the equation, we obtain

$$x^2 + y^2 - 2py + p^2 = y^2 + 2py + p^2$$
$$x^2 = 4py$$

We state this result formally.

Equation of a Parabola

An equation of the parabola having its focus at $(0, p)$ and having as its directrix the line $y = -p$ is

$$x^2 = 4py$$

In Figure 2, p is positive; p may be negative, however, because it is the directed distance $\overline{OF}$. Figure 3 shows a parabola for $p < 0$.

From Figures 2 and 3 we see that for the equation $x^2 = 4py$ the parabola opens upward if $p > 0$ and downward if $p < 0$. The line through the focus perpendicular to the directrix is called the **axis** of the parabola. The axis of the parabolas of Figures 2 and 3 is the y axis. The intersection of the parabola with its axis is called the **vertex,** which of course is midway between the focus and directrix. The vertex of the parabolas in Figures 2 and 3 is the origin.

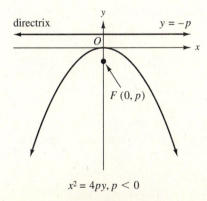

$x^2 = 4py, p < 0$

FIGURE 3

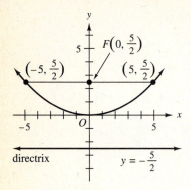

FIGURE 4

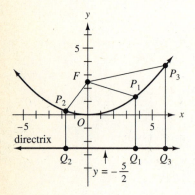

FIGURE 5

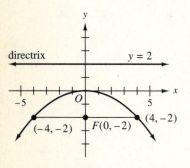

FIGURE 6

▷ **ILLUSTRATION 1**

The graph of the equation

$$x^2 = 10y$$

is a parabola whose vertex is at the origin and whose axis is the y axis. Because $4p = 10$, $p = \frac{5}{2} > 0$, and therefore the parabola opens upward. The focus is at the point $F(0, \frac{5}{2})$, and an equation of the directrix is $y = -\frac{5}{2}$. Two points on the parabola are $(5, \frac{5}{2})$ and $(-5, \frac{5}{2})$. These points are the endpoints of a chord through the focus and perpendicular to the axis of the parabola. This chord is called the **latus rectum** of the parabola. In Exercise 41 you are asked to prove that the length of the latus rectum of a parabola is $|4p|$. When sketching a parabola, it is helpful to plot the endpoints of the latus rectum. Figure 4 shows the parabola, focus, directrix, and latus rectum. ◀

The parabola of Figure 4 with three points P_1, P_2, and P_3 on it appears in Figure 5. Because the definition of a parabola states that any point on the parabola is equidistant from the focus and directrix,

$$|\overline{FP_1}| = |\overline{Q_1 P_1}| \qquad |\overline{FP_2}| = |\overline{Q_2 P_2}| \qquad |\overline{FP_3}| = |\overline{Q_3 P_3}|$$

▶ **EXAMPLE 1**   *Sketching a Parabola and Finding its Properties from its Equation*

Sketch the parabola having the equation

$$x^2 = -8y$$

and find the focus, an equation of the directrix, and the endpoints of the latus rectum.

Solution   The graph is a parabola whose vertex is at the origin and whose axis is the y axis. Because $4p = -8$, $p = -2$, and because $p < 0$, the parabola opens downward. The focus is at the point $F(0, -2)$, and an equation of the directrix is $y = 2$. The endpoints of the latus rectum are $(4, -2)$ and $(-4, -2)$. These points are obtained by substituting -2 for y in the equation of the parabola and solving for x. The parabola is sketched in Figure 6, which also shows the focus and directrix. ◀

Of course, you can check the parabola in Example 1 by plotting the graph of the equation $y = -\frac{1}{8}x^2$ on your graphics calculator.

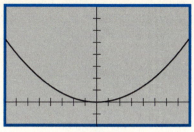

[−8, 8] by [−2, 8]

$y = \frac{1}{12}x^2$

FIGURE 7

▶ **EXAMPLE 2** *Finding an Equation of a Parabola from its Properties and Plotting its Graph*

Find an equation of the parabola having its focus at $(0, 3)$ and as its directrix the line $y = -3$. Plot the parabola.

Solution Because the focus is on the y axis and is also above the directrix, the parabola opens upward and $p = 3$. The vertex is at the origin. An equation of the parabola is of the form $x^2 = 4py$ with $4p = 12$. Thus the required equation is

$$x^2 = 12y$$

To plot the parabola, we write the equation as $y = \frac{1}{12}x^2$. See Figure 7. ◀

▶ **EXAMPLE 3** *Solving a Word Problem Having an Equation of a Parabola as a Mathematical Model*

A parabolic mirror has a depth of 12 cm at the center, and the distance across the top of the mirror is 32 cm. Find the distance from the vertex to the focus.

Solution See Figure 8. We choose the coordinate axes so that the parabola has its vertex at the origin, has its axis along the y axis, and opens upward. An equation of the parabola is, therefore, of the form

$$x^2 = 4py$$

where p centimeters is the distance from the vertex to the focus. Because the point $(16, 12)$ is on the parabola, its coordinates satisfy the equation, and we have

$$16^2 = 4p(12)$$

$$p = \frac{16}{3}$$

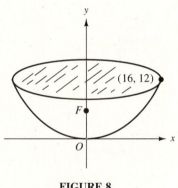

FIGURE 8

Conclusion: The distance from the vertex to the focus is $\frac{16}{3}$ cm. ◀

Some parabolas have horizontal axes. The parabola having the equation

$$y^2 = 7x$$

is an example. This equation is of the form

$$y^2 = 4px$$

which can be obtained from the equation $x^2 = 4py$ by interchanging x and y. A parabola having the equation $y^2 = 4px$ has its vertex at the origin, the

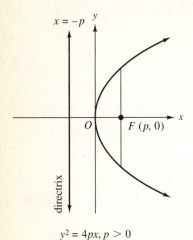

$x = -p$

O $F(p, 0)$ x

directrix

$y^2 = 4px, p > 0$

FIGURE 9

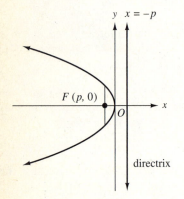

y $x = -p$

$F(p, 0)$

O x

directrix

FIGURE 10

x axis as its axis, and its focus at the point $F(p, 0)$; an equation of its directrix is $x = -p$. If $p > 0$, the parabola opens to the right, as in Figure 9, and if $p < 0$, the parabola opens to the left, as in Figure 10.

We summarize these results.

Equation of a Parabola

An equation of the parabola having its focus at $(p, 0)$ and its directrix the line $x = -p$ is

$$y^2 = 4px$$

Plotting a Parabola

To plot the graph of an equation of the form $y^2 = 4px$:

1. Solve for y by taking the square root of both sides of the equation and obtain the two equations

$$y = \sqrt{4px} \quad \text{and} \quad y = -\sqrt{4px}$$

2. The union of the graphs of these two equations gives the graph of

$$y^2 = 4px.$$

▶ **EXAMPLE 4** *Sketching a Parabola and Finding its Properties from its Equation*

Sketch the parabola having the equation

$$y^2 = 7x$$

and find the focus, an equation of the directrix, and the endpoints of the latus rectum. Check the graph by plotting it.

Solution   The given equation is of the form $y^2 = 4px$; the vertex is, therefore, at the origin, and the x axis is the axis. Because $4p = 7$, $p = \frac{7}{4} > 0$; thus the parabola opens to the right. The focus is at the point

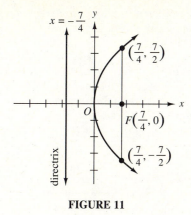

FIGURE 11

$F(\frac{7}{4}, 0)$, and an equation of the directrix is $x = -\frac{7}{4}$. To obtain the endpoints of the latus rectum, let $x = \frac{7}{4}$ in the given equation, and we have

$$y^2 = \frac{49}{4}$$

$$y = \pm\frac{7}{2}$$

The endpoints of the latus rectum are, therefore, $(\frac{7}{4}, \frac{7}{2})$ and $(\frac{7}{4}, -\frac{7}{2})$. The parabola is sketched in Figure 11, which also shows the focus and directrix.

To plot the parabola, we plot the graphs of

$$y = \sqrt{7x} \quad \text{and} \quad y = -\sqrt{7x}$$

in the same viewing rectangle.

EXERCISES 3.3

In Exercises 1 through 16, for the parabola having the given equation, find (a) the vertex, (b) the axis, (c) the focus, (d) an equation of the directrix, and (e) the endpoints of the latus rectum. Sketch the parabola.

1. $x^2 = 4y$ **2.** $x^2 = 8y$

3. $x^2 = -16y$ **4.** $x^2 = -12y$

5. $x^2 - y = 0$ **6.** $x^2 - 2y = 0$

7. $y^2 = 12x$ **8.** $y^2 = -6x$

9. $y^2 = -8x$ **10.** $y^2 = x$

11. $y^2 - 5x = 0$ **12.** $y^2 + 3x = 0$

13. $3x^2 + 8y = 0$ **14.** $2x^2 + 5y = 0$

15. $2y^2 - 9x = 0$ **16.** $3y^2 - 4x = 0$

In Exercises 17 through 22, plot the parabola having the given equation.

17. (a) $y = 4x^2$ (b) $y = -4x^2$
       (c) $x = 4y^2$ (d) $x = -4y^2$

18. (a) $y = 2x^2$ (b) $y = -2x^2$
       (c) $x = 2y^2$ (d) $x = -2y^2$

19. (a) $y = \frac{1}{4}x^2$ (b) $y = -\frac{1}{4}x^2$
       (c) $x = \frac{1}{4}y^2$ (d) $x = -\frac{1}{4}y^2$

20. (a) $y = \frac{1}{2}x^2$ (b) $y = -\frac{1}{2}x^2$
       (c) $x = \frac{1}{2}y^2$ (d) $x = -\frac{1}{2}y^2$

21. (a) $x^2 - 16y = 0$ (b) $x^2 + 16y = 0$
       (c) $y^2 - 16x = 0$ (d) $y^2 + 16x = 0$

22. (a) $4x^2 - 3y = 0$ (b) $4x^2 + 3y = 0$
       (c) $4y^2 - 3x = 0$ (d) $4y^2 + 3x = 0$

In Exercises 23 through 36, find an equation of the parabola having the given properties. Sketch the parabola and then check your graph by plotting it on your graphics calculator.

23. Focus, $(0, 4)$; directrix, $y = -4$.

24. Focus, $(0, -2)$; directrix, $y = 2$.

25. Focus, $(0, -5)$; directrix, $y - 5 = 0$.

26. Focus, $(0, -\frac{1}{2})$; directrix, $2y - 1 = 0$.

27. Focus, $(2, 0)$; directrix, $x = -2$.

28. Focus, $(1, 0)$; directrix, $x = -1$.

29. Focus, $(-\frac{5}{3}, 0)$; directrix, $5 - 3x = 0$.

30. Focus, $(-\frac{3}{2}, 0)$; directrix, $2x - 3 = 0$.

31. Vertex, the origin; opens upward; through the point $(6, 3)$.

32. Vertex, the origin; opens downward; through the point $(-4, -2)$.

33. Vertex, the origin; directrix, $2x = 3$.

34. Vertex, the origin; directrix, $2y + 5 = 0$.

35. Vertex, the origin; y axis is its axis; through the point $(-2, 4)$

36. Vertex, the origin; x axis is its axis; through the point $(-3, 3)$.

In Exercises 37 through 40, solve the word problem by finding an equation of a parabola as a mathematical model of the situation. Complete the exercise by writing a conclusion.

37. A reflecting telescope has a parabolic mirror for which the distance from the vertex to the focus is 30 ft. If the distance across the top of the mirror is 64 in., how deep is the mirror at the center?

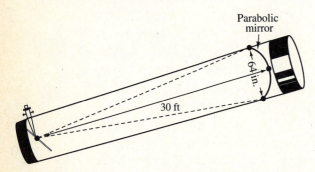

38. A parabolic arch has a height of 20 m and a width of 36 m at the base. If the vertex of the parabola is at the top of the arch, at which height above the base is it 18 m wide?

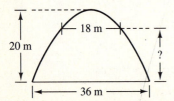

39. The cable of a suspension bridge hangs in the form of a parabola when the load is uniformly distributed horizontally. The distance between two towers is 150 m, the points of support of the cable on the towers are 22 m above the roadway, and the lowest point on the cable is 7 m above the roadway. Find the vertical distance to the cable from a point in the roadway 15 m from the foot of a tower.

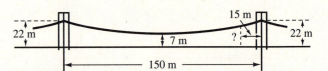

40. Assume that water issuing from the end of a horizontal pipe 25 ft above the ground describes a parabolic curve, the vertex of the parabola being at the end of the pipe. If at a point 8 ft below the line of the pipe the flow of water has curved outward 10 ft beyond a vertical line through the end of the pipe, how far beyond this vertical line will the water strike the ground?

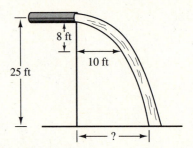

41. Prove that the length of the latus rectum of a parabola is $|4p|$.

42. Find an equation of the parabola whose vertex is at the origin and for which the endpoints of the latus rectum are at $(-8, 4)$ and $(8, 4)$.

43. The endpoints of the latus rectum of a parabola are $(5, k)$ and $(-5, k)$. If the vertex of the parabola is at the origin and the parabola opens downward, find: **(a)** the value of k; **(b)** an equation of the parabola.

44. Find all points on the parabola $y^2 = 8x$ such that the focus, the point itself, and the foot of the perpendicular drawn from the point to the directrix are vertices of an equilateral triangle.

45. Plot the graphs of $y = x^2$ and $y = x^4$. Explain why the graph of the first equation is a parabola and why the graph of the second is not. Use the definition of a parabola in your explanation.

3.4 CIRCLES

GOALS

1. **Define a circle.**
2. **Sketch circles from their equations.**
3. **Plot circles from their equations.**
4. **Learn properties of circles.**
5. **Find properties of circles from their equations.**
6. **Find equations of circles from their properties.**

You learned in the previous section that parabolas have second-degree equations involving just one second-degree term. Another curve having second-degree equations is the *circle,* but these equations have two second-degree terms, one involving x and one involving y.

DEFINITION   **A Circle**

A **circle** is the set of all points in a plane equidistant from a fixed point. The fixed point is called the **center** of the circle, and the constant equal distance is called the **radius** of the circle.

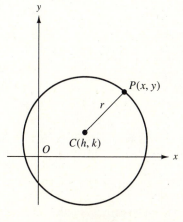

FIGURE 1

To obtain an equation of the circle having center at $C(h, k)$ and radius r, we use the distance formula. Refer to Figure 1. The point $P(x, y)$ is on the circle if and only if $|\overline{PC}| = r$; that is, if and only if

$$\sqrt{(x - h)^2 + (y - k)^2} = r$$

This equation is true if and only if

$$(x - h)^2 + (y - k)^2 = r^2 \quad (r > 0)$$

This equation is satisfied by the coordinates of those and only those points that lie on the circle, and therefore it is an equation of the circle. We state this result formally.

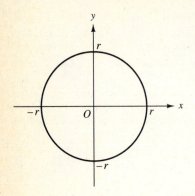

FIGURE 2

Equation of a Circle

The circle with center at the point (h, k) and radius r has as an equation

$$(x - h)^2 + (y - k)^2 = r^2$$

If the center of a circle is at the origin, then $h = 0$ and $k = 0$; therefore, its equation is

$$x^2 + y^2 = r^2$$

Such a circle appears in Figure 2. If the radius of a circle is 1, it is called a **unit circle.**

If the center and radius of a circle are known, the circle can be drawn by using a compass.

Plotting a Circle

To plot the circle having the equation

$$x^2 + y^2 = r^2$$

1. Solve this equation for y and obtain
$$y = \sqrt{r^2 - x^2} \quad \text{and} \quad y = -\sqrt{r^2 - x^2}$$
The graph of each of these equations is a semicircle.
2. Plotting these two semicircles in the same viewing rectangle gives the desired circle.

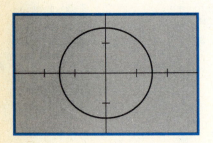

$[-7.5, 7.5]$ by $[-5, 5]$
$y = \sqrt{16 - x^2}$ and $y = -\sqrt{16 - x^2}$
FIGURE 3

▷ **ILLUSTRATION 1**

The graph of the equation

$$x^2 + y^2 = 16$$

is the circle with center at the origin and radius 4. Solving this equation for y, we obtain

$$y = \sqrt{16 - x^2} \quad \text{and} \quad y = -\sqrt{16 - x^2}$$

We plot these two semicircles in the same viewing rectangle and obtain the circle shown in Figure 3.   ◀

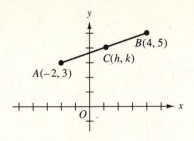

FIGURE 4

▶ **EXAMPLE 1** *Finding an Equation of a Circle Satisfying Given Conditions and Plotting its Graph*

Find an equation of the circle having a diameter with endpoints at $A(-2, 3)$ and $B(4, 5)$. Plot the circle.

Solution   The midpoint of the line segment from A to B is the center of the circle. See Figure 4. If $C(h, k)$ is the center of the circle, then

$$h = \frac{-2 + 4}{2} \qquad k = \frac{3 + 5}{2}$$
$$= 1 \qquad\qquad = 4$$

The center is at $C(1, 4)$. The radius of the circle can be computed as either $|\overline{CA}|$ or $|\overline{CB}|$. If $r = |\overline{CA}|$, then

$$r = \sqrt{(1 + 2)^2 + (4 - 3)^2}$$
$$= \sqrt{10}$$

An equation of the circle is therefore

$$(x - 1)^2 + (y - 4)^2 = 10$$
$$x^2 + y^2 - 2x - 8y + 7 = 0$$

To plot the circle, we first solve the equation for y by treating it as a quadratic equation in y:

$$y^2 - 8y + (x^2 - 2x + 7) = 0$$

From the quadratic formula where a is 1, b is -8, and c is $x^2 - 2x + 7$, we have

$$y = \frac{-b \pm \sqrt{b^2 - 4ac}}{2a}$$

$$= \frac{-(-8) \pm \sqrt{(-8)^2 - 4(1)(x^2 - 2x + 7)}}{2(1)}$$

$$= \frac{8 \pm \sqrt{64 - 4x^2 + 8x - 28}}{2}$$

$$= \frac{8 \pm \sqrt{36 + 8x - 4x^2}}{2}$$

$$= \frac{8 \pm 2\sqrt{9 + 2x - x^2}}{2}$$

$$= 4 \pm \sqrt{9 + 2x - x^2}$$

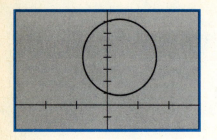

[-7.5, 7.5] by [-2, 8]
$y = 4 + \sqrt{9 + 2x - x^2}$ and
$y = 4 - \sqrt{9 + 2x - x^2}$
FIGURE 5

In the same viewing rectangle, we plot the graphs of

$$y = 4 + \sqrt{9 + 2x - x^2} \quad \text{and} \quad y = 4 - \sqrt{9 + 2x - x^2}$$

to obtain the circle appearing in Figure 5.   ◄

The equation $(x - h)^2 + (y - k)^2 = r^2$ is called the **center-radius form** of an equation of a circle. If we remove parentheses and combine like terms, we obtain

$$x^2 + y^2 - 2hx - 2ky + (h^2 + k^2 - r^2) = 0$$

By letting $D = -2h$, $E = -2k$, and $F = h^2 + k^2 - r^2$, this equation becomes

$$x^2 + y^2 + Dx + Ey + F = 0$$

which is called the **general form** of an equation of a circle. Because every circle has a center and radius, its equation can be put in the center-radius form, and hence into the general form, as we did in Example 1. If we start with an equation of a circle in the general form, we can write it in the center-radius form by completing the square. The next example shows the procedure.

► **EXAMPLE 2**   *Finding the Center and Radius of a Circle from its Equation*

Find the center and radius of the circle having the equation

$$x^2 + y^2 + 6x - 4y - 23 = 0$$

Solution   The given equation may be written as

$$(x^2 + 6x) + (y^2 - 4y) = 23$$

Completing the squares of the terms in parentheses by adding 9 and 4 on both sides of the equation, we have

$$(x^2 + 6x + 9) + (y^2 - 4y + 4) = 23 + 9 + 4$$
$$(x + 3)^2 + (y - 2)^2 = 36$$

This equation is in the center-radius form; thus it is an equation of a circle with its center at $(-3, 2)$ and radius 6.   ◄

There are equations of the form

$$x^2 + y^2 + Dx + Ey + F = 0$$

whose graphs are not circles. Suppose when we complete the squares we obtain

$$(x - h)^2 + (y - k)^2 = d \quad \text{where } d < 0$$

No real values of x and y satisfy this equation; thus the equation has no graph. In such a case we state that the graph is the empty set. See Exercise 32.

If when completing the squares, we obtain

$$(x - h)^2 + (y - k)^2 = 0$$

the only real values of x and y satisfying this equation are $x = h$ and $y = k$. Thus the graph is the single point (h, k). See Exercise 31.

 In calculus you will learn how to find an equation of the line tangent to a general curve at a point on the curve. The definition of such a tangent line requires the concept of *limit* studied in calculus. For a circle, however, the plane-geometry definition states that a tangent line at a point P on a circle is the line intersecting the circle at only one point.

▶ **EXAMPLE 3** *Finding an Equation of the Line Tangent to a Circle at a Point*

Find an equation of the tangent line to the circle

$$x^2 + y^2 - 6x - 2y - 15 = 0$$

at the point $(6, 5)$. Plot the circle and tangent line in the same viewing rectangle.

Solution   We write the equation of the circle in the center-radius form by completing the squares:

$$(x^2 - 6x) + (y^2 - 2y) = 15$$
$$(x^2 - 6x + 9) + (y^2 - 2y + 1) = 15 + 9 + 1$$
$$(x - 3)^2 + (y - 1)^2 = 25$$

From this equation, the center of the circle is at $C(3, 1)$ and the radius is 5. Figure 6 shows the circle and a piece of the tangent line at $P(6, 5)$. If m_1 is the slope of the line through C and P,

$$m_1 = \frac{5 - 1}{6 - 3}$$

$$= \frac{4}{3}$$

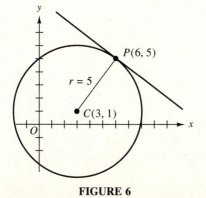

FIGURE 6

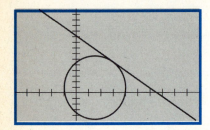

$y = 1 + \sqrt{25 - (x - 3)^2}$ and
$y = 1 - \sqrt{25 - (x - 3)^2}$ and
$$y = -\frac{3}{4}x + \frac{39}{4}$$

FIGURE 7

From plane geometry, we know that the tangent line is perpendicular to the line through C and P. Therefore, if m_2 is the slope of the tangent line,

$$m_2 m_1 = -1$$
$$m_2\left(\frac{4}{3}\right) = -1$$
$$m_2 = -\frac{3}{4}$$

Hence from the point-slope form of an equation of the line through $(5, 6)$ with slope $-\frac{3}{4}$, we have as the required equation

$$y - 6 = -\frac{3}{4}(x - 5)$$
$$4y - 24 = -3x + 15$$
$$3x + 4y - 39 = 0$$

Figure 7 shows the circle and tangent line plotted in the same viewing rectangle.   ◀

EXERCISES 3.4

In Exercises 1 through 8, sketch the graph of the equation.

1. (a) $y = \sqrt{4 - x^2}$ **(b)** $y = -\sqrt{4 - x^2}$
(c) $x^2 + y^2 = 4$

2. (a) $y = \sqrt{25 - x^2}$ **(b)** $y = -\sqrt{25 - x^2}$
(c) $x^2 + y^2 = 25$

3. $9x^2 + 9y^2 = 1$ **4.** $4x^2 + 4y^2 = 1$
5. $(x - 3)^2 + (y + 4)^2 = 16$
6. $(x + 1)^2 + (y - 5)^2 = 36$
7. $(x + 4)^2 + y^2 = 1$ **8.** $x^2 + (y - 2)^2 = 9$

In Exercises 9 through 14, plot the graph of the equation.

9. $x^2 + y^2 = 36$ **10.** $x^2 + y^2 = 16$
11. $4x^2 + 4y^2 = 81$ **12.** $9x^2 + 9y^2 = 49$
13. $(x + 2)^2 + (y - 3)^2 = 100$
14. $(x - 4)^2 + (y + 7)^2 = 64$

In Exercises 15 through 20, find an equation of the circle with center at C and radius r. Write the equation in both the center-radius form and the general form. Plot the circle.

15. $C(4, -3)$, $r = 5$ **16.** $C(0, 0)$, $r = 8$
17. $C(-5, -12)$, $r = 3$ **18.** $C(-1, 1)$, $r = 2$
19. $C(0, 7)$, $r = 1$ **20.** $C(-3, 0)$, $r = 4$

In Exercises 21 through 24, find an equation of the circle satisfying the given conditions. Plot the circle.

21. Center is at $(1, 2)$ and through the point $(3, -1)$.
22. Center is at $(-3, 4)$ and through the point $(2, 0)$.
23. Diameter has endpoints at $(3, -4)$ and $(7, 2)$.
24. Diameter has endpoints at $(-1, -5)$ and $(4, -6)$.

In Exercises 25 through 30, find the center and radius of the circle. Sketch the circle.

25. $x^2 + y^2 - 6x - 8y + 9 = 0$
26. $x^2 + y^2 - 10x - 10y + 25 = 0$
27. $x^2 + y^2 + 2x + 10y + 18 = 0$
28. $x^2 + y^2 + 6x - 1 = 0$
29. $3x^2 + 3y^2 + 4y - 7 = 0$
30. $2x^2 + 2y^2 - 2x + 2y + 7 = 0$

31. Prove that the graph of
$$x^2 + y^2 - 4x + 10y + 29 = 0$$
is a point.

32. Prove that the graph of
$$x^2 + y^2 + 8x - 6y + 30 = 0$$
is the empty set.

In Exercises 33 through 38, determine whether the graph is a circle, a point, or the empty set.

33. $x^2 + y^2 - 2x + 10y + 19 = 0$

34. $x^2 + y^2 + 2x - 4y + 5 = 0$

35. $x^2 + y^2 - 10x + 6y + 36 = 0$

36. $4x^2 + 4y^2 + 24x - 4y + 1 = 0$

37. $2x^2 + 2y^2 - 2x + 6y + 5 = 0$

38. $9x^2 + 9y^2 + 6x - 6y + 5 = 0$

In Exercises 39 through 42, find an equation of the line tangent to the circle at point P. Plot the circle and tangent line in the same viewing rectangle.

39. $x^2 + y^2 = 25$; $P(-4, 3)$

40. $16x^2 + 16y^2 = 25$; $P(\frac{3}{4}, -1)$

41. $x^2 + y^2 - 4x + 6y - 12 = 0$; $P(5, 1)$

42. $x^2 + y^2 + 14x - 8y - 35 = 0$; $P(-1, -4)$

43. Use analytic geometry to prove that an angle inscribed in a semicircle is a right angle.

44. Use analytic geometry to prove that a line from the center of any circle bisecting any chord is perpendicular to the chord.

45. What inequality involving D, E, and F is necessary for the graph of the equation

$$x^2 + y^2 + Dx + Ey + F = 0$$

to be a circle.

46. From the origin, chords of the circle

$$x^2 + y^2 + 4x = 0$$

are drawn. Prove that the set of midpoints of these chords is a circle.

47. The circumscribed circle of a triangle is the circle containing the three vertices of the triangle. Given the three vertices, explain how you can determine the center and radius of the circumscribed circle.

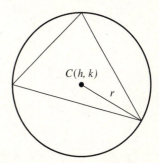

48. Use your explanation in Exercise 47 to find the center and radius of the circumscribed circle of the triangle having vertices at $(-3, 2)$, $(4, -1)$, and $(5, 2)$.

49. Describe the set of points (x, y) in R^2 for which
(a) $x^2 + y^2 \leq 1$ **(b)** $1 < x^2 + y^2 \leq 4$
(c) $x^2 + y^2 > 4$.

50. Use the fact that $ab = 0$ if and only if $a = 0$ or $b = 0$ to write an equation of each of the following graphs: **(a)** the graph consisting of all points on either of the two circles, each having its center at the origin and one having radius 2 and the other having radius 3; **(b)** the graph consisting of the origin and all points on the unit circle whose center is the origin.

3.5 TRANSLATION OF AXES

GOALS

1. **Learn equations for translating the axes.**
2. **Translate the axes to simplify an equation.**
3. **Learn standard forms of an equation of a parabola.**
4. **Find properties of a parabola from its equation in standard form and sketch the parabola from these properties.**
5. **Obtain the graph of one equation from the graph of another equation and a suitable translation of axes.**

The shape of a graph is not affected by the position of the coordinate axes, but its equation is affected. For example, a circle with a radius of 3 and having its center at the point $(4, -1)$ has the equation

$$(x - 4)^2 + (y + 1)^2 = 9$$

However, if the coordinate axes are chosen so that the origin is at the center, the same circle has the simpler equation

$$x^2 + y^2 = 9$$

If we may select the coordinate axes as we please, we generally do so in such a way that the equations will be as simple as possible. If the axes are given, however, we may wish to find a simpler equation of a particular graph relative to a different set of axes. If these different axes are chosen parallel to the given ones, we say that there has been a **translation of axes.**

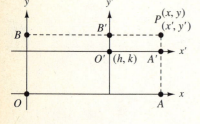

FIGURE 1

In particular, let the given x and y axes be translated to new axes x' and y' having origin (h, k) with respect to the given axes. Also assume that the positive numbers lie on the same side of the origin on the x' and y' axes as on the x and y axes. See Figure 1. A point P in the plane having coordinates (x, y) with respect to the given coordinate axes will have coordinates (x', y') with respect to the new axes. We now obtain relationships between these two sets of coordinates. We draw two lines through P, one parallel to the y and y' axes and one parallel to the x and x' axes. Let the first line intersect the x axis at the point A and the x' axis at the point A', and let the second line intersect the y axis at the point B and the y' axis at the point B'. These lines are shown in Figure 1.

With respect to the x and y axes, the coordinates of P are (x, y), the coordinates of A are $(x, 0)$, and the coordinates of A' are (x, k). Because $\overline{A'P} = \overline{AP} - \overline{AA'}$,

$$y' = y - k$$

With respect to the x and y axes, the coordinates of B are $(0, y)$, and the coordinates of B' are (h, y). Because $\overline{B'P} = \overline{BP} - \overline{BB'}$,

$$x' = x - h$$

We state these results formally.

Equations for Translating the Axes

If (x, y) represents a point P with respect to a given set of axes, and (x', y') is a representation of P after the axes are translated to a new origin having coordinates (h, k) with respect to the given axes, then

$$x' = x - h \quad \text{and} \quad y' = y - k$$

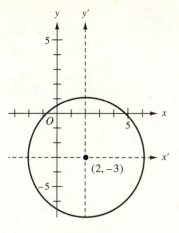

FIGURE 2

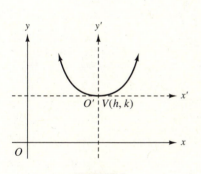

FIGURE 3

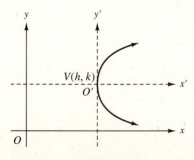

FIGURE 4

▶ **EXAMPLE 1** *Translating the Axes to Simplify an Equation*

Given the equation

$$x^2 + y^2 - 4x + 6y - 3 = 0$$

translate the axes so that the equation of the graph with respect to the x' and y' axes contains no first-degree terms.

Solution   We rewrite the given equation as

$$(x^2 - 4x) + (y^2 + 6y) = 3$$

Completing the squares of the terms in parentheses by adding 4 and 9 on both sides of the equation, we have

$$(x^2 - 4x + 4) + (y^2 + 6y + 9) = 3 + 4 + 9$$
$$(x - 2)^2 + (y + 3)^2 = 16$$

If we let $x' = x - 2$ and $y' = y - 3$, we obtain

$$x'^2 + y'^2 = 16$$

The graph of this equation with respect to the x' and y' axes is a circle with its center at the origin and radius 4. Because the substitutions of $x' = x - 2$ and $y' = y + 3$ result in a translation of axes to a new origin of $(2, -3)$, the graph of the given equation with respect to the x and y axes is a circle with center at $(2, -3)$ and radius 4. This result agrees with our discussion of the circle in Section 3.4. Figure 2 shows the circle with both sets of axes. ◀

We now apply translation of axes to find the general equation of a parabola having its vertex at the point (h, k) and either a vertical or a horizontal axis. In particular, let the axis be vertical. Let the x' and y' axes be such that the origin is at $V(h, k)$. See Figure 3. With respect to the x' and y' axes, an equation of the parabola in this figure is

$$x'^2 = 4py'$$

To obtain an equation of this parabola with respect to the x and y axes, we let $x' = x - h$ and $y' = y - k$, which gives

$$(x - h)^2 = 4p(y - k)$$

In Figure 4 the axis of the parabola is horizontal, and the vertex is at $V(h, k)$. By a similar argument, its equation with respect to the x and y axes is

$$(y - k)^2 = 4p(x - h)$$

We have obtained the standard forms of an equation of a parabola.

Standard Forms of an Equation of a Parabola

If p is the directed distance from the vertex to focus, an equation of the parabola with its vertex at (h, k) and with its axis vertical is

$$(x - h)^2 = 4p(y - k)$$

A parabola with the same vertex and with its axis horizontal has the equation

$$(y - k)^2 = 4p(x - h)$$

The graph of any quadratic equation of the form

$$y = ax^2 + bx + c \tag{1}$$

where a, b, and c are constants and $a \neq 0$ is a parabola whose axis is vertical. This statement can be proved by showing that (1) is equivalent to an equation of the form

$$(x - h)^2 = 4p(y - k)$$

You are asked to do this in Exercise 49. The equation in the following example is the special case of (1) where a is $-\frac{1}{4}$, b is 1, and c is 6.

▶ **EXAMPLE 2** *Finding Properties of a Parabola from its Equation and Sketching the Parabola from These Properties*

Given the parabola having the equation

$$y = -\frac{1}{4}x^2 + x + 6$$

find the vertex, an equation of the axis, the focus, and the endpoints of the latus rectum. Sketch the parabola from these properties, and check the graph on a graphics calculator.

Solution   The given equation is equivalent to

$$4y = -x^2 + 4x + 24$$
$$x^2 - 4x = -4y + 24$$

Completing the square on the left by adding 4 to each side, we get

$$x^2 - 4x + 4 = -4y + 24 + 4$$
$$(x - 2)^2 = -4y + 28$$
$$(x - 2)^2 = -4(y - 7)$$

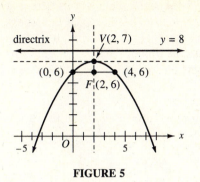

FIGURE 5

This equation is of the form

$$(x - h)^2 = 4p(y - k)$$

with $h = 2$, $k = 7$, and $p = -1$. Therefore its graph is a parabola with vertex at $(2, 7)$, and the axis is vertical. Thus the axis has the equation $x = 2$. Because $p < 0$, it opens downward. Furthermore, the focus is the point on the axis 1 unit below the vertex; thus the focus is at $(2, 6)$. Because the length of the latus rectum is $4p = 4$, its endpoints are 2 units to the right and left of the focus at $(4, 6)$ and $(0, 6)$.

Figure 5 shows the parabola sketched from these properties. A graphics calculator verifies this graph. ◀

If x and y are interchanged in (1), we have the equation

$$x = ay^2 + by + c \qquad (2)$$

The graph of any equation of this form is a parabola whose axis is horizontal. This fact can be verified by showing that (2) is equivalent to an equation of the form

$$(y - k)^2 = 4p(x - h)$$

▶ **EXAMPLE 3** *Finding Properties of a Parabola from its Equation and Sketching the Parabola from These Properties*

Follow the instructions of Example 2 for the parabola having the equation

$$x = 2y^2 + 8y + 11$$

Solution   The given equation is equivalent to

$$2y^2 + 8y = x - 11$$
$$2(y^2 + 4y) = x - 11$$

To complete the square of the expression within the parentheses on the left, we add 4 to $y^2 + 4y$. We are actually adding 8 to the left side; so we also add 8 to the right side, and we have

$$2(y^2 + 4y + 4) = x - 11 + 8$$
$$2(y + 2)^2 = x - 3$$
$$(y + 2)^2 = \frac{1}{2}(x - 3)$$

This equation is of the form

$$(y - k)^2 = 4p(x - h)$$

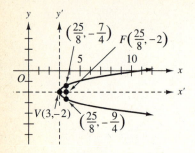

FIGURE 6

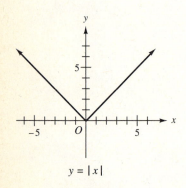

$y = |x|$

FIGURE 7

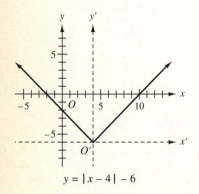

$y = |x - 4| - 6$

FIGURE 8

with $h = 3$, $k = -2$, and $p = \frac{1}{8}$. Therefore the parabola has its vertex at $(3, -2)$, its axis is the horizontal line $y = -2$, and because $p > 0$, it opens to the right. Because the focus is $\frac{1}{8}$ unit to the right of the vertex, it is at the point $(\frac{25}{8}, -2)$. The length of the latus rectum is $|4p| = \frac{1}{2}$; thus the endpoints of the latus rectum are $\frac{1}{4}$ unit above and below the focus at $(\frac{25}{8}, -\frac{7}{4})$ and $(\frac{25}{8}, -\frac{9}{4})$.

The parabola sketched from these properties appears in Figure 6. To plot the parabola on a graphics calculator, we first write the given equation as

$$2y^2 + 8y + (11 - x) = 0$$

and then solve for y in terms of x by the quadratic formula where a is 2, b is 8, and c is $(11 - x)$. We get two values for y:

$$y = -2 + \frac{1}{2}\sqrt{2x - 6} \quad \text{and} \quad y = -2 - \frac{1}{2}\sqrt{2x - 6}$$

When we plot the graphs of these two equations in the same viewing rectangle, we obtain the parabola in Figure 6.   ◀

In the next two examples, we apply translation of axes to other graphs.

▶ **EXAMPLE 4** *Obtaining the Graph of One Equation from the Graph of Another and a Suitable Translation of Axes*

From the graph of $y = |x|$ and a suitable translation of axes, obtain the graph of $y = |x - 4| - 6$.

Solution   The graph of $y = |x|$ appears in Figure 15 of Section 1.5. We reproduce it here in Figure 7. The equation

$$y = |x - 4| - 6$$

is equivalent to

$$y + 6 = |x - 4|$$

To obtain the graph of this equation, we let

$$x' = x - 4 \quad \text{and} \quad y' = y + 6$$

We have translated the axes to the new origin $(4, -6)$, and the equation becomes $y' = |x'|$. The graph of this equation with respect to the x' and y' axes is the same as the graph in Figure 7 with respect to the x and y axes. Thus we obtain the graph shown in Figure 8.   ◀

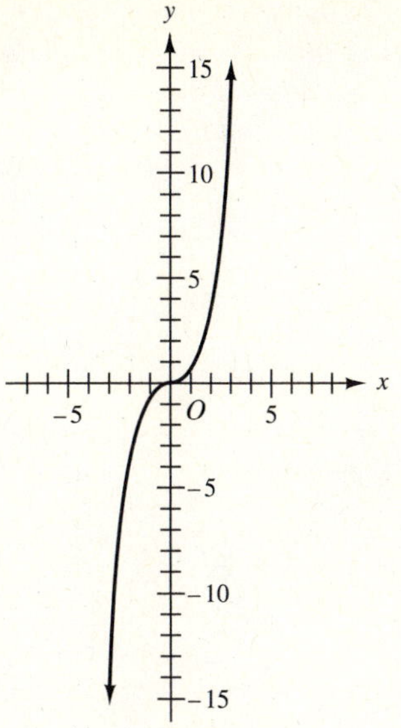

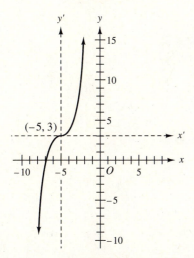

FIGURE 9

▶ **EXAMPLE 5** *Obtaining the Graph of One Equation from the Graph of Another and a Suitable Translation of Axes*

Use the graph of $y = \frac{1}{2}x^3$ from Example 4 of Section 1.5 along with a suitable translation of axes to obtain the graph of the equation

$$y = \frac{1}{2}(x + 5)^3 + 3$$

Solution   Figure 9 shows the graph of $y = \frac{1}{2}x^3$. The equation

$$y = \frac{1}{2}(x + 5)^3 + 3$$

is equivalent to

$$y - 3 = \frac{1}{2}(x + 5)^3$$

To obtain the graph of this equation, we let

$$x' = x + 5 \quad \text{and} \quad y' = y - 3$$

We have translated the axes to the new origin $(-5, 3)$, and the equation becomes $y' = \frac{1}{2}x'^3$. Figure 10 shows the graph of this equation with respect to the x' and y' axes. It is the same as the graph in Figure 9 with respect to the x and y axes.   ◀

FIGURE 10

EXERCISES 3.5

In Exercises 1 through 4, translate the axes so that an equation of the graph with respect to the new axes contains no first-degree terms. Draw the original and new axes and sketch the graph.

1. $x^2 + y^2 + 6x + 4y = 0$
2. $x^2 + y^2 - 2x - 8y + 1 = 0$
3. $x^2 + y^2 + x - 2y + 1 = 0$
4. $x^2 + y^2 - 10x + 4y + 13 = 0$

In Exercises 5 and 6, translate the axes so that an equation of the graph with respect to the new x' and y' axes contains no first-degree term in x' and no constant term. Draw the original and new axes and sketch the graph.

5. $x^2 - 4x - 8y - 28 = 0$

6. $x^2 + 4x + 2y = 0$

In Exercises 7 and 8, translate the axes so that an equation of the graph with respect to the new x' and y' axes contains no first-degree term in y' and no constant term. Draw the original and new axes and sketch the graph.

7. $y^2 + 6x + 7y + 39 = 0$

8. $2y^2 - 2x - 4y + 3 = 0$

In Exercises 9 through 24, for the given parabola, find (a) the vertex, (b) an equation of the axis, (c) the focus, (d) an equation of the directrix, and (e) the endpoints of the latus rectum. (f) Sketch the parabola from these properties and check your graph on your graphics calculator.

9. $y = x^2 - 4$ **10.** $y = x^2 + 4x$

11. $y = -x^2 + 4x - 5$ **12.** $y = x^2 + 6x - 2$

13. $x = y^2 - 6y$ **14.** $x = -y^2 + 1$

15. $x^2 - 6x - 4y + 13 = 0$

16. $x^2 - 4x + 8y + 28 = 0$

17. $y^2 + 4x + 12y = 0$

18. $y^2 - 12x - 14y + 25 = 0$

19. $y = -\frac{1}{2}x^2 + 4x - 5$ **20.** $y = \frac{1}{16}x^2 + \frac{1}{2}x$

21. $y = \frac{1}{8}x^2 - \frac{1}{2}x - \frac{3}{2}$ **22.** $x = 2y^2 + 10y + 3$

23. $x = -2y^2 - 8y - 5$ **24.** $x = -\frac{1}{4}y^2 - \frac{3}{2}y - 2$

In Exercises 25 through 28, plot the parabola having the given equation.

25. $y^2 - 4x - 2y + 9 = 0$

26. $4y^2 - x + 16y + 12 = 0$

27. $5y^2 - 4x + 10y + 17 = 0$

28. $3y^2 + 8x - 12y + 20 = 0$

In Exercises 29 through 46, do the following: (a) sketch the graph of the first equation; (b) from the graph obtained in part (a) and a suitable translation of axes, sketch the graph of the second equation. (c) Check your graphs in parts (a) and (b) by plotting them in the same viewing rectangle.

29. $y = |x|; y = |x - 2|$

30. $y = |x|; y = |x + 3|$

31. $y = |x|; y = |x| + 3$

32. $y = |x|; y = |x| - 2$

33. $y = |x|; y = |x + 4| - 5$

34. $y = |x|; y = |x - 1| + 6$

35. $y = x^3; y = (x - 4)^3$

36. $2y = -x^3; 2y + 2 = -x^3$

37. $y = x^3; y = (x + 1)^3 + 1$

38. $2y = -x^3; 2y = -(x - 4)^3 + 4$

39. $y = \sqrt{x}; y = \sqrt{x - 2} + 4$

40. $y = \sqrt{x}; y = \sqrt{x + 3} - 2$

41. $y = x^2; y = (x - 4)^2$

42. $y = x^2; y = (x + 3)^2$

43. $y = x^2; y = x^2 + 3$

44. $y = x^2; y = x^2 - 4$

45. $y = x^2; y = (x + 1)^2 - 5$

46. $y = x^2; y = (x - 2)^2 + 1$

47. Given the parabola having the equation

$$y = ax^2 + bx + c$$

with $a \neq 0$, find the coordinates of the vertex.

48. Find the coordinates of the focus of the parabola in Exercise 47.

49. Show that the equation $y = ax^2 + bx + c$ is equivalent to an equation of the form $(x - h)^2 = 4p(y - k)$ by solving the second equation for y.

50. If a parabola has its focus at the origin and the x axis as its axis, prove that it must have an equation of the form $y^2 = 4kx + 4k^2$, $k \neq 0$.

51. (a) Show that the equation $y = x^2 + bx + c$ can be written in the form $y = (x - h)^2 + k$. (b) Explain how to sketch the graph of $y = (x - h)^2 + k$ from the graph of $y = x^2$. In your explanation make up a particular example.

CHAPTER 3 REVIEW

▶ LOOKING BACK

3.1 Before finding equations of lines, we defined the slope of a line and then sketched lines and determined their slopes. We obtained two important forms of an equation of a line, the point-slope form and the slope-intercept form. We used these two forms and the theorems involving slopes of parallel lines and slopes of perpendicular lines to find equations of lines having certain properties.

3.2 We treated systems of linear equations in two unknowns and determined both graphically and algebraically if the equations were (i) consistent and independent, (ii) inconsistent, or (iii) dependent. Algebraic solutions involved both the substitution and elimination methods. We also solved systems of two equations linear in the reciprocals of the unknowns. Word problems having a system of two linear equations as a mathematical model appeared in the examples and exercises.

3.3 Our discussion of parabolas included the definition and properties of a parabola. We found properties of parabolas from their equations and equations of parabolas from their properties. From their equations, we sketched and plotted parabolas.

3.4 We followed the same format for circles as we used for parabolas. To plot a circle on a graphics calculator, we first solved the equation of the circle for y, which resulted in two equations expressing y in terms of x. The graph of each of these equations was a semicircle, and we obtained the circle by plotting the two semicircles in the same viewing rectangle.

3.5 We simplified equations by translating the axes. To obtain the standard form of an equation of a parabola, we applied translation of axes. From this equation, we found properties of the parabola and sketched the parabola from these properties. We also obtained graphs of some equations from graphs of simpler equations and suitable translations of axes.

▶ REVIEW EXERCISES

In Exercises 1 and 2, do the following: (a) sketch the line through the two points; (b) determine the slope of the line; (c) find an equation of the line.

1. $(1, -3)$ and $(4, 5)$ **2.** $(-2, -5)$ and $(6, -7)$

In Exercises 3 and 4, do the following: (a) sketch the line through the point P and having slope m; (b) find an equation of the line.

3. $P(5, -2)$; $m = -\dfrac{3}{2}$ **4.** $P(-2, 1)$; $m = \dfrac{3}{4}$

In Exercises 5 and 6, do the following: (a) find the slope and y intercept of the line having the given equation; (b) sketch the line.

5. $2x - 5y - 10 = 0$ **6.** $2x + 3y + 12 = 0$

7. Find an equation of the line through the point $(-3, -2)$ and parallel to the line whose equation is $7x - 3y - 4 = 0$. Sketch each line on the same coordinate system.

8. Find an equation of the line through the point $(-1, 6)$ and perpendicular to the line whose equation is $4x + 2y - 5 = 0$. Sketch each line on the same coordinate system.

9. Prove that the lines having the equations $2x + 5y + 20 = 0$ and $5x - 2y - 10 = 0$ are perpendicular, and sketch their graphs.

10. Prove that the lines having the equations $2x - 3y + 12 = 0$ and $4x - 6y - 3 = 0$ are parallel, and sketch their graphs.

11. Find the abscissa of the point whose ordinate is -3 and for which the line through it and the point $(2, 7)$ is parallel to the line having the equation $3x - 4y = 12$.

12. Use slopes to do Exercise 39 in the Review Exercises for Chapter 1.

In Exercises 13 and 14, use slopes to determine whether the three points lie on a line.

13. $(-2, -3), (-1, 4), (1, 17)$

14. $(0, -3), (1, 4), (2, 11)$

In Exercises 15 through 18, plot the graph of the system of equations. Classify the equations as (i) consistent and independent, (ii) inconsistent, or (iii) dependent. If the equations are consistent and independent, determine the solution set of the system from the graphs, and verify your solution by substituting into the two equations.

15. $\begin{cases} 4x + 3y = 6 \\ 2x + y = 4 \end{cases}$ **16.** $\begin{cases} 3x - 2y = 4 \\ 9x - 6y = 8 \end{cases}$

17. $\begin{cases} 2y = 4x - 6 \\ 6x = 3y + 9 \end{cases}$ **18.** $\begin{cases} 4x + 2y = 5 \\ 8x - 2y = 1 \end{cases}$

In Exercises 19 through 22, find the solution set of the system algebraically by using either the substitution or elimination method. Verify your solution by plotting the graph of the system of equations.

19. $\begin{cases} 2x + y + 1 = 0 \\ 3x + 2y + 4 = 0 \end{cases}$ **20.** $\begin{cases} 3x + 4y - 6 = 0 \\ x - 2y + 8 = 0 \end{cases}$

21. $\begin{cases} 2x - 5y = 7 \\ 6x - 15y = 14 \end{cases}$ **22.** $\begin{cases} 3x - 2y + 7 = 0 \\ 2x - 3y + 8 = 0 \end{cases}$

In Exercises 23 and 24, find the solution set of the system algebraically.

23. $\begin{cases} \dfrac{4}{x} - \dfrac{7}{y} = 4 \\ \dfrac{12}{x} + \dfrac{3}{y} = 4 \end{cases}$ **24.** $\begin{cases} \dfrac{3}{x} - \dfrac{2}{y} = 8 \\ \dfrac{9}{x} + \dfrac{4}{y} = -6 \end{cases}$

In Exercises 25 and 26, plot the graphs of the three equations in the same viewing rectangle. Are the equations consistent or inconsistent? Verify your answer algebraically: To show they are inconsistent, solve a system of two of the equations and show that no member of the solution set satisfies the third equation; to show they are consistent, find the solution set.

25. $\begin{cases} 3x + 2y = 8 \\ 2x - 3y = 14 \\ 5x + 6y = 8 \end{cases}$ **26.** $\begin{cases} 3x + 4y = 11 \\ 7x - 2y = 9 \\ 4x - 5y = 25 \end{cases}$

In Exercises 27 through 42, for the parabola having the given equation, find (a) the vertex, (b) the axis, (c) the focus, (d) an equation of the directrix, and (e) the endpoints of the latus rectum. (f) Sketch the parabola and check your graph on your graphics calculator.

27. $x^2 - 16y = 0$ **28.** $x^2 + 4y = 0$

29. $4x^2 + y = 0$ **30.** $3x^2 - 8y = 0$

31. $y^2 + 10x = 0$ **32.** $y^2 - 6x = 0$

33. $x^2 - y - 3 = 0$ **34.** $x^2 - 4y + 8 = 0$

35. $y^2 - x + 6 = 0$ **36.** $y^2 + 8x - 12 = 0$

37. $y^2 - x - 8y = 0$ **38.** $x^2 - y + 10x = 0$

39. $x^2 + 6x - 4y + 1 = 0$

40. $4y^2 - x + 16y + 21 = 0$

41. $y = -\dfrac{1}{8}x^2 + \dfrac{3}{4}x + \dfrac{7}{8}$ **42.** $x = -\dfrac{1}{6}y^2 - \dfrac{2}{3}y + \dfrac{7}{3}$

In Exercises 43 through 48, find an equation of the parabola having the given properties, and sketch the parabola. Check your graph on your graphics calculator.

43. Focus, $(0, 2)$; directrix, $y = -2$

44. Focus, $(-4, 0)$; directrix, $x = 4$

45. Vertex, the origin; opens to the left; through the point $(-3, 6)$.

46. Vertex, the origin; opens upward; through the point $(-5, 2)$.

47. Focus, the origin; directrix, $y = 6$.

48. Focus, the origin; directrix, $x = -3$.

In Exercises 49 through 56, find the center and radius of the circle. Sketch the circle, and check your graph on your graphics calculator.

49. $x^2 + y^2 = 9$ **50.** $2x^2 + 2y^2 = 1$

51. $x^2 + y^2 - 6x + 5 = 0$

52. $x^2 + y^2 - 8y = 0$

53. $x^2 + y^2 + 4x - 6y - 3 = 0$

54. $x^2 + y^2 - 8x + 4y + 8 = 0$

55. $3x^2 + 3y^2 + 4x - 4 = 0$

56. $x^2 + y^2 - 14x + 16y + 32 = 0$

In Exercises 57 and 58, determine whether the graph is a circle, a point, or the empty set.

57. $x^2 + y^2 - 2x + 2y + 4 = 0$

58. $x^2 + y^2 + 6x + 4y + 13 = 0$

In Exercises 59 through 62, find an equation of the circle satisfying the given conditions and plot the circle.

59. Center at $(3, -5)$ and radius 2.

60. Center at $(-2, 5)$ and tangent to the line $x = 7$.

61. Diameter whose endpoints are at $(2, -1)$ and $(6, -5)$.

62. Center at $(4, 3)$ and through the point $(-7, -1)$.

In Exercises 63 and 64, find an equation of the line tangent to the circle at the point P. Plot the circle and the tangent line in the same viewing rectangle.

63. $x^2 + y^2 - 2x + 6y - 15 = 0$; $P(4, -7)$

64. $x^2 + y^2 + 8x + 4y - 80 = 0$; $P(2, 6)$

In Exercises 65 through 74, plot the graph having the given equation.

65. $3x^2 + 4y = 0$ **66.** $5x^2 - 16y = 0$

67. $5y^2 - 12x = 0$ **68.** $7y^2 - 24x = 0$

69. $x^2 + y^2 = 10$ **70.** $4x^2 + 4y^2 = 25$

71. $x^2 + y^2 - 2x + 6y + 1 = 0$

72. $x^2 + y^2 + 4x - 8y + 4 = 0$

73. $3x^2 - 12x - 2y + 10 = 0$

74. $4y^2 - 3x + 16y + 4 = 0$

In Exercises 75 through 82, do the following: (a) sketch the graph of the first equation; (b) from the graph obtained in part (a) and a suitable translation of axes, sketch the graph of the second equation. (c) Check your graphs in parts (a) and (b) by plotting them in the same viewing rectangle.

75. $y = |x|$; $y = |x - 3| + 2$

76. $y = |x|$; $y = |x + 2| - 3$

77. $y = \sqrt{x}$; $y = \sqrt{x + 5} - 8$

78. $y = 2x^2$; $y = 2(x - 1)^2 + 4$

79. $y = 2x^{1/3}$; $y = 2(x - 4)^{1/3} - 6$

80. $y = \frac{1}{9}x^3$; $y = \frac{1}{9}(x + 3)^3 + 7$

81. $y = \frac{1}{4}x^4$; $y = \frac{1}{4}(x + 6)^4 + 1$

82. $y = 4x^{1/4}$; $y = 4(x - 5)^{1/4} - 2$

In Exercises 83 through 86, the points A, B, C, and D are vertices of a quadrilateral. Use slopes to determine if the quadrilateral is a rectangle, a parallelogram, or a trapezoid. Draw the quadrilateral.

83. $A(3, 1)$, $B(5, 2)$, $C(15, 5)$, $D(17, 6)$

84. $A(-8, 0)$, $B(-3, -5)$, $C(1, 4)$, $D(3, 2)$

85. $A(3, 1)$, $B(2, -2)$, $C(-1, -1)$, $D(0, 2)$

86. $A(2, 13)$, $B(-2, 5)$, $C(3, -1)$, $D(7, 7)$

87. Find an equation of the perpendicular bisector of the line segment from $(-1, 5)$ to $(3, 2)$.

88. Find an equation of the line through the point $(5, -3)$ and perpendicular to the line whose equation is $2x - 5y = 1$.

89. Find an equation of the parabola having its vertex at $(-3, 5)$ and its focus at $(-3, -1)$.

90. Find an equation of the parabola whose vertex is at $(5, 1)$, whose axis is parallel to the y axis, and through the point $(9, 3)$.

In Exercises 91 through 96, solve the word problem by finding a system of equations as a mathematical model of the situation. Complete the exercise by writing a conclusion.

91. A man placed his savings in two investments. The interest rate on investment A is 10 percent and on investment B it is 12 percent. The annual income from the two investments is $3760. If investment A were at 12 percent and investment B at 10 percent, his annual income would be $3720. What is the total amount of his savings?

92. At a supermarket that sells fruit by the pound, one person bought 3 lb of oranges and 6 lb of grapefruit for a total cost of $6, while a second person paid $6.40 for 5 lb of oranges and 4 lb of grapefruit. What is the price per pound of the oranges and the grapefruit?

93. A chemist has a solution that is 50 percent acid. By adding water, the solution is reduced to one containing 40 percent acid. By adding 500 cm³ more water, the solution then contains only 35 percent acid. Determine **(a)** how many cubic centimeters of the 50 percent acid solution the chemist had originally and **(b)** how many cubic centimeters of the 35 percent acid solution the chemist had finally.

94. An investment yields an annual interest of $7500. If $5000 more is invested and the rate is 1 percent less, the annual interest is $6500. What is the amount of the investment and the rate of interest?

95. Workers A and B can complete a particular job if they work together for 12 days. If A works alone for 20 days and then B completes the job alone in 6 more days, how long does it take each worker to do the job alone? See the hint for Exercise 33 in Exercises 3.2.

96. A woman has a certain amount of money invested at a particular rate of interest. If she had $2000 more invested at a rate 2 percent lower, she would receive the same annual interest. If she had $2000 less invested at a rate 3 percent higher, she also would receive the same annual interest. How much does she have invested and at what rate?

In Exercises 97 and 98, solve the word problem by finding an equation of a parabola as a mathematical model of the situation. Complete the exercise by writing a conclusion.

97. Any section of a parabolic mirror made by passing a plane through the axis of the mirror is a segment of a parabola. The altitude of the segment is 12 cm and the length of the base is 18 cm. A section of the mirror made by a plane perpendicular to its axis is a circle. Find the circumference of the circular plane section if the plane perpendicular to the axis is 3 cm from the vertex.

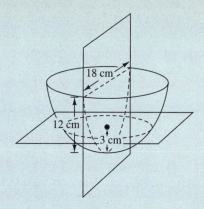

98. A wire attached 60 ft above the ground to two telephone poles, 180 ft apart, hangs in the shape of a parabola. If halfway between the poles, the wire is 50 ft above the ground, find the height of the wire 45 ft from either pole.

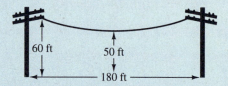

In Exercises 99 and 100, use analytic geometry to prove the given theorem from plane geometry.

99. If the diagonals of a rectangle are perpendicular, the rectangle is a square.

100. The line segment joining the midpoints of two sides of a triangle is parallel to the third side and its length is one-half the length of the third side.

Functions and Their Graphs

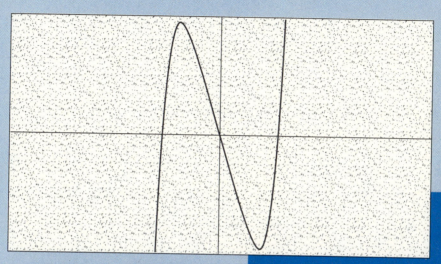

Often in practical applications the value of one quantity depends on the value of another. A person's salary may depend on the number of hours worked; the total production at a factory may depend on the number of machines used; the distance traveled by an object may depend on the time elapsed since it left a specific point; the volume of the space occupied by a gas having a constant pressure depends on the temperature of the gas; the resistance of an electrical cable of fixed length depends on its diameter; and so forth. A relationship between such quantities is often given by means of a function. The notion of a function, one of the most important concepts in mathematics, is fundamental for the study of calculus and serves as a unifying concept throughout the remainder of this text.

4.1 FUNCTIONS

GOALS

1. **Learn the intuitive concept of a function.**
2. **Determine the domain of a function.**
3. **State the formal definition of a function.**
4. **Compute function values.**
5. **Compute difference quotients.**
6. **Find the sum, difference, product, and quotient of two functions.**

 Because calculus is concerned with *functions* of real numbers, such functions are of primary interest to us. We introduce the idea of a function in an intuitive manner.

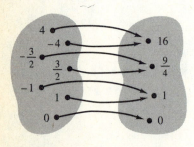

> A function can be thought of as a correspondence from a set X of real numbers x to a set Y of real numbers y, where the number y is unique for a specific value of x.

X Y

FIGURE 1

Figure 1 gives a visualization of such a correspondence where the sets X and Y consist of points in a plane region.

Stating the concept of a function another way, we intuitively consider the real number y in set Y to be a *function* of the real number x in set X if there is some rule by which a unique value of y is assigned to a value of x. This rule is often given by an equation. For example, the equation

$$y = x^2$$

defines a function for which X is the set of all real numbers and Y is the set of nonnegative numbers: The value of y in Y assigned to the value of x in X is obtained by multiplying x by itself. Table 1 gives the value of y assigned to some particular values of x, and Figure 2 visualizes the correspondence for the numbers in the table.

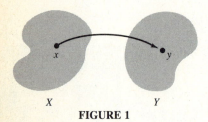

X: all real numbers Y: nonnegative numbers

FIGURE 2

Table 1

x	1	$\frac{3}{2}$	4	0	-1	$-\frac{3}{2}$	-4
$y = x^2$	1	$\frac{9}{4}$	16	0	1	$\frac{9}{4}$	16

We use symbols such as f, g, and h to denote a function. The set X of real numbers described above is the *domain* of the function, and the set Y of real numbers assigned to the values of x in X is the *range* of the function. The numbers x and y are variables. Because values are assigned to x, and because the value of y is dependent on the choice of x, x is the *independent variable* and y is the *dependent variable*.

▷ **ILLUSTRATION 1**

The equation

$$y = 2x^2 + 5$$

defines a function. Call this function f. The equation gives the rule by which a unique value of y can be determined whenever x is given; that is, multiply the number x by itself, then multiply that product by 2, and add 5. The domain of f is the set of all real numbers, and it can be denoted with interval notation as $(-\infty, +\infty)$. The smallest value that y can assume is 5 (when $x = 0$). The range of f is then the set of all positive numbers greater than or equal to 5, which is $[5, +\infty)$.   ◀

▷ **ILLUSTRATION 2**

Let g be the function defined by the equation

$$y = \sqrt{x^2 - 4}$$

Because the numbers are confined to real numbers, y is a function of x only for $x \geq 2$ or $x \leq -2$ (or simply $|x| \geq 2$); for any x satisfying either of these inequalities, a unique value of y is determined. However, if x is in the interval $(-2, 2)$, a square root of a negative number is obtained, and hence no real number y exists. We must, therefore, restrict x so that $|x| \geq 2$. The domain of g is $(-\infty, -2] \cup [2, +\infty)$, and the range is $[0, +\infty)$.   ◀

We can consider a function as a set of ordered pairs. For instance, the function defined by the equation $y = x^2$ consists of all the ordered pairs (x, y) satisfying the equation. The ordered pairs in this function given by Table 1 are $(1, 1)$, $(\frac{3}{2}, \frac{9}{4})$, $(4, 16)$, $(0, 0)$, $(-1, 1)$, $(-\frac{3}{2}, \frac{9}{4})$, and $(-4, 16)$. Of course, there is an unlimited number of ordered pairs in the function. Some others are $(2, 4)$, $(-2, 4)$, $(5, 25)$, $(-5, 25)$, $(\sqrt{3}, 3)$, and so on.

▷ **ILLUSTRATION 3**

The function f of Illustration 1 is the set of ordered pairs (x, y) for which $y = 2x^2 + 5$. Some of the ordered pairs in f are $(0, 5)$, $(1, 7)$, $(\sqrt{2}, 9)$, $(2, 13)$, $(-1, 7)$, $(-\sqrt{5}, 15)$ and so on.   ◀

▷ **ILLUSTRATION 4**

The function g of Illustration 2 is the set of ordered pairs (x, y) for which $y = \sqrt{x^2 - 4}$. Some of the ordered pairs in g are $(2, 0)$, $(5, \sqrt{21})$, $(-2, 0)$, $(3, \sqrt{5})$, $(-9, \sqrt{78})$, and so on.   ◀

We now give the formal definition of a function. Defining a function as a set of ordered pairs rather than as a rule or correspondence makes its meaning precise.

> **DEFINITION   A Function**
>
> A **function** is a set of ordered pairs of real numbers (x, y) in which no two distinct ordered pairs have the same first number. The set of all admissible values of x is called the **domain** of the function, and the set of all resulting values of y is called the **range** of the function.

In this definition, the restriction that no two distinct ordered pairs can have the same first number ensures that y is unique for a specific value of x.

If f is the function having as its domain variable x and as its range variable y, the symbol $f(x)$ (read "f of x") denotes the particular value of y that corresponds to the value of x. This notation is due to the Swiss mathematician and physicist Leonhard Euler (1707–1783).

▷ **ILLUSTRATION 5**

In Illustration 1, f is the function defined by the equation $y = 2x^2 + 5$. Thus

$$f(x) = 2x^2 + 5$$

Because when $x = 1$, $2x^2 + 5 = 7$, we have $f(1) = 7$. Similarly, $f(-2) = 13$, $f(0) = 5$, and so on.   ◀

▷ **ILLUSTRATION 6**

In Illustration 2, g is the function defined by the equation $y = \sqrt{x^2 - 4}$. Therefore

$$g(x) = \sqrt{x^2 - 4}$$

We now compute $g(x)$ for some specific values of x. $g(2) = 0$, $g(5) = \sqrt{21}$, $g(-2) = 0$, $g(-9) = \sqrt{78}$, and so on.   ◀

When defining a function, the domain of the function must be given either implicitly or explicitly. For instance, if f is defined by

$$f(x) = 3x^2 - 5x + 2$$

it is implied that x can be any real number. However, if f is defined by

$$f(x) = 3x^2 - 5x + 2 \qquad 1 \le x \le 10$$

then the domain of f consists of all real numbers between and including 1 and 10.

Similarly, if g is defined by the equation

$$g(x) = \frac{5x - 2}{x + 4}$$

it is implied that $x \ne -4$ because the quotient is undefined for $x = -4$; hence the domain of g is the set of all real numbers except -4.

If

$$h(x) = \sqrt{9 - x^2}$$

it is implied that x is in the closed interval $[-3, 3]$ because $\sqrt{9 - x^2}$ is not a real number for $x > 3$ or $x < -3$. Thus the domain of h is $[-3, 3]$, and the range is $[0, 3]$.

▶ **EXAMPLE 1** *Computing Function Values*

Given that f is the function defined by

$$f(x) = x^2 + 3x - 4$$

find: **(a)** $f(0)$; **(b)** $f(2)$; **(c)** $f(h)$; **(d)** $f(2h)$; **(e)** $f(2x)$.

Solution

(a) $f(0) = 0^2 + 3 \cdot 0 - 4$
$\qquad = -4$

(b) $f(2) = 2^2 + 3 \cdot 2 - 4$
$\qquad = 6$

(c) $f(h) = h^2 + 3h - 4$

(d) $f(2h) = (2h)^2 + 3(2h) - 4$
$\qquad\quad = 4h^2 + 6h - 4$

(e) $f(2x) = (2x)^2 + 3(2x) - 4$
$\qquad\quad = 4x^2 + 6x - 4$ ◀

▶ **EXAMPLE 2** *Computing Function Values*

For the function of Example 1, find: **(a)** $f(x + h)$; **(b)** $f(x) + f(h)$.

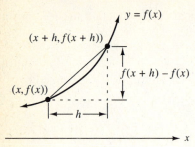

FIGURE 3

Solution

(a) $f(x + h) = (x + h)^2 + 3(x + h) - 4$
$$= x^2 + 2hx + h^2 + 3x + 3h - 4$$
$$= x^2 + (2h + 3)x + (h^2 + 3h - 4)$$

(b) $f(x) + f(h) = (x^2 + 3x - 4) + (h^2 + 3h - 4)$
$$= x^2 + 3x + (h^2 + 3h - 8) \qquad \blacktriangleleft$$

Compare the computations in Example 2. In part (a) we found $f(x + h)$, which is the function value at the sum of x and h. In part (b), where $f(x) + f(h)$ is computed, we obtain the sum of the two function values $f(x)$ and $f(h)$.

In calculus we often compute quotients of the form

$$\frac{f(x + h) - f(x)}{h}$$

called a *difference quotient*. A difference quotient arises as the slope of the line through the points $(x, f(x))$ and $(x + h, f(x + h))$ on the graph of the equation $y = f(x)$. See Figure 3.

▶ **EXAMPLE 3** *Computing a Difference Quotient*

If $f(x) = 3x^2 - 2x + 4$, and $h \neq 0$, find

$$\frac{f(x + h) - f(x)}{h}$$

Solution

$$\frac{f(x + h) - f(x)}{h} = \frac{3(x + h)^2 - 2(x + h) + 4 - (3x^2 - 2x + 4)}{h}$$

$$= \frac{3x^2 + 6hx + 3h^2 - 2x - 2h + 4 - 3x^2 + 2x - 4}{h}$$

$$= \frac{6hx - 2h + 3h^2}{h}$$

$$= 6x - 2 + 3h \qquad \blacktriangleleft$$

We now define some operations on functions. In the definition new functions are formed from given functions by adding, subtracting, multiplying, and dividing function values. Accordingly, these new functions are known as the *sum, difference, product,* and *quotient* of the original functions.

DEFINITION **The Sum, Difference, Product, and Quotient of Two Functions**

Given the two functions f and g:

(i) their **sum,** denoted by $f + g$, is the function defined by
$$(f + g)(x) = f(x) + g(x)$$

(ii) their **difference,** denoted by $f - g$, is the function defined by
$$(f - g)(x) = f(x) - g(x)$$

(iii) their **product,** denoted by $f \cdot g$, is the function defined by
$$(f \cdot g)(x) = f(x) \cdot g(x)$$

(iv) their **quotient,** denoted by f/g, is the function defined by
$$(f/g)(x) = f(x)/g(x)$$

In each case the *domain* of the resulting function consists of those values of x common to the domains of f and g, with the additional requirement in case (iv) that the values of x for which $g(x) = 0$ are excluded.

▶ **EXAMPLE 4** *Operating on Functions*

Given that f and g are the functions defined by

$$f(x) = \sqrt{x + 1} \quad \text{and} \quad g(x) = \sqrt{x - 4}$$

find: **(a)** $(f + g)(x)$; **(b)** $(f - g)(x)$; **(c)** $(f \cdot g)(x)$; **(d)** $(f/g)(x)$. In each case determine the domain of the resulting function.

Solution

(a) $(f + g)(x) = \sqrt{x + 1} + \sqrt{x - 4}$
(b) $(f - g)(x) = \sqrt{x + 1} - \sqrt{x - 4}$

(c) $(f \cdot g)(x) = \sqrt{x + 1} \cdot \sqrt{x - 4}$ **(d)** $(f/g)(x) = \dfrac{\sqrt{x + 1}}{\sqrt{x - 4}}$

The domain of f is $[-1, +\infty)$, and the domain of g is $[4, +\infty)$. So in parts (a), (b), and (c) the domain of the resulting function is $[4, +\infty)$. In part (d) the denominator is zero when $x = 4$; thus 4 is excluded from the domain, and the domain is therefore $(4, +\infty)$. ◀

EXERCISES 4.1

1. Given $f(x) = 2x - 1$, find:
   (a) $f(3)$ (b) $f(-2)$ (c) $f(0)$
   (d) $f(a + 1)$ (e) $f(x + 1)$

2. Given $f(x) = x^2 + 1$, find:
   (a) $f(2)$ (b) $f(-3)$ (c) $f(0)$
   (d) $f(a - 1)$ (e) $f(x - 1)$

3. Given $f(x) = 3x^2 - 5x + 4$, find:
   (a) $f(-1)$ (b) $f(4)$ (c) $[f(x)]^2$
   (d) $f(x^2)$

4. Given $f(x) = 8 - x^3$, find:
   (a) $f(-2)$ (b) $f(5)$ (c) $[f(x)]^2$
   (d) $f(x^2)$

5. Given $f(x) = \dfrac{3}{x}$, find:
   (a) $f(1)$ (b) $f(-3)$ (c) $f\left(\dfrac{1}{3}\right)$
   (d) $f\left(\dfrac{3}{a}\right)$ (e) $f\left(\dfrac{3}{x}\right)$ (f) $\dfrac{f(3)}{f(x)}$

6. Given $f(x) = \dfrac{2}{x + 1}$, find:
   (a) $f(7)$ (b) $f(-5)$ (c) $f\left(\dfrac{1}{2}\right)$
   (d) $f\left(\dfrac{a}{2}\right)$ (e) $f\left(\dfrac{x}{2}\right)$ (f) $\dfrac{f(x)}{f(2)}$

7. Given $f(x) = \sqrt{2x + 3}$, find:
   (a) $f(-1)$ (b) $f(4)$ (c) $f\left(\dfrac{1}{2}\right)$
   (d) $f(11)$ (e) $f(2x + 3)$

8. Given $f(x) = \sqrt{2x^2 + 1}$, find:
   (a) $f(-2)$ (b) $f(0)$ (c) $f(1)$
   (d) $f\left(\dfrac{4}{7}\right)$ (e) $f(2x^2 + 1)$

In Exercises 9 through 12, find each of the following:
(a) $2f(x)$; (b) $f(2x)$; (c) $f(x) + f(h)$; (d) $f(x + h)$.

9. The function of Exercise 1

10. The function of Exercise 2

11. The function of Exercise 5

12. The function of Exercise 6

$\iint\frac{dy}{dx}$ *In Exercises 13 through 22, compute and simplify the difference quotient*
$$\frac{f(x + h) - f(x)}{h}, h \neq 0$$

13. The function of Exercise 1

14. The function of Exercise 2

15. The function of Exercise 3

16. The function of Exercise 4

17. The function of Exercise 5

18. The function of Exercise 6

19. The function of Exercise 7; simplify by rationalizing the numerator.

20. The function of Exercise 8; simplify by rationalizing the numerator.

21. $f(x) = \dfrac{2}{\sqrt{x + 1}}$; simplify by rationalizing the numerator.

22. $f(x) = \dfrac{1}{\sqrt{x - 2}}$; simplify by rationalizing the numerator.

In Exercises 23 through 32, define the following functions and determine the domain of the resulting function:
(a) $f + g$; (b) $f - g$; (c) $f \cdot g$; (d) f/g (e) g/f.

23. $f(x) = x - 5$; $g(x) = x^2 - 1$

24. $f(x) = \sqrt{x}$; $g(x) = x^2 + 1$

25. $f(x) = \dfrac{x + 1}{x - 1}$; $g(x) = \dfrac{1}{x}$

26. $f(x) = \sqrt{x}$; $g(x) = 4 - x^2$

27. $f(x) = \sqrt{x}$; $g(x) = x^2 - 1$

28. $f(x) = |x|$; $g(x) = |x - 3|$

29. $f(x) = x^2 + 1$; $g(x) = 3x - 2$

30. $f(x) = \sqrt{x + 4}$; $g(x) = x^2 - 4$

31. $f(x) = \dfrac{1}{x + 1}$; $g(x) = \dfrac{x}{x - 2}$

32. $f(x) = x^2$; $g(x) = \dfrac{1}{\sqrt{x}}$

33. Given $H(x) = |x - 2| - |x| + 2$, express $H(x)$ without absolute-value bars if x is in the interval:
    (a) $[2, +\infty)$ (b) $(-\infty, 0)$ (c) $[0, 2)$

34. Given $f(t) = \dfrac{|3 + t| - |t| - 3}{t}$, express $f(t)$

without absolute-value bars if t is in the interval:
(a) $(0, +\infty)$ **(b)** $[-3, 0)$ **(c)** $(-\infty, -3)$

35. Given

$$f(x) = \begin{cases} \dfrac{|x|}{x} & \text{if } x \neq 0 \\ 1 & \text{if } x = 0 \end{cases}$$

find:
(a) $f(1)$ **(b)** $f(-1)$ **(c)** $f(4)$
(d) $f(-4)$ **(e)** $f(-x)$ **(f)** $f(x + 1)$
(g) $f(x^2)$ **(h)** $f(-x^2)$

36. In this section we introduced notations f and $f(x)$, pertaining to functions and having different meanings. Explain what each notation means, and in your explanation make up an equation defining a function and use that equation to differentiate between f and $f(x)$.

4.2 GRAPHS OF FUNCTIONS

GOALS

1. Define the graph of a function.
2. Find the domain and range of a function.
3. Sketch graphs of functions.
4. Define types of functions.
5. Define an even function and an odd function.
6. Determine whether a function is even, odd, or neither.

You learned in Section 4.1 that a function f can be represented by an equation of the form $y = f(x)$, which gives the set of ordered pairs (x, y) in the function. The concept of a function as a set of ordered pairs permits us to give the following definition of the *graph* of a function.

> **DEFINITION**   **The Graph of a Function**
>
> If f is a function, then the **graph** of f is the set of all points (x, y) in R^2 for which (x, y) is an ordered pair in f.

From this definition, it follows that the graph of the function f is the same as the graph of the equation $y = f(x)$.

Recall that for a function a unique value of the dependent variable exists for each value of the independent variable in the domain of the function. In geometric terms, this means:

> The graph of a function can be intersected by a vertical line in at most one point.

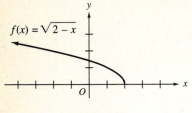

$f(x) = \sqrt{2 - x}$

FIGURE 1

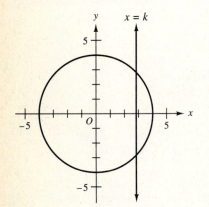

FIGURE 2

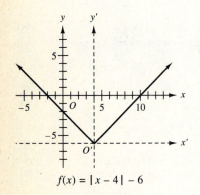

$f(x) = |x - 4| - 6$

FIGURE 3

▷ **ILLUSTRATION 1**

Let

$$f(x) = \sqrt{2 - x}$$

The graph of f is sketched in Figure 1. Observe that a vertical line having the equation $x = k$, where $k \leq 2$, intersects the graph in only one point. The domain of f is the set of all real numbers less than or equal to 2, which is the interval $(-\infty, 2]$, and the range is the set of all nonnegative real numbers, which is $[0, +\infty)$.   ◀

▷ **ILLUSTRATION 2**

Consider the set of all ordered pairs (x, y) for which

$$x^2 + y^2 = 25$$

The graph of this set is sketched in Figure 2. This set of ordered pairs is not a function because for any x in the interval $(-5, 5)$ there are two ordered pairs having x as the first number. For example, both $(3, 4)$ and $(3, -4)$ are ordered pairs in the given set. Furthermore, observe that the graph of the set is a circle with center at the origin and radius 5, and a vertical line having the equation $x = k$, where $-5 < k < 5$, intersects the circle in two points.   ◀

The domain of a function is usually apparent from the function's definition. Often the range can be determined by the graph of the function.

▷ **ILLUSTRATION 3**

The graph of the function defined by

$$f(x) = |x - 4| - 6$$

is, of course, the graph of the equation $y = |x - 4| - 6$, which we obtained in Example 4 of Section 3.5. It is reproduced here as Figure 3. The domain of f is $(-\infty, +\infty)$, and the range, apparent from the graph, is $[-6, +\infty)$.   ◀

In the next illustration, we have a *piecewise-defined function,* one that is defined by more than one equation.

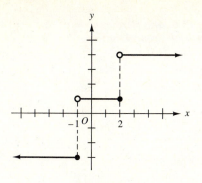

FIGURE 4

▷ **ILLUSTRATION 4**

Let the function g be defined by

$$g(x) = \begin{cases} -3 & \text{if } x \leq -1 \\ 1 & \text{if } -1 < x \leq 2 \\ 4 & \text{if } 2 < x \end{cases}$$

The graph of g is sketched in Figure 4. The solid dot at $(-1, -3)$ and the open dot at $(-1, 1)$ indicate that the value of y is -3 when $x = -1$. The domain of g is $(-\infty, +\infty)$, whereas the range consists of the three numbers -3, 1, and 4. ◄

To plot the graph of a piecewise-defined function, consult your manual for the procedure for your particular calculator.

 Piecewise-defined functions are frequently used in calculus as examples and counterexamples of functions possessing certain properties. For instance, the graph of the function in Illustration 4 has breaks at the points where $x = 1$ and $x = 2$, indicating that the function is *discontinuous* for those values of x. In the following example, we have a piecewise-defined function whose graph has no break at $x = 1$, the value of x at which the defining equations change. You will learn in calculus, however, that this graph does not have a tangent line at the point where $x = 1$.

▶ **EXAMPLE 1** *Sketching the Graph of a Piecewise-Defined Function and Determining its Domain and Range*

Sketch the graph of the function h defined by

$$h(x) = \begin{cases} 3x - 2 & \text{if } x < 1 \\ \frac{1}{2}(x^2 + 1) & \text{if } 1 \leq x \end{cases}$$

Determine the domain and range of h. Check the graph on a graphics calculator.

Solution When $x < 1$, the function values are on the line $y = 3x - 2$, and when $1 \leq x$, the function values are on the parabola $y = \frac{1}{2}(x^2 + 1)$. The graph is sketched in Figure 5. Both the domain and range are $(-\infty, +\infty)$. Our graphics calculator gives the same graph. ◄

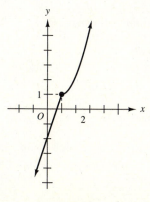

FIGURE 5

▶ **EXAMPLE 2**   *Sketching a Function's Graph Having a Hole and Determining its Domain and Range*

Sketch the graph of the function F defined by

$$F(x) = \frac{x^2 - 9}{x - 3}$$

and find its domain and range.

Solution   Because a function value is determined for each value of x except 3, the domain of F consists of all real numbers except 3. When $x = 3$, both the numerator and denominator are zero, and $\frac{0}{0}$ is undefined.

Factoring the numerator into $(x - 3)(x + 3)$, we obtain

$$F(x) = \frac{(x - 3)(x + 3)}{x - 3}$$

or $F(x) = x + 3$, provided that $x \neq 3$. In other words, the function F consists of all ordered pairs (x, y) such that

$$y = x + 3 \quad \text{and} \quad x \neq 3$$

From this definition of F it is apparent that the graph contains all points on the line $y = x + 3$ except the point $(3, 6)$. The graph is sketched in Figure 6. The range of F is the set of all real numbers except 6.   ◀

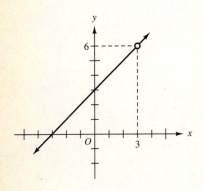

FIGURE 6

In Example 2, the graph has a "hole" at $x = 3$, where $F(3)$ is not defined. In the next example, the graph also has a hole at $x = 3$, but the function value at 3 is defined.

▶ **EXAMPLE 3**   *Sketching a Function's Graph Having a Hole and Determining its Domain and Range*

Sketch the graph of the function G defined by

$$G(x) = \begin{cases} x + 3 & \text{if } x \neq 3 \\ 2 & \text{if } x = 3 \end{cases}$$

Determine the domain and range of G.

Solution   The graph of G is sketched in Figure 7. The graph contains the point $(3, 2)$ and all points on the line $y = x + 3$ except $(3, 6)$. Function G is defined for all values of x, and therefore the domain is $(-\infty, +\infty)$. The range is the set of all real numbers except 6.   ◀

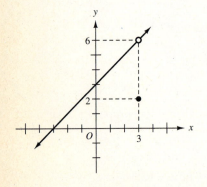

FIGURE 7

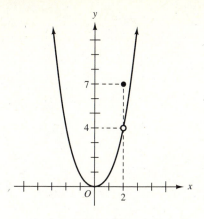

FIGURE 8

▶ **EXAMPLE 4** *Sketching a Function's Graph Having a Hole and Determining its Domain and Range*

Sketch the graph of the function f defined by

$$f(x) = \begin{cases} x^2 & \text{if } x \neq 2 \\ 7 & \text{if } x = 2 \end{cases}$$

Determine the domain and range.

Solution   The graph of f, sketched in Figure 8, consists of the point $(2, 7)$ and all points on the parabola $y = x^2$ except $(2, 4)$. Because f is defined for all real numbers, its domain is $(-\infty, +\infty)$. The range is the set of all nonnegative real numbers.   ◀

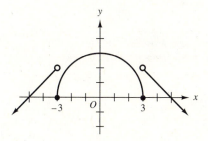

FIGURE 9

▶ **EXAMPLE 5** *Sketching the Graph of a Piecewise-defined Function and Determining its Domain and Range*

Sketch the graph of the function g defined by

$$g(x) = \begin{cases} x + 5 & \text{if } x < -3 \\ \sqrt{9 - x^2} & \text{if } -3 \leq x \leq 3 \\ 5 - x & \text{if } 3 < x \end{cases}$$

Determine the domain and range of g.

Solution   The part of the graph of g for $x < -3$ is a portion of the line $y = x + 5$. For $-3 \leq x \leq 3$, $y = \sqrt{9 - x^2}$, which is the upper semicircle of $x^2 + y^2 = 9$. The part of the graph for $3 < x$ is a portion of the line $y = 5 - x$. The graph is sketched in Figure 9. The domain of g is $(-\infty, +\infty)$ and the range is $(-\infty, 3]$.   ◀

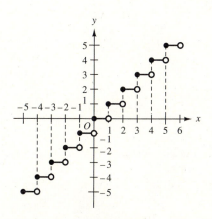

FIGURE 10

Just as the absolute-value function is built-in on a graphics calculator, so is the *greatest integer function*, whose function values are denoted by $[\![x]\!]$ defined by

$$[\![x]\!] = n \quad \text{if } n \leq x < n + 1, \quad \text{where } n \text{ is an integer.}$$

That is, $[\![x]\!]$ is the greatest integer less than or equal to n. Thus, $[\![1]\!] = 1$, $[\![1.3]\!] = 1$, $[\![\frac{1}{2}]\!] = 0$, $[\![-4.2]\!] = -5$, $[\![-8]\!] = -8$, and so on.

The graph of the greatest integer function is sketched in Figure 10. On many graphics calculators the greatest integer function is denoted by

$INT(x)$. The following values are used to obtain the graph:

$$-5 \le x < -4 \qquad [\![x]\!] = -5$$
$$-4 \le x < -3 \qquad [\![x]\!] = -4$$
$$-3 \le x < -2 \qquad [\![x]\!] = -3$$
$$-2 \le x < -1 \qquad [\![x]\!] = -2$$
$$-1 \le x < 0 \qquad [\![x]\!] = -1$$
$$0 \le x < 1 \qquad [\![x]\!] = 0$$
$$1 \le x < 2 \qquad [\![x]\!] = 1$$
$$2 \le x < 3 \qquad [\![x]\!] = 2$$
$$3 \le x < 4 \qquad [\![x]\!] = 3$$
$$4 \le x < 5 \qquad [\![x]\!] = 4$$

The domain of the greatest integer function is the set of all real numbers and its range consists of all the integers.

▶ **EXAMPLE 6** *Sketching the Graph of a Function Involving the Greatest Integer Function and Determining its Domain and Range*

Sketch the graph of the function H defined by

$$H(x) = [\![x]\!] - x$$

Determine the domain and range of H and check the graph on a graphics calculator.

Solution

If $0 \le x < 1$, $[\![x]\!] = 0$; so $[\![x]\!] - x = -x$
If $1 \le x < 2$, $[\![x]\!] = 1$; so $[\![x]\!] - x = 1 - x$
If $2 \le x < 3$, $[\![x]\!] = 2$; so $[\![x]\!] - x = 2 - x$
If $-1 \le x < 0$, $[\![x]\!] = -1$; so $[\![x]\!] - x = -1 - x$
If $-2 \le x < -1$, $[\![x]\!] = -2$; so $[\![x]\!] - x = -2 - x$

And so on.

The graph of H is sketched in Figure 11. The domain is the set of all real numbers, and the range is $(-1, 0]$.

We obtain the graph on our graphics calculator by letting

$$y = INT(x) - x \qquad \blacktriangleleft$$

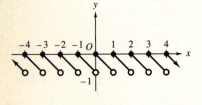

FIGURE 11

A function whose range consists of only one number is called a **constant function.** Thus, if $f(x) = c$, where c is any real number, f is a constant function. The graph of a constant function is a horizontal line at a directed distance of c units from the x axis.

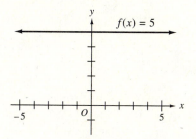

FIGURE 12

▷ **ILLUSTRATION 5**

(a) The function defined by $f(x) = 5$ is a constant function, and its graph, shown in Figure 12, is a horizontal line 5 units above the x axis.

(b) The function defined by $g(x) = -4$ is a constant function, whose graph is a horizontal line 4 units below the x axis. See Figure 13.   ◀

A **linear function** is defined by

$$f(x) = mx + b$$

where m and b are constants and $m \neq 0$. Its graph is a line having slope m and y intercept b.

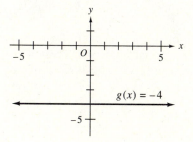

FIGURE 13

▷ **ILLUSTRATION 6**

The function defined by

$$f(x) = 3x - 2$$

is linear. Its graph is the line appearing in Figure 14.   ◀

The particular linear function defined by

$$f(x) = x$$

is called the **identity function.** Its graph, shown in Figure 15, is the line bisecting the first and third quadrants.

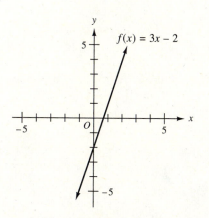

FIGURE 14

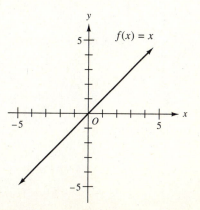

FIGURE 15

If a function f is defined by

$$f(x) = a_n x^n + a_{n-1} x^{n-1} + a_{n-2} x^{n-2} + \ldots + a_1 x + a_0$$

where $a_0, a_1, \ldots, a_n$ are real numbers ($a_n \neq 0$) and n is a nonnegative integer, then f is called a **polynomial function** of degree n. Thus the function defined by

$$f(x) = 3x^5 - x^2 + 7x - 1$$

is a polynomial function of degree 5.

A linear function is a polynomial function of degree 1. If the degree of a polynomial function is 2, it is called a **quadratic function,** and if the degree is 3, it is called a **cubic function.**

Quadratic functions are discussed in Section 4.3, and graphs of polynomial functions are treated in Section 5.3.

A function that can be expressed as the quotient of two polynomial functions is called a **rational function.** Graphs of rational functions are considered in Section 5.5.

An **algebraic function** is one formed by a finite number of algebraic operations on the identity function and the constant function. These algebraic operations include addition, subtraction, multiplication, division, raising to powers, and extracting roots. Polynomial and rational functions are particular kinds of algebraic functions. A complicated example of an algebraic function is the one defined by

$$f(x) = \frac{(x^2 - 3x + 1)^3}{\sqrt{x^4 + 1}}$$

Transcendental functions are also discussed in this text. Examples of transcendental functions are the exponential and logarithmic functions presented in Chapter 6 and the trigonometric functions introduced in Chapter 7.

A function whose graph is symmetric with respect to the y axis is an *even function,* and a function whose graph is symmetric with respect to the origin is an *odd function.* Following are the formal definitions.

DEFINITION **An Even Function and an Odd Function**

(i) A function f is said to be an **even** function if for every x in the domain of f, $f(-x) = f(x)$.

(ii) A function f is said to be an **odd** function if for every x in the domain of f, $f(-x) = -f(x)$.

In both parts (i) and (ii) it is understood that $-x$ is in the domain of f whenever x is.

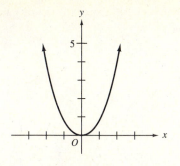

FIGURE 16

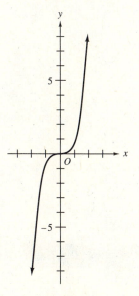

FIGURE 17

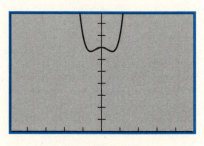

$[-5, 5]$ by $[0, 10]$
$f(x) = 3x^4 - 2x^2 + 7$
FIGURE 18

The symmetry properties of even and odd functions follow from the symmetry tests given in Section 1.5.

▷ **ILLUSTRATION 7**

(a) If $f(x) = x^2$, $f(-x) = (-x)^2$. Therefore $f(-x) = f(x)$, and f is an even function. Its graph is a parabola symmetric with respect to the y axis. See Figure 16.

(b) If $g(x) = x^3$, $g(-x) = (-x)^3$. Because $g(-x) = -g(x)$, g is an odd function. The graph of g, shown in Figure 17, is symmetric with respect to the origin. ◀

▶ **EXAMPLE 7** *Determining Graphically and Algebraically Whether a Function Is Odd, Even, or Neither*

Plot the graph of the given function and from the graph state whether the function is even, odd, or neither. Then prove the statement algebraically.

(a) $f(x) = 3x^4 - 2x^2 + 7$

(b) $g(x) = 3x^5 - 4x^3 - 9x$

(c) $h(x) = 2x^4 + 7x^3 - x^2 + 9$

Solution

(a) The graph of f, shown in Figure 18, is symmetric with respect to the y axis. The function is, therefore, even. To prove this fact algebraically, we compute $f(-x)$:

$$f(-x) = 3(-x)^4 - 2(-x)^2 + 7$$
$$= 3x^4 - 2x^2 + 7$$
$$= f(x)$$

Because $f(-x) = f(x)$, f is even.

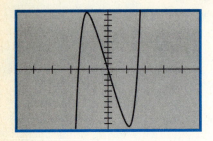

[−5, 5] by [−11, 11]
$g(x) = 3x^5 - 4x^3 - 9x$
FIGURE 19

(b) Figure 19 shows the graph of g, symmetric with respect to the origin. Thus the function is odd. We compute $g(-x)$:

$$g(-x) = 3(-x)^5 - 4(-x)^3 - 9(-x)$$
$$= -3x^5 + 4x^3 + 9x$$
$$= -(3x^5 - 4x^3 - 9x)$$
$$= -g(x)$$

Because $g(-x) = -g(x)$, we have proved algebraically that g is odd.

(c) The graph of h, appearing in Figure 20, is not symmetric with respect to either the y axis or the origin. The function is, therefore, neither even nor odd. We compute $h(-x)$:

$$h(-x) = 2(-x)^4 + 7(-x)^3 - (-x)^2 + 9$$
$$= 2x^4 - 7x^3 - x^2 + 9$$

Because $h(-x) \neq h(x)$ and $h(-x) \neq -h(x)$, h is neither even nor odd.

◀

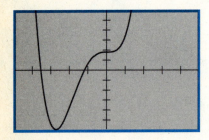

[−5, 5] by [−30, 30]
$h(x) = 2x^4 + 7x^3 - x^2 + 9$
FIGURE 20

EXERCISES 4.2

In Exercises 1 through 42, sketch the graph of the function and determine its domain and range. Check your graph on your graphics calculator.

1. $f(x) = 3x - 1$

2. $g(x) = 4 - x$

3. $F(x) = 2x^2$

4. $G(x) = x^2 + 2$

5. $g(x) = 5 - x^2$

6. $f(x) = (x - 1)^2$

7. $G(x) = \sqrt{x - 1}$

8. $F(x) = \sqrt{9 - x}$

9. $f(x) = \sqrt{x^2 - 4}$

10. $g(x) = \sqrt{4 - x^2}$

11. $g(x) = \sqrt{9 - x^2}$

12. $f(x) = \sqrt{x^2 - 1}$

13. $h(x) = |x - 3|$

14. $H(x) = |5 - x|$

15. $F(x) = |3x + 2|$

16. $G(x) = \dfrac{x^2 - 4}{x - 2}$

17. $H(x) = \dfrac{x^2 - 25}{x + 5}$

18. $f(x) = \dfrac{2x^2 + 7x + 3}{x + 3}$

19. $f(x) = \dfrac{x^2 - 4x + 3}{x - 1}$

20. $g(x) = \dfrac{(x^2 - 4)(x - 3)}{x^2 - x - 6}$

21. $f(x) = \begin{cases} -2 & \text{if } x \leq 3 \\ 2 & \text{if } 3 < x \end{cases}$

22. $g(x) = \begin{cases} -4 & \text{if } x < -2 \\ -1 & \text{if } -2 \leq x \leq 2 \\ 3 & \text{if } 2 < x \end{cases}$

23. $g(x) = \begin{cases} 2x - 1 & \text{if } x \neq 2 \\ 0 & \text{if } x = 2 \end{cases}$

24. $f(x) = \begin{cases} 3x + 2 & \text{if } x \neq 1 \\ 8 & \text{if } x = 1 \end{cases}$

25. $F(x) = \begin{cases} x^2 - 4 & \text{if } x \neq 3 \\ -2 & \text{if } x = 3 \end{cases}$

26. $G(x) = \begin{cases} 9 - x^2 & \text{if } x \neq -3 \\ 4 & \text{if } x = -3 \end{cases}$

27. $G(x) = \begin{cases} 1 - x^2 & \text{if } x < 0 \\ 3x + 1 & \text{if } 0 \leq x \end{cases}$

28. $F(x) = \begin{cases} x^2 - 4 & \text{if } x < 3 \\ 2x - 1 & \text{if } 3 \leq x \end{cases}$

29. $g(x) = \begin{cases} 6x + 7 & \text{if } x \leq -2 \\ 4 - x & \text{if } -2 < x \end{cases}$

30. $f(x) = \begin{cases} x - 2 & \text{if } x \leq 0 \\ x^2 + 1 & \text{if } 0 < x \end{cases}$

31. $h(x) = \begin{cases} x + 3 & \text{if } x < -5 \\ \sqrt{25 - x^2} & \text{if } -5 \leq x \leq 5 \\ 3 - x & \text{if } 5 < x \end{cases}$

32. $H(x) = \begin{cases} x + 2 & \text{if } x \leq -4 \\ \sqrt{16 - x^2} & \text{if } -4 < x < 4 \\ 2 - x & \text{if } 4 \leq x \end{cases}$

33. $F(x) = \dfrac{x^3 - 2x^2}{x - 2}$ **34.** $G(x) = \dfrac{x^3 + 3x^2}{x + 3}$

35. $h(x) = |x| + |x - 1|$ **36.** $H(x) = |x^2 - 1|$

37. $g(x) = |x| \cdot |x - 1|$ **38.** $F(x) = [\![x + 2]\!]$

39. $f(x) = [\![x - 4]\!]$ **40.** $H(x) = |x| + [\![x]\!]$

41. $G(x) = x - [\![x]\!]$ **42.** $h(x) = [\![x^2]\!]$

In Exercises 43 through 48, plot the graph of the function and from the graph state whether the function is even, odd, or neither. Then prove your statement algebraically.

43. (a) $f(x) = 2x^4 - 3x^2 + 1$ **(b)** $g(x) = 5x^5 + 1$

44. (a) $f(x) = x^2 + 2x + 2$ **(b)** $g(x) = x^6 - 1$

45. (a) $f(x) = 5x^3 - 7x$ **(b)** $g(x) = |x|$

46. (a) $f(x) = 4x^5 + 3x^3$ **(b)** $g(x) = x^3 + 1$

47. (a) $f(x) = \sqrt[3]{x}$ **(b)** $g(x) = 5x^4 - 4$

48. (a) $f(x) = \dfrac{|x|}{x}$ **(b)** $g(x) = 2|x| + 3$

In Exercises 49 and 50, determine algebraically whether the function is even, odd, or neither.

49. (a) $f(y) = \dfrac{y^3 - y}{y^2 + 1}$ **(b)** $g(r) = \dfrac{r^2 - 1}{r^2 + 1}$

(c) $f(x) = \dfrac{|x|}{x^2 + 1}$

50. (a) $h(x) = \dfrac{x^2 - 5}{2x^3 + x}$ **(b)** $g(z) = \dfrac{z - 1}{z + 1}$

(c) $f(x) = \begin{cases} -1, & \text{if } x < 0 \\ 1, & \text{if } 0 \leq x \end{cases}$

51. (a) Sketch the graph of the *unit step function*, denoted by U, and defined by

$$U(x) = \begin{cases} 0 & \text{if } x < 0 \\ 1 & \text{if } 0 \leq x \end{cases}$$

Find formulas defining each of the following functions and sketch their graphs:
(b) $U(x - 1)$ **(c)** $U(x) - U(x - 1)$

52. Find formulas defining each of the following functions and sketch their graphs where U is the unit step function defined in Exercise 51:
(a) $x \cdot U(x)$ **(b)** $(x + 1) \cdot U(x + 1)$
(c) $(x + 1) \cdot U(x + 1) - x \cdot U(x)$

53. (a) Sketch the graph of the *signum function* (or *sign function*), denoted by *sgn*, and defined by

$$\text{sgn } x = \begin{cases} -1 & \text{if } x < 0 \\ 0 & \text{if } x = 0 \\ 1 & \text{if } 0 < x \end{cases}$$

sgn x is read "signum of *x*." Find formulas defining each of the following functions and sketch their graphs:
(b) $x \cdot \text{sgn } x$ **(c)** $2 - x \cdot \text{sgn } x$ **(d)** $x - 2 \text{ sgn } x$

54. Find formulas defining each of the following functions and sketch their graphs where *sgn* is the signum function defined in Exercise 53:
(a) $\text{sgn}(x + 1)$ **(b)** $\text{sgn}(x - 1)$
(c) $\text{sgn}(x + 1) - \text{sgn}(x - 1)$

55. Prove that if f and g are both odd functions, then $(f + g)$ and $(f - g)$ are also odd functions, and $f \cdot g$ and f/g are both even functions.

56. There is one function, whose domain is the set of all real numbers, that is both even and odd. What is that function? Prove it is the only such function.

57. The graph of the function f in the figure resembles the letter W. Define $f(x)$ piecewise.

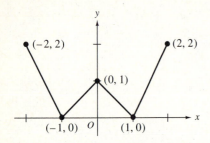

58. The graph of the function f in the figure resembles the letter M. Define $f(x)$ piecewise.

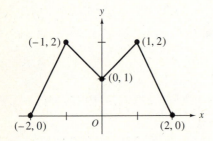

59. Graphs of the functions in Exercises 57 and 58 resemble letters of the alphabet. Define two other

functions whose graphs resemble two different letters of the alphabet.

60. In the figure, the graph resembling the letter X is the union of the graphs of two functions f_1 and f_2 plotted in the $[-1, 1]$ by $[-1, 1]$ viewing rectangle. Define $f_1(x)$ and $f_2(x)$.

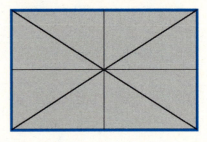

$[-1, 1]$ by $[-1, 1]$

61. There are three functions f_1, f_2, and f_3, the union of whose graphs plotted in the $[-1, 1]$ by $[-1, 1]$ viewing rectangle resembles the letter Z. Define $f_1(x)$, $f_2(x)$, and $f_3(x)$.

62. Explain why the definition of the graph of a function is consistent with the definition of a function as a set of ordered pairs. In your explanation use a specific example.

4.3 QUADRATIC FUNCTIONS

GOALS
1. **Sketch and plot graphs of quadratic functions.**
2. **Find the zeros of a quadratic function.**
3. **Find the extreme value of a quadratic function.**
4. **Solve word problems having a quadratic function as a mathematical model.**

In the previous section you learned that a quadratic function is a polynomial function of degree 2. Thus the general quadratic function is defined by

$$f(x) = ax^2 + bx + c$$

where a, b, and c are constants representing real numbers and $a \neq 0$. The graph of f is the same as the graph of the equation

$$y = ax^2 + bx + c$$

which we learned in Section 3.3 is a parabola whose axis is vertical. The properties discussed in Sections 3.3 and 3.5 are helpful when sketching parabolas.

▷ ILLUSTRATION 1

If the function f is defined by

$$f(x) = -2x^2 + 8x - 5$$

the graph of f is the same as the graph of the equation

$$y = -2x^2 + 8x - 5$$

This equation is equivalent to

$$2(x^2 - 4x) = -y - 5$$

To complete the square of the binomial within the parentheses, we add $2(4)$ to both sides of the equation, and we have

$$2(x^2 - 4x + 4) = -y - 5 + 8$$

$$(x - 2)^2 = -\frac{1}{2}(y - 3)$$

This equation is of the form

$$(x - h)^2 = 4p(y - k)$$

where (h, k) is $(2, 3)$ and $p = -\frac{1}{8}$. Therefore the vertex of the parabola is at $(2, 3)$, and the axis is the line $x = 2$. Because $p < 0$, the parabola opens downward. We locate a few more points on the parabola and sketch the graph shown in Figure 1. Verify this graph on your graphics calculator.

◀

FIGURE 1

The **zeros** of a function f are the values of x for which $f(x) = 0$.

▷ ILLUSTRATION 2

The function of Illustration 1 is defined by

$$f(x) = -2x^2 + 8x - 5$$

To find the zeros of this function, we substitute 0 for $f(x)$ and get

$$-2x^2 + 8x - 5 = 0$$

$$2x^2 - 8x + 5 = 0$$

Solving this equation by the quadratic formula where a is 2, b is -8, and c is 5, we have

$$x = \frac{-b \pm \sqrt{b^2 - 4ac}}{2a}$$

$$= \frac{8 \pm \sqrt{64 - 40}}{4}$$

$$= \frac{8 \pm \sqrt{24}}{4}$$

$$= \frac{8 \pm 2\sqrt{6}}{4}$$

$$= \frac{4 \pm \sqrt{6}}{2}$$

Hence the zeros of f are

$$\frac{4 + \sqrt{6}}{2} \approx 3.22 \qquad \text{and} \qquad \frac{4 - \sqrt{6}}{2} \approx 0.78$$

These numbers are also the x intercepts of the parabola in Figure 1. Verify these x intercepts by zooming in on your graphics calculator. ◀

In general, if

$$f(x) = ax^2 + bx + c$$

then the zeros of f are the roots of the equation

$$ax^2 + bx + c = 0$$

In Section 2.2 we learned that a quadratic equation in one variable can have two real roots, one real root (of multiplicity 2), or two imaginary roots. If a quadratic equation has two real roots, the corresponding quadratic function has two real zeros and its graph intersects the x axis at two distinct points. This situation is shown in Figure 2, where the parabola opens upward ($a > 0$), and in Figure 3, where the parabola opens downward ($a < 0$). If the function has one real zero, the graph intersects the x axis at a single point, as shown in Figures 4 ($a > 0$) and 5($a < 0$). If the function has two imaginary zeros, the graph does not intersect the x axis; this is shown in Figures 6 ($a > 0$) and 7 ($a < 0$).

The function value at the vertex of the graph of a quadratic function is called an **extreme value.** When the parabola opens upward, the function has a **minimum value** at the vertex. There is no maximum value for such a function. When the parabola opens downward, the function has a **maximum value** at the vertex; it has no minimum value.

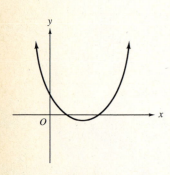

FIGURE 2

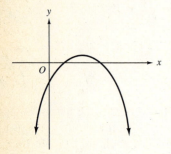

FIGURE 3

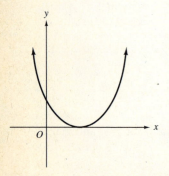

FIGURE 4

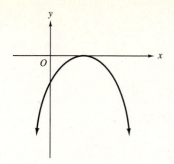

FIGURE 5

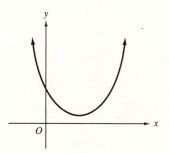

FIGURE 6

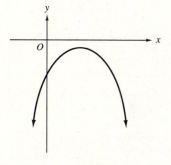

FIGURE 7

▷ **ILLUSTRATION 3**

The function of Illustrations 1 and 2 is defined by

$$f(x) = -2x^2 + 8x - 5$$

The graph of f is the parabola of Figure 1. The parabola opens downward and has its vertex at $(2, 3)$. This function, therefore, has a maximum value of 3 occurring when $x = 2$.

▶ **EXAMPLE 1** *Finding the Extreme Value of a Quadratic Function*

Find either a maximum or minimum value of the function defined by

$$f(x) = 3x^2 + 3x + 2$$

Solution   The graph of f is the parabola having the equation

$$y = 3x^2 + 3x + 2$$

We write this equation in the form $(x - h)^2 = 4p(y - k)$. The equation is equivalent to

$$3(x^2 + x) = y - 2$$

Completing the square of the binomial in parentheses by adding $\frac{3}{4}$ to both sides of the equation, we have

$$3\left(x^2 + x + \frac{1}{4}\right) = y - 2 + \frac{3}{4}$$

$$3\left(x + \frac{1}{2}\right)^2 = y - \frac{5}{4}$$

$$\left(x + \frac{1}{2}\right)^2 = \frac{1}{3}\left(y - \frac{5}{4}\right)$$

The parabola opens upward, and its vertex is at $(-\frac{1}{2}, \frac{5}{4})$. Therefore the minimum value of f is $\frac{5}{4}$, occurring at $x = -\frac{1}{2}$. ◀

Verify the result of Example 1 by plotting the parabola.

We now apply the method used in the solution of Example 1 to the general quadratic function defined by

$$f(x) = ax^2 + bx + c$$

In this equation we replace $f(x)$ by y and obtain

$$y = ax^2 + bx + c$$

which is equivalent to

$$ax^2 + bx = y - c$$

$$a\left(x^2 + \frac{b}{a}x\right) = y - c$$

We complete the square of the binomial in parentheses.

$$a\left(x^2 + \frac{b}{a}x + \frac{b^2}{4a^2}\right) = y - c + \frac{b^2}{4a}$$

$$\left(x + \frac{b}{2a}\right)^2 = \frac{1}{a}\left(y + \frac{b^2 - 4ac}{4a}\right)$$

The graph of this equation is a parabola having its vertex at the point where $x = -\dfrac{b}{2a}$. If $a > 0$, the parabola opens upward, and so f has a minimum value at the point where $x = -\dfrac{b}{2a}$. If $a < 0$, the parabola opens downward, and so f has a maximum value at the point where $x = -\dfrac{b}{2a}$. These results are given in the following theorem.

THEOREM 1

The quadratic function defined by $f(x) = ax^2 + bx + c$, where $a \neq 0$, has an extreme value at the point where $x = -\dfrac{b}{2a}$. If $a > 0$, the extreme value is a minimum value, and if $a < 0$, the extreme value is a maximum value.

▶ **EXAMPLE 2** *Finding the Extreme Value of a Quadratic Function*

Use Theorem 1 to find either a maximum or minimum value of the function g if

$$g(x) = -\frac{3}{2}x^2 + 6x - 10$$

Verify the result by plotting the graph of g.

Solution For the given quadratic function, $a = -\frac{3}{2}$ and $b = 6$. Because $a < 0$, g has a maximum value at the point where

$$x = -\frac{b}{2a}$$

$$= -\frac{6}{2\left(-\dfrac{3}{2}\right)}$$

$$= 2$$

The maximum value is

$$g(2) = -\frac{3}{2}(2)^2 + 6(2) - 10$$

$$= -4$$

The graph of g, plotted on a graphics calculator and shown in Figure 8, is a parabola with vertex at $(2, -4)$ and opening downward. This agrees with our result. ◀

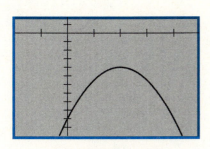

$[-2, 5]$ by $[-12, 2]$

$g(x) = -\dfrac{3}{2}x^2 + 6x - 10$

FIGURE 8

In applications of calculus, we often need to obtain a function as a mathematical model. We now give four examples involving quadratic functions as mathematical models.

▶ **EXAMPLE 3** *Solving a Word Problem Having a Quadratic Function as a Mathematical Model*

In Example 7 of Section 2.6 we had the following situation: A decorator designs and sells wall fixtures and can sell at a price of $75 each all the fixtures she produces. If x fixtures are manufactured each day, then the number of dollars in the total cost of production is $x^2 + 25x + 96$. **(a)** How many fixtures should be produced each day for the decorator to have the greatest profit? **(b)** Check the answer in part (a) on a graphics calculator.

Solution
(a) If $P(x)$ dollars is the daily profit from the sale of x fixtures, then, as in Example 7 of Section 2.6,

$$P(x) = -x^2 + 50x - 96$$

Function P is quadratic with $a = -1$ and $b = 50$. Because $a < 0$, P

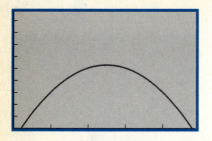

[0, 50] by [0, 1000]
$P(x) = -x^2 + 50x - 96$
FIGURE 9

has a maximum value at the point where

$$x = -\frac{b}{2a}$$
$$= 25$$

Conclusion: To have the greatest profit, the number of fixtures produced and sold each day should be 25.

(b) Figure 9 shows the graph of P plotted in the [0, 50] by [0, 1000] viewing rectangle. The x coordinate of the vertex of the parabola appears to be 25, which agrees with our answer in part (a).   ◀

▶ **EXAMPLE 4** *Solving a Word Problem Having a Quadratic Function as a Mathematical Model*

A clock manufacturer can produce a particular clock at a cost of $15 per clock. If the selling price of the clock is x dollars, then $(125 - x)$ clocks are sold per week. **(a)** Express the number of dollars in the manufacturer's weekly profit as a function of x. **(b)** From the function in part (a), determine the weekly profit if the selling price is $45 per clock. **(c)** Estimate the selling price for the manufacturer's weekly profit to be a maximum by plotting the graph of the function in part (a). **(d)** Check the estimate in part (c) algebraically.

Solution

(a) The profit can be obtained by subtracting the total cost from the total revenue. Let $R(x)$ dollars be the weekly revenue. Because the revenue is the product of the selling price and the number of clocks sold,

$$R(x) = x(125 - x)$$

Let $C(x)$ dollars be the total cost of the clocks that are sold per week. Because the total cost is the product of the cost of each clock and the number of clocks sold,

$$C(x) = 15(125 - x)$$

If $P(x)$ dollars is the weekly profit, then

$$P(x) = R(x) - C(x)$$
$$= x(125 - x) - 15(125 - x)$$
$$= (125 - x)(x - 15)$$

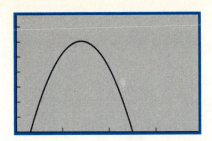

[0, 200] by [0, 4000]
$P(x) = (125 - x)(x - 15)$
FIGURE 10

(b) If the selling price is $45, the number of dollars in the weekly profit is $P(45)$. From the expression for $P(x)$ in part (a)

$$P(45) = (125 - 45)(45 - 15)$$
$$= 80 \cdot 30$$
$$= 2400$$

Conclusion: The weekly profit is $2400 when the clocks are sold at $45 each.

(c) Figure 10 shows the graph of P plotted in the [0, 200] by [0, 4000] viewing rectangle. The vertex of the parabola, opening downward, appears to be at the point where $x = 70$. Thus, for the manufacturer's weekly profit to be a maximum, we estimate the selling price per clock to be $70.

(d) From the expression for $P(x)$ in part (a), we have

$$P(x) = -x^2 + 140x - 1875$$

Function P is quadratic with $a = -1$ and $b = 140$. Because $a < 0$, P has a maximum value at the point where

$$x = -\frac{b}{2a}$$
$$= 70$$

Conclusion: The manufacturer's weekly profit will be a maximum when the selling price of the clock is $70.
This conclusion agrees with our estimate in part (c).   ◄

In the next example, we first obtain two equations involving a dependent variable and two independent variables. We then express the dependent variable as a function of a single independent variable by eliminating the other independent variable from the pair of equations.

▶ **EXAMPLE 5** *Solving a Word Problem Having a Quadratic Function as a Mathematical Model*

A rectangular field is to be fenced off along the bank of a river, and no fence is required along the river. The material for the fence costs $8 per running foot for the two ends and $12 per running foot for the side parallel to the river; $3600 worth of fence is to be used. **(a)** If x feet is the length of an end, express the number of square feet in the area of the field as a function of x. **(b)** What is the domain of the resulting function? **(c)** Find the dimensions of the field of largest possible area that can be enclosed with the $3600 worth of fence. What is the largest area?

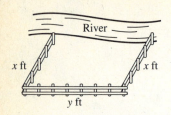

FIGURE 11

Solution

(a) Let y feet be the length of the side of the field parallel to the river and A square feet be the area of the field. See Figure 11.

$$A = xy$$

Because the cost of the material for each end is \$8 per running foot and the length of an end is x feet, the total cost for the fence for each end is $8x$ dollars. Similarly, the total cost of the fence for the third side is $12y$ dollars. We have then

$$8x + 8x + 12y = 3600 \qquad \textbf{(1)}$$

To express A in terms of a single variable, we first solve Equation (1) for y in terms of x.

$$12y = 3600 - 16x$$

$$y = 300 - \frac{4}{3}x$$

We substitute this value of y into the equation $A = xy$, yielding A as a function of x, and

$$A(x) = x\left(300 - \frac{4}{3}x\right)$$

(b) Both x and y must be nonnegative. The smallest value that x can assume is 0. The smallest value that y can assume is 0, and when $y = 0$, we obtain, from Equation (1), $x = 225$. Thus 225 is the largest value that x can assume. Hence x must be in the closed interval $[0, 225]$, and this closed interval is the domain of A.

(c) From the expression for $A(x)$ in part (a), we have

$$A(x) = -\frac{4}{3}x^2 + 300x$$

Function A is quadratic with $a = -\frac{4}{3}$ and $b = 300$. Because $a < 0$, A has a maximum value at the point where

$$x = -\frac{b}{2a}$$

$$= -\frac{300}{2\left(-\frac{4}{3}\right)}$$

$$= \frac{225}{2}$$

When $x = \frac{225}{2}$, $300 - \frac{4}{3}x = 150$. Furthermore

$$A\left(\frac{225}{2}\right) = \frac{225}{2}(150)$$
$$= 16{,}875$$

Conclusion: The largest possible area that can be enclosed for $3600 is 16,875 ft², and this is obtained when the side parallel to the river is 150 ft long and the ends are each 112.5 ft long. ◀

▶ **EXAMPLE 6** *Solving a Word Problem Having a Quadratic Function as a Mathematical Model*

The financial manager of a college newsletter determines that 1000 copies of the newsletter will be sold if the price is 50 cents and that the number of copies sold decreases by 10 for each 1 cent added to the price. What price will yield the largest gross income from sales, and what is the largest gross income?

Solution The number of cents in the gross income depends on the price per copy. Let $f(x)$ cents be the gross income when x cents is the price per copy.

The amount by which x exceeds 50 is $x - 50$. To determine the number of copies sold when x cents is the price per copy, we must subtract from 1000 the product of 10 and this excess. Hence, when x cents is the price per copy, the number of copies sold is $1000 - 10(x - 50)$.

We obtain an expression for the gross income by multiplying the number of copies sold by the price per copy. Therefore

$$f(x) = [1000 - 10(x - 50)]x$$
$$f(x) = (1500 - 10x)x$$
$$f(x) = -10x^2 + 1500x$$

For this quadratic function $a = -10$, $b = 1500$, and $c = 0$. Because $a < 0$, f has a maximum value at the point where

$$x = -\frac{b}{2a}$$
$$= -\frac{1500}{2(-10)}$$
$$= 75$$

The maximum value is

$$f(75) = -10(75)^2 + 1500(75)$$
$$= -10(5625) + 112,500$$
$$= 56,250$$

Conclusion: The price of 75 cents per copy will yield the largest gross income from sales, $562.50. ◀

EXERCISES 4.3

In Exercises 1 through 4, find the exact values of the zeros of the function. Check your answer by plotting the parabola.

1. $f(x) = x^2 - 2x - 3$ **2.** $f(x) = x^2 - 3x + 1$
3. $f(x) = 2x^2 - 2x - 1$ **4.** $f(x) = 6x^2 - 7x - 5$

In Exercises 5 through 14, sketch the graph of the function and determine from the graph which of the following statements characterizes the zeros of the function: (a) two real zeros; (b) one real zero of multiplicity 2; or (c) two imaginary zeros. Check your answer by plotting the parabola.

5. $f(x) = x^2 - 4x$ **6.** $f(x) = x^2 - 3$
7. $f(x) = -x^2 + 4$ **8.** $g(x) = x^2 - 6x + 11$
9. $g(x) = -4x^2 + 8x - 8$
10. $g(x) = 2x^2 + 4x + 1$
11. $h(x) = 9 - 6x + x^2$
12. $h(x) = 1 - 4x - x^2$
13. $f(x) = \frac{1}{8}(4x^2 + 20x + 49)$
14. $f(x) = -4x^2 + 12x - 9$

In Exercises 15 through 18, use the method of Example 1 to find either a maximum value or a minimum value of the function. Check your answer by plotting the parabola.

15. $f(x) = 4x^2 + 8x + 7$ **16.** $f(x) = 2 + 6x - x^2$
17. $g(x) = -\frac{1}{2}(x^2 + 6x + 5)$
18. $G(x) = \frac{1}{8}(x^2 - 4x - 4)$

In Exercises 19 through 22, apply Theorem 1 to find either a maximum or a minimum value of the function. Check your answer by plotting the parabola.

19. $f(x) = 2 + 4x - 3x^2$ **20.** $g(x) = 3x^2 + 6x + 9$

21. $G(x) = \frac{1}{8}(4x^2 + 12x - 9)$
22. $F(x) = -\frac{1}{2}(x^2 + 8x + 8)$

23. An object is thrown straight upward from the ground with an initial velocity of 96 ft/sec. If the height of the object is $f(t)$ feet after t seconds and if air resistance is neglected,

$$f(t) = 96t - 16t^2$$

(a) Estimate the maximum height reached by the object and how many seconds after the object is thrown it reaches its maximum height by plotting the graph of f. **(b)** Check your estimates in part (a) algebraically. **(c)** Write a conclusion.

24. A projectile is shot straight upward from a point 15 ft above the ground with an initial velocity of 176 ft/sec. If the height of the projectile is $f(t)$ feet after t seconds and if air resistance is neglected,

$$f(t) = 15 + 176t - 16t^2$$

(a) Estimate the projectile's maximum height and how long it takes the projectile to reach its maximum height by plotting the graph of f. **(b)** Check your estimates in part (a) algebraically. **(c)** Write a conclusion.

In Exercises 25 through 38, solve the word problem by finding a quadratic function as a mathematical model of the situation. Be sure to complete the exercise by writing a conclusion.

25. In Exercise 43 of Exercises 2.6 we had the following situation: A firm can sell at a price of $100 per unit all of a particular commodity it produces. If x units are produced each day, the number of dollars in the total cost of each day's production is $x^2 + 20x + 700$. **(a)** Express the number of dollars in the firm's daily profit as a function of x. **(b)** Estimate the greatest daily profit and how many units should be produced each day for the firm to have that profit by plotting the graph of your function in part (a). **(c)** Check your estimates in part (b) algebraically.

26. In Exercise 44 of Exercises 2.6, we had the following situation: A company that builds and sells desks can sell at a price of $400 per desk all the desks it produces. If x desks are built and sold each week, then the number of dollars in the total cost of the week's production is $2x^2 + 80x + 3000$. **(a)** Express the number of dollars in the company's weekly profit as a function of x. **(b)** Estimate the greatest weekly profit and how many desks should be built each week for the company to have that profit by plotting the graph of your function in part (a). **(c)** Check your estimates in part (b) algebraically.

27. A carpenter can construct bookcases at a cost of $40 each. If the carpenter sells the bookcases for x dollars each, it is estimated that $300 - 2x$ bookcases will be sold per month. **(a)** Express the number of dollars in the carpenter's monthly profit as a function of x. **(b)** Use your function in part (a) to determine the monthly profit if the selling price is $110 per bookcase. **(c)** Estimate the selling price of each bookcase that will give the carpenter the greatest monthly profit by plotting the graph of your function in part (a). **(d)** Check the estimate in part (c) algebraically.

28. A toy manufacturer can produce a particular toy at a cost of $10 per toy. If the selling price of the toy is x dollars, then $45 - x$ toys will be sold daily. **(a)** Express the number of dollars in the manufacturer's daily profit as a function of x. **(b)** Use your function in part (a) to determine the daily profit if the selling price is $30 per toy. **(c)** Estimate the selling price of each toy that will enable the manufacturer to realize the maximum daily profit by plotting the graph of your function in part (a). **(d)** Check your estimate in part (c) algebraically.

29. A rectangular field is to be enclosed with 240 m of fence. **(a)** If x meters is the length of the field, express the number of square meters in the area of the field as a function of x. **(b)** What is the domain of your function in part (a)? **(c)** Estimate the dimensions of the largest rectangular field that can be enclosed with the 240 m of fence by plotting the graph of your function in part (a). **(d)** Check your estimates in part (c) algebraically.

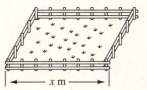

30. A rectangular garden is to be fenced off with 100 ft of fencing material. **(a)** If x feet is the length of the garden, express the number of square feet in the area of the garden as a function of x. **(b)** What is the domain of your function in part (a)? **(c)** Estimate the dimensions of the largest rectangular garden that can be fenced off with the 100 ft of fencing material by plotting the graph of your function in part (a). **(d)** Check your estimates in part (c) algebraically.

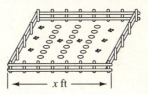

31. Do Exercise 29 if one side of the field is to have a river as a natural boundary and the fencing material is to be used for the other three sides. Let x meters be the length of the side of the field parallel to the river.

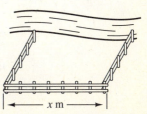

32. Do Exercise 30 if the garden is to be placed so that a side of a house serves as a boundary and the fencing material is to be used for the other three sides. Let x feet be the length of the side of the garden parallel to the house.

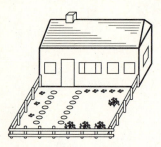

33. A rectangular plot of ground is to be enclosed by a fence and then divided down the middle by another fence. The fence down the middle costs $4 per running foot and the other fence costs $10 per running foot, and $1920 worth of fencing material is to be used. **(a)** If x feet is the length of the fence down the middle, express the number of square feet in the area of the plot as a function of x. **(b)** What is the domain of your function in part (a)? **(c)** Find the dimensions of the largest rectangular field with a fence down the middle that can be enclosed with the $1920 worth of fencing material.

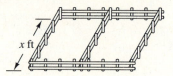

34. Find two numbers whose sum is 10 and whose product is a maximum.

35. Find two numbers whose difference is 14 and whose product is a minimum.

36. Find two positive numbers whose sum is 50 such that the sum of their squares is a minimum.

37. A travel agency offers an organization an all-inclusive tour for $800 per person if not more than 100 people take the tour. However, the cost per person will be reduced $5 for each person in excess of 100. How many people should take the tour, for the travel agency to receive the largest gross revenue, and what is this largest gross revenue?

38. A student club on a college campus charges annual membership dues of $20, less 10 cents for each member over 60. How many members would give the club the most revenue from annual dues?

4.4 FUNCTIONS AS MATHEMATICAL MODELS

GOALS

1. **Define direct variation.**
2. **Define inverse variation.**
3. **Define joint variation.**
4. **Solve word problems involving variation by using a function as a mathematical model.**
5. **Solve word problems having a cubic function as a mathematical model.**

Applications involving the dependence of one variable on another occur in business, economics, and the physical, life, and social sciences. The formulas used in these applications often determine functions. For instance, if y dollars is the simple interest for one year earned by a principal of x dollars at the rate of 12 percent per year, then

$$y = 0.12x$$

For a given nonnegative value of x there corresponds a unique value of y; thus the value of y depends on the value of x. If f is the function defined by $f(x) = 0.12x$, and the domain of f is the set of nonnegative real numbers, then the equation $y = 0.12x$ can be written as $y = f(x)$. The equation $y = 0.12x$ is an example of *direct proportion*, and y is said to be *directly proportional* to x.

DEFINITION   **Directly Proportional**

A variable y is said to be **directly proportional** to a variable x if

$$y = kx$$

where k is a nonzero constant. More generally, a variable y is said to be **directly proportional** to the nth power of x $(n > 0)$ if

$$y = kx^n$$

The constant k is called the **constant of proportionality.**

▶ **EXAMPLE 1** *Solving a Word Problem Involving a Direct Proportion by Using a Function as a Mathematical Model*

A person's approximate brain weight is directly proportional to his or her body weight, and a person weighing 150 lb has an approximate brain weight of 4 lb. **(a)** Express the number of pounds in the approximate brain weight of a person as a function of the person's body weight. **(b)** Find the approximate brain weight of a person whose body weight is 176 lb.

Solution

(a) Let $f(x)$ pounds be the approximate brain weight of a person having a body weight of x pounds. Then

$$f(x) = kx \qquad (1)$$

Because a person of body weight 150 lb has a brain weighing approximately 4 lb, we substitute 150 for x and 4 for $f(x)$ in (1), and we have

$$4 = k(150)$$

$$k = \frac{2}{75}$$

We replace k in (1) by this value and obtain

$$f(x) = \frac{2}{75}x$$

(b) Because $f(x) = \dfrac{2}{75}x$,

$$f(176) = \dfrac{2}{75}(176)$$

$$= 4.7$$

Conclusion: The approximate brain weight of a person weighing 176 lb is 4.7 lb.   ◀

DEFINITION   Inversely Proportional

A variable y is said to be **inversely proportional** to a variable x if

$$y = \dfrac{k}{x}$$

where k is a nonzero constant. More generally, a variable y is said to be **inversely proportional** to the nth power of x $(n > 0)$ if

$$y = \dfrac{k}{x^n}$$

▶ **EXAMPLE 2** *Solving a Word Problem Involving an Inverse Proportion by Using a Function as a Mathematical Model*

The illuminance from a given light source is inversely proportional to the square of the distance from it. **(a)** Express the number of luxes (lx) in the illuminance as a function of the number of meters in the distance from the light source if the illuminance is 225 lx at a distance of 5 m from the source. **(b)** Find the illuminance at a point 15 m from the source.

Solution

(a) Let $f(x)$ luxes be the illuminance from the light source at x meters from it. Then

$$f(x) = \dfrac{k}{x^2} \tag{2}$$

Because the illuminance is 225 lx at a distance of 5 m from the source, we replace x by 5 and $f(x)$ by 225 in (2) and obtain

$$225 = \dfrac{k}{5^2}$$

$$k = 5625$$

Substituting this value of k in (2), we have

$$f(x) = \frac{5625}{x^2}$$

(b) From the preceding expression for $f(x)$, we get

$$f(15) = \frac{5625}{15^2}$$

$$= \frac{5625}{225}$$

$$= 25$$

Conclusion: The illuminance at a point 15 m from the source is 25 lx.   ◄

DEFINITION   Jointly Proportional

A variable z is said to be **jointly proportional** to variables x and y if

$$z = kxy$$

where k is a nonzero constant. More generally, a variable z is said to be **jointly proportional** to the nth power of x and the mth power of $y (n > 0$ and $m > 0)$ if

$$z = kx^n y^m$$

► **EXAMPLE 3** *Solving a Word Problem Involving a Joint Proportion by Using a Quadratic Function as a Mathematical Model*

In a limited environment where A is the optimum number of bacteria supportable by the environment, the rate of bacterial growth is jointly proportional to the number present and the difference between A and the number present. Suppose 1 million bacteria is the optimum number supportable by the environment, and the rate of growth is 60 bacteria per minute when 1000 bacteria are present. **(a)** Express the rate of bacterial growth as a function of the number of bacteria present. **(b)** Find the rate of growth when 100,000 bacteria are present. **(c)** Find how many bacteria are present when the rate of growth is a maximum.

Solution

(a) Let $f(x)$ bacteria per minute be the rate of growth when there are x bacteria present. Then

$$f(x) = kx(1,000,000 - x) \tag{3}$$

Because the rate of growth is 60 bacteria per minute when there are 1000 bacteria present, we replace x by 1000 and $f(x)$ by 60 in (3), and we have

$$60 = k(1000)(1,000,000 - 1000)$$

$$k = \frac{60}{999,000,000}$$

$$= \frac{1}{16,650,000}$$

Replacing k in (3) by this value, we obtain

$$f(x) = \frac{x(1,000,000 - x)}{16,650,000}$$

(b) From the preceding expression for $f(x)$, we have

$$f(100,000) = \frac{100,000(1,000,000 - 100,000)}{16,650,000}$$

$$= \frac{100,000(900,000)}{16,650,000}$$

$$= 5405$$

Conclusion: The rate of growth is 5405 bacteria per minute when 100,000 bacteria are present.

(c) From the expression for $f(x)$ in part (a), we have

$$f(x) = -\frac{1}{16,650,000}x^2 + \frac{20}{333}x$$

Function f is quadratic with $a = -\frac{1}{16,650,000}$ and $b = \frac{20}{333}$. Because $a < 0$, f has a maximum value at the point where

$$x = -\frac{b}{2a}$$

$$= -\frac{20}{333}\left(-\frac{16,650,000}{2}\right)$$

$$= 500,000$$

Conclusion: The rate of growth is a maximum when 500,000 bacteria are present. ◄

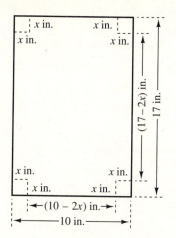

FIGURE 1

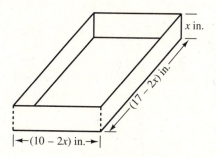

FIGURE 2

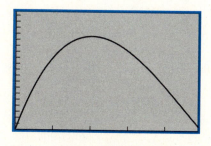

[0, 5] by [0, 200]
$V(x) = 170x - 54x^2 + 4x^3$
FIGURE 3

In Section 4.3 we determined an extreme value of a quadratic function by computing the y coordinate of the vertex of the parabola that is the graph of the function. For most nonquadratic functions, the computation of exact extreme function values requires techniques of calculus. With a graphics calculator, however, we can compute approximate extreme function values. To illustrate the procedure, we return to the box problem of Example 6 in Section 2.4 where an open box was formed by cutting equal squares from the four corners of a piece of cardboard and turning up the sides. In that example, we found the length of the side of the squares to be cut out if the box formed was to have a specific volume. We now consider the problem of determining the length of the side of the cut-out squares if the box is to have maximum volume.

▶ **EXAMPLE 4** *Solving a Word Problem Having a Cubic Function as a Mathematical Model*

A cardboard box manufacturer wishes to make open boxes from rectangular pieces of cardboard with dimensions 10 in. by 17 in. by cutting equal squares from the four corners and turning up the sides. **(a)** If x inches is the length of the side of the square to be cut out, express the number of cubic inches in the volume of the box as a function of x. **(b)** What is the domain of the function obtained in part (a)? **(c)** On a graphics calculator, find accurate to two decimal places the length of the side of the square to be cut out so that the box has the largest possible volume. What is the maximum volume?

Solution

(a) Figure 1 represents a given piece of cardboard, and Figure 2 represents the box obtained from the cardboard. The number of inches in the dimensions of the box are x, $10 - 2x$, and $17 - 2x$. The volume of the box is the product of the three dimensions. Therefore, if $V(x)$ cubic inches is the volume,

$$V(x) = x(10 - 2x)(17 - 2x)$$
$$= 170x - 54x^2 + 4x^3$$

(b) From the expression for $V(x)$ in part (a), we observe that $V(0) = 0$ and $V(5) = 0$. From conditions of the problem we know that x can be neither negative nor greater than 5. Thus the domain of V is the closed interval [0, 5].

(c) The graph of function V plotted in the viewing rectangle [0, 5] by [0, 200] appears in Figure 3. We observe that V has a maximum value on its domain. The x coordinate of the highest point on the graph gives the length of the side of the square to be cut out for maximum volume and the y coordinate gives the maximum volume. Using "zoom-in" on our graphics calculator, we determine the highest point is (2.03, 156.03).

Conclusion: The length of the side of the square to be cut out should be 2.03 in. to give a box of maximum volume 156.03 in.[3] ◀

By techniques of calculus, we can determine that the exact maximum value of the function in Example 4 occurs when

$$x = \frac{27 - \sqrt{219}}{6} \approx 2.03$$

▶ **EXAMPLE 5** *Solving a Word Problem Having a Cubic Function as a Mathematical Model*

Under a monopoly the demand equation for a certain commodity is $x^2 + p = 360$ where x units are produced daily when p dollars is the price per unit. **(a)** Express the number of dollars in the daily total revenue as a function of x. **(b)** What is the domain of the function obtained in part (a)? **(c)** On a graphics calculator, find the number of units that should be produced daily to maximize the daily total revenue. What is the maximum daily total revenue?

Solution

(a) Total revenue equals the price per unit times the number of units produced and sold. (Note: Under a monopoly the number of units produced is the same as the number of units sold.) Therefore, if $R(x)$ dollars is the daily total revenue,

$$R(x) = px$$

Solving the demand equation for p, we obtain

$$p = 360 - x^2$$

Substituting this expression for p in the equation for $R(x)$, we have

$$R(x) = (360 - x^2)x$$

(b) From the equation defining $R(x)$ in part (a), we observe that $R(0) = 0$ and $R(\sqrt{360}) = 0$. If x is either negative or greater than $\sqrt{360}$, the total revenue is negative. The domain of R is, therefore, the closed interval $[0, \sqrt{360}]$.

(c) Figure 4 shows the graph of function R plotted in the viewing rectangle $[0, 20]$ by $[0, 3000]$. We observe that R has a maximum value on its domain. Using zoom-in on our graphics calculator, we determine the highest point is $(10.95, 2629.07)$. Because x must be an integer (the number of units produced), R has a maximum value when x is 11, and $R(11) = 2629$.

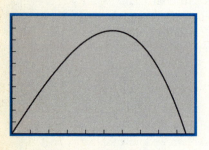

[0, 20] by [0, 3000]
$R(x) = (360 - x^2)x$
FIGURE 4

Conclusion: Eleven units should be produced daily to realize the maximum daily total revenue of $2629.   ◀

 Again from techniques of calculus, the maximum value of the function R in Example 5 occurs when

$$x = \sqrt{120} \approx 10.95$$

EXERCISES 4.4

In each exercise, solve the word problem by finding a function as a mathematical model of the situation. Be sure to write a conclusion.

1. The daily payroll for a work crew is directly proportional to the number of workers, and a crew of 12 workers earns a payroll of $540. **(a)** Express the number of dollars in the daily payroll as a function of the number of workers. **(b)** What is the daily payroll for a crew of 15 workers?

2. For a gas having a constant pressure, its volume is directly proportional to the absolute temperature, and at a temperature of 180° the gas occupies 100 m³. **(a)** Express the number of cubic meters in the volume of the gas as a function of the number of degrees in the absolute temperature. **(b)** What is the volume of the gas at a temperature of 150°?

3. The period (the time for one complete oscillation) of a pendulum is directly proportional to the square root of the length of the pendulum, and a pendulum of length 8 ft has a period of 2 sec. **(a)** Express the number of seconds in the period of a pendulum as a function of the number of feet in its length. **(b)** Find the period of a pendulum of length 2 ft.

4. For a vibrating string, the rate of vibrations is directly proportional to the square root of the tension on the string. **(a)** If a particular string vibrates 864 times per second under a tension of 24 kg, express the number of vibrations per second as a function of the number of kilograms in the tension. **(b)** Find the number of vibrations per second under a tension of 6 kg.

5. The weight of a body is inversely proportional to the square of its distance from the center of the earth. **(a)** If a body weighs 200 lb on the earth's surface, express the number of pounds in its weight as a function of the number of miles from the center of the earth. Assume that the radius of the earth is 4000 miles. **(b)** How much does the body weigh at a distance of 400 miles above the earth's surface?

6. For an electrical cable of fixed length, the resistance is inversely proportional to the square of the diameter of the cable. **(a)** If a cable having a fixed length is $\frac{1}{2}$ cm in diameter and has a resistance of 0.1 ohm, express the number of ohms in the resistance as a function of the number of centimeters in the diameter. **(b)** What is the resistance of a cable having a fixed length and a diameter of $\frac{2}{3}$ cm?

7. In a small town of population 5000 the rate of growth of an epidemic (the rate of change of the number of infected persons) is jointly proportional to the number of people infected and the number of people not infected. **(a)** If the epidemic is growing at the rate of 9 people per day when 100 people are infected, express the rate of growth of the epidemic as a function of the number of infected people. **(b)** How fast is the epidemic growing when 200 people are infected? **(c)** Determine how many people are infected when the rate of growth of the epidemic is a maximum.

8. In a community of 8000 people the rate at which a rumor spreads is jointly proportional to the number of people who have heard the rumor and the number of people who have not heard it. **(a)** If the rumor is spreading at the rate of 20 people per hour when 200 people have heard it, express the rate at which the rumor is spreading as a function of the number of people who have heard it. **(b)** How fast is the rumor spreading when 500 people have heard it? **(c)** How many people have heard the rumor when the rumor is being spread at the greatest rate?

9. A particular lake can support a maximum of 14,000 fish, and the rate of growth of the fish population is jointly proportional to the number of fish present and the difference between 14,000 and the number present. **(a)** If $f(x)$ fish per day is the rate of growth when x fish are present, write an equation defining $f(x)$. **(b)** What is the domain of function f? **(c)** What value of x makes $f(x)$ a maximum?

10. The maximum number of bacteria supportable by a particular environment is 900,000, and the rate of bacterial growth is jointly proportional to the number present and the difference between 900,000 and the number present. **(a)** If $f(x)$ bacteria per minute is the rate of growth when x bacteria are present, write an equation defining $f(x)$. **(b)** What is the domain of function f? **(c)** What value of x makes $f(x)$ a maximum?

11. A manufacturer of open tin boxes wishes to make use of pieces of tin with dimensions 8 in. by 15 in. by cutting equal squares from the four corners and turning up the sides. **(a)** If x inches is the length of the side of the square to be cut out, express the number of cubic inches in the volume of the box as a function of x. **(b)** What is the domain of your function in part (a)? **(c)** On your graphics calculator, find accurate to the nearest tenth of an inch the length of the side of the square to be cut out so that the box has the largest possible volume. What is the maximum volume to the nearest cubic inch?

12. A cardboard box manufacturer makes open boxes from square pieces of cardboard of side 12 cm by cutting equal squares from the four corners and turning up the sides. **(a)** If x centimeters is the length of the side of the square to be cut out, express the number of cubic centimeters in the volume of the box as a function of x. **(b)** What is the domain of your function in part (a)? **(c)** On your graphics calculator find accurate to the nearest centimeter the length of the side of the square to be cut out so that the volume of the box is a maximum. What is the maximum volume to the nearest cubic centimeter?

13. Do Exercise 11 if the manufacturer makes the open boxes from rectangular pieces of tin with dimensions 12 in. by 15 in. In part (c) find the length of the side of the square to be cut out accurate to two decimal places.

14. Do Exercise 12 if the manufacturer makes the open boxes from rectangular pieces of cardboard with dimensions 40 cm by 50 cm. In part (c) find the length of the side of the square to be cut out accurate to two decimal places.

15. Under a monopoly the demand equation for a particular commodity is $p - (8 - \frac{1}{100}x)^2 = 0$, where x units are produced daily when p dollars is the price per unit. **(a)** Express the number of dollars in the daily total revenue as a function of x. **(b)** What is the domain of your function in part (a)? **(c)** On your graphics calculator, find the number of units that should be produced per day to maximize the daily total revenue. What is the maximum daily total revenue?

16. Under a monopoly the demand equation for a particular commodity is $x^2 + p^2 = 36$, where $100x$ units are produced daily when p dollars is the price per unit. **(a)** Express the number of dollars in the daily total revenue as a function of x. **(b)** What is the domain of your function in part (a)? **(c)** On your graphics calculator, find the number of units that should be produced per day to maximize the daily total revenue. What is the maximum daily total revenue?

17. The demand equation for a monopolist is $(100 - x)^2 = 100p$, where x units are demanded daily when p dollars is the price per unit and x is in the closed interval $[0, 34]$. The total cost function is given by $C(x) = 55x - \frac{4}{5}x^2$, where $C(x)$ dollars is the total cost of producing x units and x is in the closed interval $[0, 34]$. **(a)** Express the number of dollars in the daily profit as a function of x. (*Hint:* Profit equals total revenue minus total cost). **(b)** On your graphics calculator find the number of units that should be produced daily to maximize the daily profit. What is the maximum daily profit?

18. A package in the shape of a rectangular box with a square cross section is to have the sum of its length and girth (the perimeter of a cross section) equal to

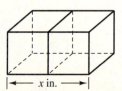

$\longleftarrow x$ in. $\longrightarrow$

100 in. **(a)** If x inches is the length of the package, express the volume of the box as a function of x. **(b)** What is the domain of your function in part (a)? **(c)** On your graphics calculator find accurate to the nearest inch the dimensions of the package having the maximum volume.

19. A Norman window consists of a rectangle surmounted by a semicircle. Suppose a particular Norman window is to have a perimeter of 200 in. Furthermore, assume that the amount of light transmitted by the window is directly proportional to the area of the window. **(a)** If r inches is the radius of the semicircle, express the amount of light transmitted by the window as a function of r. **(b)** What is the domain of your function in part (a)? **(c)** Determine the shape of such a window that will admit the most light.

20. Do Exercise 19 if the window is such that the region bounded by the semicircle transmits only one-half as much light per square inch of area as the region bounded by the rectangle.

21. A page of print is to contain 24 in.² of printed region, a margin of 1.5 in. at the top and bottom, and a margin of 1 in. at the sides. **(a)** If $A(x)$ square inches is the total area of the page when x inches is the width of the printed portion, write an equation defining $A(x)$. **(b)** What is the domain of function A? **(c)** On your graphics calculator, determine accurate to the nearest

one-hundredth of an inch the dimensions of the smallest page that satisfies these requirements.

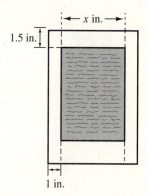

22. A one-story building having a rectangular floor space of 13,200 ft² is to be constructed where a walkway 22 ft wide is required in the front and back and a walkway 15 ft wide is required on each side. **(a)** If $A(x)$ square feet is the total area of the lot on which the building and walkways will be located when x feet is the length of the front and back of the building, write an equation defining $A(x)$. **(b)** What is the domain of function A? **(c)** On your graphics calculator, determine accurate to the nearest one-hundredth of a foot, the dimensions of the lot having the least area on which this building can be located.

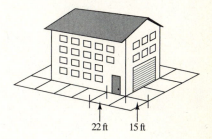

4.5 COMPOSITE FUNCTIONS

GOALS
1. **Define a composite function of two functions.**
2. **Compute composite function values.**
3. **Determine the domain of the composition of two functions.**
4. **Express a given function as the composition of two functions.**
5. **Solve word problems having a composite function as a mathematical model.**

In Section 4.1 we defined four operations on two functions: the sum, difference, product, and quotient. Obtaining the *composite function* of two given functions is another operation.

DEFINITION   **Composite Function**

Given the two functions f and g, the **composite function,** denoted by $f \circ g$, is defined by

$$(f \circ g)(x) = f(g(x))$$

and the domain of $f \circ g$ is the set of all numbers x in the domain of g such that $g(x)$ is in the domain of f.

The definition indicates that when computing $(f \circ g)(x)$, we first apply function g to x and then function f to $g(x)$. To visualize this computation see Figure 1. The function g assigns the value $g(x)$ to the number x in the domain of g. Then the function f assigns the value $f(g(x))$ to the number $g(x)$ in the domain of f. Observe in Figure 1 that the range of g is a subset of the domain of f and the range of $f \circ g$ is a subset of the range of f.

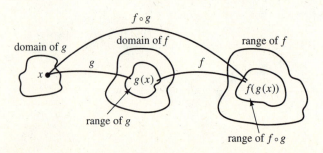

FIGURE 1

▷ **ILLUSTRATION 1**

If f and g are defined by

$$f(x) = \sqrt{x} \quad \text{and} \quad g(x) = 2x - 3$$

$$(f \circ g)(x) = f(g(x))$$
$$= f(2x - 3)$$
$$= \sqrt{2x - 3}$$

The domain of g is $(-\infty, +\infty)$, and the domain of f is $[0, +\infty)$. Therefore the domain of $f \circ g$ is the set of real numbers for which $2x - 3 \geq 0$ or, equivalently, $[\frac{3}{2}, +\infty)$.   ◀

▶ **EXAMPLE 1**   *Computing a Composite Function Value by Two Methods*

Given

$$f(x) = \frac{5}{x - 2} \qquad g(x) = 2x + 1$$

(a) Find $g(3)$ and use that number to find $f(g(3))$. **(b)** Compute $(f \circ g)(x)$ and use that value to find $(f \circ g)(3)$.

Solution

(a) $g(3) = 2(3) + 1$
$$= 7$$

Thus
$$f(g(3)) = f(7)$$
$$= \frac{5}{7 - 2}$$
$$= 1$$

(b) $(f \circ g)(x) = f(g(x))$
$$= f(2x + 1)$$
$$= \frac{5}{(2x + 1) - 2}$$
$$= \frac{5}{2x - 1}$$

Therefore
$$(f \circ g)(3) = \frac{5}{2(3) - 1}$$
$$= 1 \qquad ◀$$

▶ **EXAMPLE 2**   *Computing Composite Function Values and Determining Their Domains*

Given that f and g are defined by

$$f(x) = \sqrt{x} \quad \text{and} \quad g(x) = x^2 - 1$$

find: **(a)** $f \circ f$; **(b)** $g \circ g$; **(c)** $f \circ g$; **(d)** $g \circ f$. Also determine the domain of each composite function.

Solution   The domain of f is $[0, +\infty)$, and the domain of g is $(-\infty, +\infty)$.

(a) $(f \circ f)(x) = f(f(x))$

$= f(\sqrt{x})$

$= \sqrt{\sqrt{x}}$

$= \sqrt[4]{x}$

The domain is $[0, +\infty)$.

(b) $(g \circ g)(x) = g(g(x))$

$= g(x^2 - 1)$

$= (x^2 - 1)^2 - 1$

$= x^4 - 2x^2$

The domain is $(-\infty, +\infty)$.

(c) $(f \circ g)(x) = f(g(x))$

$= f(x^2 - 1)$

$= \sqrt{x^2 - 1}$

The domain is

$(-\infty, -1] \cup [1, +\infty)$.

(d) $(g \circ f)(x) = g(f(x))$

$= g(\sqrt{x})$

$= (\sqrt{x})^2 - 1$

$= x - 1$

The domain is $[0, +\infty)$.

In part (d) note that even though $x - 1$ is defined for all values of x, the domain of $g \circ f$, by the definition of a composite function, is the set of all numbers x in the domain of f such that $f(x)$ is in the domain of g. Thus the domain of $g \circ f$ must be a subset of the domain of f.   ◀

Observe from the results of parts (c) and (d) of the preceding example that $(f \circ g)(x)$ and $(g \circ f)(x)$ are not necessarily equal.

 An important theorem in calculus, called the *chain rule*, involves composite functions. When applying the chain rule it is necessary to think of a function as the composition of two other functions, as shown in the following illustration.

▷ **ILLUSTRATION 2**

If $h(x) = (4x^2 + 1)^3$, we can express h as the composition of the two functions f and g for which

$$f(x) = x^3 \qquad g(x) = 4x^2 + 1$$

because

$$(f \circ g)(x) = f(g(x))$$
$$= f(4x^2 + 1)$$
$$= (4x^2 + 1)^3$$
◀

The function h in Illustration 2 can be expressed as the composition of other pairs of functions. For example, if

$$F(x) = (4x + 1)^3 \qquad G(x) = x^2$$

then

$$(F \circ G)(x) = F(G(x))$$
$$= F(x^2)$$
$$= (4x^2 + 1)^3$$

 In calculus, we would use the composition in Illustration 2.

 ▶ **EXAMPLE 3** *Expressing a Given Function as the Composition of Two Other Functions*

Given

$$h(x) = \frac{1}{\sqrt{x^2 + 3}}$$

express h as the composition of two functions f and g in two ways: **(a)** the function f contains the radical; **(b)** the function g contains the radical.

Solution

(a) $f(x) = \dfrac{1}{\sqrt{x + 3}}$ **(b)** $f(x) = \dfrac{1}{x}$

$g(x) = x^2$ $g(x) = \sqrt{x^2 + 3}$

Then   Then

$$(f \circ g)(x) = f(g(x))$$ $$(f \circ g)(x) = f(g(x))$$
$$= f(x^2)$$ $$= f(\sqrt{x^2 + 3})$$
$$= \frac{1}{\sqrt{x^2 + 3}}$$ $$= \frac{1}{\sqrt{x^2 + 3}} \quad ◀$$

▶ **EXAMPLE 4** *Solving a Word Problem Having a Composite Function as a Mathematical Model*

In a forest a predator feeds on prey, and the predator population is a function f of x, the number of prey in the forest, which in turn is a function g of t, the number of weeks that have elapsed since the end of the hunting season. If

$$f(x) = \frac{1}{4}x^2 + 90 \quad \text{and} \quad g(t) = 7t + 86$$

do the following: **(a)** Express the predator population as a function of t; **(b)** find the predator population 8 weeks after the close of the hunting season.

Solution

(a) The predator population t weeks after the close of the hunting season is given by $(f \circ g)(t)$.

$$(f \circ g)(t) = f(g(t))$$
$$= f(7t + 86)$$
$$= \frac{1}{4}(7t + 86)^2 + 90$$

(b) When $t = 8$, we have

$$(f \circ g)(8) = \frac{1}{4}(7 \cdot 8 + 86)^2 + 90$$

$$= 5131$$

Conclusion: Eight weeks after the close of the hunting season the predator population is 5131.   ◀

EXERCISES   4.5

In Exercises 1 through 4, for the functions f and g and the number c, compute $(f \circ g)(c)$ by two methods: (a) Find g(c) and use that number to find f(g(c)); (b) Compute $(f \circ g)(x)$ and use that value to find $(f \circ g)(c)$.

1. $f(x) = 3x^2 - 4x; g(x) = 2x - 5; c = 4$

2. $f(x) = \sqrt{x^2 - 36}; g(x) = x^2 - 3x; c = 5$

3. $f(x) = \dfrac{1}{x - 1}; g(x) = \dfrac{2}{x^2 + 1}; c = \dfrac{1}{2}$

4. $f(x) = \dfrac{2\sqrt{x + 3}}{x}; g(x) = \dfrac{2x + 5}{x^4}; c = -2$

In Exercises 5 through 14, define the following functions and determine the domain of the composite function: (a) $f \circ g$; (b) $g \circ f$.

5. $f(x) = x - 2; g(x) = x + 7$

6. $f(x) = 3 - 2x; g(x) = 6 - 3x$

7. $f(x) = x - 5; g(x) = x^2 - 1$

8. $f(x) = \sqrt{x}; g(x) = x^2 + 1$

9. $f(x) = \sqrt{x - 2}; g(x) = x^2 - 2$

10. $f(x) = x^2 - 1; g(x) = \dfrac{1}{x}$

11. $f(x) = \dfrac{1}{x}; g(x) = \sqrt{x}$

12. $f(x) = \sqrt{x}; g(x) = -\dfrac{1}{x}$

13. $f(x) = |x|; g(x) = |x + 2|$

14. $f(x) = \sqrt{x^2 - 1}; g(x) = \sqrt{x - 1}$

In Exercises 15 through 24, define the following functions and determine the domain of the composite function: (a) $f \circ f$; (b) $g \circ g$.

15. The functions of Exercise 5

16. The functions of Exercise 6

17. The functions of Exercise 7

18. The functions of Exercise 8

19. The functions of Exercise 9

20. The functions of Exercise 10

21. The functions of Exercise 11

22. The functions of Exercise 12

23. The functions of Exercise 13

24. The functions of Exercise 14

 In Exercises 25 through 30, express h as the composition of two functions f and g in two ways.

25. $h(x) = \sqrt{x^2 - 4}$ **26.** $h(x) = (9 + x^2)^{-2}$

27. $h(x) = \left(\dfrac{1}{x - 2}\right)^3$ **28.** $h(x) = \dfrac{4}{\sqrt[3]{x^3 + 3}}$

29. $h(x) = (x^2 + 4x - 5)^4$ **30.** $h(x) = \sqrt{|x| + 4}$

In Exercises 31 through 34, solve the word problem by finding a composite function as a mathematical model of the situation. Be sure to write a conclusion.

31. The surface area of a sphere is a function of its radius. If r centimeters is the radius of a sphere, and $A(r)$ square centimeters is the surface area, then $A(r) = 4\pi r^2$. Suppose a balloon maintains the shape of a sphere as it is being inflated so that the radius is changing at a constant rate of 3 centimeters per second. If $f(t)$ centimeters is the radius of the balloon after t seconds, do the following: **(a)** Compute $(A \circ f)(t)$ and interpret your result. **(b)** Find the surface area of the balloon after 4 seconds.

32. In a lake a large fish feeds on a medium-size fish, and the population of the large fish is a function f of x, the number of medium-size fish in the lake. In turn the medium-size fish feed on small fish, and the population of the medium-size fish is a function of w, the number of small fish in the lake.
If

$$f(x) = \sqrt{2x} + 1500 \quad \text{and} \quad g(w) = \sqrt{w} + 5000$$

do the following: **(a)** Express the population of the large fish as a function of w; **(b)** find the number of large fish in the lake when there are 9 million small fish in the lake.

33. The consumer demand for a particular toy in a certain marketplace is a function f of p, the number of dollars in its price, which in turn is a function g of t, the number of months since the toy reached the market-place. If

$$f(p) = \frac{5000}{p^2} \quad \text{and} \quad g(t) = \frac{1}{20}t^2 + \frac{7}{20}t + 5$$

do the following: **(a)** Express the consumer demand as a function of t; **(b)** find the consumer demand 5 months after the toy reached the marketplace.

34. The volume of a sphere is a function of its radius. If r feet is the radius of a sphere and $V(r)$ cubic feet is the volume, then $V(r) = \frac{4}{3}\pi r^3$. Suppose a snowball, spherical in shape with a radius of 2 ft started to melt so that its radius is changing at a constant rate of 4.5 inches per minute. If $f(t)$ feet is the radius of the snowball after t minutes, do the following:
(a) Compute $(V \circ f)(t)$ and interpret your result.
(b) Find the volume of the snowball after 3 minutes.

35. Is the composition of two functions commutative; that is, if f and g are any two functions, are $(f \circ g)(x)$ and $(g \circ f)(x)$ equal? Justify your answer by giving an example.

If f and g are two functions such that $(f \circ g)(x) = x$ and $(g \circ f)(x) = x$, then f and g are inverse functions. In Exercises 36 through 40, show that f and g are inverse functions.

36. $f(x) = 2x - 3$ and $g(x) = \dfrac{x + 3}{2}$

37. $f(x) = \dfrac{1}{x + 1}$ and $g(x) = \dfrac{1 - x}{x}$

38. $f(x) = x^2, x \geq 0$, and $g(x) = \sqrt{x}$

39. $f(x) = x^2, x \leq 0$, and $g(x) = -\sqrt{x}$

40. $f(x) = (x - 1)^3$ and $g(x) = 1 + \sqrt[3]{x}$

41. Determine whether the composite function $f \circ g$ is odd or even if **(a)** f and g are both even, and **(b)** f and g are both odd.

42. Determine whether the composite function $f \circ g$ is odd or even if **(a)** f is even and g is odd, and **(b)** f is odd and g is even.

43. The unit step function U and the signum function were defined in Exercises 51 and 53, respectively, of Exercises 4.2. Find formulas for $sgn(U(x))$ and sketch the graph.

44. Find formulas for $(f \circ g)(x)$ if

$$f(x) = \begin{cases} 0 & \text{if } x < 0 \\ 2x & \text{if } 0 \le x \le 1 \\ 0 & \text{if } 1 < x \end{cases}$$

and

$$g(x) = \begin{cases} 1 & \text{if } x < 0 \\ \frac{1}{2}x & \text{if } 0 \le x \le 1 \\ 1 & \text{if } 1 < x \end{cases}$$

Sketch the graphs of f, g, and $f \circ g$.

45. Find formulas for $(g \circ f)(x)$ for the functions of Exercise 44. Sketch the graph of $g \circ f$.

46. If $f(x) = x^2 + 2x + 2$, find two functions g for which $(f \circ g)(x) = x^2 - 4x + 5$.

47. If $f(x) = x^2$, find two functions g for which $(f \circ g)(x) = 4x^2 - 12x + 9$.

48. Prove that if f and g are both linear functions, then $f \circ g$ is a linear function.

49. Let $f(x) = \dfrac{1}{x}$, $g(x) = \dfrac{1}{x}$, and $h(x) = x$. Explain why $(f \circ g)(x) = (g \circ f)(x)$ but neither $f \circ g$ nor $g \circ f$ is the same as h.

CHAPTER 4 REVIEW

▶ LOOKING BACK

4.1 Before stating the formal definition of a function, we treated a function as a correspondence between sets of real numbers. After computing some function values, we stressed evaluating difference quotients, which are important in calculus. We then showed how new functions are formed by operating on given functions.

4.2 The formal definition of a function as a set of ordered pairs of real numbers led naturally to the definition of the graph of a function. We sketched graphs of functions and also plotted some of them. We introduced piecewise-defined functions and the greatest integer function as preparation for their use in calculus, and then defined various types of algebraic functions as well as even and odd functions.

4.3 Graphs of quadratic functions, which are parabolas, were sketched and plotted. We found exact values of the zeros of a quadratic function algebraically and checked them by plotting the corresponding parabola. We showed how to find the extreme value of a quadratic function and solved word problems having a quadratic function as a mathematical model.

4.4 We solved word problems involving direct variation, inverse variation, and joint variation by using a function as a mathematical model. On a graphics calculator, we obtained approximate solutions to word problems having a cubic function as a mathematical model.

4.5 We defined a composite function, computed composite function values, and showed how to express a given function as the composition of two functions, a procedure necessary in calculus. We also solved word problems having a composite function as a mathematical model.

▶ REVIEW EXERCISES

1. Given $f(x) = 4 - x^2$, find:
(a) $f(1)$　(b) $f(-2)$　(c) $f(3)$
(d) $f(x - 1)$　(e) $f(x^2)$

2. Given $f(x) = \dfrac{3}{x - 1}$, find:
(a) $f(-2)$　(b) $f(2)$　(c) $f(4)$
(d) $f\left(\dfrac{3}{x} + 1\right)$　(e) $f\left(\dfrac{3}{x + 1}\right)$

3. Given $g(x) = \sqrt{3x + 1}$, find:
(a) $g(0)$　(b) $g(1)$　(c) $g(5)$
(d) $g(8)$　(e) $g(16)$　(f) $g(3x + 2)$

4. Given $h(x) = \dfrac{|x + 3|}{|x| + 3}$, find:
(a) $h(1)$　(b) $h(-1)$　(c) $h(3)$
(d) $h(-3)$　(e) $h(x^2)$　(f) $h(\sqrt{x})$

5. Given $f(x) = \dfrac{4}{x}$, find:

   (a) $x^2 f(x) - f\left(\dfrac{1}{x}\right)$ **(b)** $f(x) - xf(x^2)$

   (c) $f(|x|) - |f(x)|$ **(d)** $f\left(\dfrac{4}{x}\right) - \dfrac{4}{f(x)}$.

6. Given $g(x) = \dfrac{x}{x - 1}$, find:

   (a) $g\left(\dfrac{1}{x}\right) + \dfrac{g(x)}{x}$ **(b)** $\dfrac{1}{g(x) - 1} - \dfrac{x}{g(x)}$.

In Exercises 7 through 12, compute and simplify the difference quotient $\dfrac{f(x + h) - f(x)}{h}$, $h \neq 0$.

7. The function of Exercise 1

8. The function of Exercise 2

9. The function of Exercise 5

10. $f(x) = \sqrt{1 - x}$

11. $f(x) = \dfrac{1}{\sqrt{x - 2}}$

12. $f(x) = \dfrac{x + 3}{x - 4}$

In Exercises 13 through 18, define the following functions and determine the domain of the resulting function:
(a) $f + g$; (b) $f - g$; (c) $f \cdot g$; (d) f/g; (e) g/f.

13. $f(x) = x^2 - 4$; $g(x) = 4x - 3$

14. $f(x) = \sqrt{x}$; $g(x) = x^2 - 1$

15. $f(x) = \sqrt{x + 2}$; $g(x) = x^2 - 4$

16. $f(x) = x^2 - 9$; $g(x) = \sqrt{x + 5}$

17. $f(x) = \dfrac{1}{x^2}$; $g(x) = \sqrt{x}$

18. $f(x) = \dfrac{x}{x - 1}$; $g(x) = \dfrac{1}{x + 2}$

In Exercises 19 through 24, define the following functions and determine the domain of the resulting function:
(a) $f \circ g$; (b) $g \circ f$.

19. The functions of Exercise 13

20. The functions of Exercise 14

21. The functions of Exercise 15

22. The functions of Exercise 16

23. The functions of Exercise 17

24. The functions of Exercise 18

In Exercises 25 through 34, plot the graph of the function and determine its domain and range.

25. $f(x) = 4 - 2x$ **26.** $g(x) = 3x + 2$

27. $g(x) = x^2 - 4$ **28.** $f(x) = 9 - x^2$

29. $h(x) = \sqrt{x^2 - 16}$ **30.** $H(x) = \sqrt{1 - x^2}$

31. $F(x) = \sqrt{16 - x^2}$ **32.** $G(x) = \sqrt{x^2 - 1}$

33. $f(x) = |5 - x|$ **34.** $g(x) = |x + 4|$

In Exercises 35 through 44, sketch the graph of the function and determine its domain and range.

35. $g(x) = \dfrac{x^2 - 16}{x + 4}$ **36.** $f(x) = \dfrac{x^2 + x - 6}{x - 2}$

37. $G(x) = \begin{cases} x - 4 & \text{if } x \neq -4 \\ 3 & \text{if } x = -4 \end{cases}$

38. $F(x) = \begin{cases} x + 3 & \text{if } x \neq 2 \\ 1 & \text{if } x = 2 \end{cases}$

39. $F(x) = \begin{cases} 3 - x & \text{if } x < 0 \\ 3 + 2x & \text{if } 0 \leq x \end{cases}$

40. $G(x) = \begin{cases} 3x + 2 & \text{if } x \leq 0 \\ 4 - 2x & \text{if } 0 < x \end{cases}$

41. $h(x) = \begin{cases} x^2 - 1 & \text{if } x \leq 0 \\ x - 1 & \text{if } 0 < x \end{cases}$

42. $H(x) = \begin{cases} x^2 & \text{if } x < -1 \\ (x + 2)^2 & \text{if } -1 \leq x \end{cases}$

43. $f(x) = \begin{cases} x + 4 & \text{if } x < -4 \\ \sqrt{16 - x^2} & \text{if } -4 \leq x \leq 4 \\ 4 - x & \text{if } 4 < x \end{cases}$

44. $g(x) = \begin{cases} x^2 - 4 & \text{if } x \leq 2 \\ 4 - x & \text{if } 2 < x \leq 4 \\ x - 8 & \text{if } 4 < x \end{cases}$

In Exercises 45 through 48, sketch the graph of the function and determine its domain and range. Check your graph by plotting it.

45. $f(x) = [\![x - 2]\!]$ **46.** $g(x) = [\![x + 1]\!] - x$

47. $h(x) = x - [\![x]\!]$

48. $F(x) = \dfrac{1 + (-1)^n}{2}$ where $n = [\![x]\!]$

In Exercises 49 and 50, U and sgn are the unit step function and the signum function defined in Exercises 51 and 53, respectively, of Exercises 4.2. Find formulas for the given function and sketch its graph.

49. $f(x) = \text{sgn } x - U(x)$ **50.** $g(x) = \text{sgn } x \, U(x + 1)$

In Exercises 51 through 54, plot the graph of the function and determine from the graph which of the following statements characterizes the zeros of the function: (a) two real zeros; (b) one real zero of multiplicity 2; or (c) two imaginary zeros.

51. $f(x) = x^2 - 6x$ **52.** $f(x) = 2x^2 + 8x + 11$

53. $g(x) = -x^2 + 8x - 16$

54. $g(x) = -\frac{1}{2}x^2 + 2x + 2$

In Exercises 55 through 58, find either a maximum or a minimum value of the function.

55. $f(x) = 4 - 4x - 2x^2$ **56.** $g(x) = 3x^2 + 6x + 7$

57. $g(x) = \frac{1}{4}(x^2 + 2x - 11)$

58. $f(x) = \frac{1}{6}(15 + 6x - x^2)$

In Exercises 59 and 60, for the functions f and g and the number c, compute $(f \circ g)(c)$ by two methods: (a) Find $g(c)$ and use that number to find $f(g(c))$; (b) compute $(f \circ g)(x)$ and use that value to find $(f \circ g)(c)$.

59. $f(x) = \sqrt{100 - x^2}; g(x) = x^2 + x; c = -3$

60. $f(x) = \dfrac{x^4}{x^2 - 2}; g(x) = \sqrt{x - 2}; c = 6$

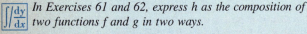

In Exercises 61 and 62, express h as the composition of two functions f and g in two ways.

61. $h(x) = (3x^2 - 2)^{1/3}$

62. $h(x) = \sqrt{9x^2 + 6x + 4}$

In Exercises 63 and 64, determine if the function is even, odd, or neither.

63. **(a)** $f(x) = 2x^3 - 3x$ **(b)** $g(x) = 5x^4 + 2x^2 - 1$
(c) $h(x) = 3x^5 - 2x^3 + x^2 - x$
(d) $F(x) = \dfrac{x^2 + 1}{x^3 - x}$

64. **(a)** $f(x) = \dfrac{x^3 - 2x}{x^2 - 1}$ **(b)** $g(x) = \dfrac{x}{|x|}$
(c) $h(x) = \dfrac{\sqrt[3]{x}}{x}$ **(d)** $F(x) = x^2 \cdot \text{sgn } x$

In Exercises 65 through 82, solve the word problem by finding a function as a mathematical model of the situation. Be sure to write a conclusion.

65. The distance a body falls from rest is directly proportional to the square of the time it has been falling, and a body falls 64 ft in 2 sec. **(a)** Express the number of feet in the distance a body falls from rest as a function of the number of seconds it has been falling. **(b)** How far will a body fall from rest in $\frac{5}{2}$ sec?

66. Boyle's law states that at a constant temperature the volume of a gas is inversely proportional to the pressure of the gas, and a gas occupies 100 m³ at a pressure of 24 kg/cm². **(a)** Express the number of cubic meters occupied by a gas as a function of the number of kilograms per square centimeter in its pressure. **(b)** What is the volume of a gas when its pressure is 16 kg/cm²?

67. If a pond can support a maximum of 10,000 fish, the rate of growth of the fish population is jointly proportional to the number of fish present and the difference between 10,000 and the number present. **(a)** If the rate of growth is 90 fish per week when 1000 fish are present, express the rate of population growth as a function of the number present. **(b)** Find the rate of population growth when 2000 fish are present.

68. Find two positive numbers whose sum is 12 such that their product is a maximum.

69. Find two positive numbers whose sum is 12 such that the sum of their squares is a minimum.

70. Show that among all the rectangles having a perimeter of 36 in., the square of side 9 in. has the greatest area.

71. A school-sponsored trip that can accommodate up to 250 students will cost each student $15 if not more than 150 students make the trip; however the cost per student will be reduced 5 cents for each student in excess of 150 until the cost reaches $10 per student. **(a)** If x students make the trip, express the number of dollars in the gross income as a function of x. **(b)** What is the domain of your function in part (a)? **(c)** How many students should make the trip for the school to receive the largest gross revenue?

72. A wholesaler offers to deliver to a dealer 300 chairs at $90 per chair and to reduce the price per chair on the entire order by 25 cents for each additional chair over 300. **(a)** If x chairs are ordered, express the number of dollars in the dealer's cost as a function of x. **(b)** What is the domain of your function in part (a)? **(c)** Find the dollar total involved in the largest possible transaction between the wholesaler and the dealer under these circumstances.

73. In a town of population 11,000 the growth rate of an epidemic is jointly proportional to the number of people infected and the number of people not infected. **(a)** If the epidemic is growing at the rate of $f(x)$ people per day when x people are infected, write an equation defining $f(x)$. **(b)** What is the domain of the function f in part (a)? **(c)** Determine the number of people infected when the epidemic is growing at a maximum rate.

74. A carpenter can sell all the end tables that are made at a price of $64 per table. If x tables are built and sold each week, then the number of dollars in the total cost of the week's production is $x^2 + 15x + 225$. **(a)** Express the number of dollars in the carpenter's weekly profit as a function of x. **(b)** How many tables should be constructed each week in order for the carpenter to have the greatest weekly profit? **(c)** What is the greatest weekly profit?

75. In a lake a predator fish feeds on a smaller fish, and the predator population at any time is a function f of x, the number of small fish in the lake, which in turn is a function g of t, the number of weeks that have elapsed since the end of the fishing season. If

$$f(x) = \tfrac{1}{4}x^2 + 80 \qquad \text{and} \qquad g(t) = 8t + 90$$

do the following: **(a)** Express the population of the predator fish as a function of t; **(b)** find the population of the predator fish 9 weeks after the close of the fishing season.

76. Under a monopoly the demand equation for a particular article of merchandise is $x^2 + p = 324$, where x units are produced daily when p dollars is the price per unit. **(a)** Express the number of dollars in the daily total revenue as a function of x. **(b)** What is the domain of your function in part (a)? **(c)** On your graphics calculator, find the number of units that should be produced daily to maximize the daily total revenue. What is the maximum daily total revenue?

77. Under the monopoly of Exercise 76, suppose it costs $80 to produce each unit so that the total cost function is C where $C(x) = 80x$. **(a)** Express the number of dollars in the daily profit as a function of x. **(b)** On your graphics calculator, find the number of units that should be produced per day to maximize the daily profit. What is the maximum daily profit?

78. An open metal pan is to be made by cutting out squares of the same size from the corners of a rectangular piece of metal 14 in. by 18 in. and turning up the sides. **(a)** If x inches is the length of the side of the square to be cut out, express the number of cubic inches in the volume of the pan as a function of x. **(b)** What is the domain of your function in part (a)? **(c)** On your graphics calculator, find accurate to the nearest one-hundredth of an inch the length of the side of the square to be cut out so that the volume of the box is a maximum. What is the maximum volume?

79. To form an open box, equal squares are cut from each corner of a square piece of tin having an area of 400 cm². **(a)** If x centimeters is the length of the side of the square cut out, express the number of cubic centimeters in the volume of the box as a function of x. **(b)** What is the domain of your function in part (a)? **(c)** On your graphics calculator, find to the nearest one-hundredth of a centimeter the length of the side of the square cut out so that the volume of the box is a maximum. What is the maximum volume?

80. A sign containing 50 m² of printed material is required to have margins of 4 m at the top and bottom and 2 m on each side. **(a)** If $A(x)$ square meters is the total area of the sign when x meters is the horizontal dimension of the region covered by the printed material, write an equation defining $A(x)$. **(b)** What is the domain of function A? **(c)** On your graphics calculator, determine to the nearest meter, the dimensions of the smallest sign that will meet these specifications.

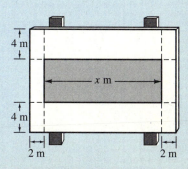

81. An open box having a square base is to have a volume of 4000 in.^{3} **(a)** If $S(x)$ square inches is the total surface area of the box when x inches is the length of a side of the square base, write an equation defining $S(x)$. **(b)** What is the domain of function S? **(c)** On your graphics calculator, determine to the nearest inch, the dimensions of the box that can be constructed with the least amount of material.

82. Do Exercise 81 if the box is to be closed.

83. Prove that any function, whose domain is the set of all real numbers, can be expressed as the sum of an even function and an odd function by writing

$$f(x) = \tfrac{1}{2}[f(x) + f(-x)] + \tfrac{1}{2}[f(x) - f(-x)]$$

and showing that the function having function values $\tfrac{1}{2}[f(x) + f(-x)]$ is an even function and the function having function values $\tfrac{1}{2}[f(x) - f(-x)]$ is an odd function.

84. The function g is defined by $g(x) = x^2$. Define a function f such that $(f \circ g)(x) = x$ if
(a) $x \geq 0$ **(b)** $x < 0$

85. Find formulas for $(f \circ g)(x)$ if

$$f(x) = \begin{cases} 0 & \text{if } x < 0 \\ x & \text{if } 0 \leq x \leq 1 \\ 0 & \text{if } 1 < x \end{cases}$$

and

$$g(x) = \begin{cases} 1 & \text{if } x < 0 \\ 2x & \text{if } 0 \leq x \leq 1 \\ 1 & \text{if } 1 < x \end{cases}$$

Sketch the graphs of f, g, and $f \circ g$.

86. Find formulas for $(g \circ f)(x)$ for the functions of Exercise 85.

87. Sketch the graph of $g \circ f$ for the functions of Exercise 85.

Polynomial and Rational Functions

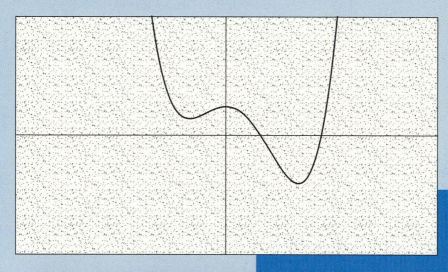

Our treatment of polynomial and rational functions is as complete as possible without using calculus. To obtain graphs of polynomial functions we will stress geometric transformations. Our coverage of graphs of rational functions will emphasize the important role asymptotes play in obtaining these graphs. We will define asymptotes only after an intuitive treatment of limits involving infinity, a concept studied more thoroughly in calculus. Another important objective of this chapter is to learn methods for finding zeros of polynomial functions both algebraically and graphically. The tools needed include the factor theorem and synthetic substitution.

5.1 GRAPHS OF POLYNOMIAL FUNCTIONS

GOALS

1. Learn the vertical translation rule.
2. Learn the horizontal translation rule.
3. Learn the vertical stretching and shrinking rule.
4. Learn the x axis reflection rule.
5. Sketch graphs of polynomial functions by geometric transformations.
6. Describe the end behavior of graphs of polynomial functions.
7. Determine the most relative extrema of a polynomial function.
8. Estimate the relative extrema of a polynomial function.
9. Plot graphs of polynomial functions.

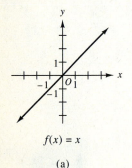

$f(x) = x$

(a)

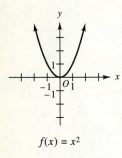

$f(x) = x^2$

(b)

In Sections 4.3 and 4.4, we showed how polynomial functions occur as mathematical models in word problems. In this section, we discuss graphs of polynomial functions.

You have already learned that the graph of a linear function is a line and the graph of a quadratic function is a parabola. Consider now the **power function** defined by

$$f(x) = x^n \qquad n \text{ is a positive integer}$$

Figure 1 shows graphs of this function for n having positive-integer values 1 through 6. All of these graphs contain the origin, which is the only intersection of the curve with either axis. If $n > 1$, the x axis is tangent to the graph at the origin. If n is a positive even integer, the graph is in the first and second quadrants and is symmetric with respect to the y axis. If n is a positive odd integer, the graph is in the first and third quadrants and is

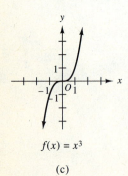

$f(x) = x^3$

(c)

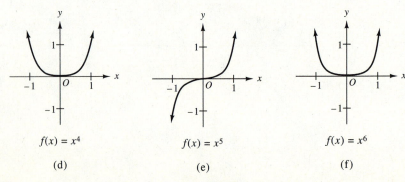

$f(x) = x^4$

(d)

$f(x) = x^5$

(e)

$f(x) = x^6$

(f)

FIGURE 1

symmetric with respect to the origin. As $|x|$ increases without bound, so does $|f(x)|$.

We now show how the graph of a function defined by an equation of the form

$$F(x) = x^n + k$$

can be obtained from the graph of $f(x) = x^n$. The procedure is based on the following *vertical translation rule* that follows from the equations for translating the axes in Section 3.5.

Vertical Translation Rule

The graph of a function defined by an equation of the form

$$F(x) = f(x) + k$$

can be obtained from the graph of the function f by a **vertical translation** (or shift) of k units in the upward direction if $k > 0$ and $|k|$ units in the downward direction if $k < 0$.

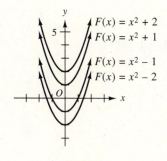

FIGURE 2

▷ **ILLUSTRATION 1**

Figure 2 shows the graphs of the functions

$$F(x) = x^2 + k$$

where k takes on the values -2, -1, 1, and 2. Observe that the graphs are obtained from the graph of $f(x) = x^2$ [Figure 1(b)] by vertical translations as follows:

$F(x) = x^2 - 2$	2 units in the downward direction
$F(x) = x^2 - 1$	1 unit in the downward direction
$F(x) = x^2 + 1$	1 unit in the upward direction
$F(x) = x^2 + 2$	2 units in the upward direction

Note that the value of k is the y intercept of the graph.   ◀

▶ **EXAMPLE 1** *Using the Vertical Translation Rule*

Sketch the graphs of each of the following functions by a vertical translation of the graph of $f(x) = x^3$:

(a) $F(x) = x^3 + 2$ **(b)** $G(x) = x^3 - 3$

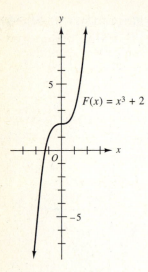

FIGURE 3

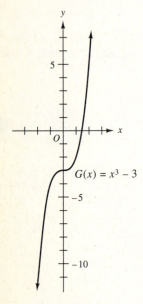

FIGURE 4

Solution

(a) The graph of $F(x) = x^3 + 2$, appearing in Figure 3, is obtained from the graph of $f(x) = x^3$ (Figure 1(c)) by a vertical translation of 2 units upward.

(b) Figure 4 shows the graph of $G(x) = x^3 - 3$, obtained from the graph of $f(x) = x^3$ by a vertical translation of 3 units downward.   ◄

The *horizontal translation rule*, which also follows from the equations for translating the axes in Section 3.5, enables us to obtain the graph of a function defined by an equation of the form

$$F(x) = (x - h)^n$$

from the graph of $f(x) = x^n$.

Horizontal Translation Rule

The graph of a function defined by an equation of the form

$$F(x) = f(x - h)$$

can be obtained from the graph of the function f by a **horizontal translation** (or shift) of h units to the right if $h > 0$ and $|h|$ units to the left if $h < 0$.

▷ **ILLUSTRATION 2**

Figure 5 shows graphs of the functions defined by

$$F(x) = (x - h)^3$$

for h having values -2, -1, 1, and 2. These graphs are obtained from the graph of $f(x) = x^3$ by horizontal translations as follows:

$$F(x) = (x + 2)^3 \qquad \text{2 units to the left}$$
$$F(x) = (x + 1)^3 \qquad \text{1 unit to the left}$$
$$F(x) = (x - 1)^3 \qquad \text{1 unit to the right}$$
$$F(x) = (x - 2)^3 \qquad \text{2 units to the right}$$

Observe that the value of h is the x intercept of the graph.   ◄

▶ **EXAMPLE 2** *Using the Horizontal Translation Rule*

Sketch the graphs of each of the following functions by a horizontal translation of the graph of $f(x) = x^4$:

(a) $F(x) = (x + 1)^4$ (b) $G(x) = (x - 2)^4$

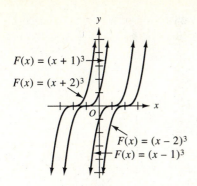

$F(x) = (x + 1)^3$

$F(x) = (x + 2)^3$

$F(x) = (x - 2)^3$

$F(x) = (x - 1)^3$

FIGURE 5

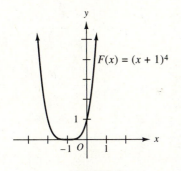

$F(x) = (x + 1)^4$

FIGURE 6

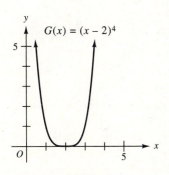

$G(x) = (x - 2)^4$

FIGURE 7

Solution

(a) The graph of $F(x) = (x + 1)^4$ is obtained from the graph of $f(x) = x^4$ (Figure 1(d)) by a horizontal translation of 1 unit to the left. The graph appears in Figure 6.

(b) Figure 7 shows the graph of $G(x) = (x - 2)^4$, obtained from the graph of $f(x) = x^4$ by a horizontal translation of 2 units to the right. ◀

To obtain the graph of a function defined by an equation of the form

$$F(x) = (x - h)^n + k$$

from the graph of $f(x) = x^n$, we utilize both a vertical and a horizontal translation as shown in the following example.

▶ **EXAMPLE 3** *Sketching the Graph of a Polynomial Function by a Vertical and Horizontal Translation*

Sketch the graph of the following function by a vertical and horizontal translation of the graph of $f(x) = x^3$:

$$F(x) = (x - 4)^3 - 2$$

Check the graph by plotting the graphs of both functions in the same viewing rectangle.

Solution
The graph of $f(x) = x^3$ appears in Figure 1(c). By a vertical translation of 2 units downward and a horizontal translation of 4 units to the right, we obtain the required graph shown in Figure 8.

Figure 9 shows the graphs of both functions plotted in the $[-10, 10]$ by $[-10, 10]$ viewing rectangle.

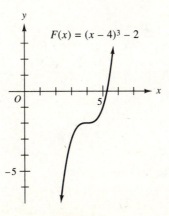

$F(x) = (x - 4)^3 - 2$

FIGURE 8

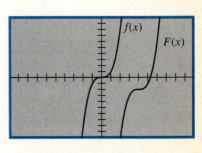

$[-10, 10]$ by $[-10, 10]$
$f(x) = x^3$ and $F(x) = (x - 4)^3 - 2$
FIGURE 9 ◀

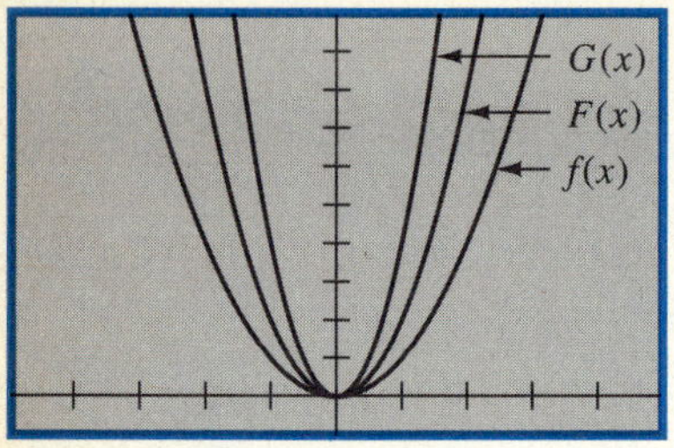

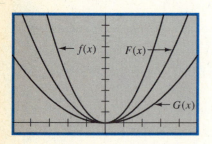

[−5, 5] by [−1, 10]
$f(x) = x^2, F(x) = 2x^2,$ and $G(x) = 4x^2$
FIGURE 10

Let us now investigate the effect on the graph of $f(x) = x^n$ caused by multiplying x^n by a positive constant.

▷ ILLUSTRATION 3

(a) Figure 10 shows the graphs of

$$f(x) = x^2 \qquad F(x) = 2x^2 \qquad G(x) = 4x^2$$

plotted in the same viewing rectangle. Notice that multiplying x^2 by the factors 2 and 4 causes a vertical "stretch" of the graph of $f(x) = x^2$ by the factor.

(b) Figure 11 shows the graphs of the functions defined by

$$f(x) = x^2 \qquad F(x) = \frac{1}{2}x^2 \qquad G(x) = \frac{1}{4}x^2$$

Here multiplying x^2 by the factors $\frac{1}{2}$ and $\frac{1}{4}$ causes the graph of $f(x) = x^2$ to shrink vertically by the factor.   ◀

[−5, 5] by [−1, 10]
$f(x) = x^2, F(x) = \frac{1}{2}x^2,$ and $G(x) = \frac{1}{4}x^2$
FIGURE 11

Illustration 3 gives a special case of the following rule.

Vertical Stretching and Shrinking Rule

The graph of a function defined by an equation of the form

$$F(x) = af(x) \qquad a > 0$$

can be obtained from the graph of f by

   (i) vertically stretching the graph by a factor of a if $a > 1$.
   (ii) vertically shrinking the graph by a factor of a if $0 < a < 1$.

▶ **EXAMPLE 4** *Using the Vertical Stretching and Shrinking Rule*

Sketch the graph of each of the following functions by either vertically stretching or shrinking the graph of $f(x) = x^4$.

(a) $F(x) = 2x^4$ (b) $G(x) = \frac{1}{4}x^4$

Solution

(a) The graph of $f(x) = x^4$ appears in Figure 1(d). The graph of F is obtained from this graph by multiplying each ordinate by 2, which has the effect of vertically stretching the graph by a factor of 2. See Figure 12.

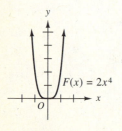

FIGURE 12

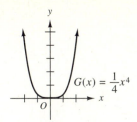

$G(x) = \frac{1}{4}x^4$

FIGURE 13

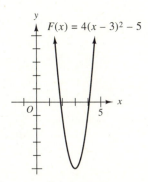

$F(x) = 4(x - 3)^2 - 5$

FIGURE 14

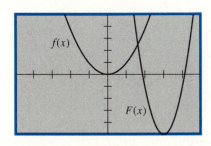

$f(x)$

$F(x)$

$[-5, 5]$ by $[-5, 5]$
$f(x) = x^2$ and $F(x) = 4(x - 3)^2 - 5$
FIGURE 15

(b) To obtain the graph of G, shown in Figure 13, we multiply each ordinate of the graph of f by $\frac{1}{4}$, which vertically shrinks the graph by a factor of $\frac{1}{4}$. ◄

We summarize our discussion so far in this section.

Geometric Transformations of a Graph

The graph of a function defined by an equation of the form

$$F(x) = af(x - h) + k \qquad a \text{ is a positive constant}$$

can be obtained from the graph of f by the following geometric transformations:

1. Vertically stretch $(a > 1)$ or shrink $(0 < a < 1)$ by a factor of a. We obtain the function defined by $af(x)$.
2. Horizontal translation h units to the right if $h > 0$ and $|h|$ units to the left if $h < 0$. We obtain the function defined by $af(x - h)$.
3. Vertical translation k units upward if $k > 0$ and $|k|$ units downward if $k < 0$. We obtain $F(x) = af(x - h) + k$.

► **EXAMPLE 5** *Sketching the Graph of a Polynomial Function by Geometric Transformations*

Sketch the graph of $F(x) = 4(x - 3)^2 - 5$ by appropriate geometric transformations of the graph of $f(x) = x^2$. Check the graph by plotting the graphs of both functions in the same viewing rectangle.

Solution We perform the following geometric transformations on the graph of f in the indicated order to obtain the graph of F shown in Figure 14.

1. Vertical stretch by a factor of 4.
2. Horizontal translation 3 units to the right.
3. Vertical translation 5 units downward.

Figure 15 shows the graphs of both functions as they appear on our graphics calculator. ◄

Consider now the power function defined by

$$f(x) = -x^n \qquad n \text{ is a positive integer}$$

Sketches of the graphs of this function for n having positive-integer values

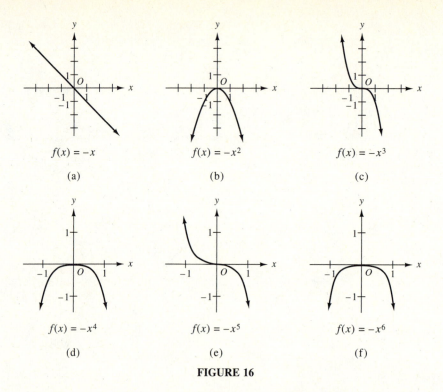

FIGURE 16

1 through 6 appear in Figure 16. Observe that they are the mirror images in the x axis of the corresponding graphs in Figure 1; that is, the graph of $f(x) = -x^n$ can be obtained by reflecting the graph of $f(x) = x^n$ through the x axis. We have the following rule.

x Axis Reflection Rule

The graph of a function defined by an equation of the form

$$F(x) = -f(x)$$

can be obtained from the graph of f by reflecting the graph through the x axis.

Using the x axis reflection rule along with the other geometric transformations in this section enables us to obtain the graph of

$$f(x) = -a(x - h)^n + k \quad a \text{ is a positive constant}$$

from the graph of $f(x) = x^n$. Be sure, however, to perform the x axis reflection first, as demonstrated in the following example.

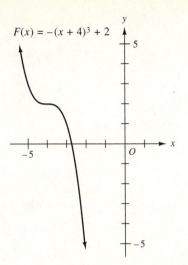

$F(x) = -(x + 4)^3 + 2$

FIGURE 17

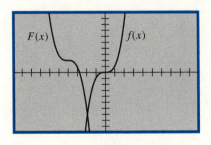

$[-10, 10]$ by $[-10, 10]$
$f(x) = x^3$ and $F(x) = -(x + 4)^3 + 2$
FIGURE 18

▶ **EXAMPLE 6** *Sketching the Graph of a Polynomial Function by Geometric Transformations*

Sketch by hand the graph of $F(x) = -(x + 4)^3 + 2$ by appropriate geometric transformations of the graph of $f(x) = x^3$. Check your graph by plotting the graphs of both functions in the same viewing rectangle.

Solution The graph of F in Figure 17 is obtained by performing the following geometric transformations on the graph of f in the indicated order.

1. Reflection through the x axis.
2. Horizontal translation 4 units to the left.
3. Vertical translation 2 units upward.

Figure 18 shows the graphs of both functions as they appear on our graphics calculator. ◀

We now discuss graphs of more general polynomial functions of degree 3 or more. If P is such a function of the nth degree,

$$P(x) = a_n x^n + a_{n-1} x^{n-1} + a_{n-2} x^{n-2} + \ldots + a_1 x + a_0$$

where $a_0, a_1, \ldots, a_n$ are real numbers with $a_n \neq 0$ and n is a nonnegative integer. Consider $a_n x^n$, the first term of the polynomial. As $|x|$ increases without bound, $|a_n x^n|$ increases without bound and will become larger than the sum of all the other terms in the polynomial. Therefore the form of the graph for large values of $|x|$ will be affected by the values of the terms $a_n x^n$. We can conclude that the shape of the graph for large values of $|x|$ will be similar to that of the graph of the power function of degree n. Let us consider as separate cases a_n positive and a_n negative.

Case 1: $a_n > 0$: The function values will be increasing for large values of x; so the graph will be going up to the right as in Figure 19(a) and (b). If n is even, the graph comes down from the left as in Figure 19(a), and if n is odd, the graph comes up from the left as in Figure 19(b).

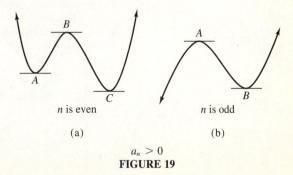

n is even n is odd

(a)   (b)

$a_n > 0$
FIGURE 19

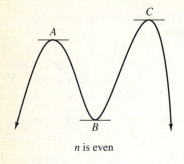

n is even

(a)

n is odd

(b)

$a_n < 0$

FIGURE 20

Case 2: $a_n < 0$: The function values will be decreasing for large values of x; thus the graph will be going down to the right as in Figure 20(a) and (b). If n is even, the graph comes up from the left as in Figure 20(a), and if n is odd, the graph comes down from the left as in Figure 20(b).

Properties of a graph at the left and right ends of the x axis are referred to as the **end behavior** of the graph. Thus the above conclusions under Cases 1 and 2 give the end behavior of graphs of polynomial functions, as summarized in Table 1.

Table 1   End Behavior of Graphs of Polynomial Functions

a_n	n	End Behavior of the Graph	Example
Positive	Even	Comes down from the left Goes up to the right	Figure 19(a)
Positive	Odd	Comes up from the left Goes up to the right	Figure 19(b)
Negative	Even	Comes up from the left Goes down to the right	Figure 20(a)
Negative	Odd	Comes down from the left Goes down to the right	Figure 20(b)

The graphs of polynomial functions are "unbroken" curves because polynomial functions are *continuous*, which implies that a small change in x causes a small change in $f(x)$. A precise definition of a continuous function requires the concept of *limit*, studied in calculus.

Significant points on graphs of polynomial functions are the points where the function has *relative extrema*, which are the *relative minimum* *and relative maximum function values*.

DEFINITION   Relative Minimum and Relative Maximum Function Values

The function value $f(a)$ is a **relative minimum value** of f if there is an open interval (c, d) containing a on which f is defined such that

$$f(a) \leq f(x) \quad \text{for all } x \text{ in } (c, d)$$

The function value $f(b)$ is a **relative maximum value** of f if there is an open interval (c, d) containing b on which f is defined such that

$$f(b) \geq f(x) \quad \text{for all } x \text{ in } (c, d)$$

The graph of a polynomial function will have a horizontal tangent line at a point where there is a relative extremum. In Figure 19(a) the function has one relative maximum value at B and two relative minimum values at A and C. Observe the horizontal tangent lines at these three points. In Figure 19(b), the function has two relative extrema, a relative maximum value at A and a relative minimum value at B. The function in Figure 20(a) has three relative extrema, and the function in Figure 20(b) has two relative extrema.

We now state without proof a theorem giving the number of possible relative extrema for a polynomial function.

THEOREM 1

A polynomial function of the n*th* degree has at most n $-$ 1 relative extrema.

From this theorem it follows that a polynomial function of the fourth degree has at most three relative extrema. Therefore the graph in Figure 19(a) could be that of a fourth-degree polynomial. The graph in Figure 19(b) could be that of a polynomial of the third degree because such a polynomial has at most two relative extrema.

In the following examples, we apply Theorem 1 as well as the end-behavior information given in Table 1.

▶ **EXAMPLE 7** *Describing and Plotting the Graph of a Polynomial Function and Estimating the Relative Extrema*

Describe the end behavior of the graph of the function defined by

$$P(x) = x^3 - 6x^2 + 9x - 1$$

Determine the most relative extrema of P. Plot the graph of P on a graphics calculator and estimate the relative extrema of P.

Solution Because the coefficient of x^3 is positive and the degree of the polynomial is odd, from Table 1 we conclude that the graph comes up from the left and goes up to the right. From Theorem 1 there are at most two relative extrema. The graph is, therefore, probably similar to that shown in Figure 19(b).

Figure 21 shows the graph of P plotted in the $[-5, 5]$ by $[-5, 5]$ viewing rectangle. Using trace and zoom-in on our calculator we estimate that P has a relative maximum value of 3 at $x = 1$ and a relative minimum value of -1 at $x = 3$. In your calculus course you will be able to show that these are the exact relative extrema. ◀

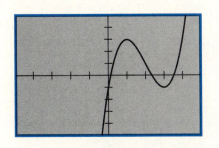

$[-5, 5]$ by $[-5, 5]$
$P(x) = x^3 - 6x^2 + 9x - 1$
FIGURE 21

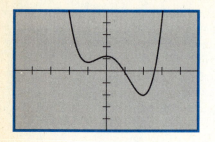

$[-5, 5]$ by $[-50, 50]$
$P(x) = 3x^4 - 4x^3 - 12x^2 + 12$
FIGURE 22

▶ **EXAMPLE 8** *Describing and Plotting the Graph of a Polynomial Function and Estimating the Relative Extrema*

Follow the instructions of Example 7 for the function defined by

$$P(x) = 3x^4 - 4x^3 - 12x^2 + 12$$

Solution   The degree of the polynomial is 4 and the coefficient of x^4 is positive. From Table 1 we, therefore, conclude that the graph comes down from the left and goes up to the right. Furthermore, from Theorem 1, there are at most three relative extrema. Thus the graph is probably something like the one shown in Figure 19(a).

We plot the graph of P in the $[-5, 5]$ by $[-50, 50]$ viewing rectangle and obtain Figure 22. By tracing and zooming-in, we estimate that P has a relative maximum value of 12 at $x = 0$ and relative minimum values of 7 and -20 occurring when x is -1 and 2, respectively. These relative extrema are exact, as you will be able to show in your calculus course.   ◀

EXERCISES 5.1

In Exercises 1 through 6, sketch the graphs of F and G on the same coordinate axes by either a vertical or horizontal translation of the graph of f.

1. $f(x) = x^2$; $F(x) = x^2 + 3$; $G(x) = x^2 - 4$

2. $f(x) = x^3$; $F(x) = x^3 - 2$; $G(x) = x^3 + 3$

3. $f(x) = x^2$; $F(x) = (x + 3)^2$; $G(x) = (x - 4)^2$

4. $f(x) = x^3$; $F(x) = (x - 2)^3$; $G(x) = (x + 3)^3$

5. $f(x) = x^3$; $F(x) = x^3 - 6$; $G(x) = (x - 6)^3$

6. $f(x) = x^4$; $F(x) = x^4 + 1$; $G(x) = (x + 1)^4$

In Exercises 7 through 10, sketch the graphs of F and G on the same coordinate axes by either stretching or shrinking the graph of f vertically.

7. $f(x) = x^4$; $F(x) = 2x^4$; $G(x) = \frac{1}{3}x^4$

8. $f(x) = x^2$; $F(x) = 3x^2$; $G(x) = \frac{1}{2}x^2$

9. $f(x) = -x^3$; $F(x) = -3x^3$; $G(x) = -\frac{1}{2}x^3$

10. $f(x) = -x^4$; $F(x) = -2x^4$; $G(x) = -\frac{1}{4}x^4$

In Exercises 11 through 16, sketch the graph of F by appropriate geometric transformations of the graph of f. Check your graph by plotting the graphs of both functions in the same viewing rectangle.

11. $f(x) = x^2$; $F(x) = 3(x + 5)^2 - 6$

12. $f(x) = x^3$; $F(x) = -2(x - 4)^3 + 1$

13. $f(x) = x^3$; $F(x) = -\frac{1}{2}(x - 1)^3 - 2$

14. $f(x) = x^2$; $F(x) = \frac{1}{3}(x + 4)^2 + 5$

15. $f(x) = x^4$; $F(x) = -2(x + 3)^4 + 1$

16. $f(x) = x^4$; $F(x) = \frac{1}{4}(x - 2)^4 - 3$

In Exercises 17 through 22, sketch the graph of the function. Check your graph on your graphics calculator.

17. (a) $f(x) = -4(x - 2)^2$ **(b)** $g(x) = 3(x + 1)^2$

18. (a) $f(x) = \frac{1}{8}(x - 1)^4$ **(b)** $g(x) = -\frac{1}{8}(x + 1)^4$

19. (a) $f(x) = (x - 2)^4 + 3$
(b) $g(x) = -(x + 2)^4 - 3$

20. (a) $f(x) = (x - 3)^3 + 2$
(b) $g(x) = -(x + 3)^3 - 2$

21. (a) $f(x) = 2(x + 4)^5 + 1$
(b) $g(x) = \frac{1}{2}(x - 4)^5 - 1$

22. (a) $f(x) = -2(x + 1)^5 + 4$
(b) $g(x) = -\frac{1}{2}(x - 1)^5 - 4$

In Exercises 23 through 36, for the given function do the following: (a) describe the end behavior of the graph; (b) determine the most relative extrema; (c) plot the graph; (d) estimate the relative extrema.

23. $P(x) = x^3 - 2x^2 - 5x + 6$

24. $P(x) = x^3 - 3x^2 - 9x + 9$

25. $F(x) = x^3 - 3x^2 + 3$

26. $G(x) = x^3 + 4x^2 + 4x$

27. $g(x) = 3x^3 - 4x^2 - 5x + 2$

28. $f(x) = 6x^3 + 29x^2 + x - 6$

29. $f(x) = x^4 - 5x^3 + 2x^2 + 8x$

30. $g(x) = x^4 - 5x^2 + 4$

31. $P(x) = x^4 + x^3 - 7x^2 - x + 6$

32. $P(x) = x^4 - 6x^3 + 11x^2 - 6x$

33. $G(x) = 3x^4 + 5x^3 - 5x^2 - 5x + 2$

34. $h(x) = 2x^4 - x^3 - 6x^2 - x + 2$

35. $f(x) = -x^4 + x^2$ **36.** $f(x) = x^5 - 4x^3$

In Exercises 37 through 40, use geometric transformations to sketch the graph of function F from the graph of the absolute value function.

37. $F(x) = 2|x - 3| + 4$

38. $F(x) = -2|x + 4| - 3$

39. $F(x) = -3|x + 1| - 5$

40. $F(x) = 3|x - 2| + 4$

41. Geometric transformations on the graph of a function f used to sketch the graph of the function F defined by

$$F(x) = cf(x - h) + k$$

where c, h, and k are constants, include stretching and shrinking, horizontal and vertical translations, and the x axis reflection rule. Is the order in which these transformations are performed important? Explain your answer.

5.2 THE FACTOR THEOREM AND SYNTHETIC SUBSTITUTION

GOALS
1. **Learn and apply the remainder theorem.**
2. **Learn and apply the factor theorem.**
3. **Learn and apply synthetic division.**
4. **Use synthetic division to factor a polynomial.**
5. **Compute polynomial function values by synthetic substitution.**
6. **Apply Horner's algorithm.**

In the next section, we will determine zeros of a polynomial function P by finding linear factors of the polynomial $P(x)$. To accomplish this, we will need to apply the techniques discussed in this section.

We begin by stating the theorem which is the familiar *division algorithm* for polynomials, where the divisor is a linear expression of the form $x - r$ and r is a real number.

THEOREM 1   *Division Algorithm for P(x) Divided by x − r*

If $P(x)$ is a polynomial and r is a real number, then when $P(x)$ is divided by $x - r$, we obtain as the quotient a unique polynomial $Q(x)$ and as the remainder a real number R such that for all values of x,

$$P(x) = (x - r)Q(x) + R$$

▷ **ILLUSTRATION 1**

Let $P(x) = 2x^3 - 5x^2 + 6x - 3$. We use long division to divide $P(x)$ by $x - 2$.

$$
\begin{array}{r}
2x^2 - x + 4 \\
x - 2 \overline{)\,2x^3 - 5x^2 + 6x - 3} \\
\underline{2x^3 - 4x^2} \\
-x^2 + 6x \\
\underline{-x^2 + 2x} \\
4x - 3 \\
\underline{4x - 8} \\
5
\end{array}
$$

Hence the quotient $Q(x)$ is $2x^2 - x + 4$, and the remainder R is 5. We can, therefore, write

$$2x^3 - 5x^2 + 6x - 3 = (x - 2)(2x^2 - x + 4) + 5$$

which is written in the form of the equation in Theorem 1.   ◀

▷ **ILLUSTRATION 2**

In Illustration 1, $P(x) = 2x^3 - 5x^2 + 6x - 3$. Thus

$$
\begin{aligned}
P(2) &= 2(2)^3 - 5(2)^2 + 6(2) - 3 \\
&= 5
\end{aligned}
$$

Observe that this value of $P(2)$ is the same number as the remainder obtained in Illustration 1 when $P(x)$ was divided by $x - 2$.   ◀

In Illustration 2 we have a special case of the following theorem, known as the *remainder theorem*.

THEOREM 2   *Remainder Theorem*

If $P(x)$ is a polynomial and r is a real number, then if $P(x)$ is divided by $x - r$, the remainder is $P(r)$.

Proof   From Theorem 1 it follows that when $P(x)$ is divided by $x - r$, we obtain a polynomial $Q(x)$ as the quotient and a real number R as the remainder such that for all values of x,

$$P(x) = (x - r)Q(x) + R$$

Because this equation is an identity, it is satisfied when $x = r$. Thus

$$P(r) = (r - r)Q(r) + R$$
$$= 0 + R$$
$$= R$$

and the theorem is proved.   ■

▶ **EXAMPLE 1** *Applying the Remainder Theorem*

Use the remainder theorem to find the remainder when $2x^3 + 3x^2 - 4x - 17$ is divided by **(a)** $x - 3$ and **(b)** $x + 2$.

Solution

(a) When $P(x)$ is divided by $x - 3$, the remainder is $P(3)$. Because
$P(x) = 2x^3 + 3x^2 - 4x - 17$

$$P(3) = 2(3)^3 + 3(3)^2 - 4(3) - 17$$
$$= 52$$

Therefore, when $P(x)$ is divided by $x - 3$, the remainder is 52.

(b) When $P(x)$ is divided by $x + 2$, the remainder is $P(-2)$.

$$P(-2) = 2(-2)^3 + 3(-2)^2 - 4(-2) - 17$$
$$= -13$$

Thus when $P(x)$ is divided by $x + 2$, the remainder is -13.   ◀

A consequence of the remainder theorem is the *factor theorem*. It enables us to determine if a specific expression of the form $x - r$ is a factor of a given polynomial.

THEOREM 3   *Factor Theorem*

If $P(x)$ is a polynomial and r is a real number, then $P(x)$ has $x - r$ as a factor if and only if $P(r) = 0$.

Proof   Because the statement of the theorem has an *if and only if* qualification, there are two parts to be proved. Part 1: $x - r$ is a factor of $P(x)$ if $P(r) = 0$; Part 2: $x - r$ is a factor of $P(x)$ only if $P(r) = 0$.

Proof of Part 1: From Theorem 1 and the remainder theorem it follows that for the polynomial $P(x)$ and the real number r there exists a unique polynomial $Q(x)$ such that

$$P(x) = (x - r)Q(x) + P(r)$$

If $P(r) = 0$, then

$$P(x) = (x - r)Q(x)$$

Therefore, $x - r$ is a factor of $P(x)$.

Proof of Part 2: We wish to prove that if $x - r$ is a factor of $P(x)$, then $P(r) = 0$. If $x - r$ is a factor of $P(x)$, then when $P(x)$ is divided by $x - r$, the remainder must be zero. Thus, from the remainder theorem, it follows that $P(r) = 0$. ∎

▶ **EXAMPLE 2** *Applying the Factor Theorem*

Use the factor theorem to show that $x - 4$ is a factor of $2x^3 - 6x^2 - 5x - 12$.

Solution   If $P(x) = 2x^3 - 6x^2 - 5x - 12$, then

$$
\begin{aligned}
P(4) &= 2(4)^3 - 6(4)^2 - 5(4) - 12 \\
&= 2(64) - 6(16) - 20 - 12 \\
&= 128 - 96 - 32 \\
&= 0
\end{aligned}
$$

Therefore, by the factor theorem, $x - 4$ is a factor of $P(x)$. ◀

When the factor theorem is used to show that a linear expression of the form $x - r$ is a factor of a polynomial $P(x)$ (as in Example 2), the other factor is obtained by dividing $P(x)$ by $x - r$. To simplify the computation of such divisions, we use a procedure called *synthetic division*, which we now explain.

In Illustration 1 we used long division to divide $2x^3 - 5x^2 + 6x - 3$ by $x - 2$. The computation is as follows:

$$
\require{enclose}
\begin{array}{r}
2x^2 - x + 4 \\
x - 2 \overline{\smash{)}2x^3 - 5x^2 + 6x - 3} \\
\underline{2x^3 - 4x^2} \\
-x^2 + 6x \\
\underline{-x^2 + 2x} \\
4x - 3 \\
\underline{4x - 8} \\
5
\end{array}
$$

The writing can be shortened by omitting the powers of x and recording only the coefficients. By doing this, the computation takes the following form:

$$
\begin{array}{r}
2 \quad -1 \quad\; 4 \\
\overline{1-2)2 \quad -5 \quad\; 6 \quad -3} \\
2 \quad -4 \\
\overline{-1 \quad\; 6 } \\
-1 \quad\; 2 \\
\overline{4 \quad -3} \\
4 \quad -8 \\
\overline{5}
\end{array}
$$

In the divisor, $x - 2$, the coefficient of x is 1. Thus the coefficient of the first term in each remainder is the same as that of the succeeding term of the quotient. Furthermore, the first term of the next partial product is the same as the coefficient of the first term in each remainder. Hence we can omit the terms of the quotient as well as the first terms of the partial products. With these terms omitted, we have

$$
\begin{array}{r}
\underline{1-2}\,|\,2 \quad -5 \quad\; 6 \quad -3 \\
-4 \\
\overline{-1 \quad\; 6 } \\
2 \\
\overline{4 \quad -3} \\
-8 \\
\overline{5}
\end{array}
$$

In synthetic division the divisor is a polynomial of the form $x - r$, and so the first coefficient in the divisor is always 1; thus we delete the coefficient 1. We can also move the numbers up so that they are arranged in three lines; doing this, we have

$$
\begin{array}{r}
\underline{-2}\,|\,2 \quad -5 \quad\; 6 \quad -3 \\
-4 \quad\; 2 \quad -8 \\
\overline{-1 \quad\; 4 \quad\;\; 5}
\end{array}
$$

We now write 2, the first coefficient in the dividend, in the first position in the bottom row, and we have

$$
\begin{array}{r}
\underline{-2}\,|\,2 \quad -5 \quad\; 6 \quad -3 \\
-4 \quad\; 2 \quad -8 \\
\overline{2 \quad -1 \quad\; 4 \quad\;\; 5}
\end{array}
$$

Notice that the first three numbers in the bottom row are the coefficients 2, -1, and 4 of the quotient; the last number in the bottom row is 5, and 5 is the remainder. The numbers in the second row are obtained by multiplying the number in the bottom row of the preceding column by -2, and the numbers in the bottom row are found by subtracting the numbers in the second row from those of the top row. If the multiplier, -2, is replaced by 2, the numbers in the second row can then be added to those of the top row to obtain the numbers in the bottom row. We make this change, and the work appears as follows:

$$\begin{array}{r|rrrr} 2 & 2 & -5 & 6 & -3 \\ & & 4 & -2 & 8 \\ \hline & 2 & -1 & 4 & 5 \end{array}$$

This arrangement of the computation is the **synthetic division** of the polynomial $2x^3 - 5x^2 + 6x - 3$ by $x - 2$, with the quotient $2x^2 - x + 4$ and the remainder 5.

In general, the following steps give the procedure for synthetic division of a polynomial $P(x)$ by $x - r$. As you read Illustrations 3 and 4, refer back to these steps.

Synthetic Division

1. Write $P(x)$ in the form $a_n x^n + a_{n-1} x^{n-1} + a_{n-2} x^{n-2} + \ldots + a_1 x + a_0$ and insert a zero coefficient for any missing term.
2. Write the coefficients of $P(x)$ in order in a horizontal row.
3. Bring down the first coefficient a_n of $P(x)$ to the bottom row.
4. Multiply a_n by r, and write the product in the second row below the coefficient a_{n-1}; then add the product to a_{n-1}, and write the sum in the bottom row.
5. Multiply this sum by r, and write the product in the second row below the coefficient a_{n-2}; add the product to a_{n-2}, and write the sum in the bottom row.
6. Continue the process of steps 4 and 5 as long as possible.
7. The last number in the bottom row is the remainder, and the preceding numbers are the coefficients of the successive terms of the quotient, which is a polynomial of degree 1 less than that of $P(x)$.

▷ ILLUSTRATION 3

We use synthetic division to divide $3x^3 - 7x^2 + x + 5$ by $x - 1$. The coefficients of the dividend are 3, -7, 1, and 5. The colored arrows indicate the order in which the numbers are found.

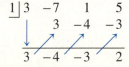

Because the bottom row consists of the numbers 3, -4, -3, and 2, the quotient is $3x^2 - 4x - 3$, and the remainder is 2. Therefore

$$3x^3 - 7x^2 + x + 5 = (x - 1)(3x^2 - 4x - 3) + 2 \qquad \blacktriangleleft$$

▷ **ILLUSTRATION 4**

We use synthetic division to find the quotient and remainder when $x^4 - 7x^2 + 2x - 6$ is divided by $x + 3$. Because we are dividing by $x + 3$ or, equivalently, $x - (-3)$, r is -3. The coefficients of $x^4 - 7x^2 + 2x - 6$ are $1, 0, -7, 2$, and -6 (we insert a zero for the coefficient of the missing term involving x^3). The computation has the following form:

$$
\begin{array}{r|rrrrr}
-3 & 1 & 0 & -7 & 2 & -6 \\
   &   & -3 & 9 & -6 & 12 \\
\hline
   & 1 & -3 & 2 & -4 & 6
\end{array}
$$

Therefore the quotient is $x^3 - 3x^2 + 2x - 4$, and the remainder is 6. Thus

$$x^4 - 7x^2 + 2x - 6 = (x + 3)(x^3 - 3x^2 + 2x - 4) + 6 \qquad \blacktriangleleft$$

▶ **EXAMPLE 3** *Applying Synthetic Division*

Use synthetic division to find the quotient and remainder when $x^5 - 3x^4 + 4x + 5$ is divided by $x - 2$.

Solution   The coefficients of $x^5 - 3x^4 + 4x + 5$ are $1, -3, 0, 0, 4$, and 5, where the two zeros represent the coefficients of the missing terms involving x^3 and x^2. Following is the computation by synthetic division:

$$
\begin{array}{r|rrrrrr}
2 & 1 & -3 & 0 & 0 & 4 & 5 \\
  &   & 2 & -2 & -4 & -8 & -8 \\
\hline
  & 1 & -1 & -2 & -4 & -4 & -3
\end{array}
$$

The quotient is $x^4 - x^3 - 2x^2 - 4x - 4$, and the remainder is -3. ◀

▶ **EXAMPLE 4** *Using Synthetic Division To Factor a Polynomial*

Given

$$P(x) = 6x^3 + x^2 - 4x + 1$$

Show that $3x - 1$ is a factor of $P(x)$ and find the other factors.

Solution   Because $3x - 1 = 3(x - \frac{1}{3})$, $3x - 1$ will be a factor of $P(x)$ if $x - \frac{1}{3}$ is a factor. We use synthetic division to divide $P(x)$ by $x - \frac{1}{3}$.

$$
\begin{array}{r|rrrr}
\frac{1}{3} & 6 & 1 & -4 & 1 \\
            &   & 2 & 1 & -1 \\
\hline
            & 6 & 3 & -3 & 0
\end{array}
$$

Because the remainder $P(\frac{1}{3}) = 0$, we conclude from the factor theorem that $x - \frac{1}{3}$ is a factor of $P(x)$. Furthermore,

$$
\begin{aligned}
P(x) &= (x - \tfrac{1}{3})(6x^2 + 3x - 3) \\
&= (x - \tfrac{1}{3})[3(2x^2 + x - 1)] \\
&= (3x - 1)(2x^2 + x - 1)
\end{aligned}
$$

Therefore, $3x - 1$ is a factor of $P(x)$. Because

$$2x^2 + x - 1 = (2x - 1)(x + 1)$$

we can write $P(x)$ in completely factored form as

$$P(x) = (3x - 1)(2x - 1)(x + 1) \qquad \blacktriangleleft$$

The remainder theorem states that for a given polynomial $P(x)$, the value of $P(r)$ is the remainder when $P(x)$ is divided by $x - r$. Synthetic division, then, provides a fast way of obtaining $P(r)$, and when this method is used we say we compute $P(r)$ by *synthetic substitution*. It is usually easier to compute $P(r)$ by synthetic substitution than by direct substitution.

▶ **EXAMPLE 5** *Computing Polynomial Function Values by Synthetic Substitution*

If $P(x) = 2x^5 + 4x^4 - 10x^3 - 20x - 10$, find $P(0)$, $P(-1)$, $P(3)$, and $P(-4)$ by either direct substitution or synthetic substitution, whichever is easier.

Solution   We obtain $P(0)$ and $P(-1)$ by direct substitution and $P(3)$ and $P(-4)$ by synthetic substitution.

$$P(0) = -10 \qquad P(-1) = -2 + 4 + 10 + 20 - 10$$
$$= 22$$

$$
\begin{array}{r|rrrrrr}
3 & 2 & 4 & -10 & 0 & -20 & -10 \\
  &   & 6 & 30 & 60 & 180 & 480 \\
\hline
  & 2 & 10 & 20 & 60 & 160 & 470
\end{array}
$$

$$
\begin{array}{r|rrrrrr}
-4 & 2 & 4 & -10 & 0 & -20 & -10 \\
   &   & -8 & 16 & -24 & 96 & -304 \\
\hline
   & 2 & -4 & 6 & -24 & 76 & -314
\end{array}
$$

Thus $P(3) = 470$ and $P(-4) = -314$. $\qquad \blacktriangleleft$

Computation of polynomial function values can be done on your calculator. On some calculators you can compute $P(r)$ by letting $y = P(x)$ and then storing r in x from which you compute the corresponding value of y. Consult your manual for the procedure to use on your particular calculator.

A process known as *Horner's algorithm*, named for William Horner (1786–1837), an English mathematician, provides a method easily adapted for computation on a calculator. To demonstrate this process, consider the polynomial of Example 4:

$$P(x) = 6x^3 + x^2 - 4x + 1 \tag{1}$$
$$= x[6x^2 + x - 4] + 1$$
$$= x[x(6x + 1) - 4] + 1$$

Hence

$$P(r) = r[r(6r + 1) - 4] + 1 \tag{2}$$

The right-hand side of (2) is easy to evaluate on a calculator by starting with the expression in the innermost set of parentheses as follows: multiply 6 (the first coefficient in (1)) by r; add 1 (the next coefficient); multiply the sum by r; add -4 (the next coefficient); multiply that sum by r; add 1 (the next coefficient).

Observe that when we compute $P(r)$ by synthetic substitution we have the following:

$$
\begin{array}{r|llll}
r & 6 & 1 & -4 & 1 \\
 & & 6r & r(6r+1) & r[r(6r+1)-4] \\
\hline
 & 6 & 6r+1 & r(6r+1)-4 & r[r(6r+1)-4]+1
\end{array}
$$

Then by the remainder theorem, $P(r) = r[r(6r + 1) - 4] + 1$ as in (2). Thus *Horner's algorithm* and *synthetic substitution* are two names for the same computation.

▶ **EXAMPLE 6** *Applying Horner's Algorithm*

Write the polynomial

$$P(x) = 3x^4 - 2x^3 + 4x^2 - x + 5$$

in the form of Horner's algorithm, and use this form to compute $P(2)$ and $P(-3)$ on a calculator.

Solution

$$P(x) = 3x^4 - 2x^3 + 4x^2 - x + 5$$
$$= x(3x^3 - 2x^2 + 4x - 1) + 5$$
$$= x(x[3x^2 - 2x + 4] - 1) + 5$$
$$= x(x[x(3x - 2) + 4] - 1) + 5$$

With this form we compute on our calculator, $P(2) = 51$ and $P(-3) = 341$. ◀

EXERCISES 5.2

In Exercises 1 through 4, use the remainder theorem to find the remainder when the polynomial is divided by the linear expression.

1. (a) $(3x^2 - 4x + 5) \div (x - 3)$
  (b) $(3x^4 + 7x^3 + x^2 + x + 9) \div (x + 1)$

2. (a) $(4x^2 + 7x - 5) \div (x - 1)$
  (b) $(x^3 - 4x^2 + 5) \div (x + 3)$

3. (a) $(x^3 + 9) \div (x + 2)$
  (b) $(3x^5 - 7x^4 - 5x^3 - 4x^2 + 1) \div (x - 3)$

4. (a) $(x^4 - 8) \div (x - 2)$
  (b) $(8x^5 + 7x^2 - 3) \div (x - \frac{1}{2})$

In Exercises 5 through 10, use the factor theorem to answer the question.

5. Is $x - 3$ a factor of $2x^3 - 6x^2 - 5x + 15$?

6. Is $x + 3$ a factor of $3x^3 - x^2 - 22x - 24$?

7. Is $x + 2$ a factor of $x^4 + 2x^3 - 12x^2 - 11x + 6$?

8. Is $x - 2$ a factor of $x^7 - 128$?

9. Is $x + 3$ a factor of $x^5 + 243$?

10. Is $x - a$ a factor of $x^8 + a^8$?

In Exercises 11 through 20, find the quotient and remainder by synthetic division.

11. $(2x^3 - x^2 + 3x + 12) \div (x - 4)$

12. $(y^3 + 4y^2 + 3y - 6) \div (y - 2)$

13. $(2x^4 + 5x^3 - 2x - 1) \div (x + 4)$

14. $(x^3 + 4x^2 - 7) \div (x + 3)$

15. $(3z^5 + z^4 - 4z^2 + 7) \div (z - 2)$

16. $(4x^6 + 21x^5 - 26x^3 + 27x) \div (x + 5)$

17. $(6x^3 - x^2 + 2x + 2) \div (x + \frac{1}{3})$

18. $(8x^3 - 6x^2 + 5x - 3) \div (x - \frac{1}{4})$

19. $(x^7 - 1) \div (x - 1)$ **20.** $(x^7 + 1) \div (x + 1)$

In Exercises 21 through 26, use synthetic substitution to find the function values.

21. If $P(x) = 4x^3 - 5x^2 - 4$, find $P(2)$ and $P(-3)$.

22. If $P(x) = 3x^3 + 4x^2 - 9$, find $P(-2)$ and $P(1)$.

23. If $P(x) = x^4 + 3x^3 - 5x^2 + 9$, find $P(-4)$ and $P(3)$.

24. If $P(x) = 2x^4 - 7x^3 - 15x + 1$, find $P(4)$ and $P(-2)$.

25. If $P(x) = 6x^3 - x^2 - 7x + 2$, find $P(-\frac{1}{3})$ and $P(\frac{3}{2})$.

26. If $P(x) = x^3 + 2x + 4$, find $P(-1.3)$ and $P(2.1)$.

In Exercises 27 through 30, show that the linear expression is a factor of the polynomial and find the other factors.

27. $x + 3;\ P(x) = 4x^3 + 9x^2 - 10x - 3$

28. $x - 5;\ P(x) = 2x^3 - 13x^2 + 16x - 5$

29. $2x - 1;\ P(x) = 6x^3 - 7x^2 + 1$
   [*Hint:* $2x - 1 = 2(x - \frac{1}{2})$]

30. $3x + 2;\ P(x) = 12x^3 + 5x^2 - 11x - 6$
   [*Hint:* $3x + 2 = 3(x + \frac{2}{3})$]

In Exercises 31 through 34, determine if the linear expression is a factor of $P(x)$.

31. $x - 4;\ P(x) = 2x^4 - 7x^3 - 14x + 8$

32. $x - 3;\ P(x) = x^4 - 6x^2 - 5x - 12$

33. $2x + 3;\ P(x) = 4x^3 - 4x^2 - 11x + 6$

34. $3x - 1;\ P(x) = 9x^3 + 3x^2 - 5x - 1$

In Exercises 35 through 38, write the polynomial in the form of Horner's algorithm and use this form to compute the indicated function values on a calculator.

35. $P(x) = 2x^3 - 6x^2 - 3x - 12;\ P(7);\ P(-5)$

36. $P(x) = 5x^4 - 8x^3 + 2x^2 - 4x + 1;\ P(6);\ P(-3)$

37. $P(x) = 4x^4 - x^3 + 2x - 1;\ P(3);\ P(-4)$

38. $P(x) = 3x^5 - 2x^3 - x^2 - 10;\ P(5);\ P(-2)$

39. Find a value of k so that $x + 2$ is a factor of $3x^3 + 5x^2 + kx - 10$.

40. Find a value of k so that $x - 5$ is a factor of $kx^3 - 17x^2 - 4kx + 5$.

41. Find values of k so that $x - 4$ is a factor of $x^3 - k^2x^2 - 8kx - 16$.

42. Find values of k so that $x + 1$ is a factor of $5x^3 + k^2x^2 + 2kx - 3$.

43. (a) Is $x - 3$ a factor of $x^{50} - 3^{50}$?
   (b) Is $x - 3$ a factor of $x^{50} + 3^{50}$?
   (c) Is $x + 3$ a factor of $x^{49} - 3^{49}$?
   (d) Is $x + 3$ a factor of $x^{49} + 3^{49}$?

44. (a) Is $x + 3$ a factor of $x^{50} - 3^{50}$?
   (b) Is $x + 3$ a factor of $x^{50} + 3^{50}$?
   (c) Is $x - 3$ a factor of $x^{49} - 3^{49}$?
   (d) Is $x - 3$ a factor of $x^{49} + 3^{49}$?

45. (a) Determine the integer values of n for which $x - y$ is a factor of $x^n - y^n$, and explain your reasoning.

(b) Determine the integer values of n for which $x + y$ is a factor of $x^n - y^n$, and explain your reasoning.

46. (a) Determine the integer values of n for which $x + y$ is a factor of $x^n + y^n$, and explain your reasoning.

(b) Explain why $x - y$ is not a factor of $x^n + y^n$ for any integer value of n.

5.3 RATIONAL ZEROS OF POLYNOMIAL FUNCTIONS

GOALS

1. **Given all but two zeros of a polynomial function, find the other two zeros.**
2. **Learn the rational zeros theorem.**
3. **Apply the rational zeros theorem to find zeros of a polynomial function.**
4. **Apply the rational zeros theorem to find roots of a polynomial equation.**
5. **Prove that a particular polynomial equation has no rational roots.**

The factor theorem and synthetic substitution can be applied to find rational zeros of polynomial functions. In this section, you will learn the procedure involved. We shall employ two theorems proved in the next section where complex zeros (including both real and imaginary numbers) of polynomial functions are treated. The first of these theorems states that if

$$P(x) = a_n x^n + a_{n-1} x^{n-1} + a_{n-2} x^{n-2} + \ldots + a_1 x + a_0 \qquad n \geq 1, a_n \neq 0$$

and if $r_1, r_2, \ldots, r_n$ are complex zeros of P, then

$$P(x) = a_n(x - r_1)(x - r_2) \ldots (x - r_n) \qquad \text{(1)}$$

If in this equation a factor $x - r_i$ occurs k times, then r_i is called a **zero of multiplicity k**. If a zero of multiplicity k is counted as k zeros, then it follows from (1) that a polynomial function P, for which $P(x)$ is of degree $n \geq 1$, has *at least n zeros*, some of which may be repeated. The second theorem proved in Section 5.4 states that such a polynomial has *exactly n zeros*.

▷ **ILLUSTRATION 1**

The function P defined by

$$P(x) = (x - 4)^3(x + 1)^2(x - 3)$$

is of degree 6 and P has six zeros; they are 4, 4, 4, -1, -1, and 3. The number 4 is a zero of multiplicity 3, and -1 is a zero of multiplicity 2. ◀

For each theorem relating to the zeros of a polynomial function, we have a statement regarding the roots of a polynomial equation. For instance, from the theorem that states that a polynomial of degree n has exactly n zeros, we have the fact that a polynomial equation of degree n has exactly n roots.

▷ **ILLUSTRATION 2**

The polynomial equation corresponding to the polynomial function of Illustration 1 is

$$(x - 4)^3(x + 1)^2(x - 3) = 0$$

This is an equation of the sixth degree, and the six roots are 4, 4, 4, -1, -1, and 3.   ◀

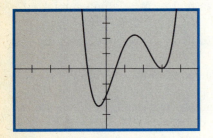

$[-5, 5]$ by $[-20, 20]$

$P(x) = 2x^4 - 11x^3 + 11x^2 + 15x - 9$

FIGURE 1

▶ **EXAMPLE 1** *Showing That a Number Is a Zero of Multiplicity 2 of a Polynomial Function and Finding the Other Zeros*

Given:

$$P(x) = 2x^4 - 11x^3 + 11x^2 + 15x - 9$$

Plot the graph of P and observe that the graph is tangent to the x axis. Then prove that the abscissa of the point of tangency is a zero of multiplicity 2 of P and find the other two zeros.

Solution   Figure 1 shows the graph of P. The point of tangency appears to be where $x = 3$. To prove that 3 is a zero of multiplicity 2 of P, we begin by dividing synthetically $P(x)$ by $x - 3$.

$$\begin{array}{r|rrrrr} 3 & 2 & -11 & 11 & 15 & -9 \\ & & 6 & -15 & -12 & 9 \\ \hline & 2 & -5 & -4 & 3 & 0 \end{array}$$

Hence

$$2x^4 - 11x^3 + 11x^2 + 15x - 9 = (x - 3)(2x^3 - 5x^2 - 4x + 3) \quad \textbf{(2)}$$

We now divide $2x^3 - 5x^2 - 4x + 3$ by $x - 3$.

$$\begin{array}{r|rrrr} 3 & 2 & -5 & -4 & 3 \\ & & 6 & 3 & -3 \\ \hline & 2 & 1 & -1 & 0 \end{array}$$

Therefore

$$2x^3 - 5x^2 - 4x + 3 = (x - 3)(2x^2 + x - 1) \quad \textbf{(3)}$$

Substituting from (3) into (2), we obtain

$$2x^4 - 11x^3 + 11x^2 + 15x - 9 = (x - 3)^2(2x^2 + x - 1)$$

The quadratic factor can now be factored into two linear factors, and we have

$$2x^4 - 11x^3 + 11x^2 + 15x - 9 = (x - 3)^2(2x - 1)(x + 1)$$

Because $2x - 1 = 2(x - \frac{1}{2})$, it follows that

$$2x^4 - 11x^3 + 11x^2 + 15x - 9 = 2(x - 3)^2(x - \frac{1}{2})(x + 1)$$

Thus the zeros of P are $3, 3, \frac{1}{2}$, and -1. Refer back to Figure 1 and observe that $\frac{1}{2}$ and -1 are intercepts of the graph, which agrees with the fact that these numbers are zeros of P. ◄

If the coefficients in the equation defining $P(x)$ are integers, then the rational zeros of P can be found by applying the following important theorem.

THEOREM 1   *Rational Zeros Theorem*

Suppose that

$$P(x) = a_n x^n + a_{n-1} x^{n-1} + a_{n-2} x^{n-2} + \ldots + a_1 x + a_0$$

where $a_0, a_1, \ldots, a_n$ are integers. If $\dfrac{p}{q}$, in lowest terms, is a rational number and a zero of P, then p is an integer factor of a_0 and q is an integer factor of a_n.

We defer the proof of the rational zeros theorem until the end of this section. Observe that the theorem does not guarantee that a polynomial function with integer coefficients has a rational zero. The theorem, however, does provide the following procedure to find any rational zeros.

To Find Rational Zeros of a Polynomial Function P:

1. Apply the rational zeros theorem to locate the possible rational zeros.
2. Plot the graph of P and by observing where the graph intersects the x axis determine which of the numbers in step 1 are likely candidates for rational zeros.
3. If x_1 is a candidate, compute $P(x_1)$ and if $P(x_1) = 0$, you have found a zero.
4. Apply synthetic division to find $Q(x)$, the quotient when $P(x)$ is divided by $x - x_1$, so that $P(x) = (x - x_1)Q(x)$.
5. Continue the procedure in steps 3 and 4 with $Q(x)$.
6. If you obtain a quadratic factor, equate the quadratic factor to zero and solve that quadratic equation.

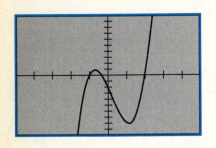

[−5, 5] by [−10, 10]
$P(x) = 3x^3 − 2x^2 − 7x − 2$
FIGURE 2

▷ **ILLUSTRATION 3**

We apply the above steps when P is defined by

$$P(x) = 3x^3 − 2x^2 − 7x − 2$$

From the rational zeros theorem, we know that any rational zero $\frac{p}{q}$ of P must be such that p is an integer factor of $−2$ and q is an integer factor of 3. Therefore the possible values of p are $1, −1, 2,$ and $−2$; and the possible values of q are $1, −1, 3,$ and $−3$. Thus the set of possible rational zeros of P is

$$\{1, −1, 2, −2, \tfrac{1}{3}, −\tfrac{1}{3}, \tfrac{2}{3}, −\tfrac{2}{3}\}$$

We plot the graph of P shown in Figure 2. Comparing the intercepts of the graph with the possible rational zeros, we suspect that $−1$ and 2 may be zeros. We compute $P(−1)$ and obtain 0, which verifies that $−1$ is indeed a zero. We now apply synthetic division.

$$
\begin{array}{r|rrrr}
-1 & 3 & -2 & -7 & -2 \\
   &   & -3 & 5 & 2 \\
\hline
   & 3 & -5 & -2 & 0
\end{array}
$$

Therefore

$$P(x) = (x + 1)(3x^2 − 5x − 2)$$

The other two zeros of P are found by equating the quadratic factor to zero and solving the equation.

$$3x^2 − 5x − 2 = 0$$
$$(x − 2)(3x + 1) = 0$$
$$x − 2 = 0 \qquad 3x + 1 = 0$$
$$x = 2 \qquad\quad x = −\tfrac{1}{3}$$

Thus the three zeros of P are $−1, 2$ and $−\tfrac{1}{3}$.
Observe that we have also factored $P(x)$.

$$P(x) = (x + 1)(x − 2)(3x + 1)$$
◀

▶ **EXAMPLE 2** *Applying the Rational Zeros Theorem To Find the Zeros of a Polynomial Function*

Given: $P(x) = 4x^3 + 14x^2 + 10x − 3$. Find any rational zeros of P and, if possible, find any irrational or imaginary zeros.

Solution   By the rational zeros theorem, any rational zero $\frac{p}{q}$ of P must be such that p is an integer factor of $−3$ and q is an integer factor of 4. Thus the possible values of p are $1, -1, 3,$ and -3; and the possible values of q are $1, -1, 2, -2, 4,$ and -4. Hence the set of possible rational zeros of P is

$$\{1, -1, 3, -3, \tfrac{1}{2}, -\tfrac{1}{2}, \tfrac{3}{2}, -\tfrac{3}{2}, \tfrac{1}{4}, -\tfrac{1}{4}, \tfrac{3}{4}, -\tfrac{3}{4}\}$$

The graph of P, plotted on our graphics calculator, appears in Figure 3. The only possible integer zeros are $1, -1, 3,$ and -3, and from Figure 3, none of these numbers is an intercept of the graph. Furthermore, $\tfrac{1}{2}, -\tfrac{1}{2},$ and $\tfrac{3}{2}$ are obviously not intercepts. However, the figure indicates that $-\tfrac{3}{2}$ is a possible intercept and, therefore, a possible zero of P. We compute $P(-\tfrac{3}{2})$ and obtain 0. Thus $-\tfrac{3}{2}$ is a zero of P. We now use synthetic division to divide $P(x)$ by $x + \tfrac{3}{2}$.

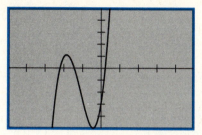

$[-5, 5]$ by $[-5, 5]$
$P(x) = 4x^3 + 14x^2 + 10x - 3$
FIGURE 3

$$
\begin{array}{r|rrrr}
-\tfrac{3}{2} & 4 & 14 & 10 & -3 \\
 & & -6 & -12 & 3 \\
\hline
 & 4 & 8 & -2 & 0 \\
\end{array}
$$

Thus

$$
\begin{aligned}
P(x) &= (x + \tfrac{3}{2})(4x^2 + 8x - 2) \\
&= 2(x + \tfrac{3}{2})(2x^2 + 4x - 1)
\end{aligned}
$$

The other two zeros can be found by setting the quadratic factor equal to zero and solving the equation.

$$2x^2 + 4x - 1 = 0$$

$$
\begin{aligned}
x &= \frac{-4 \pm \sqrt{16 + 8}}{4} \\
&= \frac{-4 \pm 2\sqrt{6}}{4} \\
&= \frac{-2 \pm \sqrt{6}}{2}
\end{aligned}
$$

Therefore the three zeros of P are $-\dfrac{3}{2}, \dfrac{-2 + \sqrt{6}}{2},$ and $\dfrac{-2 - \sqrt{6}}{2}$. ◀

A special case of the rational zeros theorem occurs when a_n, the coefficient of x_n, is 1. Then

$$P(x) = x^n + a_{n-1}x^{n-1} + a_{n-2}x^{n-2} + \ldots + a_1 x + a_0$$

where $a_0, a_1, \ldots, a_{n-1}$ are integers. For such a polynomial any rational zero of P must be an integer and, furthermore, must be an integer factor of a_0. This follows from the fact that if $\dfrac{p}{q}$ is a rational zero of P, then p must be a factor of a_0 and q must be a factor of 1, the coefficient of x^n.

▶ **EXAMPLE 3** *Applying the Rational Zeros Theorem To Find the Zeros of a Polynomial Function*

Given:

$$P(x) = x^4 + 3x^3 - 12x^2 - 13x - 15$$

Find any rational zeros of P and, if possible, find any irrational or imaginary zeros.

Solution   The possible rational zeros are integer factors of -15. These numbers are 1, -1, 3, -3, 5, -5, 15, and -15. We plot the graph of P shown in Figure 4. From the graph, the only possible rational zeros are -5 and 3. We compute $P(-5)$ and $P(3)$, and in each case we obtain 0. Therefore, -5 and 3 are two rational zeros of P. We now use synthetic division to divide $P(x)$ by $x + 5$ and then divide that quotient by $x - 3$.

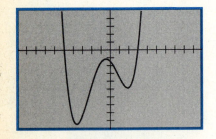

[−10, 10] by [−100, 50]
$P(x) = x^4 + 3x^3 - 12x^2 - 13x - 15$
FIGURE 4

$$
\begin{array}{r|rrrrr}
-5 & 1 & 3 & -12 & -13 & -15 \\
   &   & -5 & 10 & 10 & 15 \\
\hline
3 & 1 & -2 & -2 & -3 & 0 \\
  &   & 3 & 3 & 3 \\
\hline
  & 1 & 1 & 1 & 0
\end{array}
$$

Therefore

$$P(x) = (x + 5)(x - 3)(x^2 + x + 1)$$

We set the quadratic factor equal to zero and solve the equation.

$$x^2 + x + 1 = 0$$
$$x = \frac{-1 \pm \sqrt{1 - 4}}{2}$$
$$= \frac{-1 \pm i\sqrt{3}}{2}$$

Therefore, the zeros of P are -5, 3, $\frac{1}{2}(-1 \pm i\sqrt{3})$.   ◀

We can apply the rational zeros theorem to find the rational roots of a polynomial equation if the equation is equivalent to one in which the coefficients are integers. If all but two of the roots of such a polynomial equation are rational, then any irrational or imaginary roots can be found by the quadratic formula, as shown in Examples 2 and 3.

▶ **EXAMPLE 4** *Applying the Rational Zeros Theorem To Find the Roots of a Polynomial Equation*

Find the rational roots of the equation

$$x^3 + \frac{15}{4}x^2 + x - \frac{1}{2} = 0$$

If possible, find any irrational or imaginary roots.

Solution   To apply the rational zeros theorem the coefficients must be integers. Thus we first multiply each side of the given equation by 4. We obtain the equivalent equation

$$4x^3 + 15x^2 + 4x - 2 = 0$$

Let $P(x) = 4x^3 + 15x^2 + 4x - 2$. If $\frac{p}{q}$ is a rational root of the equation $P(x) = 0$, then p must be an integer factor of -2 and q must be an integer factor of 4. Therefore the possible values of p are $1, -1, 2,$ and -2; the possible values of q are $1, -1, 2, -2, 4,$ and -4. The set of possible rational roots of the equation is, therefore,

$$\{1, -1, 2, -2, \tfrac{1}{2}, -\tfrac{1}{2}, \tfrac{1}{4}, -\tfrac{1}{4}\}$$

On our graphics calculator we obtain the graph of P appearing in Figure 5. We observe that there are no integer x intercepts of the graph. This eliminates $1, -1, 2,$ and -2 as possible roots. The graph indicates that $\tfrac{1}{4}$ is a possible rational root. Computing $P(\tfrac{1}{4})$ by synthetic substitution, we have

$$\begin{array}{r|rrrr} \tfrac{1}{4} & 4 & 15 & 4 & -2 \\ & & 1 & 4 & 2 \\ \hline & 4 & 16 & 8 & 0 \end{array}$$

Because $P(\tfrac{1}{4}) = 0$, $\tfrac{1}{4}$ is a root. Furthermore

$$P(x) = (x - \tfrac{1}{4})(4x^2 + 16x + 8)$$
$$= 4(x - \tfrac{1}{4})(x^2 + 4x + 2)$$

We obtain the other roots by setting the quadratic factor equal to zero and solving the equation.

$$x^2 + 4x + 2 = 0$$
$$x = \frac{-4 \pm \sqrt{16 - 8}}{2}$$
$$= -2 \pm \sqrt{2}$$

The roots of the equation are, therefore, $\tfrac{1}{4}, -2 + \sqrt{2},$ and $-2 - \sqrt{2}$. ◀

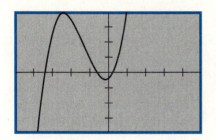

[−5, 5] by [−20, 20]
$P(x) = 4x^3 + 15x^2 + 4x - 2$
FIGURE 5

▶ **EXAMPLE 5** *Proving That a Particular Polynomial Equation Has No Rational Roots*

Prove that the equation

$$2x^3 - 2x^2 - 4x + 1 = 0$$

has no rational roots.

Solution Let $P(x) = 2x^3 - 2x^2 - 4x + 1$. If $\frac{p}{q}$ is a rational root of the equation $P(x) = 0$, then p must be an integer factor of 1 and q must be an integer factor of 2. The possible values of p are therefore 1 and -1; the possible values of q are 1, -1, 2, and -2. Thus the set of possible rational roots of the equation is $\{1, -1, \frac{1}{2}, -\frac{1}{2}\}$. We test each of these possible roots by synthetic substitution.

$$
\begin{array}{r|rrrr}
1 & 2 & -2 & -4 & 1 \\
  &   & 2 & 0 & -4 \\
\hline
  & 2 & 0 & -4 & -3
\end{array}
\qquad
\begin{array}{r|rrrr}
-1 & 2 & -2 & -4 & 1 \\
   &   & -2 & 4 & 0 \\
\hline
   & 2 & -4 & 0 & 1
\end{array}
$$

$$
\begin{array}{r|rrrr}
\frac{1}{2} & 2 & -2 & -4 & 1 \\
            &   & 1 & -\frac{1}{2} \\
\hline
            & 2 & -1 & \text{stop}
\end{array}
\qquad
\begin{array}{r|rrrr}
-\frac{1}{2} & 2 & -2 & -4 & 1 \\
             &   & -1 & \frac{3}{2} \\
\hline
             & 2 & -3 & \text{stop}
\end{array}
$$

The notation "stop" is indicated because once a fraction has appeared in the second row, each successive entry in the bottom row will also be a fraction. Thus the last number in the bottom row cannot be zero.

Because none of $P(1)$, $P(-1)$, $P(\frac{1}{2})$, and $P(-\frac{1}{2})$ is zero, the equation has no rational roots. ◀

We now prove the rational zeros theorem, its importance evinced by the previous four examples.

Proof of the Rational Zeros Theorem Because $\frac{p}{q}$ is a zero of P, it is a solution of the equation

$$a_n x^n + a_{n-1} x^{n-1} + a_{n-2} x^{n-2} + \ldots + a_1 x + a_0 = 0$$

Therefore

$$a_n \left(\frac{p}{q}\right)^n + a_{n-1}\left(\frac{p}{q}\right)^{n-1} + a_{n-2}\left(\frac{p}{q}\right)^{n-2} + \ldots + a_1\left(\frac{p}{q}\right) + a_0 = 0$$

Multiplying on each side of this equation by q^n, we obtain

$$a_n p^n + a_{n-1} p^{n-1} q + a_{n-2} p^{n-2} q^2 + \ldots + a_1 p q^{n-1} + a_0 q^n = 0 \quad \text{(4)}$$

We now add $-a_0 q^n$ on each side and factor p from each term on the resulting left side. Thus we have the equivalent equation

$$p(a_n p^{n-1} + a_{n-1} p^{n-2} q + a_{n-2} p^{n-3} q^2 + \ldots + a_1 q^{n-1}) = -a_0 q^n$$

Because $a_i (i$ is $1, 2, \ldots, n), p,$ and q are integers, and the sum and product of integers are integers, the expression in parentheses on the left side is an integer. If we represent this integer by t, we have the equation

$$pt = -a_0 q^n$$

The left side is an integer having p as a factor. Therefore p must be a factor of the right side, $-a_0 q^n$. Because $\dfrac{p}{q}$ is in lowest terms, p has no factor in common with q. Thus p must be a factor of a_0.

Equation (4) is also equivalent to the equation

$$q(a_{n-1} p^{n-1} + a_{n-2} p^{n-2} q + \ldots + a_1 p q^{n-2} + a_0 q^{n-1}) = -a_n p^n$$

Now the left side is an integer having q as a factor; hence q must be a factor of the right member, $-a_n p^n$. Because q has no factor in common with p, it follows that q must be a factor of a_n. ∎

EXERCISES 5.3

In Exercises 1 through 8, find the zeros of the polynomial function. State the multiplicity of each zero.

1. $P(x) = (x - 4)^2(x^2 - 4)$

2. $P(x) = x^3(x^2 - 5)$

3. $P(x) = x^2(x + 1)^2(x^2 - 3)$

4. $P(x) = (x^2 - 25)^2$

5. $P(x) = (x + 7)^3(x^2 - 7)^2$

6. $P(x) = (x^2 - 2)(x^2 - 4)(2x + 1)$

7. $P(x) = (3x + 4)^3(4x^2 - 9)^2(4x^2 + 12x + 9)$

8. $P(x) = (x^2 - 9)^2(5x^2 - 17x + 6)^2$

9. Show that -2 and 3 are zeros of the polynomial function defined by

$$P(x) = x^4 - 4x^3 - 7x^2 + 22x + 24$$

and find the other two zeros.

10. Show that 5 and -1 are zeros of the polynomial function defined by

$$P(x) = x^4 + x^3 - 31x^2 - x + 30$$

and find the other two zeros.

11. Show that -4 is a zero of multiplicity 2 of the polynomial function defined by

$$P(x) = x^4 + 9x^3 + 23x^2 + 8x - 16$$

and find the other two zeros.

12. Show that 3 is a zero of multiplicity 2 of the polynomial function defined by

$$P(x) = x^4 - 3x^3 - 11x^2 + 39x - 18$$

and find the other two zeros.

13. Given that -2 is a root of the equation

$$5x^3 + 3x^2 - 12x + 4 = 0$$

find the other two roots.

14. Given that $\frac{4}{3}$ is a root of the equation

$$3x^3 - 16x^2 + 28x - 16 = 0$$

find the other two roots.

15. Given that $\frac{1}{2}$ and $-\frac{2}{3}$ are roots of the equation

$$6x^4 + 25x^3 + 8x^2 - 7x - 2 = 0$$

find the other two roots.

16. Given that $\sqrt{3}$ and $-\sqrt{3}$ are roots of the equation

$$x^4 + 3x^3 - 5x^2 - 9x + 6 = 0$$

find the other two roots.

In Exercises 17 through 28, find all the rational zeros of the polynomial function. If possible, find any irrational or imaginary zeros.

17. $P(x) = x^3 - 3x^2 - x + 3$

18. $P(x) = x^3 - 4x^2 + x + 6$

19. $P(x) = x^3 - 7x - 6$

20. $P(x) = x^3 - x^2 - 8x + 12$

21. $P(x) = x^4 + 3x^3 - 12x^2 - 13x - 15$

22. $P(x) = x^4 - 3x^3 + x^2 + 7x - 30$

23. $P(x) = 3x^3 + 8x^2 - 1$

24. $P(x) = 4x^3 - 31x + 15$

25. $P(x) = 6x^4 - 37x^3 + 63x^2 - 33x + 5$

26. $P(x) = 8x^4 + 6x^3 - 13x^2 - x + 3$

27. $P(x) = x^4 - 2x^3 - 9x^2 + 20x - 4$

28. $P(x) = 2x^4 - x^3 + 2x^2 - 7x + 3$

In Exercises 29 through 40, find all the rational roots of the equation. If possible, find any irrational or imaginary roots.

29. $x^3 + 2x^2 - 7x + 4 = 0$

30. $x^3 - 3x^2 - 10x + 24 = 0$

31. $2x^3 - 13x^2 + 27x - 18 = 0$

32. $x^3 - 8x - 8 = 0$

33. $x^5 + 2x^4 - 13x^3 - 14x^2 + 24x = 0$

34. $9x^4 - 3x^3 + 7x^2 - 3x - 2 = 0$

35. $12x^4 - 5x^3 - 38x^2 + 15x + 6 = 0$

36. $2x^3 - \frac{25}{2}x^2 + \frac{7}{2}x - 3 = 0$

37. $x^3 + \frac{17}{3}x^2 - \frac{5}{3}x + 2 = 0$

38. $3x^4 + x^3 + 12x^2 - 5x - 3 = 0$

39. $\frac{1}{14}x^4 + \frac{1}{7}x^3 - \frac{1}{2}x^2 - \frac{1}{2}x + 1 = 0$

40. $18x^6 + 3x^5 - 25x^4 - 41x^3 - 15x^2 = 0$

In Exercises 41 through 44, prove that the equation has no rational roots.

41. $x^3 - 9x - 6 = 0$ **42.** $2x^3 + 6x^2 - 3 = 0$

43. $3x^4 - x^3 + 4x^2 + 2x - 2 = 0$

44. $x^4 - x^3 - 4x^2 - 16 = 0$

In Exercises 45 through 49, solve the word problem by using an equation as a mathematical model of the situation. Be sure to write a conclusion.

45. A rectangular box is to be made from a square piece of tin that measures 12 in. on a side by cutting out small squares of the same size from the four corners and turning up the sides. If the volume of the box is to be 108 in.^{3}, what should be the length of the side of the square to be cut out? There are two possible answers, one rational and one irrational. Express the irrational answer accurate to the nearest one-hundredth of an inch.

46. A rectangular box is to be made from a piece of cardboard 6 cm wide and 14 cm long by cutting out squares of the same size from the four corners and turning up the sides. If the volume of the box is to be 40 cm^3, what should be the length of the side of the square to be cut out? There are two possible answers, one rational and one irrational. Express the irrational answer accurate to the nearest one-hundredth of a centimeter.

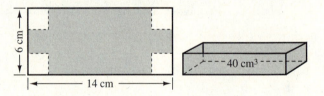

47. A slice of thickness 1 cm is cut off from one side of a cube. If the volume of the remaining figure is 180 cm^3, how long is the edge of the original cube?

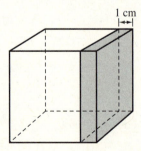

48. A rectangular box has dimensions 12 in., 4 in., and 4 in. If the first two dimensions are decreased and the other dimension is increased by the same amount, a

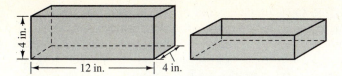

second box is formed, and its volume is five-eighths of the volume of the first box. Determine the dimensions of the second box.

49. A right-circular cone is inscribed in a sphere, and 32 times its volume is equal to 9 times the volume of the sphere. If the radius of the sphere is 2 ft, find the altitude of the cone. There are two possible answers, one rational and one irrational. Express the irrational

answer to the nearest one-hundredth of a foot. The formula for the volume of a cone is $V = \frac{1}{3}\pi r^2 h$, and the formula for the volume of a sphere is $V = \frac{4}{3}\pi r^3$.

50. Explain why the number of times the graph of a cubic polynomial crosses the x axis must be either 1 or 3 and why it cannot be 2.

5.4 COMPLEX ZEROS OF POLYNOMIAL FUNCTIONS

GOALS

1. **Find a polynomial with real coefficients having given complex numbers as zeros.**
2. **Find the other zeros of a polynomial function having a given complex number as a zero.**
3. **Learn Descartes' rule of signs.**
4. **Determine the character of the zeros of a polynomial function.**
5. **Prove that certain numbers are irrational.**

So far in this chapter our main concern has been finding real zeros of polynomials with real coefficients. We now extend our discussion to include complex numbers, which of course comprise the real numbers. We shall deal with complex zeros as well as polynomial functions having complex coefficients. Such functions arise in more advanced courses and have applications in various fields, especially engineering and physics.

You know how to find the exact zeros of linear and quadratic functions. You also know how to approximate real zeros of a polynomial function from its graph. Furthermore, you learned in Section 5.3 methods for obtaining exact values of any rational zeros of a polynomial function. A natural question that arises is: Can we find exact values of all complex zeros of any polynomial function?

For cubic and quartic (fourth degree) polynomials, this question was answered in the sixteenth century. In the book *Ars Magna* (Great Art) published in 1545, the Italian mathematician Girolamo Cardano presented methods for finding exact values of the three complex zeros of cubic polynomials and the four complex zeros of quartic polynomials. These methods are quite complicated and will not be discussed in this text. For zeros of polynomials of the fifth degree or higher, there is no general formula in terms of a finite number of operations on the coefficients. This fact was proved independently by the Italian mathematician Paolo Ruffini (1767–1822) in 1799 and by the Norwegian mathematician Niels Henrik Abel (1802–1829) in 1824. There is a theorem, however, called the *fundamental theorem of algebra*, that guarantees that every polynomial function of nonzero degree has at least one complex zero.

THEOREM   *The Fundamental Theorem of Algebra*

Every polynomial function of degree greater than zero, with complex coefficients, has at least one complex zero.

There are many proofs of this theorem, but all involve concepts beyond the level of this book. The theorem was first proved in 1799 by the German mathematician Carl Friedrich Gauss (1777–1855) in his doctorate dissertation.

The fundamental theorem of algebra and the factor theorem are used to prove the next theorem. Recall that we applied this theorem in Section 5.3.

THEOREM 1

If $P(x)$ is the polynomial with complex coefficients defined by

$$P(x) = a_n x^n + a_{n-1} x^{n-1} + a_{n-2} x^{n-2} + \ldots + a_1 x + a_0$$

where $n \geq 1$, then

$$P(x) = a_n(x - r_1)(x - r_2) \cdot \ldots \cdot (x - r_n) \qquad a_n \neq 0 \qquad (1)$$

where each r_i (i is $1, 2, \ldots, n$) is a complex zero of P.

Proof   From the fundamental theorem of algebra, the function P has at least one complex zero, r_1. That is, there exists a complex number r_1 such that $P(r_1) = 0$. Therefore, by the factor theorem, $x - r_1$ is a factor of $P(x)$. Thus

$$P(x) = (x - r_1)Q_1(x) \qquad (2)$$

where $Q_1(x)$ is the quotient obtained when $P(x)$ is divided by $x - r_1$, and $Q_1(x)$ is of degree $n - 1$. From the fundamental theorem of algebra, if $n - 1 \geq 1$, there exists a complex number r_2 such that $Q_1(r_2) = 0$. Then, by the factor theorem,

$$Q_1(x) = (x - r_2)Q_2(x) \qquad (3)$$

where $Q_2(x)$ is the quotient obtained when $Q_1(x)$ is divided by $x - r_2$. Substituting from (3) into (2), we get

$$P(x) = (x - r_1)(x - r_2)Q_2(x)$$

Because $Q_1(r_2) = 0$, it follows from (2) that $P(r_2) = 0$, and hence r_2 is a complex zero of P. We continue this procedure until the factoring has been performed n times; then we have

$$P(x) = (x - r_1)(x - r_2) \cdot \ldots \cdot (x - r_n)Q_n(x)$$

where each r_i (i is $1, 2, \ldots, n$) is a complex zero of P. Because there are n factors of the form $x - r_i$, the polynomial $Q_n(x)$ must be a constant, and that constant must be the coefficient of x^n in the expansion. Thus $Q_n(x) = a_n$, and therefore

$$P(x) = a_n(x - r_1)(x - r_2) \cdot \ldots \cdot (x - r_n)$$

where each r_i is a complex zero of P. ∎

In Section 5.3 you learned that if in Equation (1) a factor $x - r_i$ occurs exactly k times, then r_i is called a zero of multiplicity k. If such a zero is counted as k zeros, then it follows from Theorem 1 that a polynomial function P, for which $P(x)$ is of degree $n \geq 1$, has *at least n zeros*, some of which may be repeated. However, we can prove that such a polynomial function has *exactly n zeros,* and this fact is stated in the next theorem, which we also applied in Section 5.3.

THEOREM 2

If $P(x)$ is a polynomial of degree $n \geq 1$, with complex coefficients, then P has exactly n complex zeros.

Proof   From Theorem 1, P has at least n complex zeros. If we now show that P cannot have more than n zeros, the theorem will be proved.
Equation (1) is

$$P(x) = a_n(x - r_1)(x - r_2) \cdot \ldots \cdot (x - r_n) \qquad a_n \neq 0$$

where each r_i (i is $1, 2, \ldots, n$) is a complex zero of P. Let r be any number other than $r_1, r_2, \ldots, r_n$. Because Equation (1) is an identity,

$$P(r) = a_n(r - r_1)(r - r_2) \cdot \ldots \cdot (r - r_n) \qquad a_n \neq 0$$

Because $r \neq r_i$, none of the factors $r - r_i$ is zero; therefore the right side of this equation is not zero. Thus $P(r) \neq 0$; so r is not a zero of P. Hence P has exactly n complex zeros. ∎

The polynomial function with real coefficients in Example 3 of Section 5.3 has two zeros $\frac{1}{2}(-1 + i\sqrt{3})$ and $\frac{1}{2}(-1 - i\sqrt{3})$ that are conjugates of each other. Furthermore, from the quadratic formula, it follows that if a quadratic function having real coefficients has a complex zero, then the other zero is its conjugate. These two situations are special cases of the following theorem, which we present without proof. In the statement of the theorem the notation $\bar{z}$ is used to denote the conjugate of the complex number z; that is,

$$\text{if } z = a + bi \qquad \text{then} \qquad \bar{z} = a - bi$$

THEOREM 3

If $P(x)$ is a polynomial with real coefficients and if z is a complex zero of P, then the conjugate $\bar{z}$ is also a zero of P.

▶ **EXAMPLE 1** *Applying Theorem 3*

Find a polynomial $P(x)$ of the fourth degree with real coefficients if P has $1 - i$ and $-2i$ as zeros.

Solution   From Theorem 3, if $1 - i$ and $-2i$ are zeros of P, then their conjugates $1 + i$ and $2i$ are also zeros. Therefore

$$\begin{aligned} P(x) &= [x - (1 - i)][x - (1 + i)][x - (-2i)][x - 2i] \\ &= (x^2 - 2x + 2)(x^2 + 4) \\ &= x^4 - 2x^3 + 6x^2 - 8x + 8 \end{aligned}$$ ◀

In Section 5.2 we stated the factor theorem when the number r is a real number and the polynomial $P(x)$ has real coefficients. When r is complex and $P(x)$ has complex coefficients, the factor theorem is also valid.

▶ **EXAMPLE 2** *Finding the Other Zeros of a Polynomial Function Having a Given Complex Number as a Zero*

Given that i is a zero of the function defined by

$$P(x) = 2x^4 - 5x^3 + 3x^2 - 5x + 1$$

find the other zeros.

Solution   Because i is a zero of the given function, its conjugate, $-i$, is also a zero. We use synthetic division to divide $P(x)$ by $x - i$; then we divide the quotient by $x - (-i)$.

$$
\begin{array}{r|rrrrr}
i & 2 & -5 & 3 & -5 & 1 \\
& & 2i & -2-5i & 5+i & -1 \\
\hline
-i & 2 & -5+2i & 1-5i & i & 0 \\
& & -2i & 5i & -i & \\
\hline
& 2 & -5 & 1 & 0 &
\end{array}
$$

Therefore, by the factor theorem, $P(x)$ can be written as

$$P(x) = (x - i)(x + i)(2x^2 - 5x + 1)$$

Equating the quadratic factor to zero, we obtain

$$2x^2 - 5x + 1 = 0$$
$$x = \frac{5 \pm \sqrt{25 - 8}}{4}$$
$$= \frac{5 \pm \sqrt{17}}{4}$$

Thus the zeros of P are i, $-i$, $\frac{1}{4}(5 + \sqrt{17})$, and $\frac{1}{4}(5 - \sqrt{17})$.   ◀

We now present two theorems that are consequences of Theorem 3.

THEOREM 4

If $P(x)$ is a polynomial with real coefficients and the degree of P is an odd number, then P has at least one real zero.

Proof   Assume P has no real zeros. Then because the degree of P is an odd number, it has an odd number of imaginary zeros. However, the number of imaginary zeros of P must be even because by Theorem 3, for each imaginary zero its conjugate must also be a zero. Therefore we have a contradiction. Thus our assumption that P has no real zeros is false. Hence P has at least one real zero.   ∎

THEOREM 5

If $P(x)$ is a polynomial with real coefficients, then $P(x)$ can be expressed as a product of linear or quadratic polynomials with real coefficients.

Proof   If $P(x)$ is of degree n, then by Theorem 2, P has exactly n complex zeros. Denote these zeros by $c_1, c_2, \ldots, c_n$. Thus

$$P(x) = a_n(x - c_1)(x - c_2) \cdot \ldots \cdot (x - c_n)$$

For every c_i that is a real number, $x - c_i$ is a linear factor of $P(x)$. Suppose

z_i is an imaginary zero of P. Then by Theorem 3, $\overline{z}_i$ is also a zero of P and is one of the complex zeros $c_1, c_2, \ldots, c_n$. Therefore $x - z_i$ and $x - \overline{z}_i$ are factors of $P(x)$, and

$$(x - z_i)(x - \overline{z}_i) = x^2 - (z_i + \overline{z}_i)x + z_i \cdot \overline{z}_i$$

Let $z_i = a + bi$ and $\overline{z}_i = a - bi$. Then

$$\begin{aligned} z_i + \overline{z}_i &= (a + bi) + (a - bi) & z_i \cdot \overline{z}_i &= (a + bi)(a - bi) \\ &= 2a & &= a^2 - b^2 i^2 \\ & & &= a^2 + b^2 \end{aligned}$$

Hence

$$(x - z_i)(x - \overline{z}_i) = x^2 - 2ax + (a^2 + b^2)$$

Because $-2a$ and $a^2 + b^2$ are real numbers, $(x - z_i)(x - \overline{z}_i)$ is a quadratic polynomial with real coefficients.

Therefore, we conclude that $P(x)$ can be expressed as a product of linear or quadratic polynomials with real coefficients.   ■

Theorem 5 is applied in calculus when partial fractions, discussed in Section 12.3, are used as a computational technique. The following two illustrations verify the theorem for two particular polynomials.

▷ ILLUSTRATION 1

The function P in Example 3 of Section 5.3 is defined by

$$P(x) = x^4 + 3x^3 - 12x^2 - 13x - 15$$

In the solution of that example, we showed that

$$P(x) = (x + 5)(x - 3)(x^2 + x + 1)$$

which is the product of linear and quadratic polynomials.   ◀

▷ ILLUSTRATION 2

For the function P of Example 2

$$\begin{aligned} P(x) &= (x - i)(x + i)(2x^2 - 5x + 1) \\ &= (x^2 + 1)(2x^2 - 5x + 1) \end{aligned}$$

which is the product of two quadratic polynomials.   ◀

From Theorem 3, the imaginary zeros of a polynomial function with real coefficients must occur in pairs. This means that a polynomial function must always have an even number of imaginary zeros. This fact often helps us determine the character of the zeros of a polynomial function.

▷ **ILLUSTRATION 3**

In Example 5 of Section 5.3 we showed that the equation

$$2x^3 - 2x^2 - 4x + 1 = 0$$

has no rational roots or, equivalently, that the function P defined by

$$P(x) = 2x^3 - 2x^2 - 4x + 1$$

has no rational zeros. Because P is of the third degree, we know from Theorem 2 that P has exactly three complex zeros. These zeros must therefore be either irrational or imaginary. Because the number of imaginary zeros must be even, P has either three real zeros, all of which are irrational, or two imaginary zeros and one real irrational zero. ◀

In the discussion that follows we shall need the concept of *variation in sign* of a polynomial. If the terms of a polynomial with real coefficients are written in descending powers of the variable (the terms involving zero coefficients are omitted), then a **variation in sign** occurs if two successive coefficients are opposite in sign.

▷ **ILLUSTRATION 4**

If

$$Q(x) = x^4 - 6x^2 - 2x - 1$$

the coefficients have, successively, the signs $+, -, -, -$. Thus $Q(x)$ has one variation in sign. Furthermore,

$$Q(-x) = x^4 - 6x^2 + 2x - 1$$

and $Q(-x)$ has three variations in sign.

The polynomial $x^3 + 2x + 5$ has no variations in sign. ◀

Descartes' Rule of Signs

If $P(x)$ is a polynomial having real coefficients, the number of positive roots of the equation $P(x) = 0$ either is equal to the number of variations in sign of $P(x)$ or is less than this number by an even natural number. Furthermore, the number of negative roots of the equation is equal to the number of variations in sign of $P(-x)$ or is less than this number by an even natural number.

Descartes' rule of signs is named for the French mathematician René Descartes, the originator of analytic geometry. The proof of Descartes' rule of signs is beyond the scope of this book.

▷ **ILLUSTRATION 5**

Function P of Illustration 3 has two variations in sign. Therefore, by Descartes' rule of signs, P has either two or no positive zeros. Furthermore,

$$P(-x) = -2x^3 - 2x^2 + 4x + 1$$

and $P(-x)$ has one variation in sign. Thus P has one negative zero.

The three zeros of P can now be described more fully. From our conclusion in Illustration 3 and what we have just shown, either two of the zeros are positive irrational numbers and one is a negative irrational number or else two are imaginary numbers and one is a negative irrational number. ◀

The following theorem gives more information about the real zeros of a polynomial function.

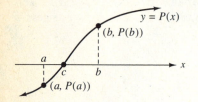

FIGURE 1

THEOREM 6
Suppose $P(x)$ is a polynomial and a and b are real numbers such that $a < b$. Then if $P(a)$ and $P(b)$ are opposite in sign, there is a real number c between a and b such that $P(c) = 0$.

The proof of Theorem 6 is omitted. However, because the graph of a polynomial function is a continuous unbroken curve, the following argument should make the theorem seem reasonable: If $P(a)$ and $P(b)$ are opposite in sign, then the points $(a, P(a))$ and $(b, P(b))$ are on opposite sides of the x axis; thus the graph of $y = P(x)$ must intersect the x axis in at least one point $(c, 0)$ where c is between a and b. We show this situation in Figures 1 and 2. In Figure 1 there is a portion of the graph of a polynomial function P from the point $(a, P(a))$ to $(b, P(b))$ where $P(a) < 0$ and $P(b) > 0$. The graph intersects the x axis at the point $(c, 0)$, where $a < c < b$. Figure 2 shows the case when $P(a) > 0$ and $P(b) < 0$.

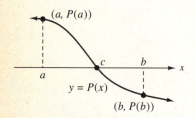

FIGURE 2

▷ **ILLUSTRATION 6**

For the function P in Illustration 5 defined by

$$P(x) = 2x^3 - 2x^2 - 4x + 1$$

we compute $P(1)$ and $P(2)$ by synthetic substitution.

$$\begin{array}{r|rrrr} 1 & 2 & -2 & -4 & 1 \\ & & 2 & 0 & -4 \\ \hline & 2 & 0 & -4 & -3 \end{array} \qquad \begin{array}{r|rrrr} 2 & 2 & -2 & -4 & 1 \\ & & 4 & 4 & 0 \\ \hline & 2 & 2 & 0 & 1 \end{array}$$

Therefore $P(1) = -3$ and $P(2) = 1$. Because $P(1)$ and $P(2)$ are opposite in sign, it follows from Theorem 6 that a real number c between 1 and 2 exists such that $P(c) = 0$. Thus P has a positive zero between 1 and 2. With this fact and the conclusion of Illustration 5, we are certain that P has two positive irrational zeros and one negative irrational zero.

We can apply Theorem 6 to locate integers between which the other two zeros lie. Because $P(0) = 1$, it follows that $P(0)$ and $P(1)$ are opposite in sign, and hence P has a positive zero between 0 and 1. We compute $P(-1)$ and $P(-2)$ by synthetic substitution.

$$\begin{array}{r|rrrr} -1 & 2 & -2 & -4 & 1 \\ & & -2 & 4 & 0 \\ \hline & 2 & -4 & 0 & 1 \end{array} \qquad \begin{array}{r|rrrr} -2 & 2 & -2 & -4 & 1 \\ & & -4 & 12 & -16 \\ \hline & 2 & -6 & 8 & -15 \end{array}$$

Thus $P(-1)$ and $P(-2)$ are opposite in sign, and consequently P has a negative zero between -2 and -1.

Figure 3 shows the graph of P plotted on a graphics calculator. The x intercepts of the graph verify our conclusions about the zeros of P. ◄

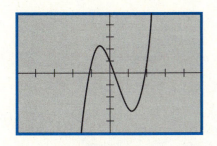

[−5, 5] by [−5, 5]
$P(x) = 2x^3 - 2x^2 - 4x + 1$
FIGURE 3

▶ **EXAMPLE 3** *Determining the Character of the Zeros of a Polynomial Function*

For each of the following functions, determine all the information you can concerning the number of positive, negative, and imaginary zeros. Then plot the graph of the function and verify the information.

(a) $F(x) = 3x^4 + x^2 + 7x + 1$ **(b)** $G(x) = x^5 + 5x^2 - 4$

Solution

(a) Because $F(x)$ has no variations in sign, F has no positive zero. $F(-x) = 3x^4 + x^2 - 7x + 1$. Because $F(-x)$ has two variations in sign, F has either two or no negative zeros. Computing $F(0)$ and $F(-1)$ we get, respectively, 1 and -2, opposite in sign. Therefore from Theorem 6, a number c between 0 and -1 exists such that $F(c) = 0$. This number c is a negative zero of F. Thus F has two negative zeros and two imaginary zeros. The graph of F, plotted in Figure 4, verifies our conclusion.

(b) $G(x)$ has one variation in sign. Therefore G has one positive zero. $G(-x) = -x^5 + 5x^2 - 4$, and $G(-x)$ has two variations in sign. Thus G has either two or no negative zeros. Because $G(-1) = 0$, -1 is a zero of G. Consequently G has one positive zero, two negative zeros, and two imaginary zeros. Figure 5 shows the graph of G plotted on our graphics calculator. The graph verifies our conclusion. ◄

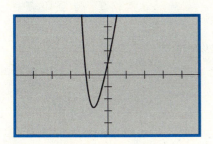

[−5, 5] by [−5, 5]
$F(x) = 3x^4 + x^2 + 7x + 1$
FIGURE 4

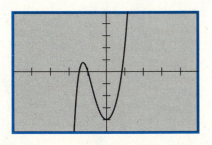

[−5, 5] by [−5, 5]
$G(x) = x^5 + 5x^2 - 4$
FIGURE 5

$\iint \dfrac{dy}{dx}$ Irrational zeros of a polynomial function can be approximated by a technique known as *Newton's method*, which involves concepts of calculus. Of course, irrational zeros can also be approximated by using a graphics calculator to zoom in on the x intercepts of the graph of the function. We conclude this chapter with an example showing how to prove that certain numbers are irrational.

▶ **EXAMPLE 4** *Proving That a Number is Irrational*

Prove that $\sqrt{3}$ is irrational.

Solution Let $P(x) = x^2 - 3$. Because $P(x)$ has one variation in sign, by Descartes' rule of signs P has one positive zero. From the rational zeros theorem, the only possible rational zeros are 1 and 3. Obviously neither is a zero of P. Thus the positive zero is an irrational number, and it is $\sqrt{3}$.

◀

EXERCISES 5.4

In Exercises 1 through 8, find a polynomial $P(x)$ of the stated degree with real coefficients for which the numbers are zeros of P.

1. Second degree; $4 + 3i$ is a zero of P.

2. Second degree; $3 - i$ is a zero of P.

3. Third degree; $2 - i\sqrt{5}$ and -4 are zeros of P.

4. Third degree; $5 + i\sqrt{3}$ and 2 are zeros of P.

5. Fourth degree; $-3i$ is a zero of multiplicity 2 of P.

6. Fourth degree; $2 + i$ and $1 - i\sqrt{2}$ are zeros of P.

7. Fifth degree; 3, $3 + i\sqrt{2}$, and $-i\sqrt{2}$ are zeros of P.

8. Fifth degree; $3 - i$ (multiplicity 2) and 1 are zeros of P.

In Exercises 9 through 12, find the solution set of the equation if the number is a root.

9. $3x^4 - 2x^3 + 2x^2 - 8x - 40 = 0$; $2i$ is a root.

10. $5x^4 - 2x^3 + 46x^2 - 18x + 9 = 0$; $-3i$ is a root.

11. $2x^4 + 6x^3 + 33x^2 - 36x + 20 = 0$; $-2 - 4i$ is a root.

12. $3x^4 + 4x^3 + 9x^2 - 6x + 4 = 0$; $-1 + i\sqrt{3}$ is a root.

In Exercises 13 through 20, find the zeros of the polynomial function and use these zeros to express the polynomial as a product of linear or quadratic polynomials.

13. $P(x) = x^3 - 4x^2 + 6x - 4$

14. $P(x) = x^3 + 2x^2 - 2x + 3$

15. $P(x) = 2x^4 - 7x^3 + 21x^2 + 17x - 13$

16. $P(x) = x^4 - 2x^3 + 2x^2 + 2x - 3$

17. $P(x) = x^5 - 2x^4 + 8x^3 - 16x^2 + 16x - 32$

18. $P(x) = 3x^5 - 18x^4 + 38x^3 - 36x^2 + 24x - 16$

19. $P(x) = 9x^4 - 42x^3 + 79x^2 - 40x + 6$

20. $P(x) = 6x^4 + 5x^3 + 4x^2 - 2x - 1$

In Exercises 21 through 32, determine all the information you can concerning the number of positive, negative, and imaginary zeros of the polynomial function. Find any rational zeros and locate any irrational zeros between two consecutive integers. Then plot the graph of the function on your graphics calculator and verify your answer.

21. $F(x) = x^3 - 4x^2 - 2$ **22.** $G(x) = 5x^3 - 3x - 7$

23. $G(x) = x^4 + 7x^3 + x - 8$

24. $F(x) = 5x^4 - 3x^3 - 2$ **25.** $H(x) = x^3 - 6x + 3$

26. $H(x) = x^3 + 3x - 20$ **27.** $F(x) = x^4 + x^2 - 1$

28. $G(x) = x^3 + 3x^2 - 2x - 5$

29. $G(x) = 4x^4 - 3x^3 + 2x - 5$

30. $F(x) = 2x^4 - 14x^3 + 24x^2 + x - 4$

31. $H(x) = 3x^4 - 21x^3 + 36x^2 + 2x - 8$

32. $H(x) = x^4 + 2x^3 - 9x^2 - 8x + 14$

In Exercises 33 through 38, prove that the number is irrational.

33. $\sqrt{5}$ **34.** $2\sqrt{7}$ **35.** $\sqrt[3]{10}$

36. $\sqrt[4]{8}$ **37.** $2 - \sqrt{5}$ **38.** $3 + 2\sqrt{3}$

In Exercises 39 and 40, determine the number of positive, negative, and imaginary zeros of the function, where n is a positive integer.

39. (a) $P(x) = x^{2n} - 1$ **(b)** $Q(x) = x^{2n-1} + 1$

40. (a) $P(x) = x^{2n} + 1$ **(b)** $Q(x) = x^{2n+1} - 1$

41. If $P(x)$ has only even powers of x with positive coefficients, explain why the function P has neither a positive nor a negative zero. *Hint:* Refer to Descartes' rule of signs.

42. If $P(x)$ has only odd powers of x with positive coefficients, explain why the function P has no real zero except the number 0. *Hint:* Refer to Descartes' rule of signs.

43. Show that $2 - i$ is a zero of the function defined by $F(x) = x^2 - 2x + 1 + 2i$, but that its conjugate is not. Explain why Theorem 3 is not contradicted.

5.5 RATIONAL FUNCTIONS

GOALS

1. Define a vertical asymptote of a graph.
2. Define a horizontal asymptote of a graph.
3. Find any vertical and horizontal asymptotes of a graph.
4. Find an oblique asymptote of a graph.
5. Sketch the graph of a rational function.
6. Solve word problems having a rational function as a mathematical model.
7. Solve rational inequalities graphically.

We begin our treatment of rational functions by considering their graphs and follow this discussion with a word problem having a rational function as a mathematical model. We conclude the section by solving rational inequalities graphically.

If P and Q are polynomial functions and f is the function defined by

$$f(x) = \frac{P(x)}{Q(x)}$$

then f is a rational function whose domain is the set of all real numbers except the zeros of Q. We shall assume that $P(x)$ and $Q(x)$ have no common factor. Knowing the behavior of $f(x)$ when x is close to a zero of Q is helpful when obtaining the graph of f. Consider, for example, the function defined by

$$f(x) = \frac{x + 2}{x - 3}$$

Table 1

x	4	$\frac{7}{2}$	$\frac{10}{3}$	$\frac{13}{4}$	$\frac{31}{10}$	$\frac{301}{100}$	$\frac{3001}{1000}$
$f(x) = \dfrac{x+2}{x-3}$	6	11	16	21	51	501	5001

The domain of f is the set of all real numbers except 3. We shall investigate the function values when x is close to 3 but not equal to 3. First let x take on the values $4, \frac{7}{2}, \frac{10}{3}, \frac{13}{4}, \frac{31}{10}, \frac{301}{100}, \frac{3001}{1000}$, and so on. We are taking values of x closer and closer to 3 but greater than 3; in other words, the variable x is approaching 3 from the right. We compute the corresponding values of $f(x)$ and enter them in Table 1. From the table we see intuitively that as x gets closer and closer to 3 from the right, $f(x)$ increases without bound. In other words, we can make $f(x)$ greater than any preassigned positive number by taking x close enough to 3 and x greater than 3. To indicate that $f(x)$ increases without bound as x approaches 3 from the right, we use the symbolism

$$f(x) \longrightarrow +\infty \qquad \text{as } x \longrightarrow 3^+$$

The symbol $+\infty$ (positive infinity) does not represent a real number; it is used to indicate the behavior of the function values $f(x)$ as x gets closer and closer to 3. The plus symbol as a superscript after the 3 indicates that x is approaching 3 from the right.

Now let the variable x approach 3 through values less than 3; that is, let x take on the values $2, \frac{5}{2}, \frac{8}{3}, \frac{11}{4}, \frac{29}{10}, \frac{299}{100}, \frac{2999}{1000}$, and so on. Refer to Table 2 for the corresponding function values. Notice that as x gets closer and closer to 3 from the left, the values of $f(x)$ decrease without bound (the values of $f(x)$ are negative numbers whose absolute values increase without bound); that is, we can make $f(x)$ less than any preassigned negative number by taking x close enough to 3 and x less than 3. We use the following notation to indicate that $f(x)$ decreases without bound as x approaches 3 from the left:

$$f(x) \longrightarrow -\infty \qquad \text{as } x \longrightarrow 3^-$$

In Figure 1 we have the graph of f showing the behavior of $f(x)$ near $x = 3$. As x gets closer and closer to 3 from either the right or left, the absolute value of $f(x)$ gets larger and larger. Observe that the graph does not

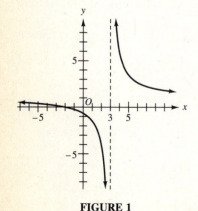

FIGURE 1

Table 2

x	2	$\frac{5}{2}$	$\frac{8}{3}$	$\frac{11}{4}$	$\frac{29}{10}$	$\frac{299}{100}$	$\frac{2999}{1000}$
$f(x) = \dfrac{x+2}{x-3}$	-4	-9	-14	-19	-49	-499	-4999

intersect the line $x = 3$, indicated with dashes in the figure. The line $x = 3$ is called a *vertical asymptote* of the graph of f.

DEFINITION   A Vertical Asymptote

The line $x = a$ is said to be a **vertical asymptote** of the graph of the function f if at least one of the following statements is true:

(i) $f(x) \to +\infty$ as $x \to a^+$
(ii) $f(x) \to -\infty$ as $x \to a^+$
(iii) $f(x) \to +\infty$ as $x \to a^-$
(iv) $f(x) \to -\infty$ as $x \to a^-$

In Figure 1 both statements (i) and (iv) of the above definition are true for the function f when a is 3.

▷ ILLUSTRATION 1

Figure 2 shows the graph of a function for which statements (i) and (iii) of the above definition are true, and for the function whose graph appears in Figure 3, statements (ii) and (iii) are true.   ◀

The following theorem can be proved from the definition of a vertical asymptote.

THEOREM 1

The graph of a rational function of the form $P(x)/Q(x)$, where $P(x)$ and $Q(x)$ have no common factors, has the line $x = a$ as a vertical asymptote if $Q(a) = 0$.

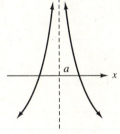

FIGURE 2

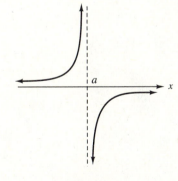

FIGURE 3

We apply Theorem 1 in the next example.

▶ EXAMPLE 1   *Sketching the Graph of a Rational Function*

Sketch the graph of the function defined by

$$f(x) = \frac{3}{x^2}$$

Check the graph on a graphics calculator.

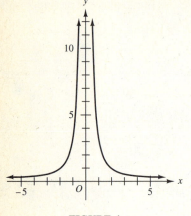

FIGURE 4

Solution   The domain of f is the set of all real numbers except 0. Thus the graph has no y intercept. Because $f(x)$ is never zero, the graph has no x intercept. Also observe from the equation that $f(x)$ is never negative. Therefore the graph is confined to the first and second quadrants. Because $f(-x) = f(x)$, f is an even function, and the graph is symmetric with respect to the y axis.

To obtain any vertical asymptotes, we use Theorem 1. We set the denominator equal to zero and get $x = 0$. Thus the y axis is a vertical asymptote.

A few points on the graph in the first quadrant are given in Table 3. We locate these points and use the symmetry property to obtain corresponding points in the second quadrant. We complete the sketch shown in Figure 4 by connecting the points in each quadrant with an unbroken curve and using the preceding information. We obtain the same graph on our graphics calculator.

Table 3

x	$\frac{1}{2}$	1	2	3	4
$f(x) = \dfrac{3}{x^2}$	12	3	$\frac{3}{4}$	$\frac{1}{3}$	$\frac{3}{16}$

◀

Observe from the graph in Figure 4 that the function values $f(x)$ approach zero as $|x|$ increases without bound. For this reason the x axis is a *horizontal asymptote* of the graph. To define a horizontal asymptote, we use the notation $f(x) \rightarrow b^+$ to mean that $f(x)$ approaches b through values greater than b, and $f(x) \rightarrow b^-$ to mean that $f(x)$ approaches b through values less than b. The notation $x \rightarrow +\infty$ indicates that x is *increasing without bound*, and $x \rightarrow -\infty$ means that x is *decreasing without bound*.

DEFINITION   A Horizontal Asymptote of the Graph of a Rational Function

The line $y = b$ is said to be a **horizontal asymptote** of the graph of the rational function f if at least one of the following statements is true:

(i) $f(x) \rightarrow b^+$ as $x \rightarrow +\infty$
(ii) $f(x) \rightarrow b^+$ as $x \rightarrow -\infty$
(iii) $f(x) \rightarrow b^-$ as $x \rightarrow +\infty$
(iv) $f(x) \rightarrow b^-$ as $x \rightarrow -\infty$

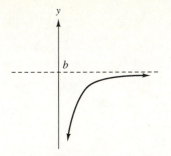

FIGURE 5

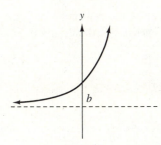

FIGURE 6

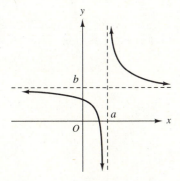

FIGURE 7

▷ **ILLUSTRATION 2**

Figure 5 shows the graph of a function for which statement (iii) of the above definition is true, and for the graph in Figure 6 statement (ii) is true. Both statements (i) and (iv) are true for the graph of the function shown in Figure 7. The graph in Figure 7 also has the line $x = a$ as a vertical asymptote because $f(x) \rightarrow +\infty$ as $x \rightarrow a^+$ and $f(x) \rightarrow -\infty$ as $x \rightarrow a^-$. ◀

THEOREM 2

The graph of a rational function of the form

$$\frac{a_n x^n + a_{n-1}x^{n-1} + \ldots + a_1 x + a_0}{b_m x^m + b_{m-1}x^{m-1} + \ldots + b_1 x + b_0}$$

has

(i) the x axis as a horizontal asymptote if $n < m$;

(ii) the line $y = \dfrac{a_n}{b_m}$ as a horizontal asymptote if $n = m$;

(iii) no horizontal asymptote if $n > m$.

The proof of Theorem 2 is omitted, but the following two illustrations should make parts (i) and (ii) plausible. Later, in Example 5, we have a function for which part (iii) applies.

▷ **ILLUSTRATION 3**

Let the function g be defined by

$$g(x) = \frac{4x}{x^2 - 25}$$

The degree of the numerator is 1, and the degree of the denominator is 2. Because $1 < 2$, g is a rational function for which part (i) of Theorem 2 should apply. To see this, let us divide the numerator and denominator by x^2. We then have

$$g(x) = \frac{\dfrac{4}{x}}{1 - \dfrac{25}{x^2}}$$

As $x \rightarrow +\infty$, $\dfrac{4}{x} \rightarrow 0^+$ and $\dfrac{25}{x^2} \rightarrow 0^+$. Therefore, as $x \rightarrow +\infty$, $g(x) \rightarrow 0^+$. As

$x \to -\infty$, $\dfrac{4}{x} \to 0^-$ and $\dfrac{25}{x^2} \to 0^+$. Therefore, as $x \to -\infty$, $g(x) \to 0^-$. Thus from the definition, the line $y = 0$ is a horizontal asymptote of the graph of g. Function g is discussed in Example 3 later in this section, and its graph appears in Figure 9. ◀

▷ **ILLUSTRATION 4**

Let the function h be defined by

$$h(x) = \frac{3x^2}{2x^2 - 32}$$

The degree of the numerator equals the degree of the denominator. Hence h is a rational function for which part (ii) of Theorem 2 should apply. If we divide the numerator and denominator by x^2, we get

$$h(x) = \frac{3}{2 - \dfrac{32}{x^2}}$$

As $x \to +\infty$ or $x \to -\infty$, $\dfrac{32}{x^2} \to 0^+$. Thus as $x \to +\infty$ or $x \to -\infty$, $h(x) \to \tfrac{3}{2}^+$.

From the definition it follows that the line $y = \tfrac{3}{2}$ is a horizontal asymptote. Function h is discussed in Example 4 later in this section, and its graph appears in Figure 10. ◀

▶ **EXAMPLE 2** *Sketching the Graph of a Rational Function*

Sketch the graph of the function defined by

$$f(x) = \frac{2x - 3}{x + 1}$$

Check the graph on a graphics calculator.

Solution   The domain of f is the set of all real numbers except -1. Because $f(0) = -3$, the y intercept is -3. The x intercept is $\tfrac{3}{2}$, which is obtained by setting the numerator equal to zero. Because f is neither even nor odd, there is no symmetry with respect to the y axis or origin.

By setting the denominator equal to zero, we obtain the line $x = -1$ as a vertical asymptote. Because the degrees of the numerator and denominator are equal, it follows from Theorem 2(ii) that a horizontal asymptote is the line $y = 2$. Some points on the graph, determined by computing $f(x)$ for selected values of x, are given by Table 4. With these points and using the asymptotes as guides, we obtain the sketch of the graph of f appearing in Figure 8.

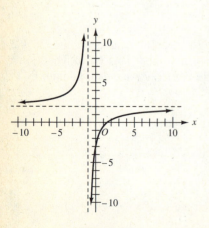

FIGURE 8

Table 4

x	0	2	4	6	8	10	-2	-4	-6	-8	-10	-12
$f(x) = \dfrac{2x-3}{x+1}$	-3	$\frac{1}{3}$	1	$\frac{9}{7}$	$\frac{13}{9}$	$\frac{17}{11}$	7	$\frac{11}{3}$	3	$\frac{19}{7}$	$\frac{23}{9}$	$\frac{25}{11}$

We check our graph on our graphics calculator. Notice that the horizontal asymptote does not appear on the display screen when you plot the graph of the function. We can, however, plot the line $y = 2$ in the same viewing rectangle to show the horizontal asymptote in the figure. ◄

▶ **EXAMPLE 3** *Sketching the Graph of a Rational Function*

Sketch the graph of the function defined by

$$g(x) = \frac{4x}{x^2 - 25}$$

Check the graph on a graphics calculator.

Solution Factoring the denominator, we have

$$g(x) = \frac{4x}{(x - 5)(x + 5)}$$

The domain of g is the set of all real numbers except 5 and -5. Because $g(0) = 0$, the graph has an intercept at the origin. Because $g(-x) = -g(x)$, g is an odd function, and the graph is symmetric with respect to the origin.

Setting the denominator equal to zero, we obtain the vertical asymptotes $x = 5$ and $x = -5$. From Theorem 2(i), the x axis is a horizontal asymptote (we also showed this fact in Illustration 3). Table 5 gives a few points on the graph. We locate these points and use the asymptotes as guides to draw the portion of the graph in the first and fourth quadrants. Using symmetry, we complete the graph in the second and third quadrants. Figure 9 shows the graph. We obtain the same graph on our graphics calculator.

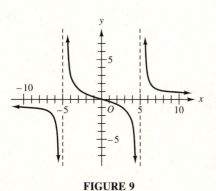

FIGURE 9

Table 5

x	0	1	2	3	4	6	8	10
$g(x) = \dfrac{4x}{x^2 - 25}$	0	$-\frac{1}{6}$	$-\frac{8}{21}$	$-\frac{3}{4}$	$-\frac{16}{9}$	$\frac{24}{11}$	$\frac{32}{39}$	$\frac{40}{75}$

◄

▶ **EXAMPLE 4** *Sketching the Graph of a Rational Function*

Sketch the graph of the function defined by

$$h(x) = \frac{3x^2}{2x^2 - 32}$$

Check the graph on a graphics calculator.

Solution We factor the denominator and obtain

$$h(x) = \frac{3x^2}{2(x - 4)(x + 4)}$$

The domain of h is the set of all real numbers except 4 and −4. Because $h(0) = 0$, the graph has an intercept at the origin. Because $h(-x) = h(x)$, h is an even function, and the graph is symmetric with respect to the y axis.

The vertical asymptotes are obtained by equating the denominator to zero. They are $x = 4$ and $x = -4$. As demonstrated in Illustration 4, the line $y = \frac{3}{2}$ is a horizontal asymptote, which also follows from Theorem 2(ii). A few points on the graph are obtained from Table 6. We locate these points and use the asymptotes as guides to draw the portion of the graph in the first and fourth quadrants. From properties of symmetry we complete the graph in the second and third quadrants. Figure 10 shows the required graph.

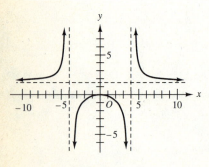

FIGURE 10

Table 6

x		1	2	3	5	6	8	10
$h(x) = \dfrac{3x^2}{2x^2 - 32}$		$-\frac{1}{10}$	$-\frac{1}{2}$	$-\frac{27}{14}$	$\frac{75}{18}$	$\frac{27}{10}$	2	$\frac{75}{42}$

On our graphics calculator we plot the graph of the function as well as the horizontal asymptote $y = \frac{3}{2}$, which agrees with Figure 10. ◀

▷ **ILLUSTRATION 5**

Let the function f be defined by

$$f(x) = \frac{x^2 - 16}{x - 3}$$

Dividing the numerator by the denominator, we obtain

$$f(x) = x + 3 - \frac{7}{x - 3}$$

Because, as $x \to +\infty$ or $x \to -\infty$, $\dfrac{7}{x - 3} \to 0$, it follows from the preceding equation that

$$\text{as } x \to +\infty \quad \text{or} \quad x \to -\infty, \quad f(x) \to x + 3$$

Because of this fact, we say that the line $y = x + 3$ is an *oblique asymptote* of the graph of f. Refer to Figure 11, which is obtained in Example 5 where the graph of f is discussed further. ◀

If a line is an asymptote of a graph but is neither horizontal nor vertical, it is called an **oblique asymptote.** The graph of any rational function of the form $P(x)/Q(x)$, where the degree of $P(x)$ is one more than the degree of $Q(x)$, will have an oblique asymptote. To find it in such a case, proceed as we did in Illustration 5: Divide the polynomial in the numerator by the polynomial in the denominator and obtain the sum of a linear function and a rational function. As $|x|$ increases without bound, the values of the original function approach the values of the linear function. The oblique asymptote is the graph of that linear function.

▶ **EXAMPLE 5** *Sketching the Graph of a Rational Function*

Sketch the graph of the function defined by

$$f(x) = \frac{x^2 - 16}{x - 3}$$

Check the graph on a graphics calculator.

Solution The domain of f is the set of all real numbers except 3. Because $f(0) = \frac{16}{3}$, the y intercept of the graph is $\frac{16}{3}$. The x intercepts of the graph are obtained by setting the numerator equal to zero. Doing this, we obtain $x = \pm 4$. Because f is neither even nor odd, it is not symmetric with respect to the y axis or the origin.

By setting the denominator equal to zero, we obtain the line $x = 3$ as a vertical asymptote. Because the degree of the numerator is greater than the degree of the denominator, it follows from Theorem 2(iii) that there are no horizontal asymptotes. There is an oblique asymptote, however, because the degree of the numerator is one more than the degree of the denominator. As seen in Illustration 5, this is the line $y = x + 3$.

Table 7 gives the results of computing $f(x)$ for various values of x. With selected points and using the asymptotes as guides, we obtain the graph of f shown in Figure 11.

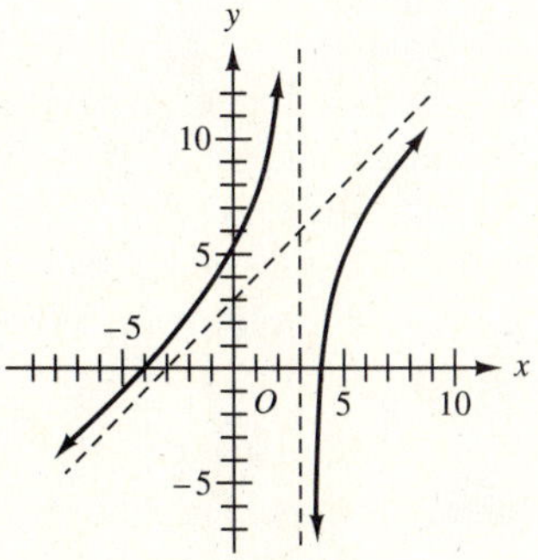

FIGURE 11

Table 7

x	-6	-4	-2	-1	0	1	2	4	5	6
$f(x) = \dfrac{x^2 - 16}{x - 3}$	$-\frac{20}{9}$	0	$\frac{12}{5}$	$\frac{15}{4}$	$\frac{16}{3}$	$\frac{15}{2}$	12	0	$\frac{9}{2}$	$\frac{20}{3}$

When we check the graph on our graphics calculator, the oblique asymptote does not appear. To show the oblique asymptote, we plot the line $y = x + 3$ in the same viewing rectangle.   ◀

 When you have learned techniques of calculus, you will be able to discuss thoroughly graphs of rational functions. At this stage, however, we recommend that you perform the following steps to sketch such graphs.

To Sketch the Graph of a Rational Function:

1. Find any intercepts.
2. Test for symmetry with respect to the y axis and the origin.
3. Find any vertical asymptotes by applying Theorem 1.
4. Find any horizontal asymptotes by applying Theorem 2.
5. If the degree of the numerator is 1 more than the degree of the denominator, find an oblique asymptote by the method of Illustration 5.
6. Plot a few points on the graph. Select as many points as are necessary to complete the sketch from the information obtained in steps 1–5.

▶ **EXAMPLE 6** *Solving a Word Problem Having a Rational Function as a Mathematical Model*

A tin can with an open top and having a volume 16π in.^{3} is to be in the form of a right-circular cylinder. **(a)** If x inches is the base radius of the cylinder, express the number of square inches in the total surface area of the cylinder as a function of x. **(b)** What is the domain of the resulting function? **(c)** Use a graphics calculator to find accurate to two decimal places what the base radius should be so that the least amount of material is used in the manufacture of the can; that is, so that the total surface area is a minimum. What is the minimum total surface area?

Solution

(a) Figure 12 shows the can having base radius x inches and height h inches. The lateral surface area of the cylinder is $2\pi xh$ square inches and the area of the bottom is πx^2 square inches. Therefore, if $S(x)$ square inches is the total surface area,

$$S(x) = 2\pi xh + \pi x^2 \tag{1}$$

To express $S(x)$ in terms of x only, we need an equation involving x and h. Because the volume of a right circular cylinder is given by $\pi x^2 h$, and the volume of the can is to be 16π in.^{3}, we have the equation

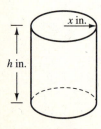

FIGURE 12

x in.

h in.

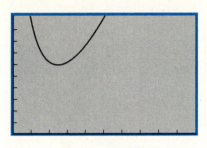

[0, 10] by [0, 100]

$$S(x) = \frac{32\pi}{x} + \pi x^2$$

FIGURE 13

$$\pi x^2 h = 16\pi$$

$$h = \frac{16}{x^2}$$

Substituting this value of h into (1) we obtain

$$S(x) = 2\pi x\left(\frac{16}{x^2}\right) + \pi x^2$$

$$S(x) = \frac{32\pi}{x} + \pi x^2$$

(b) Because x may be any positive number, the domain of S is $(0, +\infty)$.

(c) The graph of S in the viewing rectangle [0, 10] by [0, 100] appears in Figure 13. The x coordinate of the lowest point on the graph gives the base radius of the can of minimum total surface area, and the y coordinate gives the minimum total surface area. Using zoom-in on our graphics calculator, we determine the lowest point is (2.52, 59.84).

Conclusion: The base radius should be 2.52 in. to give a total surface area of 59.84 in.^{2} ◀

 In your calculus course you will be able to determine the exact value of the required base radius in Example 6. It is $2\sqrt[3]{2}$ in. ≈ 2.52 in.

In Sections 2.5 and 2.6 you learned how to solve linear and polynomial inequalities both algebraically and graphically. Certainly for linear inequalities, the algebraic method is easier than the graphical. However, a rational inequality, one that contains a rational expression involving the variable, is more easily solved on your graphics calculator than by applying algebra. Consequently, we postponed our discussion of rational inequalities until now, where we concentrate on the graphical solution.

▷ **ILLUSTRATION 6**

To solve the inequality

$$\frac{5x}{x - 1} < 4$$

we let

$$y_1 = \frac{5x}{x - 1} \qquad y_2 = 4$$

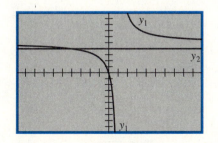

[−10, 10] by [−10, 10]

$$y_1 = \frac{5x}{x - 1} \text{ and } y_2 = 4$$

FIGURE 14

Figure 14 shows the graphs of y_1 and y_2 plotted in the [−10, 10] by [−10, 10] viewing rectangle. Observe that the line $x = 1$ is an asymptote of the graph of y_1. Because the graph of y_1 is below the graph of y_2 when x is in the open interval $(-4, 1)$, that interval is the solution set of the inequality. ◀

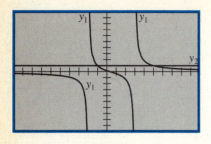

$[-10, 10]$ by $[-10, 10]$

$y_1 = \dfrac{3x - 1}{x^2 - x - 6}$ and $y_2 = 1$

FIGURE 15

▶ **EXAMPLE 7** *Solving a Rational Inequality Graphically*

Find the solution set of the inequality

$$\frac{3x - 1}{x^2 - x - 6} \le 1$$

Solution   Let

$$y_1 = \frac{3x - 1}{x^2 - x - 6} \qquad y_2 = 1$$

We plot the graphs of y_1 and y_2 in the $[-10, 10]$ by $[-10, 10]$ viewing rectangle appearing in Figure 15. The graph of y_1 has two vertical asymptotes: $x = -2$ and $x = 3$. The solution set of the inequality consists of all values of x for which the graph of y_1 either intersects the graph of y_2 or is below it. Therefore, from the figure we conclude that the solution set is $(-\infty, -2) \cup [-1, 3) \cup [5, +\infty)$.   ◀

EXERCISES 5.5

In Exercises 1 through 32, a rational function is defined. (a) Determine the domain of the function. (b) Find any intercepts of the graph of the function. (c) Test for symmetry of the graph with respect to the y axis and the origin. (d) Find any vertical and horizontal asymptotes of the graph if there are any. (e) Find an oblique asymptote if there is one. (f) Sketch the graph. (g) Check your graph on your graphics calculator.

1. $f(x) = \dfrac{1}{x}$

2. $f(x) = \dfrac{1}{x - 3}$

3. $g(x) = \dfrac{4}{x - 2}$

4. $g(x) = \dfrac{2}{x + 4}$

5. $f(x) = \dfrac{1 + x}{3 - x}$

6. $f(x) = \dfrac{x - 3}{x + 1}$

7. $h(x) = \dfrac{2x - 4}{x + 4}$

8. $g(x) = \dfrac{3x + 6}{x - 2}$

9. $f(x) = \dfrac{4}{x^2}$

10. $h(x) = -\dfrac{2}{x^2}$

11. $g(x) = -\dfrac{2}{x^3}$

12. $f(x) = \dfrac{3}{x^3}$

13. $f(x) = \dfrac{-1}{(x + 2)^2}$

14. $f(x) = \dfrac{4}{(x - 3)^2}$

15. $f(x) = \dfrac{5x}{x^2 - 4}$

16. $g(x) = \dfrac{7x}{x^2 - 9}$

17. $h(x) = \dfrac{9x}{16 - x^2}$

18. $f(x) = \dfrac{2x}{1 - x^2}$

19. $g(x) = \dfrac{2x^2}{x^2 - 9}$

20. $h(x) = \dfrac{3x^2}{x^2 - 4}$

21. $f(x) = \dfrac{x^2 + 1}{x^2 - 1}$

22. $f(x) = \dfrac{x^2 + 12}{x^2 - 16}$

23. $f(x) = \dfrac{x + 1}{x^2 + x - 6}$

24. $f(x) = \dfrac{x - 5}{x^2 - 8x + 12}$

25. $g(x) = \dfrac{x^2 - 9}{x - 2}$

26. $g(x) = \dfrac{x^2 - 25}{x - 4}$

27. $h(x) = \dfrac{x^2 + 4}{x}$

28. $h(x) = \dfrac{2x^2 + 2}{x}$

29. $f(x) = \dfrac{4}{x^2 + 4}$

30. $f(x) = \dfrac{6}{2x^2 + 3}$

31. $g(x) = \dfrac{2x^4}{x^4 + 1}$

32. $f(x) = \dfrac{3x^2}{x^2 + 4}$

In Exercises 33 through 40, find the solution set of the inequality graphically.

33. $\dfrac{1}{x - 5} > 1$

34. $\dfrac{3}{x + 4} \le 1$

35. $\dfrac{2x}{x - 2} \le 3$

36. $\dfrac{5x}{x - 4} > 6$

37. $\dfrac{2}{3 - x} > \dfrac{3}{2 + x}$

38. $\dfrac{6 - 2x}{4 + x} \le 5$

39. $\dfrac{2x - 7}{x^2 - 6x + 8} \le 1$

40. $\dfrac{5x}{x^2 + 2x - 8} \ge 3$

In Exercises 41 through 48, solve the word problem by us-
ing a rational function as a mathematical model of the sit-
uation. Be sure to write a conclusion.

41. Solve Example 6 if the tin can is closed.

42. For a closed can in the form of a right-circular cylinder
of volume 16π in.^{3}, the cost of material for the top and
bottom is 2 cents per square inch and the cost of the
material for the sides is 1 cent per square inch. **(a)** If
x inches is the base radius of the cylinder, express the
number of cents in the total cost of the material as a
function of x. **(b)** What is the domain of your function
in part (a)? **(c)** Use your graphics calculator to find
accurate to two decimal places what the base radius
should be so that the cost of the material is a minimum. What is the minimum cost of the material?

43. A closed box with a square base is to have a volume
of 2000 in.^{3} The material for the top and bottom of
the box is to cost 3 cents per square inch and the material for the sides is to cost 1.5 cents per square inch.
(a) If x inches is the length of a side of the square
base, express the number of cents in the total cost of
the material as a function of x. **(b)** What is the domain of your function in part (a)? **(c)** Use your graphics calculator to find accurate to two decimal places
what the length of a side of the square base should be
so that the cost of the material is a minimum. What is
the minimum cost of the material?

44. A rectangle has an area of 81 in.^{2} **(a)** If x inches is
the length of the rectangle, express the number of
inches in the perimeter as a function of x. **(b)** What is
the domain of your function in part (a)? **(c)** Use your
graphics calculator to find, to the nearest one-tenth of
an inch, the dimensions of the rectangle having the
least perimeter.

45. Do Exercise 44 if the rectangle has an area of 100 in.^{2}

46. In a particular community, a certain epidemic spreads
in such a way that x months after the start of the epidemic, $P(x)$ percent of the population is infected,
where

$$P(x) = \frac{30x^2}{(1 + x^2)^2}$$

Use your graphics calculator to estimate in how many
months the most people will be infected, and what
percent of the population this is?

47. It is determined that if salaries are excluded, the number of dollars in the cost per kilometer for operating a
truck is $8 + \frac{1}{300}x$, where x kilometers per hour is the
speed of the truck. The combined salary of the driver
and the driver's assistant is $27 per hour. **(a)** Express
the number of dollars in the total cost per kilometer to
operate a truck as a function of x. **(b)** What is the domain of your function in part (a)? **(c)** Use your graphics calculator to estimate to the nearest kilometer per
hour what the speed of the truck should be for the cost
per kilometer to be the least.

48. The number of dollars in the cost per hour of fuel for
a cargo ship is $\frac{1}{50}x^3$, where x knots (nautical miles per
hour) is the speed of the ship. There are additional
costs of $400 per hour. **(a)** Express the number of dol-
lars in the cost per nautical mile as a function of x.
(b) What is the domain of your function in part (a)?
(c) Use your graphics calculator to estimate to the
nearest knot what the speed of the ship should be for
the cost per nautical mile to be the least.

49. When two resistors having resistances R_1 ohms and
R_2 ohms are connected in parallel, the total resistance
R ohms is given by

$$\frac{1}{R} = \frac{1}{R_1} + \frac{1}{R_2}$$

If the total resistance of two resistors connected in
parallel must be at least 2 ohms, and one of the resistors has a resistance of 3 ohms, what must be the resistance of the other resistor? Use an inequality as a
mathematical model of the situation and be sure to
write a conclusion.

50. Explain in words only (no symbols) without using the
words *infinity* or *approaches* what we mean when we
say: "The line $x = 2$ is a vertical asymptote of the
graph of function f."

51. Explain in words only (no symbols) without using the
words *infinity* or *approaches* what we mean when we
say: "The line $y = 4$ is a horizontal asymptote of the
graph of function f."

52. If a, b, c, and d are positive numbers and $\frac{a}{b} < \frac{c}{d}$, describe how you could show that $\frac{a + c}{b + d}$ is between $\frac{a}{b}$
and $\frac{c}{d}$.

CHAPTER 5 REVIEW

▶ LOOKING BACK

5.1 Geometric transformations that we used to obtain graphs of polynomial functions included the vertical and horizontal translation rules, the vertical stretching and shrinking rule, and the x axis reflection rule. We showed how the end behavior of the graph of a polynomial function of the nth degree can be obtained from the sign of the coefficient of the nth degree term and the evenness or oddness of n. We applied but did not prove the theorem that states that a polynomial function of the nth degree has at most $n - 1$ relative extrema, and used a graphics calculator to estimate the relative extrema.

5.2 The remainder theorem tells us that when $P(x)$ is divided by $x - r$, the remainder is $P(r)$. A consequence of this theorem is the factor theorem, which states that $P(x)$ has $x - r$ as a factor if and only if $P(r) = 0$. We introduced synthetic division to enable us to apply these two theorems easily. We used the terminology synthetic substitution when $P(r)$ was computed by synthetic division. We showed how Horner's algorithm is a convenient way of performing synthetic substitution on a calculator.

5.3 The computational techniques of Section 5.2 were applied along with the rational zeros theorem to find rational zeros of polynomial functions with real coefficients.

5.4 We used the fundamental theorem of algebra to prove the crucial theorem that states that a polynomial function of degree $n \geq 1$, with complex coefficients, has exactly n complex zeros. We also stated but did not prove that if a complex number is a zero of a polynomial function, then its conjugate is also a zero. This fact permitted us to find other zeros of a polynomial function when some were known. We introduced Descartes' rule of signs that enabled us to determine the character of the zeros of a polynomial function. We also stated and proved a theorem used in calculus when working with partial fractions: A polynomial with real coefficients can be expressed as a product of linear or quadratic polynomials with real coefficients.

5.5 Vertical, horizontal, and oblique asymptotes of graphs of rational functions were defined and employed to sketch the graphs. We solved rational inequalities by plotting graphs of rational functions.

▶ REVIEW EXERCISES

In Exercises 1 and 2, use the remainder theorem to find the remainder for the division.

1. $(3x^3 + 4x^2 - 3x - 5) \div (x + 2)$

2. $(2x^4 - 5x^2 - 2x + 1) \div (x - 1)$

In Exercises 3 and 4, use the factor theorem to answer the question.

3. Is $x - 3$ a factor of $x^3 + 2x^2 - 12x - 9$?

4. Is $x + 4$ a factor of $2x^3 + 9x^2 + 6x + 8$?

5. Find a value of k so that $x - 3$ is a factor of $2kx^3 - 5x^2 + 3kx$.

6. Find a value of k so that $x + 2$ is a factor of $2x^4 + 2kx^3 - x^2 - 3kx - 8$.

In Exercises 7 through 10, find the quotient and remainder by synthetic division.

7. $(2x^4 + 7x^3 - 4x + 5) \div (x + 3)$

8. $(x^3 - 6x^2 + 8x - 5) \div (x - 4)$

9. $(x^6 - 64) \div (x - 2)$ **10.** $(x^5 + 243) \div (x + 3)$

In Exercises 11 and 12, find the function values by synthetic substitution.

11. $P(x) = 2x^4 - 8x^2 - 10x - 3$
    (a) $P(-2)$ **(b)** $P(3)$

12. $P(x) = 3x^4 + 10x^3 - 6x^2 + 1$
    (a) $P(-4)$ **(b)** $P(-\frac{1}{3})$

In Exercises 13 through 16, find the zeros of the polynomial function, and state the multiplicity of each zero.

13. $P(x) = (x^2 + 2x - 3)(2x^2 + x - 15)$

14. $P(x) = (x^2 - 1)(x^2 - 4)(x^2 + x - 2)$

15. $P(x) = (x^2 - 9)(x^2 - 4)^2(6x^2 + x - 15)$

16. $P(x) = (x - 5)^3(x^2 - 36)(x^2 + 2x - 1)^2$

In Exercises 17 and 18, show that the given numbers are zeros of the polynomial function, and find the other two zeros.

17. $P(x) = x^4 + x^3 - 8x^2 + 8$; -1 and 2

18. $P(x) = 2x^4 + 5x^3 - 6x^2 - 7x + 6$; 1 and -3

In Exercises 19 through 22, find all the rational zeros of the polynomial function. If possible, find any irrational or imaginary zeros.

19. $P(x) = x^4 - 3x^3 - 8x^2 + 26x - 12$

20. $P(x) = x^3 + 5x^2 + 5x - 3$

21. $P(x) = 3x^3 - 11x^2 + 9x - 2$

22. $P(x) = 2x^4 + x^3 - 17x^2 - 4x + 6$

In Exercises 23 through 26, find all the rational roots of the equation. If possible, find any irrational or imaginary roots.

23. $x^3 + x^2 - 15x + 9 = 0$

24. $x^4 - 3x^3 - 10x^2 + 28x - 16 = 0$

25. $6x^4 - 25x^3 - 3x^2 + 5x + 1 = 0$

26. $3x^3 - x^2 + 16x + 12 = 0$

27. Find the solution set of the equation
$$x^4 + 2x^3 - 4x^2 - 4x + 4 = 0$$
given that $\sqrt{2}$ and $-\sqrt{2}$ are roots.

28. Find the solution set of the equation
$$x^4 + x^3 - 4x^2 - 3x + 3 = 0$$
given that $\sqrt{3}$ and $-\sqrt{3}$ are roots.

In Exercises 29 through 40, sketch the graph of the function. Check your graph on your graphics calculator.

29. (a) $f(x) = x^2 - 2$ (b) $g(x) = (x - 2)^2$

30. (a) $f(x) = -2x^2 + 3$ (b) $g(x) = -2(x + 3)^2$

31. (a) $f(x) = \frac{1}{2}(x + 1)^3$ (b) $g(x) = \frac{1}{2}x^3 + 1$

32. (a) $f(x) = (x + 1)^3 - 5$ (b) $g(x) = (x + 5)^3 - 1$

33. (a) $f(x) = -(x - 1)^4 + 3$
     (b) $g(x) = -(x + 1)^4 - 3$

34. (a) $f(x) = \frac{1}{4}(x - 2)^4 - 3$
     (b) $g(x) = -\frac{1}{4}(x + 3)^4 + 2$

35. $h(x) = x^3 - 6x^2 + 9x + 6$

36. $f(x) = x^4 - 8x^2 + 9$

37. $g(x) = x^4 - 14x^2 - 24x$

38. $h(x) = x^3 + 3x^2 - 4$

39. $f(x) = x^5 - 5x^4 + 6x^3 + 2x^2 - 7x + 3$

40. $g(x) = x^4 + 4x^3 + 5x^2 + 2x$

In Exercises 41 through 52, a rational function is defined. (a) Determine the domain of the function. (b) Find any intercepts of the graph of the function. (c) Test for symmetry of the graph with respect to the y axis and the origin. (d) Find any vertical and horizontal asymptotes of the graph if there are any. (e) Find an oblique asymptote if there is one. (f) Sketch the graph. (g) Check your graph on your graphics calculator.

41. $f(x) = \dfrac{2}{x - 5}$ **42.** $f(x) = \dfrac{x + 2}{x - 3}$

43. $g(x) = \dfrac{8}{x^2}$ **44.** $g(x) = -\dfrac{4}{x^3}$

45. $f(x) = \dfrac{16}{(x + 2)^3}$ **46.** $f(x) = \dfrac{6}{(x + 1)^2}$

47. $f(x) = \dfrac{3x}{x^2 - 1}$ **48.** $g(x) = \dfrac{x^2 + 4}{x^2 - 4}$

49. $g(x) = \dfrac{4x^2}{9 - x^2}$ **50.** $f(x) = \dfrac{5x}{x^2 - 16}$

51. $h(x) = \dfrac{x^2 - 4}{x - 1}$ **52.** $h(x) = \dfrac{3x^2 - 12}{x}$

In Exercises 53 through 56, find a polynomial P(x) of the stated degree with real coefficients for which the given numbers are zeros of P.

53. Fourth degree; $1 - i$ and $1 + i\sqrt{3}$

54. Fifth degree; $2 - i\sqrt{2}$ (multiplicity 2) and -1

55. Sixth degree; $3i$ (multiplicity 2) and $2 - i\sqrt{2}$

56. Sixth degree; $2 + 3i$, $2 + i$, and $1 + 3i$

In Exercises 57 through 60, (a) find the zeros of the polynomial function and (b) use the result of part (a) to express the polynomial as a product of linear or quadratic polynomials.

57. $P(x) = 2x^3 + 3x^2 + 6x - 4$

58. $P(x) = 3x^4 + x^3 - 3x - 1$

59. $P(x) = x^4 + 2x^3 + 6x^2 + 8x + 8$; $2i$ is a zero

60. $P(x) = 2x^4 - 2x^3 + 3x^2 - 2x + 1$; $-i$ is a zero

61. Find the solution set of the equation
$$x^4 + 3x^3 + 3x^2 - 2 = 0$$
given that $-1 + i$ is a root.

62. Find the solution set of the equation
$$2x^4 - x^3 + 33x^2 - 16x + 16 = 0$$
given that $4i$ is a root.

In Exercises 63 through 66, prove that the equation has no rational roots. Then use Descartes' rule of signs to determine information concerning the number of positive, negative, and imaginary roots. Use your graphics calculator to verify your answer.

63. $x^3 - 7x^2 + x + 3 = 0$

64. $x^3 - 3x^2 - 5 = 0$

65. $x^4 - 6x - 9 = 0$

66. $x^4 + 2x^3 + 6x - 3 = 0$

In Exercises 67 through 70, determine all the information you can concerning the number of positive, negative, and imaginary zeros of the function. Find any rational zeros and locate any irrational zeros between two consecutive integers. Then plot the graph of the function and verify your answer.

67. $F(x) = x^3 - 3x^2 + 6x - 24$

68. $G(x) = x^3 + 3x^2 - 3x - 2$

69. $G(x) = x^4 - 2x^3 - 13x^2 + 33x - 14$

70. $F(x) = 3x^4 + 10x^3 - 11x^2 - 4x + 2$

In Exercises 71 through 74, find the solution set of the equation.

71. $4x^3 - 11x^2 + 26x - 15 = 0$

72. $3x^3 - x^2 + 16x + 12 = 0$

73. $6x^4 - 25x^3 - 3x^2 + 5x + 1 = 0$

74. $2x^4 - 9x^3 + 17x^2 - 3x - 7 = 0$

In Exercises 75 through 78, find the solution set of the inequality graphically.

75. $\dfrac{2x + 1}{x - 5} > 1$

76. $\dfrac{3x - 4}{2x - 3} \le 2$

77. $\dfrac{8x + 6}{x^2 + x - 12} \le 1$

78. $\dfrac{10 - 4x}{x^2 - 7x + 10} > 3$

In Exercises 79 through 81, solve the word problem by using an equation as a mathematical model of the situation. Be sure to write a conclusion.

79. The dimensions of a rectangular box are 3 in., 4 in., and 5 in. The volume of the box is doubled if each dimension is increased by the same number of inches. Determine this number.

80. The volume of a box is 504 cm³, and the numbers of centimeters in the dimensions of the box are three consecutive integers. What are the dimensions?

81. The area of a right triangle is 6 cm². Find the lengths of the sides of the triangle if the length of one of the sides is 2 cm shorter than the length of the hypotenuse.

In Exercises 82 through 85, solve the word problem by using a rational function as a mathematical model of the situation. Be sure to write a conclusion.

82. **(a)** If x feet is the length of a rectangle having an area of 90 ft², express the number of feet in the perimeter as a function of x. **(b)** What is the domain of your function in part (a)? **(c)** Use your graphics calculator to find, to the nearest one-tenth of a foot, the dimensions of the rectangle having the least perimeter.

83. A box manufacturer wishes to produce an open box of volume 288 in.³, where the base is a rectangle having a length three times its width, from the least amount of material. **(a)** If x inches is the width of the rectangular base, express the number of square inches in the total surface area of the box as a function of x. **(b)** What is the domain of your function in part (a)? **(c)** Use your graphics calculator to find accurate to the nearest one-tenth of an inch the dimensions of the box so that the total surface area is a minimum. What is the minimum total surface area?

84. Solve Exercise 83 if the box is to be closed.

85. A closed tin can having a volume of 27 in.³ is to be in the form of a right-circular cylinder. The circular top and bottom are cut from square pieces of tin. **(a)** If x inches is the radius of the cylinder, express the number of square inches in the total surface area as a function of x. Include the tin that is wasted when obtaining the surface area for the top and bottom. **(b)** What is the domain of your function in part (a)? **(c)** Use your graphics calculator to find accurate to the nearest one-tenth of an inch the radius and height of the can so that the total surface area is a minimum. What is the minimum total surface area?

86. Prove that if z_1 and z_2 are complex numbers, then

$$\overline{z_1 \cdot z_2} = \overline{z_1} \cdot \overline{z_2}$$

that is, the conjugate of the product of two complex numbers is the product of their conjugates.

Inverse Functions, Exponential Functions, and Logarithmic Functions

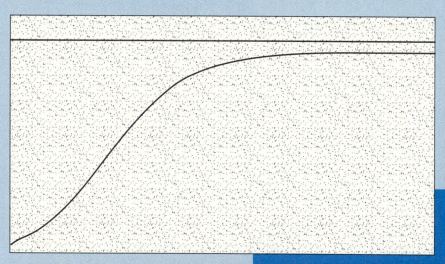

Until now we have considered only algebraic functions. Functions that are not algebraic are called *transcendental*, examples of which are the exponential and logarithmic functions, defined in this chapter. Because logarithmic and exponential functions are *inverses* of each other, we devote the first section to a discussion of inverse functions and their properties.

Applications of exponential and logarithmic functions arise in all of the sciences: physical, life, and social. They pertain to many diverse fields such as physics, chemistry, engineering, biology, business, economics, psychology, and sociology. Applications in the examples and exercises of this chapter include radioactive substances, atmospheric pressure, Newton's law of cooling, bacteria growth, interest compounded continuously, a worker's productivity on a job, the learning curve, the spread of a rumor or a disease, and the intensity of an earthquake measured on the Richter scale.

6.1 INVERSE FUNCTIONS

1. Define a one-to-one function.
2. Learn the horizontal-line test.
3. Apply the horizontal-line test to determine if a function is one-to-one.
4. Define an increasing function and a decreasing function.
5. Learn that a monotonic function is one-to-one.
6. Define the inverse of a function.
7. Learn the equations $f(f^{-1}(x)) = x$ and $f^{-1}(f(x)) = x$ where f and f^{-1} are inverse functions.
8. Determine if a given function has an inverse and, if it does, find it.
9. Learn that the graphs of a function and its inverse are reflections of each other with respect to the line $y = x$.

You are already familiar with *inverse operations*. Addition and subtraction are inverse operations; so are multiplication and division or raising to powers and extracting roots. One of a pair of inverse operations essentially "undoes" the other. For instance, if 4 is added to x, the sum is $x + 4$; if 4 is then subtracted from this sum, the difference is x. In the following illustration we use pairs of functions associated with inverse operations.

▷ **ILLUSTRATION 1**

We compute composite function values for some specific functions f and g.

(a) Let $f(x) = x + 4$ and $g(x) = x - 4$. Then

$$
\begin{aligned}
f(g(x)) &= f(x - 4) & g(f(x)) &= g(x + 4) \\
&= (x - 4) + 4 & &= (x + 4) - 4 \\
&= x & &= x
\end{aligned}
$$

(b) Let $f(x) = 2x$ and $g(x) = \dfrac{x}{2}$. Then

$$
\begin{aligned}
f(g(x)) &= f\left(\frac{x}{2}\right) & g(f(x)) &= g(2x) \\
&= 2\left(\frac{x}{2}\right) & &= \frac{2x}{2} \\
&= x & &= x
\end{aligned}
$$

(c) Let $f(x) = x^3$ and $g(x) = \sqrt[3]{x}$. Then

$$f(g(x)) = f(\sqrt[3]{x}) \qquad g(f(x)) = g(x^3)$$
$$= (\sqrt[3]{x})^3 \qquad\qquad = \sqrt[3]{x^3}$$
$$= x \qquad\qquad\qquad = x$$

◄

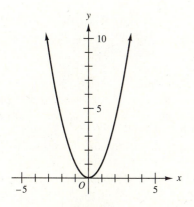

Each pair of functions f and g in Illustration 1 satisfies the following two equations:

$$f(g(x)) = x \qquad \text{for } x \text{ in the domain of } g$$

and

$$g(f(x)) = x \qquad \text{for } x \text{ in the domain of } f$$

Observe that for the functions f and g in these two equations the composite functions $f(g(x))$ and $g(f(x))$ are equal, a relationship that is not generally true for arbitrary functions f and g. You will learn in Illustration 7 that each pair of functions in Illustration 1 is a set of *inverse functions*, and that is the reason the two equations are satisfied.

We lead up to the formal definition of the *inverse of a function* by considering some more particular functions. The graph of the function defined by

$$f(x) = x^2$$

FIGURE 1

is plotted in Figure 1. The domain of f is the set of real numbers, and the range of f is the interval $[0, +\infty)$. To each value of x in the domain there corresponds one and only one number in the range. For instance, because $f(2) = 4$, the number in the range that corresponds to the number 2 in the domain is 4. However, because $f(-2) = 4$, the number corresponding to the number -2 in the domain is also the number 4 in the range. So 4 is the function value of two distinct numbers in the domain. Furthermore, every number except 0 in the range of this function is the function value of two distinct numbers in the domain. In particular, $\frac{25}{4}$ is the function value of both $\frac{5}{2}$ and $-\frac{5}{2}$, and 9 is the function value of both 3 and -3.

A different situation occurs with the function g defined by

$$g(x) = x^3 \qquad -2 \le x \le 2$$

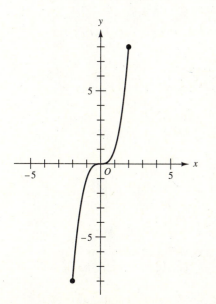

The domain of g is the closed interval $[-2, 2]$, and the range is $[-8, 8]$. The graph of g, plotted in Figure 2 is one for which a number in its range is the function value of one and only one number in the domain. Such a function is called *one-to-one*.

FIGURE 2

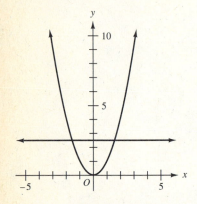

FIGURE 3

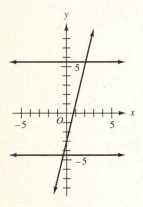

FIGURE 4

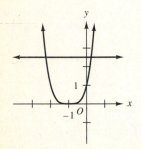

FIGURE 5

DEFINITION   A One-to-One Function

A function f is said to be **one-to-one** if and only if whenever a and b are any two distinct numbers in the domain of f, then $f(a) \neq f(b)$.

▷ ILLUSTRATION 2

The function defined by $f(x) = x^2$ does not satisfy the above definition because, for instance, 3 and -3 are two distinct numbers in the domain, yet $f(3) = f(-3)$. This function is, therefore, not one-to-one.   ◀

You learned in Section 4.2 that a vertical line can intersect the graph of a function in at most one point. For a one-to-one function, it is also true that a horizontal line can intersect the graph in at most one point. Notice that this is the situation for the one-to-one function defined by $g(x) = x^3$, where $-2 \leq x \leq 2$, whose graph appears in Figure 2. Furthermore, observe that for the function defined by $f(x) = x^2$, which is not one-to-one, any horizontal line above the x axis intersects the graph in two points (see Figure 3). Thus we have the following geometric test for determining if a function is one-to-one.

Horizontal-Line Test

A function is one-to-one if and only if every horizontal line intersects the graph of the function in not more than one point.

▶ **EXAMPLE 1**   *Determining if a Function Is One-to-One by the Horizontal-Line Test*

For each of the following functions, use the horizontal-line test to determine if it is one-to-one:

(a) $f(x) = 4x - 3$ **(b)** $f(x) = (x + 1)^4$

(c) $f(x) = |x|$ **(d)** $f(x) = \dfrac{3x + 4}{x - 2}$

Solution

(a) This function is linear, and its graph is the line in Figure 4. Because any horizontal line intersects the graph in exactly one point, the function is one-to-one.

(b) We obtained the graph of this function in Example 2(a) of Section 5.1. It is reproduced here in Figure 5, showing that any horizontal line above

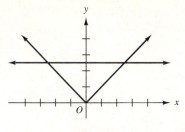

FIGURE 6

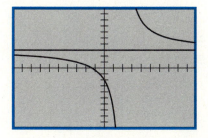

$[-10, 10]$ by $[-10, 10]$

$f(x) = \dfrac{3x + 4}{x - 2}$ and $y = 3$

FIGURE 7

the x axis intersects the graph in two points. Therefore, the function is not one-to-one.

(c) The graph of the absolute-value function appears in Figure 6. Observe that any horizontal line above the x axis intersects this graph in two points. Thus the absolute-value function is not one-to-one.

(d) Figure 7 shows the graph of the given rational function and its horizontal asymptote, the line $y = 3$, plotted in the same viewing rectangle. Any horizontal line, except the asymptote, intersects the graph in exactly one point. The function is, therefore, one-to-one. ◀

Another method of determining if a function is one-to-one relies on the following definitions.

DEFINITIONS   **An Increasing Function and a Decreasing Function**

Let f be a function. Then

(i) f is said to be an **increasing function** if

$$x_1 < x_2 \quad \text{implies} \quad f(x_1) < f(x_2)$$

where x_1 and x_2 are any numbers in the domain of f.

(ii) f is said to be a **decreasing function** if

$$x_1 < x_2 \quad \text{implies} \quad f(x_1) > f(x_2)$$

where x_1 and x_2 are any numbers in the domain of f.

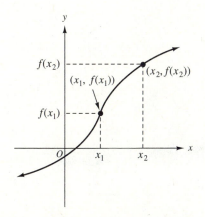

FIGURE 8

If a function is either increasing or decreasing, it is said to be **monotonic.**

The concept of an increasing function is illustrated in Figure 8, showing the graph of such a function f. For the two points $(x_1, f(x_1))$ and $(x_2, f(x_2))$ on the graph, we see that $x_1 < x_2$ and $f(x_1) < f(x_2)$.

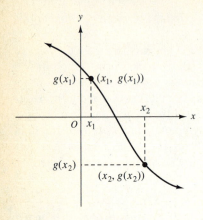

FIGURE 9

Figure 9 shows the graph of a decreasing function g. The two points $(x_1, g(x_1))$ and $(x_2, g(x_2))$ indicate that $x_1 < x_2$ and $g(x_1) > g(x_2)$.

▷ **ILLUSTRATION 3**

Consider again the function g defined by

$$g(x) = x^3 \qquad -2 \leq x \leq 2$$

whose graph is sketched in Figure 2. Because

$$x_1 < x_2 \qquad \text{implies} \qquad x_1^3 < x_2^3$$

it follows from the definition that g is an increasing function. ◄

▷ **ILLUSTRATION 4**

Consider the function f defined by

$$f(x) = 5 - 2x$$

From properties of inequalities, it follows that

$$x_1 < x_2 \qquad \text{implies} \qquad -2x_1 > -2x_2$$

If we add 5 to each member of the second inequality, we have

$$x_1 < x_2 \qquad \text{implies} \qquad 5 - 2x_1 > 5 - 2x_2$$

But $5 - 2x_1 = f(x_1)$ and $5 - 2x_2 = f(x_2)$. Therefore

$$x_1 < x_2 \qquad \text{implies} \qquad f(x_1) > f(x_2)$$

Thus, by the definition, f is a decreasing function. Refer to the graph of f in Figure 10. ◄

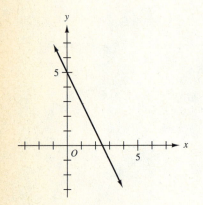

FIGURE 10

The following theorem gives a test that can be used to show that a function is one-to-one.

THEOREM 1

A monotonic function is one-to-one.

Proof Assume that f is an increasing function. If x_1 and x_2 are two numbers in the domain of f and $x_1 \neq x_2$, then either $x_1 < x_2$ or $x_2 < x_1$. If $x_1 < x_2$, it follows from the definition of an increasing function that

$f(x_1) < f(x_2)$; so $f(x_1) \neq f(x_2)$. If $x_2 < x_1$, then $f(x_2) < f(x_1)$, and so again $f(x_1) \neq f(x_2)$. Thus f is one-to-one. The proof is similar if f is a decreasing function. ∎

▷ **ILLUSTRATION 5**

In Illustration 3 we proved that the function defined by

$$g(x) = x^3 \qquad -2 \leq x \leq 2$$

is increasing. Therefore, by Theorem 1, g is one-to-one. This result agrees with our earlier demonstration by the horizontal-line test.

In the equation defining $g(x)$, if we replace $g(x)$ by y, we have

$$y = x^3 \qquad -2 \leq x \leq 2$$

Solving this equation for x, we obtain

$$x = \sqrt[3]{y} \qquad -8 \leq y \leq 8$$

which defines a function G where

$$G(y) = \sqrt[3]{y} \qquad -8 \leq y \leq 8 \qquad ◀$$

The function G of Illustration 5 is called the *inverse* of the function g. In the following formal definition of the inverse of a function, we use the notation f^{-1} to denote the inverse of f. This notation is read "f inverse," and it should not be confused with the use of -1 as an exponent.

DEFINITION **The Inverse of a Function**

If f is a one-to-one function that is the set of ordered pairs (x, y), then there is a function f^{-1}, called the **inverse of f,** where f^{-1} is the set of ordered pairs (y, x) defined by

$$x = f^{-1}(y) \qquad \text{if and only if} \qquad y = f(x)$$

The domain of f^{-1} is the range of f, and the range of f^{-1} is the domain of f.

In the preceding definition the requirement that f be a one-to-one function ensures that $f^{-1}(y)$ is unique for each value of y.

Eliminating y from the equations of the definition by writing the equation

$$f^{-1}(y) = x$$

and replacing y by $f(x)$, we obtain

$$f^{-1}(f(x)) = x \tag{1}$$

where x is in the domain of f.

Eliminating x from the same pair of equations by writing the equation

$$f(x) = y$$

and replacing x by $f^{-1}(y)$, we get

$$f(f^{-1}(y)) = y$$

where y is in the domain of f^{-1}. Because the symbol used for the independent variable is arbitrary, we can replace y by x to obtain

$$f(f^{-1}(x)) = x \tag{2}$$

where x is in the domain of f^{-1}.

From Equations (1) and (2) we see that if the inverse of the function f is the function f^{-1}, then the inverse of f^{-1} is f. We state these results formally as the following theorem.

THEOREM 2

If f is a one-to-one function having f^{-1} as its inverse, then f^{-1} is a one-to-one function having f as its inverse. Furthermore,

$$f^{-1}(f(x)) = x \qquad \text{for } x \text{ in the domain of } f$$

and

$$f(f^{-1}(x)) = x \qquad \text{for } x \text{ in the domain of } f^{-1}$$

We use the terminology *inverse functions* when referring to a function and its inverse.

▷ **ILLUSTRATION 6**

In Illustration 5 the function G defined by

$$G(y) = \sqrt[3]{y} \qquad -8 \le y \le 8$$

is the inverse of the function g defined by

$$g(x) = x^3 \qquad -2 \le x \le 2$$

Therefore g^{-1} can be written in place of G, and we have

$$g^{-1}(y) = \sqrt[3]{y} \qquad -8 \le y \le 8$$

or, equivalently, if we replace y by x,

$$g^{-1}(x) = \sqrt[3]{x} \qquad -8 \le x \le 8$$

Observe that the domain of g is $[-2, 2]$, which is the range of g^{-1}; also the range of g is $[-8, 8]$, which is the domain of g^{-1}. ◄

If a function f has an inverse, then $f^{-1}(x)$ can be found by the method used in the following illustration.

▷ ILLUSTRATION 7

Each of the functions f in Illustration 1 is one-to-one. Therefore, $f^{-1}(x)$ exists. For each function we compute $f^{-1}(x)$ from the definition of $f(x)$ by substituting y for $f(x)$ and solving the resulting equation for x. This procedure gives the equation $x = f^{-1}(y)$. We then have the definition of $f^{-1}(y)$ from which we obtain $f^{-1}(x)$.

(a) $f(x) = x + 4$ **(b)** $f(x) = 2x$ **(c)** $f(x) = x^3$

$y = x + 4$ $\qquad\qquad$ $y = 2x$ $\qquad\qquad$ $y = x^3$

$x = y - 4$ $\qquad\qquad$ $x = \dfrac{y}{2}$ $\qquad\qquad$ $x = \sqrt[3]{y}$

$f^{-1}(y) = y - 4$ $\qquad\qquad\qquad\qquad\quad$ $f^{-1}(y) = \sqrt[3]{y}$

$f^{-1}(x) = x - 4$ $\qquad\qquad$ $f^{-1}(y) = \dfrac{y}{2}$ $\qquad\quad$ $f^{-1}(x) = \sqrt[3]{x}$

$\qquad\qquad\qquad\qquad\quad$ $f^{-1}(x) = \dfrac{x}{2}$

Observe that the function f^{-1} in each part is the function g in the corresponding part of Illustration 1. ◄

▶ **EXAMPLE 2** *Finding the Inverse of a Function and Verifying the Equations of Theorem 2*

Find $f^{-1}(x)$ for the function f of Example 1(a) and verify the equations of Theorem 2 for f and f^{-1}. Plot the graphs of f and f^{-1} in the same viewing rectangle.

Solution   In Example 1(a) we showed by the horizontal-line test that the function defined by

$$f(x) = 4x - 3$$

is one-to-one. Therefore f^{-1} exists. To find $f^{-1}(x)$, we write the equation

$$y = 4x - 3$$

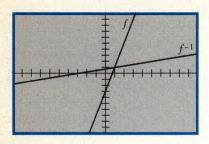

$[-10, 10]$ by $[-10, 10]$

$f(x) = 4x - 3$ and $f^{-1}(x) = \frac{1}{4}(x + 3)$

FIGURE 11

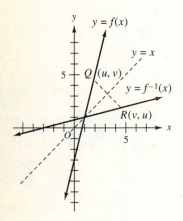

FIGURE 12

and solve for x. We obtain

$$x = \frac{y + 3}{4}$$

Therefore

$$f^{-1}(y) = \frac{y + 3}{4} \qquad f^{-1}(x) = \frac{x + 3}{4}$$

We verify the equations of Theorem 2.

$$f^{-1}(f(x)) = f^{-1}(4x - 3) \qquad f(f^{-1}(x)) = f\left(\frac{x + 3}{4}\right)$$

$$= \frac{(4x - 3) + 3}{4} \qquad\qquad = 4\left(\frac{x + 3}{4}\right) - 3$$

$$= \frac{4x}{4} \qquad\qquad\qquad = (x + 3) - 3$$

$$= x \qquad\qquad\qquad\quad = x$$

Figure 11 shows the graphs of f and f^{-1} plotted in the same viewing rectangle. ◀

Refer to Figure 12 showing the graphs of f and f^{-1} of Example 2 with the point $Q(u, v)$ on the graph of f and the point $R(v, u)$ on the graph of f^{-1}. The line segment QR in the figure is perpendicular to the line $y = x$ and is bisected by it. The point Q is a *reflection of the point R* with respect to the line $y = x$, and the point R is a *reflection of the point Q* with respect to the line $y = x$.

If x and y are interchanged in the equation $y = f(x)$, we obtain the equation $x = f(y)$, and the graph of the equation $x = f(y)$ is a *reflection of the graph* of the equation $y = f(x)$ with respect to the line $y = x$. Because the equation $x = f(y)$ is equivalent to the equation $y = f^{-1}(x)$, the graph of the equation $y = f^{-1}(x)$ is a reflection of the graph of the equation $y = f(x)$ with respect to the line $y = x$. Therefore, if a function has an inverse, the graphs of the functions are reflections of each other with respect to the line $y = x$.

▷ **ILLUSTRATION 8**

Functions g and g^{-1} of Illustration 6 are defined by

$$g(x) = x^3 \qquad -2 \leq x \leq 2$$

and

$$g^{-1}(x) = \sqrt[3]{x} \qquad -8 \leq x \leq 8$$

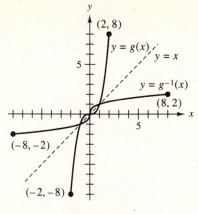

FIGURE 13

The graphs of g and g^{-1} appear in Figure 13. Observe that these graphs are reflections of each other with respect to the line $y = x$. ◀

▶ **EXAMPLE 3** *Showing That a Function is Monotonic and Finding its Inverse*

Let f be the function defined by

$$f(x) = x^2 \quad x \geq 0$$

(a) Show that f is increasing and therefore one-to-one. **(b)** Find $f^{-1}(x)$. **(c)** Plot the graphs of f, f^{-1}, and the line $y = x$ in the same viewing rectangle and observe that the graphs of f and f^{-1} are reflections of each other with respect to the line $y = x$.

Solution
(a) To show that f is an increasing function, we verify that the definition of an increasing function holds.

Let x_1 and x_2 be two numbers in the domain of f so that $x_1 < x_2$. Because the domain of f is the set of nonnegative numbers, it follows that x_1 is nonnegative and x_2 is positive. Therefore

$$x_1 < x_2 \quad \text{implies} \quad x_1^2 < x_2^2$$

But $x_1^2 = f(x_1)$ and $x_2^2 = f(x_2)$. Therefore

$$x_1 < x_2 \quad \text{implies} \quad f(x_1) < f(x_2)$$

Hence f is an increasing function and therefore one-to-one.
(b) Because f is one-to-one, it has an inverse f^{-1}. To find $f^{-1}(x)$, we replace $f(x)$ by y in the given equation, and we have

$$y = x^2 \quad x \geq 0$$

We solve this equation for x. Bearing in mind that $x \geq 0$, we obtain

$$x = \sqrt{y}$$

Thus

$$f^{-1}(y) = \sqrt{y}$$

To use x as the independent variable, we replace y by x and get

$$f^{-1}(x) = \sqrt{x}$$

(c) Figure 14 shows the graphs of f, f^{-1}, and the line $y = x$ in the same viewing rectangle. We observe that the graphs of f and f^{-1} appear to be reflections of each other with respect to the line $y = x$. ◀

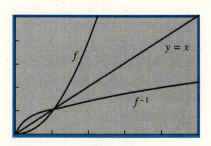

[0, 5] by [0, 5]
$f(x) = x^2, x > 0$ and $f^{-1}(x) = \sqrt{x}$ and $y = x$
FIGURE 14

▶ **EXAMPLE 4** *Finding the Inverse of a Function and Verifying the Equations of Theorem 2*

For the function of Example 1(d), find $f^{-1}(x)$ and verify the equations of Theorem 2 for f and f^{-1}.

Solution   From Example 1(d), we know that the function f defined by

$$f(x) = \frac{3x + 4}{x - 2}$$

is one-to-one and, therefore, it has an inverse. To find $f^{-1}(x)$, we solve the equation

$$y = \frac{3x + 4}{x - 2}$$

for x:

$$y(x - 2) = 3x + 4$$
$$xy - 2y = 3x + 4$$
$$xy - 3x = 2y + 4$$
$$x(y - 3) = 2y + 4$$
$$x = \frac{2y + 4}{y - 3}$$

Therefore

$$f^{-1}(y) = \frac{2y + 4}{y - 3} \qquad f^{-1}(x) = \frac{2x + 4}{x - 3}$$

We verify the equations of Theorem 2.

$$f^{-1}(f(x)) = f^{-1}\left(\frac{3x + 4}{x - 2}\right) \qquad\qquad f(f^{-1}(x)) = f\left(\frac{2x + 4}{x - 3}\right)$$

$$= \frac{2\left(\dfrac{3x + 4}{x - 2}\right) + 4}{\dfrac{3x + 4}{x - 2} - 3} \qquad\qquad = \frac{3\left(\dfrac{2x + 4}{x - 3}\right) + 4}{\dfrac{2x + 4}{x - 3} - 2}$$

$$= \frac{2(3x + 4) + 4(x - 2)}{3x + 4 - 3(x - 2)} \qquad\qquad = \frac{3(2x + 4) + 4(x - 3)}{2x + 4 - 2(x - 3)}$$

$$= \frac{6x + 8 + 4x - 8}{3x + 4 - 3x + 6} \qquad\qquad = \frac{6x + 12 + 4x - 12}{2x + 4 - 2x + 6}$$

$$= \frac{10x}{10} \qquad\qquad\qquad\qquad = \frac{10x}{10}$$

$$= x \qquad\qquad\qquad\qquad\qquad = x \qquad\qquad ◀$$

EXERCISES 6.1

In Exercises 1 through 14, sketch the graph of the function, and use the horizontal-line test to determine if it is one-to-one. Check your graph on your graphics calculator.

1. $f(x) = 2x + 3$ **2.** $g(x) = 8 - 4x$

3. $f(x) = x^2 - 6$ **4.** $f(x) = 4 - x^2$

5. $g(x) = 4 - x^3$ **6.** $h(x) = x^3 + 1$

7. $f(x) = (x - 3)^4$ **8.** $g(x) = x^4 - 3$

9. $h(x) = \sqrt{x + 3}$ **10.** $f(x) = \sqrt{1 - x^2}$

11. $f(x) = \dfrac{x + 5}{x - 4}$ **12.** $g(x) = \dfrac{3}{(x - 1)^2}$

13. $g(x) = |x - 2|$ **14.** $f(x) = 5$

In Exercises 15 through 20, do the following: (a) Sketch the graph of f and use the horizontal-line test to prove that f is one-to-one; (b) find $f^{-1}(x)$; (c) verify the equations of Theorem 2 for f and f^{-1}; (d) plot the graphs of f, f^{-1}, and the line $y = x$ in the same viewing rectangle and observe that the graphs of f and f^{-1} are reflections of each other with respect to the line $y = x$.

15. $f(x) = 3 - 4x$ **16.** $f(x) = 3x - 2$

17. $f(x) = x^3 + 2$ **18.** $f(x) = (x + 2)^3$

19. $f(x) = \dfrac{1}{x + 1}$ **20.** $f(x) = \dfrac{x + 5}{x - 2}$

In Exercises 21 through 28, (a) state the range of the one-to-one function f; (b) determine $f^{-1}(x)$ and state the domain of f^{-1}; (c) plot the graphs of f, f^{-1}, and the line $y = x$ in the same viewing rectangle and observe that the graphs of f and f^{-1} are reflections of each other with respect to the line $y = x$.

21. $f(x) = x^2 - 5, x \geq 0$ **22.** $f(x) = 2 - x^2, x \leq 0$

23. $f(x) = \sqrt{x^2 - 9}, x \geq 3$

24. $f(x) = -\sqrt{x^2 - 9}, x \geq 3$

25. $f(x) = \sqrt{x^2 - 9}, x \leq -3$

26. $f(x) = -\sqrt{x^2 - 9}, x \leq -3$

27. $f(x) = \frac{1}{8}x^3, -1 \leq x \leq 1$

28. $f(x) = (2x + 1)^3, -\frac{1}{2} \leq x \leq \frac{1}{2}$

In Exercises 29 through 36, (a) show that f is monotonic and therefore one-to-one; (b) find $f^{-1}(x)$; (c) sketch the graphs of f and f^{-1} on the same set of axes; (d) check your graphs in part (c) by plotting them in the same viewing rectangle.

29. $f(x) = 2x + 5$ **30.** $f(x) = 6 - \frac{1}{2}x$

31. $f(x) = (x + 1)^3$ **32.** $f(x) = (1 - x)^3$

33. $f(x) = (x - 2)^2, x \geq 2$

34. $f(x) = (x + 3)^2, x \leq -3$

35. $f(x) = 4 - x^2, x \leq 0$

36. $f(x) = x^2 - 9, x \geq 0$

In Exercises 37 and 38, show that function f is its own inverse.

37. $f(x) = \sqrt{16 - x^2}, 0 \leq x \leq 4$

38. $f(x) = \dfrac{x + 6}{x - 1}$

39. Find the value of k so that the one-to-one function f defined by

$$f(x) = \frac{x + 5}{x + k}$$

will be its own inverse.

In Exercises 40 and 41, show that the function is its own inverse for any constant k.

40. $f(x) = \dfrac{x + k}{x - 1}$ **41.** $f(x) = \dfrac{kx + 1}{x - k}$

42. Show that the function defined by

$$f(x) = \frac{x + h}{kx - 1}$$

is its own inverse for any values of constants h and k.

In Exercises 43 and 44, (a) show that function f is not one-to-one and hence does not have an inverse; (b) restrict the domain and obtain two one-to-one functions f_1 and f_2 each having the same range as f; (c) find $f_1^{-1}(x)$ and $f_2^{-1}(x)$ and state their domains; (d) plot the graphs of f_1 and f_1^{-1} in the same viewing rectangle; (e) plot the graphs of f_2 and f_2^{-1} in the same viewing rectangle.

43. $f(x) = x^2 + 4$ **44.** $f(x) = x^2 - 9$

45. Given

$$f(x) = \begin{cases} x & \text{if } x < 1 \\ x^2 & \text{if } 1 \leq x \leq 9 \\ 27\sqrt{x} & \text{if } 9 < x \end{cases}$$

Prove that f has an inverse function and find $f^{-1}(x)$.

6.2 EXPONENTS AND THE NUMBER e

1. Learn and apply the laws of real-number exponents.
2. Learn scientific notation.
3. Learn how to round off a numeral to k significant digits.
4. Solve word problems involving simple interest.
5. Learn the formula for computing the amount of an investment at compound interest.
6. Solve word problems involving compound interest.
7. Define the number e.
8. Solve word problems involving interest compounded continuously.

In this chapter you need to be familiar with laws of exponents. Thus you may wish to review these laws in Section 1.2 where we defined a rational exponent. In particular, 2^x has been defined for any rational value of x. For instance,

$$2^5 = 2 \cdot 2 \cdot 2 \cdot 2 \cdot 2 \qquad 2^0 = 1 \qquad 2^{-3} = \frac{1}{2^3} \qquad 2^{2/3} = \sqrt[3]{2^2}$$
$$= 32 \qquad\qquad\qquad\qquad = \frac{1}{8} \qquad\quad = \sqrt[3]{4}$$

It is not simple to define 2^x when x is irrational. For example, what is meant by $2^{\sqrt{3}}$? We can give an intuitive argument showing how $2^{\sqrt{3}}$ can be interpreted. To do this, we make use of the following theorem, which is stated without proof.

THEOREM 1

If r and s are rational numbers, then

(i) if $b > 1$, $r < s$ implies $b^r < b^s$;
(ii) if $0 < b < 1$, $r < s$ implies $b^r > b^s$.

We demonstrate this theorem in the following illustration.

▷ **ILLUSTRATION 1**

In part (i) of Theorem 1, $b > 1$. If $b = 4$, then because $2 < 3$, $4^2 < 4^3$. In part (ii), $0 < b < 1$. If $b = \frac{1}{3}$, then because $2 < 3$, $(\frac{1}{3})^2 > (\frac{1}{3})^3$. ◀

A decimal approximation for $\sqrt{3}$ can be obtained accurate to any number of decimal places desired. To four decimal places, $\sqrt{3} \approx 1.7321$. Because $1 < 1.7 < 2$, from Theorem 1(i)

$$2^1 < 2^{1.7} < 2^2 \Leftrightarrow 2 < 2^{1.7} < 4$$

Because $1.7 < 1.73 < 1.8$,

$$2^{1.7} < 2^{1.73} < 2^{1.8} \Leftrightarrow 3.2 < 2^{1.73} < 3.5$$

Because $1.73 < 1.732 < 1.74$,

$$2^{1.73} < 2^{1.732} < 2^{1.74} \Leftrightarrow 3.32 < 2^{1.732} < 3.34$$

Because $1.732 < 1.7321 < 1.733$,

$$2^{1.732} < 2^{1.7321} < 2^{1.733} \Leftrightarrow 3.322 < 2^{1.7321} < 3.324$$

and so on. In each inequality there is a power of 2 for which the exponent is a decimal approximation of the value of $\sqrt{3}$. In each successive inequality the exponent contains one more decimal place than the exponent in the previous inequality. By following this procedure indefinitely, the difference between the left member of the inequality and the right member can be made as small as we please. Hence our intuition leads us to expect that there is a value of $2^{\sqrt{3}}$ that satisfies each successive inequality as the procedure is continued indefinitely. A similar discussion could be given for any irrational power of any positive number. Furthermore, Theorem 1 is valid if _r_ and _s_ are any real numbers.

 Leonhard Euler was the first mathematician to conceive of any real number being an exponent. The definition of a real-number exponent requires a knowledge of calculus and appears in many calculus texts. The laws of rational exponents are also valid if the exponents are any real numbers. These laws for real-number exponents are summarized in the following theorem, whose proof requires techniques of calculus.

THEOREM 2

If _a_ and _b_ are any positive numbers, and _x_ and _y_ are any real numbers, then

(i) $a^x a^y = a^{x+y}$

(ii) $\dfrac{a^x}{a^y} = a^{x-y}$

(iii) $(a^x)^y = a^{xy}$

(iv) $(ab)^x = a^x b^x$

(v) $\left(\dfrac{a}{b}\right)^x = \dfrac{a^x}{b^x}$

▶ **EXAMPLE 1** *Applying Laws of Real-Number Exponents*

Simplify each of the following:

(a) $2^{\sqrt{3}} \cdot 2^{\sqrt{12}}$ **(b)** $(7^{\sqrt{5}})^{\sqrt{20}}$

Solution

(a) $2^{\sqrt{3}} \cdot 2^{\sqrt{12}} = 2^{\sqrt{3}} \cdot 2^{2\sqrt{3}}$ **(b)** $(7^{\sqrt{5}})^{\sqrt{20}} = 7^{\sqrt{5} \cdot \sqrt{20}}$

$$= 2^{\sqrt{3} + 2\sqrt{3}} \qquad\qquad = 7^{\sqrt{100}}$$

$$= 2^{3\sqrt{3}} \qquad\qquad\qquad = 7^{10} \qquad\blacktriangleleft$$

Any positive number can be written in the form

$$x = a \cdot 10^c \qquad \text{where } 1 \le a < 10 \text{ and } c \text{ is an integer}$$

A number expressed in this form is said to be written in **scientific notation.** Scientific notation, a way of expressing very large or very small numbers using powers of 10, is especially important in physics, chemistry, astronomy, and computer science.

▷ **ILLUSTRATION 2**

Each of the following numbers is written in scientific notation:

$$582 = (5.82)10^2$$
$$97{,}136 = (9.7136)10^4$$
$$92{,}900{,}000{,}000 = (9.29)10^{10}$$
$$0.627 = (6.27)10^{-1}$$
$$0.00002381 = (2.381)10^{-5}$$
$$2.04 = (2.04)10^0 \qquad\blacktriangleleft$$

To write a number in scientific notation, the first factor is obtained by placing a decimal point after the first left-hand nonzero digit. The second factor is a power of 10, where the exponent is obtained by counting the number of digits that must be passed over to move from the new position of the decimal point to the original position of the decimal point; if the movement is to the right, the exponent is positive; if the movement is to the left, the exponent is negative. You should verify this rule by applying it to the numbers in Illustration 2.

If a number is written in scientific notation, it can be written in standard form by moving the decimal point in the first factor the number of places indicated by the exponent of the power of 10; the decimal point is moved to the right if the exponent is positive and to the left if the exponent is negative. This rule is applied in the next illustration.

▷ **ILLUSTRATION 3**

$$(3.659)10^4 = 36,590$$
$$(8.007)10^2 = 800.7$$
$$(9.46)10^{12} = 9,460,000,000,000$$
$$(3.92)10^{-3} = 0.00392$$
$$(4.018)10^{-2} = 0.04018$$
$$(1.6)10^{-19} = 0.00000\ 00000\ 00000\ 00016$$
◄

Calculators often display very large or very small numbers by using scientific notation. The manual for your calculator will tell you how to enter and read in the display a number written in scientific notation.

Scientific notation affords a convenient way of indicating the significant digits in a numeral such as 83,200. For example, a measurement of 83,200 ft may be written as $(8.32)10^4$ ft to indicate that there are three significant digits, meaning the measurement is accurate to the nearest 100 ft. If we write $(8.320)10^4$ ft, then there are four significant digits, and the measurement is accurate to the nearest 10 ft. Similarly, if we write $(8.3200)10^4$ ft, there are five significant digits, and the measurement is accurate to the nearest foot.

A numeral having more than *k* significant digits is said to be **rounded off to *k* significant digits** if it is replaced by the number having *k* significant digits to which it is closest in value. For instance, the numeral 0.52368 is rounded off to four significant digits as 0.5237, while the numeral 78.142 is rounded off to four significant digits as 78.14. To round off to four significant digits a five-digit numeral whose fifth digit is 5, we adopt the following convention: If the fifth digit is 5 and the fourth digit is even, we round off to the fourth digit; if the fifth digit is 5 and the fourth digit is odd, we increase the fourth digit by 1. Hence we round off 261.85 to 261.8, and 0.0039235 is rounded off to 0.003924. A similar convention is used to round off to any number of significant digits.

▶ **EXAMPLE 2** *Computing with Scientific Notation*

Use scientific notation to perform the following computations:

(a) $(0.00002350)(56,300)$, where 56,300 has four significant digits;

(b) $\dfrac{(92,900,000,000)(0.00000262)}{(0.000310)(581)}$, where 92,900,000,000 has three significant digits.

Solution   We first write each number in scientific notation.

(a) $[(2.350)10^{-5}][(5.630)10^4] = [(2.350)(5.630)][10^{-5} \cdot 10^4]$

$$= (13.23)10^{-1}$$

$$= 1.323$$

(b) $\dfrac{[(9.29)10^{10}][(2.62)10^{-6}]}{[(3.10)10^{-4}][(5.81)10^2]} = \dfrac{(9.29)(2.62)}{(3.10)(5.81)} \cdot \dfrac{10^{10} \cdot 10^{-6}}{10^{-4} \cdot 10^2}$

$$= (1.35)10^{10-6+4-2}$$

$$= (1.35)10^6 \qquad \blacktriangleleft$$

We can apply exponents to help compute the interest on an investment. This application will lead us to the definition of e, a number that arises not only in business and economics but in the physical and life sciences as well.

If money is loaned at an interest rate of 0.12 (that is, 12 percent) per year, then the borrower's debt at the end of a year is $1.12 for each $1 borrowed. In general, if the interest rate is r (that is, $100r$ percent) per year, then for each dollar borrowed the repayment at the end of a year is $(1 + r)$ dollars. If P dollars is borrowed, then the debt is $P(1 + r)$ dollars at the end of a year.

We shall consider several types of interest. **Simple interest** is due only on the original amount that is borrowed. In this case no interest is paid on any accrued interest. For example, suppose that 10 percent simple interest on $100 is due annually. Then the lender would receive $10 at the end of each year.

Suppose now that P dollars is invested at a simple interest rate of $100r$ percent. Then the interest earned at the end of the year is Pr dollars. If no withdrawals are made for n years, the total interest earned is Pnr dollars, and if A dollars is the total amount on deposit at the end of n years,

$$A = P + Pnr$$

$$= P(1 + nr)$$

Simple interest is sometimes used for short-term investments or loans of a period of possibly 30, 60, or 90 days. In such cases, to simplify calculations, a year is considered as having 360 days, and each month is assumed to contain 30 days; then 30 days is equivalent to one-twelfth of a year.

▶ **EXAMPLE 3**   *Solving a Word Problem Involving Simple Interest*

A loan of $500 is made for a period of 90 days at a simple interest rate of 16 percent annually. Determine the amount to be repaid at the end of 90 days.

Solution   We are given $P = 500$, $r = 0.16$, and $n = \frac{90}{360}$; that is, $n = \frac{1}{4}$. Therefore, if A dollars is the amount to be repaid,

$$A = P(1 + nr)$$
$$= 500[1 + \tfrac{1}{4}(0.16)]$$
$$= 520$$

Conclusion: The amount to be repaid is $520.   ◀

Rates of interest are customarily stated as annual rates, but often the interest is computed more than once a year. When the interest for each period is added to the principal and itself earns interest, we have **compound interest.** Whenever the word *interest* is used without an adjective, it is customarily assumed to be compound interest. If the interest is compounded *m* times per year, then the annual rate must be divided by *m* to determine the interest for each period. For example, if $100 is deposited in a savings account that pays 8 percent compounded quarterly, then the number of dollars in the account at the end of the first 3-month period will be

$$100\left(1 + \frac{0.08}{4}\right) = 100(1.02)$$

For the second quarter we consider the principal to be $100(1.02)$. Therefore the number of dollars in the account at the end of the second 3-month period will be

$$[100(1.02)](1.02) = 100(1.02)^2$$

At the end of the third 3-month period the number of dollars in the account will be

$$[100(1.02)^2](1.02) = 100(1.02)^3$$

and so on. At the end of the *n*th 3-month period the number of dollars in the account will be

$$100(1.02)^n$$

More generally, we have the following theorem.

THEOREM 3

If P dollars is invested at an annual interest rate of $100r$ percent compounded m times per year, and if A_n dollars is the amount of the investment at the end of n interest periods, then

$$A_n = P\left(1 + \frac{r}{m}\right)^n$$

The proof of this theorem requires mathematical induction, which is discussed in Section 12.5.

If t is the number of years for which P dollars is invested at an interest rate of $100r$ percent compounded m times per year, then the number of interest periods n is mt. Letting A dollars be the total amount at t years, the formula of Theorem 3 can be written as

$$A = P\left(1 + \frac{r}{m}\right)^{mt} \tag{1}$$

▶ **EXAMPLE 4** *Solving a Word Problem Involving Compound Interest*

Suppose that $400 is deposited into a savings account that pays 8 percent interest per year compounded semiannually. If no withdrawals and no additional deposits are made, find the amount on deposit at the end of 3 years.

Solution   The interest is compounded twice a year; so $m = 2$. Because the time is 3 years, $t = 3$. Furthermore, $P = 400$ and $r = 0.08$. Therefore, if A dollars is the amount on deposit at the end of 3 years, we have from (1)

$$A = P\left(1 + \frac{r}{m}\right)^{mt}$$
$$= 400\left(1 + \frac{0.08}{2}\right)^{2(3)}$$
$$= 400(1.04)^6$$
$$= 506.13$$

Conclusion: The amount on deposit at the end of 3 years is $506.13. ◀

Formula (1) gives the number of dollars in the amount after t years if P dollars is invested at a rate of $100r$ percent compounded m times per year. Let us imagine the interest is continuously compounding. That is, suppose in formula (1) that the number of interest periods per year increases without bound. Thus we are concerned with the behavior of A as $m \to +\infty$. Because $mt = \dfrac{m}{r}(rt)$, we can write formula (1) as

$$A = P\left[\left(1 + \frac{r}{m}\right)^{m/r}\right]^{rt}$$

In this equation let $x = \dfrac{m}{r}$; then

$$A = P\left[\left(1 + \frac{1}{x}\right)^{x}\right]^{rt} \tag{2}$$

Because "$m \to +\infty$" is equivalent to "$x \to +\infty$," let us examine

$$\left(1 + \frac{1}{x}\right)^x \qquad \text{as } x \to +\infty$$

On our calculator we compute values of $\left(1 + \frac{1}{x}\right)^x$ as x takes on larger and larger numbers. Refer to Table 1 for some of these values.

Table 1

x	10	100	1,000	10,000	100,000
$\left(1 + \dfrac{1}{x}\right)^x$	2.5937	2.7048	2.7169	2.7181	2.7183

This table leads us to suspect that $\left(1 + \frac{1}{x}\right)^x$ probably approaches a finite number as x increases without bound. This is indeed the case, and the finite number is denoted by the letter e. The letter e was chosen by Leonhard Euler. The number e is an irrational number. Its value can be expressed to any required degree of accuracy using infinite series, which are studied in calculus. To nine decimal places, the value of e is 2.718281828. We write

$$e \approx 2.718281828$$

You can compute powers of e on your calculator with the $\boxed{e^x}$ key.

To illustrate our conclusion graphically, we plot the graph of the function defined by

$$f(x) = \left(1 + \frac{1}{x}\right)^x$$

and the line $y = e$ in the same viewing rectangle. Figure 1 shows the graphs in the viewing rectangle [0,20] by [1, 3]. The line is a horizontal asymptote of the graph of f, which means that $f(x)$ approaches the number e as x increases without bound.

Returning now to the discussion of interest compounding continuously, we have from the preceding argument and (2) the formula

$$A = Pe^{rt} \tag{3}$$

where A dollars is the amount after t years if P dollars is invested at a rate of $100r$ percent compounded continuously. This value of A is an upper bound for the amount given by (1) when interest is compounded frequently and can be used as an approximation in such a situation. This fact is demonstrated in the following illustration, where we compare the amount at the end of 1 year when interest is compounded continuously with the corre-

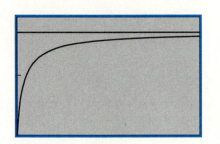

[0, 20] by [1, 3]
$f(x) = \left(1 + \dfrac{1}{x}\right)^x$ and $y = e$

FIGURE 1

sponding amounts obtained when interest is compounded monthly, semimonthly, and daily.

▷ ILLUSTRATION 4

Suppose that $5000 is borrowed at an interest rate of 12 percent compounded monthly, and the loan is to be repaid in one payment at the end of the year. If A dollars is the amount to be repaid, then from formula (1) with $P = 5000$, $r = 0.12$, $m = 12$, and $t = 1$, we have

$$A = 5000(1.01)^{12}$$
$$= 5634.13$$

If the interest rate of 12 percent is compounded semimonthly instead of monthly, then from (1) with $m = 24$, we have

$$A = 5000\left(1 + \frac{0.12}{24}\right)^{24}$$
$$= 5000(1.005)^{24}$$
$$= 5635.80$$

If the interest rate of 12 percent is compounded daily, then from (1) with $m = 365$, we have

$$A = 5000\left(1 + \frac{0.12}{365}\right)^{365}$$
$$= 5000(1.0003)^{365}$$
$$= 5637.37$$

Now suppose that the interest is compounded continuously at 12 percent. Because $P = 5000$, $r = 0.12$, $t = 1$, we have from (3)

$$A = 5000e^{0.12}$$
$$= 5637.484$$

Taking $5637.50 as the amount when interest is compounded continuously at 12 percent, this amount is the upper bound for the amount regardless of how often interest is compounded.   ◀

If in (3), $P = 1$, $r = 1$, and $t = 1$, we get

$$A = e$$

which gives a justification for the economist's interpretation of the number e as the yield on an investment of $1 for a year at an interest rate of 100 percent compounded continuously.

▶ **EXAMPLE 5** *Solving a Word Problem Involving Interest Compounded Continuously*

A bank advertises that interest on savings accounts is computed at 6 percent per year compounded daily. If \$100 is deposited into a savings account at this bank, find **(a)** an approximate amount at the end of 1 year by taking the interest rate at 6 percent compounded continuously and **(b)** the exact amount at the end of 1 year by considering an annual interest rate of 6 percent compounded 365 times per year.

Solution

(a) From (3), with $P = 100$, $r = 0.06$, and $t = 1$, we have, if A dollars is the amount,

$$A = 100e^{0.06}$$
$$= 106.18$$

Thus \$106.18 is an approximate amount on deposit at the end of 1 year.

(b) From (1) with $P = 100$, $r = 0.06$, $m = 365$, and $t = 1$, we have, if A_{365} dollars is the amount,

$$A_{365} = 100\left(1 + \frac{0.06}{365}\right)^{365}$$
$$= 100(1.0001644)^{365}$$
$$= 106.18$$

Conclusion: The exact amount on deposit at the end of 1 year is \$106.18.   ◀

EXERCISES 6.2

In Exercises 1 through 4, use a calculator to compute the power to three significant digits. In Exercises 3 and 4, express the result in scientific notation.

1. (a) $(35.7)^{2.5}$ (b) $(3.78)^{-5}$
   (c) $(0.261)^8$ (d) $(0.403)^{-3}$

2. (a) $(6.23)^{3.5}$ (b) $(15.7)^{-4}$
   (c) $(0.362)^6$ (d) $(0.916)^{-2}$

3. (a) $(4.26)^{25}$ (b) $(0.0312)^5$
   (c) $(0.00172)^{-12}$ (d) $(324)^{-10}$

4. (a) $(78.5)^{15}$ (b) $(0.00247)^8$
   (c) $(0.0311)^{-7}$ (d) $(589)^{-12}$

In Exercises 5 through 12, simplify the expression by applying laws of exponents. In Exercises 7 and 8, x is a real number.

5. (a) $3^{\sqrt{2}} \cdot 3^{\sqrt{50}}$ (b) $(e^{\sqrt{2}})^{\sqrt{50}}$

6. (a) $2^{\sqrt{12}} \cdot 2^{\sqrt{27}}$ (b) $(e^{\sqrt{12}})^{\sqrt{27}}$

7. (a) $(5^{\sqrt{15}})^{\sqrt{6}}$ (b) $5^{\sqrt[3]{x}} \cdot 5^{\sqrt[3]{x^2}}$

8. (a) $(10^{\sqrt{10}})^{\sqrt{5}}$ (b) $(10^{\sqrt{3x}})^{\sqrt{15x}}$

9. (a) $\dfrac{4^{\sqrt{32}}}{2^{\sqrt{18}}}$ (b) $\dfrac{250^{\sqrt{5}}}{10^{\sqrt{20}}}$

10. (a) $\dfrac{3^{\sqrt{45}}}{9^{\sqrt{20}}}$ **(b)** $\dfrac{14^{\sqrt{98}}}{28^{\sqrt{72}}}$

11. $\dfrac{e^2 \cdot e^{\sqrt{54}}}{e^{\sqrt{24}}}$ **12.** $\left(\dfrac{e^{\sqrt{14}}}{e^3 \cdot e^{\sqrt{7}}}\right)^2$

In Exercises 13 through 16, write the number in scientific notation.

13. (a) 52.60   **(b)** 0.0061
   (c) 172,000 (3 significant digits)
   (d) 172,000 (4 significant digits)

14. (a) 43,851   **(b)** 0.276
   (c) 3400 (2 significant digits)
   (d) 3400 (3 significant digits)

15. (a) 0.03960   **(b)** 0.0000080022
   (c) 1.723;   **(d)** 426.0

16. (a) 0.00006405 **(b)** 0.0001030
   (c) 98.0   **(d)** 7820.0

In Exercises 17 and 18, use scientific notation and a calculator to perform the operations. Assume three significant digits for each number.

17. (a) $(0.0470)(320,000)^2$

   (b) $\dfrac{(180,000)^3 \sqrt{0.0000450}}{623,000}$

18. (a) $\dfrac{(0.0000831)^2}{(140)^3}$

   (b) $\dfrac{(256,000,000)^2 \sqrt[3]{0.000712}}{(0.000348)^3 \sqrt{5100}}$

In Exercises 19 through 21, use scientific notation and a calculator.

19. Determine the distance from the earth to the sun in meters if the sun is $(9.29)10^7$ miles from the earth and 1 mile is 1.61 km.

20. Determine the distance from the earth to a star that is 7.00 light-years away if the speed of light is $(1.86)10^5$ mi/sec and 1 year is 365 days.

21. Determine the mass of the earth in tons if the number of grams in the earth's mass is $(5.97)10^{27}$ and 1 g is $(2.205)10^{-3}$ pound.

Exercises 22 through 24 are based on the following:
   In calculus we show that

   $(1 + z)^{1/z} \rightarrow e$ *as* $z \rightarrow 0$

Observe that $(1 + z)^{1/z}$ *can be obtained from* $\left(1 + \dfrac{1}{x}\right)^x$ *by letting* $x = \dfrac{1}{z}$.

22. Use a calculator to compute the values of $(1 + z)^{1/z}$ when $z = 0.001$ and $z = -0.001$. Then obtain an approximation of the number e to three decimal places by using these values to find the average value of $(1.001)^{1000}$ and $(0.999)^{-1000}$.

23. Use a calculator to compute the values of $(1 + z)^{1/z}$ when $z = 0.0001$ and $z = -0.0001$. Then obtain an approximation of the number e to four decimal places by using these values to find the average value of $(1.0001)^{10,000}$ and $(0.9999)^{-10,000}$.

24. Plot the graph of the function defined by

   $$f(x) = (1 + x)^{1/x}$$

   in the $[-0.5, 0.5]$ by $[1, 3]$ viewing rectangle. Because $f(0)$ is not defined, there is a break in the graph at $x = 0$. By using zoom-in, show that if we redefine the function at 0 so that $f(0) = e$, then there is no break in the graph at $x = 0$ (that is, function f is continuous at 0).

25. A loan of $2000 is made at a simple interest rate of 12 percent annually. Determine the amount to be repaid if the period of the loan is **(a)** 90 days; **(b)** 6 months; **(c)** 1 year.

26. Solve Exercise 25 if the loan is for $1500 and the rate is 10 percent annually.

27. A man borrowed $10,000 at an annual interest rate of 9 percent with the understanding that interest was to be paid monthly. However, the borrower did not make the monthly interest payments, and so the principal with interest at 9 percent compounded monthly was due at the end of the year. What was the amount due?

28. On his twenty-fifth birthday a man inherited $5000. If he invested this amount at 8 percent compounded annually, how much would he receive when he retires at the age of 65?

29. Determine the amount at the end of 4 years of an investment of $1000 if the annual interest rate is 8 percent and **(a)** simple interest is earned; **(b)** interest is compounded annually; **(c)** interest is compounded semiannually; **(d)** interest is compounded quarterly; **(e)** interest is compounded continuously.

30. Find the amount of an investment of $500 at the end of 2 years if the annual interest rate is 6 percent and **(a)** simple interest is earned; **(b)** interest is compounded annually; **(c)** interest is compounded semiannually; **(d)** interest is compounded monthly; **(e)** interest is compounded continuously.

31. An investment of $5000 earns interest at the rate of 16 percent per year and the interest is paid once at the end of the year. Find the interest earned during the first year if **(a)** interest is compounded quarterly and **(b)** interest is compounded continuously.

32. A loan of $2000 is repaid in one payment at the end of a year with interest at an annual rate of 12 percent. Determine the total amount repaid if **(a)** interest is compounded quarterly and **(b)** interest is compounded continuously.

33. A deposit of $1000 is made at a savings bank that advertises that interest on accounts is computed at an annual rate of 9 percent compounded daily. Find **(a)** an approximate amount at the end of 1 year by taking the interest rate as 9 percent compounded continuously and **(b)** the exact amount at the end of 1 year by considering an annual interest rate of 9 percent compounded 365 times per year.

34. Solve Exercise 33 if the bank advertises that interest is computed at an annual rate of 7 percent compounded daily, and **(a)** take the rate as 7 percent compounded continuously and **(b)** consider an annual interest rate of 7 percent compounded 365 times per year.

35. How much should be deposited in a savings account now if it is desired to have $5000 in the account at the end of 5 years if the annual interest rate is 8 percent compounded quarterly?

36. At the end of 4 years a savings account had a balance of $3000. One deposit was made at the beginning of the 4-year period and no withdrawals were made. How much was the original deposit if the interest rate is 6 percent per year compounded monthly?

37. Solve Exercise 35 if the annual interest rate is 8 percent compounded continuously.

38. Solve Exercise 36 if the interest rate is 6 percent per year compounded continuously.

39. In calculus we show that

$$\left(1 + \frac{2}{x}\right)^x$$

approaches e^2 as x increases without bound. Illustrate this statement by plotting the graph of a suitably chosen function f and its horizontal asymptote in an appropriate viewing rectangle. Define your function f and write an equation of the asymptote. Describe why your graph illustrates the statement.

40. Follow the instructions of Exercise 39 for the statement: In calculus we show that

$$\left(1 + \frac{0.5}{x}\right)^x$$

approaches $\sqrt{e}$ as x increases without bound.

41. Describe the three kinds of interest: simple, compounded quarterly, and compounded continuously. In your description, indicate what rate of simple interest and what rate of interest compounded quarterly would be necessary for an investment to yield the same annual interest as it earns at 10 percent compounded continuously.

6.3 EXPONENTIAL FUNCTIONS

GOALS
1. **Define the exponential function with base b.**
2. **Sketch the graph of the exponential function with base b.**
3. **Define the natural exponential function.**
4. **Sketch the graph of the natural exponential function.**
5. **Solve word problems involving exponential growth.**
6. **Solve word problems involving exponential decay.**
7. **Solve word problems involving bounded growth.**
8. **Solve word problems involving logistic growth.**

$\int / \frac{dy}{dx}$ In calculus you will study problems involving exponential growth and decay, which lead to mathematical models containing exponential functions with base e. We begin our discussion of exponential functions by considering such functions having any positive number as base.

If $b > 0$, then for each real number x there corresponds a unique real number b^x. Thus we can make the following definition.

DEFINITION   **The Exponential Function with Base b**

If $b > 0$ and $b \neq 1$, then the **exponential function with base b** is the function f defined by

$$f(x) = b^x$$

The domain of f is the set of real numbers, and the range is the set of positive numbers.

Note: If $b = 1$, then b^x becomes 1^x, and because for any x, $1^x = 1$, we have a constant function. For this reason, we impose the condition that $b \neq 1$ in the preceding definition.

In the following two illustrations we consider the graphs of the exponential function with bases 2 and $\frac{1}{2}$.

▷ **ILLUSTRATION 1**

The exponential function with base 2 is defined by

$$F(x) = 2^x$$

Table 1 gives some rational values of x with the corresponding function values.

Table 1

x	-3	-2	-1	0	1	2	3
2^x	$\frac{1}{8}$	$\frac{1}{4}$	$\frac{1}{2}$	1	2	4	8

The graph sketched in Figure 1 is drawn by locating the points whose coordinates are given by Table 1 and connecting these points with a curve. The function is seen to be increasing, which follows from Theorem 1(i) of Section 6.2 with r and s real numbers, and the definition of an increasing function given in Section 6.1.

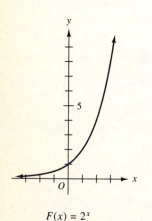

$F(x) = 2^x$

FIGURE 1

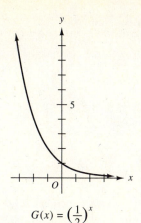

$$G(x) = \left(\tfrac{1}{2}\right)^x$$

FIGURE 2

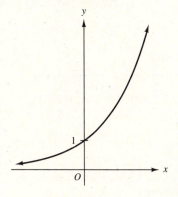

$f(x) = b^x, b > 1$

FIGURE 3

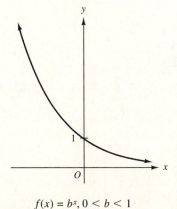

$f(x) = b^x, 0 < b < 1$

FIGURE 4

Note that

$$2^x \to 0^+ \qquad \text{as } x \to -\infty$$

that is, 2^x approaches zero through values greater than zero as x decreases without bound. Therefore, the x axis is a horizontal asymptote of the graph of F. Furthermore,

$$2^x \to +\infty \qquad \text{as } x \to +\infty$$

that is, 2^x increases without bound as x increases without bound. ◀

▷ ILLUSTRATION 2

The exponential function with base $\tfrac{1}{2}$ is the function G such that

$$G(x) = \left(\tfrac{1}{2}\right)^x$$

Table 2

x	-3	-2	-1	0	1	2	3
$\left(\tfrac{1}{2}\right)^x$	8	4	2	1	$\tfrac{1}{2}$	$\tfrac{1}{4}$	$\tfrac{1}{8}$

Table 2 gives function values for some rational values of x. By locating the points with these coordinates and connecting the points with a curve, we sketch the graph appearing in Figure 2. The function is decreasing, which follows from Theorem 1(ii) of Section 6.2, with r and s real numbers, and the definition of a decreasing function given in Section 6.1. Because

$$\left(\tfrac{1}{2}\right)^x \to 0^+ \qquad \text{as } x \to +\infty$$

the x axis is a horizontal asymptote of the graph of G. Furthermore,

$$\left(\tfrac{1}{2}\right)^x \to +\infty \qquad \text{as } x \to -\infty$$

that is, $\left(\tfrac{1}{2}\right)^x$ increases without bound as x decreases without bound. ◀

Figure 3 shows the graph of the exponential function with base b when $b > 1$. The function is increasing. In Figure 4, we have the graph of the exponential function with base b when $0 < b < 1$. This function is decreasing.

From the definition of the exponential function with base b, the base can be any positive number except 1. When the base is e, we have the *natural exponential function*.

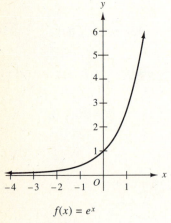

$f(x) = e^x$

FIGURE 5

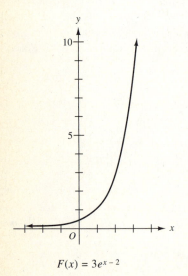

$F(x) = 3e^{x-2}$

FIGURE 6

> **DEFINITION** **The Natural Exponential Function**
>
> The **natural exponential function** is the function f defined by
>
> $$f(x) = e^x$$
>
> The domain of the natural exponential function is the set of real numbers, and its range is the set of positive numbers.

▷ ILLUSTRATION 3

Table 3 gives some values of x and the corresponding values of the natural exponential function obtained from a calculator. To sketch the graph of the natural exponential function, we locate the points whose coordinates are given in Table 3 and connect these points with a curve. We obtain the graph shown in Figure 5.

Table 3

x	0	0.5	1	1.5	2	2.5	−0.5	−1	−2
e^x	1	1.6	2.7	4.5	7.4	12.2	0.6	0.4	0.1

◀

▶ **EXAMPLE 1** *Sketching the Graph of an Exponential Function by Geometric Transformations*

Sketch the graph of each of the following functions by geometric transformations of the graph of $f(x) = e^x$.

(a) $F(x) = 3e^{x-2}$ **(b)** $G(x) = -e^x + 3$

Solution
(a) We first vertically stretch the graph of f by a factor of 3 and then make a horizontal translation of 2 units to the right. We obtain the graph of F shown in Figure 6.

(b) For the graph of G, the minus sign indicates that we first reflect the graph of f through the x axis. We then make a vertical translation 3 units upward and obtain the graph shown in Figure 7. ◀

Exponential growth and *exponential decay* give mathematical models involving powers of e. A function defined by an equation of the form

$$f(t) = Be^{kt} \qquad t \geq 0 \tag{1}$$

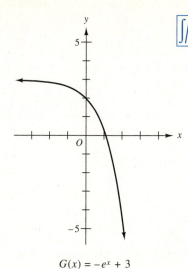

$G(x) = -e^x + 3$

FIGURE 7

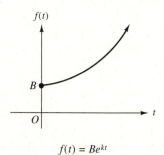

$f(t) = Be^{kt}$

FIGURE 8

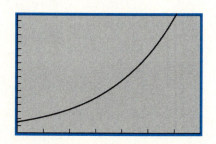

$[0, 70]$ by $[0, 17,000]$
$f(t) = 1500e^{0.04t}$
FIGURE 9

where B and k are positive constants is said to describe **exponential growth.** In calculus you will learn that such a function results when the rate of growth of a quantity is proportional to its size. To sketch the graph of $f(t) = Be^{kt}$, note that $f(0) = B$ and $f(t)$ is always positive. Furthermore,

$$Be^{kt} \to +\infty \qquad \text{as } t \to +\infty$$

Thus Be^{kt} increases without bound as t increases without bound. The graph of (1) appears in Figure 8.

The following example involves exponential growth in biology.

▶ **EXAMPLE 2** *Solving a Word Problem Involving Exponential Growth*

In a particular bacterial culture, if $f(t)$ bacteria are present at t minutes, then

$$f(t) = Be^{0.04t}$$

where B is a constant. **(a)** Compute B from the fact that 1500 bacteria are present initially. **(b)** With the value of B from part (a), plot the graph of f. **(c)** Determine algebraically how many bacteria will be present after 1 hr, and check the answer on a graphics calculator.

Solution

(a) Because 1500 bacteria are present initially, $f(0) = 1500$. From the equation defining $f(t)$

$$f(0) = Be^{0.04(0)}$$
$$1500 = Be^0$$
$$1500 = B$$

(b) With the value of B from part (a)

$$f(t) = 1500e^{0.04t}$$

The graph of f plotted in the viewing rectangle $[0, 70]$ by $[0, 17,000]$ appears in Figure 9.

(c) The number of bacteria present after 1 hr is $f(60)$, and from the equation in part (b)

$$f(60) = 1500e^{0.04(60)}$$
$$= 1500e^{2.4}$$
$$= 16,535$$

<u>Conclusion:</u>   After 1 hour, 16,535 bacteria are present in the culture. We check this answer on a graphics calculator by using the trace button and zoom-in with the graph in Figure 9.   ◀

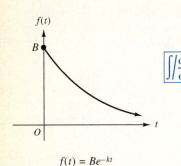

$f(t) = Be^{-kt}$

FIGURE 10

A function defined by an equation of the form

$$f(t) = Be^{-kt} \qquad t \geq 0 \tag{2}$$

where B and k are positive constants is said to describe **exponential decay.** Exponential decay occurs when the rate of decrease of a quantity is proportional to its size, as shown in calculus. For instance, it is known from experiments that the rate of decay of radium is proportional to the amount of radium present at a given instant. The graph of (2) appears in Figure 10. Observe that

$$Be^{-kt} \to 0^+ \qquad \text{as } t \to +\infty$$

The next example involves exponential decay where the value of some equipment is decreasing exponentially.

▶ **EXAMPLE 3** *Solving a Word Problem Involving Exponential Decay*

If $V(t)$ dollars is the value of a certain piece of equipment t years after its purchase, then

$$V(t) = Be^{-0.20t}$$

where B is a constant. If the equipment was purchased for $8000, what will be its value in 2 years? Check the answer on a graphics calculator.

Solution   Because the equipment was purchased for $8000, $V(0) = 8000$. From the equation defining $V(t)$

$$V(0) = Be^{-0.20(0)}$$
$$8000 = Be^0$$
$$8000 = B$$

Therefore with $B = 8000$, we have

$$V(t) = 8000e^{-0.20t}$$

The number of dollars in the value of the equipment after 2 years is $V(2)$, and from the preceding equation

$$V(2) = 8000e^{-0.20(2)}$$
$$= 8000e^{-0.40}$$
$$= 5362.56$$

Conclusion: The value of the equipment in 2 years will be $5362.56.

To check the answer on our graphics calculator, we plot the graph of $V(t) = 8000e^{-0.20t}$ in the viewing rectangle [0, 10] by [0, 10,000] shown in Figure 11. Using the trace key, move the cursor to where $t = 2$, and find the function value 5362.56.   ◀

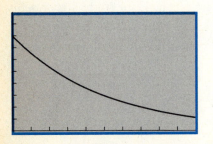

[0, 10] by [0, 10,000]
$V(t) = 8000e^{-0.2t}$
FIGURE 11

Another mathematical model involving powers of e is given by the function defined by

$$f(t) = A(1 - e^{-kt}) \qquad t \geq 0$$

where A and k are positive constants. This function describes **bounded growth.** Because

$$A(1 - e^{-kt}) = A - Ae^{-kt}$$

and $Ae^{-kt} \rightarrow 0^+$ as $t \rightarrow +\infty$, it follows that

$$A(1 - e^{-kt}) \rightarrow A^- \qquad \text{as } t \rightarrow +\infty$$

Therefore, the graph of f has as a horizontal asymptote the line A units above the t axis. Also note that

$$f(0) = A(1 - e^0)$$
$$= 0$$

From this information we obtain the graph shown in Figure 12. This graph is sometimes called a **learning curve.** The name appears appropriate when $f(t)$ represents someone's competence in performing a job. As a person's experience increases, competence increases rapidly at first and then slows down as additional experience has little effect on the skill with which the task is performed.

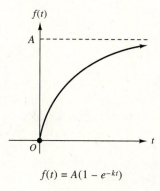

$$f(t) = A(1 - e^{-kt})$$

FIGURE 12

▶ **EXAMPLE 4** *Solving a Word Problem Involving Bounded Growth*

A typical worker at a certain factory can produce $f(t)$ units per day after t days on the job, where

$$f(t) = 50(1 - e^{-0.34t})$$

(a) How many units per day can the worker produce after 7 days on the job? **(b)** How many units per day can the worker eventually be expected to produce? **(c)** Use the answer in part (b) to determine the horizontal asymptote of the graph of f and plot this line and the graph of f in the same viewing rectangle.

Solution

(a) We wish to find $f(7)$. From the equation defining $f(t)$

$$f(7) = 50(1 - e^{-0.34(7)})$$
$$= 50(1 - e^{-2.38})$$
$$= 50(1 - 0.093)$$
$$= 45$$

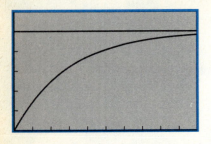

[0, 10] by [0, 60]
$f(t) = 50(1 - e^{-0.34t})$ and $g(t) = 50$

FIGURE 13

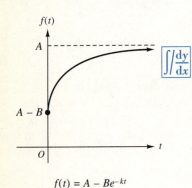

$f(t) = A - Be^{-kt}$

FIGURE 14

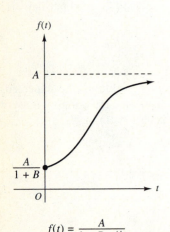

$f(t) = \dfrac{A}{1 + Be^{-Akt}}$

FIGURE 15

Conclusion: The worker can produce 45 units per day after 7 days on the job.

(b) Because $e^{-0.34t} \to 0^+$ as $t \to +\infty$, it follows that

$$50(1 - e^{-0.34t}) \to 50^- \qquad \text{as } t \to +\infty$$

Conclusion: The worker can eventually be expected to produce 50 units per day.

(c) The horizontal asymptote of the graph of f is the line $g(t) = 50$. Figure 13 shows this line and the graph of f in the viewing rectangle [0, 10] by [0, 60]. ◀

Bounded growth is also described by a function defined by

$$f(t) = A - Be^{-kt} \qquad t \geq 0$$

where A, B, and k are positive constants. For this function $f(0) = A - B$. The graph is sketched in Figure 14. In calculus we can show that bounded growth occurs when a quantity grows at a rate proportional to the difference between a fixed number A and the size of the quantity. In this case, A serves as an upper bound.

Consider now the growth of a population when the environment imposes an upper bound on its size. For instance, space or reproduction may be factors that are limited by the environment. In such cases a mathematical model of the form (1) does not apply because the population does not increase beyond a certain point. A model that takes into account environmental factors is given by the function defined by

$$f(t) = \frac{A}{1 + Be^{-Akt}} \qquad t \geq 0$$

where A, B, and k are positive constants. Figure 15 shows the graph of this function. It is called a curve of **logistic growth.** Observe that when t is small, the graph is similar to the one for exponential growth in Figure 8, and as t increases, the curve is analogous to that shown in Figure 14 for bounded growth.

An application of logistic growth in economics is the distribution of information about a particular product. Logistic growth is used by biologists to describe the spread of a disease and by sociologists to describe the spread of a rumor or a joke.

▶ **EXAMPLE 5** *Solving a Word Problem Involving Logistic Growth*

In a certain community the spread of a particular flu germ was such that t weeks after its outbreak $f(t)$ persons had contracted the flu, where

$$f(t) = \frac{45{,}000}{1 + 224e^{-0.9t}}$$

How many people had the flu (a) at the outbreak; (b) after 3 weeks; and (c) after 10 weeks? (d) If the epidemic continues indefinitely, how many people will contract the flu? (e) Use the answer in part (d) to determine the horizontal asymptote of the graph of f and plot this line and the graph of f in the same viewing rectangle.

Solution
(a) The number of people who had the flu at the outbreak is $f(0)$, and

$$f(0) = \frac{45{,}000}{1 + 224e^{-0.9(0)}}$$

$$= \frac{45{,}000}{1 + 224}$$

$$= 200$$

Conclusion: At the outbreak, 200 people had the flu.

(b) After 3 weeks the number of people who had the flu is $f(3)$; and (c) after 10 weeks the number of people who had the flu is $f(10)$.

(b) $f(3) = \dfrac{45{,}000}{1 + 224e^{-0.9(3)}}$ (c) $f(10) = \dfrac{45{,}000}{1 + 224e^{-0.9(10)}}$

$\quad\quad\quad = \dfrac{45{,}000}{1 + 224e^{-2.7}}$ $= \dfrac{45{,}000}{1 + 224e^{-9}}$

$\quad\quad\quad = 2803$ $= 43{,}790$

Conclusion: (b) After 3 weeks, 2803 people had the flu and (c) after 10 weeks 43,790 people had the flu.

(d) Because

$$224e^{-0.9t} \to 0^+ \quad \text{as } t \to +\infty$$

we conclude that

$$\frac{45{,}000}{1 + 224e^{-0.9t}} \to 45{,}000 \quad \text{as } t \to +\infty$$

Conclusion: Approximately 45,000 people will contract the flu if the epidemic continues indefinitely.

(e) The horizontal asymptote is the line $g(t) = 45{,}000$. This line and the graph of f plotted in the viewing rectangle [0, 11] by [0, 50,000] appear in Figure 16. ◀

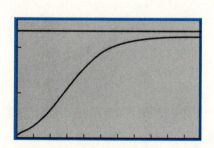

[0, 11] by [0, 50,000]

$f(t) = \dfrac{45{,}000}{1 + 224e^{-0.9t}}$ and $g(t) = 45{,}000$

FIGURE 16

EXERCISES 6.3

In Exercises 1 through 8, sketch the graph of the exponential function. Check your graph on your graphics calculator.

1. $f(x) = 3^x$

2. $g(x) = 4^x$

3. $g(x) = 3^{-x}$

4. $f(x) = 4^{-x}$

5. $F(x) = (\frac{1}{5})^x$

6. $G(x) = 10^x$

7. $G(x) = e^{-x}$

8. $F(x) = e^{2x}$

In Exercises 9 through 14, sketch the graph of function F by geometric transformations of the graph of f.

9. (a) $F(x) = 3^{x-4}, f(x) = 3^x$
   (b) $F(x) = 3^x - 4, f(x) = 3^x$

10. (a) $F(x) = 2^{x+3}, f(x) = 2^x$
   (b) $F(x) = 2^x + 3, f(x) = 2^x$

11. (a) $F(x) = -e^{x+2}, f(x) = e^x$
   (b) $F(x) = -e^x + 2, f(x) = e^x$

12. (a) $F(x) = -2e^{x-1}, f(x) = e^x$
   (b) $F(x) = -2e^x - 1, f(x) = e^x$

13. (a) $F(x) = 2^{x+1} + 3, f(x) = 2^x$
   (b) $F(x) = e^{x-3} - 4, f(x) = e^x$

14. (a) $F(x) = 3^{x-5} - 2, f(x) = 3^x$
   (b) $F(x) = e^{x+4} + 1, f(x) = e^x$

In Exercises 15 through 18, sketch the graph of the exponential function. Check your graph on your graphics calculator.

15. $f(x) = 10e^{0.2x}$

16. $g(x) = 10e^{0.1x}$

17. $g(x) = 10e^{-0.1x}$

18. $f(x) = 10e^{-0.2x}$

19. Sketch the graphs of $y = 3^x$ and $x = 3^y$ on the same coordinate axes.

20. Sketch the graphs of $y = e^x$ and $x = e^y$ on the same coordinate axes.

In Exercises 21 through 33, solve the word problem by using a mathematical model involving an exponential function. Be sure to write a conclusion.

21. The value of a particular machine t years after its purchase is $V(t)$ dollars, where $V(t) = ke^{-0.3t}$ and k is a constant. **(a)** Compute k from the fact that the machine was purchased 8 years ago for $10,000. **(b)** With the value of k from part (a), plot the graph of V. **(c)** Determine algebraically the present value of the machine, and check your answer on your graphics calculator.

22. If $f(t)$ grams of a radioactive substance are present after t seconds, then $f(t) = ke^{-0.3t}$ where k is a constant. **(a)** Compute k from the fact that 100 grams of the substance are present initially. **(b)** With the value of k from part (a), plot the graph of f. **(c)** Determine algebraically how much of the substance will be present after 5 sec, and check your answer on your graphics calculator.

23. The population of a particular town is increasing at a rate proportional to its size. If this rate is 6 percent and if the population t years from now is expected to be $P(t)$, then $P(t) = ke^{0.06t}$ where k is a constant. **(a)** Compute k from the fact that the current population is 10,000. **(b)** With the value of k from part (a) plot the graph of P. Determine algebraically the expected population **(c)** after 10 years and **(d)** after 20 years. Check your answers to parts (c) and (d) on your graphics calculator.

24. Suppose $f(t)$ is the number of bacteria present in a certain culture at t minutes and $f(t) = ke^{0.035t}$ where k is a constant. **(a)** Compute k from the fact that 5000 bacteria are present after 10 min have elapsed. **(b)** With the value of k from part (a) plot the graph of f. **(c)** Determine algebraically how many bacteria were present initially and check your answer on your graphics calculator.

25. If $P(h)$ pounds per square foot is the atmospheric pressure at a height h feet above sea level, then $P(h) = ke^{-0.00003h}$ where k is a constant. If the atmospheric pressure at sea level is 2116 $1b/ft^2$, find the atmospheric pressure outside of an airplane that is 10,000 ft high. Check your answer on your graphics calculator.

26. In 1985, it was estimated that for the succeeding 20 years the population of a particular town is expected to be $f(t)$ people t years from 1985, where $f(t) = C \cdot 10^{kt}$ where C and k are constants. If the population in 1985 was 1000 and in 1990 it was 4000, what is the expected population in the year 2000? Check your answer on your graphics calculator.

27. A historically important abstract painting was purchased in 1932 for $200, and its value has doubled every 10 years since its purchase. If $f(t)$ dollars is the value t years after its purchase, (a) define $f(t)$, and (b) what was the value of the painting in 1992?

28. After t hours of practice typing, it was determined that a certain person could type $f(t)$ words per minute, where $f(t) = 90(1 - e^{-0.03t})$. (a) How many words per minute can the person type after 30 hr of practice? (b) How many words per minute can the person eventually be expected to type? (c) Use your answer in part (b) to determine the horizontal asymptote of the graph of f and plot this line and the graph of f in the same viewing rectangle.

29. The efficiency of a typical worker at a certain factory is given by the function defined by

$$f(t) = 100 - 60e^{-0.2t}$$

where the worker can complete $f(t)$ units of work per day after being on the job for t months. (a) How many units per day can be completed by a beginning worker? (b) How many units per day can be completed by a worker having a year's experience? (c) How many units per day can the typical worker eventually be expected to complete? (d) Use your answer in part (c) to determine the horizontal asymptote of the graph of f and plot this line and the graph of f in the same viewing rectangle.

30. The resale value of a certain piece of equipment is $f(t)$ dollars t years after its purchase, where

$$f(t) = 1200 + 8000e^{-0.25t}$$

(a) What is the value of the equipment when it is purchased? (b) What is the value of the equipment 10 years after its purchase? (c) What is the anticipated scrap value of the equipment after a long period of time? (d) Use your answer in part (c) to determine the horizontal asymptote of the graph of f, and plot this line and the graph of f in the same viewing rectangle.

31. One day on a college campus when 5000 people were in attendance, a particular student heard that a certain controversial speaker was going to make an unscheduled appearance. This information was told to friends who in turn related it to others. After t minutes elapsed, $f(t)$ people had heard the rumor, where

$$f(t) = \frac{5000}{1 + 4999e^{-0.5t}}$$

How many people had heard the rumor (a) after 10

min and (b) after 20 min? (c) How many people will eventually hear the rumor? (d) Use your answer in part (c) to determine the horizontal asymptote of the graph of f and plot this line and the graph of f in the same viewing rectangle.

32. In a particular town of population A, 20 percent of the residents heard a radio announcement about a local political scandal. After t hours $f(t)$ people had heard about the scandal, where

$$f(t) = \frac{A}{1 + Be^{-Akt}}$$

If 50 percent of the population heard about the scandal after 1 hr, how long was it until 80 percent of the population heard about it?

33. In a community in which A people are susceptible to a particular virus, this virus spread in such a way that t weeks after its appearance, $f(t)$ persons had caught the virus, where

$$f(t) = \frac{A}{1 + Be^{-Akt}}$$

If 10 percent of those susceptible had the virus initially and 25 percent had been infected after 3 weeks, what percent of those susceptible had been infected after 6 weeks?

34. The function f defined by $f(x) = e^{-x^2}$ is important in statistics. Sketch the graph of f by assuming that f is continuous (the graph has no breaks) and locating the points for which x has the values, -2, $-\frac{3}{2}$, -1, $-\frac{1}{2}$, 0, $\frac{1}{2}$, 1, $\frac{3}{2}$, and 2. Use your calculator for powers of e. Check your graph on your graphics calculator.

35. Newton's law of cooling states that the rate at which a body changes temperature is proportional to the difference between its temperature and that of the surrounding medium. In calculus, we can show from this law that if a body in air of temperature $35°$ cools from $120°$ to $60°$ in 40 min, then if $f(t)$ degrees is the temperature of the body at t minutes

$$f(t) = 35 + 85\left(\frac{5}{17}\right)^{t/40}$$

Find the temperature of the body (a) after 10 min, (b) after 50 min (c) after 100 min, and (d) after 3 hr. (e) What will the temperature of the body be eventually? (f) Use your answer in part (e) to determine the horizontal asymptote of the graph of f and plot this line and the graph of f in the same viewing rectangle.

36. By observation, one of the solutions of the equation

$$x^2 = 2^x$$

is 2. **(a)** There is another positive integer solution. What is it? **(b)** There is an irrational number solution. To get a numerical approximation to this solution, plot the graphs of $y = x^2$ and $y = 2^x$ in the same viewing rectangle, and use trace and zoom-in to determine the solution to the nearest one-hundredth. From the three solutions of the equation and your graphs in part (b) determine the values of x for which **(c)** $x^2 < 2^x$ and **(d)** $x^2 > 2^x$.

 The functions S and C in Exercises 37 through 40 occur in calculus. They are called hyperbolic functions

and are defined by

$$S(x) = \tfrac{1}{2}(e^x - e^{-x}) \quad \text{and} \quad C(x) = \tfrac{1}{2}(e^x + e^{-x})$$

In Exercises 37 and 38, determine the indicated function values.

37. (a) $S(0)$ **(b)** $C(1)$ **(c)** $S(-1)$
(d) $S(3.5)$ **(e)** $C(-2)$

38. (a) $C(0)$ **(b)** $S(1)$ **(c)** $C(3)$
(d) $S(-2.5)$ **(e)** $C(-2.5)$

39. (a) Prove that S is an odd function. **(b)** Sketch the graph of S, and check your graph on your graphics calculator.

40. (a) Prove that C is an even function. **(b)** Sketch the graph of C, and check your graph on your graphics calculator.

6.4 LOGARITHMIC FUNCTIONS

GOALS

1. **Define the logarithmic function with base b.**
2. **Find the value of a logarithm.**
3. **Solve a logarithmic equation.**
4. **Sketch the graph of the logarithmic function with base b.**
5. **Define the natural logarithmic function.**
6. **Sketch the graph of the natural logarithmic function.**
7. **Solve word problems involving the natural exponential and natural logarithmic functions.**

In Example 2 of Section 6.3 we had the equation

$$f(t) = 1500e^{0.04t}$$

where $f(t)$ bacteria are present in a certain culture at t minutes. Suppose now that we wish to find in how many minutes 15,000 bacteria will be present. If T is the number of minutes to be determined,

$$15{,}000 = 1500e^{0.04T}$$

In this equation the unknown T appears in an exponent. At the present we cannot solve such an equation, but the concept of a logarithm will give us the means to do so. We will develop this concept and then return to the problem in Example 3.

When $b > 1$, the exponential function with base b is increasing, and when $0 < b < 1$, it is decreasing. Thus, from Theorem 1 of Section 6.1, it follows that the exponential function with base b is one-to-one. It therefore has an inverse function, which is given in the following definition.

> **DEFINITION**   **The Logarithmic Function with Base b**
>
> The **logarithmic function with base b** is the inverse of the exponential function with base b.

We use the notation $\log_b$ to denote the logarithmic function with base b. The function values of $\log_b$ are denoted by $\log_b(x)$ or more simply $\log_b x$ (read "logarithm with base b of x"). Therefore, because $\log_b$ and the exponential function with base b are inverse functions,

$$y = \log_b x \quad \text{if and only if} \quad x = b^y \tag{1}$$

The domain of the exponential function with base b is the set of real numbers, and its range is the set of positive numbers. The domain of $\log_b$ is, therefore, the set of positive numbers, and the range is the set of real numbers.

The two equations appearing in (1) are equivalent. We make use of this fact in the following two illustrations.

▷ **ILLUSTRATION 1**

$$3^2 = 9 \Leftrightarrow \log_3 9 = 2 \qquad 2^3 = 8 \Leftrightarrow \log_2 8 = 3$$
$$\left(\tfrac{1}{16}\right)^{1/2} = \tfrac{1}{4} \Leftrightarrow \log_{1/16} \tfrac{1}{4} = \tfrac{1}{2} \qquad 5^{-2} = \tfrac{1}{25} \Leftrightarrow \log_5 \tfrac{1}{25} = -2 \qquad ◀$$

▷ **ILLUSTRATION 2**

$$\log_{10} 10{,}000 = 4 \Leftrightarrow 10^4 = 10{,}000 \qquad \log_8 2 = \tfrac{1}{3} \Leftrightarrow 8^{1/3} = 2$$
$$\log_6 1 = 0 \Leftrightarrow 6^0 = 1 \qquad \log_9 \tfrac{1}{3} = -\tfrac{1}{2} \Leftrightarrow 9^{-1/2} = \tfrac{1}{3} \qquad ◀$$

▶ **EXAMPLE 1**   *Finding the Value of a Logarithm*

Find the value of each of the following logarithms:

(a) $\log_7 49$ **(b)** $\log_5 \sqrt{5}$ **(c)** $\log_6 \tfrac{1}{6}$ **(d)** $\log_3 81$ **(e)** $\log_{10} 0.001$

Solution   In each part we let y represent the given logarithm and obtain an equivalent equation in exponential form. We then solve for y by making use of the fact that if $b > 0$ and $b \neq 1$, then

$$b^y = b^n \quad \text{implies} \quad y = n$$

(a) Let $\log_7 49 = y$. This equation is equivalent to $7^y = 49$. Because $49 = 7^2$, we have

$$7^y = 7^2$$

Therefore $y = 2$; that is, $\log_7 49 = 2$.

(b) Let $\log_5 \sqrt{5} = y$. Therefore $5^y = \sqrt{5}$ or, equivalently,

$$5^y = 5^{1/2}$$

Hence $y = \frac{1}{2}$; that is, $\log_5 \sqrt{5} = \frac{1}{2}$.

(c) Let $\log_6 \frac{1}{6} = y$. Thus $6^y = \frac{1}{6}$ or, equivalently,

$$6^y = 6^{-1}$$

Therefore $y = -1$; that is, $\log_6 \frac{1}{6} = -1$.

(d) Let $\log_3 81 = y$. Thus $3^y = 81$ or, equivalently,

$$3^y = 3^4$$

Hence $y = 4$; that is, $\log_3 81 = 4$.

(e) Let $\log_{10} 0.001 = y$. Then $10^y = 0.001$. Because $10^{-3} = 0.001$, we have

$$10^y = 10^{-3}$$

Therefore $y = -3$; that is, $\log_{10} 0.001 = -3$. ◀

▶ **EXAMPLE 2** *Solving a Logarithmic Equation*

Solve the equation for either x or b:

(a) $\log_6 x = 2$ **(b)** $\log_{27} x = \frac{2}{3}$ **(c)** $\log_b 4 = \frac{1}{3}$ **(d)** $\log_b 81 = -2$

Solution   We write each logarithmic equation as an equivalent exponential equation.

(a) $\log_6 x = 2 \Leftrightarrow 6^2 = x$ **(b)** $\log_{27} x = \frac{2}{3} \Leftrightarrow 27^{2/3} = x$

Therefore   Hence

$$x = 36$$ $$x = (\sqrt[3]{27})^2$$
$$= 3^2$$
$$= 9$$

(c) $\log_b 4 = \frac{1}{3} \Leftrightarrow b^{1/3} = 4$

Thus

$(b^{1/3})^3 = 4^3$

$$b = 64$$

(d) $\log_b 81 = -2 \Leftrightarrow b^{-2} = 81$

Therefore

$(b^{-2})^{-1/2} = 81^{-1/2}$

$$b = \frac{1}{81^{1/2}}$$

$$= \frac{1}{9} \qquad \blacktriangleleft$$

From (1), $b^y = x$ and $y = \log_b x$ are equivalent equations. If in the first of these equations we replace y by $\log_b x$, we obtain

$$b^{\log_b x} = x \tag{2}$$

where $b > 0$, $b \neq 1$, and $x > 0$.

From (2) we observe that a *logarithm is an exponent;* that is, $\log_b x$ is the exponent to which b must be raised to yield x.

▷ ILLUSTRATION 3

From (2)

$$3^{\log_3 7} = 7 \quad \text{and} \quad 10^{\log_{10} 5} = 5 \qquad \blacktriangleleft$$

If we write the equations in (1) as $\log_b x = y$ and $x = b^y$ and replace x in the first of these equations by b^y, we obtain

$$\log_b b^y = y \tag{3}$$

where $b > 0$, $b \neq 1$, and y is any real number.

▷ ILLUSTRATION 4

From (3)

$$\log_2 2^{-5} = -5 \quad \text{and} \quad \log_{10} 10^3 = 3 \qquad \blacktriangleleft$$

Equations (2) and (3) were obtained from statement (1) as a result of the fact that the logarithmic and exponential functions are inverses of each other.

Recall from Section 6.1 that the graphs of a function and its inverse are reflections of each other with respect to the line $y = x$. We use this fact to obtain the graph of a logarithmic function from the graph of the corresponding exponential function.

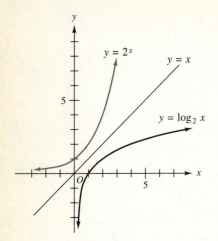

FIGURE 1

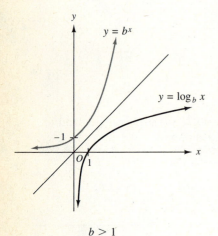

$b > 1$

FIGURE 2

▷ **ILLUSTRATION 5**

The graph of $y = 2^x$ appears as the light curve in Figure 1. The graph of $y = \log_2 x$ is darker in the figure. The two graphs are reflections of each other with respect to the line $y = x$. ◀

In Figure 1 we have a special case of the graph of the logarithmic function with base b where $b > 1$. In Figure 2 we have the general case. This figure shows the graph of

$$f(x) = \log_b x \qquad b > 1$$

This graph is symmetric, with respect to the line $y = x$, to the graph of the exponential function with base b ($b > 1$), which also appears in Figure 2.

▷ **ILLUSTRATION 6**

Figure 3 shows the graph of $y = (\frac{1}{2})^x$. The graph of $y = \log_{1/2} x$ is obtained from this graph by a reflection of $y = (\frac{1}{2})^x$ with respect to the line $y = x$. ◀

The graph in Figure 3 is a special case of the graph of the logarithmic function with base b where $0 < b < 1$. For the general case refer to Figure 4, showing the graph of

$$f(x) = \log_b x \qquad 0 < b < 1$$

Also in Figure 4 we have the graph of the exponential function with base b ($0 < b < 1$), and we observe that the two graphs are symmetric with respect to the line $y = x$.

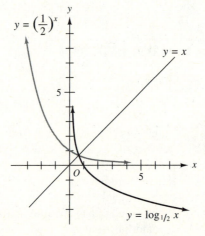

FIGURE 3

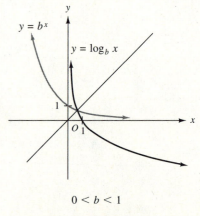

$0 < b < 1$

FIGURE 4

From the graphs of $\log_b$ in Figures 2 and 4, we note the following properties.

Properties of the Logarithmic Function with Base b

1. If $b > 1$, $\log_b$ is an increasing function. If $0 < b < 1$, $\log_b$ is a decreasing function.
2. If $b > 1$, $\log_b x$ is positive if $x > 1$, and $\log_b x$ is negative if $0 < x < 1$. If $0 < b < 1$, $\log_b x$ is negative if $x > 1$, and $\log_b x$ is positive if $0 < x < 1$. Furthermore, $\log_b x$ is not defined if x is nonpositive.
3. The only zero of the function $\log_b$ is 1; that is, $\log_b x = 0$ if and only if $x = 1$.
4. $\log_b x = 1$ if and only if $x = b$.
5. If $b > 1$, $\log_b x \to -\infty$ as $x \to 0^+$; and if $0 < b < 1$, $\log_b x \to +\infty$ as $x \to 0^+$.

Property 3 gives the equation

$$\log_b 1 = 0$$

and property 4 gives

$$\log_b b = 1$$

The logarithmic function with base e is called the *natural logarithmic function*.

DEFINITION   The Natural Logarithmic Function

The **natural logarithmic function** is the inverse of the natural exponential function.

The natural logarithmic function can be denoted by $\log_e$, but a more customary notation is *ln*. The function values of ln are denoted by $\ln x$ (read "natural logarithm of x"). Because ln and the natural exponential function are inverse functions,

$$y = \ln x \quad \text{if and only if} \quad x = e^y \tag{4}$$

The $\boxed{\text{ln}}$ key on your calculator can be used to obtain natural logarithmic function values. The graph of the natural logarithmic function is sketched in Figure 5, where the graph of the natural exponential function also appears. The two graphs are symmetric with respect to the line $y = x$. In the

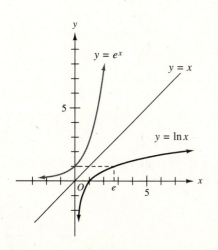

FIGURE 5

figure observe that e is the number whose natural logarithm is 1; that is,

$$\ln e = 1$$

This follows from statement (4) because

$$\ln e = 1 \quad \text{is equivalent to} \quad e = e^1$$

If in (2) $b = e$, we have

$$e^{\ln x} = x$$

and if $b = e$ in (3), we have

$$\ln e^x = x$$

▶ **EXAMPLE 3** *Solving a Word Problem Involving the Natural Exponential and Natural Logarithmic Functions*

In Example 2 of Section 6.3 we obtained the equation

$$f(t) = 1500e^{0.04t}$$

where $f(t)$ is the number of bacteria present in a certain culture at t minutes when 1500 bacteria are present initially. Use a graphics calculator to estimate how many minutes elapse until 15,000 bacteria are present. Then check the estimate algebraically.

Solution To estimate how many minutes elapse until 15,000 bacteria are present we plot the line $g(t) = 15,000$ and the graph of f in the viewing rectangle [0, 100] by [0, 20,000] as shown in Figure 6. We use the trace and zoom-in features of the calculator to determine that the line intersects the graph of f when $t = 58$. Thus our estimate is 58 minutes.

To check our estimate algebraically we let T represent the number of minutes that elapse until there are 15,000 bacteria present. Then in the equation for $f(t)$ we substitute T for t, and we have

$$f(T) = 1500e^{0.04T}$$

Because $f(T) = 15,000$

$$15,000 = 1500e^{0.04T}$$

$$10 = e^{0.04T}$$

Because the equation $x = e^y$ is equivalent to the equation $y = \ln x$, then the equation $10 = e^{0.04T}$ is equivalent to

$$0.04T = \ln 10$$

$$T = \frac{\ln 10}{0.04}$$

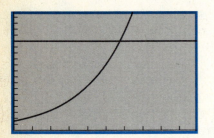

[0, 100] by [0, 20,000]
$f(t) = 1500e^{0.04t}$ and $g(t) = 15,000$
FIGURE 6

$$T = 57.56$$

This answer agrees with our estimate.

Conclusion: 58 minutes elapse until 15,000 bacteria are present. ◀

▶ **EXAMPLE 4** *Solving a Word Problem Involving the Natural Exponential and Natural Logarithmic Functions*

A deposit of $1000 is placed into a savings account that pays an annual interest rate of 6 percent compounded continuously and no withdrawals or additional deposits are made. **(a)** Express the number of dollars in the amount on deposit at t years as a function of t. **(b)** Use a graphics calculator to estimate how long it will be until $1500 is on deposit. **(c)** Check the estimate algebraically.

Solution
(a) From formula (3) of Section 6.2, if $A(t)$ dollars is the amount on deposit at t years,

$$A(t) = 1000e^{0.06t}$$

(b) Figure 7 shows the graph of A and the line $g(t) = 1500$ plotted in the viewing rectangle [0, 10] by [0, 2000]. Using the trace and zoom-in features of the calculator we determine that the line intersects the graph of A when $t = 6.758$.

(c) Let T years be how long it will be until $1500 is on deposit. In the equation for $A(t)$ we substitute T for t and we have

$$A(T) = 1000e^{0.06T}$$

Because $A(T) = 1500$, we have

$$1500 = 1000e^{0.06T}$$
$$e^{0.06T} = 1.5$$
$$0.06T = \ln 1.5$$
$$T = \frac{\ln 1.5}{0.06}$$
$$T = 6.758$$

This answer agrees with our estimate.

Conclusion: Because 6.758 years is equivalent to 6 years, 9 months, and 3 days, it will take that long until $1500 is on deposit. ◀

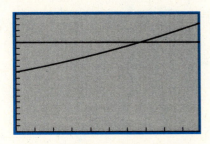

[0, 10] by [0, 2000]
$A(t) = 1000e^{0.06t}$ and $g(t) = 1500$
FIGURE 7

▶ **EXAMPLE 5** *Solving a Word Problem Involving the Natural Exponential and Natural Logarithmic Functions*

In Example 4 of Section 6.3 we had the equation

$$f(t) = 50(1 - e^{-0.34t})$$

where a typical worker at a certain factory can produce $f(t)$ units per day after t days on the job. **(a)** Use a graphics calculator to estimate how many days the worker must spend on the job until the worker produces 40 units per day. **(b)** Check the estimate algebraically.

Solution

(a) In the solution of Example 4 of Section 6.3 we obtained Figure 13 of that section that shows the graph of f and its horizontal asymptote, the line $g(t) = 50$, plotted in the viewing rectangle [0, 10] by [0, 60]. To this figure we add the line $h(t) = 40$ and we obtain Figure 8. Using trace and zoom-in, we determine that the line $h(t) = 40$ intersects the graph of f when $t = 4.7$.

(b) Let T represent the number of days on the job necessary for the worker to produce 40 units per day. Substituting T for t in the given equation, we have

$$f(T) = 50(1 - e^{-0.34T})$$

Because $f(T) = 40$,

$$40 = 50(1 - e^{-0.34T})$$
$$0.80 = 1 - e^{-0.34T}$$
$$e^{-0.34T} = 0.2$$
$$-0.34T = \ln 0.2$$
$$T = \frac{\ln 0.2}{-0.34}$$
$$T = 4.7$$

This answer agrees with our estimate.

<u>**Conclusion:**</u>   After 5 days on the job the worker will produce 40 units per day.   ◀

In problems involving exponential decay, the **half-life** of a substance is the time required for one-half of it to decay.

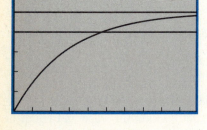

[0, 10] by [0, 60]
$f(t) = 50(1 - e^{-0.34t})$, $g(t) = 50$, and $h(t) = 40$
FIGURE 8

▶ **EXAMPLE 6** *Solving a Word Problem Involving the Natural Exponential and Logarithmic Functions*

The half-life of radium is 1690 years. Suppose that $f(t)$ milligrams of radium will be present t years from now, and

$$f(t) = Ae^{kt}$$

where k is a constant. If 60 mg of radium is present now, how much radium will be present 100 years from now?

Solution   Because 60 mg of radium is present now, $f(0) = 60$. Therefore, from the given equation

$$f(0) = Ae^{k(0)}$$
$$60 = A$$

Replacing A by 60 in the given equation, we obtain

$$f(t) = 60e^{kt}$$

Because the half-life of radium is 1690 years, $f(1690) = 30$. From the preceding equation with $t = 1690$, we have

$$f(1690) = 60e^{k(1690)}$$
$$30 = 60e^{1690k}$$
$$e^{1690k} = 0.5$$
$$1690k = \ln 0.5$$
$$k = \frac{\ln 0.5}{1690}$$
$$k = -0.000410$$

Substituting this value of k in the equation $f(t) = 60e^{kt}$, we get

$$f(t) = 60e^{-0.000410t}$$

With this equation and the fact that the number of milligrams of radium present 100 years from now is $f(100)$, we have

$$f(100) = 60e^{-0.0410}$$
$$= 57.6$$

Conclusion: There will be 57.6 mg of radium present 100 years from now.   ◀

EXERCISES 6.4

In Exercises 1 through 4, express the relationship in the equation by using logarithmic notation.

1. (a) $3^4 = 81$ **(b)** $5^3 = 125$ **(c)** $10^{-3} = 0.001$

2. (a) $2^5 = 32$ **(b)** $7^2 = 49$ **(c)** $5^{-2} = \frac{1}{25}$

3. (a) $8^{2/3} = 4$ **(b)** $625^{-3/4} = \frac{1}{125}$ **(c)** $2^0 = 1$

4. (a) $81^{3/4} = 27$ **(b)** $10^0 = 1$ **(c)** $64^{-2/3} = \frac{1}{16}$

In Exercises 5 through 8, express the relationship in the equation by using exponential notation.

5. (a) $\log_8 64 = 2$ **(b)** $\log_3 81 = 4$
   (c) $\log_2 1 = 0$

6. (a) $\log_{10} 10,000 = 4$ **(b)** $\log_5 125 = 3$
   (c) $\log_{10} 1 = 0$

7. (a) $\log_8 2 = \frac{1}{3}$ **(b)** $\log_{1/3} 9 = -2$
   (c) $\log_9 \frac{1}{3} = -\frac{1}{2}$

8. (a) $\log_{32} 2 = \frac{1}{5}$ **(b)** $\log_{1/2} 64 = -6$
   (c) $\log_{16} \frac{1}{8} = -\frac{3}{4}$

In Exercises 9 through 12, find the value of the logarithm.

9. (a) $\log_{10} 100$ **(b)** $\log_{27} 9$ **(c)** $\log_2 \frac{1}{8}$

10. (a) $\log_4 64$ **(b)** $\log_6 6$ **(c)** $\log_3 \frac{1}{81}$

11. (a) $\log_8 \frac{1}{2}$ **(b)** $\log_{27} \frac{1}{81}$ **(c)** $\ln \sqrt{e}$

12. (a) $\log_7 7$ **(b)** $\log_{1/4} \frac{1}{32}$ **(c)** $\ln e^2$

In Exercises 13 through 16, solve the equation for x.

13. (a) $\log_7 x = 3$ **(b)** $\log_{1/3} x = -4$

14. (a) $\log_4 x = 3$ **(b)** $\log_{1/4} x = -3$

15. (a) $\log_2 x = \frac{3}{2}$ **(b)** $\log_{1/4} x = \frac{7}{2}$

16. (a) $\log_3 x = \frac{2}{3}$ **(b)** $\log_{1/9} x = \frac{5}{2}$

In Exercises 17 through 20, solve the equation for b.

17. (a) $\log_b 144 = 2$ **(b)** $\log_b 6 = \frac{1}{3}$

18. (a) $\log_b 1000 = 3$ **(b)** $\log_b 3 = \frac{1}{4}$

19. (a) $\log_b 0.01 = -2$ **(b)** $\log_b \frac{1}{4} = -\frac{2}{3}$

20. (a) $\log_b 27 = -3$ **(b)** $\log_b 0.001 = -\frac{3}{2}$

In Exercises 21 through 24, simplify the expression.

21. (a) $\log_6(\log_5 5)$ **(b)** $\log_2(\log_9 81)$

22. (a) $\log_5(\log_2 32)$ **(b)** $\log_2(\log_3 81)$

23. (a) $\log_2(\log_2 256)$ **(b)** $\log_b(\log_b b), b > 0$

24. (a) $\log_3(\log_3 3)$
   (b) $\log_b(\log_a a^b), a > 0$ and $b > 0$

In Exercises 25 through 30 sketch the graph of the function.

25. $f(x) = \log_{10} x$ **26.** $f(x) = \log_3 x$

27. $g(x) = \log_{1/3} x$ **28.** $g(x) = \log_{1/10} x$

29. $F(x) = \log_3 x^2$ **30.** $G(x) = \log_2 |x|$

In Exercises 31 through 38, sketch the graph of the function by geometric transformations on the graph of the natural logarithmic function.

31. $f(x) = -\ln x$ **32.** $g(x) = 2 \ln x$

33. $g(x) = \ln(x - 1)$ **34.** $f(x) = \ln(x + 4)$

35. $F(x) = \ln x + 2$ **36.** $G(x) = \ln x - 3$

37. $f(x) = 3 \ln(x + 4) - 5$

38. $g(x) = -\ln(x - 2) + 1$

In Exercises 39 through 50, the word problem has a mathematical model involving an exponential function. Be sure to write a conclusion.

39. Refer to Exercise 21 in Exercises 6.3. The value of a machine purchased 8 years ago for $10,000 is $V(t)$ dollars t years after its purchase, where

$$V(t) = 10,000e^{-0.3t}$$

If the machine will be replaced when its value is $500, use your graphics calculator to estimate when a new machine will be purchased. Check your estimate algebraically.

40. Refer to Exercise 22 in Exercises 6.3. If $f(t)$ grams of a radioactive substance are present after t seconds and 100 grams are present initially, then

$$f(t) = 100e^{-0.3t}$$

Use your graphics calculator to estimate after how many seconds 10 grams of the substance will be present. Check your estimate algebraically.

41. Refer to Exercise 23 in Exercises 6.3. The current population of a particular town is 10,000, and it is increasing at a rate proportional to its size. If this rate is 6 percent and if the population after t years is $P(t)$, then

$$P(t) = 10,000e^{0.06t}$$

Use your graphics calculator to estimate when the population is expected to be 45,000. Check your estimate algebraically.

42. In a certain culture 2000 bacteria are present initially, and after t minutes $f(t)$ bacteria are present, where

$$f(t) = 2000e^{0.035t}$$

Use your graphics calculator to estimate when 10,000 bacteria will be present. Check your estimate algebraically.

43. (a) If $500 is invested at 9 percent compounded continuously, express the number of dollars in the value of that investment at t years as a function of t.
(b) Use your graphics calculator to estimate how long it will be until the value of the investment is $900.
(c) Check your estimate algebraically.

44. Solve Exercise 43 if money is invested at 12 percent compounded continuously.

45. Refer to Exercise 29 in Exercises 6.3. The efficiency of a typical worker at a certain factory is given by the function defined by

$$f(t) = 100 - 60e^{-0.2t}$$

where the worker can complete $f(t)$ units of work per day after being on the job for t months. Use your graphics calculator to estimate how many months the worker must spend on the job until the worker produces 70 units per day. Check your estimate algebraically.

46. Refer to Exercise 30 in Exercises 6.3. The resale value of a certain piece of equipment is $f(t)$ dollars t years after its purchase, where

$$f(t) = 1200 + 8000e^{-0.25t}$$

Use your graphics calculator to estimate how long after its purchase the resale value of the equipment will be $2000. Check your estimate algebraically.

47. If a particular radioactive substance has a half-life of 500 years and an original sample of 5 grams has decayed to 4 grams, how old is the sample? Assume

that the number of grams present t years from the time of the sample is $f(t)$ and $f(t) = Ae^{kt}$, where k is a constant.

48. Suppose that $f(t)$ units of a radioactive substance will be present t years from now and $f(t) = Ae^{kt}$, where k is a constant. If 30 percent of the substance disappears in 15 years, find its half-life.

49. How long will it take for an investment to double itself if interest is paid at the rate of 8 percent compounded continuously?

50. How long will it take for an investment to triple itself if interest is paid at the rate of 12 percent compounded continuously?

In Exercises 51 and 52, the functions S and C are the hyperbolic functions defined in Exercises 37–40 of Section 6.3 as follows:

$$S(x) = \tfrac{1}{2}(e^x - e^{-x}) \quad \text{and} \quad C(x) = \tfrac{1}{2}(e^x + e^{-x})$$

51. Find $S^{-1}(x)$. *Hint:* Let $y = S(x)$, write the equation as quadratic in e^x, and solve for e^x by the quadratic formula.

52. Find $C^{-1}(x)$. Use a hint similar to that for Exercise 51 with a restriction on the domain.

6.5 PROPERTIES OF LOGARITHMIC FUNCTIONS AND LOGARITHMIC EQUATIONS

GOALS
1. **Learn the theorem about the logarithm of a product.**
2. **Learn the theorem about the logarithm of a quotient.**
3. **Learn the theorem about the logarithm of a power.**
4. **Apply properties of logarithms.**
5. **Solve logarithmic equations.**

Before the arrival of electronic calculators, logarithms were used to perform tedious calculations involving products, quotients, powers, and roots. For this purpose logarithms to the base 10 were applied, and tables of logarithms were involved. This application of logarithms is now obsolete. Today we are concerned primarily with properties of logarithmic functions and their use in solving logarithmic equations in this section and exponential equations in Section 6.6. The computations in the illustrations, examples, and exercises in this section are presented only to demonstrate these properties and are not intended to advocate such an application of them.

The three theorems of this section concern properties of logarithms that follow from corresponding laws of exponents. After the statement of each theorem an illustration shows the law of exponents involved. In the proofs we use the fact that

$$x = b^y \quad \text{is equivalent to} \quad y = \log_b x$$

We refer to the equation $x = b^y$ as the *exponential form* of the equation $y = \log_b x$, and we refer to the equation $y = \log_b x$ as the *logarithmic form* of the equation $x = b^y$.

THEOREM 1

If $b > 0$, $b \neq 1$, and u and v are positive numbers, then

$$\log_b uv = \log_b u + \log_b v$$

▷ **ILLUSTRATION 1**

Suppose in the statement of Theorem 1 that b is 2, u is 4, and v is 8. Then

$$\log_b uv = \log_2 4 \cdot 8 \qquad\qquad \log_b u + \log_b v = \log_2 4 + \log_2 8$$
$$= \log_2 2^2 \cdot 2^3 \qquad\qquad\qquad\qquad\quad\; = \log_2 2^2 + \log_2 2^3$$
$$= \log_2 2^{2+3} \qquad\qquad\qquad\qquad\qquad\;\; = 2 + 3$$
$$= \log_2 2^5 \qquad\qquad\qquad\qquad\qquad\qquad\; = 5$$
$$= 5 \; (\text{because } \log_b b^y = y)$$

Therefore, when b is 2, u is 4, and v is 8, Theorem 1 is valid. ◀

Proof of Theorem 1   Let

$$r = \log_b u \quad \text{and} \quad s = \log_b v$$

The exponential forms of these equations are, respectively,

$$u = b^r \quad \text{and} \quad v = b^s$$

Therefore

$$uv = b^r \cdot b^s$$
$$uv = b^{r+s}$$

The logarithmic form of this equation is

$$\log_b uv = r + s$$

But $\log_b u = r$ and $\log_b v = s$. Thus

$$\log_b uv = \log_b u + \log_b v \qquad \blacksquare$$

▷ **ILLUSTRATION 2**

If we are given $\log_{10} 2 = 0.3010$ and $\log_{10} 3 = 0.4771$, we can apply Theorem 1 to find $\log_{10} 6$.

$$\begin{aligned}
\log_{10} 6 &= \log_{10}(2 \cdot 3) \\
&= \log_{10} 2 + \log_{10} 3 \\
&= 0.3010 + 0.4771 \\
&= 0.7781 \qquad \blacktriangleleft
\end{aligned}$$

Because $\log_{10} 2$, $\log_{10} 3$, and $\log_{10} 6$ are irrational numbers, the values given for them in Illustration 2 are only decimal approximations. However, in computations such as Illustration 2 we shall use the equals symbol.

THEOREM 2

If $b > 0$, $b \neq 1$, and u and v are positive numbers, then

$$\log_b \frac{u}{v} = \log_b u - \log_b v$$

▷ **ILLUSTRATION 3**

Suppose in the statement of Theorem 2 that b is 2, u is 128, and v is 16. Then

$$\begin{aligned}
\log_b \frac{u}{v} &= \log_2 \frac{128}{16} & \log_b u - \log_b v &= \log_2 128 - \log_2 16 \\
&= \log_2 \frac{2^7}{2^4} & &= \log_2 2^7 - \log_2 2^4 \\
&= \log_2 2^{7-4} & &= 7 - 4 \\
&= \log_2 2^3 & &= 3 \\
&= 3
\end{aligned}$$

Hence, when b is 2, u is 128, and v is 16, Theorem 2 holds. ◀

Proof of Theorem 2   As in the proof of Theorem 1, let

$$r = \log_b u \qquad \text{and} \qquad s = \log_b v$$

The exponential forms of these equations are, respectively,

$$u = b^r \quad \text{and} \quad v = b^s$$

Hence

$$\frac{u}{v} = \frac{b^r}{b^s}$$

$$\frac{u}{v} = b^{r-s}$$

The logarithmic form of this equation is

$$\log_b \frac{u}{v} = r - s$$

Because $\log_b u = r$ and $\log_b v = s$, we have

$$\log_b \frac{u}{v} = \log_b u - \log_b v \qquad \blacksquare$$

▷ ILLUSTRATION 4

From Theorem 2

$$\log_{10} \tfrac{3}{2} = \log_{10} 3 - \log_{10} 2$$

Substituting the values of $\log_{10} 3$ and $\log_{10} 2$ given in Illustration 2, we obtain

$$\log_{10} \tfrac{3}{2} = 0.4771 - 0.3010$$
$$= 0.1761 \qquad \blacktriangleleft$$

THEOREM 3

If $b > 0$, $b \neq 1$, n is any real number, and u is a positive number, then

$$\log_b u^n = n \log_b u$$

▷ ILLUSTRATION 5

Suppose in the statement of Theorem 3 that b is 2, n is 3, and u is 4. Then

$$\log_b u^n = \log_2 4^3 \qquad\qquad n \log_b u = 3 \log_2 4$$
$$= \log_2 (2^2)^3 \qquad\qquad\quad = 3 \log_2 2^2$$
$$= \log_2 2^{2 \cdot 3} \qquad\qquad\quad\; = 3 \cdot 2$$

$$= \log_2 2^6 \qquad\qquad = 6$$
$$= 6$$

Thus, when b is 2, n is 3, and u is 4, Theorem 3 is valid. ◄

Proof of Theorem 3   Let

$$r = \log_b u \Leftrightarrow u = b^r$$

Then

$$u^n = (b^r)^n$$
$$u^n = b^{nr}$$

The logarithmic form of this equation is

$$\log_b u^n = nr$$

Because $\log_b u = r$, we have

$$\log_b u^n = n \log_b u$$ ∎

▷ **ILLUSTRATION 6**

Because $\log_{10} 2 = 0.3010$, it follows from Theorem 3 that

$$\log_{10} 32 = \log_{10} 2^5 \qquad \log_{10} \sqrt[3]{2} = \log_{10} 2^{1/3}$$
$$= 5 \log_{10} 2 \qquad\qquad\quad = \tfrac{1}{3} \log_{10} 2$$
$$= 5(0.3010) \qquad\qquad\quad = \tfrac{1}{3}(0.3010)$$
$$= 1.5050 \qquad\qquad\qquad = 0.1003$$ ◄

► **EXAMPLE 1**   *Applying Properties of Logarithms*

Express each of the following in terms of logarithms of x, y, and z, where the variables represent positive numbers:

(a) $\log_b x^2 y^3 z^4$ **(b)** $\log_b \dfrac{x}{yz^2}$ **(c)** $\log_b \sqrt[5]{\dfrac{xy^2}{z^3}}$

Solution
(a) By Theorem 1

$$\log_b x^2 y^3 z^4 = \log_b x^2 + \log_b y^3 + \log_b z^4$$

Applying Theorem 3 to each of the logarithms on the right side, we obtain

$$\log_b x^2 y^3 z^4 = 2 \log_b x + 3 \log_b y + 4 \log_b z$$

(b) From Theorem 2

$$\log_b \frac{x}{yz^2} = \log_b x - \log_b yz^2$$

Applying Theorems 1 and 3 to the second logarithm on the right side, we have

$$\log_b \frac{x}{yz^2} = \log_b x - (\log_b y + \log_b z^2)$$

$$= \log_b x - \log_b y - 2 \log_b z$$

(c) From Theorem 3

$$\log_b \sqrt[5]{\frac{xy^2}{z^3}} = \frac{1}{5} \log_b \frac{xy^2}{z^3}$$

Applying Theorem 2 to the right side, we obtain

$$\log_b \sqrt[5]{\frac{xy^2}{z^3}} = \tfrac{1}{5}(\log_b xy^2 - \log_b z^3)$$

$$= \tfrac{1}{5}(\log_b x + \log_b y^2 - \log_b z^3)$$

$$= \tfrac{1}{5}(\log_b x + 2 \log_b y - 3 \log_b z)$$

$$= \tfrac{1}{5} \log_b x + \tfrac{2}{5} \log_b y - \tfrac{3}{5} \log_b z \qquad \blacktriangleleft$$

▶ **EXAMPLE 2** *Applying Properties of Logarithms*

Write each of the following expressions as a single logarithm with a coefficient of 1:

(a) $\log_b x + 2 \log_b y - 3 \log_b z$

(b) $\frac{1}{3}(\log_b 4 - \log_b 3 + 2 \log_b x - \log_b y)$

Solution

(a) $\log_b x + 2 \log_b y - 3 \log_b z = (\log_b x + \log_b y^2) - \log_b z^3$

$$= \log_b xy^2 - \log_b z^3$$

$$= \log_b \frac{xy^2}{z^3}$$

(b) $\frac{1}{3}(\log_b 4 - \log_b 3 + 2 \log_b x - \log_b y)$

$$= \tfrac{1}{3}[(\log_b 4 + \log_b x^2) - (\log_b 3 + \log_b y)]$$

$$= \tfrac{1}{3}[\log_b 4x^2 - \log_b 3y]$$

$$= \frac{1}{3} \log_b \frac{4x^2}{3y}$$

$$= \log_b \sqrt[3]{\frac{4x^2}{3y}} \qquad \blacktriangleleft$$

▶ **EXAMPLE 3** *Applying Properties of Logarithms*

Given $\log_{10} 2 = 0.3010$, $\log_{10} 3 = 0.4771$, and $\log_{10} 7 = 0.8451$, use the properties of logarithms from Theorems 1 through 3 to find the value of each of the following logarithms:

(a) $\log_{10} 5$ **(b)** $\log_{10} 28$ **(c)** $\log_{10} 2100$ **(d)** $\log_{10} \sqrt[3]{4.2}$

Solution In addition to the given logarithms we can easily determine the logarithm with base 10 of any integer power of 10; for instance, $\log_{10} 10 = 1$, $\log_{10} 10^2 = 2$, $\log_{10} 10^3 = 3$, $\log_{10} 10^{-1} = -1$, and so on.

(a) $\log_{10} 5 = \log_{10} \frac{10}{2}$

$\qquad = \log_{10} 10 - \log_{10} 2$

$\qquad = 1 - 0.3010$

$\qquad = 0.6990$

(b) $\log_{10} 28 = \log_{10} 2^2 \cdot 7$

$\qquad = \log_{10} 2^2 + \log_{10} 7$

$\qquad = 2 \log_{10} 2 + \log_{10} 7$

$\qquad = 2(0.3010) + 0.8451$

$\qquad = 0.6020 + 0.8451$

$\qquad = 1.4471$

(c) $\log_{10} 2100 = \log_{10} 3 \cdot 7 \cdot 10^2$

$\qquad = \log_{10} 3 + \log_{10} 7 + \log_{10} 10^2$

$\qquad = 0.4771 + 0.8451 + 2$

$\qquad = 3.3222$

(d) $\log_{10} \sqrt[3]{4.2} = \log_{10} \left(\dfrac{2 \cdot 3 \cdot 7}{10} \right)^{1/3}$

$\qquad = \tfrac{1}{3} (\log_{10} 2 + \log_{10} 3 + \log_{10} 7 - \log_{10} 10)$

$\qquad = \tfrac{1}{3} (0.3010 + 0.4771 + 0.8451 - 1)$

$\qquad = \tfrac{1}{3} (0.6232)$

$\qquad = 0.2077$ ◀

▶ **EXAMPLE 4** *Applying Properties of Logarithms*

Use the values of $\log_{10} 2$ and $\log_{10} 7$ given in Example 3 to find the value of each of the following:

(a) $\log_{10} \dfrac{7}{2}$ **(b)** $\dfrac{\log_{10} 7}{\log_{10} 2}$

Solution

(a) $\log_{10} \dfrac{7}{2} = \log_{10} 7 - \log_{10} 2$

$\qquad = 0.8451 - 0.3010$

$\qquad = 0.5441$

(b) $\dfrac{\log_{10} 7}{\log_{10} 2} = \dfrac{0.8451}{0.3010}$

$\qquad = 2.808$ ◀

Compare the computations in parts (a) and (b) of Example 4. In part (a) there is the logarithm of a quotient, that, upon applying Theorem 2, is the difference of two logarithms. In part (b) we have the quotient of the logarithms of two numbers. The computation is performed by dividing 0.8451 by 0.3010.

An equation involving logarithms is called a **logarithmic equation.** Because the domain of a logarithmic function is restricted so that logarithms of only positive numbers are obtained, you must *check* any possible solution in the given equation. The next three examples involve the solution of a logarithmic equation.

▶ **EXAMPLE 5** *Solving a Logarithmic Equation*

Find the solution set of the equation

$$\log_{10}(x + 3) = 2$$

Solution   The exponential form of the equation is

$$x + 3 = 10^2$$
$$x = 100 - 3$$
$$x = 97$$

For this value of x, the given equation becomes $\log_{10} 100 = 2$, which is true. Therefore the solution set is {97}.   ◀

▶ **EXAMPLE 6** *Solving a Logarithmic Equation*

Find the solution set of the equation

$$\log_2(x + 4) - \log_2(x - 3) = 3$$

Solution   Because the difference of two logarithms is the logarithm of a quotient, we have

$$\log_2 \frac{x + 4}{x - 3} = 3$$

Writing this equation in the equivalent exponential form, we get

$$\frac{x + 4}{x - 3} = 2^3$$
$$x + 4 = 8(x - 3)$$
$$x + 4 = 8x - 24$$
$$-7x = -28$$
$$x = 4$$

Replacing x by 4 on the left side of the given equation, we get $\log_2 8 - \log_2 1$. Because $\log_2 8 = 3$ and $\log_2 1 = 0$, the solution checks. Thus the solution set is $\{4\}$. ◀

▶ **EXAMPLE 7** *Solving a Logarithmic Equation*

Find the solution set of the equation

$$\log_3 x + \log_3(2x - 3) = 3$$

Solution   The left side of the given equation is the sum of two logarithms; writing this as the logarithm of a product, we have

$$\log_3 x(2x - 3) = 3$$

The equivalent exponential form of this equation is

$$x(2x - 3) = 3^3$$
$$2x^2 - 3x - 27 = 0$$
$$(x + 3)(2x - 9) = 0$$
$$x + 3 = 0 \qquad 2x - 9 = 0$$
$$x = -3 \qquad x = \tfrac{9}{2}$$

When $x = -3$, neither $\log_3 x$ nor $\log_3(2x - 3)$ is defined; hence we reject the root -3. When $x = \tfrac{9}{2}$, the left side of the given equation is

$$\log_3 \tfrac{9}{2} + \log_3 6 = \log_3(\tfrac{9}{2} \cdot 6)$$
$$= \log_3 27$$
$$= 3$$

Therefore the solution set is $\{\tfrac{9}{2}\}$. ◀

EXERCISES 6.5

In Exercises 1 through 8, express the logarithm in terms of logarithms of x, y, and z, where the variables represent positive numbers.

1. (a) $\log_b(5xy)$ **(b)** $\log_b\left(\dfrac{y}{z}\right)$ **(c)** $\log_b\left(\dfrac{x}{yz}\right)$

2. (a) $\log_b(3xyz)$ **(b)** $\log_b\left(\dfrac{xy}{z}\right)$ **(c)** $\log_b(x^4y^2)$

3. (a) $\log_b(xy^5)$ **(b)** $\log_b\sqrt{xy}$

4. (a) $\log_b(z^{1/3})$ **(b)** $\log_b\sqrt[3]{yz^2}$

5. (a) $\log_b(x^{1/3}z^3)$ **(b)** $\log_b\left(\dfrac{xy^{1/2}}{z^4}\right)$

6. (a) $\log_b(x^2y^3z)$ **(b)** $\log_b\left(\dfrac{y^2}{x^5z^{1/4}}\right)$

7. (a) $\log_b\sqrt[3]{\dfrac{x^2}{yz^2}}$ **(b)** $\log_b(\sqrt[3]{x^2}\sqrt{yz})$

8. (a) $\log_b\sqrt[5]{\dfrac{x^3y^4}{z^2}}$ **(b)** $\log_b(\sqrt[4]{xy^3}\sqrt{z})$

In Exercises 9 through 12, write the expression as a single logarithm with a coefficient of 1.

9. (a) $4\log_{10} x + \tfrac{1}{2}\log_{10} y$
(b) $\tfrac{3}{4}\log_b x - 6\log_b y - \tfrac{4}{5}\log_b z$

10. (a) $5 \log_{10} x + \frac{1}{2} \log_{10} y - \frac{1}{3} \log_{10} z$
   (b) $\frac{2}{3} \log_b x - 4 \log_b y + \log_b z$

11. (a) $\log_{10} g + 2 \log_{10} t - \log_{10} 2$
   (b) $\ln \pi + \ln h + 2 \ln r - \ln 3$

12. (a) $\log_{10} 4 + \log_{10} \pi + 3 \log_{10} r - \log_{10} 3$
   (b) $\ln 2 + \ln \pi + \frac{1}{2} \ln t - \frac{1}{2} \ln g$

In Exercises 13 through 20, determine the value of the given logarithm by using the following: $\log_{10} 2 = 0.3010$, $\log_{10} 3 = 0.4771$, *and* $\log_{10} 7 = 0.8451$.

13. (a) $\log_{10} 14$ **(b)** $\log_{10} 15$
14. (a) $\log_{10} 18$ **(b)** $\log_{10} 42$
15. (a) $\log_{10} 63$ **(b)** $\log_{10} 140$
16. (a) $\log_{10} 120$ **(b)** $\log_{10} 0.21$
17. (a) $\log_{10} \sqrt[3]{10.5}$ **(b)** $\log_{10}\left(\frac{\sqrt[5]{49}}{36^2}\right)$
18. (a) $\log_{10} \sqrt[3]{126}$ **(b)** $\log_{10}\left(\frac{14}{\sqrt[3]{84}}\right)$
19. (a) $\log_{10} \frac{2}{3}$ **(b)** $\frac{\log_{10} 2}{\log_{10} 3}$
20. (a) $\log_{10} \frac{7}{5}$ **(b)** $\frac{\log_{10} 7}{\log_{10} 5}$

In Exercises 21 through 24, determine the value of the logarithm by using the following: $\ln 2 = 0.6931$, $\ln 3 = 1.0986$, *and* $\ln 5 = 1.6094$.

21. (a) $\ln 300$ **(b)** $\ln 7.5$
22. (a) $\ln 90$ **(b)** $\ln 1.2$
23. (a) $\ln\left(\frac{\sqrt{10}}{\sqrt[3]{3}}\right)$ **(b)** $\ln \frac{5}{3} e^2$
24. (a) $\ln\left(\frac{\sqrt[3]{5}}{\sqrt{6}}\right)$ **(b)** $\ln \frac{2}{5} e^3$

In Exercises 25 through 32, find the solution set of the equation.

25. $\log_5(4x - 3) = 2$

26. $\log_2(2 - 3x) = -3$

27. $\log_{10} x + 3 \log_{10} 2 = 3$

28. $\log_{10} x + \log_{10}(x + 15) = 2$

29. $\log_3(x + 6) - \log_3(x - 2) = 2$

30. $\log_2(11 - x) = \log_2(x + 1) + 3$

31. $\log_2(x + 1) + \log_2(3x - 5) = \log_2(5x - 3) + 2$

32. $\log_3(2x - 1) - \log_3(5x + 2) = \log_3(x - 2) - 2$

33. The Richter scale, named for its inventor Charles Richter (1900–1985), measures an earthquake's intensity. If x is the measure of the magnitude of an earthquake, x is given by the formula

$$x = \log_{10}\left(\frac{y}{T}\right) + D$$

where y measures the energy released by the earthquake and T and D are constants that depend on the period of time of earth movement and the distance from the epicenter of the earthquake to the receiving station. **(a)** Prove that $y = k \cdot 10^x$, where k is a constant. **(b)** Use the formula in part (a) to show that an earthquake having a magnitude 7 on the Richter scale has an intensity 100 times that of an earthquake of magnitude 5 with the same epicenter. **(c)** If y_1 and y_2 are the measures of energy released by two earthquakes with the same epicenter and having magnitudes of 6.7 and 5.4, respectively, on the Richter scale, then y_1 is how many times y_2?

34. By knowing the values of $\log_{10} 2$ and $\log_{10} 3$, explain why you can compute, without a calculator, $\log_{10} 4$, $\log_{10} 5$, $\log_{10} 6$, $\log_{10} 8$, and $\log_{10} 9$, but not $\log_{10} 7$.

35. The only solution to the equation

$$\log_{10} x = \ln x$$

is $x = 1$. Explain why this is a solution and why there are no other solutions.

6.6 EXPONENTIAL EQUATIONS

GOALS
1. Define common logarithms.
2. Solve exponential equations.
3. Find the value of a logarithm to a base other than e or 10.
4. Solve word problems having exponential equations as mathematical models.

John Napier (1550–1617), a Scottish aristocrat, invented logarithms in the early seventeenth century. Shortly thereafter, the Englishman Henry Briggs (1560–1630), in close touch with Napier, first used logarithms with base 10, at one time called Briggsian logarithms but now called **common logarithms.** Leonhard Euler later related logarithms to exponents. Today in algebra courses, exponents are studied before logarithms. But in calculus courses, logarithmic functions are usually treated before exponential functions. This order of presentation allows us to define a real-number exponent.

Because $\log_b u^n = n \log_b u$, logarithms are applied to solve an exponential equation, one in which a variable occurs in an exponent. Because we express numbers in decimal notation, common logarithms are often used for this purpose. When writing common logarithms, it is customary to omit the subscript 10. Thus $\log x$ is understood to represent the number $\log_{10} x$, and the function *log* denotes the logarithmic function with base 10. Hence

$$\log x = y \Leftrightarrow 10^y = x$$

▷ **ILLUSTRATION 1**

$$\log 10 = 1 \quad \text{because } 10^1 = 10$$
$$\log 100 = 2 \quad \text{because } 10^2 = 100$$
$$\log 1000 = 3 \quad \text{because } 10^3 = 1000$$
$$\log 10,000 = 4 \quad \text{because } 10^4 = 10,000$$

and so on. Furthermore,

$$\log 1 = 0 \quad \text{because } 10^0 = 1$$
$$\log 0.1 = -1 \quad \text{because } 10^{-1} = 0.1$$
$$\log 0.01 = -2 \quad \text{because } 10^{-2} = 0.01$$
$$\log 0.001 = -3 \quad \text{because } 10^{-3} = 0.001$$
$$\log 0.0001 = -4 \quad \text{because } 10^{-4} = 0.0001$$

and so on. ◄

Common logarithms can be found on a calculator by using the key labeled ⟨log⟩.

To solve an exponential equation, consider the equivalent equation obtained by equating the common (or natural) logarithms of the two sides and then solving the resulting equation, as demonstrated in the following illustration.

▷ **ILLUSTRATION 2**

The exponential equation

$$3^x = 16$$

is equivalent to the equation

$$x = \log_3 16$$

However, this equation does not give a numerical value for x. To obtain such a value, we equate the common logarithms of the two sides of the given equation:

$$\log 3^x = \log 16$$
$$x \log 3 = \log 16$$
$$x = \frac{\log 16}{\log 3}$$
$$x = \frac{1.2041}{0.47712}$$
$$x = 2.5237$$

Therefore, the solution set is {2.5237}. Note that 2.5237 is an approximation, to five significant digits, of the value of x. The exact value of x is given by either $\log_3 16$ or $\dfrac{\log 16}{\log 3}$.

The given equation can also be solved by equating the natural logarithms of the two sides. If this is done, the solution has the following form:

$$\ln 3^x = \ln 16$$
$$x \ln 3 = \ln 16$$
$$x = \frac{\ln 16}{\ln 3}$$
$$x = \frac{2.7726}{1.0986}$$
$$x = 2.5237$$

◀

▶ **EXAMPLE 1** *Solving an Exponential Equation*

Find the solution set of the equation

$$5^{2x-1} = 8$$

Check the answer on a graphics calculator.

Solution Equating the common logarithms of both sides of the given equation, we have

$$\log 5^{2x-1} = \log 8$$
$$(2x - 1)\log 5 = \log 2^3$$
$$2x \log 5 - \log 5 = 3 \log 2$$
$$2x \log 5 = 3 \log 2 + \log 5$$
$$x = \frac{3 \log 2 + \log 5}{2 \log 5}$$
$$x = 1.1460$$

To check the answer on our graphics calculator we plot the graph of $y = 5^{2x-1} - 8$ and use trace and zoom-in to estimate the x intercept of the graph. ◀

▶ **EXAMPLE 2** *Solving an Exponential Equation*

Find the solution set of the equation

$$7^x = 3^{x+1}$$

Check the answer on a graphics calculator.

Solution We equate the common logarithms of both sides of the given equation and get

$$\log 7^x = \log 3^{x+1}$$
$$x \log 7 = (x + 1)\log 3$$
$$x \log 7 = x \log 3 + \log 3$$
$$x \log 7 - x \log 3 = \log 3$$
$$x(\log 7 - \log 3) = \log 3$$
$$x = \frac{\log 3}{\log 7 - \log 3}$$
$$x = 1.296$$

Thus the solution set is {1.296}.

We check our answer by plotting the graph of $y = 7^x - 3^{x+1}$ on our graphics calculator and estimating the x intercept of the graph by tracing and zooming-in.   ◄

As you know, values of natural logarithms and common logarithms can be found on your calculator. Values of logarithms to other bases can be found by solving exponential equations. The next example shows the method.

► **EXAMPLE 3**   *Finding the Value of a Logarithm to a Base Other Than e or 10*

Find the value of $\log_4 19$.

Solution   Let

$$y = \log_4 19$$

Writing this equation in exponential form, we have

$$4^y = 19$$

Therefore

$$\log 4^y = \log 19$$
$$y \log 4 = \log 19$$
$$y = \frac{\log 19}{\log 4}$$
$$y = 2.124$$

Thus, to four significant digits, $\log_4 19 = 2.124$.   ◄

The procedure applied in Example 3 can be used to obtain a formula relating $\log_a x$ and $\log_b x$, that is, logarithms with different bases of a given number. See Exercise 42.

The following two examples and Exercises 21 through 32 give applications of exponential equations.

► **EXAMPLE 4**   *Solving a Word Problem Having an Exponential Equation as a Mathematical Model*

Suppose on January 1, 1993, the population of a certain city was 800,000. From then until the year 2001 the population is expected to increase at the rate of 3.5 percent per year. Therefore t years after January 1, 1993, the population is expected to be $800,000(1.035)^t$ where $0 \le t \le 8$. Use a graphics calculator to predict to the nearest month when the population will be 1 million. Check the prediction algebraically.

Solution   On our graphics calculator we plot the graphs of

$$y_1 = 800,000(1.035)^t \quad \text{and} \quad y_2 = 1,000,000$$

These graphs in the viewing rectangle [0, 10] by [700,000, 1,100,000] appear in Figure 1. We use trace and zoom-in to estimate that the graphs intersect where $t = 6.5$.

Six and one-half years from January 1, 1993, is July 1, 1999, which is when we predict the population will be 1 million.

To check this prediction algebraically, we wish to determine the value of t such that

$$800,000(1.035)^t = 1,000,000$$

$$(1.035)^t = \frac{1,000,000}{800,000}$$

$$(1.035)^t = 1.25$$

Because the variable t appears in the exponent, we equate the logarithms of the two sides and obtain

$$\log(1.035)^t = \log 1.25$$

$$t \log 1.035 = \log 1.25$$

$$t = \frac{\log 1.25}{\log 1.035}$$

$$t = 6.5$$

This result checks with the value of t obtained on our calculator.

Conclusion: We predict that the population of the city will be 1 million on July 1, 1999.   ◀

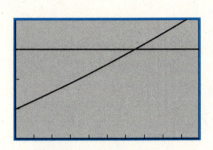

[0, 10] by [700,000, 1,100,000]
$y_1 = 800,000(1.035)^t$ and $y_2 = 1,000,000$
FIGURE 1

▶ **EXAMPLE 5** *Solving a Word Problem Having an Exponential Equation as a Mathematical Model*

If $1000 is deposited into a savings account that pays an annual interest rate of 6 percent compounded quarterly and no withdrawals or additional deposits are made, how long will it take until $1500 is on deposit?

Solution   From Theorem 3 of Section 6.2

$$A_n = P\left(1 + \frac{r}{m}\right)^n$$

where A_n dollars is the amount at the end of n interest periods of an investment of P dollars at an annual interest rate of $100r$ percent compounded m times per year. In this problem $A_n = 1500$, $P = 1000$, $r = 0.06$, and $m = 4$.

We wish to find n. From the formula

$$1500 = 1000\left(1 + \frac{0.06}{4}\right)^n$$

$$1.5 = (1.015)^n$$

$$\log 1.5 = \log(1.015)^n$$

$$\log 1.5 = n \log 1.015$$

$$n = \frac{\log 1.5}{\log 1.015}$$

$$n = 27.33$$

Conclusion: Because n is the number of interest periods and interest is compounded quarterly, it will take 28 quarters until there is $1500 on deposit.   ◄

Compare the preceding example with Example 4 in Section 6.4, which involves the same data except that interest is compounded continuously instead of quarterly.

▶ **EXAMPLE 6** *Solving an Exponential Equation*

Use a graphics calculator to estimate to four significant digits the solution of the equation

$$3^x - 3^{-x} = 4$$

Check the answer algebraically.

Solution   Figure 2 shows the graphs of

$$y_1 = 3^x - 3^{-x} \quad \text{and} \quad y_2 = 4$$

plotted in the viewing rectangle $[-5, 5]$ by $[-5, 5]$. We use trace and zoom-in to estimate that the graphs intersect at the point where $x = 1.314$.

To check the answer algebraically, we write the given equation as

$$3^x - \frac{1}{3^x} = 4$$

$$3^{2x} - 1 = 4(3^x)$$

$$(3^x)^2 - 4(3^x) - 1 = 0$$

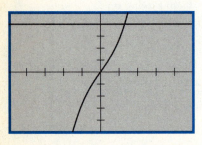

$[-5, 5]$ by $[-5, 5]$
$y_1 = 3^x - 3^{-x}$ and $y_2 = 4$
FIGURE 2

By letting $u = 3^x$, we obtain the quadratic equation

$$u^2 - 4u - 1 = 0$$

We solve this equation by the quadratic formula.

$$u = \frac{4 \pm \sqrt{16 + 4}}{2}$$

$$= \frac{4 \pm 2\sqrt{5}}{2}$$

$$= 2 \pm \sqrt{5}$$

Replacing u by 3^x, we get the two equations

$$3^x = 2 + \sqrt{5} \quad \text{and} \quad 3^x = 2 - \sqrt{5}$$

The second equation has no solution because $3^x > 0$ and $2 - \sqrt{5}$ is negative. Therefore

$$3^x = 2 + \sqrt{5}$$

Equating the common logarithms of the two sides of this equation, we have

$$\log 3^x = \log(2 + \sqrt{5})$$

$$x \log 3 = \log(2 + \sqrt{5})$$

$$x = \frac{\log(2 + \sqrt{5})}{\log 3}$$

$$x = 1.314$$

This answer agrees with our estimate from our graphics calculator.   ◀

EXERCISES 6.6

In Exercises 1 through 12, find the solution set of the equation. Express the results to four significant digits. Check your answer on your graphics calculator.

1. $4^x = 7$

2. $3^x = 25$

3. $5^{2x} = 4$

4. $100^x = 65$

5. $3^{2+x} = 5^x$

6. $10^{3x-2} = 37$

7. $3^{x+1} = 4^{x-1}$

8. $3^{2x+1} = 5^{3x-1}$

9. $(1.02)^x = 1.892$

10. $(1.04)^x = 0.932$

11. $e^{3x} = 21$

12. $10^{3x} = 57$

In Exercises 13 through 20, find the value of the logarithm to four significant digits.

13. $\log_3 12$

14. $\log_5 200$

15. $\log_2 18$

16. $\log_6 54$

17. $\log_4 155$

18. $\log_8 28$

19. $\log_{100} 75$

20. $\log_{20} 100$

In Exercises 21 through 32, solve the word problem by using a mathematical model involving an exponential function. Be sure to write a conclusion.

21. For the city of Example 4, use your graphics calculator to predict when the population will be 900,000. Check your prediction algebraically.

22. Use your graphics calculator to determine if the population of the city of Example 4 will reach 1,100,000 before the year 2001. Prove your answer algebraically.

23. How long will it take $1000 to triple itself if it is earning interest at an annual rate of 6 percent compounded semiannually?

24. How long will it take an investment to double itself if the annual interest rate is 8 percent compounded quarterly?

25. Refer to Exercise 25 of Exercises 6.3. At what height is the atmospheric pressure 500 lb/ft²? Check your answer on your graphics calculator.

26. For the town in Exercise 23 of Exercises 6.3, when is the population expected to be 21,000? Check your answer on your graphics calculator.

27. When is the value of the abstract painting in Exercise 27 of Exercises 6.3 expected to be $18,000? Check your answer on your graphics calculator.

28. What will the age of the man in Exercise 28 of Exercises 6.2 be when there is $50,000 in the investment account? Check your answer on your graphics calculator.

29. After how many minutes will the temperature of the body in Exercise 35 of Exercises 6.3 be 45°?

30. After how many hours of practice typing can the person in Exercise 28 of Exercises 6.3 type 60 words per minute?

31. In a certain speculative investment, a piece of real estate was purchased 3 years ago for $20,000 and sold today for $100,000. What is the annual rate of interest compounded monthly that has been earned?

32. A simple electrical circuit containing no condensers, a resistance of R ohms, and an inductance of L henries has the electromotive force cut off when the current is I_0 amperes. The current dies down so that at t seconds the current is i amperes, and $i = I_0 e^{-(R/L)t}$. Use natural logarithms to solve this equation for t in terms of i and the constants R, L, and I_0.

In Exercises 33 through 38, use your graphics calculator to estimate to four significant digits the solution of the equation. Check your answer algebraically.

33. $10^x - 10^{-x} = 2$ **34.** $10^x + 10^{-x} = 4$

35. $4^x - 4^{-x} = 3$ **36.** $5^x - 5^{-x} = 8$

37. $\frac{1}{2}(e^x + e^{-x}) = 4$ **38.** $\frac{1}{2}(e^x - e^{-x}) = 3$

In Exercises 39 and 40, solve for x in terms of y.

39. $y = \dfrac{10^x - 10^{-x}}{10^x + 10^{-x}}$ **40.** $y = \dfrac{e^x + e^{-x}}{e^x - e^{-x}}$

41. **(a)** Use your graphics calculator to determine to the nearest one-hundredth the value of x between 1 and 2 for which $2^x = x^5$. **(b)** Use your graphics calculator to determine to the nearest one-hundredth the value of x between 1 and 2 for which

$$\frac{\ln x}{x} = \frac{1}{5} \ln 2$$

(c) Why are the answers in parts (a) and (b) the same?

42. Use the procedure in the solution of Example 3 to prove the following formula that relates logarithms with different bases a and b of a number x:

$$\log_a x = \frac{\log_b x}{\log_b a}$$

43. **(a)** Describe how you would use your graphics calculator to plot the graph of the function defined by $f(x) = \log_x k$, where k is a positive constant.
(b) Apply your answer in part (a) to plot the graph of the function defined by $f(x) = \log_x 5$. **(c)** From your graph in part (b) determine the domain and range of function f.

44. **(a)** Describe how you would use your graphics calculator to plot the graph of the function defined by $f(x) = \log_x(x + k)$, where k is a positive constant.
(b) Apply your answer in part (a) to plot the graph of the function defined by $f(x) = \log_x(x + 2)$. **(c)** From your graph in part (b) determine the domain and range of function f.

45. Explain why we can conclude from the two equations

$$x^a = y \quad \text{and} \quad y^b = x$$

that a and b are reciprocals of each other, if x is not $\neq 1$.

CHAPTER 6 REVIEW

▶ LOOKING BACK

6.1 We commenced our discussion of inverse functions by defining a one-to-one function and stating the horizontal-line test that determines if a function is one-to-one. We defined an increasing function and a decreasing function, and proved the theorem that establishes that monotonic functions are one-to-one. Following the definition of the inverse of a function, we proved the theorem that guarantees that if a one-to-one function f has f^{-1} as its inverse, then f^{-1} is a one-to-one function having f as its inverse. We then explained how to find $f^{-1}(x)$ when $f(x)$ is known, and we demonstrated that the graphs of f and f^{-1} are reflections of each other with respect to the line $y = x$.

6.2 After laws of real-number exponents were stated, we showed how scientific notation affords a convenient way to write numerals representing very large or very small numbers as well as a way to indicate the number of significant digits in a numeral. Three kinds of investment interest were considered: simple, compound, and compounded continuously. The latter form of interest led to the introduction of the number e. We followed this discussion with an intuitive explanation of how e can be shown to be the number that the expression

$$\left(1 + \frac{1}{x}\right)^x$$

approaches as x increases without bound.

6.3 Our treatment of exponential functions began with the definition and graph of the exponential function with base b. We then discussed the natural exponential function and its graph. Exponential growth was illustrated by an increase in the number of bacteria present in a culture, while exponential decay was a model for a decrease in the value of a piece of equipment. We used the learning curve to demonstrate bounded growth. The growth of a population when the environment imposes an upper bound on its size gave us a model for logistic growth.

6.4 We defined a logarithmic function as the inverse of an exponential function. In particular, the natural logarithmic function is the inverse of the natural exponential function. We showed how to find the value of a logarithm and how to solve a logarithmic equation. Applications involved the use of logarithms to solve equations in which the variable was present in an exponent.

6.5 Properties of logarithms were in the form of theorems regarding logarithms of products, quotients, and powers. We proved these theorems from corresponding properties of exponents. Logarithmic equations provided an application of the properties of logarithms.

6.6 Common logarithms, those with base 10, were introduced. We applied these logarithms, as well as natural logarithms, to find logarithms to other bases and to solve exponential equations.

▶ REVIEW EXERCISES

In Exercises 1 through 4, sketch the graph of the function and use the horizontal-line test to determine if it is one-to-one. Check your graph on your graphics calculator.

1. $f(x) = 4 + 3x$ **2.** $g(x) = 1 - x^3$

3. $g(x) = x^2 - 1$ **4.** $f(x) = |x + 4|$

In Exercises 5 through 8, do the following: (a) Sketch the graph of f and use the horizontal-line test to prove that f is one-to-one; (b) find $f^{-1}(x)$; (c) verify that $f^{-1}(f(x)) = x$ and $f(f^{-1}(x)) = x$; (d) plot the graphs of f, f^{-1}, and the line $y = x$ in the same viewing rectangle and observe that

the graphs of f and f^{-1} are reflections of each other with respect to the line $y = x$.

5. $f(x) = x^3 - 8$ **6.** $f(x) = 5x - 10$

7. $f(x) = \dfrac{x - 1}{x + 2}$ **8.** $f(x) = \dfrac{-4}{x + 1}$

In Exercises 9 through 12, f is a one-to-one function. (a) State the range of f. (b) Determine f^{-1} and state the domain of f^{-1}. (c) Plot the graphs of f, f^{-1}, and the line $y = x$ in the same viewing rectangle.

9. $f(x) = 9 - x^2, x \geq 0$ **10.** $f(x) = x^2 - 9, x \leq 0$

11. $f(x) = \sqrt{x^2 - 1}, x \leq -1$

12. $f(x) = -\sqrt{x^2 - 1}, x \geq 1$

In Exercises 13 through 16, (a) show that f is monotonic and therefore one-to-one; (b) find $f^{-1}(x)$; (c) sketch the graphs of f and f^{-1} on the same set of axes; (d) check your graphs of f and f^{-1} by plotting them in the same viewing rectangle.

13. $f(x) = (1 - x)^3$ **14.** $f(x) = (x + 8)^3$

15. $f(x) = x^2 - 4, x \geq 0$ **16.** $f(x) = 1 - x^2, x \leq 0$

In Exercises 17 through 46, sketch the graph of the function. Check your graph on your graphics calculator.

17. $f(x) = 6^x$ **18.** $g(x) = 5^x$

19. $F(x) = 6^{-x}$ **20.** $G(x) = 5^{-x}$

21. $g(x) = 2^{x/2}$ **22.** $f(x) = 3^{x/3}$

23. $G(x) = 2^{-x/2}$ **24.** $F(x) = 3^{-x/3}$

25. $f(x) = 3e^x$ **26.** $g(x) = 2e^{-x}$

27. $g(x) = -e^{-x}$ **28.** $f(x) = -2e^{3x}$

29. $G(x) = 4e^{-2x}$ **30.** $F(x) = 5e^{-0.1x}$

31. $f(x) = e^{x-2}$ **32.** $g(x) = e^{x+1}$

33. $F(x) = e^x - 3$ **34.** $G(x) = e^x + 1$

35. $f(x) = 2 \ln x$ **36.** $g(x) = 4 \log_2 x$

37. $g(x) = -\frac{1}{2} \log_{10} x$ **38.** $f(x) = 3 \ln(-x)$

39. $f(x) = 3 - \ln x$ **40.** $g(x) = \ln(x + 2)$

41. $g(x) = 3 \ln(x - 2)$ **42.** $f(x) = 2 \ln x + 1$

43. $F(x) = -2e^{x-3} + 4$ **44.** $F(x) = \frac{1}{2}e^{x+4} - 3$

45. $f(x) = 3 \ln(x + 2) - 5$

46. $g(x) = -\ln(x - 1) + 2$

In Exercises 47 through 50, solve the equation for x, y, or b.

47. (a) $\log_5 x = 4$ (b) $\log_8 16 = y$

48. (a) $\log_9 x = -\frac{3}{2}$ (b) $\log_{27} 81 = y$

49. (a) $\log_b 4 = -\frac{1}{3}$ (b) $\ln x = -2$

50. (a) $\log_b 256 = \frac{4}{3}$ (b) $\ln x = \frac{3}{4}$

In Exercises 51 through 54, express the logarithm in terms of logarithms of x, y, and z, where the variables represent positive numbers.

51. $\log_b(x^3 y^2 \sqrt{z})$ **52.** $\log_b(\sqrt[3]{x} yz^5)$

53. $\log_b\left(\dfrac{x\sqrt[3]{y}}{z^4}\right)$ **54.** $\log_b \sqrt[5]{\dfrac{x^4 y^2}{z^3}}$

In Exercises 55 through 58, write the expression as a single logarithm with a coefficient of 1.

55. $\frac{2}{3} \log_{10} y - 4 \log_{10} x - \frac{1}{3} \log_{10} z$

56. $\ln k + 4 \ln L - \ln b - 3 \ln d$

57. $\ln 4 + \ln \pi + 2 \ln r + \ln h - \ln 3$

58. $\log_b 3 + \frac{1}{2} \log_b x + 3 \log_b z - \frac{1}{3} \log_b y - \log_b 2$

In Exercises 59 through 66, find the solution set of the equation, and express the results to four significant digits. Check your answer on your graphics calculator.

59. $5^x = 26$ **60.** $3^{x-2} = 8$

61. $(21.6)^x = 104$ **62.** $e^{-4x} = 0.231$

63. $e^{x/3} = 14.8$ **64.** $8^x - 8^{-x} = 8$

65. $2^x + 2^{-x} = 6$ **66.** $\dfrac{e^x - e^{-x}}{e^x + e^{-x}} = \dfrac{1}{2}$

In Exercises 67 through 70, find the value of the logarithm to four significant digits.

67. $\log_8 7$ **68.** $\log_7 100$

69. $\log_2 38$ **70.** $\log_5 e$

In Exercises 71 through 74, find the solution set of the equation.

71. $\log_4(2x + 3) - 2 \log_4 x = 2$

72. $\log_3(2x - 3) + \log_3(x + 3) = 4$

73. $\log_{10} x + \log_{10}(x - 200) - \log_{10} 4 = 5 - \log_{10} 5$

74. $\log_2(x + 2) - 3 = \log_2 3 - \log_2 x$

75. If $\ln i = \ln I - RT/L$, show that $i = Ie^{-RT/L}$.

In Exercises 76 through 95, solve the word problem by using a mathematical model involving an exponential function. Be sure to write a conclusion.

76. A loan of $1000 at an annual interest rate of 16 percent is repaid in one payment at the end of a year. Find the total amount repaid if (a) simple interest is earned, (b) interest is compounded quarterly, and (c) interest is compounded monthly.

77. An investment of $8000 earns interest at the rate of 12 percent per year, and the interest is paid once at the end of the year. Find the interest for the first year if (a) simple interest is earned, (b) interest is compounded semiannually, and (c) interest is compounded quarterly.

78. Work Exercise 76 if interest is compounded continuously.

79. Work Exercise 77 if interest is compounded continuously.

80. An amount of $500 is deposited into a savings account and earns interest for 7 years at a rate of 6 percent. If there are no withdrawals or additional deposits, how much is in the account after 7 years if **(a)** interest is compounded quarterly and **(b)** interest is compounded continuously.

81. A house purchased 10 years ago was sold for $100,000. If the annual interest rate was determined to be 20 percent compounded quarterly, what was the purchase price of the house to the nearest thousand dollars?

82. Interest on a savings account is computed at 8 percent per year compounded continuously. If one wishes to have $1000 in the account at the end of a year by making a single deposit now, what should the amount of the deposit be?

83. How long will it take for an investment to double itself if interest is earned at the rate of 12 percent per year compounded **(a)** quarterly and **(b)** continuously?

84. How long will it take for a deposit of $500 into a savings account to accumulate to $600 if interest is computed at 8 percent per year compounded **(a)** quarterly and **(b)** continuously?

85. If $f(t)$ milligrams of radium are present after t years, then $f(t) = ke^{-0.0004t}$, where k is a constant. **(a)** Compute k from the fact that 60 mg are present now. **(b)** With the value of k from part (a) plot the graph of f. **(c)** Determine algebraically how much radium will be present 100 years from now, and check your answer on your graphics calculator. **(d)** Use your graphics calculator to estimate how long it will take until only 50 mg of radium are present. Check your estimate algebraically.

86. In t minutes $f(t)$ bacteria will be present in a certain culture, where $f(t) = ke^{0.03t}$ and k is a constant. **(a)** Compute k from the fact that 60,000 bacteria are present initially. **(b)** With the value of k from part (a) plot the graph of f. **(c)** Determine algebraically how many bacteria will be present in 15 min, and check your answer on your graphics calculator. **(d)** Use your graphics calculator to estimate how many minutes will elapse until 200,000 bacteria will be present. Check your estimate algebraically.

87. After how many months on the job can the worker in Exercise 29 of Exercises 6.3 complete 80 units per day?

88. The population of a certain city t years from now is expected to be $f(t)$, where $f(t) = 40,000e^{kt}$ and k is a constant. If the population is expected to be 60,000 in 40 years, when is it expected to be 80,000?

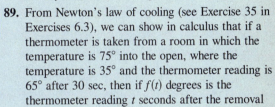

89. From Newton's law of cooling (see Exercise 35 in Exercises 6.3), we can show in calculus that if a thermometer is taken from a room in which the temperature is 75° into the open, where the temperature is 35° and the thermometer reading is 65° after 30 sec, then if $f(t)$ degrees is the thermometer reading t seconds after the removal

$$f(t) = 35 + 40\left(\frac{3}{4}\right)^{t/30}$$

What is the thermometer reading **(a)** 3 min after the removal; **(b)** 5 min after the removal; **(c)** 10 min after the removal; and **(d)** 30 min after the removal? **(e)** What will the thermometer reading be eventually? **(f)** Use your answer in part (e) to determine the horizontal asymptote of the graph of f and plot this line and the graph of f in the same viewing rectangle.

90. In a small town an epidemic spread in such a way that $f(t)$ persons had contracted the disease t weeks after its outbreak, where

$$f(t) = \frac{10,000}{1 + 599e^{-0.8t}}$$

How many people had the disease **(a)** initially; **(b)** after 6 weeks; and **(c)** after 12 weeks? **(d)** If the epidemic continues indefinitely, how many people will contract the disease? **(e)** Use your answer in part (d) to determine the horizontal asymptote of the graph of f and plot this line and the graph of f in the same viewing rectangle.

91. After how many minutes will the temperature of the body in Exercise 89 be 45°?

92. In the town of Exercise 90, after how many weeks will 5000 people, one-half of the town's population, contract the disease?

93. There was a time when a U.S. government bond sold at $74 to be redeemed 12 years later at a maturity value of $100. Determine the annual rate of interest, compounded monthly, that was earned.

94. An investment of $1000 earns $80 interest in one year at a simple interest rate of 8 percent per year. What annual interest rate would yield the same amount of interest in one year if **(a)** interest is compounded quarterly, and **(b)** interest is compounded continuously?

95. Two earthquakes having the same epicenter register magnitudes of 7.3 and 5.6 on the Richter scale. If y_1 and y_2 are respectively the measures of energy released by the two earthquakes, then y_1 is how many times y_2? *Hint:* See Exercise 33 in Exercises 6.5.

In Exercises 96 through 98, f is an exponential function; that is, $f(x) = b^x$. Use laws of exponents to prove the equality.

96. $f(x + y) = f(x) \cdot f(y)$ **97.** $f(x - y) = \dfrac{f(x)}{f(y)}$

98. $f(nx) = [f(x)]^n$

In Exercises 99 through 101, g is a logarithmic function; that is, $g(x) = \log_b x$. Use properties of logarithms to prove the equality.

99. $g(xy) = g(x) + g(y)$ **100.** $g\!\left(\dfrac{x}{y}\right) = g(x) - g(y)$

101. $g(x^n) = n\, g(x)$

102. Given

$$f(x) = \tfrac{1}{2}(e^x - e^{-x}) \quad \text{and} \quad g(x) = \ln(x + \sqrt{x^2 + 1})$$

(a) Plot the graphs of f and g and the line $y = x$ in the same viewing rectangle and observe that the graphs of f and g are reflections of each other with respect to the line $y = x$. **(b)** Prove algebraically that f and g are inverse functions.

103. (a) Use your answer in Exercise 43 in Exercises 6.6 to plot the graph of the function defined by $f(x) = \log_x 10$. **(b)** From your graph in part (a) determine the domain and range of function f.

104. Let function F be defined by $F(x) = \log_x(x - 3)$. **(a)** Compute $F(10)$ on your calculator. **(b)** Compute $F(100)$ on your calculator by using the key for $\log_{10}$. **(c)** Compute $F(1000)$ on your calculator by using the key for $\log_{10}$. **(d)** From your answers in parts (a) – (c) what does $F(x)$ appear to be approaching as x increases without bound? **(e)** Use your answer in part (d) to write an equation of the line you suspect is a horizontal asymptote of the graph of F. **(f)** Plot the line in part (e) and the graph of F in the same viewing rectangle. *Hint:* For a method to plot the graph of F, refer to your answers for Exercises 43 and 44 in Exercises 6.6. **(g)** From your graphs in part (f), determine the domain and range of F.

105. By observation, one of the solutions of the equation $x^3 = 3^x$ is 3. **(a)** There is also an irrational number solution. To get a numerical approximation to this solution, plot the graph of $y = x^3 - 3^x$ and use trace and zoom-in to determine the solution to the nearest one-hundredth. From the two solutions of the equation and your graphs in part (a), determine the values of x for which **(b)** $x^3 < 3^x$, and **(c)** $x^3 > 3^x$.

Trigonometric Functions of Real Numbers

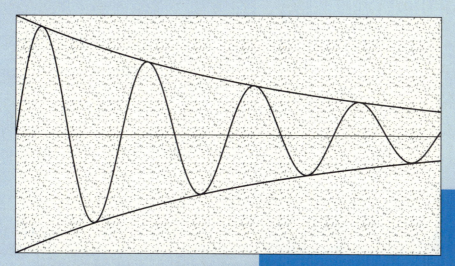

We continue the discussion of transcendental functions that began in Chapter 6 with the exponential and logarithmic functions. Here we present the trigonometric functions.

In Greek the word *trigonom* means triangle and *metron* means measure. Thus *trigonometry* means measurement of triangles, and its use in ancient times was in surveying, navigation, and astronomy to find relationships between lengths of sides of triangles and measurement of angles. It is still used for these purposes, and in such instances the trigonometric functions have angle measurements as their domains. In modern times another use of these functions involves applications in connection with periodically repetitive phenomena such as wave motion, alternating electrical current, vibrating strings, oscillating pendulums, business cycles, and biological rhythms. These applications of trigonometric functions require their domains to be sets of real numbers. Because of the importance of the periodic nature of the trigonometric functions to the study of calculus, we first define them with real number domains. Later, in Chapter 8, we discuss trigonometric functions in which the domains are sets of angle measurements.

7.1 THE SINE AND COSINE FUNCTIONS

We begin the discussion of trigonometric functions of real numbers by introducing the *length of arc on the unit circle*.

In Section 3.4 you learned that the graph of the equation

$$x^2 + y^2 = 1$$

is the unit circle, having its center at the origin and radius 1. Let us denote the unit circle by U. We shall show that a one-to-one correspondence exists between the lengths of all arcs of U, starting at the initial point $(1, 0)$, and the elements of the set R of real numbers. We start by imagining the real-number line "wrapped around" U, so that the number zero on the real-number line coincides with the point $(1, 0)$ on U. See Figure 1(a)–(c). Figure 1(a) shows U and the real-number line tangent to U at the point $(1, 0)$. In Figure 1(b), the positive side of the real-number line is wrapped around U

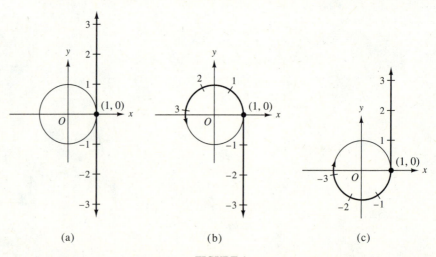

(a)   (b)   (c)

FIGURE 1

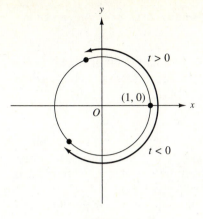

FIGURE 2

in a counterclockwise sense, and in Figure 1(c) the negative side is wrapped around U in a clockwise sense. If we consider an arc with initial point at $(1, 0)$ to have its terminal point in a counterclockwise direction from $(1, 0)$, a positive real number represents the length of arc; if the arc is considered to have its terminal point in a clockwise direction from $(1, 0)$, a negative real number represents the length of arc. See Figure 2.

Because the circumference of a circle is given by $2\pi r$, where r is the measure of the radius, the circumference of U is 2π. Thus the distance one-half of the way around U is π, the distance one-fourth of the way around is $\frac{1}{2}\pi$, the distance one-eighth of the way around is $\frac{1}{4}\pi$, and so on. Figure 3 shows some terminal points of arcs on U where the length of arc is measured in the counterclockwise direction from the point $(1, 0)$. The corresponding arc length in terms of π is indicated in the figure at the terminal point; it is a positive number. Figure 4 shows the same terminal points of arcs of U as in Figure 3, but in this case the length of arc is measured in the clockwise direction from the point $(1, 0)$; therefore, the corresponding arc length is a negative number.

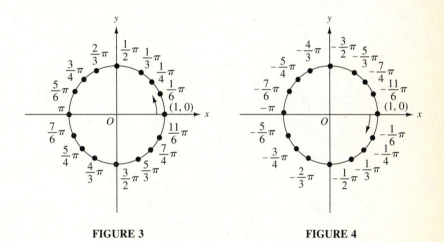

FIGURE 3   **FIGURE 4**

A length of arc of U is often given in terms of π. However, when decimals are used, we can approximate π by 3.14 and write $\pi \approx 3.14$. Thus

$$\frac{1}{2}\pi \approx 1.57 \qquad \frac{3}{2}\pi \approx 4.71 \qquad 2\pi \approx 6.28$$
$$\frac{1}{4}\pi \approx 0.79 \qquad -\frac{1}{3}\pi \approx -1.05 \qquad -\frac{3}{4}\pi \approx -2.36$$

and so on.

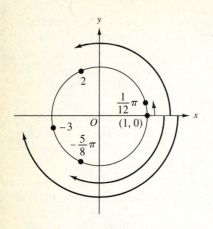

FIGURE 5

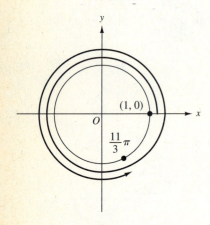

FIGURE 6

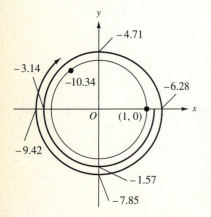

FIGURE 7

▶ **EXAMPLE 1** *Locating on U the Terminal Point of an Arc Having a Given Arc Length*

Show by a figure the location on U of the terminal point of the arc having initial point at $(1, 0)$ and having the given arc length; also state the quadrant in which the terminal point lies: **(a)** $\frac{1}{12}\pi$; **(b)** 2; **(c)** $-\frac{5}{8}\pi$; **(d)** -3.

Solution   Refer to Figure 5.

(a) Because $0 < \frac{1}{12}\pi < \frac{1}{2}\pi$, the terminal point is in the first quadrant.
(b) Because $1.57 < 2 < 3.14$, the terminal point is in the second quadrant.
(c) Because $-\pi < -\frac{5}{8}\pi < -\frac{1}{2}\pi$, the terminal point is in the third quadrant.
(d) Because $-3.14 < -3 < -1.57$, the terminal point is in the third quadrant. ◀

We have imagined the real-number line as wrapped around U. Thus, if the length of an arc from $(1, 0)$ is more than 2π or less than -2π, the wrapped part of the number line will traverse more than the circumference of U.

▷ **ILLUSTRATION 1**

(a) Figure 6 shows an arc of length $\frac{11}{3}\pi$. Because $\frac{11}{3}\pi = 2\pi + \frac{5}{3}\pi$, and $\frac{3}{2}\pi < \frac{5}{3}\pi < 2\pi$, the terminal point of this arc is in the fourth quadrant.
(b) Figure 7 shows an arc of length -10.34. Because

$$-10.34 = -6.28 + (-4.06) \quad \text{and} \quad -4.71 < -4.06 < -3.14$$

the terminal point of this arc is in the second quadrant. ◀

With t representing the length of an arc of U with initial point at $(1, 0)$, we have demonstrated that t can be any real number.

The rectangular cartesian coordinates of the terminal point of an arc on U with initial point $(1, 0)$ are functions of the arc length t. These functions are called *sine* (abbreviated *sin*) and *cosine* (abbreviated *cos*).

DEFINITION   **The Sine and Cosine Functions**

If t is a real number representing the length of arc on U with initial point $(1, 0)$ and terminal point (x, y), then

$$\sin t = y \quad \text{and} \quad \cos t = x$$

The domain of the sine and cosine functions is the set of real numbers. To determine the ranges of these functions, note that (x, y) is a point on the unit circle U. Thus

$$|y| \leq 1 \quad \text{and} \quad |x| \leq 1$$

Therefore the range of each function is the closed interval $[-1, 1]$; that is,

$$-1 \leq \sin t \leq 1 \quad \text{and} \quad -1 \leq \cos t \leq 1$$

Because $\sin t$ and $\cos t$ are coordinates of a point on a circle, the sine and cosine functions are called **circular functions.** They are also called **trigonometric functions defined on the real numbers.**

Figure 8 shows the unit circle U, and on U there are tick marks for every 0.1 unit. The circumference of U is $2\pi \approx 6.28$. For any real number t, we can approximate $\sin t$ and $\cos t$ by obtaining approximate values of the rectangular cartesian coordinates of the point whose arc length from $(1, 0)$ is t. We do this for three values of t in the following illustration.

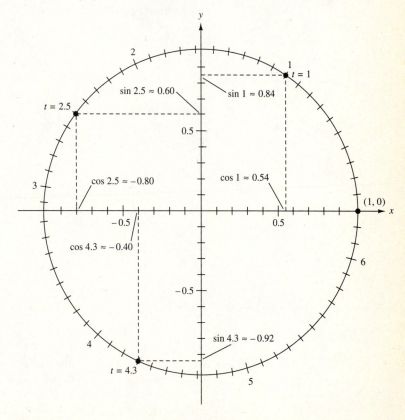

FIGURE 8

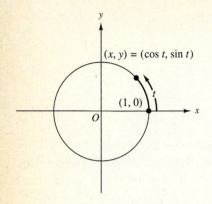

FIGURE 9

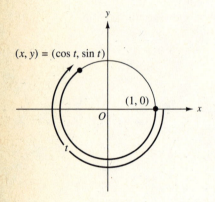

FIGURE 10

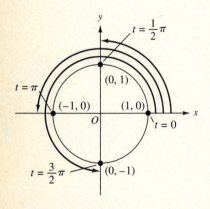

FIGURE 11

▷ **ILLUSTRATION 2**

Figure 8 shows the points on U whose arc lengths from $(1, 0)$ are $t = 1$, $t = 2.5$, and $t = 4.3$. The sine and cosine of each of these values of t are approximated by the point's y and x coordinates, respectively. From the figure we obtain

$\sin 1 \approx 0.84$	$\cos 1 \approx 0.54$
$\sin 2.5 \approx 0.60$	$\cos 2.5 \approx -0.80$
$\sin 4.3 \approx -0.92$	$\cos 4.3 \approx -0.40$	◀

▷ **ILLUSTRATION 3**

(a) Figure 9 shows U and an arc for which $t > 0$ and that has its terminal point in the first quadrant. Observe that $\sin t$ and $\cos t$ are both positive.

(b) In Figure 10 there is an arc on U for which $t < 0$ and that has its terminal point in the second quadrant. In this case $\sin t$ is positive and $\cos t$ is negative. ◀

Exact values of the sine and cosine functions for certain real numbers can be obtained by using geometry. In particular, refer to Figure 11, which shows arcs for which t is 0, $\frac{1}{2}\pi$, π, and $\frac{3}{2}\pi$. These values of t are called **quadrantal** numbers.

When $t = 0$, the arc length is zero, and so the initial and terminal points of the arc are both at $(1, 0)$. Thus

$$\sin 0 = 0 \quad \text{and} \quad \cos 0 = 1$$

When $t = \frac{1}{2}\pi$, the terminal point of the arc is at $(0, 1)$. Therefore

$$\sin \tfrac{1}{2}\pi = 1 \quad \text{and} \quad \cos \tfrac{1}{2}\pi = 0$$

When $t = \pi$, the terminal point of the arc is at $(-1, 0)$. Hence

$$\sin \pi = 0 \quad \text{and} \quad \cos \pi = -1$$

When $t = \frac{3}{2}\pi$, the terminal point of the arc is at $(0, -1)$. Thus

$$\sin \tfrac{3}{2}\pi = -1 \quad \text{and} \quad \cos \tfrac{3}{2}\pi = 0$$

We summarize these results in Table 1.

We now proceed to find the sine and cosine of $\frac{1}{4}\pi$. The distance one-eighth of the way around U is $\frac{1}{4}\pi$. Refer to Figure 12, which shows the terminal point (x, y) for an arc on U having length $\frac{1}{4}\pi$. At this point $x = y$.

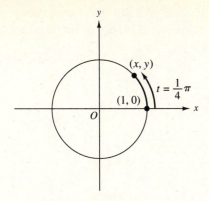

FIGURE 12

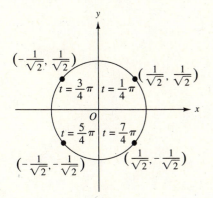

FIGURE 13

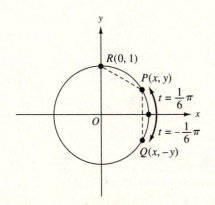

FIGURE 14

Table 1

t	(x, y)	$\sin t$	$\cos t$
0	$(1, 0)$	0	1
$\frac{1}{2}\pi$	$(0, 1)$	1	0
π	$(-1, 0)$	0	-1
$\frac{3}{2}\pi$	$(0, -1)$	-1	0

Substituting y for x in the equation of U, which is $x^2 + y^2 = 1$, we have

$$y^2 + y^2 = 1$$
$$2y^2 = 1$$
$$y^2 = \tfrac{1}{2}$$

$$y = \frac{1}{\sqrt{2}}$$

We reject the negative square root of $\frac{1}{2}$ because the point is in the first quadrant. Because $x = y$, we obtain

$$x = \frac{1}{\sqrt{2}}$$

Therefore

$$\sin \frac{1}{4}\pi = \frac{1}{\sqrt{2}} \quad \text{and} \quad \cos \frac{1}{4}\pi = \frac{1}{\sqrt{2}}$$

From these results and because U is symmetric with respect to both coordinate axes and the origin, we obtain the coordinates of the points for which t is $\frac{3}{4}\pi$, $\frac{5}{4}\pi$, and $\frac{7}{4}\pi$. See Figure 13. Thus

$$\sin \frac{3}{4}\pi = \frac{1}{\sqrt{2}} \quad \text{and} \quad \cos \frac{3}{4}\pi = -\frac{1}{\sqrt{2}}$$

$$\sin \frac{5}{4}\pi = -\frac{1}{\sqrt{2}} \quad \text{and} \quad \cos \frac{5}{4}\pi = -\frac{1}{\sqrt{2}}$$

$$\sin \frac{7}{4}\pi = -\frac{1}{\sqrt{2}} \quad \text{and} \quad \cos \frac{7}{4}\pi = \frac{1}{\sqrt{2}}$$

To obtain the sine and cosine of $\frac{1}{6}\pi$, we wish to find the terminal point $P(x, y)$ of the arc having length $\frac{1}{6}\pi$. Refer to Figure 14. Because P lies on U, (x, y) is a solution of the equation $x^2 + y^2 = 1$. Thus $(x, -y)$ is also a solution of this equation, and so the point $Q(x, -y)$ is also on U. The point Q is the terminal point of the arc having length $-\frac{1}{6}\pi$. The length of the arc of U from Q to P is $\frac{1}{6}\pi + \frac{1}{6}\pi = \frac{1}{3}\pi$. Furthermore, the length of the arc from P to $R(0, 1)$ is $\frac{1}{2}\pi - \frac{1}{6}\pi = \frac{1}{3}\pi$. Because arcs of equal lengths on a circle subtend chords of equal lengths, the distance $|\overline{QP}|$ equals the distance $|\overline{PR}|$. From the distance formula

$$|\overline{QP}| = \sqrt{(x - x)^2 + (y + y)^2} \qquad |\overline{PR}| = \sqrt{(x - 0)^2 + (y - 1)^2}$$
$$= \sqrt{4y^2} \qquad\qquad\qquad = \sqrt{x^2 + y^2 - 2y + 1}$$

Because $|\overline{QP}| = |\overline{PR}|$, $|\overline{QP}|^2 = |\overline{PR}|^2$. Therefore

$$4y^2 = x^2 + y^2 - 2y + 1$$

Because (x, y) is on U, $x^2 + y^2 = 1$. So we can substitute 1 for $x^2 + y^2$, and we obtain

$$4y^2 = 1 - 2y + 1$$
$$4y^2 + 2y - 2 = 0$$
$$2y^2 + y - 1 = 0$$
$$(2y - 1)(y + 1) = 0$$
$$y = \tfrac{1}{2} \quad \text{or} \quad y = -1$$

Because (x, y) is in the first quadrant, $y > 0$. Therefore we take $y = \frac{1}{2}$. If $\frac{1}{2}$ is substituted for y in the equation of U, we get

$$x^2 + (\tfrac{1}{2})^2 = 1$$
$$x^2 = \tfrac{3}{4}$$
$$x = \frac{\sqrt{3}}{2}$$

We reject the negative square root of $\frac{3}{4}$ because $x > 0$. Thus

$$\sin\frac{1}{6}\pi = \frac{1}{2} \quad \text{and} \quad \cos\frac{1}{6}\pi = \frac{\sqrt{3}}{2}$$

From these results and the symmetry of U, we obtain the coordinates of the points for which t is $\frac{5}{6}\pi$, $\frac{7}{6}\pi$, and $\frac{11}{6}\pi$. See Figure 15. Therefore

$$\sin\frac{5}{6}\pi = \frac{1}{2} \quad \text{and} \quad \cos\frac{5}{6}\pi = -\frac{\sqrt{3}}{2}$$
$$\sin\frac{7}{6}\pi = -\frac{1}{2} \quad \text{and} \quad \cos\frac{7}{6}\pi = -\frac{\sqrt{3}}{2}$$
$$\sin\frac{11}{6}\pi = -\frac{1}{2} \quad \text{and} \quad \cos\frac{11}{6}\pi = \frac{\sqrt{3}}{2}$$

Now let $P(x, y)$ be the terminal point of the arc having length $\frac{1}{3}\pi$. See Figure 16. The length of the arc of U from P to $R(0, 1)$ is $\frac{1}{2}\pi - \frac{1}{3}\pi = \frac{1}{6}\pi$. The length of the arc of U from $T(1, 0)$ to $S(\frac{1}{2}\sqrt{3}, \frac{1}{2})$ is $\frac{1}{6}\pi$. Therefore $|\overline{PR}| = |\overline{TS}|$. As before,

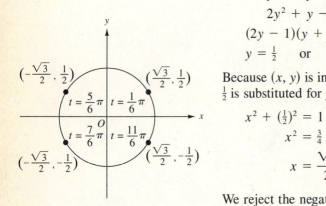

FIGURE 15

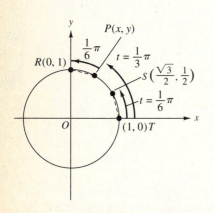

FIGURE 16

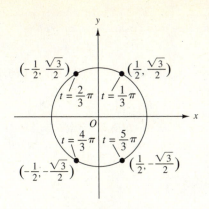

FIGURE 17

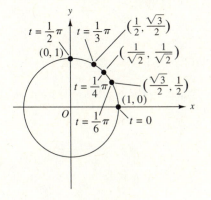

FIGURE 18

Table 2

t	$\sin t$	$\cos t$
0	0	1
$\dfrac{1}{6}\pi$	$\dfrac{1}{2}$	$\dfrac{\sqrt{3}}{2}$
$\dfrac{1}{4}\pi$	$\dfrac{1}{\sqrt{2}}$	$\dfrac{1}{\sqrt{2}}$
$\dfrac{1}{3}\pi$	$\dfrac{\sqrt{3}}{2}$	$\dfrac{1}{2}$
$\dfrac{1}{2}\pi$	1	0

$$|\overline{PR}| = \sqrt{x^2 + y^2 - 2y + 1} \qquad |\overline{TS}| = \sqrt{(\tfrac{1}{2}\sqrt{3} - 1)^2 + (\tfrac{1}{2} - 0)^2}$$

$$= \sqrt{\tfrac{3}{4} - \sqrt{3} + 1 + \tfrac{1}{4}}$$

$$= \sqrt{2 - \sqrt{3}}$$

Because $|\overline{PR}|^2 = |\overline{TS}|^2$, we have

$$x^2 + y^2 - 2y + 1 = 2 - \sqrt{3}$$

Replacing $x^2 + y^2$ by 1 because (x, y) is on U, we get

$$1 - 2y + 1 = 2 - \sqrt{3}$$
$$-2y = -\sqrt{3}$$
$$y = \frac{\sqrt{3}}{2}$$

We substitute $\tfrac{1}{2}\sqrt{3}$ for y in the equation of U and obtain

$$x^2 + \left(\frac{\sqrt{3}}{2}\right)^2 = 1$$
$$x^2 + \tfrac{3}{4} = 1$$
$$x^2 = \tfrac{1}{4}$$
$$x = \tfrac{1}{2}$$

Again we reject the negative square root because $x > 0$. Hence

$$\sin \frac{1}{3}\pi = \frac{\sqrt{3}}{2} \qquad \text{and} \qquad \cos \frac{1}{3}\pi = \frac{1}{2}$$

The coordinates of the points for which t is $\tfrac{2}{3}\pi$, $\tfrac{4}{3}\pi$, and $\tfrac{5}{3}\pi$ follow from these values and the symmetry of U, as shown in Figure 17. Thus

$$\sin \frac{2}{3}\pi = \frac{\sqrt{3}}{2} \qquad \text{and} \qquad \cos \frac{2}{3}\pi = -\frac{1}{2}$$

$$\sin \frac{4}{3}\pi = -\frac{\sqrt{3}}{2} \qquad \text{and} \qquad \cos \frac{4}{3}\pi = -\frac{1}{2}$$

$$\sin \frac{5}{3}\pi = -\frac{\sqrt{3}}{2} \qquad \text{and} \qquad \cos \frac{5}{3}\pi = \frac{1}{2}$$

In Figure 18 and Table 2 the sine and cosine are given when t is 0, $\tfrac{1}{6}\pi$, $\tfrac{1}{4}\pi$, $\tfrac{1}{3}\pi$, and $\tfrac{1}{2}\pi$. The corresponding function values for integer multiples of these numbers can be obtained from U by symmetry.

▶ **EXAMPLE 2** *Determining Exact Sine and Cosine Function Values from Points on U*

Determine the value of each of the following: **(a)** $\cos(-\frac{3}{4}\pi)$; **(b)** $\sin(-\frac{4}{3}\pi)$; **(c)** $\cos(\frac{17}{6}\pi)$; **(d)** $\sin(-\frac{5}{2}\pi)$.

Solution

(a) The point on U for $t = -\frac{3}{4}\pi$ is in the third quadrant, and it is symmetric with respect to the origin to the first quadrant point for which $t = \frac{1}{4}\pi$. Therefore

$$\cos\left(-\frac{3}{4}\pi\right) = -\frac{1}{\sqrt{2}}$$

(b) The point on U for $t = -\frac{4}{3}\pi$ is in the second quadrant, and it is symmetric with respect to the y axis to the point in the first quadrant for which $t = \frac{1}{3}\pi$. Thus

$$\sin\left(-\frac{4}{3}\pi\right) = \frac{\sqrt{3}}{2}$$

(c) Because $\frac{17}{6}\pi = 2\pi + \frac{5}{6}\pi$, the point on U for $t = \frac{17}{6}\pi$ is in the second quadrant, and it is symmetric with respect to the y axis to the point in the first quadrant for which $t = \frac{1}{6}\pi$. Hence

$$\cos\frac{17}{6}\pi = -\frac{\sqrt{3}}{2}$$

(d) Because $-\frac{5}{2}\pi = -2\pi + (-\frac{1}{2}\pi)$, $-\frac{5}{2}\pi$ is a quadrantal number and the point on U is $(0, -1)$. Therefore

$$\sin\left(-\frac{5}{2}\pi\right) = -1$$ ◀

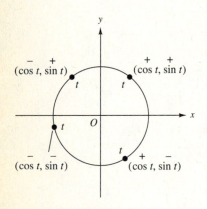

FIGURE 19

Because $\sin t$ and $\cos t$ are coordinates of a point on U, their signs depend on the quadrant in which the point lies. Refer to Figure 19. Table 3 summarizes the results in the figure. The quadrant in the table refers to the one containing the terminal point of the arc with length t.

If t is any real number, then because $(\cos t, \sin t)$ is a point on U and an equation of U is $y^2 + x^2 = 1$, we have

$$(\sin t)^2 + (\cos t)^2 = 1$$

This equation is true for all t. Instead of $(\sin t)^2$ and $(\cos t)^2$, it is customary to write $\sin^2 t$ and $\cos^2 t$. Therefore, we have the following identity:

$$\sin^2 t + \cos^2 t = 1$$

It is called the **fundamental Pythagorean identity** because the distance

Table 3

Quadrant	sin t	cos t
First	+	+
Second	+	−
Third	−	−
Fourth	−	+

formula (obtained from the Pythagorean theorem) was used to derive the equation $x^2 + y^2 = r^2$. The identity shows the relationship between the sine and cosine values and can be used to compute one of them when the other is known.

▶ **EXAMPLE 3** *Applying the Fundamental Pythagorean Identity*

If $\sin t = \frac{2}{3}$ and $\frac{1}{2}\pi < t < \pi$, find $\cos t$.

Solution   From the fundamental Pythagorean identity

$$\sin^2 t + \cos^2 t = 1$$

Because $\sin t = \frac{2}{3}$, we have

$$(\tfrac{2}{3})^2 + \cos^2 t = 1$$
$$\tfrac{4}{9} + \cos^2 t = 1$$
$$\cos^2 t = \tfrac{5}{9}$$
$$\cos t = \pm \frac{\sqrt{5}}{3}$$

We reject the positive value because $\cos t < 0$ because $\frac{1}{2}\pi < t < \pi$. Thus

$$\cos t = -\frac{\sqrt{5}}{3}$$ ◀

EXERCISES 7.1

In Exercises 1 through 8, show by a figure the location on U of the terminal point of the arc having initial point at (1, 0) and having the given arc length; also state the quadrant in which the terminal point lies.

1. (a) $\frac{1}{7}\pi$ (b) $\frac{3}{5}\pi$ **2.** (a) $\frac{1}{5}\pi$ (b) $\frac{9}{8}\pi$

3. (a) 1.23   (b) 5   **4.** (a) 3   (b) -0.25

5. (a) $\frac{17}{6}\pi$ (b) $-\frac{8}{7}\pi$ **6.** (a) $\frac{29}{8}\pi$ (b) $-\frac{8}{5}\pi$

7. (a) -2 (b) 12.2   **8.** (a) 10.6   (b) -4

In Exercises 9 through 16, use a figure similar to Figure 8 to show on the unit circle U the point whose arc length from (1, 0) is t. Then approximate the value of sin t and cos t to two decimal places.

9. $t = 2$ **10.** $t = 3$ **11.** $t = 5.2$

12. $t = 4.6$ **13.** $t = -3$ **14.** $t = -2$

15. $t = -6.1$ **16.** $t = -0.8$

In Exercises 17 through 36, determine the exact function value.

17. (a) $\sin 0$ (b) $\cos \pi$

18. (a) $\cos \frac{1}{2}\pi$ (b) $\sin \frac{3}{2}\pi$

19. (a) $\cos(-\frac{1}{2}\pi)$ (b) $\sin(-\frac{3}{2}\pi)$

20. (a) $\sin(-\pi)$ (b) $\cos(-\pi)$

21. (a) $\cos \frac{1}{6}\pi$ (b) $\sin \frac{5}{6}\pi$

22. (a) $\sin \frac{1}{4}\pi$ (b) $\cos \frac{3}{4}\pi$

23. (a) $\sin \frac{5}{4}\pi$ (b) $\cos(-\frac{1}{4}\pi)$

24. (a) $\cos \frac{7}{6}\pi$ (b) $\sin(-\frac{1}{6}\pi)$

25. (a) $\cos \frac{1}{3}\pi$ (b) $\cos(-\frac{1}{3}\pi)$

26. (a) $\sin \frac{1}{3}\pi$ (b) $\sin(-\frac{1}{3}\pi)$

27. (a) $\sin \frac{7}{6}\pi$ (b) $\cos \frac{4}{3}\pi$

28. (a) $\cos \frac{5}{6}\pi$ (b) $\sin \frac{2}{3}\pi$

29. (a) $\cos(-\frac{5}{4}\pi)$ (b) $\sin\frac{3}{4}\pi$

30. (a) $\sin(-\frac{1}{4}\pi)$ (b) $\cos\frac{7}{4}\pi$

31. (a) $\cos\frac{5}{2}\pi$ (b) $\sin(-\frac{5}{2}\pi)$

32. (a) $\cos 4\pi$ (b) $\sin 5\pi$

33. (a) $\sin\frac{8}{3}\pi$ (b) $\cos(-\frac{17}{6}\pi)$

34. (a) $\cos(-\frac{13}{6}\pi)$ (b) $\sin\frac{10}{3}\pi$

35. (a) $\cos(-\frac{25}{4}\pi)$ (b) $\cos(-\frac{15}{4}\pi)$

36. (a) $\sin\frac{19}{4}\pi$ (b) $\sin(-\frac{11}{4}\pi)$

In Exercises 37 and 38, determine the quadrant containing the terminal point of the arc of length t if the initial point is at (1, 0).

37. (a) $\sin t > 0$ and $\cos t > 0$
  (b) $\sin t < 0$ and $\cos t > 0$

38. (a) $\sin t < 0$ and $\cos t < 0$
  (b) $\sin t > 0$ and $\cos t < 0$

39. If $\cos t = \frac{3}{5}$ and $0 < t < \frac{1}{2}\pi$, find $\sin t$.

40. If $\sin t = \frac{5}{13}$ and $0 < t < \frac{1}{2}\pi$, find $\cos t$.

41. If $\sin t = -\frac{8}{17}$ and $\pi < t < \frac{3}{2}\pi$, find $\cos t$.

42. If $\cos t = -\frac{4}{5}$ and $\frac{1}{2}\pi < t < \pi$, find $\sin t$.

43. If $\cos t = \frac{12}{13}$ and $-\frac{1}{2}\pi < t < 0$, find $\sin t$.

44. If $\sin t = -\frac{15}{17}$ and $\frac{3}{2}\pi < t < 2\pi$, find $\cos t$.

45. If $\cos t = -\frac{3}{4}$ and $\frac{1}{2}\pi < t < \pi$, find $\sin t$.

46. If $\cos t = -\frac{1}{3}$ and $-\pi < t < \frac{1}{2}\pi$, find $\sin t$.

47. If $\sin t = 0$ and $\frac{1}{2}\pi < t < \frac{3}{2}\pi$, find t.

48. If $\cos t = 0$ and $\pi < t < 2\pi$, find $\sin t$.

In Exercises 49 through 52, use a calculator to determine the quadrant containing the terminal point of the arc having initial point at (1, 0) and having the given arc length.

49. (a) 52 (b) 26.74 **50.** (a) 32 (b) 48.22

51. (a) 1984 (b) −1492 **52.** (a) 2001 (b) −1776

In Exercises 53 through 56, use the fundamental Pythagorean identity and a calculator to find to four significant digits the function value from the given information.

53. Find $\sin t$ if $\cos t = 0.7816$ and $0 < t < \frac{1}{2}\pi$.

54. Find $\cos t$ if $\sin t = 0.1234$ and $0 < t < \frac{1}{2}\pi$.

55. Find $\cos t$ if $\sin t = -0.4178$ and $\pi < t < \frac{3}{2}\pi$.

56. Find $\sin t$ if $\cos t = -0.8245$ and $\frac{1}{2}\pi < t < \pi$.

In Exercises 57 and 58, $f(t) = \cos t$, $g(t) = \sin t$, $h(t) = \frac{1}{6}t$, and $\phi(t) = \frac{1}{4}t$. Compute the composite function value.

57. (a) $f(h(\pi))$ (b) $g(\phi(2\pi))$
  (c) $f(h(4\pi))$ (d) $g(\phi(5\pi))$

58. (a) $g(h(2\pi))$ (b) $f(\phi(3\pi))$
  (c) $g(h(3\pi))$ (d) $f(\phi(-\pi))$

7.2 SINE AND COSINE FUNCTION VALUES AND PERIODIC FUNCTIONS

GOALS

1. Approximate sine and cosine function values on a calculator.
2. Define a periodic function.
3. Learn that the sine and cosine functions are periodic with period 2π.
4. Find sine and cosine function values by using the concept of periodicity.
5. Find periods of functions defined by $\sin at$ and $\cos at$.
6. Solve word problems having periodic functions as mathematical models.
7. Use Taylor polynomials to find sine and cosine function values.

A calculator can be used to find sine and cosine function values of real numbers. In addition to keys for sine and cosine, your calculator can be set in either the *degree* or *radian* mode. When obtaining trigonometric function values of real numbers, the calculator should be set in the radian mode. The use of the word *radian* for this purpose will be apparent after you study Section 8.1. The number of significant digits displayed for a trigonometric function value will vary according to the calculator used. In this book, when a trigonometric function value is obtained from a calculator, we shall round off the number to four significant digits.

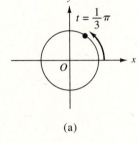

(a)

▷ ILLUSTRATION 1

To evaluate cos 1.384 on a calculator, first set it in the radian mode. If the display gives ten significant digits, you will read

$$\cos 1.384 \approx 0.1857119105$$

Rounding off the result to four significant digits, we get

$$\cos 1.384 \approx 0.1857 \qquad \blacktriangleleft$$

▶ EXAMPLE 1 *Approximating Sine and Cosine Function Values on a Calculator*

Approximate to four significant digits each of the following:

(a) sin 1.072   **(b)** $\sin \frac{3}{14}\pi$ **(c)** cos 1   **(d)** $\cos \frac{5}{12}\pi$

Solution Be sure the calculator is in the radian mode. We round off each displayed entry to four significant digits.

(a) $\sin 1.072 \approx 0.8782$ **(b)** $\sin \frac{3}{14}\pi \approx 0.6235$

(c) $\cos 1 \approx 0.5403$ **(d)** $\cos \frac{5}{12}\pi \approx 0.2588$ $\blacktriangleleft$

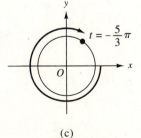

(c)

FIGURE 1

$\iint \frac{dy}{dx}$ In the introduction to this chapter, we referred to the importance of the periodic nature of the trigonometric functions to the study of calculus, which motivated our definitions of sin *t* and cos *t* as coordinates of a point on the unit circle.

Because the circumference of the unit circle is 2π, two arcs having initial point at $(1, 0)$ and differing in length by an integer multiple of 2π have the same terminal point on U. For example, refer to Figure 1. Figure 1(a) shows an arc on U of length $\frac{1}{3}\pi$. In Figure 1(b) there is an arc of length $\frac{7}{3}\pi = \frac{1}{3}\pi + 2\pi$, and this arc has the same terminal point. In Figure 1(c) there is an arc of length $-\frac{5}{3}\pi = \frac{1}{3}\pi + (-1)(2\pi)$, and this arc also has the same terminal point. As a matter of fact, any arc for which

$$t = \tfrac{1}{3}\pi + k(2\pi) \qquad k \in J$$

will have the same terminal point. Because the coordinates of the terminal point of the arc determine the sine and cosine of the arc length, it follows that

$$\sin(\tfrac{1}{3}\pi + k \cdot 2\pi) = \sin \tfrac{1}{3}\pi \qquad \text{and} \qquad \cos(\tfrac{1}{3}\pi + k \cdot 2\pi) = \cos \tfrac{1}{3}\pi \quad \textbf{(1)}$$

where k is any integer.

▷ **ILLUSTRATION 2**

$$\sin \frac{1}{3}\pi = \frac{\sqrt{3}}{2} \qquad\qquad \cos \frac{1}{3}\pi = \frac{1}{2}$$

$$\sin \frac{7}{3}\pi = \sin\left(\frac{1}{3}\pi + 2\pi\right) \qquad \cos \frac{7}{3}\pi = \cos\left(\frac{1}{3}\pi + 2\pi\right)$$

$$= \frac{\sqrt{3}}{2} \qquad\qquad\qquad = \frac{1}{2}$$

$$\sin \frac{13}{3}\pi = \sin\left(\frac{1}{3}\pi + 4\pi\right) \qquad \cos \frac{13}{3}\pi = \cos\left(\frac{1}{3}\pi + 4\pi\right)$$

$$= \frac{\sqrt{3}}{2} \qquad\qquad\qquad = \frac{1}{2}$$

$$\sin\left(-\frac{5}{3}\pi\right) = \sin\left(\frac{1}{3}\pi - 2\pi\right) \qquad \cos\left(-\frac{5}{3}\pi\right) = \cos\left(\frac{1}{3}\pi - 2\pi\right)$$

$$= \frac{\sqrt{3}}{2} \qquad\qquad\qquad = \frac{1}{2}$$

and so on.   ◀

Because an arc of length $t + k \cdot 2\pi$, where $k \in J$, has the same terminal point as an arc of length t, it follows that Equations (1) are valid if $\frac{1}{3}\pi$ is replaced by any real number t. We state this result as a theorem.

THEOREM 1

If t is a real number and k is any integer,

$$\sin(t + k \cdot 2\pi) = \sin t \qquad \text{and} \qquad \cos(t + k \cdot 2\pi) = \cos t$$

The property of sine and cosine given in Theorem 1 is called *periodicity,* which we now formally define.

DEFINITION A Periodic Function

A function f is said to be **periodic** if there exists a positive real number p such that whenever x is in the domain of f, then $x + p$ is also in the domain of f, and

$$f(x + p) = f(x)$$

The smallest such positive real number p is called the **period** of f.

Compare this definition with Theorem 1. Because 2π can be shown to be the smallest positive number p having the property that $\sin(t + p) = \sin t$ and $\cos(t + p) = \cos t$, the sine and cosine are periodic with period 2π.

▶ **EXAMPLE 2** *Applying the Concept of Periodicity to Find Sine and Cosine Function Values*

Use the values of $\sin t$ and $\cos t$ when $0 \leq t < 2\pi$ and periodicity to find each of the following values: **(a)** $\sin \frac{17}{4}\pi$; **(b)** $\cos(-\frac{7}{6}\pi)$; **(c)** $\sin \frac{15}{2}\pi$.

Solution

(a) $\sin \frac{17}{4}\pi = \sin(\frac{1}{4}\pi + 2 \cdot 2\pi)$ **(b)** $\cos(-\frac{7}{6}\pi) = \cos[\frac{5}{6}\pi + (-1)2\pi]$

$\qquad\qquad = \sin \frac{1}{4}\pi$ $\qquad\qquad\qquad\qquad = \cos \frac{5}{6}\pi$

$\qquad\qquad = \dfrac{1}{\sqrt{2}}$ $\qquad\qquad\qquad\qquad = -\dfrac{\sqrt{3}}{2}$

(c) $\sin \frac{15}{2}\pi = \sin(\frac{3}{2}\pi + 3 \cdot 2\pi)$

$\qquad\qquad = \sin \frac{3}{2}\pi$

$\qquad\qquad = -1$ ◀

From the fact that sine and cosine have period 2π, we can show that the functions defined by $\sin at$ and $\cos at$, where a is any real number, are periodic. This fact is demonstrated in the following illustration.

▷ **ILLUSTRATION 3**

(a) Let f be the function defined by

$$f(t) = \sin 4t$$

Because the sine has period 2π,

$$f(t) = \sin(4t + 2\pi)$$
$$= \sin[4(t + \tfrac{1}{2}\pi)]$$
$$= f(t + \tfrac{1}{2}\pi)$$

Thus f is periodic with period $\tfrac{1}{2}\pi$.

(b) Let g be the function defined by

$$g(t) = \cos \tfrac{1}{3}t$$

Because the cosine has period 2π,

$$g(t) = \cos(\tfrac{1}{3}t + 2\pi)$$
$$= \cos[\tfrac{1}{3}(t + 6\pi)]$$
$$= g(t + 6\pi)$$

Therefore g is periodic with period 6π.   ◄

For each function in the next illustration, the period is a rational number instead of a multiple of π.

▷ **ILLUSTRATION 4**

(a) Let f be the function defined by

$$f(t) = \sin 6\pi t$$

Because the sine has period 2π,

$$f(t) = \sin(6\pi t + 2\pi)$$
$$= \sin[6\pi(t + \tfrac{1}{3})]$$
$$= f(t + \tfrac{1}{3})$$

Therefore f is periodic with period $\tfrac{1}{3}$.

(b) Let g be the function defined by

$$g(t) = 5 \cot \tfrac{2}{3}\pi t$$

Because the cotangent has period π,

$$g(t) = 5 \cot(\tfrac{2}{3}\pi t + \pi)$$
$$= 5 \cot[(\tfrac{2}{3}\pi(t + \tfrac{3}{2})]$$
$$= g(t + \tfrac{3}{2})$$

Thus g is periodic with period $\tfrac{3}{2}$.   ◄

We now present some applications of sine and cosine functions in connection with periodically repetitive phenomena. Further applications will be given in Section 7.4.

 EXAMPLE 3 *Solving a Word Problem Having a Periodic Function as a Mathematical Model*

The electromotive force for an electric circuit with a simplified generator is $E(t)$ volts at t seconds where

$$E(t) = 50 \sin 120\pi t$$

(a) Find the period of E. Find the electromotive force at **(b)** 0.02 sec and **(c)** 0.2 sec.

Solution
(a) Because the sine has period 2π,

$$E(t) = 50 \sin(120\pi t + 2\pi)$$
$$= 50 \sin[120\pi(t + \tfrac{1}{60})]$$
$$= E(t + \tfrac{1}{60})$$

Conclusion: The period of E is $\tfrac{1}{60}$.

(b) The electromotive force at 0.02 sec is $E(0.02)$ volts, and

$$E(0.02) = 50 \sin 120\pi(0.02)$$
$$= 50 \sin 2.4\pi$$
$$= 50 \sin(0.4\pi + 2\pi)$$
$$= 50 \sin 0.4\pi$$
$$= 50(0.9511)$$
$$= 47.55$$

Conclusion: The electromotive force at 0.02 sec is 47.55 volts.

(c) The electromotive force at 0.2 sec is $E(0.2)$ volts, and

$$E(0.2) = 50 \sin 120\pi(0.2)$$
$$= 50 \sin 24\pi$$
$$= 50 \sin[0 + 12(2\pi)]$$
$$= 50 \sin 0$$
$$= 0$$

Conclusion: The electromotive force at 0.2 sec is 0.   ◀

▶ **EXAMPLE 4** *Solving a Word Problem Having a Periodic Function as a Mathematical Model*

A company that sells men's overcoats starts its fiscal year on July 1. For the fiscal year beginning July 1, 1992, the profit from sales was given by

$$P(t) = 20{,}000(1 - \cos \tfrac{1}{6}\pi t) \qquad 0 \le t \le 36$$

where $P(t)$ dollars was the monthly profit t months since July 1, 1992. **(a)** Find the period of P. Find the monthly profit on **(b)** October 1, 1992; **(c)** November 1, 1992; **(d)** December 1, 1992; **(e)** January 1, 1993; **(f)** April 1, 1993; and **(g)** July 1, 1993.

Solution
(a) Because the cosine has period 2π,

$$\begin{aligned} P(t) &= 20{,}000[1 - \cos(\tfrac{1}{6}\pi t + 2\pi)] \\ &= 20{,}000[1 - \cos \tfrac{1}{6}\pi(t + 12)] \\ &= P(t + 12) \end{aligned}$$

Conclusion: The period of P is 12.

(b)–(g) The monthly profits on October 1, 1992; November 1, 1992; December 1, 1992; January 1, 1993; April 1, 1993; and July 1, 1993, were, respectively, $P(3)$, $P(4)$, $P(5)$, $P(6)$, $P(9)$, and $P(12)$ dollars.

(b) $\begin{aligned}[t] P(3) &= 20{,}000(1 - \cos \tfrac{1}{2}\pi) \\ &= 20{,}000(1 - 0) \\ &= 20{,}000 \end{aligned}$

Conclusion: On October 1, 1992, the monthly profit was $20,000.

(c) $\begin{aligned}[t] P(4) &= 20{,}000(1 - \cos \tfrac{2}{3}\pi) \\ &= 20{,}000(1 + \tfrac{1}{2}) \\ &= 30{,}000 \end{aligned}$

Conclusion: On November 1, 1992, the monthly profit was $30,000.

(d) $\begin{aligned}[t] P(5) &= 20{,}000(1 - \cos \tfrac{5}{6}\pi) \\ &= 20{,}000\left(1 + \frac{\sqrt{3}}{2}\right) \\ &= 37{,}321 \end{aligned}$

Conclusion: On December 1, 1992, the monthly profit was $37,321.

(e) $\begin{aligned}[t] P(6) &= 20{,}000(1 - \cos \pi) \\ &= 20{,}000(1 + 1) \\ &= 40{,}000 \end{aligned}$

Conclusion: On January 1, 1993, the monthly profit was \$40,000.

(f) $P(9) = 20,000(1 - \cos \frac{3}{2}\pi)$

$\qquad\quad = 20,000(1 - 0)$

$\qquad\quad = 20,000$

Conclusion: On April 1, 1993 the monthly profit was \$20,000.

(g) $P(12) = 20,000(1 - \cos 2\pi)$

$\qquad\quad\; = 20,000(1 - 1)$

$\qquad\quad\; = 0$

Conclusion: On July 1, 1993, there was no monthly profit. ◀

In calculus, you will learn various methods of approximating a given function by polynomials. One of the most widely used methods is attributed to the English mathematician Brook Taylor (1685–1731). The Taylor polynomials for sine and cosine are

$$\sin t = t - \frac{t^3}{3!} + \frac{t^5}{5!} - \frac{t^7}{7!} + \frac{t^9}{9!} - \frac{t^{11}}{11!} + \ldots + R_n \qquad (2)$$

and

$$\cos t = 1 - \frac{t^2}{2!} + \frac{t^4}{4!} - \frac{t^6}{6!} + \frac{t^8}{8!} - \frac{t^{10}}{10!} + \ldots + R_n \qquad (3)$$

where R_n denotes a remainder after n terms, and $2! = 2 \cdot 1, 3! = 3 \cdot 2 \cdot 1,$ $4! = 4 \cdot 3 \cdot 2 \cdot 1,$ and so on. An approximation of the sine or cosine of a specific number t can be found by taking terms of the corresponding Taylor polynomial; the error that results is less than the absolute value of the next term in the polynomial.

 ▷ **ILLUSTRATION 5**

If polynomial (3) is used to compute the value of $\cos 1$ to four significant digits, we have

$$\cos 1 = 1 - \frac{1}{2!} + \frac{1}{4!} - \frac{1}{6!} + \frac{1}{8!} - \frac{1}{10!} + \ldots + R_n$$

$$= 1 - \frac{1}{2} + \frac{1}{24} - \frac{1}{720} + \frac{1}{40,320} - \frac{1}{3,628,800} + \ldots + R_n$$

$$= 1 - 0.5 + 0.04167 - 0.00139 + 0.00002 - 0.0000003 + \ldots + R_n$$

If we take the first five terms for the approximate value of $\cos 1$, the error is less than the absolute value of the sixth term. Thus the error is less than

0.0000003. Adding the first five terms, we obtain 0.54030. Rounding off to four significant digits gives

$$\cos 1 \approx 0.5403$$

◀

EXAMPLE 5 *Finding a Sine Function Value from a Taylor Polynomial*

Use Taylor polynomial (2) to find the value of sin 0.75 to four significant digits.

Solution   Replacing t by 0.75 in (2), we get

$$\sin 0.75 = 0.75 - \frac{(0.75)^3}{3!} + \frac{(0.75)^5}{5!} - \frac{(0.75)^7}{7!} + \frac{(0.75)^9}{9!} - \ldots + R_n$$

$$= 0.75 - \frac{0.42188}{6} + \frac{0.23730}{120} - \frac{0.13348}{5040} + \frac{0.07508}{362,880} - \ldots + R_n$$

$$= 0.75 - 0.07031 + 0.00198 - 0.00003 + 0.0000002 - \ldots + R_n$$

If we take the first four terms, the error is less than 0.0000002. Adding the first four terms, we get 0.68164, and rounding off to four significant digits gives

$$\sin 0.75 \approx 0.6816$$

◀

EXERCISES 7.2

In Exercises 1 through 6, use a calculator to evaluate the function value to four significant digits.

1. (a) sin 0.34   (b) cos 0.34
   (c) sin 2.85   (d) cos 2.85

2. (a) sin 1.27   (b) cos 1.27
   (c) sin 1.72   (d) cos 1.72

3. (a) sin 4.29   (b) cos 4.29
   (c) sin(−1.36)   (d) cos(−1.36)

4. (a) sin 5.08   (b) cos 5.08
   (c) sin(−2.69)   (d) cos(−2.69)

5. (a) $\sin \frac{5}{11}\pi$ (b) $\cos \frac{1}{8}\pi$
   (c) $\sin \frac{8}{9}\pi$ (d) $\cos \frac{9}{5}\pi$

6. (a) $\sin \frac{2}{7}\pi$ (b) $\cos \frac{4}{9}\pi$
   (c) $\sin \frac{7}{5}\pi$ (d) $\cos \frac{11}{7}\pi$

In Exercises 7 through 14, use the periodicity of the sine and cosine functions as well as the values of sin t and cos t when $0 \le t < 2\pi$ to find the function value.

7. (a) $\sin \frac{9}{4}\pi$ (b) $\cos \frac{9}{4}\pi$
   (c) $\sin \frac{10}{3}\pi$ (d) $\cos \frac{10}{3}\pi$

8. (a) $\sin \frac{17}{6}\pi$ (b) $\cos \frac{17}{6}\pi$
   (c) $\sin \frac{15}{4}\pi$ (d) $\cos \frac{15}{4}\pi$

9. (a) $\sin(-\frac{1}{6}\pi)$ (b) $\cos(-\frac{1}{6}\pi)$
   (c) $\sin(-\frac{5}{4}\pi)$ (d) $\cos(-\frac{5}{4}\pi)$

10. (a) $\sin(-\frac{2}{3}\pi)$ (b) $\cos(-\frac{2}{3}\pi)$
   (c) $\sin(-\frac{11}{6}\pi)$ (d) $\cos(-\frac{11}{6}\pi)$

11. (a) $\sin(-\frac{7}{3}\pi)$ (b) $\cos \frac{11}{3}\pi$
   (c) $\sin 8\pi$ (d) $\cos 10\pi$

12. (a) $\sin \frac{13}{4}\pi$ (b) $\cos(-\frac{11}{4}\pi)$
   (c) $\sin \frac{7}{2}\pi$ (d) $\cos \frac{5}{2}\pi$

13. (a) $\sin(-\frac{7}{2}\pi)$ (b) $\cos(-\frac{5}{2}\pi)$
   (c) $\sin(-\frac{11}{2}\pi)$ (d) $\cos(-\frac{9}{2}\pi)$

14. (a) $\sin(-8\pi)$ (b) $\cos(-10\pi)$
   (c) $\sin(-7\pi)$ (d) $\cos(-9\pi)$

In Exercises 15 through 20, find the period of the function.

15. **(a)** $\sin 3t$ **(b)** $3 \cos 6t$
    (c) $\cos \frac{1}{4}t$ **(d)** $2 \sin \frac{1}{5}t$

16. **(a)** $\cos 5t$ **(b)** $4 \sin 4t$
    (c) $\sin \frac{5}{6}t$ **(d)** $\frac{1}{2} \cos \frac{1}{3}t$

17. **(a)** $4 \cos 7t$ **(b)** $\sin \frac{4}{7}t$
    (c) $\cos 4\pi t$ **(d)** $\frac{1}{2} \sin \frac{1}{3}\pi t$

18. **(a)** $3 \sin 8t$ **(b)** $\cos \frac{3}{8}t$
    (c) $4 \cos 3\pi t$ **(d)** $\sin \frac{2}{3}\pi t$

19. **(a)** $10(2 - \cos \frac{1}{5}\pi t)$ **(b)** $6 + \sin \pi(1 + t)$

20. **(a)** $3 + 4 \cos \pi(\frac{1}{2} + t)$ **(b)** $3(4 - \sin \frac{1}{2}\pi t)$

$\iint \frac{dy}{dx}$ *In Exercises 21 through 26, use Taylor polynomial* (1) *or* (2) *to find the function value to four significant digits.*

21. $\cos 0.75$ **22.** $\cos 2.39$

23. $\sin 4.26$ **24.** $\sin 5.73$

25. $\sin(-0.81)$ **26.** $\cos(-1.27)$

In Exercises 27 through 34, the functions are mathematical models describing periodic phenomena discussed in Section 7.4. Be sure to write a conclusion.

27. In an electric circuit the electromotive force is $E(t)$ volts at t seconds, where $E(t) = 2 \cos 50\pi t$. **(a)** Determine the period of E. Find the electromotive force at **(b)** 0.02 sec; **(c)** 0.03 sec; **(d)** 0.04 sec; and **(e)** 0.06 sec.

28. Do Exercise 27 if $E(t) = 40 \sin 120\pi t$.

29. An alternating current of electricity is described by $I(t) = 10 \sin 2800t$ where $I(t)$ amperes is the current at t seconds. **(a)** Determine the period of I. Find the

current at **(b)** 0.001 sec; **(c)** 0.003 sec; **(d)** 0.005 sec; and **(e)** 0.01 sec.

30. Do Exercise 29 if $I(t) = 2 \sin 3000t$.

31. A weight suspended from a spring is vibrating vertically, and $f(t)$ centimeters is the directed distance of the weight from its central position (the origin) at t seconds where the positive direction is upward. If $f(t) = 2 \sin 3t$, **(a)** determine the period of f, and find the position of the weight at **(b)** 0 sec; **(c)** 1 sec; **(d)** 2 sec; and **(e)** 5 sec.

32. Do Exercise 31 if $f(t) = 6 \cos 4t$.

33. A wave produced by a simple sound has the equation $P(t) = 0.02 \sin 1500\pi t$, where $P(t)$ dynes per square centimeter is the difference between the atmospheric pressure and the air pressure at the eardrum at t seconds. **(a)** Determine the period of P. Find the difference between the atmospheric pressure and the air pressure at the eardrum at **(b)** $\frac{1}{9}$ sec; **(c)** $\frac{1}{8}$ sec; **(d)** $\frac{1}{7}$ sec; and **(e)** $\frac{1}{6}$ sec.

34. Do Exercise 33 if $P(t) = 0.003 \sin 1800\pi t$.

35. The function H of Example 6 in Section 4.2 is defined by

$$H(x) = [\![x]\!] - x$$

Show that H is periodic with period 1.

36. Do Exercise 35 for the function G of Exercise 41 in Exercises 4.2 defined by

$$G(x) = x - [\![x]\!]$$

37. Explain why a periodic function cannot be one-to-one.

7.3 GRAPHS OF THE SINE AND COSINE FUNCTIONS AND OTHER SINE WAVES

GOALS
1. Obtain the graph of the sine function.
2. Obtain the graph of the cosine function.
3. Approximate sine and cosine function values by plotting the sine and cosine curves.
4. Approximate values of t if either $\sin t$ or $\cos t$ is given by plotting the sine and cosine curves.

5. **Learn that the period of a function of the form sin bt or cos bt is $\dfrac{2\pi}{|b|}$.**

6. **Define the amplitude of a sine wave.**

7. **Define the phase shift of a sine wave.**

8. **Sketch sine waves.**

In our previous discussions of graphs, the coordinate axes were denoted by the symbols x and y. However, in our treatment of the trigonometric functions of real numbers, x and y are used for the coordinates of the terminal point of the arc of length t on the unit circle U. Therefore, in this chapter where we discuss graphs of the trigonometric functions, we shall denote the horizontal axis by t and the vertical axis by $f(t)$. Often on the t axis it is convenient to use rational multiples of π for the tick marks. With $\pi \approx 3.14$, we show in Figure 1 where these tick marks occur in relation to those for integers.

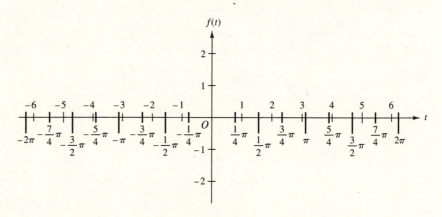

FIGURE 1

The periodicity of the sine and cosine functions plays an important part in obtaining the graphs of these functions. We first consider the graph of the sine function. Let

$$f(t) = \sin t$$

Because the sine is periodic with period 2π, it is sufficient to determine the portion of the graph for $0 \leq t \leq 2\pi$. This portion will then be repeated in intervals of length 2π on the t axis. Because $-1 \leq \sin t \leq 1$, the greatest value $\sin t$ assumes is 1, and the least value it assumes is -1. Table 1 summarizes the behavior of $\sin t$ in each of the four quadrants as t increases from 0 to 2π.

Table 1

As t increases from	$\sin t$ goes from
0 to $\frac{1}{2}\pi$	0 to 1
$\frac{1}{2}\pi$ to π	1 to 0
π to $\frac{3}{2}\pi$	0 to -1
$\frac{3}{2}\pi$ to 2π	-1 to 0

Table 2

t	$\sin t$	t	$\sin t$
0	0	$\frac{5}{4}\pi$	$-\frac{1}{\sqrt{2}}$
$\frac{1}{4}\pi$	$\frac{1}{\sqrt{2}}$	$\frac{3}{2}\pi$	-1
$\frac{1}{2}\pi$	1	$\frac{7}{4}\pi$	$-\frac{1}{\sqrt{2}}$
$\frac{3}{4}\pi$	$\frac{1}{\sqrt{2}}$	2π	0
π	0		

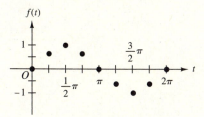

FIGURE 2

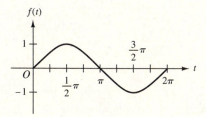

FIGURE 3

We next locate some specific points on the graph in the interval $[0, 2\pi]$. Table 2 contains the values of $\sin t$ for every $\frac{1}{4}\pi$ units, and Figure 2 shows the points whose coordinates are the number pairs $(t, \sin t)$ given in the table. In calculus we prove that the sine function is continuous, which indicates that there are no "breaks" in its graph. Therefore, we may connect the points in Figure 2 with a curve and obtain the graph shown in Figure 3. To ensure the accuracy of the graph, the points for which t is $\frac{1}{6}\pi, \frac{1}{3}\pi, \frac{2}{3}\pi$, $\frac{5}{6}\pi$, and so on can also be located. Function values from a calculator can be used to obtain still other points.

Now that we have the portion of the graph over the interval $[0, 2\pi]$, this portion is repeated for every interval on the t axis of length 2π: $[2\pi, 4\pi]$, $[4\pi, 6\pi]$, $[-2\pi, 0]$, $[-4\pi, -2\pi]$, and so on. Figure 4 shows the complete graph; it continues indefinitely to the left and right for t any real number. This graph, called the **sine curve,** is also referred to as a **sine wave.** The portion of the graph over one period is called a **cycle.**

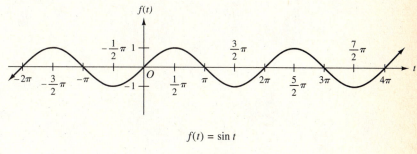

$$f(t) = \sin t$$

FIGURE 4

Because the zeros of the sine function are the values of t at which the graph intersects the t axis, they are $t = k\pi, k \in J$.

In a manner similar to that just described, we obtain the graph of the cosine function, called the **cosine curve.** Let

$$f(t) = \cos t$$

Because the period of cosine is 2π, we consider first the behavior of $\cos t$ for $t \in [0, 2\pi]$ as summarized in Table 3.

Table 3

As t increases from	$\cos t$ goes from
0 to $\frac{1}{2}\pi$	1 to 0
$\frac{1}{2}\pi$ to π	0 to -1
π to $\frac{3}{2}\pi$	-1 to 0
$\frac{3}{2}\pi$ to 2π	0 to 1

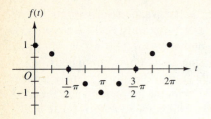

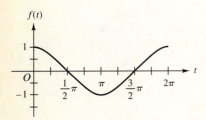

FIGURE 5

Table 4

t	0	$\frac{1}{4}\pi$	$\frac{1}{2}\pi$	$\frac{3}{4}\pi$	π	$\frac{5}{4}\pi$	$\frac{3}{2}\pi$	$\frac{7}{4}\pi$	2π
$\cos t$	1	$\dfrac{1}{\sqrt{2}}$	0	$-\dfrac{1}{\sqrt{2}}$	-1	$-\dfrac{1}{\sqrt{2}}$	0	$\dfrac{1}{\sqrt{2}}$	1

Table 4 gives values of $\cos t$ for every $\frac{1}{4}\pi$ units in the interval $[0, 2\pi]$; Figure 5 shows the points having as coordinates the number pairs $(t, \cos t)$. Because the cosine function is continuous, we connect the points to get the curve in Figure 6. By repeating this portion of the graph for every interval on the t axis of length 2π, we obtain Figure 7. As with the sine, additional points can be located using known exact values of the cosine or taking values from a calculator.

FIGURE 6

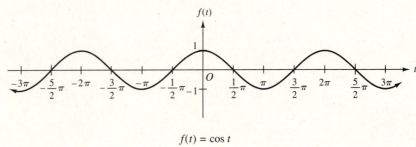

$$f(t) = \cos t$$

FIGURE 7

The t coordinates of the points at which the graph intersects the t axis are the zeros of the cosine function. They are $t = \frac{1}{2}\pi + k\pi, k \in J$.

Observe that the graph of the sine is symmetric with respect to the origin and the graph of the cosine is symmetric with respect to the $f(t)$ axis. These symmetry properties follow from the identities

$$\sin(-t) = -\sin t \tag{1}$$

and

$$\cos(-t) = \cos t \tag{2}$$

To justify these identities, refer to Figure 8. Observe that because points $P(\cos t, \sin t)$ and $\overline{P}(\cos(-t), \sin(-t))$ are symmetric with respect to the x axis, (1) and (2) are obtained. From (1) we can conclude that the sine is an odd function, and from (2) it follows that the cosine is an even function.

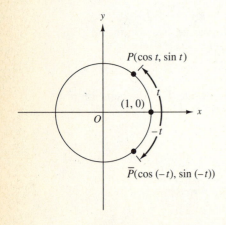

FIGURE 8

Notice that the graphs of the sine and cosine have the same shape. Actually, the graph of the cosine can be obtained by a horizontal translation of the graph of the sine a distance of $\frac{1}{2}\pi$ units to the left. Thus the graph of the cosine is also referred to as a sine wave, and again the portion of the graph for one period is a cycle.

By plotting the sine and cosine curves we can approximate sine and cosine function values.

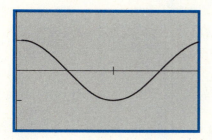

$[0, 2\pi]$ by $[-2, 2]$

$f(t) = \cos t$

FIGURE 9

▷ **ILLUSTRATION 1**

Let us find the value of cos 1.36 to three significant digits on our graphics calculator. Figure 9 shows the cosine curve in the viewing rectangle $[0, 2\pi]$ by $[-2, 2]$. Using trace and zoom-in we obtain

$$\cos 1.36 = 0.209$$ ◀

We can also use the sine and cosine curves and our graphics calculator to approximate values of t if either sin t or cos t is given.

▷ **ILLUSTRATION 2**

Suppose we wish to determine, to three significant digits, values of t in the interval $[0, 2\pi]$ for which

$$\sin t = -0.688$$

Figure 10 shows the sine curve and the line $g(t) = -0.688$ in the viewing rectangle $[0, 2\pi]$ by $[-2, 2]$. Observe that the line intersects the sine curve at two points in the interval $[0, 2\pi]$. Using trace and zoom-in, we get

$$\sin 3.90 = -0.688 \qquad \text{and} \qquad \sin 5.52 = -0.688$$ ◀

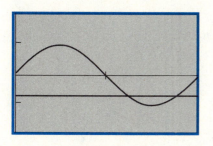

$[0, 2\pi]$ by $[-2, 2]$

$f(t) = \sin t$ and $y(t) = -0.688$

FIGURE 10

Other sine waves are obtained from functions defined by equations of the form

$$f(t) = a \sin b(t - c) \tag{3}$$

and

$$f(t) = a \cos b(t - c) \tag{4}$$

where a, b, and c are real numbers, $a \neq 0$, and $b \neq 0$. To learn how values of the constants a, b, and c affect the appearance of the sine wave defined by one of these equations, we consider special cases.

The function defined by

$$f(t) = a \sin t \tag{5}$$

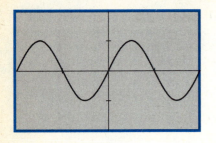

$[-2\pi, 2\pi]$ by $[-2, 2]$
$F(t) = \sin t$
(a)

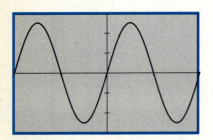

$[-2\pi, 2\pi]$ by $[-6, 6]$
$G(t) = 5 \sin t$
(b)

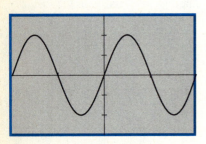

$[-2\pi, 2\pi]$ by $[-3, 3]$
$H(t) = 2 \sin t$
(c)
FIGURE 11

is the special case of (3) where $c = 0$ and $b = 1$. When $a > 0$, the factor a causes the graph of (5) to vertically stretch the sine curve by the factor. The ordinate of a point on the graph of (5) is a times the corresponding ordinate on the sine curve.

▷ **ILLUSTRATION 3**

Figure 11(a)–(c) shows graphs of the functions

$$F(t) = \sin t \qquad G(t) = 5 \sin t \qquad H(t) = 2 \sin t$$

Observe that the ordinates on the graphs of G and H are, respectively, 5 and 2 times the corresponding ordinate on the graph of F. ◀

Refer again to (5). Because

$$-1 \leq \sin t \leq 1$$

the maximum value of $f(t)$ is $|a|$, and the minimum value is $-|a|$. The number $|a|$ is called the **amplitude** of the sine wave.

▷ **ILLUSTRATION 4**

Suppose we wish to sketch the graph of

$$f(t) = 3 \sin t$$

The amplitude of the graph is 3, and each ordinate is 3 times the corresponding ordinate of the graph of $f(t) = \sin t$. In Figure 12 the graph of

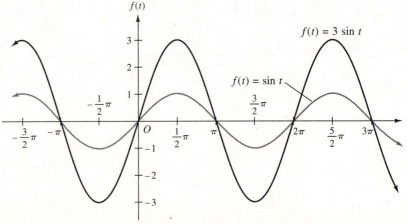

FIGURE 12

$f(t) = \sin t$ is shown by the lighter curve. By multiplying the ordinates on this curve by 3, we obtain the ordinates on the required curve.   ◀

The discussion relating to equation (5) also applies to graphs of functions defined by an equation of the form

$$f(t) = a \cos t \tag{6}$$

If $a < 0$ in either (5) or (6), the ordinates of points on the graph are the negatives of the corresponding ordinates on the graphs of $f(t) = |a| \sin t$ or $f(t) = |a| \cos t$, respectively.

▶ **EXAMPLE 1**   *Sketching Sine Waves*

Sketch the sine waves defined by the following equations and check the graphs on a graphics calculator: **(a)** $f(t) = -2 \cos t$; **(b)** $f(t) = \frac{1}{2} \cos t$.

Solution

(a) The equation $f(t) = -2 \cos t$ is of the form of (6) where $a = -2$. The amplitude of the graph is $|-2| = 2$. Each ordinate of the graph is -2 times the corresponding ordinate of the graph of $f(t) = \cos t$. The required curve appears in Figure 13, where the graph of $f(t) = \cos t$ is indicated by a lighter curve.

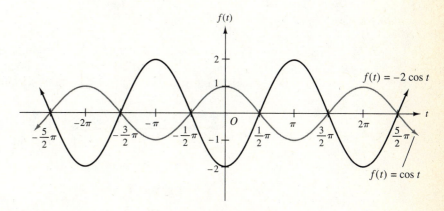

FIGURE 13

(b) The graph of the function defined by $f(t) = \frac{1}{2} \cos t$ has amplitude $\frac{1}{2}$. Each ordinate of the graph is one-half times the corresponding ordinate of the graph of $f(t) = \cos t$. Figure 14 shows the required curve, and the lighter curve denotes the graph of $f(t) = \cos t$.

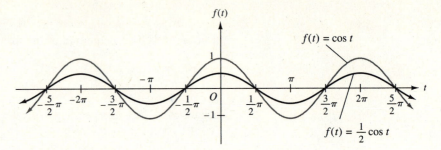

FIGURE 14

Our graphics calculator verifies the graphs in Figures 13 and 14.   ◄

Consider now an equation of the form

$$f(t) = \sin bt \qquad\qquad (7)$$

This function is the special case of $f(t) = a \sin b(t - c)$ where $c = 0$ and $a = 1$. Because the sine function has period 2π,

$$f(t) = \sin(bt + 2\pi)$$
$$= \sin\left[b\left(t + \frac{2\pi}{b}\right)\right]$$
$$= f\left(t + \frac{2\pi}{b}\right)$$

Therefore, f is a periodic function, and because, by definition, the period is positive, it is $2\pi/|b|$. The same argument applies to the function defined by $f(t) = \cos bt$. This result is stated as a theorem.

THEOREM 1

The period P of a periodic function defined by either

$$f(t) = \sin bt \qquad \text{or} \qquad f(t) = \cos bt$$

where $b \neq 0$, is given by

$$P = \frac{2\pi}{|b|}$$

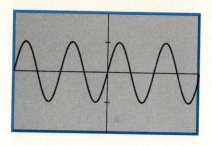

$[-\pi, \pi]$ by $[-2, 2]$
$f(t) = \sin 4t$
FIGURE 15

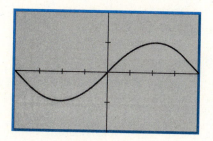

$[-4\pi, 4\pi]$ by $[-2, 2]$
$f(t) = \sin \dfrac{1}{4}t$
FIGURE 16

▷ **ILLUSTRATION 5**

From Theorem 1 the period of the function defined by

$$f(t) = \sin 4t$$

is $\frac{2\pi}{4} = \frac{1}{2}\pi$. Figure 15 shows the graph of f plotted in the viewing rectangle $[-\pi, \pi]$ by $[-2, 2]$. Observe that the graph repeats itself every interval of length $\frac{1}{2}\pi$ units. This fact is consistent with the period $\frac{1}{2}\pi$. ◄

▷ **ILLUSTRATION 6**

Figure 16 shows the graph of the function defined by

$$f(t) = \sin \tfrac{1}{4}t$$

plotted in the viewing rectangle $[-4\pi, 4\pi]$ by $[-2, 2]$. From Theorem 1, the period of this function is $\dfrac{2\pi}{\frac{1}{4}} = 8\pi$, which is consistent with Figure 16 showing one cycle of the corresponding sine wave. ◄

▷ **ILLUSTRATION 7**

To sketch the graph of

$$f(t) = \cos 2t$$

we first apply Theorem 1 to find the period P. Because $b = 2$, $P = \pi$. The amplitude is 1. The graph appears in Figure 17.

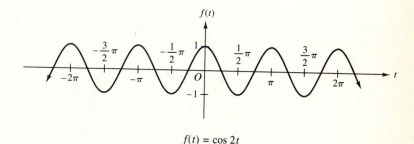

$f(t) = \cos 2t$

FIGURE 17 ◄

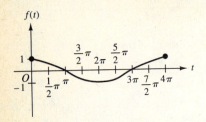

$$f(t) = \cos \frac{1}{2}t$$

FIGURE 18

▷ **ILLUSTRATION 8**

For the graph of

$$f(t) = \cos \tfrac{1}{2}t$$

the period is P, where from Theorem 1

$$P = \frac{2\pi}{\frac{1}{2}}$$
$$= 4\pi$$

The amplitude is 1. Figure 18 shows one cycle of this sine wave sketched on the interval $[0, 4\pi]$. ◀

▶ **EXAMPLE 2** *Sketching a Sine Wave*

Sketch the sine wave defined by

$$f(t) = \sin 3t$$

and check the graph on a graphics calculator.

Solution   The amplitude is 1. Because $b = 3$, from Theorem 1 the period is $\frac{2}{3}\pi$. In Figure 19 the units on the t axis are marked off on every interval of length $\frac{1}{3}\pi$. The figure shows the graph. Our graphics calculator verifies this graph.

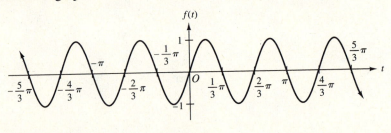

$$f(t) = \sin 3t$$

FIGURE 19 ◀

▶ **EXAMPLE 3** *Sketching One Cycle of a Sine Wave*

Sketch one cycle of the sine wave defined by

$$f(t) = \sin \tfrac{2}{3}t$$

Check the graph on a graphics calculator.

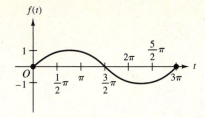

$$f(t) = \sin \frac{2}{3}t$$

FIGURE 20

Solution The amplitude is 1. We compute the period P from Theorem 1:

$$P = \frac{2\pi}{\frac{2}{3}}$$

$$= 3\pi$$

One cycle of the sine wave is, therefore, over the interval $[0, 3\pi]$. This cycle appears in Figure 20 and agrees with what we obtain on our graphics calculator. ◀

The period of the graph of $f(t) = \sin t$ is 2π. In Illustration 7 for the graph of $f(t) = \cos 2t$, the period is π, and in Example 2 the period of the graph of $f(t) = \sin 3t$ is $\frac{2}{3}\pi$. For the graphs of $f(t) = \sin bt$ and $f(t) = \cos bt$, as b increases when $b > 0$, the period decreases, and the cycles of the sine wave get closer together. Notice in Example 3 that the period of the graph of $f(t) = \sin \frac{2}{3}t$ is 3π. Furthermore, in Illustration 8 the period of the graph of $f(t) = \cos \frac{1}{2}t$ is 4π. As b decreases when $b > 0$, the period for the graphs of $f(t) = \sin bt$ and $f(t) = \cos bt$ gets larger. In particular, the graph of $f(t) = \sin \frac{1}{8}t$ has a period of 16π. Thus in the interval $[0, 16\pi]$ there is only one cycle of this sine wave.

What about the graphs of $f(t) = \sin bt$ and $f(t) = \cos bt$ when $b < 0$? To answer this question, we make use of identities (1) and (2). For example, the graph of $f(t) = \sin(-3t)$ is the same as the graph of $f(t) = -\sin 3t$, and the graph of $f(t) = \cos(-3t)$ is the same as the graph of $f(t) = \cos 3t$.

If in the equation $f(t) = a \sin b(t - c)$, $a = 1$ and $b = 1$, we have an equation of the form

$$f(t) = \sin(t - c) \tag{8}$$

The graph of this equation can be obtained from the graph of $f(t) = \sin t$ by a horizontal translation (or shift) c units to the right if $c > 0$ and $|c|$ units to the left if $c < 0$. The absolute value of c is called the **phase shift** of the graph of (8). The phase shift of a graph is also the phase shift of the corresponding function.

▷ **ILLUSTRATION 9**

Consider the graph of

$$f(t) = \sin(t - \tfrac{1}{4}\pi)$$

Here $c = \frac{1}{4}\pi$; thus the phase shift of this graph is $\frac{1}{4}\pi$. We obtain the graph by shifting the graph of $f(t) = \sin t$ a distance of $\frac{1}{4}\pi$ units to the right because $c > 0$. Figure 21 shows this graph. The graph of $f(t) = \sin t$ on the interval $[0, 2\pi]$ is indicated by a lighter curve.

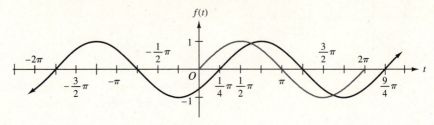

$$f(t) = \sin\left(t - \frac{1}{4}\pi\right)$$

FIGURE 21 ◄

▷ **ILLUSTRATION 10**

For the graph of

$$f(t) = \sin(t + \tfrac{1}{2}\pi)$$

$c = -\frac{1}{2}\pi$. Thus the phase shift is $\frac{1}{2}\pi$, and the graph is obtained by shifting the graph of $f(t) = \sin t$ a distance of $\frac{1}{2}\pi$ units to the left since $c < 0$. Figure 22 shows the graph as well as the graph of $f(t) = \sin t$ on the interval $[0, 2\pi]$, indicated by a lighter curve.

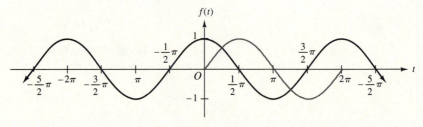

$$f(t) = \sin\left(t + \frac{1}{2}\pi\right)$$

FIGURE 22 ◄

We now apply the properties of the graphs of (5), (7), and (8) to obtain the graphs of the functions defined by equations of the form

$$f(t) = a \sin b(t - c) \quad \text{and} \quad f(t) = a \cos b(t - c)$$

▶ **EXAMPLE 4** *Sketching a Sine Wave*

Sketch the graph of

$$f(t) = \tfrac{1}{2}\sin(t + \pi)$$

Check the graph on a graphics calculator.

Solution   This equation is of the form $f(t) = a \sin b(t - c)$ where $a = \tfrac{1}{2}$, $b = 1$, and $c = -\pi$. Because $a = \tfrac{1}{2}$, the amplitude of the graph is $\tfrac{1}{2}$. Because $b = 1$, the period of the graph is 2π. Because $c = -\pi$, the phase shift is π, and the required graph is obtained by shifting the graph of $f(t) = \tfrac{1}{2}\sin t$ a distance of π units to the left. The required graph appears in Figure 23, and the graph of $f(t) = \tfrac{1}{2}\sin t$ on the interval $[0, 2\pi]$ is shown as a lighter curve. We obtain the same graph on our graphics calculator.

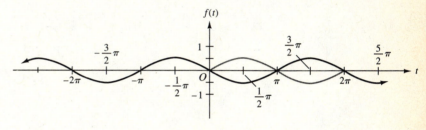

$$f(t) = \frac{1}{2}\sin(t + \pi)$$

FIGURE 23   ◀

▶ **EXAMPLE 5** *Sketching a Sine Wave*

Sketch the graph of

$$f(t) = 4\cos(2t - \tfrac{1}{2}\pi)$$

Check the graph on a graphics calculator.

Solution   We write the given equation in the form

$$f(t) = a \cos b(t - c)$$

by factoring 2 from the expression $2t - \tfrac{1}{2}\pi$. We obtain

$$f(t) = 4\cos 2(t - \tfrac{1}{4}\pi)$$

Because $a = 4$, the amplitude of the graph is 4. The period of the graph is π because $b = 2$. Since $c = \tfrac{1}{4}\pi$, the phase shift is $\tfrac{1}{4}\pi$, and the required graph is obtained by shifting the graph of $f(t) = 4\cos 2t$ a distance of $\tfrac{1}{4}\pi$

units to the right. Figure 24 shows the required graph, and the graph of $f(t) = 4 \cos 2t$ on the interval $[0, \pi]$ is represented by a lighter curve. The same graph appears on our graphics calculator.

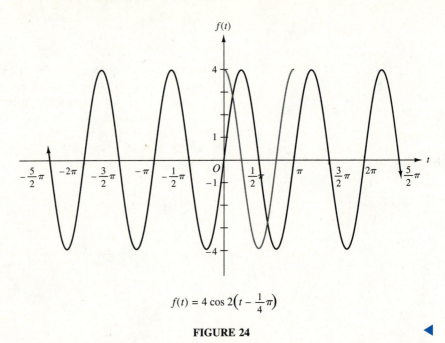

$$f(t) = 4 \cos 2\left(t - \frac{1}{4}\pi\right)$$

FIGURE 24 ◀

The function in the next example differs from those previously discussed because the period is a rational number instead of a multiple of π. In such a case, when sketching the graph, it is more convenient to use rational numbers for the tick marks on the t axis.

▶ **EXAMPLE 6** *Sketching a Sine Wave*

Sketch the graph of

$$f(t) = 2 \sin \tfrac{1}{2}\pi t$$

Check the graph on a graphics calculator.

Solution   We compare the given equation with $f(t) = a \sin bt$. Because $a = 2$, the amplitude is 2. If P is the period, then because $b = \frac{1}{2}\pi$,

$$P = \frac{2\pi}{\frac{1}{2}\pi}$$

$$= 4$$

The graph appears in Figure 25 as well as on our graphics calculator.

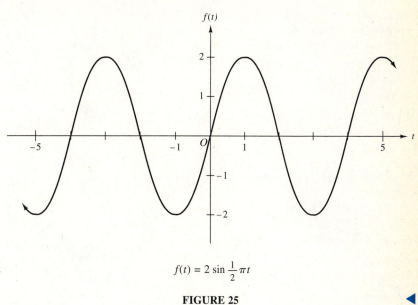

$$f(t) = 2 \sin \frac{1}{2} \pi t$$

FIGURE 25   ◀

▶ **EXAMPLE 7** *Sketching One Cycle of a Sine Wave*

Sketch one cycle of the sine wave given by the function of Example 3 in Section 7.2. Check the graph on a graphics calculator.

Solution   The function is defined by

$$E(t) = 50 \sin 120\pi t$$

and the period of E is $\frac{1}{60}$. The amplitude of E is 50. We use different scales on the two axes; on the t axis we take tick marks at every $\frac{1}{480}$ unit, and on the $E(t)$ axis we take tick marks at every 10 units. Table 5 gives values of $E(t)$ for every $\frac{1}{480}$ unit in the interval $[0, \frac{1}{60}]$. The cycle appears in Figure 26. We obtain the same curve on our graphics calculator.

Table 5

t	0	$\frac{1}{480}$	$\frac{1}{240}$	$\frac{1}{160}$	$\frac{1}{120}$	$\frac{1}{96}$	$\frac{1}{80}$	$\frac{7}{480}$	$\frac{1}{60}$
$E(t)$	0	35.4	50	35.4	0	-35.4	-50	-35.4	0

◀

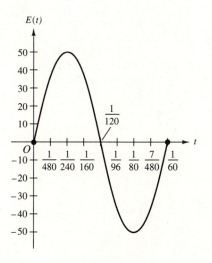

$$E(t) = 50 \sin 120\pi t$$

FIGURE 26

EXERCISES 7.3

In Exercises 1 through 4, use the method of Illustration 1 and your graphics calculator to approximate the function value to three significant digits. Then check your approximation by computing the function value by using the $\boxed{\text{sin}}$ *and* $\boxed{\text{cos}}$ *keys on your calculator.*

1. (a) sin 0.546   **(b)** cos 2.73

2. (a) sin 1.82   **(b)** cos 0.379

3. (a) sin(−1.28)   **(b)** cos 4.67

4. (a) sin 5.06   **(b)** cos(−2.41)

In Exercises 5 through 8, use the method of Illustration 2 and your graphics calculator to approximate the values of t to three significant digits such that $t \in [0, 2\pi]$ *for which the equation is satisfied.*

5. (a) sin t = 0.976   **(b)** cos t = 0.489

6. (a) sin t = 0.291   **(b)** cos t = 0.923

7. (a) sin t = −0.641   **(b)** cos t = −0.506

8. (a) sin t = −0.996   **(b)** cos t = −0.785

In Exercises 9 through 16, sketch the sine wave defined by the function and check your graph on your graphics calculator.

9. (a) $f(t) = 2 \sin t$ **(b)** $g(t) = \sin 2t$

10. (a) $f(t) = 3 \cos t$ **(b)** $g(t) = \cos 3t$

11. (a) $f(t) = \frac{1}{4} \cos t$ **(b)** $g(t) = \cos \frac{1}{4} t$

12. (a) $f(t) = \frac{1}{3} \sin t$ **(b)** $g(t) = \sin \frac{1}{3} t$

13. (a) $f(t) = -\sin t$ **(b)** $g(t) = \sin(-t)$

14. (a) $f(t) = -\cos t$ **(b)** $g(t) = \cos(-t)$

15. (a) $f(t) = 5 \cos \frac{1}{3} t$ **(b)** $g(t) = \frac{1}{3} \sin 5t$

16. (a) $f(t) = 4 \cos \frac{1}{2} t$ **(b)** $g(t) = \frac{1}{2} \sin 4t$

In Exercises 17 through 20, sketch one cycle of the sine wave defined by the function and check your graph on your graphics calculator.

17. $f(t) = \cos \frac{2}{3} t$ **18.** $f(t) = \sin \frac{2}{3} t$

19. $f(t) = 5 \sin \frac{4}{5} t$ **20.** $f(t) = 3 \cos \frac{4}{3} t$

In Exercises 21 through 42, sketch the sine wave defined by the function and check your graph on your graphics calcu- $\boxed{\iint \frac{dy}{dx}}$ *lator.*

21. $f(t) = 6 \cos \pi t$ **22.** $f(t) = 4 \sin \pi t$

23. $f(t) = 2 \cos \frac{1}{2} \pi t$ **24.** $f(t) = 3 \sin \frac{1}{3} \pi t$

25. $f(t) = \frac{1}{2} \sin 2\pi t$ **26.** $f(t) = \frac{1}{2} \cos \frac{1}{2} \pi t$

27. $f(t) = \cos(t + \frac{1}{2} \pi)$ **28.** $f(t) = \cos(t + \frac{1}{4} \pi)$

29. $f(t) = 2 \sin(t - \frac{1}{4} \pi)$ **30.** $f(t) = 3 \sin(t - \frac{1}{2} \pi)$

31. $f(t) = \cos(t - \frac{1}{3} \pi)$ **32.** $f(t) = \sin(t + \frac{1}{3} \pi)$

33. $f(t) = -3 \cos(t + \frac{1}{6} \pi)$ **34.** $f(t) = 6 \cos(t - \frac{1}{6} \pi)$

35. $f(t) = 2 \sin(\frac{3}{2} \pi - t)$ **36.** $f(t) = \frac{1}{2} \cos(\pi - t)$

37. $f(t) = 4 \sin(2t - \pi)$ **38.** $f(t) = 2 \cos(4t + \pi)$

39. $f(t) = 5 \cos(3t + \frac{1}{2} \pi)$

40. $f(t) = -2 \sin(3t - \frac{1}{2} \pi)$

41. $f(t) = 2 \sin(\frac{1}{3} \pi t + \frac{1}{2} \pi)$

42. $f(t) = 3 \cos(\frac{1}{2} \pi t + \frac{1}{3} \pi)$

In Exercises 43 through 46, sketch one cycle of the sine wave defined by the function in the indicated exercise of Exercises 7.2.

43. Exercise 27   **44.** Exercise 28

45. Exercise 29   **46.** Exercise 30

47. In a particular city on each of the dates January 15, 16, and 17, the temperature varied from −5° Celsius at 2 A.M. to 5° Celsius at 2 P.M. With the assumption that the graph is a sine wave, **(a)** write an equation defining $T(t)$ as a function of t where $T(t)$ degrees Celsius was the temperature t hours since 2 A.M., January 15, and $0 \le t \le 48$. What was the temperature at **(b)** 6 A.M., January 15; **(c)** 12 noon, January 15; **(d)** 4 P.M., January 15; and **(e)** 10 P.M., January 15? **(f)** sketch the graph of T.

48. In a certain city, at any particular time of day from October 1 through October 4, the Fahrenheit temperature was $T(t)$ degrees at t hours since midnight, September 30, where

$$T(t) = 60 - 15 \sin \tfrac{1}{12} \pi(8 - t)   0 \le t \le 96$$

(a) Determine the period of T. Find the temperature at **(b)** 8 A.M., October 1; **(c)** 12 noon, October 1; **(d)** 2 P.M., October 1; **(e)** 6 P.M., October 1; and **(f)** midnight, October 1. **(g)** Sketch the graph of T.

49. The function defined by

$$f(t) = \frac{\sin t}{t}$$

arises in calculus. The function is not defined when $t = 0$. In calculus, however, it is necessary to know the behavior of this function as t gets closer and closer to zero. Use a calculator to compute **(a)** $f(0.1)$ and $f(-0.1)$; **(b)** $f(0.06)$ and $f(-0.06)$; **(c)** $f(0.02)$ and $f(-0.02)$; **(d)** $f(0.01)$ and $f(-0.01)$; **(e)** $f(0.001)$ and $f(-0.001)$. **(f)** What value does $f(t)$ appear to be approaching as t gets closer and closer to zero? **(g)** Plot the graph of f on your graphics calculator to verify your answer in part (f).

50. The function defined by

$$g(t) = \frac{1 - \cos t}{t}$$

arises in calculus, and $g(t)$ is not defined when $t = 0$. Apply the instructions of Exercise 49 to this function.

7.4 APPLICATIONS OF SINE AND COSINE FUNCTIONS TO PERIODIC PHENOMENA

GOALS

1. Use sine and cosine functions as mathematical models describing simple harmonic motion.
2. Sketch graphs of functions describing simple harmonic motion.
3. Solve word problems involving simple harmonic motion.
4. Sketch graphs of functions describing other periodic phenomena.
5. Solve word problems involving other periodic phenomena.

The functions discussed in Section 7.3 and defined by

$$f(t) = a \sin b(t - c) \tag{1}$$

and

$$f(t) = a \cos b(t - c) \tag{2}$$

are mathematical models describing **simple harmonic motion,** either vibrating or oscillating.

An example of simple harmonic motion occurs when a weight is suspended from a spring and is vibrating vertically. Let $f(t)$ centimeters be the directed distance of the weight from its central, or rest, position after t seconds of time. See Figure 1, where a positive value of $f(t)$ indicates that the weight is above its central position. If on a rectangular cartesian coordinate system the function values $f(t)$ are plotted for specific values of t, then if friction is neglected, the resulting graph will have an equation of the form of (1) or (2). The constants a, b, and c are determined by the weight and the spring as well as by how the weight is set into motion. For instance, the further the weight is pulled down before it is released, the greater will be a, the amplitude of the motion. Furthermore, the stiffer the spring, the more rapidly the weight will vibrate and thus the smaller will be P, the period of the motion; recall that the constants b and P are related by the equation

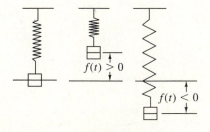

FIGURE 1

$P = 2\pi/|b|$. The **frequency** of a simple harmonic motion is the number of vibrations, or oscillations, per unit of time. Thus, if n is the frequency of the motion, $n = 1/P$.

▷ **ILLUSTRATION 1**

A weight is vibrating vertically according to the equation

$$f(t) = 8 \cos \tfrac{1}{3} \pi t \tag{3}$$

where $f(t)$ centimeters is the directed distance of the weight from its central position (the origin) at t seconds and the positive direction is upward. Equation (3) is the special case of (2) where $a = 8$, $b = \tfrac{1}{3}\pi$, and $c = 0$. Therefore, the motion is simple harmonic. Because the amplitude is 8, the maximum displacement is 8 cm. Because $b = \tfrac{1}{3}\pi$, the period P is given by

$$P = \frac{2\pi}{|b|}$$

$$= \frac{2\pi}{\tfrac{1}{3}\pi}$$

$$= 6$$

Therefore it takes 6 sec for one complete vibration of the weight. The frequency n is given by

$$n = \frac{1}{P}$$

$$= \tfrac{1}{6}$$

Thus there is $\tfrac{1}{6}$ of a vibration per second. From (3) we obtain values of $f(t)$ for the particular values of t shown in Table 1. From these values we can discuss the motion of the weight.

Table 1

t	0	$\frac{1}{2}$	1	$\frac{3}{2}$	2	$\frac{5}{2}$	3	$\frac{7}{2}$	4	$\frac{9}{2}$	5	$\frac{11}{2}$	6
$f(t)$	8	$4\sqrt{3} \approx 6.9$	4	0	-4	$-4\sqrt{3} \approx -6.9$	-8	-6.9	-4	0	4	6.9	8

Initially the weight is 8 cm above the origin, the central position. In the first $\tfrac{1}{2}$ sec the weight moves downward 1.1 cm to a point 6.9 cm above the origin. Then in the next $\tfrac{1}{2}$ sec the weight moves downward 2.9 cm to a position 4 cm above the origin. In the third $\tfrac{1}{2}$ sec the weight moves downward a distance of 4 cm to its central position. Thus the speed of the weight is increasing in the first $\tfrac{3}{2}$ sec. In the next $\tfrac{3}{2}$ sec the motion of the weight

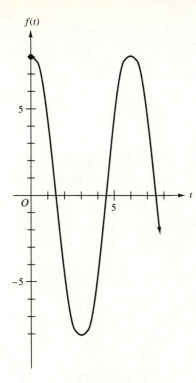

FIGURE 2

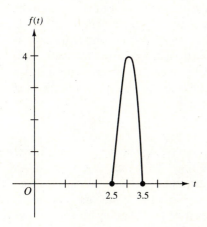

FIGURE 3

continues downward while its speed is decreasing until after a total of 3 sec the weight is 8 cm below its central position. Then the weight reverses its direction, and its speed increases until it attains its central position, following which the speed decreases until it is back to its starting position after a total of 6 sec. The weight then reverses its direction, and the motion down and up is repeated indefinitely. The graph of (3) appears in Figure 2. ◄

So far we have neglected friction, which would cause the weight eventually to come to rest. We discuss *damped harmonic motion,* for which friction is taken into account, in Section 7.5.

▶ **EXAMPLE 1** *Solving a Word Problem Involving Simple Harmonic Motion*

A weight suspended from a spring is vibrating vertically. Suppose the weight passes through its central position as it rises when $t = 2.5$, and then attains a maximum upward displacement of 4 cm, and passes through its central position as it descends when $t = 3.5$. The motion is simple harmonic described by an equation of the form

$$f(t) = a \sin b(t - c)$$

where $f(t)$ centimeters is the directed distance of the weight from its central position after t seconds and the positive direction is upward. Find this equation.

Solution The motion from $t = \frac{5}{2}$ to $t = \frac{7}{2}$ is indicated in Figure 3. Because the maximum upward displacement is 4 cm, the amplitude of the motion is 4; thus $a = 4$. One-half cycle of the motion is completed between $t = \frac{5}{2}$ and $t = \frac{7}{2}$. Therefore, if the period is P,

$$\frac{P}{2} = \frac{7}{2} - \frac{5}{2}$$

$$P = 2$$

Because $P = 2\pi/|b|$, we have, if $b > 0$,

$$\frac{2\pi}{b} = 2$$

$$b = \pi$$

The graph of the required function can be thought of as a sine wave shifted $\frac{5}{2}$ units to the right from the origin. Thus $c = \frac{5}{2}$. Therefore from the equation $f(t) = a \sin b(t - c)$ with $a = 4$, $b = \pi$, and $c = \frac{5}{2}$, we obtain

$$f(t) = 4 \sin \pi(t - \tfrac{5}{2})$$ ◄

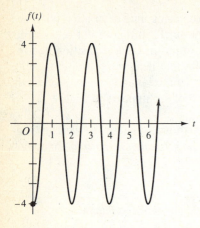

FIGURE 4

▶ **EXAMPLE 2** *Sketching the Graph of a Function Describing Simple Harmonic Motion and Discussing the Motion*

Sketch the graph of the function in Example 1 and check the graph on a graphics calculator. Discuss the motion.

Solution   The equation is

$$f(t) = 4 \sin \pi(t - \tfrac{5}{2})$$

The period is 2. Thus it takes 2 sec for one complete vibration of the weight. The frequency n is given by $n = 1/P$; thus $n = \tfrac{1}{2}$. Therefore there is $\tfrac{1}{2}$ of a vibration per second. Table 2 gives values of $f(t)$ for every $\tfrac{1}{4}$ unit when t is in the interval $[0, 2]$. The portion of the graph on this interval is drawn from these values, and the graph repeats this behavior on every interval of 2 units. See Figure 4, which agrees with the graph obtained on our graphics calculator.

Table 2

t	0	$\tfrac{1}{4}$	$\tfrac{1}{2}$	$\tfrac{3}{4}$	1	$\tfrac{5}{4}$	$\tfrac{3}{2}$	$\tfrac{7}{4}$	2
$f(t)$	-4	$-2\sqrt{2} \approx -2.8$	0	$2\sqrt{2} \approx 2.8$	4	2.8	0	-2.8	-4

Initially the weight is 4 cm below the central position. In the first $\tfrac{1}{4}$ sec the weight moves upward a distance of 1.2 cm to a point 2.8 cm below the central position. In the second $\tfrac{1}{4}$ sec the weight moves upward a distance of 2.8 cm to its central position. The speed is increasing in the first $\tfrac{1}{2}$ sec. In the next $\tfrac{1}{4}$ sec the motion of the weight continues upward while its speed is decreasing until after a total of 1 sec the weight is 4 cm above its central position. The weight then reverses its direction, and its speed increases until it attains its central position, after which the speed decreases until it has returned to its starting position after a total of 2 sec. This motion is repeated indefinitely every 2 sec. ◀

We now consider applications to other periodic phenomena.

If a wire bent in the form of a rectangle is rotated between the north and south poles of magnets, an alternating current of electricity is generated. As the wire makes one complete revolution, the current flows first in one direction and then in the opposite direction and repeats itself for each revolution. The current is periodic and can be described by an equation of the form

$$I(t) = a \sin b(t - c) \tag{4}$$

where $I(t)$ amperes is the current at t seconds produced by a generator.

▷ **ILLUSTRATION 2**

Suppose an alternating current of electricity is described by the equation

$$I(t) = 10 \sin 120\pi t$$

where $I(t)$ amperes is the current at t seconds. We compare this equation with (4). Because $a = 10$, the maximum current is 10 amperes. Because $b = 120\pi$, the frequency n is given by

$$n = \frac{120\pi}{2\pi}$$

$$= 60$$

Therefore the frequency is 60 cycles per second.   ◀

An electric current is produced by an electromotive force measured in volts. For a simple generator, if $E(t)$ volts is the electromotive force at t seconds, an equation describing $E(t)$ is

$$E(t) = a \sin bt \tag{5}$$

Such an equation occurred in Example 3 of Section 7.2.

▷ **ILLUSTRATION 3**

Suppose $E(t)$ volts is the electromotive force at t seconds and is described by an equation of the form of (5). If the maximum electromotive force is 110 volts and the frequency is 60 cycles per second, then $a = 110$, and

$$\frac{b}{2\pi} = 60$$

$$b = 120\pi$$

Thus an equation describing the electromotive force is

$$E(t) = 110 \sin 120\pi t$$ ◀

In a simple electric circuit containing only a resistance, the electromotive force and the current are related by the equation

$$E(t) = RI(t)$$

where R is a constant. This equation is known as Ohm's law, named for G. S. Ohm (1789–1854), and the resistance is R ohms.

▷ **ILLUSTRATION 4**

If the equations of Illustrations 2 and 3 pertain to the same electric circuit, then

$$E(t) = 11I(t)$$

and Ohm's law is satisfied where the resistance is 11 ohms.   ◀

Observe that in an electric circuit satisfying Ohm's law the maximum value of $E(t)$ occurs at the same time as the maximum value of $I(t)$. It is possible that these maximum values occur at different times. In such a case we say that the current and electromotive force are *out of phase*. Suppose, for instance, that

$$E(t) = a \sin bt$$

and

$$I(t) = A \sin b(t - c)$$

The graph of $I(t)$ can be obtained by shifting the graph of $f(t) = A \sin bt$ c units to the right if $c > 0$ and $|c|$ units to the left if $c < 0$. The current and electromotive force are out of phase by c units. If $c > 0$, we say that the current *lags* the electromotive force by c units, and if $c < 0$, we say the current *leads* the electromotive force by $|c|$ units.

▶ **EXAMPLE 3** *Sketching Graphs of Functions Defining Electromotive Force and Current in an Electric Circuit*

In a particular electric circuit, at t seconds, the electromotive force is $E(t)$ volts, where

$$E(t) = 150 \sin 120\pi t$$

and the current is $I(t)$ amperes, where

$$I(t) = 25 \sin 120\pi(t - \tfrac{1}{360})$$

Sketch the graphs of E and I on the same set of coordinate axes. Check the graphs on a graphics calculator. Does the current lag or lead the electromotive force and by how much?

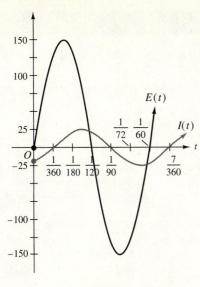

FIGURE 5

Solution   For each of the graphs the period is P, where

$$P = \frac{2\pi}{|b|}$$

$$= \frac{2\pi}{120\pi}$$

$$= \frac{1}{60}$$

We use different scales on the two axes. On the t axis we take the tick marks at every $\frac{1}{360}$ unit, and on the vertical axis we take the tick marks at every 25 units.

For the graph of E the amplitude is 150. The graph of E appears in Figure 5. For the graph of I the amplitude is 25. The graph of I, shown by the lighter curve in Figure 5, is obtained by shifting the graph of $f(t) = 25 \sin 120\pi t$ a distance of $\frac{1}{360}$ unit to the right. We obtain the same graphs on our graphics calculator.

Conclusion: The current lags the electromotive force by $\frac{1}{360}$ sec.   ◄

When a sound wave reaches the ear, there is a vibration of air particles at the eardrum. This vibration can be described by the variation of air pressure. A *simple sound* is one that produces on an oscilloscope a wave that can be described by an equation of the form

$$F(t) = a \sin bt \qquad\qquad\qquad \textbf{(6)}$$

where $F(t)$ dynes per square centimeter is the difference between the atmospheric pressure and the air pressure at the eardrum at t seconds. Positive values of $F(t)$ correspond to inward pressure on the eardrum, and negative values correspond to outward pressure. Equation (6) is an equation of the sound wave. The constant a gives the *pressure amplitude*.

▷ **ILLUSTRATION 5**

A tuning fork produces a simple sound having a pressure amplitude of 0.02 dyn/cm², and it is vibrating at 250 cycles per second. The sound wave produced has an equation of the form of (6). Because the pressure amplitude is 0.02 dyn/cm², $a = 0.02$. Because the frequency is 250 cycles per second,

$$\frac{b}{2\pi} = 250$$

$$b = 500\pi$$

Therefore an equation of the sound wave is

$$F(t) = 0.02 \sin 500\pi t \qquad\qquad ◄$$

EXERCISES 7.4

In Exercises 1 through 6, the equation describes the simple harmonic motion of a weight suspended from a spring and vibrating vertically, where f(t) centimeters is the directed distance of the weight from its central position (the origin) at t seconds and the positive direction is upward. For each motion, determine (a) the amplitude; (b) the period; (c) the frequency; and (d) the positions of the weight at t_1, t_2, t_3, and t_4 seconds.

1. $f(t) = 3 \cos \frac{1}{6} \pi t$; $t_1 = 0$, $t_2 = 2$, $t_3 = 4$, $t_4 = 6$
2. $f(t) = 2 \sin \frac{1}{2} \pi t$; $t_1 = 1$, $t_2 = 2$, $t_3 = 3$, $t_4 = 4$
3. $f(t) = 5 \sin 2t$; $t_1 = 0$, $t_2 = \frac{1}{4} \pi$, $t_3 = \frac{1}{2} \pi$, $t_4 = \frac{3}{4} \pi$
4. $f(t) = 6 \cos 3t$; $t_1 = 0$, $t_2 = \frac{1}{6} \pi$, $t_3 = \frac{1}{3} \pi$, $t_4 = \frac{1}{2} \pi$
5. $f(t) = 8 \cos \pi(2t - \frac{1}{3})$; $t_1 = 0$, $t_2 = \frac{1}{6}$, $t_3 = \frac{1}{3}$, $t_4 = \frac{1}{2}$
6. $f(t) = 3 \cos \pi(3t + \frac{1}{2})$; $t_1 = 0$, $t_2 = \frac{1}{6}$, $t_3 = \frac{1}{3}$, $t_4 = \frac{1}{2}$

In Exercises 7 through 10, sketch the graph of the function in the indicated exercise and check your graph on your graphics calculator. Discuss the motion.

7. Exercise 1
8. Exercise 2
9. Exercise 5
10. Exercise 6

In Exercises 11 through 14, the equation describes an alternating current of electricity, where I(t) amperes is the current at t seconds. In each exercise, (a) plot the graph of I for $t_1 \le t \le t_4$; determine (b) the maximum current; (c) the frequency; and (d) the current at t_1, t_2, t_3, and t_4 seconds.

11. $I(t) = 5 \sin 30\pi t$; $t_1 = 0.05$, $t_2 = 0.1$, $t_3 = 0.15$, $t_4 = 0.2$
12. $I(t) = 20 \sin 240\pi t$; $t_1 = 0.01$, $t_2 = 0.03$, $t_3 = 0.05$, $t_4 = 0.07$
13. $I(t) = 0.6 \sin 400t$; $t_1 = 1$, $t_2 = 3.5$, $t_3 = 8$, $t_4 = 10$
14. $I(t) = 0.3 \sin 650t$; $t_1 = 1$, $t_2 = 4$, $t_3 = 7$, $t_4 = 10$

In Exercises 15 through 18, E(t) volts is the electromotive force at t seconds. In each exercise, (a) plot the graph of E for $t_1 \le t \le t_4$; determine (b) the maximum electromotive force; (c) the frequency; and (d) the electromotive force at t_1, t_2, t_3, and t_4 seconds.

15. $E(t) = 220 \sin 120\pi t$; $t_1 = 0.01$, $t_2 = 0.05$, $t_3 = 0.06$, $t_4 = 0.1$

16. $E(t) = 110 \sin 120\pi t$; $t_1 = 0.02$, $t_2 = 0.03$, $t_3 = 0.04$, $t_4 = 0.05$
17. $E(t) = 8 \sin 332t$; $t_1 = 0.05$, $t_2 = 0.15$, $t_3 = 0.25$, $t_4 = 0.35$
18. $E(t) = 40 \sin 510t$; $t_1 = 1.2$, $t_2 = 2$, $t_3 = 2.8$, $t_4 = 3.6$

In Exercises 19 through 22, the equation describes a wave produced by a simple sound where F(t) dynes per square centimeter is the difference between the atmospheric pressure and the air pressure at the eardrum at t seconds. In each exercise, (a) plot the graph of F for $t_1 \le t \le t_4$; determine (b) the pressure amplitude; (c) the frequency; and (d) the difference between the atmospheric pressure and the air pressure at the eardrum at t_1, t_2, t_3, and t_4 seconds.

19. $F(t) = 0.005 \sin 880\pi t$; $t_1 = 0.0025$, $t_2 = 0.005$, $t_3 = 0.0075$, $t_4 = 0.01$
20. $F(t) = 0.04 \sin 200\pi t$; $t_1 = 0.001$, $t_2 = 0.002$, $t_3 = 0.003$, $t_4 = 0.004$
21. $F(t) = 0.02 \sin 600t$; $t_1 = 0.002$, $t_2 = 0.006$, $t_3 = 0.018$, $t_4 = 0.054$
22. $F(t) = 0.006 \sin 2400t$; $t_1 = 0.005$, $t_2 = 0.010$, $t_3 = 0.015$, $t_4 = 0.020$.

23. A weight suspended from a spring is lifted up to a point 2 cm above its central position and then released. It takes $\frac{1}{2}$ sec for the weight to complete one vibration. **(a)** Write an equation defining $f(t)$, where $f(t)$ centimeters is the directed distance of the weight from its central position t seconds after the start of the motion and the positive direction is upward. **(b)** Plot the graph of f. Use the graph to estimate the position of the weight **(c)** $\frac{1}{12}$ sec after the start of the motion and **(d)** $\frac{1}{4}$ sec after the start of the motion. Check your estimates in parts (c) and (d) by computing $f(\frac{1}{12})$ and $f(\frac{1}{4})$, respectively.

24. A weight suspended from a spring is set into vibratory motion by pulling it down 4 cm from its central position and then releasing it. It takes 1.5 sec for the weight to complete one vibration. **(a)** Write an equation defining $f(t)$, where $f(t)$ centimeters is the directed distance of the weight from its central position t seconds after the start of the motion and the positive direction is upward. **(b)** Plot the graph of f.

Use the graph to estimate the position of the weight **(c)** 1 sec after the start of the motion and **(d)** 2 sec after the start of the motion. Check your estimates in parts (c) and (d) by computing $f(1)$ and $f(2)$, respectively.

25. A weight suspended from a spring is vibrating vertically. The weight passes through its central position as it rises when $t = 2$, attains a maximum displacement of 9 cm, and passes through its central position as it descends when $t = 3.2$. The motion is simple harmonic described by an equation of the form $f(t) = a \cos b(t - c)$ where $f(t)$ centimeters is the directed distance of the weight from its central position after t seconds and the positive direction is upward. Find this equation and plot the graph of f.

26. A weight suspended from a spring is vibrating vertically. Suppose the weight passes through its central position at 3 sec and 7 sec. Between these times the weight attains twice a maximum displacement of 10 cm above its central position and attains once a maximum displacement of 10 cm below its central position. The motion is simple harmonic described by an equation of the form $f(t) = a \sin b(t - c)$ where $f(t)$ centimeters is the directed distance of the weight from its central position at t seconds and the positive direction is upward. Find this equation and plot the graph of f.

27. A 60-cycle alternating current is described by an equation of the form $I(t) = a \sin b(t - c)$, where $I(t)$ amperes is the current at t seconds. The maximum current is 20 amperes, and the current is 10 amperes for the first time when $t = \frac{1}{360}$. Write the equation and plot the graph of I.

28. An electric generator produces a 30-cycle alternating current reaching a maximum of 50 amperes and described by an equation of the form $I(t) = a \sin b(t - c)$, where $I(t)$ amperes is the current at t seconds and the current is 25 amperes for the first time at 0.01 sec. Write the equation and plot the graph of I.

29. In a certain electric circuit the electromotive force at t seconds is $E(t)$ volts, where $E(t) = 110 \sin 120\pi t$, and the current is $I(t)$ amperes, where $I(t) = 55 \sin 120\pi(t - \frac{1}{240})$. Sketch the graphs of E and I on the same set of coordinate axes. Check your graphs on your graphics calculator in the same viewing rectangle. Does the current lag or lead the electromotive force and by how much?

30. Do Exercise 29 if $E(t) = 220 \sin 60\pi t$ and $I(t) = 44 \sin 60\pi(t + \frac{1}{180})$.

31. The number of dynes per square centimeter in the difference between the atmospheric pressure and the air pressure at the eardrum at t seconds produced by a particular sound wave is described by the equation $F(t) = 5 \sin 100\pi(t - \frac{1}{400})$. **(a)** Plot the graph of F. Determine **(b)** the pressure amplitude and **(c)** the frequency. Find the difference between the atmospheric pressure and the air pressure at the eardrum at **(d)** 0 sec; **(e)** 0.01 sec; and **(f)** 0.015 sec.

32. Do Exercise 31 if $F(t) = 3 \sin 880\pi(t + \frac{1}{100})$.

 33. Suppose the motion of a particle along a straight line is simple harmonic and is described by an equation of the form $S(t) = a \sin b(t - c)$ where $S(t)$ centimeters is the displacement of the particle from a fixed point (the origin) at t seconds. Then, it is proved in calculus, if $V(t)$ centimeters per second is the velocity of the particle at t seconds, $V(t) = ab \cos b(t - c)$, and if $A(t)$ centimeters per second per second is the acceleration of the particle at t seconds, $A(t) = -ab^2 \sin b(t - c)$. If an equation describing the motion is $S(t) = 2 \sin \frac{1}{3}\pi(t - 1)$, determine the particle's position, velocity, and acceleration at **(a)** 0 sec; **(b)** 1 sec; **(c)** 2 sec; **(d)** 3 sec; and **(e)** 4 sec.

 34. Do Exercise 33 if $S(t) = 4 \sin \frac{1}{6}\pi(t + 2)$.

35. Discuss the simple harmonic motion of a weight described by the equation of Exercise 1, as we did in the concluding paragraph of Illustration 1.

36. Follow the instructions of Exercise 35 for the simple harmonic motion of a weight described by the equation of Exercise 2.

7.5 OTHER GRAPHS INVOLVING SINE AND COSINE FUNCTIONS

GOALS
1. Determine properties of the graph of a function describing damped harmonic motion.
2. Determine properties of the graph of a function describing resonance.
3. Sketch the graph of the function defined by $f(t) = \dfrac{\sin t}{t}$.

In Section 7.4, you learned that simple harmonic motion continues indefinitely, repeating a cycle every interval of length a period. For instance, in Example 1 of that section, the weight suspended from a spring moves vertically upward and downward, and one complete oscillation occurs every interval of 2 sec. In practice, however, friction would cause the amplitude of the motion to decrease until the weight finally came to rest. This is the case of **damped harmonic motion,** which can be described by the product of a sine function and a nonconstant function called a **damping factor.** The damping factor causes the decrease in amplitude. An important damping factor is an exponential function whose values approach zero as the independent variable increases without bound. The following example illustrates the effect of this factor.

▶ **EXAMPLE 1** *Determining Properties of the Graph of a Function Describing Damped Harmonic Motion*

The three functions $F, f,$ and G are defined by

$$F(t) = -e^{-t/4} \qquad f(t) = e^{-t/4} \sin 4t \qquad G(t) = e^{-t/4}$$

where $t \geq 0$. The function f is a mathematical model describing damped harmonic motion.

(a) Show that $F(t) \leq f(t) \leq G(t)$.
(b) Plot the graphs of the three functions in the viewing rectangle $[0, 2\pi]$ by $[-1, 1]$.
(c) Determine algebraically the values of t at all points of intersection of the graph of f with the graphs of F and G. Which of these values are in the interval $[0, 2\pi]$?
(d) Determine algebraically all t intercepts of the graph of f. Which of these intercepts are in the interval $[0, 2\pi]$?

Solution

(a) Function f is the product of two functions. Because $|\sin 4t| \le 1$ and $e^{-t/4} > 0$ for all t,

$$|f(t)| \le e^{-t/4} \qquad \text{for all } t$$

Thus

$$-e^{-t/4} \le f(t) \le e^{-t/4} \text{ for all } t$$

That is,

$$F(t) \le f(t) \le G(t)$$

(b) Figure 1 shows the required graphs. Observe that the graph of f is between the graphs of F and G.

(c) The graph of f intersects the graph of G when $\sin 4t = 1$. Because $\sin x = 1$ when $x = \frac{1}{2}\pi + 2k\pi$, $k \in J$, $\sin 4t = 1$ when

$$4t = \tfrac{1}{2}\pi + 2k\pi \quad k \in J$$
$$t = \tfrac{1}{8}\pi + \tfrac{1}{2}k\pi \quad k \in J$$

For values of t in $[0, 2\pi]$, we let k equal 0, 1, 2, and 3, and we obtain, respectively,

$$t = \tfrac{1}{8}\pi + 0\pi \qquad t = \tfrac{1}{8}\pi + \tfrac{1}{2}\pi \qquad t = \tfrac{1}{8}\pi + \pi \qquad t = \tfrac{1}{8}\pi + \tfrac{3}{2}\pi$$
$$= \tfrac{1}{8}\pi \qquad\qquad = \tfrac{5}{8}\pi \qquad\qquad = \tfrac{9}{8}\pi \qquad\qquad = \tfrac{13}{8}\pi$$

The graph of f intersects the graph of F when $\sin 4t = -1$. Because $\sin x = -1$ when $x = \frac{3}{2} + 2k\pi$, $k \in J$, $\sin 4t = -1$ when

$$4t = \tfrac{3}{2}\pi + 2k\pi \quad k \in J$$
$$t = \tfrac{3}{8}\pi + \tfrac{1}{2}k\pi \quad k \in J$$

For values of t in $[0, 2\pi]$, we let k equal 0, 1, 2, and 3, and we obtain, respectively,

$$t = \tfrac{3}{8}\pi + 0\pi \qquad t = \tfrac{3}{8}\pi + \tfrac{1}{2}\pi \qquad t = \tfrac{3}{8}\pi + \pi \qquad t = \tfrac{3}{8}\pi + \tfrac{3}{2}\pi$$
$$= \tfrac{3}{8}\pi \qquad\qquad = \tfrac{7}{8}\pi \qquad\qquad = \tfrac{11}{8}\pi \qquad\qquad = \tfrac{15}{8}\pi$$

(d) The graph of f intersects the t axis when $\sin 4t = 0$. Because $\sin x = 0$ when $x = k\pi$, $k \in J$, $\sin 4t = 0$ when

$$4t = k\pi \qquad k \in J$$
$$t = \tfrac{1}{4}k\pi \qquad k \in J$$

For values of t in $[0, 2\pi]$, we let k equal 0 through 8, and we obtain, respectively, the following values of t: 0, $\frac{1}{4}\pi$, $\frac{1}{2}\pi$, $\frac{3}{4}\pi$, π, $\frac{5}{4}\pi$, $\frac{3}{2}\pi$, $\frac{7}{4}\pi$, 2π.

Observe that the values of t in the interval $[0, 2\pi]$ obtained in parts (c) and (d) are consistent with those shown on the graph in part (b). ◀

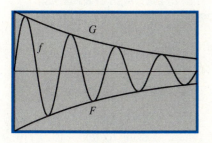

$[0, 2\pi]$ by $[-1, 1]$
$F(t) = -e^{-t/4}$, $f(t) = e^{-t/4} \sin 4t$,
and $G(t) = e^{-t/4}$

FIGURE 1

Damped harmonic motion is important in the design of buildings, bridges, and vehicles. For instance, shock absorbers are used to damp the oscillations when an automobile encounters a bump in the road.

For damped harmonic motion, the amplitude decreases to zero as time increases. If the amplitude increases without bound as time increases, then **resonance** occurs. The following example gives a mathematical model describing resonance.

▶ **EXAMPLE 2** *Determining Properties of the Graph of a Function Describing Resonance*

The three functions F, f, and G are defined by

$$F(t) = -2^t \qquad f(t) = 2^t \cos 4t \qquad G(t) = 2^t$$

where $t \geq 0$. The function f is a mathematical model describing resonance.

(a) Show that $F(t) \leq f(t) \leq G(t)$.
(b) Plot the graphs of the three functions in the viewing rectangle $[0, \pi]$ by $[-10, 10]$.
(c) Determine algebraically the values of t at all points of intersection of the graph of f with the graphs of F and G. Which of these values are in the interval $[0, \pi]$?
(d) Determine algebraically all t intercepts of the graph of f. Which of these intercepts are in the interval $[0, \pi]$?

Solution
(a) Because $|\cos 4t| \leq 1$ and $2^t > 0$ for all t,

$$|f(t)| \leq 2^t \qquad \text{for all } t$$

Hence

$$-2^t \leq f(t) \leq 2^t \qquad \text{for all } t$$

That is,

$$F(t) \leq f(t) \leq G(t)$$

(b) Figure 2 shows the graphs of f, F, and G plotted on our graphics calculator, where the graph of f is between the graphs of F and G.
(c) The graph of f intersects the graph of G when $\cos 4t = 1$. Because $\cos x = 1$ when $x = 2k\pi$, $k \in J$, $\cos 4t = 1$ when

$$4t = 2k\pi \qquad k \in J$$
$$t = \tfrac{1}{2}k\pi \qquad k \in J$$

For values of t in $[0, \pi]$, we let k equal 0, 1, and 2, and we obtain,

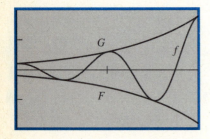

$[0, \pi]$ by $[-10, 10]$
$F(t) = -2^t$, $f(t) = 2^t \cos 4t$, and $G(t) = 2^t$

FIGURE 2

respectively, the following values of t: $0, \frac{1}{2}\pi, \pi$. The graph of f intersects the graph of F when $\cos 4t = -1$. Because $\cos x = -1$ when $x = (2k + 1)\pi$, $k \in J$, $\cos 4t = -1$ when

$$4t = (2k + 1)\pi \qquad k \in J$$
$$t = (\tfrac{1}{2}k + \tfrac{1}{4})\pi \qquad k \in J$$

For values of t in $[0, \pi]$, we let k equal 0 and 1, and we obtain, respectively, $t = \frac{1}{4}\pi$ and $t = \frac{3}{4}\pi$.

(d) The graph of f intersects the t axis when $\cos 4t = 0$. Because $\cos x = 0$ when $x = \frac{1}{2}\pi + k\pi$, $k \in J$, $\cos 4t = 0$ when

$$4t = \tfrac{1}{2}\pi + k\pi \qquad k \in J$$
$$t = \tfrac{1}{8}\pi + \tfrac{1}{4}k\pi \qquad k \in J$$

For values of t in $[0, \pi]$, we let k equal 0, 1, 2, and 3, and we obtain, respectively,

$$t = \tfrac{1}{8}\pi + 0\pi \qquad t = \tfrac{1}{8}\pi + \tfrac{1}{4}\pi \qquad t = \tfrac{1}{8}\pi + \tfrac{1}{2}\pi \qquad t = \tfrac{1}{8}\pi + \tfrac{3}{4}\pi$$
$$= \tfrac{1}{8}\pi \qquad\qquad = \tfrac{3}{8}\pi \qquad\qquad = \tfrac{5}{8}\pi \qquad\qquad = \tfrac{7}{8}\pi$$

Notice that the values of t in the interval $[0, \pi]$ obtained in parts (c) and (d) are consistent with those shown on the graph in part (b). ◀

In Exercise 49 of Exercises 7.3 you were given the function defined by

$$f(t) = \frac{\sin t}{t}$$

 and you were asked to compute values of $f(t)$ for small values of t. These function values suggested that $f(t)$ approaches 1 as t approaches 0, a fact with important consequences in calculus. In that exercise you were also asked to plot the graph of f on your graphics calculator. In the following example we show how to sketch the graph of the function by hand.

 ▶ **EXAMPLE 3**   *Sketching the Graph of a Function Important in Calculus*

Sketch the graph of the function defined by

$$f(t) = \frac{\sin t}{t}$$

for $t \in [-\pi, 0) \cup (0, \pi]$.

Solution Observe that $f(0)$ does not exist. Thus the graph does not intersect the vertical axis. Furthermore, because $\sin(-t) = -\sin t$, we have

$$f(-t) = \frac{\sin(-t)}{-t}$$

$$= \frac{-\sin t}{-t}$$

$$= \frac{\sin t}{t}$$

$$= f(t)$$

Hence f is an even function; so its graph is symmetric with respect to the vertical axis. We shall first obtain the portion of the graph for $t \in (0, \pi]$. If $t > 0$, then because $|\sin t| \leq 1$,

$$-\frac{1}{t} \leq \frac{\sin t}{t} \leq \frac{1}{t}$$

Let

$$F(t) = -\frac{1}{t} \quad \text{and} \quad G(t) = \frac{1}{t}$$

Then if $t > 0$,

$$F(t) \leq f(t) \leq G(t)$$

and for positive values of t the graph of f lies between the graphs of F and G. Because $\sin \frac{1}{2}\pi = 1$, the graph of f intersects the graph of G at $t = \frac{1}{2}\pi$. Because $\sin \pi = 0$, the graph of f intersects the t axis at $t = \pi$. Also, when $0 < t < \pi$, $\sin t > 0$. Therefore, on the interval $(0, \pi)$ the graph lies above the t axis. Values for $f(t)$ on $(0, \pi)$ are found and points are located. From this information the graph of f on $(0, \pi]$ is drawn as shown in Figure 3. The portion of the graph on $[-\pi, 0)$ follows from properties of symmetry. The open dot on the vertical axis indicates that $f(0)$ is not defined. ◄

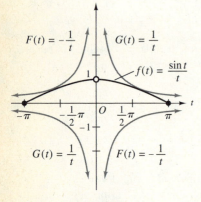

FIGURE 3

EXERCISES 7.5

In Exercises 1 through 8, the function f is a mathematical model describing damped harmonic motion. In each exercise do the following: (a) Show that $F(t) \leq f(t) \leq G(t)$. (b) Plot the graphs of the three functions in the viewing rectangle $[0, 2\pi]$ by $[-1, 1]$. (c) Determine algebraically the values of t at all points of intersection of the graph of f with the graphs of F and G. Which of these values are in the interval $[0, 2\pi]$? (d) Determine algebraically all t intercepts of the graph of f. Which of these intercepts are in the interval $[0, 2\pi]$?

1. $f(t) = e^{-t/2} \sin 2t$; $F(t) = -e^{-t/2}$; $G(t) = e^{-t/2}$
2. $f(t) = e^{-t/2} \cos 4t$; $F(t) = -e^{-t/2}$; $G(t) = e^{-t/2}$
3. $f(t) = 2^{-t/4} \cos 4t$; $F(t) = -2^{-t/4}$; $G(t) = 2^{-t/4}$
4. $f(t) = 2^{-t/4} \sin 2t$; $F(t) = -2^{-t/4}$; $G(t) = 2^{-t/4}$
5. $f(t) = e^{-t/8} \sin 3t$; $F(t) = -e^{-t/8}$; $G(t) = e^{-t/8}$
6. $f(t) = e^{-t/6} \cos 3t$; $F(t) = -e^{-t/6}$; $G(t) = e^{-t/6}$
7. $f(t) = e^{-t/4} \cos 8t$; $F(t) = -e^{-t/4}$; $G(t) = e^{-t/4}$
8. $f(t) = 3^{-t/3} \sin 6t$; $F(t) = -3^{-t/3}$; $G(t) = 3^{-t/3}$

In Exercises 9 through 16, the function f is a mathematical model describing resonance. In each exercise do the following: (a) Show that $F(t) \le f(t) \le G(t)$. (b) Plot the graphs of the three functions in the indicated viewing rectangle. (c) Determine algebraically the values of t at all points of intersection of the graph of f with the graphs of F and G. Which of these values of t give intersection points appearing on your graph in part (b)? (d) Determine algebraically all t intercepts of the graph of f. Which of these intercepts appear on your graph in part (b)?

9. $f(t) = 2^t \sin 4t$; $F(t) = -2^t$; $G(t) = 2^t$; $[0, \pi]$ by $[-10, 10]$

10. $f(t) = 2^t \cos 5t$; $F(t) = -2^t$; $G(t) = 2^t$; $[0, \pi]$ by $[-10, 10]$

11. $f(t) = e^t \cos 6t$; $F(t) = -e^t$; $G(t) = e^t$; $[0, \pi]$ by $[-30, 30]$

12. $f(t) = e^{t/2} \sin 6t$; $F(t) = -e^{t/2}$; $G(t) = e^{t/2}$; $[0, \pi]$ by $[-10, 10]$

13. $f(t) = t \sin 2t$; $F(t) = -t$; $G(t) = t$; $[0, 2\pi]$ by $[-10, 10]$

14. $f(t) = t \cos 3t$; $F(t) = -t$; $G(t) = t$; $[0, 2\pi]$ by $[-10, 10]$

15. $f(t) = 4^{t/6} \cos 6t$; $F(t) = -4^{t/6}$; $G(t) = 4^{t/6}$; $[0, 2\pi]$ by $[-5, 5]$

16. $f(t) = 3^{t/3} \sin 8t$; $F(t) = -3^{t/3}$; $G(t) = 3^{t/3}$; $[0, 2\pi]$ by $[-10, 10]$

17. The function defined by

$$f(t) = \frac{1 - \cos t}{t}$$

is important in calculus. Observe that $f(0)$ is not defined. In Exercise 50 of Exercises 7.3, you were asked to compute values of $f(t)$ for small values of t. These values suggested that $f(t)$ approaches 0 as t approaches 0. In that exercise you were asked to plot the graph of f on your graphics calculator. Now sketch by hand the graph of f for $t \in [-\pi, 0) \cup (0, \pi]$.

In Exercises 18 through 22, (a) plot the graph of the function for $t \in [-\pi, 0) \cup (0, \pi]$. (b) The function is not defined at 0, but what value does $f(t)$ appear to be approaching as t approaches 0? Compute the following values on your calculator: (c) $f(0.1)$ and $f(-0.1)$; (d) $f(0.01)$ and $f(-0.01)$; (e) $f(0.001)$ and $f(-0.001)$. (f) Do the function values in parts (c)–(e) agree with your answer in part (b)?

18. $f(t) = \dfrac{1 - \cos t}{\sin t}$

19. $f(t) = \dfrac{1 - \cos t}{t^2}$

20. $f(t) = \dfrac{\sin 2t}{t}$

21. $f(t) = \dfrac{2 \sin t}{t \cos t}$

22. $f(t) = \dfrac{1 - \cos 4t}{t^2}$

23. Let $f(t) = \cos^2 t$ and $g(t) = \frac{1}{2} + \frac{1}{2} \cos 2t$. **(a)** Plot the graph of f in the viewing rectangle $[-2\pi, 2\pi]$ by $[-2, 2]$ and sketch what you see. **(b)** Plot the graph of g in the viewing rectangle $[-2\pi, 2\pi]$ by $[-2, 2]$ and sketch what you see. **(c)** How do your graphs in (a) and (b) compare? Check by plotting the graphs of f and g in the same viewing rectangle.

24. Follow the instructions of Exercise 23 if $f(t) = \sin^2 t$ and $g(t) = \frac{1}{2} - \frac{1}{2} \cos 2t$.

7.6 THE TANGENT, COTANGENT, SECANT, AND COSECANT FUNCTIONS

GOALS

1. **Define the tangent function.**
2. **Define the cotangent function.**
3. **Define the secant and cosecant functions.**
4. **Compute exact tangent, cotangent, secant, and cosecant function values of quadrantal numbers when they are defined.**
5. **Compute exact tangent, cotangent, secant, and cosecant function values of $\frac{1}{4}\pi$, $\frac{1}{6}\pi$, and $\frac{1}{3}\pi$.**
6. **Approximate tangent, cotangent, secant, and cosecant function values on a calculator.**

7. Determine the quadrant containing the terminal point of an arc from signs of trigonometric functions of the arc's length.

8. Learn the eight fundamental trigonometric identities.

9. Find the other five trigonometric functions of a number when one function of the number is given.

10. Write a trigonometric expression in terms of sine and cosine functions.

Four other trigonometric (or circular) functions are defined in terms of the sine and cosine functions. The first of these is the *tangent* (abbreviated *tan*), the quotient of the sine and cosine.

DEFINITION **The Tangent Function**

If t is a real number and $\cos t \neq 0$, then

$$\tan t = \frac{\sin t}{\cos t}$$

The values of t for which $\cos t = 0$ are $\frac{1}{2}\pi + k\pi$, where $k \in J$. In particular, when $k = 0$, $\frac{1}{2}\pi + k\pi = \frac{1}{2}\pi$; when $k = 1$, $\frac{1}{2}\pi + k\pi = \frac{3}{2}\pi$; when $k = -1$, $\frac{1}{2}\pi + k\pi = -\frac{1}{2}\pi$; when $k = 2$, $\frac{1}{2}\pi + k\pi = \frac{5}{2}\pi$; and so on. These numbers are excluded from the domain of the tangent function because zero cannot be used as a divisor. Thus

domain of tangent is $\{t \mid t \neq \frac{1}{2}\pi + k\pi, k \in J\}$

Because $-1 \leq \cos t \leq 1$ and $-1 \leq \sin t \leq 1$, $|\tan t|$ can be made as large as we please by taking values of t so that $\cos t$ is sufficiently close to zero. In particular, if t is close to $\frac{1}{2}\pi$, $\sin t$ is close to 1 and $|\cos t|$ is close to zero. Thus $|\tan t|$ is large. When t is close to $\frac{1}{2}\pi$ and less than $\frac{1}{2}\pi$, both $\sin t$ and $\cos t$ are positive, and so $\tan t$ is positive. When t is close to $\frac{1}{2}\pi$ and greater than $\frac{1}{2}\pi$, $\sin t$ is positive and $\cos t$ is negative, and so $\tan t$ is negative. It follows that the range of the tangent function is the set of real numbers.

By using the definition of $\tan t$ and values of $\sin t$ and $\cos t$ given in Table 1 of Section 7.1, we obtain $\tan t$ at the quadrantal numbers $0, \frac{1}{2}\pi, \pi,$ and $\frac{3}{2}\pi$. We have

$$\tan 0 = \frac{\sin 0}{\cos 0} \qquad \tan \pi = \frac{\sin \pi}{\cos \pi}$$

$$= \frac{0}{1} \qquad\qquad = \frac{0}{-1}$$

$$= 0 \qquad\qquad\quad = 0$$

Table 1

t	$\sin t$	$\cos t$	$\tan t$
0	0	1	0
$\frac{1}{2}\pi$	1	0	undefined
π	0	-1	0
$\frac{3}{2}\pi$	-1	0	undefined

Table 2

t	$\tan t$
0	0
$\frac{1}{6}\pi$	$\dfrac{1}{\sqrt{3}}$
$\frac{1}{4}\pi$	1
$\frac{1}{3}\pi$	$\sqrt{3}$
$\frac{1}{2}\pi$	Undefined

Because $\cos \frac{1}{2}\pi$ and $\cos \frac{3}{2}\pi$ are zero, $\tan \frac{1}{2}\pi$ and $\tan \frac{3}{2}\pi$ are undefined. These results are summarized in Table 1.

We can determine $\tan t$ when t is $\frac{1}{6}\pi$, $\frac{1}{4}\pi$, and $\frac{1}{3}\pi$ from the sine and cosine of these numbers given in Table 2 of Section 7.1. Therefore

$$\tan \tfrac{1}{6}\pi = \frac{\sin \frac{1}{6}\pi}{\cos \frac{1}{6}\pi} \qquad \tan \tfrac{1}{4}\pi = \frac{\sin \frac{1}{4}\pi}{\cos \frac{1}{4}\pi} \qquad \tan \tfrac{1}{3}\pi = \frac{\sin \frac{1}{3}\pi}{\cos \frac{1}{3}\pi}$$

$$= \frac{\frac{1}{2}}{\frac{\sqrt{3}}{2}} \qquad\qquad = \frac{\frac{1}{\sqrt{2}}}{\frac{1}{\sqrt{2}}} \qquad\qquad = \frac{\frac{\sqrt{3}}{2}}{\frac{1}{2}}$$

$$= \frac{1}{\sqrt{3}} \qquad\qquad = 1 \qquad\qquad = \sqrt{3}$$

These values as well as those for $\tan 0$ and $\tan \frac{1}{2}\pi$ are summarized in Table 2.

Because $(\cos t, \sin t)$ is a point (x, y) on U,

$$\tan t = \frac{y}{x} \qquad \text{if } x \neq 0$$

Thus the value of $\tan t$ can be found by determining the coordinates of the terminal point of the arc having length t.

▷ ILLUSTRATION 1

Figure 1 shows the three points on U for which $t = \frac{2}{3}\pi$, $t = -\frac{1}{6}\pi$, and $t = -\frac{3}{4}\pi$. These points are symmetric to the points in the first quadrant for which $t = \frac{1}{3}\pi$, $t = \frac{1}{6}\pi$, and $t = \frac{1}{4}\pi$, respectively. From the coordinates of these points, we obtain

$$\tan \frac{2}{3}\pi = \frac{y}{x} \qquad \tan\left(-\frac{1}{6}\pi\right) = \frac{y}{x} \qquad \tan\left(-\frac{3}{4}\pi\right) = \frac{y}{x}$$

$$= \frac{\frac{\sqrt{3}}{2}}{-\frac{1}{2}} \qquad\qquad = \frac{-\frac{1}{2}}{\frac{\sqrt{3}}{2}} \qquad\qquad = \frac{-\frac{1}{\sqrt{2}}}{-\frac{1}{\sqrt{2}}}$$

$$= -\sqrt{3} \qquad\qquad = -\frac{1}{\sqrt{3}} \qquad\qquad = 1 \quad ◀$$

FIGURE 1

The *cotangent* (abbreviated *cot*) function is defined as the quotient of the cosine and sine.

DEFINITION   **The Cotangent Function**

If t is a real number and $\sin t \neq 0$, then

$$\cot t = \frac{\cos t}{\sin t}$$

The values of t for which $\sin t = 0$ are $k\pi$, where $k \in J$, and these numbers are excluded from the domain of the cotangent function. Hence

domain of cotangent is $\{t \mid t \neq k\pi, k \in J\}$

As was the case of the tangent, the range of the cotangent is the set of real numbers.

The function values for cotangent when t is $0, \frac{1}{6}\pi, \frac{1}{4}\pi, \frac{1}{3}\pi, \frac{1}{2}\pi, \pi$, and $\frac{3}{2}\pi$ can be found by using the definition and the function values for sine and cosine given in Tables 1 and 2 of Section 7.1.

▷ **ILLUSTRATION 2**

(a) $\cot \frac{1}{6}\pi = \dfrac{\cos \frac{1}{6}\pi}{\sin \frac{1}{6}\pi}$ **(b)** $\cot \frac{1}{4}\pi = \dfrac{\cos \frac{1}{4}\pi}{\sin \frac{1}{4}\pi}$ **(c)** $\cot \frac{1}{2}\pi = \dfrac{\cos \frac{1}{2}\pi}{\sin \frac{1}{2}\pi}$

$$= \dfrac{\dfrac{\sqrt{3}}{2}}{\dfrac{1}{2}} \qquad\qquad = \dfrac{\dfrac{1}{\sqrt{2}}}{\dfrac{1}{\sqrt{2}}} \qquad\qquad = \dfrac{0}{1}$$

$$= \sqrt{3} \qquad\qquad\qquad = 1 \qquad\qquad\qquad = 0$$

Because $\sin 0$ and $\sin \pi$ are zero, $\cot 0$ and $\cot \pi$ are undefined.   ◀

When neither $\sin t$ nor $\cos t$ is zero,

$$\frac{\cos t}{\sin t} = \frac{1}{\dfrac{\sin t}{\cos t}}$$

Therefore

$$\cot t = \frac{1}{\tan t} \qquad \text{if } t \neq \tfrac{1}{2}k\pi, k \in J$$

This equation is an alternate definition of the cotangent, and because of the relationship in it, the tangent and cotangent are **reciprocal functions.** The sine and cosine also have reciprocal functions, and they are the *cosecant* (abbreviated *csc*) and *secant* (abbreviated *sec*), respectively.

DEFINITION   The Secant and Cosecant Functions

If t is a real number, then

(i) $\sec t = \dfrac{1}{\cos t}$ if $t \neq \frac{1}{2}\pi + k\pi, k \in J$

(ii) $\csc t = \dfrac{1}{\sin t}$ if $t \neq k\pi, k \in J$

Observe from the definition that

domain of secant is $\{t \mid t \neq \frac{1}{2}\pi + k\pi, k \in J\}$

and

domain of cosecant is $\{t \mid t \neq k\pi, k \in J\}$

Because $|\cos t| \leq 1$ for all real numbers t, then $|\sec t| \geq 1$ for all t in the domain of the secant. Thus the range of the secant is $(-\infty, -1] \cup [1, +\infty)$. Similarly, because $|\sin t| \leq 1$ for all real numbers t, then $|\csc t| \geq 1$ for all t in the domain of the cosecant, and so the range of the cosecant is also the set $(-\infty, -1] \cup [1, +\infty)$.

▷ **ILLUSTRATION 3**

(a) $\sec 0 = \dfrac{1}{\cos 0}$

$= \dfrac{1}{1}$

$= 1$

(b) $\sec \frac{1}{3}\pi = \dfrac{1}{\cos \frac{1}{3}\pi}$

$= \dfrac{1}{\frac{1}{2}}$

$= 2$

(c) $\csc \frac{3}{2}\pi = \dfrac{1}{\sin \frac{3}{2}\pi}$

$= \dfrac{1}{-1}$

$= -1$

(d) $\csc\left(-\frac{5}{4}\pi\right) = \dfrac{1}{\sin\left(-\frac{5}{4}\pi\right)}$

$= \dfrac{1}{\dfrac{1}{\sqrt{2}}}$

$= \sqrt{2}$

(e) $\sec(-\frac{1}{6}\pi) = \dfrac{1}{\cos(-\frac{1}{6}\pi)}$

$= \dfrac{1}{\dfrac{\sqrt{3}}{2}}$

$= \dfrac{2}{\sqrt{3}}$

(f) $\csc(-\frac{1}{3}\pi) = \dfrac{1}{\sin(-\frac{1}{3}\pi)}$

$= \dfrac{1}{-\dfrac{\sqrt{3}}{2}}$

$= -\dfrac{2}{\sqrt{3}}$

Because $\cos \frac{1}{2}\pi$ and $\cos \frac{3}{2}\pi$ are zero, $\sec \frac{1}{2}\pi$ and $\sec \frac{3}{2}\pi$ are undefined. Furthermore, because $\sin 0$ and $\sin \pi$ are zero, $\csc 0$ and $\csc \pi$ are undefined. ◀

Again using the fact that $(\cos t, \sin t)$ is a point (x, y) on U,

$$\cot t = \frac{x}{y} \ \text{ if } y \neq 0 \qquad \sec t = \frac{1}{x} \ \text{ if } x \neq 0 \qquad \csc t = \frac{1}{y} \ \text{ if } y \neq 0$$

Therefore the values of these functions can be found by determining the coordinates of the terminal point of the arc having length t.

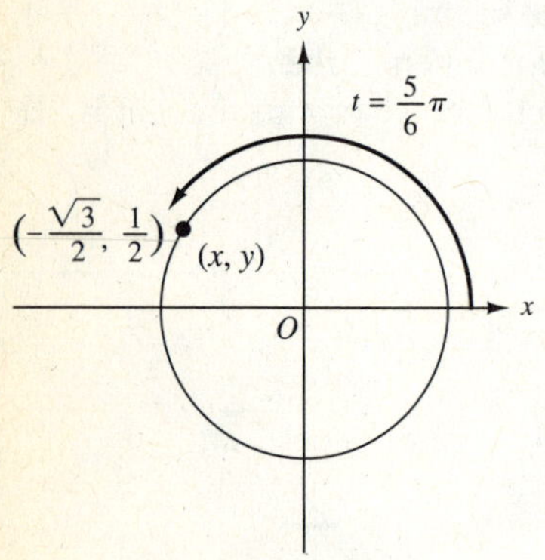

▶ **EXAMPLE 1**   *Finding Exact Trigonometric Function Values from the Coordinates of a Point on U*

Find the tangent, cotangent, secant, and cosecant of $\frac{5}{6}\pi$ by determining the coordinates of the terminal point of the arc having length $\frac{5}{6}\pi$.

Solution   Figure 2 shows the point on U for which $t = \frac{5}{6}\pi$, and for this point, (x, y) is $(-\frac{\sqrt{3}}{2}, \frac{1}{2})$. Therefore

FIGURE 2

$$\tan \frac{5}{6}\pi = \frac{y}{x} \qquad \cot \frac{5}{6}\pi = \frac{x}{y} \qquad \sec \frac{5}{6}\pi = \frac{1}{x} \qquad \csc \frac{5}{6}\pi = \frac{1}{y}$$

$$= \frac{\dfrac{1}{2}}{-\dfrac{\sqrt{3}}{2}} \qquad\qquad = \frac{-\dfrac{\sqrt{3}}{2}}{\dfrac{1}{2}} \qquad\qquad = \frac{1}{-\dfrac{\sqrt{3}}{2}} \qquad\qquad = \frac{1}{\dfrac{1}{2}}$$

$$= -\frac{1}{\sqrt{3}} \qquad\qquad = -\sqrt{3} \qquad\qquad = -\frac{2}{\sqrt{3}} \qquad\qquad = 2$$

◀

▷ **ILLUSTRATION 4**

To evaluate $\tan 1.205$ on a calculator in the radian mode, we apply the $\boxed{\tan}$

key and read

$$\tan 1.205 \approx 2.610727883$$

Rounding off the result to four significant digits, we get

$$\tan 1.205 \approx 2.611 \qquad \blacktriangleleft$$

Calculators generally do not have keys for cotangent, secant, and cosecant. To obtain values of $\cot t$, $\sec t$, and $\csc t$, we use the identities

$$\cot t = \frac{1}{\tan t} \qquad \sec t = \frac{1}{\cos t} \qquad \csc t = \frac{1}{\sin t}$$

▷ **ILLUSTRATION 5**

To find $\csc 0.562$, we apply the equality

$$\csc 0.562 = \frac{1}{\sin 0.562}$$

and from our calculator in the radian mode, we have

$$\csc 0.562 \approx 1.876596357$$

Rounding off the result to four significant digits gives

$$\csc 0.562 \approx 1.877 \qquad \blacktriangleleft$$

▶ **EXAMPLE 2**   *Approximating Trigonometric Function Values on a Calculator*

Approximate to four significant digits each of the following: **(a)** $\tan 2$; **(b)** $\cot \frac{2}{7}\pi$; **(c)** $\sec 4.391$; **(d)** $\csc(-3.672)$.

Solution   Be sure the calculator is in the radian mode. We round off each displayed entry to four significant digits.

(a) $\tan 2 \approx -2.185$

(b) $\cot \frac{2}{7}\pi = \dfrac{1}{\tan \frac{2}{7}\pi}$

$\qquad\qquad \approx 0.7975$

(c) $\sec 4.391 = \dfrac{1}{\cos 4.391}$

$\qquad\qquad \approx -3.166$

(d) $\csc(-3.672) = \dfrac{1}{\sin(-3.672)}$

$\qquad\qquad\qquad \approx 1.977 \qquad \blacktriangleleft$

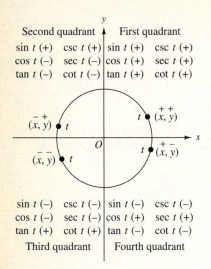

FIGURE 3

The sign ($+$ or $-$) of each of the six trigonometric functions of the number t is determined by the quadrant containing the terminal point of the arc of length t. Figure 3 shows a point on U in each of the four quadrants. Above the variables x and y are their signs, determined by the quadrant in which the point lies. Also indicated in the figure is the sign of the function in the particular quadrant. Table 3 summarizes the results given in the figure for the signs. Observe that when the sign of a function is determined for a specific quadrant, its reciprocal function will have the same sign in that quadrant.

Table 3

Quadrant	sin t	cos t	tan t	cot t	sec t	csc t
First	$+$	$+$	$+$	$+$	$+$	$+$
Second	$+$	$-$	$-$	$-$	$-$	$+$
Third	$-$	$-$	$+$	$+$	$-$	$-$
Fourth	$-$	$+$	$-$	$-$	$+$	$-$

▶ **EXAMPLE 3**   *Determining the Quadrant Containing the Terminal Point of an Arc*

From the signs of the indicated trigonometric functions of t, determine the quadrant containing the terminal point of the arc of length t: **(a)** sin $t < 0$ and tan $t > 0$; **(b)** cos $t > 0$ and csc $t < 0$.

Solution
(a) From Table 3, sin $t < 0$ for the third and fourth quadrants, and tan $t > 0$ for the first and third quadrants. For both conditions to hold, the point must be in the third quadrant.
(b) From the table, cos $t > 0$ for the first and fourth quadrants, and csc $t < 0$ for the third and fourth quadrants. Thus the point is in the fourth quadrant.   ◀

In Section 7.1 we had the fundamental Pythagorean identity
$$\sin^2 t + \cos^2 t = 1$$
If cos $t \neq 0$, we can divide on each side by $\cos^2 t$ and obtain
$$\frac{\sin^2 t}{\cos^2 t} + \frac{\cos^2 t}{\cos^2 t} = \frac{1}{\cos^2 t}$$
$$\left(\frac{\sin t}{\cos t}\right)^2 + 1 = \left(\frac{1}{\cos t}\right)^2$$
$$\tan^2 t + 1 = \sec^2 t$$

We can divide on each side of the fundamental Pythagorean identity by $\sin^2 t$ provided that $\sin t \neq 0$. Doing this, we get

$$\frac{\sin^2 t}{\sin^2 t} + \frac{\cos^2 t}{\sin^2 t} = \frac{1}{\sin^2 t}$$

$$1 + \left(\frac{\cos t}{\sin t}\right)^2 = \left(\frac{1}{\sin t}\right)^2$$

$$1 + \cot^2 t = \csc^2 t$$

Therefore, we have the following two identities that are valid when the functions are defined:

$$\tan^2 t + 1 = \sec^2 t$$

$$1 + \cot^2 t = \csc^2 t$$

These identities are also called Pythagorean identities.

The relationships given by the Pythagorean identities and the identities in the definitions of this section constitute the eight fundamental trigonometric identities. These identities are often used to write an expression involving trigonometric functions in an equivalent form. Because of their importance, we list them together. They should be memorized and recognized whenever they occur.

The Eight Fundamental Trigonometric Identities

I $\sin t \csc t = 1 \leftrightarrow \csc t = \dfrac{1}{\sin t} \leftrightarrow \sin t = \dfrac{1}{\csc t}$

II $\cos t \sec t = 1 \leftrightarrow \sec t = \dfrac{1}{\cos t} \leftrightarrow \cos t = \dfrac{1}{\sec t}$

III $\tan t \cot t = 1 \leftrightarrow \cot t = \dfrac{1}{\tan t} \leftrightarrow \tan t = \dfrac{1}{\cot t}$

IV $\tan t = \dfrac{\sin t}{\cos t}$

V $\cot t = \dfrac{\cos t}{\sin t}$

VI $\sin^2 t + \cos^2 t = 1 \leftrightarrow \sin^2 t = 1 - \cos^2 t \leftrightarrow \cos^2 t = 1 - \sin^2 t$

VII $\tan^2 t + 1 = \sec^2 t \leftrightarrow \tan^2 t = \sec^2 t - 1 \leftrightarrow \sec^2 t - \tan^2 t = 1$

VIII $1 + \cot^2 t = \csc^2 t \leftrightarrow \cot^2 t = \csc^2 t - 1 \leftrightarrow \csc^2 t - \cot^2 t = 1$

▶ **EXAMPLE 4** *Finding Other Trigonometric Functions of a Number When One Function is Given*

If $\sin t = -\frac{3}{5}$ and $\cos t > 0$, find $\cos t$, $\tan t$, $\cot t$, $\sec t$, and $\csc t$.

Solution   Because $\sin t < 0$ and $\cos t > 0$, the terminal point of the arc of length t is in the fourth quadrant.
From fundamental identity VI and because $\cos t > 0$,

$$
\begin{aligned}
\cos t &= \sqrt{1 - \sin^2 t} \\
&= \sqrt{1 - (-\tfrac{3}{5})^2} \\
&= \sqrt{\tfrac{16}{25}} \\
&= \tfrac{4}{5}
\end{aligned}
$$

From fundamental identity IV,

$$
\begin{aligned}
\tan t &= \frac{\sin t}{\cos t} \\
&= \frac{-\frac{3}{5}}{\frac{4}{5}} \\
&= -\tfrac{3}{4}
\end{aligned}
$$

From fundamental identities I through III,

$$
\cot t = \frac{1}{\tan t} \qquad \sec t = \frac{1}{\cos t} \qquad \csc t = \frac{1}{\sin t}
$$
$$
= -\tfrac{4}{3} \qquad\qquad = \tfrac{5}{4} \qquad\qquad = -\tfrac{5}{3} \qquad ◀
$$

Observe that in the eight fundamental identities, the sine and cosine functions appear more often than the other four. Identities I, II, IV, and V enable us to express the other four functions in terms of either the sine or cosine or both.

▶ **EXAMPLE 5** *Writing a Trigonometric Expression in Terms of Sine and Cosine*

Write expression **(a)** in terms of $\sin t$ and expression **(b)** in terms of $\cos t$ and simplify: **(a)** $\tan^2 t \csc t$; **(b)** $\dfrac{\sec^2 t - 1}{\sin^2 t}$.

Solution

(a) $\tan^2 t \csc t = \left(\dfrac{\sin t}{\cos t}\right)^2 \cdot \dfrac{1}{\sin t}$

$$= \dfrac{\sin^2 t}{\cos^2 t \sin t}$$

$$= \dfrac{\sin t}{\cos^2 t}$$

$$= \dfrac{\sin t}{1 - \sin^2 t}$$

(b) $\dfrac{\sec^2 t - 1}{\sin^2 t} = \dfrac{\tan^2 t}{\sin^2 t}$

$$= \dfrac{\left(\dfrac{\sin t}{\cos t}\right)^2}{\sin^2 t}$$

$$= \dfrac{\sin^2 t}{\cos^2 t} \cdot \dfrac{1}{\sin^2 t}$$

$$= \dfrac{1}{\cos^2 t} \quad \blacktriangleleft$$

EXERCISES 7.6

In Exercises 1 through 10, find the exact values of (a) tan t, (b) cot t, (c) sec t, and (d) csc t from the given information.

1. $\sin t = \dfrac{\sqrt{3}}{2}$ and $\cos t = \dfrac{1}{2}$

2. $\sin t = \dfrac{1}{\sqrt{2}}$ and $\cos t = \dfrac{1}{\sqrt{2}}$

3. $\sin t = \dfrac{2}{3}$ and $\cos t = -\dfrac{\sqrt{5}}{3}$

4. $\sin t = -\dfrac{3}{5}$ and $\cos t = \dfrac{4}{5}$

5. $\sin t = -\dfrac{12}{13}$ and $\cos t = -\dfrac{5}{13}$

6. $\sin t = -\dfrac{8}{17}$ and $\cos t = -\dfrac{15}{17}$

7. $\sin t = -\dfrac{3}{\sqrt{34}}$ and $\cos t = \dfrac{5}{\sqrt{34}}$

8. $\sin t = \dfrac{3}{\sqrt{13}}$ and $\cos t = -\dfrac{2}{\sqrt{13}}$

9. $\sin t = -1$ and $\cos t = 0$

10. $\sin t = 0$ and $\cos t = -1$

In Exercises 11 through 16, find the six trigonometric functions of the number by determining the coordinates of the terminal point of the arc whose length is the given number and whose initial point is at (1, 0).

11. $\dfrac{7}{6}\pi$ **12.** $\dfrac{11}{6}\pi$ **13.** $-\dfrac{1}{4}\pi$

14. $\dfrac{3}{4}\pi$ **15.** $\dfrac{2}{3}\pi$ **16.** $-\dfrac{2}{3}\pi$

In Exercises 17 through 24, find the exact function value.

17. (a) $\tan \frac{7}{4}\pi$ **(b)** $\cot(-\frac{3}{4}\pi)$

18. (a) $\tan \frac{5}{4}\pi$ **(b)** $\cot(-\frac{7}{4}\pi)$

19. (a) $\csc \frac{4}{3}\pi$ **(b)** $\sec(-\frac{7}{6}\pi)$

20. (a) $\sec(-\frac{1}{3}\pi)$ **(b)** $\csc \frac{7}{6}\pi$

21. (a) $\tan \frac{13}{6}\pi$ **(b)** $\sec(-\frac{7}{3}\pi)$

22. (a) $\cot \frac{7}{3}\pi$ **(b)** $\csc(-\frac{13}{6}\pi)$

23. (a) $\csc \frac{3}{2}\pi$ **(b)** $\cot(-\frac{7}{2}\pi)$

24. (a) $\sec(-\pi)$ **(b)** $\tan 3\pi$

In Exercises 25 through 34, use a calculator to evaluate to four significant digits the function value.

25. (a) $\tan 1.26$ **(b)** $\cot 1.26$
   (c) $\sec 0.34$ **(d)** $\csc 0.34$

26. (a) $\tan 0.57$ **(b)** $\cot 0.57$
   (c) $\sec 1.42$ **(d)** $\csc 1.42$

27. (a) $\tan 1.05$ **(b)** $\cot 1.05$
   (c) $\sec 0.21$ **(d)** $\csc 0.21$

28. (a) $\tan 0.18$ **(b)** $\cot 0.18$
   (c) $\sec 1.33$ **(d)** $\csc 1.33$

29. (a) $\tan 1.84$ **(b)** $\csc 1.84$
   (c) $\cot 3.62$ **(d)** $\sec 3.62$

30. (a) $\tan 4.06$ **(b)** $\sec 4.06$
   (c) $\cot 2.75$ **(d)** $\csc 2.75$

31. (a) $\tan \frac{5}{11}\pi$ **(b)** $\sec(-\frac{1}{8}\pi)$ **(c)** $\csc(-\frac{1}{8}\pi)$

32. (a) $\tan(-\frac{3}{5}\pi)$ **(b)** $\sec(-\frac{3}{5}\pi)$ **(c)** $\csc\frac{7}{5}\pi$

33. (a) $\cot\frac{4}{5}\pi$ **(b)** $\sec(-\frac{4}{7}\pi)$ **(c)** $\csc\frac{17}{10}\pi$

34. (a) $\cot\frac{5}{12}\pi$ **(b)** $\sec\frac{23}{14}\pi$ **(c)** $\csc(-\frac{6}{5}\pi)$

In Exercises 35 through 38, determine the quadrant containing the terminal point of the arc of length t if the initial point is at (1, 0).

35. (a) $\cos t > 0$ and $\tan t < 0$
   (b) $\sin t < 0$ and $\cot t > 0$

36. (a) $\cos t < 0$ and $\tan t > 0$
   (b) $\sin t > 0$ and $\cot t < 0$

37. (a) $\tan t < 0$ and $\csc t > 0$
   (b) $\cot t > 0$ and $\sec t < 0$

38. (a) $\tan t > 0$ and $\csc t < 0$
   (b) $\cot t < 0$ and $\sec t > 0$

39. If $\cos t = -\frac{4}{5}$ and $\sin t > 0$, find the exact values of $\sin t$, $\tan t$, $\cot t$, $\sec t$, and $\csc t$.

40. If $\sin t = -\frac{12}{13}$ and $\cos t > 0$, find the exact values of $\cos t$, $\tan t$, $\cot t$, $\sec t$, and $\csc t$.

41. If $\tan t = \frac{15}{8}$ and $\sec t < 0$, find the exact values of $\sin t$, $\cos t$, $\cot t$, $\sec t$, and $\csc t$.

42. If $\cot t = \frac{4}{3}$ and $\csc t < 0$, find the exact values of $\sin t$, $\cos t$, $\tan t$, $\sec t$, and $\csc t$.

43. If $\cot t = -\frac{5}{12}$ and $\csc t < 0$, find the exact values of $\sin t$, $\cos t$, $\tan t$, $\sec t$, and $\csc t$.

44. If $\tan t = -\frac{8}{15}$ and $\sec t < 0$, find the exact values of $\sin t$, $\cos t$, $\cot t$, $\sec t$, and $\csc t$.

In Exercises 45 through 50, write expression (a) in terms of sin t and expression (b) in terms of cos t.

45. (a) $\cot^2 t \csc t$ **(b)** $\tan t \csc t$

46. (a) $\cot t \sec t$ **(b)** $\cot^2 t \sec t$

47. (a) $\dfrac{\sin^3 t}{\tan^2 t}$ **(b)** $\dfrac{1 - \csc^2 t}{\csc^2 t}$

48. (a) $\dfrac{1 - \sec^2 t}{\sec^2 t}$ **(b)** $\dfrac{\cos^3 t}{\cot^2 t}$

49. (a) $\dfrac{\csc^2 t - 1}{\cos t \cot t}$ **(b)** $\dfrac{\cot^2 t + 1}{\sin t}$

50. (a) $\dfrac{\tan^2 t + 1}{\cos t}$ **(b)** $\dfrac{\sec^2 t - 1}{\sin t \tan t}$

51. Express $\dfrac{\tan^2 t - \csc^2 t}{\sec^2 t}$ in terms of $\sin t$.

52. Express $\dfrac{\sec^2 t - \cot^2 t}{\csc^2 t}$ in terms of $\cos t$.

53. In Section 2.1 we defined an identity as an equation whose solution set is the same as the domain of the unknown. Consequently a trigonometric identity in the variable t is valid for all t for which the functions are defined. State each of the eight fundamental trigonometric identities, and for each identity indicate all values of t for which the identity is valid.

7.7 PERIODICITY AND GRAPHS OF THE TANGENT, COTANGENT, SECANT, AND COSECANT FUNCTIONS

GOALS

1. **Learn that the tangent and cotangent functions are periodic with period π.**

2. **Learn that the secant and cosecant functions are periodic with period 2π.**

3. **Find trigonometric function values by using the periodicity of the functions.**

4. **Obtain the graph of the tangent function.**

5. **Obtain the graph of the cotangent function.**

6. Learn that the period of a function of the form **tan** *bt* or **cot** *bt*

   is $\dfrac{\pi}{|b|}$.

7. **Sketch graphs of tangent and cotangent functions.**
8. **Obtain the graph of the secant function.**
9. **Obtain the graph of the cosecant function.**
10. **Learn that the period of a function of the form sec** *bt* **or csc** *bt*

    is $\dfrac{2\pi}{|b|}$.

11. **Sketch graphs of secant and cosecant functions.**

Prior to discussing the graphs of the tangent, cotangent, secant, and cosecant functions, we show that these functions are periodic. The concept of periodicity will help us to obtain the graphs.

Consider point $P(\cos t, \sin t)$ and point $\bar{P}(-\cos t, -\sin t)$ on the unit circle U. These are endpoints of a diameter. See Figure 1, showing the point P in each of the four quadrants.

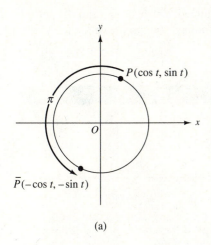

(a)

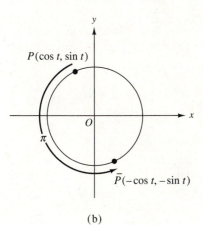

(b)

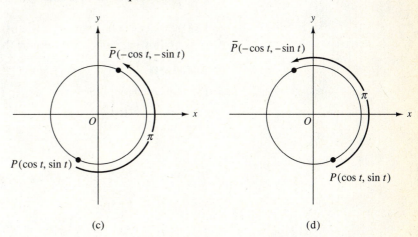

(c)   (d)

FIGURE 1

From the definitions of tangent and cotangent

$$\tan t = \frac{\sin t}{\cos t} \quad \text{if } \cos t \neq 0 \qquad \text{and} \qquad \cot t = \frac{\cos t}{\sin t} \quad \text{if } \sin t \neq 0$$

Because $\bar{P}(-\cos t, -\sin t)$ is the terminal point of an arc of length $t + \pi$, then

$$\tan (t + \pi) = \frac{-\sin t}{-\cos t} \qquad\qquad \cot(t + \pi) = \frac{-\cos t}{-\sin t}$$

$$= \frac{\sin t}{\cos t} \quad \text{if } \cos t \neq 0 \qquad\qquad = \frac{\cos t}{\sin t} \quad \text{if } \sin t \neq 0$$

Thus

$$\tan(t + \pi) = \tan t \quad \text{if } \cos t \neq 0 \quad \text{and} \quad \cot(t + \pi) = \cot t \quad \text{if } \sin t \neq 0$$

From these formulas the tangent and cotangent are periodic, and the period is π. By repeated applications of the formulas, we can prove the following theorem.

THEOREM 1

If t is any real number and k is any integer,

$$\tan(t + k\pi) = \tan t \quad \text{if } t \neq \tfrac{1}{2}\pi + k\pi$$
$$\cot(t + k\pi) = \cot t \quad \text{if } t \neq k\pi$$

▶ **EXAMPLE 1** *Applying the Concept of Periodicity to Find Tangent and Cotangent Function Values*

Use the values of $\tan t$ and $\cot t$ when $0 \leq t < \pi$ and periodicity to find each of the following values: **(a)** $\tan \tfrac{5}{4}\pi$; **(b)** $\cot \tfrac{5}{3}\pi$; **(c)** $\cot \tfrac{7}{2}\pi$; **(d)** $\tan\left(-\tfrac{19}{6}\pi\right)$.

Solution

(a) $\tan \tfrac{5}{4}\pi = \tan(\tfrac{1}{4}\pi + \pi)$ **(b)** $\cot \tfrac{5}{3}\pi = \cot(\tfrac{2}{3}\pi + \pi)$

$\qquad\qquad = \tan \tfrac{1}{4}\pi$ $\qquad\qquad\qquad\qquad = \cot \tfrac{2}{3}\pi$

$\qquad\qquad = 1$ $\qquad\qquad\qquad\qquad\qquad = -\dfrac{1}{\sqrt{3}}$

(c) $\cot \tfrac{7}{2}\pi = \cot(\tfrac{1}{2}\pi + 3\pi)$ **(d)** $\tan(-\tfrac{19}{6}\pi) = \tan(\tfrac{5}{6}\pi + (-4)\pi)$

$\qquad\qquad = \cot \tfrac{1}{2}\pi$ $\qquad\qquad\qquad\qquad\qquad = \tan \tfrac{5}{6}\pi$

$\qquad\qquad = 0$ $\qquad\qquad\qquad\qquad\qquad\quad = -\dfrac{1}{\sqrt{3}}$ ◀

The periodicity of secant and cosecant is related to the periodicity of cosine and sine. Because

$$\sec t = \frac{1}{\cos t} \quad \text{if } \cos t \neq 0 \quad \text{and} \quad \csc t = \frac{1}{\sin t} \quad \text{if } \sin t \neq 0$$

then

$$\sec(t + k \cdot 2\pi) = \frac{1}{\cos(t + k \cdot 2\pi)} \quad \text{if } \cos(t + k \cdot 2\pi) \neq 0$$

Table 1

t	$\tan t$
0	0
$\dfrac{1}{6}\pi$	$\dfrac{1}{\sqrt{3}}$
$\dfrac{1}{4}\pi$	1
$\dfrac{1}{3}\pi$	$\sqrt{3}$
$\dfrac{1}{2}\pi$	undefined
$\dfrac{2}{3}\pi$	$-\sqrt{3}$
$\dfrac{3}{4}\pi$	-1
$\dfrac{5}{6}\pi$	$\dfrac{1}{\sqrt{3}}$
π	0

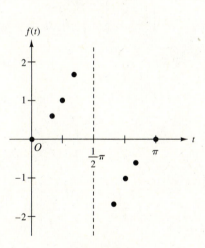

FIGURE 2

and

$$\csc(t + k \cdot 2\pi) = \frac{1}{\sin(t + k \cdot 2\pi)} \quad \text{if } \sin(t + k \cdot 2\pi) \neq 0$$

But $\cos(t + k \cdot 2\pi) = \cos t$ and $\sin(t + k \cdot 2\pi) = \sin t$. Thus

$$\sec(t + k \cdot 2\pi) = \frac{1}{\cos t} \quad \text{if } \cos t \neq 0$$

and

$$\csc(t + k \cdot 2\pi) = \frac{1}{\sin t} \quad \text{if } \sin t \neq 0$$

Replacing $1/(\cos t)$ by $\sec t$ and $1/(\sin t)$ by $\csc t$, we have the following theorem.

THEOREM 2

If t is a real number and k is any integer,

$$\sec(t + k \cdot 2\pi) = \sec t \quad \text{if } t \neq \tfrac{1}{2}\pi + k\pi$$
$$\csc(t + k \cdot 2\pi) = \csc t \quad \text{if } t \neq k\pi$$

From Theorem 2, the secant and cosecant are periodic, and the period is 2π.

We now concentrate on the graphs. The procedure for obtaining the graph of the tangent function is similar to that used for the graphs of the sine and cosine. Because the period of the tangent is π, we first determine the portion of the graph on the interval $[0, \pi]$. To get some specific points on the graph, we use some special values of $\tan t$ listed in Table 1. Figure 2 shows the points having as coordinates the number pairs $(t, \tan t)$ given in the table, where we use $\sqrt{3} \approx 1.7$ and $\frac{1}{\sqrt{3}} \approx 0.6$.

Because $\tan \frac{1}{2}\pi$ is not defined, there is no point on the graph for $t = \frac{1}{2}\pi$. But what happens as t approaches $\frac{1}{2}\pi$? To answer this question, we use the identity

$$\tan t = \frac{\sin t}{\cos t}$$

As t approaches $\frac{1}{2}\pi$, $\sin t$ approaches 1 and $\cos t$ approaches zero. More specifically, as t approaches $\frac{1}{2}\pi$ through values less than $\frac{1}{2}\pi$ (that is, $t \to \frac{1}{2}\pi^-$), $\cos t$ approaches zero through positive values; thus $\tan t$ increases without bound. So we write

$$\tan t \to +\infty \quad \text{as } t \to \tfrac{1}{2}\pi^-$$

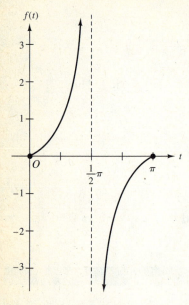

FIGURE 3

As t approaches $\frac{1}{2}\pi$ through values greater than $\frac{1}{2}\pi$ (that is, $t \to \frac{1}{2}\pi^+$), cos t approaches zero through negative values; hence tan t decreases without bound. We write

$$\tan t \to -\infty \qquad \text{as } t \to \tfrac{1}{2}\pi^+$$

Therefore the line $t = \frac{1}{2}\pi$ is a vertical asymptote of the graph. With this information we connect the points in Figure 2 with a curve and obtain the graph shown in Figure 3. As an aid in drawing the graph, additional points can be located from values of tan t obtained from a calculator.

The portion of the curve over the interval $[0, \pi]$ is repeated for every interval on the t axis of length π: $[\pi, 2\pi], [2\pi, 3\pi], [-\pi, 0], [-2\pi, -\pi]$, and so on. Figure 4 shows the complete graph. The vertical asymptotes are lines having the equations

$$t = \tfrac{1}{2}\pi + k\pi$$
$$\Leftrightarrow \qquad t = (2k+1)\tfrac{1}{2}\pi$$

where $k \in J$. The values of t at which the graph intersects the t axis are the zeros of the tangent function. They are $t = k\pi$, where $k \in J$.

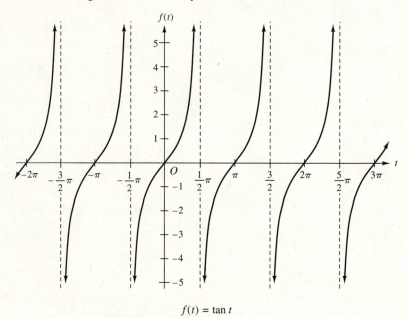

$$f(t) = \tan t$$

FIGURE 4

Observe that the graph of the tangent function is symmetric with respect to the origin because the tangent is an odd function; that is,

$$\tan(-t) = -\tan t$$

This identity can be obtained from identities (1) and (2) in Section 7.3 as follows:

$$\tan(-t) = \frac{\sin(-t)}{\cos(-t)}$$

$$= \frac{-\sin t}{\cos t}$$

$$= -\tan t$$

The graph of the cotangent function can be obtained by locating points as we did for the tangent. However, an easier method is one that makes use of the identity

$$\cot t = \frac{1}{\tan t}$$

if $t \neq \frac{1}{2}k\pi$, where $k \in J$. At a value of t in the domains of both functions, an ordinate on the graph of the cotangent can be found by taking the reciprocal of the corresponding ordinate on the graph of the tangent.

If $t = \frac{1}{2}k\pi$, where k is an even integer, then $\tan t = 0$, and $\cot t$ is undefined. Therefore, there is no point on the graph of the cotangent for these values of t. When $t = \frac{1}{2}k\pi$, where k is an odd integer, $\cot t = 0$. Thus the graph of the cotangent intersects the t axis for these values of t. We use the graph of the tangent as an aid by first sketching it with a light curve as indicated in Figure 5. Then by taking reciprocals of the ordinates, we obtain

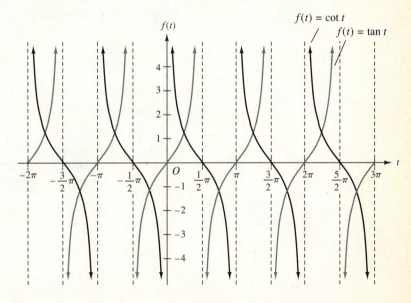

$$f(t) = \cot t$$

FIGURE 5

points on the graph of the cotangent shown in the figure. Observe that the vertical asymptotes of the graph of the cotangent are the lines having the equations $t = k\pi$, where $k \in J$, and the zeros of the cotangent function are $t = \frac{1}{2}\pi + k\pi$, where $k \in J$.

From their graphs it is apparent that the range of the tangent and cotangent functions is the set R of real numbers.

Theorem 1 of Section 7.3 stated that a function defined by either of the equations $f(t) = \sin bt$ or $f(t) = \cos bt$, where $b \neq 0$, has period $2\pi/|b|$. The proof was based on the fact that the sine and cosine have period 2π. A similar proof can be given for the following theorem. The period of a function defined by either $f(t) = \tan bt$ or $f(t) = \cot bt$ is $\pi/|b|$ because the period of the tangent and cotangent is π.

THEOREM 3

The period P of a periodic function defined by either

$$f(t) = \tan bt \qquad \text{or} \qquad f(t) = \cot bt$$

where $b \neq 0$, is given by

$$P = \frac{\pi}{|b|}$$

▶ **EXAMPLE 2** *Sketching the Graph of a Tangent Function*

Sketch the graph of

$$f(t) = \tan 3t$$

where t is any number in the interval $[0, \pi]$ at which the function is defined. Write equations of the asymptotes of the graph on the interval. Check the graph on a graphics calculator.

Solution From Theorem 3, with $b = 3$, the period of f is $\frac{1}{3}\pi$. To find the zeros of f, we solve the equation $\tan 3t = 0$ and obtain $3t = k\pi$, where $k \in J$. Therefore, on $[0, \pi]$ the intersections of the graph with the t axis are at $0, \frac{1}{3}\pi, \frac{2}{3}\pi$, and π.

To find equations of the asymptotes, we observe that $\tan 3t$ is not defined when

$$3t = \frac{1}{2}\pi + k\pi$$

where $k \in J$. Because t is in $[0, \pi]$, $3t$ is in $[0, 3\pi]$. Thus equations of the asymptotes on $[0, \pi]$ are given by

$$3t = \tfrac{1}{2}\pi \qquad 3t = \tfrac{3}{2}\pi \qquad 3t = \tfrac{5}{2}\pi$$
$$\Leftrightarrow \quad t = \tfrac{1}{6}\pi \qquad \quad t = \tfrac{1}{2}\pi \qquad \quad t = \tfrac{5}{6}\pi$$

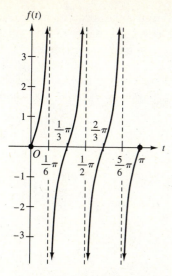

$f(t) = \tan 3t$

FIGURE 6

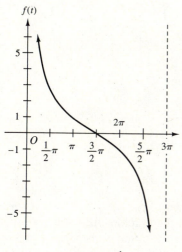

$f(t) = 2 \cot \dfrac{1}{3}t$

FIGURE 7

With this information and by locating a few points, we obtain the graph shown in Figure 6. For this graph the interval $[0, \pi]$ on the t axis is marked off every $\frac{1}{6}\pi$ units. We obtain the same graph on our graphics calculator. ◀

▶ **EXAMPLE 3** *Sketching the Graph of a Cotangent Function*

Sketch the graph of

$$f(t) = 2 \cot \tfrac{1}{3}t$$

over an interval whose length is the period. Check the graph on a graphics calculator.

Solution   From Theorem 3, with $b = \frac{1}{3}$, the period of f is $\dfrac{\pi}{\frac{1}{3}} = 3\pi$. Therefore, we obtain the graph over the interval $[0, 3\pi]$.

The graph intersects the t axis when $\cot \frac{1}{3}t = 0$, that is, when

$$\tfrac{1}{3}t = \tfrac{1}{2}\pi + k\pi$$

where $k \in J$. Because $\frac{1}{3}t$ is in $[0, \pi]$, the only point of intersection with the t axis is given by

$$\tfrac{1}{3}t = \tfrac{1}{2}\pi$$
$$t = \tfrac{3}{2}\pi$$

For the asymptotes, we determine where $\cot \frac{1}{3}t$ is not defined. This happens when

$$\tfrac{1}{3}t = k\pi$$

where k is any integer. On the interval $[0, 3\pi]$, the asymptotes are given by

$$\tfrac{1}{3}t = 0 \qquad \tfrac{1}{3}t = \pi$$
$$\Leftrightarrow \qquad t = 0 \qquad t = 3\pi$$

From the preceding information and a few points we obtain the required graph appearing in Figure 7, which checks with the graph plotted on our graphics calculator. ◀

The graphs of the secant and cosecant function are obtained in a manner similar to that used for the cotangent. For the identity

$$\sec t = \frac{1}{\cos t}$$

if $t \neq \frac{1}{2}\pi + k\pi$, where $k \in J$. On the graph of $f(t)$
is the reciprocal of the corresponding ordinate of th

except for values of t for which $\cos t = 0$. Because the secant function has no zeros, the graph does not intersect the t axis. As a matter of fact, since $|\sec t| \geq 1$, there are no ordinates of the graph between -1 and 1.

The graph of $f(t) = \sec t$ appears in Figure 8 where the graph of $f(t) = \cos t$ is indicated by a lighter curve. The vertical asymptotes of the graph of the secant are the lines having the equations $t = \frac{1}{2}\pi + k\pi$, where $k \in J$.

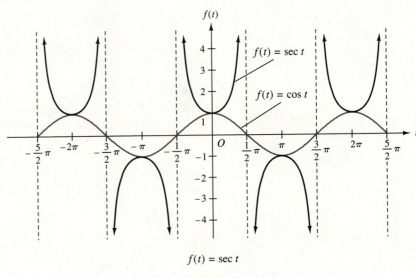

$$f(t) = \sec t$$

FIGURE 8

For the graph of the cosecant, we use the identity

$$\csc t = \frac{1}{\sin t}$$

if $t \neq k\pi$, where $k \in J$. Each ordinate of the graph of the cosecant is the reciprocal of the corresponding ordinate of the graph of the sine except for values of t for which $\sin t = 0$. As with the secant, there are no ordinates of the graph between -1 and 1.

Figure 9 shows the graph of the cosecant function and the graph of the sine represented by a lighter curve. The vertical asymptotes of the graph of the cosecant are the lines having the equations $t = k\pi$, where $k \in J$.

Because the secant and cosecant are reciprocals of the cosine and sine, respectively, we have the following theorem, corresponding to Theorem 1 in Section 7.3.

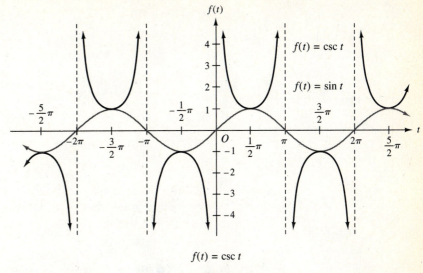

$$f(t) = \csc t$$

FIGURE 9

THEOREM 4

The period P of a periodic function defined by either

$$f(t) = \sec bt \quad \text{or} \quad f(t) = \csc bt$$

where $b \neq 0$, is given by

$$P = \frac{2\pi}{|b|}$$

▶ **EXAMPLE 4** *Sketching the Graph of a Cosecant Function*

Sketch the graph of

$$f(t) = -2 \csc \tfrac{1}{2}t$$

over an interval whose length is the period. Check the graph on a graphics calculator.

Solution   From Theorem 4, with $b = \tfrac{1}{2}$, the period of f is $\dfrac{2\pi}{\frac{1}{2}} = 4\pi$. Thus we obtain the graph over the interval $[0, 4\pi]$.

To find the vertical asymptotes, we determine where $\csc \tfrac{1}{2}t$ is not defined. We therefore let $\tfrac{1}{2}t = k\pi$, where $k \in J$. On the interval $[0, 4\pi]$, the

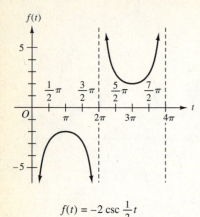

$$f(t) = -2 \csc \frac{1}{2} t$$

FIGURE 10

asymptotes are

$$\tfrac{1}{2}t = 0 \qquad \tfrac{1}{2}t = \pi \qquad \tfrac{1}{2}t = 2\pi$$

$$\Leftrightarrow \quad t = 0 \qquad t = 2\pi \qquad t = 4\pi$$

Because $\left| \csc \tfrac{1}{2}t \right| \geq 1$, $f(t) \leq -2$ or $f(t) \geq 2$. Furthermore, $f(t) = -2$ when $\csc \tfrac{1}{2}t = 1$, that is, when

$$\tfrac{1}{2}t = \tfrac{1}{2}\pi \Leftrightarrow t = \pi$$

And $f(t) = 2$ when $\csc \tfrac{1}{2}t = -1$, that is, when

$$\tfrac{1}{2}t = \tfrac{3}{2}\pi \Leftrightarrow t = 3\pi$$

With the preceding information and by locating a few points, we have the required graph shown in Figure 10. We obtain the same graph on our graphics calculator. ◀

▶ **EXAMPLE 5** *Sketching the Graph of a Secant Function*

Sketch the graph of

$$f(t) = \sec(t - \tfrac{1}{2}\pi)$$

Check the graph on a graphics calculator.

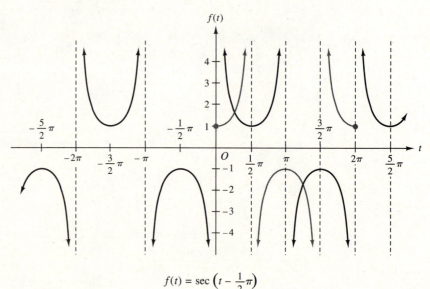

$$f(t) = \sec \left(t - \frac{1}{2}\pi \right)$$

FIGURE 11

Solution   The graph of the given function can be obtained from the graph of the function defined by $g(t) = \sec t$ by a horizontal translation of $\frac{1}{2}\pi$ units to the right to get the required graph appearing in Figure 11, which includes the graph of $f(t) = \sec t$ on the interval $[0, 2\pi]$ represented by the lighter curve. Our graphics calculator shows the same graph.   ◀

As we did with the sine and cosine curves in Section 7.3, we can use the tangent, cotangent, secant, and cosecant curves plotted on a graphics calculator to approximate particular values of these trigonometric functions. We can also use the curves to approximate values of t if one of these trigonometric functions of t is given. See Exercises 21 through 24.

EXERCISES 7.7

In Exercises 1 through 8, use the periodicity of the tangent and cotangent functions as well as the values of $\tan t$ *and* $\cot t$ *when* $0 \le t < \pi$ *to find the function value.*

1. (a) $\tan \frac{7}{4}\pi$ **(b)** $\cot \frac{7}{4}\pi$
(c) $\tan(-\frac{1}{4}\pi)$ **(d)** $\cot(-\frac{1}{4}\pi)$

2. (a) $\tan \frac{4}{3}\pi$ **(b)** $\cot \frac{4}{3}\pi$
(c) $\tan(-\frac{2}{3}\pi)$ **(d)** $\cot(-\frac{2}{3}\pi)$

3. (a) $\tan \frac{7}{6}\pi$ **(b)** $\cot \frac{7}{6}\pi$
(c) $\tan(-\frac{5}{6}\pi)$ **(d)** $\cot(-\frac{5}{6}\pi)$

4. (a) $\tan \frac{11}{6}\pi$ **(b)** $\cot \frac{11}{6}\pi$
(c) $\tan(-\frac{1}{6}\pi)$ **(d)** $\cot(-\frac{1}{6}\pi)$

5. (a) $\tan \frac{11}{3}\pi$ **(b)** $\cot \frac{11}{3}\pi$
(c) $\tan(-\frac{7}{3}\pi)$ **(d)** $\cot(-\frac{7}{3}\pi)$

6. (a) $\tan \frac{13}{4}\pi$ **(b)** $\cot \frac{13}{4}\pi$
(c) $\tan(-\frac{11}{4}\pi)$ **(d)** $\cot(-\frac{11}{4}\pi)$

7. (a) $\tan 5\pi$ **(b)** $\cot 6\pi$
(c) $\tan(-\frac{11}{2}\pi)$ **(d)** $\cot(-\frac{9}{2}\pi)$

8. (a) $\tan \frac{11}{2}\pi$ **(b)** $\cot \frac{9}{2}\pi$
(c) $\tan(-5\pi)$ **(d)** $\cot(-6\pi)$

In Exercises 9 through 14, use the periodicity of the secant and cosecant functions as well as the values of $\sec t$ *and* $\csc t$ *when* $0 \le t < 2\pi$ *to find the function value.*

9. (a) $\sec \frac{9}{4}\pi$ **(b)** $\csc \frac{9}{4}\pi$
(c) $\sec \frac{10}{3}\pi$ **(d)** $\csc \frac{10}{3}\pi$

10. (a) $\sec \frac{17}{6}\pi$ **(b)** $\csc \frac{17}{6}\pi$
(c) $\sec \frac{15}{4}\pi$ **(d)** $\csc \frac{15}{4}\pi$

11. (a) $\sec(-\frac{1}{6}\pi)$ **(b)** $\csc(-\frac{1}{6}\pi)$
(c) $\sec(-\frac{5}{4}\pi)$ **(d)** $\csc(-\frac{5}{4}\pi)$

12. (a) $\sec(-\frac{2}{3}\pi)$ **(b)** $\csc(-\frac{2}{3}\pi)$
(c) $\sec(-\frac{11}{6}\pi)$ **(d)** $\csc(-\frac{11}{6}\pi)$

13. (a) $\sec 7\pi$ **(b)** $\csc 9\pi$
(c) $\sec(-\frac{11}{2}\pi)$ **(d)** $\csc(-\frac{9}{2}\pi)$

14. (a) $\sec \frac{11}{2}\pi$ **(b)** $\csc \frac{9}{2}\pi$
(c) $\sec(-7\pi)$ **(d)** $\csc(-9\pi)$

In Exercises 15 through 20, find the period of the function.

15. (a) $2 \tan 3t$ **(b)** $\cot 6t$
(c) $\tan \frac{1}{5}t$ **(d)** $\frac{1}{3}\cot \frac{1}{4}t$

16. (a) $3 \cot 5t$ **(b)** $\tan 4t$
(c) $\tan \frac{2}{3}t$ **(d)** $5 \cot \frac{1}{6}t$

17. (a) $4 \sec 7t$ **(b)** $\csc \frac{4}{7}t$
(c) $\frac{1}{4}\tan 7t$ **(d)** $\cot \frac{7}{4}t$

18. (a) $3 \csc 8t$ **(b)** $\sec \frac{3}{8}t$
(c) $\frac{1}{3}\tan 8t$ **(d)** $3 \cot \frac{1}{8}t$

19. (a) $\sec 5\pi t$ **(b)** $\frac{1}{3}\csc \frac{1}{2}\pi t$
(c) $\tan 5\pi t$ **(d)** $\cot \frac{3}{4}\pi t$

20. (a) $3 \sec 4\pi t$ **(b)** $\csc \frac{3}{2}\pi t$
(c) $\tan \frac{5}{4}\pi t$ **(d)** $\cot 10\pi t$

In Exercises 21 and 22, use the graph of the given function plotted on your graphics calculator to approximate the function value to three significant digits. Then check your approximation by computing the function value by using the appropriate key, or keys, on your calculator.

21. (a) $\tan 0.894$ **(b)** $\cot 2.53$
(c) $\tan 5.67$ **(d)** $\cot(-1.98)$
(e) $\sec 1.84$ **(f)** $\csc(-4.65)$

22. (a) $\tan 1.35$ (b) $\cot 4.06$
    (c) $\tan(-2.29)$ (d) $\cot 0.374$
    (e) $\sec(-3.07)$ (f) $\csc 6.16$

In Exercises 23 and 24, use the graph of the given function plotted on your graphics calculator to approximate the values of t to three significant digits such that $0 \le t < 2\pi$ for which the equation is satisfied.

23. (a) $\tan t = 0.775$ (b) $\cot t = -0.775$
    (c) $\tan t = -3.60$ (d) $\cot t = 4.07$
    (e) $\sec t = -12.4$ (f) $\csc t = 2.79$

24. (a) $\tan t = -1.67$ (b) $\cot t = 1.67$
    (c) $\tan t = 0.690$ (d) $\cot t = -0.484$
    (e) $\sec t = 1.39$ (f) $\csc t = -3.92$

In Exercises 25 through 40, (a) sketch the graph of the function where t is any number in the indicated interval at which the function is defined. (b) Write equations of the asymptotes of the graph on the interval. (c) Check your graph on your graphics calculator.

25. $f(t) = 2 \tan t$; $[0, 2\pi]$
26. $f(t) = 3 \cot t$; $(0, 2\pi)$
27. $f(t) = \cot 2t$; $(0, \pi)$
28. $f(t) = \tan 3t$; $[0, \pi]$
29. $f(t) = \tan \frac{1}{2}t$; $[-2\pi, 2\pi]$
30. $f(t) = \frac{1}{2} \cot 4t$; $(-\frac{1}{2}\pi, \frac{1}{2}\pi)$
31. $f(t) = -2 \cot 3t$; $(0, \pi)$
32. $f(t) = -3 \tan \frac{1}{3}t$; $[0, 3\pi]$
33. $f(t) = 3 \sec t$; $[0, 2\pi]$
34. $f(t) = 2 \csc t$; $(0, 2\pi)$

35. $f(t) = \csc 3t$; $(0, 2\pi)$
36. $f(t) = \sec 2t$; $[-\pi, \pi]$
37. $f(t) = \frac{1}{2} \cot \pi t$; $(0, 2)$
38. $f(t) = \tan \frac{1}{2} \pi t$; $[0, 4]$
39. $f(t) = 2 \sec \frac{1}{2} \pi t$; $[0, 2]$
40. $f(t) = 3 \csc \frac{1}{3} \pi t$; $(-3, 3)$

In Exercises 41 through 48, sketch the graph of the function over an interval whose length is the period. Check your graph on your graphics calculator.

41. $f(t) = 3 \tan 2t$ **42.** $f(t) = 2 \cot 3t$
43. $f(t) = 2 \cot \frac{1}{2}t$ **44.** $f(t) = -\tan \frac{1}{4}t$
45. $f(t) = -\sec 4t$ **46.** $f(t) = 3 \csc \frac{1}{3}t$
47. $f(t) = \frac{1}{2} \tan \frac{1}{4} \pi t$ **48.** $f(t) = \frac{1}{2} \sec 3\pi t$

In Exercises 49 through 58, sketch the graph of the function. Check your graph on your graphics calculator.

49. $f(t) = \tan(t + \frac{1}{4}\pi)$ **50.** $f(t) = \tan(t - \frac{1}{4}\pi)$
51. $f(t) = \cot(t - \frac{1}{6}\pi)$ **52.** $f(t) = \cot(t + \frac{1}{3}\pi)$
53. $f(t) = 2 \sec(t - \pi)$ **54.** $f(t) = \frac{1}{2} \csc(t - \frac{1}{2}\pi)$
55. $f(t) = \frac{1}{3} \csc(t + \frac{1}{2}\pi)$ **56.** $f(t) = 3 \sec(t + \frac{1}{6}\pi)$
57. $f(t) = \frac{1}{4} \tan(\frac{1}{4}\pi - t)$ **58.** $f(t) = 2 \cot(\frac{1}{2}\pi - t)$

59. Example 3 involved the graph of a function defined by an equation of the form $f(t) = a \cot bt$. If g is the function defined by $g(t) = \cot bt$, how can the graph of f be obtained from the graph of g? Does the graph of f have an amplitude? Explain.

CHAPTER 7 REVIEW

▶ LOOKING BACK

7.1 We began the study of trigonometry with trigonometric functions of real numbers because they are essential for calculus. Our discussion of geometric concepts involving the unit circle and the real-number line "wrapped around" it laid the groundwork for definitions of the sine and cosine of a real number t as coordinates of the terminal point of an arc of length t on the unit circle. We used the definitions to compute exact function values of the sine and cosine of quadrantal numbers and $\frac{1}{4}\pi, \frac{1}{6}\pi$, and $\frac{1}{3}\pi$. The fundamental Pythagorean identity was introduced to find either the sine or cosine of a number when the other was given.

7.2 After computing sine and cosine function values on a calculator, we demonstrated the periodicity of sine and cosine by showing that these functions have period 2π. We used this periodicity property to find function values of other numbers from the function value of a given number. We also solved word problems having periodic functions as mathematical models.

7.3 Periodicity played an important part in obtaining graphs of the sine and cosine functions and other sine waves. You learned how to sketch these graphs. You also learned how to use the graphs plotted on a graphics calculator to approximate sine and cosine function values as well as to approximate numbers having given sine and cosine function values.

7.4 The applications of sine and cosine functions to periodic phenomena presented here included simple harmonic motion of a weight suspended from a spring and vibrating vertically, alternating current and electromotive force in electric circuits, and sound waves produced by simple sounds.

7.5 Damped harmonic motion was illustrated by a mathematical model involving the product of a sine function and an exponential function of t whose values approach zero as t increases without bound. We illustrated resonance by a mathematical model involving the product of a cosine function and an exponential function of t whose values increase without bound as t increases without bound. We gave an example where

we sketched the graph of the important function in calculus defined by $\dfrac{\sin t}{t}$, which approaches 1 as t approaches zero.

7.6 We defined the tangent, cotangent, secant, and cosecant functions as quotients of the sine and cosine functions and then computed exact values of these functions of quadrantal numbers, and $\frac{1}{4}\pi$, $\frac{1}{6}\pi$, and $\frac{1}{3}\pi$. We also discussed the signs of the trigonometric functions in the various quadrants. After presenting the other seven fundamental trigonometric identities, we applied all eight of them to write a trigonometric expression in terms of sine and cosine functions and to find five trigonometric functions of a number when the sixth function of the number was given.

7.7 We used the periodicity of the tangent, cotangent, secant, and cosecant functions to find particular values of these functions. Then, just as we did in Section 7.3 with the sine and cosine functions, we applied the periodicity of the other four trigonometric functions to obtain their graphs.

▶ REVIEW EXERCISES

In Exercises 1 through 4, determine the exact function value. Do not use a calculator.

1. (a) $\cos 0$ (b) $\tan \pi$ (c) $\sin \frac{3}{2}\pi$
    (d) $\sin \frac{1}{6}\pi$ (e) $\cos \frac{2}{3}\pi$ (f) $\tan \frac{3}{4}\pi$

2. (a) $\sin \frac{1}{2}\pi$ (b) $\cos \pi$ (c) $\cot \frac{3}{2}\pi$
    (d) $\tan \frac{1}{4}\pi$ (e) $\sin \frac{5}{6}\pi$ (f) $\cos \frac{4}{3}\pi$

3. (a) $\sin \frac{1}{4}\pi$ (b) $\cos \frac{3}{4}\pi$ (c) $\tan \frac{1}{3}\pi$
    (d) $\cot \frac{7}{6}\pi$ (e) $\sec(-\frac{1}{3}\pi)$ (f) $\csc(-\frac{5}{6}\pi)$

4. (a) $\sin \frac{1}{3}\pi$ (b) $\cos \frac{5}{4}\pi$ (c) $\tan \frac{5}{6}\pi$
    (d) $\cot(-\frac{2}{3}\pi)$ (e) $\sec(-\frac{7}{6}\pi)$ (f) $\csc(-\frac{5}{4}\pi)$

In Exercises 5 through 14, either sin t, cos t, tan t, cot t, sec t, or csc t is given. Find the exact values of the other five without using a calculator.

5. $\sin t = \frac{5}{13}$ and $0 < t < \frac{1}{2}\pi$

6. $\cos t = -\frac{3}{5}$ and $\frac{1}{2}\pi < t < \pi$

7. $\tan t = -\frac{15}{8}$ and $\frac{1}{2}\pi < t < \pi$

8. $\cot t = \frac{5}{12}$ and $\pi < t < \frac{3}{2}\pi$

9. $\sec t = -\frac{5}{4}$ and $-\pi < t < -\frac{1}{2}\pi$

10. $\csc t = -\frac{17}{15}$ and $-\frac{1}{2}\pi < t < 0$

11. $\cos t = \frac{12}{13}$ and $\sin t < 0$

12. $\tan t = \frac{4}{3}$ and $\cos t > 0$

13. $\csc t = -\frac{17}{8}$ and $\tan t > 0$

14. $\sin t = -\frac{5}{13}$ and $\cot t < 0$

In Exercises 15 through 20, write expression (a) in terms of sin t and expression (b) in terms of cos t.

15. (a) $\tan t \csc t$ (b) $\cot t \csc t$

16. (a) $\cot t \sec t$ (b) $\tan t \sec t$

17. (a) $\dfrac{1 - \tan^2 t}{\cot^2 t}$ (b) $\dfrac{1 - \sec^2 t}{1 + \cot^2 t}$

18. (a) $\dfrac{1 - \csc^2 t}{1 + \tan^2 t}$ (b) $\dfrac{1 - \cot^2 t}{\tan^2 t}$

19. (a) $\dfrac{\cos t}{\sec t - \tan t}$ (b) $\dfrac{1}{1 + \sin t} + \dfrac{1}{1 - \sin t}$

20. (a) $\dfrac{1}{1 + \cos t} + \dfrac{1}{1 - \cos t}$ (b) $\dfrac{\sin t}{\csc t - \cot t}$

In Exercises 21 through 26, use a calculator to evaluate to four significant digits the function value.

21. (a) $\sin 2.42$ (b) $\cos 2.42$ (c) $\tan 3.57$
    (d) $\sec 3.57$ (e) $\cot 5.38$ (f) $\csc 5.38$

22. (a) $\sin 4.26$ (b) $\cos 4.26$ (c) $\tan 5.84$
    (d) $\sec 5.84$ (e) $\cot 1.97$ (f) $\csc 1.97$

23. (a) $\cos 18.31$ (b) $\sin(-14.86)$ (c) $\tan 56.00$

24. (a) $\cos(-11.43)$ (b) $\sin 33.00$ (c) $\tan(-10.53)$

25. (a) $\cos 3.49$ (b) $\sec 3.49$ (c) $\tan \frac{12}{5}\pi$
    (d) $\cot(-\frac{8}{3}\pi)$

26. (a) $\sin \frac{7}{8}\pi$ (b) $\csc \frac{7}{8}\pi$ (c) $\cos(-7.32)$
    (d) $\tan 12.28$

In Exercises 27 through 30, find the period of the function.

27. (a) $\sin 5t$ (b) $\cos \frac{1}{5}t$
    (c) $3 \tan \frac{5}{4}t$ (d) $5 \sec \frac{3}{4}t$

28. (a) $\cos 4t$ (b) $\sin \frac{1}{4}t$
    (c) $7 \cot \frac{2}{3}t$ (d) $2 \csc \frac{7}{3}t$

29. (a) $2 \cos 6\pi t$ (b) $\sin \frac{3}{5}\pi t$
    (c) $5 \tan 2\pi t$ (d) $\frac{1}{2} \cot \frac{1}{3}\pi t$

30. (a) $4 \sin 5\pi t$ (b) $\frac{2}{3} \cos \frac{3}{2}\pi t$
    (c) $\frac{1}{4} \tan \pi t$ (d) $3 \cot \frac{3}{4}\pi t$

In Exercises 31 and 32, use the periodicity of the sine, cosine, secant, and cosecant functions as well as values of $\sin t$, $\cos t$, $\sec t$, *and* $\csc t$, *when* $0 \leq t < 2\pi$ *to find the function value.*

31. (a) $\sin \frac{21}{4}\pi$ (b) $\cos \frac{14}{3}\pi$
    (c) $\sec(-\frac{11}{6}\pi)$ (d) $\sin(-5\pi)$
    (e) $\cos \frac{11}{2}\pi$ (f) $\csc \frac{13}{2}\pi$

32. (a) $\sin \frac{8}{3}\pi$ (b) $\cos \frac{13}{6}\pi$
    (c) $\csc(-\frac{17}{4}\pi)$ (d) $\sin 7\pi$
    (e) $\cos(-\frac{9}{2}\pi)$ (f) $\sec(-4\pi)$

In Exercises 33 and 34, use the periodicity of the tangent and cotangent functions as well as the values of $\tan t$ *and* $\cot t$ *when* $0 \leq t < \pi$ *to find the function value.*

33. (a) $\tan \frac{5}{4}\pi$ (b) $\cot \frac{5}{3}\pi$
    (c) $\tan(-\frac{1}{2}\pi)$ (d) $\tan \frac{14}{3}\pi$
    (e) $\cot(-\frac{11}{6}\pi)$ (f) $\cot(-6\pi)$

34. (a) $\tan \frac{13}{6}\pi$ (b) $\cot(-\frac{1}{4}\pi)$
    (c) $\cot(-\pi)$ (d) $\tan(-\frac{4}{3}\pi)$
    (e) $\cot \frac{17}{4}\pi$ (f) $\tan(-\frac{7}{2}\pi)$

In Exercises 35 through 38, use the graph of the given function plotted on your graphics calculator to approximate the function value to three significant digits. Then check your approximation by computing the function value by using the appropriate key, or keys, on your calculator.

35. (a) $\sin 1.27$ (b) $\cos(-0.703)$

36. (a) $\sin(-0.384)$ (b) $\cos 2.39$

37. (a) $\tan 1.65$ (b) $\cot 4.11$

38. (a) $\sec 0.602$ (b) $\csc 4.85$

In Exercises 39 through 42, use the graph of the given function plotted on your graphics calculator to approximate the values of t to three significant digits such that $0 \leq t < 2\pi$ *for which the equation is satisfied.*

39. (a) $\sin t = 0.542$ (b) $\cos t = -0.287$

40. (a) $\sin t = -0.964$ (b) $\cos t = 0.750$

41. (a) $\tan t = 8.99$ (b) $\csc t = -1.03$

42. (a) $\tan t = -0.389$ (b) $\sec t = 5.56$

43. A weight suspended from a spring is vibrating vertically, and $f(t)$ centimeters is the directed distance of the weight from its central position (the origin) at t seconds with the positive direction upward. If $f(t) = 2 \sin 3t$, (a) determine the period of f. (b) Sketch the graph of f and check your graph on your graphics calculator. Use your graph on your graphics calculator to estimate the position of the weight at (c) 0 sec; (d) 1 sec; (e) 2 sec; and (f) 6 sec. Check your estimates by computing $f(0)$, $f(1)$, $f(2)$, and $f(6)$.

44. Do Exercise 43 if $f(t) = 3 \cos 4t$.

In Exercises 45 through 68, sketch the graph of the function. Check your graph on your graphics calculator.

45. $f(t) = \cos 2t$ 46. $f(t) = \sin 3t$

47. $g(t) = 2 \cos t$ 48. $g(t) = 3 \sin t$

49. $f(t) = \sin 2\pi t$ 50. $f(t) = \cos \frac{1}{2}\pi t$

51. $g(t) = 2 \sin \pi t$ 52. $g(t) = \frac{1}{2} \cos \pi t$

53. $f(t) = \sin(t - \frac{1}{2}\pi t)$ 54. $f(t) = \cos(t - \pi)$

55. $g(t) = \cos(t + \frac{1}{3}\pi)$ 56. $g(t) = \sin(t + \frac{1}{6}\pi)$

57. $f(t) = \tan 2t$ 58. $f(t) = \cot \frac{1}{2}t$

59. $g(t) = \frac{1}{2} \cot t$ 60. $g(t) = 2 \tan t$

61. $f(t) = \tan 3\pi t$ 62. $f(t) = \cot 2\pi t$

63. $f(t) = 3 \csc t$ 64. $f(t) = 2 \sec t$

65. $g(t) = 2 \sec \frac{1}{3}\pi t$ 66. $g(t) = 3 \csc \frac{1}{2}\pi t$

67. $f(t) = \cot(t + \frac{1}{3}\pi)$ 68. $f(t) = \frac{1}{2} \tan(t - \frac{1}{4}\pi)$

In Exercises 69 and 70, the function f is a mathematical model describing damped harmonic motion. In each exercise do the following: (a) Show that $F(t) \leq f(t) \leq G(t)$. (b) Plot the graphs of the three functions in the viewing rectangle $[0, 2\pi]$ by $[-1, 1]$. (c) Determine algebraically the values of t at all points of intersection of the graph of f with the graphs of F and G. Which of these values are in the interval $[0, 2\pi]$? (d) Determine algebraically all t intercepts of the graph of f. Which of these intercepts are in the interval $[0, 2\pi]$?

69. $f(t) = e^{-t/4} \sin 6t$; $F(t) = -e^{-t/4}$; $G(t) = e^{-t/4}$

70. $f(t) = 2^{-t/8} \cos 3t$; $F(t) = -2^{-t/8}$; $G(t) = 2^{-t/8}$

In Exercises 71 and 72, the function f is a mathematical model describing resonance. In each exercise do the following: (a) Show that $F(t) \leq f(t) \leq G(t)$. (b) Plot the graphs of the three functions in the viewing rectangle $[0, \pi]$, by $[-10, 10]$. (c) Determine algebraically the values of t at all points of intersection of the graph of f with the graphs of F and G. Which of these values are in the interval $[0, \pi]$? (d) Determine algebraically all t intercepts of the graph of f. Which of these intercepts are in the interval $[0, \pi]$?

71. $f(t) = 3^{t/2} \cos 4t$; $F(t) = -3^{t/2}$; $G(t) = 3^{t/2}$

72. $f(t) = e^{t/2} \sin 6t$; $F(t) = -e^{t/2}$; $G(t) = e^{t/2}$

73. A weight suspended from a spring is set into vibratory motion by pulling it down 6 cm from its central position and then releasing it. It takes 1.8 sec for the weight to complete one vibration. **(a)** Write an equation defining $f(t)$, where $f(t)$ centimeters is the directed distance of the weight from its central position t seconds after the start of the motion and the positive direction is upward. **(b)** Plot the graph of f. Use the graph to estimate the position of the weight when it has been in motion **(c)** 1 sec and **(d)** 4 sec. Check your estimates in parts (c) and (d) by computing $f(1)$ and $f(4)$, respectively.

74. A weight suspended from a spring is vibrating vertically. Suppose the weight passes through its central position at 4 sec and 8 sec. Between these times the weight attains once a maximum displacement of 12 cm above its central position and attains twice a maximum displacement of 12 cm below its central position. The motion is simple harmonic

and is described by an equation of the form $f(t) = a \sin b(t - c)$ where $f(t)$ centimeters is the directed distance of the weight from its central position at t seconds and the positive direction is upward. Find this equation and plot the graph of f.

75. In an electric circuit the electromotive force is $E(t)$ volts at t seconds where $E(t) = 4 \cos 100\pi t$. **(a)** Plot the graph of E for $0.01 \leq t \leq 0.08$. Determine **(b)** the maximum electromotive force; **(c)** the period of E; and **(d)** the frequency of the motion. Find the electromotive force at **(e)** 0.01 sec; **(f)** 0.02 sec; **(g)** 0.05 sec; and **(h)** 0.08 sec.

76. Do Exercise 75 if $E(t) = 6 \sin 50\pi t$.

77. A wave produced by a simple sound is described by the equation $F(t) = 0.04 \sin 1000t$ where $F(t)$ dynes per square centimeter is the difference between the atmospheric pressure and the air pressure at the eardrum at t seconds. **(a)** Plot the graph of F for $0.001 \leq t \leq 0.05$. Determine **(b)** the pressure amplitude and **(c)** the frequency. Find the difference between the atmospheric pressure and the air pressure at the eardrum at **(d)** 0.001 sec; **(e)** 0.01 sec; **(f)** 0.02 sec; and **(g)** 0.05 sec.

78. Do Exercise 77 if $F(t) = 0.008 \sin 600t$.

79. A 30-cycle alternating current is described by an equation of the form $I(t) = a \sin b(t - c)$ where $I(t)$ amperes is the current at t seconds. The maximum current is 40 amperes, and the current is 20 amperes at 0.2 sec. Write the equation and plot the graph of I.

80. In a particular electric circuit the electromotive force at t seconds is $E(t)$ volts, where $E(t) = 200 \sin 120\pi t$, and the current is $I(t)$ amperes, where $I(t) = 50 \sin 120\pi(t - \frac{1}{180})$. Sketch the graphs of E and I on the same set of coordinate axes. Check your graphs on your graphics calculator in the same viewing rectangle. Does the current lag or lead the electromotive force and by how much?

∫∫ dy/dx In Exercises 81 and 82, use Taylor polynomial (2) or (3) of Section 7.2 to find the function value to four significant digits.

81. (a) $\sin 0.49$ **(b)** $\cos 2.63$

82. (a) $\sin 3.58$ **(b)** $\cos(-1.07)$

In Exercises 83 and 84, (a) plot the graph of the function for $t \in [-\frac{1}{4}\pi, 0) \cup (0, \frac{1}{4}\pi]$. (b) The function is not defined at 0, but what value does $f(t)$ appear to be approaching as t approaches 0? Compute the following values on a calculator: (c) $f(0.1)$ and $f(-0.1)$; (d) $f(0.01)$ and $f(-0.01)$; (e) $f(0.001)$ and $f(-0.001)$. (f) Do the function values in parts (c)–(e) agree with your answer in part (b)?

83. $f(t) = \dfrac{t^2 + 3t}{\sin t}$

84. $f(t) = \dfrac{\sin 6t}{\sin 3t}$

 In Exercises 85 and 86, (a) plot the graph of the function for $t \in [0, \frac{1}{2}\pi) \cup (\frac{1}{2}\pi, \pi]$. (b) The function is not defined at $\frac{1}{2}\pi$, but what value does $f(t)$ appear to be approaching as t approaches $\frac{1}{2}\pi$? Compute the following values on a calculator: (c) $f(1.56)$ and $f(1.58)$; (d) $f(1.569)$ and $f(1.571)$; (e) $f(1.5707)$ and $f(1.5709)$. (f) Do the function values in parts (c)–(e) agree with your answer in part (b)?

85. $f(t) = \dfrac{1 - \sin t}{\frac{1}{2}\pi - t}$

86. $f(t) = \dfrac{\frac{1}{2}\pi - t}{\cos t}$

Trigonometric Functions of Angles

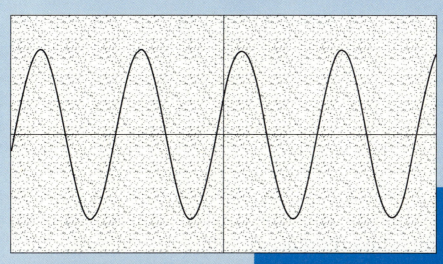

In geometry an angle is thought of as the union of two rays, called the sides, having a common endpoint called the vertex. We define an angle in Section 8.1 as a rotation, and then in the next section define the trigonometric functions so that their domains are measurements of angles. We put these functions to use to solve right triangles as well as oblique triangles, those not containing a right angle. This traditional approach to the trigonometric functions leads to applications in many fields including surveying, navigation, astronomy, and physics.

8.1 ANGLES AND THEIR MEASUREMENT

An **angle** in the plane is obtained by rotating a ray about its endpoint, called the **vertex.** The original position of the ray is called the **initial side.** An angle having its vertex at the origin and its initial side lying on the positive side of the *x* axis is said to be in **standard position.** Any angle is congruent to some angle in standard position. Figure 1 shows an angle *AOB* in standard position with *OA* as the initial side. The other side, *OB*, is called the **terminal side.** The angle *AOB* can be formed by rotating the side *OA* to the side *OB*, and under such a rotation the point *A* moves to the point *B* along the circumference of a circle having its center at *O* and radius $|\overline{OA}|$. The angle is **positive** if *OA* is rotated in a counterclockwise direction to *OB* and **negative** if it is rotated in a clockwise direction.

Greek letters are often used to represent angles, and the direction of rotation is indicated by an arc with an arrow at its endpoint. Figure 2 shows a positive angle α and a negative angle β. Figure 3 shows a positive angle γ and a negative angle θ.

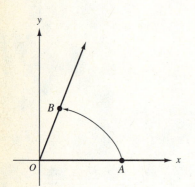

FIGURE 1

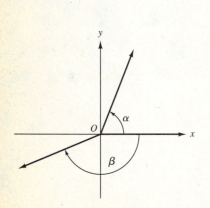

FIGURE 2

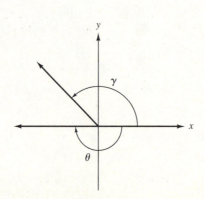

FIGURE 3

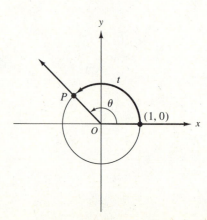

FIGURE 4

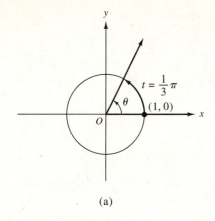

(a)

We consider an angle to be in the quadrant containing its terminal side. However, if the terminal side lies on an axis, the angle is said to be **quadrantal.** In Figure 2, α is in the first quadrant and β is in the third quadrant. In Figure 3, γ is in the second quadrant and θ is a quadrantal angle.

Consider an angle θ in standard position, and let its terminal side intersect the unit circle U at P, the terminal point of the arc of length t measured along U from the point $(1, 0)$. See Figure 4. Because t is a real number, there is a one-to-one correspondence between the set R of real numbers and all angles θ in standard position. Figure 5 shows the angles when t is $\frac{1}{3}\pi, \frac{1}{2}\pi, -\frac{3}{4}\pi$, and $-\frac{3}{2}\pi$. The number t corresponding to the angle θ is a measure of the size of the angle. Furthermore, when t is positive, the angle has a counterclockwise rotation, and when t is negative, the angle's rotation is clockwise. The measurement of the angle for which $t = 1$ is called a *radian*. See Figure 6.

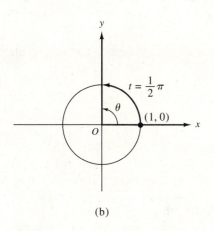

(b)

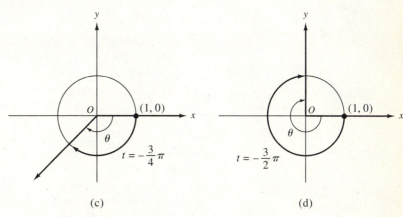

(c)   (d)

FIGURE 5

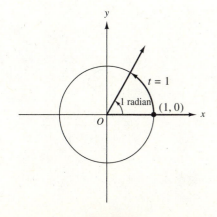

FIGURE 6

DEFINITION   **Radian**

If an angle has its vertex at the center of the unit circle U and intercepts on U an arc of length 1, it has a measurement of 1 **radian.**

▷ **ILLUSTRATION 1**

The radian measures of the angles in Figure 5 are $\frac{1}{3}\pi, \frac{1}{2}\pi, -\frac{3}{4}\pi$, and $-\frac{3}{2}\pi$.

◀

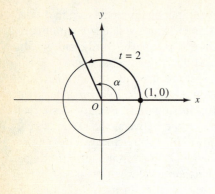

$$m^R(\alpha) = 2$$

FIGURE 7

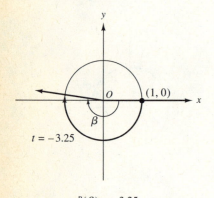

$$m^R(\beta) = -3.25$$

FIGURE 8

If the measurement of an angle θ is t radians, we write

$$m^R(\theta) = t$$

This equality is sometimes read as "the radian measure of angle θ is t."

▷ **ILLUSTRATION 2**

(a) If an angle is denoted by α, and we have

$$m^R(\alpha) = 2$$

the equation means that the radian measure of α is 2. Figure 7 shows this angle α that intercepts on U an arc of length 2, and the rotation for α is counterclockwise.

(b) If β denotes an angle and

$$m^R(\beta) = -3.25$$

then the radian measure of β is -3.25, β intercepts on U an arc of length -3.25, and the rotation for β is clockwise. See Figure 8. ◄

Because the circumference of the unit circle is 2π, one complete revolution of the terminal side from the initial side in the counterclockwise direction generates an angle of radian measure 2π. See Figure 9, where we have designated the angle by writing its radian measure. More than one complete revolution in the counterclockwise direction generates an angle of radian measure greater than 2π. For instance, in Figure 10 is an angle having radian measure $\frac{9}{4}\pi$. Observe that the angle of radian measure $\frac{1}{4}\pi$ has the same terminal side. Another angle having this same terminal side is the one for which the radian measure is $-\frac{7}{4}\pi$. In fact, there are an unlimited number of angles having this terminal side. These angles are called *coterminal*.

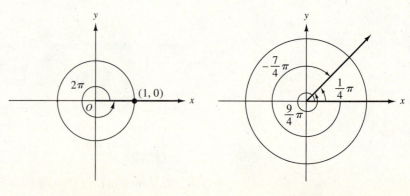

FIGURE 9   **FIGURE 10**

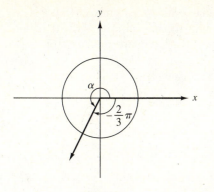

FIGURE 11

In general, **coterminal** angles are those in standard position that have the same terminal side.

▶ **EXAMPLE 1** *Finding the Radian Measure of an Angle Coterminal with a Given Angle*

Find the radian measure of the smallest positive angle coterminal with the angle having the given radian measure and sketch both angles: **(a)** $-\frac{2}{3}\pi$; **(b)** $\frac{11}{4}\pi$; **(c)** 7.15; **(d)** -0.54.

Solution

(a) An angle of radian measure $-\frac{2}{3}\pi$ appears in Figure 11, where the angle is designated by its radian measure. In the figure the required angle is designated by α, and $m^R(\alpha) = 2\pi - \frac{2}{3}\pi$. Therefore $m^R(\alpha) = \frac{4}{3}\pi$.

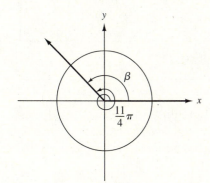

FIGURE 12

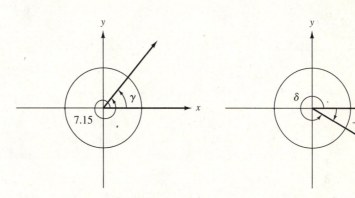

FIGURE 13   **FIGURE 14**

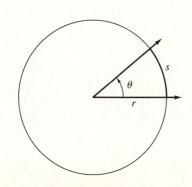

FIGURE 15

(b) Figure 12 shows an angle of radian measure $\frac{11}{4}\pi$. If β is the required angle, $m^R(\beta) = \frac{11}{4}\pi - 2\pi$. Hence $m^R(\beta) = \frac{3}{4}\pi$.

(c) Let γ be the required angle. See Figure 13. To the nearest one-hundredth, 2π is 6.28. Thus $m^R(\gamma) = 7.15 - 6.28$, and so $m^R(\gamma) = 0.87$.

(d) See Figure 14. Let δ be the required angle. Then $m^R(\delta) = 6.28 - 0.54$. Therefore $m^R(\delta) = 5.74$.   ◄

A **central angle** of a circle is one whose vertex is at the center of the circle. Figure 15 shows a circle of radius r and a central angle θ. The angle intercepts on the circle an arc of length s. We now proceed to obtain a formula relating r, s, and t, the radian measure of θ.

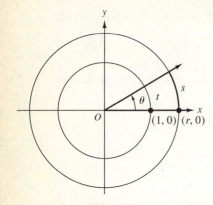

FIGURE 16

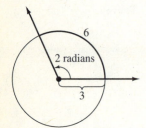

FIGURE 17

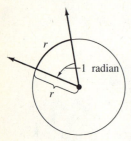

FIGURE 18

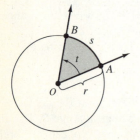

FIGURE 19

We construct a rectangular cartesian coordinate system such that θ is in standard position. Assume $r > 1$, so that the unit circle U is within the given circle. See Figure 16. The length of the arc intercepted by θ on U is t. From a theorem in geometry, the ratio of the arc lengths t and s is equal to the ratio of the radii of the two circles. Thus

$$\frac{t}{s} = \frac{1}{r}$$

$$rt = s$$

We have proved the following theorem.

THEOREM 1

If r is the radius of a circle and t is the radian measure of a central angle that intercepts on the circle an arc of length s, then

$$s = rt$$

▷ **ILLUSTRATION 3**

If in the formula of Theorem 1, $r = 3$ and $t = 2$, then $s = 6$. This result states that on a circle of radius 3 units, a central angle of 2 radians intercepts on the circle an arc of length 6 units. Refer to Figure 17.   ◀

Observe that if $r = 1$ in the formula of Theorem 1, then $s = t$. Therefore on the unit circle U the length of the intercepted arc is the radian measure of the central angle. Also observe that if $t = 1$, the formula becomes $s = r$. This equation states that an arc of a circle equal in length to the radius subtends a central angle of 1 radian. See Figure 18.

A **sector** of a circle is the region bounded by an arc of the circle and the sides of a central angle. For the circle in Figure 19, r units is the radius, t is the radian measure of the central angle, s units is the length of the intercepted arc, and the shaded region is the sector AOB. From geometry, the ratio of the area of the sector to the area of the circle (given by πr^2) is equal to the ratio of the length of the intercepted arc to the circumference of the circle (given by $2\pi r$). Thus, if K square units is the area of the sector, then

$$\frac{K}{\pi r^2} = \frac{s}{2\pi r}$$

$$K = \tfrac{1}{2} rs$$

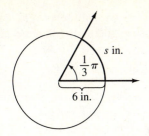

$\frac{1}{3}\pi$ s in.

6 in.

FIGURE 20

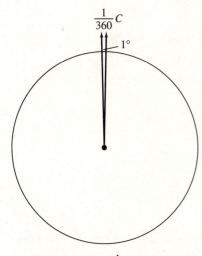

$\frac{1}{360}C$

$1°$

Length of arc is $\frac{1}{360}C$ where C

is the circumference of circle

FIGURE 21

Replacing s by rt (from Theorem 1), we obtain

$$K = \tfrac{1}{2}r^2t$$

▶ **EXAMPLE 2** *Finding the Arc Length and Area of a Sector of a Circle*

A circle of radius 6 in. has a sector whose central angle has radian measure $\frac{1}{3}\pi$. Find the arc length and area of the sector.

Solution   Refer to Figure 20. We are given $r = 6$ and $t = \frac{1}{3}\pi$. With these values

$$s = rt$$
$$= 6(\tfrac{1}{3}\pi)$$
$$= 2\pi$$

Thus the arc length is 2π in.

If K square inches is the area of the sector, then with the given values of r and t

$$K = \tfrac{1}{2}r^2t$$
$$= \tfrac{1}{2}(6)^2(\tfrac{1}{3}\pi)$$
$$= 6\pi$$

Hence the area of the sector is 6π in.^{2} ◀

The *degree* is another unit of angle measurement. If a circle has a central angle subtended by an arc whose length is $\frac{1}{360}$ of the circumference of the circle, the angle is said to have **degree measure** 1. A measurement of 1 degree is written $1°$. See Figure 21. Observe that the definition of degree measure is independent of the radius of the circle. Figure 22 shows some angles in standard position and their measurements in degrees.

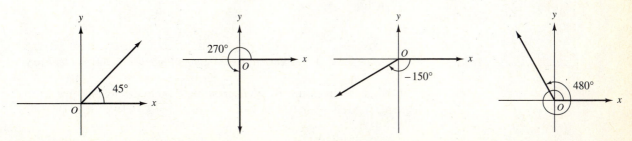

FIGURE 22

An angle formed by one complete revolution, so that the terminal side coincides with the initial side, has degree measure 360 and radian measure 2π. It follows that

$$180° \sim \pi \text{ radians}$$

where the symbol $\sim$ (read "corresponds to") indicates that the given measurements are for congruent angles. We thus have the following correspondence between angle measurements in degrees and those in radians:

$$1° \sim \frac{\pi}{180} \text{ radian} \quad \text{and} \quad 1 \text{ radian} \sim \frac{180°}{\pi}$$

If 3.1416 is taken as an approximation for π, we obtain

$$1° \sim 0.017453 \text{ radian} \quad \text{and} \quad 1 \text{ radian} \sim 57.296°$$

From this correspondence between degrees and radians, the measurement of an angle can be converted from one system of units to the other.

▷ **ILLUSTRATION 4**

(a) $162° \sim 162 \cdot \dfrac{\pi}{180}$ radians

$162° \sim \dfrac{9\pi}{10}$ radians

(b) $\dfrac{5\pi}{12}$ radians $\sim \dfrac{5\pi}{12} \cdot \dfrac{180°}{\pi}$

$\dfrac{5\pi}{12}$ radians $\sim 75°$ ◀

Table 1 gives the corresponding degree and radian measures for certain angles.

We use the notation $m°(\theta)$ to indicate the degree measure of an angle θ. It follows from the preceding discussion that $m°(\theta)$ and $m^R(\theta)$ are related by the equation

$$m°(\theta) = \frac{180}{\pi} m^R(\theta)$$

$$\Leftrightarrow \quad m^R(\theta) = \frac{\pi}{180} m°(\theta)$$

Table 1

Degree Measure	Radian Measure
30	$\frac{1}{6}\pi$
45	$\frac{1}{4}\pi$
60	$\frac{1}{3}\pi$
90	$\frac{1}{2}\pi$
120	$\frac{2}{3}\pi$
135	$\frac{3}{4}\pi$
150	$\frac{5}{6}\pi$
180	π
270	$\frac{3}{2}\pi$
360	2π

▶ **EXAMPLE 3**   *Converting from Radian Measure to Degree Measure*

Find the degree measure to the nearest one-hundredth of a degree for the angle having the given radian measure (let $\pi \approx 3.1416$):

(a) $m^R(\alpha) = \dfrac{5}{7}\pi$

(b) $m^R(\beta) = 0.3826$

Solution

(a) $m°(\alpha) = \dfrac{180}{\pi} \cdot \dfrac{5\pi}{7}$

≈ 128.57

(b) $m°(\beta) = \dfrac{180}{\pi}(0.3826)$

≈ 21.92

Therefore angle α has a measurement of $128.57°$ and angle β has a measurement of $21.92°$. ◀

As in Example 3, we have been using decimals for measurements of less than 1 degree. Another way of dealing with such measurements is to use minutes and seconds. One minute is $\frac{1}{60}$ of a degree; that is, 60 minutes is equivalent to 1 degree. Also 1 second is $\frac{1}{60}$ of a minute; thus 60 seconds is equivalent to 1 minute. Obviously, 1 second is $\frac{1}{3600}$ of a degree, and 3600 seconds is equivalent to 1 degree. The symbol for minutes is ′, and the symbol for seconds is ″. Thus $26°14′46″$ is a measurement of 26 degrees, 14 minutes, and 46 seconds.

We will express angle measurements of less than 1 degree by decimals. If you are given an angle measurement in minutes and seconds, you can convert it easily to a form using decimals. For example, to the nearest one-hundredth of a degree,

$$26°14′46″ = (26 + \tfrac{14}{60} + \tfrac{46}{3600})°$$
$$= (26 + 0.233 + 0.013)°$$
$$= 26.25°$$

From time to time we write equations such as

$$\theta = 45°$$

where the symbol for an angle, θ, is equated to the measurement of the angle, $45°$. It is understood that such an equation indicates that the degree measure of the angle is 45. We also write inequalities such as

$$0° \leq \theta < 360°$$

which indicates that the interval containing the degree measure of θ is $[0, 360)$.

▶ **EXAMPLE 4** *Solving a Word Problem Involving Arc Length*

Find the distance on the surface of the earth from a point having latitude $38.40°$ N to the closest point on the equator. Assume that the earth is a sphere of radius 3960 miles.

Solution Refer to Figure 23, where C is at the center of the earth, P is

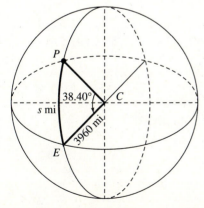

FIGURE 23

the given point, and E is the point on the equator closest to P. We wish to find the length of the arc from E to P on the circle with center at C and radius 3960 miles. Let s miles be this length. From Theorem 1 $s = rt$, where $r = 3960$ and t is the radian measure of the angle at C. We first compute t by converting 38.40° to radians.

$$38.40° \sim 38.40\left(\frac{\pi}{180}\right)$$

$$\approx 0.6702$$

Thus

$$s = rt$$
$$= 3960(0.6702)$$
$$= 2654$$

Conclusion: The distance is 2654 miles.   ◀

EXERCISES 8.1

In Exercises 1 through 6, show by a diagram the angle having the given radian measure.

1. (a) $\frac{1}{6}\pi$ **(b)** $\frac{2}{3}\pi$ **(c)** π **(d)** $\frac{5}{4}\pi$ **(e)** $\frac{11}{6}\pi$

2. (a) $\frac{1}{4}\pi$ **(b)** $\frac{5}{6}\pi$ **(c)** $\frac{4}{3}\pi$ **(d)** $\frac{3}{2}\pi$ **(e)** $\frac{7}{4}\pi$

3. (a) $-\frac{1}{4}\pi$ **(b)** $-\frac{1}{2}\pi$ **(c)** $-\frac{5}{6}\pi$ **(d)** $-\frac{5}{4}\pi$ **(e)** $-\frac{5}{3}\pi$

4. (a) $-\frac{1}{6}\pi$ **(b)** $-\frac{3}{4}\pi$ **(c)** $-\pi$ **(d)** $-\frac{4}{3}\pi$ **(e)** $-\frac{11}{6}\pi$

5. (a) 0.78 **(b)** 3   **(c)** −2   **(d)** 5.20   **(e)** −4.35

6. (a) 2.35 **(b)** −1   **(c)** 4   **(d)** −1.80 **(e)** −6.18

In Exercises 7 through 10, find the radian measure of the smallest positive angle coterminal with the angle having the given radian measure and sketch both angles.

7. (a) $-\frac{3}{4}\pi$ **(b)** $\frac{19}{6}\pi$ **(c)** $\frac{7}{2}\pi$ **(d)** $-\frac{7}{3}\pi$

8. (a) $-\frac{7}{6}\pi$ **(b)** $\frac{8}{3}\pi$ **(c)** $\frac{11}{4}\pi$ **(d)** $-\frac{7}{2}\pi$

9. (a) 7.28 **(b)** 9   **(c)** −4.25   **(d)** −11

10. (a) 8.28 **(b)** −2.63   **(c)** −14   **(d)** 10

In Exercises 11 through 14, find the equivalent radian measurement for the angle having the given degree measurement.

11. (a) 60°   **(b)** 135°   **(c)** 210°   **(d)** −150°

12. (a) 45°   **(b)** 120°   **(c)** 240°   **(d)** −225°

13. (a) 20°   **(b)** 450°   **(c)** −75°   **(d)** 100°

14. (a) 15°   **(b)** 540°   **(c)** −48°   **(d)** 2°

In Exercises 15 through 18, find the equivalent degree measurement for the angle having the given radian measurement.

15. (a) $\frac{1}{4}\pi$ **(b)** $\frac{2}{3}\pi$ **(c)** $\frac{11}{6}\pi$ **(d)** $-\frac{1}{2}\pi$

16. (a) $\frac{1}{6}\pi$ **(b)** $\frac{4}{3}\pi$ **(c)** $\frac{3}{4}\pi$ **(d)** -5π

17. (a) $\frac{1}{2}$ **(b)** −2   **(c)** 4.78   **(d)** 0.23

18. (a) $\frac{1}{3}$ **(b)** 0.2   **(c)** −2.75   **(d)** 5.66

In Exercises 19 and 20, convert the angle measurement to a form using decimals to the nearest one-hundredth of a degree and then find the equivalent radian measurement.

19. (a) 35°22′12″   **(b)** 102°31′27″

20. (a) 68°53′48″   **(b)** 251°8′14″

In Exercises 21 through 24, find the degree measurement of the smallest positive angle coterminal with the angle having the given degree measurement and sketch both angles.

21. (a) −45° **(b)** 510°   **(c)** −540°   **(d)** −120°

22. (a) $-30°$ **(b)** $585°$ **(c)** $-240°$ **(d)** $-630°$

23. (a) $382.56°$ **(b)** $-118.24°$ **(c)** $-253.85°$ **(d)** $302.36°$

24. (a) $496.58°$ **(b)** $-28.16°$ **(c)** $-342.15°$ **(d)** $197.74°$

In Exercises 25 through 30, find (a) the arc length and (b) the area of the sector of the circle having the radius and central angle α with the given measurement.

25. radius is 9 in.; $m^R(\alpha) = \frac{2}{3}\pi$

26. radius is 8 cm; $m^R(\alpha) = \frac{1}{4}\pi$

27. radius is 6 cm; $m°(\alpha) = 135$

28. radius is 12 in.; $m°(\alpha) = 120$

29. radius is 4.72 in.; $m°(\alpha) = 22.14$

30. radius is 8.53 cm; $m°(\alpha) = 80.35$

In Exercises 31 through 33, assume the radius of the earth is 3960 miles.

31. A point P on the surface of the earth is 1500 miles north of the point on the equator closest to it. Find the latitude of the point P in degree measurement.

32. One nautical mile can be defined as the distance on the surface of the earth from a point having latitude $1'$ N to the closest point on the equator. Show that 1 nautical mile is approximately 1.15 ordinary miles.

33. Two points A and B on the surface of the earth are on the same circle which is a meridian, having center at C

where C is the center of the earth. If A has latitude $10°$ N and B has latitude $4.6°$ S, find the distance between A and B.

34. The end of a pendulum of length 40 cm travels an arc length of 5 cm as it swings through an angle α. Find **(a)** $m^R(\alpha)$ and **(b)** $m°(\alpha)$.

35. An automobile tire has diameter of 36 in. How many revolutions will the wheel make as the automobile travels 1 mile (5280 ft)?

36. If the minute hand of a clock has a length of 6 in., how far does its tip travel in 18 min?

37. If the hour hand of a clock has a length of 4 in., how far does its tip travel in 1 hr and 20 min?

38. A pulley having a diameter of 36 cm is turned by a belt that moves at the rate of 5 m/sec. How many revolutions does the pulley make per second?

39. One *mil* is the measurement of the central angle of a circle that intercepts on the circle an arc equal in length to $\frac{1}{6400}$ of the circumference of the circle. Determine the number of mils in angle α if **(a)** $m°(\alpha) = 34.4$ and **(b)** $m^R(\alpha) = 2.3$.

40. If the measurement of angle α is 34 mils (see Exercise 39), find **(a)** $m°(\alpha)$ and **(b)** $m^R(\alpha)$.

41. The two most common units for measuring angles are radians and degrees. Define these units of measurement and explain their relationship.

8.2 TRIGONOMETRIC FUNCTIONS OF ANGLES

GOALS

1. **Define the six trigonometric functions of angles.**
2. **Find exact trigonometric function values of certain angles.**
3. **Find trigonometric function values of an angle when a point on its terminal side is given.**
4. **Learn the theorem that expresses the trigonometric functions of an acute angle in a right triangle as ratios of lengths of the sides of the triangle.**
5. **Learn exact values of trigonometric functions of quadrantal angles.**
6. **Learn exact values of trigonometric functions of 30°, 45°, and 60°.**
7. **Approximate trigonometric function values of angles on a calculator.**

8. **Define a reference angle.**

9. **Express the trigonometric function value of any angle θ in terms of the corresponding function value of the reference angle associated with θ.**

10. **Determine an exact or approximate trigonometric function value by using a reference angle.**

11. **Find an angle when a trigonometric function value of that angle is given.**

In Chapter 7, we defined trigonometric functions with real-number domains. We now consider trigonometric functions whose domains are angle measurements. A trigonometric function of an angle θ is the corresponding function of the real number t where t is the radian measure of θ.

DEFINITION   The Trigonometric Functions of Angles

If θ is an angle having radian measure t, then

$$\sin \theta = \sin t \qquad \cos \theta = \cos t \qquad \tan \theta = \tan t$$
$$\cot \theta = \cot t \qquad \sec \theta = \sec t \qquad \csc \theta = \csc t$$

When considering a trigonometric function of an angle θ, often the measurement of the angle is used in place of θ. For instance, if $m°(\theta) = 60$ or, equivalently, $m^R(\theta) = \frac{1}{3}\pi$, then in place of $\sin \theta$ we could write $\sin 60°$ or $\sin \frac{1}{3}\pi$. Notice that when the measurement of the angle is in degrees, the degree symbol is written. However, when there is no symbol attached, the measurement of the angle is in radians. For instance, $\cos 2°$ means the cosine of an angle having degree measure 2, while $\cos 2$ means the cosine of an angle having radian measure 2. This is consistent with our discussion in Section 7.1, where $\cos 2$ refers to the cosine of the real number 2 because, by definition, the cosine of an angle having radian measure 2 is equal to the cosine of the real number 2.

▷ **ILLUSTRATION 1**

Figure 1 shows an angle θ for which $m°(\theta) = 60$ and $m^R(\theta) = \frac{1}{3}\pi$. The angle can be designated by the symbol θ or by its measurement in either degrees or radians. By definition, $\cos \theta = \cos \frac{1}{3}\pi$. Thus we can write

$$\cos \theta = \tfrac{1}{2}$$

We could also write

$$\cos 60° = \tfrac{1}{2} \quad \text{and} \quad \cos \tfrac{1}{3}\pi = \tfrac{1}{2} \qquad \blacktriangleleft$$

FIGURE 1

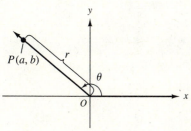

(a)

(b)

> **EXAMPLE 1** *Finding Exact Trigonometric Function Values of Angles*

Sketch the angle and find the function value for each of the following:
(a) sin 45°; **(b)** cos 150°; **(c)** tan(−135°); **(d)** cot 270°.

Solution   The angles are shown in Figure 2. From the definition and the results in Chapter 7, we compute the function values:

(a) $\sin 45° = \sin \frac{1}{4}\pi$

$$= \frac{1}{\sqrt{2}}$$

(b) $\cos 150° = \cos \frac{5}{6}\pi$

$$= -\frac{\sqrt{3}}{2}$$

(c) $\tan(-135°) = \tan(-\frac{3}{4}\pi)$

$$= 1$$

(d) $\cot 270° = \cot \frac{3}{2}\pi$

$$= 0$$

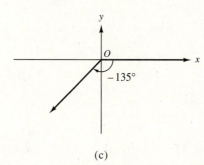

(c)

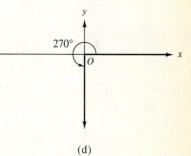

(d)

FIGURE 2   ◀

We now show how trigonometric functions of an angle can be expressed as ratios of real numbers. Figure 3 shows an angle θ whose terminal side is in the second quadrant. The point $P(a, b)$ is any point other than the origin on the terminal side of θ, and the number r is the distance $|\overline{OP}|$; therefore

$$r = \sqrt{a^2 + b^2}$$

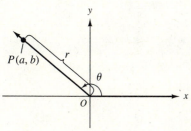

FIGURE 3

FIGURE 4

Let us assume that the point P is not on the unit circle U. Then $r \neq 1$. In Figure 4, where $r > 1$, $Q(x, y)$ is the point on the terminal side of θ that is on U. Thus, $|\overline{OQ}| = 1$. In Figure 4 vertical line segments from P and Q intersect the x axis at the points $M(a, 0)$ and $N(x, 0)$, respectively. The two right triangles ONQ and OMP are similar. Because in similar triangles the ratios of the measures of corresponding sides are equal, we have

$$\frac{|\overline{ON}|}{|\overline{OQ}|} = \frac{|\overline{OM}|}{|\overline{OP}|} \quad \text{and} \quad \frac{|\overline{NQ}|}{|\overline{OQ}|} = \frac{|\overline{MP}|}{|\overline{OP}|}$$

Therefore

$$\frac{|x|}{1} = \frac{|a|}{r} \quad \text{and} \quad \frac{y}{1} = \frac{b}{r}$$

Because $x < 0$ and $a < 0$,

$$x = \frac{a}{r} \quad \text{and} \quad y = \frac{b}{r}$$

Replacing x by $\cos \theta$ and y by $\sin \theta$, we obtain

$$\cos \theta = \frac{a}{r} \quad \text{and} \quad \sin \theta = \frac{b}{r} \tag{1}$$

From a fundamental identity

$$\tan \theta = \frac{\sin \theta}{\cos \theta} \quad \text{if } \cos \theta \neq 0$$

Therefore

$$\tan \theta = \frac{\dfrac{b}{r}}{\dfrac{a}{r}}$$

$$\tan \theta = \frac{b}{a} \quad \text{if } a \neq 0 \tag{2}$$

The same results as in (1) and (2) are obtained no matter which quadrant contains the terminal side of θ. From (1) and (2) and because the cotangent, secant, and cosecant are reciprocals of the tangent, cosine, and sine, respectively, we have the following theorem.

THEOREM 1

If θ is an angle in standard position, $P(a, b)$ is any point other than the origin on the terminal side of θ, and $r = \sqrt{a^2 + b^2}$, then

$$\sin \theta = \frac{b}{r} \qquad\qquad \csc \theta = \frac{r}{b} \quad \text{if } b \neq 0$$

$$\cos \theta = \frac{a}{r} \qquad\qquad \sec \theta = \frac{r}{a} \quad \text{if } a \neq 0$$

$$\tan \theta = \frac{b}{a} \quad \text{if } a \neq 0 \qquad \cot \theta = \frac{a}{b} \quad \text{if } b \neq 0$$

Suppose in the discussion preceding the statement of Theorem 1 we had selected the point $\overline{P}(\overline{a}, \overline{b})$ on the terminal side of θ instead of the point $P(a, b)$. See Figure 5. Because triangles OMP and $O\overline{M}\overline{P}$ are similar, it

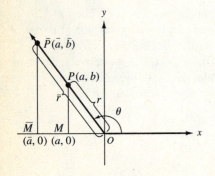

FIGURE 5

follows that

$$\frac{b}{r} = \frac{\bar{b}}{\bar{r}} \qquad \frac{a}{r} = \frac{\bar{a}}{\bar{r}} \qquad \frac{b}{a} = \frac{\bar{b}}{\bar{a}} \qquad \frac{r}{b} = \frac{\bar{r}}{\bar{b}} \qquad \frac{r}{a} = \frac{\bar{r}}{\bar{a}} \qquad \frac{a}{b} = \frac{\bar{a}}{\bar{b}}$$

Therefore we can conclude that the formulas in Theorem 1 are determined only by the position of the terminal side of angle θ and not by the particular point $P(a, b)$ selected on the terminal side.

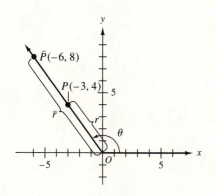

FIGURE 6

▷ **ILLUSTRATION 2**

Let $P(-3, 4)$ and $\bar{P}(-6, 8)$ be two points on the terminal side of angle θ. See Figure 6. If the point P is selected to compute the trigonometric functions of θ, $r = \sqrt{(-3)^2 + 4^2}$; thus $r = 5$, and from Theorem 1 $\sin \theta = \frac{4}{5}$, $\cos \theta = -\frac{3}{5}$, $\tan \theta = -\frac{4}{3}$, $\cot \theta = -\frac{3}{4}$, $\sec \theta = -\frac{5}{3}$, and $\csc \theta = \frac{5}{4}$. If $\bar{P}$ is selected, then $\bar{r} = \sqrt{(-6)^2 + 8^2}$, that is, $\bar{r} = 10$, and from Theorem 1 $\sin \theta = \frac{8}{10}$, $\cos \theta = -\frac{6}{10}$, $\tan \theta = -\frac{8}{6}$, $\cot \theta = -\frac{6}{8}$, $\sec \theta = -\frac{10}{6}$, and $\csc \theta = \frac{10}{8}$, which are the same values.   ◀

▶ **EXAMPLE 2** *Finding Trigonometric Function Values of an Angle When a Point on its Terminal Side Is Given*

If θ is an angle in standard position and the point $P(5, -12)$ is on the terminal side of θ, find the values of the trigonometric functions of θ.

Solution   Figure 7 shows angle θ with the point $P(5, -12)$ on its terminal side. The distance $|\overline{OP}|$ is r, which is $\sqrt{5^2 + (-12)^2} = \sqrt{169}$; thus $r = 13$. Therefore, from Theorem 1, with $a = 5$, $b = -12$, and $r = 13$, we obtain

$$\sin \theta = -\tfrac{12}{13} \qquad \csc \theta = -\tfrac{13}{12}$$

$$\cos \theta = \tfrac{5}{13} \qquad \sec \theta = \tfrac{13}{5}$$

$$\tan \theta = -\tfrac{12}{5} \qquad \cot \theta = -\tfrac{5}{12}$$

◀

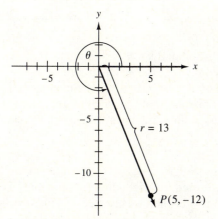

FIGURE 7

When discussing Theorem 1, we did not consider the possibility that the terminal side of θ could lie on either the x axis or the y axis, in which case θ is a quadrantal angle. The formulas of Theorem 1 are also valid for these angles, as shown in the following illustration.

▷ **ILLUSTRATION 3**

(a) Suppose we have an angle of 90°. We can choose any point P on the terminal side of this angle; in particular choose $P(0, 1)$. Then $a = 0$, $b = 1$, and $r = 1$. From the formulas of Theorem 1

$$\sin 90° = \tfrac{1}{1} \qquad \cos 90° = \tfrac{0}{1} \qquad \cot 90° = \tfrac{0}{1} \qquad \csc 90° = \tfrac{1}{1}$$
$$= 1 \qquad\qquad = 0 \qquad\qquad = 0 \qquad\qquad = 1$$

Neither tan 90° nor sec 90° is defined because when using the formulas for tangent and secant, we obtain 0 in the denominator.

(b) If we have an angle of 180°, choose point $P(-1, 0)$ on the terminal side. Then $a = -1$, $b = 0$, and $r = 1$. From the formulas of Theorem 1

$$\sin 180° = \tfrac{0}{1} \qquad \cos 180° = \tfrac{-1}{1} \qquad \tan 180° = \tfrac{0}{-1} \qquad \sec 180° = \tfrac{1}{-1}$$
$$= 0 \qquad\qquad = -1 \qquad\qquad = 0 \qquad\qquad = -1$$

Because $b = 0$, neither cot 180° nor csc 180° is defined.   ◀

In Section 7.6 we determined the sign of a trigonometric function of a real number t by considering the quadrant containing the terminal point of the arc on U of length t. The signs of the trigonometric functions for the four quadrants are given in Table 3 of that section. This table can also be used to determine the sign of a trigonometric function of an angle because the sign depends on the quadrant containing the terminal side of the angle.

▷ **ILLUSTRATION 4**

Suppose θ is an angle in standard position and $\sin \theta > 0$ and $\cos \theta < 0$. We wish to determine the quadrant containing the terminal side of θ. Because $\sin \theta > 0$, the terminal side of θ is in either the first or second quadrant. Because $\cos \theta < 0$, the terminal side of θ is in either the second or third quadrant. For both conditions to hold, the terminal side of θ must be in the second quadrant.   ◀

Because an acute angle is a positive angle of degree measure less than 90, it can be an angle in a right triangle, and the trigonometric functions of an acute angle can be expressed as ratios of the measures of the sides of a right triangle. Figure 8 shows a right triangle having an acute angle θ and a rectangular cartesian coordinate system placed so that θ is in standard position. The following theorem is the result of applying Theorem 1 where

a is the measure of the side adjacent to θ (abbreviated *adj*)
b is the measure of the side opposite θ (abbreviated *opp*)
r is the measure of the hypotenuse (abbreviated *hyp*)

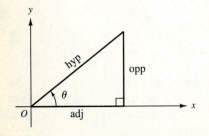

FIGURE 8

If θ is an acute angle in a right triangle,

$$\sin \theta = \frac{\text{opp}}{\text{hyp}} \qquad \csc \theta = \frac{\text{hyp}}{\text{opp}}$$

$$\cos \theta = \frac{\text{adj}}{\text{hyp}} \qquad \sec \theta = \frac{\text{hyp}}{\text{adj}}$$

$$\tan \theta = \frac{\text{opp}}{\text{adj}} \qquad \cot \theta = \frac{\text{adj}}{\text{opp}}$$

▷ **ILLUSTRATION 5**

Figure 9 shows a right triangle with an acute angle θ. The length of the hypotenuse is 8 units, and the length of the side adjacent to θ is 5 units. We can find the values of the six trigonometric functions of θ by the formulas of Theorem 2, but first we must compute the length of the side opposite θ. If z is this length, from the Pythagorean theorem we have

$$z^2 + 5^2 = 8^2$$
$$z^2 + 25 = 64$$
$$z^2 = 39$$
$$z = \sqrt{39}$$

FIGURE 9

Therefore

$$\sin \theta = \frac{\text{opp}}{\text{hyp}} \qquad \csc \theta = \frac{\text{hyp}}{\text{opp}}$$
$$= \frac{\sqrt{39}}{8} \qquad\qquad = \frac{8}{\sqrt{39}}$$

$$\cos \theta = \frac{\text{adj}}{\text{hyp}} \qquad \sec \theta = \frac{\text{hyp}}{\text{adj}}$$
$$= \frac{5}{8} \qquad\qquad = \frac{8}{5}$$

$$\tan \theta = \frac{\text{opp}}{\text{adj}} \qquad \cot \theta = \frac{\text{adj}}{\text{opp}}$$
$$= \frac{\sqrt{39}}{5} \qquad\qquad = \frac{5}{\sqrt{39}}$$

◀

The formulas of Theorem 2 are used in trigonometric applications involving right triangles. This topic is discussed in Section 8.3. The formulas can also be used to determine trigonometric functions of angles of degree measurements 30°, 60°, and 45° as shown in the next two illustrations.

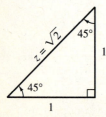

FIGURE 10

▷ **ILLUSTRATION 6**

A theorem from geometry states that in a right triangle having acute angles of measurements 30° and 60°, the length of the side opposite the 30° angle is one-half the length of the hypotenuse. Figure 10 shows such a right triangle with a hypotenuse of length 2 units. Thus the length of the side opposite the 30° angle is 1 unit. To compute the length of the other side, let it be z units, and find z by applying the Pythagorean theorem. The computation is as follows:

$$z^2 + 1^2 = 2^2$$
$$z^2 + 1 = 4$$
$$z^2 = 3$$
$$z = \sqrt{3}$$

From the triangle, we obtain

$$\sin 30° = \frac{1}{2} \qquad \csc 30° = 2$$

$$\cos 30° = \frac{\sqrt{3}}{2} \qquad \sec 30° = \frac{2}{\sqrt{3}}$$

$$\tan 30° = \frac{1}{\sqrt{3}} \qquad \cot 30° = \sqrt{3}$$

and

$$\sin 60° = \frac{\sqrt{3}}{2} \qquad \csc 60° = \frac{2}{\sqrt{3}}$$

$$\cos 60° = \frac{1}{2} \qquad \sec 60° = 2$$

$$\tan 60° = \sqrt{3} \qquad \cot 60° = \frac{1}{\sqrt{3}}$$

◀

FIGURE 11

▷ **ILLUSTRATION 7**

To obtain the trigonometric functions of an angle having measurement 45°, we consider an isosceles right triangle with two sides of length 1 unit. See Figure 11. If z units is the length of the hypotenuse,

$$z^2 = 1^2 + 1^2$$
$$z^2 = 2$$
$$z = \sqrt{2}$$

Thus, from the triangle, we have

$$\sin 45° = \frac{1}{\sqrt{2}} \qquad \csc 45° = \sqrt{2}$$

$$\cos 45° = \frac{1}{\sqrt{2}} \qquad \sec 45° = \sqrt{2}$$

$$\tan 45° = 1 \qquad \cot 45° = 1 \qquad\blacktriangleleft$$

Illustrations 6 and 7 provide a practical method for obtaining the values of the trigonometric functions of 30°, 60°, and 45° or, equivalently, $\frac{1}{6}\pi$, $\frac{1}{3}\pi$, and $\frac{1}{4}\pi$, respectively. Table 1 summarizes these values as well as those for 0 and $\frac{1}{2}\pi$. These values occur often and are exact. You should be able to give these exact values without using a calculator. It is not necessary to memorize them if you recall the methods of obtaining them from right triangles and points on the coordinate axes.

Table 1

t (Radians)	θ (Degrees)	$\sin t$ and $\sin \theta$	$\cos t$ and $\cos \theta$	$\tan t$ and $\tan \theta$	$\cot t$ and $\cot \theta$	$\sec t$ and $\sec \theta$	$\csc t$ and $\csc \theta$
0	$0°$	0	1	0	undefined	1	undefined
$\frac{1}{6}\pi$	$30°$	$\frac{1}{2}$	$\frac{\sqrt{3}}{2}$	$\frac{1}{\sqrt{3}}$	$\sqrt{3}$	$\frac{2}{\sqrt{3}}$	2
$\frac{1}{4}\pi$	$45°$	$\frac{1}{\sqrt{2}}$	$\frac{1}{\sqrt{2}}$	1	1	$\sqrt{2}$	$\sqrt{2}$
$\frac{1}{3}\pi$	$60°$	$\frac{\sqrt{3}}{2}$	$\frac{1}{2}$	$\sqrt{3}$	$\frac{1}{\sqrt{3}}$	2	$\frac{2}{\sqrt{3}}$
$\frac{1}{2}\pi$	$90°$	1	0	undefined	0	undefined	1

Trigonometric function values of angles can be found on a calculator set in the correct mode, either radian or degree, dependent on how the angle is measured.

▶ **EXAMPLE 3** *Approximating Trigonometric Function Values of Angles on a Calculator*

Use a calculator to approximate the function value: **(a)** sin 16.72°; **(b)** cos 154.38°; **(c)** $\tan(-\frac{3}{5}\pi)$; **(d)** sin 5.27; **(e)** sec(−1.26); **(f)** cot 319.4°.

Solution In parts (a), (b), and (f) we set the calculator in degree mode and in parts (c), (d), and (e) in radian mode.

(a) $\sin 16.72° = 0.2877$ **(b)** $\cos 154.38° = -0.9017$

(c) $\tan(-\frac{3}{5}\pi) = 3.0777$ **(d)** $\sin 5.27 = -0.8485$

(e) $\sec(-1.26) = \dfrac{1}{\cos(-1.26)}$ **(f)** $\cot 319.4° = \dfrac{1}{\tan 319.4°}$

$\qquad\qquad\quad = 3.270$ $\qquad\qquad\quad = -1.1667$ ◄

Trigonometric function values of any angle can be determined by knowing function values of acute angles (an angle having positive radian measure less than $\frac{1}{2}\pi$). The procedure involves the concept of a *reference angle*.

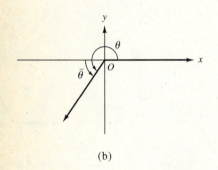

DEFINITION   Reference Angle

The reference angle associated with a nonquadrantal angle θ in standard position is the acute angle $\bar{\theta}$ formed by the terminal side of θ and the x axis.

Four angles in standard position, each having its terminal side in a different quadrant, appear in Figure 12. The reference angle $\bar{\theta}$ associated with angle θ is indicated in the figure.

(a)

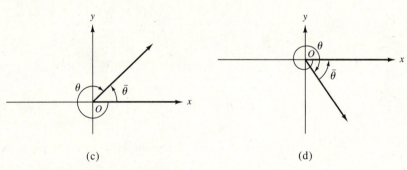

(b)   (c)   (d)

FIGURE 12

▷ **ILLUSTRATION 8**

Figure 13 shows four angles in standard position with $\theta_1 = \frac{2}{3}\pi$, $\theta_2 = 225°$, $\theta_3 = 5.25$, and $\theta_4 = 432°$, and their associated reference angles $\bar{\theta}_1$, $\bar{\theta}_2$, $\bar{\theta}_3$, and $\bar{\theta}_4$, respectively. The computation of the measurements of these reference angles is as follows.

(a) $\bar{\theta}_1 = \pi - \frac{2}{3}\pi$ **(b)** $\bar{\theta}_2 = 225° - 180°$

$\qquad = \frac{1}{3}\pi$ $\qquad\qquad = 45°$

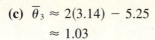

(c) $\bar{\theta}_3 \approx 2(3.14) - 5.25$

≈ 1.03

(d) $\bar{\theta}_4 = 432° - 360°$

$= 72°$

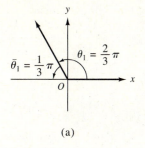

(a)

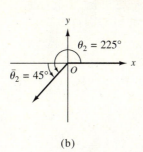

(b)

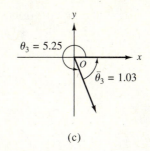

(c)

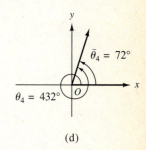

(d)

FIGURE 13   ◀

The next illustration shows how to obtain the reference angle associated with a negative angle.

▷ **ILLUSTRATION 9**

Figure 14 shows the four negative angles $\theta_1 = -150°$, $\theta_2 = -\frac{5}{3}\pi$, $\theta_3 = -202.4°$, and $\theta_4 = -7.36$. The computations of the respective reference angles are as follows.

(a) $\bar{\theta}_1 = 180° - 150°$

$= 30°$

(b) $\bar{\theta}_2 = 2\pi - \frac{5}{3}\pi$

$= \frac{1}{3}\pi$

(c) $\bar{\theta}_3 = 202.4° - 180°$

$= 22.4°$

(d) $\bar{\theta}_4 \approx 7.36 - 2(3.14)$

≈ 1.08

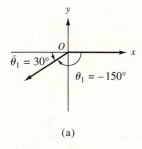

(a)

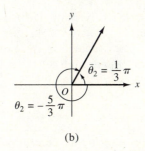

(b)

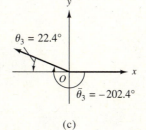

(c)

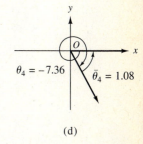

(d)

FIGURE 14   ◀

We now show how reference angles are used to find values of the trigonometric functions. Consider an angle θ in standard position and choose on the terminal side of θ the point $P(x, y)$ where $r = |\overline{OP}|$. The associated reference angle is $\bar{\theta}$. Construct the angle $\bar{\theta}$ in standard position;

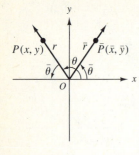

(a)

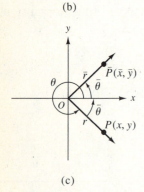

(b)

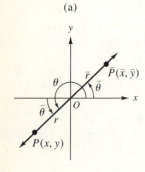

(c)

FIGURE 15

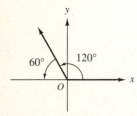

FIGURE 16

this is a first-quadrant angle. Select on the terminal side of $\bar{\theta}$ the point $\bar{P}(\bar{x}, \bar{y})$ so that $\bar{r} = r$. Figure 15 shows angle θ in the second, third, and fourth quadrants. Because of the symmetry of points P and $\bar{P}$ with respect to either the origin or one of the coordinate axes, we have for each position of $P(x, y)$

$$|x| = \bar{x} \quad \text{and} \quad |y| = \bar{y}$$

Therefore

$$|\sin \theta| = \frac{|y|}{r} \qquad |\cos \theta| = \frac{|x|}{r} \qquad |\tan \theta| = \frac{|y|}{|x|}$$

$$= \frac{\bar{y}}{\bar{r}} \qquad\qquad = \frac{\bar{x}}{\bar{r}} \qquad\qquad = \frac{\bar{y}}{\bar{x}}$$

$$= \sin \bar{\theta} \qquad\qquad = \cos \bar{\theta} \qquad\qquad = \tan \bar{\theta}$$

In a similar manner

$$|\csc \theta| = \csc \bar{\theta} \qquad |\sec \theta| = \sec \bar{\theta} \qquad |\cot \theta| = \cot \bar{\theta}$$

From these relationships, we conclude that a trigonometric function value of any angle θ can be found by determining the corresponding function value of the reference angle $\bar{\theta}$ and prefixing the appropriate algebraic sign ($+$ or $-$). The procedure can be summarized as follows:

Using Reference Angles to Find a Trigonometric Function Value of a Nonacute Angle θ

1. Determine the reference angle $\bar{\theta}$ associated with θ.
2. Find the value of the corresponding trigonometric function of $\bar{\theta}$. This can be an exact value if θ is 45°, 30°, or 60°, or it can be an approximate value obtained from a calculator.
3. Affix the proper algebraic sign for the particular function by noting which quadrant contains θ.

▶ **EXAMPLE 4**   *Determining an Exact Trigonometric Function Value by Using a Reference Angle*

Find the exact value of each of the following: **(a)** cos 120°; **(b)** $\cot(-\frac{2}{3}\pi)$.

Solution

(a) An angle of measurement 120° appears in Figure 16. The reference

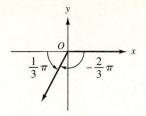

FIGURE 17

angle has measurement 60°. Therefore

$$\cos 120° = -\cos 60°$$
$$= -\tfrac{1}{2}$$

(b) Refer to Figure 17 showing an angle of radian measure $-\tfrac{2}{3}\pi$ and its reference angle of radian measure $\tfrac{1}{3}\pi$. Therefore

$$\cot(-\tfrac{2}{3}\pi) = \cot \tfrac{1}{3}\pi$$
$$= \frac{1}{\sqrt{3}}$$

It is not necessary to use reference angles when trigonometric function values are obtained from a calculator. However, because of the importance of reference angles in later work, you need practice using them. The following example and Exercises 33 through 38 give that practice.

▶ **EXAMPLE 5** *Approximating a Trigonometric Function Value by Using a Reference Angle*

Find an approximate function value by expressing it in terms of a function of the associated reference angle and then using a calculator: **(a)** sin 116.4°; **(b)** tan(−16.8°).

Solution   We set the calculator in degree mode.

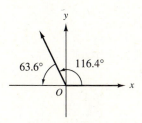

FIGURE 18

(a) Figure 18 shows an angle of measurement 116.4°. The reference angle has measurement 180° − 116.4° = 63.6°.

$$\sin 116.4° = \sin 63.6°$$
$$= 0.8957$$

(b) An angle of measurement −16.8° appears in Figure 19. The reference angle has measurement 16.8°.

$$\tan(-16.8°) = -\tan 16.8°$$
$$= -0.3019$$

◀

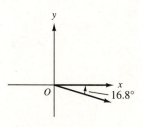

FIGURE 19

The problem of finding an angle from a trigonometric function value of the angle is the *inverse* of determining the function value of a given angle. When a calculator is used, the inverse process is performed by using the $\boxed{\sin^{-1}}$, $\boxed{\cos^{-1}}$, and $\boxed{\tan^{-1}}$ keys. These keys refer to *inverse trigonometric functions*, discussed in Section 9.5.

▷ **ILLUSTRATION 10**

To find the positive angle θ less than 90° for which

$$\cos \theta = 0.8367$$

we set our calculator in degree mode, press the $\boxed{\cos^{-1}}$ key followed by 0.8367, and obtain

$$\theta = 33.21°$$ ◄

▷ **ILLUSTRATION 11**

Suppose we wish to find the positive angle θ less than 90° for which

$$\cot \theta = 2.906$$

To use our calculator and the $\boxed{\tan^{-1}}$ key we write the equation as

$$\tan \theta = \frac{1}{2.906}$$

Then with the calculator in degree mode we press the $\boxed{\tan^{-1}}$ key followed by $\frac{1}{2.906}$, and we obtain

$$\theta = 18.99°$$ ◄

▷ **ILLUSTRATION 12**

Let us determine all θ such that $0° \leq \theta < 360°$ for which

$$\sin \theta = \frac{1}{\sqrt{2}}$$

Because $\sin \theta > 0$, θ may be in either the first or second quadrant. Therefore, there is a θ_1 such that

$$\sin \theta_1 = \frac{1}{\sqrt{2}} \qquad 0° < \theta_1 < 90°$$

and a θ_2 such that

$$\sin \theta_2 = \frac{1}{\sqrt{2}} \qquad 90° < \theta_2 < 180°$$

Figure 20(a) shows that

$$\theta_1 = 45°$$

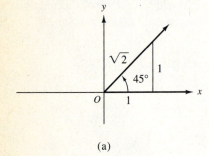

(a)

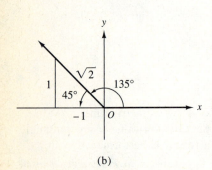

(b)

FIGURE 20

The angle θ_2 is a second-quadrant angle having a reference angle of 45°. See Figure 20(b), which shows that

$$\theta_2 = 180° - 45°$$
$$= 135°$$

◀

▶ **EXAMPLE 6** *Finding Angles Having a Given Trigonometric Function Value*

Find all θ such that $0° \leq \theta < 360°$ for which the given equation is satisfied:
(a) $\tan \theta = -\sqrt{3}$; **(b)** $\cos \theta = -0.4493$.

Solution
(a) Because $\tan \theta < 0$, θ may be in either the second or fourth quadrant. Therefore there is a θ_1 such that

$$\tan \theta_1 = -\sqrt{3} \qquad 90° < \theta_1 < 180°$$

and a θ_2 such that

$$\tan \theta_2 = -\sqrt{3} \qquad 270° < \theta_2 < 360°$$

We first determine the reference angle $\bar{\theta}$ for which $\tan \bar{\theta} = \sqrt{3}$. See Figure 21(a) and observe that $\bar{\theta} = 60°$. The angle θ_1 is a second-quadrant angle having a reference angle of 60°. Refer to Figure 21(b), which shows that

$$\theta_1 = 180° - 60°$$
$$= 120°$$

The angle θ_2 is in the fourth quadrant and has a reference angle of 60°.

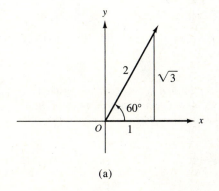

(a)

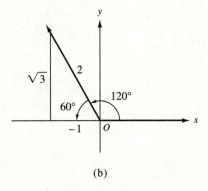

(b)

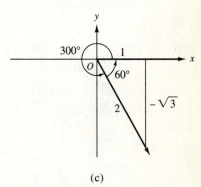

(c)

FIGURE 21

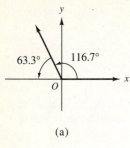

(a)

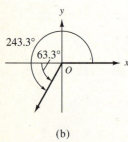

(b)

FIGURE 22

Figure 21(c) shows this angle, and from the figure we see that

$$\theta_2 = 360° - 60°$$
$$= 300°$$

(b) Because $\cos \theta < 0$, θ may be in either the second or third quadrant. Therefore there is a θ_1 such that

$$\cos \theta_1 = -0.4493 \qquad 90° < \theta_1 < 180°$$

and a θ_2 such that

$$\cos \theta_2 = -0.4493 \qquad 180° < \theta_2 < 270°$$

To find the reference angle $\bar{\theta}$ for which $\cos \bar{\theta} = 0.4493$, we use the $\boxed{\cos^{-1}}$ key on our calculator in degree mode, followed by 0.4493, and we obtain $\bar{\theta} = 63.3°$. Thus θ_1 is a second-quadrant angle having a reference angle of 63.3°. Refer to Figure 22(a) and observe that

$$\theta_1 = 180° - 63.3°$$
$$= 116.7°$$

The angle θ_2 is in the third quadrant and has a reference angle of 63.3°. This angle appears in Figure 22(b) and

$$\theta_2 = 180° + 63.3°$$
$$= 243.3°$$ ◀

EXERCISES 8.2

In Exercises 1 through 6, sketch the angle and find the exact function value.

1. (a) $\sin 60°$ **(b)** $\cos 30°$ **(c)** $\tan 45°$

2. (a) $\sin 30°$ **(b)** $\cos 45°$ **(c)** $\tan 60°$

3. (a) $\cos 135°$ **(b)** $\sin 210°$ **(c)** $\tan 300°$

4. (a) $\cos 240°$ **(b)** $\sin 150°$ **(c)** $\tan 315°$

5. (a) $\sin(-30°)$ **(b)** $\cos(-225°)$ **(c)** $\tan(-120°)$

6. (a) $\sin(-135°)$ **(b)** $\cos(-60°)$ **(c)** $\tan(-210°)$

In Exercises 7 through 10, find the four trigonometric function values of the given quadrantal angle that are defined. Which two function values are not defined?

7. $0°$ **8.** $-180°$ **9.** $-90°$ **10.** $270°$

In Exercises 11 through 18, θ is an angle in standard position and the point P is on the terminal side of θ. Find the values of the six trigonometric functions of θ.

11. $P(3, 4)$ **12.** $P(8, 6)$ **13.** $P(-15, 8)$

14. $P(-12, -5)$ **15.** $P(6, -3)$ **16.** $P(-6, 2)$

17. $P(-1, -4)$ **18.** $P(8, -15)$

In Exercises 19 through 22, find the exact values of the six trigonometric functions of the acute angle θ shown in the right triangle.

19.

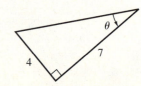

20.

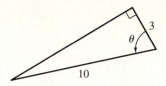

21.

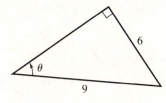

22.

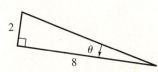

In Exercises 23 through 28, use a calculator to approximate the function value.

23. (a) sin 32.4°   **(b)** cos 57.6°   **(c)** tan 73.7°
    (d) cot 16.3°   **(e)** sec 44.5°   **(f)** csc 45.5°

24. (a) sin 81.1°   **(b)** cos 8.9°   **(c)** tan 25.8°
    (d) cot 64.2°   **(e)** sec 52.3°   **(f)** csc 37.7°

25. (a) sin 75.4°   **(b)** cos 22.3°   **(c)** tan 34.8°
    (d) cot 88.1°   **(e)** sec 52.6°   **(f)** csc 2.7°

26. (a) sin 1.5°   **(b)** cos 48.2°   **(c)** tan 76.9°
    (d) cot 14.6°   **(e)** sec 33.7°   **(f)** csc 80.4°

27. (a) $\sin \frac{3}{8}\pi$ **(b)** cos 0.457   **(c)** tan 1.264
    (d) $\cot \frac{2}{5}\pi$

28. (a) sin 1.302   **(b)** $\cos \frac{5}{11}\pi$ **(c)** $\tan \frac{4}{9}\pi$
    (d) cot 0.839

In Exercises 29 through 32, express the function value in terms of a function value of the associated reference angle. Then determine the exact value. Do not use a calculator.

29. (a) sin 135°   **(b)** cos 210°   **(c)** tan(−240°)
    (d) cot 330°   **(e)** sec 180°   **(f)** csc(−120°)

30. (a) sin 315°   **(b)** cos 120°   **(c)** tan 210°
    (d) cot(−225°)   **(e)** sec 240°   **(f)** csc(−90°)

31. (a) $\sin(-\frac{2}{3}\pi)$ **(b)** $\cos \frac{7}{4}\pi$ **(c)** $\tan \frac{5}{4}\pi$
    (d) $\cot \frac{3}{2}\pi$ **(e)** $\sec(-\frac{7}{6}\pi)$ **(f)** $\csc(-\frac{5}{6}\pi)$

32. (a) $\sin \pi$ **(b)** $\cos \frac{11}{6}\pi$ **(c)** $\tan \frac{2}{3}\pi$
    (d) $\cot(-\frac{5}{6}\pi)$ **(e)** $\sec(-\frac{3}{4}\pi)$ **(f)** $\csc(-\frac{7}{4}\pi)$

In Exercises 33 through 38, express the function value in terms of a function value of the associated reference angle. Then approximate the value on a calculator.

33. (a) sin 124°18′   **(b)** cos 243°36′
    (c) tan(−15°6′)

34. (a) cos 151°12′   **(b)** sin(−98°24′)
    (c) tan 196°42′

35. (a) cos(−172.4°)   **(b)** sin(−263.8°)
    (c) tan 200.2°

36. (a) sin(−193.7°)   **(b)** cos(−10.9°)
    (c) tan 108.3°

37. (a) cot(−169°30′)   **(b)** sec 292.6°
    (c) csc 175.5°

38. (a) cot 348.1°   **(b)** sec(−304.9°)
    (c) csc(−126.5°)

In Exercises 39 and 40, find all θ such that 0° ≤ θ < 360° for which the equation is satisfied.

39. (a) $\sin \theta = \frac{1}{2}$ **(b)** $\cos \theta = -\frac{1}{2}$ **(c)** $\tan \theta = 1$

40. (a) $\sin \theta = \dfrac{\sqrt{3}}{2}$ **(b)** $\cos \theta = \dfrac{1}{\sqrt{2}}$
    (c) $\tan \theta = -\dfrac{1}{\sqrt{3}}$

In Exercises 41 and 42, find all θ such that 0 ≤ θ < 2π for which the equation is satisfied.

41. (a) $\sin \theta = -\dfrac{1}{\sqrt{2}}$ **(b)** $\cos \theta = \dfrac{\sqrt{3}}{2}$
    (c) $\tan \theta = 0$

42. (a) $\sin \theta = -\frac{1}{2}$ **(b)** $\cos \theta = 0$
    (c) $\tan \theta = \sqrt{3}$

In Exercises 43 and 44, find all θ to the nearest one-tenth of a degree such that 0° ≤ θ < 360° for which the equation is satisfied.

43. (a) sin θ = 0.5165   **(b)** cos θ = 0.1685
    (c) tan θ = 2.367   **(d)** cot θ = 1.053

44. (a) sin θ = 0.9078   **(b)** cos θ = 0.7638
    (c) tan θ = 0.2035   **(d)** cot θ = 7.595

In Exercises 45 and 46, find all θ to four significant digits such that 0 ≤ θ < 2π for which the equation is satisfied.

45. (a) sin θ = 0.4207   **(b)** cos θ = −0.3586
    (c) tan θ = 2.190   **(d)** cot θ = −4.371

46. (a) sin θ = −0.8236   **(b)** cos θ = 0.5017
    (c) tan θ = −6.495   **(d)** cot θ = 0.9721

47. In the figure below, the terminal side of angle θ intersects the unit circle at point P. The point T is the intersection of the line through O and P and the tangent line to the unit circle at point $A(1, 0)$. The tangent line to the unit circle at $B(0, 1)$ intersects the line through O and P at point S. Show that $\tan \theta$ is equal to the length of line segment AT and $\cot \theta$ is equal to the length of line segment BS.

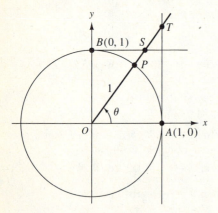

48. The figure below is obtained from the figure for Exercise 47 by drawing line segment CP, which is the

perpendicular from point P to the x axis. Determine the line segment whose length is **(a)** $\sin \theta$, **(b)** $\cos \theta$, **(c)** $\sec \theta$, and **(d)** $\csc \theta$.

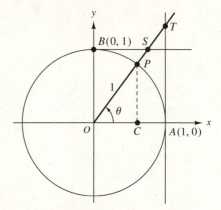

49. Before the advent of electronic calculators, trigonometric function values were obtained from tables of values for positive angles less than 90°. Explain how trigonometric function values of other angles were obtained at that time. As illustrations in your discussion, use a positive angle in the second quadrant and a negative angle in the third quadrant.

8.3 SOLUTIONS OF RIGHT TRIANGLES

GOALS

1. **Learn that any trigonometric function of an acute angle is the cofunction of the angle's complement.**
2. **Solve a right triangle when an acute angle and a side are given.**
3. **Solve a right triangle when two sides are given.**
4. **Solve word problems involving right triangles.**

In Section 8.2 you learned that the trigonometric functions of an acute angle can be expressed as ratios of the measures of the sides of a right triangle. This interpretation of trigonometric functions can be used to *solve a right triangle*, which means to find the measures of its sides and acute angles. Solving right triangles has applications in various fields, especially surveying and navigation, as you will see later in this section.

The vertices of a right triangle are usually denoted by A, B, and C, where C is used for the vertex at which the 90° angle appears. The measures of the sides opposite A, B, and C, are designated by a, b, and c, respectively; thus

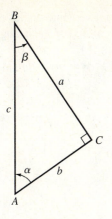

FIGURE 1

c represents the measure of the hypotenuse, and a and b represent the measures of the legs. The acute angles at vertices A and B are denoted by α and β, respectively. See Figure 1. Using these symbols for the sides and angles in a right triangle, we have the following theorem which follows from Theorem 2 of Section 8.2.

THEOREM 1

If α and β are the two acute angles of a right triangle, a and b are, respectively, the measures of the two sides opposite these angles, and c is the measure of the hypotenuse; then

$$\sin \alpha = \frac{a}{c} \qquad \sin \beta = \frac{b}{c}$$

$$\cos \alpha = \frac{b}{c} \qquad \cos \beta = \frac{a}{c}$$

$$\tan \alpha = \frac{a}{b} \qquad \tan \beta = \frac{b}{a}$$

$$\cot \alpha = \frac{b}{a} \qquad \cot \beta = \frac{a}{b}$$

$$\sec \alpha = \frac{c}{b} \qquad \sec \beta = \frac{c}{a}$$

$$\csc \alpha = \frac{c}{a} \qquad \csc \beta = \frac{c}{b}$$

From the formulas of Theorem 1, observe that

$$\sin \alpha = \cos \beta \qquad \text{and} \qquad \cos \alpha = \sin \beta \qquad (1)$$
$$\tan \alpha = \cot \beta \qquad \text{and} \qquad \cot \alpha = \tan \beta \qquad (2)$$
$$\sec \alpha = \csc \beta \qquad \text{and} \qquad \csc \alpha = \sec \beta \qquad (3)$$

Because $\alpha + \beta = 90°$,

$$\beta = 90° - \alpha \qquad (4)$$

Substituting from (4) into (1), (2), and (3), we have

$$\sin \alpha = \cos(90° - \alpha) \qquad \text{and} \qquad \cos \alpha = \sin(90° - \alpha) \qquad (5)$$
$$\tan \alpha = \cot(90° - \alpha) \qquad \text{and} \qquad \cot \alpha = \tan(90° - \alpha) \qquad (6)$$
$$\sec \alpha = \csc(90° - \alpha) \qquad \text{and} \qquad \csc \alpha = \sec(90° - \alpha) \qquad (7)$$

Because of Equations (5), we say that the sine and cosine are **cofunctions** of each other. Furthermore, because of (6), the tangent and cotangent are cofunctions of each other, and because of (7), the secant and cosecant are cofunctions of each other. When the sum of two acute angles is 90°, we say that the two angles are **complementary** and that each angle is the **complement** of the other. Therefore Equations (5), (6), and (7) state that *any trigonometric function of an acute angle is the cofunction of the angle's complement.*

▷ **ILLUSTRATION 1**

(a) $\cos(90° - 23.4°) = \cos 66.6°$

$= 0.3971$

$= \sin 23.4°$

(b) $\tan(90° - 39.8°) = \tan 50.2°$

$= 1.2002$

$= \cot 39.8°$

(c) $\csc(90° - 81.3°) = \csc 8.7°$

$= 6.6111$

$= \sec 81.3°$ ◀

▶ **EXAMPLE 1** *Expressing a Trigonometric Function Value as a Function Value of an Acute Angle Less than 45°*

Write each of the following as the function value of a positive angle less than 45°: **(a)** $\sin 72.1°$; **(b)** $\cos 45.5°$; **(c)** $\cot 89.7°$.

Solution

(a) $\sin 72.1° = \cos(90° - 72.1°)$

$= \cos 17.9°$

(b) $\cos 45.5° = \sin(90° - 45.5°)$

$= \sin 44.5°$

(c) $\cot 89.7° = \tan(90° - 89.7°)$

$= \tan 0.3°$ ◀

We now apply the formulas of Theorem 1 to solve a right triangle. In computations with these formulas, the accuracy of the given data determines the accuracy of the result. Table 1 gives the relationship between the accuracies of the measures of the sides and the measurements of the acute angles in degrees. Although we are dealing with approximate numbers, we

Table 1

Number of Significant Digits in Measures of Sides	Measurements of Acute Angles
4	To nearest 0.01°
3	To nearest 0.1°
2	To nearest degree

shall use the equals symbol with the understanding that the equality is valid for only the number of significant digits warranted by Table 1.

When solving a triangle, we recommend that you sketch the triangle approximately to scale to gain an understanding of the problem and to enable you to notice an error if the computed value is unreasonably large or much too small.

▶ **EXAMPLE 2** *Solving a Right Triangle*

Solve the right triangle for which $\alpha = 24.2°$ and $c = 16.3$.

Solution   Figure 2 shows the triangle. We wish to determine β, a, and b. Because $\alpha + \beta = 90°$,

$$\beta = 90° - \alpha$$
$$= 90° - 24.2°$$
$$= 65.8°$$

To find a, we need a formula containing a and the given values of c and α. The formulas for $\sin \alpha$ and $\csc \alpha$ involve these quantities. If we use the sine, we have

$$\sin 24.2° = \frac{a}{16.3}$$
$$a = 16.3 \sin 24.2°$$
$$a = 6.68$$

To find b, we wish to use a formula involving α, c, and b. From the formula for cosine, we have

$$\cos 24.2° = \frac{b}{16.3}$$
$$b = 16.3 \cos 24.2°$$
$$b = 14.9$$

We could have obtained the value for b by using the computed value of a and $\cot \alpha$. By this method we have

$$\cot 24.2° = \frac{b}{6.68}$$
$$b = 6.68 \cot 24.2°$$
$$b = 14.9$$

Observe that by computing b by two methods, we have a check on the work. ◀

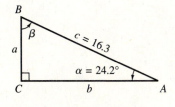

FIGURE 2

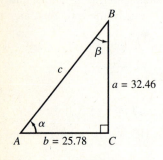

FIGURE 3

▶ **EXAMPLE 3** *Solving a Right Triangle*

Solve the right triangle for which $a = 32.46$ and $b = 25.78$.

Solution   The triangle appears in Figure 3. The unknowns are α, β, and c. We first determine one of the angles. To solve for α, we use the tangent function and have

$$\tan \alpha = \frac{32.46}{25.78}$$

$$\tan \alpha = 1.259$$

$$\alpha = 51.54°$$

To find β, we also use the tangent function.

$$\tan \beta = \frac{25.78}{32.46}$$

$$\tan \beta = 0.7942$$

$$\beta = 38.46°$$

Because $\alpha + \beta = 90°$, we could have computed β by subtracting the value of α from 90°. However, by finding β from the given information we have a check on our work by showing that

$$\alpha + \beta = 51.54° + 38.46°$$

$$= 90°$$

To solve for c, we can use any of the functions of α and β. Using $\sin \alpha$, we have

$$\sin 51.54° = \frac{32.46}{c}$$

$$c = \frac{32.46}{\sin 51.54°}$$

$$c = 41.45$$

Of course c can also be obtained from $c^2 = a^2 + b^2$. This equation can be used as a check for our computed value of c.   ◀

The area of a triangle can be found from the formula

$$K = \tfrac{1}{2}bh$$

where K square units is the area, b units is the length of the base, and h units is the length of the altitude. In a right triangle the sides (legs) opposite the acute angles are a base and altitude.

▷ **ILLUSTRATION 2**

(a) If K square units is the area of the triangle of Example 2, then

$$K = \tfrac{1}{2}ba$$
$$= \tfrac{1}{2}(14.9)(6.68)$$
$$= 49.8$$

(b) If K square units is the area of the triangle of Example 3, then

$$K = \tfrac{1}{2}ba$$
$$= \tfrac{1}{2}(25.78)(32.46)$$
$$= 418.4 \qquad \blacktriangleleft$$

We now discuss some applications involving the solution of right triangles. In the following example, the length of a line segment is given as 50.00 m, which indicates that the measurement is accurate to the nearest one-hundredth of a meter. That is, there are four significant digits.

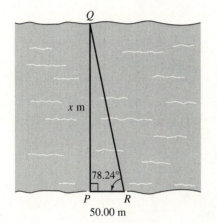

FIGURE 4

▶ **EXAMPLE 4** *Solving a Word Problem Involving a Right Triangle*

To determine the distance across a lake, a surveyor selected two points P and Q, one on each shore and directly opposite each other. On the shore containing P, another point R was chosen, 50.00 m from P, so that the line segment PR was perpendicular to the line segment PQ. The angle having the sides PR and RQ was measured to be 78.24°. What is the distance across the lake?

Solution Refer to Figure 4, showing the right triangle having its right angle at P. The distance across the lake is x meters. Then

$$\tan 78.24° = \frac{x}{50.00}$$
$$x = 50.00 \tan 78.24°$$
$$x = 240.2$$

Conclusion: The distance across the lake is 240.2 m. ◀

A line segment from an observation point O to a point P being observed is called the **line of sight** of P. The angle, having its vertex at O, made by a horizontal ray and the line of sight is called the **angle of elevation** of P or the **angle of depression** of P, according to whether P is above or below the point O. See Figure 5.

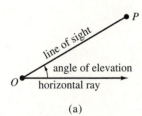

(a)

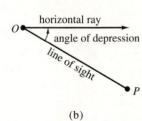

(b)

FIGURE 5

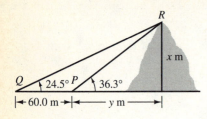

FIGURE 6

(a)

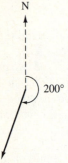

(b)

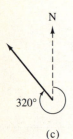

(c)

FIGURE 7

▶ **EXAMPLE 5** *Solving a Word Problem Involving a Right Triangle*

At a point P the angle of elevation of the top of a hill is 36.3°. At a point Q on the same horizontal line as P and the foot of the hill and 60.0 m from P, the angle of elevation is 24.5°. Find the height of the hill.

Solution   See Figure 6, where point R is at the top of the hill. The height of the hill is x meters. Let y meters be the distance from P to the foot of the hill. There are two right triangles. From the right triangle having a vertex at P, we have

$$\cot 36.3° = \frac{y}{x}$$

$$y = x \cot 36.3°$$

From the right triangle having a vertex at Q, we obtain

$$\cot 24.5° = \frac{60.0 + y}{x}$$

$$60.0 + y = x \cot 24.5°$$

We replace y in this equation by $x \cot 36.3°$ and obtain

$$60.0 + x \cot 36.3° = x \cot 24.5°$$

$$60.0 = x(\cot 24.5° - \cot 36.3°)$$

$$x = \frac{60.0}{\cot 24.5° - \cot 36.3°}$$

$$x = 72.0$$

Conclusion: The height of the hill is 72.0 m.   ◀

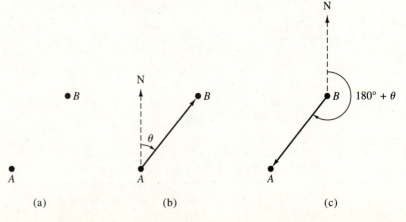

(a)   (b)   (c)

FIGURE 8

In navigation, the **course** of a ship or airplane is the angle measured in degrees clockwise from the north to the direction in which the carrier is traveling. The angle is considered positive even though it is in the clockwise sense. Figure 7 shows courses of 30°, 200°, and 320°. The **bearing** of a particular location P from an observer at O is the angle measured in degrees clockwise from the north to the line segment OP. Refer to Figure 8(a), which shows two points A and B. The bearing from A to B is the angle shown in Figure 8(b), and the bearing from B to A is the angle shown in Figure 8(c). Be sure to distinguish the two situations. In this case, θ is the bearing from A to B, and $180° + \theta$ is the bearing from B to A.

▶ **EXAMPLE 6** *Solving a Word Problem Involving a Right Triangle*

A navigator on a ship, sailing on a course of 338° at 12 knots (nautical miles per hour), observes a lighthouse due north of the ship. Fifteen minutes later the lighthouse is due east of the ship. How far is the lighthouse from the ship at that time?

Solution   Refer to Figure 9, where x nautical miles is the distance to be determined, C is the position of the lighthouse, A is the ship's position at the time of the initial observation, and B is the ship's position 15 min later. Because the ship is traveling at 12 knots, it covers a distance of 3 nautical miles in 15 min. Because the course is 338°, the angle at A in the triangle is $360° - 338°$, or 22°. From the right triangle, we have

$$\sin 22° = \frac{x}{3}$$

$$x = 3 \sin 22°$$

$$x = 1.1$$

Conclusion: The lighthouse is 1.1 nautical miles from the ship at the indicated time.   ◀

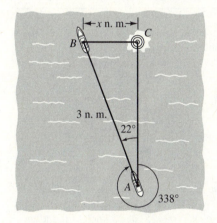

FIGURE 9

EXERCISES 8.3

In Exercises 1 through 4, express the function value as the function value of a positive angle not greater than 45°.

1. (a) sin 60°   **(b)** cos 84.3°   **(c)** tan 49.8°
  (d) cot 52.1°   **(e)** csc 67.5°

2. (a) sin 79.6°   **(b)** cos 46°   **(c)** tan 64.3°
  (d) cot 88.8°   **(e)** sec 59.7°

3. (a) sin 47°18′   **(b)** cos 71°42′   **(c)** tan 46°
  (d) cot 55°6′   **(e)** sec 64°30′

4. (a) sin 62°12′   **(b)** cos 80°54′   **(c)** tan 75°24′
  (d) cot 60°   **(e)** csc 49°36′

In Exercises 5 through 8, solve the right triangle where the right angle is at vertex C. For the trigonometric function values, do not use a calculator. Express the results to the number of significant digits justified by the given information.

5. $\alpha = 60°$, $c = 2.7$ **6.** $\alpha = 45°$, $a = 34$

7. $\beta = 45°$, $a = 56$ **8.** $\beta = 30°$, $c = 8.2$

In Exercises 9 through 30, solve the right triangle where the right angle is at vertex C. Express the results to the number of significant digits justified by the given information.

9. $\alpha = 24°$, $a = 16$

10. $\alpha = 65°$, $b = 6.3$

11. $\beta = 71°$, $c = 44$

12. $\beta = 10°$, $b = 22$

13. $b = 26$, $c = 38$

14. $a = 3.4$, $b = 5.7$

15. $\beta = 37.4°$, $a = 4.18$

16. $\alpha = 52.3°$, $c = 48.5$

17. $\alpha = 16.9°$, $a = 136$

18. $\beta = 29.7°$, $a = 0.534$

19. $a = 63.6$, $b = 58.1$

20. $a = 154$, $c = 393$

21. $\alpha = 58.43°$, $c = 625.3$

22. $\beta = 18.63°$, $c = 52.10$

23. $\beta = 40.92°$, $b = 36.72$

24. $\alpha = 70.25°$, $a = 6584$

25. $a = 312.7$, $c = 809.0$

26. $b = 4.218$, $c = 6.759$

27. $\beta = 65°18'$, $c = 39.2$

28. $\alpha = 35°48'$, $c = 25.0$

29. $\alpha = 29°36'$, $b = 287$

30. $\beta = 81°12'$, $a = 43.6$

In Exercises 31 through 34, find the area of the triangle of the indicated exercise.

31. Exercise 11

32. Exercise 18

33. Exercise 25

34. Exercise 28

In Exercises 35 through 44, the solution of the word problem involves a right triangle. Be sure to write a conclusion.

35. From the top of a cliff 126 m high, the angle of depression of a boat is 20.7°. How far is the boat from the foot of the cliff?

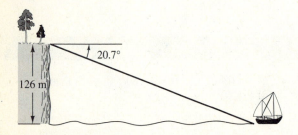

36. A tower 135 ft high stands on the water's edge of a lake. From the top of the tower, the angle of depression of an object on the water's edge on the other side of the lake is 36.3°. What is the distance across the lake?

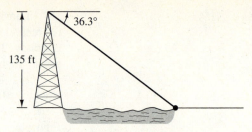

37. The Empire State Building is 1250 ft tall. What is the angle of elevation of the top from a point on the ground 1 mile (5280 ft) from the base of the building?

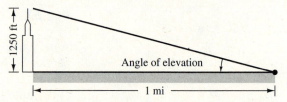

38. If the angle of elevation of the sun is 42°, what is the length of the shadow on the ground of a man who is 6.1 ft tall?

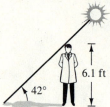

39. Two ships leave a port at the same time. The first ship sails on a course of 35° at 15 knots while the second ship sails on a course of 125° at 20 knots. Find after 2 hr **(a)** the distance between the ships, **(b)** the bearing from the first ship to the second, and **(c)** the bearing from the second ship to the first.

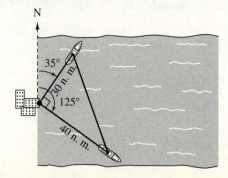

40. A ship leaves a port and sails for 4 hr on a course of 78° at 18 knots. Then the ship changes its course to 168° and sails for 6 hr at 16 knots. After the 10 hr **(a)** what is the distance of the ship from the port and **(b)** what is the bearing from the port to the ship?

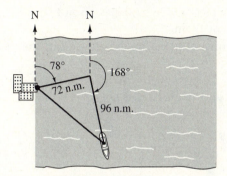

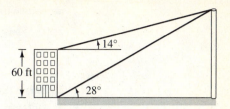

41. Points A and B are on the same horizontal line with the foot of a hill, and the angles of depression of these points from the top of the hill are 30.2° and 22.5°, respectively. If the distance between A and B is 75.0 m, what is the height of the hill?

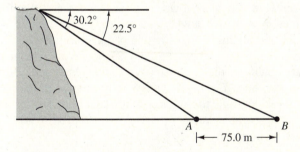

42. From the top of a building 60 ft high, the angle of elevation of the top of a pole is 14°. At the bottom of the building the angle of elevation of the top of the pole is 28°. Find **(a)** the height of the pole and **(b)** the distance of the pole from the building.

43. From the top of a mountain 532 m higher than a nearby river, the angle of depression of a point P on the closer bank of the river is 52.6°, and the angle of depression of a point Q directly opposite P on the other side is 34.5°. Points P and Q and the foot of the mountain are on the same horizontal line. Find the distance across the river from P to Q.

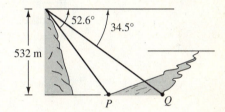

44. Point T is at the top of a mountain. From a point P on the ground, the angle of elevation of T is 16.3°. From a point Q on the same horizontal line with P and the foot of the mountain, the angle of elevation of T is 28.7°. Find the height of the mountain if the distance between P and Q is 125 m.

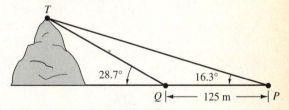

45. Refer to a table of trigonometric function values that appears in almost any pre-1990 trigonometry textbook. Explain how the construction of the table utilizes the fact that *any trigonometric function of an acute angle is the cofunction of the angle's complement.*

8.4 THE LAW OF SINES

In many practical situations we wish to find sides and angles in an **oblique triangle,** one that does not contain a right angle. In this section and the next, we solve such triangles.

As with a right triangle, the vertices of oblique triangles are labeled A, B, and C, and the measures of the sides opposite them are designated by a, b, and c, respectively. The angles at the vertices A, B, and C are denoted by α, β, and γ, respectively. Figure 1 shows an oblique triangle having all acute angles. An oblique triangle having at vertex A an obtuse angle (an angle whose degree measure is between 90 and 180) appears in Figure 2.

To solve an oblique triangle, we must know the measure of one side and any two other measures. In this section we consider two cases. In the first we are given the measures of two angles and a side. In the second the measures of two sides and an angle opposite one of them are known.

For each of the triangles in Figures 1 and 2 choose a rectangular cartesian coordinate system so that the origin is at the vertex A and the positive x axis is along the side AB. The triangles and the coordinate axes are shown in Figure 3. In each triangle the angle α is in standard position. Also for each triangle, a line segment is drawn through C parallel to the y axis and intersecting the x axis at D. Let $h = |\overline{DC}|$. In either case

$$\sin \alpha = \frac{h}{b}$$

$$h = b \sin \alpha \tag{1}$$

Also from right triangle BDC, in either case

$$\sin \beta = \frac{h}{a}$$

$$h = a \sin \beta \tag{2}$$

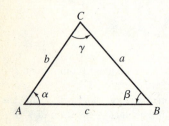

FIGURE 1

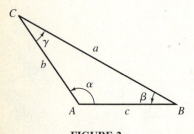

FIGURE 2

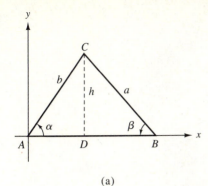

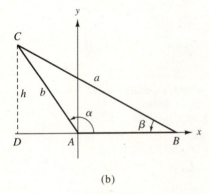

(a)

(b)

FIGURE 3

Substituting from (2) into (1), we have

$$a \sin \beta = b \sin \alpha$$

$$\frac{a}{\sin \alpha} = \frac{b}{\sin \beta} \tag{3}$$

If the coordinate axes are chosen so that the origin is at A and the positive x axis is along the side AC, then by a similar argument, we obtain

$$\frac{a}{\sin \alpha} = \frac{c}{\sin \gamma} \tag{4}$$

Observe that (4) holds if the triangle is a right triangle. That is, if γ is 90°, then because $\sin 90° = 1$, (4) becomes

$$\frac{a}{\sin \alpha} = \frac{c}{1}$$

$$\sin \alpha = \frac{a}{c}$$

which is the first formula of Theorem 1 in Section 8.3.

From (3) and (4) we have the following theorem, known as the *law of sines*.

THEOREM 1   The Law of Sines

If α, β, and γ are the angles of any triangle and a, b, and c are, respectively, the measures of the sides opposite these angles, then

$$\frac{a}{\sin \alpha} = \frac{b}{\sin \beta} = \frac{c}{\sin \gamma}$$

▶ **EXAMPLE 1** *Solving an Oblique Triangle Given Two Angles and a Side*

Solve the triangle for which $\alpha = 51.2°$, $\beta = 48.6°$, and $a = 23.5$.

Solution   Figure 4 shows the triangle. Because $\alpha + \beta + \gamma = 180°$,

$$\gamma = 180° - \alpha - \beta$$
$$= 180° - 51.2° - 48.6°$$
$$= 80.2°$$

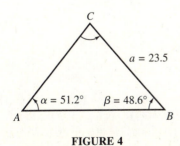

FIGURE 4

From the law of sines

$$\frac{23.5}{\sin 51.2°} = \frac{b}{\sin 48.6°}$$

$$b = \frac{23.5(\sin 48.6°)}{\sin 51.2°}$$

$$b = 22.6$$

Also from the law of sines

$$\frac{23.5}{\sin 51.2°} = \frac{c}{\sin 80.2°}$$

$$c = \frac{23.5(\sin 80.2°)}{\sin 51.2°}$$

$$c = 29.7$$ ◄

▶ **EXAMPLE 2**   *Solving a Word Problem Involving an Oblique*
Triangle and the Law of Sines

On a hill inclined at an angle of 14.2° with the horizontal, stands a vertical tower. At a point P that is 62.5 m down the hill from the foot of the tower, the angle of elevation of the top of the tower is 43.6°. How tall is the tower?

Solution   See Figure 5, where x meters is the height of the tower, the top of the tower is denoted by T, and F represents the point at the foot of the tower. We have an oblique triangle with vertices at P, T, and F. The angle at P in the triangle is found by computing 43.6° − 14.2°, which is 29.4°. The angle at T in the triangle is found by computing 90° − 43.6°, which is 46.4°. Thus we know the measures of two angles and a side of the triangle. From the law of sines

$$\frac{x}{\sin 29.4°} = \frac{62.5}{\sin 46.4°}$$

$$x = \frac{62.5(\sin 29.4°)}{\sin 46.4°}$$

$$x = 42.4$$

Conclusion: The tower is 42.4 m tall.   ◄

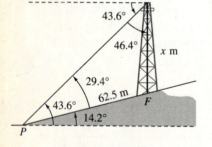

FIGURE 5

The law of sines can also be used when the measures of two sides and the angle opposite one of them are given. However, in such a case we do not always have a unique triangle. Suppose, for instance, we are given a, b, and α, where α is an acute angle. To construct a triangle having the given

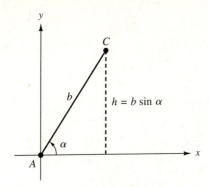

FIGURE 6

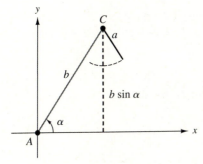

α is acute and $a < b \sin \alpha$
no triangle

FIGURE 7

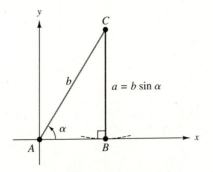

α is acute and $a = b \sin \alpha$
one right triangle

FIGURE 8

measurements, let the angle be in standard position on a rectangular coordinate system so that the vertex A is at the origin. See Figure 6. Because b is given, a line segment AC of length b units is marked off on the terminal side of α. Thus the position of vertex C is determined. The side BC of length a units is to be opposite vertex A and the vertex B should be on the x axis. To locate the possible position of B, first draw the perpendicular line segment from C to the x axis. If this line segment has length h units, then

$$\sin \alpha = \frac{h}{b}$$

$$h = b \sin \alpha$$

The position of the vertex B will depend upon the relationship between a and $b \sin \alpha$. There are four possibilities: $a < b \sin \alpha$; $a = b \sin \alpha$; $b \sin \alpha < a < b$; and $a \geq b$. We consider each possibility separately and follow each discussion with an illustration involving a particular set of values of a, b, and α.

Possibility 1: $a < b \sin \alpha$. See Figure 7. A side BC of length a units does not intersect the x axis. Therefore there is no triangle possible.

▷ **ILLUSTRATION 1**

Let $a = 2.3$, $b = 4.5$, and $\alpha = 42°$. From the law of sines

$$\frac{a}{\sin \alpha} = \frac{b}{\sin \beta}$$

$$\frac{2.3}{\sin 42°} = \frac{4.5}{\sin \beta}$$

$$\sin \beta = \frac{4.5(\sin 42°)}{2.3}$$

$$\sin \beta = 1.309$$

Because $|\sin \beta|$ cannot be greater than 1, this equation has no solution. Thus there is no triangle satisfying the given information.

Observe that this set of values satisfies Possibility 1 because

$$b \sin \alpha = 4.5(\sin 42°)$$

$$= 3.01$$

Furthermore, $a = 2.3$ and $2.3 < 3.01$.   ◄

Possibility 2: $a = b \sin \alpha$. See Figure 8. The perpendicular distance from C to the x axis is a units, and so at vertex B there is a right angle. Hence there is one right triangle.

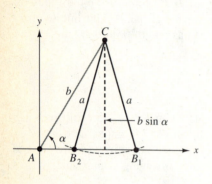

α is acute and $b \sin \alpha < a < b$
two triangles

FIGURE 9

▷ **ILLUSTRATION 2**

Suppose $a = 2.0$, $b = 4.0$, and $\alpha = 30°$. Then from the law of sines

$$\frac{a}{\sin \alpha} = \frac{b}{\sin \beta}$$

$$\frac{2.0}{\sin 30°} = \frac{4.0}{\sin \beta}$$

$$\sin \beta = \frac{4.0(\sin 30°)}{2.0}$$

$$\sin \beta = \frac{4.0(\frac{1}{2})}{2.0}$$

$$\sin \beta = 1$$

Therefore $\beta = 90°$ and the triangle is a right triangle.

Possibility 2 holds because for this set of values, $b \sin \alpha = 4.0(\sin 30°)$; that is, $b \sin \alpha = 2$ and $a = 2$. ◀

Possibility 3: $b \sin \alpha < a < b$. See Figure 9. There are two possible positions of vertex B on the x axis shown in the figure as B_1 and B_2. Thus there are two triangles possible.

▷ **ILLUSTRATION 3**

Let $a = 25.2$, $b = 30.5$, and $\alpha = 54.2°$. From the law of sines

$$\frac{a}{\sin \alpha} = \frac{b}{\sin \beta}$$

$$\frac{25.2}{\sin 54.2°} = \frac{30.5}{\sin \beta}$$

$$\sin \beta = \frac{30.5(\sin 54.2°)}{25.2}$$

$$\sin \beta = 0.9817$$

There are two angles β having degree measure between 0 and 180 for which $\sin \beta = 0.9817$. Let the acute angle be β_1. From a calculator

$$\beta_1 = 79.0°$$

If the obtuse angle is β_2, then because β_1 is the reference angle of β_2, we have

$$\beta_2 = 180° - 79.0°$$
$$= 101.0°$$

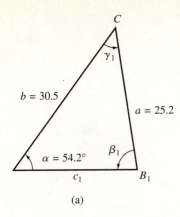

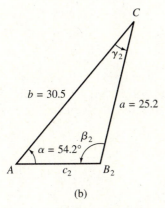

FIGURE 10

Therefore there are two triangles. Figure 10(a) shows the triangle for which $a = 25.2$, $b = 30.5$, $\alpha = 54.2°$, and $\beta_1 = 79.0°$. Figure 10(b) shows the triangle for which $a = 25.2$, $b = 30.5$, $\alpha = 54.2°$, and $\beta_2 = 101.0°$. We have two triangles to solve.

For the triangle of Figure 10(a), we compute γ_1 and c_1.

$$\gamma_1 = 180° - \alpha - \beta_1$$
$$= 180° - 54.2° - 79.0°$$
$$= 46.8°$$

From the law of sines

$$\frac{c_1}{\sin \gamma_1} = \frac{b}{\sin \beta_1}$$

$$\frac{c_1}{\sin 46.8°} = \frac{30.5}{\sin 79.0°}$$

$$c_1 = \frac{30.5(\sin 46.8°)}{\sin 79.0°}$$

$$c_1 = \frac{30.5(0.7290)}{0.9816}$$

$$c_1 = 22.7$$

We now compute γ_2 and c_2 for the triangle of Figure 10(b).

$$\gamma_2 = 180° - \alpha - \beta_2$$
$$= 180° - 54.2° - 101.0°$$
$$= 24.8°$$

From the law of sines

$$\frac{c_2}{\sin \gamma_2} = \frac{b}{\sin \beta_2}$$

$$\frac{c_2}{\sin 24.8°} = \frac{30.5}{\sin 101.0°}$$

$$c_2 = \frac{30.5(\sin 24.8°)}{\sin 101.0°}$$

$$c_2 = \frac{30.5(0.4195)}{0.9816}$$

$$c_2 = 13.0$$

The given set of values satisfies Possibility 3 because

$$b \sin \alpha = 30.5 \sin 54.2°$$
$$= 24.7$$

Futhermore, $a = 25.2$, $b = 30.5$, and $24.7 < 25.2 < 30.5$. ◄

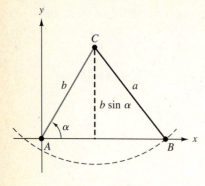

α is acute and $a \geq b$
one triangle

FIGURE 11

Possibility 4: $a \geq b$. See Figure 11. There is only one potential position of B on the x axis. Therefore there is one triangle possible. If $a = b$, the triangle is isosceles.

▷ ILLUSTRATION 4

Suppose $a = 5.21$, $b = 3.06$, and $\alpha = 47.6°$. From the law of sines

$$\frac{a}{\sin \alpha} = \frac{b}{\sin \beta}$$

$$\frac{5.21}{\sin 47.6°} = \frac{3.06}{\sin \beta}$$

$$\sin \beta = \frac{3.06(\sin 47.6°)}{5.21}$$

$$\sin \beta = 0.4337$$

There are two angles, having degree measure between 0 and 180, whose sine has a value 0.4337. But because $a > b$, it follows that $\alpha > \beta$; therefore $47.6° > \beta$. Thus we have only one value for β. From a calculator,

$$\beta = 25.7°$$

We compute γ and c for this triangle.

$$\gamma = 180° - \alpha - \beta$$
$$= 180° - 47.6° - 25.7°$$
$$= 106.7°$$

From the law of sines

$$\frac{c}{\sin \gamma} = \frac{a}{\sin \alpha}$$

$$\frac{c}{\sin 106.7°} = \frac{5.21}{\sin 47.6°}$$

$$c = \frac{5.21(\sin 106.7°)}{\sin 47.6°}$$

$$c = 6.76$$

It is apparent that the given set of values satisfies Possibility 4 because $a = 5.21$, $b = 3.06$, and $5.21 > 3.06$. ◀

If an angle of a triangle is obtuse, the measure of the side opposite it must be greater than the measures of the other sides. Therefore, if a, b, and α are given and α is obtuse, then one triangle is possible if and only if

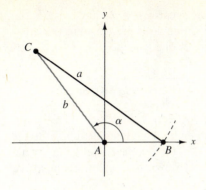

α is obtuse and $a > b$
one triangle

(a)

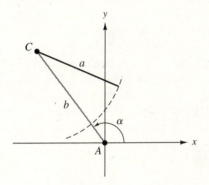

α is obtuse and $a < b$
no triangle

(b)

FIGURE 12

$a > b$. Figure 12(a) shows one triangle where α is obtuse and $a > b$. Figure 12(b) indicates there is no triangle when α is obtuse and $a < b$.

Because of the various situations that can occur, the case when two sides and the angle opposite one of them are given is called the *ambiguous case*. The same results are obtained if the known sides and the angle opposite one of them are represented by other symbols such as c, b, and γ; b, a, and β; and so on.

▶ **EXAMPLE 3** *Determining the Number of Possible Triangles Given Two Sides and the Angle Opposite One of Them*

In parts (a)–(d), we are given the measures of two sides of a triangle and the angle opposite one of them; thus we have the ambiguous case. Determine the number of possible triangles. **(a)** $c = 2.0$, $b = 6.0$, $\gamma = 30°$; **(b)** $b = 32.4$, $a = 20.6$, $\beta = 52.1°$; **(c)** $a = 10.3$, $c = 16.5$, $\alpha = 23.8°$; **(d)** $b = 32$, $c = 25$, $\beta = 114°$.

Solution

(a) The given angle is γ and $c < b$. The measure of the side opposite the given angle is less than the other given measure of a side. Therefore we first compute $b \sin \gamma$.

$$b \sin \gamma = 6.0 \sin 30°$$
$$= 6.0(\tfrac{1}{2})$$
$$= 3.0$$

Because $c = 2.0$,

$$c < b \sin \gamma$$

Thus we have Possibility 1. Therefore there is no triangle.

(b) The given angle is β and $b > a$. Here the measure of the side opposite the given angle is greater than the other given measure of a side. Therefore, from Possibility 4, there is one triangle.

(c) The given angle is α and $a < c$. The measure of the side opposite the given angle is less than the other given measure of a side. So we compute $c \sin \alpha$.

$$c \sin \alpha = 16.5 \sin 23.8°$$
$$= 16.5(0.4035)$$
$$= 6.66$$

Because $a = 10.3$, $c = 16.5$, and $6.66 < 10.3 < 16.5$,

$$c \sin \alpha < a < c$$

Therefore, from Possibility 3, there are two triangles.

(d) The given angle is β, which is obtuse. Because $b = 32$ and $c = 25$, the measure of the side opposite the given angle is greater than the other given measure of a side. Hence there is one triangle. ◀

▶ **EXAMPLE 4** *Solving a Word Problem Involving an Oblique Triangle and the Law of Sines*

A ladder 35.4 ft long is leaning against an embankment inclined 62.5° to the horizontal. If the bottom of the ladder is 10.2 ft from the embankment, what is the distance from the top of the ladder down the embankment to the ground?

Solution   See Figure 13, where x feet is the distance to be determined, B is the point at the top of the ladder, A is the point at the bottom of the ladder, and P is the point at the bottom of the embankment. The angle at P in the triangle is found by computing $180° - 62.5°$, which is $117.5°$. We have an oblique triangle for which the measures of two sides and the angle opposite one of them are known. Thus we have the ambiguous case. Because the known angle is obtuse and the measure of the side opposite this angle is greater than the other given measure of a side, there is one triangle.

Before we can use the law of sines to determine x, we must know the measurement of the angle at A in the triangle. Let α be this angle and let β be the angle at B in the triangle. We first find β from the law of sines.

$$\frac{35.4}{\sin 117.5°} = \frac{10.2}{\sin \beta}$$

$$\sin \beta = \frac{10.2(\sin 117.5°)}{35.4}$$

$$\sin \beta = 0.2556$$

$$\beta = 14.8°$$

Because $\alpha + \beta + 117.5° = 180°$,

$$\alpha = 180° - \beta - 117.5°$$
$$= 180° - 14.8° - 117.5°$$
$$= 47.7°$$

From the law of sines

$$\frac{x}{\sin 47.7°} = \frac{35.4}{\sin 117.5°}$$

$$x = \frac{35.4(\sin 47.7°)}{\sin 117.5°}$$

$$x = 29.5$$

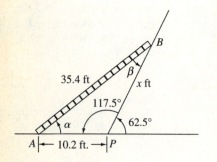

FIGURE 13

Conclusion: The distance from the top of the ladder down the embankment to the ground is 29.5 ft. ◀

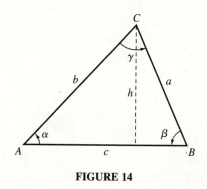

FIGURE 14

Figure 14 shows an oblique triangle with the customary symbols denoting the vertices, angles, and measures of the sides. If the base is considered to be side AB and h units is the length of the altitude, then if K square units is the area of the triangle,

$$K = \tfrac{1}{2}ch$$

Because $h = b \sin \alpha$, we have

$$K = \tfrac{1}{2}c(b \sin \alpha)$$
$$K = \tfrac{1}{2}bc \sin \alpha \tag{5}$$

Another equation giving h is $h = a \sin \beta$. Substituting this value of h in $K = \tfrac{1}{2}ch$, we get

$$K = \tfrac{1}{2}c(a \sin \beta)$$
$$K = \tfrac{1}{2}ac \sin \beta \tag{6}$$

If the base is considered to be side AC, then by a similar argument, we obtain

$$K = \tfrac{1}{2}ab \sin \gamma \tag{7}$$

From formulas (5), (6), and (7) we have the following theorem.

THEOREM 2   *Area of a Triangle*
The measure of the area of a triangle is one-half the product of the measures of two sides and the sine of the angle included between the two sides.

▷ **ILLUSTRATION 5**

We compute the area of the triangle in Example 1 by each of the three formulas (5), (6), and (7). If K square units is the area of the triangle,

$$K = \tfrac{1}{2}bc \sin \alpha \qquad\qquad K = \tfrac{1}{2}ac \sin \beta$$
$$= \tfrac{1}{2}(22.6)(29.7)(\sin 51.2°) \qquad = \tfrac{1}{2}(23.5)(29.7)(\sin 48.6°)$$
$$= 262 \qquad\qquad\qquad\qquad = 262$$
$$K = \tfrac{1}{2}ab \sin \gamma$$
$$= \tfrac{1}{2}(23.5)(22.6)(\sin 80.2°)$$
$$= 262$$

◀

EXERCISES 8.4

In Exercises 1 through 4, two angles of a triangle and the measure of the side opposite one of them are given. Find the measure of the side opposite the other given angle, but do not use a calculator for the trigonometric function values. Express the results to the number of significant digits justified by the given information.

1. $a = 4.6$, $\alpha = 45°$, $\beta = 60°$

2. $b = 23$, $\beta = 45°$, $\gamma = 30°$

3. $c = 88$, $\alpha = 30°$, $\gamma = 120°$

4. $a = 9.5$, $\alpha = 135°$, $\beta = 30°$

In Exercises 5 through 12, solve the triangle. Express the results to the number of significant digits justified by the given information.

5. $\alpha = 34°$, $\beta = 71°$, $a = 24$

6. $\alpha = 62°$, $\gamma = 55°$, $a = 8.3$

7. $\beta = 48.6°$, $\gamma = 61.4°$, $c = 53.2$

8. $\alpha = 26.5°$, $\beta = 32.7°$, $b = 187$

9. $\alpha = 73.2°$, $\gamma = 23.8°$, $b = 2.30$

10. $\beta = 84.6°$, $\gamma = 51.9°$, $a = 46.4$

11. $\alpha = 52°42'$, $\beta = 75°36'$, $b = 408$

12. $\beta = 101°6'$, $\gamma = 23°24'$, $c = 0.149$

In Exercises 13 through 24, the measures of two sides of a triangle and the angle opposite one of them are given. Therefore it is the ambiguous case. Determine the number of triangles that satisfy the given set of conditions and solve each triangle. Express the results to the number of significant digits justified by the given information.

13. $a = 6.4$, $b = 4.7$, $\alpha = 42°$

14. $b = 27$, $a = 46$, $\beta = 38°$

15. $b = 17$, $c = 34$, $\beta = 30°$

16. $c = 18.3$, $b = 12.5$, $\gamma = 58.3°$

17. $c = 42.5$, $a = 68.0$, $\gamma = 35.2°$

18. $a = 245$, $b = 302$, $\alpha = 136.4°$

19. $b = 846$, $a = 431$, $\beta = 116.4°$

20. $a = 40.2$, $b = 52.4$, $\alpha = 41.5°$

21. $a = 54.0$, $c = 83.7$, $\alpha = 43.6°$

22. $c = 9.04$, $a = 3.52$, $\gamma = 128.1°$

23. $b = 3.562$, $c = 4.210$, $\beta = 50.23°$

24. $b = 5649$, $a = 6382$, $\beta = 59.43°$

In Exercises 25 through 32, find the area of the triangle of the indicated exercise.

25. Exercise 5   **26.** Exercise 6

27. Exercise 9   **28.** Exercise 10

29. Exercise 13   **30.** Exercise 16

31. Exercise 19   **32.** Exercise 22

In Exercises 33 through 38, the solution of the word problem involves an oblique triangle. Be sure to write a conclusion.

33. A building is located at the end of a street that is inclined at an angle of 8.4° with the horizontal. At a point P that is 210 m down the street from the building, the angle subtended by the building is 15.6°. How tall is the building?

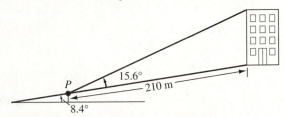

34. A flagpole is situated at the top of a building 115 ft tall. From a point in the same horizontal plane as the base of the building the angles of elevation of the top and bottom of the flagpole are 63.2° and 58.6°, respectively. How tall is the flagpole?

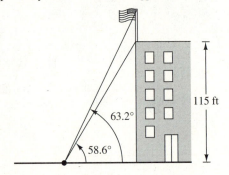

35. To determine the distance across a straight river, a surveyor chooses two points P and Q on the bank, where the distance between P and Q is 200 m. At each of these points a point R on the opposite bank is sighted. The angle having sides PQ and PR is measured to be 63.1°, and the angle having sides PQ

and QR is measured to be 80.4°. What is the distance across the river?

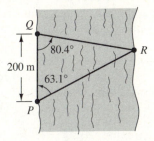

36. A triangular parcel of land with vertices at R, S, and T was to be enclosed by a fence, but it was discovered that the surveyor's mark at S was missing. From a deed to the property, it was learned that the distance from T to R is 324 m, the distance from T to S is 506 m, and the angle at R in the triangle is 125.4°. Determine the location of S by finding the distance from R to S.

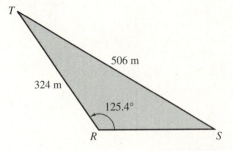

37. A ramp is inclined at an angle of 41.3° with the ground. One end of a board 20.6 ft in length is located on the ground at a point P that is 12.2 ft from the

base Q of the ramp, and the other end rests on the ramp at point R. Find the distance from point Q up the ramp to point R.

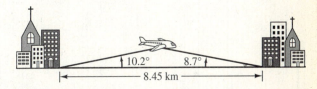

38. At a particular instant, when an airplane was directly above a straight-line road connecting two small towns, the angles of depression of these towns were 10.2° and 8.7°. **(a)** Find the straight-line distances from the airplane to each of the towns at this instant if the towns are 8.45 km apart. **(b)** Determine the height of the airplane at this instant.

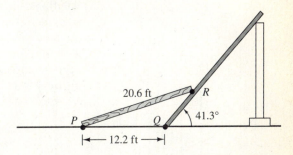

39. State the law of sines in words only. Do not use any symbols for the measures of the sides and angles of a triangle.

40. Suppose you are solving a triangle when the measures of two sides and the angle opposite one of them are known. Explain why we refer to this situation as the *ambiguous case*.

8.5 THE LAW OF COSINES

GOALS
1. **Learn the law of cosines.**
2. **Solve an oblique triangle given two sides and the included angle.**
3. **Solve an oblique triangle given the three sides.**
4. **Find the area of a parallelogram.**
5. **Solve word problems involving oblique triangles and the law of cosines.**

A unique triangle is determined if the measures of two sides and the angle included between them are known. We should therefore be able to solve a triangle when we are given a, b, and γ; or a, c, and β; or b, c, and α. The solution cannot be obtained by using the law of sines exclusively. Nevertheless, there is a theorem, called the *law of cosines*, that can be applied, and we now discuss it.

Suppose a, b, and γ are known. Figure 1 shows a triangle with a rectangular coordinate system chosen so that γ is in standard position. In the figure γ is an obtuse angle, but the discussion is also valid if γ is acute. The vertex B is at $(a, 0)$. To determine the coordinates of A, let the point be (x, y). Then

$$\cos \gamma = \frac{x}{b} \quad \text{and} \quad \sin \gamma = \frac{y}{b}$$

Therefore

$$x = b \cos \gamma \quad \text{and} \quad y = b \sin \gamma$$

From the distance formula applied to side BA,

$$c^2 = (x - a)^2 + (y - 0)^2$$

Substituting $b \cos \gamma$ for x and $b \sin \gamma$ for y, we get

$$c^2 = (b \cos \gamma - a)^2 + (b \sin \gamma - 0)^2$$
$$= b^2 \cos^2 \gamma - 2ab \cos \gamma + a^2 + b^2 \sin^2 \gamma$$
$$= b^2(\cos^2 \gamma + \sin^2 \gamma) - 2ab \cos \gamma + a^2$$

Because $\cos^2 \gamma + \sin^2 \gamma = 1$, we have

$$c^2 = a^2 + b^2 - 2ab \cos \gamma$$

This equation gives one form of the law of cosines. Two other forms are obtained in a similar manner by having either β or α in standard position on a rectangular coordinate system. We now state the law formally.

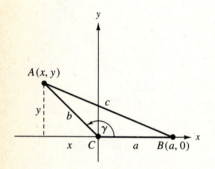

FIGURE 1

THEOREM 1   *The Law of Cosines*

If α, β, and γ are the angles of any triangle and a, b, and c are, respectively, the measures of the sides opposite these angles, then

$$c^2 = a^2 + b^2 - 2ab \cos \gamma$$
$$b^2 = a^2 + c^2 - 2ac \cos \beta$$
$$a^2 = b^2 + c^2 - 2bc \cos \alpha$$

Observe that if $\gamma = 90°$, we have a right triangle, and from the law of cosines

$$c^2 = a^2 + b^2 - 2ab \cos 90°$$
$$= a^2 + b^2 - 2ab(0)$$
$$= a^2 + b^2$$

which is the Pythagorean theorem. Rather than memorizing the separate forms of the law of cosines, think of it as a generalized version of the Pythagorean theorem that states

The square of the measure of any side of a triangle is equal to the sum of the squares of the measures of the other two sides minus twice the product of the measures of the other two sides and the cosine of the angle between them.

▶ **EXAMPLE 1** *Solving an Oblique Triangle Given Two Sides and the Included Angle*

Solve the triangle for which $a = 24.0$, $c = 32.0$, and $\beta = 64.0°$.

Solution   The triangle appears in Figure 2. From the law of cosines

$$b^2 = a^2 + c^2 - 2ac \cos \beta$$
$$b^2 = (24.0)^2 + (32.0)^2 - 2(24.0)(32.0)\cos 64.0°$$
$$b^2 = 926.6$$
$$b = 30.4$$

Because we now have values for b and β, we can use the law of sines.

$$\frac{a}{\sin \alpha} = \frac{b}{\sin \beta}$$

$$\frac{24.0}{\sin \alpha} = \frac{30.4}{\sin 64.0°}$$

$$\sin \alpha = \frac{24.0(\sin 64.0°)}{30.4}$$

$$\sin \alpha = 0.7096$$

There are two angles in a triangle for which the sine is 0.7096: 45.2° and 134.8°. However, because $a < b$, $\alpha < \beta$. Therefore, we reject 134.8° and

$$\alpha = 45.2°$$

Because $\alpha + \beta + \gamma = 180°$,

$$\gamma = 180° - 45.2° - 64.0°$$
$$= 70.8°$$

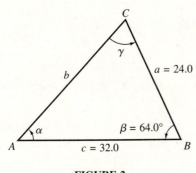

FIGURE 2

The law of sines can be applied to find γ. If this method is used, there is a check on the work by verifying that $\alpha + \beta + \gamma = 180°$.   ◄

▶ **EXAMPLE 2**   *Solving a Word Problem Involving an Oblique Triangle by Using the Law of Cosines and the Law of Sines*

Two ships leave the same port at the same time. One ship sails on a course of 125° at 18 knots while the other sails on a course of 230° at 24 knots. Find, after 3 hr, **(a)** the distance between the ships and **(b)** the bearing from the first ship to the second.

Solution

(a) Refer to Figure 3. The port is at point P. After 3 hr the first ship is at point A and the second ship is at point B. Because the first ship is traveling at 18 knots, the distance from P to A is 54 nautical miles. The second ship is traveling at 24 knots, and so the distance from P to B is 72 nautical miles. The angle at P in the triangle is $230° - 125° = 105°$. Let the distance between the two ships after 3 hr be x nautical miles. From the law of cosines

$$x^2 = (72)^2 + (54)^2 - 2(72)(54)\cos 105°$$
$$x^2 = 10{,}112$$
$$x = 100$$

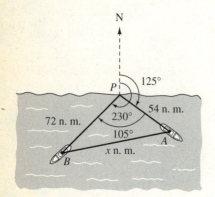

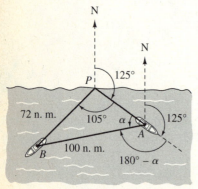

FIGURE 3

Conclusion: After 3 hr the distance between the two ships is 100 nautical miles.

(b) The bearing from the first ship to the second ship after 3 hr is the bearing from A to B. See Figure 4. To find the bearing, we must first determine α, the angle at A in the triangle. From the law of sines

$$\frac{72}{\sin \alpha} = \frac{100}{\sin 105°}$$

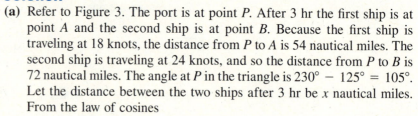

$$\sin \alpha = \frac{72(\sin 105°)}{100}$$

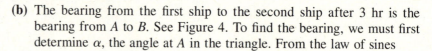

$$\sin \alpha = 0.6954$$
$$\alpha = 44°$$

From Figure 4 we observe that the bearing from A to B is $125° + (180° - \alpha)$, which is $125° + (180° - 44°)$, or 261°.

Conclusion: After 3 hr, the bearing from the first ship to the second is 261°.   ◄

FIGURE 4

Because the law of cosines involves the measures of the three sides and one angle of any triangle, it can be used to find an angle of a triangle when the measures of the three sides are known. For instance, one form of the law of cosines is

$$c^2 = a^2 + b^2 - 2ab \cos \gamma$$

If this equation is solved for $\cos \gamma$, we obtain

$$2ab \cos \gamma = a^2 + b^2 - c^2$$

$$\cos \gamma = \frac{a^2 + b^2 - c^2}{2ab}$$

Similarly, the other two forms of the law of cosines can be used to solve for $\cos \beta$ and $\cos \alpha$, and we have

$$\cos \beta = \frac{a^2 + c^2 - b^2}{2ac} \qquad \cos \alpha = \frac{b^2 + c^2 - a^2}{2bc}$$

If you wish to determine the three angles of a triangle when the measures of the three sides are given, first you should use the law of cosines to find the largest angle, which is the one opposite the longest side. This is so that you can determine if the angle is obtuse (when its cosine is negative) or acute (when its cosine is positive). In either case the other two angles will be acute and can be found from the law of sines. You cannot assume that the other two angles will be acute if the first angle found is not the largest.

▶ **EXAMPLE 3** *Solving an Oblique Triangle Given the Three Sides*

Solve the triangle for which $a = 28.4$, $b = 40.3$, and $c = 25.7$.

Solution   Figure 5 shows the triangle. We wish to find α, β, and γ. Because β is opposite the longest side, we find it first. From the law of cosines

$$\cos \beta = \frac{a^2 + c^2 - b^2}{2ac}$$

$$\cos \beta = \frac{(28.4)^2 + (25.7)^2 - (40.3)^2}{2(28.4)(25.7)}$$

$$\cos \beta = -0.1075$$

$$\beta = 96.2°$$

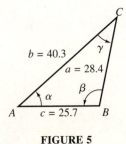

$b = 40.3$

$a = 28.4$

γ

β

α

A $c = 25.7$ B

C

FIGURE 5

We use the law of sines to compute α and γ.

$$\frac{\sin \alpha}{28.4} = \frac{\sin 96.2°}{40.3} \qquad \frac{\sin \gamma}{25.7} = \frac{\sin 96.2°}{40.3}$$

$$\sin \alpha = \frac{28.4(\sin 96.2°)}{40.3} \qquad \sin \gamma = \frac{25.7(\sin 96.2°)}{40.3}$$

$$\sin \alpha = 0.7006 \qquad \sin \gamma = 0.6340$$

$$\alpha = 44.5° \qquad \gamma = 39.3°$$

Of course, we could have computed just one of these angles by the law of sines and then found the other angle from the fact that the sum of the degree measures is 180. However, by computing them separately, we have a check:

$$\alpha + \beta + \gamma = 44.5° + 96.2° + 39.3°$$
$$= 180° \qquad \blacktriangleleft$$

▶ **EXAMPLE 4** *Finding the Area of a Parallelogram*

The lengths of two sides of a parallelogram are 7.4 cm and 9.2 cm, and one of the diagonals has a length of 6.2 cm. Find the area of the parallelogram.

Solution The parallelogram appears in Figure 6. Because the diagonal divides the parallelogram into two congruent triangles, we first find the area of one of these triangles. Recall from Section 8.4 that the measure of the area of a triangle is one-half the product of the measures of two sides and the sine of the angle included between the two sides. To use this fact, we find the angle opposite the diagonal of length 6.2 cm. If θ is this angle, then from the law of cosines

$$\cos \theta = \frac{(9.2)^2 + (7.4)^2 - (6.2)^2}{2(9.2)(7.4)}$$

$$\cos \theta = 0.742$$

$$\theta = 42°$$

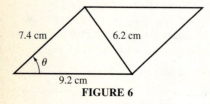

7.4 cm   6.2 cm

θ

9.2 cm

FIGURE 6

We now can find the area of the triangle formed by two sides and the given diagonal. If K square centimeters is the area of this triangle,

$$K = \tfrac{1}{2}(7.4)(9.2)\sin 42°$$
$$= 23$$

Thus

$$2K = 46$$

Conclusion: The area of the parallelogram is 46 cm². ◀

EXERCISES 8.5

In Exercises 1 through 4, measures of two sides of a triangle and the angle included between them are given. Find the measure of the third side, but do not use a calculator for the trigonometric function values. Express the results to the number of significant digits justified by the given information.

1. $a = 4.5$, $b = 6.3$, $\gamma = 60°$
2. $b = 26$, $c = 37$, $\alpha = 45°$
3. $a = 15$, $c = 22$, $\beta = 135°$
4. $a = 1.4$, $b = 2.1$, $\gamma = 120°$

In Exercises 5 through 12, solve the triangle. Express the results to the number of significant digits justified by the given information.

5. $b = 3.4$, $c = 2.8$, $\alpha = 82°$
6. $a = 43$, $c = 32$, $\beta = 59°$
7. $a = 11.2$, $b = 15.3$, $\gamma = 116.4°$
8. $a = 40.2$, $b = 45.3$, $\gamma = 72.2°$
9. $a = 2045$, $c = 3126$, $\beta = 10.52°$
10. $b = 182.4$, $c = 245.1$, $\alpha = 126.81°$
11. $a = 5.26$, $b = 3.74$, $\gamma = 135°12'$
12. $a = 325$, $c = 108$, $\beta = 18°36'$

In Exercises 13 through 18, find the area of the triangle of the indicated exercise.

13. Exercise 1
14. Exercise 4
15. Exercise 7
16. Exercise 8
17. Exercise 9
18. Exercise 10

In Exercises 19 through 26, solve the triangle. Express the results to the number of significant digits justified by the given information.

19. $a = 5.2$, $b = 7.1$, $c = 3.5$
20. $a = 8.4$, $b = 2.7$, $c = 7.3$
21. $a = 20.7$, $b = 10.2$, $c = 24.3$
22. $a = 1.24$, $b = 1.56$, $c = 1.38$
23. $a = 408$, $b = 256$, $c = 283$
24. $a = 11.3$, $b = 25.0$, $c = 27.6$
25. $a = 66.92$, $b = 53.46$, $c = 15.78$
26. $a = 718.5$, $b = 634.2$, $c = 528.4$

27. A triangle has sides of lengths 34 cm, 23 cm, and 42 cm. **(a)** Find the measurement of the smallest angle. **(b)** Determine the area of the triangle.

28. A triangle has sides of lengths 2.8 in., 3.2 in., and 4.1 in. What is **(a)** the measurement of the largest angle and **(b)** the area of the triangle?

29. A parallelogram has sides of lengths 10.3 cm and 23.2 cm, and one of the angles is 54.2°. What is **(a)** the length of the longer diagonal and **(b)** the area of the parallelogram?

30. The sides of a parallelogram have lengths of 15.6 cm and 33.0 cm. If one of the angles is 42.6°, find **(a)** the length of the shorter diagonal and **(b)** the area of the parallelogram.

In Exercises 31 through 40, the solution of the word problem involves an oblique triangle. Be sure to write a conclusion.

31. At 9 A.M. a boat leaves a pier on a course of 63.2° at 8 knots. At 10 A.M. another boat leaves the same pier on a course of 108.4° at 10 knots. At 12 noon **(a)** what is the distance between the boats and **(b)** what is the bearing from the first boat to the second?

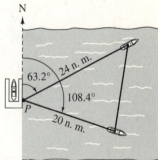

32. A point P is 1.4 km from one end of a lake and 2.2 km from the other end. If at P the lake subtends an angle of 54°, what is the length of the lake?

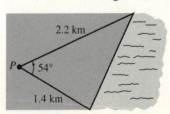

33. Two points P and Q are on opposite sides of a building. To determine the distance between these points, a third point R is selected where the distance from P to R is 50.2 m and the distance from Q to R is 61.4 m. The angle formed by the line segments PR and QR is measured as 62.5°. Determine the distance from P to Q.

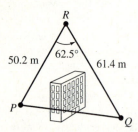

34. Two straight roads intersect at a point P and make an angle of 42.6° there. At a point R on one road is a building that is 368 m from P and at a point S on the other road is a building that is 426 m from P. Determine the direct distance from R to S.

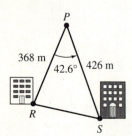

35. A tower 23.5 m tall makes an angle of 110.2° with the inclined road on which it is located. Determine the angle subtended by the tower at a point down the road 28.2 m from its foot.

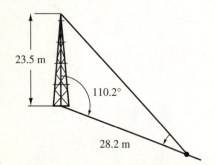

36. A triangular field has sides of lengths 212 m, 255 m, and 168 m. Determine the area of the field.

37. A ladder 24 ft long is leaning against a sloping embankment. The foot of the ladder is 11 ft from the base of the embankment, and the distance from the top of the ladder down the embankment to the ground is 16 ft. What is the angle at which the embankment is inclined to the horizontal?

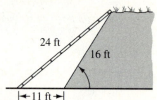

38. Two straight-line flight patterns intersect each other at an angle of 50.6°. At a particular time an airplane on one flight pattern is 53.4 miles from the intersection and a plane on the other pattern is 63.9 miles from the intersection. What is the distance between the planes at this time? There are two solutions.

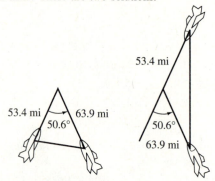

39. An airplane leaves an airport on a course of 310°. After flying 150 miles, it must return to the airport. Because of a navigational error, the plane flies 150 miles on a course of 115°. After flying the 300 miles, **(a)** how far is the plane from the airport and **(b)** what is the bearing from the plane to the airport?

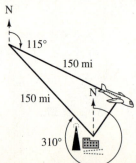

40. On a particular day the distance from the earth at E to the sun at S was $(9.2)10^7$ miles, and the distance from Mars at M to the sun at S was $(1.4)10^8$ miles. If the angle between line segments ES and MS was $59°$, what was the distance between the earth and Mars on that day?

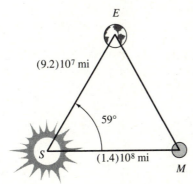

41. From the law of cosines, $\cos \alpha = \dfrac{b^2 + c^2 - a^2}{2bc}$. Use this equation to prove that

$$1 + \cos \alpha = \frac{(b + c + a)(b + c - a)}{2bc}$$

and

$$1 - \cos \alpha = \frac{(a - b + c)(a + b - c)}{2bc}$$

42. If K square units is the area of a triangle having angle α included between the sides of lengths b units and c units, then $K = \frac{1}{2}bc \sin \alpha$. Use this equation to prove that

$$K = \sqrt{\tfrac{1}{2}bc(1 + \cos \alpha) \cdot \tfrac{1}{2}bc(1 - \cos \alpha)}$$

43. Use the results of Exercises 41 and 42 to prove *Heron's formula:*

$$K = \sqrt{s(s - a)(s - b)(s - c)}$$

where $s = \frac{1}{2}(a + b + c)$. Heron's formula is used to compute the area of a triangle when only the lengths of the three sides are known.

In Exercises 44 through 46, use Heron's formula from Exercise 43 to find the area of the triangle for the given values of a, b, and c.

44. $a = 18.7$, $b = 12.6$, $c = 17.9$

45. $a = 325$, $b = 236$, $c = 411$

46. $a = 1.847$, $b = 2.112$, $c = 1.903$

In Exercises 47 and 48, use Heron's formula from Exercise 43 to find the area of the triangle having vertices at the given points.

47. $(-2, 1)$, $(2, -3)$, and $(5, 4)$

48. $(0, -3)$, $(2, 4)$, and $(5, 2)$

49. State the law of cosines in words only. Do not use any symbols for the measures of the sides and angles of a triangle. Then explain why the law of cosines can be considered as a generalization of the Pythagorean theorem.

50. Suppose you are solving a triangle when the measures of the three sides are known and you compute the first angle by the law of cosines and the other two angles by the law of sines. Explain why you should compute the largest angle first.

CHAPTER 8   REVIEW

▶ *LOOKING BACK*

8.1 Angle measurement followed our definition of an angle as a rotation. We defined a radian and learned how to convert from an angle's radian measure to its degree measure and from its degree measure back to radian measure. Formulas for computing the length of an arc of a circle and the area of a sector of a circle were applied.

8.2 We defined the six trigonometric functions of an angle having radian measure t as the corresponding functions of the real number t. Given a point on the terminal side of an angle, we computed trigonometric function values of the angle. The theorem that expresses the trigonometric functions of an acute angle in a right triangle as ratios of lengths of the sides of the triangle was presented. We found exact trigonometric function values of quadrantal angles and certain other angles, and showed how reference angles allow us to express the function value of any angle θ in terms of the corresponding function value of the reference angle associated with θ. Reference angles were applied to approximate certain trigonometric function values as well as to find angles having a given trigonometric function value.

8.3 The theorem pertaining to trigonometric functions of an acute angle in a right triangle was utilized to show that any trigonometric function of an acute angle is the cofunction of the angle's complement. We then applied the theorem to solve right triangles and word problems involving right triangles.

8.4 The law of sines was employed to solve oblique triangles given (i) two angles and a side or (ii) two sides and the angle opposite one of them. Situation (ii), the ambiguous case, presented three possibilities: no triangle, one triangle, or two triangles. We solved word problems involving oblique triangles and the law of sines.

8.5 Oblique triangles given two sides and the included angle or given the three sides were solved by the law of cosines. Applications in the word problems of this section, as well as of the previous two sections, concerned surveying, sea and air navigation, astronomy, and measurements of geometrical figures.

▶ REVIEW EXERCISES

In Exercises 1 through 4, show by a diagram an angle in standard position that has the degree measurement. Also find the equivalent radian measurement of the angle.

1. (a) 30° **(b)** 225° **2. (a)** 15° **(b)** 330°

3. (a) $-120°$ **(b)** 100° **4. (a)** $-135°$ **(b)** 250°

In Exercises 5 through 8, show by a diagram an angle in standard position that has the radian measurement. Also find the equivalent degree measurement of the angle.

5. (a) $\frac{1}{3}\pi$ **(b)** $\frac{7}{4}\pi$ **6. (a)** $\frac{5}{6}\pi$ **(b)** $\frac{5}{4}\pi$

7. (a) $-\frac{1}{3}$ **(b)** 2.36 **8. (a)** $\frac{5}{6}$ **(b)** -1.48

In Exercises 9 and 10, find (a) the arc length and (b) the area of the sector of the circle having the radius and central angle α with the given measurement.

9. radius is 12 cm; $m°(\alpha) = 150$

10. radius is 6.48 in.; $m°(\alpha) = 27.25$

In Exercises 11 through 14, if θ is an angle in standard position and the point P is on the terminal side of θ, find the six trigonometric functions of θ.

11. $P(-8, 15)$ **12.** $P(3, -4)$

13. $P(-5, 12)$ **14.** $P(-1, 0)$

In Exercises 15 through 22, sketch the angle in standard position and find the exact values of the six trigonometric functions of the angle.

15. 45° **16.** 60° **17.** $\frac{2}{3}\pi$

18. $\frac{3}{4}\pi$ **19.** $-150°$ **20.** $-135°$

21. $-\frac{9}{4}\pi$ **22.** $-\frac{1}{6}\pi$

In Exercises 23 through 26, sketch the quadrantal angle in standard position. Find the exact values of the four trigonometric functions that are defined. Which two trigonometric functions are not defined?

23. 270° **24.** 90° **25.** $-\pi$ **26.** 2π

In Exercises 27 and 28, find the reference angle for the given angle.

27. (a) $\theta = 315°$ **(b)** $\theta = 218°$ **(c)** $\theta = \frac{13}{6}\pi$
(d) $\theta = -4.27$ **(e)** $\theta = -\frac{5}{9}\pi$ **(f)** $\theta = 11.23$

28. (a) $\theta = 150°$ **(b)** $\theta = 291°$ **(c)** $\theta = -\frac{3}{4}\pi$
(d) $\theta = 7.65$ **(e)** $\theta = \frac{11}{10}\pi$ **(f)** $\theta = -216.3°$

In Exercises 29 and 30, express the function value in terms of a function of the associated reference angle; then determine the exact value.

29. (a) $\sin 150°$ **(b)** $\cos 225°$ **(c)** $\tan(-240°)$
(d) $\cot \frac{7}{4}\pi$ **(e)** $\sec(-\frac{2}{3}\pi)$ **(f)** $\csc \frac{11}{6}\pi$

30. (a) $\sin 240°$ **(b)** $\cos(-45°)$ **(c)** $\tan 120°$
(d) $\cot(-\frac{3}{4}\pi)$ **(e)** $\sec(-\frac{11}{6}\pi)$ **(f)** $\csc \frac{2}{3}\pi$

In Exercises 31 and 32, express the function value in terms of a function value of the associated reference angle. Then approximate the value on a calculator.

31. (a) $\sin 136.4°$ **(b)** $\cos(-10.8°)$ **(c)** $\tan 327.1°$
(d) $\cot(-205.7°)$ **(e)** $\sec 194.8°$ **(f)** $\csc(-98.6°)$

32. (a) $\sin 263.2°$ **(b)** $\cos 348.9°$ **(c)** $\tan(-123.6°)$
(d) $\cot(-19.7°)$ **(e)** $\sec 165.3°$ **(f)** $\csc(-244.5°)$

In Exercises 33 and 34, find all θ such that $0° \leq \theta < 360°$ for which the given equation is satisfied.

33. (a) $\cos \theta = \frac{1}{2}$ (b) $\tan \theta = -1$

34. (a) $\sin \theta = \dfrac{1}{\sqrt{2}}$ (b) $\cot \theta = -\sqrt{3}$

In Exercises 35 and 36, find all θ such that $0 \leq \theta < 2\pi$ for which the given equation is satisfied.

35. (a) $\sin \theta = -\dfrac{\sqrt{3}}{2}$ (b) $\sec \theta = -1$

36. (a) $\cos \theta = -\frac{1}{2}$ (b) $\csc \theta = 1$

In Exercises 37 and 38, find all θ to the nearest one-tenth of a degree such that $0° \leq \theta < 360°$ for which the equation is satisfied.

37. (a) $\sin \theta = 0.3217$ (b) $\cos \theta = 0.4806$
   (c) $\tan \theta = 2.953$ (d) $\cot \theta = -4.715$

38. (a) $\sin \theta = 0.8066$ (b) $\cos \theta = 0.7315$
   (c) $\tan \theta = -0.9424$ (d) $\cot \theta = 6.019$

In Exercises 39 and 40, find all θ to four significant digits such that $0 \leq \theta < 2\pi$ for which the equation is satisfied.

39. (a) $\sin \theta = -0.7049$ (b) $\cos \theta = 0.2185$
   (c) $\tan \theta = -1.356$ (d) $\cot \theta = 0.8462$

40. (a) $\sin \theta = 0.4728$ (b) $\cos \theta = -0.2093$
   (c) $\tan \theta = 5.614$ (d) $\cot \theta = -9.762$

In Exercises 41 through 46, measures of sides and angles of a triangle are given. Find the measure of the indicated side but do not use a calculator for the trigonometric function values. Express the results to the number of significant digits justified by the given information.

41. $\gamma = 90°$, $\beta = 30°$, $b = 16$; find c.

42. $\gamma = 90°$, $\alpha = 45°$, $c = 7.4$; find b.

43. $\alpha = 30°$, $\gamma = 45°$, $c = 5.3$; find a.

44. $\beta = 135°$, $a = 29$, $c = 14$; find b.

45. $\alpha = 120°$, $b = 35$, $c = 46$; find a.

46. $\beta = 120°$, $\gamma = 30°$, $b = 7.8$; find c.

In Exercises 47 through 54, solve the triangle. Express the results to the number of significant digits justified by the given information.

47. $\gamma = 90°$, $a = 4.8$, $b = 3.2$

48. $a = 60.4$, $b = 72.3$, $c = 54.7$

49. $\alpha = 43.2°$, $\beta = 61.4°$, $b = 26.8$

50. $\gamma = 90°$, $\alpha = 23.2°$, $a = 12.2$

51. $\gamma = 105.3°$, $a = 21.6$, $b = 32.4$

52. $\alpha = 114°$, $\gamma = 32°$, $a = 85$

53. $a = 518.2$, $b = 439.7$, $c = 630.4$

54. $\alpha = 29.42°$, $b = 7134$, $c = 6024$

In Exercises 55 through 58, find the area of the triangle of the indicated exercise.

55. Exercise 49   56. Exercise 52

57. Exercise 53   58. Exercise 54

In Exercises 59 through 62, the measures of two sides of a triangle and the angle opposite one of them are given. Therefore it is the ambiguous case. Determine the number of triangles that satisfy the given set of conditions and solve each triangle. Express the results to the number of significant digits justified by the given information.

59. $\alpha = 54.4°$, $a = 112$, $b = 131$

60. $\gamma = 32.4°$, $b = 50.3$, $c = 25.1$

61. $\beta = 39.7°$, $b = 12.8$, $c = 10.8$

62. $\beta = 42.5°$, $a = 12.4$, $b = 10.1$

63. An automobile tire has diameter of 30 in. How many revolutions per minute will the wheel make when the automobile maintains a speed of 30 mi/hr?

In Exercises 64 through 72, the solution of the word problem involves a triangle. Be sure to write a conclusion.

64. A tower is 150 ft high and from its top the angle of depression of an object on the ground is 36.4°.
(a) Determine the distance from the base of the tower to the object. (b) How far is the object from the top of the tower?

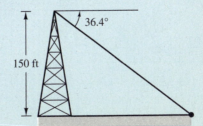

65. From the roof of a building 60 ft high, the angle of elevation of the top of a pole is 11.2°. From the bottom of the building the angle of elevation of the top of the pole is 23.4°. Find **(a)** the height of the pole and **(b)** the distance from the building to the pole.

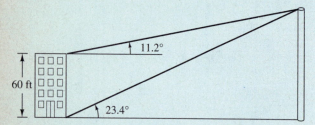

66. At a point P south of a building, the angle of elevation of the top of the building is 58°. At a point Q, 250 ft west of P, the angle of elevation is 27°. Find the height of the building.

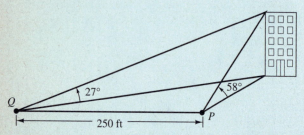

67. A tree is situated on a hill, and at a point P that is 23 m down the hill from the tree, the angle subtended by the tree is 18.5°. If the height of the tree is 36.5 m, at what angle is the hill inclined with the horizontal?

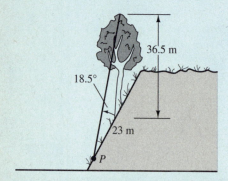

68. A triangular lot has sides of lengths 242 ft, 160 ft, and 184 ft. If the cost per square foot of land is appraised at $40, what is the appraised value of the lot?

69. A pilot going from city A to city B must avoid a particular mountain range. The pilot first flies a course of 52° for a distance of 160 miles and then alters the course to 105° and arrives at B after flying another 108 miles. **(a)** What is the direct distance from A to B? **(b)** What is the bearing from A to B?

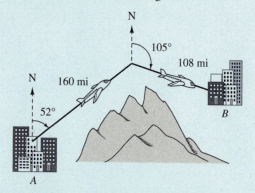

70. A hill is inclined at an angle of 16.2° with the horizontal, and a tunnel running through the side of the hill is inclined at an angle of 11.3° with the horizontal. From a point 256 ft down the tunnel, what is the vertical distance to the surface of the hill?

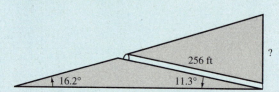

71. A vacant lot in the form of a parallelogram is situated on the corner of two streets that intersect at an angle of 98.3°, and the street frontages of the lot are 76.7 ft and 91.4 ft. If instead of going around the street sides of the lot, a girl decides to cross the lot along a diagonal from one corner to another, what distance does she save?

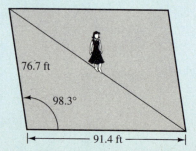

72. To determine the distance between two points P and Q on opposite sides of a building, a third point R is chosen such that the distance from P to R is 120 m and the distance from Q to R is 140 m. If the angle formed by the line segments PR and QR is 72.3°, what is the distance from P to Q?

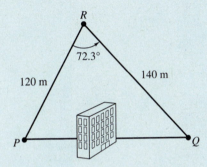

73. If one acre is equivalent to 4840 yd², use Heron's formula in Exercise 43 of Exercises 8.5 to find the number of acres in the area of the triangular field having sides of lengths 453 yd, 592 yd, and 700 yd.

74. If r units is the radius of the inscribed circle of a triangle having sides of lengths a units, b units, and c units, use Heron's formula in Exercise 43 of Exercises 8.5 to show that

$$r = \sqrt{\frac{(s - a)(s - b)(s - c)}{s}}$$

where $s = \frac{1}{2}(a + b + c)$.

In Exercises 75 and 76, use the formula of Exercise 74 to find the radius of the inscribed circle of the triangle for the given values of a, b, and c.

75. $a = 325, b = 236, c = 411$

76. $a = 1.847, b = 2.112, c = 1.903$

Analytic Trigonometry

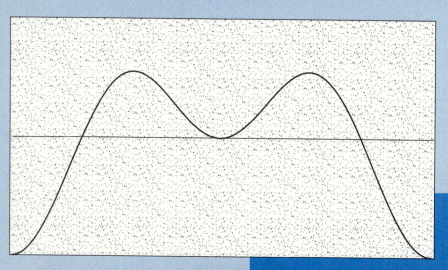

In calculus it is often necessary to convert a trigonometric expression from one form to an equivalent simpler one. To perfect that skill we apply the fundamental identities and algebraic manipulations to verify other identities. All the primary identities are listed for reference on the endpapers in the front and back of the book.

Some of the basic identities are used in our discussion of the inverse trigonometric functions. These functions, as well as familiarity with trigonometric identities, enable us to solve trigonometric equations.

9.1 TRIGONOMETRIC IDENTITIES

GOALS
1. **Apply the eight fundamental identities to prove other identities.**
2. **Learn suggestions for proving identities.**
3. **Learn how to show an equation is not an identity.**

Recall from Section 2.1 that an identity is an equation for which the solution set is the same as the domain of the variable. For trigonometric identities we will use symbols such as s, t, u, and v to represent real numbers and symbols such as α, β, γ, and θ to represent angles. Symbols such as x, y, and z will be used to represent real numbers or angles.

In Section 7.6 we discussed the eight fundamental trigonometric identities and listed them. In that section t represented a real number. We repeat the identities here for reference as Table 1, where x represents either a real number or an angle.

Table 1

The Eight Fundamental Trigonometric Identities

$$\text{I} \quad \sin x \csc x = 1 \leftrightarrow \csc x = \frac{1}{\sin x} \leftrightarrow \sin x = \frac{1}{\csc x}, \, x \neq k\pi, \, k \in J$$

$$\text{II} \quad \cos x \sec x = 1 \leftrightarrow \sec x = \frac{1}{\cos x} \leftrightarrow \cos x = \frac{1}{\sec x}, \, x \neq \tfrac{1}{2}\pi + k\pi, \, k \in J$$

$$\text{III} \quad \tan x \cot x = 1 \leftrightarrow \cot x = \frac{1}{\tan x} \leftrightarrow \tan x = \frac{1}{\cot x}, \, x \neq \tfrac{1}{2}k\pi, \, k \in J$$

$$\text{IV} \quad \tan x = \frac{\sin x}{\cos x}, \, x \neq \tfrac{1}{2}\pi + k\pi, \, k \in J$$

$$\text{V} \quad \cot x = \frac{\cos x}{\sin x}, \, x \neq k\pi, \, k \in J$$

$$\text{VI} \quad \sin^2 x + \cos^2 x = 1 \leftrightarrow \sin^2 x = 1 - \cos^2 x \leftrightarrow \cos^2 x = 1 - \sin^2 x$$

$$\text{VII} \quad 1 + \tan^2 x = \sec^2 x \leftrightarrow \tan^2 x = \sec^2 x - 1 \leftrightarrow \sec^2 x - \tan^2 x = 1$$

$$\text{VIII} \quad 1 + \cot^2 x = \csc^2 x \leftrightarrow \cot^2 x = \csc^2 x - 1 \leftrightarrow \csc^2 x - \cot^2 x = 1$$

We now verify, or prove, other trigonometric identities. No general method can be applied. As we present the examples, we shall give suggestions helpful in determining the best approach for a proof. Familiarity with the eight fundamental identities in their various forms is crucial.

If one side of an identity is in a more complicated form than the other, you may wish to start with it and transform it to the simpler form on the other side. Bear this simpler form in mind as you proceed. It represents your objective.

Before beginning the proof of an identity, you may wish to plot the graph of the function on each side. The fact that the graphs are the same at all points for which the functions are defined will convince you that you have a valid identity.

▶ **EXAMPLE 1** *Proving an Identity*

Prove the identity

$$\frac{1 + \sin x}{\cos x} = \sec x + \tan x$$

Solution Because the left side is a fraction, we consider it as more complicated than the right side. We therefore begin with the left side, which we write as the sum of two fractions. We then apply two fundamental identities to obtain the right side.

$$\frac{1 + \sin x}{\cos x} = \frac{1}{\cos x} + \frac{\sin x}{\cos x}$$
$$= \sec x + \tan x \qquad ◀$$

You will often find it expedient to convert an expression to one containing only sine and cosine.

▶ **EXAMPLE 2** *Proving an Identity*

Prove the identity

$$\frac{\csc x + \sec x}{1 + \tan x} = \csc x$$

Solution Because the left side is the more complicated one, we start with it. We use fundamental identities to express it in terms of the sine and cosine, and then apply algebraic processes to get the right side. In the second step, we multiply numerator and denominator by $\sin x \cos x$ (the LCD of all the fractions).

$$\frac{\csc x + \sec x}{1 + \tan x} = \frac{\dfrac{1}{\sin x} + \dfrac{1}{\cos x}}{1 + \dfrac{\sin x}{\cos x}}$$

$$= \frac{\sin x \cos x \left(\dfrac{1}{\sin x} + \dfrac{1}{\cos x}\right)}{\sin x \cos x \left(1 + \dfrac{\sin x}{\cos x}\right)}$$

$$= \frac{\cos x + \sin x}{\sin x \cos x + \sin^2 x}$$

$$= \frac{\cos x + \sin x}{\sin x (\cos x + \sin x)}$$

$$= \frac{1}{\sin x}$$

$$= \csc x \qquad \blacktriangleleft$$

▶ **EXAMPLE 3** *Proving an Identity*

Prove the identity

$$(1 + \sec \theta)(1 - \cos \theta) = \tan \theta \sin \theta$$

Solution   The left side is the more complicated one. We start with it and begin by performing the indicated multiplication.

$$(1 + \sec \theta)(1 - \cos \theta) = 1 + \sec \theta - \cos \theta - \sec \theta \cos \theta$$

$$= 1 + \frac{1}{\cos \theta} - \cos \theta - 1$$

$$= \frac{1 - \cos^2 \theta}{\cos \theta}$$

$$= \frac{\sin^2 \theta}{\cos \theta}$$

$$= \frac{\sin \theta}{\cos \theta}(\sin \theta)$$

$$= \tan \theta \sin \theta \qquad \blacktriangleleft$$

The verification of the identity in the next example involves a procedure different from those used in the preceding examples. We transform each side separately into the same equivalent form.

▶ **EXAMPLE 4** *Proving an Identity*

Prove the identity

$$\frac{1 + \cot y}{\csc y} = \frac{1 + \tan y}{\sec y}$$

Solution   We start with the left side and express $\cot y$ and $\csc y$ in terms of $\sin y$ and $\cos y$.

$$\frac{1 + \cot y}{\csc y} = \frac{1 + \dfrac{\cos y}{\sin y}}{\dfrac{1}{\sin y}}$$

$$= \frac{\sin y \left(1 + \dfrac{\cos y}{\sin y}\right)}{\sin y \left(\dfrac{1}{\sin y}\right)}$$

$$= \sin y + \cos y$$

We now express the right side in terms of $\sin y$ and $\cos y$.

$$\frac{1 + \tan y}{\sec y} = \frac{1 + \dfrac{\sin y}{\cos y}}{\dfrac{1}{\cos y}}$$

$$= \frac{\cos y \left(1 + \dfrac{\sin y}{\cos y}\right)}{\cos y \left(\dfrac{1}{\cos y}\right)}$$

$$= \cos y + \sin y$$

Because each side is equal to $\sin y + \cos y$, we can conclude that

$$\frac{1 + \cot y}{\csc y} = \frac{1 + \tan y}{\sec y}$$ ◀

As shown in the following example, sometimes it may be advantageous to convert an expression to one involving only a single function.

▶ **EXAMPLE 5** *Proving an Identity*

Prove the identity

$$(1 - \tan \beta)^3 = \frac{\cot \beta - 1}{\cot \beta}(\sec^2 \beta - 2 \tan \beta)$$

Solution   The right side is more complicated. We start with it, and because the left side involves only the tangent, we express the right side in terms of $\tan \beta$.

$$\frac{\cot \beta - 1}{\cot \beta}(\sec^2 \beta - 2 \tan \beta)$$

$$= \frac{\dfrac{1}{\tan \beta} - 1}{\dfrac{1}{\tan \beta}}(1 + \tan^2 \beta - 2 \tan \beta)$$

$$= \frac{\tan \beta\left(\dfrac{1}{\tan \beta} - 1\right)}{\tan \beta\left(\dfrac{1}{\tan \beta}\right)}(1 - 2 \tan \beta + \tan^2 \beta)$$

$$= \frac{1 - \tan \beta}{1}(1 - \tan \beta)^2$$

$$= (1 - \tan \beta)^3 \qquad\qquad ◀$$

▶ **EXAMPLE 6** *Proving an Identity*

Prove the identity

$$\frac{\sin x}{1 + \cos x} = \frac{1 - \cos x}{\sin x}$$

Solution   Each side of the identity involves a fraction containing only the sine and cosine. On the left the binomial $1 + \cos x$ appears in the denominator, and on the right the binomial $1 - \cos x$ appears in the numerator. We can start with the left side and obtain a factor of $1 - \cos x$ in the numerator by multiplying the numerator and denominator by $1 - \cos x$. This operation is equivalent to multiplying the fraction by 1. Therefore, we have

$$\frac{\sin x}{1 + \cos x} = \frac{\sin x(1 - \cos x)}{(1 + \cos x)(1 - \cos x)}$$

$$= \frac{\sin x(1 - \cos x)}{1 - \cos^2 x}$$

$$= \frac{\sin x(1 - \cos x)}{\sin^2 x}$$

$$= \frac{1 - \cos x}{\sin x} \qquad \blacktriangleleft$$

We now summarize the suggestions given for proving an identity. Review them before doing the exercises.

Suggestions for Proving an Identity

1. Start with the more complicated side and transform it to the simpler form on the other side. See Examples 1, 2, 3, and 5.
2. Instead of suggestion 1, it may be more convenient to transform each side into the same equivalent form. See Example 4.
3. Often it is desirable to convert an expression to one containing only the sine and cosine. See Examples 2 and 4.
4. Instead of suggestion 3, it may be advantageous to convert an expression to one involving only a single function provided no radicals are introduced. See Example 5.
5. Consider the possibilities of applying algebraic processes such as multiplying, factoring, combining fractions into a single fraction, writing a single fraction having more than one term in the numerator into a sum of fractions, and simplifying a complex fraction. See Examples 1, 2, 3, and 5.
6. To obtain a particular factor in the numerator or denominator of a fraction, you may multiply the numerator and denominator by this desired factor. See Example 6.

If you suspect an equation is not an identity, first plot the graph of the function on each side of the equation to verify that the graphs are different. Then to show algebraically that the equation is not an identity, you need only find one number in the domain of the variable for which the equality is not true. By doing this, you have found a *counterexample*.

▶ **EXAMPLE 7** *Showing an Equation Is Not an Identity*

Demonstrate graphically and algebraically that the equation

$$\sqrt{1 - \cos^2 x} = \sin x$$

is not an identity.

Solution   When you plot the graph of the function defined by $\sqrt{1 - \cos^2 x}$, you obtain only the top half of the sine curve. To show algebraically that the equation is not an identity, observe that if x is any number for which $\sin x < 0$, the right side is negative and the left side is nonnegative. In particular, if $x = \frac{7}{6}\pi$, the left side is

$$\sqrt{1 - \cos^2 \frac{7}{6}\pi} = \sqrt{1 - \left(-\frac{\sqrt{3}}{2}\right)^2}$$
$$= \sqrt{1 - \frac{3}{4}}$$
$$= \sqrt{\frac{1}{4}}$$
$$= \frac{1}{2}$$

and the right side is

$$\sin \frac{7}{6}\pi = -\frac{1}{2}$$ ◀

▶ **EXAMPLE 8** *Showing an Equation Is Not an Identity*

Demonstrate graphically and algebraically that the equation

$$4 \sin t \cos t = \sec t$$

is not an identity.

Solution   The graph of the function defined by the left side of the equation appears in Figure 1. It is not the secant curve. Thus the equation is not an identity. To show this fact algebraically, let $t = 0$. The left side is

$$4 \sin 0 \cos 0 = 4(0)(1)$$
$$= 0$$

and the right side is

$$\sec 0 = 1$$

Therefore the equation is not an identity. ◀

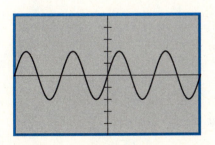

$[-2\pi, 2\pi]$ by $[-5, 5]$
$f(t) = 4 \sin t \cos t$
FIGURE 1

EXERCISES 9.1

In Exercises 1 through 6, write the expression in terms of either sin x or cos x, or both, in a simplified form.

1. (a) $\cos x \tan x$ **(b)** $\dfrac{\cos x}{\sec x}$

2. (a) $\sin x \cot x$ **(b)** $\dfrac{\sin x}{\csc x}$

3. (a) $\sec x \cot x$ **(b)** $\dfrac{\cot^2 x}{\csc^2 x}$

4. (a) $\csc x \tan x$ **(b)** $\dfrac{\tan^2 x}{\sec^2 x}$

5. (a) $\tan x + \cot x$ **(b)** $\sin x + \cos^2 x \csc x$

6. (a) $\sec^2 x + \csc^2 x$ **(b)** $\cos x + \sin^2 x \sec x$

In Exercises 7 through 12, prove that the first expression is equivalent to the second expression for all values of the variable for which both expressions are defined.

7. (a) $\cos \theta \csc \theta, \cot \theta$ **(b)** $\sin \theta \sec \theta \cot \theta, 1$

8. (a) $\sin \theta \sec \theta, \tan \theta$ **(b)** $\cos \theta \csc \theta \tan \theta, 1$

9. (a) $\dfrac{\sec \alpha}{\csc \alpha}, \tan \alpha$ **(b)** $\dfrac{\tan \alpha}{\sin \alpha}, \sec \alpha$

10. (a) $\dfrac{\csc \alpha}{\sec \alpha}, \cot \alpha$ **(b)** $\dfrac{\cot \alpha}{\cos \alpha}, \csc \alpha$

11. (a) $\dfrac{1 + \cot^2 \beta}{\sec^2 \beta}, \cot^2 \beta$

   (b) $(1 + \sin \beta)(1 - \sin \beta), \cos^2 \beta$

12. (a) $\dfrac{1 + \tan^2 \beta}{\csc^2 \beta}, \tan^2 \beta$

   (b) $(1 + \cos \beta)(1 - \cos \beta), \sin^2 \beta$

In Exercises 13 through 42, prove the identity.

13. $\sec^2 x + \csc^2 x = \sec^2 x \cdot \csc^2 x$

14. $(\tan x + \cot x)^2 = \sec^2 x \cdot \csc^2 x$

15. $\tan^2 x - \sin^2 x = \tan^2 x \cdot \sin^2 x$

16. $\cot^2 x - \cos^2 x = \cos^2 x \cdot \cot^2 x$

17. $\dfrac{\sec \theta + 1}{\sec \theta - 1} = \dfrac{1 + \cos \theta}{1 - \cos \theta}$

18. $\dfrac{\csc \theta + 1}{\csc \theta - 1} = \dfrac{1 + \sin \theta}{1 - \sin \theta}$

19. $\dfrac{1 - \cot^2 \theta}{1 + \cot^2 \theta} = \sin^2 \theta - \cos^2 \theta$

20. $\dfrac{1}{\tan \theta + \cot \theta} = \sin \theta \cos \theta$

21. $\dfrac{\csc^2 \alpha - 1}{\sec^2 \alpha - 1} = \cot^4 \alpha$

22. $\dfrac{1}{\csc \alpha - 1} - \dfrac{1}{\csc \alpha + 1} = 2 \tan^2 \alpha$

23. $\dfrac{1}{1 + \sin \alpha} + \dfrac{1}{1 - \sin \alpha} = 2 \sec^2 \alpha$

24. $\dfrac{1}{1 + \cos \alpha} + \dfrac{1}{1 - \cos \alpha} = 2 \csc^2 \alpha$

25. $\dfrac{\sin \beta}{1 + \cos \beta} + \dfrac{1 + \cos \beta}{\sin \beta} = 2 \csc \beta$

26. $\dfrac{\cos \beta}{1 + \sin \beta} + \dfrac{\cos \beta}{1 - \sin \beta} = 2 \sec \beta$

27. $\dfrac{\cos \beta}{1 + \sin \beta} = \dfrac{1 - \sin \beta}{\cos \beta}$

28. $\dfrac{\sec \beta + 1}{\tan \beta} = \dfrac{\tan \beta}{\sec \beta - 1}$

29. $\dfrac{\cos t}{\sec t - \tan t} = 1 + \sin t$

30. $\sin t \tan t = \sec t - \cos t$

31. $\dfrac{\tan y}{\tan^2 y - 1} = \dfrac{1}{\tan y - \cot y}$

32. $\cot y \csc y = \dfrac{1}{\sec y - \cos y}$

33. $(\tan x + \cot x)^2 = \sec^2 x + \csc^2 x$

34. $\sec x + \tan x = \dfrac{1}{\sec x - \tan x}$

35. $\csc^4 \theta - \cot^4 \theta = \csc^2 \theta + \cot^2 \theta$

36. $\cos^4 \theta - \sin^4 \theta = \cos^2 \theta - \sin^2 \theta$

37. $\tan^4 \alpha + \tan^2 \alpha = \sec^4 \alpha - \sec^2 \alpha$

38. $\csc^4 \alpha - \csc^2 \alpha = \cot^4 \alpha + \cot^2 \alpha$

39. $\sin^3 t + \cos^3 t + \sin t \cos^2 t + \sin^2 t \cos t$
$$= \sin t + \cos t$$

40. $\dfrac{\sin^3 t + \cos^3 t}{\sin t + \cos t} = 1 - \sin t \cos t$

41. $\dfrac{\tan^3 x + \sin x \sec x - \sin x \cos x}{\sec x - \cos x}$

$$= \tan x \sec x + \sin x$$

42. $\dfrac{2 \tan x}{1 - \tan^2 x} + \dfrac{1}{2 \cos^2 x - 1} = \dfrac{\cos x + \sin x}{\cos x - \sin x}$

In Exercises 43 through 46, show graphically and algebraically that the equation is not an identity.

43. $\sin t \cos t + 1 = \sin t + \cos t$

44. $\sin t \sec t + 1 = 2 \sin t + \sec t$

45. $\sec y = 1 + \tan^2 y$

46. $\tan^2 x + \tan x = 2 \tan^3 x$

In Exercises 47 through 54, to determine if the equation is an identity, plot the graph of the function on each side. If the equation is an identity, prove this fact algebraically. If it is not an identity, find a counterexample.

47. $\dfrac{1}{1 + \cos \theta} - \dfrac{1}{1 - \cos \theta} = \dfrac{2}{\sec \theta - \cos \theta}$

48. $\dfrac{1}{1 - \sin \theta} - \dfrac{1}{1 + \sin \theta} = \dfrac{2}{\tan \theta \cos \theta}$

49. $\dfrac{1}{1 - \cos \theta} - \dfrac{1}{1 + \cos \theta} = 2 \cot \theta \csc \theta$

50. $\tan \theta \sec \theta - \sin \theta = \tan \theta$

51. $1 - \tan^3 x = (2 \tan x - \sec^2 x)(\tan x - 1)$

52. $\csc^4 x(1 - \cos^4 x) = 1 + 2 \cot^2 x$

53. $\dfrac{1 + \cos x}{1 - \cos x} + \dfrac{1 + \sin x}{1 - \sin x} = \dfrac{2(\cos x - \csc x)}{\cot x - \cos x - \csc x + 1}$

54. $\dfrac{1 + \cos x}{1 - \cos x} + \dfrac{1 - \sin x}{1 + \sin x} = \dfrac{2(\sin x + \sec x)}{\tan x - \sin x + \sec x - 1}$

55. The fundamental Pythagorean identity is

$$\sin^2 x + \cos^2 x = 1$$

Explain in both graphical and algebraic terms why neither

$$\sin x = \sqrt{1 - \cos^2 x} \quad \text{nor} \quad \cos x = \sqrt{1 - \sin^2 x}$$

is an identity.

56. Explain why the equations

$$\sec^2 x - \tan^2 x = 1 \quad \text{and} \quad \csc^2 x - \cot^2 x = 1$$

are identities even though the left hand sides are not defined for certain values of x. Include in your explanation the values of x for which the left hand sides are not defined.

9.2 SUM AND DIFFERENCE IDENTITIES

GOALS

1. Learn and apply the cosine difference identity.
2. Learn and apply the cosine sum identity.
3. Learn and apply the cofunction identities.
4. Learn and apply the sine sum identity.
5. Learn and apply the sine difference identity.
6. Learn and apply the tangent sum identity.
7. Learn and apply the tangent difference identity.
8. Find exact function values from the sum and difference identities.
9. Use the sum and difference identities to prove other identities.
10. Write an expression of the form $A \sin bt + B \cos bt$ as $a \sin (bt + c)$.

We now obtain identities that express trigonometric function values of the sum or difference of two real numbers (or angles) in terms of trigonometric function values of each real number (or angle).

We begin by deriving a formula for $\cos(u - v)$ where u and v are real numbers. In our discussion we take $\frac{1}{2}\pi < u < \pi$ and $0 < v < \frac{1}{2}\pi$. A similar argument can be used for u and v in any quadrants. Figure 1 shows the unit circle U and the point $A(1, 0)$. Point P_1 is the terminal point of the arc having initial point at A and length u, and P_2 is the terminal point of the arc having initial point at A and length v. By the definition of the sine and cosine of a real number, P_1 is the point $(\cos u, \sin u)$ and P_2 is the point $(\cos v, \sin v)$. The length of the arc from P_2 to P_1 is $u - v$. Let P_3 on the unit circle U be the terminal point of the arc having initial point at A and length $u - v$. Thus P_3 is the point $(\cos(u - v), \sin(u - v))$. Because the length of the arc from P_2 to P_1 is the same as the length of the arc from A to P_3, it follows from a theorem in geometry that $|\overline{P_2 P_1}| = |\overline{AP_3}|$; hence

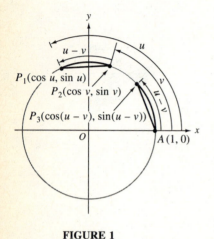

FIGURE 1

$$|\overline{P_2 P_1}|^2 = |\overline{AP_3}|^2 \tag{1}$$

From the distance formula

$$|\overline{P_2 P_1}|^2$$
$$= (\cos u - \cos v)^2 + (\sin u - \sin v)^2$$
$$= \cos^2 u - 2\cos u \cos v + \cos^2 v + \sin^2 u - 2\sin u \sin v + \sin^2 v$$
$$= (\cos^2 u + \sin^2 u) + (\cos^2 v + \sin^2 v) - 2(\cos u \cos v + \sin u \sin v)$$
$$= 1 + 1 - 2(\cos u \cos v + \sin u \sin v)$$
$$= 2 - 2(\cos u \cos v + \sin u \sin v) \tag{2}$$

Also from the distance formula

$$|\overline{AP_3}|^2 = [\cos(u - v) - 1]^2 + [\sin(u - v) - 0]^2$$
$$= \cos^2(u - v) - 2\cos(u - v) + 1 + \sin^2(u - v)$$
$$= [\cos^2(u - v) + \sin^2(u - v)] + 1 - 2\cos(u - v)$$
$$= 1 + 1 - 2\cos(u - v)$$
$$= 2 - 2\cos(u - v) \tag{3}$$

Substituting from (2) and (3) into (1), we have

$$2 - 2(\cos u \cos v + \sin u \sin v) = 2 - 2\cos(u - v)$$
$$\cos(u - v) = \cos u \cos v + \sin u \sin v$$

This formula is an identity because it is true for all real numbers. It is also true for all angles. It is called the cosine difference identity, and we write it now in terms of x and y.

Cosine Difference Identity

$$\cos(x - y) = \cos x \cos y + \sin x \sin y$$

If in this identity $-y$ is substituted for y, we have

$$\cos[x - (-y)] = \cos x \cos(-y) + \sin x \sin(-y)$$

Because $\sin(-y) = -\sin y$ and $\cos(-y) = \cos y$, we obtain the cosine sum identity.

Cosine Sum Identity

$$\cos(x + y) = \cos x \cos y - \sin x \sin y$$

In the following example we apply the sum and difference identities to compute certain exact function values. This particular application of the identities is not important in itself because you can obtain approximate function values on your calculator. The purpose of the example, and exercises like it, is to give practice in using the formulas so that you gain familiarity with them.

▶ **EXAMPLE 1** *Finding Exact Function Values from the Sum and Difference Identities*

Determine the exact value of **(a)** $\cos \frac{1}{12} \pi$ and **(b)** $\cos 105°$.

Solution

(a) Because $\frac{1}{12} \pi = \frac{1}{4} \pi - \frac{1}{6} \pi$ and we know the exact function values of $\frac{1}{4} \pi$ and $\frac{1}{6} \pi$, we use the cosine difference identity.

$$\begin{aligned}
\cos \tfrac{1}{12} \pi &= \cos(\tfrac{1}{4} \pi - \tfrac{1}{6} \pi) \\
&= \cos \tfrac{1}{4} \pi \cos \tfrac{1}{6} \pi + \sin \tfrac{1}{4} \pi \sin \tfrac{1}{6} \pi \\
&= \frac{1}{\sqrt{2}} \cdot \frac{\sqrt{3}}{2} + \frac{1}{\sqrt{2}} \cdot \frac{1}{2} \\
&= \frac{\sqrt{3} + 1}{2\sqrt{2}}
\end{aligned}$$

(b) Because $105° = 60° + 45°$ and the exact function values of $60°$ and $45°$ are known, we apply the cosine sum identity.

$$\cos 105° = \cos(60° + 45°)$$
$$= \cos 60° \cos 45° - \sin 60° \sin 45°$$
$$= \frac{1}{2} \cdot \frac{1}{\sqrt{2}} - \frac{\sqrt{3}}{2} \cdot \frac{1}{\sqrt{2}}$$
$$= \frac{1 - \sqrt{3}}{2\sqrt{2}} \qquad \blacktriangleleft$$

The cosine sum and difference identities can be used to obtain other identities. If in the cosine difference identity we let $x = \frac{1}{2}\pi$, we get

$$\cos(\tfrac{1}{2}\pi - y) = \cos \tfrac{1}{2}\pi \cos y + \sin \tfrac{1}{2}\pi \sin y$$
$$= (0)\cos y + (1)\sin y$$
$$= \sin y$$

Now in this formula let $y = \frac{1}{2}\pi - x$. We have then

$$\cos[\tfrac{1}{2}\pi - (\tfrac{1}{2}\pi - x)] = \sin(\tfrac{1}{2}\pi - x)$$
$$\cos x = \sin(\tfrac{1}{2}\pi - x)$$

From a fundamental identity

$$\tan(\tfrac{1}{2}\pi - x) = \frac{\sin(\tfrac{1}{2}\pi - x)}{\cos(\tfrac{1}{2}\pi - x)}$$

Replacing $\sin(\frac{1}{2}\pi - x)$ by $\cos x$ and $\cos(\frac{1}{2}\pi - x)$ by $\sin x$, we have

$$\tan(\tfrac{1}{2}\pi - x) = \frac{\cos x}{\sin x}$$
$$= \cot x$$

We have obtained the following three identities, called the cofunction identities.

Cofunction Identities

$$\cos(\tfrac{1}{2}\pi - x) = \sin x$$
$$\sin(\tfrac{1}{2}\pi - x) = \cos x$$
$$\tan(\tfrac{1}{2}\pi - x) = \cot x$$

Observe that these formulas for angles appear in Equations (5) and (6) of Section 8.3. However, there the angle was restricted to being acute.

We now derive the sine sum and difference identities. From the first cofunction identity, with x replaced by $x + y$, we have

$$\sin(x + y) = \cos[\tfrac{1}{2}\pi - (x + y)]$$
$$= \cos[(\tfrac{1}{2}\pi - x) - y]$$

We now apply the cosine difference identity and get

$$\sin(x + y) = \cos(\tfrac{1}{2}\pi - x)\cos y + \sin(\tfrac{1}{2}\pi - x)\sin y$$

Using cofunction identities on the right side, we obtain the following sine sum identity.

Sine Sum Identity

$$\sin(x + y) = \sin x \cos y + \cos x \sin y$$

If in this identity we replace y by $-y$, we get

$$\sin[x + (-y)] = \sin x \cos(-y) + \cos x \sin(-y)$$

Replacing $\cos(-y)$ by $\cos y$ and $\sin(-y)$ by $-\sin y$, we have the following sine difference identity.

Sine Difference Identity

$$\sin(x - y) = \sin x \cos y - \cos x \sin y$$

▶ **EXAMPLE 2** *Finding Exact Function Values from the Sum Identities*

If $\sin \alpha = \frac{24}{25}$, where α is in the first quadrant, and $\sin \beta = \frac{4}{5}$, where β is in the second quadrant, find **(a)** $\sin(\alpha + \beta)$, **(b)** $\cos(\alpha + \beta)$, and **(c)** the quadrant containing $\alpha + \beta$.

Solution   To use the sine sum identity and the cosine sum identity, we need to determine $\cos \alpha$ and $\cos \beta$. To find $\cos \alpha$, we use the fundamental identity $\sin^2 \alpha + \cos^2 \alpha = 1$ with $\sin \alpha = \frac{24}{25}$ and $\cos \alpha > 0$ because α is in the first quadrant.

$$(\tfrac{24}{25})^2 + \cos^2 \alpha = 1$$
$$\cos^2 \alpha = 1 - \tfrac{576}{625}$$
$$\cos^2 \alpha = \tfrac{49}{625}$$
$$\cos \alpha = \tfrac{7}{25}$$

To find $\cos \beta$, we use the identity $\sin^2 \beta + \cos^2 \beta = 1$ with $\sin \beta = \frac{4}{5}$ and $\cos \beta < 0$ because β is in the second quadrant.

$$(\tfrac{4}{5})^2 + \cos^2 \beta = 1$$
$$\cos^2 \beta = 1 - \tfrac{16}{25}$$
$$\cos^2 \beta = \tfrac{9}{25}$$
$$\cos \beta = -\tfrac{3}{5}$$

(a) From the sine sum identity

$$\sin(\alpha + \beta) = \sin \alpha \cos \beta + \cos \alpha \sin \beta$$
$$= (\tfrac{24}{25})(-\tfrac{3}{5}) + (\tfrac{7}{25})(\tfrac{4}{5})$$
$$= -\tfrac{72}{125} + \tfrac{28}{125}$$
$$= -\tfrac{44}{125}$$

(b) From the cosine sum identity

$$\cos(\alpha + \beta) = \cos \alpha \cos \beta - \sin \alpha \sin \beta$$
$$= (\tfrac{7}{25})(-\tfrac{3}{5}) - (\tfrac{24}{25})(\tfrac{4}{5})$$
$$= -\tfrac{21}{125} - \tfrac{96}{125}$$
$$= -\tfrac{117}{125}$$

(c) Because $\sin(\alpha + \beta) < 0$ and $\cos(\alpha + \beta) < 0$, we conclude that $\alpha + \beta$ is in the third quadrant. ◀

Observe in Example 2 that $\sin(\alpha + \beta)$ and $\cos(\alpha + \beta)$ were obtained without finding the actual values of α and β.

To express $\tan(x + y)$ in terms of $\tan x$ and $\tan y$, we begin with the fundamental identity that states the tangent is the quotient of the sine and cosine. We then use the sine and cosine sum identities. We have

$$\tan(x + y) = \frac{\sin(x + y)}{\cos(x + y)}$$

$$= \frac{\sin x \cos y + \cos x \sin y}{\cos x \cos y - \sin x \sin y}$$

So that the identity involves $\tan x$ and $\tan y$, we divide the numerator and denominator by $\cos x \cos y$, with the assumption that $\cos x \cos y \neq 0$. Thus

$$\tan(x + y) = \frac{\dfrac{\sin x}{\cos x} \cdot \dfrac{\cos y}{\cos y} + \dfrac{\cos x}{\cos x} \cdot \dfrac{\sin y}{\cos y}}{\dfrac{\cos x}{\cos x} \cdot \dfrac{\cos y}{\cos y} - \dfrac{\sin x}{\cos x} \cdot \dfrac{\sin y}{\cos y}}$$

$$= \frac{\tan x \cdot 1 + 1 \cdot \tan y}{1 \cdot 1 - \tan x \cdot \tan y}$$

We have therefore obtained the following tangent sum identity.

Tangent Sum Identity

$$\tan(x + y) = \frac{\tan x + \tan y}{1 - \tan x \tan y}$$

If in this identity y is replaced by $-y$, we get

$$\tan[x + (-y)] = \frac{\tan x + \tan(-y)}{1 - \tan x \tan(-y)}$$

In Section 7.7, we proved that $\tan(-y) = -\tan y$. Making this substitution in the preceding equation, we have

$$\tan(x - y) = \frac{\tan x + (-\tan y)}{1 - \tan x(-\tan y)}$$

from which the following tangent difference identity is obtained.

Tangent Difference Identity

$$\tan(x - y) = \frac{\tan x - \tan y}{1 + \tan x \tan y}$$

In the derivation of the tangent sum identity, the restriction that $\cos x \cos y \neq 0$ indicates that the identity does not hold if either x or y is $\frac{1}{2}\pi + k\pi$, where k is any integer. Observe that if you attempt to apply either the tangent sum identity or the tangent difference identity when x or y has a value of $\frac{1}{2}\pi + k\pi$, the right side involves $\tan(\frac{1}{2}\pi + k\pi)$, which is not defined.

▶ **EXAMPLE 3** *Proving an Identity*

Prove the identity

$$\tan(x - \tfrac{1}{4}\pi) = \frac{\tan x - 1}{\tan x + 1}$$

Solution   From the tangent difference identity

$$\tan(x - \tfrac{1}{4}\pi) = \frac{\tan x - \tan \tfrac{1}{4}\pi}{1 + \tan x \tan \tfrac{1}{4}\pi}$$

Because $\tan \frac{1}{4}\pi = 1$, we get

$$\tan(x - \tfrac{1}{4}\pi) = \frac{\tan x - 1}{1 + \tan x(1)}$$

$$= \frac{\tan x - 1}{\tan x + 1} \qquad \blacktriangleleft$$

Identities that express a trigonometric function of $\frac{1}{2}k\pi \pm x$, where k is any integer, in terms of a function of x are called **reduction formulas.** In particular, the cofunction identities are reduction formulas. Some others are

$$\sin(\tfrac{1}{2}\pi + x) = \cos x \qquad \cos(\tfrac{1}{2}\pi + x) = -\sin x \qquad \tan(\tfrac{1}{2}\pi + x) = -\cot x$$

$$\sin(\pi - x) = \sin x \qquad \cos(\pi - x) = -\cos x \qquad \tan(\pi - x) = -\tan x$$

$$\sin(\pi + x) = -\sin x \qquad \cos(\pi + x) = -\cos x \qquad \tan(\pi + x) = \tan x$$

$$\sin(\tfrac{3}{2}\pi - x) = -\cos x \qquad \cos(\tfrac{3}{2}\pi - x) = -\sin x \qquad \tan(\tfrac{3}{2}\pi - x) = \cot x$$

$$\sin(\tfrac{3}{2}\pi + x) = -\cos x \qquad \cos(\tfrac{3}{2}\pi + x) = \sin x \qquad \tan(\tfrac{3}{2}\pi + x) = -\cot x$$

$$\sin(2\pi - x) = -\sin x \qquad \cos(2\pi - x) = \cos x \qquad \tan(2\pi - x) = -\tan x$$

and so on. The following illustrations give the proofs of some of these formulas. The other formulas are proved in a similar way. You are asked to supply some of these proofs in Exercises 21 through 24.

▷ ILLUSTRATION 1

From the sine difference and sine sum identities

$$\sin(\pi - x) = \sin \pi \cos x - \cos \pi \sin x$$

$$= (0)\cos x - (-1)\sin x$$

$$= \sin x$$

$$\sin(\tfrac{3}{2}\pi + x) = \sin \tfrac{3}{2}\pi \cos x + \cos \tfrac{3}{2}\pi \sin x$$

$$= (-1)\cos x + (0)\sin x$$

$$= -\cos x \qquad \blacktriangleleft$$

▷ ILLUSTRATION 2

From the cosine difference and cosine sum identities

$$\cos(\pi - x) = \cos \pi \cos x + \sin \pi \sin x$$

$$= (-1)\cos x + (0)\sin x$$

$$= -\cos x$$

$$\cos(\tfrac{3}{2}\pi + x) = \cos \tfrac{3}{2}\pi \cos x - \sin \tfrac{3}{2}\pi \sin x$$

$$= (0)\cos x - (-1)\sin x$$

$$= \sin x \qquad \blacktriangleleft$$

▷ **ILLUSTRATION 3**

From the tangent difference identity

$$\tan(\pi - x) = \frac{\tan \pi - \tan x}{1 + \tan \pi \tan x}$$

$$= \frac{0 - \tan x}{1 + (0)\tan x}$$

$$= -\tan x$$

From a fundamental identity and the results of Illustrations 1 and 2

$$\tan(\tfrac{3}{2}\pi + x) = \frac{\sin(\tfrac{3}{2}\pi + x)}{\cos(\tfrac{3}{2}\pi + x)}$$

$$= \frac{-\cos x}{\sin x}$$

$$= -\cot x$$

We cannot use the tangent sum identity to find $\tan(\tfrac{3}{2}\pi + x)$. An attempt to use it would result in an expression containing $\tan \tfrac{3}{2}\pi$, which is not defined.

◀

In physics, when discussing electricity, heat, and dynamics, it is sometimes necessary to write an expression of the form

$$A \sin bt + B \cos bt$$

in the form

$$a \sin(bt + c)$$

so that the amplitude, period, frequency, phase shift, and graph of the corresponding function can be obtained more easily. We shall now obtain a theorem used in such a case. Suppose the function f is defined by the equation

$$f(t) = A \sin bt + B \cos bt \tag{4}$$

where A, B, and b are constants and $A \neq 0$. We can write $f(t)$ in the form

$$f(t) = \sqrt{A^2 + B^2} \left(\frac{A}{\sqrt{A^2 + B^2}} \sin bt + \frac{B}{\sqrt{A^2 + B^2}} \cos bt \right) \tag{5}$$

Let c be a real number such that

$$\tan c = \frac{B}{A} \tag{6}$$

and

$$\cos c = \frac{A}{\sqrt{A^2 + B^2}} \quad \text{and} \quad \sin c = \frac{B}{\sqrt{A^2 + B^2}} \tag{7}$$

You are asked to prove that Equation (6) follows from (7) in Exercise 42. Substituting from (7) into (5), we have

$$f(t) = \sqrt{A^2 + B^2}\,(\cos c \sin bt + \sin c \cos bt) \tag{8}$$

Observe that the expression in parentheses is $\sin(bt + c)$. If we let

$$a = \sqrt{A^2 + B^2}$$

then (8) can be written in the form

$$f(t) = a \sin(bt + c) \tag{9}$$

By equating the right sides of (4) and (9), we have the following theorem.

THEOREM 1

If t is any real number, A, B, and b are constants, and $A \neq 0$,

$$A \sin bt + B \cos bt = a \sin(bt + c)$$

where $a = \sqrt{A^2 + B^2}$ and c satisfies (6) and (7)

▶ **EXAMPLE 4**   *Applying Theorem 1*

Given

$$f(t) = 2 \sin t + 2\sqrt{3} \cos t$$

(a) Define $f(t)$ by an equation of the form $f(t) = a \sin(t + c)$.
(b) Determine the amplitude, period, and phase shift of f, and sketch the graph of f. **(c)** Check the answers by plotting the graphs of both the given equation and the equation in part (b).

Solution
(a) From Theorem 1

$$2 \sin t + 2\sqrt{3} \cos t = a \sin(t + c)$$

where $a = \sqrt{2^2 + (2\sqrt{3})^2}$; that is, $a = 4$; furthermore, $\tan c = \dfrac{2\sqrt{3}}{2}$; that is, $\tan c = \sqrt{3}$. Thus we take $c = \frac{1}{3}\pi$. Hence

$$f(t) = 4 \sin(t + \tfrac{1}{3}\pi)$$

(b) The amplitude of f is 4. Because the coefficient of t is 1, the period is $2\pi/1 = 2\pi$. The phase shift is $\frac{1}{3}\pi$, and the graph of $f(t) = 4 \sin t$ is

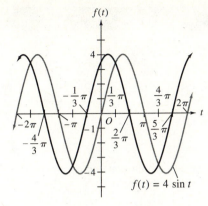

$f(t) = 4 \sin t$

$$f(t) = 2 \sin t + 2\sqrt{3} \cos t$$
$$= 4 \sin\left(t + \frac{1}{3}\pi\right)$$

FIGURE 2

shifted $\frac{1}{3}\pi$ units to the left to obtain the required graph. Figure 2 shows this graph in a black curve, and the graph of the function defined by the equation $f(t) = 4 \sin t$ is indicated by a lighter curve.

(c) On our graphics calculator the graphs of both equations are the same as the graph in Figure 2.   ◀

From Theorem 1 we can show that the sum of two sine functions having the same period is a sine function having that common period. This fact has important applications in physics in the fields of sound and electricity. The next example gives a particular case.

▶ **EXAMPLE 5** *Applying Theorem 1*

At a particular point in space two atmospheric waves produce pressures of $F(t)$ dyn/cm^2 and $G(t)$ dyn/cm^2 at t seconds, where

$$F(t) = 0.021 \sin 400\pi t \quad \text{and} \quad G(t) = 0.043 \sin\left(400\pi t + \tfrac{1}{5}\pi\right)$$

(a) Define the sum of these two pressures by an equation of the form $f(t) = a \sin(bt + c)$. **(b)** Show that all three functions F, G, and f have the same period. **(c)** Check the answers by plotting the graphs of $F + G$ and f.

Solution

(a) If f is the sum of F and G,

$$f(t) = F(t) + G(t)$$
$$= 0.021 \sin 400\pi t + 0.043 \sin\left(400\pi t + \tfrac{1}{5}\pi\right)$$
$$= 0.021 \sin 400\pi t + 0.043[\sin 400\pi t \cos \tfrac{1}{5}\pi + \cos 400\pi t \sin \tfrac{1}{5}\pi]$$
$$= 0.021 \sin 400\pi t + (0.043 \cos \tfrac{1}{5}\pi) \sin 400\pi t$$
$$\qquad\qquad\qquad\qquad + (0.043 \sin \tfrac{1}{5}\pi) \cos 400\pi t$$
$$= 0.021 \sin 400\pi t + 0.035 \sin 400\pi t + 0.025 \cos 400\pi t$$
$$= 0.056 \sin 400\pi t + 0.025 \cos 400\pi t$$

From Theorem 1, $f(t) = a \sin(400\pi t + c)$, where

$$a = \sqrt{(0.056)^2 + (0.025)^2} \quad \text{and} \quad \tan c = \frac{0.025}{0.056}$$
$$= 0.061 \qquad\qquad\qquad\qquad \tan c = 0.446$$
$$\qquad\qquad\qquad\qquad\qquad\qquad c = 0.42$$

Therefore

$$f(t) = 0.061 \sin(400\pi t + 0.42)$$

(b) The period of each of the functions F, G, and f is $2\pi/400\pi = \frac{1}{200}$.

(c) In the viewing rectangle $[-0.01, 0.01]$ by $[-0.1, 0.1]$ we obtain the graph shown in Figure 3 for both $F + G$ and f.   ◀

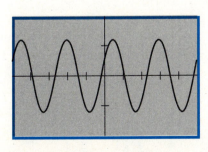

$[-0.01, 0.01]$ by $[-0.1, 0.1]$
$f(t) = 0.061 \sin(400\pi t + 0.42)$

FIGURE 3

EXERCISES 9.2

In Exercises 1 through 12, determine the exact value. Do not use a calculator.

1. (a) $\sin 15°$ **(b)** $\cos 165°$

2. (a) $\cos 75°$ **(b)** $\sin 105°$

3. (a) $\sin(-\frac{7}{12}\pi)$ **(b)** $\cos \frac{23}{12}\pi$

4. (a) $\sin \frac{11}{12}\pi$ **(b)** $\cos(-\frac{11}{12}\pi)$

5. (a) $\tan 105°$ **(b)** $\tan \frac{13}{12}\pi$

6. (a) $\tan 345°$ **(b)** $\tan \frac{5}{12}\pi$

7. (a) $\sin 32° \cos 58° + \cos 32° \sin 58°$
   (b) $\cos 116° \cos 64° - \sin 116° \sin 64°$

8. (a) $\sin 110° \cos 20° - \cos 110° \sin 20°$
   (b) $\sin 110° \sin 70° - \cos 110° \cos 70°$

9. (a) $\sin 85° \sin 25° + \cos 85° \cos 25°$
   (b) $\sin 70° \cos 205° - \cos 70° \sin 205°$

10. (a) $\cos 50° \cos 275° + \sin 50° \sin 275°$
   (b) $\cos 50° \sin 100° + \sin 50° \cos 100°$

11. (a) $\dfrac{\tan 20° + \tan 25°}{1 - \tan 20° \tan 25°}$

   (b) $\dfrac{\tan 200° - \tan 80°}{1 + \tan 200° \tan 80°}$

12. (a) $\dfrac{\tan 110° + \tan 100°}{1 - \tan 110° \tan 100°}$

   (b) $\dfrac{\tan 65° - \tan 110°}{1 + \tan 65° \tan 110°}$

In Exercises 13 through 16, without finding α and β, find (a) $\sin(\alpha + \beta)$, (b) $\cos(\alpha + \beta)$, (c) $\sin(\alpha - \beta)$, (d) $\cos(\alpha - \beta)$, (e) the quadrant containing $\alpha + \beta$, and (f) the quadrant containing $\alpha - \beta$.

13. $\sin \alpha = \frac{12}{13}$, α in the first quadrant; $\sin \beta = \frac{7}{25}$, β in the first quadrant.

14. $\cos \alpha = \frac{24}{25}$, α in the first quadrant; $\cos \beta = \frac{4}{5}$, β in the first quadrant.

15. $\sin \alpha = -\frac{3}{5}$, α in the fourth quadrant; $\cos \beta = -\frac{12}{13}$, β in the third quadrant.

16. $\cos \alpha = -\frac{5}{13}$, α in the second quadrant; $\sin \beta = -\frac{7}{25}$, β in the fourth quadrant.

In Exercises 17 through 20, without finding x and y, find (a) $\tan(x + y)$, (b) $\tan(x - y)$, (c) the quadrant containing $x + y$, and (d) the quadrant containing $x - y$.

17. $\tan x = \frac{7}{3}$, $0 < x < \frac{1}{2}\pi$; $\tan y = \frac{3}{4}$, $0 < y < \frac{1}{2}\pi$

18. $\tan x = \frac{2}{3}$, $0 < x < \frac{1}{2}\pi$; $\tan y = \frac{5}{6}$, $0 < y < \frac{1}{2}\pi$

19. $\tan x = \frac{3}{10}$, $\pi < x < \frac{3}{2}\pi$; $\tan y = -\frac{5}{4}$, $\frac{1}{2}\pi < y < \pi$

20. $\tan x = -\frac{7}{2}$, $\frac{3}{2}\pi < x < 2\pi$; $\tan y = \frac{4}{5}$, $\pi < y < \frac{3}{2}\pi$

In Exercises 21 through 24, use the sum and difference identities to prove the reduction formula.

21. (a) $\sin(\frac{1}{2}\pi + x) = \cos x$
   (b) $\sin(\pi + x) = -\sin x$
   (c) $\sin(\frac{3}{2}\pi - x) = -\cos x$

22. (a) $\cos(\frac{1}{2}\pi + x) = -\sin x$
   (b) $\cos(\pi + x) = -\cos x$
   (c) $\cos(\frac{3}{2}\pi - x) = -\sin x$

23. (a) $\sin(2\pi - x) = -\sin x$
   (b) $\tan(\pi + x) = \tan x$
   (c) $\tan(\frac{1}{2}\pi + x) = -\cot x$

24. (a) $\cos(2\pi - x) = \cos x$
   (b) $\tan(2\pi - x) = -\tan x$
   (c) $\tan(\frac{3}{2}\pi - x) = \cot x$

In Exercises 25 through 30, prove the identity.

25. $\cot(\alpha + \beta) = \dfrac{\cot \alpha \cot \beta - 1}{\cot \beta + \cot \alpha}$

26. $\cot(\alpha - \beta) = \dfrac{\cot \alpha \cot \beta + 1}{\cot \beta - \cot \alpha}$

27. $\sec(\alpha + \beta) = \dfrac{\sec \alpha \sec \beta}{1 - \tan \alpha \tan \beta}$

28. $\csc(\alpha + \beta) = \dfrac{\csc \alpha \csc \beta}{\cot \beta + \cot \alpha}$

29. $\tan(\frac{1}{4}\pi + x) = \dfrac{1 + \tan x}{1 - \tan x}$

30. $\tan(x - \frac{3}{4}\pi) = \dfrac{1 + \tan x}{1 - \tan x}$

In Exercises 31 through 36, simplify the expression by writing it as $\pm\sin kx$, $\pm\cos kx$, or $\pm\tan kx$, where k is a positive integer.

31. (a) $\cos 8x \cos x + \sin 8x \sin x$
   (b) $\sin 5x \cos 2x + \cos 5x \sin 2x$

32. (a) $\cos 3x \cos 2x - \sin 3x \sin 2x$
(b) $\sin 5x \cos 4x - \cos 5x \sin 4x$

33. (a) $\sin 4x \sin 6x - \cos 4x \cos 6x$
(b) $\sin 3x \cos 4x - \cos 3x \sin 4x$

34. (a) $\sin x \cos 3x + \cos x \sin 3x$
(b) $\sin x \sin 5x + \cos x \cos 5x$

35. (a) $\dfrac{\tan 8x + \tan 2x}{1 - \tan 8x \tan 2x}$ **(b)** $\dfrac{\tan \frac{1}{3}x - \tan \frac{7}{3}x}{1 + \tan \frac{1}{3}x \tan \frac{7}{3}x}$

36. (a) $\dfrac{\tan \frac{1}{4}x + \tan \frac{3}{4}x}{1 - \tan \frac{1}{4}x \tan \frac{3}{4}x}$ **(b)** $\dfrac{\tan 4x - \tan 5x}{1 + \tan 4x \tan 5x}$

In Exercises 37 through 40, prove the identity.

37. $\dfrac{\sin(u + v) + \sin(u - v)}{\cos(u + v) + \cos(u - v)} = \tan u$

38. $\dfrac{\sin(u - v)}{\sin u \sin v} = \cot v - \cot u$

39. $\cos(u + v)\cos(u - v) = \cos^2 u - \sin^2 v$

40. $\cos u \sin(u + v) - \sin u \cos(u + v) = \sin v$

41. Prove that if α, β, and γ are angles in a triangle, then
(a) $\sin \alpha \cos \beta + \cos \alpha \sin \beta = \sin \gamma$
and
(b) $\cos \alpha \cos \beta - \sin \alpha \sin \beta = -\cos \gamma$

42. Prove that Equation (6) follows from (7).

In Exercises 43 through 48, (a) express f(t) in the form $a \sin(bt + c)$; (b) determine the amplitude, period, and phase shift of f; (c) sketch the graph of f; and (d) check your graph in part (c) on your graphics calculator.

43. $f(t) = \sqrt{3} \sin t + \cos t$

44. $f(t) = 2 \sin t + 2 \cos t$

45. $f(t) = 3 \sin t - 3 \cos t$

46. $f(t) = 3 \sin t - 3\sqrt{3} \cos t$

47. $f(t) = 2.65 \sin 4t + 1.28 \cos 4t$

48. $f(t) = 0.37 \sin 6t - 0.42 \cos 6t$

49. A weight suspended from a spring is vibrating vertically according to the equation $f(t) = 3 \sin 6t + 4 \cos 6t$, where $f(t)$ centimeters is the directed distance of the weight from its central position t seconds after the start of the motion. **(a)** Define $f(t)$ by an equation of the form $f(t) = a \sin(bt + c)$. **(b)** Determine the amplitude, period, and frequency of f. **(c)** Sketch the graph of f. **(d)** Check your graph in part (c) on your graphics calculator.

50. Do Exercise 49 if $f(t) = -6 \sin 4t - 8 \cos 4t$.

51. Two atmospheric waves at a given point in space produce pressures of $F(t)$ dyn/cm^2 and $G(t)$ dyn/cm^2 at t seconds, where

$$F(t) = 0.06 \sin 2\pi nt$$

and

$$G(t) = 0.04 \sin(2\pi nt - \tfrac{3}{4}\pi)$$

Define the sum of F and G by an equation of the form

$$f(t) = a \sin(2\pi nt + c)$$

Check your answer by plotting the graphs of $F + G$ and f.

52. In an electric circuit the electromotive force is $E(t)$ volts, where

$$E(t) = 50 + 6 \sin 60\pi t + 3.2 \cos 60\pi t$$
$$- 0.6 \sin 120\pi t + 0.8 \cos 120\pi t$$

Define $E(t)$ by an equation of the form

$$E(t) = a_0 + a_1 \sin(60\pi t + c_1) + a_2 \sin(120\pi t + c_2)$$

53. A particle is moving along a straight line according to the equation of motion

$$s = \sin(4t + \tfrac{1}{3}\pi) + \sin(4t + \tfrac{1}{6}\pi)$$

where s centimeters is the directed distance of the particle from the origin at t seconds. **(a)** Show that the motion is simple harmonic by defining s by an equation of the form $s = a \sin(bt + c)$. **(b)** Find the amplitude and frequency of the motion.

54. Do Exercise 53 if $s = 6 \sin(t + \tfrac{1}{3}\pi) + 4 \sin(t - \tfrac{1}{6}\pi)$.

55. A flagpole 15 ft high is situated on top of a building 10 ft high. On the ground, how far from the base of the building will the flagpole and the building subtend equal angles? Be sure to write a conclusion.

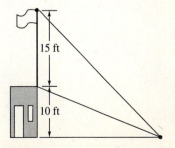

56. Give an explanation of why $\sin(a + b)$ and $\sin a + \sin b$ are not equal for arbitrary values of a and b by discussing how the graphs of

$$F(x) = \sin(x + 2) \quad \text{and} \quad G(x) = \sin x + \sin 2$$

can be obtained from the sine curve.

 57. The quotient

$$\frac{\sin(x + h) - \sin x}{h} \tag{10}$$

arises in calculus and is not defined when $h = 0$. However, in calculus it is necessary to know the behavior of this quotient as h gets closer and closer to zero. **(a)** Show that

$$\frac{\sin(x + h) - \sin x}{h} = \cos x\left(\frac{\sin h}{h}\right) - \sin x\left(\frac{1 - \cos h}{h}\right)$$

(b) From your conclusions in Exercises 49 and 50 of Exercises 7.3 and the identity in part (a), what value does quotient (10) appear to be approaching as h gets closer and closer to zero?

9.3 MULTIPLE-MEASURE IDENTITIES

GOALS

1. **Learn and apply the sine double-measure identity.**
2. **Learn and apply the cosine double-measure identity.**
3. **Learn and apply alternate forms of the cosine double-measure identity.**
4. **Learn and apply the tangent double-measure identity.**
5. **Learn and apply identities for $\sin^2 x$, $\cos^2 x$, and $\tan^2 x$ in terms of $\cos 2x$.**
6. **Learn and apply the half-measure identities.**
7. **Find exact function values from multiple-measure identities.**
8. **Prove other identities from the multiple-measure identities.**
9. **Express even powers of sine and cosine in terms of cosine values with no exponent greater than 1.**
10. **Express $\sin x$ and $\cos x$ as rational functions of z by the substitution $z = \tan \frac{1}{2}x$.**

The double-measure identities are special cases of the sine sum, cosine sum, and tangent sum identities obtained in Section 9.2. In the sine sum identity

$$\sin(x + y) = \sin x \cos y + \cos x \sin y$$

let $y = x$, and we have

$$\sin(x + x) = \sin x \cos x + \cos x \sin x$$

from which we have the following sine double-measure identity.

Sine Double-Measure Identity

$$\sin 2x = 2 \sin x \cos x$$

We proceed in a similar fashion with the cosine sum identity.

$$\cos(x + y) = \cos x \cos y - \sin x \sin y$$
$$\cos(x + x) = \cos x \cos x - \sin x \sin x$$

We have then the cosine double-measure identity.

Cosine Double-Measure Identity

$$\cos 2x = \cos^2 x - \sin^2 x$$

If x is an angle, the double-measure identities are referred to as *double-angle identities*.

There are two other forms for the cosine double-measure identity. If $1 - \sin^2 x$ is substituted for $\cos^2 x$ in the formula for $\cos 2x$, we get

$$\cos 2x = (1 - \sin^2 x) - \sin^2 x$$
$$= 1 - 2 \sin^2 x$$

Furthermore, if $1 - \cos^2 x$ is substituted for $\sin^2 x$ in the cosine double-measure identity, we have

$$\cos 2x = \cos^2 x - (1 - \cos^2 x)$$
$$= 2 \cos^2 x - 1$$

We state these two results formally.

Alternate Forms of Cosine Double-Measure Identity

$$\cos 2x = 1 - 2 \sin^2 x$$
$$\cos 2x = 2 \cos^2 x - 1$$

▶ **EXAMPLE 1** *Using the Sine and Cosine Double-Measure Identities*

If $\sin t = \frac{3}{5}$ and $\frac{1}{2}\pi < t < \pi$, find $\sin 2t$ and $\cos 2t$.

Solution We first find $\cos t$ from the fundamental identity

$$\sin^2 t + \cos^2 t = 1$$

Because $\sin t = \frac{3}{5}$, we have

$$(\tfrac{3}{5})^2 + \cos^2 t = 1$$
$$\cos^2 t = 1 - \tfrac{9}{25}$$
$$\cos^2 t = \tfrac{16}{25}$$

Because $\frac{1}{2}\pi < t < \pi$, $\cos t < 0$. Therefore, when we solve for $\cos t$ by taking the square root, we reject the positive value. Thus

$$\cos t = -\tfrac{4}{5}$$

Therefore

$$\sin 2t = 2 \sin t \cos t \qquad \cos 2t = \cos^2 t - \sin^2 t$$
$$= 2(\tfrac{3}{5})(-\tfrac{4}{5}) \qquad\qquad = (-\tfrac{4}{5})^2 - (\tfrac{3}{5})^2$$
$$= -\tfrac{24}{25} \qquad\qquad\qquad = \tfrac{7}{25} \qquad\qquad \blacktriangleleft$$

The next example shows how another multiple-measure identity follows from the double-measure identities and the sine sum identity.

▶ **EXAMPLE 2** *Obtaining an Identity for sin 3x from Other Identities*

Obtain an identity for $\sin 3x$ in terms of $\sin x$.

Solution

$$\sin 3x = \sin(2x + x)$$
$$= \sin 2x \cos x + \cos 2x \sin x$$
$$= (2 \sin x \cos x)\cos x + (1 - 2 \sin^2 x)\sin x$$
$$= 2 \sin x \cos^2 x + \sin x - 2 \sin^3 x$$
$$= 2 \sin x(1 - \sin^2 x) + \sin x - 2 \sin^3 x$$
$$= 2 \sin x - 2 \sin^3 x + \sin x - 2 \sin^3 x$$
$$= 3 \sin x - 4 \sin^3 x \qquad\qquad \blacktriangleleft$$

To derive the formula expressing $\tan 2x$ in terms of $\tan x$, we start with the tangent sum identity.

$$\tan(x + y) = \frac{\tan x + \tan y}{1 - \tan x \tan y}$$

We let $y = x$ and get

$$\tan(x + x) = \frac{\tan x + \tan x}{1 - \tan x \tan x}$$

from which we have the following tangent double-measure identity.

Tangent Double-Measure Identity

$$\tan 2x = \frac{2 \tan x}{1 - \tan^2 x}$$

This identity does not hold if $x = \frac{1}{4}\pi + \frac{1}{2}k\pi$, where k is an integer, because for these values of x the denominator is zero. The identity also does not hold if $x = \frac{1}{2}\pi + k\pi$, where k is an integer, because for these values $\tan x$ does not exist. Of course, the equation is still an identity because it is valid for all values of x for which both sides are defined.

▶ **EXAMPLE 3** *Using the Tangent Double-Measure Identity*

Obtain an identity for $\tan 4\theta$ in terms of $\tan \theta$.

Solution From the tangent double-measure identity with $x = 2\theta$, we have

$$\tan 2(2\theta) = \frac{2 \tan 2\theta}{1 - \tan^2 2\theta}$$

On the right side of this identity we apply the tangent double-measure identity with $x = \theta$, and we have

$$\tan 4\theta = \frac{2\left(\dfrac{2 \tan \theta}{1 - \tan^2 \theta}\right)}{1 - \left(\dfrac{2 \tan \theta}{1 - \tan^2 \theta}\right)^2}$$

$$= \frac{\dfrac{4 \tan \theta}{1 - \tan^2 \theta}}{\dfrac{(1 - \tan^2 \theta)^2 - 4 \tan^2 \theta}{(1 - \tan^2 \theta)^2}}$$

$$= \frac{(1 - \tan^2 \theta)^2 \left(\dfrac{4 \tan \theta}{1 - \tan^2 \theta}\right)}{(1 - \tan^2 \theta)^2 \left[\dfrac{(1 - 2 \tan^2 \theta + \tan^4 \theta) - 4 \tan^2 \theta}{(1 - \tan^2 \theta)^2}\right]}$$

$$= \frac{4 \tan \theta(1 - \tan^2 \theta)}{1 - 6 \tan^2 \theta + \tan^4 \theta}$$ ◀

Useful identities expressing $\sin^2 x$, $\cos^2 x$, and $\tan^2 x$ in terms of $\cos 2x$ are obtained from the alternate forms of the cosine double-measure identity.

$$\cos 2x = 1 - 2 \sin^2 x \qquad \cos 2x = 2 \cos^2 x - 1$$

$$2 \sin^2 x = 1 - \cos 2x \qquad 2 \cos^2 x = 1 + \cos 2x$$

$$\sin^2 x = \frac{1 - \cos 2x}{2} \qquad \cos^2 x = \frac{1 + \cos 2x}{2}$$

$$\tan^2 x = \frac{\sin^2 x}{\cos^2 x}$$

$$= \frac{\dfrac{1 - \cos 2x}{2}}{\dfrac{1 + \cos 2x}{2}}$$

$$= \frac{1 - \cos 2x}{1 + \cos 2x}$$

We have then the following three identities.

> **Identities for sin² x, cos² x, and tan² x in Terms of cos 2x**
>
> $$\sin^2 x = \frac{1 - \cos 2x}{2}$$
>
> $$\cos^2 x = \frac{1 + \cos 2x}{2}$$
>
> $$\tan^2 x = \frac{1 - \cos 2x}{1 + \cos 2x}$$

In the first two of these identities, x can be any real number or angle. In the third identity, because $1 + \cos 2x$ cannot be zero, $x \neq \frac{1}{2}\pi + k\pi$, where k is any integer.

In calculus it is sometimes necessary to express an even power of a trigonometric function in terms of trigonometric function values with no exponent greater than 1. The preceding identities are useful for this purpose, as shown in the following example.

▶ **EXAMPLE 4** *Expressing an Even Power of Sine in Terms of Cosine Values with No Exponent Greater than 1*

Express $\sin^4 t$ in terms of cosine values with no exponent greater than 1.

Solution   Because $\sin^4 t = (\sin^2 t)^2$, we have from the first of the preceding formulas

$$\sin^4 t = \left(\frac{1 - \cos 2t}{2} \right)^2$$

$$\sin^4 t = \tfrac{1}{4}(1 - 2 \cos 2t + \cos^2 2t)$$

To obtain a substitution for $\cos^2 2t$ on the right side, we apply the formula for $\cos^2 x$ with $x = 2t$, and we get

$$\sin^4 t = \tfrac{1}{4}\left(1 - 2 \cos 2t + \frac{1 + \cos 4t}{2} \right)$$

$$= \tfrac{1}{8}(2 - 4 \cos 2t + 1 + \cos 4t)$$

$$= \tfrac{3}{8} - \tfrac{1}{2} \cos 2t + \tfrac{1}{8} \cos 4t \qquad \blacktriangleleft$$

If in the identities for $\sin^2 x$, $\cos^2 x$, and $\tan^2 x$ we let $x = \tfrac{1}{2}y$, then $2x = y$, and we obtain the following half-measure identities.

Half-Measure Identities

$$\sin^2 \tfrac{1}{2}y = \frac{1 - \cos y}{2}$$

$$\cos^2 \tfrac{1}{2}y = \frac{1 + \cos y}{2}$$

$$\tan^2 \tfrac{1}{2}y = \frac{1 - \cos y}{1 + \cos y}$$

When applying these identities to find $\sin \tfrac{1}{2}y$, $\cos \tfrac{1}{2}y$, and $\tan \tfrac{1}{2}y$, the sign ($+$ or $-$) is determined by the value of $\tfrac{1}{2}y$.

▶ **EXAMPLE 5** *Finding an Exact Function Value from a Half-Measure Identity*

Find the exact value of $\cos \tfrac{1}{8}\pi$.

Solution   We use the identity for $\cos^2 \tfrac{1}{2}y$ with $y = \tfrac{1}{4}\pi$.

$$\cos^2 \tfrac{1}{8}\pi = \frac{1 + \cos \tfrac{1}{4}\pi}{2}$$

$$\cos^2 \tfrac{1}{8}\pi = \frac{1 + \dfrac{\sqrt{2}}{2}}{2}$$

$$\cos^2 \tfrac{1}{8}\pi = \frac{2 + \sqrt{2}}{4}$$

Because $0 < \frac{1}{8}\pi < \frac{1}{2}\pi$, $\cos\frac{1}{8}\pi > 0$. Therefore

$$\cos\tfrac{1}{8}\pi = \frac{\sqrt{2 + \sqrt{2}}}{2} \qquad \blacktriangleleft$$

There are two other identities for $\tan\frac{1}{2}y$. One of them is derived by starting with the identity

$$\tan\tfrac{1}{2}y = \frac{\sin\frac{1}{2}y}{\cos\frac{1}{2}y}$$

and multiplying the numerator and denominator by $2\sin\frac{1}{2}y$. We then have

$$\tan\tfrac{1}{2}y = \frac{2\sin^2\frac{1}{2}y}{2\sin\frac{1}{2}y\cos\frac{1}{2}y} \tag{1}$$

From the identity for $\sin^2\frac{1}{2}y$, we obtain

$$2\sin^2\tfrac{1}{2}y = 1 - \cos y \tag{2}$$

and from the sine double-measure identity with $x = \frac{1}{2}y$, we have

$$2\sin\tfrac{1}{2}y\cos\tfrac{1}{2}y = \sin y \tag{3}$$

Substituting from (2) and (3) in (1), we get

$$\tan\tfrac{1}{2}y = \frac{1 - \cos y}{\sin y}$$

Another identity involving $\tan\frac{1}{2}y$ is obtained from this one by multiplying the numerator and denominator by $1 + \cos y$:

$$\begin{aligned}
\tan\tfrac{1}{2}y &= \frac{(1 - \cos y)(1 + \cos y)}{\sin y(1 + \cos y)} \\
&= \frac{1 - \cos^2 y}{\sin y(1 + \cos y)} \\
&= \frac{\sin^2 y}{\sin y(1 + \cos y)} \\
&= \frac{\sin y}{1 + \cos y}
\end{aligned}$$

We state these two identities formally.

Tangent Half-Measure Identities

$$\tan\tfrac{1}{2}y = \frac{1 - \cos y}{\sin y}$$

$$\tan\tfrac{1}{2}y = \frac{\sin y}{1 + \cos y}$$

In the first of these identities y can be any real number or angle except $k\pi$, where k is any integer. In the second, y can be any real number or angle except $(2k + 1)\pi$, where k is any integer. Note that $(2k + 1)\pi$ is an odd multiple of π.

When y represents an angle in the half-measure identities, they are also referred to as *half-angle identities*.

▶ **EXAMPLE 6** *Finding Exact Function Values from the Half-Measure Identities*

Use the half-measure identities to find the exact value of **(a)** tan 22.5° and **(b)** cos 105°.

Solution
(a) From the identity

$$\tan \tfrac{1}{2}y = \frac{1 - \cos y}{\sin y}$$

with $y = 45°$, we have

$$\tan 22.5° = \frac{1 - \cos 45°}{\sin 45°}$$

$$= \frac{1 - \dfrac{1}{\sqrt{2}}}{\dfrac{1}{\sqrt{2}}}$$

$$= \sqrt{2} - 1$$

(b) From the identity

$$\cos^2 \tfrac{1}{2}y = \frac{1 + \cos y}{2}$$

with $y = 210°$, we have

$$\cos^2 105° = \frac{1 + \cos 210°}{2}$$

Because 105° is a second-quadrant angle, cos 105° is negative. Therefore, with $\cos 210° = -\tfrac{1}{2}\sqrt{3}$, we have

$$\cos 105° = -\sqrt{\frac{1 - \dfrac{\sqrt{3}}{2}}{2}}$$

$$= -\frac{\sqrt{2 - \sqrt{3}}}{2}$$

◀

Compare the value of cos 105° computed from the cosine half-measure identity in the above example with the value of cos 105° computed from the cosine sum identity in Example 1(b) of Section 9.2. In Exercise 21 you are asked to verify on your calculator that these two values are equal and then to prove algebraically that they are equal.

▶ **EXAMPLE 7** *Finding* $\tan \frac{1}{2}\theta$ *When* $\tan \theta$ *is Known*

If $\tan \theta = -\frac{7}{24}$ and $270° < \theta < 360°$, find $\tan \frac{1}{2}\theta$.

Solution   Figure 1 shows angle θ with the point $(24, -7)$ chosen on its terminal side. From the Pythagorean theorem, $r = 25$. Then

$$\sin \theta = -\frac{7}{25} \quad \text{and} \quad \cos \theta = \frac{24}{25}$$

Therefore

$$\tan \frac{1}{2}\theta = \frac{1 - \cos \theta}{\sin \theta}$$

$$= \frac{1 - \frac{24}{25}}{-\frac{7}{25}}$$

$$= -\frac{1}{7} \qquad ◀$$

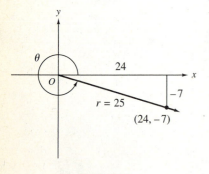

FIGURE 1

    In calculus we sometimes find it necessary to reduce a rational function of $\sin x$ and $\cos x$ to a rational function of z by the substitution

$$z = \tan \frac{1}{2}x$$

From the identity

$$\cos 2y = 2 \cos^2 y - 1$$

with $y = \frac{1}{2}x$,

$$\cos x = 2 \cos^2 \tfrac{1}{2}x - 1$$

$$= \frac{2}{\sec^2 \frac{1}{2}x} - 1$$

$$= \frac{2}{1 + \tan^2 \frac{1}{2}x} - 1$$

$$= \frac{2}{1 + z^2} - 1$$

$$= \frac{2 - (1 + z^2)}{1 + z^2}$$

Thus

$$\cos x = \frac{1 - z^2}{1 + z^2} \qquad \textbf{(4)}$$

In a similar manner, from the identity $\sin 2y = 2 \sin y \cos y$, with $y = \frac{1}{2}x$,

$$\sin x = 2 \sin \tfrac{1}{2}x \cos \tfrac{1}{2}x$$

$$= 2 \cdot \frac{\sin \tfrac{1}{2}x}{\cos \tfrac{1}{2}x} \cdot \cos^2 \tfrac{1}{2}x$$

$$= 2 \tan \tfrac{1}{2}x \cdot \frac{1}{\sec^2 \tfrac{1}{2}x}$$

$$= 2 \tan \tfrac{1}{2}x \cdot \frac{1}{1 + \tan^2 \tfrac{1}{2}x}$$

Therefore

$$\sin x = \frac{2z}{1 + z^2} \qquad (5)$$

By using (4) and (5), $\cos x$ and $\sin x$ can be expressed as rational functions of z. The procedure is shown in the following example.

▶ **EXAMPLE 8** *Reducing a Rational Function of sin x and cos x to a Rational Function of z by the Substitution z = tan $\frac{1}{2}$x*

Write

$$3 \sin x - 2 \cos x$$

as a rational function of z by letting $z = \tan \frac{1}{2}x$.

Solution   With $z = \tan \frac{1}{2}x$, we have from (4) and (5)

$$3 \sin x - 2 \cos x = 3\left(\frac{2z}{1 + z^2}\right) - 2\left(\frac{1 - z^2}{1 + z^2}\right)$$

$$= \frac{6z}{1 + z^2} + \frac{-2 + 2z^2}{1 + z^2}$$

$$= \frac{2z^2 + 6z - 2}{1 + z^2} \qquad \blacktriangleleft$$

The other four trigonometric functions can be expressed as rational functions of z by applying (4) and (5) and the identities

$$\tan x = \frac{\sin x}{\cos x} \qquad \cot x = \frac{\cos x}{\sin x} \qquad \sec x = \frac{1}{\cos x} \qquad \csc x = \frac{1}{\sin x}$$

EXERCISES 9.3

In Exercises 1 through 8, find (a) $\sin 2t$, *(b)* $\cos 2t$, *and (c)* $\tan 2t$.

1. $\sin t = \frac{4}{5}$ and $0 < t < \frac{1}{2}\pi$

2. $\cos t = \frac{7}{25}$ and $0 < t < \frac{1}{2}\pi$

3. $\cos t = -\frac{5}{13}$ and $\frac{1}{2}\pi < t < \pi$

4. $\sin t = \frac{3}{5}$ and $\frac{1}{2}\pi < t < \pi$

5. $\tan t = \frac{8}{15}$ and $\sin t < 0$

6. $\tan t = -\frac{12}{5}$ and $\sin t < 0$

7. $\sin t = -\frac{7}{25}$ and $\cos t > 0$

8. $\cos t = -\frac{15}{17}$ and $\tan t > 0$

In Exercises 9 through 16, obtain an identity for the first function value in terms of the second function value.

9. $\cos 3x$; $\cos x$

10. $\cos 4x$; $\cos x$

11. $\sin 4x$; $\sin x$

12. $\sin 5x$; $\sin x$

13. $\tan 3x$; $\tan x$

14. $\cot 2x$; $\cot x$

15. $\sec 2x$; $\sec x$

16. $\csc 2x$; $\csc x$

$\iint \frac{dy}{dx}$ *In Exercises 17 through 20, write the expression in terms of cosine values with no exponent greater than 1.*

17. $\cos^4 t$

18. $\sin^4 2t$

19. $\sin^2 3t \cos^2 3t$

20. $\cos^4 4t \sin^2 4t$

21. In Example 6(b) with the cosine half-measure identity, we obtained $\cos 105° = -\dfrac{\sqrt{2 - \sqrt{3}}}{2}$; and in Example 1(b) of Section 9.2 with the cosine sum identity, we obtained $\cos 105° = \dfrac{1 - \sqrt{3}}{2\sqrt{2}}$. Verify on your calculator that these two values are equal and then prove algebraically that they are equal.

22. (a) Use the cosine half-measure identity to find the exact value of $\cos \frac{1}{12}\pi$. **(b)** Verify on your calculator that the value obtained in part (a) is equal to the value of $\cos \frac{1}{12}\pi$ obtained in Example 1(a) of Section 9.2 with the cosine difference identity. Then prove algebraically that these two values are equal.

23. (a) Use the sine half-measure identity to find the exact value of $\sin 15°$. **(b)** Verify on your calculator that the value obtained in part (a) is equal to the value of $\sin 15°$ obtained in Exercise 1(a) of Exercises 9.2 with

the sine difference identity. Then prove algebraically that these two values are equal.

24. (a) Use the tangent half-measure identity to find the exact value of $\tan \frac{5}{12}\pi$. **(b)** Verify on your calculator that the value obtained in part (a) is equal to the value of $\tan \frac{5}{12}\pi$ obtained in Exercise 6(b) of Exercises 9.2 with the tangent sum identity. Then prove algebraically that these two values are equal.

In Exercises 25 through 28, use the half-measure identities to determine the exact function value.

25. $\cos \frac{1}{8}\pi$

26. $\sin 165°$

27. $\tan 112.5°$

28. $\tan \frac{3}{8}\pi$

In Exercises 29 through 34, use a half-measure identity to find the function value.

29. $\cos t = \frac{1}{3}$ and $0 < t < \frac{1}{2}\pi$; find $\sin \frac{1}{2}t$.

30. $\cos t = \frac{1}{2}$ and $0 < t < \frac{1}{2}\pi$; find $\cos \frac{1}{2}t$.

31. $\sin t = \frac{24}{25}$ and $\frac{1}{2}\pi < t < \pi$; find $\cos \frac{1}{2}t$.

32. $\sin t = \frac{4}{5}$ and $\cos t < 0$; find $\tan \frac{1}{2}t$.

33. $\tan t = -\frac{5}{12}$ and $\sin t > 0$; find $\tan \frac{1}{2}t$.

34. $\tan t = \frac{24}{7}$ and $\sin t < 0$; find $\cos \frac{1}{2}t$.

In Exercises 35 through 44, simplify the expression by writing it as a $\sin kx$, *a* $\cos kx$, *or a* $\tan kx$, *where a is an integer and k is a positive integer.*

35. (a) $\cos^2 x - \sin^2 x$ 　　**(b)** $\cos^2 2x - \sin^2 2x$
　　　(c) $2 \sin \frac{3}{2}x \cos \frac{3}{2}x$

36. (a) $1 - 2 \sin^2 x$ 　　**(b)** $1 - 2 \sin^2 3x$
　　　(c) $1 - 2 \cos^2 x$

37. (a) $2 \sin x \cos x$ 　　**(b)** $4 \sin 4x \cos 4x$
　　　(c) $2 \sin \frac{1}{2}x \cos \frac{1}{2}x$

38. (a) $6 \sin 2x \cos 2x$ 　　**(b)** $2 \sin 6x \cos 6x$
　　　(c) $2 \sin \frac{3}{2}x \cos \frac{3}{2}x$

39. (a) $\dfrac{2 \tan x}{1 - \tan^2 x}$ 　　**(b)** $\dfrac{4 \tan 2x}{\tan^2 2x - 1}$
　　　(c) $\dfrac{6 \tan \frac{3}{2}x}{1 - \tan^2 \frac{3}{2}x}$

40. (a) $\dfrac{2 \tan \frac{1}{2}x}{1 - \tan^2 \frac{1}{2}x}$ 　　**(b)** $\dfrac{4 \tan 4x}{1 - \tan^2 4x}$
　　　(c) $\dfrac{8 \tan 3x}{\tan^2 3x - 1}$

41. (a) $\dfrac{\sin 2t}{1 + \cos 2t}$ (b) $\dfrac{\sin 4t}{\cos 4t + 1}$

    (c) $\dfrac{\cos 4t - 1}{\sin 4t}$

42. (a) $\dfrac{1 - \cos 2t}{\sin 2t}$ (b) $\dfrac{1 - \cos 6t}{\sin 6t}$

    (c) $\dfrac{\cos 8t - 1}{\sin 8t}$

43. (a) $\dfrac{\cos 2t \sin 4t}{1 + \cos 4t}$ (b) $\pm\sqrt{\dfrac{1 - \cos 8t}{2}}$

    (c) $\pm\sqrt{2 - 2\cos 8t}$

44. (a) $\dfrac{\cos 3t(1 - \cos 6t)}{\sin 6t}$ (b) $\pm\sqrt{\dfrac{1 + \cos 4t}{2}}$

    (c) $\pm\sqrt{2 + 2\cos 4t}$

In Exercises 45 through 52, prove the identity.

45. $\dfrac{2 \tan x}{1 + \tan^2 x} = \sin 2x$

46. $\dfrac{\tan x + \cot x}{\cot x - \tan x} = \sec 2x$

47. $\dfrac{2}{1 + \cos 2t} = \sec^2 t$

48. $\dfrac{1 + \cos 2t}{1 - \cos 2t} = \cot^2 t$

49. $\cos^4 \theta - \sin^4 \theta = \cos 2\theta$

50. $\dfrac{1 - \tan^2 \theta}{1 + \tan^2 \theta} = \cos 2\theta$

51. $\dfrac{1 + \sin 2\beta + \cos 2\beta}{1 + \sin 2\beta - \cos 2\beta} = \cot \beta$

52. $\dfrac{1 - \cos 8x}{8} = \sin^2 2x \cos^2 2x$

53. If α and β are the acute angles in a right triangle, verify that $\sin 2\alpha = \sin 2\beta$.

54. If K square units is the area of a right triangle, c units is the length of the hypotenuse, and α is an acute angle, verify that $K = \frac{1}{4}c^2 \sin 2\alpha$.

In Exercises 55 through 60, use the substitution $z = \tan \frac{1}{2}x$ to write the expression as a rational function of z.

55. $1 - \sin x + \cos x$ **56.** $\sin x - \cos x + 2$

57. $\sin 2x + \tan x$ **58.** $\sec x + \tan x$

59. $\sec^2 x + \tan^2 x$ **60.** $\dfrac{2 - \cos 2x}{\sin x}$

61. A particle is moving along a straight line according to the equation of motion $s = 12 \cos^2 4t - 6$, where s centimeters is the directed distance of the particle from the origin at t seconds. **(a)** Show that the motion is simple harmonic by defining s by an equation of the form $s = a \sin(bt + c)$. **(b)** Find the amplitude and frequency of the motion.

62. Do Exercise 61 if $s = 5 - 10 \sin^2 2t$.

63. A pendulum of length 10 cm has swung so that θ is the radian measure of the angle formed by the pendulum and a vertical line. Show that the number of centimeters in the vertical height of the end of the pendulum above its lowest position is $20 \sin^2 \frac{1}{2}\theta$.

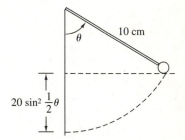

9.4 IDENTITIES FOR THE PRODUCT, SUM, AND DIFFERENCE OF SINE AND COSINE

GOALS
1. Learn and apply the product sine and cosine identities.
2. Learn and apply the sum and difference sine and cosine identities.
3. Find exact function values from identities for the product, sum, and difference of sine and cosine.

4. Prove other identities from identities for the product, sum, and difference of sine and cosine.

5. Express the sum of two sine functions having the same amplitude as the product of sine and cosine functions.

 In certain computations in calculus it is necessary to write an expression involving the product of sine and cosine functions as a sum or difference. The tools for doing this are provided by the product sine and cosine identities, which follow from the sine and cosine sum and difference identities.

The sine sum and difference identities are

$$\sin(x + y) = \sin x \cos y + \cos x \sin y$$
$$\sin(x - y) = \sin x \cos y - \cos x \sin y$$

If we add corresponding terms of these two equations, we obtain

$$\sin(x + y) + \sin(x - y) = 2 \sin x \cos y$$

and if we subtract terms of the second equation from corresponding terms of the first, we get

$$\sin(x + y) - \sin(x - y) = 2 \cos x \sin y$$

These results give the following identities.

Product Sine and Cosine Identities

$$\sin x \cos y = \tfrac{1}{2}[\sin(x + y) + \sin(x - y)]$$
$$\cos x \sin y = \tfrac{1}{2}[\sin(x + y) - \sin(x - y)]$$

These two identities express the product of a sine and cosine function as the sum or difference of two sine functions. They are valid for all real numbers and angles x and y.

▷ **ILLUSTRATION 1**

(a) From the identity $\sin x \cos y = \tfrac{1}{2}[\sin(x + y) + \sin(x - y)]$,

$$\sin 5t \cos 3t = \tfrac{1}{2}[\sin(5t + 3t) + \sin(5t - 3t)]$$
$$= \tfrac{1}{2}(\sin 8t + \sin 2t)$$

(b) From the identity $\cos x \sin y = \tfrac{1}{2}[\sin(x + y) - \sin(x - y)]$,

$$\cos 3t \sin 5t = \tfrac{1}{2}[\sin(3t + 5t) - \sin(3t - 5t)]$$
$$= \tfrac{1}{2}[\sin 8t - \sin(-2t)]$$
$$= \tfrac{1}{2}[\sin 8t - (-\sin 2t)]$$
$$= \tfrac{1}{2}(\sin 8t + \sin 2t)$$ ◀

The cosine sum and difference identities are

$$\cos(x + y) = \cos x \cos y - \sin x \sin y$$
$$\cos(x - y) = \cos x \cos y + \sin x \sin y$$

Adding corresponding terms of these two equations, we have

$$\cos(x + y) + \cos(x - y) = 2 \cos x \cos y$$

and subtracting terms of the first equation from corresponding terms of the second, we get

$$\cos(x - y) - \cos(x + y) = 2 \sin x \sin y$$

From these two results we have the following identities.

Product Sine and Cosine Identities

$$\cos x \cos y = \tfrac{1}{2}[\cos(x + y) + \cos(x - y)]$$
$$\sin x \sin y = \tfrac{1}{2}[\cos(x - y) - \cos(x + y)]$$

The first identity expresses the product of two cosine functions as the sum of two cosine functions, and the second identity expresses the product of two sine functions as the difference of two cosine functions. They are both valid for all real numbers and all angles x and y.

▷ **ILLUSTRATION 2**

(a) From the identity $\cos x \cos y = \tfrac{1}{2}[\cos(x + y) + \cos(x - y)]$,

$$\cos 4\theta \cos 2\theta = \tfrac{1}{2}[\cos(4\theta + 2\theta) + \cos(4\theta - 2\theta)]$$
$$= \tfrac{1}{2}(\cos 6\theta + \cos 2\theta)$$

(b) From the identity $\sin x \sin y = \tfrac{1}{2}[\cos(x - y) - \cos(x + y)]$,

$$\sin 4\theta \sin 2\theta = \tfrac{1}{2}[\cos(4\theta - 2\theta) - \cos(4\theta + 2\theta)]$$
$$= \tfrac{1}{2}(\cos 2\theta - \cos 6\theta)$$

◀

▶ **EXAMPLE 1** *Finding the Exact Value of the Product of Sine and Cosine Function Values*

Find the exact value of

$$\sin \tfrac{25}{24} \pi \cos \tfrac{5}{24} \pi$$

Solution   From the identity $\sin x \cos y = \frac{1}{2}[\sin(x + y) + \sin(x - y)]$,

$$
\begin{aligned}
\sin \tfrac{25}{24} \pi \cos \tfrac{5}{24} \pi &= \tfrac{1}{2}[\sin(\tfrac{25}{24} \pi + \tfrac{5}{24} \pi) + \sin(\tfrac{25}{24} \pi - \tfrac{5}{24} \pi)] \\
&= \tfrac{1}{2}(\sin \tfrac{30}{24} \pi + \sin \tfrac{20}{24} \pi) \\
&= \tfrac{1}{2}(\sin \tfrac{5}{4} \pi + \sin \tfrac{5}{6} \pi) \\
&= \frac{1}{2}\left(-\frac{\sqrt{2}}{2} + \frac{1}{2} \right) \\
&= \frac{1 - \sqrt{2}}{4}
\end{aligned}
$$

◀

▶ **EXAMPLE 2** *Proving an Identity*

Prove the identity

$$(\sin 2t)(1 + 2 \cos t) = \sin t + \sin 2t + \sin 3t$$

Solution   We start with the left side. After removing parentheses, we apply the product sine and cosine identity for $\sin x \cos y$.

$$
\begin{aligned}
(\sin 2t)(1 + 2 \cos t) &= \sin 2t + 2 \sin 2t \cos t \\
&= \sin 2t + 2 \cdot \tfrac{1}{2}[\sin(2t + t) + \sin(2t - t)] \\
&= \sin 2t + (\sin 3t + \sin t) \\
&= \sin t + \sin 2t + \sin 3t
\end{aligned}
$$

◀

The product sine and cosine identities can be used to write a sum or difference of sine and cosine functions as a product. We make the substitutions

$$x + y = w \quad \text{and} \quad x - y = z \tag{1}$$

Then

$$
\begin{aligned}
(x + y) + (x - y) &= w + z \\
x &= \frac{w + z}{2}
\end{aligned}
\tag{2}
$$

Also

$$
\begin{aligned}
(x + y) - (x - y) &= w - z \\
y &= \frac{w - z}{2}
\end{aligned}
\tag{3}
$$

Substituting from (1), (2), and (3) in the four product sine and cosine identities, we obtain the following identities.

Sum and Difference Sine and Cosine Identities

$$\sin w + \sin z = 2 \sin\left(\frac{w+z}{2}\right)\cos\left(\frac{w-z}{2}\right)$$

$$\sin w - \sin z = 2 \cos\left(\frac{w+z}{2}\right)\sin\left(\frac{w-z}{2}\right)$$

$$\cos w + \cos z = 2 \cos\left(\frac{w+z}{2}\right)\cos\left(\frac{w-z}{2}\right)$$

$$\cos w - \cos z = -2 \sin\left(\frac{w+z}{2}\right)\sin\left(\frac{w-z}{2}\right)$$

The first and third of these identities are called the *sum sine* and *sum cosine identities*, respectively. The second and fourth are called the *difference sine* and *difference cosine identities*, respectively. They are valid for all real numbers and angles w and z.

▷ **ILLUSTRATION 3**

To write $\sin 8x + \sin 4x$ as a product, we apply the sum sine identity and obtain

$$\sin 8x + \sin 4x = 2 \sin\left(\frac{8x+4x}{2}\right)\cos\left(\frac{8x-4x}{2}\right)$$

$$= 2 \sin 6x \cos 2x \qquad \blacktriangleleft$$

▶ **EXAMPLE 3** *Proving an Identity*

Prove the identity

$$\frac{\cos 2y + \cos 4y}{\sin 4y - \sin 2y} = \cot y$$

Solution   We begin with the left side and apply the sum cosine identity to the numerator and the difference sine identity to the denominator.

$$\frac{\cos 2y + \cos 4y}{\sin 4y - \sin 2y} = \frac{2 \cos\left(\dfrac{2y + 4y}{2}\right)\cos\left(\dfrac{2y - 4y}{2}\right)}{2 \cos\left(\dfrac{4y + 2y}{2}\right)\sin\left(\dfrac{4y - 2y}{2}\right)}$$

$$= \frac{2 \cos 3y \cos(-y)}{2 \cos 3y \sin y}$$

$$= \frac{\cos(-y)}{\sin y}$$

$$= \frac{\cos y}{\sin y}$$

$$= \cot y \qquad\qquad \blacktriangleleft$$

In Example 5 of Section 9.2 we showed that the sum of two sine functions having the same period is a sine function having that common period. The following example involves the addition of two sine functions having different periods but the same amplitude.

▶ **EXAMPLE 4** *Expressing the Sum of Two Sine Functions as the Product of Sine and Cosine Functions and Sketching the Graph*

Given

$$f(t) = \sin 44\pi t + \sin 36\pi t$$

(a) Express $f(t)$ as the product of sine and cosine functions. **(b)** Sketch the graph of f over one period of the cosine function found in part (a) and check the graph on a graphics calculator.

Solution

(a) Applying the sum sine identity on the right side of the given equation, we have

$$f(t) = 2 \sin\left(\frac{44\pi t + 36\pi t}{2}\right)\cos\left(\frac{44\pi t - 36\pi t}{2}\right)$$

$$= 2 \sin 40\pi t \cos 4\pi t$$

(b) The period of $\cos 4\pi t$ is $\dfrac{2\pi}{4\pi}$, or $\frac{1}{2}$. Therefore we wish to sketch the graph of f on $[0, \frac{1}{2}]$.

Because $-1 \le \sin 40\pi t \le 1$, the graph of f lies between the graphs of

$$g(t) = 2 \cos 4\pi t \qquad \text{and} \qquad h(t) = -2 \cos 4\pi t$$

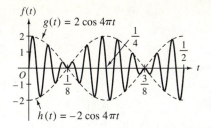

$f(t) = \sin 44\pi t + \sin 36\pi t$
$\quad = 2 \sin 40\pi t \cos 4\pi t$

FIGURE 1

shown as dashed curves in Figure 1. The graph of f intersects the graph of g or h when $\sin 40\pi t = \pm 1$; that is, when $40\pi t = \frac{1}{2}\pi + k \cdot \pi$, $k \in J$, or, equivalently, when $t = \frac{1}{80} + k \cdot \frac{1}{40}$, $k \in J$. With this information, we sketch the graph of f on $[0, \frac{1}{2}]$ as shown in Figure 1. We obtain the same graph on our graphics calculator.   ◀

The graph of the function in Example 4 demonstrates the *principle of superposition*. This kind of behavior occurs whenever we have the sum of two sine or cosine functions of the same amplitude and for which the difference of the frequencies is small compared with the sum of the frequencies. For instance, in Example 4, if we let

$$F(t) = \sin 44\pi t \qquad G(t) = \sin 36\pi t$$
$$\quad = \sin 2\pi(22)t \qquad \quad = \sin 2\pi(18)t$$

the frequencies of F and G are $n_1 = 22$ and $n_2 = 18$, respectively; $n_1 - n_2 = 4$ and $n_1 + n_2 = 40$. Thus the graph of $F + G$ has the appearance shown in Figure 1, where the amplitude is changing according to a specific pattern.

The principle of superposition of sound waves produce what physicists call *beats*. The concept of the superposition of two or more waves has applications in the fields of television, radio, and telephone communications as well as electricity involving current with a high frequency.

EXERCISES 9.4

In Exercises 1 through 4, write the product as a sum or difference of function values.

1. (a) $\sin 4x \cos x$ **(b)** $\sin 4x \sin x$
(c) $\cos 4x \cos x$

2. (a) $\sin 7x \cos 3x$ **(b)** $\sin 7x \sin 3x$
(c) $\cos 7x \cos 3x$

3. (a) $\sin 2x \cos 6x$ **(b)** $\sin 2x \sin 6x$
(c) $\cos 2x \cos 6x$

4. (a) $\sin 4x \cos 5x$ **(b)** $\sin 4x \sin 5x$
(c) $\cos 4x \cos 5x$

In Exercises 5 through 8, find the exact value.

5. (a) $\sin \frac{7}{24}\pi \cos \frac{1}{24}\pi$ **(b)** $\cos \frac{1}{8}\pi \cos \frac{7}{8}\pi$

6. (a) $\sin \frac{3}{8}\pi \cos \frac{1}{8}\pi$ **(b)** $\sin \frac{1}{24}\pi \sin \frac{5}{24}\pi$

7. (a) $\cos \frac{9}{8}\pi \sin \frac{3}{8}\pi$ **(b)** $\sin \frac{5}{24}\pi \sin \frac{25}{24}\pi$

8. (a) $\cos \frac{31}{24}\pi \sin \frac{13}{24}\pi$ **(b)** $\cos \frac{3}{8}\pi \cos \frac{11}{8}\pi$

In Exercises 9 through 12, write the sum or difference as a product of function values.

9. (a) $\sin 6t + \sin 2t$ **(b)** $\sin 6t - \sin 2t$
(c) $\cos 6t + \cos 2t$ **(d)** $\cos 6t - \cos 2t$

10. (a) $\sin 5t + \sin 3t$ **(b)** $\sin 5t - \sin 3t$
(c) $\cos 5t + \cos 3t$ **(d)** $\cos 5t - \cos 3t$

11. (a) $\sin 3\theta + \sin 7\theta$ **(b)** $\sin 3\theta - \sin 7\theta$
(c) $\cos 3\theta + \cos 7\theta$ **(d)** $\cos 3\theta - \cos 7\theta$

12. (a) $\sin 4\theta + \sin 6\theta$ **(b)** $\sin 4\theta - \sin 6\theta$
(c) $\cos 4\theta + \cos 6\theta$ **(d)** $\cos 4\theta - \cos 6\theta$

In Exercises 13 through 16, find the exact value.

13. (a) $\sin 75° + \sin 15°$ **(b)** $\cos 75° - \cos 15°$

14. (a) $\cos 105° + \cos 15°$ **(b)** $\sin 105° - \sin 15°$

15. (a) $\sin 165° + \cos 195°$ **(b)** $\sin 165° - \cos 195°$

16. (a) $\sin 195° + \cos 345°$ **(b)** $\sin 195° - \cos 345°$

In Exercises 17 through 20, verify the equality.

17. $\dfrac{\sin 37° + \sin 23°}{\cos 37° + \cos 23°} = \dfrac{1}{\sqrt{3}}$

18. $\dfrac{\cos 62° - \cos 28°}{\sin 62° - \sin 28°} = -1$

19. $\dfrac{\sin 144° - \sin 126°}{\cos 144° - \cos 126°} = 1$

20. $\dfrac{\sin 140° - \sin 20°}{\cos 140° + \cos 20°} = \sqrt{3}$

In Exercises 21 through 34, prove the identity.

21. $\dfrac{\sin 3x + \sin 7x}{\cos 3x + \cos 7x} = \tan 5x$

22. $\dfrac{\cos 6x - \cos 2x}{\sin 6x - \sin 2x} = -\tan 4x$

23. $\dfrac{\sin 2\alpha - \sin 2\beta}{\sin 2\alpha + \sin 2\beta} = \dfrac{\tan(\alpha - \beta)}{\tan(\alpha + \beta)}$

24. $\dfrac{\sin 2\alpha + \sin 2\beta}{\cos 2\alpha + \cos 2\beta} = \tan(\alpha + \beta)$

25. $\dfrac{\cos t - \cos 5t}{\cos t \sin t} = 4 \sin 3t$

26. $\dfrac{\sin t - \sin 3t}{\sin^2 t - \cos^2 t} = 2 \sin t$

27. $\dfrac{\sin \theta - \sin 7\theta}{2 \sin^2 2\theta - 1} = 2 \sin 3\theta$

28. $\dfrac{\sin 6\theta + \sin 4\theta}{\sin \theta \sin 5\theta} = 2 \cot \theta$

29. $\dfrac{\cos 2y - \cos 3y}{\sin 3y + \sin 2y} = \dfrac{\sin y}{1 + \cos y}$

30. $\dfrac{\sin 5y - \sin 4y}{\cos 5y + \cos 4y} = \dfrac{1 - \cos y}{\sin y}$

31. $\dfrac{\sin x + \sin 2x + \sin 3x}{\cos x + \cos 2x + \cos 3x} = \tan 2x$

32. $\dfrac{\sin x + \sin 3x + \sin 5x + \sin 7x}{\cos x + \cos 3x + \cos 5x + \cos 7x} = \tan 4x$

33. $\sin^2 t \cos^4 t = \dfrac{2 + \cos 2t - 2 \cos 4t - \cos 6t}{32}$

34. $\sin^3 t \cos^3 t = \dfrac{3 \sin 2t - \sin 6t}{32}$

In Exercises 35 through 42, use the method of Example 4 to sketch the graph of the function on the indicated interval. Check your graph on your graphics calculator.

35. $f(t) = \sin 21\pi t + \sin 19\pi t$; $[0, 2]$

36. $f(t) = \cos 9\pi t + \cos 11\pi t$; $[0, 2]$

37. $f(t) = \cos 9\pi t - \cos 11\pi t$; $[0, 2]$

38. $f(t) = \sin 21\pi t - \sin 19\pi t$; $[0, 2]$

39. $f(t) = 2 \cos 80t + 2 \cos 100\pi t$; $[0, \frac{1}{5}]$

40. $f(t) = 2 \sin 80\pi t + 2 \sin 100\pi t$; $[0, \frac{1}{5}]$

41. $f(t) = 3 \sin 95t - 3 \sin 105t$; $[0, \frac{2}{5}]$

42. $f(t) = 3 \cos 95t - 3 \cos 105t$; $[0, \frac{2}{5}]$

In Exercises 43 through 46, verify the equality if α, β, and γ are angles in a triangle.

43. $\sin 2\alpha + \sin 2\beta + \sin 2\gamma = 4 \sin \alpha \sin \beta \sin \gamma$

44. $\cos 2\alpha + \cos 2\beta + \cos 2\gamma = -1 - 4 \cos \alpha \cos \beta \cos \gamma$

45. $\tan \alpha + \tan \beta + \tan \gamma = \tan \alpha \tan \beta \tan \gamma$

46. $\sin \alpha + \sin \beta - \sin \gamma = 4 \sin \frac{1}{2}\alpha \sin \frac{1}{2}\beta \cos \frac{1}{2}\gamma$

47. Two tuning forks are struck at the same instant, and each produces a sound with the same pressure amplitude of 0.02 dyn/cm². If one fork is vibrating at 252 cycles per second and the other is vibrating at 248 cycles per second, an equation of the sound wave produced is

$$f(t) = 0.02 \sin 504\pi t + 0.02 \sin 496\pi t$$

where $f(t)$ dynes per square centimeter is the difference between the atmospheric pressure and the air pressure at the eardrum at t seconds. Write this equation as the product of a sine and cosine function. Check your answer by verifying on your graphics calculator that the graphs of both equations are the same.

48. A note produced by a certain musical instrument is such that the pure tone has a frequency of 250 cycles per second and the only significant overtone is the first. If the pressure amplitudes of the pure tone and the overtone are each 0.04 dyn/cm², an equation of the sound wave produced by this instrument for this note is

$$f(t) = 0.04 \sin 500\pi t + 0.04 \sin 1000\pi t$$

where $f(t)$ dynes per square centimeter is the difference between the atmospheric pressure and the air pressure at the eardrum at t seconds. Write this equation as the product of a sine and cosine function. Check your answer by verifying on your graphics calculator that the graphs of both equations are the same.

9.5 INVERSE TRIGONOMETRIC FUNCTIONS

GOALS

1. Define the inverse sine function.
2. Define the inverse cosine function.
3. Define the inverse tangent function.
4. Define the inverse cotangent function.
5. Define the inverse secant function.
6. Define the inverse cosecant function.
7. Sketch graphs of inverse trigonometric functions.
8. Compute inverse trigonometric function values on a calculator.
9. Find exact function values involving inverse trigonometric functions.
10. Prove identities involving inverse trigonometric functions.
11. Solve word problems having mathematical models involving inverse trigonometric functions.

Before beginning our study of inverse trigonometric functions, you may wish to review Section 6.1 where we introduced inverse functions. We showed there that a function must be one-to-one to have an inverse, and we applied the horizontal-line test to determine if a function is one-to-one.

Figure 1 shows the graph of the sine function. This function is not one-to-one because every number in its range is the function value of more

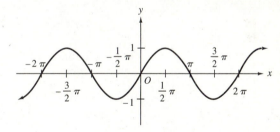

$$y = \sin x$$

FIGURE 1

than one number in its domain. Therefore the sine function does not have an inverse. However, observe in Figure 1 that on the interval $[-\frac{1}{2}\pi, \frac{1}{2}\pi]$, every horizontal line intersects this portion of the graph in no more than one point. Thus from the horizontal-line test, the function F for which

$$F(x) = \sin x \qquad \text{and} \qquad -\tfrac{1}{2}\pi \le x \le \tfrac{1}{2}\pi \tag{1}$$

is one-to-one and therefore has an inverse function. The domain of F is the closed interval $[-\frac{1}{2}\pi, \frac{1}{2}\pi]$, and its range is the closed interval $[-1, 1]$. The graph of F appears in Figure 2. The inverse of the function defined by (1)

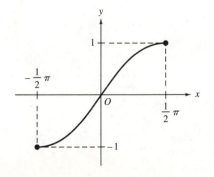

FIGURE 2

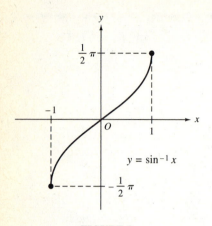

FIGURE 3

is called the *inverse sine function* and is denoted by the symbol $\sin^{-1}$. Following is the formal definition.

DEFINITION **The Inverse Sine Function**

The **inverse sine function,** denoted by $\sin^{-1}$, is defined as follows:

$$y = \sin^{-1} x \quad \text{if and only if} \quad x = \sin y \quad \text{and} \quad -\tfrac{1}{2}\pi \le y \le \tfrac{1}{2}\pi$$

The domain of $\sin^{-1}$ is the closed interval $[-1, 1]$, and the range is the closed interval $[-\tfrac{1}{2}\pi, \tfrac{1}{2}\pi]$.

To sketch the graph of the inverse sine function, let

$$f(x) = \sin^{-1} x$$

Table 1 gives values of $f(x)$ for some specific values of x. The graph appears in Figure 3. You can verify this graph on your graphics calculator.

Table 1

x	-1	$-\dfrac{\sqrt{3}}{2}$	$-\dfrac{1}{2}$	0	$\dfrac{1}{2}$	$\dfrac{\sqrt{3}}{2}$	1
$\sin^{-1} x$	$-\dfrac{1}{2}\pi$	$-\dfrac{1}{3}\pi$	$-\dfrac{1}{6}\pi$	0	$\dfrac{1}{6}\pi$	$\dfrac{1}{3}\pi$	$\dfrac{1}{2}\pi$

▶ **EXAMPLE 1** *Sketching Graphs of Functions Involving Inverse Sine*

Find the domain and range of the function defined by the equation and sketch the graph of the function. Check the graph on a graphics calculator.

(a) $f(x) = \sin^{-1} \tfrac{1}{2} x$ **(b)** $g(x) = 2 \sin^{-1} 3x$

Solution
(a) Because the domain of the inverse sine function is $[-1, 1]$, we can obtain the domain of f by writing the inequality $-1 \le \tfrac{1}{2} x \le 1$ as $-2 \le x \le 2$. Thus the domain is $[-2, 2]$, and the range is $[-\tfrac{1}{2}\pi, \tfrac{1}{2}\pi]$. With this information and by locating a few points, we obtain the graph, which agrees with what we get on our graphics calculator in Figure 4.
(b) For the domain of g, we write the inequality $-1 \le 3x \le 1$ as $-\tfrac{1}{3} \le x \le \tfrac{1}{3}$. The domain is, therefore, $[-\tfrac{1}{3}, \tfrac{1}{3}]$. Because the range of the inverse sine function is $[-\tfrac{1}{2}\pi, \tfrac{1}{2}\pi]$, the range of g is $[-\pi, \pi]$. We locate a few points and obtain the graph, which agrees with the graph on our graphics calculator shown in Figure 5. ◀

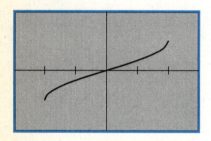

$[-3, 3]$ by $[-\pi, \pi]$
$f(x) = \sin^{-1} \dfrac{1}{2} x$
FIGURE 4

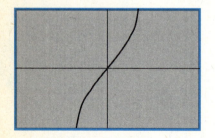

$[-1, 1]$ by $[-\pi, \pi]$
$g(x) = 2 \sin^{-1} 3x$
FIGURE 5

The use of the symbol -1 to represent the inverse sine function makes it necessary to denote the reciprocal of $\sin x$ by $(\sin x)^{-1}$ to avoid confusion. A similar convention is applied when using any negative exponent with a trigonometric function. For instance,

$$\frac{1}{\sin x} = (\sin x)^{-1} \qquad \frac{1}{\sin^2 x} = (\sin x)^{-2} \qquad \frac{1}{\cos^3 x} = (\cos x)^{-3}$$

and so on.

The terminology **arc sine** is sometimes used in place of inverse sine, and the notation arc sin x can be used instead of $\sin^{-1} x$. This notation probably comes from the fact that if $t = $ arc sin u, then $\sin t = u$ and t units is the length of the arc on the unit circle for which the sine is u.

▷ **ILLUSTRATION 1**

(a) $\sin^{-1}\dfrac{1}{\sqrt{2}} = \dfrac{1}{4}\pi$

(b) arc sin $\dfrac{1}{\sqrt{2}} = \dfrac{1}{4}\pi$

(c) $\sin^{-1}\left(-\dfrac{1}{\sqrt{2}}\right) = -\dfrac{1}{4}\pi$

(d) arc sin$\left(-\dfrac{1}{\sqrt{2}}\right) = -\dfrac{1}{4}\pi$ ◄

Approximate inverse sine function values can be obtained on your calculator by using the $\boxed{\sin^{-1}}$ key.

▷ **ILLUSTRATION 2**

To determine $\sin^{-1} 0.8724$ on your calculator, first set the calculator in the radian mode. Then with the $\boxed{\sin^{-1}}$ key and by entering 0.8724, you get

$$\sin^{-1} 0.8724 \approx 1.060$$ ◄

From the definition of the inverse sine function

$$\sin(\sin^{-1} x) = x \qquad \text{for } x \text{ in } [-1, 1]$$
$$\sin^{-1}(\sin y) = y \qquad \text{for } y \text{ in } [-\tfrac{1}{2}\pi, \tfrac{1}{2}\pi]$$

Observe that $\sin^{-1}(\sin y) \neq y$ if y is not in the interval $[-\tfrac{1}{2}\pi, \tfrac{1}{2}\pi]$. For example,

$$\sin^{-1}\left(\sin \frac{3}{4}\pi\right) = \sin^{-1}\frac{1}{\sqrt{2}} \qquad \text{and} \qquad \sin^{-1}\left(\sin \frac{7}{4}\pi\right) = \sin^{-1}\left(-\frac{1}{\sqrt{2}}\right)$$

$$= \frac{1}{4}\pi \qquad\qquad\qquad\qquad = -\frac{1}{4}\pi$$

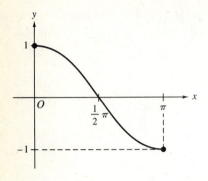

FIGURE 6

Table 2

x	$\cos^{-1} x$
-1	π
$-\dfrac{\sqrt{3}}{2}$	$\dfrac{5}{6}\pi$
$-\dfrac{1}{2}$	$\dfrac{2}{3}\pi$
0	$\dfrac{1}{2}\pi$
$\dfrac{1}{2}$	$\dfrac{1}{3}\pi$
$\dfrac{\sqrt{3}}{2}$	$\dfrac{1}{6}\pi$
1	0

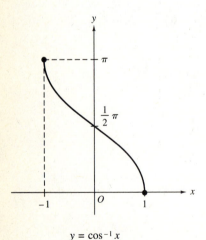

$y = \cos^{-1} x$

FIGURE 7

▶ **EXAMPLE 2** *Finding Function Values Involving Inverse Sine*

Find: **(a)** $\cos[\sin^{-1}(-\frac{1}{2})]$; **(b)** $\sin^{-1}(\cos\frac{2}{3}\pi)$.

Solution   Because the range of the inverse sine function is $[-\frac{1}{2}\pi, \frac{1}{2}\pi]$, $\sin^{-1}(-\frac{1}{2}) = -\frac{1}{6}\pi$.

(a) $\cos\left[\sin^{-1}\left(-\frac{1}{2}\right)\right] = \cos\left(-\frac{1}{6}\pi\right)$ **(b)** $\sin^{-1}\left(\cos\frac{2}{3}\pi\right) = \sin^{-1}\left(-\frac{1}{2}\right)$

$$= \frac{\sqrt{3}}{2} \qquad\qquad\qquad = -\frac{1}{6}\pi \quad ◀$$

The cosine does not have an inverse function because it also is not one-to-one. To define the inverse cosine function, we restrict the cosine to the interval $[0, \pi]$. Consider the function G defined by

$$G(x) = \cos x \qquad \text{and} \qquad 0 \le x \le \pi$$

The domain of G is the closed interval $[0, \pi]$, and the range is the closed interval $[-1, 1]$. The graph of G is shown in Figure 6. From the horizontal-line test, we observe that G is one-to-one. Therefore it has an inverse function, called the *inverse cosine function* and denoted by $\cos^{-1}$.

DEFINITION   The Inverse Cosine Function

The **inverse cosine function,** denoted by $\cos^{-1}$, is defined as follows:

$$y = \cos^{-1} x \quad \text{if and only if} \quad x = \cos y \quad \text{and} \quad 0 \le y \le \pi$$

The domain of $\cos^{-1}$ is the closed interval $[-1, 1]$, and the range is the closed interval $[0, \pi]$.

Table 2 gives values of $\cos^{-1} x$ for some particular values of x. From them we obtain the graph of the inverse cosine function shown in Figure 7. Check this graph on your graphics calculator.

The inverse cosine function is also called the **arc cosine** function, and the notation arc cos x can be used in place of $\cos^{-1} x$.

▷ **ILLUSTRATION 3**

(a) $\cos^{-1}\dfrac{1}{\sqrt{2}} = \dfrac{1}{4}\pi$ **(b)** arc cos $\dfrac{1}{\sqrt{2}} = \dfrac{1}{4}\pi$

(c) $\cos^{-1}\left(-\dfrac{1}{\sqrt{2}}\right) = \dfrac{3}{4}\pi$ **(d)** arc $\cos\left(-\dfrac{1}{\sqrt{2}}\right) = \dfrac{3}{4}\pi$ ◀

From the definition of the inverse cosine function

$$\cos(\cos^{-1} x) = x \qquad \text{for } x \text{ in } [-1, 1]$$
$$\cos^{-1}(\cos y) = y \qquad \text{for } y \text{ in } [0, \pi]$$

Notice there is again a restriction on y in order to have the equality $\cos^{-1}(\cos y) = y$. For example, because $\frac{3}{4}\pi$ is in $[0, \pi]$

$$\cos^{-1}\left(\cos \frac{3}{4}\pi\right) = \frac{3}{4}\pi$$

However,

$$\cos^{-1}\left(\cos \frac{5}{4}\pi\right) = \cos^{-1}\left(-\frac{1}{\sqrt{2}}\right) \qquad \text{and} \qquad \cos^{-1}\left(\frac{7}{4}\pi\right) = \cos^{-1}\left(\frac{1}{\sqrt{2}}\right)$$

$$= \frac{3}{4}\pi \qquad\qquad\qquad\qquad\qquad = \frac{1}{4}\pi$$

▶ **EXAMPLE 3** *Finding an Exact Function Value Involving Inverse Cosine*

Find the exact value of $\sin[2 \cos^{-1}(-\frac{3}{5})]$.

Solution Because we wish to obtain trigonometric functions of the number $\cos^{-1}(-\frac{3}{5})$, we shall let t represent this number.

$$t = \cos^{-1}(-\tfrac{3}{5})$$

Because the range of the inverse cosine function is $[0, \pi]$ and $\cos t$ is negative, t is in the second quadrant. Thus

$$\cos t = -\tfrac{3}{5} \qquad \text{and} \qquad \tfrac{1}{2}\pi < t < \pi$$

We wish to find the exact value of $\sin 2t$. From the sine double-measure identity, $\sin 2t = 2 \sin t \cos t$. Thus we need to compute $\sin t$. From the identity $\sin^2 t + \cos^2 t = 1$, and because $\sin t > 0$ since t is in $(\frac{1}{2}\pi, \pi)$, $\sin t = \sqrt{1 - \cos^2 t}$. Thus

$$\sin t = \sqrt{1 - (-\tfrac{3}{5})^2}$$
$$= \tfrac{4}{5}$$

Therefore

$$\sin 2t = 2 \sin t \cos t$$
$$= 2(\tfrac{4}{5})(-\tfrac{3}{5})$$
$$= -\tfrac{24}{25}$$

from which we conclude that

$$\sin[2 \cos^{-1}(-\tfrac{3}{5})] = -\tfrac{24}{25}$$

◀

▶ **EXAMPLE 4** *Proving an Identity*

Prove the identity

$$\cos^{-1} x = \tfrac{1}{2}\pi - \sin^{-1} x \qquad \text{for } |x| \leq 1$$

Solution   Let x be in $[-1, 1]$, and let

$$t = \cos(\tfrac{1}{2}\pi - \sin^{-1} x) \qquad \qquad (2)$$

Applying the reduction formula $\cos(\tfrac{1}{2}\pi - v) = \sin v$ with $v = \sin^{-1} x$ on the right side of (2), we get

$$t = \sin(\sin^{-1} x)$$

Because x is in $[-1, 1]$, $\sin(\sin^{-1} x) = x$; therefore

$$t = x$$

Replacing t by x in (2) gives

$$x = \cos(\tfrac{1}{2}\pi - \sin^{-1} x) \qquad \qquad (3)$$

Because $-\tfrac{1}{2}\pi \leq \sin^{-1} x \leq \tfrac{1}{2}\pi$, by adding $-\tfrac{1}{2}\pi$ to each member, we have

$$-\pi \leq -\tfrac{1}{2}\pi + \sin^{-1} x \leq 0$$

Multiplying each member of this inequality by -1 and reversing the direction of the inequality signs gives

$$0 \leq \tfrac{1}{2}\pi - \sin^{-1} x \leq \pi \qquad \qquad (4)$$

From (3), (4), and the definition of $\cos^{-1}$, it follows that

$$\cos^{-1} x = \tfrac{1}{2}\pi - \sin^{-1} x \qquad \text{for } |x| \leq 1$$

which is what we wished to prove.   ◀

Observe in the solution of Example 4 that the identity depends on our choosing the range of the inverse cosine function to be $[0, \pi]$.

To obtain the inverse tangent function, we first restrict the tangent function to the open interval $(-\tfrac{1}{2}\pi, \tfrac{1}{2}\pi)$. We let H be the function defined by

$$H(x) = \tan x \qquad \text{and} \qquad -\tfrac{1}{2}\pi < x < \tfrac{1}{2}\pi$$

The domain of H is the open interval $(-\tfrac{1}{2}\pi, \tfrac{1}{2}\pi)$, and the range is the set R of real numbers. Its graph appears in Figure 8. From the horizontal-line test, H is one-to-one. Therefore it has an inverse function, called the *inverse tangent function* denoted by $\tan^{-1}$.

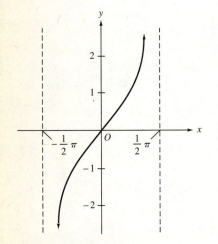

FIGURE 8

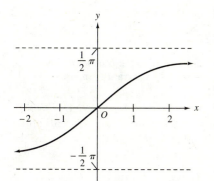

$y = \tan^{-1} x$

FIGURE 9

DEFINITION **The Inverse Tangent Function**

The **inverse tangent function,** denoted by $\tan^{-1}$, is defined as follows:

$y = \tan^{-1} x$ if and only if $x = \tan y$ and $-\tfrac{1}{2}\pi < y < \tfrac{1}{2}\pi$

The domain of $\tan^{-1}$ is the set R of real numbers, and the range is the open interval $(-\tfrac{1}{2}\pi, \tfrac{1}{2}\pi)$.

Figure 9 shows the graph of $\tan^{-1}$. Check it on your graphics calculator.

The inverse tangent function is sometimes referred to as the arc tangent function, and then arc tan x is used instead of $\tan^{-1} x$.

▷ **ILLUSTRATION 4**

(a) $\tan^{-1} \sqrt{3} = \dfrac{1}{3}\pi$ **(b)** arc $\tan\left(-\dfrac{1}{\sqrt{3}}\right) = -\dfrac{1}{6}\pi$ **(c)** $\tan^{-1} 0 = 0$ ◀

From the definition of the inverse tangent function

$\tan(\tan^{-1} x) = x$ for x in R

$\tan^{-1}(\tan y) = y$ for y in $(-\tfrac{1}{2}\pi, \tfrac{1}{2}\pi)$

▷ **ILLUSTRATION 5**

$\tan^{-1}(\tan \tfrac{1}{4}\pi) = \tfrac{1}{4}\pi$ and $\tan^{-1}[\tan(-\tfrac{1}{4}\pi)] = -\tfrac{1}{4}\pi$

However

$\tan^{-1}(\tan \tfrac{3}{4}\pi) = \tan^{-1}(-1)$ and $\tan^{-1}(\tan \tfrac{5}{4}\pi) = \tan^{-1} 1$

$\qquad\qquad\qquad = -\tfrac{1}{4}\pi \qquad\qquad\qquad\qquad\qquad = \tfrac{1}{4}\pi$ ◀

▶ **EXAMPLE 5** *Finding an Exact Function Value Involving Arc Tangent*

Find the exact value of sec[arc $\tan(-3)$].

Solution   We shall do this problem by letting arc $\tan(-3)$ be an angle. Let

$\theta = $ arc $\tan(-3)$

Because the range of the arc tangent function is $(-\tfrac{1}{2}\pi, \tfrac{1}{2}\pi)$, and because $\tan \theta$ is negative, $-\tfrac{1}{2}\pi < \theta < 0$. Thus

$\tan \theta = -3$ and $-\tfrac{1}{2}\pi < \theta < 0$

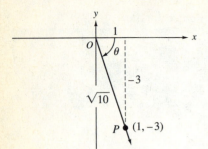

FIGURE 10

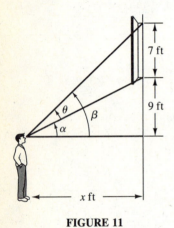

FIGURE 11

Figure 10 shows an angle θ that satisfies these requirements. Observe that the point P selected on the terminal side of θ is $(1, -3)$. From the Pythagorean theorem, r is $\sqrt{1^2 + (-3)^2} = \sqrt{10}$. Therefore $\sec \theta = \sqrt{10}$. Hence

$$\sec[\text{arc tan}(-3)] = \sqrt{10} \qquad \blacktriangleleft$$

▶ **EXAMPLE 6**   *Solving a Word Problem Having a Mathematical Model Involving the Inverse Tangent Function*

A picture 7 ft high is placed on a wall with its base 9 ft above the level of the eye of an observer. Suppose the observer is x feet from the wall and θ is the radian measure of the angle subtended at the observer's eye by the picture. **(a)** Define θ as a function of x. Find θ when **(b)** $x = 10$; **(c)** $x = 12$; and **(d)** $x = 15$.

Solution

(a) In Figure 11, α is the radian measure of the angle subtended at the observer's eye by the portion of the wall above eye level and below the picture. Furthermore,

$$\alpha + \theta = \beta$$

So

$$\theta = \beta - \alpha$$

From the tangent difference identity,

$$\tan \theta = \frac{\tan \beta - \tan \alpha}{1 + \tan \beta \tan \alpha}$$

Observe from Figure 11 that

$$\tan \alpha = \frac{9}{x} \quad \text{and} \quad \tan \beta = \frac{16}{x}$$

Substituting these values into the expression for $\tan \theta$, we obtain

$$\tan \theta = \frac{\dfrac{16}{x} - \dfrac{9}{x}}{1 + \dfrac{16}{x} \cdot \dfrac{9}{x}}$$

$$= \frac{16x - 9x}{x^2 + 144}$$

$$= \frac{7x}{x^2 + 144}$$

Therefore

$$\theta = \tan^{-1} \frac{7x}{x^2 + 144}$$

(b) When $x = 10$,

$$\theta = \tan^{-1} \frac{70}{100 + 144}$$

$$= 0.2794$$

(c) When $x = 12$,

$$\theta = \tan^{-1} \frac{84}{144 + 144}$$

$$= 0.2838$$

(d) When $x = 15$,

$$\theta = \tan^{-1} \frac{105}{225 + 144}$$

$$= 0.2772$$

◀

In Example 6 when x is large (that is, when the observer is far away from the wall), θ is small. As the observer gets closer to the wall, θ increases until it reaches a maximum value. Then, as the observer gets even closer to the wall, θ gets smaller. In calculus we can find the value of x that will make θ a maximum. So we can determine how far from the wall the observer should stand in order for the angle subtended at the observer's eye by the picture to be the greatest. When θ is a maximum, the observer has the "best view" of the picture.

▶ **EXAMPLE 7** *Solving a Word Problem Having a Mathematical Model Involving the Inverse Tangent Function*

Use a graphics calculator to estimate, to the nearest foot, how far from the wall the observer in Example 6 should stand to have the "best view" of the picture.

Solution   Figure 12 shows the graph of

$$\theta = \tan^{-1} \frac{7x}{x^2 + 144}$$

in the viewing rectangle [0, 20] by [0, 0.4]. By using trace and zoom-in we estimate that the maximum value of θ occurs when $x = 12$.

Conclusion: To the nearest foot, the observer should stand 12 ft from the wall to have the best view of the picture.

◀

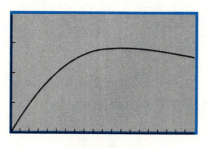

[0, 20] by [0, 0.4]
$\theta = \tan^{-1} \dfrac{7x}{x^2 + 144}$
FIGURE 12

It turns out that our estimate in Example 7 is the exact value of x obtained by using techniques of calculus. That is, the best view of the picture occurs when the observer is exactly 12 ft from the wall.

Before defining the inverse cotangent function, we refer back to Example 4 in which we proved the identity

$$\cos^{-1} x = \tfrac{1}{2}\pi - \sin^{-1} x \qquad \text{for } |x| \le 1$$

This identity can be used to define the inverse cosine function, and then it can be proved that the range of $\cos^{-1}$ is $[0, \pi]$. We use this kind of procedure in discussing the *inverse cotangent function*.

DEFINITION **The Inverse Cotangent Function**

The **inverse cotangent function,** denoted by $\cot^{-1}$, is defined by

$$\cot^{-1} x = \tfrac{1}{2}\pi - \tan^{-1} x \qquad \text{where } x \text{ is any real number}$$

From the definition, the domain of $\cot^{-1}$ is the set of all real numbers. To obtain the range, we write the equation in the definition as

$$\tan^{-1} x = \tfrac{1}{2}\pi - \cot^{-1} x \qquad\qquad\qquad (5)$$

Because

$$-\tfrac{1}{2}\pi < \tan^{-1} x < \tfrac{1}{2}\pi$$

by substituting from (5) into this inequality we get

$$-\tfrac{1}{2}\pi < \tfrac{1}{2}\pi - \cot^{-1} x < \tfrac{1}{2}\pi$$

Subtracting $\tfrac{1}{2}\pi$ from each member, we get

$$-\pi < -\cot^{-1} x < 0$$

Now multiplying each member by -1 and reversing the direction of the inequality signs, we obtain

$$0 < \cot^{-1} x < \pi$$

Therefore the range of the inverse cotangent function is the open interval $(0, \pi)$. Its graph appears in Figure 13.

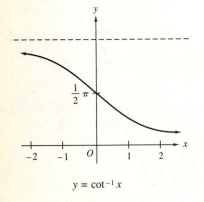

$y = \cot^{-1} x$

FIGURE 13

▷ **ILLUSTRATION 6**

(a) $\tan^{-1} 1 = \tfrac{1}{4}\pi$

(b) $\tan^{-1}(-1) = -\tfrac{1}{4}\pi$

(c) $\cot^{-1} 1 = \tfrac{1}{2}\pi - \tan^{-1} 1$
$\qquad\quad = \tfrac{1}{2}\pi - \tfrac{1}{4}\pi$
$\qquad\quad = \tfrac{1}{4}\pi$

(d) $\cot^{-1}(-1) = \tfrac{1}{2}\pi - \tan^{-1}(-1)$
$\qquad\qquad\quad = \tfrac{1}{2}\pi - (-\tfrac{1}{4}\pi)$
$\qquad\qquad\quad = \tfrac{3}{4}\pi$ ◀

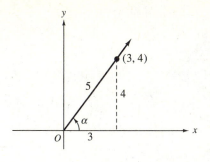

FIGURE 14

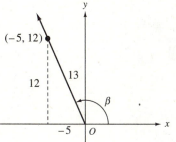

FIGURE 15

▶ **EXAMPLE 8** *Finding an Exact Function Value Involving Inverse Cotangent*

Find the exact value of $\cos[\cot^{-1}\frac{3}{4} + \cot^{-1}(-\frac{5}{12})]$.

Solution Let $\alpha = \cot^{-1}\frac{3}{4}$ and $\beta = \cot^{-1}(-\frac{5}{12})$. Then

$$\cot \alpha = \tfrac{3}{4} \qquad \text{and} \qquad 0 < \alpha < \tfrac{1}{2}\pi$$
$$\cot \beta = -\tfrac{5}{12} \qquad \text{and} \qquad \tfrac{1}{2}\pi < \beta < \pi$$

We wish to find $\cos(\alpha + \beta)$. From the cosine sum identity,

$$\cos(\alpha + \beta) = \cos \alpha \cos \beta - \sin \alpha \sin \beta \qquad (6)$$

To determine $\sin \alpha$ and $\cos \alpha$, refer to Figure 14, which shows a first-quadrant angle α for which $\cot \alpha = \frac{3}{4}$. From the figure,

$$\sin \alpha = \tfrac{4}{5} \qquad \cos \alpha = \tfrac{3}{5} \qquad (7)$$

Figure 15 shows a second-quadrant angle β for which $\cot \beta = -\frac{5}{12}$. From the figure,

$$\sin \beta = \tfrac{12}{13} \qquad \cos \beta = -\tfrac{5}{13} \qquad (8)$$

Substituting from (7) and (8) into (6), we have

$$\cos(\alpha + \beta) = \tfrac{3}{5}(-\tfrac{5}{13}) - \tfrac{4}{5}(\tfrac{12}{13})$$
$$= -\tfrac{63}{65} \qquad \blacktriangleleft$$

No universal agreement prevails regarding the ranges of the inverse secant and cosecant functions. In calculus, however, certain computations are simplified if the *inverse secant function* is defined so that if $x \leq -1$, then $\pi \leq \sec^{-1} x < \frac{3}{2}\pi$ (see Exercise 67). Thus we give the following definition.

DEFINITION **The Inverse Secant Function**

The **inverse secant function,** denoted by $\sec^{-1}$, is defined as follows:

$$y = \sec^{-1} x \quad \text{if and only if} \quad x = \sec y \quad \text{and} \quad \begin{cases} 0 \leq y < \tfrac{1}{2}\pi & \text{if } x \geq 1 \\ \pi \leq y < \tfrac{3}{2}\pi & \text{if } x \leq -1 \end{cases}$$

The domain of $\sec^{-1}$ is $(-\infty, -1] \cup [1, +\infty)$, and the range is the set $[0, \frac{1}{2}\pi) \cup [\pi, \frac{3}{2}\pi)$. Figure 16 shows the graph of $\sec^{-1}$.

From the definition of the inverse secant function

$$\sec(\sec^{-1} x) = x \qquad \text{for } x \text{ in } (-\infty, -1] \cup [1, +\infty)$$
$$\sec^{-1}(\sec y) = y \qquad \text{for } y \text{ in } [0, \tfrac{1}{2}\pi) \cup [\pi, \tfrac{3}{2}\pi)$$

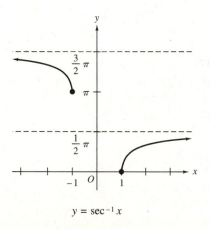

$y = \sec^{-1} x$

FIGURE 16

We now define the *inverse cosecant function* in terms of the inverse secant function.

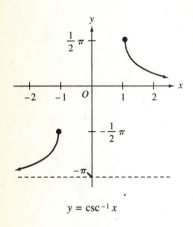

$y = \csc^{-1} x$

FIGURE 17

> **DEFINITION   The Inverse Cosecant Function**
>
> The **inverse cosecant function,** denoted by $\csc^{-1}$, is defined by
> $$\csc^{-1} x = \tfrac{1}{2}\pi - \sec^{-1} x \qquad \text{for } |x| \geq 1$$

From the definition the domain of $\csc^{-1}$ is $(-\infty, -1] \cup [1, +\infty)$. The range of $\csc^{-1}$ can be found in a manner similar to that used to determine the range of $\cot^{-1}$. The range of $\csc^{-1}$ is $(-\pi, -\tfrac{1}{2}\pi] \cup (0, \tfrac{1}{2}\pi]$, and you are asked to show this in Exercise 68. The graph of $\csc^{-1}$ appears in Figure 17.

▷ **ILLUSTRATION 7**

(a) $\sec^{-1} 2 = \tfrac{1}{3}\pi$

(b) $\sec^{-1}(-2) = \tfrac{4}{3}\pi$

(c) $\csc^{-1} 2 = \tfrac{1}{2}\pi - \sec^{-1} 2$
$\qquad = \tfrac{1}{2}\pi - \tfrac{1}{3}\pi$
$\qquad = \tfrac{1}{6}\pi$

(d) $\csc^{-1}(-2) = \tfrac{1}{2}\pi - \sec^{-1}(-2)$
$\qquad = \tfrac{1}{2}\pi - \tfrac{4}{3}\pi$
$\qquad = -\tfrac{5}{6}\pi$ ◀

EXERCISES 9.5

In Exercises 1 through 6, determine the exact function value. Do not use a calculator.

1. (a) $\sin^{-1}\tfrac{1}{2}$ **(b)** $\sin^{-1}(-\tfrac{1}{2})$
(c) $\cos^{-1}\tfrac{1}{2}$ **(d)** $\cos^{-1}(-\tfrac{1}{2})$

2. (a) $\sin^{-1}\dfrac{\sqrt{3}}{2}$ **(b)** $\sin^{-1}\left(-\dfrac{\sqrt{3}}{2}\right)$

(c) $\cos^{-1}\dfrac{\sqrt{3}}{2}$ **(d)** $\cos^{-1}\left(-\dfrac{\sqrt{3}}{2}\right)$

3. (a) $\tan^{-1}\dfrac{1}{\sqrt{3}}$ **(b)** $\tan^{-1}(-\sqrt{3})$

(c) $\sec^{-1}\dfrac{2}{\sqrt{3}}$ **(d)** $\sec^{-1}\left(-\dfrac{2}{\sqrt{3}}\right)$

4. (a) $\cot^{-1}\dfrac{1}{\sqrt{3}}$ **(b)** $\cot^{-1}(-\sqrt{3})$

(c) $\csc^{-1}\dfrac{2}{\sqrt{3}}$ **(d)** $\csc^{-1}\left(-\dfrac{2}{\sqrt{3}}\right)$

5. (a) $\sin^{-1} 1$ **(b)** $\sin^{-1}(-1)$ **(c)** $\csc^{-1} 1$
(d) $\csc^{-1}(-1)$ **(e)** $\sin^{-1} 0$

6. (a) $\cos^{-1} 1$ **(b)** $\cos^{-1}(-1)$ **(c)** $\sec^{-1} 1$
(d) $\sec^{-1}(-1)$ **(e)** $\cos^{-1} 0$

In Exercises 7 through 12, use a calculator to approximate the function value.

7. (a) $\sin^{-1} 0.4882$ **(b)** $\sin^{-1}(-0.4882)$
(c) $\cos^{-1} 0.4882$ **(d)** $\cos^{-1}(-0.4882)$

8. (a) $\sin^{-1} 0.2764$ **(b)** $\sin^{-1}(-0.2764)$
(c) $\cos^{-1} 0.2764$ **(d)** $\cos^{-1}(-0.2764)$

9. (a) $\tan^{-1} 0.4346$ **(b)** $\tan^{-1}(-0.4346)$
(c) $\cot^{-1} 0.4346$ **(d)** $\cot^{-1}(-0.4346)$

10. (a) $\tan^{-1} 2.733$ **(b)** $\tan^{-1}(-2.733)$
(c) $\cot^{-1} 2.733$ **(d)** $\cot^{-1}(-2.733)$

11. (a) $\sec^{-1} 2.083$ **(b)** $\sec^{1}(-2.083)$
(c) $\csc^{-1} 2.083$ **(d)** $\csc^{-1}(-2.083)$

12. (a) $\sec^{-1} 1.256$ **(b)** $\sec^{-1}(-1.256)$
(c) $\csc^{-1} 1.256$ **(d)** $\csc^{-1}(-1.256)$

13. Given $x = \arc\sin\tfrac{1}{3}$, find the exact value of each of the following: **(a)** $\cos x$; **(b)** $\tan x$; **(c)** $\cot x$; **(d)** $\sec x$; **(e)** $\csc x$.

14. Given $x = \text{arc cos }\frac{2}{3}$, find the exact value of each of the following: **(a)** $\sin x$; **(b)** $\tan x$; **(c)** $\cot x$; **(d)** $\sec x$; **(e)** $\csc x$.

15. Do Exercise 13 if $x = \text{arc sin}(-\frac{1}{3})$.

16. Do Exercise 14 if $x = \text{arc cos}(-\frac{2}{3})$.

17. Given $y = \tan^{-1}(-2)$, find the exact value of each of the following: **(a)** $\sin y$; **(b)** $\cos y$; **(c)** $\cot y$; **(d)** $\sec y$; **(e)** $\csc y$.

18. Given $y = \cot^{-1}(-\frac{1}{2})$, find the exact value of each of the following: **(a)** $\sin y$; **(b)** $\cos y$; **(c)** $\tan y$; **(d)** $\sec y$; **(e)** $\csc y$.

19. Given $t = \csc^{-1}(-\frac{3}{2})$, find the exact value of each of the following: **(a)** $\sin t$; **(b)** $\cos t$; **(c)** $\tan t$; **(d)** $\cot t$; **(e)** $\sec t$.

20. Given $t = \sec^{-1}(-3)$, find the exact value of each of the following: **(a)** $\sin t$; **(b)** $\cos t$; **(c)** $\tan t$; **(d)** $\cot t$; **(e)** $\csc t$.

In Exercises 21 through 46, find the exact function value.

21. (a) $\sin^{-1}(\sin \frac{1}{6}\pi)$ **(b)** $\sin^{-1}[\sin(-\frac{1}{6}\pi)]$
(c) $\sin^{-1}(\sin \frac{5}{6}\pi)$ **(d)** $\sin^{-1}(\sin \frac{11}{6}\pi)$

22. (a) $\sin^{-1}(\sin \frac{1}{3}\pi)$ **(b)** $\sin^{-1}[\sin(-\frac{1}{3}\pi)]$
(c) $\sin^{-1}(\sin \frac{2}{3}\pi)$ **(d)** $\sin^{-1}(\sin \frac{5}{3}\pi)$

23. (a) $\cos^{-1}(\cos \frac{1}{3}\pi)$ **(b)** $\cos^{-1}[\cos(-\frac{1}{3}\pi)]$
(c) $\cos^{-1}(\cos \frac{2}{3}\pi)$ **(d)** $\cos^{-1}(\cos \frac{4}{3}\pi)$

24. (a) $\cos^{-1}(\cos \frac{1}{4}\pi)$ **(b)** $\cos^{-1}[\cos(-\frac{1}{4}\pi)]$
(c) $\cos^{-1}(\cos \frac{3}{4}\pi)$ **(d)** $\cos^{-1}(\cos \frac{5}{4}\pi)$

25. (a) $\tan^{-1}(\tan \frac{1}{6}\pi)$ **(b)** $\tan^{-1}[\tan(-\frac{1}{3}\pi)]$
(c) $\tan^{-1}(\tan \frac{7}{6}\pi)$ **(d)** $\tan^{-1}[\tan (-\frac{4}{3}\pi)]$

26. (a) $\tan^{-1}(\tan \frac{1}{3}\pi)$ **(b)** $\tan^{-1}[\tan(-\frac{1}{6}\pi)]$
(c) $\tan^{-1}(\tan \frac{4}{3}\pi)$ **(d)** $\tan^{-1}[\tan(-\frac{7}{6}\pi)]$

27. (a) $\cot^{-1}(\cot \frac{1}{6}\pi)$ **(b)** $\cot^{-1}[\cot(-\frac{1}{3}\pi)]$
(c) $\cot^{-1}(\cot \frac{7}{6}\pi)$ **(d)** $\cot^{-1}[\cot(-\frac{4}{3}\pi)]$

28. (a) $\cot^{-1}(\cot \frac{1}{3}\pi)$ **(b)** $\cot^{-1}[\cot(-\frac{1}{6}\pi)]$
(c) $\cot^{-1}(\cot \frac{4}{3}\pi)$ **(d)** $\cot^{-1}[\cot(-\frac{7}{6}\pi)]$

29. (a) $\sec^{-1}(\sec \frac{1}{3}\pi)$ **(b)** $\sec^{-1}[\sec(-\frac{1}{3}\pi)]$
(c) $\sec^{-1}(\sec \frac{2}{3}\pi)$ **(d)** $\sec^{-1}(\sec \frac{4}{3}\pi)$

30. (a) $\sec^{-1}(\sec \frac{1}{4}\pi)$ **(b)** $\sec^{-1}[\sec(-\frac{1}{4}\pi)]$
(c) $\sec^{-1}(\sec \frac{3}{4}\pi)$ **(d)** $\sec^{-1}(\sec \frac{5}{4}\pi)$

31. (a) $\csc^{-1}(\csc \frac{1}{6}\pi)$ **(b)** $\csc^{-1}[\csc(-\frac{1}{6}\pi)]$
(c) $\csc^{-1}(\csc \frac{5}{6}\pi)$ **(d)** $\csc^{-1}(\csc \frac{11}{6}\pi)$

32. (a) $\csc^{-1}(\csc \frac{1}{3}\pi)$ **(b)** $\csc^{-1}[\csc(-\frac{1}{3}\pi)]$
(c) $\csc^{-1}(\csc \frac{2}{3}\pi)$ **(d)** $\csc^{-1}(\csc \frac{5}{3}\pi)$

33. (a) $\tan[\sin^{-1}\frac{1}{2}\sqrt{3}]$ **(b)** $\sin[\tan^{-1}\frac{1}{2}\sqrt{3}]$

34. (a) $\cos[\tan^{-1}(-3)]$ **(b)** $\tan[\sec^{-1}(-3)]$

35. (a) $\cos[\text{arc sin}(-\frac{1}{2})]$ **(b)** $\sin[\text{arc cos}(-\frac{1}{2})]$

36. (a) $\tan[\text{arc cot}(-1)]$ **(b)** $\cot[\text{arc tan}(-1)]$

37. $\cos[2 \sin^{-1}(-\frac{5}{13})]$ **38.** $\tan[2 \sec^{-1}(-\frac{5}{4})]$

39. $\sin[\text{arc sin }\frac{2}{3} + \text{arc cos }\frac{1}{3}]$

40. $\cos[\text{arc sin}(-\frac{1}{2}) + \text{arc sin }\frac{1}{4}]$

41. $\cos[\sin^{-1}\frac{2}{3} + 2 \sin^{-1}(-\frac{1}{3})]$

42. $\tan[\tan^{-1}(-\frac{2}{5}) - \cos^{-1}(-\frac{1}{2}\sqrt{2})]$

43. $\tan(\text{arc tan }\frac{3}{4} - \text{arc sin }\frac{1}{2})$

44. $\tan[\text{arc sec }\frac{5}{3} + \text{arc csc}(-\frac{13}{12})]$

45. $\cos(\sin^{-1}\frac{1}{3} - \tan^{-1}\frac{1}{2})$

46. $\sin[\cos^{-1}(-\frac{2}{3}) + 2 \sin^{-1}(-\frac{1}{3})]$

47. Prove: $\cos^{-1}\dfrac{3}{\sqrt{10}} + \cos^{-1}\dfrac{2}{\sqrt{5}} = \dfrac{1}{4}\pi$

48. Prove: $2 \tan^{-1}\frac{1}{3} - \tan^{-1}(-\frac{1}{7}) = \frac{1}{4}\pi$

In Exercises 49 through 56, sketch the graph of the function. Check your graph on your graphics calculator.

49. $f(x) = 2 \sin^{-1} x$ **50.** $g(x) = \text{arc sin } 2x$

51. $g(x) = \text{arc tan }\frac{1}{2}x$ **52.** $f(x) = \frac{1}{2} \tan^{-1} x$

53. $h(x) = \cos^{-1} 3x$ **54.** $h(x) = 2 \cos^{-1}\frac{1}{2}x$

55. $f(x) = \frac{1}{2} \text{arc sin}(x + 3)$ **56.** $g(x) = 3 \sin^{-1}(x - 2)$

57. (a) Sketch the graph of $f(x) = \sin(\sin^{-1} x)$, and check your graph on your graphics calculator. What is **(b)** the domain of f and **(c)** the range of f?

58. Do Exercise 57 if $f(x) = \cos(\cos^{-1} x)$.

59. Do Exercise 57 if $f(x) = \sin^{-1}(\sin x)$.

60. Do Exercise 57 if $f(x) = \cos^{-1}(\cos x)$.

$\boxed{\int\!/\frac{dy}{dx}}$ **61.** Solve this problem by obtaining a mathematical model involving the inverse tangent function, and be sure to write a conclusion: A sign 3 ft high is placed on a wall with its base 2 ft above the eye level of a woman attempting to read it. **(a)** If the woman is x feet from the wall and θ is the radian measure of the angle subtended at the woman's eye, define θ as a function of x. Find θ when **(b)** $x = 2$; **(c)** $x = 3$; and **(d)** $x = 4$. **(e)** Use your graphics calculator to estimate, to the nearest one-hundredth of a foot, how far the woman should stand so that she has the "best view" of the sign, that is, so that the angle subtended at the woman's eye by the sign is a maximum.

62. (a) In Example 6, show that another equation defining θ in terms of x is

$$\theta = \cot^{-1}\frac{x}{16} - \cot^{-1}\frac{x}{9}$$

Use the equation in part (a) to find θ when
(b) $x = 10$; **(c)** $x = 12$; and **(d)** $x = 15$.

63. A picture w feet high is placed on a wall with its base z feet above the level of the eye of an observer. If the observer is x feet from the wall and θ is the radian measure of the angle subtended at the observer's eye

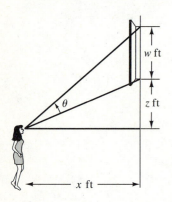

by the picture, show that

$$\theta = \cot^{-1}\frac{x}{w + z} - \cot^{-1}\frac{x}{z}$$

64. In Exercise 63, show that another equation defining θ in terms of w, z, and x is

$$\theta = \tan^{-1}\left(\frac{wx}{x^2 + wz + z^2}\right)$$

In Exercises 65 through 67, an algebraic expression in the variable x is put in the form of a trigonometric expression in the variable θ by a substitution involving an inverse trigonometric function. This kind of substitution is sometimes required for a computational technique in calculus, called indefinite integration.

65. Show that the substitution $\theta = \sin^{-1}(\frac{1}{3}x)$ in the expression $\sqrt{9 - x^2}$ yields $3\cos\theta$ and explain how the domain of θ is applied.

66. Show that the substitution $\theta = \tan^{-1}(\frac{1}{2}x)$ in the expression $\sqrt{x^2 + 4}$ yields $2\sec\theta$ and explain how the domain of θ is applied.

67. Show that the substitution $\theta = \sec^{-1}(\frac{1}{5}x)$ in the expression $\sqrt{x^2 - 25}$ yields $5\tan\theta$ and explain how the domain of θ is applied.

68. Prove that the range of $\csc^{-1}$ is $(-\pi, -\frac{1}{2}\pi] \cup (0, \frac{1}{2}\pi]$.

9.6 TRIGONOMETRIC EQUATIONS

GOALS

1. **Solve trigonometric equations involving trigonometric functions of real numbers.**
2. **Solve trigonometric equations involving trigonometric functions of angles.**
3. **Solve word problems having a trigonometric equation as a mathematical model.**

In the first four sections of this chapter we were concerned with trigonometric identities, which are equations satisfied by all real numbers, or angles, for which each member of the equation is defined. We now discuss conditional trigonometric equations, which are satisfied by only particular values of the variable.

The methods used to find solutions of trigonometric equations are similar to those used to solve algebraic equations. However, here we first solve for a particular trigonometric function value. As with algebraic equations, we obtain a succession of equivalent equations until we have one for which the trigonometric function value is apparent. Trigonometric identities are helpful in securing equivalent equations.

▷ ILLUSTRATION 1

To solve the equation

$$2 \sin x - 1 = 0$$

for $0 \leq x \leq \frac{1}{2}\pi$, we first solve for $\sin x$:

$$2 \sin x = 1$$
$$\sin x = \tfrac{1}{2}$$
$$x = \sin^{-1} \tfrac{1}{2}$$
$$x = \tfrac{1}{6}\pi$$ ◀

▶ EXAMPLE 1 *Solving a Trigonometric Equation*

Find the solution of the equation if $0 \leq x \leq \frac{1}{2}\pi$.

(a) $\tan^2 x - 3 = 0$ **(b)** $2 \cos^2 x - 1 = 0$

Solution

(a) $\tan^2 x - 3 = 0$
$$\tan^2 x = 3$$

Because $0 \leq x \leq \frac{1}{2}\pi$, $\tan x > 0$; thus

$$\tan x = \sqrt{3}$$
$$x = \tan^{-1} \sqrt{3}$$
$$x = \tfrac{1}{3}\pi$$

(b) $2 \cos^2 x - 1 = 0$
$$\cos^2 x = \tfrac{1}{2}$$

Because $0 \leq x \leq \frac{1}{2}\pi$, $\cos x > 0$; thus

$$\cos x = \frac{1}{\sqrt{2}}$$
$$x = \cos^{-1} \frac{1}{\sqrt{2}}$$
$$x = \tfrac{1}{4}\pi$$ ◀

▶ EXAMPLE 2 *Solving a Trigonometric Equation*

Solve the equation

$$\cot^2 x - 1 = 0$$

if $0 \leq x \leq \pi$.

Solution   We first solve for $\cot x$.

$$\cot^2 x = 1$$
$$\cot x = \pm 1$$

The value of x in $[0, \pi]$ for which $\cot x = 1$ is $\frac{1}{4}\pi$. The value of x in $[0, \pi]$ for which $\cot x = -1$ is $\frac{3}{4}\pi$. Therefore the solution set of the given equation is $\{\frac{1}{4}\pi, \frac{3}{4}\pi\}$.   ◀

As with algebraic equations, check your answers on your graphics calculator by first writing the equation in the form $f(x) = 0$ and then plotting the graph of f. The x intercepts of the graph are the solutions of the equation.

▶ **EXAMPLE 3**   *Solving a Trigonometric Equation*

Solve the equation

$$2 \sin^2 t - \cos t - 1 = 0$$

if $0 \le t < 2\pi$. Check on a graphics calculator.

Solution   We first replace $\sin^2 t$ by $1 - \cos^2 t$ and then solve the resulting quadratic equation by factoring the left side.

$$2(1 - \cos^2 t) - \cos t - 1 = 0$$
$$2 - 2 \cos^2 t - \cos t - 1 = 0$$
$$2 \cos^2 t + \cos t - 1 = 0$$
$$(2 \cos t - 1)(\cos t + 1) = 0$$

$$2 \cos t - 1 = 0 \qquad \cos t + 1 = 0$$
$$\cos t = \tfrac{1}{2} \qquad\qquad \cos t = -1$$

The values of t in $[0, 2\pi)$ for which $\cos t = \frac{1}{2}$ are $\frac{1}{3}\pi$ and $\frac{5}{3}\pi$. The value of t in $[0, 2\pi)$ for which $\cos t = -1$ is π. Therefore the solution set of the given equation is $\{\frac{1}{3}\pi, \pi, \frac{5}{3}\pi\}$.

Figure 1 shows the graph of

$$f(t) = 2 \sin^2 t - \cos t - 1$$

in the $[0, 2\pi]$ by $[-2, 2]$ viewing rectangle on our graphics calculator. We check our solutions by using trace and zoom-in.   ◀

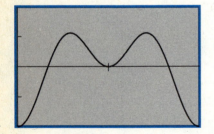

$[0, 2\pi]$ by $[-2, 2]$
$f(t) = 2 \sin^2 t - \cos t - 1$
FIGURE 1

▶ **EXAMPLE 4**   *Solving a Trigonometric Equation*

Solve the equation

$$8 \sin^2 \theta + 6 \sin \theta - 9 = 0$$

if $0° \le \theta < 360°$. Check on a graphics calculator.

Solution   We factor the left side and equate each factor to zero.

$$(2 \sin \theta + 3)(4 \sin \theta - 3) = 0$$

$$2 \sin \theta + 3 = 0 \qquad 4 \sin \theta - 3 = 0$$

$$\sin \theta = -\tfrac{3}{2} \qquad \sin \theta = \tfrac{3}{4}$$

$$\sin \theta = 0.75$$

The solution set of $\sin \theta = -\tfrac{3}{2}$ is $\varnothing$ because $|\sin \theta| \leq 1$. For the equation $\sin \theta = 0.75$, we have both a first-quadrant and a second-quadrant angle. The first-quadrant angle is $\sin^{-1} 0.75 \approx 48.6°$, obtained from a calculator. The second-quadrant angle is $180° - 48.6° = 131.4°$. Thus the solution set is $\{48.6°, 131.4°\}$.

On our graphics calculator, in the degree mode, we plot the graph of

$$f(\theta) = 8 \sin^2 \theta + 6 \sin \theta - 9$$

in the $[0°, 150°]$ by $[-10, 10]$ viewing rectangle and obtain the graph shown in Figure 2. Using trace and zoom-in, we check our solutions.   ◀

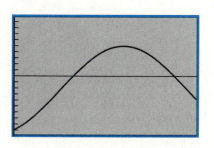

$[0°, 150°]$ by $[-10, 10]$
$f(\theta) = 8 \sin^2 \theta + 6 \sin \theta - 9$
FIGURE 2

▶ **EXAMPLE 5** *Solving a Trigonometric Equation*

Solve the equation

$$\sec^2 x - \tan x = 1$$

if $0 \leq x < 2\pi$.

Solution   We use the identity $\sec^2 x = 1 + \tan^2 x$ and solve the resulting equation by factoring.

$$(1 + \tan^2 x) - \tan x = 1$$

$$\tan^2 x - \tan x = 0$$

$$\tan x (\tan x - 1) = 0$$

$$\tan x = 0 \qquad \qquad \tan x - 1 = 0$$

$$x = 0 \quad x = \pi \qquad \qquad \tan x = 1$$

$$x = \tfrac{1}{4}\pi \qquad x = \tfrac{5}{4}\pi$$

The solution set is $\{0, \tfrac{1}{4}\pi, \pi, \tfrac{5}{4}\pi\}$.   ◀

▷ **ILLUSTRATION 2**

Suppose we wish to determine all real-number solutions of the equation of Example 3. That is, we have

$$2 \sin^2 t - \cos t - 1 = 0 \qquad t \in R$$

We proceed as in Example 3, and for $t \in [0, 2\pi)$ the solutions are $\tfrac{1}{3}\pi, \tfrac{5}{3}\pi$, and π. Because the period of cosine is 2π, all real-number solutions are obtained by adding to each of these numbers $k \cdot 2\pi$, where k is any integer. Thus the solution set for $t \in R$ is

$$\{t \mid t = \tfrac{1}{3}\pi + k \cdot 2\pi\} \cup \{t \mid t = \tfrac{5}{3}\pi + k \cdot 2\pi\} \cup \{t \mid t = \pi + k \cdot 2\pi\}$$

where $k \in J$.   ◄

▷ **ILLUSTRATION 3**

To obtain all angles θ that are solutions of the equation of Example 4, we first find those θ for which

$$8 \sin^2 \theta + 6 \sin \theta - 9 = 0$$

and $0° \le \theta < 360°$. In Example 4 they were found to be $48.6°$ and $131.4°$. Thus if θ can be any angle, the solution set is

$$\{\theta \mid \theta \approx 48.6° + k \cdot 360°\} \cup \{\theta \mid \theta \approx 131.4° + k \cdot 360°\} \quad k \in J$$

◄

▷ **ILLUSTRATION 4**

The equation of Example 5 is

$$\sec^2 x - \tan x = 1$$

and if $x \in [0, 2\pi)$, the solution set is $\{0, \tfrac{1}{4}\pi, \pi, \tfrac{5}{4}\pi\}$. If there is no restriction on x, then because the period of the tangent is π, all the solutions of this equation are in the set

$$\{x \mid x = k\pi\} \cup \{x \mid x = \tfrac{1}{4}\pi + k\pi\} \quad k \in J$$ ◄

▶ **EXAMPLE 6** *Finding All Solutions of a Trigonometric Equation*

Find all solutions of the equation

$$4 \sin 2\theta - 3 \cos \theta = 0$$

if θ is any angle. Express the solutions in degree measurement.

Solution   We begin by applying the sine double-measure identity.

$$4(2 \sin \theta \cos \theta) - 3 \cos \theta = 0$$
$$\cos \theta(8 \sin \theta - 3) = 0$$
$$\cos \theta = 0 \qquad 8 \sin \theta - 3 = 0$$
$$\sin \theta = \tfrac{3}{8}$$
$$\sin \theta = 0.375$$

The solution set of $\cos \theta = 0$ is $\{\theta \mid \theta = 90° + k \cdot 180°\}$, $k \in J$. We obtain a first-quadrant angle for which $\sin \theta = 0.375$ on a calculator and get $\theta \approx 22.0°$. A second-quadrant angle is $180° - 22.0°$, or $158.0°$. Therefore the solution set of $\sin \theta = 0.375$ is

$$\{\theta \mid \theta \approx 22.0° + k \cdot 360°\} \cup \{\theta \mid \theta \approx 158.0° + k \cdot 360°\} \qquad k \in J$$

The solution set of the given equation is then

$$\{\theta \mid \theta = 90° + k \cdot 180°\} \cup \{\theta \mid \theta \approx 22.0° + k \cdot 360°\} \cup \{\theta \mid \theta \approx 158.0° + k \cdot 360°\}$$

where $k \in J$. ◀

▶ **EXAMPLE 7** *Solving a Word Problem Having a Trigonometric Equation as a Mathematical Model*

An electric generator produces a 30-cycle alternating current described by the equation

$$I(t) = 40 \sin 60\pi(t - \tfrac{7}{72})$$

where $I(t)$ amperes is the current at t seconds. Find the smallest positive value of t for which the current is 20 amperes and check the answer on a graphics calculator.

Solution If $I(t)$ is replaced by 20 in the equation, we have

$$40 \sin 60\pi(t - \tfrac{7}{72}) = 20$$
$$\sin 60\pi(t - \tfrac{7}{72}) = \tfrac{1}{2}$$

$$60\pi(t - \tfrac{7}{72}) = \tfrac{1}{6}\pi + k \cdot 2\pi \quad \text{or} \quad 60\pi(t - \tfrac{7}{72}) = \tfrac{5}{6}\pi + k \cdot 2\pi \quad k \in J$$

$$t - \tfrac{7}{72} = \tfrac{1}{360} + k \cdot \tfrac{1}{30} \qquad\qquad t - \tfrac{7}{72} = \tfrac{1}{72} + k \cdot \tfrac{1}{30}$$

$$t = \tfrac{1}{10} + k \cdot \tfrac{1}{30} \qquad\qquad\qquad t = \tfrac{1}{9} + k \cdot \tfrac{1}{30}$$

$$t = \tfrac{3}{30} + k \cdot \tfrac{1}{30} \qquad\qquad\qquad t = \tfrac{10}{90} + k \cdot \tfrac{3}{90}$$

The smallest positive value of t obtained from $t = \tfrac{3}{30} + k \cdot \tfrac{1}{30}$ occurs when $k = -2$; this value of t is $\tfrac{1}{30}$. The smallest positive value of t obtained from $t = \tfrac{10}{90} + k \cdot \tfrac{3}{90}$ occurs when $k = -3$; this value of t is $\tfrac{1}{90}$.

Conclusion: The smallest positive value of t for which the current is 20 amperes is $\tfrac{1}{90}$.

Figure 3 shows the graph of the given equation as well as the line $f(t) = 20$ in the viewing rectangle $[0, 0.1]$ by $[-50, 50]$ on our graphics calculator. By using trace and zoom-in, the t coordinate of the point of intersection of the curve and the line is 0.011 to two significant digits. This result agrees with our answer $\tfrac{1}{90}$. ◀

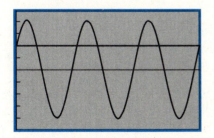

$[0, 0.1]$ by $[-50, 50]$

$I(t) = 40 \sin 60\pi\left(t - \dfrac{7}{72}\right)$ and $f(t) = 20$

FIGURE 3

▶ **EXAMPLE 8** *Solving a Trigonometric Equation by Applying the Quadratic Formula*

Find the solutions of the equation

$$3 \tan t - \cot t - 5 = 0$$

if $0 \leq t < 2\pi$.

Solution By making the substitution $\cot t = \dfrac{1}{\tan t}$, we obtain an equivalent equation containing only one function.

$$3 \tan t - \cot t - 5 = 0$$

$$3 \tan t - \frac{1}{\tan t} - 5 = 0$$

$$3 \tan^2 t - 1 - 5 \tan t = 0$$

$$3 \tan^2 t - 5 \tan t - 1 = 0$$

This is a quadratic equation, which we solve by applying the quadratic formula.

$$\tan t = \frac{-b \pm \sqrt{b^2 - 4ac}}{2a}$$

$$= \frac{-(-5) \pm \sqrt{(-5)^2 - 4(3)(-1)}}{2(3)}$$

$$= \frac{5 \pm \sqrt{37}}{6}$$

$$\approx \frac{5 \pm 6.083}{6}$$

$$\tan t \approx \frac{5 + 6.083}{6} \qquad \tan t \approx \frac{5 - 6.083}{6}$$

$$\approx 1.847 \qquad\qquad \approx -0.180$$

For $\tan t \approx 1.847$ we find, from a calculator, $\tan^{-1} 1.847 \approx 1.07$. Thus one solution is $t \approx 1.07$. Because t is in $[0, 2\pi)$, another solution is $t \approx 3.14 + 1.07$; that is, $t \approx 4.21$. For $\tan t \approx -0.180$, we first determine from a calculator a $\bar{t}$ for which $\bar{t} \approx \tan^{-1} 0.180$; it is $\bar{t} \approx 0.18$. There are two values of t in $[0, 2\pi)$ for which $\tan t \approx -0.180$; they are $t \approx 3.14 - 0.18$, or $t \approx 2.96$, and $t \approx 6.28 - 0.18$, or $t \approx 6.10$. Thus the solution set of the given equation is $\{1.07, 2.96, 4.21, 6.10\}$. ◀

The equation in the next example involves a trigonometric function of a multiple measure.

▶ **EXAMPLE 9** *Solving a Trigonometric Equation Involving a Function of a Multiple Measure*

Find all values of x in $[0, 2\pi)$ for which

$$\tan 3x = 1$$

Solution   If $0 \le x < 2\pi$, then $0 \le 3x < 6\pi$. Therefore, to find all the values of x in $[0, 2\pi)$ that are solutions of the equation, we must first determine all the values of $3x$ in $[0, 6\pi)$ for which

$$\tan 3x = 1$$

They are

$$3x = \tfrac{1}{4}\pi \quad 3x = \tfrac{5}{4}\pi \quad 3x = \tfrac{9}{4}\pi \quad 3x = \tfrac{13}{4}\pi \quad 3x = \tfrac{17}{4}\pi \quad 3x = \tfrac{21}{4}\pi$$

Dividing by 3 on both sides of each of these equations, we obtain all the solutions of the given equation in $[0, 2\pi)$. They are

$$x = \tfrac{1}{12}\pi \quad x = \tfrac{5}{12}\pi \quad x = \tfrac{3}{4}\pi \quad x = \tfrac{13}{12}\pi \quad x = \tfrac{17}{12}\pi \quad x = \tfrac{7}{4}\pi \quad ◀$$

▷ **ILLUSTRATION 5**

To determine all solutions of the equation in Example 9, we first obtain the value of $3x$ in $[0, \pi)$ that satisfies the equation. It is

$$3x = \tfrac{1}{4}\pi$$

Because the period of the tangent function is π, all solutions of the equation are given by

$$3x = \tfrac{1}{4}\pi + k \cdot \pi \quad k \in J$$
$$\Leftrightarrow \quad x = \tfrac{1}{12}\pi + k \cdot \tfrac{1}{3}\pi \quad k \in J \quad ◀$$

▶ **EXAMPLE 10** *Solving a Trigonometric Equation Involving Functions of Multiple Measure*

Find the solutions of the equation

$$2 \sin 2\theta \cos 3\theta + \cos 3\theta = 0$$

if $0° \le \theta < 360°$.

Solution   Factoring out $\cos 3\theta$ on the left side of the equation, we have

$$\cos 3\theta(2 \sin 2\theta + 1) = 0$$
$$\cos 3\theta = 0 \quad 2 \sin 2\theta + 1 = 0$$
$$\sin 2\theta = -\tfrac{1}{2}$$

We first find the solutions of the equation $\cos 3\theta = 0$. Because $0° \leq \theta < 360°$, then $0° \leq 3\theta < 1080°$. The solutions are

$$3\theta = 90° \quad 3\theta = 270° \quad 3\theta = 450° \quad 3\theta = 630° \quad 3\theta = 810° \quad 3\theta = 990°$$

Therefore

$$\theta = 30° \quad \theta = 90° \quad \theta = 150° \quad \theta = 210° \quad \theta = 270° \quad \theta = 330°$$

We now find the solutions of the equation $\sin 2\theta = -\frac{1}{2}$. Because $0° \leq \theta < 360°$, then $0° \leq 2\theta < 720°$. The solutions are

$$2\theta = 210° \quad 2\theta = 330° \quad 2\theta = 570° \quad 2\theta = 690°$$

Hence

$$\theta = 105° \quad \theta = 165° \quad \theta = 285° \quad \theta = 345°$$

The solution set of the given equation then is

$$\{30°, 90°, 105°, 150°, 165°, 210°, 270°, 285°, 330°, 345°\} \qquad \blacktriangleleft$$

EXERCISES 9.6

In Exercises 1 through 10, find the solution of the equation if $0 \leq x \leq \frac{1}{2}\pi$.

1. (a) $\sin x - 1 = 0$ **(b)** $2 \cos x - 1 = 0$

2. (a) $\cos x - 1 = 0$ **(b)** $\tan x - 1 = 0$

3. (a) $2 \sin^2 x - 1 = 0$ **(b)** $3 \cot^2 x - 1 = 0$

4. (a) $\cot^2 x - 3 = 0$ **(b)** $4 \cos^2 x - 3 = 0$

5. (a) $\sec^2 x - 1 = 0$ **(b)** $\csc^2 x - 2 = 0$

6. (a) $\csc^2 x - 1 = 0$ **(b)** $\sec^2 x - 2 = 0$

7. (a) $\sin x \cos x = 0$ **(b)** $\cos x \cot x = 0$

8. (a) $\tan x \sec x = 0$ **(b)** $\sin x \tan x = 0$

9. (a) $4 \sin^3 x - 3 \sin x = 0$
     (b) $\tan^3 x - \tan x = 0$

10. (a) $4 \cos^3 x - \cos x = 0$
     (b) $3 \csc^3 x - 4 \csc x = 0$

In Exercises 11 through 18, find the solutions of the equation if $0 \leq t < 2\pi$.

11. (a) $4 \sin^2 t - 1 = 0$ **(b)** $\tan^2 t - 1 = 0$

12. (a) $\sec^2 t - 4 = 0$ **(b)** $3 \tan^2 t - 1 = 0$

13. (a) $2 \cos^2 t + 3 \cos t + 1 = 0$
     (b) $2 \sin^2 t - 5 \sin t - 3 = 0$

14. (a) $2 \sin^2 t + \sin t - 1 = 0$
     (b) $2 \cos^2 t + 3 \cos t - 2 = 0$

15. (a) $\sec^2 t - 2 \tan t = 0$ **(b)** $\tan^2 t - \sec t = 1$

16. (a) $\tan t + \cot t + 2 = 0$ **(b)** $\cot^2 t + \csc t = 1$

17. (a) $\sin t + \cos t = 0$ **(b)** $2 \sin t + \sec t = 0$

18. (a) $\tan t - \cot t = 0$ **(b)** $\tan t + \sec t = 0$

In Exercises 19 through 22, find the solutions of the equation if $0 \leq t < 2\pi$. Check on your graphics calculator.

19. $\sin^2 t + 5 \cos t + 2 = 0$

20. $3 \sec^2 t + \tan t - 5 = 0$

21. $3 \tan^2 t - \tan t - 3 = 0$

22. $10 \cos^2 t - 4 \cos t - 5 = 0$

In Exercises 23 through 28, find all angles θ that are solutions of the equation if $0° \leq \theta < 360°$. Check on your graphics calculator.

23. $9 \cos^2 \theta + 6 \cos \theta - 8 = 0$

24. $5 \sin^2 \theta - 11 \sin \theta + 2 = 0$

25. $2 \sin 2\theta - 3 \sin \theta = 0$

26. $2 \cos 2\theta + 3 \cos \theta + 1 = 0$

27. $\tan \theta - 3 \cot \theta = 2$

28. $2 \tan^2 \theta - \sec \theta = 1$

In Exercises 29 through 34, find all real-number solutions of the equations of the indicated exercise.

29. Exercise 11   **30.** Exercise 14

31. Exercise 15   **32.** Exercise 16

33. Exercise 19   **34.** Exercise 20

In Exercises 35 through 40, find all angles θ that are solutions of the equation of the indicated exercise. Express the solutions in degree measurement.

35. Exercise 23   **36.** Exercise 24

37. Exercise 25   **38.** Exercise 26

39. Exercise 27   **40.** Exercise 28

In Exercises 41 through 48, find all values of x in $[0, 2\pi)$ that are solutions of the equation.

41. (a) $\tan 3x = -1$ **(b)** $\sin 3x = \frac{1}{2}$

42. (a) $\cot 3x = \sqrt{3}$ **(b)** $\cos 3x = -\frac{1}{2}$

43. (a) $\cot 4x = -\sqrt{3}$ **(b)** $\sec 5x = 2$

44. (a) $\cot 4x = -1$ **(b)** $\csc 5x = \sqrt{2}$

45. (a) $\sin^2 2x = 1$ **(b)** $\cos \frac{1}{2}x = -\frac{1}{2}$

46. (a) $\cos^2 2x = \frac{1}{4}$ **(b)** $\sin \frac{1}{3}x = 1$

47. (a) $\tan^2 \frac{1}{2}x = 3$ **(b)** $\csc \frac{1}{3}x = 2$

48. (a) $\sec^2 \frac{1}{2}x = 2$ **(b)** $\cot \frac{1}{2}x = -1$

In Exercises 49 through 52, find all real-number solutions of the equations of the indicated exercise.

49. Exercise 41   **50.** Exercise 42

51. Exercise 45   **52.** Exercise 46

In Exercises 53 through 56, find all angles θ that are solutions of the equation if $0° \leq \theta < 360°$.

53. $2 \cos 3\theta \sin 2\theta - \sin 2\theta = 0$

54. $2 \sin 3\theta \cos 3\theta = -1$

55. $\tan 2\theta + 5 = 3 \sec^2 2\theta$

56. $\cot^2 4\theta - 1 = \csc 4\theta$

In Exercises 57 through 62, do the following: (a) Plot the graph of the function on each side of the equation and determine if the equation is an identity. (b) If the equation is an identity, prove it algebraically. If the equation is not an identity, solve it for all values of x in $[0, 2\pi)$ and check your solutions on your graphics calculator.

57. $\cot x + \tan x = \csc x \sec x$

58. $1 - \tan^2 x = \tan x \sec x$

59. $\sin 3x - \sin x = \cos 2x$

60. $\cos 4x + \cos 2x = 2 \cos x \cos 3x$

61. $\sin^2 4x + \cos^2 2x = 1$

62. $\tan 2x - \tan x = \tan x \sec 2x$

Exercises 63 through 68 are word problems having a trigonometric equation as a mathematical model. Be sure to write a conclusion.

63. In an electric circuit, the electromotive force is $E(t)$ volts, where $E(t) = 2 \cos 50\pi t$. Find the smallest positive value of t for which the electromotive force is **(a)** 2 volts and **(b)** -2 volts. **(c)** Check your answers for parts (a) and (b) on your graphics calculator.

64. Do Exercise 63 if $E(t) = 4 \sin 120\pi t$.

65. A weight suspended from a spring is vibrating vertically according to the equation

$$f(t) = 10 \sin \tfrac{3}{4} \pi(t - 3)$$

where $f(t)$ centimeters is the directed distance of the weight from its central position at t seconds and the positive direction is upward. Determine the smallest positive value of t for which the displacement of the weight above its central position is **(a)** 5 cm and **(b)** 6 cm. **(c)** Check your answers for parts (a) and (b) on your graphics calculator.

66. Do Exercise 65 if $f(t) = 10 \cos \frac{5}{6} \pi (t - \frac{2}{5})$.

67. A weight suspended from a spring is vibrating vertically according to the equation

$$y = 2 \sin 4\pi(t + \tfrac{1}{8})$$

where y centimeters is the directed distance of the weight from its central position t seconds after the start of the motion and the positive direction is upward. **(a)** Solve the equation for t. **(b)** Use the equation in part (a) to determine the smallest three positive values of t for which the weight is 1 cm above its central position. **(c)** Check your answers for part (b) on your graphics calculator.

68. A 60-cycle alternating current is described by the equation $x = 20 \sin 120\pi(t - \frac{11}{720})$ where x amperes is the current at t seconds. **(a)** Solve the equation for t. **(b)** Use the equation in part (a) to determine the smallest three positive values of t for which the current is 10 amperes. **(c)** Check your answers for part (b) on your graphics calculator.

69. Explain why the equation $\sin^2 x - \frac{1}{4} = 0$ has four solutions in the interval $[-\pi, \pi]$ whereas the equation $x^2 - (\sin^{-1} \frac{1}{2})^2 = 0$ has only two solutions in that interval. Include the solutions in your explanation.

CHAPTER 9 REVIEW

▶ LOOKING BACK

9.1 A step-by-step summary of suggestions for proving trigonometric identities was highlighted and applied to the examples. Certain exercises and examples required determining whether a particular equation was an identity.

9.2 We first obtained the cosine difference identity, derived by using the distance formula applied to points on the unit circle. The cosine sum identity followed from this identity. The sine sum and sine difference identities were derived from the corresponding ones for the cosine by applying cofunction identities. The tangent sum and difference identities then followed by using the fundamental identity expressing the tangent as the ratio of sine to cosine. We applied the sum and difference identities to compute exact function values and to prove other identities. We also proved the theorem that enables us to write an expression of the form $A \sin bt + B \cos bt$ as $a \sin(bt + c)$.

9.3 As special cases of the sine, cosine, and tangent sum identities, we derived the double-measure identities. We then used these identities to derive other multiple-measure identities. We showed how the identities for $\sin^2 x$ and $\cos^2 x$ in terms of $\cos 2x$ are used to express an even power of sine or cosine in terms of cosine function values with no exponent greater than 1, a procedure useful in calculus. Another substitution helpful in calculus, mentioned here, is the substitution $z = \tan \frac{1}{2}x$, which reduces a rational function of $\sin x$ and $\cos x$ to a rational function of z.

9.4 Identities for the product of sine and cosine were discussed. Although not used as often as those in Sections 9.2 and 9.3, they are sometimes necessary in calculus

to write a trigonometric expression involving the product of sine and cosine functions as a sum or difference. Identities for the sum and difference of sine and cosine were also treated. We applied the sum sine identity in an example to express the sum of two sine functions, having different periods but the same amplitude, as the product of sine and cosine functions. The graph in that example demonstrated the principle of superposition having applications in various fields including television, radio, and telephone communications.

9.5 Definitions and graphs of the inverse trigonometric functions were featured. We found exact function values involving these functions and proved identities pertaining to them. We stressed that the unusual choice of the range of the inverse secant function was made so that certain computations in calculus are simplified. An example showed how the inverse tangent function can be used to determine the position of an observer that gives the "best view" of an object.

9.6 To solve trigonometric equations involving both real numbers and angles, we applied the inverse trigonometric functions. We showed how to obtain solutions of trigonometric equations when the variable is restricted to a particular interval as well as how to obtain all solutions of an equation when there is no restriction on the domain of the unknown. We included some equations involving trigonometric functions of multiple measure. The discussions in this section, as well as in other sections of the chapter were augmented by applications in the examples and exercises. These applications included periodic phenomena, solving triangles, and calculus-related problems.

▶ REVIEW EXERCISES

In Exercises 1 through 28, prove the identity.

1. $\sin \theta \sec \theta = \tan \theta$

2. $\dfrac{\sin x}{\tan x} = \cos x$

3. $\cos^2 y(\tan^2 y + 1) = 1$

4. $\sec^2 \alpha \cot^2 \alpha = 1 + \cot^2 \alpha$

5. $\sin x(\csc x - \sin x) = \cos^2 x$

6. $\csc x(\csc x - \sin x) = \cot^2 x$

7. $\dfrac{1 - \tan^2 x}{1 + \tan^2 x} = 1 - 2 \sin^2 x$

8. $\dfrac{1 - \cot^2 \theta}{1 + \cot^2 \theta} = 2 \cos^2 \theta - 1$

9. $\dfrac{\sin \alpha}{\csc \alpha - \cot \alpha} = 1 + \cos \alpha$

10. $\dfrac{\cos y}{\sec y - \cos y} = \cot^2 y$

11. $\dfrac{\sec t - 1}{\cos t} = \dfrac{\tan^2 t}{1 + \cos t}$

12. $\dfrac{\tan x}{\sec x - \cos x} = \dfrac{\sec x}{\tan x}$

13. $\sec^2 \theta - \csc^2 \theta = \tan^2 \theta - \cot^2 \theta$

14. $\sec^4 \beta - \tan^4 \beta = 1 + 2 \tan^2 \beta$

15. $\dfrac{\sin x}{1 + \cos x} = \csc x - \cot x$

16. $\dfrac{\sin x}{1 - \cos x} = \csc x + \cot x$

17. $\dfrac{\cos y}{1 - \tan y} + \dfrac{\sin y}{1 - \cot y} = \sin y + \cos y$

18. $\dfrac{\sec^2 \theta + \csc^2 \theta}{\sec^2 \theta - \csc^2 \theta} = \dfrac{\tan^2 \theta + 1}{\tan^2 \theta - 1}$

19. $\dfrac{\sin(\alpha + \beta)}{\sin(\alpha - \beta)} = \dfrac{\tan \alpha + \tan \beta}{\tan \alpha - \tan \beta}$

20. $\dfrac{\cos(\alpha + \beta)}{\cos(\alpha - \beta)} = \dfrac{\cot \alpha - \tan \beta}{\cot \alpha + \tan \beta}$

21. $\tan(\theta - \tfrac{1}{4}\pi) = \dfrac{\tan \theta - 1}{\tan \theta + 1}$

22. $\sin(\tfrac{1}{4}\pi + x)\sin(\tfrac{1}{4}\pi - x) = \tfrac{1}{2} \cos 2x$

23. $\tan \tfrac{1}{2}(x - y) = \dfrac{\sin x - \sin y}{\cos x + \cos y}$

24. $\tan \tfrac{1}{2}(x + y) = \dfrac{\sin x + \sin y}{\cos x + \cos y}$

25. $2 \cos 2\theta \cos \theta - \cos 3\theta = \cos \theta$

26. $2 \cos 2\theta \sin \theta - \sin 3\theta = -\sin \theta$

27. $\dfrac{\sin x + \sin 3x + \sin 5x}{\cos x + \cos 3x + \cos 5x} = \tan 3x$

28. $\dfrac{\sin x + \sin 2x + \sin 3x}{\cos x + \cos 2x + \cos 3x} = \tan 2x$

29. Compute the exact value of $\cos 75°$ in two ways: **(a)** use the cosine sum identity; **(b)** use a half-angle identity.

30. Compute the exact value of $\sin 105°$ in two ways: **(a)** use the sine sum identity; **(b)** use a half-angle identity.

31. Compute the exact value of $\tan \tfrac{7}{12}\pi$ in two ways: **(a)** use the tangent sum identity; **(b)** use a half-measure identity.

32. If $\tan x = \tfrac{4}{3}$, $0 < x < \tfrac{1}{2}\pi$, and $\tan y = -\tfrac{2}{3}$, $\tfrac{1}{2}\pi < y < \pi$, find; **(a)** $\tan(x + y)$; **(b)** $\tan(x - y)$; **(c)** the quadrant containing $x + y$; **(d)** the quadrant containing $x - y$.

33. If $\sin \alpha = \tfrac{5}{13}$ with α in the second quadrant and $\cos \beta = -\tfrac{7}{25}$ with β in the third quadrant, find: **(a)** $\sin(\alpha + \beta)$; **(b)** $\cos(\alpha + \beta)$; **(c)** $\sin(\alpha - \beta)$; **(d)** $\cos(\alpha - \beta)$; **(e)** the quadrant containing $\alpha + \beta$; **(f)** the quadrant containing $\alpha - \beta$.

34. If $\tan t = \tfrac{24}{7}$ and $\cos t > 0$, find $\sin \tfrac{1}{2}t$.

35. If $\cos t = -\tfrac{3}{5}$ and $\sin t < 0$, find $\tan \tfrac{1}{2}t$.

36. Find the exact value: **(a)** $\sin \tfrac{9}{8}\pi \cos \tfrac{3}{8}\pi$; **(b)** $\sin 195° - \cos 165°$.

 In Exercises 37 and 38, write the expression in terms of cosine values with no exponent greater than 1.

37. $\cos^2 2t \sin^4 2t$ 38. $\sin^4 3t$

 In Exercises 39 and 40, use the substitution $z = \tan \tfrac{1}{2}x$ to write the expression as a rational function of z.

39. $3 \cos x - 2 \sin x + 3$ 40. $\dfrac{2 + \cos x}{\sin 2x}$

In Exercises 41 through 44, sketch the graph of the function. Check your graph on your graphics calculator.

41. $f(x) = \tfrac{1}{2} \cos^{-1} 2x$ 42. $g(x) = 2 \sin^{-1} \tfrac{1}{2}x$

43. $g(x) = \tan^{-1} 3x$ 44. $f(x) = 3 \cot^{-1} x$

In Exercises 45 through 54, find the solutions of the equation if $0 \le t < 2\pi$.

45. **(a)** $2 \cos t - 1 = 0$ **(b)** $\tan^2 t - 1 = 0$

46. **(a)** $\cot t + 1 = 0$ **(b)** $2 \sin^2 t - 1 = 0$

47. $2 \sin^2 t + 3 \sin t - 2 = 0$

48. $\tan^2 t - 3 \sec t + 3 = 0$

49. $\tan t - 2 = 3 \cot t$

50. $6 \cos^2 t - \sin t - 4 = 0$

51. $4 \cos^2 3t - 3 = 0$ 52. $4 \sin^2 2t - 1 = 0$

53. $\cot^2 2t = 4$ 54. $\tan^2 3t = 2$

In Exercises 55 through 58, find all real-number solutions of the equations of the indicated exercise.

55. Exercise 45   56. Exercise 46

57. Exercise 49   58. Exercise 50

In Exercises 59 through 64, find all angles θ that are solutions of the equation if $0° \leq \theta < 360°$.

59. $\tan \theta - \cot \theta = 0$

60. $2 \cos \theta + \sec \theta - 3 = 0$

61. $2 \sin^2 \frac{1}{2} \theta = 1$

62. $\cot^2 \frac{1}{2} \theta = 3$

63. $\tan^2 \theta + 3 \sec \theta + 3 = 0$

64. $2 \sin 2\theta + 1 = \csc 2\theta$

In Exercises 65 through 68, find all angles θ that are solutions of the equation of the indicated exercise. Express the solutions in degree measurement.

65. Exercise 59

66. Exercise 60

67. Exercise 61

68. Exercise 62

In Exercises 69 through 72, do the following: (a) Plot the graph of the function on each side of the equation and determine if the equation is an identity. (b) If the equation is an identity, prove it algebraically. If the equation is not an identity, solve it for all values of x in $[0, 2\pi)$ and check your solutions on your graphics calculator.

69. $\sin 2x - \cos 2x \tan x = \tan x$

70. $\cos 2x = 2 \sin x \cos x$

71. $\cos^2 2x + 3 \sin 2x = 3$

72. $\cot x - \tan x = 2 \cot 2x$

In Exercises 73 and 74, use a calculator to approximate the function value.

73. (a) $\sin^{-1}(0.6032)$ (b) $\sin^{-1}(-0.6032)$
(c) $\cos^{-1}(0.6032)$ (d) $\cos^{-1}(-0.6032)$
(e) $\tan^{-1}(0.6032)$ (f) $\tan^{-1}(-0.6032)$

74. (a) $\sin^{-1}(0.4833)$ (b) $\sin^{-1}(-0.4833)$
(c) $\cos^{-1}(0.4833)$ (d) $\cos^{-1}(-0.4833)$
(e) $\tan^{-1}(0.4833)$ (f) $\tan^{-1}(-0.4833)$

75. Given $x = \arcsin \frac{3}{5}$ and $y = \arccos(-\frac{4}{5})$, find the exact value of each of the following:
(a) $\cos x$ (b) $\sin y$ (c) $\tan x$ (d) $\tan y$

76. Given $x = \cos^{-1} \frac{7}{25}$ and $y = \tan^{-1}(-\frac{7}{24})$, find the exact value of each of the following:
(a) $\sin x$ (b) $\sin y$ (c) $\cos y$ (d) $\tan x$

In Exercises 77 through 82, find the exact function value.

77. (a) $\sin^{-1}(\cos \frac{1}{4} \pi)$ (b) $\sin^{-1}[\cos(-\frac{1}{4} \pi)]$
(c) $\cos^{-1}[\sin(-\frac{1}{4} \pi)]$ (d) $\sin^{-1}(\cos \frac{3}{4} \pi)$
(e) $\cos^{-1}(\sin \frac{3}{4} \pi)$

78. (a) $\cos^{-1}(\sin \frac{1}{6} \pi)$ (b) $\cos^{-1}[\sin(-\frac{1}{6} \pi)]$
(c) $\sin^{-1}[\cos(-\frac{1}{6} \pi)]$ (d) $\cos^{-1}(\sin \frac{5}{6} \pi)$
(e) $\sin^{-1}(\cos \frac{5}{6} \pi)$

79. (a) $\tan^{-1}(\tan \frac{5}{6} \pi)$ (b) $\tan^{-1}(\cot \frac{5}{6} \pi)$
(c) $\tan^{-1}[\tan(-\frac{5}{6} \pi)]$ (d) $\tan^{-1}[\cot(-\frac{5}{6} \pi)]$

80. (a) $\tan^{-1}(\tan \frac{2}{3} \pi)$ (b) $\tan^{-1}(\cot \frac{2}{3} \pi)$
(c) $\tan^{-1}[\tan(-\frac{2}{3} \pi)]$ (d) $\tan^{-1}[\cot(-\frac{2}{3} \pi)]$

81. (a) $\sin[2 \cos^{-1}(-\frac{12}{13})]$
(b) $\tan[\cos^{-1} \frac{3}{5} + \sin^{-1}(-\frac{7}{25})]$

82. (a) $\tan[2 \sin^{-1}(-\frac{24}{25})]$
(b) $\cos[\tan^{-1} \frac{4}{3} - \sin^{-1}(-\frac{5}{13})]$

In Exercises 83 and 84, (a) express f (t) in the form a sin(bt + c), (b) determine the amplitude, period, and phase shift of f, (c) sketch the graph of f, and (d) check your graph in part (c) on your graphics calculator.

83. $f(t) = 1.47 \sin 2t + 2.65 \cos 2t$

84. $f(t) = 2 \sin 4t - 2\sqrt{3} \cos 4t$

In Exercises 85 and 86, use the method of Example 4 in Section 9.4 to sketch the graph of the function on the indicated interval. Check your graph on your graphics calculator.

85. $f(t) = \sin 10\pi t + \sin 12\pi t$; $[0, 2]$

86. $f(t) = 2 \cos 30\pi t - 2 \cos 50\pi t$; $[0, \frac{1}{5}]$

87. Given $f(t) = \sin 41\pi t + \sin 39\pi t$. (a) Express $f(t)$ as the product of sine and cosine functions.
(b) Sketch the graph of f over one-half period of the cosine function found in part (a) and check your graph on your graphics calculator.

88. At a particular point in space two atmospheric waves produce pressures of $F(t)$ dynes per square centimeter and $G(t)$ dynes per square centimeter at t seconds where $F(t) = 0.02 \sin(200\pi t + \frac{2}{3} \pi)$ and $G(t) = 0.04 \sin(200\pi t - \frac{1}{3} \pi)$. Define the sum of F and G by an equation of the form

$$f(t) = a \sin(200\pi t + c)$$

Check your answer by plotting the graphs of $F + G$ and f.

89. Do Exercise 88 if $F(t) = 0.037 \sin(200\pi t + 0.26)$ and $G(t) = 0.024 \sin 200\pi t$.

90. A weight suspended from a spring is vibrating vertically according to the equation

$$f(t) = -4 \sin 10t - 3 \cos 10t$$

where $f(t)$ centimeters is the directed distance of the weight from its central position t seconds after the start of the motion and the positive direction is upward. **(a)** Define $f(t)$ by an equation of the form $f(t) = a \sin(bt + c)$. **(b)** Determine the amplitude, period, and frequency of f. **(c)** Sketch the graph of f. **(d)** Check your graph in part (c) on your graphics calculator.

91. A particle is moving along a straight line according to the equation of motion

$$s = \sin(6t - \tfrac{1}{3}\pi) + \sin(6t + \tfrac{1}{6}\pi)$$

where s centimeters is the directed distance of the particle from the origin at t seconds. **(a)** Show that the motion is simple harmonic by defining s by an equation of the form $s = a \sin(bt + c)$. **(b)** Determine the amplitude and frequency of the motion.

92. Do Exercise 91 if $s = 8 \cos^2 6t - 4$.

93. A weight suspended from a spring is vibrating vertically according to the equation

$$y = 4 \sin 2\pi(t + \tfrac{1}{6})$$

where y centimeters is the directed distance of the weight from its central position t seconds after the start of the motion and the positive direction is upward. **(a)** Solve the equation for t. **(b)** Use the equation in part (a) to determine the smallest two positive values of t for which the displacement of the weight above its central position is 3 cm. **(c)** Check your answers for part (b) on your graphics calculator.

94. In an electric circuit, the electromotive force is $E(t)$ volts where $E(t) = 20 \cos 120\pi t$. Find the smallest positive value of t for which the electromotive force is **(a)** 5 volts and **(b)** -5 volts. **(c)** Check your answers for parts (a) and (b) on your graphics calculator.

95. A picture 5 ft high is placed on a wall with its base 4 ft above the eye level of an observer. Let θ be the radian measure of the angle subtended at the

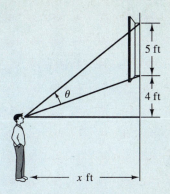

observer's eye by the picture when the observer is x feet from the wall. **(a)** Define θ as a function of x. Find θ when **(b)** $x = 5$, **(c)** $x = 6$, and **(d)** $x = 7$. **(e)** Use your graphics calculator to estimate, to the nearest one-hundredth of a foot, how far the observer should stand to have the "best view" of the picture, that is, so that the angle subtended at the observer's eye by the picture is a maximum.

96. If t is any real number, A, B, and b are constants, and $A > 0$, prove that

$$A \cos bt + B \sin bt = a \cos(bt - c)$$

where $a = \sqrt{A^2 + B^2}$ and $c = \tan^{-1}\left(\dfrac{B}{A}\right)$.

97. Given $f(t) = \sqrt{3} \cos 2t - \sin 2t$. **(a)** Use the formula of Exercise 96 to define $f(t)$ by an equation of the form $f(t) = a \cos(bt - c)$. **(b)** Determine the amplitude, period, and phase shift of f. **(c)** Sketch the graph of f. **(d)** Check your answers by plotting the graphs of both the given equation and your equation in part (a).

98. Do Exercise 97 if $f(t) = 4.83 \cos 4\pi t + 5.07 \sin 4\pi t$.

99. Prove the identity

$$\sin 2x + \sin 2y - \sin 2(x + y) = 4 \sin x \sin y \sin(x + y)$$

100. Prove that $\tan^{-1} x + \tan^{-1}\left(\dfrac{1}{x}\right) = \tfrac{1}{2}\pi$, if $x > 0$.

[*Hint:* Use the reduction formula $\cot u = \tan(\tfrac{1}{2}\pi - u)$.]

Vectors, Parametric Equations, Polar Coordinates, and Complex Numbers

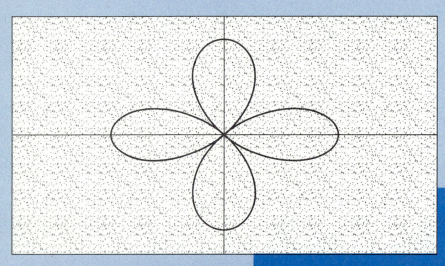

We now apply the trigonometry you learned in the previous three chapters to vectors and some algebraic concepts. Our treatment of vectors is a brief introduction because you will study vectors thoroughly in your calculus course, where vector-valued functions play a prominent role. Graphs of vector-valued functions can be represented by parametric equations, which we also introduce in this chapter.

LOOKING AHEAD

10.1 VECTORS

GOALS

1. **Define a vector.**
2. **Compute the magnitude of a vector.**
3. **Compute the direction angle of a vector.**
4. **Find the sum of two given vectors.**
5. **Find the difference of two given vectors.**
6. **Find the product of a scalar and a vector.**
7. **Express a vector in terms of its magnitude and direction angle.**
8. **Solve word problems involving vectors.**

Applications of mathematics often involve quantities that possess both magnitude and direction. An example of such a quantity is *velocity*. For instance, an airplane's velocity has magnitude (the speed of the airplane) and direction, which determines the course of the airplane. Other examples of such quantities are *force, displacement,* and *acceleration.* Physicists and engineers refer to a directed line segment as a *vector,* and the quantities that have both magnitude and direction are called **vector quantities.** In contrast, a quantity that has magnitude but not direction is called a **scalar quantity.** Examples of scalar quantities are length, area, volume, and speed. The study of vectors is called **vector analysis.**

The approach to vector analysis can be on either a geometric or an analytic basis. If the geometric approach is taken, we first define a directed line segment as a line segment from a point P to a point Q and denote this directed line segment by $\overrightarrow{PQ}$. The point P is called the **initial point,** and the point Q is called the **terminal point.** Two directed line segments $\overrightarrow{PQ}$ and $\overrightarrow{RS}$ are said to be **equal** if they have the same *length* and *direction,* and we write $\overrightarrow{PQ} = \overrightarrow{RS}$ (see Figure 1). The directed line segment $\overrightarrow{PQ}$ is called the **vector** from P to Q. A vector is denoted by a single letter, set in boldface type, such as **A.** In some books, a letter in lightface type, with an arrow above it, is used to indicate a vector, for example $\vec{A}$. When doing your work, you may use that notation or $\underline{A}$ to distinguish the symbol for a vector from the symbol for a real number.

$\overrightarrow{PQ} = \overrightarrow{RS}$

FIGURE 1

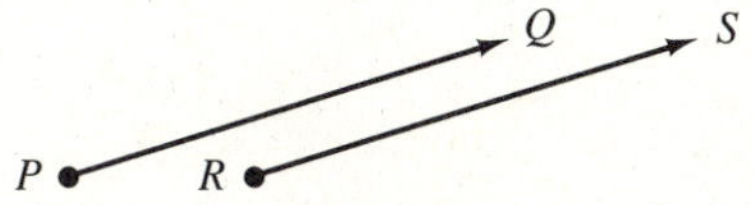

Continuing with the geometric approach to vector analysis, note that if the directed line segment $\overrightarrow{PQ}$ is the vector **A**, and $\overrightarrow{PQ} = \overrightarrow{RS}$, the directed line segment $\overrightarrow{RS}$ is also the vector **A.** Then a vector is considered to remain unchanged if it is moved parallel to itself. With this interpretation of a vector, we can assume for convenience that every vector has its initial point at some fixed reference point. By taking this point as the origin of a rectangular cartesian coordinate system, a vector can be defined analytically in terms of real numbers. Such a definition permits the study of vector analysis

from a purely analytic viewpoint, and it is this approach that is used in calculus. A vector in the plane is denoted by an ordered pair of real numbers. The notation $\langle x, y \rangle$ is used instead of (x, y) to avoid confusing a vector with a point. Following is the formal definition.

DEFINITION **Vector**

A vector in the plane is an ordered pair of real numbers $\langle x, y \rangle$. The numbers x and y are called the **components** of the vector.

Two vectors are said to be **equal** if and only if they have the same components.

There is a one-to-one correspondence between the vectors $\langle x, y \rangle$ in the plane and the points (x, y) in the plane. Let the vector **A** be the ordered pair of real numbers $\langle a_1, a_2 \rangle$. If A is the point (a_1, a_2), then the vector **A** may be represented geometrically by the directed line segment $\overrightarrow{OA}$. Such a directed line segment is called a **representation** of vector **A**. Any directed line segment that is equal to $\overrightarrow{OA}$ is also a representation of vector **A**. The particular representation of a vector that has its initial point at the origin is called the **position representation** of the vector.

▷ ILLUSTRATION 1

The vector $\langle 2, 3 \rangle$ has as its position representation the directed line segment from the origin to the point $(2, 3)$. The representation of $\langle 2, 3 \rangle$ whose initial point is (h, k) has as its terminal point $(h + 2, k + 3)$; refer to Figure 2.   ◀

The vector $\langle 0, 0 \rangle$ is called the **zero vector,** and it is denoted by **0;** that is,

$$\mathbf{0} = \langle 0, 0 \rangle$$

Any point is a representation of the zero vector.

The **magnitude** of a vector **A** is the length of any of its representations and is denoted by $\|\mathbf{A}\|$. The **direction** of a nonzero vector is the direction of any of its representations.

THEOREM 1

If **A** is the vector $\langle a_1, a_2 \rangle$, then

$$\|\mathbf{A}\| = \sqrt{a_1{}^2 + a_2{}^2}$$

Proof   Because $\|\mathbf{A}\|$ is the length of any of the representations of **A,** it will be the length of the position representation, which is the distance from the origin to the point (a_1, a_2). From the distance formula

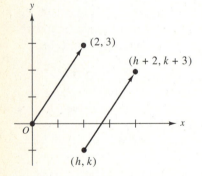

FIGURE 2

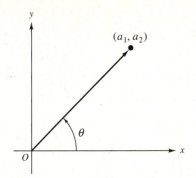

FIGURE 3

$$\|\mathbf{A}\| = \sqrt{(a_1 - 0)^2 + (a_2 - 0)^2}$$
$$= \sqrt{a_1{}^2 + a_2{}^2}$$

∎

Observe that $\|\mathbf{A}\|$ is a nonnegative number and is not a vector. From Theorem 1, it follows that

$$\|\mathbf{0}\| = 0$$

that is, the magnitude of the zero vector is 0.

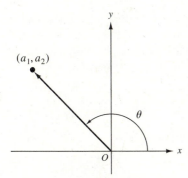

FIGURE 4

▷ **ILLUSTRATION 2**

If $\mathbf{A} = \langle -3, 5 \rangle$, then
$$\|\mathbf{A}\| = \sqrt{(-3)^2 + 5^2}$$
$$= \sqrt{34}$$

◀

The **direction angle** of any nonzero vector is the angle θ measured from the positive side of the x axis counterclockwise to the position representation of the vector. If θ is measured in radians, $0 \le \theta < 2\pi$. If $\mathbf{A} = \langle a_1, a_2 \rangle$, then

$$\tan \theta = \frac{a_2}{a_1} \quad \text{if } a_1 \ne 0$$

If $a_1 = 0$ and $a_2 > 0$, then $\theta = \frac{1}{2}\pi$; if $a_1 = 0$ and $a_2 < 0$, then $\theta = \frac{3}{2}\pi$. Figures 3 through 5 show the direction angle θ for specific vectors whose position representations are drawn.

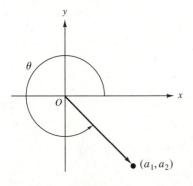

FIGURE 5

▶ **EXAMPLE 1** *Computing the Direction Angle of a Vector*

Find the radian measure of the direction angle of each of the following vectors: **(a)** $\langle -1, 1 \rangle$; **(b)** $\langle 0, -5 \rangle$; **(c)** $\langle 1, -2 \rangle$.

Solution The position representations of the vectors in (a), (b), and (c) appear in Figures 6, 7, and 8, respectively, on the next page.

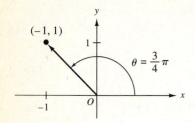

FIGURE 6

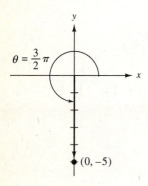

FIGURE 7

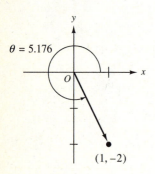

FIGURE 8

(a) $\tan \theta = \frac{1}{-1}$

$\qquad = -1$

Because $\frac{1}{2}\pi < \theta < \pi$, $\theta = \frac{3}{4}\pi$.

(b) Because $a_1 = 0$, $\tan \theta$ does not exist. Therefore, because $a_2 < 0$, $\theta = \frac{3}{2}\pi$.

(c) $\tan \theta = \frac{-2}{1}$

$\qquad = -2$

Because $\frac{3}{2}\pi < \theta < 2\pi$, $\theta = 5.176$. ◄

Observe that if $\mathbf{A} = \langle a_1, a_2 \rangle$ and θ is the direction angle of $\mathbf{A}$, then

$$a_1 = \|\mathbf{A}\|\cos\theta \qquad \text{and} \qquad a_2 = \|\mathbf{A}\|\sin\theta \qquad (1)$$

If the vector $\mathbf{A} = \langle a_1, a_2 \rangle$, then the representation of $\mathbf{A}$ whose initial point is (x, y) has as its endpoint $(x + a_1, y + a_2)$. Figure 9 illustrates five representations of the vector $\mathbf{A} = \langle a_1, a_2 \rangle$. In each case $\mathbf{A}$ translates the point (x_i, y_i) into the point $(x_i + a_1, y_i + a_2)$.

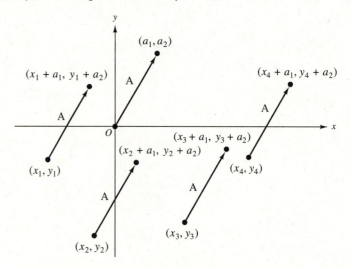

FIGURE 9

The following definition gives the method for adding two vectors.

DEFINITION **The Sum of Two Vectors**

The **sum** of two vectors $\mathbf{A} = \langle a_1, a_2 \rangle$ and $\mathbf{B} = \langle b_1, b_2 \rangle$ is the vector $\mathbf{A} + \mathbf{B}$ defined by

$$\mathbf{A} + \mathbf{B} = \langle a_1 + b_1, a_2 + b_2 \rangle$$

▷ **ILLUSTRATION 3**

If $\mathbf{A} = \langle 3, -1 \rangle$ and $\mathbf{B} = \langle -4, 5 \rangle$, then

$$\mathbf{A} + \mathbf{B} = \langle 3 + (-4), -1 + 5 \rangle$$
$$= \langle -1, 4 \rangle \qquad \blacktriangleleft$$

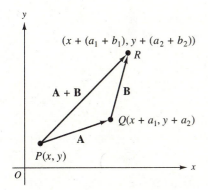

FIGURE 10

The geometric interpretation of the sum of two vectors appears in Figure 10. Let $\mathbf{A} = \langle a_1, a_2 \rangle$, $\mathbf{B} = \langle b_1, b_2 \rangle$, and P be the point (x, y). Then $\mathbf{A}$ translates the point P into the point $(x + a_1, y + a_2) = Q$. The vector $\mathbf{B}$ translates the point Q into the point $((x + a_1) + b_1, (y + a_2) + b_2)$ or, equivalently, $(x + (a_1 + b_1), y + (a_2 + b_2)) = R$. Furthermore,

$$\mathbf{A} + \mathbf{B} = \langle a_1 + b_1, a_2 + b_2 \rangle$$

Therefore $\mathbf{A} + \mathbf{B}$ translates P into $(x + (a_1 + b_1), y + (a_2 + b_2)) = R$. Thus, in Figure 10 $\overrightarrow{PQ}$ is a representation of the vector $\mathbf{A}$, $\overrightarrow{QR}$ is a representation of the vector $\mathbf{B}$, and $\overrightarrow{PR}$ is a representation of the vector $\mathbf{A} + \mathbf{B}$. The representations of the vectors $\mathbf{A}$ and $\mathbf{B}$ are adjacent sides of a parallelogram, and the representation of the vector $\mathbf{A} + \mathbf{B}$ is a diagonal of the parallelogram. This diagonal is called the **resultant** of the vectors $\mathbf{A}$ and $\mathbf{B}$. The rule for the addition of vectors is sometimes referred to as the **parallelogram law.**

Force is a vector quantity where the magnitude is expressed in force units and the direction angle is determined by the direction of the force. It is shown in physics that two forces applied to an object at a particular point can be replaced by an equivalent force that is their resultant.

In the following example, we refer to the **angle between two vectors,** which is the angle θ determined by the position representations of the two vectors such that $0° \leq \theta \leq 180°$.

▶ **EXAMPLE 2** *Solving a Word Problem Involving Vectors*

Two forces of magnitudes 200 lb and 250 lb make an angle of 60° with each other and are applied to an object at the same point. Find **(a)** the magnitude of the resultant force and **(b)** the angle it makes with the force of 200 lb.

Solution Refer to Figure 11, where the axes are chosen so that the position representation of the force of 200 lb is along the positive side of the x axis. The vector $\mathbf{A}$ represents this force and $\mathbf{A} = \langle 200, 0 \rangle$. The vector $\mathbf{B}$ represents the force of 250 lb. From Formulas (1), if $\mathbf{B} = \langle b_1, b_2 \rangle$, then

$$b_1 = 250 \cos 60° \qquad b_2 = 250 \sin 60°$$
$$= 125 \qquad\qquad = 216.5$$

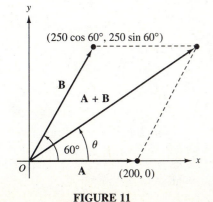

FIGURE 11

Thus, $\mathbf{B} = \langle 125, 216.5 \rangle$. The resultant force is $\mathbf{A} + \mathbf{B}$, and

$$\mathbf{A} + \mathbf{B} = \langle 200, 0 \rangle + \langle 125, 216.5 \rangle$$
$$= \langle 325, 216.5 \rangle$$

(a) $\|\mathbf{A} + \mathbf{B}\| = \sqrt{(325)^2 + (216.5)^2}$
$$= 390.5$$

Conclusion: The magnitude of the resultant force is 390.5 1b.

(b) If θ is the angle the vector $\mathbf{A} + \mathbf{B}$ makes with $\mathbf{A}$, then

$$\tan \theta = \frac{216.5}{325}$$
$$\tan \theta = 0.6662$$
$$\theta = 33.67°$$

Conclusion: The resultant force makes an angle of 33.67° with the force of 200 lb. ◄

The following illustration gives an alternative solution for Example 2.

▷ **ILLUSTRATION 4**

From Figure 11 we have the triangle shown in Figure 12. Applying the law of cosines to this triangle, we obtain

$$\|\mathbf{A} + \mathbf{B}\|^2 = (200)^2 + (250)^2 - 2(200)(250)\cos 120°$$
$$\|\mathbf{A} + \mathbf{B}\|^2 = 40,000 + 62,500 - 100,000(-\tfrac{1}{2})$$
$$\|\mathbf{A} + \mathbf{B}\|^2 = 152,500$$
$$\|\mathbf{A} + \mathbf{B}\| = \sqrt{152,500}$$
$$\|\mathbf{A} + \mathbf{B}\| = 390.5$$

We can compute θ by applying the law of sines to the triangle of Figure 12.

$$\frac{\sin \theta}{250} = \frac{\sin 120°}{390.5}$$
$$\sin \theta = \frac{250 \sin 120°}{390.5}$$
$$\sin \theta = 0.5544$$
$$\theta = 33.67°$$ ◄

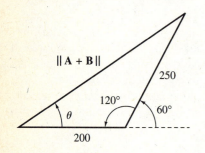

FIGURE 12

If $\mathbf{V}$ is a velocity vector, then $\|\mathbf{V}\|$ is a speed, a scalar quantity. In the next example, pertaining to marine navigation, we refer to a boat's velocity relative to the water and the velocity of the current. The resultant of these two velocities is the velocity of the boat relative to the land.

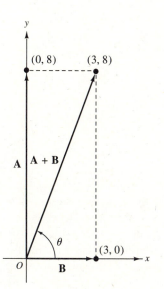

FIGURE 13

▶ **EXAMPLE 3** *Solving a Word Problem Involving Vectors*

A boat leaves the south bank of a river with a compass heading of north and traveling 8 mi/hr relative to the water. If the velocity of the current is 3 mi/hr toward the east, what is the speed of the boat relative to the land and what is its course?

Solution   See Figure 13, showing the position representations of vectors **A, B,** and **A + B**. The vector **A** represents the velocity of the boat relative to the water. Because **A** has a magnitude of 8 and a direction angle of 90°, **A** = $\langle 0, 8 \rangle$. The vector **B** represents the velocity of the current relative to the land, which has a magnitude of 3 and a direction angle of 0°. Thus **B** = $\langle 3, 0 \rangle$. The resultant of **A** and **B** is **A + B,** which is the velocity of the boat relative to the land.

$$\mathbf{A} + \mathbf{B} = \langle 0, 8 \rangle + \langle 3, 0 \rangle$$
$$= \langle 3, 8 \rangle$$
$$\|\mathbf{A} + \mathbf{B}\| = \sqrt{3^2 + 8^2}$$
$$= \sqrt{73}$$
$$\approx 8.54$$

If θ is the direction angle of **A + B,** then

$$\tan \theta = \tfrac{8}{3}$$
$$\tan \theta = 2.667$$
$$\theta = 69.4°$$

Conclusion: The boat is traveling at a speed of 8.54 mi/hr relative to the land in the direction of 69.4° with respect to the south bank or, equivalently, a course of 20.6°.   ◀

If **A** is the vector $\langle a_1, a_2 \rangle$, then the **negative** of **A,** denoted by $-\mathbf{A}$, is the vector $\langle -a_1, -a_2 \rangle$. If the directed line segment $\overrightarrow{PQ}$ is a representation of the vector **A,** then the directed line segment $\overrightarrow{QP}$ is a representation of $-\mathbf{A}$. Any directed line segment that is parallel to $\overrightarrow{PQ}$, has the same length as $\overrightarrow{PQ}$, and has a direction opposite to that of $\overrightarrow{PQ}$ is also a representation of $-\mathbf{A}$. See Figure 14.

We now define subtraction of two vectors.

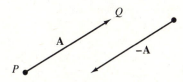

FIGURE 14

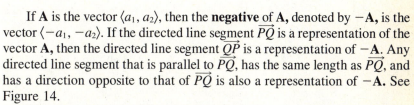

DEFINITION   **The Difference of Two Vectors**

The **difference** of the two vectors **A** and **B,** denoted by **A − B,** is the vector obtained by adding **A** to the negative of **B;** that is,

$$\mathbf{A} - \mathbf{B} = \mathbf{A} + (-\mathbf{B})$$

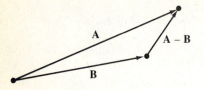

FIGURE 15

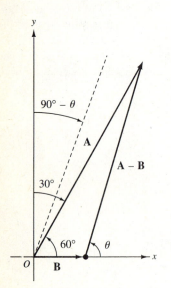

FIGURE 16

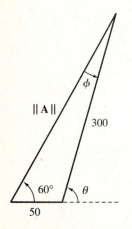

FIGURE 17

If $\mathbf{A} = \langle a_1, a_2 \rangle$ and $\mathbf{B} = \langle b_1, b_2 \rangle$, then $-\mathbf{B} = \langle -b_1, -b_2 \rangle$, and from the definition

$$\mathbf{A} - \mathbf{B} = \langle a_1 - b_1, a_2 - b_2 \rangle$$

▷ **ILLUSTRATION 5**

If $\mathbf{A} = \langle 4, -2 \rangle$ and $\mathbf{B} = \langle 6, -3 \rangle$, then

$$\begin{aligned} \mathbf{A} - \mathbf{B} &= \langle 4, -2 \rangle - \langle 6, -3 \rangle \\ &= \langle 4, -2 \rangle + \langle -6, 3 \rangle \\ &= \langle -2, 1 \rangle \end{aligned}$$

◀

To interpret the difference of two vectors geometrically, let the representations of the vectors $\mathbf{A}$ and $\mathbf{B}$ have the same initial point. Then the directed line segment from the endpoint of the representation of $\mathbf{B}$ to the endpoint of the representation of $\mathbf{A}$ is a representation of the vector $\mathbf{A} - \mathbf{B}$. This obeys the parallelogram law $\mathbf{B} + (\mathbf{A} - \mathbf{B}) = \mathbf{A}$. See Figure 15.

The following example, involving the difference of two vectors, is concerned with air navigation. The *air speed* of a plane refers to its speed relative to the air, and the *ground speed* is its speed relative to the ground. When there is a wind, the velocity of the plane relative to the ground is the resultant of the vector representing the wind's velocity and the vector representing the velocity of the plane relative to the air.

▶ **EXAMPLE 4** *Solving a Word Problem Involving Vectors*

An airplane can fly at an air speed of 300 mi/hr. If there is a wind blowing toward the east at 50 mi/hr, what should be the plane's compass heading in order for its course to be 30°? What will be the plane's ground speed if it flies this course?

Solution Refer to Figure 16, showing position representations of the vectors $\mathbf{A}$ and $\mathbf{B}$ as well as a representation of $\mathbf{A} - \mathbf{B}$. The vector $\mathbf{A}$ represents the velocity of the plane relative to the ground on a course of 30°. The direction angle of $\mathbf{A}$ is 60°, which is $90° - 30°$. The vector $\mathbf{B}$ represents the velocity of the wind. Because $\mathbf{B}$ has a magnitude of 50 and a direction angle of 0°, $\mathbf{B} = \langle 50, 0 \rangle$. The vector $\mathbf{A} - \mathbf{B}$ represents the velocity of the plane relative to the air; thus $\| \mathbf{A} - \mathbf{B} \| = 300$. Let θ be the direction angle of $\mathbf{A} - \mathbf{B}$. From Figure 16 we obtain the triangle shown in Figure 17.

Applying the law of sines to this triangle, we get

$$\frac{\sin \phi}{50} = \frac{\sin 60°}{300}$$

$$\sin \phi = \frac{50 \sin 60°}{300}$$

$$\sin \phi = 0.1443$$

$$\phi = 8.3°$$

Therefore

$$\theta = 60° + 8.3°$$

$$= 68.3°$$

Again applying the law of sines to the triangle in Figure 17, we have

$$\frac{\|\mathbf{A}\|}{\sin(180° - \theta)} = \frac{300}{\sin 60°}$$

$$\|\mathbf{A}\| = \frac{300 \sin 111.7°}{\sin 60°}$$

$$\|\mathbf{A}\| = 322$$

Conclusion: The plane's compass heading should be $90° - \theta$, which is $21.7°$, and if the plane flies this course, its ground speed will be 322 mi/hr.

◀

Scalar multiplication is another operation with vectors. We now give the definition of the multiplication of a vector by a scalar.

DEFINITION   Scalar Multiplication

If c is a scalar and $\mathbf{A}$ is the vector $\langle a_1, a_2 \rangle$, then the **product** of c and $\mathbf{A}$, denoted by $c\mathbf{A}$, is a vector and is given by

$$c\mathbf{A} = c\langle a_1, a_2 \rangle$$

$$= \langle ca_1, ca_2 \rangle$$

▷ **ILLUSTRATION 6**

If $\mathbf{A} = \langle 4, -5 \rangle$, then

$$3\mathbf{A} = 3\langle 4, -5 \rangle$$

$$= \langle 12, -15 \rangle$$

◀

▶ **EXAMPLE 5** *Multiplying a Vector by the Scalar 0 and Multiplying the Zero Vector by a Scalar*

If **A** is any vector and c is any scalar show that

(a) $0(\mathbf{A}) = \mathbf{0}$ **(b)** $c(\mathbf{0}) = \mathbf{0}$

Solution   We apply the definition of scalar multiplication.

(a) $0(\mathbf{A}) = 0\langle a_1, a_2 \rangle$ **(b)** $c(\mathbf{0}) = c\langle 0, 0 \rangle$

$\phantom{0(\mathbf{A})} = \langle 0, 0 \rangle$ $\phantom{c(\mathbf{0})} = \langle 0, 0 \rangle$

$\phantom{0(\mathbf{A})} = \mathbf{0}$ $\phantom{c(\mathbf{0})} = \mathbf{0}$ ◀

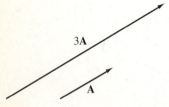

3A

A

FIGURE 18

We compute the magnitude of vector $c\mathbf{A}$ as follows:

$$\|c\mathbf{A}\| = \sqrt{(ca_1)^2 + (ca_2)^2}$$
$$= \sqrt{c^2(a_1{}^2 + a_2{}^2)}$$
$$= \sqrt{c^2}\,\sqrt{a_1{}^2 + a_2{}^2}$$
$$= |c|\|\mathbf{A}\|$$

Therefore the magnitude of $c\mathbf{A}$ is the absolute value of c times the magnitude of **A**.

A

$-\frac{1}{2}\mathbf{A}$

FIGURE 19

Figures 18 and 19 show the geometric interpretation of the vector $c\mathbf{A}$. If $c > 0$, then $c\mathbf{A}$ is a vector whose representation has a length c times the magnitude of **A** and the same direction as **A**; an example of this appears in Figure 18, where $c = 3$. If $c < 0$, then $c\mathbf{A}$ is a vector whose representation has a length that is $|c|$ times the magnitude of **A** and a direction opposite to that of **A**. This is shown in Figure 19, where $c = -\frac{1}{2}$.

We now take an arbitrary vector and write it in a special form.

$$\langle a_1, a_2 \rangle = \langle a_1, 0 \rangle + \langle 0, a_2 \rangle$$
$$\langle a_1, a_2 \rangle = a_1 \langle 1, 0 \rangle + a_2 \langle 0, 1 \rangle \tag{2}$$

Because the magnitude of each of the two vectors $\langle 1, 0 \rangle$ and $\langle 0, 1 \rangle$ is one unit, they are called **unit vectors.** We introduce the following notations for these two unit vectors:

$$\mathbf{i} = \langle 1, 0 \rangle \qquad \mathbf{j} = \langle 0, 1 \rangle$$

With these notations, we have from (2)

$$\langle a_1, a_2 \rangle = a_1 \mathbf{i} + a_2 \mathbf{j} \tag{3}$$

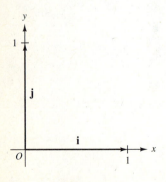

FIGURE 20

The position representations of the vectors **i** and **j** appear in Figure 20.

▷ **ILLUSTRATION 7**

From (3),

$$\langle 3, -4 \rangle = 3\mathbf{i} - 4\mathbf{j}$$ ◀

Let **A** be the vector $\langle a_1, a_2 \rangle$ and θ the direction angle of **A**. See Figure 21, where the point (a_1, a_2) is in the second quadrant and the position representation of **A** is shown. Because $\mathbf{A} = a_1\mathbf{i} + a_2\mathbf{j}, a_1 = \|\mathbf{A}\| \cos \theta$, and $a_2 = \|\mathbf{A}\| \sin \theta$, we can write

$$\mathbf{A} = \|\mathbf{A}\| \cos \theta\mathbf{i} + \|\mathbf{A}\| \sin \theta\mathbf{j}$$

$$\mathbf{A} = \|\mathbf{A}\|(\cos \theta\mathbf{i} + \sin \theta\mathbf{j}) \qquad \textbf{(4)}$$

This equation expresses the vector **A** in terms of its magnitude, direction angle, and the unit vectors **i** and **j**.

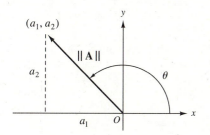

FIGURE 21

▶ **EXAMPLE 6** *Expressing a Vector in Terms of Its Magnitude and the Sine and Cosine of Its Direction Angle*

Express the vector $\langle -5, -2 \rangle$ in the form of (4).

Solution   Refer to Figure 22, which shows the position representation of the vector $\langle -5, -2 \rangle$.

$$\|\langle -5, -2 \rangle\| = \sqrt{(-5)^2 + (-2)^2} \qquad \cos \theta = -\frac{5}{\sqrt{29}} \qquad \sin \theta = -\frac{2}{\sqrt{29}}$$

$$= \sqrt{29}$$

Therefore from (4) we have

$$\langle -5, -2 \rangle = \sqrt{29}\left(-\frac{5}{\sqrt{29}}\mathbf{i} - \frac{2}{\sqrt{29}}\mathbf{j}\right)$$ ◀

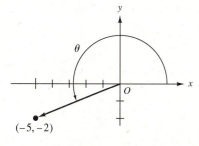

FIGURE 22

EXERCISES 10.1

*In Exercises 1 and 2, draw the position representation of the vector **A** and also the particular representation through the point P; find the magnitude of **A**.*

1. (a) $\mathbf{A} = \langle 3, 4 \rangle$, $P = (2, 1)$
   (b) $\mathbf{A} = \langle 0, -2 \rangle$, $P = (-3, 4)$

2. (a) $\mathbf{A} = \langle -2, 5 \rangle$, $P = (3, -4)$
   (b) $\mathbf{A} = \langle 4, 0 \rangle$, $P = (2, 6)$

*In Exercises 3 and 4, find the vector **A** having $\overrightarrow{PQ}$ as a representation. Draw $\overrightarrow{PQ}$ and the position representation of **A**.*

3. (a) $P = (3, 7)$, $Q = (5, 4)$
   (b) $P = (-3, 5)$, $Q = (-5, -2)$

4. (a) $P = (5, 4)$, $Q = (3, 7)$
   (b) $P = (-5, -3)$, $Q = (0, 3)$

In Exercises 5 and 6, find the point S so that $\overrightarrow{PQ}$ and $\overrightarrow{RS}$ are each representations of the same vector.

5. (a) $P = (2, 5)$, $Q = (1, 6)$, $R = (-3, 2)$
 (b) $P = (0, 3)$, $Q = (5, -2)$, $R = (7, 0)$

6. (a) $P = (-1, 4)$, $Q = (2, -3)$, $R = (-5, -2)$
 (b) $P = (-2, 0)$, $Q = (-3, -4)$, $R = (4, 2)$

In Exercises 7 and 8, find the sum of the pairs of vectors and illustrate geometrically.

7. (a) $\langle 2, 4 \rangle$, $\langle -3, 5 \rangle$ **(b)** $\langle -3, 0 \rangle$, $\langle 4, -5 \rangle$
8. (a) $\langle 0, 3 \rangle$, $\langle -2, 3 \rangle$ **(b)** $\langle 2, 5 \rangle$, $\langle 2, 5 \rangle$

In Exercises 9 and 10, subtract the second vector from the first and illustrate geometrically.

9. (a) $\langle 4, 5 \rangle$, $\langle -3, 2 \rangle$ **(b)** $\langle -3, -4 \rangle$, $\langle 6, 0 \rangle$
10. (a) $\langle 0, 5 \rangle$, $\langle 2, 8 \rangle$ **(b)** $\langle 3, 7 \rangle$, $\langle 3, 7 \rangle$

In Exercises 11 and 12, find the vector or scalar if $A = \langle 2, 4 \rangle$, $B = \langle 4, -3 \rangle$, and $C = \langle -3, 2 \rangle$.

11. (a) $A + B$ **(b)** $\|C - B\|$ **(c)** $\|7A - B\|$
12. (a) $A - B$ **(b)** $\|C\|$ **(c)** $\|2A + 3B\|$

In Exercises 13 through 16, find the given vector or scalar if $A = 2i + 3j$ and $B = 4i - j$.

13. (a) $5A$ **(b)** $-6B$
 (c) $A + B$ **(d)** $\|A + B\|$
14. (a) $-2A$ **(b)** $3B$
 (c) $A - B$ **(d)** $\|A - B\|$
15. (a) $\|A\| + \|B\|$ **(b)** $5A - 6B$
 (c) $\|5A - 6B\|$ **(d)** $\|5A\| - \|6B\|$
16. (a) $\|A\| - \|B\|$ **(b)** $3B - 2A$
 (c) $\|3B - 2A\|$ **(d)** $\|3B\| - \|2A\|$

In Exercises 17 through 20, find the components of the vector having the given magnitude and direction angle.

17. $18; \frac{1}{5}\pi$ **18.** $24; 41.2°$
19. $35; 250°$ **20.** $110; \frac{1}{3}\pi$

In Exercises 21 through 24, write the given vector in the form $r(\cos\theta i + \sin\theta j)$, where r is the magnitude and θ is the direction angle.

21. (a) $3i - 4j$ **(b)** $2i + 2j$
22. (a) $8i + 6j$ **(b)** $2\sqrt{5}i + 4j$
23. (a) $-4i + 4\sqrt{3}j$ **(b)** $-16i$
24. (a) $3i - 3j$ **(b)** $2j$

Exercises 25 through 35 are word problems involving vectors. Be sure to write a conclusion.

25. Two forces of magnitudes 60 lb and 80 lb make an angle of 30° with each other and are applied to an object at the same point. Find **(a)** the magnitude of the resultant force and **(b)** to the nearest degree the angle it makes with the force of 60 lb. Use the method of Example 2.

26. Two forces of magnitudes 340 lb and 475 lb make an angle of 34.6° with each other and are applied to an object at the same point. Find **(a)** the magnitude of the resultant force and **(b)** to the nearest tenth of a degree the angle it makes with the force of 475 lb. Use the method of Example 2.

27. Do Exercise 25 by the method of Illustration 4.

28. Do Exercise 26 by the method of Illustration 4.

29. A force of magnitude 112 lb and one of 84 lb are applied to an object at the same point, and the resultant force has a magnitude of 162 lb. Find to the nearest tenth of a degree the angle made by the resultant force with the force of 112 lb.

30. A force of 22 lb and a force of 34 lb are applied to an object at the same point and make an angle of θ with each other. If the resultant force has a magnitude of 46 lb, find θ to the nearest degree.

31. A swimmer who can swim at a speed of 1.5 mi/hr relative to the water leaves the south bank of a river and is headed north directly across the river. If the river's current is toward the east at 0.8 mi/hr, **(a)** in what direction is the swimmer going? **(b)** What is the swimmer's speed relative to the land? **(c)** If the distance across the river is 1 mile, how far down the river does the swimmer reach the north bank?

32. Suppose the swimmer in Exercise 31 wishes to reach the point directly north across the river. **(a)** In what direction should the swimmer head? **(b)** What will be the swimmer's speed relative to the land if this direction is taken?

33. In an airplane that has an air speed of 250 mi/hr, a pilot wishes to fly due north. If there is a wind blowing at 60 mi/hr toward the east, **(a)** what should be the plane's compass heading? **(b)** What will be the plane's ground speed if it flies this course?

34. A plane has an air speed of 350 mi/hr. For the actual course of the plane to be due north, the compass heading is 340°. If the wind is blowing from the west, **(a)** what is the magnitude of its velocity? **(b)** What is the plane's ground speed?

35. A boat can travel 15 knots relative to the water. On a river whose current is 3 knots toward the west the boat has a compass heading of south. What is the speed of the boat relative to the land and what is its course?

36. Let $\overrightarrow{PQ}$ be a representation of vector **A**, $\overrightarrow{QR}$ be a representation of vector **B**, and $\overrightarrow{RS}$ be a representation of vector **C**. Prove that if $\overrightarrow{PQ}$, $\overrightarrow{QR}$, and $\overrightarrow{RS}$ are sides of a triangle, then $\mathbf{A} + \mathbf{B} + \mathbf{C} = \mathbf{0}$.

37. Explain the difference between a vector quantity and a scalar quantity. In your explanation use velocity and speed as examples.

10.2 VECTOR-VALUED FUNCTIONS AND PARAMETRIC EQUATIONS

GOALS

1. Define a vector-valued function.
2. Find the domain of a vector-valued function.
3. Learn about parametric equations.
4. Sketch the graph of a vector equation or its equivalent pair of parametric equations.
5. Find a cartesian equation of a graph from its vector equation or the equivalent pair of parametric equations.
6. Plot the graph of a vector equation or its equivalent pair of parametric equations.
7. Solve word problems involving motion of a projectile.
8. Derive parametric equations of a cycloid.

Suppose a particle moves so that the coordinates (x, y) of its position at any time t are given by the equations $x = f(t)$ and $y = g(t)$. Then for every number t in the domain common to f and g there is a vector $f(t)\mathbf{i} + g(t)\mathbf{j}$, and the endpoints of the position representations of these vectors trace a curve C traveled by the particle. This leads us to consider a function whose domain is a set of real numbers and whose range is a set of vectors. Such a function is called a *vector-valued function*.

DEFINITION **Vector-Valued Function**

Let f and g be two real-valued functions of a real variable t. Then for every number t in the domain common to f and g there is a vector **R** defined by

$$\mathbf{R}(t) = f(t)\mathbf{i} + g(t)\mathbf{j}$$

and **R** is called a **vector-valued function.**

▶ **EXAMPLE 1** *Finding the Domain of a Vector-Valued Function*

Determine the domain of the vector-valued function **R** defined by

$$\mathbf{R}(t) = \sqrt{t - 2}\,\mathbf{i} + (t - 3)^{-1}\mathbf{j}$$

Solution   Let

$$f(t) = \sqrt{t - 2} \quad \text{and} \quad g(t) = (t - 3)^{-1}$$

The domain of **R** is the set of values of t for which both $f(t)$ and $g(t)$ are defined. Because $f(t)$ is defined for $t \geq 2$ and $g(t)$ is defined for all real numbers except 3, the domain of **R** is $\{t \mid t \geq 2, t \neq 3\}$.   ◀

The equation $\mathbf{R}(t) = f(t)\mathbf{i} + g(t)\mathbf{j}$ is called a **vector equation,** and it defines a curve C. The same curve C is also defined by the equations

$$x = f(t) \quad \text{and} \quad y = g(t) \tag{1}$$

which are called **parametric equations** of C. The variable t is a **parameter.** The curve C is also called a **graph;** that is, the set of all points (x, y) satisfying (1) is the graph of the vector-valued function **R.**

A vector equation of a curve, as well as parametric equations of a curve, gives the curve a direction at each point. That is, if we think of the curve as being traced by a particle, we can consider the positive direction along a curve as the direction in which the particle moves as the parameter t increases. In such a case as this, t may be taken to be the measure of time, and the vector $\mathbf{R}(t)$ is called the **position vector.** Sometimes $\mathbf{R}(t)$ is referred to as the **radius vector.**

If the parameter t is eliminated from the pair of Equations (1), we obtain one equation in x and y, called a **cartesian equation** of C.

▶ **EXAMPLE 2** *Sketching the Graph of a Vector Equation and Finding a Cartesian Equation of the Graph*

Given the vector equation

$$\mathbf{R}(t) = 2 \cos t\,\mathbf{i} + 2 \sin t\,\mathbf{j}$$

(a) sketch the graph of this equation, and **(b)** find a cartesian equation of the graph.

Solution

(a) The domain of **R** is the set of all real numbers. The values of x and y for particular values of t can be tabulated. We find the magnitude of the position vector. For every t,

$$\begin{aligned}
\|\mathbf{R}(t)\| &= \sqrt{4 \cos^2 t + 4 \sin^2 t} \\
&= 2\sqrt{\cos^2 t + \sin^2 t} \\
&= 2
\end{aligned}$$

Therefore the endpoint of the position representation of each vector $\mathbf{R}(t)$ is two units from the origin. By letting t take on all numbers in the closed interval $[0, 2\pi]$, we obtain a circle having its center at the origin and radius 2. This is the entire graph because any value of t will give a point on this circle. The circle appears in Figure 1. Parametric equations of the circle are

$$x = 2 \cos t \quad \text{and} \quad y = 2 \sin t$$

(b) We can find a cartesian equation of the graph by eliminating t from the two parametric equations, which when squaring on both sides of each equation and adding gives

$$x^2 + y^2 = (2 \cos t)^2 + (2 \sin t)^2$$
$$x^2 + y^2 = 4(\cos^2 t + \sin^2 t)$$
$$x^2 + y^2 = 4 \qquad \blacktriangleleft$$

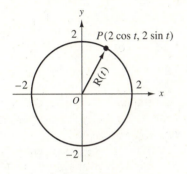

FIGURE 1

In Example 2 for the graph of the given vector equation, or the equivalent pair of parametric equations, we eliminated the parameter t and obtained a cartesian equation of the graph. Often we cannot eliminate the parameter, and we must, therefore, obtain the graph directly from the parametric equations.

To sketch the graph of a pair of parametric equations, it is helpful to find horizontal and vertical tangent lines of the graph. In general we cannot do this without techniques of calculus. So here we are confined to locating points whose x and y coordinates are found from values of the parameter and to determining certain characteristics of the graph from the form of the parametric equations.

To plot the graph of a pair of parametric equations on your graphics calculator, you should consult the user's manual for your particular calculator. Some calculators require you to set the calculator in parametric mode. Other calculators may utilize a parametric plot format. In any case, you will need to enter into the calculator equations defining $x(t)$ and $y(t)$. Furthermore, you must indicate the smallest and largest values of t under consideration. We shall denote these values by t min and t max.

▷ ILLUSTRATION 1

To plot the graph of the parametric equations of Example 2 we enter

$$x(t) = 2 \cos t \quad \text{and} \quad y(t) = 2 \sin t$$

and let t min $= 0$ and t max $= 2\pi$. In the viewing rectangle $[-4.5, 4.5]$ by $[-3, 3]$ we obtain the circle shown in Figure 2. $\qquad \blacktriangleleft$

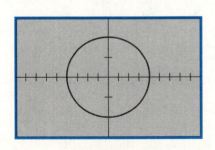

$[-4.5, 4.5]$ by $[-3, 3]$
$x(t) = 2 \cos t$ and $y(t) = 2 \sin t$
FIGURE 2

Table 1

t	x	y
0	0	0
$\frac{1}{2}$	$\frac{3}{4}$	$\frac{1}{2}$
1	3	4
2	12	32
$-\frac{1}{2}$	$\frac{3}{4}$	$-\frac{1}{2}$
-1	3	-4
-2	12	-32

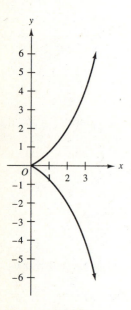

FIGURE 3

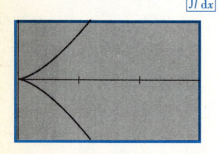

$[0, 3]$ by $[-1, 1]$
$x(t) = 3t^2$ and $y(t) = 4t^3$
FIGURE 4

▶ **EXAMPLE 3** *Sketching the Graph of a Pair of Parametric Equations and Finding a Cartesian Equation of the Graph*

(a) Sketch the graph of the parametric equations

$$x = 3t^2 \quad \text{and} \quad y = 4t^3$$

and check the graph on a graphics calculator. **(b)** Find a cartesian equation of the graph.

Solution

(a) We observe that x is nonnegative. Thus the graph is restricted to the first and fourth quadrants. Table 1 gives values of x and y for particular values of t. We locate the points having the corresponding x and y coordinates and obtain the graph in Figure 3. Locating points close to the origin leads us to believe that the tangent line at the origin is horizontal, but we cannot be certain that this is the case without techniques of calculus.

Figure 4 shows the graph plotted in the viewing rectangle $[0, 3]$ by $[-1, 1]$ with t min $= -2$ and t max $= 2$.

(b) From the two parametric equations, we get $x^3 = 27t^6$ and $y^2 = 16t^6$. Solving each of these equations for t, we have

$$t^6 = \frac{x^3}{27} \quad \text{and} \quad t^6 = \frac{y^2}{16}$$

Equating these two values for t^6, we obtain

$$\frac{x^3}{27} = \frac{y^2}{16}$$
$$16x^3 = 27y^2$$

which is the cartesian equation desired. ◀

$\boxed{\int / \frac{dy}{dx}}$ In calculus, vectors are used to derive equations of motion of a projectile under the assumption that the projectile is moving in a vertical plane and that the only force acting on the projectile is due to gravity. We neglect the force attributed to air resistance, which for heavy bodies traveling at small speeds has no noticeable effect. We cannot give a complete discussion here, but we will introduce you to this topic.

The positive direction is taken vertically upward and horizontally to the right, and the coordinate axes are set up so that the gun is located at the origin. Refer to Figure 5. If the projectile is shot from a gun having an angle of elevation of radian measure α, and the number of feet per second in the initial speed, or *muzzle speed,* is denoted by v_0, the initial velocity vector, $\mathbf{V}_0$, of the projectile is given by

$$\mathbf{V}_0 = v_0 \cos \alpha \mathbf{i} + v_0 \sin \alpha \mathbf{j} \tag{2}$$

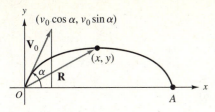

FIGURE 5

Let t seconds be the time elapsed since the firing of the gun, x feet be the horizontal distance of the projectile from the starting point, and y feet be the vertical distance of the projectile at t seconds. Then if $\mathbf{R}(t)$ is the position vector of the projectile at t seconds, we show in calculus that

$$\mathbf{R}(t) = -\tfrac{1}{2} gt^2 \mathbf{j} + \mathbf{V}_0 t \tag{3}$$

where g ft/sec^2 is the acceleration constant due to gravity. An approximate value for g near sea level is 32. Substituting the value of $\mathbf{V}_0$ from (2) into (3) we obtain

$$\mathbf{R}(t) = -\tfrac{1}{2} gt^2 \mathbf{j} + (v_0 \cos \alpha \mathbf{i} + v_0 \sin \alpha \mathbf{j})t$$
$$\mathbf{R}(t) = tv_0 \cos \alpha \mathbf{i} + (tv_0 \sin \alpha - \tfrac{1}{2} gt^2)\mathbf{j} \tag{4}$$

Equation (4) gives the position vector of the projectile at t seconds. From this equation we can discuss the motion of the projectile. We are usually concerned with the following questions:

1. What is the range of the projectile? The range is the distance $|OA|$ along the x axis (see Figure 5).
2. What is the total time of flight, that is, the time it takes the projectile to go from O to A?
3. What is the maximum height of the projectile?
4. What is a cartesian equation of the path of the projectile?
5. What is the velocity vector of the projectile at impact?

We answer the first four questions in the following example. Techniques of calculus are needed to answer the fifth question.

▶ **EXAMPLE 4** *Solving a Word Problem Involving Motion of a Projectile*

A projectile is shot from a gun at an angle of elevation of radian measure $\tfrac{1}{6}\pi$. Its muzzle speed is 480 ft/sec. Find: **(a)** the position vector of the projectile at any time; **(b)** parametric equations of the projectile's position at any time; **(c)** the total time of flight; **(d)** the range of the projectile; **(e)** the maximum height attained by the projectile; **(f)** a cartesian equation of the path of the projectile. **(g)** Plot the path of the projectile.

Solution From (2) with $v_0 = 480$ and $\alpha = \tfrac{1}{6}\pi$, the initial velocity vector is

$$\mathbf{V}_0 = 480 \cos \tfrac{1}{6}\pi \mathbf{i} + 480 \sin \tfrac{1}{6}\pi \mathbf{j}$$
$$= 240\sqrt{3}\mathbf{i} + 240\mathbf{j}$$

(a) We can obtain the position vector at t seconds by applying (4); we get

$$\mathbf{R}(t) = 240\sqrt{3}t\mathbf{i} + (240t - \tfrac{1}{2}gt^2)\mathbf{j}$$

By letting $g = 32$ we have

$$\mathbf{R}(t) = 240\sqrt{3}t\,\mathbf{i} + (240t - 16t^2)\mathbf{j} \tag{5}$$

(b) If (x, y) is the position of the projectile at t seconds, parametric equations of the projectile's position are

$$x = 240\sqrt{3}t \quad \text{and} \quad y = 240t - 16t^2 \tag{6}$$

(c) To determine the time of flight, we must find t when $y = 0$. Setting $y = 0$ in the second equation of (6), we have

$$240t - 16t^2 = 0$$
$$t(240 - 16t) = 0$$
$$t = 0 \quad t = 15$$

The value $t = 0$ occurs when the projectile is fired. The value $t = 15$ gives the total time of flight.

Conclusion: The total time of flight is 15 sec.

(d) To find the range, we determine x when $t = 15$. From the first equation of (6) with $t = 15$, we obtain $x = 3600\sqrt{3}$.

Conclusion: The range is $3600\sqrt{3}$ ft ≈ 6235 ft.

(e) The maximum height is attained when y has its maximum value. From the second equation of (6)

$$y = 240t - 16t^2$$

Because the graph of this equation is a parabola opening downward, the maximum value of y occurs at the vertex, which from Theorem 1 in Section 4.3 is at $t = -\frac{240}{2(-16)}$ or at $t = \frac{15}{2}$, which is one-half the total time of flight. When $t = \frac{15}{2}$, $y = 900$.

Conclusion: The maximum height attained is 900 ft.

(f) To find a cartesian equation of the path of the projectile, we eliminate t between parametric equations (6). From the first of these equations, $t = x/(240\sqrt{3})$. Substituting this value of t into the second equation, we have

$$y = 240\left(\frac{x}{240\sqrt{3}}\right) - 16\left(\frac{x}{240\sqrt{3}}\right)^2$$

$$y = \frac{1}{\sqrt{3}}x - \frac{1}{10{,}800}x^2$$

which is an equation of a parabola. Since t is restricted to the interval $[0, 15]$, the path of the projectile is the portion of the parabola in that interval.

(g) We plot the path of the projectile by using parametric equations (6) with t min $= 0$ and t max $= 15$. Because the maximum value of x is 6235

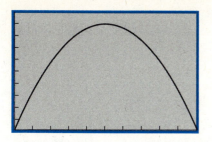

$[0, 6240]$ by $[0, 1000]$
$x(t) = 240\sqrt{3}\,t$ and $y(t) = 240t - 16t^2$
FIGURE 6

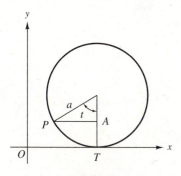

FIGURE 7

and the maximum value of y is 900 we choose the viewing rectangle $[0, 6240]$ by $[0, 1000]$. We obtain Figure 6. ◀

We now show how parametric equations can be used to define a curve that is described by a physical motion. The curve we consider is a **cycloid,** which is the curve traced by a point on the circumference of a circle as the circle rolls along a straight line. Suppose the circle has radius a. Let the fixed straight line on which the circle rolls be the x axis, and let the origin be one of the points at which the given point P comes in contact with the x axis. See Figure 7, showing the circle after it has rolled through an angle of t radians. From Figure 7, where $\mathbf{V}(\overrightarrow{OT})$, $\mathbf{V}(\overrightarrow{TA})$, $\mathbf{V}(\overrightarrow{AP})$, and $\mathbf{V}(\overrightarrow{OP})$ are the vectors having the indicated directed line segments as representations, we have

$$\mathbf{V}(\overrightarrow{OT}) + \mathbf{V}(\overrightarrow{TA}) + \mathbf{V}(\overrightarrow{AP}) = \mathbf{V}(\overrightarrow{OP}) \qquad (7)$$

The length of the arc PT is $\| \mathbf{V}(\overrightarrow{OT}) \| = at$. Because the direction of $\mathbf{V}(\overrightarrow{OT})$ is along the positive x axis,

$$\mathbf{V}(\overrightarrow{OT}) = at\mathbf{i} \qquad (8)$$

Also, $\| \mathbf{V}(\overrightarrow{TA}) \| = a - a \cos t$. And because the direction of $\mathbf{V}(\overrightarrow{TA})$ is the same as the direction of $\mathbf{j},$

$$\mathbf{V}(\overrightarrow{TA}) = a(1 - \cos t)\mathbf{j} \qquad (9)$$

$\| \mathbf{V}(\overrightarrow{AP}) \| = a \sin t$, and the direction of $\mathbf{V}(\overrightarrow{AP})$ is the same as the direction of $-\mathbf{i};$ thus

$$\mathbf{V}(\overrightarrow{AP}) = -a \sin t\mathbf{i}$$

Substituting from this equation, (8), and (9) into (7) we obtain

$$at\mathbf{i} + a(1 - \cos t)\mathbf{j} - a \sin t\mathbf{i} = \mathbf{V}(\overrightarrow{OP})$$

$$\Leftrightarrow \qquad \mathbf{V}(\overrightarrow{OP}) = a(t - \sin t)\mathbf{i} + a(1 - \cos t)\mathbf{j}$$

This is a vector equation of the cycloid. So parametric equations of the cycloid are

$$x = a(t - \sin t) \qquad \text{and} \qquad y = a(1 - \cos t) \qquad (10)$$

where t is any real number. A portion of the cycloid appears in Figure 8.

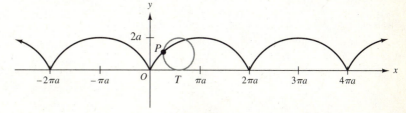

FIGURE 8

EXERCISES 10.2

In Exercises 1 through 10, find the domain of the vector-valued function.

1. $R(t) = (1/t)i + \sqrt{4 - t}\,j$

2. $R(t) = (t^2 - 1)^{-1}i + \sqrt{t^2 + 1}\,j$

3. $R(t) = \ln t\,i + e^t j$

4. $R(t) = (e^t + 1)^{-1}i + (e^t - 1)^{-1}j$

5. $R(t) = \sin t\,i + \tan t\,j$

6. $R(t) = \sec t\,i + \csc t\,j$

7. $R(t) = (\sin^{-1} t)i + (\cos^{-1} t)j$

8. $R(t) = \ln(t + 1)i + (\tan^{-1} t)j$

9. $R(t) = \sqrt{t^2 - 9}\,i + \sqrt{t^2 + 2t - 8}\,j$

10. $R(t) = \sqrt{t - 4}\,i + \sqrt{4 - t}\,j$

In Exercises 11 through 16, sketch the graph of the vector equation and find a cartesian equation of the graph.

11. $R(t) = \cos t\,i + \sin t\,j$

12. $R(t) = 4 \cos t\,i + 4 \sin t\,j$

13. $R(t) = 4 \cos t\,i + 4 \sin t\,j \qquad t \in [0, \pi]$

14. $R(t) = 4 \cos t\,i + 4 \sin t\,j \qquad t \in [-\frac{1}{2}\pi, \frac{1}{2}\pi]$

15. $R(t) = 3t\,i + 2t^2 j$

16. $R(t) = (1 - t^2)i + (1 + t)j$

In Exercises 17 through 20, (a) sketch the graph of the parametric equations and check your graph on your graphics calculator; (b) find a cartesian equation of the graph.

17. $x = 3 - 2t, y = 4 + t$

18. $x = 2t - 5, y = t + 1$

19. $x = 2t^3, y = 4t^2$

20. $x = 2t^2, y = 3t^3$

In Exercises 21 and 22, do the following: (a) Plot the graph of the parametric equations in an appropriate viewing rectangle for t in the indicated interval; (b) Find a cartesian equation of the graph. (Hint: Apply the fundamental Pythagorean identity.)

21. $x = 9 \cos t$ and $y = 4 \sin t$; $t \in [0, 2\pi]$

22. $x = 4 \cos t$ and $y = 25 \sin t$; $t \in [0, 2\pi]$

 In Exercises 23 and 24, a projectile moves so that the coordinates of its position at t seconds are given by the parametric equations. Find: (a) the total time of flight; (b) the

range of the projectile; (c) the maximum height attained by the projectile; (d) a cartesian equation of the path of the projectile. (e) Plot the path of the projectile.

23. $x = 60t, y = 80t - 16t^2$

24. $x = 40t, y = 56t - 16t^2$

 25. A projectile is shot from a gun at an angle of elevation of radian measure $\frac{1}{4}\pi$, and its muzzle speed is 2500 ft/sec. Find **(a)** the position vector of the projectile at any time; **(b)** parametric equations of the projectile's position at any time; **(c)** the total time of flight; **(d)** the range of the projectile; **(e)** the maximum height attained by the projectile; **(f)** a cartesian equation of the path of the projectile. **(g)** Plot the path of the projectile.

 26. Do Exercise 25 if the angle of elevation has radian measure $\frac{1}{3}\pi$ and the muzzle speed is 160 ft/sec.

 27. At a baseball game, a ball is batted with an initial velocity of 98 ft/sec at an angle of elevation of 58°. Determine **(a)** the horizontal distance traveled and **(b)** the maximum height reached by the ball. **(c)** Plot the curve traveled by the ball. **(d)** Find a cartesian equation of the curve traveled by the ball.

 28. The muzzle speed of a gun is 160 ft/sec. At what angle of elevation should the gun be fired so that a projectile will hit an object on the same level as the gun and a distance of 400 ft from it?

29. A cycloid is generated by the path of a point on the circumference of a circle of radius 4 units rolling along the x axis. **(a)** Write parametric equations of the cycloid where the parameter t is the number of radians in the angle through which the circle has rolled. **(b)** Sketch the cycloid for $-2\pi \leq t \leq 4\pi$.

30. Do Exercise 29 if the radius of the circle is 3 units.

31. A *hypocycloid* is the curve traced by a point P on a circle of radius b rolling inside a fixed circle of radius a, where $a > b$. If the origin is at the center of the fixed circle, $A(a, 0)$ is one of the points at which the point P comes in contact with the fixed circle, B is the moving point of tangency of the two circles, and the parameter t is the number of radians in the angle AOB, prove that parametric equations of the hypocycloid are

$$x = (a - b)\cos t + b \cos \frac{a - b}{b}t$$

and

$$y = (a - b)\sin t - b \sin \frac{a - b}{b} t$$

Plot the hypocycloid if $a = 6$ and $b = 2$, and $-2\pi \leq t \leq 2\pi$.

32. If $a = 4b$ in Exercise 31, we have a *hypocycloid of four cusps.* Show that parametric equations of this curve are

$$x = a \cos^3 t \qquad \text{and} \qquad y = a \sin^3 t$$

Plot the hypocycloid of four cusps if $a = 3$ and $-2\pi \leq t \leq 2\pi$.

33. (a) Plot the graph of the parametric equations

$$x = \tfrac{1}{2}(e^t + e^{-t}) \qquad \text{and} \qquad y = \tfrac{1}{2}(e^t - e^{-t})$$

for $-3 \leq t \leq 3$ in an appropriate viewing rectangle. (b) Follow the directions of part (a) for the parametric equations

$$x = -\tfrac{1}{2}(e^t + e^{-t}) \qquad \text{and} \qquad y = \tfrac{1}{2}(e^t - e^{-t})$$

(c) Plot the graphs of the two sets of parametric equations given in parts (a) and (b) in the same viewing rectangle. (d) Describe the graphs obtained in parts (a), (b), and (c) and discuss their relationship. The graph obtained in part (c) is called an *equilateral hyperbola* and is treated in Section 11.2

10.3 POLAR COORDINATES

GOALS
1. **Define polar coordinates of a point.**
2. **Locate a point from its polar coordinates and find other sets of polar coordinates of the point.**
3. **Obtain the equations relating the rectangular cartesian coordinates and polar coordinates of a point.**
4. **Find polar coordinates of a point from its rectangular cartesian coordinates.**
5. **Find a cartesian equation of a graph from a polar equation.**
6. **Find a polar equation of a graph from a cartesian equation.**

Until now we have located a point in a plane by its rectangular cartesian coordinates. Other coordinate systems, however, give the position of a point in a plane. The polar coordinate system is one of them, and it is important because certain curves have simpler equations in that system.

Cartesian coordinates are numbers, the abscissa and ordinate, and these numbers are directed distances from two fixed lines. Polar coordinates consist of a directed distance and the measure of an angle, which is taken relative to a fixed point and a fixed ray (or half line). The fixed point is called the **pole** (or origin), designated by the letter O. The fixed ray is called the **polar axis** (or polar line), which we label OA. The ray OA is usually drawn horizontally and to the right, and it extends indefinitely. See Figure 1.

Let P be any point in the plane distinct from O. Let θ be the radian measure of a directed angle AOP, positive when measured counterclockwise and negative when measured clockwise, having as its initial side the ray OA and as its terminal side the ray OP. Then if r is the undirected distance from O to P (that is, $r = |\overline{OP}|$), one set of polar coordinates of P is given by r and θ, and we write these coordinates as (r, θ).

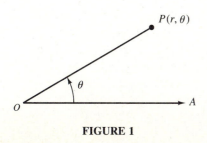

FIGURE 1

▶ **EXAMPLE 1**   *Locating Points from Their Polar Coordinates*

Locate each of the following points having the given set of polar coordinates:
(a) $(2, \frac{1}{4}\pi)$; **(b)** $(5, \frac{1}{2}\pi)$; **(c)** $(1, \frac{2}{3}\pi)$; **(d)** $(3, \frac{7}{6}\pi)$; **(e)** $(4, -\frac{1}{3}\pi)$; **(f)** $(\frac{5}{2}, -\pi)$.

Solution

(a) The point $(2, \frac{1}{4}\pi)$ is determined by first drawing the angle with radian measure $\frac{1}{4}\pi$, having its vertex at the pole and its initial side along the polar axis. The point on the terminal side that is 2 units from the pole is the point $(2, \frac{1}{4}\pi)$. See Figure 2(a). In a similar manner we obtain the points appearing in Figure 2(b)–(f).   ◀

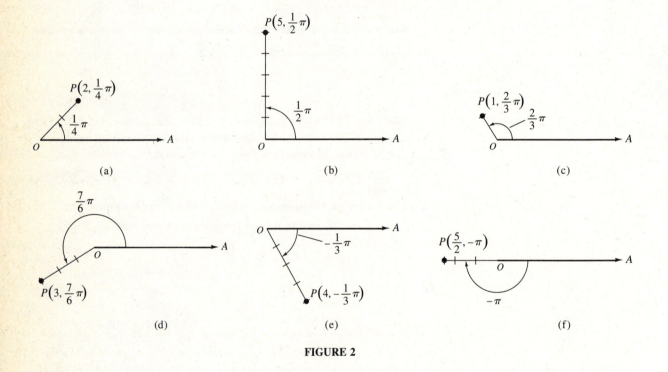

FIGURE 2

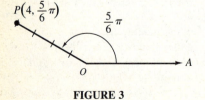

FIGURE 3

▷ **ILLUSTRATION 1**

Figure 3 shows the point $(4, \frac{5}{6}\pi)$. Another set of polar coordinates for this point is $(4, -\frac{7}{6}\pi)$; see Figure 4. Furthermore, the polar coordinates $(4, \frac{17}{6}\pi)$ also yield the same point, as shown in Figure 5.   ◀

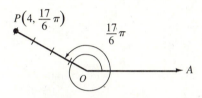

$P\left(4, -\frac{7}{6}\pi\right)$

$-\frac{7}{6}\pi$

FIGURE 4

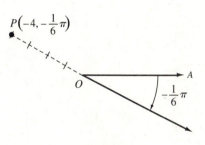

$P\left(4, \frac{17}{6}\pi\right)$

$\frac{17}{6}\pi$

FIGURE 5

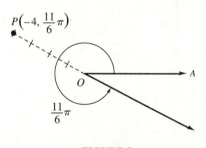

$P\left(-4, -\frac{1}{6}\pi\right)$

$-\frac{1}{6}\pi$

FIGURE 6

$P\left(-4, \frac{11}{6}\pi\right)$

$\frac{11}{6}\pi$

FIGURE 7

Actually the coordinates $(4, \frac{5}{6}\pi + 2k\pi)$, where k is any integer, give the same point as $(4, \frac{5}{6}\pi)$. Thus a given point has an unlimited number of sets of polar coordinates. This is unlike the rectangular cartesian coordinate system in which there is a one-to-one correspondence between the coordinates and the position of points in the plane. There is no such one-to-one correspondence between the polar coordinates and the position of points in the plane. A further example is obtained by considering sets of polar coordinates for the pole. If $r = 0$ and θ is any real number, we have the pole, which is therefore designated by $(0, \theta)$.

We now consider polar coordinates for which r is negative. In this case, instead of being on the terminal side of the angle, the point is on the extension of the terminal side, which is the ray from the pole extending in the direction opposite to the terminal side. Hence if P is on the extension of the terminal side of the angle of radian measure θ, a set of polar coordinates of P is (r, θ), where $r = -|\overline{OP}|$.

▷ **ILLUSTRATION 2**

The point $(-4, -\frac{1}{6}\pi)$ shown in Figure 6 is the same point as $(4, \frac{5}{6}\pi)$, $(4, -\frac{7}{6}\pi)$, and $(4, \frac{17}{6}\pi)$ in Illustration 1. Still another set of polar coordinates for this point is $(-4, \frac{11}{6}\pi)$; see Figure 7. ◄

The angle is usually measured in radians; thus a set of polar coordinates of a point is an ordered pair of real numbers. For each ordered pair of real numbers there is a unique point having this set of polar coordinates. However, we have seen that a particular point can be given by an unlimited number of ordered pairs of real numbers. If the point P is not the pole, and r and θ are restricted so that $r > 0$ and $0 \le \theta < 2\pi$, then there is a unique set of polar coordinates for P.

▶ **EXAMPLE 2** *Locating a Point from Its Polar Coordinates and Finding Other Sets of Polar Coordinates for the Point*

(a) Locate the point having polar coordinates $(3, -\frac{2}{3}\pi)$. Find another set of polar coordinates of this point for which **(b)** $r < 0$ and $0 < \theta < 2\pi$; **(c)** $r > 0$ and $0 < \theta < 2\pi$; **(d)** $r < 0$ and $-2\pi < \theta < 0$.

Solution
(a) The point is located by drawing the angle of radian measure $-\frac{2}{3}\pi$ in a clockwise direction from the polar axis. Because $r > 0$, P is on the

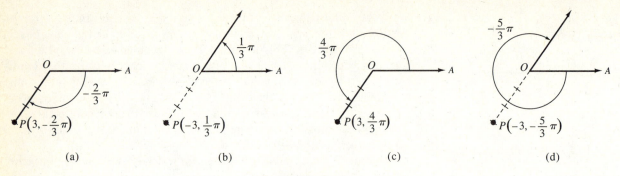

FIGURE 8

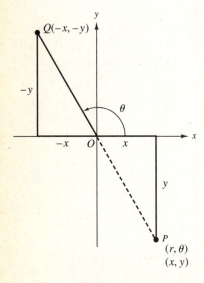

FIGURE 9

FIGURE 10

terminal side of the angle, three units from the pole; see Figure 8(a).

The answers to (b), (c), and (d) are, respectively, $(-3, \frac{1}{3}\pi)$, $(3, \frac{4}{3}\pi)$, and $(-3, -\frac{5}{3}\pi)$. They are illustrated in Figure 8(b)–(d). ◀

Often we wish to refer to both the rectangular cartesian coordinates and the polar coordinates of a point. To do this, we take the origin of the first system and the pole of the second system coincident, the polar axis as the positive side of the x axis, and the ray for which $\theta = \frac{1}{2}\pi$ as the positive side of the y axis.

Suppose P is a point whose representation in the rectangular cartesian coordinate system is (x, y) and (r, θ) is a polar coordinate representation of P. We distinguish two cases: $r > 0$ and $r < 0$. In the first case, if $r > 0$, then the point P is on the terminal side of the angle of radian measure θ, and $r = |\overline{OP}|$. Figure 9 shows such a case. Then

$$\cos \theta = \frac{x}{|\overline{OP}|} \qquad \sin \theta = \frac{y}{|\overline{OP}|}$$

$$= \frac{x}{r} \qquad\qquad = \frac{y}{r}$$

Thus

$$x = r \cos \theta \qquad \text{and} \qquad y = r \sin \theta \qquad\qquad (1)$$

In the second case, if $r < 0$, then the point P is on the extension of the terminal side and $r = -|\overline{OP}|$. See Figure 10. Then if Q is the point $(-x, -y)$,

$$\cos \theta = \frac{-x}{|\overline{OQ}|} \qquad \sin \theta = \frac{-y}{|\overline{OQ}|}$$

$$= \frac{-x}{|\overline{OP}|} \qquad\qquad = \frac{-y}{|\overline{OP}|}$$

$$= \frac{-x}{-r} \qquad\qquad = \frac{-y}{-r}$$

$$= \frac{x}{r} \qquad\qquad = \frac{y}{r}$$

Hence

$$x = r \cos \theta \quad \text{and} \quad y = r \sin \theta$$

These equations are the same as Equations (1); thus they hold in all cases.

From Equations (1) we can obtain the rectangular cartesian coordinates of a point when its polar coordinates are known. Also, from the equations we can find a polar equation of a curve if a rectangular cartesian equation is given.

To obtain equations that give a set of polar coordinates of a point when its rectangular cartesian coordinates are known, we square on both sides of each equation in (1) and obtain

$$x^2 = r^2 \cos^2 \theta \quad \text{and} \quad y^2 = r^2 \sin^2 \theta$$

Equating the sum of the left members of these equations to the sum of the right members, we have

$$x^2 + y^2 = r^2 \cos^2 \theta + r^2 \sin^2 \theta$$
$$x^2 + y^2 = r^2(\sin^2 \theta + \cos^2 \theta)$$
$$x^2 + y^2 = r^2$$
$$\Leftrightarrow \qquad r = \pm \sqrt{x^2 + y^2} \qquad (2)$$

From the equations in (1) and dividing, we have

$$\frac{r \sin \theta}{r \cos \theta} = \frac{y}{x}$$

$$\Leftrightarrow \qquad \tan \theta = \frac{y}{x} \qquad (3)$$

▷ **ILLUSTRATION 3**

The point whose polar coordinates are $(-6, \frac{7}{4}\pi)$ is located in Figure 11. We find its rectangular cartesian coordinates. From (1),

$$
\begin{aligned}
x &= r \cos \theta & y &= r \sin \theta \\
&= -6 \cos \tfrac{7}{4}\pi & &= -6 \sin \tfrac{7}{4}\pi \\
&= -6 \cdot \frac{\sqrt{2}}{2} & &= -6\left(-\frac{\sqrt{2}}{2}\right) \\
&= -3\sqrt{2} & &= 3\sqrt{2}
\end{aligned}
$$

So the point is $(-3\sqrt{2}, 3\sqrt{2})$. ◀

The graph of an equation in polar coordinates r and θ consists of all those points and only those points having at least one pair of coordinates that satisfy the equation. If an equation of a graph is given in polar coordinates, it is called a **polar equation** to distinguish it from a cartesian equa-

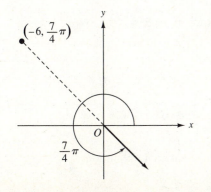

FIGURE 11

tion. We discuss methods of obtaining the graph of a polar equation in Section 10.4.

▶ **EXAMPLE 3**   *Finding a Cartesian Equation of a Graph from a Polar Equation*

Find a cartesian equation of a graph having the polar equation

$$r^2 = 4 \sin 2\theta$$

Solution   Because $\sin 2\theta = 2 \sin \theta \cos \theta$ we have $\sin 2\theta = 2(y/r)(x/r)$. With this substitution and $r^2 = x^2 + y^2$, we obtain from the given polar equation

$$x^2 + y^2 = 4(2)\frac{y}{r} \cdot \frac{x}{r}$$

$$x^2 + y^2 = \frac{8xy}{r^2}$$

$$x^2 + y^2 = \frac{8xy}{x^2 + y^2}$$

$$(x^2 + y^2)^2 = 8xy \qquad \blacktriangleleft$$

▶ **EXAMPLE 4**   *Finding Polar Coordinates of a Point from Its Rectangular Cartesian Coordinates*

Find (r, θ) if $r > 0$ and $0 \le \theta < 2\pi$, for the point whose rectangular cartesian coordinate representation is $(-\sqrt{3}, -1)$.

Solution   The point $(-\sqrt{3}, -1)$ is located in Figure 12. From (2), because $r > 0$,

$$r = \sqrt{3 + 1}$$

$$= 2$$

From (3), $\tan \theta = -1/(-\sqrt{3})$, and since $\pi < \theta < \tfrac{3}{2}\pi$,

$$\theta = \tfrac{7}{6}\pi$$

So the point is $(2, \tfrac{7}{6}\pi)$. $\qquad \blacktriangleleft$

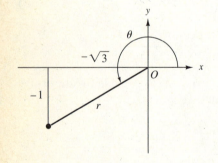

FIGURE 12

▶ **EXAMPLE 5**   *Finding a Polar Equation of a Graph from a Cartesian Equation*

Find a polar equation of the graph whose cartesian equation is

$$x^2 + y^2 - 4x = 0$$

Solution   Substituting $x = r \cos \theta$ and $y = r \sin \theta$ in

$$x^2 + y^2 - 4x = 0$$

we have

$$r^2 \cos^2 \theta + r^2 \sin^2 \theta - 4r \cos \theta = 0$$
$$r^2 - 4r \cos \theta = 0$$
$$r(r - 4 \cos \theta) = 0$$

Therefore

$$r = 0 \quad \text{or} \quad r - 4 \cos \theta = 0$$

The graph of $r = 0$ is the pole. However, the pole is a point on the graph of $r - 4 \cos \theta = 0$ because $r = 0$ when $\theta = \frac{1}{2}\pi$. Therefore a polar equation of the graph is

$$r = 4 \cos \theta$$

The graph of $x^2 + y^2 - 4x = 0$ is a circle. The equation may be written in the form

$$(x - 2)^2 + y^2 = 4$$

which is an equation of the circle with center at $(2, 0)$ and radius 2.   ◀

EXERCISES 10.3

In Exercises 1 through 4, locate the point having the given set of polar coordinates.

1. (a) $(3, \frac{1}{6}\pi)$ (b) $(2, \frac{2}{3}\pi)$ (c) $(1, \pi)$
     (d) $(4, \frac{5}{4}\pi)$ (e) $(5, \frac{11}{6}\pi)$

2. (a) $(4, \frac{1}{4}\pi)$ (b) $(3, \frac{3}{4}\pi)$ (c) $(1, \frac{7}{6}\pi)$
     (d) $(2, \frac{3}{2}\pi)$ (e) $(5, \frac{5}{3}\pi)$

3. (a) $(1, -\frac{1}{4}\pi)$ (b) $(3, -\frac{5}{6}\pi)$ (c) $(-1, \frac{1}{4}\pi)$
     (d) $(-3, \frac{5}{6}\pi)$ (e) $(-2, -\frac{1}{2}\pi)$

4. (a) $(5, -\frac{2}{3}\pi)$ (b) $(2, -\frac{7}{6}\pi)$ (c) $(-5, \frac{2}{3}\pi)$
     (d) $(-2, \frac{7}{6}\pi)$ (e) $(-4, -\frac{5}{4}\pi)$

In Exercises 5 through 10, locate the point having the given set of polar coordinates; then find another set of polar coordinates for the same point for which (a) $r < 0$ and $0 \le \theta < 2\pi$; (b) $r > 0$ and $-2\pi < \theta \le 0$; (c) $r < 0$ and $-2\pi < \theta \le 0$.

5. $(4, \frac{1}{4}\pi)$ **6.** $(3, \frac{5}{6}\pi)$ **7.** $(2, \frac{1}{2}\pi)$

8. $(3, \frac{3}{2}\pi)$ **9.** $(\sqrt{2}, \frac{7}{4}\pi)$ **10.** $(2, \frac{4}{3}\pi)$

11. Locate the point having the polar coordinates $(2, -\frac{1}{4}\pi)$. Find another set of polar coordinates for

this point for which (a) $r < 0$ and $0 \le \theta < 2\pi$; (b) $r < 0$ and $-2\pi < \theta \le 0$; (c) $r > 0$ and $2\pi \le \theta < 4\pi$.

12. Locate the point having the polar coordinates $(-3, -\frac{2}{3}\pi)$. Find another set of polar coordinates for this point for which (a) $r > 0$ and $0 \le \theta < 2\pi$; (b) $r > 0$ and $-2\pi < \theta \le 0$; (c) $r < 0$ and $2\pi \le \theta < 4\pi$.

In Exercises 13 through 20, locate the point having the given set of polar coordinates; then write two other sets of polar coordinates of the same point, one with the same value of r and one with an r having opposite sign.

13. $(3, -\frac{2}{3}\pi)$ **14.** $(\sqrt{2}, -\frac{1}{4}\pi)$

15. $(-4, \frac{5}{6}\pi)$ **16.** $(-2, \frac{4}{3}\pi)$

17. $(-2, -\frac{5}{4}\pi)$ **18.** $(-3, -\pi)$

19. $(2, 6)$ **20.** $(5, \frac{1}{6}\pi)$

In Exercises 21 and 22, find the rectangular cartesian coordinates of the points whose polar coordinates are given.

21. (a) $(3, \pi)$ **(b)** $(\sqrt{2}, -\frac{3}{4}\pi)$
    (c) $(-4, \frac{2}{3}\pi)$ **(d)** $(-1, -\frac{7}{6}\pi)$

22. (a) $(-2, -\frac{1}{2}\pi)$ **(b)** $(-1, \frac{1}{4}\pi)$
    (c) $(2, -\frac{7}{6}\pi)$ **(d)** $(2, \frac{7}{4}\pi)$

In Exercises 23 and 24, find a set of polar coordinates of the points whose rectangular cartesian coordinates are given. Take $r > 0$ and $0 \leq \theta < 2\pi$.

23. (a) $(1, -1)$ **(b)** $(-\sqrt{3}, 1)$
    (c) $(2, 2)$ **(d)** $(-5, 0)$

24. (a) $(3, -3)$ **(b)** $(-1, \sqrt{3})$
    (c) $(0, -2)$ **(d)** $(-2, -2\sqrt{3})$

In Exercises 25 through 34, find a polar equation of the graph having the given cartesian equation.

25. $x^2 + y^2 = a^2$ **26.** $x + y = 1$

27. $y^2 = 4(x + 1)$ **28.** $x^3 = 4y^2$

29. $x^2 = 6y - y^2$ **30.** $x^2 - y^2 = 16$

31. $(x^2 + y^2)^2 = 4(x^2 - y^2)$ **32.** $2xy = a^2$

33. $x^3 + y^3 - 3axy = 0$ **34.** $y = \dfrac{2x}{x^2 + 1}$

In Exercises 35 through 44, find a cartesian equation of the graph having the given polar equation.

35. $r^2 = 2 \sin 2\theta$ **36.** $r^2 \cos 2\theta = 10$

37. $r^2 = \cos \theta$ **38.** $r^2 = 4 \cos 2\theta$

39. $r^2 = \theta$ **40.** $r = 2 \sin 3\theta$

41. $r \cos \theta = -1$ **42.** $r^6 = r^2 \cos^2 \theta$

43. $r = \dfrac{6}{2 - 3 \sin \theta}$ **44.** $r = \dfrac{4}{3 - 2 \cos \theta}$

45. Explain why a one-to-one correspondence exists between the position of a point in the plane and its rectangular cartesian coordinates but no such correspondence exists for the point's polar coordinates. In your explanation give two points as examples: one in the first quadrant and one in the second quadrant.

10.4 GRAPHS OF POLAR EQUATIONS

GOALS

1. **Find polar equations of lines.**
2. **Find polar equations of circles.**
3. **Learn symmetry tests for polar graphs.**
4. **Describe properties of limaçons and roses from their polar equations.**
5. **Sketch and plot polar graphs.**

In Section 10.3, you learned that the graph of a polar equation, called a **polar graph,** consists of those points, and only those points, having at least one pair of polar coordinates that satisfy the equation. In this section, we discuss properties of such graphs and obtain them by hand and on a graphics calculator.

The equation

$$\theta = C$$

where C is a constant, is satisfied by all points having polar coordinates (r, C) whatever the value of r. Therefore, the graph of this equation is a line containing the pole and making an angle of radian measure C with the polar axis. See Figure 1. The same line is given by the equation

$$\theta = C \pm k\pi$$

where k is any integer.

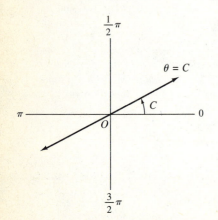

FIGURE 1

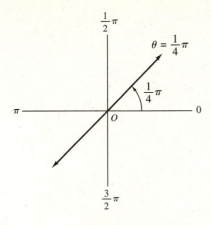

FIGURE 2

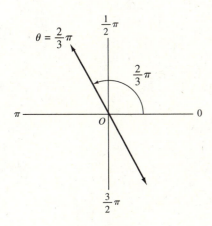

FIGURE 3

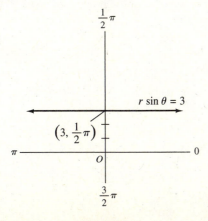

FIGURE 4

▷ **ILLUSTRATION 1**

(a) The graph of the equation

$$\theta = \tfrac{1}{4}\pi$$

appears in Figure 2. It is the line containing the pole and making an angle of radian measure $\tfrac{1}{4}\pi$ with the polar axis. The same line is given by the equations

$$\theta = \tfrac{5}{4}\pi \qquad \theta = \tfrac{9}{4}\pi \qquad \theta = -\tfrac{3}{4}\pi \qquad \theta = -\tfrac{7}{4}\pi$$

and so on.

(b) Figure 3 shows the graph of the equation

$$\theta = \tfrac{2}{3}\pi$$

It is the line passing through the pole and making an angle of radian measure $\tfrac{2}{3}\pi$ with the polar axis. Other equations of this line are

$$\theta = \tfrac{5}{3}\pi \qquad \theta = \tfrac{8}{3}\pi \qquad \theta = -\tfrac{1}{3}\pi \qquad \theta = -\tfrac{4}{3}\pi$$

and so on.   ◀

In general, the polar form of an equation of a line is not as simple as the cartesian form. However, if the line is parallel to either the polar axis or the $\tfrac{1}{2}\pi$ axis, the equation is fairly simple.

If a line is parallel to the polar axis and contains the point B whose cartesian coordinates are $(0, b)$, so that the polar coordinates of B are $(b, \tfrac{1}{2}\pi)$, then a cartesian equation is $y = b$. If we replace y by $r \sin \theta$, we have

$$r \sin \theta = b$$

which is a polar equation of any line parallel to the polar axis. If b is positive, the line is above the polar axis. If b is negative, it is below the polar axis.

▷ **ILLUSTRATION 2**

In Figure 4 we have the graph of the equation

$$r \sin \theta = 3$$

and in Figure 5 we have the graph of the equation

$$r \sin \theta = -3$$ ◀

Now consider a line parallel to the $\tfrac{1}{2}\pi$ axis or, equivalently, perpendicular to the polar axis. If the line goes through the point A whose cartesian coordinates are $(a, 0)$ and polar coordinates are $(a, 0)$, a cartesian equation is $x = a$. Replacing x by $r \cos \theta$, we obtain

$$r \cos \theta = a$$

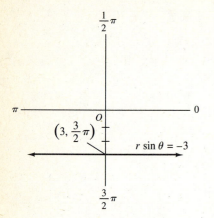

FIGURE 5

which is an equation of any line parallel to the $\frac{1}{2}\pi$ axis. If a is positive, the line is to the right of the $\frac{1}{2}\pi$ axis. If a is negative, the line is to the left of the $\frac{1}{2}\pi$ axis.

▷ **ILLUSTRATION 3**

Figure 6 shows the graph of the equation

$$r \cos \theta = 3$$

and Figure 7 shows the graph of the equation

$$r \cos \theta = -3$$

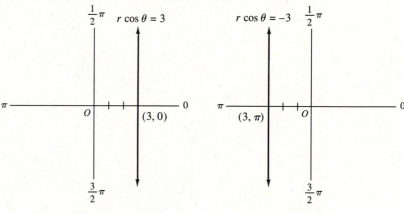

FIGURE 6   **FIGURE 7**   ◀

The graph of the equation

$$r = C$$

where C is any constant, is a circle whose center is at the pole and radius is $|C|$. The same circle is given by the equation

$$r = -C$$

▷ **ILLUSTRATION 4**

In Figure 8 we have the graph of the equation

$$r = 4$$

It is a circle with center at the pole and radius 4. The same circle is given by the equation

$$r = -4$$

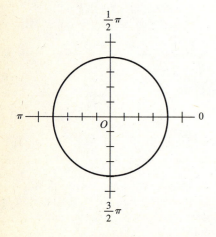

FIGURE 8

although the use of such an equation is uncommon.   ◀

As with the line, the general polar equation of a circle is not as simple as the cartesian form. However, further special cases of an equation of a circle are worth considering in polar form.

If a circle contains the origin (the pole) and has its center at the point having cartesian coordinates (a, b), then a cartesian equation of the circle is

$$x^2 + y^2 - 2ax - 2by = 0$$

A polar equation of this circle is

$$(r \cos \theta)^2 + (r \sin \theta)^2 - 2a(r \cos \theta) - 2b(r \sin \theta) = 0$$
$$r^2(\cos^2 \theta + \sin^2 \theta) - 2ar \cos \theta - 2br \sin \theta = 0$$
$$r^2 - 2ar \cos \theta - 2br \sin \theta = 0$$
$$r(r - 2a \cos \theta - 2b \sin \theta) = 0$$
$$r = 0 \qquad r - 2a \cos \theta - 2b \sin \theta = 0$$

Because the graph of the equation $r = 0$ is the pole and the pole ($r = 0$ when $\theta = \tan^{-1}(-a/b)$) is on the graph of $r - 2a \cos \theta - 2b \sin \theta = 0$, a polar equation of the circle is

$$r = 2a \cos \theta + 2b \sin \theta$$

When $b = 0$ in this equation, we have

$$r = 2a \cos \theta$$

This is a polar equation of the circle of radius $|a|$ units, tangent to the $\frac{1}{2}\pi$ axis, and with its center on the polar axis or its extension. If $a > 0$, the circle is to the right of the pole as in Figure 9, and if $a < 0$, the circle is to the left of the pole.

If $a = 0$ in the equation $r = 2a \cos \theta + 2b \sin \theta$, we have

$$r = 2b \sin \theta$$

which is a polar equation of the circle of radius $|b|$ units, tangent to the polar axis, and with its center on the $\frac{1}{2}\pi$ axis or its extension. If $b > 0$, the circle is above the pole, and if $b < 0$, the circle is below the pole.

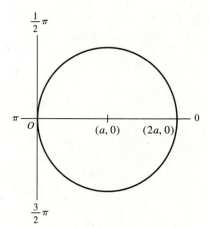

FIGURE 9

▶ **EXAMPLE 1** *Sketching Polar Graphs*

Sketch the graph of each of the following equations: **(a)** $r = 5 \cos \theta$; **(b)** $r = -6 \sin \theta$.

Solution

(a) The equation

$$r = 5 \cos \theta$$

is of the form $r = 2a \cos \theta$ with $a = \frac{5}{2}$. Thus the graph is a circle with center at the point having polar coordinates $(\frac{5}{2}, 0)$ and tangent to the $\frac{1}{2}\pi$ axis. The graph appears in Figure 10.

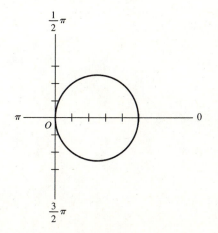

FIGURE 10

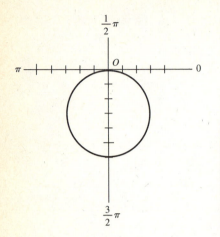

FIGURE 11

(b) The equation

$$r = -6 \sin \theta$$

is of the form $r = 2b \sin \theta$ with $b = -3$. The graph is the circle with center at the point having polar coordinates $(3, \frac{3}{2}\pi)$ and tangent to the polar axis. Figure 11 shows the graph.   ◀

Summary of Polar Equations of Lines and Circles

C, a, and b are constants

$\theta = C$	Line containing pole; making angle of radian measure C with polar axis.
$r \sin \theta = b$	Line parallel to polar axis; above polar axis if $b > 0$; below polar axis if $b < 0$.
$r \cos \theta = a$	Line parallel to $\frac{1}{2}\pi$ axis; to right of $\frac{1}{2}\pi$ axis if $b > 0$; to left of $\frac{1}{2}\pi$ axis if $b < 0$.
$r = C$	Circle; center at pole; radius is C.
$r = 2a \cos \theta$	Circle; radius is $|a|$; tangent to $\frac{1}{2}\pi$ axis; center on polar axis or its extension.
$r = 2b \sin \theta$	Circle; radius is $|b|$; tangent to polar axis; center on $\frac{1}{2}\pi$ axis or its extension.

Before discussing other polar graphs, we consider properties of symmetry. The following symmetry tests can be proved from the definition of symmetry of a graph given in Section 1.5.

Symmetry Tests

A polar graph is

(i) symmetric with respect to the polar axis if an equivalent equation is obtained when (r, θ) is replaced by either $(r, -\theta)$ or $(-r, \pi - \theta)$;

(ii) symmetric with respect to the $\frac{1}{2}\pi$ axis if an equivalent equation is obtained when (r, θ) is replaced by either $(r, \pi - \theta)$ or $(-r, -\theta)$;

(iii) symmetric with respect to the pole if an equivalent equation is obtained when (r, θ) is replaced by either $(-r, \theta)$ or $(r, \pi + \theta)$.

▷ **ILLUSTRATION 5**

For the graph of the equation

$$r = 4 \cos 2\theta$$

we test for symmetry with respect to the polar axis, the $\frac{1}{2}\pi$ axis, and the pole.

To test for symmetry with respect to the polar axis, we replace (r, θ) by $(r, -\theta)$ and obtain $r = 4 \cos(-2\theta)$, which is equivalent to $r = 4 \cos 2\theta$. So the graph is symmetric with respect to the polar axis.

To test for symmetry with respect to the $\frac{1}{2}\pi$ axis, we replace (r, θ) by $(r, \pi - \theta)$ in the given equation and get $r = 4 \cos(2(\pi - \theta))$ or, equivalently, $r = 4 \cos(2\pi - 2\theta)$, which is equivalent to $r = 4 \cos 2\theta$. Therefore the graph is symmetric with respect to the $\frac{1}{2}\pi$ axis.

To test for symmetry with respect to the pole, we replace (r, θ) by $(-r, \theta)$ and obtain the equation $-r = 4 \cos 2\theta$, which is not equivalent to the given equation. But we must also determine if the other set of coordinates works. We replace (r, θ) by $(r, \pi + \theta)$ and obtain $r = 4 \cos 2(\pi + \theta)$ or, equivalently, $r = 4 \cos(2\pi + 2\theta)$, which is equivalent to the equation $r = 4 \cos 2\theta$. Therefore the graph is symmetric with respect to the pole. ◀

When sketching a polar graph, we determine if it contains the pole by substituting 0 for r and solving for θ.

▶ **EXAMPLE 2** *Sketching Polar Graphs*

Sketch the graph of the equation

$$r = 1 - 2 \cos \theta$$

Solution   Because we obtain an equivalent equation when (r, θ) is replaced by $(r, -\theta)$, the graph is symmetric with respect to the polar axis.

If $r = 0$, we obtain $\cos \theta = \frac{1}{2}$, and if $0 \leq \theta \leq \pi$, then $\theta = \frac{1}{3}\pi$. Thus the point $(0, \frac{1}{3}\pi)$, the pole, is on the graph. Table 1 gives coordinates of some other points on the graph. From these points we sketch one-half of the graph; the remainder is sketched from its symmetry with respect to the polar axis. The graph appears in Figure 12. ◀

In Figure 12, the graph is sketched on a polar coordinate system. You are asked to sketch some polar graphs on a polar coordinate system in Exercises 39 through 48.

Table 1

θ	r
0	-1
$\frac{1}{6}\pi$	$1 - \sqrt{3}$
$\frac{1}{3}\pi$	0
$\frac{1}{2}\pi$	1
$\frac{2}{3}\pi$	2
$\frac{5}{6}\pi$	$1 + \sqrt{3}$
π	3

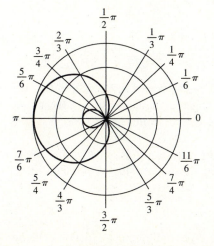

FIGURE 12

Parametric equations can be used to plot polar graphs on a graphics calculator. The following theorem shows how to define a polar graph by a pair of parametric equations.

THEOREM 1

The graph of the polar equation $r = f(\theta)$ is defined by the parametric equations

$$x = f(t) \cos t \qquad \text{and} \qquad y = f(t) \sin t$$

Proof   Let (x, y) be the cartesian representation of a point P whose polar representation is (r, θ). Then

$$x = r \cos \theta \qquad \text{and} \qquad y = r \sin \theta$$

Because $r = f(\theta)$, we have

$$x = f(\theta) \cos \theta \qquad \text{and} \qquad y = f(\theta) \sin \theta$$

Replacing θ by t so that the parameter is t, we have

$$x = f(t) \cos t \qquad \text{and} \qquad y = f(t) \sin t \qquad\qquad \blacksquare$$

▷ **ILLUSTRATION 6**

To plot the polar graph of Example 2 on our graphics calculator, we use the parametric equations of Theorem 1:

$$x = f(t) \cos t \qquad \text{and} \qquad y = f(t) \sin t$$

where $f(\theta) = 1 - 2 \cos \theta$; that is,

$$x = (1 - 2 \cos t) \cos t \qquad \text{and} \qquad y = (1 - 2 \cos t) \sin t$$

With our calculator in parametric and radian mode and $0 \le t \le 2\pi$, we select $[-3, 3]$ by $[-2, 2]$ as our viewing rectangle and obtain the graph shown in Figure 13, which agrees with the curve in Figure 12.   ◄

The polar graph in Example 2 and Illustration 6 is called a *limaçon*, a French word from the Latin *limax* meaning snail or slug. A **limaçon** is the graph of an equation of the form

$$r = a \pm b \cos \theta \qquad \text{or} \qquad r = a \pm b \sin \theta$$

where $a > 0$ and $b > 0$. There are four types of limaçons, depending on the ratio a/b.

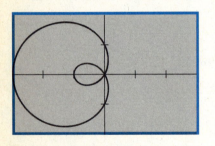

$[-3, 3]$ by $[-2, 2]$
$r = 1 - 2 \cos \theta$
FIGURE 13

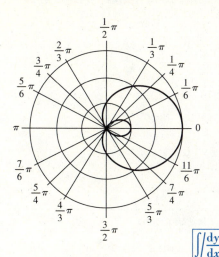

$$0 < \frac{a}{b} < 1$$

limaçon with a loop

(a)

Types of Limaçons

From the equation $r = a + b \cos \theta$ where $a > 0$ and $b > 0$.

1. $0 < \dfrac{a}{b} < 1$ **Limaçon with a loop.** See Figure 14(a).

2. $\dfrac{a}{b} = 1$ **Cardioid** (heart-shaped). See Figure 14(b).

3. $1 < \dfrac{a}{b} < 2$ **Limaçon with a dent.** See Figure 14(c).

4. $2 \le \dfrac{a}{b}$ **Convex limaçon** (no dent). See Figure 14(d).

$\iint \boxed{\frac{dy}{dx}}$ When you study calculus, where horizontal and vertical tangent lines of polar graphs are discussed, the reason that limaçons of type 3 have a dent and those of type 4 have no dent will be apparent.

From the equation of a limaçon, we can also determine its symmetry and the direction in which it points.

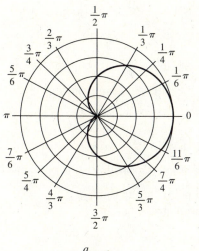

$$\frac{a}{b} = 1$$

cardioid

(b)

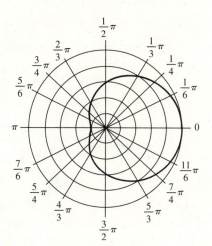

$$1 < \frac{a}{b} < 2$$

limaçon with a dent

(c)

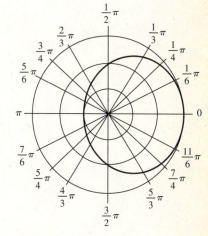

$$2 \le \frac{a}{b}$$

convex limaçon

(d)

FIGURE 14

Symmetry and Direction of a Limaçon

$a > 0$ and $b > 0$

$r = a + b \cos \theta$	Symmetry with respect to polar axis; points to right.
$r = a - b \cos \theta$	Symmetry with respect to polar axis; points to left.
$r = a + b \sin \theta$	Symmetry with respect to the $\frac{1}{2}\pi$ axis; points upward.
$r = a - b \sin \theta$	Symmetry with respect to the $\frac{1}{2}\pi$ axis; points downward.

▶ **EXAMPLE 3** *Describing and Plotting Limaçons*

For each of the following limaçons, determine the type, its symmetry, and the direction in which it points. Plot the limaçon: **(a)** $r = 3 + 2 \sin \theta$; **(b)** $r = 2 + 2 \cos \theta$; **(c)** $r = 2 - \sin \theta$.

Solution

(a) The equation $r = 3 + 2 \sin \theta$ is of the form $r = a + b \sin \theta$ with $a = 3$ and $b = 2$. Because $\dfrac{a}{b} = \dfrac{3}{2}$ and $1 < \dfrac{3}{2} < 2$, the graph is a limaçon with a dent. It is symmetric with respect to the $\frac{1}{2}\pi$ axis and points upward.

We plot the graph of the parametric equations

$$x = (3 + 2 \sin t)\cos t \quad \text{and} \quad y = (3 + 2 \sin t)\sin t$$

for $0 \le t \le 2\pi$ in the viewing rectangle $[-9, 9]$ by $[-6, 6]$ and obtain the limaçon in Figure 15.

(b) The equation $r = 2 + 2 \cos \theta$ is of the form $r = a + b \cos \theta$ with $a = 2$ and $b = 2$. Because $\dfrac{a}{b} = 1$, the graph is a cardioid. It is symmetric with respect to the polar axis and points to the right.

We plot the cardioid from the parametric equations

$$x = (2 + 2 \cos t)\cos t \quad \text{and} \quad y = (2 + 2 \cos t)\sin t$$

for $0 \le t \le 2\pi$ in the viewing rectangle $[-7.5, 7.5]$ by $[-5, 5]$ as shown in Figure 16.

(c) The equation $r = 2 - \sin \theta$ is of the form $r = a - b \sin \theta$ with $a = 2$ and $b = 1$. Because $\dfrac{a}{b} = 2$, the graph is a convex limaçon. It is symmetric with respect to the $\frac{1}{2}\pi$ axis and points downward.

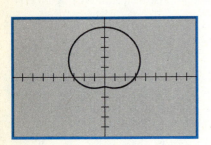

$[-9, 9]$ by $[-6, 6]$
$r = 3 + 2 \sin \theta$
FIGURE 15

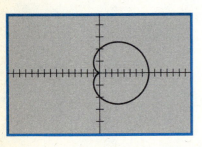

$[-7.5, 7.5]$ by $[-5, 5]$
$r = 2 + 2 \cos \theta$
FIGURE 16

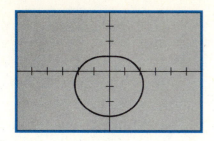

[−6, 6] by [−4, 4]
$r = 2 - \sin \theta$
FIGURE 17

The graph appears in Figure 17, plotted from the parametric equations

$$x = (2 - \sin t)\cos t \quad \text{and} \quad y = (2 - \sin t)\sin t$$

for $0 \leq t \leq 2\pi$ in the viewing rectangle $[−6, 6]$ by $[−4, 4]$. ◄

The graph of an equation of the form

$$r = a \cos n\theta \quad \text{or} \quad r = a \sin n\theta$$

is a **rose,** having n leaves if n is odd and $2n$ leaves if n is even. The length of a leaf is $|a|$.

► **EXAMPLE 4** *Describing and Plotting a Polar Graph*

Describe and plot the graph of the equation

$$r = 4 \cos 2\theta$$

Solution The equation is of the form $r = a \cos n\theta$ where n is 2. Because n is even, the graph is a four-leafed rose. The length of a leaf is 4. In Illustration 5, we proved that the graph is symmetric with respect to the polar axis, the $\frac{1}{2}\pi$ axis, and the pole. The graph contains the pole because when $r = 0$ we have

$$\cos 2\theta = 0$$

from which we obtain, for $0 \leq \theta \leq 2\pi$,

$$\theta = \tfrac{1}{4}\pi \quad \theta = \tfrac{3}{4}\pi \quad \theta = \tfrac{5}{4}\pi \quad \theta = \tfrac{7}{4}\pi$$

We plot the graph of the parametric equations

$$x = 4 \cos 2t \cos t \quad \text{and} \quad y = 4 \cos 2t \sin t$$

for $0 \leq t \leq 2\pi$ in the viewing rectangle $[−7.5, 7.5]$ by $[−5, 5]$ as shown in Figure 18. The graph agrees with our description. ◄

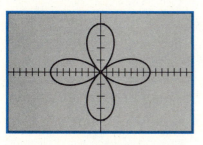

[−7.5, 7.5] by [−5, 5]
$r = 4 \cos 2\theta$
FIGURE 18

Observe that if in the equations for a rose we take $n = 1$, we get

$$r = a \cos \theta \quad \text{or} \quad r = a \sin \theta$$

which are equations for a circle. A circle can, therefore, be considered as a one-leafed rose.

Other polar graphs that occur frequently are *spirals* (see Exercises 25 through 28) and *lemniscates* (see Exercises 29 through 32). The graph in the next example is called *a spiral of Archimedes.*

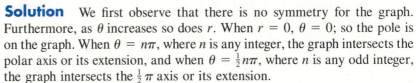

▶ **EXAMPLE 5** *Describing and Plotting a Polar Graph*

Describe and plot the graph of the equation

$$r = \theta \qquad \theta \geq 0$$

Solution We first observe that there is no symmetry for the graph. Furthermore, as θ increases so does r. When $r = 0$, $\theta = 0$; so the pole is on the graph. When $\theta = n\pi$, where n is any integer, the graph intersects the polar axis or its extension, and when $\theta = \frac{1}{2}n\pi$, where n is any odd integer, the graph intersects the $\frac{1}{2}\pi$ axis or its extension.

We plot the graph of the parametric equations

$$x = t \cos t \quad \text{and} \quad y = t \sin t$$

for $t \geq 0$ in the viewing rectangle $[-42, 42]$ by $[-28, 28]$. We obtain the graph in Figure 19, which agrees with our description. ◀

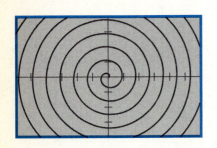

$[-42, 42]$ by $[-28, 28]$
$r = \theta \quad \theta \geq 0$
FIGURE 19

EXERCISES 10.4

In Exercises 1 through 8, sketch the graph of the equation.

1. (a) $\theta = \frac{1}{3}\pi$ **(b)** $r = \frac{1}{3}\pi$

2. (a) $\theta = \frac{3}{4}\pi$ **(b)** $r = \frac{3}{4}\pi$

3. (a) $\theta = 2$ **(b)** $r = 2$

4. (a) $\theta = -3$ **(b)** $r = -3$

5. (a) $r \cos \theta = 4$ **(b)** $r = 4 \cos \theta$

6. (a) $r \sin \theta = 2$ **(b)** $r = 2 \sin \theta$

7. (a) $r \sin \theta = -4$ **(b)** $r = -4 \sin \theta$

8. (a) $r \cos \theta = -5$ **(b)** $r = -5 \cos \theta$

In Exercises 9 through 18, determine the type of limaçon, its symmetry, and the direction in which it points. Plot the limaçon.

9. $r = 4(1 - \cos \theta)$ **10.** $r = 3(1 - \sin \theta)$

11. $r = 2(1 + \sin \theta)$ **12.** $r = 3(1 + \cos \theta)$

13. $r = 2 - 3 \sin \theta$ **14.** $r = 4 - 3 \sin \theta$

15. $r = 3 - 2 \cos \theta$ **16.** $r = 3 - 4 \cos \theta$

17. $r = 4 + 2 \sin \theta$ **18.** $r = 6 + 2 \cos \theta$

In Exercises 19 through 38, describe and plot the graph of the equation.

19. $r = 2 \sin 3\theta$ **20.** $r = 4 \sin 5\theta$

21. $r = 2 \cos 4\theta$ **22.** $r = 3 \cos 2\theta$

23. $r = 4 \sin 2\theta$ **24.** $r = 3 \cos 3\theta$

25. $r = e^{\theta}$ (logarithmic spiral)

26. $r = e^{\theta/3}$ (logarithmic spiral)

27. $r = \dfrac{1}{\theta}$ (reciprocal spiral)

28. $r = 2\theta$ (spiral of Archimedes)

29. $r^2 = 9 \sin 2\theta$ (lemniscate)

30. $r^2 = 16 \cos 2\theta$ (lemniscate)

31. $r^2 = -25 \cos 2\theta$ (lemniscate)

32. $r^2 = -4 \sin 2\theta$ (lemniscate)

33. $r = 2 \sin \theta \tan \theta$ (cissoid)

34. $r^2 = 8\theta$ (Fermat's spiral)

35. $r = 2 \sec \theta - 1$ (conchoid of Nicomedes)

36. $r = 2 \csc \theta + 3$ (conchoid of Nicomedes)

37. $r = |\sin 2\theta|$ **38.** $r = 2|\cos \theta|$

39. Polar graphs can be sketched on polar graph paper using a polar coordinate system such as in Figure 12. Construct such a system with ruler, compass, and protractor. On this system sketch the graph of $r = 1 + 4 \sin \theta$.

In Exercises 40 through 48, follow the instructions of Exercise 39 to sketch the graph of the polar equation.

40. $r = 5 - 3 \sin \theta$ **41.** $r = 4 - \cos \theta$

42. $r = 3 + 5 \cos \theta$ **43.** $r = 2(1 - \sin \theta)$

44. $r = 6(1 - \cos \theta)$ **45.** $r = 4 + 3 \cos \theta$

46. $r = 3 + \sin \theta$ **47.** $r = 3 \cos 5\theta$

48. $r = 2 \sin 4\theta$

In Exercises 49 through 52, plot the graphs of the two equations in the same viewing rectangle. Then use trace and zoom-in to approximate to two significant digits the rectangular cartesian coordinates of the points of intersection of the graphs.

49. $\begin{cases} r = 3 \\ r = 2(1 + \cos \theta) \end{cases}$ **50.** $\begin{cases} r = 5 \cos \theta \\ r = 5 \sin \theta \end{cases}$

51. $\begin{cases} r = 2 \cos 2\theta \\ r = 2 \sin \theta \end{cases}$ **52.** $\begin{cases} r = 2 \sin 3\theta \\ r = 4 \sin \theta \end{cases}$

In Exercises 53 through 56, find algebraically the polar coordinates of the points of intersection of the graphs of the indicated exercise and compare your answers with the rectangular cartesian coordinates found graphically. (Hint: Equate the two values of r in terms of θ and solve the resulting trigonometric equation.)

53. Exercise 49   **54.** Exercise 50

55. Exercise 51   **56.** Exercise 52

10.5 POLAR FORM OF COMPLEX NUMBERS

GOALS
1. Represent complex numbers as points in the complex plane.
2. Find the absolute value of a complex number.
3. Learn the polar form of a complex number.
4. Express a complex number in standard polar form.
5. Obtain formulas for the product and quotient of two complex numbers expressed in polar form.
6. Express in cartesian form the product of two complex numbers given in polar form.
7. Express in cartesian form the quotient of two complex numbers given in polar form.

In previous chapters, you have seen how complex numbers can be put to "real" use when they arise naturally in problem solving. If you continue your study of mathematics to courses beyond calculus, you will learn that complex numbers have significance in both theoretical and applied aspects of the subject. In this section we give a geometric representation of complex numbers and show how these numbers can be expressed in terms of polar coordinates. Then, in Section 10.6, we apply the *polar form* to compute powers and roots of complex numbers.

The set of complex numbers can be represented by points in a rectangular coordinate system. In such a representation, the horizontal axis is called the **real axis,** and the vertical axis is called the **imaginary axis.** The points of the plane, called the **complex plane,** are then placed in one-to-one correspondence with the complex numbers. The geometric representation

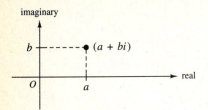

FIGURE 1

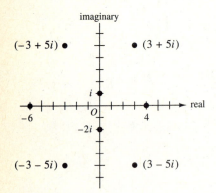

FIGURE 2

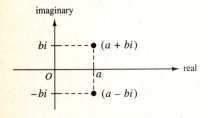

FIGURE 3

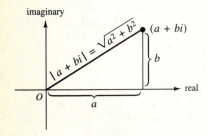

FIGURE 4

of the complex number $a + bi$ is the point $P(a, b)$ in the complex plane, and the point is called the **graph** of the number. Refer to Figure 1.

▶ **EXAMPLE 1** *Representing Complex Numbers as Points in the Complex Plane*

Show the geometric representation of each of the following complex numbers as a point in the complex plane: $3 + 5i$; $-3 + 5i$; $-3 - 5i$; $3 - 5i$; i; $-2i$; 4; and -6.

Solution   The points are shown in Figure 2.   ◀

Observe that any real number is represented by a point on the real axis and any pure imaginary number is represented by a point on the imaginary axis. The geometric representations of a complex number $a + bi$ and its conjugate $a - bi$ are points that are symmetric with respect to the real axis. See Figure 3.

If $a + bi$ is an arbitrary complex number, then the distance in the complex plane from the origin to the graph of $a + bi$ is $\sqrt{a^2 + b^2}$. Refer to Figure 4. This number is called the *absolute value* or *modulus* of $a + bi$.

DEFINITION   **The Absolute Value of a Complex Number**

The **absolute value,** or **modulus,** of the complex number $a + bi$, denoted by $|a + bi|$, is given by

$$|a + bi| = \sqrt{a^2 + b^2}$$

▶ **EXAMPLE 2** *Finding the Absolute Value of a Complex Number*

Write the following numbers without absolute-value bars: **(a)** $|6 - i|$; **(b)** $|-3 + 2i|$; **(c)** $|-4 - 3i|$; **(d)** $|5i|$.

Solution

(a) $|6 - i| = \sqrt{6^2 + (-1)^2}$ **(b)** $|-3 + 2i| = \sqrt{(-3)^2 + 2^2}$
$= \sqrt{37}$ $= \sqrt{13}$

(c) $|-4 - 3i| = \sqrt{(-4)^2 + (-3)^2}$ **(d)** $|5i| = \sqrt{0^2 + 5^2}$
$= \sqrt{25}$ $= \sqrt{25}$
$= 5$ $= 5$ ◀

We do not refer to one complex number with nonzero imaginary part being greater than, or less than, another complex number. However, we can

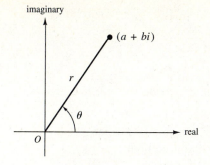

FIGURE 5

use the symbols $<$ and $>$ with the absolute values of complex numbers because they are real numbers. If $|z_1| < |z_2|$, then the point in the complex plane representing z_1 is closer to the origin than the point representing z_2. Furthermore, if $|z_1| = |z_2|$, the points representing z_1 and z_2 are at the same distance from the origin of the complex plane; that is, they lie on a circle with center at the origin.

If the point (a, b) representing $a + bi$ is expressed in polar coordinates with $r \geq 0$, as indicated in Figure 5, then

$$a = r \cos \theta \quad \text{and} \quad b = r \sin \theta$$

Therefore

$$a + bi = r \cos \theta + (r \sin \theta)i$$

$$a + bi = r(\cos \theta + i \sin \theta)$$

The right side of this equation is called the **polar form** (or **trigonometric form**) of the complex number $a + bi$. It is sometimes abbreviated as $r \operatorname{cis} \theta$, where *cis* comes from cosine, *i*, and sine. In contrast, $a + bi$ is called the **cartesian form** (or **algebraic form**).

Because we required r to be nonnegative,

$$r = |a + bi|$$

The angle θ is called an **argument** (or **amplitude**) of $a + bi$. The argument may be measured in either degrees or radians. Because

$$\sin(\theta + 2k\pi) = \sin \theta \quad \text{and} \quad \cos(\theta + 2k\pi) = \cos \theta$$

where $k \in J$, then if θ is an argument of $a + bi$, so is an angle $\theta + 2k\pi$, $k \in J$. The argument θ, for which $0 \leq \theta < 2\pi$, is called the **principal argument.** When the principal argument is used, the complex number is said to be in **standard polar form.**

▷ ILLUSTRATION 1

Let us express the complex number $\sqrt{3} + i$ in standard polar form. Figure 6 shows the geometric representation of this number. We first determine r and θ from the formulas

$$r = |a + bi| \qquad \cos \theta = \frac{a}{r} \qquad \sin \theta = \frac{b}{r}$$

$$= \sqrt{a^2 + b^2}$$

Because $a = \sqrt{3}$ and $b = 1$, we have

$$r = |\sqrt{3} + i| \qquad \cos \theta = \frac{\sqrt{3}}{2} \qquad \sin \theta = \frac{1}{2}$$

$$= \sqrt{(\sqrt{3})^2 + 1^2}$$

$$= 2$$

FIGURE 6

From the values of $\sin \theta$ and $\cos \theta$, we obtain $\theta = \frac{1}{6}\pi$. Therefore

$$\sqrt{3} + i = r(\cos \theta + i \sin \theta)$$
$$= 2(\cos \tfrac{1}{6}\pi + i \sin \tfrac{1}{6}\pi) \qquad \blacktriangleleft$$

▷ **ILLUSTRATION 2**

To express the complex number $-3 + 3i$ in standard polar form, we first determine r and θ with $a = -3$ and $b = 3$. See Figure 7.

imaginary

$(-3 + 3i)$

θ

real

O

FIGURE 7

$$r = \sqrt{a^2 + b^2} \qquad \cos \theta = \frac{a}{r} \qquad \sin \theta = \frac{b}{r}$$

$$= \sqrt{(-3)^2 + 3^2}$$

$$= 3\sqrt{2} \qquad\qquad = \frac{-3}{3\sqrt{2}} \qquad = \frac{3}{3\sqrt{2}}$$

$$= -\frac{1}{\sqrt{2}} \qquad = \frac{1}{\sqrt{2}}$$

From the values of $\cos \theta$ and $\sin \theta$, it follows that $\theta = \frac{3}{4}\pi$. Thus

$$-3 + 3i = r(\cos \theta + i \sin \theta)$$
$$= 3\sqrt{2}(\cos \tfrac{3}{4}\pi + i \sin \tfrac{3}{4}\pi) \qquad \blacktriangleleft$$

If a complex number is real, then in cartesian form it is $a + 0i$. For this number, $r = |a|$. If $a > 0$, $\cos \theta = 1$, $\sin \theta = 0$, and so $\theta = 0$; if $a < 0$, $\cos \theta = -1$, $\sin \theta = 0$, and so $\theta = \pi$. Therefore, in polar form

$$a = a(\cos 0 + i \sin 0) \qquad \text{if } a > 0$$
$$a = |a|(\cos \pi + i \sin \pi) \quad \text{if } a < 0$$

For the real number 0, the polar form is $0(\cos \theta + i \sin \theta)$ where θ is any angle.

For the pure imaginary number $0 + bi$, $r = |b|$. If $b > 0$, $\cos \theta = 0$, $\sin \theta = 1$, and so $\theta = \frac{1}{2}\pi$; if $b < 0$, $\cos \theta = 0$, $\sin \theta = -1$, and so $\theta = \frac{3}{2}\pi$. Hence in polar form

$$bi = b(\cos \tfrac{1}{2}\pi + i \sin \tfrac{1}{2}\pi) \qquad \text{if } b > 0$$
$$bi = |b|(\cos \tfrac{3}{2}\pi + i \sin \tfrac{3}{2}\pi) \qquad \text{if } b < 0$$

▶ **EXAMPLE 3** *Expressing Complex Numbers in Standard Polar Form*

Write the following numbers in standard polar form: **(a)** $4 - 4\sqrt{3}i$; **(b)** $2 + 5i$.

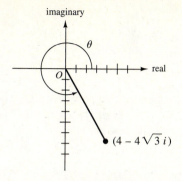

FIGURE 8

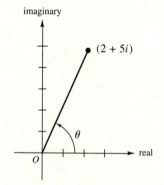

FIGURE 9

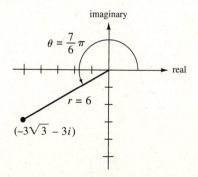

FIGURE 10

Solution We use the formulas

$$r = \sqrt{a^2 + b^2} \qquad \cos \theta = \frac{a}{r} \qquad \sin \theta = \frac{b}{r}$$

(a) If $a + bi = 4 - 4\sqrt{3}i$, then $a = 4$ and $b = -4\sqrt{3}$. The point representing the complex number is shown in Figure 8. From the formulas we get

$$r = \sqrt{16 + 48} \qquad \cos \theta = \frac{4}{8} \qquad \sin \theta = \frac{-4\sqrt{3}}{8}$$
$$= 8 \qquad\qquad = \frac{1}{2} \qquad\qquad = -\frac{\sqrt{3}}{2}$$

From the values of $\cos \theta$ and $\sin \theta$ we determine that $\theta = \frac{5}{3}\pi$. Therefore

$$4 - 4\sqrt{3}i = r(\cos \theta + i \sin \theta)$$
$$= 8(\cos \tfrac{5}{3}\pi + i \sin \tfrac{5}{3}\pi)$$

(b) If $a + bi = 2 + 5i$, then $a = 2$ and $b = 5$. See Figure 9. We compute r, $\cos \theta$, and $\sin \theta$ from the formulas and obtain

$$r = \sqrt{4 + 25} \qquad \cos \theta = \frac{2}{\sqrt{29}} \qquad \sin \theta = \frac{5}{\sqrt{29}}$$
$$= \sqrt{29}$$

Because $\cos \theta > 0$ and $\sin \theta > 0$, θ is in the first quadrant. Furthermore, because $\tan \theta = \dfrac{b}{a}$, $\tan \theta = \dfrac{5}{2}$. Thus $\theta = \arctan \dfrac{5}{2}$. Therefore

$$2 + 5i = r(\cos \theta + i \sin \theta)$$
$$= \sqrt{29}[\cos(\arctan \tfrac{5}{2}) + i \sin(\arctan \tfrac{5}{2})]$$

Because $\arctan \tfrac{5}{2} \approx 1.19$ (or 68.2°), we can write

$$2 + 5i \approx \sqrt{29}(\cos 1.19 + i \sin 1.19)$$

or

$$2 + 5i \approx \sqrt{29}(\cos 68.2° + i \sin 68.2°) \qquad \blacktriangleleft$$

The next illustration shows the conversion of the polar form of a complex number to the equivalent cartesian form.

▷ **ILLUSTRATION 3**

The complex number $6(\cos \tfrac{7}{6}\pi + i \sin \tfrac{7}{6}\pi)$ appears in Figure 10.

$$6\left(\cos\frac{7}{6}\pi + i\sin\frac{7}{6}\pi\right) = 6\left[-\frac{\sqrt{3}}{2} + i\left(-\frac{1}{2}\right)\right]$$
$$= -3\sqrt{3} - 3i \qquad \blacktriangleleft$$

To find formulas for the product and quotient of two complex numbers when the numbers are expressed in polar form, we apply trigonometric identities. The following theorem summarizes the results.

THEOREM 1

If

$$z_1 = r_1(\cos\theta_1 + i\sin\theta_1) \qquad \text{and} \qquad z_2 = r_2(\cos\theta_2 + i\sin\theta_2)$$

then

(i) $z_1 \cdot z_2 = r_1 r_2[\cos(\theta_1 + \theta_2) + i\sin(\theta_1 + \theta_2)]$

(ii) $\dfrac{z_1}{z_2} = \dfrac{r_1}{r_2}[\cos(\theta_1 - \theta_2) + i\sin(\theta_1 - \theta_2)]$

Proof of (i)

$z_1 \cdot z_2$

$= r_1(\cos\theta_1 + i\sin\theta_1) \cdot r_2(\cos\theta_2 + i\sin\theta_2)$

$= r_1 r_2[\cos\theta_1 \cdot \cos\theta_2 + \cos\theta_1(i\sin\theta_2) + (i\sin\theta_1)\cos\theta_2 + (i\sin\theta_1)(i\sin\theta_2)]$

$= r_1 r_2[\cos\theta_1\cos\theta_2 + i\cos\theta_1\sin\theta_2 + i\sin\theta_1\cos\theta_2 + i^2\sin\theta_1\sin\theta_2]$

$= r_1 r_2[(\cos\theta_1\cos\theta_2 - \sin\theta_1\sin\theta_2) + i(\cos\theta_1\sin\theta_2 + \sin\theta_1\cos\theta_2)]$

$= r_1 r_2[\cos(\theta_1 + \theta_2) + i\sin(\theta_1 + \theta_2)] \qquad \blacksquare$

The proof of part (ii) is similar and is left as an exercise. See Exercise 50.

▷ **ILLUSTRATION 4**

Let $z_1 = -3\sqrt{3} - 3i$ and $z_2 = 4 - 4\sqrt{3}i$. From Illustration 3

$$z_1 = 6(\cos\tfrac{7}{6}\pi + i\sin\tfrac{7}{6}\pi)$$

and from Example 3(a)

$$z_2 = 8(\cos \tfrac{5}{3}\pi + i \sin \tfrac{5}{3}\pi)$$

Applying Theorem 1(i), we obtain

$$z_1 \cdot z_2 = 6 \cdot 8[\cos(\tfrac{7}{6}\pi + \tfrac{5}{3}\pi) + i \sin(\tfrac{7}{6}\pi + \tfrac{5}{3}\pi)]$$

$$= 48(\cos \tfrac{17}{6}\pi + i \sin \tfrac{17}{6}\pi)$$

$$= 48\left(-\frac{\sqrt{3}}{2} + \frac{1}{2}i\right)$$

$$= -24\sqrt{3} + 24i$$

We can check this result by using the procedure of Section 1.3. We have

$$z_1 \cdot z_2 = (-3\sqrt{3} - 3i)(4 - 4\sqrt{3}i)$$

$$= -12\sqrt{3} + 36i - 12i + 12\sqrt{3}i^2$$

$$= -24\sqrt{3} + 24i$$

From Theorem 1(ii) we get

$$\frac{z_1}{z_2} = \frac{6}{8}\left[\cos\left(\frac{7}{6}\pi - \frac{5}{3}\pi\right) + i \sin\left(\frac{7}{6}\pi - \frac{5}{3}\pi\right)\right]$$

$$= \tfrac{3}{4}[\cos(-\tfrac{1}{2}\pi) + i \sin(-\tfrac{1}{2}\pi)]$$

$$= \tfrac{3}{4}[0 + i(-1)]$$

$$= -\tfrac{3}{4}i$$

As a check, we have

$$\frac{z_1}{z_2} = \frac{-3\sqrt{3} - 3i}{4 - 4\sqrt{3}i}$$

$$= \frac{(-3\sqrt{3} - 3i)(4 + 4\sqrt{3}i)}{(4 - 4\sqrt{3}i)(4 + 4\sqrt{3}i)}$$

$$= \frac{-12\sqrt{3} - 36i - 12i - 12\sqrt{3}i^2}{16 - 48i^2}$$

$$= \frac{-48i}{64}$$

$$= -\frac{3}{4}i$$

◀

▶ **EXAMPLE 4** *Expressing in Cartesian Form the Product of Two Complex Numbers Given in Polar Form*

Write

$$\tfrac{2}{3}(\cos \tfrac{2}{9}\pi + i\sin \tfrac{2}{9}\pi)\cdot 6(\cos \tfrac{19}{36}\pi + i\sin \tfrac{19}{36}\pi)$$

in the form $a + bi$.

Solution   From Theorem 1(i)

$$\tfrac{2}{3}(\cos \tfrac{2}{9}\pi + i\sin \tfrac{2}{9}\pi)\cdot 6(\cos \tfrac{19}{36}\pi + i\sin \tfrac{19}{36}\pi)$$
$$= \tfrac{2}{3}\cdot 6[\cos(\tfrac{2}{9}\pi + \tfrac{19}{36}\pi) + i\sin(\tfrac{2}{9}\pi + \tfrac{19}{36}\pi)]$$
$$= 4(\cos \tfrac{3}{4}\pi + i\sin \tfrac{3}{4}\pi)$$
$$= 4\left[-\frac{\sqrt{2}}{2} + i\left(\frac{\sqrt{2}}{2}\right)\right]$$
$$= -2\sqrt{2} + 2\sqrt{2}i$$ ◀

▶ **EXAMPLE 5** *Expressing in Cartesian Form the Quotient of Two Complex Numbers Given in Polar Form*

Write the following quotient in the form $a + bi$:

$$\frac{4(\cos 345° + i\sin 345°)}{5(\cos 105° + i\sin 105°)}$$

Solution   From Theorem 1(ii)

$$\frac{4(\cos 345° + i\sin 345°)}{5(\cos 105° + i\sin 105°)} = \frac{4}{5}[\cos(345° - 105°) + i\sin(345° - 105°)]$$
$$= \frac{4}{5}[\cos 240° + i\sin 240°]$$
$$= \frac{4}{5}\left[-\frac{1}{2} + i\left(-\frac{\sqrt{3}}{2}\right)\right]$$
$$= -\frac{2}{5} - \frac{2\sqrt{3}}{5}i$$ ◀

EXERCISES 10.5

In Exercises 1 through 8, show the geometric representation of the complex number as a point in the complex plane.

1. (a) $4 + 5i$ **(b)** $7 - 8i$

2. (a) $7 + i$ **(b)** $-4 + 9i$

3. (a) $-1 + 6i$ **(b)** $-3 - i$

4. (a) $2 - 6i$ **(b)** $-5 - 3i$

5. (a) 2 **(b)** $-6i$

6. (a) -4 **(b)** $3i$

7. (a) $2 - 6i$ **(b)** $2 + 6i$

8. (a) $-4 + 3i$ **(b)** $-4 - 3i$

In Exercises 9 through 16, write the number without absolute-value bars.

9. (a) $|5 + 2i|$ **(b)** $|-1 + 2i|$

10. (a) $|8 - 3i|$ **(b)** $|-2 - 5i|$

11. (a) $|3|$ **(b)** $|3i|$

12. (a) $|5|$ **(b)** $|5i|$

13. (a) $|-2|$ **(b)** $|-2i|$

14. (a) $|-7|$ **(b)** $|-7i|$

15. (a) $|-6 + 8i|$ **(b)** $|-6 - 8i|$

16. (a) $|6 + i|$ **(b)** $|6 - i|$

In Exercises 17 through 22, show the geometric representation of the complex number in the complex plane and write it in cartesian form.

17. (a) $3(\cos \frac{1}{3}\pi + i \sin \frac{1}{3}\pi)$
    (b) $3(\cos \frac{4}{3}\pi + i \sin \frac{4}{3}\pi)$

18. (a) $4(\cos \frac{1}{4}\pi + i \sin \frac{1}{4}\pi)$
    (b) $4(\cos \frac{3}{4}\pi + i \sin \frac{3}{4}\pi)$

19. (a) $6(\cos 150° + i \sin 150°)$
    (b) $6(\cos 330° + i \sin 330°)$

20. (a) $5(\cos 210° + i \sin 210°)$
    (b) $5(\cos 300° + i \sin 300°)$

21. (a) $2(\cos \frac{1}{2}\pi + i \sin \frac{1}{2}\pi)$
    (b) $\frac{2}{3}(\cos 180° + i \sin 180°)$

22. (a) $3(\cos 0° + i \sin 0°)$ **(b)** $\frac{1}{2}(\cos \frac{3}{2}\pi + i \sin \frac{3}{2}\pi)$

In Exercises 23 through 26, express the complex number in standard polar form.

23. (a) $4 - 4i$ **(b)** 6 **(c)** i

24. (a) $-5 + 5i$ **(b)** 0 **(c)** $-4i$

25. (a) $-3\sqrt{3} + 3i$ **(b)** -7 **(c)** $-7i$

26. (a) $2 - 2i$ **(b)** 5 **(c)** $5i$

In Exercises 27 through 30, express the complex number in standard polar form where the argument is written with (a) inverse function notation, (b) radian measurement to two decimal places, and (c) degree measurement to one-tenth of a degree.

27. $3 + 4i$ **28.** $5 + i$

29. $1 - 2i$ **30.** $-3 + 2i$

In Exercises 31 through 38, express the product in cartesian form.

31. $2(\cos 20° + i \sin 20°) \cdot 5(\cos 70° + i \sin 70°)$

32. $(\cos 70° + i \sin 70°) \cdot 4(\cos 110° + i \sin 110°)$

33. $3(\cos \frac{1}{18}\pi + i \sin \frac{1}{18}\pi) \cdot \frac{2}{3}(\cos \frac{5}{18}\pi + i \sin \frac{5}{18}\pi)$

34. $10(\cos \frac{1}{6}\pi + i \sin \frac{1}{6}\pi) \cdot \frac{2}{5}(\cos \frac{1}{12}\pi + i \sin \frac{1}{12}\pi)$

35. $4(\cos \frac{5}{9}\pi + i \sin \frac{5}{9}\pi) \cdot (\cos \frac{7}{36}\pi + i \sin \frac{7}{36}\pi)$

36. $\frac{3}{4}(\cos \frac{7}{18}\pi + i \sin \frac{7}{18}\pi) \cdot \frac{8}{3}(\cos \frac{4}{9}\pi + i \sin \frac{4}{9}\pi)$

37. $5(\cos 75° + i \sin 75°) \cdot \frac{4}{5}(\cos 255° + i \sin 255°)$

38. $2(\cos 145° + i \sin 145°) \cdot 3(\cos 95° + i \sin 95°)$

In Exercises 39 through 46, express the quotient in cartesian form.

39. $6(\cos 70° + i \sin 70°) \div 3(\cos 40° + i \sin 40°)$

40. $4(\cos 65° + i \sin 65°) \div 2(\cos 20° + i \sin 20°)$

41. $8(\cos \frac{4}{3}\pi + i \sin \frac{4}{3}\pi) \div 2(\cos \frac{7}{12}\pi + i \sin \frac{7}{12}\pi)$

42. $2(\cos \frac{17}{12}\pi + i \sin \frac{17}{12}\pi) \div 6(\cos \frac{3}{4}\pi + i \sin \frac{3}{4}\pi)$

43. $5(\cos \frac{5}{18}\pi + i \sin \frac{5}{18}\pi) \div (\cos \frac{11}{18}\pi + i \sin \frac{11}{18}\pi)$

44. $6(\cos \frac{5}{18}\pi + i \sin \frac{5}{18}\pi) \div \frac{3}{2}(\cos \frac{7}{9}\pi + i \sin \frac{7}{9}\pi)$

45. $\frac{2}{3}(\cos 350° + i \sin 350°) \div (\cos 80° + i \sin 80°)$

46. $3(\cos 310° + i \sin 310°) \div \frac{1}{2}(\cos 85° + i \sin 85°)$

47. Prove that if z is a complex number such that $|z| = 1$, then $\frac{1}{z} = \bar{z}$. *Hint:* Show that $\cos(-\theta) + i \sin(-\theta)$ is both the conjugate and the reciprocal of $\cos \theta + i \sin \theta$.

48. Prove that if $z = r(\cos \theta + i \sin \theta)$, then

$$z^2 = r^2(\cos 2\theta + i \sin 2\theta)$$

and

$$z^3 = r^3(\cos 3\theta + i \sin 3\theta)$$

49. Compare the definition of the absolute value of a real number with the definition of the absolute value of a complex number, and explain why the first definition is a special case of the second.

50. Prove Theorem 1(ii).

10.6 POWERS AND ROOTS OF COMPLEX NUMBERS AND DE MOIVRE'S THEOREM

GOALS

1. Learn De Moivre's theorem for positive integers.
2. Find positive-integer powers of complex numbers by De Moivre's theorem.
3. Define zero and negative exponents of complex numbers.
4. Learn De Moivre's theorem for all integers.
5. Find negative-integer powers of complex numbers by De Moivre's theorem.
6. Define an nth root of a complex number.
7. Learn the theorem giving the n nth roots of a complex number.
8. Find n nth roots of unity and show their geometric representations.
9. Find n nth roots of a number by using the n nth roots of unity.
10. Express the n nth roots of a complex number in cartesian form.

The polar form of complex numbers can be applied to calculate their powers and roots. The procedure is provided by a theorem called De Moivre's theorem, named for the mathematician Abraham De Moivre (1667–1754), who was born in France but lived most of his life in London. He was a friend of Sir Isaac Newton, one of the inventors of calculus. To lead up to the theorem, we consider the complex number $z = r(\cos \theta + i \sin \theta)$ and compute some positive-integer powers of z. Certainly

$$z^1 = r(\cos \theta + i \sin \theta) \tag{1}$$

From Theorem 1(i) of Section 10.5

$$
\begin{aligned}
z^2 &= [r(\cos \theta + i \sin \theta)][r(\cos \theta + i \sin \theta)] \\
&= r^2[\cos(\theta + \theta) + i \sin(\theta + \theta)] \\
&= r^2(\cos 2\theta + i \sin 2\theta) \tag{2}
\end{aligned}
$$

$$z^3 = [r(\cos \theta + i \sin \theta)][r(\cos \theta + i \sin \theta)]^2$$

Substituting from (2) in the right side, we get

$$
\begin{aligned}
z^3 &= [r(\cos \theta + i \sin \theta)][r^2(\cos 2\theta + i \sin 2\theta)] \\
&= r^3[\cos(\theta + 2\theta) + i \sin(\theta + 2\theta)] \\
&= r^3(\cos 3\theta + i \sin 3\theta) \tag{3}
\end{aligned}
$$

▷ **ILLUSTRATION 1**

(a) From (2)

$$[4(\cos 60° + i \sin 60°)]^2 = 4^2[\cos(2 \cdot 60°) + i \sin(2 \cdot 60°)]$$
$$= 16(\cos 120° + i \sin 120°)$$
$$= 16\left[-\frac{1}{2} + i\left(\frac{\sqrt{3}}{2} \right) \right]$$
$$= -8 + 8\sqrt{3}i$$

(b) From (3)

$$[4(\cos 60° + i \sin 60°)]^3 = 4^3[\cos(3 \cdot 60°) + i \sin(3 \cdot 60°)]$$
$$= 64(\cos 180° + i \sin 180°)$$
$$= 64[-1 + i(0)]$$
$$= -64 \qquad ◀$$

Equations (1), (2), and (3) are the special cases of De Moivre's theorem when n is 1, 2, and 3. We now state this theorem for n any positive integer.

THEOREM 1   *De Moivre's Theorem for Positive Integers*

If n is any positive integer,

$$[r(\cos \theta + i \sin \theta)]^n = r^n(\cos n\theta + i \sin n\theta)$$

The proof of this theorem requires mathematical induction, the topic of Section 12.5. We give the proof in Example 6 of that section.

▶ **EXAMPLE 1**   *Finding Powers of Complex Numbers with De Moivre's Theorem*

Use De Moivre's theorem to find **(a)** $(1 + i)^5$; **(b)** $(\sqrt{3} - i)^6$.

Solution

(a) We first write $1 + i$ in polar form. If $1 + i = a + bi$, then $a = 1$ and $b = 1$. Thus

$$r = \sqrt{a^2 + b^2} \qquad \cos \theta = \frac{a}{r} \qquad \sin \theta = \frac{b}{r}$$
$$= \sqrt{1^2 + (1)^2} \qquad\qquad = \frac{1}{\sqrt{2}} \qquad\qquad = \frac{1}{\sqrt{2}}$$
$$= \sqrt{2}$$

Hence $\theta = \frac{1}{4}\pi$. Therefore

$$1 + i = \sqrt{2}(\cos\tfrac{1}{4}\pi + i\sin\tfrac{1}{4}\pi)$$

From De Moivre's theorem

$$(1 + i)^5 = (\sqrt{2})^5(\cos\tfrac{5}{4}\pi + i\sin\tfrac{5}{4}\pi)$$

$$= 4\sqrt{2}\left[-\frac{1}{\sqrt{2}} + i\left(-\frac{1}{\sqrt{2}}\right)\right]$$

$$= -4 - 4i$$

(b) If $\sqrt{3} - i = a + bi$, then $a = \sqrt{3}$ and $b = -1$. Hence

$$r = \sqrt{(\sqrt{3})^2 + (-1)^2} \qquad \cos\theta = \frac{a}{r} \qquad \sin\theta = \frac{b}{r}$$

$$= 2 \qquad\qquad\qquad = \frac{\sqrt{3}}{2} \qquad\quad = -\frac{1}{2}$$

Thus $\theta = \frac{11}{6}\pi$. Therefore

$$\sqrt{3} - i = 2(\cos\tfrac{11}{6}\pi + i\sin\tfrac{11}{6}\pi)$$

From De Moivre's theorem

$$(\sqrt{3} - i)^6 = 2^6[\cos(6\cdot\tfrac{11}{6}\pi) + i\sin(6\cdot\tfrac{11}{6}\pi)]$$

$$= 64(\cos 11\pi + i\sin 11\pi)$$

$$= 64(-1 + i\cdot 0)$$

$$= -64 \qquad\qquad\qquad\qquad\qquad\blacktriangleleft$$

We wish to define zero and negative-integer exponents of complex numbers in such a way that De Moivre's theorem holds for these integers. If the theorem is to be valid for $n = 0$, then if $z = r(\cos\theta + i\sin\theta)$,

$$z^0 = r^0(\cos 0\cdot\theta + i\sin 0\cdot\theta)$$

$$= 1(\cos 0 + i\sin 0)$$

$$= 1(1 + i\cdot 0)$$

$$= 1$$

If the theorem is to be valid for $-n$ where $n > 0$, then if

$$z = r(\cos\theta + i\sin\theta)$$

$$z^{-n} = r^{-n}[\cos(-n\theta) + i\sin(-n\theta)]$$

Because $\cos(-n\theta) = \cos n\theta$ and $\sin(-n\theta) = -\sin n\theta$, this equation becomes

$$z^{-n} = \frac{\cos n\theta - i \sin n\theta}{r^n}$$

We multiply the numerator and denominator by the conjugate of the numerator, and we have

$$z^{-n} = \frac{(\cos n\theta - i \sin n\theta)(\cos n\theta + i \sin n\theta)}{r^n(\cos n\theta + i \sin n\theta)}$$

$$= \frac{\cos^2 n\theta - i^2 \sin^2 n\theta}{r^n(\cos n\theta + i \sin n\theta)}$$

$$= \frac{\cos^2 n\theta + \sin^2 n\theta}{r^n(\cos n\theta + i \sin n\theta)}$$

$$= \frac{1}{r^n(\cos n\theta + i \sin n\theta)}$$

$$= \frac{1}{z^n}$$

Thus in order for De Moivre's theorem to hold for zero and negative-integer exponents, we make the following definition, which is consistent with the definition of zero and negative-integer exponents of real numbers.

DEFINITION **Zero and Negative Exponents of Complex Numbers**

If z is a complex number and $z \neq 0 + 0i$, and n is a positive integer,

(i) $z^0 = 1$

(ii) $z^{-n} = \dfrac{1}{z^n}$

From De Moivre's theorem for positive integers and the preceding definition we have De Moivre's theorem for all integers.

THEOREM 2 *De Moivre's Theorem for All Integers*

If k is any integer,

$$[r(\cos \theta + i \sin \theta)]^k = r^k(\cos k\theta + i \sin k\theta)$$

▶ **EXAMPLE 2** *Finding Negative-Integer Powers of Complex Numbers with De Moivre's Theorem*

Use De Moivre's theorem to find **(a)** $(-1 + \sqrt{3}i)^{-4}$; **(b)** $(-1 - i)^{-14}$.

Solution

(a) If $-1 + \sqrt{3}i = a + bi$, then $a = -1$ and $b = \sqrt{3}$. Hence

$$r = \sqrt{(-1)^2 + (\sqrt{3})^2} \qquad \cos\theta = \frac{a}{r} \qquad \sin\theta = \frac{b}{r}$$

$$= 2$$

$$= -\frac{1}{2} \qquad\qquad = \frac{\sqrt{3}}{2}$$

Thus $\theta = \frac{2}{3}\pi$. Therefore

$$-1 + \sqrt{3}i = 2\left(\cos\frac{2}{3}\pi + i\sin\frac{2}{3}\pi\right)$$

From De Moivre's theorem

$$(-1 + \sqrt{3}i)^{-4} = 2^{-4}\left[\cos\left(-4\cdot\frac{2}{3}\pi\right) + i\sin\left(-4\cdot\frac{2}{3}\pi\right)\right]$$

$$= \frac{1}{16}\left[\cos\left(-\frac{8}{3}\pi\right) + i\sin\left(-\frac{8}{3}\pi\right)\right]$$

$$= \frac{1}{16}\left[-\frac{1}{2} + i\left(-\frac{\sqrt{3}}{2}\right)\right]$$

$$= -\frac{1}{32} - \frac{\sqrt{3}}{32}i$$

(b) If $-1 - i = a + bi$, then $a = -1$ and $b = -1$. Thus

$$r = \sqrt{(-1)^2 + (-1)^2} \qquad \cos\theta = \frac{a}{r} \qquad \sin\theta = \frac{b}{r}$$

$$= \sqrt{2}$$

$$= -\frac{1}{\sqrt{2}} \qquad\qquad = -\frac{1}{\sqrt{2}}$$

Therefore $\theta = \frac{5}{4}\pi$. Hence

$$-1 - i = \sqrt{2}\left(\cos\tfrac{5}{4}\pi + i\sin\tfrac{5}{4}\pi\right)$$

From De Moivre's theorem

$$(-1 - i)^{-14} = (\sqrt{2})^{-14}\left[\cos\left(-14\cdot\tfrac{5}{4}\pi\right) + i\sin\left(-14\cdot\tfrac{5}{4}\pi\right)\right]$$

$$= \frac{1}{2^7}\left[\cos\left(-\frac{35}{2}\pi\right) + i\sin\left(-\frac{35}{2}\pi\right)\right]$$

$$= \frac{1}{128}(0 + i\cdot 1)$$

$$= \frac{1}{128}i$$

◀

For the treatment of roots of complex numbers, we first define an nth root.

DEFINITION   An nth Root of a Complex Number

If z is a nonzero complex number, and n is a positive integer, the number w is said to be an **nth root** of z if

$$w^n = z$$

To find a formula for computing nth roots of complex numbers, assume $w^n = z$ and let

$$z = r(\cos \theta + i \sin \theta) \qquad \text{and} \qquad w = s(\cos \phi + i \sin \phi)$$

We wish to obtain formulas for s and ϕ in terms of r and θ. Because $w^n = z$,

$$[s(\cos \phi + i \sin \phi)]^n = r(\cos \theta + i \sin \theta)$$

Applying De Moivre's theorem to the left side, we have

$$s^n(\cos n\phi + i \sin n\phi) = r(\cos \theta + i \sin \theta) \tag{4}$$

This equation is an equality of two complex numbers. Therefore the absolute values of these complex numbers are equal; thus

$$s^n = r$$
$$s = r^{1/n}$$

Because $s^n = r$, then from (4)

$$\cos n\phi + i \sin n\phi = \cos \theta + i \sin \theta$$

From this equation and the definition of equality of two complex numbers,

$$\cos n\phi = \cos \theta \qquad \text{and} \qquad \sin n\phi = \sin \theta$$

Because the period of both the sine and cosine is 2π, these equations are valid if and only if

$$n\phi = \theta + k \cdot 2\pi \qquad k \in J$$

Therefore

$$\phi = \frac{\theta + k \cdot 2\pi}{n} \qquad k \in J$$

With this value of ϕ and $s = r^{1/n}$ and because $w = s(\cos \phi + i \sin \phi)$, we obtain

$$w = r^{1/n}\left[\cos\left(\frac{\theta + k \cdot 2\pi}{n}\right) + i \sin\left(\frac{\theta + k \cdot 2\pi}{n}\right)\right] \qquad k \in J$$

If in this equation k is replaced successively by $0, 1, \ldots, n - 1$, we obtain n distinct values of w that are all nth roots of z. Observe that if k is replaced by an integer greater than $n - 1$ or less than 0, we get no new value of w.

For instance, if $k = n$ in the preceding equation, the right side is

$$r^{1/n}\left[\cos\left(\frac{\theta + n \cdot 2\pi}{n}\right) + i\sin\left(\frac{\theta + n \cdot 2\pi}{n}\right)\right]$$

$$= r^{1/n}\left[\cos\left(\frac{\theta}{n} + 2\pi\right) + i\sin\left(\frac{\theta}{n} + 2\pi\right)\right]$$

which gives the same value as $k = 0$. We have proved the following theorem.

THEOREM 3   *nth Roots of Complex Numbers*

If z is a nonzero complex number where

$$z = r(\cos\theta + i\sin\theta)$$

and n is a positive integer, then z has exactly n distinct nth roots given by

$$r^{1/n}\left[\cos\left(\frac{\theta + k \cdot 2\pi}{n}\right) + i\sin\left(\frac{\theta + k \cdot 2\pi}{n}\right)\right] \quad k \in J$$

where k is $0, 1, \ldots, n - 1$.

If θ is measured in degrees, then the formula of Theorem 3 is written as

$$r^{1/n}\left[\cos\left(\frac{\theta + k \cdot 360°}{n}\right) + i\sin\left(\frac{\theta + k \cdot 360°}{n}\right)\right] \quad k \in J$$

The absolute value of each of the nth roots of the complex number z is

$$\sqrt{(r^{1/n})^2\left[\cos^2\left(\frac{\theta + k \cdot 2\pi}{n}\right) + \sin^2\left(\frac{\theta + k \cdot 2\pi}{n}\right)\right]}$$

$$= \sqrt{(r^{1/n})^2(1)}$$

$$= r^{1/n} \quad \text{(because } r > 0\text{)}$$

Thus the geometric representations of the nth roots lie on a circle with center at the origin and radius $\sqrt[n]{r}$. Furthermore, because the difference of the arguments of successive nth roots is $2\pi/n$, they are equally spaced around the circle.

▷ ILLUSTRATION 2

To find the four fourth roots of $-8 + 8\sqrt{3}i$, we first write this number in polar form. If $-8 + 8\sqrt{3}i = a + bi$, then $a = -8$ and $b = 8\sqrt{3}$.

Therefore

$$r = \sqrt{(-8)^2 + (8\sqrt{3})^2} \qquad \cos\theta = \frac{a}{r} \qquad \sin\theta = \frac{b}{r}$$
$$= \sqrt{64 + 64(3)}$$
$$= 8\sqrt{4} \qquad\qquad\qquad = \frac{-8}{16} \qquad\qquad = \frac{8\sqrt{3}}{16}$$
$$= 16 \qquad\qquad\qquad\qquad = -\frac{1}{2} \qquad\qquad = \frac{\sqrt{3}}{2}$$

Hence $\theta = 120°$. Thus

$$-8 + 8\sqrt{3}i = 16(\cos 120° + i \sin 120°)$$

From Theorem 3, the four fourth roots are given by

$$16^{1/4}\left[\cos\left(\frac{120° + k \cdot 360°}{4}\right) + i \sin\left(\frac{120° + k \cdot 360°}{4}\right)\right]$$
$$\Leftrightarrow \quad 2[\cos(30° + k \cdot 90°) + i \sin(30° + k \cdot 90°)]$$

where k is 0, 1, 2, 3. Replacing k successively by 0, 1, 2, and 3, we obtain

$$2(\cos 30° + i \sin 30°) = \sqrt{3} + i$$
$$2(\cos 120° + i \sin 120°) = -1 + \sqrt{3}i$$
$$2(\cos 210° + i \sin 210°) = -\sqrt{3} - i$$
$$2(\cos 300° + i \sin 300°) = 1 - \sqrt{3}i$$

Observe that if k is replaced by any other integer, one of these four complex numbers is obtained. For instance, if $k = 4$, we get

$$2(\cos 390° + i \sin 390°) = \sqrt{3} + i$$

and if $k = -1$, we get

$$2[\cos(-60°) + i \sin(-60°)] = 1 - \sqrt{3}i$$

The geometric representations of these roots appear in Figure 1. All four lie on the circle having center at the origin and radius 2, and they are equally spaced around the circle. Note that the four fourth roots occur in pairs where each member of a pair is the opposite (additive inverse) of the other member: $\sqrt{3} + i$ and $-\sqrt{3} - i$; $-1 + \sqrt{3}i$ and $1 - \sqrt{3}i$. ◄

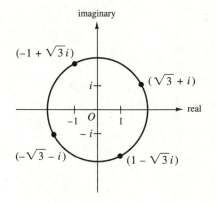

FIGURE 1

▶ **EXAMPLE 3** *Finding n nth Roots of 1 and Showing Their Geometric Representations*

Find the six sixth roots of 1 and show their geometric representations.

Solution   If $1 = a + bi$, then $a = 1$ and $b = 0$. Therefore

$$r = \sqrt{1^2 + 0^2} \qquad \cos\theta = \frac{a}{r} \qquad \sin\theta = \frac{b}{r}$$

$$= 1 \qquad\qquad\quad = 1 \qquad\qquad = 0$$

Thus $\theta = 0$. Therefore

$$1 = 1(\cos 0 + i \sin 0)$$

The six sixth roots of 1 are given by

$$1^{1/6}\left[\cos\left(\frac{0 + k \cdot 2\pi}{6}\right) + i \sin\left(\frac{0 + k \cdot 2\pi}{6}\right)\right]$$

$$\Leftrightarrow \qquad \cos\tfrac{1}{3}k\pi + i \sin\tfrac{1}{3}k\pi$$

where k is 0, 1, 2, 3, 4, 5. Replacing k successively by 0, 1, 2, 3, 4, and 5, we obtain

$$\cos 0 + i \sin 0 = 1$$

$$\cos\frac{1}{3}\pi + i \sin\frac{1}{3}\pi = \frac{1}{2} + \frac{\sqrt{3}}{2}i$$

$$\cos\frac{2}{3}\pi + i \sin\frac{2}{3}\pi = -\frac{1}{2} + \frac{\sqrt{3}}{2}i$$

$$\cos\pi + i \sin\pi = -1$$

$$\cos\frac{4}{3}\pi + i \sin\frac{4}{3}\pi = -\frac{1}{2} - \frac{\sqrt{3}}{2}i$$

$$\cos\frac{5}{3}\pi + i \sin\frac{5}{3}\pi = \frac{1}{2} - \frac{\sqrt{3}}{2}i$$

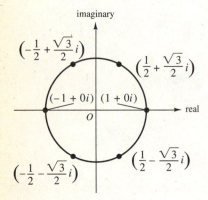

FIGURE 2

Figure 2 shows the geometric representations of the six roots. They all lie on the unit circle and are equally spaced around it.   ◀

The complex roots of 1 are called **roots of unity.** Because

$$1 = 1(\cos 0 + i \sin 0)$$

the n nth roots of unity are given by

$$\cos\frac{k \cdot 2\pi}{n} + i \sin\frac{k \cdot 2\pi}{n}$$

where k is 0, 1, 2, . . . , $n - 1$.

▷ **ILLUSTRATION 3**

The four fourth roots of unity are given by

$$\cos\frac{k \cdot 2\pi}{4} + i \sin\frac{k \cdot 2\pi}{4}$$

where k is 0, 1, 2, 3. Replacing k successively by 0, 1, 2, and 3, we get

$$\cos 0 + i \sin 0 = 1$$
$$\cos \tfrac{1}{2}\pi + i \sin \tfrac{1}{2}\pi = i$$
$$\cos \pi + i \sin \pi = -1$$
$$\cos \tfrac{3}{2}\pi + i \sin \tfrac{3}{2}\pi = -i \qquad \blacktriangleleft$$

The roots of unity can be used to obtain the roots of other real numbers as shown in the following illustration and example.

▷ ILLUSTRATION 4

We use the four fourth roots of unity to find the four fourth roots of 81. A real fourth root of 81 is 3. If each of the four fourth roots of unity found in Illustration 3 are multiplied by 3, we obtain

$$3(1) = 3 \qquad 3(i) = 3i \qquad 3(-1) = -3 \qquad 3(-i) = -3i$$

Thus the four fourth roots of 81 are 3, -3, $3i$, and $-3i$. $\qquad \blacktriangleleft$

▶ **EXAMPLE 4** *Finding n nth Roots of a Number by Using the n nth Roots of Unity*

Use the six sixth roots of unity found in Example 3 to obtain the six sixth roots of 64.

Solution A real sixth root of 64 is 2. If each of the six sixth roots of unity is multiplied by 2, we get the six sixth roots of 64:

$$2(1) = 2$$
$$2\left(\frac{1}{2} + \frac{\sqrt{3}}{2}i\right) = 1 + \sqrt{3}i$$
$$2\left(-\frac{1}{2} + \frac{\sqrt{3}}{2}i\right) = -1 + \sqrt{3}i$$
$$2(-1) = -2$$
$$2\left(-\frac{1}{2} - \frac{\sqrt{3}}{2}i\right) = -1 - \sqrt{3}i$$
$$2\left(\frac{1}{2} - \frac{\sqrt{3}}{2}i\right) = 1 - \sqrt{3}i \qquad \blacktriangleleft$$

You can visualize the n nth roots of unity on your graphics calculator. See Exercises 39 and 40 for the procedure to follow.

Observe that the four fourth roots of 81, found in Illustration 4, and the six sixth roots of 64, obtained in Example 4, can also be determined algebraically without De Moivre's theorem by solving the equations $x^4 - 81 = 0$ and $x^6 - 64 = 0$, respectively (see Exercises 43 and 44 in Exercises 2.4). However, to find the seven seventh roots of 128, for instance, De Moivre's theorem is required (see Exercise 23). The theorem is also required in the following example.

▶ **EXAMPLE 5** *Finding n nth Roots of a Complex Number*

Find the two square roots of i.

Solution We write the number i in polar form. If $i = a + bi$, then $a = 0$ and $b = 1$. Thus

$$r = \sqrt{0^2 + 1^2} \qquad \cos \theta = \frac{a}{r} \qquad \sin \theta = \frac{b}{r}$$
$$= 1 \qquad\qquad\qquad = \frac{0}{1} \qquad\qquad = \frac{1}{1}$$
$$\qquad\qquad\qquad\qquad = 0 \qquad\qquad\quad = 1$$

Therefore $\theta = \frac{1}{2}\pi$. Thus

$$i = 1(\cos \tfrac{1}{2}\pi + i \sin \tfrac{1}{2}\pi)$$

From Theorem 3, the two square roots of i are given by

$$1^{1/2}\left[\cos\left(\frac{\frac{1}{2}\pi + k \cdot 2\pi}{2}\right) + i \sin\left(\frac{\frac{1}{2}\pi + k \cdot 2\pi}{2}\right)\right]$$

where $k = 0, 1$. Replacing k by 0 and 1, we obtain, respectively,

$$1\left(\cos \frac{1}{4}\pi + i \sin \frac{1}{4}\pi\right) = \frac{1}{\sqrt{2}} + \frac{1}{\sqrt{2}}i$$
$$1\left(\cos \frac{5}{4}\pi + i \sin \frac{5}{4}\pi\right) = -\frac{1}{\sqrt{2}} - \frac{1}{\sqrt{2}}i \qquad ◀$$

▶ **EXAMPLE 6** *Expressing the n nth Roots of a Complex Number in Cartesian Form*

Express each of the three cube roots of $3 + 4i$ in the form $a + bi$. Approximate the values of a and b to two decimal places.

Solution If $3 + 4i = a + bi$, then $a = 3$ and $b = 4$. Thus

$$r = \sqrt{3^2 + 4^2} \qquad \cos \theta = \frac{a}{r} \qquad \sin \theta = \frac{b}{r} \qquad \tan \theta = \frac{b}{a}$$
$$= 5 \qquad\qquad\qquad = \frac{3}{5} \qquad\qquad = \frac{4}{5} \qquad\qquad = \frac{4}{3}$$

Hence $\theta = \arctan \frac{4}{3}$. Therefore

$$3 + 4i = 5[\cos(\arctan \tfrac{4}{3}) + i \sin(\arctan \tfrac{4}{3})]$$

Thus the three cube roots of $3 + 4i$ are given by

$$5^{1/3}\left[\cos\left(\frac{\arctan \frac{4}{3} + k \cdot 2\pi}{3} \right) + i \sin\left(\frac{\arctan \frac{4}{3} + k \cdot 2\pi}{3} \right) \right]$$

where k is 0, 1, 2. Replacing k successively by 0, 1, and 2 and using a calculator to compute the approximate values, we obtain

$$5^{1/3}[\cos \tfrac{1}{3}(\arctan \tfrac{4}{3})] + i \sin \tfrac{1}{3}(\arctan \tfrac{4}{3})]$$
$$\approx 1.710[0.953 + i(0.304)] \approx 1.63 + 0.52i$$
$$5^{1/3}[\cos \tfrac{1}{3}(\arctan \tfrac{4}{3} + 2\pi) + i \sin \tfrac{1}{3}(\arctan \tfrac{4}{3} + 2\pi)]$$
$$\approx 1.710[-0.740 + i(0.673)] \approx -1.26 + 1.15i$$
$$5^{1/3}[\cos \tfrac{1}{3}(\arctan \tfrac{4}{3} + 4\pi) + i \sin \tfrac{1}{3}(\arctan \tfrac{4}{3} + 4\pi)]$$
$$\approx 1.710[-0.213 + i(-0.977)] \approx -0.36 - 1.67i$$

◀

EXERCISES 10.6

In Exercises 1 through 20, use De Moivre's theorem to find the indicated power. Write the answer in cartesian form.

1. $[2(\cos 30° + i \sin 30°)]^2$

2. $[4(\cos 120° + i \sin 120°)]^2$

3. $[4(\cos 120° + i \sin 120°)]^3$

4. $[2(\cos 30° + i \sin 30°)]^3$

5. $(\cos \tfrac{1}{4}\pi + i \sin \tfrac{1}{4} \pi)^4$

6. $[3(\cos \tfrac{3}{4} \pi + i \sin \tfrac{3}{4} \pi)]^5$

7. $[2(\cos 48° + i \sin 48°)]^5$

8. $[(\cos 50° + i \sin 50°)]^6$

9. $(1 + i)^3$

10. $(-1 + i)^3$

11. $(-1 - \sqrt{3}i)^6$

12. $(\sqrt{3} - i)^6$

13. $(-4 + 4i)^5$

14. $\left(\dfrac{1}{4} + \dfrac{\sqrt{3}}{4}i \right)^5$

15. $\left(\dfrac{1}{2} - \dfrac{\sqrt{3}}{2}i \right)^{-3}$

16. $(-3 - 3i)^{-4}$

17. $\left(\dfrac{\sqrt{2}}{2} + \dfrac{\sqrt{2}}{2}i \right)^{-8}$

18. $(2 + 2\sqrt{3}i)^{-5}$

19. $(-3\sqrt{3} + 3i)^{30}$

20. $(4\sqrt{2} - 4\sqrt{2}i)^{40}$

21. **(a)** Find the three cube roots of unity and show their geometric representations. **(b)** Use the result of part (a) to find the three cube roots of 27.

22. **(a)** Find the five fifth roots of unity and show their geometric representations. **(b)** Use the result of part (a) to find the five fifth roots of 243.

23. **(a)** Find the seven seventh roots of unity and show their geometric representations. **(b)** Use the result of part (a) to find the seven seventh roots of 128.

24. **(a)** Find the eight eighth roots of unity and show their geometric representations. **(b)** Use the result of part (a) to find the eight eighth roots of 256.

In Exercises 25 through 28, solve the equation in two ways: (a) by De Moivre's theorem; (b) algebraically without De Moivre's theorem.

25. $x^4 - 16 = 0$

26. $x^3 - 125 = 0$

27. $x^3 + i = 0$ [*Hint* for part (b): Write $x^3 + i$ as $x^3 - i^3$]

28. $x^3 - i = 0$ [*Hint* for part (b): Write $x^3 - i$ as $x^3 + i^3$]

In Exercises 29 through 38, find the indicated roots and write them in cartesian form. In Exercises 33 through 38, in the cartesian form representation approximate the values of a and b to two decimal places.

29. The two square roots of $-4i$

30. The two square roots of $25i$

31. The three cube roots of i

32. The three cube roots of $-i$

33. The three cube roots of $4 - 4\sqrt{3}i$

34. The three cube roots of $-\dfrac{\sqrt{3}}{2} + \dfrac{1}{2}i$

35. The four fourth roots of $-\dfrac{1}{2} + \dfrac{\sqrt{3}}{2}i$

36. The four fourth roots of $-8 - 8\sqrt{3}i$

37. The five fifth roots of $-4 + 3i$

38. The five fifth roots of $2 - i$

39. The geometric representations of the n nth roots of unity are the vertices of a regular polygon of n sides inscribed in the unit circle. That is, the three cube roots of unity are the vertices of an inscribed equilateral triangle, the four fourth roots of unity are the vertices of an inscribed square, the five fifth roots of unity are the vertices of an inscribed regular pentagon, and so on. Explain why this is so.

40. The statement of Exercise 39 provides the following method of visualizing the n nth roots of unity on your

graphics calculator. (i) Set the calculator in parametric and degree modes. (ii) Let $x(t) = \cos t$ and $y(t) = \sin t$. (iii) Choose: t min $= 0$, t max $= 360$, t step $= 360/n$. (iv) Choose the viewing rectangle $[-3, 3]$ by $[-2, 2]$. (v) Use the trace button to see each nth root of unity. Apply this procedure to visualize: **(a)** the three cube roots of unity; **(b)** the four fourth roots of unity; **(c)** the five fifth roots of unity; **(d)** the six sixth roots of unity.

41. Obtain the cosine and sine double-measure identities by equating the real and imaginary parts of the two expressions for $(\cos \theta + i \sin \theta)^2$ found in the following two ways: (1) use De Moivre's theorem; (2) multiply $\cos \theta + i \sin \theta$ by itself.

In Exercises 42 through 45, use the method of Exercise 41 to obtain the indicated identity.

42. Identity for $\cos 3\theta$ in terms of $\cos \theta$

43. Identity for $\sin 3\theta$ in terms of $\sin \theta$

44. Identity for $\sin 4\theta$ in terms of $\sin \theta$

45. Identity for $\cos 4\theta$ in terms of $\cos \theta$

CHAPTER 10 REVIEW

▶ LOOKING BACK

10.1 A vector was defined as an ordered pair of real numbers and represented geometrically by a directed line segment. The position representation was used to illustrate geometrically important facts about vectors including the definitions of magnitude and direction angle, addition and subtraction of vectors, and the product of a scalar and a vector. We expressed a vector in terms of its magnitude and direction angle and the unit vectors **i** and **j**. The word problems involving vectors pertained to force, air and marine navigation, and geometry.

10.2 As a prelude to the use of vectors in calculus, we discussed vector-valued functions. We stated that a curve defined by a vector equation of the form $\mathbf{R}(t) = f(t)\mathbf{i} + g(t)\mathbf{j}$ is also defined by parametric equations $x = f(t)$ and $y = g(t)$. We found a cartesian equation of a graph from its vector equation, or the equivalent pair of parametric equations, and obtained the graph by hand and on a graphics calcula-

tor. We defined a cycloid and derived its parametric equations. The word problems in this section involved motion of a projectile.

10.3 We defined polar coordinates of a point and then located a point from its polar coordinates and found other sets of polar coordinates of the point. Equations relating the rectangular cartesian coordinates and polar coordinates of a point were used to find a cartesian equation of a graph from a polar equation and a polar equation of a graph from a cartesian equation.

10.4 Polar graphs discussed in this section included lines, circles, limaçons, roses, and spirals. We presented the theorem that shows how a polar graph can be defined by a pair of parametric equations and applied the theorem to plot polar graphs.

10.5 Before stating the polar form of a complex number, we represented complex numbers as points in the complex plane and defined the absolute value of a complex number. We obtained formulas for the product and quotient of two complex numbers given in polar form and expressed the product and quotient in cartesian form.

10.6 The polar form of a complex number appeared in the statement of De Moivre's theorem. We first presented the theorem for positive integers and used it to find positive-integer powers of complex numbers. After defining zero and negative exponents, we stated De

Moivre's theorem for all integers and found negative-integer powers of complex numbers. We then defined an nth root of a complex number and proved the theorem giving the n nth roots of a complex number. The n nth roots of unity were used to find the n nth roots of a complex number. We showed how to express these roots in cartesian form.

▶ REVIEW EXERCISES

In Exercises 1 and 2, find the sum of the pairs of vectors and illustrate geometrically.

1. (a) $\langle 3, 7 \rangle$, $\langle 2, -4 \rangle$ **(b)** $-\mathbf{i} + 5\mathbf{j}$, $-2\mathbf{j}$

2. (a) $\langle 4, 0 \rangle$, $\langle -6, -1 \rangle$ **(b)** $5\mathbf{i} + 3\mathbf{j}$, $3\mathbf{i} - 5\mathbf{j}$

In Exercises 3 and 4, subtract the second vector from the first and illustrate geometrically.

3. (a) $\langle 6, 4 \rangle$, $\langle 0, -3 \rangle$ **(b)** $-2\mathbf{i} + 8\mathbf{j}$, $4\mathbf{i} - 5\mathbf{j}$

4. (a) $\langle 9, -7 \rangle$, $\langle -2, 5 \rangle$ **(b)** $6\mathbf{i} - 4\mathbf{j}$, $-\mathbf{i} + 8\mathbf{j}$

In Exercises 5 and 6, $\mathbf{A} = 4\mathbf{i} - 6\mathbf{j}$, $\mathbf{B} = \mathbf{i} + 7\mathbf{j}$, and $\mathbf{C} = 9\mathbf{i} - 5\mathbf{j}$. Find the indicated vector or scalar.

5. (a) $3\mathbf{B} - 7\mathbf{A}$ **(b)** $\|3\mathbf{B} - 7\mathbf{A}\|$
    (c) $\|3\mathbf{B}\| - \|7\mathbf{A}\|$

6. (a) $5\mathbf{B} - 3\mathbf{C}$ **(b)** $\|5\mathbf{B} - 3\mathbf{C}\|$
    (c) $\|5\mathbf{B}\| - \|3\mathbf{C}\|$

In Exercises 7 and 8, find the components of the vector having the given magnitude and direction angle.

7. (a) $12, \frac{1}{6}\pi$ **(b)** $36, 112°$

8. (a) $25, \frac{3}{4}\pi$ **(b)** $130, 335.2°$

In Exercises 9 and 10, write the given vector in the form $r(\cos\theta\mathbf{i} + \sin\theta\mathbf{j})$, where r is the magnitude and θ is the direction angle.

9. (a) $6\mathbf{i} + 6\mathbf{j}$ **(b)** $-4\mathbf{i} + 2\sqrt{5}\mathbf{j}$

10. (a) $2\sqrt{3}\mathbf{i} + 2\mathbf{j}$ **(b)** $-3\mathbf{i} - 4\mathbf{j}$

In Exercises 11 through 14, find the domain of the vector-valued function.

11. $\mathbf{R}(t) = \dfrac{1}{t-2}\mathbf{i} + \ln t\,\mathbf{j}$

12. $\mathbf{R}(t) = \sqrt{t+1}\,\mathbf{i} + \dfrac{1}{t-3}\mathbf{j}$

13. $\mathbf{R}(t) = e^t\mathbf{i} + \tan t\,\mathbf{j}$

14. $\mathbf{R}(t) = \sin^{-1}(t-1)\mathbf{i} + \cos^{-1}(t-1)\mathbf{j}$

In Exercises 15 and 16, sketch the graph of the vector equation and find a cartesian equation of the graph.

15. $\mathbf{R}(t) = 9\cos t\,\mathbf{i} + 9\sin t\,\mathbf{j}$

16. $\mathbf{R}(t) = (4t - 1)\mathbf{i} + (4t^2 + 2)\mathbf{j}$

In Exercises 17 through 20, (a) sketch the graph of the parametric equations and check your graph on your graphics calculator; (b) find a cartesian equation of the graph.

17. $x = 2 - t$, $y = 2t$

18. $x = t^2$, $y = t^3 + 1$

19. $x = 2t^3$, $y = 3t^2 - 4$

20. $x = 4\cos t - 2$, $y = 4\sin t + 3$

In Exercises 21 and 22, locate the point having the given set of polar coordinates; then give two other sets of polar coordinates of the same point, one with the same value of r and one with an r having opposite sign.

21. (a) $\left(2, \frac{3}{4}\pi\right)$ **(b)** $\left(-3, \frac{7}{6}\pi\right)$

22. (a) $\left(4, -\frac{1}{3}\pi\right)$ **(b)** $\left(-1, \frac{1}{4}\pi\right)$

In Exercises 23 and 24, find the rectangular cartesian coordinates of the point whose polar coordinates are given.

23. (a) $\left(1, \frac{1}{2}\pi\right)$ **(b)** $\left(2, -\frac{1}{3}\pi\right)$
    (c) $\left(4, \frac{5}{4}\pi\right)$ **(d)** $\left(-3, \frac{1}{6}\pi\right)$

24. (a) $(5, \pi)$ **(b)** $\left(-2, \frac{5}{6}\pi\right)$
    (c) $\left(-\sqrt{2}, \frac{1}{4}\pi\right)$ **(d)** $\left(1, \frac{4}{3}\pi\right)$

In Exercises 25 and 26, find a set of polar coordinates of the point whose rectangular cartesian coordinates are given. Take $r > 0$ and $0 \le \theta < 2\pi$.

25. (a) $(-4, 4)$ **(b)** $(1, -\sqrt{3})$
    (c) $(0, 6)$ **(d)** $(-2\sqrt{3}, -2)$

26. (a) $(-4, 0)$ **(b)** $(\sqrt{3}, 1)$
    (c) $(-2, -2)$ **(d)** $(3, -3\sqrt{3})$

In Exercises 27 through 30, find a polar equation of the graph having the given cartesian equation.

27. $4x^2 - 9y^2 = 36$ **28.** $2xy = 1$

29. $x^2 + y^2 - 9x + 8y = 0$ **30.** $y^4 = x^2(a^2 - y^2)$

In Exercises 31 through 34, find a cartesian equation of the graph having the given polar equation.

31. $r^2 \sin 2\theta = 4$ **32.** $r(1 - \cos \theta) = 2$

33. $r^2 = \sin^2 \theta$ **34.** $r = a \tan^2 \theta$

In Exercises 35 through 38, sketch the graph of the equation.

35. (a) $\theta = \frac{1}{4}\pi$ (b) $r = 4$

36. (a) $\theta = \frac{2}{3}$ (b) $r = \frac{3}{2}$

37. (a) $r \cos \theta = 3$ (b) $r = 3 \cos \theta$

38. (a) $r \sin \theta = 6$ (b) $r = 6 \sin \theta$

In Exercises 39 through 44, determine the type of limaçon, its symmetry, and the direction in which it points. Plot the limaçon.

39. $r = 3 + 2 \cos \theta$ **40.** $r = 2 + 3 \sin \theta$

41. $r = 2(1 - \cos \theta)$ **42.** $r = 3(1 + \sin \theta)$

43. $r = 1 - 2 \sin \theta$ **44.** $r = 2 - \cos \theta$

In Exercises 45 through 50, describe and plot the graph of the equation.

45. $r = 3 \sin 2\theta$ **46.** $r = 3 \cos 2\theta$

47. $r = \sqrt{|\cos \theta|}$ **48.** $r = |\sin 2\theta|$

49. $r^2 = -\sin 2\theta$ **50.** $r^2 = 16 \cos \theta$

51. Describe and plot the graph of each of the following equations: (a) $r\theta = 3$ (reciprocal spiral); (b) $3r = \theta$ (spiral of Archimedes).

52. Show by hand that the equations $r = 1 + \sin \theta$ and $r = \sin \theta - 1$ have the same graph. Then check your graphs on your graphics calculator.

In Exercises 53 and 54, show the geometric representation of the complex number as a point in the complex plane.

53. (a) $3 + 7i$ (b) $6 - 3i$

 (c) $-1 + 5i$ (d) $-4i$

54. (a) $-3 + 2i$ (b) $-7 - i$

 (c) -4 (d) $3i$

In Exercises 55 and 56, write the number without absolute-value bars.

55. (a) $|4 - 3i|$ (b) $|-6 + 2i|$

56. (a) $|5 + 12i|$ (b) $|-6i|$

In Exercises 57 and 58, express the complex number in standard polar form.

57. (a) $-3 + \sqrt{3}i$ (b) $2\sqrt{3} + 2i$

 (c) $-4i$ (d) $-1 - i$

58. (a) $\frac{1}{2} + \frac{1}{2}i$ (b) $\sqrt{3} - i$

 (c) $6i$ (d) -6

In Exercises 59 through 62, show the geometric representation of the complex number in the complex plane and write it in cartesian form.

59. $2(\cos \frac{1}{6}\pi + i \sin \frac{1}{6}\pi)$

60. $6(\cos \frac{2}{3}\pi + i \sin \frac{2}{3}\pi)$

61. $4(\cos 135° + i \sin 135°)$

62. $8(\cos 330° + i \sin 330°)$

In Exercises 63 and 64, express the product in cartesian form.

63. $4(\cos 50° + i \sin 50°) \cdot \frac{3}{4}(\cos 70° + i \sin 70°)$

64. $5(\cos \frac{7}{12}\pi + i \sin \frac{7}{12}\pi) \cdot 2(\cos \frac{2}{3}\pi + i \sin \frac{2}{3}\pi)$

In Exercises 65 and 66, express the quotient in cartesian form.

65. $10(\cos \frac{4}{9}\pi + i \sin \frac{4}{9}\pi) \div 5(\cos \frac{43}{36}\pi + i \sin \frac{43}{36}\pi)$

66. $2(\cos 200° + i \sin 200°) \div 3(\cos 50° + i \sin 50°)$

In Exercises 67 through 70, express the power in cartesian form.

67. $(\sqrt{3} + i)^3$ **68.** $(4 - 4i)^4$

69. $(-2 + 2i)^{-6}$ **70.** $\left(\frac{1}{2} + \frac{\sqrt{3}}{2}i\right)^{-5}$

In Exercises 71 through 76, find the indicated roots, write them in cartesian form, and show their geometric representations.

71. The three cube roots of 8

72. The three cube roots of -27

73. The four fourth roots of $2 + i$

74. The three cube roots of $8i$

75. The five fifth roots of $2 - 5i$

76. The four fourth roots of $\dfrac{\sqrt{3}}{2} - \dfrac{1}{2}i$

77. Two forces of magnitudes 50 lb and 70 lb make an angle of 60° with each other and are applied to an object at the same point. Find **(a)** the magnitude of the resultant force and **(b)** to the nearest degree the angle it makes with the force of 50 lb.

78. Determine the angle between two forces of 112 lb and 136 lb applied to an object at the same point if the resultant force has a magnitude of 168 lb.

79. The compass heading of an airplane is 107°, and its air speed is 210 mi/hr. If there is a wind blowing from the west at 36 mi/hr, what is **(a)** the plane's ground speed and **(b)** its course?

80. A particle is moving along a curve having the vector equation $\mathbf{R}(t) = 3t\,\mathbf{i} + (4t - t^2)\mathbf{j}$. Find **(a)** parametric equations and **(b)** a cartesian equation of the path of the particle. **(c)** Plot the path of the particle.

 81. A projectile moves so that the coordinates of its position at any time t seconds are given by the parametric equations

$$x = 30t \quad \text{and} \quad y = 40t - 16t^2$$

Find: **(a)** the total time of flight; **(b)** the range of the projectile; **(c)** the maximum height attained by the projectile; **(d)** a cartesian equation of the path of the projectile. **(e)** Plot the path of the projectile.

 82. A projectile shot from a gun at an angle of elevation of radian measure $\frac{1}{6}\pi$ has a muzzle speed of 220 ft/sec. Find **(a)** the position vector of the projectile at any time; **(b)** parametric equations of the projectile's position at any time; **(c)** the total time of flight; **(d)** the range of the projectile; **(e)** the maximum height attained by the projectile; **(f)** a cartesian equation of the curve traveled by the projectile. **(g)** Plot the path of the projectile.

83. A golf ball, hit at point A with an initial velocity of 140 ft/sec at an angle of elevation of 40° strikes the ground at point P. **(a)** Find the horizontal distance from A to P. **(b)** How long does it take the ball to go from A to P? **(c)** Determine the maximum height reached by the ball. **(d)** Plot the curve traveled by the ball from A to P. **(e)** Find a cartesian equation of the curve traveled by the ball.

84. An *epicycloid* is the curve traced by a point P on the circumference of a circle of radius b rolling externally on a fixed circle of radius a. If the origin is at the center of the fixed circle, $A(a, 0)$ is one of the points at which the given point P comes in contact with the fixed circle, B is the moving point of tangency of the two circles, and the parameter t is the radian measure of the angle AOB, prove that parametric equations of the epicycloid are

$$x = (a + b)\cos t - b \cos \frac{a + b}{b}t$$

and

$$y = (a + b)\sin t - b \sin \frac{a + b}{b}t$$

Conic Sections

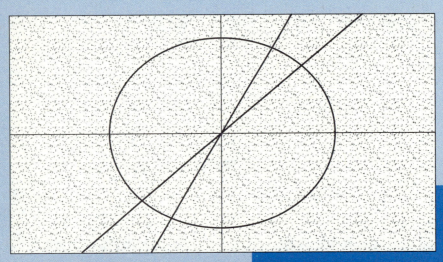

LOOKING AHEAD

Conic sections (or conics) are curves of intersection of a plane with a right circular cone, and three types of curves of intersection that occur are the *ellipse*, the *hyperbola*, and the *parabola*. The Greek mathematician Appolonius studied conic sections by using this geometrical concept. In Section 3.3, we defined a parabola as a set of points in a plane, and in the first two sections of this chapter, we also define an ellipse and a hyperbola as sets of points in a plane. We prove that the definition of an ellipse as a set of points is a consequence of its definition as a conic section; variations of this proof can be given for a hyperbola and a parabola. In the final section of the chapter, we give a unified treatment of conic sections by defining a conic in terms of its eccentricity.

11.1 ELLIPSES

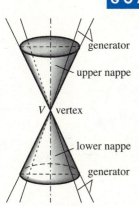

FIGURE 1

ellipse
FIGURE 2

circle
FIGURE 3

To consider the geometry of conic sections, we regard a cone as having two nappes, each extending indefinitely far. A portion of a right-circular cone of two nappes appears in Figure 1. A line lying in the cone is called a **generator** (or **element**) of the cone. All generators of a cone contain the point V called the **vertex**.

An *ellipse* is obtained as a conic section if the cutting plane is parallel to no generator, in which case the cutting plane intersects each generator as in Figure 2. A special case of the ellipse is a circle, which is formed if the cutting plane intersecting each generator is also perpendicular to the axis of the cone. See Figure 3. We now define an ellipse as a set of points in a plane. At the end of this section we prove that this definition is a consequence of the definition of an ellipse as a section of a cone.

> **DEFINITION**   **An Ellipse**
>
> An **ellipse** is the set of points in a plane the sum of whose distances from two fixed points is a constant. Each fixed point is called a **focus**.

Let the undirected distance between the foci (the plural of focus) be $2c$ where $c > 0$. To obtain an equation of an ellipse, we select the x axis as the line through F and F', and we choose the origin as the midpoint of the

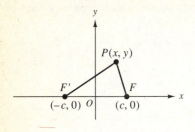

FIGURE 4

segment between F and F'. See Figure 4. The foci F and F' have coordinates $(c, 0)$ and $(-c, 0)$, respectively. Let the constant sum referred to in the definition be $2a$. Then $a > c$, and the point $P(x, y)$ in Figure 4 is any point on the ellipse if and only if

$$|\overline{FP}| + |\overline{F'P}| = 2a$$

Because

$$|\overline{FP}| = \sqrt{(x - c)^2 + y^2} \quad \text{and} \quad |\overline{F'P}| = \sqrt{(x + c)^2 + y^2}$$

P is on the ellipse if and only if

$$\sqrt{(x - c)^2 + y^2} + \sqrt{(x + c)^2 + y^2} = 2a$$

To simplify this equation requires eliminating the radicals and performing some algebraic manipulations, which you are asked to do in Exercise 35. When this is done, we obtain

$$\frac{x^2}{a^2} + \frac{y^2}{b^2} = 1$$

where $b^2 = a^2 - c^2$. We state this result formally.

Equation of an Ellipse

If $2a$ is the constant referred to in the definition of an ellipse, if the foci are at $(c, 0)$ and $(-c, 0)$, and if $b^2 = a^2 - c^2$, then an equation of the ellipse is

$$\frac{x^2}{a^2} + \frac{y^2}{b^2} = 1$$

To sketch this ellipse, first observe from the equation that the graph is symmetric with respect to both the x and y axes. If we replace y by 0 in the equation, we get $x = \pm a$, and if we replace x by 0, we obtain $y = \pm b$. Therefore the graph intersects the x axis at $(a, 0)$ and $(-a, 0)$, and it intersects the y axis at $(0, b)$ and $(0, -b)$. Because $b^2 = a^2 - c^2$, it follows that $a > b$. See Figure 5 and refer to it as you read the next paragraph.

The line through the foci is called the **principal axis.** For this ellipse the x axis is the principal axis. The points of intersection of the ellipse and its principal axis are called the **vertices.** Thus for this ellipse the vertices are at $V(a, 0)$ and $V'(-a, 0)$. The point on the principal axis that lies halfway between the two vertices is called the **center.** The origin is the center of this ellipse. The segment of the principal axis between the two vertices is called the **major axis,** and its length is $2a$ units. For this ellipse the segment of the

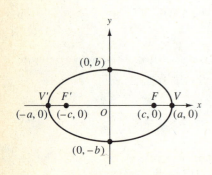

FIGURE 5

y axis between the points $(0, b)$ and $(0, -b)$ is called the **minor axis.** Its length is $2b$ units.

An ellipse is called a **central conic** in contrast to a parabola, which has no center because it has only one vertex.

▶ **EXAMPLE 1** *Finding Properties of an Ellipse and Sketching Its Graph*

For the ellipse having the equation

$$\frac{x^2}{25} + \frac{y^2}{16} = 1$$

find the vertices, endpoints of the minor axis, and foci. Sketch the ellipse and show the foci.

Solution Because the form of the equation is

$$\frac{x^2}{a^2} + \frac{y^2}{b^2} = 1$$

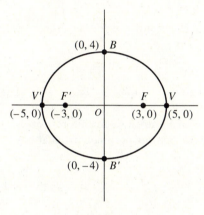

FIGURE 6

the center of the ellipse is at the origin and the principal axis is the x axis. Because $a^2 = 25$ and $b^2 = 16$, $a = 5$ and $b = 4$. Therefore the vertices are at $V(5, 0)$ and $V'(-5, 0)$, and the endpoints of the minor axis are at $B(0, 4)$ and $B'(0, -4)$.

To find the foci, we solve for c from the equation $b^2 = a^2 - c^2$ with $a^2 = 25$ and $b^2 = 16$. Thus, because $c > 0$,

$$16 = 25 - c^2$$
$$c^2 = 9$$
$$c = 3$$

Therefore the foci are at $F(3, 0)$ and $F'(-3, 0)$.

As an aid in sketching the ellipse, we find a point on it in the first quadrant by substituting 3 for x in the equation and solving for y. (Of course, any other value of x between 0 and 5 can be used.) By symmetry we have corresponding points in the other three quadrants. Figure 6 shows the ellipse and the foci. ◀

Observe from the definition of an ellipse that if P is any point on the ellipse of Example 1, then $|\overline{FP}| + |\overline{F'P}| = 10$. In Figure 7 we have taken P in the second quadrant.

To plot an ellipse on a graphics calculator, we can do what we did in Section 3.4 for graphs of circles. That is, we treat the equation of the ellipse as quadratic in y and solve it to obtain two equations defining y as two functions of x.

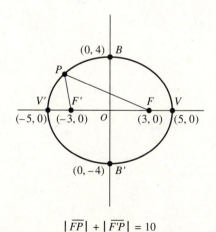

$$|\overline{FP}| + |\overline{F'P}| = 10$$

FIGURE 7

▷ **ILLUSTRATION 1**

Solving the equation of the ellipse in Example 1 for y, by first multiplying both sides of the equation by 400, we have

$$16x^2 + 25y^2 = 400$$
$$25y^2 = 400 - 16x^2$$
$$25y^2 = 16(25 - x^2)$$
$$y^2 = \frac{16}{25}(25 - x^2)$$
$$y = \pm\frac{4}{5}\sqrt{25 - x^2}$$

In the same viewing rectangle on a graphics calculator, we plot the graphs of

$$y_1 = \frac{4}{5}\sqrt{25 - x^2} \quad \text{and} \quad y_2 = -\frac{4}{5}\sqrt{25 - x^2}$$

to obtain the ellipse that appears in Figure 6.   ◀

Another method for plotting an ellipse on a graphics calculator involves parametric equations of the ellipse. To show that the parametric equations

$$x = a\cos t \quad \text{and} \quad y = b\sin t$$

represent an ellipse, we eliminate t from the equations. We first write the equations as

$$\frac{x}{a} = \cos t \quad \text{and} \quad \frac{y}{b} = \sin t$$

Squaring on both sides of each equation and adding gives

$$\frac{x^2}{a^2} + \frac{y^2}{b^2} = \cos^2 t + \sin^2 t$$
$$\frac{x^2}{a^2} + \frac{y^2}{b^2} = 1$$

which is a standard form of an equation of an ellipse.

▷ **ILLUSTRATION 2**

Parametric equations of the ellipse of Example 1 are

$$x = 5\cos t \quad \text{and} \quad y = 4\sin t$$

On our graphics calculator in parametric mode we let t take on all numbers in the closed interval $[0, 2\pi]$ to obtain the ellipse in Figure 6.   ◀

The paths of many comets and the orbits of planets and satellites are ellipses. Arches of bridges are sometimes elliptical in shape, and ellipses are used in making machine gears. An application of the ellipse in architecture for so-called whispering galleries uses its reflective property. In whispering galleries the ceilings have cross sections that are arcs of ellipses with common foci. A person located at one focus F can hear another person whispering at the other focus F' because the sound waves originating from the whisperer at F' hit the ceiling and are reflected by the ceiling to the listener at F. A famous example of a whispering gallery is under the dome of the Capitol in Washington, D.C. Another is at the Mormon Tabernacle in Salt Lake City.

▶ **EXAMPLE 2** *Solving a Word Problem Having an Equation of an Ellipse as a Mathematical Model*

An arch in the form of a semiellipse is 48 ft wide at the base and has a height of 20 ft. How wide is the arch at a height of 10 ft above the base?

Solution Figure 8 shows a sketch of the arch and the coordinate axes chosen so that the x axis is along the base and the origin is at the midpoint of the base. Then the ellipse has its principal axis on the x axis, its center at the origin, $a = 24$, and $b = 20$. Thus an equation of the ellipse is

$$\frac{x^2}{576} + \frac{y^2}{400} = 1$$

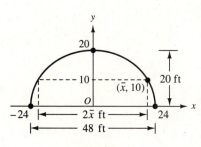

FIGURE 8

Let $2\bar{x}$ be the number of feet in the width of the arch at a height of 10 ft above the base. Therefore the point $(\bar{x}, 10)$ is on the ellipse. Thus

$$\frac{\bar{x}^2}{576} + \frac{100}{400} = 1$$

$$\bar{x}^2 = 432$$

$$\bar{x} = 12\sqrt{3}$$

Conclusion: At a height of 10 ft above the base the width of the arch is $24\sqrt{3}$ ft. ◀

If an ellipse has its center at the origin and principal axis on the y axis, then an equation of the ellipse is of the form

$$\frac{y^2}{a^2} + \frac{x^2}{b^2} = 1$$

This equation is obtained by interchanging x and y in the equation

$$\frac{x^2}{a^2} + \frac{y^2}{b^2} = 1$$

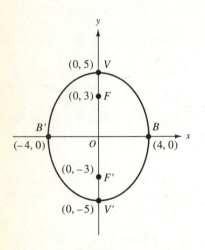

FIGURE 9

▷ **ILLUSTRATION 3**

Because for an ellipse $a > b$, it follows that the ellipse having the equation

$$\frac{x^2}{16} + \frac{y^2}{25} = 1$$

has its principal axis on the y axis. This ellipse has the same shape as the ellipse of Example 1. The vertices are at $(0, 5)$ and $(0, -5)$, the endpoints of the minor axis are at $(4, 0)$ and $(-4, 0)$, and the foci are at $(0, 3)$ and $(0, -3)$. Figure 9 shows this ellipse. You can obtain it on your graphics calculator by plotting the graph of the parametric equations

$$x = 4 \cos t \quad \text{and} \quad y = 5 \sin t \qquad ◀$$

Suppose the center of an ellipse is at the point (h, k) rather than at the origin, and the principal axis is parallel to one of the coordinate axes. Then by a translation of axes so that the point (h, k) is the new origin, an equation of the ellipse is

$$\frac{x'^2}{a^2} + \frac{y'^2}{b^2} = 1$$

if the principal axis is horizontal, and

$$\frac{y'^2}{a^2} + \frac{x'^2}{b^2} = 1$$

if the principal axis is vertical. Because $x' = x - h$ and $y' = y - k$, we obtain the following standard forms of an equation of an ellipse.

Standard Forms of an Equation of an Ellipse

If the center of an ellipse is at (h, k) and the distance between the vertices is $2a$, then an equation of the ellipse is of the form

$$\frac{(x - h)^2}{a^2} + \frac{(y - k)^2}{b^2} = 1 \qquad (a > b) \tag{1}$$

if the principal axis is horizontal, and

$$\frac{(y - k)^2}{a^2} + \frac{(x - h)^2}{b^2} = 1 \qquad (a > b) \tag{2}$$

if the principal axis is vertical.

To plot an ellipse whose principal axis is horizontal on a graphics calculator, use the parametric equations

$$x = a \cos t + h \qquad \text{and} \qquad y = b \sin t + k$$

If the principal axis is vertical, use the parametric equations

$$x = b \cos t + h \qquad \text{and} \qquad y = a \sin t + k$$

By expanding $(x - h)^2$ and $(y - k)^2$ and simplifying, we can write each of Equations (1) and (2) in the form

$$Ax^2 + Cy^2 + Dx + Ey + F = 0 \tag{3}$$

where A and C have the same sign. In the following example, we start with an equation in this form and complete squares to write it in a standard form of an equation of an ellipse.

▶ **EXAMPLE 3** *Finding Properties of an Ellipse and Obtaining Its Graph*

Show that the graph of the equation

$$25x^2 + 16y^2 + 150x - 128y - 1119 = 0$$

is an ellipse. Find the center, an equation of the principal axis, the vertices, the endpoints of the minor axis, and the foci. Sketch the ellipse and check the graph on a graphics calculator.

Solution To write this equation in one of the standard forms, we begin by completing the squares in x and y. We have then

$$25(x^2 + 6x) + 16(y^2 - 8y) = 1119$$
$$25(x^2 + 6x + 9) + 16(y^2 - 8y + 16) = 1119 + 225 + 256$$
$$25(x + 3)^2 + 16(y - 4)^2 = 1600$$
$$\frac{25(x + 3)^2}{1600} + \frac{16(y - 4)^2}{1600} = 1$$
$$\frac{(x + 3)^2}{64} + \frac{(y - 4)^2}{100} = 1$$

This equation is of the form

$$\frac{(y - k)^2}{a^2} + \frac{(x - h)^2}{b^2} = 1 \qquad (a > b)$$

where (h, k) is $(-3, 4)$, $a^2 = 100$, and $b^2 = 64$. Therefore the graph is an ellipse whose center is at $(-3, 4)$ and whose principal axis is the vertical line having the equation $x = -3$. Because $a = 10$ and $b = 8$, the vertices are at $V(-3, 14)$ and $V'(-3, -6)$ and the endpoints of the minor axis are at $B(5, 4)$ and $B'(-11, 4)$. To find the foci, we use the equation $b^2 = a^2 - c^2$

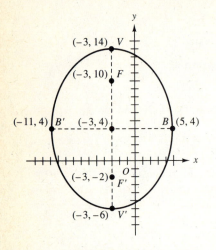

FIGURE 10

with $c > 0$ and obtain

$$64 = 100 - c^2$$
$$c^2 = 36$$
$$c = 6$$

Thus the foci are at $F(-3, 10)$ and $F'(-3, -2)$. By plotting a few more points (in particular where the ellipse intersects the x and y axes), we get the ellipse appearing in Figure 10.

To plot the ellipse on our graphics calculator, we use the parametric equations

$$x = 8 \cos t - 3 \quad \text{and} \quad y = 10 \sin t + 4 \quad \blacktriangleleft$$

In the next two illustrations, we again have equations in form (3).

▷ **ILLUSTRATION 4**

Suppose that (3) is

$$6x^2 + 9y^2 - 24x - 54y + 115 = 0$$

which can be written as

$$6(x^2 - 4x) + 9(y^2 - 6y) = -115$$

Completing the squares in x and y, we get

$$6(x^2 - 4x + 4) + 9(y^2 - 6y + 9) = -115 + 24 + 81$$
$$6(x - 2)^2 + 9(y - 3)^2 = -10$$

Because the right side of this equation is negative and the left side is nonnegative for all points (x, y), the graph is the empty set. ◀

▷ **ILLUSTRATION 5**

Because the equation

$$6x^2 + 9y^2 - 24x - 54y + 105 = 0$$

can be written as

$$6(x - 2)^2 + 9(y - 3)^2 = 0$$

its graph is the point $(2, 3)$. ◀

We can prove in general that the graph of any equation of the form (3) is either an ellipse, as in Example 3, a point, or the empty set. When the graph is a point or the empty set, as in Illustrations 4 and 5, it is said to be **degenerate**.

Observe that (3) is the special case of the general equation of the second degree in two variables,

$$Ax^2 + Bxy + Cy^2 + Dx + Ey + F = 0 \tag{4}$$

where $B = 0$ and $AC > 0$ (that is, A and C have the same sign).

The conclusions in the preceding discussion are summarized in the following theorem.

> ### THEOREM 1
>
> If in the general second-degree equation (4), $B = 0$ and $AC > 0$, then the graph is either an ellipse, a point, or the empty set.

The degenerate case of an ellipse, a point, is obtained as a conic section if the cutting plane contains the vertex of the cone but does not contain a generator. See Figure 11.

If $A = C$ in (3), the equation becomes

$$Ax^2 + Ay^2 + Dx + Ey + F = 0$$

which when dividing by A gives

$$x^2 + y^2 + \frac{D}{A}x + \frac{E}{A}y + \frac{F}{A} = 0$$

In Section 3.4 we learned that the graph of this equation is either a circle, a point, or the empty set. This statement agrees with Theorem 1 because a circle is a limiting form of an ellipse. This fact can be shown by considering the equation relating a, b, and c for an ellipse:

$$b^2 = a^2 - c^2$$

From this equation we see that if $c = 0$, $b^2 = a^2$, and then the standard forms of an equation of an ellipse become

$$\frac{(x - h)^2}{a^2} + \frac{(y - k)^2}{a^2} = 1$$

$$\Leftrightarrow \quad (x - h)^2 + (y - k)^2 = a^2$$

which is an equation of a circle having its center at (h, k) and radius a. Furthermore, when $c = 0$, the foci are coincident at the center of the circle.

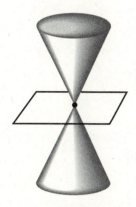

point

FIGURE 11

▶ **EXAMPLE 4** *Finding an Equation of an Ellipse from Its Properties and Obtaining Its Graph*

Find an equation of the ellipse having foci at $(-8, 2)$ and $(4, 2)$ and for which the constant referred to in the definition is 18. Sketch the ellipse and check the graph on a graphics calculator.

Solution The center of the ellipse is halfway between the foci and is the point $(-2, 2)$. The distance between the foci of an ellipse is $2c$, and the distance between $(-8, 2)$ and $(4, 2)$ is 12. Therefore $c = 6$. The constant referred to in the definition is $2a$; thus $2a = 18$ and $a = 9$. Because $b^2 = a^2 - c^2$,

$$b^2 = 81 - 36$$
$$b^2 = 45$$
$$b = 3\sqrt{5}$$

The principal axis is parallel to the x axis; hence an equation of the ellipse is of the form

$$\frac{(x - h)^2}{a^2} + \frac{(y - k)^2}{b^2} = 1$$

Because (h, k) is the point $(-2, 2)$, $a = 9$, and $b = 3\sqrt{5}$, the required equation is

$$\frac{(x + 2)^2}{81} + \frac{(y - 2)^2}{45} = 1$$

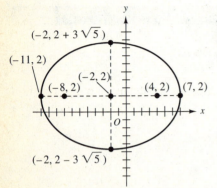

FIGURE 12

This ellipse appears in Figure 12. To plot the ellipse on our graphics calculator, we use the parametric equations

$$x = 9 \cos t - 2 \quad \text{and} \quad y = 3\sqrt{5} \sin t + 2 \qquad \blacktriangleleft$$

Some ellipses are almost circular, which happens when the foci are close together. Some ellipses are "flat," which occurs when the foci and vertices are near each other. The shape of an ellipse (its "roundness" or "flatness") is given by the *eccentricity*, which we now formally define.

DEFINITION **Eccentricity**

The **eccentricity** e of an ellipse is the ratio of the undirected distance between the foci to the undirected distance between the vertices; that is,

$$e = \frac{c}{a}$$

Because $c^2 = a^2 - b^2$, then $c < a$; therefore, $0 < e < 1$. When the foci are close together, e is close to zero, and the shape of the ellipse is close to that of a circle. See Figure 13(a) showing an ellipse for which $e = 0.3$. If a remains fixed, then as e increases the flatness of the ellipse increases. Figures 13(b) and 13(c) show ellipses with eccentricities of 0.7 and 0.95, respectively, each with the same value of a as in Figure 13(a). The limiting

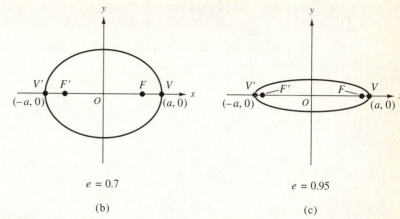

e = 0.3

(a)

e = 0.7

(b)

e = 0.95

(c)

FIGURE 13

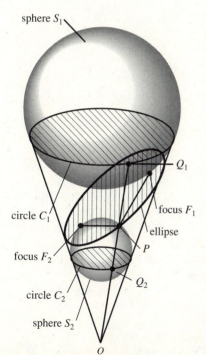

sphere S_1

Q_1

circle C_1

focus F_1

ellipse

focus F_2

P

circle C_2

Q_2

sphere S_2

O

FIGURE 14

forms of the ellipse are a circle of diameter $2a$ and a line segment of length $2a$.

As promised, we now prove that the definition of an ellipse as a set of points in a plane follows from the definition of an ellipse as a conic section. This proof, sometimes referred to as the "ice cream cone proof," was presented in 1822 by the Belgian mathematician G. P. Dandelin (1794–1847). Refer to Figure 14, which shows one nappe of a cone having vertex at O and a cutting plane intersecting the cone in an ellipse. Two spheres S_1 and S_2 are inscribed in the cone. Sphere S_1 is tangent to the cone along the circle C_1 and tangent to the cutting plane at point F_1. Sphere S_2 is tangent to the cone along the circle C_2 and tangent to the cutting plane at point F_2. The planes of the circles C_1 and C_2 are parallel. We shall prove that F_1 and F_2 are the foci of the ellipse by showing that if P is any point on the ellipse, $|\overline{PF_1}| + |\overline{PF_2}|$ is a constant. To demonstrate this, we draw the line through points O and P on the surface of the cone. Points Q_1 and Q_2 are the intersections of this line with the circles C_1 and C_2, respectively. Because PF_1 and PQ_1 are two tangent lines to sphere S_1 from the point P, it follows that

$$|\overline{PF_1}| = |\overline{PQ_1}|$$

Also PF_2 and PQ_2 are two tangent lines to sphere S_2 from the point P. Thus

$$|\overline{PF_2}| = |\overline{PQ_2}|$$

Therefore

$$|\overline{PF_1}| + |\overline{PF_2}| = |\overline{PQ_1}| + |\overline{PQ_2}|$$

Observe that $|\overline{PQ_1}| + |\overline{PQ_2}| = |\overline{Q_1Q_2}|$, which is the distance measured along the surface of the cone between the parallel planes of the circles C_1 and C_2. This distance will be the same for any choice of point P on the ellipse. Therefore $|\overline{PF_1}| + |\overline{PF_2}|$ is a constant, and F_1 and F_2 are the foci of the ellipse.

EXERCISES 11.1

In Exercises 1 through 16, for the ellipse having the given equation, find (a) the center, (b) the principal axis, (c) the vertices, (d) the endpoints of the minor axis, and (e) the foci. (f) Sketch the ellipse and show the foci. Check your graph on your graphics calculator.

1. $\dfrac{x^2}{25} + \dfrac{y^2}{9} = 1$

2. $\dfrac{x^2}{100} + \dfrac{y^2}{64} = 1$

3. $\dfrac{x^2}{4} + \dfrac{y^2}{16} = 1$

4. $\dfrac{x^2}{25} + \dfrac{y^2}{169} = 1$

5. $9x^2 + 25y^2 = 900$

6. $4x^2 + 9y^2 = 36$

7. $9x^2 + y^2 = 9$

8. $25x^2 + 4y^2 = 100$

9. $4x^2 + 9y^2 - 16x - 18y - 11 = 0$

10. $x^2 + 4y^2 - 6x + 8y - 3 = 0$

11. $4x^2 + y^2 + 8x - 4y - 92 = 0$

12. $2x^2 + 2y^2 - 2x + 18y + 33 = 0$

13. $4x^2 + 4y^2 + 20x - 32y + 89 = 0$

14. $25x^2 + y^2 - 4y - 21 = 0$

15. $x^2 + 3y^2 - 4x - 23 = 0$

16. $2x^2 + 3y^2 - 4x + 12y + 2 = 0$

In Exercises 17 and 18, determine whether the graph of the equation is an ellipse, a point, or the empty set.

17. $4x^2 + y^2 - 8x + 2y + 5 = 0$

18. $2x^2 + 3y^2 + 8x - 6y + 20 = 0$

In Exercises 19 through 28, find an equation of the ellipse having the given properties and sketch the ellipse. Check your graph on your graphics calculator.

19. Vertices at $(-\frac{5}{2}, 0)$ and $(\frac{5}{2}, 0)$ and one focus at $(\frac{3}{2}, 0)$.

20. Foci at $(-5, 0)$ and $(5, 0)$ and for which the constant referred to in the definition is 20.

21. Foci at $(0, 3)$ and $(0, -3)$ and for which the constant referred to in the definition is $6\sqrt{3}$.

22. Center at the origin, its foci on the x axis, the length of the major axis equal to 3 times the length of the minor axis, and passing through the point $(3, 3)$.

23. Vertices at $(2, 0)$ and $(-2, 0)$ and through the point $(-1, \frac{1}{2}\sqrt{3})$.

24. Vertices at $(0, 5)$ and $(0, -5)$ and through the point $(2, -\frac{5}{3}\sqrt{5})$.

25. Center at $(4, -2)$, a vertex at $(9, -2)$, and one focus at $(0, -2)$.

26. A focus at $(2, -3)$, a vertex at $(2, 4)$, and center on the x axis.

27. Foci at $(-1, -1)$ and $(-1, 7)$ and the semimajor axis of length 8 units.

28. Foci at $(2, 3)$ and $(2, -7)$ and the length of the semi-minor axis is two-thirds of the length of the semimajor axis.

In Exercises 29 through 32, solve the word problem and be sure to write a conclusion.

29. The ceiling in a hallway 10 m wide is in the shape of a semiellipse and is 9 m high in the center and 6 m high at the side walls. Find the height of the ceiling 2 m from either wall.

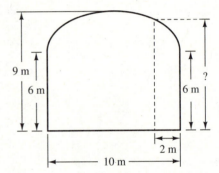

30. The orbit of the earth around the sun is elliptical in shape with the sun at one focus and a semimajor axis of length 92.96 million miles. If the eccentricity of the ellipse is 0.0167, find **(a)** how close the earth gets to the sun and **(b)** the greatest possible distance between the earth and the sun.

31. Suppose that the orbit of a planet is in the shape of an ellipse with a major axis whose length is 500 million km. If the distance between the foci is 400 million km, find an equation of the orbit.

32. The arch of a bridge is in the shape of a semiellipse having a horizontal span of 40 m and a height of 16 m at its center. How high is the arch 9 m to the right or left of the center?

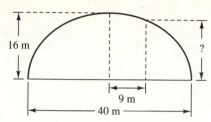

33. To trace the ellipse defined by the equation $4x^2 + 9y^2 = 36$, use the following procedure and explain why it works: First determine the points of intersection of the ellipse with the coordinate axes. Obtain the foci on the x axis by using a compass with its center at one of the points of intersection with the y axis and with radius of 3. Then fasten thumbtacks at each focus. Take a piece of string of length 6 and attach one end at one thumbtack and the other end at the other thumbtack. Place a pencil against the string and make it tight. Move the pencil against the string and trace the ellipse.

34. Use a procedure similar to that of Exercise 33 to trace the ellipse having the equation $16x^2 + 9y^2 = 144$. Explain why your procedure works.

35. Show that the equation

$$\sqrt{(x - c)^2 + y^2} + \sqrt{(x + c)^2 + y^2} = 2a$$

can be simplified to

$$\frac{x^2}{a^2} + \frac{y^2}{b^2} = 1$$

where $b^2 = a^2 - c^2$.

36. For the ellipse whose equation is

$$\frac{(x - h)^2}{a^2} + \frac{(y - k)^2}{b^2} = 1$$

where $a > b > 0$, find the coordinates of the foci in terms of h, k, a, and b.

11.2 HYPERBOLAS

GOALS

1. **Define a hyperbola.**
2. **Obtain standard forms of an equation of a hyperbola.**
3. **Sketch a hyperbola from its equation.**
4. **Find properties of a hyperbola from its equation.**
5. **Find an equation of a hyperbola from its properties.**
6. **Find equations of the asymptotes of a hyperbola.**
7. **Discuss degenerate cases of a hyperbola.**
8. **Define eccentricity of a hyperbola and show its relationship to the shape of the hyperbola.**
9. **Define the hyperbolic cosine and hyperbolic sine functions.**
10. **Plot a hyperbola on a graphics calculator by using parametric equations of the hyperbola.**

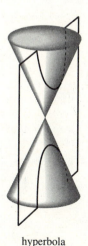

hyperbola

FIGURE 1

When a cutting plane intersects both nappes of a cone and is parallel to two generators, the conic section obtained is a *hyperbola*, shown in Figure 1. The following definition of a hyperbola as a set of points in a plane can be proved from its definition as a conic section. The proof, similar to that used for an ellipse in Section 11.1, involves a sphere in each nappe of the cone.

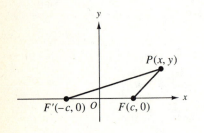

FIGURE 2

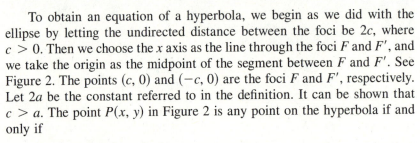

DEFINITION   A Hyperbola

A **hyperbola** is the set of points in a plane, the absolute value of the difference of whose distances from two fixed points is a constant. The two fixed points are called the **foci.**

To obtain an equation of a hyperbola, we begin as we did with the ellipse by letting the undirected distance between the foci be $2c$, where $c > 0$. Then we choose the x axis as the line through the foci F and F', and we take the origin as the midpoint of the segment between F and F'. See Figure 2. The points $(c, 0)$ and $(-c, 0)$ are the foci F and F', respectively. Let $2a$ be the constant referred to in the definition. It can be shown that $c > a$. The point $P(x, y)$ in Figure 2 is any point on the hyperbola if and only if

$$\left| |\overline{FP}| - |\overline{F'P}| \right| = 2a$$

Because

$$|\overline{FP}| = \sqrt{(x - c)^2 + y^2} \quad \text{and} \quad |\overline{F'P}| = \sqrt{(x + c)^2 + y^2}$$

P is on the hyperbola if and only if

$$\left| \sqrt{(x - c)^2 + y^2} - \sqrt{(x + c)^2 + y^2} \right| = 2a$$

or, equivalently, without absolute-value bars,

$$\sqrt{(x - c)^2 + y^2} - \sqrt{(x + c)^2 + y^2} = \pm 2a$$

This equation can be simplified by eliminating the radicals and performing some algebraic manipulations. You are asked to do this in Exercise 43. The resulting equation is

$$\frac{x^2}{a^2} - \frac{y^2}{b^2} = 1$$

where $b^2 = c^2 - a^2$. We have then the following statement.

Equation of a Hyperbola

If $2a$ is the constant referred to in the definition of a hyperbola, if the foci are at $(c, 0)$ and $(-c, 0)$, and if $b^2 = c^2 - a^2$, then an equation of the hyperbola is

$$\frac{x^2}{a^2} - \frac{y^2}{b^2} = 1$$

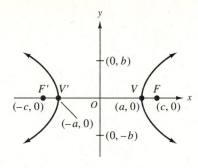

FIGURE 3

We now show how to sketch this hyperbola, which appears in Figure 3. Observe from the equation that the graph is symmetric with respect to both the x and y axes. As with the ellipse, the line through the foci is called the **principal axis.** Thus for this hyperbola the x axis is the principal axis. The points where the hyperbola intersects the principal axis are called the **vertices,** and the point that is halfway between the vertices is called the **center.** For this hyperbola the vertices are at $V(a, 0)$ and $V'(-a, 0)$ and the center is at the origin. The segment $V'V$ of the principal axis is called the **transverse axis,** and its length is $2a$ units.

Substituting 0 for x in the equation of the hyperbola, we get $y^2 = -b^2$, which has no real solutions. Consequently, the hyperbola does not intersect the y axis. However, the line segment having extremities at the points $(0, -b)$ and $(0, b)$ is called the **conjugate axis,** and its length is $2b$ units. If we solve the equation of the hyperbola for y in terms of x, we have

$$y = \pm \frac{b}{a}\sqrt{x^2 - a^2}$$

We conclude from this equation that if $|x| < a$, there is no real value of y. Thus there are no points (x, y) on the hyperbola for which $-a < x < a$. We also observe that if $|x| > a$, then y has two real values. Thus the hyperbola has two *branches.* One branch contains the vertex $V(a, 0)$ and extends indefinitely to the right of V. The other branch contains the vertex $V'(-a, 0)$ and extends indefinitely to the left of V'.

As was the case with an ellipse, because the hyperbola has a center it is called a **central conic.**

▶ **EXAMPLE 1** *Finding Properties of a Hyperbola from Its Equation and Sketching Its Graph*

Find the vertices and foci of the hyperbola having the equation

$$\frac{x^2}{9} - \frac{y^2}{16} = 1$$

Sketch the hyperbola and show the foci.

Solution Because the equation is of the form

$$\frac{x^2}{a^2} - \frac{y^2}{b^2} = 1$$

the center of the hyperbola is at the origin and the principal axis is the x axis. Because $a^2 = 9$ and $b^2 = 16$, $a = 3$ and $b = 4$. The vertices are therefore at $V(3, 0)$ and $V'(-3, 0)$. The number of units in the length of the transverse axis is $2a = 6$, and the number of units in the length of the conjugate axis

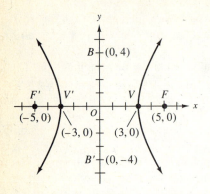

FIGURE 4

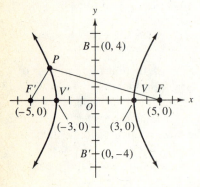

$$|\overline{FP}| - |\overline{F'P}| = 6$$

(a)

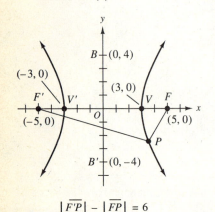

$$|\overline{F'P}| - |\overline{FP}| = 6$$

(b)

FIGURE 5

is $2b = 8$. Because $b^2 = c^2 - a^2$, with $c > 0$, we have

$$16 = c^2 - 9$$
$$c^2 = 16 + 9$$
$$c^2 = 25$$
$$c = 5$$

Hence the foci are at $F(5, 0)$ and $F'(-5, 0)$. A sketch of the hyperbola with its foci appears in Figure 4. ◀

From the definition of a hyperbola, if P is any point on the hyperbola of Example 1, $\left\|\,\overline{FP}\,\right| - \left|\overline{F'P}\,\right\| = 6$. See Figure 5(a) and (b); in (a) P is in the second quadrant and $|\overline{FP}| - |\overline{F'P}| = 6$; in (b) P is in the fourth quadrant and $|\overline{F'P}| - |\overline{FP}| = 6$.

▶ **EXAMPLE 2** *Finding an Equation of a Hyperbola Having Specific Properties*

Find an equation of the hyperbola having a focus at $(5, 0)$ and the ends of its conjugate axis at $(0, 2)$ and $(0, -2)$.

Solution Because the ends of the conjugate axis are at $(0, 2)$ and $(0, -2)$, $b = 2$, the principal axis is on the x axis, and the center is at the origin. Hence an equation is of the form

$$\frac{x^2}{a^2} - \frac{y^2}{b^2} = 1$$

Because a focus is at $(5, 0)$, $c = 5$, and because $b^2 = c^2 - a^2$, $a^2 = 25 - 4$. Thus $a = \sqrt{21}$, and an equation of the hyperbola is

$$\frac{x^2}{21} - \frac{y^2}{4} = 1$$ ◀

If in the equation

$$\frac{x^2}{a^2} - \frac{y^2}{b^2} = 1$$

x and y are interchanged, we obtain

$$\frac{y^2}{a^2} - \frac{x^2}{b^2} = 1$$

which is an equation of a hyperbola having its center at the origin and its principal axis on the y axis.

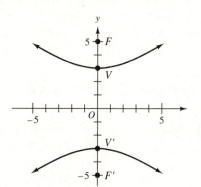

FIGURE 6

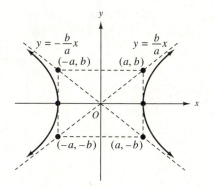

FIGURE 7

$$\iint \frac{dy}{dx}$$

▷ **ILLUSTRATION 1**

The equation

$$\frac{y^2}{9} - \frac{x^2}{16} = 1$$

can be obtained from the one in Example 1 by interchanging x and y. The graph of this equation is a hyperbola having its center at the origin, the y axis as its principal axis, its vertices at $V(0, 3)$ and $V'(0, -3)$, and its foci at $F(0, 5)$ and $F'(0, -5)$. Figure 6 shows the hyperbola and its foci. ◀

As we did with circles in Section 3.4 and with an ellipse in Illustration 1 of Section 11.1, we can plot a hyperbola on a graphics calculator by first defining y as two functions of x obtained by solving the equation of the hyperbola for y. As with an ellipse, however, it is easier to plot the graph from parametric equations of the hyperbola. These parametric equations involve two functions involving combinations of e^x and e^{-x}, called *hyperbolic functions.* We define these functions at the end of this section and show how they are used to plot hyperbolas.

In the standard equation of an ellipse, we know that $a < b$. For a hyperbola, however, there is no general inequality involving a and b. For instance, in Example 1 where $a = 3$ and $b = 4$, $a < b$; but in Example 2 where $a = \sqrt{21}$ and $b = 2$, $a > b$. Furthermore a may equal b, in which case the hyperbola is **equilateral.** The equilateral hyperbola having the equation

$$x^2 - y^2 = 1$$

is called the **unit hyperbola.**

Refer now to Figure 7, showing the hyperbola having the equation

$$\frac{x^2}{a^2} - \frac{y^2}{b^2} = 1$$

The dashed lines in the figure are asymptotes of the hyperbola. In Section 5.5 we discussed vertical, horizontal, and oblique asymptotes of rational functions. A rigorous definition of an asymptote of a more general function requires the concept of "limit," studied in calculus. Intuitively, however, we can state that if the undirected distance between a graph and a line gets smaller and smaller (but not zero) as either $|x|$ or $|y|$ gets larger and larger, then the line is an asymptote of the graph.

Observe in Figure 7 that the diagonals of the rectangle having vertices at (a, b), $(a, -b)$ $(-a, b)$, and $(-a, -b)$ are on the asymptotes of the hyperbola. This rectangle is called the **auxiliary rectangle;** its sides have lengths $2a$ and $2b$. The vertices of the hyperbola are the points of intersection of the principal axis and the auxiliary rectangle. A fairly good graph of a hyperbola can be sketched by first drawing the auxiliary rectangle. By extending the diagonals of the rectangle, we have the asymptotes. Through each vertex we draw a branch of the hyperbola by using the asymptotes as guides. Observe that because $a^2 + b^2 = c^2$, the circle having its center at the origin and passing through the vertices of the auxiliary rectangle also passes through the foci of the hyperbola.

▶ **EXAMPLE 3** *Finding Properties of a Hyperbola from Its Equation and Sketching Its Graph*

Find the vertices of the hyperbola having the equation

$$x^2 - 4y^2 = 16$$

Sketch the hyperbola and show the auxiliary rectangle and asymptotes.

Solution   The given equation is equivalent to

$$\frac{x^2}{16} - \frac{y^2}{4} = 1$$

Therefore the hyperbola has its center at the origin, and its principal axis is the x axis. Because $a^2 = 16$ and $b^2 = 4$, $a = 4$ and $b = 2$. The vertices are at $V(4, 0)$ and $V'(-4, 0)$, and the sides of the auxiliary rectangle have lengths $2a = 8$ and $2b = 4$. Figure 8 shows the auxiliary rectangle and the asymptotes. These asymptotes are used as guides to sketch the hyperbola appearing in the figure.   ◀

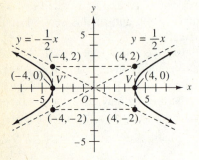

FIGURE 8

A convenient device can be used to obtain equations of the asymptotes of a hyperbola. For instance, for the hyperbola having the equation $\frac{x^2}{a^2} - \frac{y^2}{b^2} = 1$, we replace the right side by zero and obtain

$$\frac{x^2}{a^2} - \frac{y^2}{b^2} = 0$$

Upon factoring, this equation becomes

$$\left(\frac{x}{a} - \frac{y}{b}\right)\left(\frac{x}{a} + \frac{y}{b}\right) = 0$$

which is equivalent to the two equations

$$\frac{x}{a} - \frac{y}{b} = 0 \qquad \text{and} \qquad \frac{x}{a} + \frac{y}{b} = 0$$

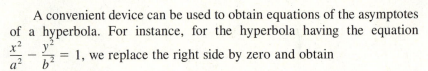

$$\Leftrightarrow \qquad y = \frac{b}{a}x \qquad \text{and} \qquad y = -\frac{b}{a}x$$

which are equations of the asymptotes of the given hyperbola.

▷ **ILLUSTRATION 2**

An equation of the hyperbola of Example 3 is

$$\frac{x^2}{16} - \frac{y^2}{4} = 1$$

To obtain equations of the asymptotes we replace the right side by zero, and we have

$$\frac{x^2}{16} - \frac{y^2}{4} = 0$$

$$\left(\frac{x}{4} - \frac{y}{2}\right)\left(\frac{x}{4} + \frac{y}{2}\right) = 0$$

$$\frac{x}{4} - \frac{y}{2} = 0 \qquad\qquad \frac{x}{4} + \frac{y}{2} = 0$$

$$y = \tfrac{1}{2}x \qquad \text{and} \qquad y = -\tfrac{1}{2}x \qquad\qquad ◀$$

Suppose the center of a hyperbola is at (h, k) and its principal axis is parallel to one of the coordinate axes. Then by a translation of axes so that the point (h, k) is the new origin, an equation of the hyperbola is

$$\frac{x'^2}{a^2} - \frac{y'^2}{b^2} = 1$$

if the principal axis is horizontal, and

$$\frac{y'^2}{a^2} - \frac{x'^2}{b^2} = 1$$

if the principal axis is vertical. If we replace x' by $x - h$ and y' by $y - k$, we obtain the following standard forms of an equation of a hyperbola.

Standard Forms of an Equation of a Hyperbola

If the center of a hyperbola is at (h, k) and the distance between the vertices is $2a$, then an equation of the hyperbola is of the form

$$\frac{(x - h)^2}{a^2} - \frac{(y - k)^2}{b^2} = 1$$

if the principal axis is horizontal, and

$$\frac{(y - k)^2}{a^2} - \frac{(x - h)^2}{b^2} = 1$$

if the principal axis is vertical.

By expanding $(x - h)^2$ and $(y - k)^2$ and simplifying, we can write each of these equations in the form

$$Ax^2 + Cy^2 + Dx + Ey + F = 0 \qquad \textbf{(1)}$$

where A and C have opposite signs. The next example involves an equation of this form.

▶ **EXAMPLE 4**   *Finding Properties of a Hyperbola and Sketching Its Graph*

Show that the graph of the equation

$$9x^2 - 4y^2 - 18x - 16y + 29 = 0$$

is a hyperbola. Find the center, an equation of the principal axis, and the vertices. Sketch the hyperbola and show the auxiliary rectangle and asymptotes.

Solution   We begin by completing the squares in x and y. We have

$$9(x^2 - 2x) - 4(y^2 + 4y) = -29$$
$$9(x^2 - 2x + 1) - 4(y^2 + 4y + 4) = -29 + 9 - 16$$
$$9(x - 1)^2 - 4(y + 2)^2 = -36$$
$$\frac{(y + 2)^2}{9} - \frac{(x - 1)^2}{4} = 1$$

This equation is that of a hyperbola whose center is at $(1, -2)$ and whose principal axis is the vertical line having the equation $x = 1$. Because $a^2 = 9$ and $b^2 = 4$, $a = 3$ and $b = 2$. The vertices are on the principal axis and 3 units above and below the center; thus they are at $V(1, 1)$ and $V'(1, -5)$. The auxiliary rectangle has sides of lengths $2a = 6$ and $2b = 4$; it appears in Figure 9 along with the asymptotes and the hyperbola.   ◀

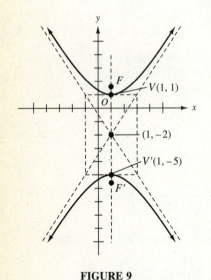

FIGURE 9

In the following illustration we have another equation in the form of (1).

▷ **ILLUSTRATION 3**

The equation

$$4x^2 - 12y^2 + 24x + 96y - 156 = 0$$

can be written as

$$4(x^2 + 6x) - 12(y^2 - 8y) = 156$$

and upon completing the squares in x and y we have

$$4(x^2 + 6x + 9) - 12(y^2 - 8y + 16) = 156 + 36 - 192$$
$$4(x + 3)^2 - 12(y - 4)^2 = 0$$
$$(x + 3)^2 - 3(y - 4)^2 = 0$$
$$[(x + 3) - \sqrt{3}(y - 4)][(x + 3) + \sqrt{3}(y - 4)] = 0$$
$$x + 3 - \sqrt{3}(y - 4) = 0 \quad \text{and} \quad x + 3 + \sqrt{3}(y - 4) = 0$$

which are equations of two lines through the point $(-3, 4)$. ◄

We can prove in general that the graph of any equation of the form (1) is either a hyperbola or two intersecting lines. The results of Example 4 and Illustration 3 are particular cases of this fact.

Equation (1) is the special case of the general equation of the second degree in two variables,

$$Ax^2 + Bxy + Cy^2 + Dx + Ey + F = 0 \tag{2}$$

where $B = 0$ and $AC < 0$ (that is, A and C have opposite signs).

The following theorem summarizes the conclusions in the preceding discussion.

> **THEOREM 1**
>
> If in the general second-degree equation (2), $B = 0$ and $AC < 0$, then the graph is either a hyperbola or two intersecting lines.

The degenerate case of a hyperbola, two intersecting lines, is obtained as a conic section if the cutting plane contains the vertex of the cone and two generators, as shown in Figure 10.

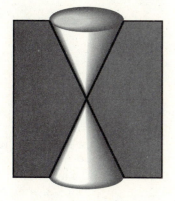

two intersecting lines

FIGURE 10

► **EXAMPLE 5** *Finding Equations of a Hyperbola and Its Asymptotes from Properties of the Hyperbola and Sketching the Graph*

The vertices of a hyperbola are at $(-5, -3)$ and $(-5, -1)$ and the endpoints of the conjugate axis are at $(-7, -2)$ and $(-3, -2)$. Find an equation of the hyperbola and equations of the asymptotes. Sketch the hyperbola and the asymptotes.

Solution   The distance between the vertices is $2a$; hence $2a = 2$ and $a = 1$. The length of the conjugate axis is $2b$; thus $2b = 4$ and $b = 2$. Because the principal axis is vertical, an equation of the hyperbola is of the

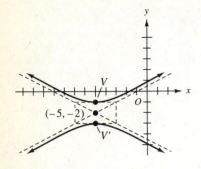

FIGURE 11

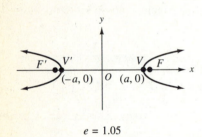

$e = 1.05$

(a)

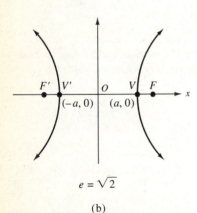

$e = \sqrt{2}$

(b)

FIGURE 12

form

$$\frac{(y - k)^2}{a^2} - \frac{(x - h)^2}{b^2} = 1$$

The center (h, k) is halfway between the vertices and is therefore at the point $(-5, -2)$. Hence an equation of the hyperbola is

$$\frac{(y + 2)^2}{1} - \frac{(x + 5)^2}{4} = 1$$

Replacing the right side by zero to obtain equations of the asymptotes, we have

$$\left(\frac{y + 2}{1} - \frac{x + 5}{2}\right)\left(\frac{y + 2}{1} + \frac{x + 5}{2}\right) = 0$$

$$y + 2 = \tfrac{1}{2}(x + 5) \qquad \text{and} \qquad y + 2 = -\tfrac{1}{2}(x + 5)$$

The hyperbola and the asymptotes appear in Figure 11.   ◀

As with an ellipse, an indication of the shape of a hyperbola is given by its *eccentricity*, defined exactly the same as for an ellipse; that is, if e is the **eccentricity** of a hyperbola,

$$e = \frac{c}{a}$$

For a hyperbola, however, $e > 1$. This follows from the fact that $c > a$ because for a hyperbola

$$c^2 = a^2 + b^2 \tag{3}$$

From this equation, when $a = b$, we obtain $c = \sqrt{2}a$. Thus the eccentricity of an equilateral hyperbola is $\sqrt{2}$. See Figure 12(b). If e approaches 1 and a remains fixed, then c approaches a and from Equation (3) b approaches 0, so that the shape of the hyperbola becomes "thin" around its principal axis. Figure 12(a) shows a hyperbola with $e = 1.05$ and the same value of a as in Figure 12(b). If e increases as a remains fixed, then c increases and b increases, and the hyperbola becomes "fat" around its principal axis. See Figure 12(c) for a hyperbola with $e = 2$ and the same value of a as in Figures 12(a) and 12(b).

The property of the hyperbola given in its definition forms the basis of several important navigational systems. These systems involve a network of pairs of radio transmitters at fixed positions at a known distance from one another. The transmitters send out radio signals that are received by a navigator. The difference in arrival time of the two signals determines the difference $2a$ of the distances from the navigator. Thus the navigator's position is known to be somewhere along one arc of a hyperbola having foci at the locations of the two transmitters. One arc, rather than both, is deter-

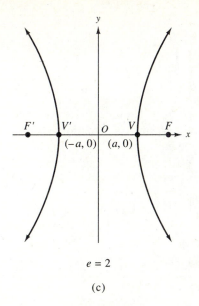

$e = 2$

(c)

FIGURE 12

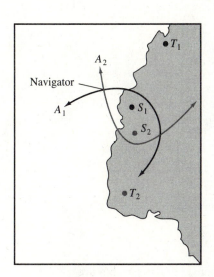

FIGURE 13

mined because of the signal delay between the two transmitters that is built into the system. The procedure is then repeated for a different pair of radio transmitters, and another arc of a hyperbola that contains the navigator's position is determined. The point of intersection of the two hyperbolic arcs is the actual position. For example, in Figure 13 suppose a pair of transmitters is located at points T_1 and S_1 and the signals from this pair determine the hyperbolic arc A_1. Another pair of transmitters is located at points T_2 and S_2 and hyperbolic arc A_2 is determined from their signals. Then the intersection of A_1 and A_2 is the position of the navigator.

The hyperbola has a reflective property used in the design of certain telescopes. Hyperbolas are also used in combat to locate the position of enemy guns by the sound of their firing, a practice called *sound ranging*. Some comets move in hyperbolic orbits. If a quantity varies inversely as another quantity, such as pressure and volume in Boyle's law for a perfect gas ($PV = k$), the graph is a hyperbola as you will learn in Section 11.3.

We now define the *hyperbolic cosine* (abbreviated cosh) and *hyperbolic sine* (abbreviated sinh) functions, which we will apply to obtain parametric equations of a hyperbola.

> **DEFINITION   The Hyperbolic Cosine and Hyperbolic Sine Functions**
>
> $$\cosh t = \frac{e^t + e^{-t}}{2} \qquad \text{and} \qquad \sinh t = \frac{e^t - e^{-t}}{2}$$
>
> The domains of the two functions are the set R of real numbers. The range of cosh is the interval $[1, +\infty)$ and the range of sinh is the set R of real numbers.

We now obtain a fundamental identity involving cosh and sinh from which parametric equations of a hyperbola follow. From the definitions of $\cosh t$ and $\sinh t$, we have

$$
\begin{aligned}
\cosh^2 t - \sinh^2 t &= \left(\frac{e^t + e^{-t}}{2}\right)^2 - \left(\frac{e^t - e^{-t}}{2}\right)^2 \\
&= \frac{e^{2t} + 2e^0 + e^{-2t}}{4} - \frac{e^{2t} - 2e^0 + e^{-2t}}{4} \\
&= \frac{e^{2t} + 2 + e^{-2t} - e^{2t} + 2 - e^{-2t}}{4} \\
&= \frac{4}{4} \\
&= 1
\end{aligned}
$$

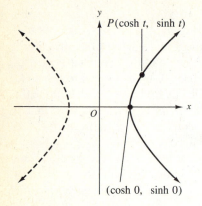

FIGURE 14

We have proved the identity

$$\cosh^2 t - \sinh^2 t = 1$$

From this identity it follows that the parametric equations

$$x = \cosh t \quad \text{and} \quad y = \sinh t \tag{4}$$

define points on the unit hyperbola $x^2 - y^2 = 1$. However, because $\cosh t$ is never less than 1, the curve defined by the parametric equations consists of only the points on the right branch of the unit hyperbola. See Figure 14. The "dashed" curve in the figure is the left branch of the unit hyperbola, which is not defined by parametric equations (4). We can, however, define the left branch by the parametric equations

$$x = -\cosh t \quad \text{and} \quad y = \sinh t$$

More generally, the parametric equations

$$x = a \cosh t \quad \text{and} \quad y = b \sinh t$$

define the right branch of the hyperbola

$$\frac{x^2}{a^2} - \frac{y^2}{b^2} = 1$$

and the parametric equations

$$x = -a \cosh t \quad \text{and} \quad y = b \sinh t$$

define the left branch. You are asked to prove this in Exercise 44. In the following two illustrations we obtain the parametric equations used to plot the hyperbolas of Examples 1 and 4 on a graphics calculator. Consult your manual to learn how to enter hyperbolic functions on your particular calculator.

▷ **ILLUSTRATION 4**

The hyperbola of Example 1 has the cartesian equation

$$\frac{x^2}{9} - \frac{y^2}{16} = 1$$

Parametric equations of the right branch of this hyperbola are

$$x = 3 \cosh t \quad \text{and} \quad y = 4 \sinh t$$

and parametric equations of the left branch are

$$x = -3 \cosh t \quad \text{and} \quad y = 4 \sinh t$$

We plot the graphs of the two sets of parametric equations in the same viewing rectangle, and we obtain the hyperbola shown in Figure 4. ◄

▷ **ILLUSTRATION 5**

The hyperbola of Example 4 has the cartesian equation

$$\frac{(y + 2)^2}{9} - \frac{(x - 1)^2}{4} = 1$$

Parametric equations of the upper branch of this hyperbola are

$$x = 2 \sinh t + 1 \quad \text{and} \quad y = 3 \cosh t - 2$$

and parametric equations of the lower branch are

$$x = 2 \sinh t + 1 \quad \text{and} \quad y = -3 \cosh t - 2$$

We plot the graphs of these two sets of parametric equations in the same viewing rectangle to obtain the hyperbola appearing in Figure 9.   ◀

An alternative method for plotting the graph of the hyperbola of Illustration 5 utilizes the parametric equations.

$$x = 2 \tan t + 1 \quad \text{and} \quad y = 3 \sec t - 2$$

where $0 \leq t \leq 2\pi$, based on the identity $\sec^2 t - \tan^2 t = 1$. See Exercise 55.

 The hyperbolic cosine and hyperbolic sine functions, as well as other hyperbolic functions, are treated in more detail in a calculus course where most of their applications are to engineering and physics. In particular, the catenary, a curve formed by a flexible cable of uniform density hanging from two points under its own weight, has an equation involving cosh.

EXERCISES 11.2

In Exercises 1 through 6, for the hyperbola having the equation, find (a) the center, (b) the principal axis, (c) the vertices, and (d) the foci. (e) Sketch the hyperbola and show the foci.

1. $\dfrac{x^2}{64} - \dfrac{y^2}{36} = 1$

2. $\dfrac{x^2}{4} - \dfrac{y^2}{4} = 1$

3. $\dfrac{y^2}{25} - \dfrac{x^2}{144} = 1$

4. $\dfrac{y^2}{16} - \dfrac{x^2}{9} = 1$

5. $9x^2 - 4y^2 = 36$

6. $25y^2 - 4x^2 = 100$

In Exercises 7 through 20, for the hyperbola having the equation, find (a) the center, (b) the principal axis, and (c) the vertices. (d) Sketch the hyperbola and show the auxiliary rectangle and asymptotes.

7. $\dfrac{x^2}{25} - \dfrac{y^2}{16} = 1$

8. $\dfrac{x^2}{9} - \dfrac{y^2}{25} = 1$

9. $\dfrac{y^2}{4} - \dfrac{x^2}{16} = 1$

10. $\dfrac{y^2}{100} - \dfrac{x^2}{49} = 1$

11. $25y^2 - 36x^2 = 900$ **12.** $4x^2 - 9y^2 = 144$

13. $x^2 - y^2 + 6x - 4y - 4 = 0$

14. $9y^2 - 4x^2 + 32x - 36y - 64 = 0$

15. $9x^2 - 16y^2 + 54x - 32y - 79 = 0$

16. $9y^2 - 25x^2 - 50x - 72y - 106 = 0$

17. $3y^2 - 4x^2 - 8x - 24y - 40 = 0$

18. $2x^2 - y^2 + 12x + 8y - 6 = 0$

19. $4y^2 - 9x^2 + 16y + 18x = 29$

20. $4x^2 - y^2 + 56x + 2y + 195 = 0$

In Exercises 21 through 26, find equations of the asymptotes of the hyperbola of the given exercise.

21. Exercise 7   **22.** Exercise 10

23. Exercise 13   **24.** Exercise 16

25. Exercise 19   **26.** Exercise 18

In Exercises 27 through 36, find an equation of the hyperbola satisfying the conditions and sketch the hyperbola.

27. Vertices at $(-2, 0)$ and $(2, 0)$ and a conjugate axis of length 6.

28. Foci at $(0, 5)$ and $(0, -5)$ and a vertex at $(0, 4)$.

29. Center at the origin, its foci on the y axis, and passing through the points $(-2, 4)$ and $(-6, 7)$.

30. Endpoints of its conjugate axis at $(0, -3)$ and $(0, 3)$ and one focus at $(5, 0)$.

31. One focus at $(26, 0)$ and asymptotes the lines $12y = \pm 5x$.

32. Center at $(3, -5)$, a vertex at $(7, -5)$, and a focus at $(8, -5)$.

33. Center at $(-2, -1)$, a vertex at $(-2, 11)$, and a focus at $(-2, 14)$.

34. Foci at $(3, 6)$ and $(3, 0)$ and passing through the point $(5, 3 + \frac{6}{5}\sqrt{5})$.

35. Foci at $(-1, 4)$ and $(7, 4)$ and length of the transverse axis is $\frac{8}{3}$.

36. One focus at $(-3 - 3\sqrt{13}, 1)$, asymptotes intersecting at $(-3, 1)$, and one asymptote passing through the point $(1, 7)$.

37. The vertices of a hyperbola are at $(-3, -1)$ and $(-1, -1)$ and the distance between the foci is $2\sqrt{5}$. Find **(a)** an equation of the hyperbola and **(b)** equations of the asymptotes.

38. The foci of a hyperbola are at $(2, 7)$ and $(2, -7)$ and the distance between the vertices is $8\sqrt{3}$. Find **(a)** an equation of the hyperbola and **(b)** equations of the asymptotes.

39. Find an equation of the hyperbola whose foci are the vertices of the ellipse $7x^2 + 11y^2 = 77$ and whose vertices are the foci of this ellipse.

40. Find an equation of the ellipse whose foci are the vertices of the hyperbola $11x^2 - 7y^2 = 77$ and whose vertices are the foci of this hyperbola.

41. The cost of production of a commodity is $12 less per unit at a point A than it is at a point B, and the distance between A and B is 100 km. Assuming that the route of delivery of the commodity is along a straight line and that the delivery cost is 20 cents per unit per kilometer, find the curve at any point of which the commodity can be supplied from either A or B at the same total cost. (*Hint:* Take points A and B at $(-50, 0)$ and $(50, 0)$, respectively.)

42. Two LORAN (long-range navigation) stations A and B lie on a line running east and west, and A is 80 miles due east of B. An airplane is traveling east on a straight-line course that is 60 miles north of the line through A and B. Signals are sent at the same time from A and B, and the signal from A reaches the plane 350 μsec (microseconds) before the one from B. If the signals travel at the rate of 0.2 mi/μsec, locate the position of the plane by the definition of a hyperbola.

43. Show that the equation
$$\sqrt{(x - c)^2 + y^2} - \sqrt{(x + c)^2 + y^2} = \pm 2a$$
can be simplified to
$$\frac{x^2}{a^2} - \frac{y^2}{b^2} = 1$$
where $b^2 = c^2 - a^2$.

44. Prove that the hyperbola having the cartesian equation $\frac{x^2}{a^2} - \frac{y^2}{b^2} = 1$ is defined by the following two sets of parametric equations:
$$x = a \cosh t \quad \text{and} \quad y = b \sinh t$$
and
$$x = -a \cosh t \quad \text{and} \quad y = b \sinh t$$

In Exercises 45 through 52, write parametric equations defining the hyperbola of the given exercise, and use them to plot the hyperbola.

45. Exercise 1   **46.** Exercise 2

47. Exercise 3   **48.** Exercise 4

49. Exercise 13   **50.** Exercise 14

51. Exercise 17   **52.** Exercise 18

53. For a hyperbola the eccentricity e is greater than 1, and for an ellipse $0 < e < 1$. Explain why the eccentricity of a parabola is equal to 1.

54. Explain how the hyperbolic cosine and hyperbolic sine functions are related to the coordinates of points on the unit hyperbola in a manner similar to that in which the corresponding trigonometric functions (cosine and sine) are related to the coordinates of points on the unit circle.

55. Explain why the hyperbola having the equation

$$\frac{(y - k)^2}{a^2} - \frac{(x - h)^2}{b^2} = 1$$

can be plotted by utilizing the parametric equations

$$x = b \tan t + h \quad \text{and} \quad y = a \sec t + k$$

56. Refer to Exercise 55. Write a set of parametric equations involving the tangent and secant functions that you could use to plot the hyperbola having the equation

$$\frac{(x - h)^2}{a^2} - \frac{(y - k)^2}{b^2} = 1$$

and explain why the procedure works.

11.3 THE GENERAL EQUATION OF THE SECOND DEGREE IN TWO VARIABLES AND ROTATION OF AXES

GOALS

1. **Use the product of the coefficients of x^2 and y^2 to identify the graph of a second-degree equation in two variables having no xy term.**
2. **Learn the formulas for rotation of axes.**
3. **Find an equation of a graph with respect to the new axes after a rotation of axes.**
4. **Eliminate the xy term in a second-degree equation by a rotation of axes.**
5. **Sketch the graph of a second-degree equation having an xy term.**
6. **Identify the graph of a second-degree equation in two variables by examining the discriminant of the equation.**
7. **Plot the graph of a second-degree equation in two variables on a graphics calculator.**

You learned in Sections 11.1 and 11.2 that the graph of the general second-degree equation in two variables,

$$Ax^2 + Bxy + Cy^2 + Dx + Ey + F = 0 \tag{1}$$

is an ellipse or a degenerate case if $B = 0$ and $AC > 0$, and a hyperbola or the degenerate case if $B = 0$ and $AC < 0$.

We now consider (1) where $B = 0$ and $AC = 0$. In such a case either $A = 0$ or $C = 0$, but not both, for if the three numbers A, B, and C are all zero, (1) is not a second-degree equation. Suppose in (1) that $B = 0$, $A = 0$, and $C \neq 0$. Then (1) becomes

$$Cy^2 + Dx + Ey + F = 0 \tag{2}$$

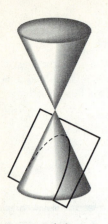

parabola

FIGURE 1

If $D \neq 0$, you learned in Section 3.5 that the graph of this equation is a parabola, the third conic section. The parabola is obtained as a conic section if the cutting plane is parallel to one and only one generator of a conic. See Figure 1.

In Section 3.3, we defined a parabola as a set of points in a plane. The proof that this definition follows from its definition as a conic section is similar to that for an ellipse. For a parabola, however, you need only one sphere tangent to the cutting plane at the focus and tangent to the cone along a circle. The intersection of the plane of the circle with the cutting plane is the directrix of the parabola.

The degenerate cases of a parabola, occurring if $D = 0$ in (2), are two parallel lines, one line, or the empty set.

$\triangleright$ **ILLUSTRATION 1**

The graph of the equation

$$4y^2 - 9 = 0$$

is two parallel lines: $2y - 3 = 0$ and $2y + 3 = 0$. The graph of the equation

$$9y^2 + 6y + 1 = 0$$

is one line because the equation is equivalent to $(3y + 1)^2 = 0$. Because the equation

$$2y^2 + y + 1 = 0$$

has no real solutions, its graph is the empty set. $\blacktriangleleft$

A discussion similar to the above can be given if, in (1), $B = 0$, $C = 0$, and $A \neq 0$. We summarize the results in the following theorem.

THEOREM 1

In the general second-degree Equation (1), if $B = 0$ and either $A = 0$ and $C \neq 0$ or $C = 0$ and $A \neq 0$, then the graph is one of the following: a parabola, two parallel lines, one line, or the empty set.

line

FIGURE 2

The degenerate case of a parabola, one line, is obtained as a conic section if the cutting plane contains the vertex of the cone and only one generator, as in Figure 2. The degenerate parabola consisting of two parallel lines cannot be obtained as a plane section of a cone unless we consider a circular cylinder as a degenerate cone with its vertex at infinity. Then a plane parallel to the elements of the cylinder and cutting two distinct elements produces the two parallel lines.

From the theorems of Sections 11.1 and 11.2 and the preceding theorem in this section, we may conclude that the graph of the general second-degree equation in two variables when $B = 0$ is either a conic or a degenerate conic. The type of conic can be determined from the product of A and C. We have the following theorem.

THEOREM 2

The graph of the equation

$$Ax^2 + Cy^2 + Dx + Ey + F = 0$$

where A and C are not both zero, is either a conic or a degenerate conic. If it is a conic, then the graph is

   (i) a *parabola* if either $A = 0$ or $C = 0$, that is, if $AC = 0$;
  (ii) an *ellipse* if A and C have the same sign, that is, if $AC > 0$;
 (iii) a *hyperbola* if A and C have opposite signs, that is, if $AC < 0$.

▶ **EXAMPLE 1** *Identifying the Graph of a Second-Degree Equation in Two Variables*

Identify the graph of each of the following equations as the type of conic or degenerate conic:

(a) $9x^2 + y^2 - 18x + 4y + 4 = 0$ **(b)** $x^2 + 4y^2 = 0$
(c) $2x^2 + 12x - 5y + 28 = 0$ **(d)** $x^2 - 4 = 0$
(e) $3x^2 - 2y^2 + 12x - 4y - 2 = 0$ **(f)** $x^2 - 4y^2 = 0$

Solution   In each part, we have a second-degree equation in two variables. The graph is, therefore, a conic or else it degenerates. From Theorem 2, the product AC determines the type of conic.

(a) Because $A = 9$ and $C = 1$, $AC = 9 > 0$. Therefore, the graph is an ellipse or else degenerates. By completing squares, the equation can be written as

$$(x - 1)^2 + \frac{(y + 2)^2}{9} = 1$$

Thus the graph is an ellipse.
(b) Because the only ordered pair satisfying this equation is $(0, 0)$, the graph is the origin, a degenerate ellipse.
(c) Because $C = 0$, $AC = 0$. So the graph is a parabola or degenerates. By completing squares, we can write the equation as

$$(x + 3)^2 = \tfrac{5}{2}(y - 2)$$

which is an equation of a parabola.

(d) Because the equation $x^2 - 4 = 0$ is equivalent to the equation $(x - 2)(x + 2) = 0$, its graph consists of the two parallel lines $x = 2$ and $x = -2$, a degenerate parabola.

(e) Because $A = 3$ and $C = -2$, $AC = -6 < 0$. The graph is, therefore, a hyperbola, or else it degenerates. The equation is equivalent to

$$\frac{(x + 2)^2}{4} - \frac{(y + 1)^2}{6} = 1$$

whose graph is a hyperbola.

(f) The equation is equivalent to $(x - 2y)(x + 2y) = 0$; thus its graph consists of the two intersecting lines $x = 2y$ and $x = -2y$, a degenerate hyperbola. ◀

We now discuss the graph of the general second-degree equation in two variables where $B \neq 0$, that is, an equation having an xy term. We transform such an equation into one having no xy term by rotating the coordinate axes. While a translation of axes gives a new coordinate system whose axes are parallel to the original x and y axes, a rotation of axes gives a coordinate system that will in general have axes that are *not* parallel to the original ones.

Suppose we have two rectangular cartesian coordinate systems with the same origin. Let one system be the xy system and the other the $\overline{x}\,\overline{y}$ system. Suppose further that the $\overline{x}$ axis makes an angle α with the x axis. See Figure 3. Of course, the $\overline{y}$ axis then makes an angle α with the y axis. In such a case we state that the xy system of coordinates is *rotated* through an angle α to form the $\overline{x}\,\overline{y}$ system of coordinates. A point P having coordinates (x, y) with respect to the original coordinate system will have coordinates $(\overline{x}, \overline{y})$ with respect to the new one. We now obtain relationships between these two sets of coordinates.

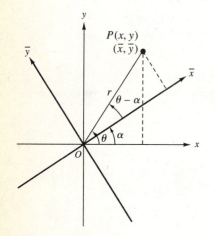

FIGURE 3

In Figure 3, let r denote the undirected distance $|\overline{OP}|$ and let θ be the angle measured from the x axis to the line segment OP. From the figure we observe that

$$x = r \cos \theta \quad \text{and} \quad y = r \sin \theta \tag{3}$$

Also from Figure 3

$$\overline{x} = r \cos(\theta - \alpha) \quad \text{and} \quad \overline{y} = r \sin(\theta - \alpha)$$

With the cosine and sine difference identities these two equations become

$$\overline{x} = r \cos \theta \cos \alpha + r \sin \theta \sin \alpha$$

and

$$\overline{y} = r \sin \theta \cos \alpha - r \cos \theta \sin \alpha$$

Substituting from Equations (3) into the preceding equations, we get

$$\overline{x} = x \cos \alpha + y \sin \alpha \quad \text{and} \quad \overline{y} = -x \sin \alpha + y \cos \alpha \tag{4}$$

Solving Equations (4) simultaneously for x and y in terms of $\bar{x}$ and $\bar{y}$ (see Exercise 38), we obtain

$$x = \bar{x} \cos \alpha - \bar{y} \sin \alpha \quad \text{and} \quad y = \bar{x} \sin \alpha + \bar{y} \cos \alpha \qquad \textbf{(5)}$$

We state these results formally.

Formulas for Rotation of Axes

If (x, y) represents a point P with respect to a given set of axes and $(\bar{x}, \bar{y})$ is a representation of P after the axes have been rotated through an angle α, then

(i) $x = \bar{x} \cos \alpha - \bar{y} \sin \alpha \quad$ and $\quad y = \bar{x} \sin \alpha + \bar{y} \cos \alpha$
(ii) $\bar{x} = x \cos \alpha + y \sin \alpha \quad$ and $\quad \bar{y} = -x \sin \alpha + y \cos \alpha$

▶ **EXAMPLE 2** *Obtaining an Equation of a Graph After a Rotation of Axes and Sketching the Graph*

Given the equation

$$xy = 1$$

(a) Find an equation of the graph with respect to the x and y axes after a rotation of axes through an angle of radian measure $\frac{1}{4}\pi$. **(b)** Sketch the graph and show both sets of axes.

Solution

(a) With $\alpha = \frac{1}{4}\pi$ in (i) of the Formulas for Rotation of Axes, we obtain

$$x = \frac{1}{\sqrt{2}}\bar{x} - \frac{1}{\sqrt{2}}\bar{y} \quad \text{and} \quad y = \frac{1}{\sqrt{2}}\bar{x} + \frac{1}{\sqrt{2}}\bar{y}$$

Substituting these expressions for x and y in the equation $xy = 1$, we get

$$\left(\frac{1}{\sqrt{2}}\bar{x} - \frac{1}{\sqrt{2}}\bar{y}\right)\left(\frac{1}{\sqrt{2}}\bar{x} + \frac{1}{\sqrt{2}}\bar{y}\right) = 1$$

$$\frac{\bar{x}^2}{2} - \frac{\bar{y}^2}{2} = 1$$

(b) This is an equation of an equilateral hyperbola whose asymptotes are the bisectors of the quadrants in the $\bar{x}\bar{y}$ system. Thus the graph of the equation $xy = 1$ is an equilateral hyperbola lying in the first and third quadrants and the asymptotes are the x and y axes. See Figure 4 for the required graph. ◀

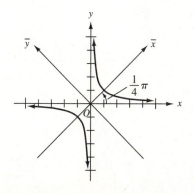

FIGURE 4

From Theorem 1 we know that when $B = 0$ and A and C are not both zero, the graph of (1), the general second-degree equation in two variables, is either a conic or a degenerate conic. We now show that if $B = 0$, then any equation of the form (1) can be transformed by a suitable rotation of axes into an equation of the form

$$\bar{A}\bar{x}^2 + \bar{C}\bar{y}^2 + \bar{D}\bar{x} + \bar{E}\bar{y} + \bar{F} = 0 \qquad (6)$$

where $\bar{A}$ and $\bar{C}$ are not both zero.

If the xy system is rotated through an angle α, then to obtain an equation of the graph of (1) with respect to the $\bar{x}\bar{y}$ system, we replace x by $\bar{x} \cos \alpha - \bar{y} \sin \alpha$ and y by $\bar{x} \sin \alpha + \bar{y} \cos \alpha$. We get

$$\bar{A}\bar{x}^2 + \bar{B}\bar{x}\bar{y} + \bar{C}\bar{y}^2 + \bar{D}\bar{x} + \bar{E}\bar{y} + \bar{F} = 0 \qquad (7)$$

where

$$\bar{A} = A \cos^2 \alpha + B \sin \alpha \cos \alpha + C \sin^2 \alpha$$
$$\bar{B} = -2A \sin \alpha \cos \alpha + B(\cos^2 \alpha - \sin^2 \alpha) + 2C \sin \alpha \cos \alpha$$
$$\bar{C} = A \sin^2 \alpha - B \sin \alpha \cos \alpha + C \cos^2 \alpha$$

We wish to find an α so that the rotation transforms (1) into an equation of the form (6). Setting the expression for $\bar{B}$ equal to zero, we have

$$B(\cos^2 \alpha - \sin^2 \alpha) + (C - A)(2 \sin \alpha \cos \alpha) = 0$$

or, equivalently, with trigonometric identities,

$$B \cos 2\alpha + (C - A) \sin 2\alpha = 0$$

Because $B \neq 0$, this gives

$$\cot 2\alpha = \frac{A - C}{B}$$

We have shown that a rotation of axes through an angle α satisfying this equation will transform (1), the general second-degree equation in two variables, where $B \neq 0$, to an equation of the form (6). We now wish to show that $\bar{A}$ and $\bar{C}$ in (6) are not both zero. To prove this, notice that (7) is obtained from (1) by rotating the axes through the angle α. Also (1) can be obtained from (7) by rotating the axes back through the angle $-\alpha$. If $\bar{A}$ and $\bar{C}$ in (7) were both zero then the substitutions

$$\bar{x} = x \cos \alpha + y \sin \alpha \quad \text{and} \quad \bar{y} = -x \sin \alpha + y \cos \alpha$$

in that equation would result in the equation

$$\bar{D}(x \cos \alpha + y \sin \alpha) + \bar{E}(-x \sin \alpha + y \cos \alpha) + \bar{F} = 0$$

This equation is of the first degree and hence different from (1) because we have assumed that at least $B \neq 0$. We have, therefore, proved the following theorem.

THEOREM 3

If $B \neq 0$, the equation

$$Ax^2 + Bxy + Cy^2 + Dx + Ey + F = 0$$

can be transformed into the equation

$$\bar{A}\bar{x}^2 + \bar{C}\bar{y}^2 + \bar{D}\bar{x} + \bar{E}\bar{y} + \bar{F} = 0$$

where $\bar{A}$ and $\bar{C}$ are not both zero, by a rotation of axes through an angle α for which

$$\cot 2\alpha = \frac{A - C}{B}$$

By Theorems 2 and 3, it follows that the graph of the general second-degree equation in two variables is either a conic or a degenerate conic. To determine which type of conic is the graph of a particular equation, we examine the expression $B^2 - 4AC$.

We use the fact that A, B, and C of (1) and $\bar{A}$, $\bar{B}$, and $\bar{C}$ of (7) satisfy the equation

$$B^2 - 4AC = \bar{B}^2 - 4\bar{A}\bar{C} \tag{8}$$

which can be proved by substituting the expressions for $\bar{A}$, $\bar{B}$, and $\bar{C}$ given after Equation (7) in the right side. You are asked to do this in Exercise 37.

The expression $B^2 - 4AC$ is called the **discriminant** and Equation (8) states that the discriminant of the general quadratic equation in two variables is **invariant** under a rotation of axes.

If the angle of rotation is chosen so that $\bar{B} = 0$, then (8) becomes

$$B^2 - 4AC = -4\bar{A}\bar{C} \tag{9}$$

Except for degenerate cases, by applying Theorem 2, the graph of the equation

$$\bar{A}\bar{x}^2 + \bar{C}\bar{y}^2 + \bar{D}\bar{x} + \bar{E}\bar{y} + \bar{F} = 0$$

is a parabola if $\bar{A}\bar{C} = 0$, an ellipse if $\bar{A}\bar{C} > 0$, and a hyperbola if $\bar{A}\bar{C} < 0$; or, equivalently, a parabola if $-4\bar{A}\bar{C} = 0$, an ellipse if $-4\bar{A}\bar{C} < 0$, and a hyperbola if $-4\bar{A}\bar{C} > 0$. From these facts and Equation (9) it follows that, except for degenerate cases, the graph of (1), the general second-degree equation, is a parabola, an ellipse, or a hyperbola depending on whether the discriminant $B^2 - 4AC$ is zero, negative, or positive. We have proved the following theorem.

THEOREM 4

The graph of the equation

$$Ax^2 + Bxy + Cy^2 + Dx + Ey + F = 0$$

is either a conic or a degenerate conic. If the graph is a conic, then it is

(i) a *parabola* if $B^2 - 4AC = 0$;
(ii) an *ellipse* if $B^2 - 4AC < 0$;
(iii) a *hyperbola* if $B^2 - 4AC > 0$.

▶ **EXAMPLE 3** *Simplifying an Equation by a Rotation of Axes and Sketching Its Graph*

(a) Identify the graph of the equation

$$17x^2 - 12xy + 8y^2 - 80 = 0$$

(b) Simplify the equation by a rotation of axes. **(c)** Sketch the graph of the equation and show both sets of axes.

Solution

(a) From the equation, $A = 17$, $B = -12$, and $C = 8$.

$$B^2 - 4AC = (-12)^2 - 4(17)(8)$$

Because $B^2 - 4AC < 0$, from Theorem 4 the graph is an ellipse or else it degenerates.

(b) To eliminate the xy term by a rotation of axes, we must choose an α such that

$$\cot 2\alpha = \frac{A - C}{B}$$

$$= \frac{17 - 8}{-12}$$

$$= -\tfrac{3}{4}$$

There is a 2α in the interval $(0, \pi)$ for which $\cot 2\alpha = -\tfrac{3}{4}$. Therefore α is in the interval $(0, \tfrac{1}{2}\pi)$. To apply the formulas for rotation of axes, it is not necessary to find α so long as we find $\cos \alpha$ and $\sin \alpha$. These functions can be found from the value of $\cot 2\alpha$ by the trigonometric identities

$$\cos \alpha = \sqrt{\frac{1 + \cos 2\alpha}{2}} \quad \text{and} \quad \sin \alpha = \sqrt{\frac{1 - \cos 2\alpha}{2}}$$

Because $\cot 2\alpha = -\frac{3}{4}$ and $0 < \alpha < \frac{1}{2}\pi$, it follows that $\cos 2\alpha = -\frac{3}{5}$. So

$$\cos \alpha = \sqrt{\frac{1 - \frac{3}{5}}{2}} \quad \text{and} \quad \sin \alpha = \sqrt{\frac{1 + \frac{3}{5}}{2}}$$

$$= \frac{1}{\sqrt{5}} \qquad\qquad = \frac{2}{\sqrt{5}}$$

Substituting $x = \bar{x}/\sqrt{5} - 2\bar{y}/\sqrt{5}$ and $y = 2\bar{x}/\sqrt{5} + \bar{y}/\sqrt{5}$ in the given equation, we obtain

$$17\left(\frac{\bar{x}^2 - 4\bar{x}\bar{y} + 4\bar{y}^2}{5}\right) - 12\left(\frac{2\bar{x}^2 - 3\bar{x}\bar{y} - 2\bar{y}^2}{5}\right) + 8\left(\frac{4\bar{x}^2 + 4\bar{x}\bar{y} + \bar{y}^2}{5}\right) - 80 = 0$$

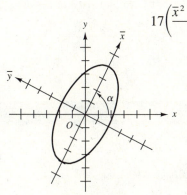

Upon simplification this equation becomes

$$\bar{x}^2 + 4\bar{y}^2 = 16$$

$$\frac{\bar{x}^2}{16} + \frac{\bar{y}^2}{4} = 1$$

The graph is, therefore, an ellipse whose major axis is 8 units long and whose minor axis is 4 units long.

(c) We apply the information obtained in part (b) to sketch the ellipse. Figure 5 shows this ellipse and both sets of axes.   ◄

FIGURE 5

As you can see from the above example, sketching the graph of a second-degree equation, having an xy term, by a rotation of axes often requires tedious computations. As shown in the next example, plotting such a graph on a graphics calculator can also entail complicated calculations when you first express y as two functions of x.

▶ **EXAMPLE 4** *Plotting the Graph of a Second-Degree Equation in Two Variables*

Plot the graph of the equation of Example 3.

Solution   The equation defines y as two functions of x. To determine these functions we treat the equation as quadratic in y and write it as

$$8y^2 - 12xy + (17x^2 - 80) = 0$$

From the quadratic formula with $a = 8$, $b = -12x$, and $c = 17x^2 - 80$, we have

$$y' = \frac{-b \pm \sqrt{b^2 - 4ac}}{2a}$$

$$= \frac{-(-12x) \pm \sqrt{(-12x)^2 - 4(8)(17x^2 - 80)}}{2(8)}$$

$$= \frac{12x \pm 4\sqrt{160 - 25x^2}}{16}$$

$$= \frac{3x \pm \sqrt{160 - 25x^2}}{4}$$

In the viewing rectangle $[-7.5, 7.5]$ by $[-5, 5]$ we plot the graphs of

$$y_1 = \frac{3x + \sqrt{160 - 25x^2}}{4} \quad \text{and} \quad y_2 = \frac{3x - \sqrt{160 - 25x^2}}{4}$$

to obtain the ellipse in Figure 5. ◄

EXERCISES 11.3

In Exercises 1 through 4, identify the graph of the equation as the type of conic or degenerate conic.

1. (a) $x^2 - 4y^2 - 6x - 24y - 31 = 0$
   (b) $4x^2 + y^2 + 8x - 14y - 47 = 0$
   (c) $y^2 + 4x - 8y + 4 = 0$
   (d) $x^2 - 16y^2 = 0$

2. (a) $2x^2 + y^2 + 8x - 2y - 9 = 0$
   (b) $2x^2 - 3y^2 + 16x + 12y + 38 = 0$
   (c) $16y^2 + 24y + 9 = 0$
   (d) $4x^2 + 16x - 3y + 19 = 0$

3. (a) $3x^2 + 5y^2 + 6x - 20y + 23 = 0$
   (b) $4x^2 - 5y^2 + 16x + 10y + 111 = 0$
   (c) $2x^2 - 16x - 5y + 22 = 0$
   (d) $9x^2 + 30x + 29 = 0$

4. (a) $5y^2 + 4x + 10y - 3 = 0$
   (b) $3x^2 + 7y^2 - 6x + 28y + 37 = 0$
   (c) $2x^2 - 3y^2 - 12x - 6y + 15 = 0$
   (d) $4x^2 - 9y^2 - 40x - 54y + 55 = 0$

In Exercises 5 through 8, (a) identify the graph of the equation, (b) find an equation of the graph with respect to the x and y axes after a rotation of axes through an angle of radian measure $\frac{1}{4}\pi$, and (c) sketch the graph and show both sets of axes.

5. $xy = 8$
6. $xy = -4$
7. $x^2 - y^2 = 8$
8. $y^2 - x^2 = 16$

In Exercises 9 through 16, (a) identify the graph of the equation, (b) remove the xy term by a rotation of axes, and (c) sketch the graph and show both sets of axes.

9. $24xy - 7y^2 + 36 = 0$
10. $4xy + 3x^2 = 4$
11. $x^2 + 2xy + y^2 - 8x + 8y = 0$
12. $x^2 + xy + y^2 = 3$
13. $xy + 16 = 0$
14. $5x^2 + 6xy + 5y^2 = 9$
15. $31x^2 + 10\sqrt{3}xy + 21y^2 = 144$
16. $6x^2 + 20\sqrt{3}xy + 26y^2 = 324$

In Exercises 17 through 26, (a) identify the graph of the equation, (b) simplify the equation by a rotation and translation of axes, and (c) sketch the graph and show the three sets of axes.

17. $x^2 + xy + y^2 - 3y - 6 = 0$
18. $19x^2 + 6xy + 11y^2 - 26x + 38y + 31 = 0$
19. $17x^2 - 12xy + 8y^2 - 68x + 24y - 12 = 0$
20. $x^2 - 10xy + y^2 + x + y + 1 = 0$
21. $x^2 + 2xy + y^2 + x - y - 4 = 0$
22. $16x^2 - 24xy + 9y^2 - 60x - 80y + 400 = 0$
23. $11x^2 - 24xy + 4y^2 + 30x + 40y - 45 = 0$

24. $3x^2 - 4xy + 8x - 1 = 0$

25. $4x^2 + 4xy + y^2 - 6x + 12 = 0$

26. $x^2 + 2xy + y^2 - x - 3y = 0$

In Exercises 27 through 34, plot the graph of the equation of the indicated exercise.

27. Exercise 9

28. Exercise 10

29. Exercise 11

30. Exercise 12

31. Exercise 17

32. Exercise 18

33. Exercise 21

34. Exercise 22

35. Show that the graph of $\sqrt{x} + \sqrt{y} = 1$ is part of a parabola by rotating the axes through an angle of radian measure $\frac{1}{4}\pi$. *Hint:* Eliminate the radicals in the equation before applying the formulas for rotation of axes.

36. Given the equation $(a^2 + b^2)xy = 1$, where $a > 0$ and $b > 0$, find an equation of the graph with respect to the $\bar{x}$ and $\bar{y}$ axes after a rotation of the axes through an angle of radian measure $\tan^{-1}(b/a)$.

37. Show that for the general second-degree equation in two variables, the discriminant $B^2 - 4AC$ is invariant under a rotation of axes.

38. Derive Equations (5) by solving Equations (4) for x and y in terms of $\bar{x}$ and $\bar{y}$. *Hint:* To solve for x, multiply both sides of the first equation by $\cos \alpha$ and both sides of the second equation by $\sin \alpha$ and then subtract corresponding members of the resulting equations. Use a similar procedure to solve for y.

39. Rotation of axes makes neither a change in the graph nor a change in the position of the graph in the plane. Explain when rotation of axes is an advantage to sketching the graph of a second-degree equation in two variables and when it is a disadvantage.

11.4 SYSTEMS INVOLVING QUADRATIC EQUATIONS

GOALS

1. **Use graphs to determine the number of ordered pairs of real numbers that are solutions of a system of equations.**
2. **Solve a system of two equations in two unknowns where one is linear and the other is quadratic.**
3. **Solve a system of two quadratic equations in two unknowns.**
4. **Solve word problems having as a mathematical model a system involving quadratic equations.**

In Section 3.2, we confined our discussion of systems of equations to linear systems. A number of applications, however, lead to nonlinear systems as illustrated in Exercises 23 through 34. The word problems in these exercises use concepts previously presented, but the resulting systems involve at least one quadratic equation. In this section we discuss methods of solving such systems of two equations in two unknowns.

By plotting the graphs of the two equations in the same viewing rectangle, we can determine the number of points of intersection of the graphs. The coordinates of these points give us the ordered pairs of real numbers that are solutions of the system.

We consider first a system that contains a linear equation and a quadratic equation. On our graphics calculator, we obtain a line that intersects the graph of the quadratic equation in at most two points. We solve the system algebraically by the substitution method. From the linear equation, we express one variable in terms of the other, and we substitute the resulting expression into the quadratic equation.

▶ **EXAMPLE 1** *Solving a System Containing a Linear Equation and a Quadratic Equation*

Plot the graphs of the equations of the following system and determine the number of ordered pairs of real numbers that are solutions of the system. Then find the solution set of the system algebraically.

$$\begin{cases} y^2 = 4x \\ x + y = 3 \end{cases} \qquad \textbf{(I)}$$

Solution   Figure 1 shows the graphs of the two equations in the same viewing rectangle. The line intersects the parabola in two points. The solution of the system, therefore, consists of two ordered pairs of real numbers.

To find the solution set algebraically, we first solve the second equation for x and obtain the equivalent system

$$\begin{cases} y^2 = 4x \\ x = 3 - y \end{cases}$$

Replacing x in the first equation by its equal from the second, we have the equivalent system

$$\begin{cases} y^2 = 4(3 - y) \\ x = 3 - y \end{cases}$$

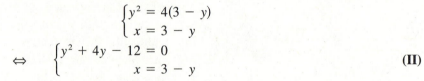

$$\Leftrightarrow \quad \begin{cases} y^2 + 4y - 12 = 0 \\ x = 3 - y \end{cases} \qquad \textbf{(II)}$$

We now solve the first equation.

$$(y - 2)(y + 6) = 0$$
$$y - 2 = 0 \qquad y + 6 = 0$$
$$y = 2 \qquad\qquad y = -6$$

Because the first equation of system (II) is equivalent to the two equations $y = 2$ and $y = -6$, system (II) is equivalent to the two systems

$$\begin{cases} y = 2 \\ x = 3 - y \end{cases} \quad \text{and} \quad \begin{cases} y = -6 \\ x = 3 - y \end{cases}$$

In each of the latter two systems we substitute into the second equation the value of y from the first, and we have

$$\begin{cases} y = 2 \\ x = 1 \end{cases} \quad \text{and} \quad \begin{cases} y = -6 \\ x = 9 \end{cases}$$

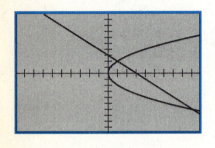

[−10, 10] by [−10, 10]
$y^2 = 4x$ and $x + y = 3$
FIGURE 1

These two systems are equivalent to system (I). Thus the solution set of (I) is $\{(1, 2), (9, -6)\}$.

The two ordered pairs give the coordinates of the points of intersection of the graphs in Figure 1. ◀

▶ **EXAMPLE 2** *Solving a System Containing a Linear Equation and a Quadratic Equation*

Follow the instructions of Example 1 for the system

$$\begin{cases} x^2 + y^2 = 25 \\ 3x + 4y = 25 \end{cases} \qquad \textbf{(III)}$$

Solution   The graphs of the two equations are plotted in Figure 2. The line appears tangent to the circle; if that is the case, we shall obtain one ordered pair of real numbers as a solution of the system. Solving the system algebraically, we express y in terms of x in the second equation, and replace y in the first equation by the resulting expression. We have the equivalent system

$$\begin{cases} x^2 + \left(\dfrac{25 - 3x}{4}\right)^2 = 25 \\ \\ y = \dfrac{25 - 3x}{4} \end{cases} \qquad \textbf{(IV)}$$

We solve the first equation by first multiplying each side by 16.

$$16x^2 + (625 - 150x + 9x^2) = 400$$
$$25x^2 - 150x + 225 = 0$$
$$x^2 - 6x + 9 = 0$$
$$(x - 3)^2 = 0$$

Hence the roots of this quadratic equation are 3 and 3; that is, 3 is a double root. Therefore system (IV) is equivalent to the system

$$\begin{cases} x = 3 \\ y = \dfrac{25 - 3x}{4} \end{cases}$$

Substituting 3 for x in the second equation, we obtain

$$\begin{cases} x = 3 \\ y = 4 \end{cases}$$

Thus the solution set of the given system (III) is $\{(3, 4)\}$.

We conclude that the line in Figure 2 is indeed tangent to the circle at the point $(3, 4)$; this is the geometric significance of the double root. The

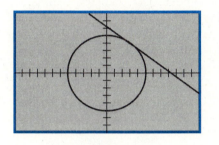

[−12, 12] by [−8, 8]
$x^2 + y^2 = 25$ and $3x + 4y = 25$
FIGURE 2

point of tangency can be considered as two intersections of the line and the circle.   ◀

▶ **EXAMPLE 3** *Solving a System Containing a Linear Equation and a Quadratic Equation*

Follow the instructions of Example 1 for the system

$$\begin{cases} x^2 + y^2 = 2 \\ x - y = 4 \end{cases} \tag{V}$$

Solution   Figure 3 shows the graphs of the two equations, a line and a circle. Because the line and the circle do not intersect, there are no ordered pairs of real numbers that are solutions of the system.

We solve the system algebraically by expressing x in terms of y in the second equation and replacing x in the first equation by the resulting expression. We obtain the equivalent system

$$\begin{cases} (y + 4)^2 + y^2 = 2 \\ x = y + 4 \end{cases} \tag{VI}$$

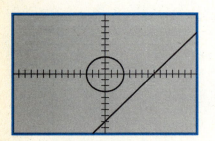

$[-7.5, 7.5]$ by $[-5, 5]$
$x^2 + y^2 = 2$ and $x - y = 4$
FIGURE 3

We solve the first equation for y.

$$y^2 + 8y + 16 + y^2 = 2$$
$$2y^2 + 8y + 14 = 0$$
$$y^2 + 4y + 7 = 0$$
$$y = \frac{-4 \pm \sqrt{4^2 - 4(1)(7)}}{2(1)}$$
$$= \frac{-4 \pm \sqrt{-12}}{2}$$
$$= \frac{-4 \pm 2i\sqrt{3}}{2}$$
$$= -2 \pm i\sqrt{3}$$

Hence the first equation of system (VI) is equivalent to the two equations $y = -2 + i\sqrt{3}$ and $y = -2 - i\sqrt{3}$. Therefore, (VI) is equivalent to the two systems

$$\begin{cases} y = -2 + i\sqrt{3} \\ x = y + 4 \end{cases} \quad \text{and} \quad \begin{cases} y = -2 - i\sqrt{3} \\ x = y + 4 \end{cases}$$

In each of these systems we substitute the value of y from the first equation into the second and we have

$$\begin{cases} y = -2 + i\sqrt{3} \\ x = 2 + i\sqrt{3} \end{cases} \quad \text{and} \quad \begin{cases} y = -2 - i\sqrt{3} \\ x = 2 - i\sqrt{3} \end{cases}$$

These two systems are equivalent to the given system (V). Thus the solution set of (V) is

$$\{(2 + i\sqrt{3}, -2 + i\sqrt{3}), (2 - i\sqrt{3}, -2 - i\sqrt{3})\}$$

Because the solutions are ordered pairs of imaginary numbers, there are no points of intersection of the graphs that correspond to the solutions. Remember that the coordinates of points in the real plane are real numbers.

◄

In Section 3.2, we introduced the elimination method to solve a system of linear equations. We apply this method in the following example involving a system of two quadratic equations.

► **EXAMPLE 4** *Solving a System of Two Quadratic Equations*

Follow the instructions of Example 1 for the system

$$\begin{cases} 2x^2 - 3y^2 = 6 \\ 6x^2 + y^2 = 58 \end{cases} \qquad \textbf{(VII)}$$

Solution   The graph of the first equation is a hyperbola, and the graph of the second is an ellipse. They appear in Figure 4 and intersect at four points. The system, therefore, has four ordered pairs of real numbers as solutions.

To solve the system algebraically, we wish to replace the given system by one that has an equation containing only one variable. We can eliminate y between the two equations by adding the first equation and 3 times the second as follows:

$$\begin{array}{r} 2x^2 - 3y^2 = 6 \\ 18x^2 + 3y^2 = 174 \\ \hline 20x^2 = 180 \end{array}$$

The following system involving the first equation of system (VII) and the preceding equation is equivalent to (VII):

$$\begin{cases} 2x^2 - 3y^2 = 6 \\ 20x^2 = 180 \end{cases}$$

$$\Leftrightarrow \quad \begin{cases} 2x^2 - 3y^2 = 6 \\ x^2 = 9 \end{cases}$$

The second equation of this system is equivalent to the two equations $x = 3$ and $x = -3$. Therefore this system is equivalent to the two systems

$$\begin{cases} 2x^2 - 3y^2 = 6 \\ x = 3 \end{cases} \quad \text{and} \quad \begin{cases} 2x^2 - 3y^2 = 6 \\ x = -3 \end{cases}$$

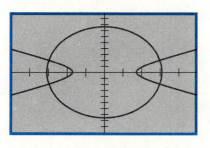

$[-5, 5]$ by $[-10, 10]$
$2x^2 - 3y^2 = 6$ and $6x^2 + y^2 = 58$
FIGURE 4

In each of these two systems we substitute into the first equation the value of x from the second, and we have

$$\begin{cases} 2(3)^2 - 3y^2 = 6 \\ \qquad\quad x = 3 \end{cases} \text{ and } \begin{cases} 2(-3)^2 - 3y^2 = 6 \\ \qquad\qquad x = -3 \end{cases}$$

$$\Leftrightarrow \quad \begin{cases} 18 - 3y^2 = 6 \\ \qquad x = 3 \end{cases} \text{ and } \begin{cases} 18 - 3y^2 = 6 \\ \qquad x = -3 \end{cases}$$

$$\Leftrightarrow \quad \begin{cases} y^2 = 4 \\ x = 3 \end{cases} \text{ and } \begin{cases} y^2 = 4 \\ x = -3 \end{cases}$$

The first equation in each of these two systems is equivalent to the two equations $y = 2$ and $y = -2$. Hence the two systems are equivalent to the four systems

$$\begin{cases} y = 2 \\ x = 3 \end{cases} \quad \begin{cases} y = -2 \\ x = 3 \end{cases} \quad \begin{cases} y = 2 \\ x = -3 \end{cases} \quad \begin{cases} y = -2 \\ x = -3 \end{cases}$$

These four systems are equivalent to system (VII). The solution set of (VII) is, therefore, $\{(3, 2), (3, -2), (-3, 2), (-3, -2)\}$.

The ordered pairs in the solution set give the coordinates of the points of intersection in Figure 4. ◀

In the next example, we have a system of two quadratic equations in which the second equation involves three second-degree terms. Because the right side of the second equation is zero and the left side can be factored, the second equation is equivalent to two linear equations. The given system is, therefore, equivalent to two systems, each consisting of a quadratic equation and a linear equation.

▶ **EXAMPLE 5** *Solving a System of Two Quadratic Equations*

Follow the instructions of Example 1 for the system

$$\begin{cases} \qquad\quad x^2 + y^2 = 16 \\ 2x^2 - 3xy + y^2 = 0 \end{cases}$$

Solution The graph of the first equation is the circle with center at the origin and radius 4. To obtain the graph of the second equation, we factor the left side and obtain

$$(2x - y)(x - y) = 0$$

This equation is equivalent to the two equations $2x - y = 0$ and $x - y = 0$, each of which has a line as its graph. The circle and the two lines appear in Figure 5 showing four points of intersection. The solution set of the system, therefore, has four ordered pairs of real numbers.

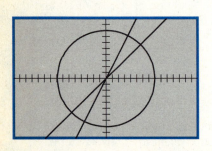

$[-7.5, 7.5]$ by $[-5, 5]$
$x^2 + y^2 = 16$ and $2x^2 - 3xy + y^2 = 0$
FIGURE 5

To find the solution set algebraically, we note that the given system is equivalent to the system

$$\begin{cases} x^2 + y^2 = 16 \\ (2x - y)(x - y) = 0 \end{cases}$$

which in turn is equivalent to the two systems

$$\begin{cases} x^2 + y^2 = 16 \\ 2x - y = 0 \end{cases} \quad \text{and} \quad \begin{cases} x^2 + y^2 = 16 \\ x - y = 0 \end{cases}$$

$$\Leftrightarrow \quad \begin{cases} x^2 + y^2 = 16 \\ y = 2x \end{cases} \quad \text{and} \quad \begin{cases} x^2 + y^2 = 16 \\ y = x \end{cases}$$

In each of the latter two systems we substitute into the first equation the value of y from the second, and we have

$$\begin{cases} x^2 + 4x^2 = 16 \\ y = 2x \end{cases} \quad \text{and} \quad \begin{cases} x^2 + x^2 = 16 \\ y = x \end{cases}$$

$$\Leftrightarrow \quad \begin{cases} x^2 = \frac{16}{5} \\ y = 2x \end{cases} \quad \text{and} \quad \begin{cases} x^2 = 8 \\ y = x \end{cases}$$

The equation $x^2 = \frac{16}{5}$ is equivalent to the two equations $x = \frac{4}{5}\sqrt{5}$ and $x = -\frac{4}{5}\sqrt{5}$. The equation $x^2 = 8$ is equivalent to the two equations $x = 2\sqrt{2}$ and $x = -2\sqrt{2}$. Therefore the preceding two systems are equivalent to the four systems

$$\begin{cases} x = \frac{4}{5}\sqrt{5} \\ y = 2x \end{cases} \quad \begin{cases} x = -\frac{4}{5}\sqrt{5} \\ y = 2x \end{cases} \quad \begin{cases} x = 2\sqrt{2} \\ y = x \end{cases} \quad \begin{cases} x = -2\sqrt{2} \\ y = x \end{cases}$$

In each of these latter systems we substitute the value of x from the first equation into the second equation, and we have

$$\begin{cases} x = \frac{4}{5}\sqrt{5} \\ y = \frac{8}{5}\sqrt{5} \end{cases} \quad \begin{cases} x = -\frac{4}{5}\sqrt{5} \\ y = -\frac{8}{5}\sqrt{5} \end{cases} \quad \begin{cases} x = 2\sqrt{2} \\ y = 2\sqrt{2} \end{cases} \quad \begin{cases} x = -2\sqrt{2} \\ y = -2\sqrt{2} \end{cases}$$

Thus the solution set of the given system is

$$\{(\tfrac{4}{5}\sqrt{5}, \tfrac{8}{5}\sqrt{5}), (-\tfrac{4}{5}\sqrt{5}, -\tfrac{8}{5}\sqrt{5}), (2\sqrt{2}, 2\sqrt{2}), (-2\sqrt{2}, -2\sqrt{2})\}$$

Observe in Figure 5 that each line intersects the circle at two points, the coordinates of which are the ordered pairs in the solution set. With $\sqrt{5} \approx 2.2$ and $\sqrt{2} \approx 1.4$, these points are $(1.8, 3.6)$, $(-1.8, -3.6)$, $(2.8, 2.8)$, and $(-2.8, -2.8)$. ◀

The system in the next example consists of two quadratic equations in which all the terms containing unknowns are of the second degree; that is, there are no first-degree terms. If one of the equations is replaced by a combination of the two equations in which no constant term appears, the system that results can be solved by the method used in Example 5.

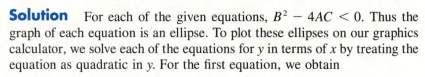

▶ **EXAMPLE 6**   *Solving a System of Two Quadratic Equations*

Follow the instructions of Example 1 for the system

$$\begin{cases} 4x^2 + xy + y^2 = 6 \\ 2x^2 - xy + y^2 = 8 \end{cases} \qquad \textbf{(VIII)}$$

Solution   For each of the given equations, $B^2 - 4AC < 0$. Thus the graph of each equation is an ellipse. To plot these ellipses on our graphics calculator, we solve each of the equations for y in terms of x by treating the equation as quadratic in y. For the first equation, we obtain

$$y = \tfrac{1}{2}(-x \pm \sqrt{24 - 15x^2})$$

and for the second, we get

$$y = \tfrac{1}{2}(x \pm \sqrt{32 - 7x^2})$$

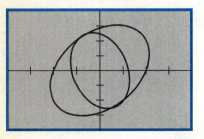

$[-4, 4]$ by $[-4, 4]$

$4x^2 - xy + y^2 = 6$ and $2x^2 - xy + y^2 = 8$

FIGURE 6

The two ellipses appear in Figure 6 where they intersect at four points.

To solve the system algebraically, we first obtain an equation having zero on the right side by adding 4 times the first equation and -3 times the second as follows:

$$\begin{array}{rcl} 16x^2 + 4xy + 4y^2 &=& 24 \\ -6x^2 + 3xy - 3y^2 &=& -24 \\ \hline 10x^2 + 7xy + y^2 &=& 0 \end{array}$$

The following system, involving the first equation of system (VIII) and the preceding equation, is equivalent to (VIII):

$$\begin{cases} 4x^2 + xy + y^2 = 6 \\ 10x^2 + 7xy + y^2 = 0 \end{cases}$$

Factoring the left member of the second equation, we obtain

$$\begin{cases} 4x^2 + xy + y^2 = 6 \\ (5x + y)(2x + y) = 0 \end{cases}$$

The second equation is equivalent to the two equations $5x + y = 0$ and $2x + y = 0$. Hence this system is equivalent to the two systems

$$\begin{cases} 4x^2 + xy + y^2 = 6 \\ \qquad\qquad\quad y = -5x \end{cases} \quad \text{and} \quad \begin{cases} 4x^2 + xy + y^2 = 6 \\ \qquad\qquad\quad y = -2x \end{cases}$$

In each of these two systems, if we substitute into the first equation the value of y from the second equation, we have

$$\begin{cases} 4x^2 + x(-5x) + (-5x)^2 = 6 \\ \qquad\qquad\qquad\qquad y = -5x \end{cases} \text{ and } \begin{cases} 4x^2 + x(-2x) + (-2x)^2 = 6 \\ \qquad\qquad\qquad\qquad y = -2x \end{cases}$$

$$\Leftrightarrow \quad \begin{cases} 4x^2 - 5x^2 + 25x^2 = 6 \\ \qquad\qquad\qquad y = -5x \end{cases} \text{ and } \begin{cases} 4x^2 - 2x^2 + 4x^2 = 6 \\ \qquad\qquad\qquad y = -2x \end{cases}$$

$$\Leftrightarrow \quad \begin{cases} x^2 = \frac{1}{4} \\ y = -5x \end{cases} \text{ and } \begin{cases} x^2 = 1 \\ y = -2x \end{cases}$$

The equation $x^2 = \frac{1}{4}$ is equivalent to the two equations $x = \frac{1}{2}$ and $x = -\frac{1}{2}$, and the equation $x^2 = 1$ is equivalent to the two equations $x = 1$ and $x = -1$. Thus the preceding two systems are equivalent to the four systems

$$\begin{cases} x = \frac{1}{2} \\ y = -5x \end{cases} \quad \begin{cases} x = -\frac{1}{2} \\ y = -5x \end{cases} \quad \begin{cases} x = 1 \\ y = -2x \end{cases} \quad \begin{cases} x = -1 \\ y = -2x \end{cases}$$

If in each of these equations we substitute in the second the value of x from the first, we have the equivalent systems

$$\begin{cases} x = \frac{1}{2} \\ y = -\frac{5}{2} \end{cases} \quad \begin{cases} x = -\frac{1}{2} \\ y = \frac{5}{2} \end{cases} \quad \begin{cases} x = 1 \\ y = -2 \end{cases} \quad \begin{cases} x = -1 \\ y = 2 \end{cases}$$

These four systems are equivalent to the given system. Therefore the solution set is $\{(\frac{1}{2}, -\frac{5}{2}), (-\frac{1}{2}, \frac{5}{2}), (1, -2), (-1, 2)\}$.

The two ellipses in Figure 6 intersect at the points whose coordinates are the ordered pairs of real numbers in the solution set. ◄

EXERCISES 11.4

In Exercises 1 through 22, plot the graphs of the equations of the system on your graphics calculator and determine the number of ordered pairs of real numbers that are solutions of the system. Then find the solution set of the system algebraically.

1. $\begin{cases} x^2 + y^2 = 25 \\ x - y + 1 = 0 \end{cases}$

2. $\begin{cases} x^2 + y^2 = 25 \\ x - 2y = -2 \end{cases}$

3. $\begin{cases} x^2 - y = 1 \\ x - 2y = -1 \end{cases}$

4. $\begin{cases} x^2 - y^2 = 9 \\ 2x + y = 6 \end{cases}$

5. $\begin{cases} x^2 - y^2 = 9 \\ x + y - 5 = 0 \end{cases}$

6. $\begin{cases} 4x^2 + y^2 = 25 \\ 2x + y + 1 = 0 \end{cases}$

7. $\begin{cases} x^2 - y - 4 = 0 \\ x - y - 3 = 0 \end{cases}$

8. $\begin{cases} 4x^2 + y - 3 = 0 \\ 8x + y - 7 = 0 \end{cases}$

9. $\begin{cases} x^2 - 2y^2 = 2 \\ x + 2y = 2 \end{cases}$

10. $\begin{cases} 4x^2 + y^2 = 17 \\ x^2 + y = 5 \end{cases}$

11. $\begin{cases} x^2 - y^2 = 15 \\ xy = 4 \end{cases}$

12. $\begin{cases} x^2 + y^2 = 25 \\ xy = 12 \end{cases}$

13. $\begin{cases} x^2 + y^2 = 4 \\ x^2 + 2y = 4 \end{cases}$

14. $\begin{cases} x^2 + xy + y^2 = 3 \\ x + y + 1 = 0 \end{cases}$

15. $\begin{cases} x^2 + y^2 = 25 \\ x^2 + 4y^2 = 64 \end{cases}$

16. $\begin{cases} 3x^2 + 2y^2 = 59 \\ 2x^2 + y^2 = 34 \end{cases}$

17. $\begin{cases} x^2 - y^2 = 9 \\ y^2 - 2x = 6 \end{cases}$

18. $\begin{cases} x^2 + y^2 = 16 \\ 9x^2 - 4y^2 = 36 \end{cases}$

19. $\begin{cases} x^2 - xy + 4 = 0 \\ 2x^2 - 2xy + y^2 = 8 \end{cases}$

20. $\begin{cases} 2x^2 - xy - y^2 = 0 \\ xy = 9 \end{cases}$

21. $\begin{cases} 10x^2 - xy + 4y^2 = 28 \\ 2x^2 - 3xy - 2y^2 = 0 \end{cases}$

22. $\begin{cases} 4x^2 - 5xy + 3y^2 = 24 \\ 2x^2 - 3xy + 2y^2 = 16 \end{cases}$

In Exercises 23 through 34, solve the word problem by finding a system of equations as a mathematical model of the situation. Be sure to write a conclusion.

23. The sum of the reciprocals of two numbers is $\frac{4}{15}$ and their product is 60. What are the numbers?

24. The sum of the squares of two numbers is $\frac{5}{18}$ and the sum of 6 times the smaller number and 4 times the larger number is 3. What are the numbers?

25. The length of the hypotenuse of a right triangle is 37 cm and its area is 210 cm². Find the lengths of the legs of the triangle.

26. Determine the dimensions of a rectangle of area 60 in.² that is inscribed in a circle of radius 6.5 in.

27. A rectangular lot has a perimeter of 40 m and an area of 96 m². What are its dimensions?

28. Find an equation of the common chord of the two circles
$$x^2 + y^2 - 4x - 1 = 0 \text{ and } x^2 + y^2 - 2y - 9 = 0$$

29. A group of students planned a field trip and agreed to contribute equal amounts toward the transportation costs of $150. Later five more students decided to go on the trip, and the transportation cost for each student was reduced by $1.50. Find the number of students who actually made the trip and the amount each paid for transportation.

30. An investment yields an annual interest of $1500. If $500 more is invested and the rate is 2 percent less, the annual interest is $1300. What is the amount of the investment and the rate of interest?

31. A piece of tin is in the form of a rectangle whose area is 486 cm². A square of side 3 cm is cut from each corner, and an open box is made by turning up the ends and sides. If the volume of the box is 504 cm³, what are the dimensions of the piece of tin?

32. A closed rectangular box, having a square base, has a total surface area of 16 ft². If a main diagonal of the box has a length of 3 ft, what are the dimensions of the box?

33. An open rectangular box, with a square base, has a surface area of 128 ft². If the cost per square foot of material was $1 for the sides and $1.20 for the bottom and the total cost of the material was $131.20, find the dimensions of the box.

34. Three listening posts are located at the points $A(0, 0)$, $B(0, \frac{21}{4})$, and $C(\frac{25}{3}, 0)$, the unit being 1 km. Microphones located at these points show that a gun is $\frac{5}{3}$ km closer to A than to C and $\frac{7}{4}$ km closer to B than to A. Determine the position of the gun by use of the definition of a hyperbola in Section 11.2.

In Exercises 35 and 36, find the solution set of the system of equations.

35. $\begin{cases} \dfrac{7}{x^2} - \dfrac{8}{y^2} = 5 \\ \dfrac{3}{x^2} - \dfrac{4}{y^2} = 2 \end{cases}$

36. $\begin{cases} \dfrac{3}{x^2} + \dfrac{1}{y^2} = 7 \\ \dfrac{5}{x^2} - \dfrac{2}{y^2} = -3 \end{cases}$

37. A parabola and a circle are the graphs pertaining to a system of two quadratic equations. Draw a figure showing an example of a parabola and a circle if the number of ordered pairs of real numbers that are solutions of the system is (a) four, (b) three, (c) two, (d) one, (e) zero. Explain why your figure indicates the specified number of such solutions.

38. Do Exercise 37 if the graphs are a hyperbola and an ellipse.

11.5 A UNIFIED TREATMENT OF CONIC SECTIONS AND POLAR EQUATIONS OF CONICS

GOALS

1. State and prove the theorem that defines a conic section in terms of its eccentricity.
2. State and prove the theorem giving the eccentricity, foci, and corresponding directrices of a central conic when its cartesian equation is known.
3. Find the eccentricity and directrices of a central conic from its cartesian equation.
4. State and prove the theorem giving polar equations of conics.
5. Find a polar equation of a conic from its properties.
6. Identify a conic from its polar equation.
7. Find properties of a conic from its polar equation.

In our treatment of conics, we have so far defined each of the three types of conic sections separately. An alternative approach is to start with a definition based on a common property of conics and then introduce each of the conics as a special case of the general definition. We state this definition in the following theorem, in which the common property is the eccentricity of the conic, denoted by e.

THEOREM 1

A conic section can be defined as the set of all points P in a plane such that the ratio of the undirected distance of P from a fixed point to the undirected distance of P from a fixed line that does not contain the fixed point is a positive constant e. Furthermore, if $e = 1$, the conic is a parabola; if $0 < e < 1$, it is an ellipse; and if $e > 1$, it is a hyperbola.

Proof   If $e = 1$, we see by comparing the definition of a parabola in Section 3.3 with the statement of the theorem that the set is a parabola having the fixed point as its focus and the fixed line as its directrix.

Suppose now that $e \neq 1$. We first obtain a polar equation of the set of points described. Let F denote the fixed point and l denote the fixed line. We take the pole at F and the polar axis and its extension perpendicular to l. We first consider the situation when the line l is to the left of the point F. Let D be the point of intersection of l with the extension of the polar axis, and let d denote the undirected distance from F to l. Refer to Figure 1. Let $P(r, \theta)$ be any point in the set to the right of l and on the terminal side of the angle of measure θ. Draw perpendiculars PQ and PR to the polar axis and line l,

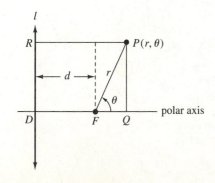

FIGURE 1

respectively. The point P is in the set described if and only if

$$|\overline{FP}| = e|\overline{RP}| \tag{1}$$

Because P is to the right of l, $\overline{RP} > 0$; thus $|\overline{RP}| = \overline{RP}$. Furthermore, $|\overline{FP}| = r$ because $r > 0$. Thus from (1),

$$r = e(\overline{RP}) \tag{2}$$

However, $\overline{RP} = \overline{DQ}$, and because $\overline{DQ} = \overline{DF} + \overline{FQ}$, we have

$$\overline{RP} = d + r \cos \theta$$

Substituting this expression for $\overline{RP}$ in (2) we get

$$r = e(d + r \cos \theta)$$

Solving for r gives

$$r = \frac{ed}{1 - e \cos \theta} \tag{3}$$

We obtain a cartesian representation of this equation by first replacing $\cos \theta$ by x/r. We have

$$r = \frac{ed}{1 - \dfrac{ex}{r}}$$

$$r = \frac{edr}{r - ex}$$

$$r - ex = ed$$

$$r = e(x + d)$$

We now replace r by $\pm\sqrt{x^2 + y^2}$ and get

$$\pm\sqrt{x^2 + y^2} = e(x + d)$$

Squaring on both sides of this equation gives

$$x^2 + y^2 = e^2x^2 + 2e^2dx + e^2d^2$$

$$y^2 + x^2(1 - e^2) = 2e^2dx + e^2d^2$$

Because $e \neq 1$, we can divide on both sides of this equation by $1 - e^2$ and obtain

$$x^2 - \frac{2e^2d}{1 - e^2}x + \frac{1}{1 - e^2}y^2 = \frac{e^2d^2}{1 - e^2}$$

Completing the square for the terms involving x by adding $e^4d^2/(1 - e^2)^2$ on both sides of the above equation we get

$$\left(x - \frac{e^2d}{1 - e^2}\right)^2 + \frac{1}{1 - e^2}y^2 = \frac{e^2d^2}{(1 - e^2)^2}$$

Dividing both sides of this equation by $e^2d^2/(1 - e^2)^2$ gives us an equation of the form

$$\frac{(x - h)^2}{\dfrac{e^2d^2}{(1 - e^2)^2}} + \frac{y^2}{\dfrac{e^2d^2}{1 - e^2}} = 1 \tag{4}$$

where

$$h = \frac{e^2d}{1 - e^2} \tag{5}$$

Now let

$$\frac{e^2d^2}{(1 - e^2)^2} = a^2 \quad \text{where } a > 0 \tag{6}$$

Then (4) can be written as

$$\frac{(x - h)^2}{a^2} + \frac{y^2}{a^2(1 - e^2)} = 1 \tag{7}$$

If $0 < e < 1$, then $a^2(1 - e^2) > 0$ and we can let

$$b^2 = a^2(1 - e^2) \quad \text{where } 0 < e < 1 \tag{8}$$

Substituting from (8) in (7), we get

$$\frac{(x - h)^2}{a^2} + \frac{y^2}{b^2} = 1$$

which is an equation of an ellipse having its principal axis on the x axis and its center at $(h, 0)$, where $h > 0$.

If $e > 1$, then $a^2(e^2 - 1) > 0$, and we can let

$$b^2 = a^2(e^2 - 1) \quad \text{where } e > 1 \tag{9}$$

Substituting from this equation in (7), we obtain

$$\frac{(x - h)^2}{a^2} - \frac{y^2}{b^2} = 1$$

which is an equation of a hyperbola having its principal axis on the x axis and its center at $(h, 0)$, where $h < 0$.

In a similar manner we can derive an equation of a central conic (an ellipse or hyperbola) from (1) when $e \neq 1$ if the line l is to the right of the point F at the pole. In this case instead of Equation (3) we have

$$r = \frac{ed}{1 + e \cos \theta} \tag{10}$$

The derivation of (10) is left as an exercise (see Exercise 31).

We can also derive an equation of a central conic from (1) when $e \neq 1$ if the line l is parallel to the polar axis and the point F is at the pole. In this case instead of Equation (3) we obtain

$$r = \frac{ed}{1 \pm e \sin \theta} \tag{11}$$

where e and d are, respectively, the eccentricity and undirected distance between F and l. The plus sign is taken when l is above F, and the minus sign is taken when it is below F. The derivations of (11) are left as exercises (see Exercises 32 and 33).

We can reverse the steps in going from (1) to (7). Thus if P is any point on a central conic, Equation (1) is satisfied.

Therefore we conclude that a conic can be defined by the described set of points. ■

In Sections 11.1 and 11.2, we defined the eccentricity e of a central conic by the equation

$$e = \frac{c}{a} \quad \Leftrightarrow \quad c = ae$$

To show that the number e in the statement of Theorem 1 satisfies this equation for an ellipse, we substitute from (8) in the equation $c^2 = a^2 - b^2$; to show that the same equation is satisfied for a hyperbola, we substitute from (9) in the equation $c^2 = a^2 + b^2$. For an ellipse, we have

$$c^2 = a^2 - a^2(1 - e^2)$$

and for a hyperbola, we have

$$c^2 = a^2 + a^2(e^2 - 1)$$

In both cases, we obtain

$$c^2 = a^2 e^2$$
$$c = ae$$

In the proof of Theorem 1 we showed that when the conic is a parabola, the fixed point F mentioned in the theorem is the focus of the parabola and the fixed line is the directrix. In Exercise 34, you are asked to show that the point F is one of the foci when we have a central conic. If (7) is an equation of an ellipse, the point F is the left-hand focus, and if (7) is an equation of a hyperbola, F is the right-hand focus.

Consider now the standard form of a cartesian equation of a central conic having its principal axis on the x axis and its center at the origin:

$$\frac{x^2}{a^2} + \frac{y^2}{a^2(1 - e^2)} = 1 \tag{12}$$

The fixed line *l* mentioned in Theorem 1 is the directrix of the central conic corresponding to the focus at *F*. When the conic defined by equation (12) is an ellipse, the directrix corresponding to the focus at $(-c, 0)$ or, equivalently, $(-ae, 0)$ has the equation

$$x = -ae - d$$

From (6) when $0 < e < 1$, $d = a(1 - e^2)/e$, so this equation becomes

$$x = -ae - \frac{a(1 - e^2)}{e}$$

$$x = -\frac{a}{e}$$

Similarly, when the conic defined by (12) is a hyperbola, the directrix corresponding to the focus at $(c, 0)$ or, equivalently, $(ae, 0)$ has the equation

$$x = ae - d$$

Again from (6), when $e > 1$, $d = a(e^2 - 1)/e$, so the above equation of the directrix can be written as

$$x = \frac{a}{e}$$

Hence we have shown that if (12) is an equation of an ellipse, a focus and its corresponding directrix are $(-ae, 0)$ and $x = -a/e$; and if (12) is an equation of a hyperbola, a focus and its corresponding directrix are $(ae, 0)$ and $x = a/e$.

Because (12) contains only even powers of *x* and *y*, its graph is symmetric with respect to both the *x* and *y* axes. Therefore, if there is a focus at $(-ae, 0)$ having a corresponding directrix of $x = -a/e$, by symmetry there is also a focus at $(ae, 0)$ having a corresponding directrix of $x = a/e$. Similarly, for a focus at $(ae, 0)$ and a corresponding directrix of $x = a/e$, there is also a focus at $(-ae, 0)$ and a corresponding directrix of $x = -a/e$. These results are summarized in the following theorem.

THEOREM 2

The central conic having the equation

$$\frac{x^2}{a^2} + \frac{y^2}{a^2(1 - e^2)} = 1$$

where $a > 0$, has a focus at $(-ae, 0,)$ whose corresponding directrix is $x = -a/e$, and a focus at $(ae, 0)$, whose corresponding directrix is $x = a/e$.

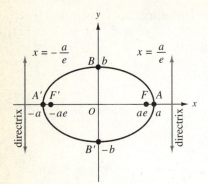

FIGURE 2

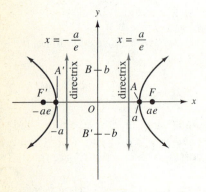

FIGURE 3

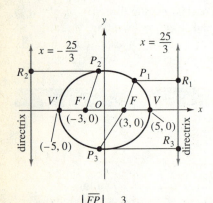

$$\frac{|\overline{FP}|}{|\overline{RP}|} = \frac{3}{5}$$

FIGURE 4

Figures 2 and 3 show the graphs of (12) together with the foci and directrices in the respective case of an ellipse and a hyperbola.

▶ **EXAMPLE 1** *Finding the Eccentricity and Directrices of an Ellipse from Its Cartesian Equation*

The ellipse of Example 1 in Section 11.1 has the equation

$$\frac{x^2}{25} + \frac{y^2}{16} = 1$$

(a) Find the eccentricity and directrices of this ellipse. **(b)** Sketch the ellipse and show the directrices and foci. Also choose any three points P on the ellipse and draw the line segments whose lengths are the undirected distances from P to a focus and its corresponding directrix. Observe that the ratio of these distances is e.

Solution

(a) From the equation of the ellipse, $a = 5$ and $b = 4$. We showed in Example 1 of Section 11.1 that $c = 3$. Because $e = c/a$, $e = \frac{3}{5}$. Because $a/e = 25/3$, it follows from Theorem 2 that the directrix corresponding to the focus at $(3, 0)$ has the equation $x = \frac{25}{3}$, and the directrix corresponding to the focus at $(-3, 0)$ has the equation $x = -\frac{25}{3}$.

(b) Figure 4 shows the ellipse, the directrices, and the foci, as well as three points P_1, P_2, and P_3 on the ellipse. For each of these points,

$$\frac{|\overline{FP}|}{|\overline{RP}|} = \frac{3}{5}$$ ◀

▶ **EXAMPLE 2** *Finding the Eccentricity and Directrices of a Hyperbola from Its Cartesian Equation*

The hyperbola of Example 1 in Section 11.2 has the equation

$$\frac{x^2}{9} - \frac{y^2}{16} = 1$$

(a) Find the eccentricity and directrices of this hyperbola. **(b)** Sketch the hyperbola and show the directrices and foci. Also choose any three points P on the hyperbola and draw the line segments whose lengths are the undirected distances from P to a focus and its corresponding directrix. Observe that the ratio of these distances is e.

Solution

(a) From the equation of the hyperbola, $a = 3$ and $b = 4$. In Example 1 of Section 11.2 we showed that $c = 5$. Because $e = c/a$, $e = \frac{5}{3}$. Because

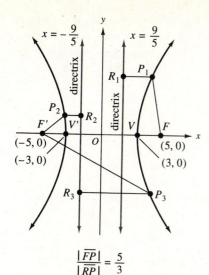

$x = -\frac{9}{5}$ $x = \frac{9}{5}$

$$\frac{|\overline{FP}|}{|\overline{RP}|} = \frac{5}{3}$$

FIGURE 5

$a/e = 9/5$, we conclude from Theorem 2 that the directrix corresponding to the focus at $(5, 0)$ has the equation $x = \frac{9}{5}$, and the directrix corresponding to the focus at $(-5, 0)$ has the equation $x = -\frac{9}{5}$.

(b) Figure 5 shows the hyperbola, the directrices, and the foci, as well as three points P_1, P_2, and P_3 on the hyperbola. For each of these points

$$\frac{|\overline{FP}|}{|\overline{RP}|} = \frac{5}{3}$$ ◀

In the proof of Theorem 1 we learned that all three types of conics have polar equations of the same form. When a focus is at the pole and the corresponding directrix is either perpendicular or parallel to the polar axis, an equation of the conic has the form of (3), (10), or (11). We have, therefore, the following theorem.

THEOREM 3

Suppose we have a conic for which e and d are, respectively, the eccentricity and the undirected distance between the focus and the corresponding directrix.

(i) If a focus of the conic is at the pole and the corresponding directrix is perpendicular to the polar axis, then an equation of the conic is

$$r = \frac{ed}{1 \pm e \cos \theta} \tag{13}$$

where the plus sign is taken when the directrix corresponding to the focus at the pole is to the right of the focus and the minus sign is taken when it is to the left of the focus.

(ii) If a focus of the conic is at the pole and the corresponding directrix is parallel to the polar axis, then an equation of the conic is

$$r = \frac{ed}{1 \pm e \sin \theta} \tag{14}$$

where the plus sign is taken when the directrix corresponding to the focus at the pole is above the focus, and the minus sign is taken when it is below the focus.

▶ **EXAMPLE 3** *Finding a Polar Equation of a Conic from Its Properties*

A parabola has its focus at the pole and its vertex at $(4, \pi)$. Find a polar equation of the parabola and an equation of the directrix. Sketch the parabola and show the directrix. Check the graph on a graphics calculator.

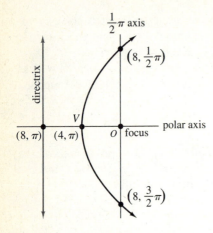

FIGURE 6

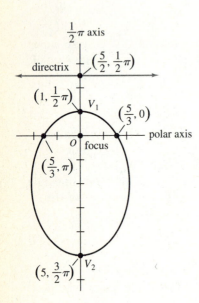

FIGURE 7

Solution   Because the focus is at the pole and the vertex is at $(4, \pi)$, the polar axis and its extension are along the axis of the parabola. Furthermore, the vertex is to the left of the focus; so the directrix is also to the left of the focus. Hence an equation of the parabola is of the form of (13) with the minus sign. Because the vertex is at $(4, \pi)$, $\frac{1}{2}d = 4$; thus $d = 8$. The eccentricity $e = 1$, and therefore we obtain the equation

$$r = \frac{8}{1 - \cos \theta}$$

An equation of the directrix is given by $r \cos \theta = -d$, and because $d = 8$, then $r \cos \theta = -8$.

Figure 6 shows the parabola and the directrix. We obtain the same graph on our graphics calculator.   ◄

▶ **EXAMPLE 4**   *Identifying and Finding Properties of a Conic from Its Polar Equation*

An equation of a conic is

$$r = \frac{5}{3 + 2 \sin \theta}$$

Identify the conic, find the eccentricity, write an equation of the directrix corresponding to the focus at the pole, and find the vertices. Sketch the curve and check the graph on a graphics calculator.

Solution   Dividing the numerator and denominator of the fraction in the given equation by 3 we obtain

$$r = \frac{\frac{5}{3}}{1 + \frac{2}{3} \sin \theta}$$

which is of the form of (14) with the plus sign. The eccentricity $e = \frac{2}{3}$. Because $e < 1$, the conic is an ellipse. Because $ed = \frac{5}{3}$, $d = \frac{5}{3} \div \frac{2}{3}$; thus $d = \frac{5}{2}$. The $\frac{1}{2}\pi$ axis and its extension are along the principal axis. The directrix corresponding to the focus at the pole is above the focus, and an equation of it is $r \sin \theta = \frac{5}{2}$. When $\theta = \frac{1}{2}\pi$, $r = 1$; and when $\theta = \frac{3}{2}\pi$, $r = 5$. The vertices are therefore at $(1, \frac{1}{2}\pi)$ and $(5, \frac{3}{2}\pi)$.

The ellipse, as sketched, appears in Figure 7. We obtain the same curve on our graphics calculator.   ◄

▶ **EXAMPLE 5**   *Finding a Polar Equation of a Conic and Other Properties of the Conic from Given Properties*

The polar axis and its extension are along the principal axis of a hyperbola having a focus at the pole. The corresponding directrix is to the left of the focus. If the hyperbola contains the point $(1, \frac{2}{3}\pi)$ and $e = 2$, find **(a)** a polar equation of the hyperbola, **(b)** the vertices, **(c)** the center, **(d)** an equation of the directrix corresponding to the focus at the pole. **(e)** Sketch the hyperbola and check the graph on a graphics calculator.

Solution   An equation of the hyperbola is of the form of (13) with the minus sign, where $e = 2$. We have, then,

$$r = \frac{2d}{1 - 2 \cos \theta}$$

(a) Because the point $(1, \frac{2}{3}\pi)$ lies on the hyperbola, its coordinates satisfy the equation. Therefore

$$1 = \frac{2d}{1 - 2(-\frac{1}{2})}$$

from which we obtain $d = 1$. Hence an equation of the hyperbola is

$$r = \frac{2}{1 - 2 \cos \theta} \tag{15}$$

(b) The vertices are the points on the hyperbola for which $\theta = 0$ and $\theta = \pi$. From (15), when $\theta = 0$, $r = -2$; and when $\theta = \pi$, $r = \frac{2}{3}$. Consequently, the left vertex V_1 is at the point $(-2, 0)$, and the right vertex V_2 is at the point $(\frac{2}{3}, \pi)$.

(c) The center C of the hyperbola is the point on the principal axis halfway between the two vertices. This is the point $(\frac{4}{3}, \pi)$.

(d) An equation of the directrix corresponding to the focus at the pole is given by $r \cos \theta = -d$. Because $d = 1$, this equation is $r \cos \theta = -1$.

(e) As an aid in sketching the hyperbola, we first draw the two asymptotes. These are lines through the center of the hyperbola that are parallel to the lines $\theta = \theta_1$ and $\theta = \theta_2$, where θ_1 and θ_2 are the values of θ in the interval $[0, 2\pi)$ for which r is not defined. From (15), r is not defined when $1 - 2 \cos \theta = 0$. Therefore $\theta_1 = \frac{1}{3}\pi$ and $\theta_2 = \frac{5}{3}\pi$. Figure 8 shows the hyperbola, as well as the two asymptotes and the directrix corresponding to the focus at the pole. Our graphics calculator gives the same graph.   ◀

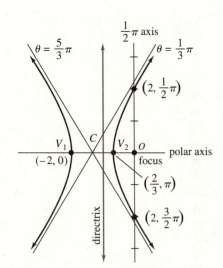

FIGURE 8

EXERCISES 11.5

In Exercises 1 through 8, (a) find the eccentricity, foci, and directrices of the central conic. (b) Sketch the conic and show the foci and directrices. Also choose any three points P (in different quadrants) on the conic and draw the line segments whose lengths are the undirected distances from P to a focus and its corresponding directrix. Observe that the ratio of these distances is e.

1. $4x^2 + 9y^2 = 36$
2. $4x^2 + 9y^2 = 4$
3. $25x^2 + 4y^2 = 100$
4. $16x^2 + 9y^2 = 144$
5. $4x^2 - 25y^2 = 100$
6. $x^2 - 9y^2 = 9$
7. $16x^2 - 9y^2 = 144$
8. $4y^2 - x^2 = 16$

In Exercises 9 and 10, the polar equation represents a conic having a focus at the pole. Identify the conic.

9. (a) $r = \dfrac{3}{1 - \cos \theta}$ **(b)** $r = \dfrac{6}{4 + 5 \sin \theta}$

(c) $r = \dfrac{5}{4 - \cos \theta}$ **(d)** $r = \dfrac{4}{1 + \sin \pi}$

10. (a) $r = \dfrac{1}{1 - \sin \theta}$ **(b)** $r = \dfrac{2}{3 + \sin \theta}$

(c) $r = \dfrac{3}{2 + 4 \cos \theta}$ **(d)** $r = \dfrac{5}{1 - \cos \pi}$

In Exercises 11 through 22, the graph of the equation is a conic having a focus at the pole. (a) Find the eccentricity; (b) identify the conic; (c) write an equation of the directrix corresponding to the focus at the pole; and (d) sketch the curve and check your graph on your graphics calculator.

11. $r = \dfrac{2}{1 - \cos \theta}$ **12.** $r = \dfrac{4}{1 + \cos \theta}$

13. $r = \dfrac{5}{2 + \sin \theta}$ **14.** $r = \dfrac{4}{1 - 3 \cos \theta}$

15. $r = \dfrac{6}{3 - 2 \cos \theta}$ **16.** $r = \dfrac{1}{2 + \sin \theta}$

17. $r = \dfrac{9}{5 - 6 \sin \theta}$ **18.** $r = \dfrac{1}{1 - 2 \sin \theta}$

19. $r = \dfrac{10}{7 - 2 \sin \theta}$ **20.** $r = \dfrac{7}{3 + 4 \cos \theta}$

21. $r = \dfrac{10}{4 + 5 \cos \theta}$ **22.** $r = \dfrac{1}{5 - 3 \sin \theta}$

In Exercises 23 through 28, find a polar equation of the conic having a focus at the pole and satisfying the given conditions.

23. Parabola; vertex at $(4, \tfrac{3}{2} \pi)$.

24. Ellipse; $e = \tfrac{1}{2}$; corresponding vertex at $(4, \pi)$.

25. Hyperbola; $e = \tfrac{4}{3}$; $r \cos \theta = 9$ is the directrix corresponding to the focus at the pole.

26. Hyperbola; vertices at $(1, \tfrac{1}{2} \pi)$ and $(3, \tfrac{1}{2} \pi)$.

27. Ellipse; vertices at $(3, 0)$ and $(1, \pi)$.

28. Parabola; vertex at $(6, \tfrac{1}{2} \pi)$.

29. (a) Find a polar equation of the hyperbola having a focus at the pole and the corresponding directrix to the

left of the focus if the point $(2, \tfrac{4}{3} \pi)$ is on the hyperbola and $e = 3$. **(b)** Write an equation of the directrix that corresponds to the focus at the pole.

30. (a) Find a polar equation of the hyperbola for which $e = 3$ and which has the line $r \sin \theta = 3$ as the directrix corresponding to a focus at the pole. **(b)** Find the polar equations of the two lines through the pole that are parallel to the asymptotes of the hyperbola.

31. Show that an equation of a conic having its principal axis along the polar axis and its extension, a focus at the pole, and the corresponding directrix to the right of the focus is

$$r = \frac{ed}{1 + e \cos \theta}$$

32. Show that an equation of a conic having its principal axis along the $\tfrac{1}{2} \pi$ axis and its extension, a focus at the pole, and the corresponding directrix above the focus is

$$r = \frac{ed}{1 + e \sin \theta}$$

33. Show that an equation of a conic having its principal axis along the $\tfrac{1}{2} \pi$ axis and its extension, a focus at the pole, and the corresponding directrix below the focus is

$$r = \frac{ed}{1 - e \sin \theta}$$

34. Show that the point F mentioned in the proof of Theorem 1 is one of the foci when the conic is either an ellipse or a hyperbola. *Hint:* Use (7), which is the cartesian equation of a central conic having its center at $(h, 0)$, the value of h from Equation (5), the value of a from Equation (6), and the fact that $c = ae$.

35. A comet is moving in a parabolic orbit around the sun at the focus of the parabola. When the comet is 80 million miles from the sun, the line segment from the sun to the comet makes an angle of $\tfrac{1}{3} \pi$ radians with the axis of the orbit. **(a)** Find an equation of the comet's orbit. **(b)** How close does the comet come to the sun?

36. A satellite is traveling around the earth in an elliptical orbit having the center of the earth at one focus and an eccentricity of $\tfrac{1}{3}$. The closest distance that the satellite gets to the earth is 300 mi. Find the farthest distance that the satellite gets from the earth. Assume the earth's radius is 4000 mi.

37. Show that the equation $r = k \csc^2 \frac{1}{2}\theta$, where k is a constant, is a polar equation of a parabola.

38. Show that if α is the angle between the asymptotes of a hyperbola of eccentricity e then $\alpha = 2 \tan^{-1}\sqrt{e^2 - 1}$.

39. Describe how the shape of a conic changes as the eccentricity takes on the following values: 0.01, 0.10, 0.50, 0.99, 1.00, 1.01, $\sqrt{2}$, 1.50, 2.00.

CHAPTER 11 REVIEW

▶ LOOKING BACK

11.1 Cartesian equations of an ellipse were obtained from the definition of an ellipse as a set of points in a plane. Later in the section, we presented Dandelin's proof that this definition is a consequence of the definition of an ellipse as a conic section. We found properties of an ellipse from its equation and used these properties to sketch the ellipse. We also obtained an equation of an ellipse from its properties. Parametric equations of an ellipse were introduced so we could use them to plot ellipses.

11.2 The structure of this section on hyperbolas was similar to that of the preceding section on ellipses. We defined a hyperbola as a set of points in a plane and then obtained a cartesian equation of the hyperbola from that definition. We sketched a hyperbola from properties obtained from its equation. The asymptotes of a hyperbola aided us in sketching its graph. We defined the hyperbolic cosine and hyperbolic sine functions so they could be applied to obtain parametric equations of a hyperbola used to plot the hyperbola.

11.3 To determine the graph of a second-degree equation

in x and y with an xy term, we transformed the equation into one without an xy term by rotating the coordinate axes through a suitably chosen angle. We then proved that the graph of such an equation is either a conic or a degenerate conic, and the type of conic is determined by the sign of the discriminant $B^2 - 4AC$. We sketched and plotted graphs of second-degree equations.

11.4 Before solving a system involving quadratic equations, we ascertained the number of real solutions by plotting the graphs of the equations and counting the number of points of intersection of the graphs. Algebraic solutions of the systems involved both the substitution and elimination methods.

11.5 All three types of conics have polar equations of the same form. This fact is part of the proof of the theorem that defines a conic section in terms of its eccentricity. We found properties of a conic from its polar equation. We also found the eccentricity, foci, and corresponding directrices of a central conic from its cartesian equation.

▶ REVIEW EXERCISES

In Exercises 1 through 20, sketch the graph of the equation. If the graph is an ellipse, find (a) the center, (b) an equation of the principal axis, (c) the vertices, (d) the endpoints of the minor axis, and (e) the foci. If the graph is a hyperbola, find (a) the center, (b) an equation of the principal axis, and (c) the vertices. Also for a hyperbola show the auxiliary rectangle and the asymptotes.

1. $\dfrac{x^2}{100} + \dfrac{y^2}{36} = 1$

2. $\dfrac{x^2}{16} + \dfrac{y^2}{64} = 1$

3. $\dfrac{y^2}{9} - \dfrac{x^2}{16} = 1$

4. $\dfrac{x^2}{36} - \dfrac{y^2}{64} = 1$

5. $9x^2 - 25y^2 = 225$

6. $4y^2 - 25x^2 = 400$

7. $25x^2 + 9y^2 = 225$

8. $4x^2 + 25y^2 = 400$

9. $4x^2 + 25y^2 = 100$

10. $25x^2 - 4y^2 = 100$

11. $x^2 - 9y^2 = 144$

12. $9x^2 + 16y^2 = 144$

13. $25x^2 + y^2 - 100x + 8y + 91 = 0$

14. $16y^2 - 9x^2 - 64y - 80 = 0$

15. $4x^2 - 4y^2 - 32x + 16y + 39 = 0$

16. $x^2 + 2y^2 + 12x - 12y + 38 = 0$

17. $4x^2 + y^2 + 24x - 16y + 84 = 0$

18. $4x^2 + 9y^2 + 32x - 18y + 37 = 0$

19. $25x^2 - y^2 + 50x + 6y - 9 = 0$

20. $3x^2 - 2y^2 + 6x - 8y + 11 = 0$

In Exercises 21 through 24, find equations of the asymptotes of the hyperbola in the given exercise.

21. Exercise 3

22. Exercise 6

23. Exercise 15

24. Exercise 20

In Exercises 25 through 32, define the graph of the given exercise by a pair of parametric equations, and use the equations to plot the graph.

25. Exercise 1

26. Exercise 2

27. Exercise 3

28. Exercise 4

29. Exercise 13

30. Exercise 14

31. Exercise 19

32. Exercise 18

In Exercises 33 through 36, find an equation of the ellipse having the given properties and sketch the graph.

33. The sum of the distances from the points $(-4, 0)$ and $(4, 0)$ to any point on the ellipse is 10.

34. Vertices at $(-4, 0)$ and $(4, 0)$ and through the point $(2, \sqrt{3})$.

35. Vertices at $(0, 8)$ and $(0, -4)$ and one focus at $(0, 6)$.

36. Foci at $(-3, -2)$ and $(-3, 6)$ and the length of the semiminor axis is three-fourths the length of the semimajor axis.

In Exercises 37 through 40, find an equation of the hyperbola having the given properties and sketch the graph.

37. Vertices at $(0, 2\sqrt{2})$ and $(0, -2\sqrt{2})$ and through the point $(2, 4)$.

38. The absolute value of the difference of the distances from the points $(0, -5)$ and $(0, 5)$ to any point on the hyperbola is 8.

39. Foci at $(-5, 1)$ and $(1, 1)$ and one vertex at $(-4, 1)$.

40. Center at $(-5, 2)$, a vertex at $(-1, 2)$, and a focus at $(0, 2)$.

In Exercises 41 through 44, (a) identify the graph of the equation; (b) remove the xy term from the equation by a rotation of axes; and (c) sketch the graph and show both sets of axes.

41. $5x^2 - 6xy + 5y^2 = 8$

42. $2x^2 + 4\sqrt{3}xy - 2y^2 = 4$

43. $x^2 + 2xy + y^2 + 8x - 8y = 0$

44. $16x^2 - 24xy + 9y^2 - 240x - 320y = 0$

In Exercises 45 through 48, (a) identify the graph of the equation; (b) simplify the equation by a rotation and translation of axes; and (c) sketch the graph and show the three sets of axes.

45. $3x^2 - 3xy - y^2 - 6y = 0$

46. $4x^2 + 3xy + y^2 - 6x + 12y = 0$

47. $3x^2 + 4xy + 16x - 8y + 19 = 0$

48. $x^2 - 4xy + 4y^2 + 36x + 28y + 24 = 0$

In Exercises 49 through 52, plot the graph of the equation of the indicated exercise.

49. Exercise 41

50. Exercise 42

51. Exercise 45

52. Exercise 46

In Exercises 53 through 60, plot the graphs of the equations of the system and determine the number of ordered pairs of real numbers that are solutions of the system. Then find the solution set of the system algebraically.

53. $\begin{cases} x^2 + y^2 = 9 \\ 3x - 2y = 6 \end{cases}$

54. $\begin{cases} 7x + 3y = 9 \\ x^2 + 2y = 1 \end{cases}$

55. $\begin{cases} 3x + 2y = -5 \\ x^2 - 3y = 3 \end{cases}$

56. $\begin{cases} x^2 + y^2 = 50 \\ 3x - 4y = 0 \end{cases}$

57. $\begin{cases} 3x^2 + 2y^2 = 7 \\ 5x^2 - y^2 = 3 \end{cases}$

58. $\begin{cases} x^2 + y^2 = 25 \\ 4x^2 - xy - 3y^2 = 0 \end{cases}$

59. $\begin{cases} 2x^2 - xy + y^2 = 8 \\ 4x^2 + xy + y^2 = 6 \end{cases}$

60. $\begin{cases} 4x^2 - 3y^2 = -8 \\ y^2 + 2xy = 8 \end{cases}$

In Exercises 61 through 64, the graph of the equation is a conic having a focus at the pole. (a) Find the eccentricity; (b) identify the conic; (c) write an equation of the directrix that corresponds to the focus at the pole; (d) sketch the curve and check your graph on your graphics calculator.

61. $r = \dfrac{2}{2 - \sin \theta}$

62. $r = \dfrac{5}{3 + 3 \sin \theta}$

63. $r = \dfrac{4}{2 + 3 \cos \theta}$

64. $r = \dfrac{4}{3 - 2 \cos \theta}$

In Exercises 65 through 68, find a polar equation of the conic satisfying the conditions. Sketch the conic and check your graph on your graphics calculator.

65. A focus at the pole; vertices at $(2, \pi)$ and $(4, \pi)$.

66. A focus at the pole; corresponding vertex at $(6, \frac{1}{2}\pi)$; $e = \frac{3}{4}$.

67. A focus at the pole; a vertex at $(3, \frac{3}{2}\pi)$; $e = 1$.

68. The line $r \sin \theta = 6$ is the directrix corresponding to the focus at the pole and $e = \frac{5}{3}$.

In Exercises 69 and 70, (a) find the eccentricity, foci, and directrices of the central conic. (b) Sketch the conic and show the foci and directrices.

69. $25x^2 - 9y^2 = 225$ **70.** $4x^2 + y^2 = 64$

71. Show that the hyperbola $x^2 - y^2 = 4$ has the same foci as the ellipse $x^2 + 9y^2 = 9$.

72. The orbit of the planet Mercury around the sun is elliptical in shape with the sun at one focus, a semi-major axis of length 36.0 million miles, and an eccentricity of 0.206. Find **(a)** how close Mercury gets to the sun and **(b)** the greatest possible distance between Mercury and the sun.

73. The arch of a bridge is in the shape of a semiellipse having a horizontal span of 60 m and a height of 20 m at its center. How high is the arch 10 m to the right or left of the center?

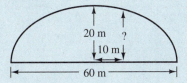

74. The area of a right triangle is 84 cm^2 and the length of the hypotenuse is 25 cm. Find the lengths of the legs of the triangle.

75. An open box is constructed from a rectangular piece of cardboard having an area of 120 in.^{2} by cutting a square of side 2 in. from each corner and turning up the ends and sides. If the volume of the box is 96 in.^{3}, what are the dimensions of the original piece of cardboard?

76. If the distance between the two directrices of an ellipse is three times the distance between the foci, find the eccentricity.

77. Find a polar equation of the parabola containing the point $(2, \frac{1}{3}\pi)$, whose focus is at the pole and whose vertex is on the extension of the polar axis.

78. Points A and B are 1000 m apart, and it is determined from the sound of an explosion heard at these points at different times that the location of the explosion is 600 m closer to A than to B. Show that the location of the explosion is restricted to a particular curve, and find an equation of it.

79. A focal chord of a conic is divided into two segments by the focus. Prove that the sum of the reciprocals of the measures of the lengths of the two segments is the same, regardless of what chord is taken. (*Hint:* Use polar coordinates.)

80. A focal chord of a conic is a line segment passing through a focus and having its endpoints on the conic. Prove that if two focal chords of a parabola are perpendicular, the sum of the reciprocals of the measures of their lengths is a constant. (*Hint:* Use polar coordinates.)

81. Show that $\sqrt{x} \pm \sqrt{y} = \pm\sqrt{a}$ represents a one-parameter family of conics, and determine the type of conic. Sketch the conics for $a = 1$, $a = 2$, and $a = 4$.

82. The graph of the equation

$$(1 - e^2)x^2 + y^2 - 2px + p^2 = 0$$

is a central conic having eccentricity e and a focus at $(p, 0)$. Simplify the equation by a translation of axes. Find the new origin with respect to the x and y axes.

Topics in Algebra

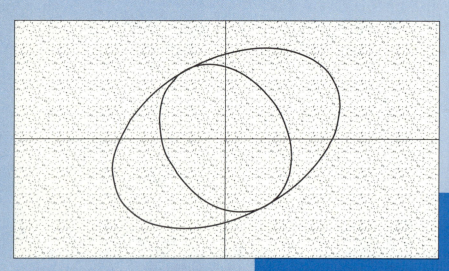

The topics in this chapter will help you develop solid skills in algebra that are crucial for the study of calculus. We apply matrices to solve systems of linear equations, initially by the Gaussian reduction method and then by matrix inverses. Our treatment of partial fractions, necessary for certain computational work in calculus, is self-contained and can be studied at any time. We provide the rudiments of sequences and series, which constitute a major part of the calculus syllabus. You will need to use sigma notation throughout your calculus course to write summations. You will also encounter mathematical induction, a technique used to prove particular calculus theorems involving positive integers. We apply mathematical induction to prove some properties of arithmetic and geometric sequences and series and to prove the binomial theorem for the expansion of $(a + b)^n$ for all positive-integer values of n. The final section in this text is devoted to an introduction to infinite series that is related to the notion of *limit*, probably the single most important concept in calculus.

12.1 SYSTEMS OF LINEAR EQUATIONS AND MATRICES

In Section 3.2 we discussed systems of two linear equations in two unknowns. You may wish to review that section at this time. In this section we introduce systems of linear equations in three unknowns.

Consider the equation

$$2x - y + 4z = 10$$

for which the replacement set of each of the three variables x, y, and z is the set R of real numbers. This equation is linear in the three variables. A solution of a linear equation in the three variables x, y, and z is the ordered triple of real numbers (r, s, t) such that if x is replaced by r, y by s, and z by t, the resulting statement is true. The set of all solutions is the solution set of the equation.

▷ **ILLUSTRATION 1**

For the equation

$$2x - y + 4z = 10$$

the ordered triple (3, 4, 2) is a solution because

$$2(3) - 4 + 4(2) = 10$$

Some other ordered triples that satisfy this equation are $(-1, 8, 5)$, $(2, -6, 0)$, $(1, 0, 2)$, $(5, 0, 0)$, $(0, -6, 1)$, $(8, 2, -1)$, and $(7, 2, -\frac{1}{2})$. It appears that the solution set is infinite.   ◀

The graph of an equation in three variables is a set of points represented by ordered triples of real numbers that appear in a three-dimensional coordinate system. When you study solid analytic geometry, you will learn that the graph of a linear equation in three variables is a plane.

Suppose we have the following system of linear equations in the variables x, y, and z:

$$\begin{cases} a_1x + b_1y + c_1z = d_1 \\ a_2x + b_2y + c_2z = d_2 \\ a_3x + b_3y + c_3z = d_3 \end{cases}$$

The solution set of this system is the intersection of the solution sets of the three equations. Because the graph of each equation is a plane, the solution set can be interpreted geometrically as the intersection of three planes. When this intersection consists of a single point as in Figure 1, the equations of the system are said to be consistent and independent. As we proceed with the discussion, we shall show other possible relative positions of three planes.

Algebraic methods for finding the solution set of a system of three linear equations in three unknowns are analogous to those used to solve linear systems in two unknowns. The following example shows the elimination method.

FIGURE 1

▶ **EXAMPLE 1**   *Solving a System of Three Linear Equations by the Elimination Method*

Find the solution set of the system

$$\begin{cases} 4x - 2y - 3z = 8 \\ 5x + 3y - 4z = 4 \\ 6x - 4y - 5z = 12 \end{cases} \tag{I}$$

Solution   We first obtain an equivalent system in which the second and third equations do not involve the variable x. To eliminate x between the first two equations, we multiply the first equation by 5 and the second by -4 and add them:

$$\begin{array}{rcr} 20x - 10y - 15z = & 40 \\ -20x - 12y + 16z = & -16 \\ \hline -22y + z = & 24 \end{array}$$

To eliminate x between the first and third equations of the given system, we multiply the first equation by 6 and the third by -4 and add them:

$$\begin{array}{rcr} 24x - 12y - 18z = & 48 \\ -24x + 16y + 20z = & -48 \\ \hline 4y + 2z = & 0 \end{array}$$

The following system (II) is equivalent to the given system (I). Its first equation is the same as the first equation in (I), and its second and third equations are those obtained above.

$$\begin{cases} 4x - 2y - 3z = 8 \\ \quad\;\; -22y + z = 24 \\ \quad\quad\;\; 4y + 2z = 0 \end{cases} \tag{II}$$

Because the coefficient of y is -22 in the second equation and 4 in the third equation, we can obtain an equation not involving y by computing the sum of 2 times the second equation and 11 times the third equation:

$$\begin{array}{r} -44y + 2z = 48 \\ 44y + 22z = \;\; 0 \\ \hline 24z = 48 \end{array}$$

Dividing both sides of this equation by 24, we have

$$z = 2$$

We now replace the third equation of system (II) by this equation. We have the following equivalent system:

$$\begin{cases} 4x - 2y - 3z = 8 \\ \quad\;\; -22y + z = 24 \\ \quad\quad\quad\quad z = 2 \end{cases} \tag{III}$$

Substituting the value of z from the third equation into the second, we obtain

$$\begin{aligned} -22y + 2 &= 24 \\ -22y &= 22 \\ y &= -1 \end{aligned}$$

Replacing the second equation of system (III) by this equation, we have the equivalent system

$$\begin{cases} 4x - 2y - 3z = 8 \\ \quad\quad\quad\quad y = -1 \\ \quad\quad\quad\quad z = 2 \end{cases} \tag{IV}$$

Substituting the values of y and z from the second and third equations into the first, we get

$$\begin{aligned} 4x - 2(-1) - 3(2) &= 8 \\ 4x + 2 - 6 &= 8 \\ 4x &= 12 \\ x &= 3 \end{aligned}$$

We replace the first equation of system (IV) by this equation and obtain the equivalent system

$$\begin{cases} x = 3 \\ y = -1 \\ z = 2 \end{cases} \qquad\qquad \textbf{(V)}$$

Because systems (V) and (I) are equivalent, the required solution set is $\{(3, -1, 2)\}$.   ◀

The solution set in the preceding example could have been found by working with other combinations of equations. We followed a methodical procedure that relied on obtaining system (III) called **triangular form.** With the system in triangular form we readily found first the value of y and then the value of x by substituting known values of the variables back into the equations. This process is called **back substitution.** When a system is in triangular form, it is a simple matter to use back substitution to obtain an equivalent system such as (V), which is the eventual goal.

The procedure we followed in Example 1 leads to the solution of a system of linear equations by *matrices.* The matrix method is of special importance because most computer and calculator solutions of linear equations depend on matrices. We lead up to the matrix method by referring back to the system of equations in Example 1.

By using a zero as the coefficient of a variable that does not appear in an equation, the triangular form (III) of the system can be written as

$$\begin{cases} 4x - 2y - 3z = 8 \\ 0x - 22y + z = 24 \\ 0x + 0y + z = 2 \end{cases} \qquad\qquad \textbf{(VI)}$$

The procedure used to obtain this system involves operations on the equations of the given system and a series of equivalent systems until a system in triangular form is found. These operations result in changes in the coefficients of the variables and changes in the constant terms in the right members of the equations. Thus we are concerned essentially with these numbers (the coefficients and constant terms) that appear in each of the equivalent systems. Therefore, to simplify the calculations, we introduce a notation for recording the coefficients and constant terms so that we do not have to write the variables. For system (I) the coefficients of the variables are listed as follows:

$$\begin{bmatrix} 4 & -2 & -3 \\ 5 & 3 & -4 \\ 6 & -4 & -5 \end{bmatrix} \qquad\qquad \textbf{(VII)}$$

If we also list the constant terms appearing on the right-hand sides of the equations of system (I), we have

$$\begin{bmatrix} 4 & -2 & -3 & 8 \\ 5 & 3 & -4 & 4 \\ 6 & -4 & -5 & 12 \end{bmatrix}$$ **(VIII)**

where the vertical line separates the coefficients from the constant terms. The numbers involved in system (VI) are listed as

$$\begin{bmatrix} 4 & -2 & -3 & 8 \\ 0 & -22 & 1 & 24 \\ 0 & 0 & 1 & 2 \end{bmatrix}$$ **(IX)**

Each of the arrays (VII), (VIII), and (IX) is called a **matrix.** The numbers in the matrix are called **elements.** The elements that appear next to each other horizontally form a **row,** and those that appear vertically form a **column.** In matrix (VIII) the elements in the first row are 4, -2, -3, and 8, and the elements in the second column are -2, 3, and -4. If there are m rows and n columns in a matrix, then its **order** is $m \times n$ (read "m by n"). Notice that the number of rows is stated first. If $m = n$, we have a **square matrix of order n.** Matrices (VIII) and (IX) are of order 3×4, and matrix (VII) is a square matrix of order 3. The **main diagonal** of a matrix contains the elements on the diagonal line starting at the upper left-hand corner. In matrix (VII) the elements on the main diagonal are 4, 3, and -5, while in matrix (IX) they are 4, -22, and 1.

Suppose we have a system of linear equations where the constants are on the right side and the terms involving the variables, appearing in the same order in each equation, are on the left side. System (I) is such a system. The matrix whose only elements are the coefficients of the variables, listed as they appear in the equations, is called the **coefficient matrix.** Thus matrix (VII) is the coefficient matrix of system (I). The matrix obtained from the coefficient matrix by listing the constants in the additional column on the right, where the coefficients are separated from the constants by a vertical line, is called the **augmented matrix.** For system (I) the augmented matrix is matrix (VIII).

To solve a system of linear equations by using matrices, we start with the augmented matrix and perform operations on the rows to obtain a matrix of an equivalent system. The process is continued until we obtain a matrix of a system in triangular form, that is, a matrix that has zeros everywhere below the main diagonal. Such a matrix is said to be in **echelon form.** For example, (VI) is the system in triangular form that is equivalent to system (I). Furthermore, matrix (IX) is the augmented matrix of system (VI) and is in echelon form.

The rules for performing operations on matrices are called **elementary row operations,** and they are given in the following theorem.

> **THEOREM 1**
>
> If we have an augmented matrix of a system of linear equations, each of the following operations produces a matrix of an equivalent system of linear equations:
>
> **(i)** Interchanging any two rows.
> **(ii)** Multiplying each element of a row by the same nonzero number.
> **(iii)** Replacing a given row by a new row whose elements are the sum of k_1 times the elements of the given row and k_2 times the corresponding elements of any other row, where k_1 and k_2 are real numbers and $k_1 \neq 0$.

The proof of Theorem 1 utilizes the corresponding operations on the equations of the system and is omitted. Observe that the theorem involves *operations on rows only*. Do not make the mistake of applying the operations to columns, as doing so does not produce a matrix of an equivalent system.

In part (ii) of Theorem 1, it is understood that to multiply a row by a number means to multiply each element of the row by the number. In a similar manner, when applying part (iii) by adding one row to another row, we are actually adding corresponding elements of the two rows.

The following illustration shows how Theorem 1 is used to solve system (I). You should compare the computation with that of Example 1.

▷ **ILLUSTRATION 2**

System (I) is

$$\begin{cases} 4x - 2y - 3z = 8 \\ 5x + 3y - 4z = 4 \\ 6x - 4y - 5z = 12 \end{cases}$$

and the augmented matrix is

$$\begin{bmatrix} 4 & -2 & -3 & | & 8 \\ 5 & 3 & -4 & | & 4 \\ 6 & -4 & -5 & | & 12 \end{bmatrix}$$

To obtain 0 as the first element in the second row, we replace the second row by the sum of 5 times the first row and −4 times the second row; and to obtain 0 as the first element in the third row, we replace the third row by the sum of 6 times the first row and −4 times the third row. Thus we have the matrix

$$\begin{bmatrix} 4 & -2 & -3 & | & 8 \\ 0 & -22 & 1 & | & 24 \\ 0 & 4 & 2 & | & 0 \end{bmatrix}$$

From this matrix we obtain 0 as the second element in the third row by replacing the third row by the sum of 2 times the second row and 11 times the third row. Doing this gives

$$\begin{bmatrix} 4 & -2 & -3 & | & 8 \\ 0 & -22 & 1 & | & 24 \\ 0 & 0 & 24 & | & 48 \end{bmatrix}$$

Multiplying the third row by $\frac{1}{24}$, we get

$$\begin{bmatrix} 4 & -2 & -3 & | & 8 \\ 0 & -22 & 1 & | & 24 \\ 0 & 0 & 1 & | & 2 \end{bmatrix}$$

which is matrix (IX) and is in echelon form; it is the augmented matrix for system (VI), which is in triangular form. The solution set is then found by back substitution as in Example 1. ◀

The procedure demonstrated in Illustration 2 for solving a system of linear equations by matrices is called the Gaussian reduction method, named after the German mathematician and scientist Karl Friedrich Gauss (1777–1855). The method requires reducing a matrix to echelon form. In linear algebra, a more advanced course, a formal step-by-step process for doing this is given. For the fairly simple systems in this text, the echelon form can be obtained by a sequence of elementary row operations determined by observation and trial and error.

In the following example a system of four linear equations in four variables is solved by the Gaussian reduction method. By a solution of a system of four equations in w, x, y, and z, we mean an ordered four-tuple (r, s, t, u) such that each of the equations is satisfied if w, x, y, and z are replaced by r, s, t, and u, respectively.

▶ **EXAMPLE 2** *Solving a System of Four Linear Equations by the Gaussian Reduction Method*

Use the Gaussian reduction method to find the solution set of the system

$$\begin{cases} w + x + y + z = 5 \\ 3z + x = w - 2 \\ 2x + 2y = 3w + 2 \\ y = w + z \end{cases}$$

Solution We replace each equation by an equivalent one in which the terms involving variables are on the left side and the constant terms are on the right side. We have the following equivalent system in which the equations are written so that terms containing the same variable are in a vertical column.

$$\begin{cases} w + x + y + z = 5 \\ -w + x + 3z = -2 \\ -3w + 2x + 2y = 2 \\ -w + y - z = 0 \end{cases}$$

The augmented matrix of this system is

$$\left[\begin{array}{cccc|c} 1 & 1 & 1 & 1 & 5 \\ -1 & 1 & 0 & 3 & -2 \\ -3 & 2 & 2 & 0 & 2 \\ -1 & 0 & 1 & -1 & 0 \end{array}\right]$$

Initially we obtain zeros as the first elements in the second, third, and fourth rows. We replace the second row by the sum of the first and second rows; we replace the third row by the sum of 3 times the first row and the third row; we replace the fourth row by the sum of the first and fourth rows. We then have the matrix

$$\left[\begin{array}{cccc|c} 1 & 1 & 1 & 1 & 5 \\ 0 & 2 & 1 & 4 & 3 \\ 0 & 5 & 5 & 3 & 17 \\ 0 & 1 & 2 & 0 & 5 \end{array}\right]$$

We now interchange the second and fourth rows because then we will have 1 as an element in the second row and second column; this will make it easier to obtain zeros in the third and fourth rows of the second column. We then have the matrix

$$\left[\begin{array}{cccc|c} 1 & 1 & 1 & 1 & 5 \\ 0 & 1 & 2 & 0 & 5 \\ 0 & 5 & 5 & 3 & 17 \\ 0 & 2 & 1 & 4 & 3 \end{array}\right]$$

Replacing the third row by the sum of the third row and −5 times the second row and replacing the fourth row by the sum of the fourth row and −2 times the second row, we obtain the matrix

$$\left[\begin{array}{cccc|c} 1 & 1 & 1 & 1 & 5 \\ 0 & 1 & 2 & 0 & 5 \\ 0 & 0 & -5 & 3 & -8 \\ 0 & 0 & -3 & 4 & -7 \end{array}\right]$$

Replacing the fourth row by the sum of 3 times the third row and -5 times the fourth row, we get the matrix

$$\begin{bmatrix} 1 & 1 & 1 & 1 & | & 5 \\ 0 & 1 & 2 & 0 & | & 5 \\ 0 & 0 & -5 & 3 & | & -8 \\ 0 & 0 & 0 & -11 & | & 11 \end{bmatrix}$$

We now multiply the fourth row by $-\frac{1}{11}$ and obtain

$$\begin{bmatrix} 1 & 1 & 1 & 1 & | & 5 \\ 0 & 1 & 2 & 0 & | & 5 \\ 0 & 0 & -5 & 3 & | & -8 \\ 0 & 0 & 0 & 1 & | & -1 \end{bmatrix}$$

This matrix is in echelon form and is the augmented matrix of the system in triangular form,

$$\begin{cases} w + x + y + z = 5 \\ x + 2y = 5 \\ -5y + 3z = -8 \\ z = -1 \end{cases}$$

We back substitute into the third equation the value of z from the fourth equation and obtain $y = 1$; then back substituting 1 for y in the second equation, we have $x = 3$. By back substituting in the first equation 3 for x, 1 for y, and -1 for z, we obtain $w = 2$. We have then the following system, which is equivalent to the given system:

$$\begin{cases} w = 2 \\ x = 3 \\ y = 1 \\ z = -1 \end{cases}$$

Thus the solution set of the given system is $\{(2, 3, 1, -1)\}$. ◄

In Section 3.2 we showed that when a system of two linear equations in two variables is inconsistent, the graphs of the two equations are parallel lines. For a system of three inconsistent linear equations in three variables, the graphs of the three equations are planes that have no common intersection. The various possibilities are shown in Figure 2(a)–(d).

In (a) the three planes are parallel. In (b) two of the planes are the same plane, and the third plane is parallel to it. In (c) two of the planes are parallel, the intersection of each of these planes with the third plane is a line, and the lines are parallel. In (d) no two planes are parallel, but two of the planes intersect in a line that is parallel to the third plane.

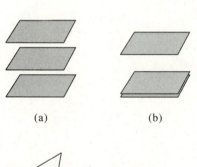

(a)　　　　(b)

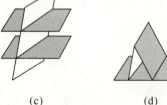

(c)　　　　(d)

FIGURE 2

In the following illustration we show what happens when the Gaussian reduction method is applied to a system of inconsistent equations.

▷ ILLUSTRATION 3

For the system

$$\begin{cases} 2x + y - z = 2 \\ x + 2y + 4z = 1 \\ 5x + y - 7z = 4 \end{cases}$$

the augmented matrix is

$$\begin{bmatrix} 2 & 1 & -1 & | & 2 \\ 1 & 2 & 4 & | & 1 \\ 5 & 1 & -7 & | & 4 \end{bmatrix}$$

Replacing the second row by the sum of 1 times the first row and -2 times the second row and replacing the third row by the sum of 5 times the first row and -2 times the third row, we get

$$\begin{bmatrix} 2 & 1 & -1 & | & 2 \\ 0 & -3 & -9 & | & 0 \\ 0 & 3 & 9 & | & 2 \end{bmatrix}$$

We replace the third row by the sum of the second and third rows and obtain

$$\begin{bmatrix} 2 & 1 & -1 & | & 2 \\ 0 & -3 & -9 & | & 0 \\ 0 & 0 & 0 & | & 2 \end{bmatrix}$$

This matrix is in echelon form and is the augmented matrix of the system

$$\begin{cases} 2x + y - z = 2 \\ -3y - 9z = 0 \\ 0 = 2 \end{cases}$$

for which the equations are inconsistent, as indicated by the third equation.

◀

Observe in Illustration 3 that in the last row of the matrix in echelon form there is a nonzero element only in the last position. Whenever this situation occurs, you can conclude that the equations of the system are inconsistent.

▶ **EXAMPLE 3** *Solving a System of Three Linear Equations by the Gaussian Reduction Method*

Use the Gaussian reduction method to find the solution set of the system

$$\begin{cases} 2x + y - 3z = 0 \\ 3x + 2y - 4z = 2 \\ x - y - 3z = -6 \end{cases}$$

Solution   The augmented matrix of the system is

$$\begin{bmatrix} 2 & 1 & -3 & | & 0 \\ 3 & 2 & -4 & | & 2 \\ 1 & -1 & -3 & | & -6 \end{bmatrix}$$

Replacing the second row by the sum of 3 times the first row and -2 times the second row and replacing the third row by the sum of the first row and -2 times the third row, we obtain

$$\begin{bmatrix} 2 & 1 & -3 & | & 0 \\ 0 & -1 & -1 & | & -4 \\ 0 & 3 & 3 & | & 12 \end{bmatrix}$$

We now replace the third row by the sum of 3 times the second row and the third row and get

$$\begin{bmatrix} 2 & 1 & -3 & | & 0 \\ 0 & -1 & -1 & | & -4 \\ 0 & 0 & 0 & | & 0 \end{bmatrix}$$

This matrix is in echelon form and is the augmented matrix of the system

$$\begin{cases} 2x + y - 3z = 0 \\ -y - z = -4 \\ 0 = 0 \end{cases}$$

From the third equation we observe that the equations of the system are dependent. The third equation is an identity for any ordered triple (x, y, z) and in particular for $(0, 0, t)$. Thus we replace the equation $0 = 0$ by the equation $z = t$, and we have the equivalent system in triangle form,

$$\begin{cases} 2x + y - 3z = 0 \\ -y - z = -4 \\ z = t \end{cases}$$

We now back substitute the value of z from the third equation into the second, solve for y, and then back substitute the values of y and z from the

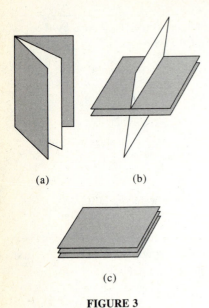

(a)   (b)

(c)

FIGURE 3

second and third equations into the first. We obtain

$$\begin{cases} x = 2t - 2 \\ y = 4 - t \\ z = t \end{cases}$$

The solution set of the given system is, therefore, $\{(2t - 2, 4 - t, t)\}$. ◀

In Example 3, observe that when the Gaussian reduction method is used to solve a system of dependent equations, the matrix in echelon form has all zeros in the last row.

The graphs of three dependent linear equations in three variables are either three planes with a line in common or else the planes are identical. Figure 3(a)–(c) shows the various possibilities. In (a) the graphs are three distinct planes with a line in common. In (b) two identical planes intersect the third plane in a line. In (c) the three planes are identical.

The Gaussian reduction method can be extended to systems of n linear equations in n unknowns, and calculators can be used to perform the matrix operations. Consult your manual for the matrix features, and how to use them, on your particular calculator.

In the solution of the word problem in the next example we obtain a system of three equations that turn out to be dependent. The word problem, however, has a finite number of solutions.

▶ **EXAMPLE 4** *Solving a Word Problem Having a System of Three Dependent Linear Equations as a Mathematical Model*

A group of 14 people spent $56 for admission tickets to a matinee performance in Cinema One, charging $5 for adults, $3 for students, and $2 for children. If the same people had attended a matinee performance in Cinema Two, charging $4 for adults, $2 for students, and $1 for children, they would have spent $42 for admission tickets. How many adults, how many students, and how many children were in the group?

Solution   The unknown quantities are the number of adults, the number of students, and the number of children in the group. We represent these numbers by a, s, and c, respectively. Then because 14 people are in the group,

$$a + s + c = 14$$

Because Cinema One charges $5 for adults, $3 for students, and $2 for children and the total for admission tickets to Cinema One was $56, we have the equation

$$5a + 3s + 2c = 56$$

Because Cinema Two charges $4 for adults, $2 for students, and $1 for children, and the total for admission tickets to Cinema Two was $42, we have the equation

$$4a + 2s + c = 42$$

We have then the system of equations

$$\begin{cases} a + s + c = 14 \\ 5a + 3s + 2c = 56 \\ 4a + 2s + c = 42 \end{cases} \qquad \textbf{(X)}$$

The augmented matrix of this system is

$$\begin{bmatrix} 1 & 1 & 1 & \bigm| & 14 \\ 5 & 3 & 2 & \bigm| & 56 \\ 4 & 2 & 1 & \bigm| & 42 \end{bmatrix}$$

To obtain zeros as the first element in the second and third rows, we replace the second row by the sum of -5 times the first row and the second row, and we replace the third row by the sum of -4 times the first row and the third row. We then have

$$\begin{bmatrix} 1 & 1 & 1 & \bigm| & 14 \\ 0 & -2 & -3 & \bigm| & -14 \\ 0 & -2 & -3 & \bigm| & -14 \end{bmatrix}$$

Replacing the third row by the sum of -1 times the second row and the third row, we obtain

$$\begin{bmatrix} 1 & 1 & 1 & \bigm| & 14 \\ 0 & -2 & -3 & \bigm| & -14 \\ 0 & 0 & 0 & \bigm| & 0 \end{bmatrix}$$

This matrix, in echelon form, is the augmented matrix of the system

$$\begin{cases} a + s + c = 14 \\ -2s - 3c = -14 \\ 0 = 0 \end{cases}$$

The third equation of this system indicates that the equations are dependent. Replacing this equation by the equation $c = t$, we have the equivalent system in triangular form,

$$\begin{cases} a + s + c = 14 \\ 2s + 3c = 14 \\ c = t \end{cases}$$

Substituting the value of c from the third equation into the second, solving for s, and substituting the values of s and c into the first equation, we have

$$\begin{cases} a = 7 + \frac{1}{2}t \\ s = 7 - \frac{3}{2}t \\ c = t \end{cases}$$

This system is equivalent to system (X). Thus a solution of (X) is an ordered triple of the form $\{(7 + \frac{1}{2}t, 7 - \frac{3}{2}t, t)\}$.

Because a, s, and c must represent nonnegative integers, each of the numbers t, $7 - \frac{3}{2}t$, and $7 + \frac{1}{2}t$ must be nonnegative integers. If $t = 0$, $7 - \frac{3}{2}t = 7$ and $7 + \frac{1}{2}t = 7$. Therefore, $(7, 7, 0)$ is a solution. If $t = 1$, $7 - \frac{3}{2}t = \frac{11}{2}$ and $7 + \frac{1}{2}t = \frac{15}{2}$; thus $t = 1$ does not give a solution to the problem. If $t = 2$, $7 - \frac{3}{2}t = 4$ and $7 + \frac{1}{2}t = 8$. Therefore $(8, 4, 2)$ is a solution. If $t = 3$, both $7 - \frac{3}{2}t$ and $7 + \frac{1}{2}t$ are not integers, and so $t = 3$ does not give a solution. If $t = 4$, $7 - \frac{3}{2}t = 1$ and $7 + \frac{1}{2}t = 9$. Hence $(9, 1, 4)$ is a solution. If t is an integer greater than 4, $7 - \frac{3}{2}t$ is a negative number. Therefore the solution set is $\{(7, 7, 0), (8, 4, 2), (9, 1, 4)\}$.

Conclusion: There are three possible combinations of people in the group: seven adults, seven students, and no children; eight adults, four students, and two children; or nine adults, one student, and four children.

Check: If seven adults, seven students, and no children are in the group, the number of dollars in the total cost of admission tickets to Cinema One is $5(7) + 3(7) = 56$, and the number of dollars in the total cost of admission tickets to Cinema Two is $(4)(7) + (2)(7) = 42$. If eight adults, four students, and two children are in the group, the number of dollars in the total cost of admission tickets to Cinema One is $5(8) + 3(4) + 2(2) = 56$, and the number of dollars in the total cost of admission tickets to Cinema Two is $(4)(8) + (2)(4) + (1)(2) = 42$. If nine adults, one student, and four children are in the group, the number of dollars in the total cost of admission tickets to Cinema One is $5(9) + 3(1) + 2(4) = 56$, and the number of dollars in the total cost of admission tickets to Cinema Two is $(4)(9) + (2)(1) + (1)(4) = 42$. ◀

The matrix method can be applied to a system of any number of linear equations in any number of variables. When there are as many equations as variables, the system has a unique solution unless, as we have seen, the equations are dependent or inconsistent. When the number of variables exceeds the number of equations, more often than not the system will have an infinite number of solutions. However, the equations could be inconsistent, which is the case for a system of two linear equations in three variables when the graphs are parallel planes. When there are more equations than variables, it is quite common for the equations to be inconsistent, although this is not necessarily the case. In Exercise 23 the system has more variables than equations, and you are asked to show that the system has an infinite number of solutions. In Exercise 24 the system has more equations than variables, and you are asked to show that the equations are inconsistent.

EXERCISES 12.1

In Exercises 1 through 10, find the solution set of the system of equations by the elimination method. If the equations are either inconsistent or dependent, then indicate that this is the case.

1. $\begin{cases} 4x + 3y + z = 15 \\ x - y - 2z = 2 \\ 2x - 2y + z = 4 \end{cases}$

2. $\begin{cases} 2x + 3y + z = 8 \\ 5x + 2y - 3z = -13 \\ x - 2y + 5z = 15 \end{cases}$

3. $\begin{cases} x - y + 3z = 2 \\ 2x + 2y - z = 5 \\ 5x + 2z = 7 \end{cases}$

4. $\begin{cases} 3x - 2y + 4z = 4 \\ 7x + 5y - z = 9 \\ x + 9y - 9z = 1 \end{cases}$

5. $\begin{cases} 2x - 3y - 5z = 4 \\ x + 7y + 6z = -7 \\ 7x - 2y - 9z = 6 \end{cases}$

6. $\begin{cases} 3x - 5y + 2z = -2 \\ 2x + 3z = -3 \\ 4y - 3z = 8 \end{cases}$

7. $\begin{cases} x - y = 2 \\ 3y + z = 1 \\ x - 2z = 7 \end{cases}$

8. $\begin{cases} 5x - 4y + 5z = 6 \\ 6x + y - 2z = 4 \\ 4x - 9y + 12z = 5 \end{cases}$

9. $\begin{cases} \dfrac{1}{x} + \dfrac{1}{y} - \dfrac{1}{z} = 5 \\ \dfrac{3}{x} - \dfrac{1}{y} + \dfrac{2}{z} = 12 \\ \dfrac{1}{x} + \dfrac{2}{y} + \dfrac{1}{z} = 9 \end{cases}$

10. $\begin{cases} \dfrac{3}{x} - \dfrac{3}{y} + \dfrac{1}{z} = -1 \\ \dfrac{2}{x} + \dfrac{1}{y} - \dfrac{4}{z} = 0 \\ \dfrac{1}{x} + \dfrac{4}{y} + \dfrac{1}{z} = 5 \end{cases}$

In Exercises 11 through 22, use the Gaussian reduction method to find the solution set of the system. If the equations are either inconsistent or dependent, then indicate that this is the case.

11. The system of Exercise 1

12. The system of Exercise 2

13. $\begin{cases} x - 2y - z = 3 \\ x + y + z = 4 \\ x - 3y - z = 4 \end{cases}$

14. $\begin{cases} x + y - 2z = 5 \\ 3x + 2y = 4 \\ 2x + z = 2 \end{cases}$

15. $\begin{cases} \dfrac{x}{2} + \dfrac{y}{3} + \dfrac{z}{4} = 2 \\ \dfrac{x}{4} - \dfrac{2y}{3} - \dfrac{z}{4} = \dfrac{1}{2} \\ \dfrac{x}{6} + \dfrac{y}{9} + \dfrac{z}{2} = 4 \end{cases}$

16. $\begin{cases} \dfrac{x}{3} + \dfrac{y}{2} - \dfrac{z}{2} = 0 \\ \dfrac{x}{6} + \dfrac{3y}{4} + \dfrac{z}{3} = \dfrac{5}{8} \\ \dfrac{3x}{2} + \dfrac{y}{4} + \dfrac{z}{6} = -\dfrac{1}{8} \end{cases}$

17. $\begin{cases} 2x + 3y - 4z = 4 \\ x + 2y - 5z = 6 \\ 4x + 5y - 2z = 0 \end{cases}$

18. $\begin{cases} w + 3x + y + z = 3 \\ -w - 3y - z = 0 \\ -2w + 3x - 4y + z = 0 \\ -w - 6x - 2y + 2z = -4 \end{cases}$

19. $\begin{cases} 2w + x - 3y - 3z = 4 \\ x + 2y + z = -3 \\ 2w - x + 3z = -3 \\ 5w - y + z = -6 \end{cases}$

20. $\begin{cases} 2w - 4x + y - 2z = 3 \\ 3w + x + 2y + 3z = 12 \\ w - 4x + 2y - 6z = 1 \\ 5w + x + 11z = 16 \end{cases}$

21. $\begin{cases} 2w + 3x - 4y - z = 3 \\ 3w + x + y + 2z = 1 \\ w - 2x + 3y - z = 0 \\ w - 2x - y - 9z = 5 \end{cases}$

22. $\begin{cases} 4w + x - 2y + z = 4 \\ 2w + 3x + 4y - 3z = -2 \\ 3w - 2x + 5y + 3z = 2 \\ w + 4x + 3y + z = 5 \end{cases}$

23. Use the Gaussian reduction method to show that the following system has an infinite number of solutions:

$\begin{cases} w - x + y + z = 4 \\ 3w - 2y + z = 2 \\ w + 2x - 3y - 5z = 3 \end{cases}$

24. Use the Gaussian reduction method to show that the following system is inconsistent:

$$\begin{cases} x + 4y - 3z = 5 \\ 2x + y + 5z = 2 \\ 3x - 2y - 4z = 4 \\ 4x - 5y + 4z = -3 \end{cases}$$

In Exercises 25 through 28, the system has more variables than equations. Use the Gaussian reduction method to find the solution set of the system.

25. $\begin{cases} 5x - 4y + 3z = 1 \\ 3x - 5y + 7z = 11 \end{cases}$

26. $\begin{cases} 3x - 2y - 5z = 5 \\ 2x - y - 4z = 3 \end{cases}$

27. $\begin{cases} 5w - x - 2y + 5z = 7 \\ 3w + 2x - 3y + 2z = 4 \\ w - 8x + 5y + 4z = 3 \end{cases}$

28. $\begin{cases} w + 2x - 3y - z = -2 \\ 3w - 4x + y - 3z = 1 \\ 4w + x + 2y - 2z = 3 \end{cases}$

In Exercises 29 through 36, the system has more equations than variables. Use the Gaussian reduction method to determine if the equations are consistent or inconsistent. If they are consistent, find the solution set.

29. The system of Exercise 39 in Exercises 3.2

30. The system of Exercise 40 in Exercises 3.2

31. The system of Exercise 41 in Exercises 3.2

32. The system of Exercise 42 in Exercises 3.2

33. $\begin{cases} 2x + y + 3z = -4 \\ x - 4y - 2z = 3 \\ 4x - 2y + z = 4 \\ 5x + 3y + 4z = 5 \end{cases}$

34. $\begin{cases} 2x + 3y - z = 5 \\ 4x - 3y + 3z = 5 \\ 3x + y + 4z = 2 \\ x - 2y + z = 1 \end{cases}$

35. $\begin{cases} 2x + 4y + 3z = 5 \\ x - 4y - 2z = 7 \\ 4x - 3y + 5z = 2 \\ 3x + 2y + 4z = 8 \end{cases}$

36. $\begin{cases} x - y - 4z = 0 \\ x - y + 2z = 6 \\ 3x + y - 5z = -1 \\ x - 2y + z = 7 \end{cases}$

In Exercises 37 and 38, find an equation of the circle containing the given points. Hint: Recall that an equation of a circle is of the form $x^2 + y^2 + Dx + Ey + F = 0$.

37. $(-2, 8)$, $(2, 6)$, and $(-7, 3)$

38. $(5, 4)$, $(-2, 3)$, and $(-4, 1)$

39. Find an equation of the parabola, whose axis is parallel to the y axis, if it contains the points $(1, -1)$, $(2, 3)$, and $(3, 15)$. *Hint:* An equation of such a parabola is of the form $y = ax^2 + bx + c$.

40. Find an equation of the parabola, whose axis is parallel to the x axis, if it contains the points $(3, 0)$, $(1, -4)$, and $(-1, 6)$. *Hint:* An equation of such a parabola is of the form $x = ay^2 + by + c$.

In Exercises 41 through 48, solve the word problem by finding a system of linear equations as a mathematical model. Be sure to write a conclusion.

41. Part of $25,000 is invested at 10 percent, another part is invested at 12 percent, and a third part is invested at 16 percent. The total yearly income from these three investments is $3200. Furthermore, the income from the 16 percent investment yields the same amount as the sum of the incomes from the other two investments. How much is invested at each rate?

42. If t degrees Celsius is the temperature at which water boils at a height of h feet above sea level, then $h = a + bt + ct^2$. If water boils at 100° Celsius at sea level, 95° Celsius at a height of 7400 ft above sea level, and 90° Celsius at a height of 14,550 ft above sea level, find a, b, and c.

43. In 20 oz of one alloy there are 6 oz of copper, 4 oz of zinc, and 10 oz of lead. In 20 oz of a second alloy there are 12 oz of copper, 5 oz of zinc, and 3 oz of lead, while in 20 oz of a third alloy there are 8 oz of copper, 6 oz of zinc, and 6 oz of lead. How many ounces of each alloy should be combined to make a new alloy containing 34 oz of copper, 17 oz of zinc, and 19 oz of lead?

44. A total of $50,000 is invested in three different securities for which the annual dividends are computed at 8 percent, 10 percent, and 12 percent. If the total annual income from the three securities is $5320, and the income from the 12 percent security is $1080 more than that from the 10 percent security, determine the amount invested in each security.

45. On a store counter, there was a supply of three sizes of Christmas cards. The large cards cost $1 each; the medium cards cost 80 cents each; and the small cards cost 60 cents each. A woman purchased 10 cards, which consisted of one-fourth of the available large cards, one-third of the available medium cards, and one-half of the available small cards. The total cost of her cards was $8.20. If there were 21 cards remaining on the counter after her purchase, how many of each kind of card did she buy?

46. Food *A* has 560 calories per pound and 80 units of vitamins per pound, food *B* has 240 calories per pound and 400 units of vitamins per pound, and food *C* has 480 calories per pound and 160 units of vitamins per pound. It is desired to have a 10-lb mixture of foods *A*, *B*, and *C* to contain a total of 2000 calories and 1200 units of vitamins. Show that these requirements lead to an inconsistent system of equations, and therefore the mixture is not possible.

47. Suppose in Exercise 46, instead of a 10-lb mixture, it is desired to have a 5-lb mixture of foods *A*, *B*, and *C* to contain a total of 2000 calories and 1200 units of vitamins. **(a)** Show that these conditions lead to a dependent system of equations. **(b)** If *x* pounds of *A*, *y* pounds of *B*, and *z* pounds of *C* are to be used to make up the 5 lb, what are the restrictions on *x*, *y*, and *z*?

48. A brochure promoting exhibitions at an art gallery is sent each month to the persons whose names appear on a mailing list, and employees *A*, *B*, and *C* assist in the preparation of the mailing. When all three work together, it takes them 2 hr and 55 min to complete the job. Last month employee *C* was away and employees *A* and *B* together took 5 hr to prepare the mailing. For this month's mailing each employee started at a different time: Employee *A* began at 9 A.M.; employee *B* joined *A* at 10 A.M.; and employee *C* joined *A* and *B* at 10:54 A.M. They finished the work at 1 P.M. How long does it take each employee working alone to prepare the mailing? See the hint for Exercise 33 in Exercises 3.2.

49. Explain why interchanging two rows of an augmented matrix of a system of linear equations produces a matrix of an equivalent system but interchanging two columns does not.

50. Explain why multiplying each element of a row of an augmented matrix of a system of linear equations by the same nonzero number produces a matrix of an equivalent system but performing the same computation on each element of a column does not.

12.2 PROPERTIES OF MATRICES AND SOLVING LINEAR SYSTEMS BY MATRIX INVERSES

GOALS

1. **Evaluate a second-order determinant.**
2. **Evaluate a third-order determinant.**
3. **Compute the product of a scalar and a matrix.**
4. **Compute the product of two matrices.**
5. **Define the multiplicative identity for a set of square matrices.**
6. **Learn that a matrix has an inverse if and only if its determinant is not zero.**
7. **Compute by hand the inverse of a square matrix of order two if it exists.**
8. **Solve systems of linear equations by matrix inverses.**

In Section 12.1 we showed how matrices are used to solve a system of linear equations by the Gaussian reduction method. In this section we present another method for solving linear systems by matrices when the number of equations in the system is the same as the number of variables. We begin the discussion by defining some terminology.

Associated with each square matrix is a number called the *determinant*.

DEFINITION **Second-Order Determinant**

If H is the square matrix of order two

$$\begin{bmatrix} a_1 & b_1 \\ a_2 & b_2 \end{bmatrix}$$

then the **determinant** of H, denoted by either det H or $\begin{vmatrix} a_1 & b_1 \\ a_2 & b_2 \end{vmatrix}$, is defined by

$$\begin{vmatrix} a_1 & b_1 \\ a_2 & b_2 \end{vmatrix} = a_1 b_2 - a_2 b_1$$

and is of the **second order**.

▶ **EXAMPLE 1** *Evaluating a Second-Order Determinant*

Evaluate the determinant of the matrix

$$\begin{bmatrix} 3 & -2 \\ 4 & -1 \end{bmatrix}$$

Solution

$$\begin{vmatrix} 3 & -2 \\ 4 & -1 \end{vmatrix} = 3(-1) - 4(-2)$$
$$= 5$$ ◀

DEFINITION **Third-Order Determinant**

If H is the square matrix of order three

$$\begin{bmatrix} a_1 & b_1 & c_1 \\ a_2 & b_2 & c_2 \\ a_3 & b_3 & c_3 \end{bmatrix}$$

then the **determinant** of H, denoted by either det H or $\begin{vmatrix} a_1 & b_1 & c_1 \\ a_2 & b_2 & c_2 \\ a_3 & b_3 & c_3 \end{vmatrix}$, is defined by

$$\begin{vmatrix} a_1 & b_1 & c_1 \\ a_2 & b_2 & c_2 \\ a_3 & b_3 & c_3 \end{vmatrix} = a_1 \begin{vmatrix} b_2 & c_2 \\ b_3 & c_3 \end{vmatrix} - b_1 \begin{vmatrix} a_2 & c_2 \\ a_3 & c_3 \end{vmatrix} + c_1 \begin{vmatrix} a_2 & b_2 \\ a_3 & b_3 \end{vmatrix}$$

and is of the **third order.**

Notice that the terms on the right-hand side of the equation defining a third-order determinant consist of the elements of the first row multiplied by the second-order determinant obtained by crossing out the row and column in which the element occurs. That is, a_1 is multiplied by the determinant obtained by crossing out the first row and first column, b_1 (with a minus sign preceding) is multiplied by the determinant obtained by crossing out the first row and second column, and c_1 is multiplied by the determinant obtained by crossing out the first row and third column.

▶ **EXAMPLE 2** *Evaluating a Third-Order Determinant*

Evaluate the determinant of the matrix

$$\begin{bmatrix} 2 & 3 & -1 \\ -4 & 1 & -2 \\ 5 & 0 & -3 \end{bmatrix}$$

Solution

$$\begin{vmatrix} 2 & 3 & -1 \\ -4 & 1 & -2 \\ 5 & 0 & -3 \end{vmatrix} = 2 \begin{vmatrix} 1 & -2 \\ 0 & -3 \end{vmatrix} - 3 \begin{vmatrix} -4 & -2 \\ 5 & -3 \end{vmatrix} + (-1) \begin{vmatrix} -4 & 1 \\ 5 & 0 \end{vmatrix}$$

$$= 2(-3 - 0) - 3(12 + 10) + (-1)(0 - 5)$$
$$= 2(-3) - 3(22) - 1(-5)$$
$$= -67 \qquad \blacktriangleleft$$

The German mathematician Gottfried Wilhelm Leibniz (1646–1716), one of the inventors of calculus, is often said to be the inventor of determinants. In fact, however, the Japanese mathematician Seki Kowa (1642–1708) used them first.

We have introduced determinants so that we can apply them later in this section to the solution of linear systems. We do not give a formal definition of a determinant of order higher than three because computation by hand for such determinants is cumbersome. You can, however, compute the determinant of a matrix on a calculator having a matrix capability. Refer to your manual for the procedure on your particular calculator.

In Section 12.1 we defined a matrix of order $m \times n$ as one having m rows and n columns. If a matrix has only one column, it is called a **column matrix,** and if it has only one row it is called a **row matrix.**

▷ **ILLUSTRATION 1**

The following matrices have the order indicated below the matrix:

$$\begin{bmatrix} 5 & 4 \\ -2 & 7 \\ 0 & -4 \end{bmatrix} \qquad \begin{bmatrix} 9 & -1 & 6 \\ 2 & -10 & -8 \end{bmatrix} \qquad \begin{bmatrix} 7 \\ -3 \\ 0 \\ 4 \end{bmatrix} \qquad [0 \quad 2 \quad -5] \qquad \begin{bmatrix} -4 & 1 & 8 \\ 0 & -2 & 11 \\ 6 & -3 & 0 \end{bmatrix}$$

3×2	2×3	4×1	1×3	3×3
		Column matrix	Row matrix	Square matrix

◀

Two matrices are said to be **equal** if and only if their orders are equal and the corresponding elements are equal. That is,

$$\begin{bmatrix} r & s \\ t & u \end{bmatrix} = \begin{bmatrix} a & b \\ c & d \end{bmatrix}$$

if and only if

$$r = a \qquad s = b \qquad t = c \qquad u = d$$

▷ **ILLUSTRATION 2**

If

$$A = \begin{bmatrix} 4 & -3 & 0 \\ 1 & 3 & 2 \end{bmatrix} \qquad B = \begin{bmatrix} 1 & 3 & 2 \\ 4 & -3 & 0 \end{bmatrix} \qquad C = \begin{bmatrix} 2 \times 2 & -3 & 0 \\ 3 - 2 & \frac{6}{2} & 2 \\ & & 1 \end{bmatrix}$$

then $A \neq B$ and $B \neq C$, but $A = C$.   ◀

We now define *multiplication* of a matrix by a *scalar* where, for our purposes, a **scalar** denotes a real number.

DEFINITION   Product of a Scalar and a Matrix

The **product** of a scalar k and a matrix H, denoted by kH, is the matrix in which each element is obtained by multiplying k by the corresponding element of H.

▷ **ILLUSTRATION 3**

$$2\begin{bmatrix} 4 & -1 & 6 \\ -3 & 0 & \dfrac{1}{2} \end{bmatrix} = \begin{bmatrix} 8 & -2 & 12 \\ -6 & 0 & 1 \end{bmatrix}$$

and

$$-3\begin{bmatrix} -5 & -3 \\ 2 & 1 \\ 0 & -4 \end{bmatrix} = \begin{bmatrix} 15 & 9 \\ -6 & -3 \\ 0 & 12 \end{bmatrix}$$ ◀

The definition of the product of two matrices may be peculiar to you. However, there are applications that warrant such a definition. Before giving the formal definition, we consider some special cases and begin with the following illustration involving a row matrix and a column matrix.

▷ **ILLUSTRATION 4**

Let

$$R = \begin{bmatrix} 2 & 3 & -5 \end{bmatrix} \quad \text{and} \quad C = \begin{bmatrix} 1 \\ 4 \\ 2 \end{bmatrix}$$

The product of R and C, denoted by RC, is the matrix containing a single element that is the sum of the products formed by multiplying each element in R by the corresponding element in C:

$$\begin{aligned}
RC &= \begin{bmatrix} 2 & 3 & -5 \end{bmatrix} \begin{bmatrix} 1 \\ 4 \\ 2 \end{bmatrix} \\
&= [(2)(1) + (3)(4) + (-5)(2)] \\
&= [2 + 12 - 10] \\
&= [4]
\end{aligned}$$ ◀

Observe that the method of matrix multiplication in Illustration 4 allows us to write the linear equation

$$ax + by + cz = d$$

in matrix form as

$$\begin{bmatrix} a & b & c \end{bmatrix} \begin{bmatrix} x \\ y \\ z \end{bmatrix} = [d]$$

▷ **ILLUSTRATION 5**

Consider the product of the matrices C and D, where

$$C = \begin{bmatrix} 3 & 0 & -4 \\ -2 & 2 & -1 \end{bmatrix} \quad \text{and} \quad D = \begin{bmatrix} 2 & -3 \\ 4 & -1 \\ 1 & 5 \end{bmatrix}$$

To obtain the element in row 1, column 1 of CD we multiply the elements in row 1 of C by the corresponding elements in column 1 of D and add the products as follows:

$$(3)(2) + (0)(4) + (-4)(1) = 2$$

For the element in row 1, column 2 of CD we multiply the elements in row 1 of C by the corresponding elements in column 2 of D and add the products:

$$(3)(-3) + (0)(-1) + (-4)(5) = -29$$

For the element in row 2, column 1 of CD we add the products obtained by multiplying the elements in row 2 of C by the corresponding elements in column 1 of D:

$$(-2)(2) + (2)(4) + (-1)(1) = 3$$

For the element in row 2, column 2 of CD we add the products obtained by multiplying the elements in row 2 of C by the corresponding elements in column 2 of D:

$$(-2)(-3) + (2)(-1) + (-1)(5) = -1$$

We have therefore obtained the following product:

$$\begin{bmatrix} 3 & 0 & -4 \\ -2 & 2 & -1 \end{bmatrix} \begin{bmatrix} 2 & -3 \\ 4 & -1 \\ 1 & 5 \end{bmatrix} = \begin{bmatrix} 2 & -29 \\ 3 & -1 \end{bmatrix} \qquad ◀$$

We now give the formal definition of the product of two matrices A and B. It is necessary that the number of columns of the first matrix A be the same as the number of rows of the second matrix B.

DEFINITION   Product of Two Matrices

Suppose that A is a matrix of order $m \times p$ and B is a matrix of order $p \times n$. Then the **product** of A and B, denoted by AB, is the $m \times n$ matrix for which the element in the ith row and jth column is the sum of the products formed by multiplying each element in the ith row of A by the corresponding element in the jth column of B.

▶ **EXAMPLE 3** *Computing the Product of Two Matrices*

Apply the definition to find the product DC for the matrices of Illustration 5.

Solution

$$
DC = \begin{bmatrix} 2 & -3 \\ 4 & -1 \\ 1 & 5 \end{bmatrix} \begin{bmatrix} 3 & 0 & -4 \\ -2 & 2 & -1 \end{bmatrix}
$$

$$
= \begin{bmatrix} (2)(3) + (-3)(-2) & (2)(0) + (-3)(2) & (2)(-4) + (-3)(-1) \\ (4)(3) + (-1)(-2) & (4)(0) + (-1)(2) & (4)(-4) + (-1)(-1) \\ (1)(3) + (5)(-2) & (1)(0) + (5)(2) & (1)(-4) + (5)(-1) \end{bmatrix}
$$

$$
= \begin{bmatrix} 12 & -6 & -5 \\ 14 & -2 & -15 \\ -7 & 10 & -9 \end{bmatrix}
$$
◀

Again we stress that to obtain the product of matrices A and B, the number of columns in A must be the same as the number of rows in B; otherwise the product is not defined. When AB is defined, this product has as many rows as A and as many columns as B. Refer back to Illustrations 4 and 5 and Example 3 and observe the following:

Illustration 4: 1×3 matrix times 3×1 matrix yields 1×1 matrix
Illustration 5: 2×3 matrix times 3×2 matrix yields 2×2 matrix
Example 3: 3×2 matrix times 2×3 matrix yields 3×3 matrix

Another observation from Illustration 5 and Example 3 is that CD is not equal to DC; that is, matrix multiplication is *not commutative*. The following theorem, however, states that matrix multiplication is *associative* when the products exist.

THEOREM 1

If A, B, and C are matrices such that the following products exist, then

$A(BC) = (AB)C$

The proof of Theorem 1 is omitted. However, you are asked to verify the theorem for particular matrices in Exercises 21 and 22.

The method of matrix multiplication can be used to write a system of linear equations in matrix form. For instance, the system

$$
\begin{cases} a_1 x + b_1 y + c_1 z = d_1 \\ a_2 x + b_2 y + c_2 z = d_2 \\ a_3 x + b_3 y + c_3 z = d_3 \end{cases}
$$

can be written as

$$\begin{bmatrix} a_1 & b_1 & c_1 \\ a_2 & b_2 & c_2 \\ a_3 & b_3 & c_3 \end{bmatrix} \begin{bmatrix} x \\ y \\ z \end{bmatrix} = \begin{bmatrix} d_1 \\ d_2 \\ d_3 \end{bmatrix}$$

Before using this matrix form to solve a system of linear equations we need two more concepts: the *multiplicative identity* for a set of square matrices and the *multiplicative inverse* of a square matrix.

The number 1 is the multiplicative identity for real numbers such that for any real number a

$$a \cdot 1 = a$$

Does the set of matrices of a given order have a multiplicative identity? The answer, in general, is no. However, there is a multiplicative identity for the set of all square matrices of a given order.

DEFINITION   Multiplicative Identity

The **multiplicative identity** for the set of square matrices of order n, denoted by I_n, is the square matrix of order n whose elements on the main diagonal are ones and whose other elements are all zeros.

▷ **ILLUSTRATION 6**

The multiplicative identities for the sets of square matrices of orders 2, 3, and 4, are, respectively, I_2, I_3, and I_4, where

$$I_2 = \begin{bmatrix} 1 & 0 \\ 0 & 1 \end{bmatrix} \qquad I_3 = \begin{bmatrix} 1 & 0 & 0 \\ 0 & 1 & 0 \\ 0 & 0 & 1 \end{bmatrix} \qquad I_4 = \begin{bmatrix} 1 & 0 & 0 & 0 \\ 0 & 1 & 0 & 0 \\ 0 & 0 & 1 & 0 \\ 0 & 0 & 0 & 1 \end{bmatrix}$$

◀

▷ **ILLUSTRATION 7**

Let H be any square matrix of order three:

$$H = \begin{bmatrix} a_1 & b_1 & c_1 \\ a_2 & b_2 & c_2 \\ a_3 & b_3 & c_3 \end{bmatrix}$$

Then

$$HI_3 = \begin{bmatrix} a_1 & b_1 & c_1 \\ a_2 & b_2 & c_2 \\ a_3 & b_3 & c_3 \end{bmatrix} \begin{bmatrix} 1 & 0 & 0 \\ 0 & 1 & 0 \\ 0 & 0 & 1 \end{bmatrix} \qquad I_3 H = \begin{bmatrix} 1 & 0 & 0 \\ 0 & 1 & 0 \\ 0 & 0 & 1 \end{bmatrix} \begin{bmatrix} a_1 & b_1 & c_1 \\ a_2 & b_2 & c_2 \\ a_3 & b_3 & c_3 \end{bmatrix}$$

$$= \begin{bmatrix} a_1 & b_1 & c_1 \\ a_2 & b_2 & c_2 \\ a_3 & b_3 & c_3 \end{bmatrix} \qquad\qquad = \begin{bmatrix} a_1 & b_1 & c_1 \\ a_2 & b_2 & c_2 \\ a_3 & b_3 & c_3 \end{bmatrix}$$

$$= H \qquad\qquad\qquad\qquad = H \qquad \blacktriangleleft$$

In Illustration 7 we have proved the following theorem for square matrices of order three. A similar proof can be given for square matrices of any order.

THEOREM 2

If H is a square matrix of order n and if I_n is the multiplicative identity of order n, then

$$HI_n = H \qquad \text{and} \qquad I_n H = H$$

We know that for every real number a, except 0, there exists a real number a^{-1}, called the multiplicative inverse of a, such that

$$a \cdot a^{-1} = 1 \qquad \text{and} \qquad a^{-1} \cdot a = 1$$

If a square matrix H of order n is to have a *multiplicative inverse H^{-1}*, it is necessary that

$$HH^{-1} = I_n \qquad \text{and} \qquad H^{-1}H = I_n$$

We now give a theorem that states that such a matrix H^{-1}, called the **multiplicative inverse** of H, exists if and only if det $H \neq 0$; that is, a determinant *determines* whether a matrix has an inverse.

THEOREM 3

If H is a square matrix of order n, there exists a matrix H^{-1} such that

$$HH^{-1} = I_n \qquad \text{and} \qquad H^{-1}H = I_n$$

if and only if det $H \neq 0$.

We omit the proof of Theorem 3.

If a matrix H has an inverse (that is, if det $H \neq 0$), H is said to be **invertible,** or **nonsingular.** If a matrix H does not have an inverse (that is, if det $H = 0$), then H is said to be **singular.**

Theorem 3 is called an existence theorem because it gives a condition for which the inverse of a matrix exists. Note that the theorem does not indicate how to find the inverse. However, we can compute the inverse of an invertible square matrix of order two by applying the following theorem.

THEOREM 4

If H is the matrix

$$\begin{bmatrix} a & b \\ c & d \end{bmatrix}$$

and if det $H \neq 0$, then

$$H^{-1} = \frac{1}{\det H} \begin{bmatrix} d & -b \\ -c & a \end{bmatrix}$$

Proof We shall show that $H^{-1}H = I_2$. The proof that $HH^{-1} = I_2$ is similar.

$$H^{-1}H = \left(\frac{1}{\det H} \begin{bmatrix} d & -b \\ -c & a \end{bmatrix} \right) \begin{bmatrix} a & b \\ c & d \end{bmatrix}$$

$$= \frac{1}{\det H} \begin{bmatrix} ad - bc & bd - bd \\ -ac + ac & -bc + ad \end{bmatrix}$$

$$= \frac{1}{\det H} \begin{bmatrix} \det H & 0 \\ 0 & \det H \end{bmatrix}$$

$$= \begin{bmatrix} 1 & 0 \\ 0 & 1 \end{bmatrix}$$

$$= I_2 \qquad\qquad \blacksquare$$

▶ **EXAMPLE 4** *Computing the Inverse of a Square Matrix of Order Two*

If

$$H = \begin{bmatrix} 2 & -4 \\ 3 & -5 \end{bmatrix}$$

prove that H is invertible and compute H^{-1}.

Solution

$$\det H = 2(-5) - 3(-4)$$
$$= 2$$

Because $\det H \neq 0$, H is invertible. From Theorem 4

$$H^{-1} = \frac{1}{\det H}\begin{bmatrix} -5 & -(-4) \\ -3 & 2 \end{bmatrix}$$

$$= \frac{1}{2}\begin{bmatrix} -5 & 4 \\ -3 & 2 \end{bmatrix}$$

$$= \begin{bmatrix} -\dfrac{5}{2} & 2 \\ -\dfrac{3}{2} & 1 \end{bmatrix}$$

◀

In general, hand computation of matrix inverses is tedious. Fortunately, a calculator having a matrix capability will compute the inverse of an invertible matrix for you. Consult your users manual for the procedure. Remember, first verify that the matrix is indeed invertible by showing that its determinant is not zero.

▶ **EXAMPLE 5** *Determining that a Square Matrix of Order Three Is Invertible*

If

$$H = \begin{bmatrix} 1 & 2 & 1 \\ 1 & 3 & 0 \\ 4 & 0 & 2 \end{bmatrix}$$

show that H is invertible.

Solution From Theorem 3, H is invertible if and only if $\det H \neq 0$. From the definition of a third-order determinant

$$\det H = 1\begin{vmatrix} 3 & 0 \\ 0 & 2 \end{vmatrix} - 2\begin{vmatrix} 1 & 0 \\ 4 & 2 \end{vmatrix} + 1\begin{vmatrix} 1 & 3 \\ 4 & 0 \end{vmatrix}$$

$$= 1(6 - 0) - 2(2 - 0) + 1(0 - 12)$$

$$= -10$$

Because $\det H \neq 0$, H is invertible. ◀

▶ **EXAMPLE 6** *Multiplying a Matrix by Its Inverse*

On a calculator compute H^{-1}, the inverse of matrix H of Example 5, and show that $HH^{-1} = I_3$.

Solution   Computation of H^{-1} on a calculator yields

$$H^{-1} = \begin{bmatrix} -\dfrac{3}{5} & \dfrac{2}{5} & \dfrac{3}{10} \\[2mm] \dfrac{1}{5} & \dfrac{1}{5} & -\dfrac{1}{10} \\[2mm] \dfrac{6}{5} & -\dfrac{4}{5} & -\dfrac{1}{10} \end{bmatrix}$$

Thus

$$HH^{-1} = \begin{bmatrix} 1 & 2 & 1 \\ 1 & 3 & 0 \\ 4 & 0 & 2 \end{bmatrix} \begin{bmatrix} -\dfrac{3}{5} & \dfrac{2}{5} & \dfrac{3}{10} \\[2mm] \dfrac{1}{5} & \dfrac{1}{5} & -\dfrac{1}{10} \\[2mm] \dfrac{6}{5} & -\dfrac{4}{5} & -\dfrac{1}{10} \end{bmatrix}$$

$$= \begin{bmatrix} -\dfrac{3}{5} + \dfrac{2}{5} + \dfrac{6}{5} & \dfrac{2}{5} + \dfrac{2}{5} - \dfrac{4}{5} & \dfrac{3}{10} - \dfrac{2}{10} - \dfrac{1}{10} \\[2mm] -\dfrac{3}{5} + \dfrac{3}{5} + 0 & \dfrac{2}{5} + \dfrac{3}{5} + 0 & \dfrac{3}{10} - \dfrac{3}{10} + 0 \\[2mm] -\dfrac{12}{5} + 0 + \dfrac{12}{5} & \dfrac{8}{5} + 0 - \dfrac{8}{5} & \dfrac{12}{10} + 0 - \dfrac{2}{10} \end{bmatrix}$$

$$= \begin{bmatrix} 1 & 0 & 0 \\ 0 & 1 & 0 \\ 0 & 0 & 1 \end{bmatrix}$$

$$= I_3$$ ◀

We now have the tools to solve systems of linear equations by matrix inverses.

▷ **ILLUSTRATION 8**

To solve the system

$$\begin{cases} 2x - 4y = 1 \\ 3x - 5y = 2 \end{cases}$$

we first write the system in matrix form as

$$\begin{bmatrix} 2 & -4 \\ 3 & -5 \end{bmatrix} \begin{bmatrix} x \\ y \end{bmatrix} = \begin{bmatrix} 1 \\ 2 \end{bmatrix}$$

The first matrix on the left is called the *matrix of coefficients* for the system; denoting this matrix by H, we have

$$H = \begin{bmatrix} 2 & -4 \\ 3 & -5 \end{bmatrix}$$

The matrix form of the system is then

$$H \begin{bmatrix} x \\ y \end{bmatrix} = \begin{bmatrix} 1 \\ 2 \end{bmatrix}$$

Because H is the matrix of Example 4, we know that H^{-1} exists. We can therefore multiply both sides of this equation on the left by H^{-1} and we obtain

$$\begin{bmatrix} x \\ y \end{bmatrix} = H^{-1} \begin{bmatrix} 1 \\ 2 \end{bmatrix}$$

Using the result of Example 4 for H^{-1}, we have

$$\begin{bmatrix} x \\ y \end{bmatrix} = \begin{bmatrix} -\dfrac{5}{2} & 2 \\ -\dfrac{3}{2} & 1 \end{bmatrix} \begin{bmatrix} 1 \\ 2 \end{bmatrix}$$

$$= \begin{bmatrix} -\dfrac{5}{2} + 4 \\ -\dfrac{3}{2} + 2 \end{bmatrix}$$

$$= \begin{bmatrix} \dfrac{3}{2} \\ \dfrac{1}{2} \end{bmatrix}$$

Therefore, $x = \frac{3}{2}$ and $y = \frac{1}{2}$. ◀

▶ **EXAMPLE 7** *Solving a System of Linear Equations by the Inverse of a Matrix*

Use the inverse of a matrix to solve the system

$$\begin{cases} x + 2y + z = 2 \\ x + 3y = 1 \\ 4x + 2z = -4 \end{cases}$$

Solution   If

$$H = \begin{bmatrix} 1 & 2 & 1 \\ 1 & 3 & 0 \\ 4 & 0 & 2 \end{bmatrix}$$

then, in matrix form, the system is

$$H \begin{bmatrix} x \\ y \\ z \end{bmatrix} = \begin{bmatrix} 2 \\ 1 \\ -4 \end{bmatrix}$$

Because H is the matrix of Example 5, we know that H^{-1} exists. Therefore we can multiply both sides of this equation on the left by H^{-1} and we obtain

$$\begin{bmatrix} x \\ y \\ z \end{bmatrix} = H^{-1} \begin{bmatrix} 2 \\ 1 \\ -4 \end{bmatrix}$$

Using the result of Example 6 for H^{-1}, we have

$$\begin{bmatrix} x \\ y \\ z \end{bmatrix} = \begin{bmatrix} -\dfrac{3}{5} & \dfrac{2}{5} & \dfrac{3}{10} \\ \dfrac{1}{5} & \dfrac{1}{5} & -\dfrac{1}{10} \\ \dfrac{6}{5} & -\dfrac{4}{5} & -\dfrac{1}{10} \end{bmatrix} \begin{bmatrix} 2 \\ 1 \\ -4 \end{bmatrix}$$

$$= \begin{bmatrix} -\dfrac{6}{5} + \dfrac{2}{5} - \dfrac{6}{5} \\ \dfrac{2}{5} + \dfrac{1}{5} + \dfrac{2}{5} \\ \dfrac{12}{5} - \dfrac{4}{5} + \dfrac{2}{5} \end{bmatrix}$$

$$= \begin{bmatrix} -2 \\ 1 \\ 2 \end{bmatrix}$$

Therefore $x = -2$, $y = 1$, and $z = 2$.   ◀

EXERCISES 12.2

In Exercises 1 through 4, evaluate the determinant.

1. (a) $\begin{vmatrix} 3 & -1 \\ 2 & 5 \end{vmatrix}$ **(b)** $\begin{vmatrix} 6 & 5 \\ -2 & 1 \end{vmatrix}$

2. (a) $\begin{vmatrix} -2 & 4 \\ 3 & 7 \end{vmatrix}$ **(b)** $\begin{vmatrix} 8 & 5 \\ 5 & 8 \end{vmatrix}$

3. $\begin{vmatrix} -1 & 6 & 4 \\ -1 & 2 & 1 \\ -2 & 7 & 4 \end{vmatrix}$ **4.** $\begin{vmatrix} -6 & -3 & 2 \\ 3 & 2 & 0 \\ 4 & 2 & -1 \end{vmatrix}$

5. Given the matrices

$$A = \begin{bmatrix} 3 & -2 & 7 & 0 \\ 4 & 5 & -1 & 8 \end{bmatrix} \qquad B = \begin{bmatrix} -6 \\ 2 \\ 5 \end{bmatrix}$$

what is each of the following: **(a)** the order of A; **(b)** the order of B; **(c)** the element in the first row and second column of A; **(d)** the product of 3 and A; **(e)** the product of -2 and B?

6. Given the matrices

$$A = \begin{bmatrix} 5 & -1 & -3 & 0 \\ 1 & 2 & -1 & -4 \\ 0 & -6 & 2 & 3 \\ 4 & -1 & 7 & 2 \end{bmatrix} \qquad B = [2 \ -1 \ 5 \ -6]$$

what is each of the following: **(a)** the order of A; **(b)** the order of B; **(c)** the element in the third row and second column of A; **(d)** the product of 2 and A; **(e)** the product of -3 and B?

In Exercises 7 through 16, find the product.

7. (a) $4\begin{bmatrix} -1 & 2 \\ 3 & 5 \\ -6 & 1 \end{bmatrix}$ **(b)** $-5\begin{bmatrix} -7 & 0 \\ -4 & 6 \end{bmatrix}$

8. (a) $-2\begin{bmatrix} -3 & 6 & -1 \\ 4 & -2 & 0 \\ -5 & 1 & -8 \end{bmatrix}$ **(b)** $3\begin{bmatrix} 4 & 3 & -3 \\ 1 & -2 & 7 \end{bmatrix}$

9. (a) $\begin{bmatrix} 2 & 3 \\ -1 & 5 \end{bmatrix}\begin{bmatrix} 2 & -1 \\ 0 & 3 \end{bmatrix}$

(b) $\begin{bmatrix} 1 & 2 & 3 \\ 4 & 5 & 7 \end{bmatrix}\begin{bmatrix} 1 & -1 \\ 2 & 0 \\ -1 & 1 \end{bmatrix}$

10. (a) $\begin{bmatrix} 4 & -5 \\ 7 & 3 \end{bmatrix}\begin{bmatrix} 5 & -1 \\ -2 & 7 \end{bmatrix}$

(b) $[-1 \ \ 3 \ \ -2]\begin{bmatrix} 2 & 0 \\ 1 & 4 \\ -1 & -2 \end{bmatrix}$

11. (a) $[1 \ \ 2 \ \ -3]\begin{bmatrix} 2 \\ 1 \\ -1 \end{bmatrix}$ **(b)** $\begin{bmatrix} 2 \\ 1 \\ -1 \end{bmatrix}[1 \ \ 2 \ \ -3]$

12. (a) $[2 \ \ 1]\begin{bmatrix} 3 \\ 2 \end{bmatrix}$ **(b)** $\begin{bmatrix} 3 \\ 2 \end{bmatrix}[2 \ \ 1]$

13. $\begin{bmatrix} 4 & -1 \\ 0 & 2 \\ 5 & 1 \end{bmatrix}\begin{bmatrix} -2 & 3 \\ 0 & -3 \end{bmatrix}$

14. $\begin{bmatrix} 2 & 0 & -2 \\ 3 & -1 & 0 \end{bmatrix}\begin{bmatrix} 5 & 2 & -1 \\ 0 & -3 & 1 \\ -2 & 6 & 0 \end{bmatrix}$

15. $\begin{bmatrix} 1 & -4 & 0 & -1 \\ 2 & 0 & 3 & -2 \end{bmatrix}\begin{bmatrix} 1 & 0 & -2 \\ -1 & 6 & 0 \\ 2 & 1 & 3 \\ 0 & 2 & -4 \end{bmatrix}$

16. $\begin{bmatrix} 1 & -1 & 0 & 3 \\ 2 & 0 & -3 & -2 \\ 0 & -4 & 5 & 2 \end{bmatrix}\begin{bmatrix} -2 & 6 \\ 0 & 1 \\ -1 & 3 \\ 2 & -3 \end{bmatrix}$

In Exercises 17 through 22, verify the statement where

$$A = \begin{bmatrix} -2 & 3 \\ 2 & -1 \end{bmatrix} \quad B = \begin{bmatrix} 2 & -1 \\ 3 & -2 \end{bmatrix} \quad C = \begin{bmatrix} -3 & 0 \\ 1 & -2 \end{bmatrix}$$

and I is the multiplicative identity of order 2.

17. $AI = A$ and $IA = A$ **18.** $BI = B$ and $IB = B$

19. $AB \neq BA$ **20.** $BC \neq CB$

21. $A(BC) = (AB)C$ **22.** $C(BA) = (CB)A$

In Exercises 23 and 24, the given matrix is H. Prove that H is invertible, find H^{-1}, and show that $HH^{-1} = I_2$.

23. $\begin{bmatrix} 7 & 2 \\ 2 & 1 \end{bmatrix}$ **24.** $\begin{bmatrix} 2 & -1 \\ -4 & 3 \end{bmatrix}$

In Exercises 25 through 28, determine if the matrix is invertible.

25. $\begin{bmatrix} 3 & 3 & 1 \\ 1 & 4 & 1 \\ 2 & 3 & 1 \end{bmatrix}$ **26.** $\begin{bmatrix} 0 & 1 & 2 \\ -3 & 4 & 0 \\ 0 & -2 & -1 \end{bmatrix}$

27. $\begin{bmatrix} 4 & -1 & 2 \\ -2 & 0 & -2 \\ 3 & 1 & 5 \end{bmatrix}$ **28.** $\begin{bmatrix} 1 & -2 & 1 \\ 2 & 2 & -1 \\ 1 & 1 & 0 \end{bmatrix}$

In Exercises 29 through 32, solve the system of equations by a matrix inverse. For Exercises 29 and 30, first verify on your calculator that

$$\begin{bmatrix} 2 & -1 \\ -5 & 3 \end{bmatrix}^{-1} = \begin{bmatrix} 3 & 1 \\ 5 & 2 \end{bmatrix}$$

and for Exercises 31 and 32, first verify on your calculator that

$$\begin{bmatrix} 1 & 0 & 2 \\ 0 & -1 & 1 \\ 1 & 3 & 0 \end{bmatrix}^{-1} = \begin{bmatrix} 3 & -6 & -2 \\ -1 & 2 & 1 \\ -1 & 3 & 1 \end{bmatrix}$$

29. $\begin{cases} 2x - y = -2 \\ -5x + 3y = 7 \end{cases}$ **30.** $\begin{cases} 2x - y = 1 \\ -5x + 3y = -5 \end{cases}$

31. $\begin{cases} x + 2z = 5 \\ -y + z = 3 \\ x + 3y = 0 \end{cases}$ **32.** $\begin{cases} x + 2z = 12 \\ -y + z = 7 \\ x + 3y = -4 \end{cases}$

In Exercises 33 through 36, solve the system of equations by a matrix inverse. Compute the inverse of the matrix of coefficients on your calculator.

33. $\begin{cases} 3x + 2y = -1 \\ 4x + 3y = 0 \end{cases}$ **34.** $\begin{cases} 3x + 2y = 3 \\ 6x - 6y = 1 \end{cases}$

35. $\begin{cases} 4x + 2y + 3z = 4 \\ 5x - y + 4z = 12 \\ x - 3y + 2z = 0 \end{cases}$ **36.** $\begin{cases} 2x + y + 3z = 5 \\ x + y + z = 0 \\ 4x + y - 2z = -15 \end{cases}$

In Exercises 37 and 38, find H^{-1} for the given matrix H without using a calculator or Theorem 4. (Hint: Let $H^{-1} = \begin{bmatrix} a & b \\ c & d \end{bmatrix}$, and from the fact that $HH^{-1} = I_2$, obtain four equations in a, b, c, and d.)

37. $\begin{bmatrix} 3 & -4 \\ -2 & 3 \end{bmatrix}$ **38.** $\begin{bmatrix} -3 & 1 \\ 2 & -2 \end{bmatrix}$

In Exercises 39 through 48, verify the equality if

$$M = \begin{bmatrix} 3 & 1 \\ 7 & 3 \end{bmatrix} \quad \text{and} \quad N = \begin{bmatrix} -1 & 1 \\ -3 & 4 \end{bmatrix}$$

39. $\det(MN) = (\det M)(\det N)$

40. $\det(NM) = (\det N)(\det M)$

41. $(MN)^{-1} = N^{-1}M^{-1}$ **42.** $(NM)^{-1} = M^{-1}N^{-1}$

43. $(M^{-1})^{-1} = M$ **44.** $(N^{-1})^{-1} = N$

45. $(M^{-1})^2 = (M^2)^{-1}$ **46.** $(N^{-1})^2 = (N^2)^{-1}$

47. $\det(M^{-1}) = \dfrac{1}{\det M}$ **48.** $\det(N^{-1}) = \dfrac{1}{\det N}$

49. Explain in words only (no symbols) the meaning of the equalities in Exercises 41 and 43.

50. Explain in words only (no symbols) the meaning of the equalities in Exercises 45 and 47.

12.3 PARTIAL FRACTIONS

GOALS

1. Decompose a fraction into partial fractions when the factors of the denominator are all linear and none is repeated.
2. Decompose a fraction into partial fractions when the factors of the denominator are all linear and some are repeated.
3. Decompose a fraction into partial fractions when the factors of the denominator are linear and quadratic and none is repeated.
4. Decompose a fraction into partial fractions when the factors of the denominator are linear and quadratic and some of the quadratic factors are repeated.

You already know how to combine two or more rational expressions into one rational expression by addition or subtraction. For example,

$$\frac{3}{x + 2} + \frac{4}{x - 3} = \frac{7x - 1}{(x + 2)(x - 3)}$$

 It is sometimes necessary to do the reverse, that is, to express a single rational expression as a sum of two or more simpler quotients, called **partial fractions.** We need to do this in calculus to perform the operation of integration on some rational functions. Often systems of equations are used to decompose a rational expression into partial fractions.

Consider a rational function H defined by

$$H(x) = \frac{P(x)}{Q(x)}$$

where $P(x)$ and $Q(x)$ are polynomials. We shall assume that we have a **proper fraction,** that is, one for which the degree of $P(x)$ is less than the degree of $Q(x)$. If we have a rational function for which the degree of the numerator is not less than the degree of the denominator, then we have an **improper fraction,** and in that case we divide the numerator by the denominator until a proper fraction is obtained. For instance,

$$\frac{x^4 - 10x^2 + 3x + 1}{x^2 - 4} = x^2 - 6 + \frac{3x - 23}{x^2 - 4}$$

In general, then, we are concerned with a method of decomposing a proper fraction of the form $P(x)/Q(x)$ into two or more partial fractions. The denominators of the partial fractions are obtained by factoring $Q(x)$ into a product of linear and quadratic factors. Sometimes it may be difficult to find these factors. However, recall Theorem 5 from Section 5.4, which states that a polynomial with real coefficients can be expressed as a product of linear or quadratic polynomials with real coefficients.

After $Q(x)$ has been factored into products of linear or quadratic factors, the method of determining the partial fractions depends on the nature of these factors. We consider various cases separately.

Case 1: The factors of $Q(x)$ are all linear, and none is repeated. That is,

$$Q(x) = (a_1x + b_1)(a_2x + b_2) \cdot \ldots \cdot (a_nx + b_n)$$

where no two of the factors are identical. In this case we write

$$\frac{P(x)}{Q(x)} = \frac{A_1}{a_1x + b_1} + \frac{A_2}{a_2x + b_2} + \ldots + \frac{A_n}{a_nx + b_n}$$

where $A_1, A_2, \ldots, A_n$ are constants to be determined. Observe that this equation is an identity because it is true for each value of x for which a denominator is not zero. The following illustration shows a method for determining the values of A_i.

▷ **ILLUSTRATION 1**

To decompose the fraction

$$\frac{7x - 1}{x^2 - x - 6}$$

into partial fractions, we factor the denominator and obtain

$$\frac{7x - 1}{x^2 - x - 6} = \frac{7x - 1}{(x + 2)(x - 3)}$$

Therefore we have

$$\frac{7x - 1}{(x + 2)(x - 3)} = \frac{A}{x + 2} + \frac{B}{x - 3} \tag{1}$$

Equation (1) is an identity for all x except -2 and 3. By multiplying on both sides of the equation by the LCD, we obtain

$$7x - 1 = A(x - 3) + B(x + 2)$$

This equation is an identity. It is true for all values of x including -2 and 3. We now find the constants A and B. Substituting 3 for x in the preceding equation, we get

$$20 = 5B \Leftrightarrow B = 4$$

Substituting -2 for x in the same equation, we obtain

$$-15 = -5A \Leftrightarrow A = 3$$

With these values for A and B, we have from (1)

$$\frac{7x - 1}{(x + 2)(x - 3)} = \frac{3}{x + 2} + \frac{4}{x - 3}$$

Observe that this equation is equivalent to the one at the beginning of this section. ◀

▶ **EXAMPLE 1** *Decomposing a Fraction into Partial Fractions when the Factors of the Denominator Are All Linear and None Is Repeated*

Decompose the fraction

$$\frac{x - 1}{x^3 - x^2 - 2x}$$

into partial fractions.

Solution We factor the denominator and have

$$\frac{x - 1}{x^3 - x^2 - 2x} = \frac{x - 1}{x(x - 2)(x + 1)}$$

Thus we have

$$\frac{x - 1}{x(x - 2)(x + 1)} = \frac{A}{x} + \frac{B}{x - 2} + \frac{C}{x + 1} \tag{2}$$

Equation (2) is an identity for all x except 0, 2 and -1. We multiply on both sides of the equation by the LCD and get

$$x - 1 = A(x - 2)(x + 1) + Bx(x + 1) + Cx(x - 2)$$

This equation is an identity that is true for all values of x including 0, 2, and -1. To find the constants, we first substitute 0 for x and obtain

$$-1 = -2A \Leftrightarrow A = \tfrac{1}{2}$$

Substituting 2 for x, we get

$$1 = 6B \Leftrightarrow B = \tfrac{1}{6}$$

Substituting -1 for x, we obtain

$$-2 = 3C \Leftrightarrow C = -\tfrac{2}{3}$$

With these values for A, B, and C we have from (2)

$$\frac{x - 1}{x(x - 2)(x + 1)} = \frac{\tfrac{1}{2}}{x} + \frac{\tfrac{1}{6}}{x - 2} + \frac{-\tfrac{2}{3}}{x + 1}$$

$$\Leftrightarrow \quad \frac{x - 1}{x(x - 2)(x + 1)} = \frac{1}{2x} + \frac{1}{6(x - 2)} - \frac{2}{3(x + 1)} \quad ◀$$

Case 2: The factors of $Q(x)$ are all linear, and some are repeated.

Suppose that $(ax + b)^p$ occurs as a factor of $Q(x)$. Then $ax + b$ is said to be a p-fold factor of $Q(x)$, and corresponding to this factor there will be the sum of p partial fractions

$$\frac{A_1}{ax + b} + \frac{A_2}{(ax + b)^2} + \cdots + \frac{A_{p-1}}{(ax + b)^{p-1}} + \frac{A_p}{(ax + b)^p}$$

where $A_1, A_2, \ldots, A_p$ are constants to be determined. Example 2 illustrates this case and the method of determining each A_i.

▶ **EXAMPLE 2** *Decomposing a Fraction into Partial Fractions when the Factors of the Denominator Are All Linear and Some Are Repeated*

Decompose the fraction

$$\frac{x^4 + x^2 + 16x - 12}{x^3(x - 2)^2}$$

into partial fractions.

Solution We write the given fraction as a sum of partial fractions as follows:

$$\frac{x^4 + x^2 + 16x - 12}{x^3(x - 2)^2} = \frac{A}{x} + \frac{B}{x^2} + \frac{C}{x^3} + \frac{D}{x - 2} + \frac{E}{(x - 2)^2} \qquad \textbf{(3)}$$

Multiplying on both sides of (3) by the LCD, we get

$$x^4 + x^2 + 16x - 12 = Ax^2(x - 2)^2 + Bx(x - 2)^2 + C(x - 2)^2 + Dx^3(x - 2) + Ex^3 \qquad \textbf{(4)}$$

We substitute 0 for x in this equation and obtain

$$-12 = 4C \Leftrightarrow C = -3$$

Substituting 2 for x in (4), we get

$$40 = 8E \Leftrightarrow E = 5$$

With these values for C and E in (4) and expanding the powers of the binomials, we have

$$x^4 + x^2 + 16x - 12 = Ax^2(x^2 - 4x + 4) + Bx(x^2 - 4x + 4) - 3(x^2 - 4x + 4) + Dx^3(x - 2) + 5x^3$$
$$x^4 + x^2 + 16x - 12 = (A + D)x^4 + (-4A + B - 2D + 5)x^3 + (4A - 4B - 3)x^2 + (4B + 12)x - 12$$

Because this equation is an identity, the coefficients on the left must equal the corresponding coefficients on the right. Therefore we have the following system of equations:

$$\begin{cases} A + D = 1 \\ -4A + B - 2D + 5 = 0 \\ 4A - 4B - 3 = 1 \\ 4B + 12 = 16 \end{cases}$$

From the fourth equation, $B = 1$. Replacing B by 1 in the third equation and solving for A, we get $A = 2$. With $A = 2$ in the first equation, we obtain $D = -1$. We use the second equation as a check:

$$-4A + B - 2D + 5 = -4(2) + 1 - 2(-1) + 5$$
$$= 0$$

Thus the values of the constants are as follows:

$$A = 2 \quad B = 1 \quad C = -3 \quad D = -1 \quad E = 5$$

With these values we have from (3)

$$\frac{x^4 + x^2 + 16x - 12}{x^3(x - 2)^2} = \frac{2}{x} + \frac{1}{x^2} - \frac{3}{x^3} - \frac{1}{x - 2} + \frac{5}{(x - 2)^2} \quad \blacktriangleleft$$

Case 3:   The factors of $Q(x)$ are linear and quadratic, and none of the quadratic factors is repeated.

Corresponding to the quadratic factor $ax^2 + bx + c$ in the denominator is the partial fraction of the form

$$\frac{Ax + B}{ax^2 + bx + c}$$

▶ **EXAMPLE 3** *Decomposing a Fraction into Partial Fractions when the Factors of the Denominator Are Linear and Quadratic and None Is Repeated*

Decompose the fraction

$$\frac{x^2 - x - 5}{x^3 + x^2 - 2}$$

into partial fractions.

Solution   We attempt to factor the denominator by using synthetic division to divide $x^3 + x^2 - 2$ by linear expressions of the form $x - r$, where r is an integer factor of -2. The division by $x - 1$ is as follows:

$$\begin{array}{r|rrrr} 1 & 1 & 1 & 0 & -2 \\ & & 1 & 2 & 2 \\ \hline & 1 & 2 & 2 & 0 \end{array}$$

Therefore $x^3 + x^2 - 2 = (x - 1)(x^2 + 2x + 2)$. The given fraction is written as a sum of partial fractions in the following way:

$$\frac{x^2 - x - 5}{(x - 1)(x^2 + 2x + 2)} = \frac{Ax + B}{x^2 + 2x + 2} + \frac{C}{x - 1} \tag{5}$$

Multiplying on both sides by the LCD, we have

$$x^2 - x - 5 = (Ax + B)(x - 1) + C(x^2 + 2x + 2) \tag{6}$$

We compute C by substituting 1 for x in (6), and we get

$$-5 = 5C \Leftrightarrow C = -1$$

We replace C by -1 in (6) and multiply on the right side to obtain

$$x^2 - x - 5 = (A - 1)x^2 + (B - A - 2)x + (-B - 2)$$

Equating coefficients of like powers of x gives the system

$$\begin{cases} A - 1 = 1 \\ B - A - 2 = -1 \\ -B - 2 = -5 \end{cases}$$

Therefore

$$A = 2 \quad B = 3$$

Substituting the values of A, B, and C in (5), we obtain

$$\frac{x^2 - x - 5}{(x - 1)(x^2 + 2x + 2)} = \frac{2x + 3}{x^2 + 2x + 2} - \frac{1}{x - 1} \qquad \blacktriangleleft$$

Case 4:   The factors of $Q(x)$ are linear and quadratic, and some of the quadratic factors are repeated.

If $ax^2 + bx + c$ is a p-fold quadratic factor of $Q(x)$, then corresponding to this factor $(ax^2 + bx + c)^p$, we have the sum of the following p partial fractions:

$$\frac{A_1 x + B_1}{ax^2 + bx + c} + \frac{A_2 x + B_2}{(ax^2 + bx + c)^2} + \cdots + \frac{A_p x + B_p}{(ax^2 + bx + c)^p}$$

▷ **ILLUSTRATION 2**

If the denominator contains the factor $(x^2 - 5x + 2)^3$, we have, corresponding to this factor, the sum of partial fractions

$$\frac{Ax + B}{x^2 - 5x + 2} + \frac{Cx + D}{(x^2 - 5x + 2)^2} + \frac{Ex + F}{(x^2 - 5x + 2)^3} \qquad \blacktriangleleft$$

▶ **EXAMPLE 4** *Decomposing a Fraction into Partial Fractions when the Factors of the Denominator Are Linear and Quadratic and the Quadratic Factor Is Repeated*

Decompose the fraction

$$\frac{3x^4 - 12x^3 + 4x^2 + 11x + 4}{x(x^2 - 3x - 2)^2}$$

into partial fractions.

Solution   The given fraction is written as a sum of partial fractions as follows:

$$\frac{3x^4 - 12x^3 + 4x^2 + 11x + 4}{x(x^2 - 3x - 2)^2} = \frac{Ax + B}{x^2 - 3x - 2} + \frac{Cx + D}{(x^2 - 3x - 2)^2} + \frac{E}{x} \qquad (7)$$

We multiply on both sides by the LCD and get

$$3x^4 - 12x^3 + 4x^2 + 11x + 4 = x(Ax + B)(x^2 - 3x - 2) + x(Cx + D) + E(x^2 - 3x - 2)^2 \qquad (8)$$

Substituting 0 for x in this equation, we obtain

$$4 = 4E \Leftrightarrow E = 1$$

With $E = 1$ in (8) and multiplying the polynomials, we get

$$3x^4 - 12x^3 + 4x^2 + 11x + 4$$
$$= Ax^4 - 3Ax^3 - 2Ax^2 + Bx^3 - 3Bx^2 - 2Bx + Cx^2 + Dx + x^4 + 9x^2 + 4 - 6x^3 - 4x^2 + 12x$$
$$= (A + 1)x^4 + (-3A + B - 6)x^3 + (-2A - 3B + C + 5)x^2 + (-2B + D + 12)x + 4$$

We equate the coefficients of corresponding powers of x and obtain the system

$$\begin{cases} A + 1 = 3 \\ -3A + B - 6 = -12 \\ -2A - 3B + C + 5 = 4 \\ -2B + D + 12 = 11 \end{cases}$$

We solve this system to obtain

$$A = 2 \quad B = 0 \quad C = 3 \quad D = -1 \quad E = 1$$

Substituting these values in (7), we have

$$\frac{3x^4 - 12x^3 + 4x^2 + 11x + 4}{x(x^2 - 3x - 2)^2} = \frac{2x}{x^2 - 3x - 2} + \frac{3x - 1}{(x^2 - 3x - 2)^2} + \frac{1}{x} \qquad ◀$$

Observe that in each of the examples, the number of constants to be determined is equal to the degree of the denominator of the original fraction being decomposed into partial fractions. This situation is always the case.

EXERCISES 12.3

In Exercises 1 through 10, decompose the fraction into partial fractions.

1. $\dfrac{12}{x^2 - 4}$

2. $\dfrac{1}{2x^2 - x}$

3. $\dfrac{x - 1}{x^2 + x}$

4. $\dfrac{x + 15}{x^2 - 9}$

5. $\dfrac{x + 5}{x^2 - 4x + 3}$

6. $\dfrac{3x}{x^2 + x - 2}$

7. $\dfrac{x + 12}{3x^2 - 5x - 2}$

8. $\dfrac{3x - 7}{4x^2 + 3x - 1}$

9. $\dfrac{3x^2 + 3x - 12}{6x^3 + 5x^2 - 6x}$

10. $\dfrac{2x^2 - 11x - 9}{x^3 - 2x^2 - 3x}$

In Exercises 11 through 14, express the improper fraction as the sum of a polynomial and partial fractions.

11. $\dfrac{2x^3 + 4}{x^2 - 4}$

12. $\dfrac{x^3 + 5}{x^2 - 1}$

13. $\dfrac{4x^3 - 8x^2 - 10x + 30}{2x^2 + x - 6}$

14. $\dfrac{6x^3 + x^2 - 5x - 7}{3x^2 - x - 2}$

In Exercises 15 through 38, decompose the fraction into partial fractions.

15. $\dfrac{3x^2 + 13x - 10}{x^3 - 2x^2}$

16. $\dfrac{x^2 + x + 1}{x^4 - x^3}$

17. $\dfrac{x^2 - 11x + 6}{(x + 2)(x^2 - 4x + 4)}$

18. $\dfrac{x^2 + 11}{(x - 5)(x^2 + 2x + 1)}$

19. $\dfrac{3x + 15}{(2x^2 - x - 1)^2}$

20. $\dfrac{9x^3 - 8x^2 - 4x + 48}{(x^2 - 4)^2}$

21. $\dfrac{x^3 + 6x - 4}{(x - 2)^3}$

22. $\dfrac{x^2 + 2}{(x - 3)^3}$

23. $\dfrac{3x^2 - x + 4}{x^3 + x^2 + x}$

24. $\dfrac{3x + 8}{x^3 + 4x}$

25. $\dfrac{3x^2 + 2x - 4}{x^3 - 8}$

26. $\dfrac{x^2 - 6x + 2}{x^3 + 1}$

27. $\dfrac{2x^2 - 7x + 1}{x^3 - x^2 + x - 1}$

28. $\dfrac{3x^2 - 9x + 8}{x^3 + x^2 + 3x + 3}$

29. $\dfrac{11x^2 + 11x + 8}{2x^3 + 8x^2 + 3x + 12}$

30. $\dfrac{3x^2 + 2x + 3}{x^4 + x^3 + x^2 + x}$

31. $\dfrac{3x^2 - 4x}{(x^2 + 1)(x^2 - x - 1)}$

32. $\dfrac{4x - 3}{x^4 + 2x^3 + 3x^2}$

33. $\dfrac{x + 6}{x^4 + 2x^3 + 3x^2}$

34. $\dfrac{3x^4 + 4}{x^4 + 4x^2 + 4}$

35. $\dfrac{x^3 - x^2}{x^4 + 2x^2 + 1}$

36. $\dfrac{x^4 + 2x^2 - 2x - 4}{(x^2 + 3)^3}$

37. $\dfrac{x^4 + x^3 - 5x^2 - 14x - 1}{x^5 - x^4 + 4x^3 - 4x^2 + 4x - 4}$

38. $\dfrac{11x - 28}{x^5 + 2x^4 + 2x^3 + 4x^2 + x + 2}$

12.4 SEQUENCES, SERIES, AND SIGMA NOTATION

GOALS

1. Define a finite-sequence function.
2. Define an infinite-sequence function.
3. Write elements of a sequence when the general element is given.
4. Learn sigma notation.
5. Write a sum with sigma notation.
6. Write a series with sigma notation.

Sequences of numbers are often encountered in mathematics. For instance, the numbers

$$2, 4, 6, 8, 10$$

form a sequence. This sequence is called *finite* because there is a last number. If a set of numbers forming a sequence does not have a last number, the sequence is said to be *infinite*. For example,

$$1, \tfrac{1}{2}, \tfrac{1}{3}, \tfrac{1}{4}, \tfrac{1}{5}, \ldots$$

is an infinite sequence because the three dots with no number following indicate that there is no last number.

Before defining a sequence, we define a **sequence function** as a function whose domain is a subset of positive integers. A **finite-sequence function** is one whose domain is the set $\{1, 2, 3, \ldots, n\}$ of the first n positive integers. An **infinite-sequence function** is one whose domain is the set of all positive integers. The numbers in the range of a sequence function are called **elements**. A **sequence** consists of the elements of a sequence function listed in order.

$\triangleright$ **ILLUSTRATION 1**

Let f be the function defined by

$$f(n) = 2n \qquad n \in \{1, 2, 3, 4, 5\}$$

Then f is a finite-sequence function, and

$$f(1) = 2 \quad f(2) = 4 \quad f(3) = 6 \quad f(4) = 8 \quad f(5) = 10$$

The elements of the sequence defined by f are then 2, 4, 6, 8, and 10, and the sequence is

$$2, 4, 6, 8, 10$$

The ordered pairs in f are $(1, 2)$, $(2, 4)$, $(3, 6)$, $(4, 8)$, and $(5, 10)$. The graph of f, shown in Figure 1, consists of the five points whose coordinates are the ordered pairs in f. ◄

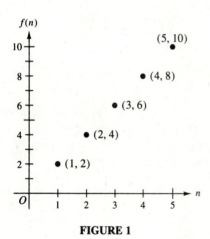

FIGURE 1

$\triangleright$ **ILLUSTRATION 2**

Let h be the function defined by

$$h(n) = \frac{2n - 1}{n^2} \qquad n \in \{1, 2, 3, \ldots\}$$

The function h is an infinite-sequence function. The elements of h are

$$h(1) = \tfrac{1}{1} \quad h(2) = \tfrac{3}{4} \quad h(3) = \tfrac{5}{9} \quad h(4) = \tfrac{7}{16}$$

and so on. The sequence defined by h is therefore

$$1, \tfrac{3}{4}, \tfrac{5}{9}, \tfrac{7}{16}, \ldots$$

Some of the ordered pairs in h are $(1, 1)$, $(2, \tfrac{3}{4})$, $(3, \tfrac{5}{9})$, $(4, \tfrac{7}{16})$, and $(5, \tfrac{9}{25})$.   ◀

We denote the first element of a sequence by a_1, the second element by a_2, the third element by a_3, and so on. The nth element is a_n, called the **general element** of the sequence.

For the sequence of Illustration 1, $a_n = f(n)$; that is, $a_n = 2n$. Therefore, $a_1 = 2$, $a_2 = 4$, $a_3 = 6$, $a_4 = 8$, and $a_5 = 10$. For the sequence of Illustration 2, $a_n = h(n)$; that is,

$$a_n = \frac{2n - 1}{n^2}$$

Hence $a_1 = \tfrac{1}{1}$, $a_2 = \tfrac{3}{4}$, $a_3 = \tfrac{5}{9}$, and so on. Sometimes we state the general element of a sequence when we list the elements in order. Thus for the elements of the sequence of Illustration 2, we would write

$$\frac{1}{1}, \frac{3}{4}, \frac{5}{9}, \frac{7}{16}, \ldots, \frac{2n - 1}{n^2}, \ldots$$

A sequence

$$a_1, a_2, a_3, \ldots, a_n, \ldots$$

is said to be **equal** to a sequence

$$b_1, b_2, b_3, \ldots, b_n, \ldots$$

if and only if $a_i = b_i$ for every positive integer i. Remember that a sequence consists of an ordering of the elements of a sequence function. It is, therefore, possible for two sequences to have the same elements and be unequal. This situation occurs in the following illustration.

▷ **ILLUSTRATION 3**

The sequence for which $a_n = \dfrac{1}{n}$, has as its elements the reciprocals of the positive integers:

$$1, \frac{1}{2}, \frac{1}{3}, \frac{1}{4}, \ldots, \frac{1}{n}, \ldots \tag{1}$$

The sequence for which

$$a_n = \begin{cases} 1 & \text{if } n \text{ is odd} \\ \dfrac{2}{n + 2} & \text{if } n \text{ is even} \end{cases}$$

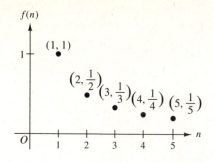

FIGURE 2

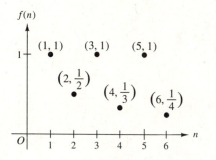

FIGURE 3

has as its elements

$$a_1 = 1 \quad a_2 = \tfrac{1}{2} \quad a_3 = 1 \quad a_4 = \tfrac{1}{3} \quad a_5 = 1 \quad a_6 = \tfrac{1}{4}$$

and so on. The sequence is

$$1, \tfrac{1}{2}, 1, \tfrac{1}{3}, 1, \tfrac{1}{4}, \ldots \qquad (2)$$

The elements of the sequences

$$1, \frac{1}{2}, \frac{1}{3}, \frac{1}{4}, \ldots, \frac{1}{n}, \ldots \quad \text{and} \quad 1, \frac{1}{2}, 1, \frac{1}{3}, 1, \frac{1}{4}, \ldots$$

are the same, but the sequences are not equal. The graphs of the sequence functions for sequences (1) and (2) appear in Figures 2 and 3, respectively. ◄

You should realize that several elements of a sequence do not determine a unique general element, as shown in the following illustration.

▷ **ILLUSTRATION 4**

(a) The sequence for which $a_n = 2n$ is

$$2, 4, 6, 8, 10, 12, \ldots, 2n, \ldots$$

(b) The sequence for which

$$a_n = \begin{cases} 2n & \text{if } n \text{ is odd} \\ 2a_{n-1} & \text{if } n \text{ is even} \end{cases}$$

is

$$2, 4, 6, 12, 10, 20, \ldots, a_n, \ldots$$

(c) The sequence for which $a_n = 2n + (n-1)(n-2)(n-3)$ is

$$2, 4, 6, 14, 34, 72, \ldots, 2n + (n-1)(n-2)(n-3), \ldots \quad ◄$$

Observe that all three sequences in Illustration 4 have 2, 4, and 6 as their first three elements, but each sequence, and thus its general element, is different. To determine a sequence uniquely, we must have a method for obtaining any element. One way is to find an equation that defines the general element, but that is not always possible. For instance, the sequence of prime numbers can be written as

$$2, 3, 5, 7, 11, 13, 17, 19, \ldots, a_n, \ldots$$

where a_n is the nth prime number. We cannot write an equation that defines a_n. However, a_n can be determined (theoretically) for every positive integer n.

▶ **EXAMPLE 1** *Writing Elements of a Sequence when the General Element is Given*

Write the first five elements of the sequence for each of the following general elements.

(a) $a_n = \dfrac{n + 2}{n(n + 1)}$ **(b)** $a_n = (-1)^n \dfrac{1}{3^{n-1}}$ **(c)** $a_n = (-1)^{n-1} x^{2n+1}$

Solution

(a) $a_1 = \dfrac{3}{1 \cdot 2}$ $a_2 = \dfrac{4}{2 \cdot 3}$ $a_3 = \dfrac{5}{3 \cdot 4}$ $a_4 = \dfrac{6}{4 \cdot 5}$ $a_5 = \dfrac{7}{5 \cdot 6}$

$\quad\quad = \dfrac{3}{2}$ $\quad = \dfrac{2}{3}$ $\quad = \dfrac{5}{12}$ $\quad = \dfrac{3}{10}$ $\quad = \dfrac{7}{30}$

Hence the first five elements are

$$\dfrac{3}{2}, \dfrac{2}{3}, \dfrac{5}{12}, \dfrac{3}{10}, \dfrac{7}{30}$$

(b) $a_1 = (-1)^1 \dfrac{1}{3^0}$ $a_2 = (-1)^2 \dfrac{1}{3^1}$ $a_3 = (-1)^3 \dfrac{1}{3^2}$

$a_4 = (-1)^4 \dfrac{1}{3^3}$ $a_5 = (-1)^5 \dfrac{1}{3^4}$

Therefore the first five elements are

$$-1, \dfrac{1}{3}, -\dfrac{1}{9}, \dfrac{1}{27}, -\dfrac{1}{81}$$

(c) $a_1 = (-1)^0 x^3$; $a_2 = (-1)^1 x^5$; $a_3 = (-1)^2 x^7$; $a_4 = (-1)^3 x^9$; $a_5 = (-1)^4 x^{11}$.

Thus the first five elements are

$$x^3, -x^5, x^7, -x^9, x^{11}$$ ◀

▶ **EXAMPLE 2** *Writing Elements of a Sequence when the General Element Is Given*

Write the first twelve elements of the sequence for which

$$a_n = \begin{cases} n & \text{if } n \text{ is odd} \\ n & \text{if } n \text{ is even and not exactly divisible by 4} \\ \frac{1}{2}(n + a_{n-2}) & \text{if } n \text{ is even and exactly divisible by 4} \end{cases}$$

Solution $a_1 = 1$; $a_2 = 2$; $a_3 = 3$; $a_4 = \frac{1}{2}(4 + 2)$; $a_5 = 5$; $a_6 = 6$; $a_7 = 7$; $a_8 = \frac{1}{2}(8 + 6)$; $a_9 = 9$; $a_{10} = 10$; $a_{11} = 11$; $a_{12} = \frac{1}{2}(12 + 10)$. Therefore the first twelve elements are

$$1, 2, 3, 3, 5, 6, 7, 7, 9, 10, 11, 11 \qquad \blacktriangleleft$$

We introduce the **sigma notation** to facilitate writing the sum of the elements of a sequence. This notation involves the use of the symbol Σ, the capital sigma of the Greek alphabet, which corresponds to the letter S. The symbol i, called the **index of summation,** is also involved. Prior to the formal definition of the sigma notation, we give some examples of it in the following illustration.

▷ **ILLUSTRATION 5**

(a) $\displaystyle\sum_{i=1}^{5} i^2 = 1^2 + 2^2 + 3^2 + 4^2 + 5^2$

Observe that $i = 1$ appears under the sigma symbol. This indicates that the first term on the right-hand side is the value of i^2 when $i = 1$. The next four terms are the values of i^2 when i is 2, 3, 4, and 5. We stop there because the number 5 appears above the sigma symbol.

(b) $\displaystyle\sum_{i=-2}^{3} (4i + 1) = [4(-2) + 1] + [4(-1) + 1] + [4(0) + 1]$
$$+ [4(1) + 1] + [4(2) + 1] + [4(3) + 1]$$
$$= (-7) + (-3) + 1 + 5 + 9 + 13$$

Observe that $i = -2$ appears under the sigma symbol and 3 appears above it. Therefore the first term on the right-hand side is the value of $4i + 1$ when $i = -2$, and the remaining terms are the values of $4i + 1$ when i is -1, 0, 1, 2, and 3.

(c) $\displaystyle\sum_{k=3}^{10} \frac{1}{k} = \frac{1}{3} + \frac{1}{4} + \frac{1}{5} + \frac{1}{6} + \frac{1}{7} + \frac{1}{8} + \frac{1}{9} + \frac{1}{10}$

(d) $\displaystyle\sum_{j=1}^{n} j^3 = 1^3 + 2^3 + 3^3 + \ldots + n^3 \qquad \blacktriangleleft$

The sigma notation can be defined by the equation

$$\sum_{i=m}^{n} F(i) = F(m) + F(m + 1) + F(m + 2) + \ldots + F(n)$$

where m and n are integers and $m \le n$. The right-hand side of this equation consists of the sum of $n - m + 1$ terms, the first of which is obtained by replacing i by m in $F(i)$, the second by replacing i by $m + 1$ in $F(i)$, and so on, until the last term is obtained by replacing i by n in $F(i)$. The number m is called the **lower limit** of the sum and n is called the **upper limit.** The symbol i is a "dummy" symbol because any other symbol could be used without changing the right-hand side. For example,

$$\sum_{i=4}^{6} i^3 = 4^3 + 5^3 + 6^3$$

can be written also as

$$\sum_{j=4}^{6} j^3 = 4^3 + 5^3 + 6^3$$

Sometimes the terms of a sum involve subscripts. For instance, the sum

$$a_1 + a_2 + a_3 + \ldots + a_n$$

can be written with sigma notation as

$$\sum_{i=1}^{n} a_i$$

▶ **EXAMPLE 3** *Writing a Sum with Sigma Notation*

Write the following sums with sigma notation:

(a) $2 + 4 + 6 + 8$ (b) $1 + 3 + 5 + 7 + 9$

(c) $-3a_3 + 4a_4 - 5a_5 + 6a_6 - 7a_7 + 8a_8$

(d) $x^2 - x^4 + x^6 - x^8 + x^{10} - x^{12} + x^{14}$

Solution

(a) The numbers 2, 4, 6, and 8 are the first four positive even integers, and they can be written as $2i$ where i is 1, 2, 3, and 4. Therefore

$$2 + 4 + 6 + 8 = \sum_{i=1}^{4} 2i$$

(b) Observe that 1, 3, 5, 7, and 9 are the first five positive odd integers, and they can be written as $2i - 1$, where i is 1, 2, 3, 4, and 5. Thus

$$1 + 3 + 5 + 7 + 9 = \sum_{i=1}^{5} (2i - 1)$$

(c) Notice in the given summation that the odd-numbered terms are preceded by a minus sign and the even-numbered terms are preceded by a plus sign. So that the odd-numbered terms contain an odd power of -1 and the even-numbered terms contain an even power of -1, we write the factor $(-1)^i$ in the sigma notation. Thus we have

$$-3a_3 + 4a_4 - 5a_5 + 6a_6 - 7a_7 + 8a_8 = \sum_{i=3}^{8} (-1)^i \, i a_i$$

(d) In the given summation the odd-numbered terms are preceded by a plus sign and the even-numbered terms are preceded by a minus sign. Therefore, the odd-numbered terms require an even power of -1 and the even-numbered terms require an odd power of -1; so we write the factor $(-1)^{i-1}$ in the sigma notation. We have

$$x^2 - x^4 + x^6 - x^8 + x^{10} - x^{12} + x^{14} = \sum_{i=1}^{7} (-1)^{i-1} x^{2i} \qquad \blacktriangleleft$$

Observe that we can write a sum an unlimited number of ways with sigma notation. For instance, in Example 3 the sum in part (b) can also be written as

$$\sum_{i=2}^{6} (2i - 3) \qquad \sum_{i=0}^{4} (2i + 1) \qquad \sum_{i=3}^{7} (2i - 5)$$

and so on.

The sum of the elements of a sequence is a **series.** For example, associated with the sequence 3, 6, 9, 12, 15, 18 is the series

$$3 + 6 + 9 + 12 + 15 + 18 = \sum_{i=1}^{6} 3i$$

Associated with the sequence

$$a_1, a_2, a_3, \ldots, a_n$$

is the series

$$a_1 + a_2 + a_3 + \ldots + a_n = \sum_{i=1}^{n} a_i$$

The terms in a series are the same as the corresponding elements in the associated sequence. The **general term** of a series is the general element of the associated sequence.

▶ **EXAMPLE 4**   *Writing a Series with Sigma Notation*

Write the following series with sigma notation:

(a) $1 + \frac{1}{2} + \frac{1}{3} + \frac{1}{4} + \frac{1}{5} + \frac{1}{6}$ **(b)** $-1 + 5 - 9 + 13 - 17$

(c) $\frac{1}{3}x^3 - \frac{1}{9}x^5 + \frac{1}{27}x^7 - \frac{1}{81}x^9$

Solution

(a) The first term can also be written as $\frac{1}{1}$. Thus each term is a fraction whose numerator is 1 and whose denominator is the number of the term. Therefore

$$1 + \frac{1}{2} + \frac{1}{3} + \frac{1}{4} + \frac{1}{5} + \frac{1}{6} = \sum_{i=1}^{6} \frac{1}{i}$$

(b) Except for the first term, each of the numbers 1, 5, 9, 13, and 17 is four more than the preceding one. This suggests $4i$ in the sigma notation. Because the first term is 1, we need to write $4i - 3$ to obtain 1 when $i = 1$. Then observe that $4i - 3 = 5$ when $i = 2$, $4i - 3 = 9$ when $i = 3$, and so on. Because the odd-numbered terms in the given series are preceded by a minus sign and the even-numbered terms are preceded by a plus sign, we need a factor of $(-1)^i$. Thus we have

$$-1 + 5 - 9 + 13 - 17 = \sum_{i=1}^{5} (-1)^i (4i - 3)$$

(c) The exponents of x are the odd integers 3, 5, 7, and 9, which can be written as $2i + 1$, where i is 1, 2, 3, and 4. Thus in the sigma notation we must have a factor of x^{2i+1}. Because the numerical coefficients are the reciprocals of successive powers of 3, there will also be a factor of $\frac{1}{3^i}$. Furthermore, because the odd-numbered terms are preceded by a plus sign and the even-numbered terms are preceded by a minus sign, we need a factor of $(-1)^{i-1}$. Hence we have

$$\frac{1}{3}x^3 - \frac{1}{9}x^5 + \frac{1}{27}x^7 - \frac{1}{81}x^9 = \sum_{i=1}^{4} (-1)^{i-1} \frac{1}{3^i} x^{2i+1} \qquad \blacktriangleleft$$

EXERCISES 12.4

In Exercises 1 through 16, write the first eight elements of the sequence whose general element is given.

1. $a_n = 2n + 3$

2. $a_n = \dfrac{3n - 1}{2}$

3. $a_n = \dfrac{n^2 + 1}{n}$

4. $a_n = \dfrac{1}{n^2 + 2}$

5. $a_n = (-1)^{n-1} \dfrac{n + 1}{2n - 1}$

6. $a_n = (-1)^{n+1} \dfrac{n}{2^n}$

7. $a_n = (-1)^n \dfrac{2^n}{1 + 2^n}$

8. $a_n = \dfrac{(-1)^{n-1}}{n(n + 1)}$

9. $a_n = \dfrac{(-1)^{n+1}}{n + 2} x^n$

10. $a_n = \dfrac{(-1)^n}{n^2} x^{2n - 1}$

11. $a_n = n + (-1)^n n$

12. $a_n = \dfrac{1}{2n} - \dfrac{1}{3n}$

13. $a_n = \begin{cases} \dfrac{2}{n + 1} & \text{if } n \text{ is odd} \\ 2 & \text{if } n \text{ is even} \end{cases}$

14. $a_n = \begin{cases} 1 & \text{if } n \text{ is odd} \\ \dfrac{4}{(n + 2)^2} & \text{if } n \text{ is even} \end{cases}$

15. $a_n = \begin{cases} \dfrac{n + 1}{2} & \text{if } n \text{ is odd} \\ a_{n-1} & \text{if } n \text{ is even} \end{cases}$

16. $a_n = \begin{cases} n & \text{if } n \text{ is odd} \\ \frac{1}{2}n & \text{if } n \text{ is even and} \\ & \text{not exactly divisible by 4} \\ \frac{1}{2}(a_{n-2} + a_{n-1}) & \text{if } n \text{ is even and} \\ & \text{exactly divisible by 4} \end{cases}$

In Exercises 17 through 30, write the series. In Exercises 17 through 24, find the sum of the series.

17. $\displaystyle\sum_{i=1}^{5} (4i - 3)$

18. $\displaystyle\sum_{i=1}^{7} (i + 1)^2$

19. $\displaystyle\sum_{j=2}^{6} \frac{j}{j - 1}$

20. $\displaystyle\sum_{k=1}^{4} \frac{(-1)^{k+1}}{k}$

21. $\displaystyle\sum_{i=1}^{100} 5$

22. $\displaystyle\sum_{i=1}^{8} \frac{3i - 6}{2}$

23. $\displaystyle\sum_{k=0}^{5} \frac{1}{2^k}$

24. $\displaystyle\sum_{j=0}^{3} \frac{(-1)^j}{2^j + 1}$

25. $\displaystyle\sum_{i=1}^{8} (-1)^{i-1} x^{2i-1}$

26. $\displaystyle\sum_{i=1}^{5} 2^i x^{3i}$

27. $\displaystyle\sum_{i=0}^{7} (-1)^i (i + 1) a_i$

28. $\displaystyle\sum_{i=0}^{6} a_i x^{2i+1}$

29. $\displaystyle\sum_{i=1}^{n} f(x_{i-1})$

30. $\displaystyle\sum_{i=0}^{n} f(x_{i+1}) h$

In Exercises 31 through 38, write the series with sigma notation (there is no unique solution).

31. $1 + 3 + 5 + 7 + 9 + 11$

32. $2 + 4 + 6 + 8 + 10$

33. $4 - 7 + 10 - 13 + 16$

34. $1 + \frac{1}{3} + \frac{1}{9} + \frac{1}{27} + \frac{1}{81} + \frac{1}{243}$

35. $1 + \frac{3}{4} + \frac{5}{9} + \frac{7}{16} + \frac{9}{25} + \frac{11}{36}$

36. $1 - \frac{1}{4} + \frac{1}{16} - \frac{1}{64}$

37. $\frac{1}{2} - \frac{x^2}{4} + \frac{x^4}{6} - \frac{x^6}{8} + \frac{x^8}{10}$

38. $x - \frac{1}{2}x^3 + \frac{1}{3}x^5 - \frac{1}{4}x^7 + \frac{1}{5}x^9$

39. Write the general element and the first six elements of three different sequences each having as the first three elements 1, 3, and 5.

40. Write the general element of a sequence whose first three elements are 2, 4, and 6 and whose fourth element is x, where x can be any real number.

41. Explain why two sequences having the same elements are not necessarily equal.

42. Explain the difference between a sequence and a series.

12.5 MATHEMATICAL INDUCTION

GOALS
1. State the principle of mathematical induction.
2. Prove a formula by mathematical induction.
3. Prove an inequality by mathematical induction.
4. Prove a statement by mathematical induction.
5. Prove a theorem by mathematical induction.

The Italian mathematician Francesco Maurolycus (1494–1575) is often credited with the introduction of **mathematical induction** as a method of proof. He used it in his *Arithmetica* of 1575 to prove that the sum of the first n positive odd integers is n^2, that is,

$$1 + 3 + 5 + \ldots + (2n - 1) = n^2$$

$$\Leftrightarrow \qquad \sum_{i=1}^{n} (2i - 1) = n^2$$

Let us verify this formula for some values of n.

$$n = 1: \qquad\qquad\qquad\qquad 1 = 1 \Leftrightarrow 1 = 1^2$$
$$n = 2: \qquad\qquad\qquad 1 + 3 = 4 \Leftrightarrow 1 + 3 = 2^2$$
$$n = 3: \qquad\qquad 1 + 3 + 5 = 9 \Leftrightarrow 1 + 3 + 5 = 3^2$$
$$n = 4: \qquad 1 + 3 + 5 + 7 = 16 \Leftrightarrow 1 + 3 + 5 + 7 = 4^2$$
$$n = 5: \quad 1 + 3 + 5 + 7 + 9 = 25 \Leftrightarrow 1 + 3 + 5 + 7 + 9 = 5^2$$

From these calculations we are convinced that the formula holds for values of n from 1 to 5. We could continue with more values of n and we would observe that the formula is still true. Of course, such a procedure is *not a proof* of the formula for all positive-integer values of n. The technique of mathematical induction provides us with a proof. We now state the *principle of mathematical induction* that provides the basis for this method of proof, which is presented in Illustration 1.

Principle of Mathematical Induction

A statement P_n, involving the positive integer n, is true for all positive integer values of n if the following two conditions are satisfied:

(i) P_1 is true; that is, the statement is true for $n = 1$.
(ii) If k is an arbitrary positive integer for which P_k is true, then P_{k+1} is also true; that is, whenever the statement is true for $n = k$, it is also true for $n = k + 1$, where k is an arbitrary positive integer.

We do not prove the principle of mathematical induction. Its use involves the following reasoning:

From (i) the statement P_n is true for $n = 1$. Because P_n is true for $n = 1$, it follows from (ii) that P_n is true for $n = 1 + 1$, or 2. Because P_n is true for $n = 2$, it is true for $n = 2 + 1$, or 3. Because P_n is true for $n = 3$, it is true for $n = 3 + 1$, or 4. And so on. Thus P_n is true for all positive-integer values of n.

A proof by mathematical induction consists of two parts verifying the two conditions of the principle of mathematical induction. We complete the proof by writing a conclusion.

▷ ILLUSTRATION 1

We shall use mathematical induction to prove that

$$1 + 3 + 5 + 7 + \ldots + (2n - 1) = n^2$$

for all positive-integer values of n.

Part 1:   We first verify that the formula is true for $n = 1$. If $n = 1$, the formula becomes

$$1 = 1^2$$

which is true.

Part 2:   We now show that if the formula is true for $n = k$, it is also true for $n = k + 1$, where k is an arbitrary positive integer. That is, we assume

$$1 + 3 + 5 + \ldots + (2k - 1) = k^2 \tag{1}$$

We wish to prove that if Equation (1) is true, then

$$1 + 3 + 5 + \ldots + (2k - 1) + [2(k + 1) - 1] = (k + 1)^2 \tag{2}$$

is also true. The last term on the left side of Equation (2) can be written as $2k + 1$. We add $2(k + 1) - 1$ to the left side of Equation (1) and its equivalent $2k + 1$ to the right side, and we obtain

$$1 + 3 + 5 + \ldots + (2k - 1) + [2(k + 1) - 1] = k^2 + (2k + 1)$$
$$1 + 3 + 5 + \ldots + (2k - 1) + [2(k + 1) - 1] = (k + 1)^2$$

which is Equation (2).

Conclusion: In Part 1 we proved that the formula is true for $n = 1$. In Part 2 we proved that when the formula is true for $n = k$, it is also true for $n = k + 1$. Therefore, by the principle of mathematical induction, the formula holds for all positive-integer values of n.   ◀

In the following example, we prove that the sum of the first n positive even integers is given by the formula

$$2 + 4 + 6 + \ldots + 2n = n(n + 1)$$

▶   **EXAMPLE 1** *Proving a Formula by Mathematical Induction*

Use mathematical induction to prove

$$\sum_{i=1}^{n} 2i = n(n + 1)$$

Solution

Part 1:   First the formula is verified for $n = 1$. When $n = 1$, the left side is

$$\sum_{i=1}^{1} 2i = 2$$

and the right side is

$$1(1 + 1) = 2$$

Therefore the formula is true when $n = 1$.

Part 2: We assume that the formula is true when $n = k$, where k is any positive integer:

$$\sum_{i=1}^{k} 2i = k(k + 1) \tag{3}$$

With this assumption we wish to prove that the formula is also true when $n = k + 1$. Thus we wish to prove

$$\sum_{i=1}^{k+1} 2i = (k + 1)[(k + 1) + 1]$$

$$\Leftrightarrow \quad \sum_{i=1}^{k+1} 2i = (k + 1)(k + 2) \tag{4}$$

When $n = k + 1$, we have

$$\sum_{i=1}^{k+1} 2i = 2 + 4 + 6 + \ldots + 2k + 2(k + 1)$$

$$= \sum_{i=1}^{k} 2i + (2k + 2)$$

$$= k(k + 1) + (2k + 2) \qquad \text{[by applying (3)]}$$

$$= k^2 + k + 2k + 2$$

$$= k^2 + 3k + 2$$

$$= (k + 1)(k + 2)$$

which is (4).

Conclusion: We have proved that the formula is true when $n = 1$; and we have also proved that when the formula is true for $n = k$, it is also true for $n = k + 1$. Therefore, by the principle of mathematical induction, the formula is true when n is any positive integer. ◀

In the next example we prove the formula that gives the sum of the squares of the first n positive integers.

▶ **EXAMPLE 2** *Proving a Formula by Mathematical Induction*

Use mathematical induction to prove

$$\sum_{i=1}^{n} i^2 = \frac{n(n + 1)(2n + 1)}{6}$$

Solution

Part 1: We first verify the formula for $n = 1$. With this value of n the left side is

$$\sum_{i=1}^{1} i^2 = 1^2$$

$$= 1$$

and the right side is

$$\frac{1(1 + 1)(2 + 1)}{6} = \frac{1 \cdot 2 \cdot 3}{6}$$

$$= 1$$

Thus the formula is true when $n = 1$.

Part 2: We assume that the formula is true when $n = k$, where k is any positive integer, or

$$\sum_{i=1}^{k} i^2 = \frac{k(k + 1)(2k + 1)}{6} \tag{5}$$

With this assumption we wish to prove that the formula is also true when $n = k + 1$; that is, we wish to prove

$$\sum_{i=1}^{k+1} i^2 = \frac{(k + 1)[(k + 1) + 1][2(k + 1) + 1]}{6} \tag{6}$$

When $n = k + 1$, we have

$$\sum_{i=1}^{k+1} i^2 = 1^2 + 2^2 + 3^2 + \ldots + k^2 + (k + 1)^2$$

$$= \sum_{i=1}^{k} i^2 + (k + 1)^2$$

$$= \frac{k(k + 1)(2k + 1)}{6} + (k + 1)^2 \quad \text{[by applying (5)]}$$

$$= \frac{k(k + 1)(2k + 1) + 6(k + 1)^2}{6}$$

$$= \frac{(k + 1)[k(2k + 1) + 6(k + 1)]}{6}$$

$$= \frac{(k + 1)(2k^2 + 7k + 6)}{6}$$

$$= \frac{(k + 1)(k + 2)(2k + 3)}{6}$$

$$= \frac{(k + 1)[(k + 1) + 1][2(k + 1) + 1]}{6}$$

which is (6).

Conclusion: We have proved that the formula is true when $n = 1$; and we have also proved that when the formula is true for $n = k$, it is also true for $n = k + 1$. Therefore, by the principle of mathematical induction, the formula is true when n is any positive integer.   ◄

▶   **EXAMPLE 3** *Proving an Inequality by Mathematical Induction*

Use mathematical induction to prove

$$2^n \geq 2n$$

For all positive-integer values of n.

Solution

Part 1:   We verify the inequality for $n = 1$. When $n = 1$, the left side is 2 and the right side is 2. Because $2 \geq 2$, the inequality is true when $n = 1$.

Part 2:   We assume that the inequality is true when $n = k$, where k is any positive integer; that is, we assume

$$2^k \geq 2k \tag{7}$$

With this assumption we wish to prove that the inequality is true when $n = k + 1$; that is, we wish to prove

$$2^{k+1} \geq 2(k + 1) \tag{8}$$

On both sides of (7) we multiply by 2 and obtain

$$2 \cdot 2^k \geq 2 \cdot 2k$$
$$2^{k+1} \geq 2k + 2k \tag{9}$$

Because $k \geq 1$, then $2k \geq 2$. Thus

$$2k + 2k \geq 2k + 2 \tag{10}$$

From (9) and (10)

$$2^{k+1} \geq 2k + 2$$
$$2^{k+1} \geq 2(k + 1)$$

which is (8).

Conclusion: Because the inequality is true when $n = 1$, and we have proved that when it is true for $n = k$, it is also true for $n = k + 1$, it follows from the principle of mathematical induction that the inequality is true when n is any positive integer.   ◄

In Example 3 we proved that the nonstrict inequality

$$2^n \geq 2n$$

is valid for all positive-integer values of n. By a similar procedure, as indicated in the following illustration, we can prove the corresponding strict inequality

$$2^n > 2n \qquad \text{when } n > 2 \tag{11}$$

and n is a positive integer. Because 3 is the smallest value of n for which the inequality is true, Part 1 of the proof by mathematical induction is to verify that P_3 (rather than P_1) is true. In this case we are applying an extension of the principle of mathematical induction.

▷ ILLUSTRATION 2

To prove (11) by mathematical induction, Part 1 is as follows:
   We verify (11) for $n = 3$. When $n = 3$, the left side of the inequality is 8 and the right side is 6. Because $8 > 6$, inequality (11) is true when $n = 3$.
   Part 2 of the proof of (11) is the same as that in Part 2 in the solution of Example 3 except that the inequality symbol $\geq$ is replaced by $>$.
   Then the conclusion states that because the inequality is true when $n = 3$, and in Part 2 we proved that when it is true for $n = k$, it is also true for $n = k + 1$, it follows from the principle of mathematical induction that inequality (11) is true when n is a positive integer and $n > 2$.   ◀

▶ EXAMPLE 4 *Proving a Statement by Mathematical Induction*

Prove the following statement by mathematical induction: $x - y$ is a factor of $x^n - y^n$ for all positive-integer values of n.

Solution

Part 1:   When $n = 1$, $x^n - y^n$ becomes $x - y$, which certainly has $x - y$ as a factor.

Part 2:   We assume that $x - y$ is a factor of $x^k - y^k$, where k is any positive integer; and with this assumption we wish to prove that $x - y$ is also a factor of $x^{k+1} - y^{k+1}$. If we subtract and add xy^k to $x^{k+1} - y^{k+1}$, we obtain

$$x^{k+1} - y^{k+1} = x^{k+1} - xy^k + xy^k - y^{k+1}$$
$$\Leftrightarrow \qquad x^{k+1} - y^{k+1} = x(x^k - y^k) + y^k(x - y) \tag{12}$$

We have assumed that $x - y$ is a factor of $x^k - y^k$; furthermore, $x - y$ is a factor of $y^k(x - y)$. Hence $x - y$ is a factor of each of the two terms in the right member of (12). Therefore $x - y$ is a factor of $x^{k+1} - y^{k+1}$.

Conclusion: We have shown that the statement is true when $n = 1$ and we have proved that when the statement is true for $n = k$, it is also true for $n = k + 1$. Therefore, by the principle of mathematical induction, the statement is true for all positive-integer values of n. ◀

Certain theorems involving laws of exponents can be proved by mathematical induction. In Example 5 we give such a proof, which utilizes the following definition of positive-integer exponents:

Let a be any real number. Then

$$a^1 = a \tag{13}$$

If k is any positive integer such that a^k is defined, let

$$a^{k+1} = a^k \cdot a \tag{14}$$

▶ **EXAMPLE 5** *Proving a Theorem by Mathematical Induction*

Prove that if m and n are positive integers and a is a real number, then

$$a^m \cdot a^n = a^{m+n} \tag{15}$$

Solution Let m be an arbitrary positive integer. We wish to prove that (15) is true for all positive-integer values of n.

Part 1: We verify that (15) is true when $n = 1$. From (13)

$$a^m \cdot a^1 = a^m \cdot a$$

Applying (14) on the right side of this equation, we have

$$a^m \cdot a^1 = a^{m+1}$$

Hence (15) is true when $n = 1$.

Part 2: We assume that (15) is true when $n = k$ where k is any positive integer:

$$a^m \cdot a^k = a^{m+k} \tag{16}$$

With this assumption we wish to prove that

$$a^m \cdot a^{k+1} = a^{m+(k+1)}$$

To prove this, we start with the left side and replace a^{k+1} by $a^k \cdot a$, which follows from (14). Thus

$$
\begin{aligned}
a^m \cdot a^{k+1} &= a^m \cdot (a^k \cdot a) \\
&= (a^m \cdot a^k) \cdot a \quad \text{(from the associative law for multiplication)} \\
&= a^{m+k} \cdot a \quad \text{[from (16)]} \\
&= a^{(m+k)+1} \quad \text{[from (14)]} \\
&= a^{m+(k+1)} \quad \text{(from the associative law for addition)}
\end{aligned}
$$

which is what we wished to prove.

Conclusion: From Part 1 we know that (15) is true when $n = 1$. From Part 2 we know that when (15) is true for $n = k$, it is also true for $n = k + 1$, where k is any positive integer. Therefore, by the principle of mathematical induction, (15) is true when n is any positive integer.   ◀

As promised in Section 10.6, we now prove Theorem 1 of that section by mathematical induction.

▶ **EXAMPLE 6** *Proving a Theorem by Mathematical Induction*

Prove by mathematical induction De Moivre's theorem for positive integers: If n is any positive integer,

$$[r(\cos \theta + i \sin \theta)]^n = r^n(\cos n\theta + i \sin n\theta)$$

Solution

Part 1:   When $n = 1$ both the left side and right side become $r(\cos \theta + i \sin \theta)$; so the theorem is valid for $n = 1$.

Part 2:   We assume the theorem is true for $n = k$, where k is a positive integer; that is, we assume

$$[r(\cos \theta + i \sin \theta)]^k = r^k(\cos k\theta + i \sin k\theta) \tag{17}$$

We wish to show that the theorem is true for $n = k + 1$; that is, we wish to prove

$$[r(\cos \theta + i \sin \theta)]^{k+1} = r^{k+1}[\cos(k + 1)\theta + i \sin(k + 1)\theta] \tag{18}$$

We have

$$[r(\cos \theta + i \sin \theta)]^{k+1} = [r(\cos \theta + i \sin \theta)][r(\cos \theta + i \sin \theta)]^k$$

Substituting from (17) in the right side of this equation, we obtain

$$[r(\cos \theta + i \sin \theta)]^{k+1} = [r(\cos \theta + i \sin \theta)][r^k(\cos k\theta + i \sin k\theta)]$$

In the right side of this equation, we apply Theorem 1(i) of Section 10.5, and we get

$$[r(\cos \theta + i \sin \theta)]^{k+1} = r^{k+1}[\cos(\theta + k\theta) + i \sin(\theta + k\theta)]$$
$$[r(\cos \theta + i \sin \theta)]^{k+1} = r^{k+1}[\cos(k + 1)\theta + i \sin(k + 1)\theta]$$

which is (18).

Conclusion: From Part 1 we know that the theorem is true when $n = 1$. From Part 2, we know that when the theorem is true for $n = k$, it is also true for $n = k + 1$, where k is any positive integer. Therefore, by the principle of mathematical induction, the theorem is true when n is any positive integer.

◀

EXERCISES 12.5

In Exercises 1 through 33, prove the formula, inequality, statement, or theorem by mathematical induction. Be sure to complete your proof by writing a conclusion.

In Exercises 1 through 14, prove that the formula is true for all positive-integer values of n.

1. $\displaystyle\sum_{i=1}^{n} i = \frac{n(n + 1)}{2}$ 2. $\displaystyle\sum_{i=1}^{n} 4i = 2n(n + 1)$

3. $\displaystyle\sum_{i=1}^{n} (3i - 2) = \frac{n(3n - 1)}{2}$

4. $\displaystyle\sum_{i=1}^{n} (3i - 1) = \frac{n(3n + 1)}{2}$

5. $\displaystyle\sum_{i=1}^{n} \frac{i(i + 1)}{2} = \frac{n(n + 1)(n + 2)}{6}$

6. $\displaystyle\sum_{i=1}^{n} (2i - 1)^2 = \frac{n(2n - 1)(2n + 1)}{3}$

7. $\displaystyle\sum_{i=1}^{n} 2^i = 2(2^n - 1)$

8. $\displaystyle\sum_{i=1}^{n} 3^i = \tfrac{3}{2}(3^n - 1)$

9. $\displaystyle\sum_{i=1}^{n} i^3 = \frac{n^2(n + 1)^2}{4}$

10. $\displaystyle\sum_{i=1}^{n} (2i - 1)^3 = n^2(2n^2 - 1)$

11. $\displaystyle\sum_{i=1}^{n} \frac{1}{i(i + 1)} = \frac{n}{n + 1}$

12. $\displaystyle\sum_{i=1}^{n} \frac{1}{(2i - 1)(2i + 1)} = \frac{n}{2n + 1}$

13. $\displaystyle\sum_{i=1}^{n} \frac{1}{(3i - 1)(3i + 2)} = \frac{n}{2(3n + 2)}$

14. $\displaystyle\sum_{i=1}^{n} ar^{i-1} = \frac{a - ar^n}{1 - r}, r \neq 1$

In Exercises 15 through 18, prove that the inequality is true for all positive-integer values of n.

15. $3^n \geq 3n$ 16. $2^n > n$

17. $a^n > 1$ if a is a real number and $a > 1$.

18. $0 < a^n < 1$ if a is a real number and $0 < a < 1$.

In Exercises 19 and 20, prove that the inequality is true for the indicated positive-integer values of n.

19. $2^n > n^2$ if $n > 4$

20. $3^n > 2^n + 10n$ if $n > 3$

In Exercises 21 through 26, prove that the statement is true for all positive-integer values of n.

21. 2 is a factor of $n^2 + n$.

22. 2 is a factor of $n^2 - n + 2$.

23. 6 is a factor of $n^3 + 3n^2 + 2n$.

24. 3 is a factor of $4^n - 1$.

25. $x + y$ is a factor of $x^{2n} - y^{2n}$.

26. $x + y$ is a factor of $x^{2n-1} + y^{2n-1}$.

In Exercises 27 and 28, prove that the formula is true for all positive-integer values of n.

27. $\sin(x + n\pi) = (-1)^n \sin x$

28. $\cos(x + n\pi) = (-1)^n \cos x$

In Exercises 29 through 33, prove the theorem.

29. If m and n are positive integers and a is a real number, then
$$(a^n)^m = a^{nm}$$

30. If n is a positive integer and a and b are real numbers, then
$$(ab)^n = a^n b^n$$

31. If n is a positive integer, a and b are real numbers, and $b \neq 0$, then
$$\left(\frac{a}{b}\right)^n = \frac{a^n}{b^n}$$

32. If P dollars is invested at an annual interest rate of $100i$ percent compounded m times per year, and if A_n dollars is the amount of the investment at the end of n interest periods, then
$$A_n = P\left(1 + \frac{i}{m}\right)^n$$

33. If $n \geq 3$, the sum of the interior angles of an n-sided polygon is $(n - 2)180°$. *Hint:* Choose one vertex and form $n - 2$ triangles by drawing $n - 3$ lines through the selected vertex and each of $n - 3$ other vertices.

34. Explain why the idea behind the principle of mathematical induction can be compared to the following concept: Suppose that infinitely many dominoes are arranged in an unending line and that when one domino falls it knocks down the next one in line; then all of the dominoes would fall if the first domino is pushed to knock down the second domino.

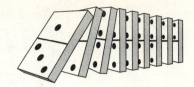

12.6 ARITHMETIC AND GEOMETRIC SEQUENCES AND SERIES

GOALS

1. Define an arithmetic sequence.
2. Learn and apply the formula for the Nth element of an arithmetic sequence.
3. Define and find the arithmetic means between two numbers.
4. Define and find the arithmetic mean of a set of numbers.
5. Learn and apply the formula for the sum of an arithmetic series.
6. Define a geometric sequence.
7. Learn and apply the formula for the Nth element of a geometric sequence.
8. Define and find the geometric mean between two numbers.
9. Define and find the geometric mean of a set of numbers.
10. Learn and apply the formula for the sum of a geometric series.
11. Solve word problems having arithmetic and geometric sequences and series as mathematical models.

Arithmetic and geometric sequences are two particular kinds of sequences that have many applications. An example of an arithmetic sequence is

$$2, 5, 8, 11, 14, 17, 20$$

where each element, except the first, is three more than the one preceding. The sequence

$$1, 2, 4, 8, 16, 32, 64, 128$$

is an example of a geometric sequence. Each element, except the first, can be obtained by multiplying the preceding element by 2.

We first discuss arithmetic sequences.

> **DEFINITION   An Arithmetic Sequence**
>
> An **arithmetic sequence** is a sequence for which any element, except the first, can be obtained by adding a constant to the preceding element.

An arithmetic sequence is sometimes called an **arithmetic progression.**

The constant addend in an arithmetic sequence is called the **common difference** and is denoted by d. We can ascertain if a given sequence is an arithmetic sequence by subtracting each element from the succeeding one.

▷ ILLUSTRATION 1

For the sequence

$$9, 5, 1, -3, -7, -11$$

note that $5 - 9 = -4$; $1 - 5 = -4$; $-3 - 1 = -4$; $-7 - (-3) = -4$; and $-11 - (-7) = -4$. Therefore we have an arithmetic sequence where the common difference d is -4. ◀

In an arithmetic sequence the number of elements is denoted by N, the first element is denoted by a_1, and the last element is denoted by a_N. In the sequence of Illustration 1, $N = 6$, $a_1 = 9$, and $a_6 = -11$.

The definition of an arithmetic sequence can be stated symbolically by giving the value of the first element a_1, the number of elements N, and the formula

$$a_{n+1} = a_n + d$$

from which every element after the first can be obtained from the preceding one. This formula is called a **recursive formula.** Recursive formulas are used in computer programming because the repeated application of a single formula is often involved.

▷ ILLUSTRATION 2

If $a_1 = 4$, $N = 8$, and

$$a_{n+1} = a_n + 3$$

the arithmetic sequence is

$$4, 7, 10, 13, 16, 19, 22, 25$$ ◀

From the recursive formula we can write the general arithmetic sequence, for which the first element is a_1, the common difference is d, and the number of elements is N. We start with the element a_1, and each successive element is obtained from the preceding one by adding d to it. Hence we have

$$a_1, a_1 + d, a_1 + 2d, a_1 + 3d, a_1 + 4d, \ldots, a_N$$

Refer to the first five elements and observe that each element is a_1 plus a multiple of d, where the coefficient of d is one less than the number of the element. Intuitively, it appears that $a_N = a_1 + (N - 1)d$. We state this formally and prove it by mathematical induction.

THEOREM 1

The Nth element of an arithmetic sequence is given by

$$a_N = a_1 + (N - 1)d$$

Proof

Part 1:   We show that the formula is true if $N = 1$ by substituting 1 for N.

$$a_1 = a_1 + (1 - 1)d$$
$$= a_1$$

Part 2:   We now assume that the formula is true if $N = k$, that is,

$$a_k = a_1 + (k - 1)d$$

We wish to show that the formula is true if $N = k + 1$, that is,

$$a_{k+1} = a_1 + [(k + 1) - 1]d$$

By the definition of an arithmetic sequence,

$$a_{k+1} = a_k + d$$

Replacing a_k by $a_1 + (k - 1)d$, we have

$$a_{k+1} = [a_1 + (k - 1)d] + d$$
$$= a_1 + kd - d + d$$
$$= a_1 + [(k + 1) - 1]d$$

which is what we wished to show.

Conclusion: We have proved that the formula is true when $N = 1$, and we have also proved that when the formula is true for $N = k$, it is also true for $N = k + 1$. Therefore, by the principle of mathematical induction, it is valid for all natural numbers.   ■

▶ **EXAMPLE 1**   *Finding a Particular Element of an Arithmetic Sequence*

Find the thirtieth element of the arithmetic sequence for which the first element is 5 and the second element is 9.

Solution   Let a_{30} be the thirtieth element of the arithmetic sequence

$$5, 9, \ldots, a_{30}$$

Then $d = 9 - 5$, or 4, $a_1 = 5$, and $N = 30$. From Theorem 1

$$a_{30} = a_1 + (30 - 1)d$$
$$= 5 + 29 \cdot 4$$
$$= 121$$
◀

DEFINITION   Arithmetic Means

If $a, c_1, c_2, \ldots, c_k, b$ is an arithmetic sequence, then the numbers $c_1, c_2, \ldots, c_k$ are the k **arithmetic means** between a and b.

▷ **ILLUSTRATION 3**

Because

$$2, 5, 8, 11, 14, 17, 20$$

is an arithmetic sequence, it follows that 5, 8, 11, 14, and 17 are the five arithmetic means between 2 and 20.
◀

▶ **EXAMPLE 2**   *Finding Arithmetic Means Between Two Numbers*

Insert three arithmetic means between 11 and 14.

Solution   If c_1, c_2, and c_3 are the three arithmetic means, then

$$11, c_1, c_2, c_3, 14$$

is an arithmetic sequence. With $N = 5$ in Theorem 1

$$a_5 = a_1 + (5 - 1)d$$

Because $a_1 = 11$ and $a_5 = 14$,

$$14 = 11 + 4d$$
$$d = \tfrac{3}{4}$$

Thus

$$c_1 = 11 + \tfrac{3}{4} \qquad c_2 = 11\tfrac{3}{4} + \tfrac{3}{4} \qquad c_3 = 12\tfrac{1}{2} + \tfrac{3}{4}$$
$$= 11\tfrac{3}{4} \qquad\qquad = 12\tfrac{1}{2} \qquad\qquad = 13\tfrac{1}{4}$$

The three arithmetic means are therefore $11\tfrac{3}{4}$, $12\tfrac{1}{2}$, and $13\tfrac{1}{4}$. ◀

▷ **ILLUSTRATION 4**

To insert one arithmetic mean between the numbers x and y, let M be the arithmetic mean, and we have the arithmetic sequence

x, M, y

The common difference can be represented by either $M - x$ or $y - M$. Therefore

$$M - x = y - M$$
$$2M = x + y$$
$$M = \frac{x + y}{2}$$

◀

The number M obtained in Illustration 4 is called the *arithmetic mean* (or *average*) of the numbers x and y. We can generalize this concept and refer to the *arithmetic mean* of a set of numbers.

DEFINITION   Arithmetic Mean

(i) The **arithmetic mean** (or **average**) of the numbers x and y is the number

$$\frac{x + y}{2}$$

(ii) The **arithmetic mean** (or **average**) of a set of numbers x_1, x_2, x_3, . . . , x_n is the number

$$\frac{x_1 + x_2 + x_3 + \ldots + x_n}{n}$$

▶ **EXAMPLE 3** *Finding the Arithmetic Mean of a Set of Numbers*

On five separate examinations a student received the following test scores: 78, 89, 62, 75, and 84. Find the arithmetic mean of these scores.

Solution   If M is the arithmetic mean, we have from the definition

$$M = \frac{78 + 89 + 62 + 75 + 84}{5}$$

$$= \frac{388}{5}$$

$$= 77.6 \qquad \blacktriangleleft$$

An **arithmetic series** is the indicated sum of the elements of an arithmetic sequence.

▷ **ILLUSTRATION 5**

The arithmetic series associated with the arithmetic sequence of Illustration 2 is

$$4 + 7 + 10 + 13 + 16 + 19 + 22 + 25$$

This arithmetic series can be written with the sigma notation as

$$\sum_{i=1}^{8} (3i + 1) \qquad \blacktriangleleft$$

The arithmetic series associated with the general arithmetic sequence is

$$a_1 + (a_1 + d) + (a_1 + 2d) + \ldots + [a_1 + (N - 1)d]$$

If we denote this sum by S_N, we have

$$S_N = a_1 + (a_1 + d) + (a_1 + 2d) + \ldots + [a_1 + (N - 1)d]$$

The series on the right can be written in the reverse order with the Nth term being written as a_N, the $(N - 1)$th term being written as $a_N - d$, and so on, until the first term is written as $a_N - (N - 1)d$. Therefore

$$S_N = a_N + (a_N - d) + (a_N - 2d) + \ldots + [a_N - (N - 1)d]$$

If we add term by term the two equations defining S_N, we obtain

$$S_N + S_N = (a_1 + a_N) + (a_1 + a_N) + (a_1 + a_N) + \ldots + (a_1 + a_N)$$

where on the right side the term $a_1 + a_N$ occurs N times. Hence

$$2S_N = N(a_1 + a_N)$$

$$S_N = \frac{N}{2}(a_1 + a_N)$$

If we substitute into this equation the value of a_N from the formula of Theorem 1, we obtain

$$S_N = \frac{N}{2}(a_1 + [a_1 + (N-1)d])$$

$$S_N = \frac{N}{2}[2a_1 + (N-1)d]$$

We have proved the following theorem.

THEOREM 2

If $a_1, a_2, a_3, \ldots, a_N$ is an arithmetic sequence with common difference d, and

$$S_N = a_1 + a_2 + a_3 + \ldots + a_N$$

then

$$S_N = \frac{N}{2}(a_1 + a_N)$$

and

$$S_N = \frac{N}{2}[2a_1 + (N-1)d]$$

The first formula in Theorem 2 can be written as

$$S_N = N\left(\frac{a_1 + a_N}{2}\right)$$

Thus S_N is the product of the number of terms and the arithmetic mean of the first and last terms.

▶ **EXAMPLE 4** *Finding the Sum of an Arithmetic Series*

Find the sum of the positive even integers less than 100.

Solution   The positive even integers less than 100 form the arithmetic sequence

$$2, 4, 6, \ldots, 96, 98$$

We wish to find the sum of the associated arithmetic series, which is

$$2 + 4 + 6 + \ldots + 96 + 98$$

For this series $a_1 = 2$, $d = 2$, $N = 49$, and $a_{49} = 98$. From Theorem 2

$$S_{49} = \frac{49}{2}(a_1 + a_{49})$$

$$= \frac{49}{2}(2 + 98)$$

$$= 2450 \qquad \blacktriangleleft$$

▶ **EXAMPLE 5** *Solving a Word Problem Having an Arithmetic Series as a Mathematical Model*

The seller of a certain piece of real estate received the following two offers from prospective purchasers:

Offer 1: The payment for the first year is $24,000, and for nine years thereafter there is an annual increase of $1800 in the payments.

Offer 2: The payment for the first six months is $12,000, and for the second six months it is $12,450. For nine years thereafter there is a semiannual increase of $450 in the payments.

Which offer will give the seller more money over a ten-year period and how much more?

Solution According to offer 1, the number of dollars received by the seller over a period of ten years is the sum of the following arithmetic series of ten terms:

$$24,000 + 25,800 + 27,600 + \ldots + a_{10}$$

Let S_{10} be this sum. Then from the second formula of Theorem 2

$$S_{10} = \tfrac{10}{2}[2a_1 + (10 - 1)d]$$

Because $a_1 = 24,000$ and $d = 1800$,

$$S_{10} = 5[2(24,000) + 9(1800)]$$

$$= 321,000$$

According to offer 2, the number of dollars received by the seller over a period of ten years is the sum of the following arithmetic series of twenty terms:

$$12,000 + 12,450 + 12,900 + \ldots + a_{20}$$

Denoting this sum by S_{20} and from Theorem 2 with $a_1 = 12,000$ and $d = 450$, we have

$$S_{20} = \tfrac{20}{2}[2a_1 + (20 - 1)d]$$

$$= 10[2(12,000) + 19(450)]$$

$$= 325,500$$

Conclusion: Offer 2 will give the seller $4500 more money over a ten-year period.   ◄

DEFINITION   A Geometric Sequence

A **geometric sequence** is a sequence such that any element after the first can be obtained by multiplying the preceding element by a constant.

A geometric sequence is also called a **geometric progression.**

The constant multiplier in a geometric sequence is called the **common ratio** and is denoted by r. We may compute r by dividing any term after the first by the preceding one.

▷ **ILLUSTRATION 6**

For the sequence

1, 2, 4, 8, 16, 32, 64, 128

we have $\frac{2}{1} = 2$, $\frac{4}{2} = 2$, $\frac{8}{4} = 2$, $\frac{16}{8} = 2$, $\frac{32}{16} = 2$, $\frac{64}{32} = 2$, and $\frac{128}{64} = 2$. The common ratio $r = 2$.   ◄

As with an arithmetic sequence, the number of elements in a geometric sequence is denoted by N, the first element is denoted by a_1, and the last element is denoted by a_N. In the sequence of Illustration 6, $N = 8$, $a_1 = 1$, and $a_8 = 128$.

A geometric sequence can be defined by giving the values of a_1 and N and a recursive formula

$$a_{n+1} = a_n r$$

from which every element after the first can be obtained from the preceding one.

▷ **ILLUSTRATION 7**

Consider the geometric sequence for which $a_1 = 128$, $N = 5$, and $a_{n+1} = a_n(-\frac{1}{4})$. Then

$$a_2 = 128(-\tfrac{1}{4}) \qquad a_3 = -32(-\tfrac{1}{4}) \qquad a_4 = 8(-\tfrac{1}{4}) \qquad a_5 = -2(-\tfrac{1}{4})$$
$$= -32 \qquad\qquad = 8 \qquad\qquad\quad = -2 \qquad\qquad = \tfrac{1}{2}$$

Therefore, the sequence is

128, -32, 8, -2, $\tfrac{1}{2}$ ◄

The general geometric sequence, for which the first element is a_1, the common ratio is r, and the number of elements is N, can be obtained by applying the recursive formula. Starting with the element a_1, we obtain each successive element by multiplying the preceding one by r. Doing this, we have

$$a_1, a_1r, a_1r^2, a_1r^3, a_1r^4, \ldots, a_N$$

In the first five elements we observe that each element is the product of a_1 and a power of r, where the exponent of r is one less than the number of the element. Therefore our intuition suggests that the Nth (last) element is $a_N = a_1r^{N-1}$.

THEOREM 3

The Nth element of a geometric sequence is given by

$$a_N = a_1r^{N-1}$$

The proof of this theorem is by mathematical induction and is left as an exercise. See Exercise 52.

▶ **EXAMPLE 6** *Solving a Word Problem Having a Geometric Sequence as a Mathematical Model*

A city has a current population of 100,000. If the population is expected to increase 10 percent every five years, what is the expected population forty years from now?

Solution   The population at the end of five years is expected to be

$$100{,}000 + 0.10(100{,}000) = (1.10)(100{,}000)$$

The expected population at the end of each successive five-year period is 1.10 times the population at the end of the preceding five-year period. Hence we have the geometric sequence of nine elements

$$100{,}000, (1.10)(100{,}000), (1.10)^2(100{,}000), \ldots, a_9$$

where a_9 is the expected population at the end of forty years. From Theorem 3 with $N = 9$, $a_1 = 100{,}000$, and $r = 1.10$,

$$a_9 = a_1r^{9-1}$$
$$= 100{,}000(1.10)^8$$
$$\approx (2.144)10^5$$

Conclusion: Forty years from now, the population is expected to be 214,400 to the nearest hundred.   ◀

DEFINITION **Geometric Means**

If $a, c_1, c_2, \ldots, c_k, b$ is a geometric sequence, then the numbers $c_1, c_2, \ldots, c_k$ are a set of k **geometric means** between a and b.

▷ **ILLUSTRATION 8**

The sequence

2, 6, 18, 54, 162

is a geometric sequence with $r = 3$. From the definition the numbers 6, 18, and 54 form a set of three geometric means between 2 and 162.
Because the sequence

2, −6, 18, −54, 162

is also a geometric sequence ($r = -3$), the numbers −6, 18, and −54 form another set of three geometric means between 2 and 162. ◄

If m is a geometric mean between two numbers x and y, then

x, m, y

is a geometric sequence. Therefore

$$\frac{m}{x} = \frac{y}{m}$$

$$m^2 = xy$$

This equation implies that either both x and y are positive or both x and y are negative. Furthermore, the equation has two solutions: $m = \sqrt{xy}$ and $m = -\sqrt{xy}$. Because we want the geometric mean to be between the numbers x and y, we choose for the value of m the number having the same sign as x and y. Therefore we have the following definition.

DEFINITION **Geometric Mean**

The **geometric mean** between the numbers x and y is

$\sqrt{xy}$ if x and y are positive

$-\sqrt{xy}$ if x and y are negative

▶ **EXAMPLE 7** *Finding the Geometric Mean Between Two Numbers*

Find the geometric mean between each set of numbers: **(a)** 4 and 9;
(b) $-\frac{3}{10}$ and $-\frac{5}{6}$.

Solution In each part let m be the geometric mean. We compute m by
applying the definition.

(a) $m = \sqrt{4 \cdot 9}$ **(b)** $m = -\sqrt{(-\frac{3}{10})(-\frac{5}{6})}$

$\quad\quad = \sqrt{36}$ $\quad\quad\quad = -\sqrt{\frac{15}{60}}$

$\quad\quad = 6$ $\quad\quad\quad = -\sqrt{\frac{1}{4}}$

$\quad\quad\quad = -\frac{1}{2}$ ◀

By a generalization of the definition of the geometric mean between two
numbers, we define the **geometric mean** of a set of numbers $x_1, x_2, x_3, \ldots ,$
x_n to be the number $\sqrt[n]{x_1 x_2 x_3 \ldots x_n}$.

▷ **ILLUSTRATION 9**

The geometric mean of the numbers 4, 10, and 25 is

$$\sqrt[3]{(4)(10)(25)} = \sqrt[3]{1000}$$
$$= 10 \quad\quad ◀$$

With any geometric sequence, there is an associated **geometric series,**
which is the indicated sum of the elements of the geometric sequence.

▷ **ILLUSTRATION 10**

The geometric sequence of Illustration 7 is

$$128, -32, 8, -2, \tfrac{1}{2}$$

Associated with this sequence is the geometric series

$$128 - 32 + 8 - 2 + \tfrac{1}{2}$$

which can be written with sigma notation as

$$\sum_{i=1}^{5} 128(-\tfrac{1}{4})^{i-1} \quad\quad ◀$$

Let S_N be the sum of N terms of a geometric series. Then

$$S_N = a_1 + a_1 r + a_1 r^2 + a_1 r^3 + \ldots + a_1 r^{N-2} + a_1 r^{N-1}$$

If we multiply both sides of this equation by r, we have

$$rS_N = a_1r + a_1r^2 + a_1r^3 + a_1r^4 + \ldots + a_1r^{N-1} + a_1r^N$$

The sum of the first equation and -1 times the second gives

$$S_N - rS_N = a_1 - a_1r^N$$
$$(1 - r)S_N = a_1 - a_1r^N$$

If $1 - r \neq 0$, we can divide each side of this equation by $1 - r$ and obtain

$$S_N = \frac{a_1 - a_1r^N}{1 - r} \qquad \text{if } r \neq 1$$

$$S_N = \frac{a_1(1 - r^N)}{1 - r} \qquad \text{if } r \neq 1$$

A formula for S_N in terms of a_1, r, and a_N is found by expressing a_1r^N as $r(a_1r^{N-1})$. Doing this, we have

$$S_N = \frac{a_1 - r(a_1r^{N-1})}{1 - r} \qquad \text{if } r \neq 1$$

From Theorem 3, $a_1r^{N-1} = a_N$. Thus

$$S_N = \frac{a_1 - ra_N}{1 - r} \qquad \text{if } r \neq 1$$

We have proved the following theorem.

THEOREM 4

If $a_1, a_2, a_3, \ldots, a_N$ is a geometric sequence with common ratio r, and

$$S_N = a_1 + a_2 + a_3 + \ldots + a_N$$

then

(i) $S_N = \dfrac{a_1(1 - r^N)}{1 - r} \qquad \text{if } r \neq 1$

and

(ii) $S_N = \dfrac{a_1 - ra_N}{1 - r} \qquad \text{if } r \neq 1$

▶ **EXAMPLE 8** *Finding the Sum of a Geometric Series*

Find the sum of the geometric series

$$\sum_{i=1}^{5} 2\left(\tfrac{1}{3}\right)^{i-1}$$

Solution   For the given series, $a_1 = 2$, $r = \frac{1}{3}$, and $N = 5$. Thus, from Theorem 4(i)

$$S_5 = \frac{a_1(1 - r^5)}{1 - r}$$

$$= \frac{2[1 - (\frac{1}{3})^5]}{1 - \frac{1}{3}}$$

$$= \frac{2(1 - \frac{1}{243})}{\frac{2}{3}}$$

$$= 3 - \frac{1}{81}$$

$$= 2\frac{80}{81}$$ ◀

In the next example we use Theorem 3 of Section 6.2, which states that if P dollars is invested at an interest rate of $100i$ percent compounded m times per year and if A_n dollars is the amount of the investment at the end of n interest periods, then

$$A_n = P\left(1 + \frac{i}{m}\right)^n$$

▶ **EXAMPLE 9**   *Solving a Word Problem Having a Geometric Series as a Mathematical Model*

To create a sinking fund that will provide capital to purchase some new equipment, a company deposits $25,000 into an account on January 1 every year for ten years. If the account earns 12 percent interest compounded annually, how much is in the sinking fund immediately after the tenth deposit is made?

Solution   Immediately after the tenth deposit is made, the tenth payment has earned no interest; the ninth payment has earned interest for one year; the eighth payment has earned interest for two years; and so on; and the first payment has earned interest for nine years. To find the number of dollars in the fund immediately after the tenth payment, we apply the above formula for A_n with $P = 25,000$, $i = 0.12$, and $m = 1$ to find the dollar amount from each payment. The results are as follows:

10th payment: 25,000   (no interest)
9th payment: $25,000(1.12)^1$ (interest for one year; $n = 1$)
8th payment: $25,000(1.12)^2$ (interest for two years; $n = 2$)
.
.
.
1st payment: $25,000(1.12)^9$ (interest for nine years; $n = 9$)

If x dollars is the total amount in the sinking fund immediately after the tenth deposit is made, then

$$x = 25{,}000 + 25{,}000(1.12)^1 + 25{,}000(1.12)^2 + \ldots + 25{,}000(1.12)^9$$

We observe that x is the sum of a geometric series for which $N = 10$, $r = 1.12$, and $a_1 = 25{,}000$. From Theorem 4(i)

$$x = \frac{25{,}000[1 - (1.12)^{10}]}{1 - 1.12}$$

$$\approx (4.39)10^5$$

Conclusion: Immediately after the tenth deposit is made, the amount in the sinking fund is $439,000 to the nearest thousand dollars. ◀

EXERCISES 12.6

In Exercises 1 through 4, write the first five elements of an arithmetic sequence whose first element is a and whose common difference is d.

1. (a) $a = 5, d = 3$ **(b)** $a = 10, d = -4$

2. (a) $a = -3, d = 2$ **(b)** $a = 16, d = -5$

3. (a) $a = -5, d = -7$ **(b)** $a = x, d = 2y$

4. (a) $a = 20, d = 10$ **(b)** $a = u + v, d = -3v$

In Exercises 5 through 8, write the first five elements of a geometric sequence whose first element is a and whose common ratio is r.

5. (a) $a = 5, r = 3$ **(b)** $a = 8, r = -\frac{1}{2}$

6. (a) $a = 3, r = 2$ **(b)** $a = 2, r = \sqrt{2}$

7. (a) $a = -\dfrac{9}{16}, r = -\dfrac{2}{3}$ **(b)** $a = \dfrac{x}{y}, r = -\dfrac{y}{x}$

8. (a) $a = -81, r = \dfrac{1}{3}$ **(b)** $a = \dfrac{s}{t}, r = \dfrac{1}{u}$

In Exercises 9 through 12, determine if the elements form an arithmetic sequence or a geometric sequence. If they do, write the next two elements.

9. (a) $3, -1, -5$ **(b)** $1, 3, 9$
     (c) $\sqrt{2}, \sqrt{6}, 3\sqrt{2}$ **(d)** $-6, 10, -14$

10. (a) $12, 7, 2$ **(b)** $2, -4, 8$
     (c) $-1, -\frac{1}{3}, \frac{1}{3}$ **(d)** $\frac{1}{2}, \frac{1}{3}, \frac{1}{4}$

11. (a) $3.33, 2.22, 1.11$ **(b)** $\frac{1}{3}, \frac{1}{4}, \frac{1}{6}$
     (c) $-6, 2, -\frac{2}{3}$ **(d)** $x, 2x + y, 3x + 2y$

12. (a) $\frac{1}{2}, \frac{3}{4}, \frac{7}{8}$ **(b)** $3^{-2}, 3^0, 3^2$
     (c) $\sqrt[3]{3}, \sqrt[6]{3}, 1$ **(d)** $s, t, 2t - s$

13. Find the twelfth element of an arithmetic sequence whose first element is 2 and whose second element is 5.

14. Find the tenth element of an arithmetic sequence whose first element is 8 and whose third element is 2.

15. Find the first element of an arithmetic sequence whose eighth element is 2 and whose common difference is -2.

16. The ninth element of an arithmetic sequence is 28 and the twenty-first element is 100. What is the fifteenth element?

17. The first three elements of an arithmetic sequence are 20, 16, and 12. Which element is -96?

18. In the arithmetic sequence whose first three elements are $\frac{1}{6}, \frac{1}{4}$, and $\frac{1}{3}$, which element is 4?

19. Find the third element of a geometric sequence whose fifth element is 81 and whose ninth element is 16.

20. If the first element of a geometric sequence is $\frac{1}{8}$ and the eighth element is -16, find the sixth element.

21. Find the common ratio of a geometric sequence whose third element is -2 and whose sixth element is 54.

22. If the first element of a geometric sequence is 1 and the common ratio is 3, determine the smallest four-digit numeral that represents an element of this geometric sequence.

23. In the geometric sequence whose first element is 0.0003 and whose common ratio is 10, which element is 3,000,000?

24. In the geometric sequence whose first three elements are 27, -18, and 12, which element is $-\frac{512}{729}$?

25. (a) Insert four arithmetic means between 5 and 6;
(b) insert five geometric means between 1 and 64.

26. (a) Insert seven arithmetic means between 3 and 9;
(b) insert three geometric means between 162 and 2.

27. (a) Find the geometric mean between $-\frac{2}{3}$ and -6;
(b) find the geometric mean of the numbers 9, 21, and 49.

28. (a) Find the geometric mean between 16 and 25;
(b) find the geometric mean of the numbers $\frac{1}{3}$, 4, and 6.

In Exercises 29 through 34, find the sum of the series.

29. (a) $\displaystyle\sum_{i=1}^{8} (3i - 1)$ (b) $\displaystyle\sum_{i=1}^{8} 2^i$

30. (a) $\displaystyle\sum_{i=1}^{18} \frac{2i - 1}{3}$ (b) $\displaystyle\sum_{i=2}^{7} (-3)^i$

31. (a) $\displaystyle\sum_{i=2}^{12} (8 - 2i)$ (b) $\displaystyle\sum_{j=3}^{9} 5\left(\frac{1}{3}\right)^{j-3}$

32. (a) $\displaystyle\sum_{k=1}^{50} (2k - 1)$ (b) $\displaystyle\sum_{i=1}^{10} (1.02)^i$

33. (a) $\displaystyle\sum_{i=1}^{12} (1.02)^{i-1}$ (b) $\displaystyle\sum_{k=1}^{20} (5k - 1)$

34. (a) $\displaystyle\sum_{j=1}^{6} \left(\frac{2}{3}\right)^j$ (b) $\displaystyle\sum_{i=3}^{12} \left(\frac{1}{2}i - 5\right)$

35. Find the arithmetic mean of the following set of test scores: 72, 53, 85, 74, 62, and 83.

36. Find the sum of all the positive integers less than 100.

37. Find the sum of all the positive even integers consisting of two digits.

38. Find the sum of all the integer multiples of 8 between 9 and 199.

In Exercises 39 through 47, solve the word problem by using as a mathematical model a sequence or series, either arithmetic or geometric. Be sure to write a conclusion.

39. The sum of $1000 is distributed among four people so that each person after the first receives $20 less than the preceding person. How much does each person receive?

40. A student's grade on the first of 12 quizzes in her algebra course was 45. However, on each successive quiz her score was 5 more than on the preceding one. What was the arithmetic mean (average) of the twelve scores?

41. In a display window a grocer wishes to place boxes of detergent in pyramid form so that the bottom row contains 15 boxes, the next row contains 14 boxes, the next row contains 13 boxes, and so on, with 1 box on top. How many boxes of detergent are necessary for the pyramid?

42. To dig a well, a company charges $80 for the first foot, $100 for the second foot, $120 for the third foot, and so on; the cost of each foot is $20 more than the cost of the preceding foot. What is the depth of a well that costs $23,400 to dig?

43. From a barrel filled with 1 gal of wine, 1 pint is withdrawn and then the barrel is filled with water. If this procedure is followed six times, what fractional part of the original contents is in the barrel?

44. If a town having a population of 5000 in 1971 has a 20 percent increase every five years, what is its expected population in the year 2001?

45. A contractor who does not meet the deadline on the construction of a building is fined $800 per day for each of the first ten days of extra time, and for each additional day thereafter the fine is increased by $160 each day. If the contractor is fined $20,160, by how many extra days was the construction time delayed?

46. Some logs should be piled in layers so that the top layer contains one log, the next layer contains two logs, the next layer contains three logs, and so on, each layer containing one more log than the layer on top of it. If there are 190 logs, determine if all the logs can be used in such a grouping, and if so, how many logs are in the bottom layer.

47. Payments of $1000 are deposited into a sinking fund every six months and the account earns 10 percent interest compounded semiannually. How much is in the fund immediately after the twentieth payment is made?

48. Find a sequence of four numbers, the first of which is 6 and the fourth of which is 16, if the first three numbers form an arithmetic sequence and the last three numbers form a geometric sequence.

49. Three numbers form an arithmetic sequence having a common difference of 4. If the first number is increased by 2, the second number is increased by 3, and the third number is increased by 5, the resulting numbers form a geometric sequence. Find the original numbers.

50. Three numbers whose sum is 3 form an arithmetic sequence, and their squares form a geometric sequence. What are the numbers?

51. Explain why the geometric mean of two unequal positive numbers is less than their arithmetic mean.

52. Prove Theorem 3 by mathematical induction.

12.7 THE BINOMIAL THEOREM

G O A L S

1. **Learn and apply factorial notation.**
2. **Learn and apply binomial coefficient notation.**
3. **Learn and apply the binomial theorem.**
4. **Prove the binomial theorem by mathematical induction.**
5. **Learn and apply the formula for the *r*th term in a binomial expansion.**
6. **Use Pascal's triangle to find the coefficients in a binomial expansion.**

A power of a binomial is a special kind of series called a **binomial expansion.** In this section we present the *binomial theorem,* which gives the expansion of

$$(a + b)^n$$

where n is any positive integer. In our discussion of the binomial theorem we use two notations with which you need to be familiar. One is the **factorial notation** $n!$, read "n factorial," which is defined by

$$n! = n(n - 1)(n - 2) \cdot \ldots \cdot 2 \cdot 1$$

▷ **ILLUSTRATION 1**

$$1! = 1 \quad 2! = 2 \cdot 1 \quad 3! = 3 \cdot 2 \cdot 1 \quad 4! = 4 \cdot 3 \cdot 2 \cdot 1$$
$$= 2 \qquad\quad = 6 \qquad\qquad = 24 \qquad\quad ◀$$

Because

$$n! = n(n - 1)(n - 2) \cdot \ldots \cdot 3 \cdot 2 \cdot 1$$

and

$$(n - 1)! = (n - 1)(n - 2) \cdot \ldots \cdot 3 \cdot 2 \cdot 1$$

it follows that

$$n! = n(n - 1)! \tag{1}$$

In particular,

$$5! = 5 \cdot 4! \quad \text{and} \quad 26! = 26 \cdot 25!$$

Furthermore, if we substitute 1 for n in formula (1), we obtain

$$1! = 1(1 - 1)!$$

or, equivalently,

$$1! = 1 \cdot 0!$$

Therefore we define

$$0! = 1$$

▷ ILLUSTRATION 2

To simplify $\frac{10!}{7!}$, we first write $10! = 10 \cdot 9 \cdot 8 \cdot (7!)$ and then divide numerator and denominator by $7!$.

$$\frac{10!}{7!} = \frac{10 \cdot 9 \cdot 8 \cdot (7!)}{7!}$$

$$= 10 \cdot 9 \cdot 8$$

$$= 720 \qquad \blacktriangleleft$$

We also need the **binomial coefficient notation** $\begin{pmatrix} n \\ r \end{pmatrix}$, read "$n$ above r" and defined as follows: If n and r are positive integers and $r \leq n$,

$$\begin{pmatrix} n \\ r \end{pmatrix} = \frac{n(n - 1)(n - 2) \cdot \ldots \cdot (n - r + 1)}{r!}$$

Note that a binomial coefficient is an integer. Furthermore,

$$\begin{pmatrix} n \\ 0 \end{pmatrix} = 1$$

▷ ILLUSTRATION 3

From the definition

$$\begin{pmatrix} 7 \\ 3 \end{pmatrix} = \frac{7 \cdot 6 \cdot 5}{3 \cdot 2 \cdot 1} \qquad \begin{pmatrix} 5 \\ 5 \end{pmatrix} = \frac{5 \cdot 4 \cdot 3 \cdot 2 \cdot 1}{5 \cdot 4 \cdot 3 \cdot 2 \cdot 1} \qquad \begin{pmatrix} 9 \\ 0 \end{pmatrix} = 1$$

$$= 35 \qquad\qquad\qquad = 1 \qquad\qquad\qquad \blacktriangleleft$$

We have just seen that $\begin{pmatrix} 5 \\ 5 \end{pmatrix} = 1$. More generally, it follows from the definition that

$$\begin{pmatrix} n \\ n \end{pmatrix} = 1$$

The following theorem gives another formula for computing $\begin{pmatrix} n \\ r \end{pmatrix}$.

THEOREM 1

If n is a positive integer, r is a nonnegative integer, and $r \leq n$,

$$\binom{n}{r} = \frac{n!}{r!(n-r)!}$$

Proof From the definition, if r is a positive integer,

$$\binom{n}{r} = \frac{n(n-1)(n-2) \cdot \ldots \cdot (n-r+1)}{r!}$$

$$= \frac{n(n-1)(n-2) \cdot \ldots \cdot (n-r+1)}{r!} \cdot \frac{(n-r)!}{(n-r)!}$$

$$= \frac{n!}{r!(n-r)!}$$

The formula is also valid if $r = 0$ because $\binom{n}{0} = 1$ and

$$\frac{n!}{0!(n-0)!} = \frac{n!}{0!n!}$$

$$= \frac{n!}{(1)n!}$$

$$= 1$$

■

▶ **EXAMPLE 1** *Computing a Binomial Coefficient*

Compute the given binomial coefficient by two methods: First use the definition and second apply Theorem 1.

(a) $\binom{5}{3}$ (b) $\binom{6}{6}$ (c) $\binom{4}{1}$

Solution
(a) From the definition   From Theorem 1

$$\binom{5}{3} = \frac{5 \cdot 4 \cdot 3}{3 \cdot 2 \cdot 1}$$ $$\binom{5}{3} = \frac{5!}{3!2!}$$

$$= 10$$ $$= \frac{5 \cdot 4 \cdot 3 \cdot 2 \cdot 1}{3 \cdot 2 \cdot 1 \cdot 2 \cdot 1}$$

$$= 10$$

(b) From the definition

$$\binom{6}{6} = \frac{6 \cdot 5 \cdot 4 \cdot 3 \cdot 2 \cdot 1}{6 \cdot 5 \cdot 4 \cdot 3 \cdot 2 \cdot 1}$$

$$= 1$$

From Theorem 1

$$\binom{6}{6} = \frac{6!}{6!0!}$$

$$= \frac{6!}{6!(1)}$$

$$= 1$$

(c) From the definition

$$\binom{4}{1} = \frac{4}{1}$$

$$= 4$$

From Theorem 1

$$\binom{4}{1} = \frac{4!}{1!3!}$$

$$= \frac{4 \cdot 3 \cdot 2 \cdot 1}{1 \cdot 3 \cdot 2 \cdot 1}$$

$$= 4 \qquad ◀$$

The next theorem is needed in the proof of the binomial theorem.

THEOREM 2

$$\binom{n}{r} + \binom{n}{r-1} = \binom{n+1}{r}$$

Proof　From Theorem 1

$$\binom{n}{r} + \binom{n}{r-1} = \frac{n!}{r!(n-r)!} + \frac{n!}{(r-1)![n-(r-1)]!}$$

$$= \frac{n!}{r!(n-r)!} \cdot \frac{n-r+1}{n-r+1} + \frac{n!}{(r-1)!(n-r+1)!} \cdot \frac{r}{r}$$

$$= \frac{(n-r+1)n!}{r!(n-r+1)!} + \frac{(r)n!}{r!(n-r+1)!}$$

$$= \frac{[(n-r+1)+r]n!}{r!(n-r+1)!}$$

$$= \frac{(n+1)n!}{r!(n+1-r)!}$$

$$= \frac{(n+1)!}{r!(n+1-r)!}$$

$$= \binom{n+1}{r} \qquad ∎$$

We now consider the binomial expansion of

$$(a + b)^n$$

for specific values of n.

$n = 1:$ $\quad (a + b)^1 = a + b$

$n = 2:$ $\quad (a + b)^2 = a^2 + 2ab + b^2$

$n = 3:$ $\quad (a + b)^3 = a^3 + 3a^2b + 3ab^2 + b^3$

$n = 4:$ $\quad (a + b)^4 = a^4 + 4a^3b + 6a^2b^2 + 4ab^3 + b^4$

$n = 5:$ $\quad (a + b)^5 = a^5 + 5a^4b + 10a^3b^2 + 10a^2b^3 + 5ab^4 + b^5$

$n = 6:$ $\quad (a + b)^6 = a^6 + 6a^5b + 15a^4b^2 + 20a^3b^3 + 15a^2b^4 + 6ab^5 + b^6$

Each equation after the first is obtained by multiplying both sides of the previous equation by $a + b$. On the right side of each equation the first term can be written with a factor b^0 and the last term with a factor a^0. Therefore, each term contains nonnegative integer powers of a and b. We also note the following properties of each of the six expansions:

1. There are $n + 1$ terms in the expansion.
2. The sum of the exponents of a and b in any term is n; the exponent of a decreases by 1 and the exponent of b increases by 1 from each term to the next.
3. **(i)** The first term in the expansion is

$$a^n = \binom{n}{0} a^n$$

(ii) The second term is

$$\frac{n}{1} a^{n-1}b = \binom{n}{1} a^{n-1}b$$

(iii) The third term is

$$\frac{n(n-1)}{2 \cdot 1} a^{n-2}b^2 = \binom{n}{2} a^{n-2}b^2$$

(iv) The fourth term is

$$\frac{n(n-1)(n-2)}{3 \cdot 2 \cdot 1} a^{n-3}b^3 = \binom{n}{3} a^{n-3}b^3$$

(v) The fifth term is

$$\frac{n(n-1)(n-2)(n-3)}{4 \cdot 3 \cdot 2 \cdot 1} a^{n-4}b^4 = \binom{n}{4} a^{n-4}b^4$$

(vi) The term involving b^r is

$$\frac{n(n-1)(n-2) \cdot \ldots \cdot (n-r+1)}{r!} a^{n-r}b^r = \binom{n}{r}a^{n-r}b^r$$

(vii) The last term is

$$b^n = \binom{n}{n}b^n$$

▷ **ILLUSTRATION 4**

The binomial expansion of $(a + b)^n$ when $n = 6$ is

$$(a + b)^6 = a^6 + 6a^5b + 15a^4b^2 + 20a^3b^3 + 15a^2b^4 + 6ab^5 + b^6$$

We show that the preceding properties apply to this expansion.

1. There are seven terms in the expansion.
2. The sum of the exponents of a and b in any term is 6; the exponent of a decreases by 1 and the exponent of b increases by 1 from each term to the next.
3. **(i)** The first term in the expansion is

$$a^6 = \binom{6}{0}a^6$$

   (ii) The second term is

$$\frac{6}{1}a^{6-1}b = \binom{6}{1}a^5b$$

   (iii) The third term is

$$\frac{6 \cdot 5}{2 \cdot 1}a^{6-2}b^2 = \binom{6}{2}a^4b^2$$

   (iv) The fourth term is

$$\frac{6 \cdot 5 \cdot 4}{3 \cdot 2 \cdot 1}a^{6-3}b^3 = \binom{6}{3}a^3b^3$$

   (v) The fifth term is

$$\frac{6 \cdot 5 \cdot 4 \cdot 3}{4 \cdot 3 \cdot 2 \cdot 1}a^{6-4}b^4 = \binom{6}{4}a^2b^4$$

   (vi) The sixth term is

$$\frac{6 \cdot 5 \cdot 4 \cdot 3 \cdot 2}{5 \cdot 4 \cdot 3 \cdot 2 \cdot 1}a^{6-5}b^5 = \binom{6}{5}ab^5$$

(vii) The last term is

$$b^6 = \binom{6}{6}b^6$$

◄

Illustration 4 and the discussion preceding it suggest a similar expression for the expansion of $(a + b)^n$, where n is any positive integer. We state this in the binomial theorem.

THEOREM   *The Binomial Theorem*

If n is any positive integer,

$$(a + b)^n = \binom{n}{0}a^n + \binom{n}{1}a^{n-1}b + \binom{n}{2}a^{n-2}b^2 + \ldots$$

$$+ \binom{n}{r}a^{n-r}b^r + \ldots + \binom{n}{n-1}ab^{n-1} + \binom{n}{n}b^n$$

Proof   We use mathematical induction.

Part 1:   We first verify that the theorem is true for $n = 1$. We have

$$(a + b)^1 = \binom{1}{0}a + \binom{1}{1}b$$

$$= 1 \cdot a + 1 \cdot b$$

$$= a + b$$

Part 2:   We now assume that the theorem is true for $n = k$; that is,

$$(a + b)^k = \binom{k}{0}a^k + \binom{k}{1}a^{k-1}b + \binom{k}{2}a^{k-2}b^2 + \ldots$$

$$+ \binom{k}{k-1}ab^{k-1} + \binom{k}{k}b^k \qquad (2)$$

We wish to show that the theorem is true for $n = k + 1$, so that

$$(a + b)^{k+1} = \binom{k+1}{0}a^{k+1} + \binom{k+1}{1}a^k b + \binom{k+1}{2}a^{k-1}b^2 + \ldots$$

$$+ \binom{k+1}{k}ab^k + \binom{k+1}{k+1}b^{k+1} \qquad (3)$$

We first write

$$(a + b)^{k+1} = (a + b)(a + b)^k$$

To obtain $(a + b)(a + b)^k$, we multiply each term on the right side of (2) by

a and then by b and add the results. Therefore

$$(a + b)^{k+1} = \left[\binom{k}{0}a^{k+1} + \binom{k}{1}a^k b + \binom{k}{2}a^{k-1}b^2 + \ldots + \binom{k}{k}ab^k \right]$$

$$+ \left[\binom{k}{0}a^k b + \binom{k}{1}a^{k-1}b^2 + \ldots + \binom{k}{k-1}ab^k + \binom{k}{k}b^{k+1} \right]$$

$$(a + b)^{k+1} = \binom{k}{0}a^{k+1} + \left[\binom{k}{1} + \binom{k}{0} \right]a^k b + \left[\binom{k}{2} + \binom{k}{1} \right]a^{k-1}b^2 + \ldots$$

$$+ \left[\binom{k}{k} + \binom{k}{k-1} \right]ab^k + \binom{k}{k}b^{k+1} \qquad (4)$$

From Theorem 2

$$\binom{k}{1} + \binom{k}{0} = \binom{k+1}{1}; \quad \binom{k}{2} + \binom{k}{1} = \binom{k+1}{2}; \quad \ldots ; \quad \binom{k}{k} + \binom{k}{k-1} = \binom{k+1}{k} \qquad (5)$$

Furthermore, because $\binom{n}{0} = 1$ and $\binom{n}{n} = 1$, we have

$$\binom{k}{0} = \binom{k+1}{0} \quad \text{and} \quad \binom{k}{k} = \binom{k+1}{k+1} \qquad (6)$$

Substituting from (5) and (6) in (4), we obtain

$$(a + b)^{k+1} = \binom{k+1}{0}a^{k+1} + \binom{k+1}{1}a^k b + \binom{k+1}{2}a^{k-1}b^2 + \ldots$$

$$+ \binom{k+1}{k}ab^k + \binom{k+1}{k+1}b^{k+1}$$

which is (3).

Conclusion: We have proved that the theorem is true when $n = 1$, and we have also proved that when the theorem is true for $n = k$, it is also true for $n = k + 1$. Therefore, by the principle of mathematical induction, the theorem is valid for all positive-integer values of n.   ∎

▶ **EXAMPLE 2** *Applying the Binomial Theorem*

Expand and simplify by the binomial theorem $(x^2 + 3y)^5$.

Solution   Applying the binomial theorem where a is x^2, b is $3y$, and n is

5, we have

$$(x^2 + 3y)^5 = \binom{5}{0}(x^2)^5 + \binom{5}{1}(x^2)^4(3y)^1 + \binom{5}{2}(x^2)^3(3y)^2$$

$$+ \binom{5}{3}(x^2)^2(3y)^3 + \binom{5}{4}(x^2)^1(3y)^4 + \binom{5}{5}(3y)^5$$

$$= 1 \cdot x^{10} + \frac{5}{1}x^8(3y) + \frac{5 \cdot 4}{2 \cdot 1}x^6(9y^2) + \frac{5 \cdot 4 \cdot 3}{3 \cdot 2 \cdot 1}x^4(27y^3)$$

$$+ \frac{5 \cdot 4 \cdot 3 \cdot 2}{4 \cdot 3 \cdot 2 \cdot 1}x^2(81y^4) + 1(243y^5)$$

$$= x^{10} + 15x^8y + 90x^6y^2 + 270x^4y^3 + 405x^2y^4 + 243y^5 \blacktriangleleft$$

▶ **EXAMPLE 3** *Applying the Binomial Theorem*

Expand and simplify

$$\left(2\sqrt{t} - \frac{1}{t}\right)^4$$

Solution   We use the binomial theorem where a is $2\sqrt{t}$, b is $-1/t$, and n is 4.

$$\left(2\sqrt{t} - \frac{1}{t}\right)^4 = \binom{4}{0}(2\sqrt{t})^4 + \binom{4}{1}(2\sqrt{t})^3\left(-\frac{1}{t}\right)^1 + \binom{4}{2}(2\sqrt{t})^2\left(-\frac{1}{t}\right)^2$$

$$+ \binom{4}{3}(2\sqrt{t})^1\left(-\frac{1}{t}\right)^3 + \binom{4}{4}\left(-\frac{1}{t}\right)^4$$

$$= 1(16t^2) + \frac{4}{1}(8t^{3/2})\left(-\frac{1}{t}\right) + \frac{4 \cdot 3}{2 \cdot 1}(4t)\left(\frac{1}{t^2}\right)$$

$$+ \frac{4 \cdot 3 \cdot 2}{3 \cdot 2 \cdot 1}(2t^{1/2})\left(-\frac{1}{t^3}\right) + 1\left(\frac{1}{t^4}\right)$$

$$= 16t^2 - 32t^{1/2} + \frac{24}{t} - \frac{8}{t^{5/2}} + \frac{1}{t^4} \blacktriangleleft$$

From the binomial theorem, the term involving b^r in the expansion of $(a + b)^n$ is the $(r + 1)$st term, which is

$$\binom{n}{r}a^{n-r}b^r$$

The rth term in the expansion of $(a + b)^n$ is obtained from this expression by replacing r by $r - 1$. Thus

the rth term in $(a + b)^n$ is $\left(\begin{array}{c} n \\ r - 1 \end{array} \right) a^{n-r+1} b^{r-1}$

Observe that the exponent of b is one less than the number of the term, and the sum of the exponents of a and b is n.

▶ **EXAMPLE 4** *Finding a Particular Term in a Binomial Expansion*

Find the seventh term of the expansion of $(2u^3 - \frac{1}{4}v^4)^{10}$.

Solution   Applying the formula for the rth term where r is 7, n is 10, a is $2u^3$, and b is $-\frac{1}{4}v^4$, we have

$$\left(\begin{array}{c} 10 \\ 6 \end{array} \right)(2u^3)^4\left(-\frac{1}{4}v^4 \right)^6 = \frac{10 \cdot 9 \cdot 8 \cdot 7 \cdot 6 \cdot 5}{6 \cdot 5 \cdot 4 \cdot 3 \cdot 2 \cdot 1}(2^4 u^{12})\left(\frac{1}{2^{12}}v^{24} \right)$$

$$= \frac{210}{2^8}u^{12}v^{24}$$

$$= \frac{105}{128}u^{12}v^{24}$$ ◀

▶ **EXAMPLE 5** *Finding a Particular Term in a Binomial Expansion*

Find the term involving x^3 in the expansion of $(x - 3x^{-1})^9$.

Solution   From the formula for the rth term, where a is x, b is $-3x^{-1}$, and n is 9, the rth term has the factors

$$x^{10-r}(-3x^{-1})^{r-1} = (-3)^{r-1}x^{11-2r}$$

The term involving x^3 is the one for which the exponent of x is 3; hence we solve the equation

$$11 - 2r = 3$$
$$r = 4$$

Thus the fourth term is the desired term. It is

$$\left(\begin{array}{c} 9 \\ 3 \end{array} \right)x^6(-3x^{-1})^3 = \frac{9 \cdot 8 \cdot 7}{3 \cdot 2 \cdot 1}(-27)x^3$$

$$= -2268x^3$$ ◀

There is an interesting pattern for the coefficients in the binomial expansion. The coefficients can be written in the following triangular arrangement:

$$
\begin{array}{ccccccccccc}
 & & & & 1 & & 1 & & & & \\
 & & & 1 & & 2 & & 1 & & & \\
 & & 1 & & 3 & & 3 & & 1 & & \\
 & 1 & & 4 & & 6 & & 4 & & 1 & \\
1 & & 5 & & 10 & & 10 & & 5 & & 1
\end{array}
$$

The first row contains the coefficients for $(a + b)^1$; the second row contains the coefficients in the expansion of $(a + b)^2$; in the third row are the coefficients for $(a + b)^3$; and so on. Observe that each row begins and ends with the number 1. Also observe that each of the other numbers is the sum of the two numbers in the previous row, one to the left and one to the right of the number. For instance, the numbers in the fifth row are the coefficients for $(a + b)^5$. The first and last numbers are 1. The second number 5 is the sum of 1 and 4; then 10 is the sum of 4 and 6; 10 is the sum of 6 and 4; and 5 is the sum of 4 and 1. This triangular array is called **Pascal's triangle**, named in honor of the French mathematician Blaise Pascal (1623–1662), who used it in his work with probability, although it was known before his time.

▶ **EXAMPLE 6** *Using Pascal's Triangle To Find the Coefficients in a Binomial Expansion*

Expand $(a + b)^7$ by first finding the coefficients from Pascal's triangle.

Solution We write the fifth row, and then obtain the sixth and seventh rows by the procedure described above.

$$
\begin{array}{ccccccccccccc}
 & 1 & & 5 & & 10 & & 10 & & 5 & & 1 & \\
 1 & & 6 & & 15 & & 20 & & 15 & & 6 & & 1 \\
1 & & 7 & & 21 & & 35 & & 35 & & 21 & & 7 & & 1
\end{array}
$$

The coefficients for $(a + b)^7$ are in the seventh row. Therefore

$$(a + b)^7 = a^7 + 7a^6b + 21a^5b^2 + 35a^4b^3 + 35a^3b^4 + 21a^2b^5 + 7ab^6 + b^7 \quad ◀$$

When n is small, the use of Pascal's triangle for the coefficients of $(a + b)^n$ is advantageous. However, if n is large or a specific term is desired, you will want to use the binomial theorem or the formula for the rth term.

EXERCISES 12.7

In Exercises 1 through 4, simplify the rational expression.

1. (a) $\dfrac{6!}{9!}$ (b) $\dfrac{32!}{28!}$ **2.** (a) $\dfrac{8!}{4!}$ (b) $\dfrac{49!}{51!}$

3. (a) $\dfrac{12!}{8!\,6!}$ (b) $\dfrac{3! + 4!}{7!}$ **4.** (a) $\dfrac{2!\,5!}{7!}$ (b) $\dfrac{8!}{6! + 7!}$

In Exercises 5 through 8, find the number.

5. (a) $\dbinom{7}{3}$ (b) $\dbinom{6}{6}$ **6.** (a) $\dbinom{9}{4}$ (b) $\dbinom{5}{1}$

7. (a) $\dbinom{12}{9}$ (b) $\dbinom{50}{48}$ **8.** (a) $\dbinom{20}{17}$ (b) $\dbinom{15}{15}$

In Exercises 9 through 20, expand the power of the binomial.

9. $(a + b)^8$ **10.** $(a + b)^9$

11. $(x - y)^9$ **12.** $(x - y)^{10}$

13. $(x + 3y)^5$ **14.** $(2x - y)^6$

15. $(4 - ab)^6$ **16.** $(2t + s^2)^7$

17. $(3e^x - 2e^{-x})^5$ **18.** $(2u^{-1} - 3u^2)^5$

19. $(a^{1/2} + b^{1/2})^8$ **20.** $(xy^{-1} - x^{-1}y)^7$

In Exercises 21 through 26, find the first four terms in the expansion of the power of the binomial and simplify each term.

21. $(2x^2 + y^2)^{12}$ **22.** $(a^3 - 2b^2)^{14}$

23. $(u^{-1} - 3v^2)^{11}$ **24.** $(e^{x/2} - e^{-x/2})^{20}$

25. $(a^{1/3} - b^{1/3})^9$ **26.** $(\frac{2}{5}a^{2/3} + b^{3/2})^{11}$

27. Find the seventh term of the expansion of $(a + b)^{14}$.

28. Find the sixth term of the expansion of $(\frac{1}{2}a - b)^{13}$.

29. Find the sixth term of the expansion of $(2x - 3)^9$.

30. Find the tenth term of the expansion of $(\sqrt{t} - t^{-1/2})^{15}$.

31. Find the middle term of the expansion of $(1 - x^3y^{-2})^{12}$.

32. Find the middle term of the expansion of $(\frac{1}{3}y + \sqrt{y})^{10}$.

33. Find the term involving a^6 in the expansion of $(\frac{1}{2} + a)^{12}$.

34. Find the term involving x^{12} in the expansion of $(x^2 - \frac{1}{2})^{11}$.

35. Find the term that does not contain x in the expansion of $(x^2 - 2x^{-2})^{10}$.

36. Find the term involving t^{-4} in the expansion of $(\frac{1}{5}t^2 - t^{-1})^{13}$.

In Exercises 37 through 40, write the binomial expansion by first finding the coefficients from Pascal's triangle.

37. $(a + b)^9$ **38.** $(x - y)^8$

39. $(r - t)^{12}$ **40.** $(u + v)^{11}$

In Exercises 41 through 44, use the following interpretation of the binomial coefficient: $\dbinom{n}{r}$ is the number of combinations of n elements taken r at a time; that is, it is the number of ways you can select r elements from a set of n elements.

41. A football conference consists of eight teams. If each team plays every other team, how many conference games are played?

42. How many different committees of 3 persons each can be chosen from a group of 12 persons?

43. A student has 10 posters to pin up on the walls of her room, but there is space for only 7. **(a)** In how many ways can she choose the posters to be pinned up? **(b)** In how many ways can she select the 3 posters not to be pinned up?

44. A student is to answer any 8 questions on a test containing 12 questions. **(a)** In how many different ways can the student choose the 8 questions to be answered? **(b)** In how many different ways can the student choose the 4 questions not to be answered?

45. Prove that $\dbinom{n}{r} = \dbinom{n}{n - r}$.

46. Explain the equality in Exercise 45 in terms of the questions asked in Exercises 43 and 44.

47. **(a)** Expand $(2x - 3)^5$ by the binomial theorem. **(b)** Verify your expansion on your graphics calculator and explain how you did it.

48. Prove by mathematical induction that the following formula is true for all positive-integer values of n:

$$\sum_{i=1}^{n}\binom{i + 1}{2} = \binom{n + 2}{3}$$

12.8 INTRODUCTION TO INFINITE SERIES

GOALS

1. Describe an infinite series.
2. Learn and apply the formula for the sum of an infinite geometric series.
3. Express a nonterminating repeating decimal as a fraction.
4. Solve word problems having an infinite geometric series as a mathematical model.
5. Learn and apply the formula for a binomial series.
6. Compute with a binomial series.
7. Apply the infinite series for $\sin x$, $\cos x$, and e^x to compute approximate values of irrational numbers.

Here, in the final section of this text, we give you a preview of what to expect when you study infinite series in your calculus course.

Our discussion of series so far has been restricted to those associated with finite sequences. An **infinite series,** denoted by

$$a_1 + a_2 + a_3 + \ldots + a_n + \ldots$$

is the series associated with the infinite sequence

$$a_1, a_2, a_3, \ldots, a_n, \ldots$$

But what do we mean by the "sum" of an infinite number of terms, and under what circumstances does such a sum exist? The answers to these questions depend on the concept of *limit*, which is studied in calculus. However, for certain infinite series we can give an intuitive idea of how to interpret such a sum.

Suppose a piece of string of length 2 ft is cut in half. One of these halves of length 1 ft is set aside and the other piece is cut in half again. One of the resulting pieces of length $\frac{1}{2}$ ft is set aside, and the other piece is cut in half so that two pieces, each of length $\frac{1}{4}$ ft, are obtained. One of the pieces of length $\frac{1}{4}$ ft is set aside and then the other piece is cut in half; so two pieces, each of length $\frac{1}{8}$ ft, are obtained. Again one of the pieces is set aside and the other is cut in half. If this procedure is continued indefinitely, the number of feet in the sum of the lengths of the pieces set aside can be considered as the infinite series

$$1 + \frac{1}{2} + \frac{1}{4} + \frac{1}{8} + \frac{1}{16} + \ldots + \frac{1}{2^{n-1}} + \ldots \tag{1}$$

This series is an example of an *infinite geometric series*—one that you will meet in calculus. Series (1) is an infinite geometric series with $r = \frac{1}{2}$. Because we started with a piece of string 2 ft in length, your intuition indicates that the sum of series (1) should be 2. We can demonstrate this situation by applying our knowledge of geometric series. From Theorem 4 of Section 12.6, if S_N is the sum of N terms of a geometric series,

$$S_N = \frac{a_1(1 - r^N)}{1 - r}$$

Therefore, for a finite number N of terms of series (1) with $a_1 = 1$ and $r = \frac{1}{2}$,

$$S_N = \frac{(1)[1 - (\frac{1}{2})^N]}{1 - \frac{1}{2}}$$

$$= \frac{1 - (\frac{1}{2})^N}{\frac{1}{2}}$$

$$\Leftrightarrow \quad S_N = 2[1 - (\tfrac{1}{2})^N]$$

Applying this formula to successive values of N, we obtain

$$
\begin{aligned}
S_1 &= 2[1 - (\tfrac{1}{2})^1] & S_2 &= 2[1 - (\tfrac{1}{2})^2] & S_3 &= 2[1 - (\tfrac{1}{2})^3] \\
&= 2 - 2(\tfrac{1}{2}) & &= 2 - 2(\tfrac{1}{4}) & &= 2 - 2(\tfrac{1}{8}) \\
&= 2 - 1 & &= 2 - \tfrac{1}{2} & &= 2 - \tfrac{1}{4} \\
S_4 &= 2[1 - (\tfrac{1}{2})^4] & S_5 &= 2[1 - (\tfrac{1}{2})^5] & S_6 &= 2[1 - (\tfrac{1}{2})^6] \\
&= 2 - 2(\tfrac{1}{16}) & &= 2 - 2(\tfrac{1}{32}) & &= 2 - 2(\tfrac{1}{64}) \\
&= 2 - \tfrac{1}{8} & &= 2 - \tfrac{1}{16} & &= 2 - \tfrac{1}{32}
\end{aligned}
$$

$$
\begin{aligned}
\vdots \\
S_{10} &= 2[1 - (\tfrac{1}{2})^{10}] \\
&= 2 - 2(\tfrac{1}{1024}) \\
&= 2 - \tfrac{1}{512}
\end{aligned}
$$

and so on. We intuitively see that we can make the value of S_N as close to 2 as we please by taking N large enough. In other words, we can make the difference between 2 and S_N as small as we please by taking N sufficiently large. Therefore we state that S_N approaches 2 as N increases without bound, and we write

$$S_N \to 2 \quad \text{as} \quad N \to +\infty \tag{2}$$

Consider now the general **infinite geometric series:**

$$a_1 + a_1 r + a_1 r^2 + a_1 r^3 + \ldots + a_1 r^{n-1} + \ldots \quad |r| < 1$$

The sum of the first N terms of this series is given by

$$S_N = \frac{a_1}{1 - r}(1 - r^N) \tag{3}$$

Let us consider what happens to r^N as N increases without bound, when $|r| < 1$. For instance,

$$\left(\frac{1}{2}\right)^1 = \frac{1}{2}, \left(\frac{1}{2}\right)^2 = \frac{1}{4}, \left(\frac{1}{2}\right)^3 = \frac{1}{8}, \ldots, \left(\frac{1}{2}\right)^{10} = \frac{1}{1024}, \ldots$$

$$\left(\frac{1}{3}\right)^1 = \frac{1}{3}, \left(\frac{1}{3}\right)^2 = \frac{1}{9}, \left(\frac{1}{3}\right)^3 = \frac{1}{27}, \ldots, \left(\frac{1}{3}\right)^{10} = \frac{1}{59,049}, \ldots$$

$$\left(\frac{2}{3}\right)^1 = \frac{2}{3}, \left(\frac{2}{3}\right)^2 = \frac{4}{9}, \left(\frac{2}{3}\right)^3 = \frac{8}{27}, \ldots, \left(\frac{2}{3}\right)^{10} = \frac{1024}{59,049}, \ldots$$

More generally, for any r for which $|r| < 1$, when N increases without bound, $|r^N|$ gets smaller and smaller; that is,

$$r^N \to 0 \quad \text{as} \quad N \to +\infty$$

Therefore, from (3)

$$S_N \to \frac{a_1}{1 - r} \quad \text{as} \quad N \to +\infty$$

This statement leads to the following definition.

> **DEFINITION**   **The Sum of an Infinite Geometric Series**
>
> The **sum** S of an infinite geometric series, for which $|r| < 1$, is given by
>
> $$S = \frac{a_1}{1 - r}$$

▷ ILLUSTRATION 1

For series (1), $a_1 = 1$ and $r = \frac{1}{2}$. Therefore, if S is the sum of this series, we have from the definition

$$S = \frac{a_1}{1 - r}$$

$$= \frac{1}{1 - \frac{1}{2}}$$

$$= 2$$

This result agrees with statement (2).   ◄

Observe that the sum of an infinite geometric series is defined only when $|r| < 1$. If we have an infinite geometric series for which $|r| > 1$, then as N increases without bound, r^N will not approach a finite number; so by referring to formula (3), it is apparent that the infinite series does not have a sum. If $r = 1$, then $S_N = Na_1$, and as N increases without bound, $|S_N|$ also increases without bound.

▶ **EXAMPLE 1** *Finding the Sum of an Infinite Geometric Series*

Find the sum of the infinite geometric series

$$6 + 4 + \tfrac{8}{3} + \ldots + 6(\tfrac{2}{3})^{n-1} + \ldots$$

Solution From the definition with $a_1 = 6$ and $r = \tfrac{2}{3}$

$$S = \frac{a_1}{1 - r}$$

$$= \frac{6}{1 - \tfrac{2}{3}}$$

$$= 18 \qquad\qquad\qquad ◀$$

An application of infinite geometric series is to express a given nonterminating repeating decimal as a fraction. We can thus show that such a decimal numeral represents a rational number. To indicate a nonterminating repeating decimal, we write a bar over the repeated digits. Then $0.33\overline{3}$ indicates $0.3333 \ldots$ and $4.024\overline{24}$ indicates $4.024242424 \ldots$.

▷ **ILLUSTRATION 2**

The nonterminating repeating decimal $0.33\overline{3}$ can be written as

$$0.3 + 0.03 + 0.003 + 0.0003 + \ldots$$

$$\Leftrightarrow \quad \frac{3}{10} + \frac{3}{100} + \frac{3}{1000} + \frac{3}{10,000} + \ldots + \frac{3}{10^n} + \ldots$$

which is an infinite geometric series with $a_1 = \tfrac{3}{10}$ and $r = \tfrac{1}{10}$. If S is the sum of this series,

$$S = \frac{a_1}{1 - r}$$

$$= \frac{\tfrac{3}{10}}{1 - \tfrac{1}{10}}$$

$$= \tfrac{1}{3}$$

Therefore the nonterminating repeating decimal $0.33\overline{3}$ and the fraction $\tfrac{1}{3}$ are representations for the same rational number. ◀

▶ **EXAMPLE 2** *Expressing a Nonterminating Repeating Decimal as a Fraction*

Express the nonterminating repeating decimal $4.0242\overline{424}$ as a fraction.

Solution   The given decimal can be written as

$$4 + \left[\frac{24}{1000} + \frac{24}{100,000} + \frac{24}{10,000,000} + \dots + \frac{24}{10^{2n+1}} + \dots \right]$$

The series in brackets is an infinite geometric series for which $a_1 = \frac{24}{1000}$ and $r = \frac{1}{100}$. If S is the sum of this series, then

$$S = \frac{a_1}{1 - r}$$

$$= \frac{\frac{24}{1000}}{1 - \frac{1}{100}}$$

$$= \frac{\frac{24}{1000}}{\frac{99}{100}}$$

$$= \frac{24}{990}$$

$$= \frac{4}{165}$$

Thus

$$4.0242\overline{424} = 4 + \tfrac{4}{165}$$

$$= \tfrac{664}{165}$$ ◀

▶ **EXAMPLE 3** *Solving a Word Problem Having an Infinite Geometric Series as a Mathematical Model*

A ball is dropped from a height of 36 m, and each time it strikes the ground it rebounds to a height of two-thirds of the distance from which it fell. Find the total distance traveled by the ball before it comes to rest.

Solution   Let d meters be the total distance traveled by the ball. To obtain d, we must add the distances it falls as well as the distances it rebounds. Thus

$$d = 36 + [(36)(\tfrac{2}{3}) + (36)(\tfrac{2}{3}) + (36)(\tfrac{2}{3})^2 + (36)(\tfrac{2}{3})^2 + \dots]$$

$$= 36 + 2[(36)(\tfrac{2}{3}) + (36)(\tfrac{2}{3})^2 + \dots + (36)(\tfrac{2}{3})^n + \dots]$$

The series in brackets is an infinite geometric series with $a_1 = (36)(\tfrac{2}{3})$ and

$r = \frac{2}{3}$. If S is the sum of this series, then

$$S = \frac{a_1}{1 - r}$$

$$= \frac{24}{1 - \frac{2}{3}}$$

$$= 72$$

Therefore

$$d = 36 + 2(72)$$

$$= 180$$

Conclusion: The ball travels 180 m before coming to rest.   ◄

In Section 12.7 we proved the binomial theorem for the expansion of $(a + b)^n$ when n is a positive integer. The following theorem states that the binomial $(1 + b)^n$ has an expansion when n is any real number and $|b| < 1$. The expansion, called a **binomial series,** is another infinite series you will encounter in your calculus course.

THEOREM 1

If n is any real number and $|b| < 1$,

$$(1 + b)^n = 1 + nb + \frac{n(n - 1)}{2!}b^2 + \frac{n(n - 1)(n - 2)}{3!}b^3 + \ldots$$

$$+ \frac{n(n - 1)(n - 2) \cdot \ldots \cdot (n - r + 1)}{r!}b^r + \ldots$$

We omit the proof of Theorem 1 because it depends on concepts studied in calculus.

 ▶ **EXAMPLE 4** *Writing a Binomial Series*

Write a binomial series for $(1 + \sqrt{x})^{-1}$ if $\sqrt{x} < 1$.

Solution   From Theorem 1, where n is -1 and b is $\sqrt{x}$,

$$(1 + \sqrt{x})^{-1} = 1 + (-1)x^{1/2} + \frac{(-1)(-2)}{2!}(x^{1/2})^2 + \frac{(-1)(-2)(-3)}{3!}(x^{1/2})^3$$

$$+ \ldots + \frac{(-1)(-2)(-3) \cdot \ldots \cdot (-r)}{r!}(x^{1/2})^r + \ldots$$

$$\Leftrightarrow \quad (1 + x^{1/2})^{-1} = 1 - x^{1/2} + x - x^{3/2} + \ldots + (-1)^r x^{r/2} + \ldots \quad ◄$$

Observe that the infinite series obtained in Example 4 is an infinite geometric series where the first term is 1 and the common ratio is $-x^{1/2}$. We defined the sum of an infinite geometric series for which the first term is a_1 and the common ratio is r, with $|r| < 1$, to be

$$\frac{a_1}{1 - r}$$

If $a_1 = 1$ and $r = -x^{1/2}$, then

$$\frac{a_1}{1 - r} = \frac{1}{1 - (-x^{1/2})}$$

$$= \frac{1}{1 + x^{1/2}}$$

$$= (1 + x^{1/2})^{-1}$$

in agreement with the result of Example 4.

Infinite series are applied to approximate irrational numbers. In the following two examples we show the procedure by computing approximate values of $\sqrt[3]{25}$ and $\sin 0.3$ by infinite series. In Exercises 31 through 42 you are asked to find approximate values of other irrational numbers by infinite series.

 ▶ **EXAMPLE 5** *Computing with a Binomial Series*

Find the value of $\sqrt[3]{25}$ accurate to three decimal places by using the binomial series for $(1 + x)^{1/3}$. Check the answer on a calculator.

Solution   From Theorem 1 we have, if $|x| < 1$,

$$(1 + x)^{1/3} = 1 + \frac{1}{3}x + \left(\frac{1}{3}\right)\left(-\frac{2}{3}\right)\frac{x^2}{2!} + \left(\frac{1}{3}\right)\left(-\frac{2}{3}\right)\left(-\frac{5}{3}\right)\frac{x^3}{3!} + \ldots$$

Because

$$\sqrt[3]{25} = \sqrt[3]{27} \, \sqrt[3]{\tfrac{25}{27}}$$

$$= 3 \sqrt[3]{1 - \tfrac{2}{27}}$$

then

$$\sqrt[3]{25} = 3(1 - \tfrac{2}{27})^{1/3} \tag{4}$$

From the binomial series for $(1 + x)^{1/3}$ with $x = -\frac{2}{27}$, we have

$$\left(1 - \frac{2}{27}\right)^{1/3} = 1 + \frac{1}{3}\left(-\frac{2}{27}\right) - \frac{2}{3^2 \cdot 2!}\left(-\frac{2}{27}\right)^2 + \frac{2 \cdot 5}{3^3 \cdot 3!}\left(-\frac{2}{27}\right)^3 + \ldots$$

$$= 1 - 0.0247 - 0.0006 - 0.00003 - \ldots$$

The fourth term has a zero in the fourth decimal place, and so has each successive term. It can be proved that no term after the third term affects the first four decimal places. Using the first three terms of the series, we obtain

$$(1 - \tfrac{2}{27})^{1/3} \approx 0.9747$$

Substituting into (4), we have

$$\sqrt[3]{25} \approx 3(0.9747)$$
$$= 2.9241$$

Rounding off to three decimal places gives $\sqrt[3]{25} \approx 2.924$. Our calculator gives the same result. ◀

 In calculus you will learn how to express many different kinds of functions as infinite series. Three such series that represent the function for all values of x are

$$\sin x = x - \frac{x^3}{3!} + \frac{x^5}{5!} - \frac{x^7}{7!} + \ldots + \frac{(-1)^{n-1}x^{2n-1}}{(2n-1)!} + \ldots \qquad (5)$$

$$\cos x = 1 - \frac{x^2}{2!} + \frac{x^4}{4!} - \frac{x^6}{6!} + \ldots + \frac{(-1)^{n-1}x^{2n-2}}{(2n-2)!} + \ldots \qquad (6)$$

$$e^x = 1 + x + \frac{x^2}{2!} + \frac{x^3}{3!} + \frac{x^4}{4!} + \ldots + \frac{x^{n-1}}{(n-1)!} + \ldots \qquad (7)$$

In the following example we use series (5) to approximate a particular sine function value. You are asked to use series (5), (6), and (7) in Exercises 37 through 42.

 ▶ **EXAMPLE 6** *Using an Infinite Series to Approximate a Sine Function Value*

Find the value of sin 0.3 accurate to four decimal places by using series (5) for sin x. Check the answer on a calculator.

Solution From series (5) with $x = 0.3$, we have

$$\sin 0.3 = 0.3 - \frac{(0.3)^3}{3!} + \frac{(0.3)^5}{5!} - \frac{(0.3)^7}{7!} + \ldots$$
$$= 0.3 - 0.0045 + 0.00002 - 0.00000004 + \ldots$$

No term after the third affects the first five decimal places. From the first three terms of this series, we have

$$\sin 0.3 \approx 0.29552$$

Rounding off to four decimal places gives sin 0.3 $\approx$ 0.2955. We obtain the same result on our calculator. ◀

EXERCISES 12.8

In Exercises 1 through 10, find the sum of the infinite geometric series.

1. $16 + 4 + 1 + \ldots$

2. $\frac{1}{3} + \frac{1}{9} + \frac{1}{27} + \ldots$

3. $60 - 6 + 0.6 - \ldots$

4. $4 - 1.6 + 0.64 - \ldots$

5. $\frac{2}{3} + \frac{1}{9} + \frac{1}{54} + \ldots$

6. $1 + (1.04)^{-1} + (1.04)^{-2} + \ldots$

7. $\frac{4}{3} - 1 + \frac{3}{4} - \ldots$

8. $-2 - \frac{1}{2} - \frac{1}{8} - \ldots$

9. $3 + \sqrt{3} + 1 + \ldots$

10. $(2 + \sqrt{3}) + 1 + (2 - \sqrt{3}) + \ldots$

In Exercises 11 through 22, express the nonterminating repeating decimal as a fraction.

11. $0.66\overline{6}$

12. $0.272\overline{72}$

13. $0.8181\overline{81}$

14. $0.252252\overline{252}$

15. $2.99\overline{9}$

16. $3.141614161\overline{4161}$

17. $1.234234\overline{234}$

18. $7.99\overline{9}$

19. $0.465346534\overline{653}$

20. $2.045045\overline{045}$

21. $3.22544\overline{4}$

22. $6.50711\overline{1}$

23. Express the nonterminating repeating decimal $2.464\overline{646}$ as a fraction by two methods: **(a)** Consider $2.464\overline{646}$ as $2 + 0.46 + 0.0046 + 0.000046 + \ldots$ and **(b)** consider $2.464\overline{646}$ as $2.4 + 0.064 + 0.00064 + 0.0000064 + \ldots$.

24. Express the nonterminating repeating decimal $5.1696\overline{969}$ as a fraction by two methods: **(a)** Consider $5.1696\overline{969}$ as $5.1 + 0.069 + 0.00069 + 0.0000069 + \ldots$ and **(b)** consider $5.1696\overline{969}$ as $5.16 + 0.0096 + 0.000096 + 0.00000096 + \ldots$.

 In Exercises 25 through 30, find the first four terms of the binomial series for the given expression.

25. $(1 + x^2)^{-1}$ $x^2 < 1$

26. $(1 - x)^{1/3}$ $|x| < 1$

27. $(1 - 2x)^{1/2}$ $|x| < \frac{1}{2}$

28. $(1 - x^2)^{-1/2}$ $x^2 < 1$

29. $(8 + x)^{1/3}$ $|x| < 8$

30. $(3 + x)^{-1}$ $|x| < 3$

 In Exercises 31 through 36, compute the value accurate to three decimal places by using a binomial series. Check your answer on your calculator.

31. $\sqrt{1.04}$ **32.** $\sqrt[3]{0.99}$ **33.** $\sqrt[3]{63}$

34. $\sqrt{38}$ **35.** $\sqrt[4]{620}$ **36.** $\dfrac{1}{\sqrt[4]{15}}$

 37. Compute the value of e accurate to four decimal places by using series (7).

38. Compute the value of $1/e$ accurate to four decimal places by using series (7). Check your answer on your calculator.

In Exercises 39 through 42 use either series (5) or (6) to compute the function value accurate to four decimal places. Check your answer on your calculator.

39. $\cos 0.4$ **40.** $\sin 1.2$

41. $\sin 22°$ **42.** $\cos 17°$

In Exercises 43 through 47, solve the word problem by using an infinite geometric series as a mathematical model. Be sure to write a conclusion.

43. A ball is dropped from a height of 12 m. Each time it strikes the ground it bounces back to a height of three-fourths of the distance from which it fell. Determine the total distance traveled by the ball before it comes to rest.

44. What is the total distance traveled by a tennis ball before coming to rest if it is dropped from a height of 100 m and if, after each fall, it rebounds $\frac{11}{20}$ of the distance from which it fell?

45. The path of each swing, after the first, of a pendulum bob is 0.93 as long as the path of the previous swing from one side to the other side. If the path of the first swing is 28 cm long, and air resistance eventually brings the pendulum to rest, how far does the bob travel before it comes to rest?

46. An equilateral triangle has sides of length 4 units; therefore its perimeter is 12 units. Another equilateral triangle is constructed by drawing line segments through the midpoints of the sides of the first triangle. This triangle has sides of length 2 units, and its perimeter is 6 units. If this procedure can be repeated an unlimited number of times, what is the total perimeter of all the triangles that are formed?

47. After a woman riding a bicycle removes her feet from the pedals, the front wheel rotates 100 times during the first 10 sec. Then in each succeeding 10-sec time period the wheel rotates four-fifths as many times as it did the previous period. Determine the number of rotations of the wheel before the bicycle stops.

48. Find an infinite geometric series whose sum is 6 and such that each term is four times the sum of all the terms that follow it.

 49. *Euler's formula* named for Leonhard Euler is

$$e^{it} = \cos t + i \sin t$$

where $i = \sqrt{-1}$. This formula gives a definition of e^{it}. **(a)** Show that Euler's formula is consistent with series (5), (6), and (7) by doing the following: obtain a series for e^{it} by substituting it for x in series (7) and simplifying powers of i; substitute t for x in series (5)

and (6); add the series for $\cos t$ to i times the series for $\sin t$ and obtain the series for e^{it}. **(b)** Use Euler's formula to show that

$$e^{i\pi} + 1 = 0$$

which is a surprising equation involving the five important constants: e, i, π, 1, and 0.

50. The infinite series

$$1 + \frac{1}{2} + \frac{1}{3} + \frac{1}{4} + \frac{1}{5} + \ldots + \frac{1}{n} + \ldots$$

whose terms are the reciprocals of the positive integers is called the *harmonic series*. How many terms of the harmonic series are required before the sum is greater than **(a)** 2, **(b)** 3, **(c)** 4, and **(d)** 8? **(e)** Do you think the harmonic series has a finite sum? Explain how you arrived at your answer.

CHAPTER 12   REVIEW

▶ LOOKING BACK

12.1 To solve a system of linear equations we first wrote the system in triangular form and then used back substitution to solve for the three variables. This procedure led to the Gaussian reduction method for solving a linear system by matrices, where we applied the theorem that gives rules for performing elementary row operations on the elements of a matrix. We showed how the method can be used to determine if a system of linear equations is consistent or inconsistent as well as independent or dependent.

12.2 Properties of matrices included the determinant of a matrix, the product of a scalar and a matrix, the product of two matrices, the multiplicative identity matrix of a set of square matrices of a particular order, and the multiplicative inverse of a matrix. These properties were applied to solve linear systems by matrix inverses.

12.3 When using systems of linear equations to decompose a rational expression into partial fractions, we considered four cases regarding the factors of the denominator: (i) all are linear and none is repeated; (ii) all are linear and some are repeated; (iii) all are linear and quadratic and none is repeated; (iv) all are linear and

quadratic and some of the quadratic factors are repeated.

12.4 Our treatment of sequences and series was based on the function concept. We introduced sigma notation and used it to write a series in abbreviated form.

12.5 The principle of mathematical induction was the basis of our mathematical-induction proofs of formulas, inequalities, statements, and theorems.

12.6 For both arithmetic and geometric sequences, we obtained formulas for the Nth element and for the sums of the associated series. Mathematical induction was used for the proofs of the formulas for the Nth element. We then solved word problems having these sequences and series as mathematical models.

12.7 The binomial theorem, proved by mathematical induction, was applied to expand $(a + b)^n$ for positive-integer values of n, and we presented the formula for finding a particular term in such expansions.

12.8 We applied the formula for the sum of an infinite geometric series and showed how it can be used to express a nonterminating repeating decimal as a fraction. We also solved word problems having an infinite geometric series as a mathematical model. Another infinite series, the binomial series, enabled us to expand $(1 + b)^n$ when n is any real number and $|b| < 1$. The binomial series and infinite series for $\sin x$, $\cos x$, and e^x were used to compute approximate values of irrational numbers.

▶ REVIEW EXERCISES

In Exercises 1 through 4, find the solution set of the system by the elimination method. If the equations are either inconsistent or dependent, then indicate that this is the case.

1. $\begin{cases} x + 2y + 2z = 1 \\ x - 3y - 2z = 4 \\ 6x + y - z = 21 \end{cases}$
2. $\begin{cases} 6x + 4y + 5z = 14 \\ 4x - 3y - z = 2 \\ 14x - 10y - 9z = 10 \end{cases}$

3. $\begin{cases} 4x - 3y - 6z = 7 \\ 2x - y - 4z = 3 \\ 3x - 2y - 5z = 5 \end{cases}$
4. $\begin{cases} 3x + 3y - 5z = 4 \\ 6x + 2y - 3z = 7 \\ 3x - 5y + 9z = 5 \end{cases}$

In Exercises 5 through 8, use the Gaussian reduction method to find the solution set of the system. If the equations are either inconsistent or dependent, then indicate that this is the case.

5. $\begin{cases} 3x - 5y + 2z = 4 \\ 4x + 2y + 7z = 1 \\ 5x - 9y - 3z = -11 \end{cases}$

6. $\begin{cases} 3x - 2y - 2z = 3 \\ 6x + 4y + 3z = 3 \\ 3x - 6y + z = -2 \end{cases}$

7. $\begin{cases} w + x + y + 3z = 3 \\ 3w + x - y = 0 \\ 2w - 2x - y + 6z = 4 \\ 4w - x - 2y - 3z = 0 \end{cases}$

8. $\begin{cases} 2w + 2x - z = 3 \\ w + 2x - 2z = 2 \\ w + y + z = 0 \\ w - x - 2z = -4 \end{cases}$

In Exercises 9 through 11, the system has more variables than equations. Use the Gaussian reduction method to find the solution set of the system.

9. $\begin{cases} 2x - 3y + 2z = -1 \\ -3x + 4y - 4z = 3 \end{cases}$

10. $\begin{cases} 2x + 3y + 4z = 5 \\ 5x - y - 3z = -1 \end{cases}$

11. $\begin{cases} 3w + x + 2y - 2z = 6 \\ 2w - 3x - 4y + 3z = 1 \\ w + 2x + 3y - 5z = 6 \end{cases}$

In Exercises 12 through 14, the system has more equations than variables. Use the Gaussian reduction method to determine if the equations are consistent or inconsistent. If they are consistent, find the solution set.

12. $\begin{cases} 3x + 4y = 5 \\ 2x - 3y = 9 \\ 5x + 8y = 7 \end{cases}$

13. $\begin{cases} 2x + 4y + 3z = 8 \\ x - 2y + 3z = 1 \\ 2x - 6y - 6z = 9 \\ x + 2y - 6z = 9 \end{cases}$

14. $\begin{cases} 4x - 4y + z = 5 \\ 2x + y + 3z = 6 \\ 6x + 3y + 4z = 8 \\ 3x - 2y + 3z = 5 \end{cases}$

In Exercises 15 and 16, evaluate the determinant.

15. (a) $\begin{vmatrix} 5 & -7 \\ 3 & -4 \end{vmatrix}$ (b) $\begin{vmatrix} 2 & 3 & -1 \\ 3 & 1 & 5 \\ 2 & 3 & 4 \end{vmatrix}$

16. (a) $\begin{vmatrix} 6 & 3 \\ -2 & -4 \end{vmatrix}$ (b) $\begin{vmatrix} 1 & 1 & -1 \\ 0 & 1 & 1 \\ -4 & 2 & -3 \end{vmatrix}$

In Exercises 17 through 20, find the product.

17. (a) $2\begin{bmatrix} -4 & 0 \\ 1 & -3 \\ 8 & -2 \end{bmatrix}$ (b) $-4\begin{bmatrix} 2 & -1 & 0 \\ 3 & 1 & -2 \end{bmatrix}$

18. (a) $-3\begin{bmatrix} -2 & 2 \\ 3 & 5 \\ -1 & -3 \end{bmatrix}$ (b) $3\begin{bmatrix} -2 & 1 & -1 \\ -3 & 0 & 2 \end{bmatrix}$

19. (a) $\begin{bmatrix} 1 & 1 \\ 0 & 1 \end{bmatrix}\begin{bmatrix} 1 & 1 \\ -1 & 1 \end{bmatrix}$ (b) $\begin{bmatrix} 1 & 2 \\ -1 & 3 \end{bmatrix}\begin{bmatrix} 2 & 1 & 3 \\ 0 & -1 & 0 \end{bmatrix}$

20. (a) $[1 \quad -1]\begin{bmatrix} 1 \\ -1 \end{bmatrix}$ **(b)** $[-1 \quad 2 \quad 0]\begin{bmatrix} -2 & 1 & 0 \\ 0 & -1 & 1 \\ 3 & 0 & 2 \end{bmatrix}$

In Exercises 21 and 22, the given matrix is H. Prove that H is invertible, find H^{-1}, and show that $HH^{-1} = I_2$.

21. $\begin{bmatrix} 2 & -1 \\ 3 & 4 \end{bmatrix}$ **22.** $\begin{bmatrix} -2 & 1 \\ 4 & -3 \end{bmatrix}$

In Exercises 23 and 24, the given matrix is H. Determine if H is invertible. If H is invertible, compute H^{-1} on your calculator and show that $HH^{-1} = I_3$.

23. $\begin{bmatrix} 1 & 6 & 0 \\ 4 & 1 & 2 \\ -3 & 0 & 5 \end{bmatrix}$ **24.** $\begin{bmatrix} 1 & -2 & -1 \\ 0 & -4 & 2 \\ 3 & 1 & 7 \end{bmatrix}$

In Exercises 25 through 28, solve the system of equations by a matrix inverse. Compute the inverse of the matrix of coefficients on your calculator.

25. $\begin{cases} x + y - z = 0 \\ 2x - y + 3z = -1 \\ 2y - 3z = 1 \end{cases}$ **26.** $\begin{cases} x + 2z = 2 \\ 3x + y = 5 \\ 2y - 3z = -5 \end{cases}$

27. $\begin{cases} 2x + y = 0 \\ 4x - 3z = -2 \\ 2y + 3z = 2 \end{cases}$ **28.** $\begin{cases} 4x + 2y - 3z = 10 \\ 2x - 3y - 4z = 8 \\ 6x - 5y - 2z = 6 \end{cases}$

In Exercises 29 through 34, decompose the fraction into partial fractions.

29. $\dfrac{2x^2 - 15x - 32}{(x-1)(x^2 + 6x + 8)}$ **30.** $\dfrac{x^3 - 6x^2 + 8x - 1}{x^2 - 6x + 9}$

31. $\dfrac{x^2 + 2x - 2}{x^3 - x^2 - x - 2}$ **32.** $\dfrac{x^2 - 3x + 6}{2x^3 - 3x^2 - 2x + 3}$

33. $\dfrac{x^4 + x^3 + x - 4}{x^4 + 2x^2 + 1}$ **34.** $\dfrac{x^4 - x^2 + 9}{x(x^2 + 3x + 3)^2}$

In Exercises 35 and 36, write the first six elements of the sequence whose general element is given.

35. $a_n = (-1)^{n-1}\dfrac{2n - 1}{3^n}$

36. $a_n = (-1)^n \dfrac{n^2 + 1}{2n}x^{2n-1}$

In Exercises 37 and 38, write the series and find the sum of the series.

37. $\displaystyle\sum_{i=2}^{7} \frac{i + 1}{i - 1}$ **38.** $\displaystyle\sum_{k=1}^{4} (-1)^{k+1}\frac{3k}{2^k}$

In Exercises 39 and 40, write the series with sigma notation (there is no unique solution).

39. $\frac{1}{2}x^2 - \frac{1}{4}x^4 + \frac{1}{8}x^6 - \frac{1}{16}x^8$

40. $-x + \frac{1}{4}x^3 - \frac{1}{7}x^5 + \frac{1}{10}x^7 - \frac{1}{13}x^9 + \frac{1}{16}x^{11}$

In Exercises 41 through 44, find the sum of the series.

41. $\displaystyle\sum_{j=1}^{10} 3\left(\frac{1}{2}\right)^j$ **42.** $\displaystyle\sum_{k=1}^{30} (3k + 1)$

43. $\displaystyle\sum_{i=1}^{20} \left(\frac{1}{3}i + 3\right)$ **44.** $\displaystyle\sum_{i=1}^{10} (1.01)^i$

In Exercises 45 through 48, find the number.

45. (a) $\dfrac{12!}{4!\,9!}$ **(b)** $\dfrac{4! + 5!}{3! + 5!}$

46. (a) $\dfrac{3!\,7!}{9!}$ **(b)** $\dfrac{4! + 5!}{6!}$

47. (a) $\dbinom{9}{3}$ **(b)** $\dbinom{40}{36}$

48. (a) $\dbinom{10}{4}$ **(b)** $\dbinom{18}{16}$

In Exercises 49 through 52, determine if the elements form an arithmetic or geometric sequence. If they do, write the next two elements of the sequence.

49. (a) 12, 16, 20 **(b)** 27, −9, 3

50. (a) 10, 4, −2 **(b)** 1, 5, 25

51. (a) $\sqrt{2}, \sqrt[4]{2}, 1$ **(b)** $\frac{2}{3}, 2, 3$

52. (a) 5.55, 4.44, 3.33 **(b)** $\frac{1}{9}, \frac{1}{6}, \frac{1}{3}$

53. Find x so that the numbers $\frac{1}{16}, \frac{1}{4}$, and x form an arithmetic sequence.

54. Find x so that the numbers 25, x, and 9 form a geometric sequence.

55. Find the first element of a geometric sequence whose fourth element is −3 and whose eighth element is −243.

56. Find the sum of the positive odd integers between 10 and 100.

57. In the arithmetic sequence whose first three elements are −8, −5, and −2, which element is 52?

58. Insert three arithmetic means between -1 and 15.

59. Insert three arithmetic means between $\frac{1}{2}$ and $\frac{2}{3}$.

60. Insert seven geometric means between $\frac{1}{4}$ and 64.

61. Insert five geometric means between 192 and 3.

62. (a) Find the arithmetic mean of 2 and 18. **(b)** Find the geometric mean between 2 and 18.

63. (a) Find the arithmetic mean of 32 and 8. **(b)** Find the geometric mean between -32 and -8.

64. Find the arithmetic mean of the following set of test scores: 83, 91, 62, 75, 96, 84, 70, and 89.

65. Find the thirtieth element of the arithmetic sequence whose seventeenth element is 7 and whose forty-seventh element is 31.

66. Find the eighth element of the geometric sequence whose third element is 192 and whose seventh element is 12.

67. (a) How many numbers between 100 and 500 are divisible by 8? **(b)** What is their sum?

68. Show that the reciprocals of the elements of a geometric sequence also form a geometric sequence.

In Exercises 69 and 70, find the sum of the infinite geometric series.

69. $0.4 + 0.02 + 0.001 + \ldots$

70. $2 + \sqrt{2} + 1 + \ldots$

In Exercises 71 and 72, express the nonterminating repeating decimal as a fraction.

71. $0.72\overline{72}$

72. $4.66363\overline{63}$

In Exercises 73 through 76, use mathematical induction to prove that the formula is true for all positive-integer values of n.

73. $\displaystyle\sum_{i=1}^{n} (4i + 1) = n(2n + 3)$

74. $\displaystyle\sum_{i=1}^{n} 4^i = \frac{4(4^n - 1)}{3}$

75. $\displaystyle\sum_{i=1}^{n} \frac{1}{(3i - 2)(3i + 1)} = \frac{n}{3n + 1}$

76. $\displaystyle\sum_{i=1}^{n} i(2i + 1) = \frac{4n^3 + 9n^2 + 5n}{6}$

In Exercises 77 and 78, expand the power of the binomial.

77. $(2x - y)^7$

78. $(a + 3b)^8$

In Exercises 79 and 80, write and simplify the first four terms in the expansion of the binomial.

79. $(x + 2x^{-1})^{20}$

80. $(4w^{-1} - \frac{1}{2}w^2)^{15}$

81. Find the tenth term of the expansion of $(t^{1/2} - t^{-1/2})^{15}$.

82. Find the middle term of the expansion of $(x^2 + 3y)^6$.

83. Find the term involving z^{12} in the expansion of $(z^2 - \frac{1}{2})^{11}$.

84. Find the term involving u^{40} in the expansion of $\left(2u^4 - \dfrac{1}{2u}\right)^{15}$.

In Exercises 85 and 86, compute the function value accurate to three decimal places by using a binomial series. Check your answer on your calculator.

85. $\sqrt[3]{25}$

86. $\sqrt[4]{17}$

In Exercises 87 through 90, compute the function value accurate to four decimal places by using series (5), (6), or (7) of Section 12.8. Check your answer on your calculator.

87. $\sin 0.6$

88. $\cos 1.3$

89. $\sqrt{e}$

90. $\sqrt[3]{e}$

91. Find an equation of the parabola whose axis is parallel to the y axis if it contains the points $(2, 3)$, $(1, 0)$, and $(5, 36)$. See the hint for Exercise 39 in Exercises 12.1.

92. Find an equation of the circle containing the points $(-2, -1)$, $(5, 0)$, and $(2, 1)$. See the hint for Exercises 37 and 38 in Exercises 12.1.

In Exercises 93 and 94, solve the word problem by using a system of linear equations as a mathematical model. Be sure to write a conclusion.

93. A man has a total of \$15,000 in three investments: bonds that pay 6 percent annual interest; a savings account that pays 5 percent annual interest; and a business. Two years ago the business lost 3 percent and his net income from the three investments was \$550. Last year the business earned 9 percent and his net income from the three investments was \$910. How much does he have in each of the investments?

94. Determine the degree measurements of the angles of a triangle if the measurement of one angle is one-half the measurement of the second, and the measurement of the third is twice the sum of the measurements of the first two.

In Exercises 95 through 98, solve the word problem by using as a mathematical model a sequence or series, either arithmetic or geometric. Be sure to write a conclusion.

95. A pile of logs has 30 logs in the bottom layer, 29 logs in the next to bottom layer, and so on, and the top layer contains 5 logs; each layer except the last contains 1 less log than the layer beneath it. How many logs are in the pile?

96. In a certain culture the number of bacteria increases 20 percent every 30 min. If there are 1000 bacteria present initially, find a formula for determining the number of bacteria in the culture at the end of t hours. How many bacteria are in the culture at the end of 5 hr?

97. **(a)** How many ancestors, to the nearest thousand, did you have 20 generations ago under the assumption that each ancestor appears only once in your family tree? **(b)** What is the total number of ancestors, to the nearest thousand, in all 20 generations?

98. A man borrows $20,000 and places a mortgage on his home. He agrees that at the end of each year for ten years he will repay $2000 of the principal together with interest at the rate of 15 percent per year on the amount outstanding during the year. What is the total amount to be paid in ten years?

In Exercises 99 and 100, use the interpretation of the binomial coefficient given for Exercises 41 through 44 in Exercises 12.7.

99. In how many ways can a tenor choose three operatic arias from a selection of ten to be sung at an audition?

100. Twelve people are qualified to operate a machine that requires four persons at one time. How many different groups of four people can operate the machine?

In Exercises 101 and 102, solve the word problem by using an infinite geometric series as a mathematical model. Be sure to write a conclusion.

101. A sheet of paper is torn in half and then each half is again torn in half, and so on, until the tearing-in-half is done 30 times. Then each of the pieces of paper is placed in a pile with one piece on top of another. If the original sheet of paper has a thickness of 0.01 cm, what will be the height of the pile to the nearest kilometer?

102. The path of each swing, after the first, of a pendulum bob is 80 percent as long as the path of the previous swing from one side to the other side. If the path of the first swing is 18 in. long, and air resistance eventually brings the pendulum to rest, how far does the bob travel before it comes to rest?

103. Three numbers whose sum is 35 form a geometric sequence. If 1 is subtracted from the first number, 2 is subtracted from the second number, and 8 is subtracted from the third number, the resulting differences form an arithmetic sequence. What are the numbers?

104. Prove by mathematical induction that the following inequality is true for all positive-integer values of n: $2^{n+3} < (n + 3)!$.

105. Prove by mathematical induction that the following formula is true for all positive-integer values of n:

$$\sum_{i=0}^{n} \binom{i + 3}{3} = \binom{n + 4}{4}$$

106. Prove by mathematical induction that $x - y$ is a factor of $x^{2n} - y^{2n}$ for all positive-integer values of n.

Appendix

A.1 PROPERTIES OF REAL NUMBERS

A **binary** operation on the set of real numbers is a rule that assigns to any two real numbers a and b, taken in a definite order, a number c. **Addition** is a binary operation on R because addition assigns to the real numbers a and b a number, denoted by $a + b$, called the **sum** of a and b. The numbers a and b are called **addends** (or **terms**). **Multiplication** is also a binary operation on R because multiplication assigns to the real numbers a and b a number denoted by ab, called the **product** of a and b. The numbers a and b are called **factors.** The following seven axioms give laws governing the operations of addition and multiplication on the set R. In the illustrations that follow the axioms, we use elements of the set of natural numbers, and we assume that the sum and product of natural numbers are known.

AXIOM 1   *Closure Laws*

If a and b are real numbers, then $a + b$ and ab are unique real numbers

---

Axiom 1 guarantees that whenever the operations of addition and multiplication are performed on two real numbers, the sum and product are real numbers. The axiom is called *closure laws* because a set is said to be **closed** with respect to an operation if, whenever the operation is performed on elements of the set, an element of the set is obtained.

▷ ILLUSTRATION 1

(a) The set $\{1, 2, 3, 4\}$ is not closed with respect to addition because whenever the operation of addition is performed on elements of this set we do not necessarily obtain an element of the set. For instance, $2 + 3 = 5$, but 5 is not an element of the set.

(b) The set of even natural numbers is closed with respect to both addition and multiplication because whenever either addition or multiplication is performed on two even natural numbers, the sum and product are even natural numbers. For instance, 6 and 8 are even natural numbers, and $6 + 8 = 14$, and $6 \cdot 8 = 48$; 14 and 48 are even natural numbers. ◀

AXIOM 2   *Commutative Laws*

If a and b are real numbers,

$$a + b = b + a \quad \text{and} \quad ab = ba$$

Axiom 2 states that the sum and product of two real numbers are not affected by the order of the numbers.

▷ ILLUSTRATION 2

(a) By the commutative law for addition, $4 + 9 = 9 + 4$.

(b) By the commutative law for multiplication, $5 \cdot 7 = 7 \cdot 5$.   ◀

AXIOM 3   *Associative Laws*
If a, b, and c are real numbers, $\qquad a + (b + c) = (a + b) + c \qquad$ and $\qquad a(bc) = (ab)c$

Although addition and multiplication are binary operations, the expressions $a + b + c$ and abc are meaningful because the grouping symbols (parentheses and brackets) can be inserted in any possible way without affecting the results. From Axiom 3 it follows that the sum of three real numbers can be obtained by grouping the addends in either of two ways and that the product of three real numbers can be found by grouping the factors in either of two ways.

▷ ILLUSTRATION 3

(a)
$$\begin{aligned} 2 + 3 + 4 &= 2 + (3 + 4) & 2 + 3 + 4 &= (2 + 3) + 4 \\ &= 2 + 7 & &= 5 + 4 \\ &= 9 & &= 9 \end{aligned}$$

(b)
$$\begin{aligned} 3 \cdot 5 \cdot 6 &= 3 \cdot (5 \cdot 6) & 3 \cdot 5 \cdot 6 &= (3 \cdot 5) \cdot 6 \\ &= 3 \cdot 30 & &= 15 \cdot 6 \\ &= 90 & &= 90 \end{aligned}$$
◀

AXIOM 4   *Distributive Law*
If a, b, and c are real numbers, $\qquad a(b + c) = ab + ac$

▷ ILLUSTRATION 4

(a) When evaluating $4(7 + 2)$, the parentheses are used to indicate that we should first perform the operation of addition of $7 + 2$ and then the sum is multiplied by 4. Therefore we have

$$4(7 + 2) = 4 \cdot 9$$
$$= 36$$

(b) When computing $(4 \cdot 7) + (4 \cdot 2)$, the parentheses indicate that each of the multiplications is performed and then the sum of the two products is obtained. We have then

$$(4 \cdot 7) + (4 \cdot 2) = 28 + 8$$
$$= 36$$

Observe that from parts (a) and (b) we can conclude that

$$4(7 + 2) = (4 \cdot 7) + (4 \cdot 2)$$

and this equality is a special case of Axiom 4.   ◄

AXIOM 5   *Existence of Identity Elements*

There exist two distinct real numbers 0 and 1, called the **additive identity** and **multiplicative identity,** respectively, such that for any real number a,

$$a + 0 = a \quad \text{and} \quad a \cdot 1 = a$$

▷ **ILLUSTRATION 5**

$$8 + 0 = 8 \quad \text{and} \quad 8 \cdot 1 = 8$$ ◄

AXIOM 6   *Existence of Additive Inverse*

For every real number a, there exists a real number, called the **opposite of a** (or **additive inverse of a**), denoted by $-a$, such that

$$a + (-a) = 0$$

▷ **ILLUSTRATION 6**

The opposite of 4 is denoted by -4 and

$$4 + (-4) = 0$$ ◄

> **AXIOM 7**   *Existence of Multiplicative Inverse*
>
> For every real number a, except 0, there exists a real number, called the **reciprocal of a** (or **multiplicative inverse of a**), denoted by $\dfrac{1}{a}$, such that
>
> $$a \cdot \frac{1}{a} = 1$$

▷ ILLUSTRATION 7

The reciprocal of 9 is $\frac{1}{9}$ and

$$9 \cdot \frac{1}{9} = 1 \qquad\qquad\qquad\blacktriangleleft$$

Axioms 1 through 7 are called **field axioms,** and if these axioms are satisfied by a set of elements, then the set is called a **field** under the two binary operations involved. Thus the set R is a field under addition and multiplication. For the set J of integers, each of Axioms 1 through 6 is satisfied, but Axiom 7 is not satisfied; for instance, the integer 2 has no multiplicative inverse in J. Therefore the set of integers is not a field under addition and multiplication.

The field axioms do not imply any order of the real numbers. That is, by means of the field axioms alone we cannot state that 2 is greater than 1, 3 is greater than 2, and so on. However, we introduce an axiom giving the concept of a real number being positive. It is called the *order axiom* because the notion of a positive number is used in Section 1.1 to define what is meant by one real number being *greater than* or *less than* another.

> **AXIOM 8**   *Order Axiom*
>
> In the set of real numbers there exists a subset called the **positive numbers** such that
>
> **(i)** if a is a real number, exactly one of the following three statements holds:
>
> $$a = 0 \qquad a \text{ is positive} \qquad -a \text{ is positive}$$
>
> **(ii)** The sum of two positive numbers is positive.
> **(iii)** The product of two positive numbers is positive.

Because the set R of real numbers satisfies the order axiom and the field axioms, we say that R is an **ordered field.**

The opposites of the elements of the set of positive numbers form the set of *negative numbers,* as given in the following definition.

DEFINITION   **Negative Number**

The real number a is **negative** if and only if $-a$ is positive.

We use the terminology *negative* when referring to a negative number. For instance, -5 can be read "negative 5," as well as "the opposite of 5." The symbolism $-(-5)$ is read "the opposite of negative 5." In summary, then, we are using the symbol $-$ in two different ways: (1) to denote the opposite of a real number; and (2) to denote a negative number. If x can be any real number, $-x$ denotes the opposite of x. Observe that $-x$ is not necessarily a negative number; if x is -3, for instance, then $-x = 3$.

From Axiom 8 and the preceding definition, it follows that a real number is either positive, negative, or zero.

▷ **ILLUSTRATION 8**

In each of the following, there is a statement that is an immediate consequence of one of the field axioms; the axiom that applies is indicated.

(a) $5 + (-5) = 0$; additive inverse axiom (Axiom 6)
(b) $1 \cdot y = y$; identity element for multiplication (Axiom 5)
(c) $x + (y + 2) = (x + y) + 2$; associative law for addition (Axiom 3)
(d) $3 + (x + 4) = (x + 4) + 3$; commutative law for addition (Axiom 2)
(e) $2(5x) = (2 \cdot 5)x$; associative law for multiplication (Axiom 3)
(f) $8(u + 12) = 8u + 8 \cdot 12$; distributive law (Axiom 4)   ◀

EXERCISES A.1

In Exercises 1 through 24, the given equality follows immediately from one of the field axioms. Indicate which axiom applies. Assume that each variable is a real number.

1. $4 \cdot 5 = 5 \cdot 4$

2. $(6 + 2) + 4 = 6 + (2 + 4)$

3. $8 + 0 = 8$

4. $1 \cdot y = y$

5. $(5 + 2) + 4 = (2 + 5) + 4$

6. $17 + 41 = 41 + 17$

7. $3(xy) = (3x)y$

8. $(7a)b = b(7a)$

9. $\pi + (-\pi) = 0$

10. $x + (y + x) = (y + x) + x$

11. $7 + (8 + 11) = 7 + (11 + 8)$

12. $b + (-b) = 0$

13. $4 \cdot \frac{1}{4} = 1$

14. $x + 0 = x$

15. $3(a + b) = (a + b)3$

16. $11 + (y + 7) = (y + 7) + 11$

17. $a(b + 0) = ab$ **18.** $4(x + y) = 4x + 4y$

19. $0 \cdot 1 = 0$ **20.** $0 + 0 = 0$

21. $w + x(y + z) = w + (xy + xz)$

22. $(r + s)u + t = (s + r)u + t$

23. $(r + s) + (t + u) = r + [s + (t + u)]$

24. $(w + x) + (y + z) = [(w + x) + y] + z$

In Exercises 25 through 36, state whether the given set is closed under the indicated operation and give an example to illustrate your answer.

25. The set N of natural numbers; addition

26. The set of negative numbers; addition

27. The set of odd natural numbers; multiplication

28. The set of odd natural numbers; addition

29. $\{0\}$; addition   **30.** $\{0\}$; multiplication

31. $\{1\}$; multiplication   **32.** $\{1\}$; addition

33. $\{1, 2\}$; addition   **34.** $\{1, 2\}$; multiplication

35. $\{0, 1\}$; multiplication   **36.** $\{0, 1\}$; addition

Answers to Odd-Numbered Exercises

EXERCISES 1.1 (PAGE 13)

1. (a) $\in$; (b) $\notin$; (c) $\in$; (d) $\notin$ **3.** (a) $N \subseteq Q$; (b) $Q \subseteq R$; (c) $N \subseteq J$; (d) $J \subseteq R$ **5.** (a) $\subseteq$; (b) $\not\subseteq$;
(c) $\subseteq$; (d) $\subseteq$ **7.** (a) Q; (b) R; (c) J; (d) $\emptyset$ **9.** (a) $\{12, 571\}$; (b) $\{12, \frac{5}{3}, 0, -38, 571, -\frac{1}{10}, 0.666, \ldots, 16.34\}$;
(c) $\{\sqrt{7}, -\sqrt{2}, \pi\}$; (d) $\{12, 0, -38, 571\}$ **11.** $\{-10, -7, -\sqrt{5}, -2, -\frac{7}{4}, -\frac{5}{3}, -1, 0, \frac{2}{3}, \frac{3}{4}, \sqrt{2}, 3, 5, 21\}$
13. (a) $\{x \mid -9 < x < 8\}$; (b) $\{y \mid -12 < y < -3\}$; (c) $\{z \mid 4z - 5 < 0\}$ **15.** (a) $\{x \mid 2x + 4 \geq 0\}$; (b) $\{r \mid 2 \leq r < 8\}$;
(c) $\{a \mid -5 < a - 2 \leq 7\}$ **17.** (a) $(2, +\infty)$; (b) $(-4, 4]$ **19.** (a) $(2, 12)$; (b) $(-\infty, -4] \cup (4, +\infty)$ **21.** (a) $(2, 12)$;
(b) $(-\infty, -4] \cup (4, +\infty)$ **23.** (a) $(-4, 0]$; (b) $(-\infty, 7]$ **25.** (a) $\{x \mid 2 < x < 7\}$; (b) $\{x \mid -3 \leq x \leq 6\}$;
(c) $\{x \mid -5 < x \leq 4\}$; (d) $\{x \mid -10 \leq x < -2\}$ **27.** (a) $\{x \mid x \geq 3\}$; (b) $\{x \mid x < 0\}$; (c) $\{x \mid x > -4\}$,
(d) the set R of real numbers   **29.** (a) 7; (b) $\frac{3}{4}$; (c) $3 - \sqrt{3}$; (d) $3 - \sqrt{3}$ **31.** (a) 6; (b) 10; (c) 10; (d) 6
33. (a) t; (b) $-t$

EXERCISES 1.2 (PAGE 28)

1. $\dfrac{-1}{x(x + h)}$ **3.** $\dfrac{-3}{(3x + 3h + 2)(3x + 2)}$ **5.** $-\dfrac{8 + h}{16(4 + h)^2}$ **7.** $-\dfrac{3x^2 + 3hx + h^2}{x^3(x + h)^3}$ **9.** (a) 6561; (b) 7776;
(c) $\frac{49}{169}$; (d) $\frac{373,248}{125}$;   **11.** (a) $-\frac{1}{16}$; (b) $\frac{1}{36}$; (c) 6; (d) 9   **13.** (a) 4; (b) -0.064; (c) $\frac{1}{56}$; (d) $-\frac{255}{16}$
15. (a) $\frac{34}{57}$; (b) 9   **17.** (a) 27; (b) 4; (c) 81   **19.** (a) $|x|^{3/2} y^2$; (b) $2s^2 |t|^5$ **21.** (a) $9y^2 |y - 2|$;
(b) $4(x - 2)^2 |2 - y|$ **23.** $4(u + 1)^2 |u - 4|$ **25.** $\dfrac{x + 2}{2(x + 1)^{3/2}}$ **27.** $\dfrac{5x^3 + 15x^2}{(2x + 5)^{3/2}}$ **29.** $\dfrac{9x^5 + 4x^3}{(3x^2 + 1)^{3/2}}$ **31.** $\dfrac{1}{\sqrt{4 + h} + 2}$
33. $\dfrac{2}{\sqrt{2(x + h) + 1} + \sqrt{2x + 1}}$ **35.** $-\dfrac{1}{\sqrt{x}\sqrt{x + h}(\sqrt{x} + \sqrt{x + h})}$ **37.** $\dfrac{h}{\sqrt{h + 1}(h + 2 + 2\sqrt{h + 1})}$
39. $\dfrac{1}{(x + h)^{2/3} + x^{1/3}(x + h)^{1/3} + x^{2/3}}$ **41.** (a) $|x + 3| - |x - 3|$; (b) $x \geq 3$

EXERCISES 1.3 (PAGE 38)

1. (a) $5 + 0i$; (b) $0 + 7i$; (c) $3 + 5i$; (d) $3 - 5i$ **3.** (a) $8 - 5i$; (b) $-8 + 5i$; (c) $-6 + 6i$; (d) $\frac{1}{3} - \frac{3}{5}\sqrt{5}i$
5. (a) $12 + 3i$; (b) $1 - 2i$ **7.** (a) $-6 + 0i$; (b) $-5 + 8i$ **9.** $8 + 26i$ **11.** $-9 + \sqrt{5}i$ **13.** (a) $-15 + 0i$;
(b) $-4 + 0i$ **15.** $0 - 72i$ **17.** $(8 - 18\sqrt{2}) + 0i$ **19.** $53 + 0i$ **21.** $-24 + 5\sqrt{3}i$ **23.** $-18 + 18\sqrt{3}i$
25. $0 + 5i$ **27.** $-\frac{3}{13} - \frac{2}{13}i$ **29.** $\frac{4}{5} + \frac{7}{5}i$ **31.** $\frac{1}{2} - \frac{3}{4}i$ **33.** $\dfrac{3}{11} - \dfrac{\sqrt{2}}{11}i$ **35.** $\frac{5}{169} - \frac{12}{169}i$ **37.** (a) $-i$; (b) i; (c) -1
39. (a) $-i$; (b) i; (c) -1 **41.** $8 - 6i$ **43.** $16 - 6i$ **45.** 0   **47.** 0   **49.** (a) $10, 10$; (b) $-10, 10$; (c) $10i, 10i$;
(d) a and b are not both negative

EXERCISES 1.4 (PAGE 46)

1. (a) first quadrant; (b) third quadrant; (c) fourth quadrant; (d) second quadrant
3. (a) $(1, 2)$; (b) $(-1, -2)$; (c) $(-1, 2)$; (d) $(-2, 1)$ **5.** (a) $(2, -2)$; (b) $(-2, 2)$; (c) $(-2, -2)$; (d) does not apply
7. (a) $(-1, 3)$; (b) $(1, -3)$; (c) $(1, 3)$; (d) $(-3, -1)$ **9.** (a) horizontal line 4 units above the x axis;
(b) vertical line 2 units to the left of the y axis; (c) points to the right of the y axis; (d) points on and below the x axis
11. (a) the y axis; (b) points to the left of the vertical line 4 units to the right of the y axis; (c) points on or above the horizontal line
2 units below the x axis; (d) points in either the first or third quadrants   **13.** (a) 7; (b) -7 **15.** (a) -4; (b) 4
17. (a) -10; (b) 6   **19.** (b) 5; (c) $(-\frac{1}{2}, 5)$ **21.** (b) 13; (c) $(\frac{11}{2}, -1)$ **23.** (a) $(-\frac{3}{2}, 2)$ **25.** (a) $(\frac{5}{2}, \frac{3}{2})$
27. $|\overline{AB}| = 10$; $|\overline{BC}| = \sqrt{17}$; $|\overline{CA}| = 13$ **29.** $\sqrt{26}$; $\frac{1}{2}\sqrt{89}$; $\frac{1}{2}\sqrt{53}$ **31.** $|\overline{AB}| = \sqrt{41}$, $|\overline{AC}| = \sqrt{41}$, $|\overline{BC}| = \sqrt{82}$, and
$|\overline{AB}|^2 + |\overline{AC}|^2 = |\overline{BC}|^2$ **37.** $17\sqrt{2}$ **39.** $(-8, 12)$ **41.** -2 or 8

EXERCISES 1.5 (PAGE 55)

9. Plot using: **(a)** $[-20, 20]$ by $[-1, 10]$;   **(b)** $[-1, 30]$ by $[-1, 10]$;   **(c)** $[-20, 20]$ by $[-1, 10]$
11. Plot using: **(a)** $[-10, 10]$ by $[-100, 100]$;   **(b)** $[-20, 20]$ by $[-5, 1]$;   **(c)** $[-100, 100]$ by $[-5, 40]$
13. Plot using: **(a)** $[-20, 20]$ by $[-1, 10]$; **(b)** $[-1, 30]$ by $[-1, 10]$;   **(c)** $[-20, 20]$ by $[-1, 10]$
15. Plot using: **(a)** $[-2, 21]$ by $[-2, 21]$;   **(b)** $[-2, 21]$ by $[-21, 2]$;   **(c)** $[-2, 21]$ by $[-2, 21]$

In Exercises 17-23, the graph of (i) is shown. The solid part is the graph of (ii); the dashed part is that of (iii).

17. Symmetric with respect to x axis

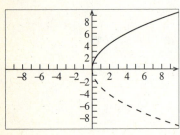

19. Symmetric with respect to x axis

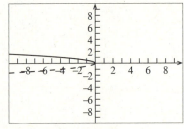

21. Symmetric with respect to x axis, y axis, and origin

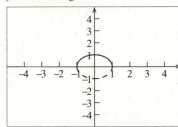

23. Symmetric with respect to x axis, y axis, and origin

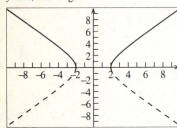

25. (a)

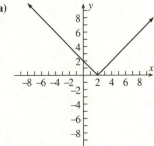

25. (b)

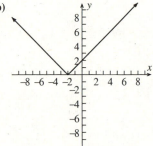

25. (c)

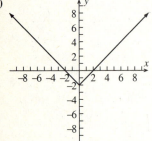

25. (d)

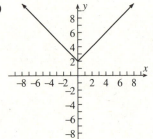

27. All are symmetric with respect to the origin.
29. All are symmetric with respect to the y axis.

31.

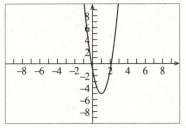

33.

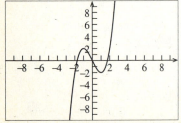

35.

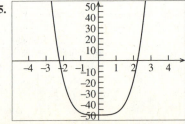

37.

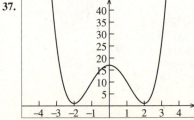

REVIEW EXERCISES FOR CHAPTER 1 (PAGE 57)

1. (a) Q; **(b)** $\varnothing$; **(c)** N; **(d)** Q **3. (a)** $\{15\}$; **(b)** $\{-4, 15, 0\}$; **(c)** $\{-\frac{1}{3}, 2, -5, 0.75, \frac{1}{2}\}$; **(d)** $\{-\sqrt{3}, \frac{2}{3}\pi\}$

5. (a) $(-\infty, 4]$; **(b)** $(2, 8)$; **(c)** $(-\infty, 0] \cup (2, +\infty)$; **(d)** $(1, 5]$ **7. (a)** 7; **(b)** $\sqrt{5}$; **(c)** $2 - \sqrt{3}$; **(d)** $2 - \sqrt{3}$

9. (a) $\{x \mid -4 < x < 6\}$; **(b)** $\{x \mid 3 \le x \le 11\}$; **(c)** $\{x \mid -10 \le x < 10\}$; **(d)** $\{x \mid x < 0\}$ **11. (a)** 64; **(b)** 125

13. (a) $\frac{125}{729}$; **(b)** 81   **15.** $\frac{1}{32}$ **17.** $\frac{1}{12}$ **19.** $\frac{22}{9}$ **21.** $\frac{3}{5}$ **23. (a)** $2|x|y^2|z|^3$; **(b)** $|y + 1|(x^2 + 4)$

25. (a) $18 + i$; **(b)** $\frac{1}{4} - \frac{2}{3}i$ **27. (a)** $-21 + 0i$; **(b)** $10 - 10i$ **29.** $-\frac{14}{25} + \frac{23}{25}i$ **31. (a)** i; **(b)** $-i$; **(c)** -1

33. (b) 5; **(c)** $(5, \frac{1}{2})$ **35. (a)** $(1, -2)$ **37.** $|\overline{AB}| = \sqrt{37}$; $|\overline{BC}| = 2\sqrt{5}$; $|\overline{CA}| = 5$ **39.** Area is 15 square units   **41.** No

47. $\dfrac{2}{7(7 + h)}$ **49.** $\dfrac{-2}{(2x + 2h + 1)(2x + 1)}$ **51.** $\dfrac{3x(3x + 2)}{(2x + 1)^{3/2}}$ **53.** $\dfrac{2}{\sqrt{9 + 2h} + 3}$ **55.** $\dfrac{1}{\sqrt{x + h - 2} + \sqrt{x - 2}}$

57. $\dfrac{4h - 3}{2\sqrt{3h + 4}(2h + 6 + 3\sqrt{3h + 4})}$ **59.** Plot using standard view.

61. (a) Plot using standard view; **(b)** Plot using $[-5, 5]$ by $[-20, 20]$ **63.** Plot using $[-5, 5]$ by $[-10, 10]$

65. Plot using $[-10, 10]$ by $[-25, 25]$ **67.** Plot using $[-10, 10]$ by $[-25, 25]$ **69.** Plot using $[-0.5, 1.5]$ by $[-0.25, 1.25]$

In Exercises 71-75, the graph of (i) is shown. The solid part is the graph of (ii); the dashed part is that of (iii).

71. Symmetric with respect to x axis   **73.** Symmetric with respect to x axis   **75.** Symmetric with respect to x axis, y axis, and origin

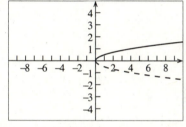 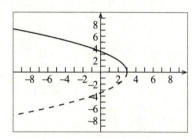 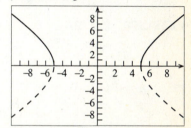

77. (a)   **(b)**   **(c)**   **(d)**

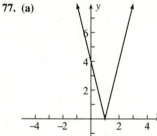 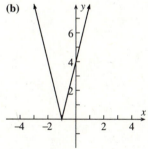 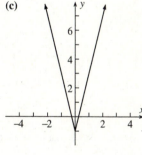 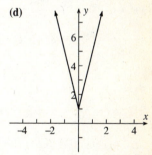

79. (a)   **(b), (c)**

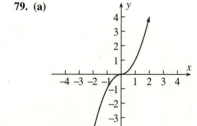 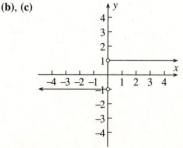

81. (b)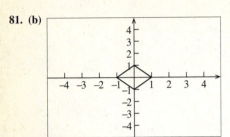

83. (b)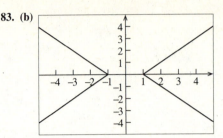

(c) a rhombus

EXERCISES 2.1 (PAGE 68)

1. $\{-\frac{7}{2}\}$ **3.** $\{\frac{8}{5}\}$ **5.** $\{3\}$ **7.** $\{2\}$ **9.** $\{3\}$ **11.** $\varnothing$ **13.** $\{-\frac{13}{49}\}$ **15.** $\{2\}$ **17.** $\{2\}$ **19.** $\{2\}$ **21.** $\{-1\}$

23. $\{3\}$ **27.** $x = \frac{3}{4}b, a \ne 0$ **29.** $y = a + 3b, a \ne 2b$ **31.** $x = 3a - 5b, a \ne -b$ **33.** $h = \dfrac{2A}{a + b}$ **35.** $r = \dfrac{E - IR}{I}$

37. $p = \dfrac{fq}{q - f}$ **39.** $r = \dfrac{a - s}{l - s}$ **41. (a)** 11;   **(b)** 6;   **(c)** 1;   **(d)** $y = 16 - \dfrac{x}{20}$;   **(e)** 4 gallons;   **(f)** 240 miles;   **(g)** 320 miles

45. (a) $\{x \mid x \in R, x \ne 0\}$;   **(b)** $\varnothing$;   **(c)** $\varnothing$

EXERCISES 2.2 (PAGE 81)

1. $\{-7, 7\}$ **3.** $\{-\frac{2}{3}\sqrt{15}, \frac{2}{3}\sqrt{15}\}$ **5.** $\{0, \frac{1}{4}\}$ **7.** $\{3, 5\}$ **9.** $\{-\frac{3}{2}, \frac{1}{4}\}$ **11.** $\{-\frac{6}{7}\}$ **13.** $\{-1, 4\}$ **15.** $\{1 \pm \sqrt{3}\}$

17. $\left\{\dfrac{2 \pm \sqrt{14}}{5}\right\}$ **19.** $\left\{\dfrac{1 \pm i}{2}\right\}$ **21.** $\{2 \pm \sqrt{3}i\}$ **23. (a)** 361, roots are real and unequal;   **(b)** 52, roots are real and unequal

25. (a) 0, roots are real and unequal;   **(b)** -31, roots are imaginary and complex conjugates of each other   **27.** $\{4, -1\}$

29. $\{-2, \frac{-5}{3}\}$ **31. (a)** $x^2 + 6x + 3^2 = (x + 3)^2$;   **(b)** $x^2 - 5x + (\frac{5}{2})^2 = (x - \frac{5}{2})^2$ **33. (a)** $x^2 - \frac{2}{3}x + (\frac{1}{3})^2 = (x - \frac{1}{3})^2$;

(b) $x^2 + \frac{3}{5}x + (\frac{3}{10})^2 = (x + \frac{3}{10})^2$ **35.** $(x - 2)^2 + (y + 3)^2 = (\sqrt{26})^2$ **37.** $(x - 1)^2 = 4 \cdot \frac{5}{12}(y - 2)$

39. $\dfrac{(x - 2)^2}{3^2} + \dfrac{(y - 1)^2}{2^2} = 1$ **41.** $\dfrac{(x + 2)^2}{1^2} - \dfrac{(y + 5)^2}{(\frac{5}{2})^2}$ **43.** $\sqrt{(x + 1)^2 - 2^2}$ **45.** $2\sqrt{(\frac{1}{2})^2 - (x - 3)^2}$ **47.** $r = \sqrt{\dfrac{3V}{\pi h}}$

49. $v = \sqrt{\dfrac{rF}{kM}}$ **51.** If $a = 0$, then $x = 0$; else $x = \dfrac{3 \pm \sqrt{9 + 40a^2}}{10a}$ **53.** $x = y + 2 \pm 2\sqrt{y^2 + y + 1}$

55. $x = \dfrac{-b \pm \sqrt{b^2 - 4a(c - y)}}{2a}$ **57. (a)** $s = -16t^2 + 128t$;   **(b)** 192 ft;   **(c)** 6;   **(d)** 8

59. (a) $s = -16t^2 + 76t + 68$;   **(c)** 156 ft;   **(d)** 2.75;   **(e)** 4.75;   **(f)** Approximately 5.52

EXERCISES 2.3 (PAGE 91)

1. $\frac{15}{2}, \frac{3}{2}$ **3.** 12 cm, 8 cm   **5.** 423 adults, 387 students   **7.** $16,250 in first, $8,750 in second

9. 24 grams of 80%-gold alloy, 16 grams of 55%-gold alloy   **11.** 6 liters   **13.** $\frac{25}{3}$ m³   **15.** approximately 10.85 m

17. 650 in Fundamentals, 590 in Composition   **19.** 8 and 10

EXERCISES 2.4 (PAGE 102)

1. $\{64\}$ **3.** $\varnothing$ **5.** $\{\frac{7}{2}\}$ **7.** $\{4\}$ **9.** $\{3\}$ **11.** $\{4\}$ **13.** $\emptyset$ **15.** $\{0, 3\}$ **17.** $\{7\}$ **19.** $\{-7\}$ **21.** $\{3, \frac{5}{3}\}$

23. $\{0, -3\}$ **25.** $\{\pm 1, \pm 2\}$ **27.** $\{\pm\sqrt{2}, \pm\sqrt{3}\}$ **29.** $\left\{\pm\dfrac{\sqrt{5}}{2}, \pm\dfrac{\sqrt{2}}{2}i\right\}$ **31.** $\{-5, \frac{1}{2}\}$

33. $\left\{2, 3, -1 \pm \sqrt{3}i, -\dfrac{3}{2} \pm \dfrac{3\sqrt{3}}{2}i\right\}$ **35.** $\{1\}$ **39.** $1, -\dfrac{1}{2} \pm \dfrac{\sqrt{3}}{2}i$ **41.** $-2, 1 \pm \sqrt{3}i$ **43.** $\pm 3, \pm 3i$ **45. (a)** -1.14;

(b) 0.23   **47. (a)** 0.73;   **(b)** 3.08   **49.** 4 ft   **51.** 3.92 in.   **53.** 40 in. and 30 in.   **55.** 1.21 in. or 2.16 in.   **57.** 0.98 in.

EXERCISES 2.5 (PAGE 111)

1. $[5, +\infty)$ **3.** $(-\infty, \frac{7}{2})$ **5.** $(-5, +\infty)$ **7.** $(-\infty, \frac{17}{2}]$ **9.** $(-\infty, -\frac{17}{3})$ **11.** $(-8, +\infty)$ **13.** $(-\infty, 4]$ **15.** $(-\infty, 5)$

17. $[4, 8)$ **19.** $(-4, 1)$ **21.** $(1, 4)$ **23.** $[-6, -4]$ **25.** $(-\frac{5}{3}, \frac{4}{3}]$ **27.** $[-3, \frac{21}{5})$ **41.** $\{F \mid 50 < F < 68\}$

43. At least $6,000   **45.** At least 49   **47.** At most 20 g and at least 14 g

EXERCISES 2.6 (PAGE 120)

1. $(-\infty, -3) \cup (3, +\infty)$ **3.** $(-3, 4)$ **5.** $(-\infty, -\frac{1}{2}) \cup (\frac{7}{2}, +\infty)$ **7.** $[1, 3]$ **9.** $[-4, 1]$ **11.** $(-\infty, -4) \cup (2, +\infty)$
13. $[-2, \frac{3}{2}]$ **15.** $(-\infty, +\infty)$ **17.** $(-\infty, \frac{5}{3}) \cup (2, +\infty)$ **19.** $(-3, 1) \cup (4, +\infty)$ **21.** $(-\infty, -4] \cup [0, 4]$
23. $\left(\frac{1}{2} - \frac{\sqrt{5}}{2}, \frac{1}{2} + \frac{\sqrt{5}}{2}\right)$ **25.** $(-\infty, +\infty)$ **27.** $(-3, -2) \cup (2, 3)$ **29.** $[-3, -1] \cup [2, 3] \cup [5, +\infty)$ **31.** $(1.15, +\infty)$
33. $(-\infty, -3.11) \cup (-0.75, 0.86)$ **35.** $(-\infty, -0.38) \cup (0.44, 2.94) \cup (4, +\infty)$ **37.** $(-\infty, -3.58) \cup (-1.71, 1) \cup (2.29, +\infty)$
39. between 0.4 and 7.6 seconds after being thrown   **41. (a)** before 0.4 seconds and after 4.3 seconds from being thrown;
(b) after 5.3 seconds from being thrown   **43.** more than 10 and less than 70   **45.** If x ft is the length of any side, $100 \le x \le 120$
47. If x meters is the width of the terrace, $8.8 \le x \le 13.5$ **49.** If x in. is the length of the side of the cutout squares, $0.9 \le x \le 1.3$
or $2.1 \le x \le 2.5$

EXERCISES 2.7 (PAGE 129)

1. $\{1, 9\}$ **3.** $\{\frac{4}{3}, 4\}$ **5.** $\{-5, \frac{5}{2}\}$ **7.** $\{-1, 8\}$ **9.** $\{1, 3\}$ **11.** $\{\frac{9}{4}, 4\}$ **13.** $\left\{1, 2, \dfrac{-3 \pm \sqrt{17}}{2}\right\}$ **15.** $[-5, 5]$
17. $(-\infty, -6) \cup (8, +\infty)$ **19.** $[2, 8]$ **21.** $(-1, 8)$ **23.** $(-\infty, -1) \cup (8, +\infty)$ **25.** $(-\infty, -\frac{16}{3}) \cup (-\frac{8}{3}, +\infty)$ **27.** $[1, \frac{9}{3}]$
29. $(-\infty, \frac{1}{2}) \cup (1, +\infty)$ **37.** $(-3, -1) \cup (1, 3)$ **39.** $[-1, 2] \cup [3, 6]$ **41.** $(-\infty, -5] \cup [-3, 3] \cup [5, +\infty)$
43. $x = -1$ or $x = 2$ **45.** $x = -1$ or $x = \frac{1}{2}$

REVIEW EXERCISES FOR CHAPTER 2 (PAGE 131)

1. $\{\frac{11}{4}\}$ **3.** $x = 3$ **5.** $x = \frac{4}{3}$ **7.** $\{\pm\frac{8}{7}\}$ **9.** $\{5, -2\}$ **11.** $\{\frac{1}{2}, -\frac{6}{5}\}$ **13.** $\left\{\dfrac{2 \pm \sqrt{14}}{2}\right\}$ **15.** $\left\{\dfrac{1 \pm \sqrt{5}i}{3}\right\}$
17. $\{3\}$ **19.** $\varnothing$ **21.** $\{5, -3\}$ **23.** $\{5, -3\}$ **25.** $\{2\}$ **27.** $\{-2\}$ **29.** $\{2, -1 \pm \sqrt{3}i\}$ **31.** $\{\pm\frac{1}{3}, \pm\frac{1}{2}\}$
33. $\left\{-1, -2, \dfrac{-3 + \sqrt{29}}{2}\right\}$ **35.** $\{6\}$ **37.** $\{-\frac{1}{27}, \frac{1}{8}\}$ **39.** $\{-6, 1\}$ **41.** $\{\frac{1}{2}, 2\}$ **43.** $x = -\dfrac{By + C}{A}, A \ne 0$
45. $x = b \pm a$ **47.** $x = \frac{1}{2}$ or $x = \dfrac{y - 1}{3}$ **49. (a)** $x^2 - 8x + 4^2 = (x - 4)^2$;   **(b)** $y^2 + 3y + (\frac{3}{2})^2 = (y + \frac{3}{2})^2$;
(c) $x^2 + \frac{5}{3}x + (\frac{5}{6})^2 = (x + \frac{5}{6})^2$ **51. (a)** 0; roots are real, rational, and equal;   **(b)** 196; roots are real, rational, and unequal
(c) -23; roots are imaginary and complex conjugates of each other   **53.** 1.75   **55.** 1.76   **57.** $(-\infty, 4]$ **59.** $(-\infty, -\frac{5}{4})$
61. $[12, +\infty)$ **63.** $(-2, 1)$ **65.** $(-4, 5]$ **67.** $(-\infty, -3] \cup [1, +\infty)$ **69.** $(-1, 6)$ **71.** $(\infty, \frac{1}{6}) \cup (\frac{17}{6}, +\infty)$
73. $(-5, 5)$ **75.** $(-\infty, -1] \cup [\frac{5}{2}, +\infty)$ **77.** $(-\infty, 0)$ **79.** $(-2, -1) \cup (4, 5)$ **81.** $(x - 3)^2 = 4(\frac{1}{16})(y + 68)$
83. $\dfrac{(x - 4)^2}{4^2} + \dfrac{(y + 2)^2}{2^2} = 1$ **85.** $\dfrac{(x + 1)^2}{5^2} - \dfrac{(y + 2)^2}{(\sqrt{10})^2} = 1$ **87.** $4\sqrt{(x + 1)^2 - (\frac{1}{2})^2}$ **91. (a)** \$195.60;
(b) \$200.80;   **(c)** \$232;   **(d)** $y = 180 + 0.52x$;   **(f)** \$223.16;   **(h)** Approximately 48 miles   **93.** \$18   **95.** $\frac{4}{3}$ liters
97. 18 cm   **99.** 1.33 cm or 5.87 cm   **101.** 1.53 ft   **103.** not greater than 7 cm

EXERCISES 3.1 (PAGE 147)

1. (a) slope is $\frac{1}{5}$ **(b)** slope is -1 **3. (a)** slope is $-\frac{3}{4}$ **(b)** slope $= -\frac{1}{7}$

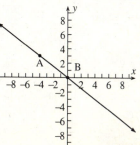

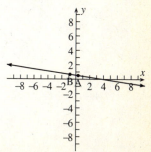

5. (a) slope is 0   **(b)** slope is $\frac{1}{4}$ **7. (a)**   **(b)**

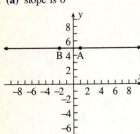 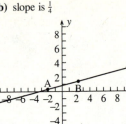 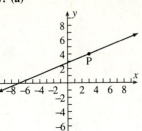 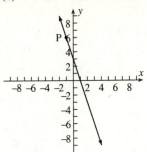

9. (a) $4x - y - 11 = 0$;   **(b)** $11x - 4y - 9 = 0$ **11. (a)** $2x + 3y - 3 = 0$;   **(b)** $6x - 3y + 8 = 0$
13. (a) $y = -7$;   **(b)** $x = 2$ **15. (a)** $4x - 3y + 12 = 0$;   **(b)** $x - y = 0$ **17.** $2x - y + 7 = 0$ **19.** $5x + y - 14 = 0$

21. (a) slope is $-\frac{1}{3}$, y intercept is 2   **(b)** slope is 0, y intercept is $\frac{9}{4}$ **23. (a)** slope is $\frac{7}{8}$, y intercept is 0   **(b)** slope is $-\frac{1}{2}$, y intercept is 3

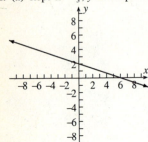 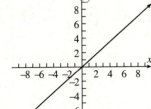 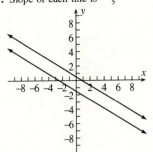 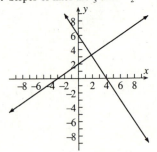

25. $y = -5x + 8$ **27.** Slope of each line is $-\frac{3}{5}$ **29.** Slopes of lines are $\frac{2}{3}$ and $-\frac{3}{2}$

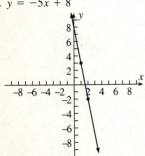

31. $\pm\frac{2}{3}$ **33. (a)** collinear;   **(b)** not collinear   **35. (a)** not collinear;   **(b)** collinear   **37.** slopes of two sides are $-\frac{1}{2}$;
slopes of other two sides are $\frac{3}{5}$ **39.** area is 5 square units   **41. (a)** $y = 25x + 3000$ **43. (a)** \$600;   **(b)** $y = 30x + 600$
45. 8   **49.** from $(3, -2)$: $y = -\frac{2}{5}(x - 3) - 2$; from $(2, 4)$: $y = \frac{9}{2}(x - 2) + 4$; from $(-1, 1)$: $y = 1$

EXERCISES 3.2 (PAGE 158)

1. (i); $\{(3, -5)\}$

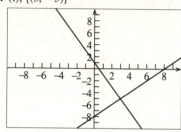

3. (i); $\{(\frac{9}{4}, \frac{3}{2})\}$

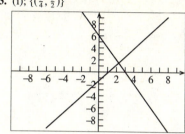

5. (iii)

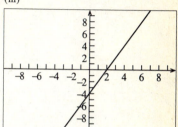

7. (ii)

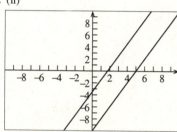

9. (i); $\{(-1.5, -1.3)\}$

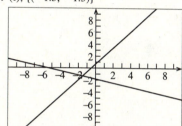

11. $\{(1, 0)\}$ **13.** $\{(0, -2)\}$ **15.** $\{(\frac{1}{2}, \frac{1}{6})\}$ **17.** $\{(\frac{2}{3}, \frac{1}{2})\}$ **19.** $\{(-4, 7)\}$ **21.** $\{(-\frac{12}{17}, \frac{42}{17})\}$ **23.** $\{(-\frac{15}{4}, -\frac{15}{2})\}$
25. $\{(\frac{3}{4}, -\frac{1}{5})\}$ **27.** \$6.30 per pound of tea; \$2.60 per pound of coffee   **29.** 20; \$120
31. $\frac{400}{3}$ cm³ of the 15%-acid solution; $\frac{800}{3}$ cm³ of the 6%-acid solution   **33.** girl, 20 minutes; brother, 25 minutes

35. any fraction of the form $\dfrac{t}{2t - 4}$, where $t \neq 2$ and $t \neq 0$ **39.** inconsistent   **41.** consistent; $\{(-3, -2)\}$

EXERCISES 3.3 (PAGE 165)

1. **(a)** $(0, 0)$; **(b)** $x = 0$;
  (c) $(0, 1)$; **(d)** $y = -1$;
  (e) $(-2, 1), (2, 1)$

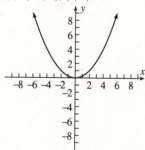

3. **(a)** $(0, 0)$; **(b)** $x = 0$;
  (c) $(0, -4)$; **(d)** $y = 4$;
  (e) $(-8, -4), (8, -4)$

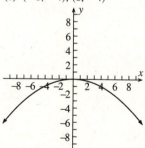

5. **(a)** $(0, 0)$; **(b)** $x = 0$;
  (c) $(0, \frac{1}{4})$; **(d)** $y = -\frac{1}{4}$;
  (e) $(-\frac{1}{2}, \frac{1}{4}), (\frac{1}{2}, \frac{1}{4})$

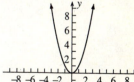
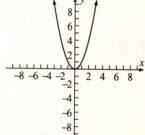

7. (a) $(0, 0)$; **(b)** $y = 0$;
(c) $(3, 0)$; **(d)** $x = -3$;
(e) $(3, -6), (3, 6)$

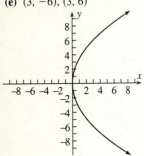

9. (a) $(0, 0)$; **(b)** $y = 0$;
(c) $(-2, 0)$; **(d)** $x = 2$;
(e) $(-2, -4); (-2, 4)$

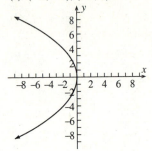

11. (a) $(0, 0)$; **(b)** $y = 0$;
(c) $(\frac{5}{4}, 0)$; **(d)** $x = -\frac{5}{4}$;
(e) $(\frac{5}{4}, -\frac{5}{2}), (\frac{5}{4}, \frac{5}{2})$

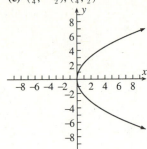

13. (a) $(0, 0)$; **(b)** $x = 0$;
(c) $(0, -\frac{2}{3})$; **(d)** $y = \frac{2}{3}$;
(e) $(-\frac{4}{3}, -\frac{2}{3}), (\frac{4}{3}, -\frac{2}{3})$

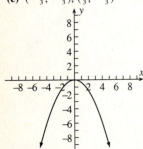

15. (a) $(0, 0)$; **(b)** $y = 0$;
(c) $(\frac{9}{8}, 0)$; **(d)** $x = -\frac{9}{8}$;
(e) $(\frac{9}{8}, -\frac{9}{4}), (\frac{9}{8}, \frac{9}{4})$

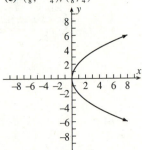

23. $x^2 = 16y$

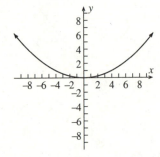

25. $x^2 = -20y$

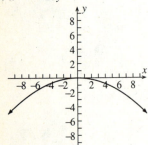

27. $y^2 = 8x$

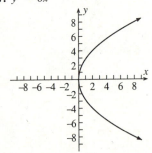

29. $3y^2 = -20x$

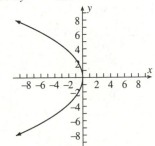

31. $x^2 = 12y$

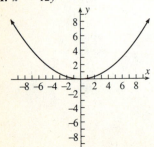

33. $3y^2 = -8x$

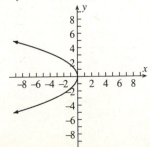

35. $x^2 = y$

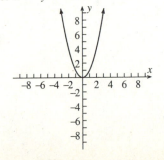

37. $\frac{32}{45}$ in.   **39.** 16.6 m   **43. (a)** $-\frac{5}{2}$;   **(b)** $x^2 = -10y$

EXERCISES 3.4 (PAGE 172)

1. The graph of (c) is shown. The solid part is the graph of (a); the dashed part is that of (b).

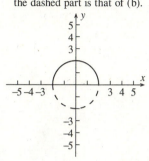

3.

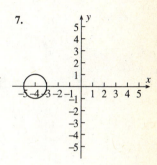

5.

7.

15. $(x - 4)^2 + (y + 3)^2 = 25$; $x^2 + y^2 - 8x + 6y = 0$ **17.** $(x + 5)^2 + (y + 12)^2 = 9$; $x^2 + y^2 + 10x + 24y + 160 = 0$
19. $x^2 + (y - 7)^2 = 1$; $x^2 + y^2 - 14y + 48 = 0$ **21.** $(x - 1)^2 + (y - 2)^2 = 13$ **23.** $(x - 5)^2 + (y + 1)^2 = 13$
25. $(3, 4)$; 4   **27.** $(-1, -5)$; $2\sqrt{2}$ **29.** $(0, -\frac{2}{3})$; $\frac{5}{3}$ **33.** circle   **35.** the empty set   **37.** point $(\frac{1}{2}, -\frac{3}{2})$
39. $y = \frac{4}{3}(x + 4) + 3$ **41.** $y = -\frac{3}{4}(x - 5) + 1$ **45.** $D^2 + E^2 > 4F$

EXERCISES 3.5 (PAGE 179)

1. $x'^2 + y'^2 = 13$; $x' = x + 3$, $y' = y + 2$

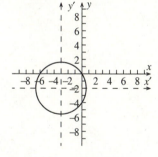

3. $x'^2 + y'^2 = \frac{1}{4}$; $x' = x + \frac{1}{2}$; $y' = y - 1$

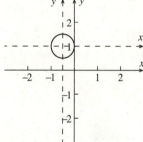

5. $x'^2 = 8y'$; $x' = x - 2$, $y' = y + 4$

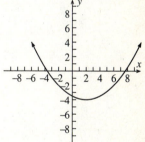
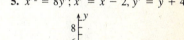

7. $y'^2 = -6x'$; $x' = x + 5$, $y' = y + 3$

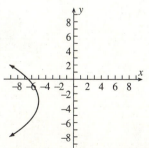

9. (a) $(0, -4)$;   **(b)** $x = 0$;
(c) $(0, -\frac{15}{4})$;   **(d)** $y = -\frac{17}{4}$;
(e) $(\frac{1}{2}, -\frac{15}{4})$, $(-\frac{1}{2}, -\frac{15}{4})$;
(f)

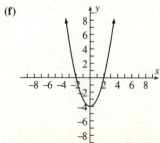

11. (a) $(2, -1)$;   **(b)** $x = 2$;
(c) $(2, -\frac{5}{4})$;   **(d)** $y = -\frac{3}{4}$;
(e) $(\frac{5}{2}, -\frac{5}{4})$, $(\frac{3}{2}, -\frac{5}{4})$;
(f)

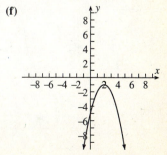

13. (a) $(-9, 3)$;   **(b)** $y = 3$;
(c) $(-\frac{35}{4}, 3)$;   **(d)** $x = -\frac{37}{4}$;
(e) $(-\frac{35}{4}, \frac{7}{2})$, $(-\frac{35}{4}, \frac{5}{2})$;

(f)

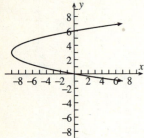

15. (a) $(3, 1)$;   **(b)** $x = 3$;
(c) $(3, 2)$;   **(d)** $y = 0$;
(e) $(5, 2)$, $(1, 2)$;

(f)

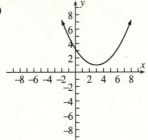

17. (a) $(9, -6)$;   **(b)** $y = -6$;
(c) $(8, -6)$;   **(d)** $x = 10$;
(e) $(8, -4)$, $(8, -8)$;

(f)

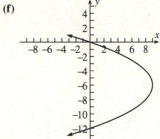

19. (a) $(4, 3)$;   **(b)** $x = 4$;
(c) $(4, \frac{5}{2})$;   **(d)** $y = \frac{7}{2}$;
(e) $(5, \frac{5}{2})$, $(3, \frac{5}{2})$;

(f)

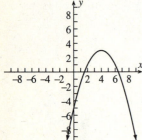

21. (a) $(2, -2)$;   **(b)** $x = 2$;
(c) $(2, 0)$;   **(d)** $y = -4$;
(e) $(6, 0)$, $(-2, 0)$;

(f)

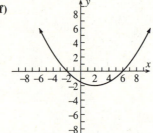

23. (a) $(3, -2)$;   **(b)** $y = -2$;
(c) $(\frac{23}{8}, -2)$;   **(d)** $x = \frac{25}{8}$;
(e) $(\frac{23}{8}, -\frac{7}{4})$, $(\frac{23}{8}, -\frac{9}{4})$;

(f)

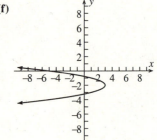

In Exercises 29-45, the graph of the first equation is dashed; that of the second is solid.

29.

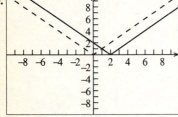

31.

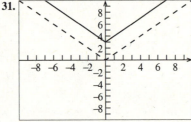

33.

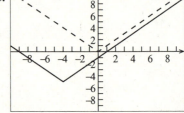

35.

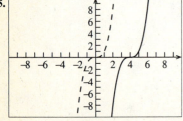

37.

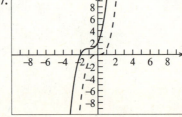

39.

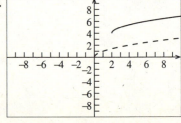

41.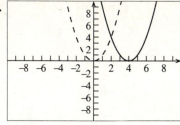

43.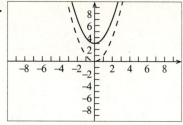

45.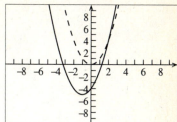

47. $\left(-\dfrac{b}{2a}, -\dfrac{b^2 - 4ac}{4a}\right)$

REVIEW EXERCISES FOR CHAPTER 3 (PAGE 181)

1. slope is $\frac{8}{3}$; $8x - 3y - 17 = 0$

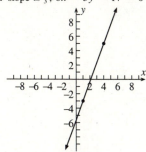

3. $3x + 2y - 11 = 0$

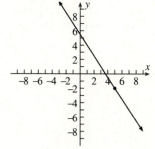

5. slope is $\frac{2}{5}$, y intercept is -2

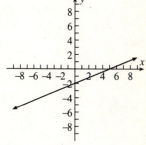

7. $7x - 3y + 15 = 0$

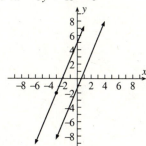

11. $-\frac{34}{3}$ **13.** no   **15.** (i); $\{(3, -2)\}$ **17.** (iii)   **19.** $\{(2, -5)\}$

21. $\varnothing$ **23.** $\{(\frac{12}{5}, -3)\}$ **25.** $\{(4, -2)\}$

27. (a) $(0, 0)$;   **(b)** $x = 0$;
(c) $(0, 4)$;   **(d)** $y = -4$;
(e) $(8, 4), (-8, 4)$;

(f)

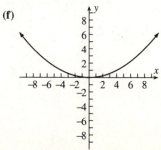

29. (a) $(0, 0)$;   **(b)** $x = 0$;
(c) $(0, -\frac{1}{16})$;   **(d)** $y = \frac{1}{16}$;
(e) $(\frac{1}{8}, -\frac{1}{16}), (-\frac{1}{8}, -\frac{1}{16})$;

(f)

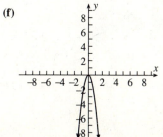

31. (a) $(0, 0)$;   **(b)** $y = 0$;
(c) $(-\frac{5}{2}, 0)$;   **(d)** $x = \frac{5}{2}$;
(e) $(-\frac{5}{2}, 5), (-\frac{5}{2}, -5)$;

(f)

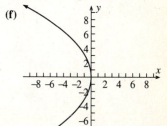

33. (a) $(0, -3)$; **(b)** $x = 0$;
(c) $(0, -\frac{11}{4})$; **(d)** $y = -\frac{13}{4}$;
(e) $(\frac{1}{2}, -\frac{11}{4})$, $(-\frac{1}{2}, -\frac{11}{4})$;

(f)

35. (a) $(6, 0)$; **(b)** $y = 0$;
(c) $(\frac{25}{4}, 0)$; **(d)** $x = \frac{23}{4}$;
(e) $(\frac{25}{4}, \frac{1}{2})$, $(\frac{25}{4}, -\frac{1}{2})$;

(f)

37. (a) $(-16, 4)$; **(b)** $y = 4$;
(c) $(-\frac{63}{4}, 4)$; **(d)** $x = -\frac{65}{4}$;
(e) $(-\frac{63}{4}, \frac{9}{2})$, $(-\frac{63}{4}, \frac{7}{2})$;

(f)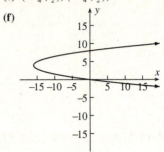

39. (a) $(-3, -2)$; **(b)** $x = -3$;
(c) $(-3, -1)$; **(d)** $y = -3$;
(e) $(-1, -1)$, $(-5, -1)$;

(f)

41. (a) $(3, 2)$; **(b)** $x = 3$;
(c) $(3, 0)$; **(d)** $y = 4$;
(e) $(7, 0)$, $(-1, 0)$;

(f)

43. $x^2 = 8y$

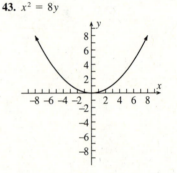

45. $y^2 = -12x$

47. $x^2 = -12(y - 3)$

49. $(0, 0)$; 3

51. $(3, 0)$; 2

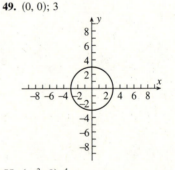

53. $(-2, 3)$; 4

55. $(-\frac{2}{3}, 0)$; $\frac{4}{3}$

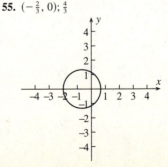

57. the empty set   **59.** $(x - 3)^2 + (y + 5)^2 = 4$ **61.** $(x - 4)^2 + (y + 3)^2 = 8$
63. $y = \frac{3}{4}(x - 4) - 7$

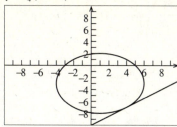

In Exercises 75-81, the graph of the first equation is dashed; that of the second is solid.

75. 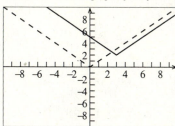   **77.** 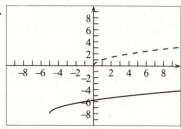   **79.**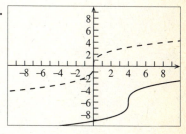

81. 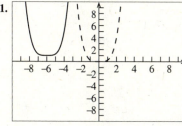   **83.** parallelogram 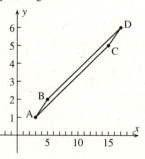   **85.** square rectangle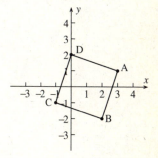

87. $y = \frac{4}{3}x + \frac{13}{6}$ **89.** $(x + 3)^2 = -24(y - 5)$ **91.** \$34,000   **93. (a)** 2800;   **(b)** 4000
95. 28 days for A and 21 days for B   **97.** 9π cm

EXERCISES 4.1 (PAGE 192)

1. (a) 5;   **(b)** -5;   **(c)** -1;   **(d)** $2a + 1$;   **(e)** $2x + 1$ **3. (a)** 12;   **(b)** 32;   **(c)** $9x^4 - 30x^3 + 49x^2 - 40x + 16$;
(d) $3x^4 - 5x^2 + 4$ **5. (a)** 3;   **(b)** -1;   **(c)** 9;   **(d)** $a, a \neq 0$;   **(e)** $x, x \neq 0$;   **(f)** $\frac{x}{3}, x \neq 0$ **7. (a)** 1;   **(b)** $\sqrt{11}$;   **(c)** 2;
(d) 5;   **(e)** $\sqrt{4x + 9}$ **9. (a)** $4x - 2$;   **(b)** $4x - 1$;   **(c)** $2x + 2h - 2$;   **(d)** $2x + 2h - 1$ **11. (a)** $\frac{6}{x}$;   **(b)** $\frac{3}{2x}$;
(c) $\frac{3}{x} + \frac{3}{h}$;   **(d)** $\frac{3}{x + h}$ **13.** 2   **15.** $6x + 3h - 5$ **17.** $\frac{-3}{x(x + h)}$ **19.** $\frac{2}{\sqrt{2x + 2h + 3} + \sqrt{2x + 3}}$

21. $\dfrac{-2}{\sqrt{x+h+1}\sqrt{x+1}(\sqrt{x+1}+\sqrt{x+h+1})}$ **23. (a)** $x^2 + x - 6$, domain: $(-\infty, +\infty)$;

(b) $-x^2 + x - 4$, domain: $(-\infty, +\infty)$;   **(c)** $x^3 - 5x^2 - x + 5$, domain: $(-\infty, +\infty)$;   **(d)** $\dfrac{x-5}{x^2-1}$, domain: $\{x \mid x \neq -1, x \neq 1\}$;

(e) $\dfrac{x^2-1}{x-5}$, domain: $\{x \mid x \neq 5\}$ **25. (a)** $\dfrac{x^2+2x-1}{x^2-x}$, domain: $\{x \mid x \neq 0, x \neq 1\}$;   **(b)** $\dfrac{x^2+1}{x^2-x}$, domain: $\{x \mid x \neq 0, x \neq 1\}$;

(c) $\dfrac{x+1}{x^2-x}$, domain: $\{x \mid x \neq 0, x \neq 1\}$;   **(d)** $\dfrac{x^2+x}{x-1}$, domain: $\{x \mid x \neq 0, x \neq 1\}$;   **(e)** $\dfrac{x-1}{x^2+x}$, domain: $\{x \mid x \neq -1, x \neq 0, x \neq 1\}$

27. (a) $\sqrt{x} + x^2 - 1$, domain: $[0, +\infty)$;   **(b)** $\sqrt{x} - x^2 + 1$, domain: $[0, +\infty)$;   **(c)** $\sqrt{x}(x^2 - 1)$, domain: $[0, +\infty)$;

(d) $\dfrac{\sqrt{x}}{x^2-1}$, domain: $[0, 1) \cup (1, +\infty)$;   **(e)** $\dfrac{x^2-1}{\sqrt{x}}$, domain: $(0, +\infty)$ **29. (a)** $x^2 + 3x - 1$, domain: $(-\infty, +\infty)$;

(b) $x^2 - 3x + 3$, domain: $(-\infty, +\infty)$;   **(c)** $3x^3 - 2x^2 + 3x - 2$, domain: $(-\infty, +\infty)$;   **(d)** $\dfrac{x^2+1}{3x-2}$, domain: $\{x \mid x \neq \frac{2}{3}\}$;

(e) $\dfrac{3x-2}{x^2+1}$, domain: $(-\infty, +\infty)$ **31. (a)** $\dfrac{x^2+2x-2}{x^2-x-2}$, domain: $\{x \mid x \neq -1, x \neq 2\}$;

(b) $\dfrac{-x^2-2}{x^2-x-2}$, domain: $\{x \mid x \neq -1, x \neq 2\}$;   **(c)** $\dfrac{x}{x^2-x-2}$, domain: $\{x \mid x \neq -1, x \neq 2\}$;

(d) $\dfrac{x-2}{x^2+x}$, domain: $\{x \mid x \neq -1, x \neq 0, x \neq 2\}$;   **(e)** $\dfrac{x^2+x}{x-2}$, domain: $\{x \mid x \neq -1, x \neq 2\}$ **33. (a)** 0;   **(b)** 4;   **(c)** $4 - 2x$

35. (a) 1;   **(b)** -1;   **(c)** 1;   **(d)** -1,   **(e)** 1 if $x \leq 0$, -1 if $x > 0$;   **(f)** 1 if $x \geq -1$, -1 if $x < -1$;   **(g)** 1;

(h) -1 if $x \neq 0$, 1 if $x = 0$

EXERCISES 4.2 (PAGE 202)

1. domain: $(-\infty, +\infty)$;
range: $(-\infty, +\infty)$

3. domain: $(-\infty, +\infty)$;
range: $[0, +\infty)$

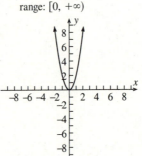

5. domain: $(-\infty, +\infty)$;
range: $(-\infty, 5]$

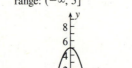

7. domain: $[1, +\infty)$;
range: $[0, +\infty)$

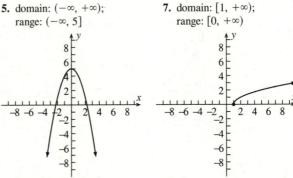

9. domain: $(-\infty, -2] \cup [2, +\infty)$;
range: $[0, +\infty)$

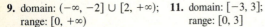

11. domain: $[-3, 3]$;
range: $[0, 3]$

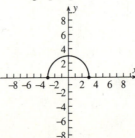

13. domain: $(-\infty, +\infty)$;
range: $[0, +\infty)$

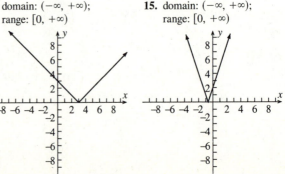

15. domain: $(-\infty, +\infty)$;
range: $[0, +\infty)$

17. domain: $\{x \mid x \neq -5\}$;
   range: $\{y \mid y \neq -10\}$

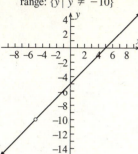

19. domain: $\{x \mid x \neq 1\}$;
   range: $\{y \mid y \neq -2\}$

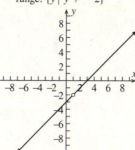

21. domain: $(-\infty, +\infty)$;
   range: $\{-2, 2\}$

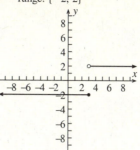

23. domain: $(-\infty, +\infty)$;
   range: $\{y \mid y \neq 3\}$

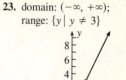

25. domain: $(-\infty, +\infty)$;
   range: $[-4, +\infty)$

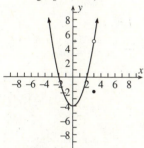

27. domain: $(-\infty, +\infty)$;
   range: $(-\infty, +\infty)$

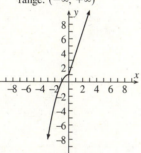

29. domain: $(-\infty, +\infty)$;
   range: $(-\infty, 6)$

31. domain: $(-\infty, +\infty)$;
   range: $(-\infty, -2) \cup [0, 5]$

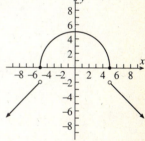

33. domain: $\{x \mid x \neq 2\}$;
   range: $[0, +\infty)$

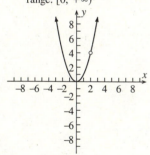

35. domain: $(-\infty, +\infty)$;
   range: $[1, +\infty)$

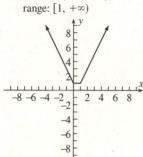

37. domain: $(-\infty, +\infty)$;
   range: $[0, +\infty)$

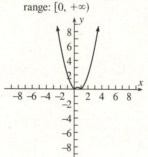

39. domain: $(-\infty, +\infty)$;
   range: $\{\text{integers}\}$

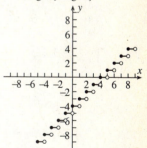

41. domain: $(-\infty, +\infty)$;
   range: $[0, 1)$

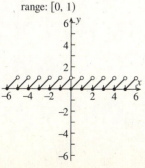

43. **(a)** even;   **(b)** neither   **45.** **(a)** odd;   **(b)** even

47. **(a)** odd;   **(b)** even **49.** **(a)** odd;   **(b)** even;   **(c)** even

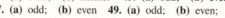
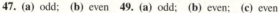

51. (a)

(b) $U(x - 1)$
$$= \begin{cases} 0 & \text{if } x < 1 \\ 1 & \text{if } 1 \le x \end{cases}$$

(c) $U(x) - U(x - 1)$
$$= \begin{cases} 0 & \text{if } x < 0 \\ 1 & \text{if } 0 \le x < 1 \\ 0 & \text{if } 1 \le x \end{cases}$$

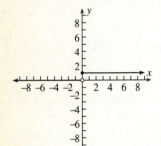

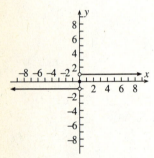

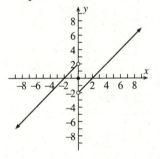

53. (a)

(b) $x \operatorname{sgn} x$
$$= \begin{cases} -x & \text{if } x < 0 \\ 0 & \text{if } x = 0 \\ x & \text{if } 0 < x \end{cases} = |x|$$

(c) $2 - x \operatorname{sgn} x$
$$= \begin{cases} 2 + x & \text{if } x < 0 \\ 2 & \text{if } x = 0 \\ 2 - x & \text{if } 0 < x \end{cases}$$

(d) $x - 2 \operatorname{sgn} x$
$$= \begin{cases} x + 2 & \text{if } x < 0 \\ 0 & \text{if } x = 0 \\ x - 2 & \text{if } 0 < x \end{cases}$$

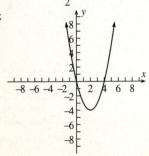

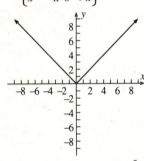

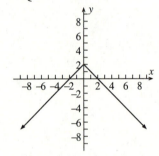

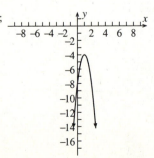

57. $f(x) = \begin{cases} -2x - 2 & \text{if } -2 \le x < -1 \\ x + 1 & \text{if } -1 \le x < 0 \\ -x + 1 & \text{if } 0 \le x < 1 \\ 2x - 2 & \text{if } 1 \le x \le 2 \end{cases}$

59. $N: f(x) = \begin{cases} x & \text{if } 0 \le x < 1 \\ 2 - x & \text{if } 1 \le x < 2 \\ x - 2 & \text{if } 2 \le x \le 3 \end{cases}$; $V: f(x) = |x|$ if $-1 \le x \le 1$

61. $f_1(x) = 1, f_2(x) = x, f_3(x) = -1$

EXERCISES 4.3 (PAGE 214)

1. $-1, 3$ **3.** $\dfrac{1 \pm \sqrt{3}}{2}$

5. (a);

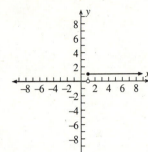

7. (a);

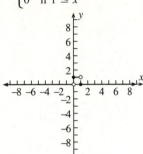

9. (c);

11. (b);

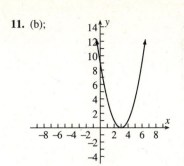

13. (c);

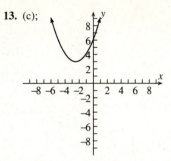

15. 3 is a minimum value
17. 2 is a maximum value
19. $\frac{10}{3}$ is a maximum value
21. $-\frac{9}{4}$ is a minimum value
23. 144 ft; 3 sec
25. (a) $P(x) = -x^2 + 80x - 700$;
  (b) \$900 for 40 units
27. (a) $P(x) = -2x^2 + 380x - 12000$;
  (b) \$5600; (c) \$95
29. (a) $A(x) = 120x - x^2$; (b) [0, 120];
  (c) 60 m by 60 m

31. (a) $A(x) = 120x - \frac{1}{2}x^2$; (b) [0, 240]; (c) 60 m by 120 m   **33.** (a) $A(x) = 96x - \frac{6}{5}x^2$; (b) [0, 80];
(c) The sides parallel to the dividing fence are 40 ft; the others are 48 ft   **35.** -7 and 7   **37.** 130; \$84,500

EXERCISES 4.4 (PAGE 223)

1. (a) $P(x) = 45x$; (b) \$675   **3.** (a) $P(x) = \sqrt{\dfrac{x}{2}}$; (b) 1 sec   **5.** (a) $W(x) = \dfrac{3.2 \times 10^9}{x^2}$; (b) 165 lb

7. (a) $f(x) = \dfrac{9x}{490,000}(5000 - x)$; (b) 17.6 people per day; (c) 2500   **9.** (a) $f(x) = kx(14,000 - x)$;

(b) $\{0, 1, 2, 3, \ldots 14,000\}$; (c) 7000   **11.** (a) $V(x) = 4x^3 - 46x^2 + 120x$; (b) [0, 4]; (c) 1.7 in., 91 in.^{3}
13. (a) $V(x) = 4x^3 - 54x^2 + 180x$; (b) [0, 6]; (c) 2.21 in., 177 in.^{3} **15.** (a) $R = (8 - \frac{1}{100}x)^2x$; (b) [0, 800];
(c) 267, \$7585   **17.** (a) $P = \frac{1}{100}(100 - x)^2x - 55x + \frac{4}{5}x^2$; (b) 30, \$540   **19.** (a) $L(r) = kr[200 - r(2 + \frac{1}{2}\pi)]$;

(b) $\left[0, \dfrac{200}{2 + \pi}\right]$; (c) radius: $\dfrac{200}{4 + \pi}$ in. ≈ 28 in., width: $\dfrac{400}{4 + \pi}$ in. ≈ 56 in., length: $\dfrac{200}{4 + \pi}$ in.

21. (a) $A(x) = 3x + \dfrac{48}{x} + 30$; (b) $(0, +\infty)$; (c) 6.00 in. by 9.00 in.

EXERCISES 4.5 (PAGE 230)

1. 15   **3.** $\frac{5}{3}$ **5.** (a) $x + 5$, domain: $(-\infty, +\infty)$; (b) $x + 5$, domain: $(-\infty, +\infty)$ **7.** (a) $x^2 - 6$, domain: $(-\infty, +\infty)$;
(b) $x^2 - 10x + 24$, domain: $(-\infty, +\infty)$ **9.** (a) $\sqrt{x^2 - 4}$, domain: $(-\infty, -2] \cup [2, +\infty)$; (b) $x - 4$, domain: $[2, +\infty)$

11. (a) $\dfrac{1}{\sqrt{x}}$, domain: $(0, +\infty)$; (b) $\dfrac{1}{\sqrt{x}}$, domain: $(0, +\infty)$ **13.** (a) $|x + 2|$, domain: $(-\infty, +\infty)$; (b) $|x| + 2$, domain: $(-\infty, +\infty)$

15. $x - 4$, domain: $(-\infty, +\infty)$; (b) $x + 14$, domain: $(-\infty, +\infty)$ **17.** $x - 10$, domain: $(-\infty, +\infty)$; (b) $x^4 - 2x^2$, domain: $(-\infty, +\infty)$

19. (a) $\sqrt{\sqrt{x - 2} - 2}$, domain: $[6, +\infty)$; (b) $x^4 - 4x^2 + 2$, domain: $(-\infty, +\infty)$ **21.** (a) x, domain: $\{x \mid x \neq 0\}$;
(b) $\sqrt[4]{x}$, domain: $[0, +\infty)$ **23.** (a) $|x|$, domain: $(-\infty, +\infty)$; (b) $|x + 2| + 2$, domain: $(-\infty, +\infty)$

25. $f(x) = \sqrt{x - 4}$, $g(x) = x^2$; $f(x) = \sqrt{x}$, $g(x) = x^2 - 4$ **27.** $f(x) = x^3$, $g(x) = \dfrac{1}{x - 2}$; $f(x) = \left(\dfrac{1}{x}\right)^3$, $g(x) = x - 2$

29. $f(x) = x^4$, $g(x) = x^2 + 4x - 5$; $f(x) = (x - 5)^4$, $g(x) = x^2 + 4x$ **31.** (a) $(A \circ f)(t) = 36\pi t^2$; (b) 576π cm^2

33. (a) $(f \circ g)(t) = \dfrac{2,000,000}{(t^2 + 7t + 100)^2}$; (b) 78.125   **35.** No   **41.** (a) even; (b) odd

43. $\text{sgn}(U(x)) = \begin{cases} 0 & \text{if } x < 0 \\ 1 & \text{if } 0 \le x \end{cases}$

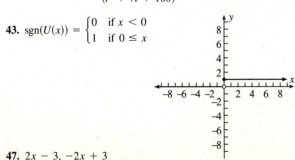

45. $(g \circ f)(x) = \begin{cases} 0 & \text{if } x < 0 \\ x & \text{if } 0 \le x \le \frac{1}{2} \\ 1 & \text{if } \frac{1}{2} < x \le 1 \\ 0 & \text{if } 1 < x \end{cases}$

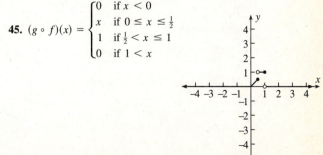

47. $2x - 3$, $-2x + 3$

REVIEW EXERCISES FOR CHAPTER 4 (PAGE 232)

1. (a) 3; (b) 0; (c) -5; (d) $-x^2 + 2x + 3$; (e) $-x^4 + 4$ **3.** (a) 1; (b) 2; (c) 4; (d) 5; (e) 7; (f) $\sqrt{9x + 7}$

5. (a) 0; (b) 0; (c) 0; (d) 0 **7.** $-2x - h$ **9.** $\dfrac{-4}{x(x + h)}$ **11.** $\dfrac{-1}{(\sqrt{x} - 2)(\sqrt{x + h} - 2)(\sqrt{x} + \sqrt{x + h})}$

13. (a) $x^2 + 4x - 7$, domain: $(-\infty, +\infty)$; (b) $x^2 - 4x - 1$, domain: $(-\infty, +\infty)$; (c) $4x^3 - 3x^2 - 16x + 12$, domain: $(-\infty, +\infty)$;

(d) $\dfrac{x^2 - 4}{4x - 3}$, domain: $\{x \mid x \neq \frac{3}{4}\}$; (e) $\dfrac{4x - 3}{x^2 - 4}$, domain: $\{x \mid x \neq \pm 2\}$ **15.** (a) $\sqrt{x + 2} + x^2 - 4$, domain: $[-2, +\infty)$;

(b) $\sqrt{x + 2} - x^2 + 4$, domain: $[-2, +\infty)$; (c) $\sqrt{x + 2}(x^2 - 4)$, domain: $[-2, +\infty)$; (d) $\dfrac{\sqrt{x + 2}}{x^2 - 4}$, domain: $(-2, 2) \cup (2, +\infty)$;

(e) $\dfrac{x^2 - 4}{\sqrt{x + 2}}$, domain: $(-2, +\infty)$ **17.** (a) $\dfrac{1}{x^2} + \sqrt{x}$, domain: $(0, +\infty)$; (b) $\dfrac{1}{x^2} - \sqrt{x}$, domain: $(0, +\infty)$; (c) $\dfrac{1}{\sqrt{x^3}}$, domain: $(0, +\infty)$;

(d) $\dfrac{1}{\sqrt{x^5}}$, domain: $(0, +\infty)$; (e) $\sqrt{x^5}$, domain: $(0, +\infty)$ **19.** (a) $16x^2 - 24x + 5$, domain: $(-\infty, +\infty)$; (b) $4x^2 - 19$, domain:

$(-\infty, +\infty)$ **21.** (a) $\sqrt{x^2 - 2}$, domain: $(-\infty, -\sqrt{2}] \cup [\sqrt{2}, +\infty)$; (b) $x - 2$, domain: $[-2, +\infty)$ **23.** (a) $\dfrac{1}{x}$, domain: $(0, +\infty)$;

(b) $\dfrac{1}{|x|}$, domain: $\{x \mid x \neq 0\}$ **25.** domain: $(-\infty, +\infty)$; range: $(-\infty, +\infty)$ **27.** domain: $(-\infty, +\infty)$; range: $[-4, +\infty)$

29. domain: $(-\infty, -4] \cup [4, +\infty)$; range: $[0, +\infty)$ **31.** domain: $[-4, 4]$; range: $[0, 4]$ **33.** domain: $(-\infty, +\infty)$; range $[0, +\infty)$

35. domain: $\{x \mid x \neq -4\}$; range: $\{y \mid y \neq -8\}$

37. domain: $(-\infty, +\infty)$; range: $\{y \mid y \neq -8\}$

39. domain: $(-\infty, +\infty)$; range: $[3, +\infty)$

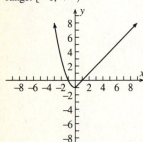 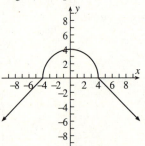 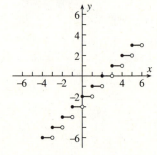

41. domain: $(-\infty, +\infty)$; range: $[-1, +\infty)$

43. domain: $(-\infty, +\infty)$; range: $(-\infty, 4]$

45. domain: $(-\infty, +\infty)$; range: {integers}

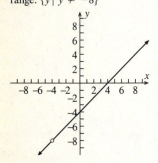 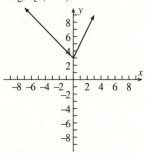

47. domain: $(-\infty, +\infty)$;
range: $[0, 1)$

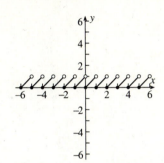

49. $\text{sgn}(x) - U(x)$
$$= \begin{cases} -1 & \text{if } x \leq 0 \\ 0 & \text{if } 0 < x \end{cases}$$

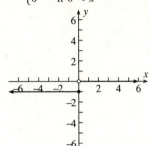

51. (a)

53. (b)

55. 6 is a maximum value,
occurring at $x = -1$

57. -3 is a minimum value,
occurring at $x = -1$

59. 8

61. $f(x) = x^{1/3}$, $g(x) = 3x^2 - 2$; $f(x) = (x - 2)^{1/3}$, $g(x) = 3x^2$ **63.** (a) odd;   (b) even;   (c) neither;   (d) odd

65. (a) $f(t) = 16t^2$;   (b) 100 ft   **67.** (a) $f(x) = \dfrac{x(10{,}000 - x)}{100{,}000}$;   (b) 160 fish per week   **69.** 6 and 6

71. (a) $f(x) = \begin{cases} 15x & \text{if } 0 \leq x \leq 150 \\ 22.5x - 0.05x^2 & \text{if } 150 < x \leq 250 \end{cases}$;   (b) $[0, 250]$;   (c) 225   **73.** (a) $f(x) = kx(11{,}000 - x)$;

(b) $[0, 11000]$;   (c) 5500   **75.** (a) $(f \circ g)(t) = 16t^2 + 360t + 2105$;   (b) 6641   **77.** (a) $P(x) = 244x - x^3$;

(b) 9, \$1467   **79.** (a) $V = (20 - 2x)^2 x$;   (b) $[0, 10]$;   (c) 3.33 cm, 592.59 cm³   **81.** (a) $S(x) = x^2 + \dfrac{16000}{x}$;

(b) $(0, +\infty)$;   (c) 20 in. by 20 in. by 10 in.

85. $(f \circ g)(x)$
$$= \begin{cases} 1 & \text{if } x < 0 \\ 2x & \text{if } 0 \leq x \leq \frac{1}{2} \\ 0 & \text{if } \frac{1}{2} < x \leq 1 \\ 1 & \text{if } 1 < x \end{cases}$$

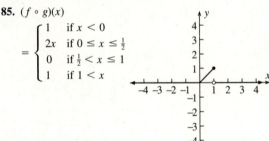

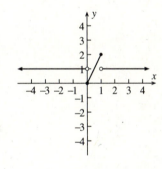

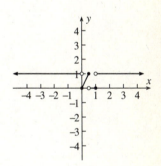

87.

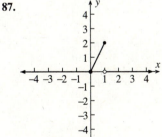

EXERCISES 5.1 (PAGE 248)

1.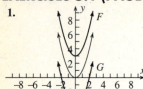

3.

5.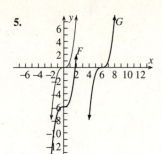

7.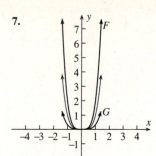

9.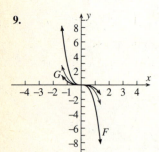

11.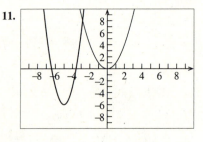

13.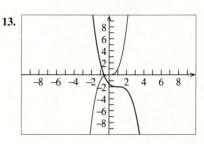

15.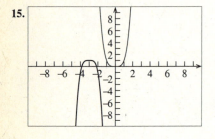

17. (a)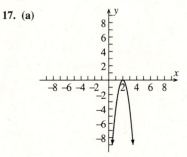

17. (b)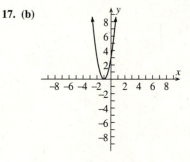

19. (a)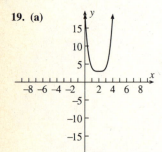

19. (b)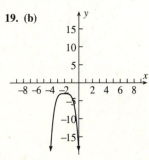

21. (a)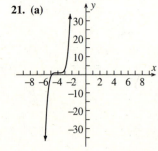

21. (b)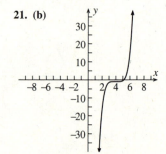

23. (a) up from left, up to right; **(b)** 2;

(c)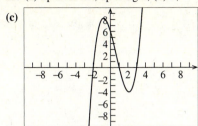

(d) rel. max. at $(-0.786, 8.209)$
rel. min. at $(2.120, -4.061)$

25. (a) up from left, up to right; **(b)** 2;

(c)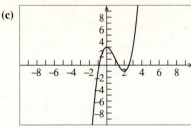

(d) rel. max. at $(0, 3)$
rel. min. at $(2, -1)$

27. (a) up from left, up to right; **(b)** 2;

(c)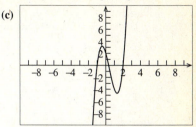

(d) rel. max. at $(-0.423, 3.172)$
rel. min. at $(1.312, -4.670)$

29. (a) down from left, up to right; **(b)** 3;

(c)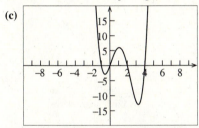

(d) rel. min. at $(-0.574, -2.879)$
rel. max. at $(1.070, 6.035)$
rel. min. at $(3.253, -12.94)$

31. (a) down from left, up to right; **(b)** 3;

(c)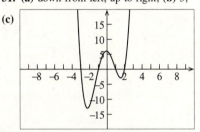

(d) rel. min. at $(-2.25, -12.95)$
rel. max. at $(-0.70, 6.035)$
rel. min. at $(1.5742, -2.879)$

33. (a) down from left, up to right; **(b)** 3;

(c)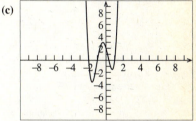

(d) rel. min. at $(-1.607, -3.620)$
rel. max. at $(-0.361, 2.969)$
rel. min. at $(0.718, -1.520)$

35. (a) up from left, down to right, **(b)** 3;

(c)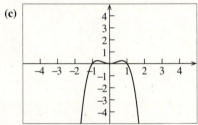

(d) rel. max. at $(\pm 0.707, 0.25)$
rel. min. at $(0, 0)$

37.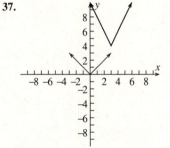

39.

EXERCISES 5.2 (PAGE 258)

1. (a) 20; **(b)** 5   **3. (a)** 1; **(b)** -8 **5.** Yes   **7.** No   **9.** Yes   **11.** The quotient is $2x^2 + 7x + 31$, and the remainder is 136   **13.** The quotient is $2x^3 - 3x^2 + 12x - 50$, and the remainder is 199
15. The quotient is $3z^4 + 7z^3 + 14z^2 + 24z + 48$, and the remainder is 103   **17.** The quotient is $6x^2 - 3x + 3$, and the remainder is 1   **19.** The quotient is $x^6 + x^5 + x^4 + x^3 + x^2 + x + 1$, and the remainder is 0   **21.** 8, -157
23. -7, 126   **25.** 4, $\frac{19}{2}$ **27.** $P(x) = (x + 3)(x - 1)(4x + 1)$ **29.** $P(x) = (2x - 1)(x - 1)(3x + 1)$ **31.** No
33. Yes   **35.** $((2x - 6)x - 3)x - 12$; 359; -397 **37.** $(((4x - 1)x + 2)x - 1$; 302; 1079   **39.** -7 **41.** -3 and 1
43. (a) Yes; **(b)** No; **(c)** No; **(d)** Yes   **45. (a)** all positive integers; **(b)** all positive even integers

EXERCISES 5.3 (PAGE 267)

1. 4, multiplicity two; -2; 2 **3.** 0, multiplicity two; -1, multiplicity two; $\sqrt{3}$; $-\sqrt{3}$ **5.** -7, multiplicity three; $\sqrt{7}$, multiplicity two; $-\sqrt{7}$, multiplicity two **7.** $-\frac{4}{3}$, multiplicity three; $\frac{3}{2}$, multiplicity two; $-\frac{3}{2}$, multiplicity four **9.** $-1, 4$

11. $\dfrac{-1 \pm \sqrt{5}}{2}$ **13.** $\frac{2}{5}, 1$ **15.** $-2 \pm \sqrt{3}$ **17.** $-1, 1, 3$ **19.** $-2, -1, 3$ **21.** $-5, 3, \frac{1}{2}(-1 \pm i\sqrt{3})$ **23.** $\frac{1}{3}, \dfrac{-3 \pm \sqrt{5}}{2}$

25. $\frac{1}{2}, \frac{5}{3}, 2 \pm \sqrt{3}$ **27.** 2 is the only rational zero **29.** -4; 1, multiplicity two **31.** $\frac{3}{2}, 2, 3$ **33.** $-4, -2, 0, 1, 3$

35. $-\frac{1}{4}, \frac{2}{3}, \pm\sqrt{3}$ **37.** $-6, \frac{1}{6}(1 \pm i\sqrt{11})$ **39.** -2 is the only rational root **45.** 3 in. or $\dfrac{9 - 3\sqrt{5}}{2}$ in. ≈ 1.15 in. **47.** 6 cm

49. 3 ft or $\dfrac{1 + \sqrt{13}}{2}$ ft ≈ 2.3 ft

EXERCISES 5.4 (PAGE 278)

1. $x^2 - 8x + 25$ **3.** $x^3 - 7x + 36$ **5.** $x^4 + 18x^2 + 81$ **7.** $x^5 - 9x^4 + 31x^3 - 51x^2 + 58x - 66$

9. $\left\{\dfrac{1 \pm \sqrt{31}}{3}, -2i, 2i\right\}$ **11.** $\left\{\dfrac{1 \pm i}{2}, -2 - 4i, -2 + 4i\right\}$ **13.** $2, 1 \pm i; (x - 2)(x^2 - 2x + 2)$

15. $-1, \frac{1}{2}, 2 \pm 3i; 2(x + 1)(x - \frac{1}{2})(x^2 - 4x + 13)$ **17.** $2, \pm 2i$ (multiplicity 2); $(x - 2)(x^2 + 4)^2$

19. $\frac{1}{3}$ (multiplicity 2), $2 \pm \sqrt{2}i; 9(x - \frac{1}{3})^2(x^2 - 4x + 6)$ **21.** One positive irrational zero between 4 and 5; two imaginary zeros

23. One negative irrational zero between -8 and -7; one positive irrational zero between 0 and 1; two imaginary zeros

25. One negative irrational zero between -3 and -2; one positive irrational zero between 0 and 1; one positive irrational zero between 2 and 3 **27.** One negative irrational zero between -1 and 0; one positive irrational zero between 0 and 1; two imaginary zeros

29. -1 is a zero; one positive irrational zero between 1 and 2; two imaginary zeros

31. 4 is a zero; one negative irrational zero between -1 and 0; one positive irrational zero between 0 and 1; one positive irrational zero between 2 and 3 **39. (a)** 1 positive, 1 negative, $2n - 2$ imaginary; **(b)** 0 positive, 1 negative, $2n - 2$ imaginary

EXERCISES 5.5 (PAGE 290)

1. (a) $\{x \mid x \neq 0\}$; **(b)** no intercepts;
(c) symmetry with respect to the origin;
(d) $x = 0, y = 0$;
(f)

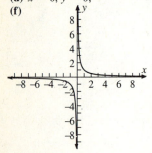

3. (a) $\{x \mid x \neq 2\}$; **(b)** $(0, -2)$;
(c) neither symmetry;
(d) $x = 2, y = 0$;
(f)

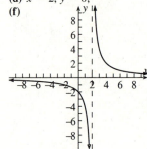

5. (a) $\{x \mid x \neq 3\}$; **(b)** $(-1, 0), (0, \frac{1}{3})$;
(c) neither symmetry;
(d) $x = 3, y = -1$;
(f)

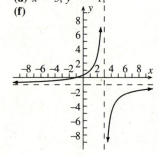

7. (a) $\{x \mid x \neq -4\}$; **(b)** $(2, 0), (0, -1)$;
(c) neither symmetry;
(d) $x = -4, y = 2$;
(f)

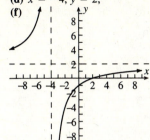

9. (a) $\{x \mid x \neq 0\}$; **(b)** no intercepts;
(c) symmetry with respect to the y axis;
(d) $x = 0, y = 0$;
(f)

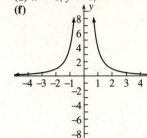

11. (a) $\{x \mid x \neq 0\}$; **(b)** no intercepts;
(c) symmetry with respect to the origin;
(d) $x = 0, y = 0$;
(f)

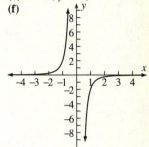

13. (a) $\{x \mid x \neq -2\}$; **(b)** $(0, -\frac{1}{4})$;
(c) neither symmetry;
(d) $x = -2, y = 0$;
(f)

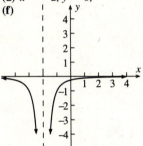

15. (a) $\{x \mid x \neq \pm 2\}$; **(b)** $(0, 0)$;
(c) symmetry with respect to the origin;
(d) $x = -2, x = 2, y = 0$;
(f)

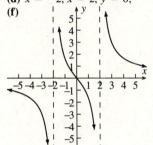

17. (a) $\{x \mid x \neq \pm 4\}$; **(b)** $(0, 0)$;
(c) symmetry with respect to the origin;
(d) $x = -4, x = 4, y = 0$;
(f)

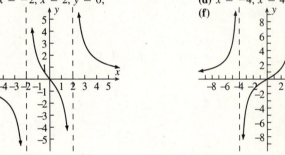

19. (a) $\{x \mid x \neq \pm 3\}$; **(b)** $(0, 0)$;
(c) symmetry with respect to the y axis;
(d) $x = -3, x = 3, y = 2$;
(f)

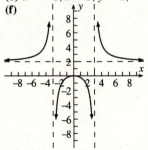

21. (a) $\{x \mid x \neq \pm 1\}$; **(b)** $(0, -1)$;
(c) symmetry with respect to the y axis;
(d) $x = -1, x = 1, y = 1$;
(f)

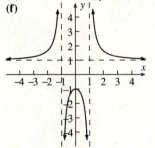

23. (a) $\{x \mid x \neq -3, x \neq 2\}$;
(b) $(-1, 0), (0, -\frac{1}{6})$;
(c) neither symmetry;
(d) $x = -3, x = 2, y = 0$;
(f)

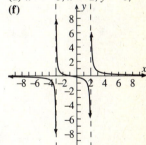

25. (a) $\{x \mid x \neq 2\}$;   (b) $(-3, 0), (3, 0), (0, \frac{9}{2})$;
(c) neither symmetry;   (d) $x = 2$;
(e) $y = x + 2$;
(f)

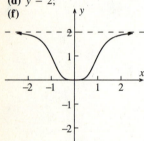

27. (a) $\{x \mid x \neq 0\}$;   (b) no intercepts;
(c) symmetry with respect to the origin;
(d) $x = 0$;   (e) $y = x$;
(f)

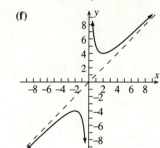

29. (a) $(-\infty, +\infty)$;   (b) $(0, 1)$;
(c) symmetry with respect to the
   y axis;
(d) $y = 0$;
(f)

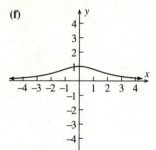

31. (a) $(-\infty, +\infty)$;   (b) $(0, 0)$;
(c) symmetry with respect to the y axis;
(d) $y = 2$;
(f)

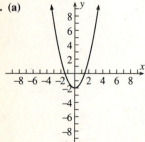

33. $(5, 6)$ **35.** $(-\infty, 2) \cup [6, +\infty)$ **37.** $(-\infty, -2) \cup (1, 3)$

39. $(-\infty, 2) \cup [3, 4) \cup [5, +\infty)$ **41.** (a) $S(x) = \dfrac{32\pi}{x} + 2\pi x^2$;   (b) $(0, +\infty)$

(c) 2.0 in., 75.40 in.^{2} **43.** (a) $C(x) = 6x^2 + \dfrac{12000}{x}$;   (b) $(0, +\infty)$;   (c) 10.0 in., $18.00

45. (a) $P(x) = 2x + \dfrac{200}{x}$;   (b) $(0, +\infty)$;   (c) 10.0 in. by 10.0 in.

47. (a) $C(x) = 8 + \dfrac{x}{300} + \dfrac{27}{x}$;   (b) $(0, +\infty)$;   (c) 90 km/hr   **49.** at least 6 ohms

REVIEW EXERCISES FOR CHAPTER 5 (PAGE 292)

1. -7 **3.** Yes   **5.** $\frac{5}{7}$ **7.** $2x^3 + x^2 - 3x + 5, -10$ **9.** $x^5 + 2x^4 + 4x^3 + 8x^2 + 16x + 32, 0$ **11.** (a) 17;   (b) 57

13. -3, multiplicity two; $1; \frac{5}{2}$ **15.** -2, multiplicity two; 2, multiplicity two; $-3; -\frac{5}{3}; \frac{3}{2}; 3$ **17.** $-1 \pm \sqrt{5}$ **19.** $-3, 2, 2 \pm \sqrt{2}$

21. $\frac{2}{3}, \dfrac{3 \pm \sqrt{5}}{2}$ **23.** $3, -2 \pm \sqrt{7}$ **25.** $-\frac{1}{3}, \frac{1}{2}, 2 \pm \sqrt{5}$ **27.** $\{\pm\sqrt{2}, -1 \pm \sqrt{3}\}$

29. (a)

29. (b)

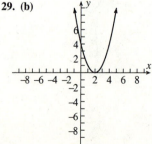

31. (a)

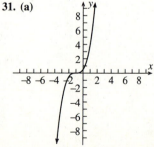

31. (b)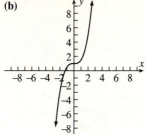

33. (a)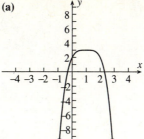

33. (b)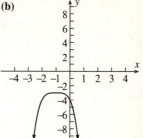

35.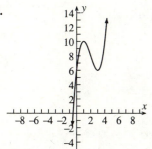

37.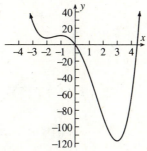

39.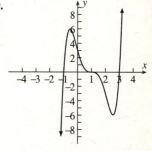

41. (a) $\{x \mid x \neq 5\}$;   **(b)** $(0, -\frac{2}{5})$;
(c) neither symmetry;
(d) $x = 5, y = 0$;
(f)

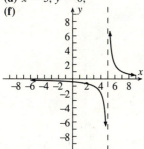

43. (a) $\{x \mid x \neq 0\}$;   **(b)** no intercepts;
(c) symmetry with respect to the y axis;
(d) $x = 0, y = 0$;
(f)

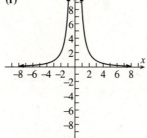

45. (a) $\{x \mid x \neq -2\}$;   **(b)** $(0, 2)$;
(c) neither symmetry;
(d) $x = -2, y = 0$;
(f)

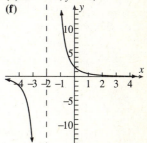

47. (a) $\{x \mid x \neq \pm 1\}$;   **(b)** $(0, 0)$;
(c) symmetry with respect to the origin;
(d) $x = -1, x = 1, y = 0$;
(f)

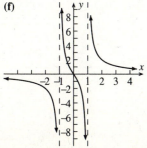

49. (a) $\{x \mid x \neq \pm 3\}$;   **(b)** $(0, 0)$;
(c) symmetry with respect to the y axis;
(d) $x = -3, x = 3, y = -4$;
(f)

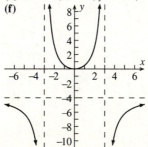

51. (a) $\{x \mid x \neq 1\}$;   **(b)** $(-2, 0), (2, 0), (0, 4)$;
(c) neither symmetry;
(d) $x = 1$;   **(e)** $y = x + 1$;
(f)

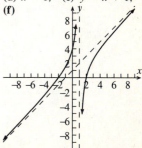

53. $x^4 - 4x^3 + 10x^2 - 12x + 8$ **55.** $x^6 - 4x^5 + 24x^4 - 72x^3 + 189x^2 - 324x + 486$

57. (a) $\frac{1}{2}, -1 \pm \sqrt{3}i$; **(b)** $(2x - 1)(x^2 + 2x + 4)$ **59. (a)** $\pm 2i, -1 \pm i$; **(b)** $(x^2 + 4)(x^2 + 2x + 2)$

61. $\left\{ -1 \pm i, \dfrac{-1 \pm \sqrt{5}}{2} \right\}$ **63.** two positive, one negative, and no imaginary; or no positive, one negative, and two imaginary

65. one positive, one negative, and two imaginary

67. one positive irrational zero between 3 and 4;
two imaginary zeros

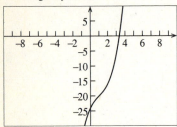

69. 2 is a zero; one negative irrational zero between -4 and -3;
one positive irrational zero between 0 and 1;
one positive irrational zero between 3 and 4

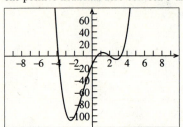

71. $\left\{\frac{3}{4}, 1 \pm 2i\right\}$ **73.** $\left\{-\frac{1}{3}, \frac{1}{2}, 2 \pm \sqrt{5}\right\}$ **75.** $(-\infty, -6) \cup (5, +\infty)$ **77.** $(-\infty, -4) \cup [-2, 3) \cup [9, +\infty)$ **79.** 1

81. 3 cm, 4 cm, 5 cm   **83. (a)** $S(x) = 3x^2 + \dfrac{768}{x}$; **(b)** $(0, +\infty)$; **(c)** 5.0 in. by 15.1 in. by 3.8 in. high; 228.6 in.²

85. (a) $S(x) = 8x^2 + \dfrac{54}{x}$; **(b)** $(0, +\infty)$; **(c)** radius is 1.5 in., height is 3.82 in.; 54 in.²

EXERCISES 6.1 (PAGE 307)

1. one-to-one

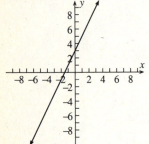

3. not one-to-one

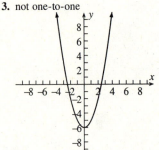

5. one-to-one

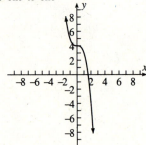

7. not one-to-one

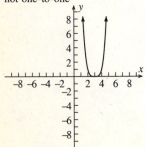

9. one-to-one

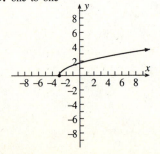

11. one-to-one

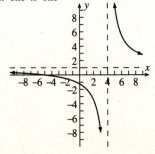

13. not one-to-one

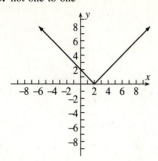

In Exercises 15-35, the graph of f is shown lighter than that of f⁻¹.

15. (b) $f^{-1}(x) = \dfrac{3 - x}{4}$;

(d)

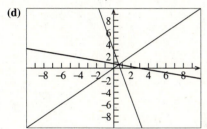

17. (b) $f^{-1}(x) = \sqrt[3]{x - 2}$;

(d)

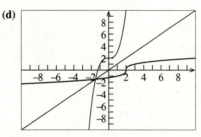

19. (b) $f^{-1}(x) = \dfrac{1 - x}{x}$;

(d)

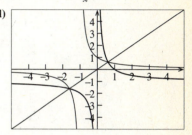

21. (a) $[-5, +\infty)$;
  (b) $f^{-1}(x) = \sqrt{x + 5}, [-5, +\infty)$;
  (c)

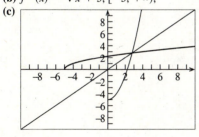

23. (a) $[0, +\infty)$;
  (b) $f^{-1}(x) = \sqrt{x^2 + 9}, [0, +\infty)$;
  (c)

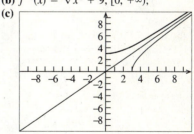

25. (a) $[0, +\infty)$;
  (b) $f^{-1}(x) = -\sqrt{x^2 + 9}, [0, +\infty)$;
  (c)

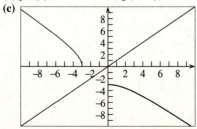

27. (a) $\left[-\frac{1}{8}, \frac{1}{8}\right]$;
  (b) $f^{-1}(x) = 2\sqrt[3]{x}, \left[-\frac{1}{8}, \frac{1}{8}\right]$;
  (c)

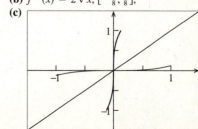

29. (b) $f^{-1}(x) = \frac{1}{2}(x - 5);$
(c)

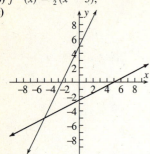

31. (b) $f^{-1}(x) = \sqrt[3]{x} - 1;$
(c)

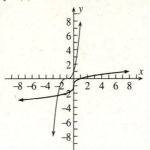

33. (b) $f^{-1}(x) = \sqrt{x} + 2;$
(c)

35. (b) $f^{-1}(x) = -\sqrt{4 - x};$
(c)

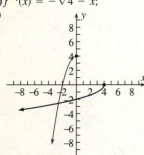

39. -1

45. $f^{-1}(x) = \begin{cases} x & \text{if } x < 1 \\ \sqrt{x} & \text{if } 1 \le x \le 81 \\ \left(\dfrac{x}{27}\right)^2 & \text{if } 81 < x \end{cases}$

43. (b) $f_1(x) = x^2 + 4, x \ge 0;$
$f_2(x) = x^2 + 4, x \le 0;$
(c) $f_1^{-1}(x) = \sqrt{x - 4}, x \ge 4;$
$f_2^{-1}(x) = -\sqrt{x - 4}, x \ge 4$

EXERCISES 6.2 (PAGE 317)

1. (a) 7620; **(b)** 0.00130; **(c)** 0.0000215; **(d)** 15.3   **3. (a)** $(5.44)10^{15}$; **(b)** $(2.96)10^{-8}$; **(c)** $(1.49)10^{33}$; **(d)** $(7.84)10^{-26}$
5. (a) $3^{6\sqrt{2}}$; **(b)** e^{10} **7. (a)** $5^{3\sqrt{10}}$; **(b)** $5^{\sqrt[3]{x} + \sqrt[3]{x^2}}$ **9. (a)** $2^{5\sqrt{2}}$; **(b)** $\left(\frac{5}{2}\right)^{\sqrt{5}}$ **11.** $e^{2+\sqrt{6}}$ **13. (a)** $(5.260)10^1$;
(b) $(6.1)10^{-3}$; **(c)** $(1.72)10^5$; **(d)** $(1.720)10^5$ **15. (a)** $(3.960)10^{-2}$; **(b)** $(8.0022)10^{-6}$; **(c)** $(1.723)10^0$; **(d)** $(4.260)10^2$
17. (a) $(4.81)10^9$; **(b)** $(6.28)10^7$ **19.** $(1.50)10^{11}$ m   **21.** $(6.58)10^{21}$ tons
23. $(1.001)^{10,000} \approx 2.71815$; $(0.9999)^{-10,000} \approx 2.71842$; $e \approx 2.7183$ **25. (a)** \$2060; **(b)** \$2120; **(c)** \$2240   **27.** \$10,938.07
29. (a) \$1320; **(b)** \$1360.49; **(c)** \$1368.57; **(d)** \$1372.79; **(e)** \$1377.13   **31. (a)** \$849.29; **(b)** \$867.55
33. (a) \$1094.17; **(b)** \$1094.16   **35.** \$3364.86   **37.** \$3351.60

39. The figure shows the graphs of $f(x) = \left(1 + \dfrac{2}{x}\right)^x$
and its asymptote, $y = e^2$,
in the viewing rectangle [0, 60] by [0, 10].

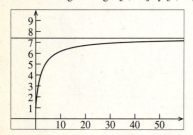

41. 10.52% simple interest; 10.13% interest compounded quarterly

EXERCISES 6.3 (PAGE 328)

1. **3.** **5.** **7.**

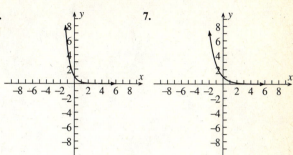

In Exercises 9-13, the graph of F is shown, along with that of f, which is shown relatively light.

9. (a) **11. (a)** **13. (a)**

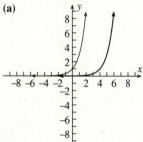 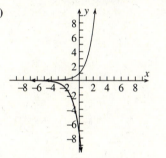 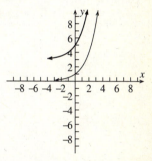

(b) **(b)** **(b)**

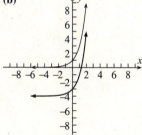 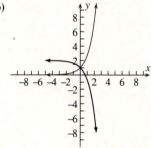 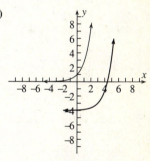

15. **17.** **19.**

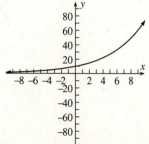 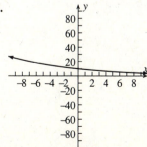 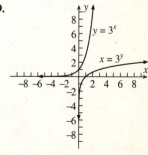

21. (a) 10,000;
(b)

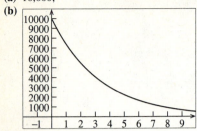

(c) $907.18

23. (a) 10,000;
(b)

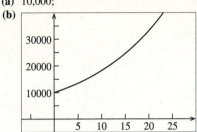

(c) 18,221;   **(d)** 33,201

25. 1568 lb/ft^2
27. (a) $f(t) = 200 \cdot 2^{t/10}$;
(b) $12,800

29. (a) 40;   **(b)** 94;   **(c)** 100;
(d) horizontal asymptote: $y = 100$

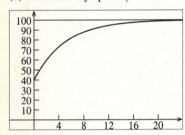

31. (a) 144;   **(b)** 4075;   **(c)** 5000
(d) horizontal asymptote: $y = 5000$

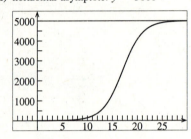

33. 50 percent

35. (a) 97.60°;   **(b)** 53.41°;   **(c)** 38.99°;
(d) 35.34°;   **(e)** 35°;
(f) horizontal asymptote: $y = 35$

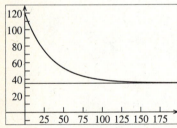

37. (a) 0;   **(b)** 1.543;   **(c)** −1.175;
(d) 16.543;   **(e)** 3.762
39. (b)

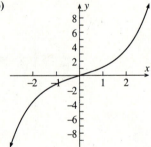

EXERCISES 6.4 (PAGE 339)

1. (a) $\log_3 81 = 4$;   **(b)** $\log_5 125 = 3$;   **(c)** $\log_{10}(0.001) = -3$ **3. (a)** $\log_8 4 = \frac{2}{3}$;   **(b)** $\log_{625}(\frac{1}{125}) = -\frac{3}{4}$;   **(c)** $\log_2 1 = 0$
5. (a) $8^2 = 64$;   **(b)** $3^4 = 81$;   **(c)** $2^0 = 1$ **7. (a)** $8^{1/3} = 2$;   **(b)** $(\frac{1}{3})^{-2} = 9$;   **(c)** $9^{-1/2} = \frac{1}{3}$ **9. (a)** 2;   **(b)** $\frac{2}{3}$;   **(c)** −3
11. (a) $-\frac{1}{3}$;   **(b)** $-\frac{4}{3}$;   **(c)** $\frac{1}{2}$ **13. (a)** 343;   **(b)** 81   **15. (a)** $2\sqrt{2}$;   **(b)** $\frac{1}{128}$ **17. (a)** 12;   **(b)** 216   **19. (a)** 10;   **(b)** 8
21. (a) 0;   **(b)** 1   **23. (a)** 3;   **(b)** 0

In Exercises 31-37, the graph of the given function is shown darker than that of the natural logarithmic function.

31.   **33.**   **35.**   **37.**

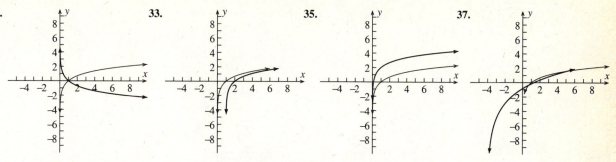

39. 2 years from now   **41.** 25 years from now   **43. (a)** $A(t) = 500e^{.09t}$;   **(b)** 6.53 years $\approx$ 6 years, 193 days   **45.** 3.47

47. 161 years   **49.** 8.66 years $\approx$ 8 years, 242 days   **51.** $S^{-1}(x) = \ln\left(x + \sqrt{x^3 + 1}\right)$

EXERCISES 6.5 (PAGE 349)

1. (a) $\log_b 5 + \log_b x + \log_b y$;   **(b)** $\log_b y - \log_b z$;   **(c)** $\log_b x - \log_b y - \log_b z$ **3. (a)** $\log_b x + 5 \log_b y$;   **(b)** $\frac{1}{2}(\log_b x + \log_b y)$

5. (a) $\frac{1}{3}\log_b x + 3 \log_b z$;   **(b)** $\log_b x + \frac{1}{2}\log_b y - 4 \log_b z$ **7. (a)** $\frac{1}{3}(2 \log_b x - \log_b y - 2 \log_b z)$;   **(b)** $\frac{2}{3}\log_b x + \frac{1}{2}(\log_b y + \log_b z)$

9. (a) $\log_{10} x^4 \sqrt{y}$ **(b)** $\log_b \dfrac{\sqrt[4]{x^3}}{y^6 \sqrt[5]{z^4}}$ **11. (a)** $\log_{10} \frac{1}{2} gt^2$;   **(b)** $\ln \frac{1}{3}\pi h r^2$ **13. (a)** 1.1461;   **(b)** 1.1761

15. (a) 1.7993;   **(b)** 2.1461   **17. (a)** 0.3404;   **(b)** -2.7744 **19. (a)** -0.1761;   **(b)** 0.6309   **21. (a)** 5.7036;   **(b)** 2.0149

23. (a) 0.7851;   **(b)** 2.5108   **25.** {7}   **27.** {125}   **29.** {3}   **31.** {7}   **33. (c)** $10^{1.3} \approx 19.95$

EXERCISES 6.6 (PAGE 357)

1. {1.404}   **3.** {0.4307}   **5.** {4.301}   **7.** {8.638}   **9.** {32.20}   **11.** {1.015}   **13.** 2.262   **15.** 4.170   **17.** 3.638

19. 0.9375   **21.** 3.42 years from January 1, 1993 (near the beginning of June 1996)   **23.** 18.58 years

25. 48089.16 ft above sea level   **27.** 1997   **29.** 69.95   **31.** 54.87%   **33.** {0.3828}   **35.** {0.8618}   **37.** {$-2.063, 2.063$}

39. $x = \frac{1}{2}\log \dfrac{1 + y}{1 - y}$ **41. (a-b)** $x \approx 1.18$ **43. (c)** domain: $(0, 1) \cup (1, +\infty)$, range: $\{y \mid y \neq 0\}$

REVIEW EXERCISES FOR CHAPTER 6 (PAGE 359)

1. one-to-one   **3.** not one-to-one

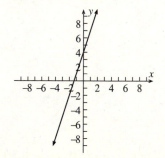   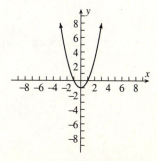

In Exercises 5-15, the graph of f is shown lighter than that of f^{-1}.

5. (b) $f^{-1}(x) = \sqrt[3]{x + 8}$;

(d)

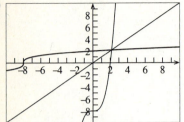

7. (b) $f^{-1}(x) = \dfrac{1 + 2x}{1 - x}$;

(d)

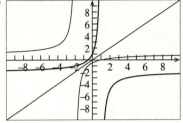

9. (a) $(-\infty, 9]$;

(b) $f^{-1}(x) = \sqrt{9 - x}$, $(-\infty, 9]$;

(c)

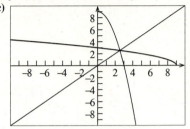

11. (a) $[0, +\infty)$;

(b) $f^{-1}(x) = -\sqrt{x^2 + 1}$, $[0, +\infty)$;

(c)

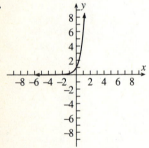

13. (a) decreasing function;

(b) $f^{-1}(x) = 1 - \sqrt[3]{x}$;

(c)

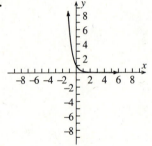

15. (a) increasing function;

(b) $f^{-1}(x) = \sqrt{x + 4}$;

(c)

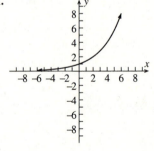

17.

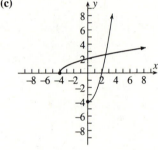

19.

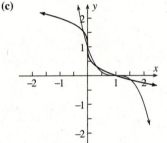

21.

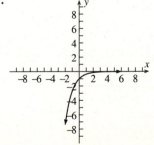

23.

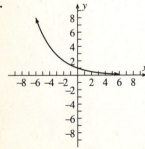

25.

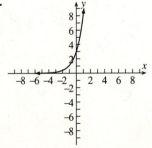

27.

29.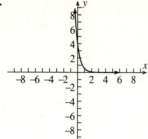

31.

33.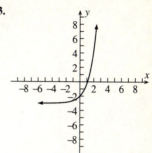

35.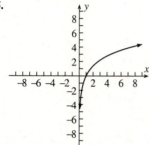

37.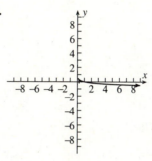

39.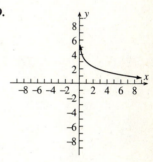

41.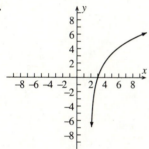

43.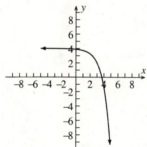

45.

47. (a) 625; **(b)** $\frac{4}{3}$ **49. (a)** $\frac{1}{64}$; **(b)** $\dfrac{1}{e^2}$ **51.** $3 \log_b x + 2 \log_b y + \frac{1}{2} \log_b z$ **53.** $\log_b x + \frac{1}{3} \log_b y - 4 \log_b z$

55. $\log_{10}\left(\dfrac{\sqrt[3]{y^2}}{x^4 \sqrt[3]{z}}\right)$ **57.** $\ln(\frac{4}{3}\pi r^2 h)$ **59.** $\{2.024\}$ **61.** $\{1.512\}$ **63.** $\{8.084\}$ **65.** $\{-2.543, 2.543\}$ **67.** 0.9358

69. 5.248   **71.** $\{\frac{1}{2}\}$ **73.** $\{400\}$ **77. (a)** \$960; **(b)** \$988.80; **(c)** \$1004.07   **79.** \$1019.97   **81.** \$14,000

83. (a) 5.86 years;   **(b)** 5.78 years

85. (a) 60;

(b)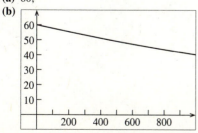

(c) 57.65 mg;   **(d)** 455.80 years

87. 5.49

89. (a) 73.87°;   **(b)** 73.13°;   **(c)** 71.34°;   **(d)** 65°;   **(e)** 35°;
(f) horizontal asymptote: $y = 35$

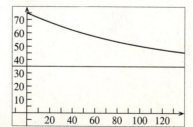

91. 144.57   **93.** 2.512%   **95.** 50.12

103. (a)

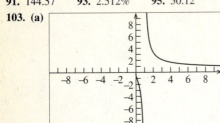

(b) domain: $(0, 1) \cup (1, +\infty)$, range: $\{y \mid y \neq 0\}$

105.

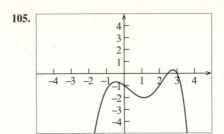

(a) $x = 2.48$;

(b) $(-\infty, 2.48) \cup (3, +\infty)$;   **(c)** $(2.48, 3)$

EXERCISES 7.1 (PAGE 373)

1. (a) first quadrant;   **(b)** second quadrant   **3. (a)** first quadrant;   **(b)** fourth quadrant
5. (a) second quadrant;   **(b)** second quadrant   **7. (a)** third quadrant;   **(b)** fourth quadrant
9. $\sin 2 \approx 0.91$; $\cos 2 \approx -0.42$ **11.** $\sin 5.2 \approx -0.88$; $\cos 5.2 \approx 0.47$ **13.** $\sin(-3) \approx -0.14$; $\cos(-3) \approx -0.99$

15. $\sin(-6.1) \approx 0.18$; $\cos(-6.1) \approx 0.98$ **17. (a)** 0;   **(b)** -1 **19. (a)** 0;   **(b)** 1   **21. (a)** $\dfrac{\sqrt{3}}{2}$;   **(b)** $\dfrac{1}{2}$

23. (a) $-\dfrac{1}{\sqrt{2}}$;   **(b)** $\dfrac{1}{\sqrt{2}}$ **25. (a)** $\frac{1}{2}$;   **(b)** $\frac{1}{2}$ **27. (a)** $-\frac{1}{2}$;   **(b)** $-\frac{1}{2}$ **29. (a)** $-\dfrac{1}{\sqrt{2}}$;   **(b)** $\dfrac{1}{\sqrt{2}}$ **31. (a)** 0;   **(b)** -1

33. (a) $\dfrac{\sqrt{3}}{2}$;   **(b)** $-\dfrac{\sqrt{3}}{2}$ **35. (a)** $\dfrac{1}{\sqrt{2}}$;   **(b)** $\dfrac{1}{\sqrt{2}}$ **37. (a)** first quadrant;   **(b)** fourth quadrant   **39.** $\frac{4}{5}$ **41.** $-\frac{15}{17}$

43. $-\frac{5}{13}$ **45.** $\dfrac{\sqrt{7}}{4}$ **47.** -1 **49. (a)** second quadrant;   **(b)** second quadrant

51. (a) fourth quadrant;   **(b)** third quadrant   **53.** 0.6238   **55.** -0.9085 **57. (a)** $\dfrac{\sqrt{3}}{2}$;   **(b)** 1;   **(c)** $-\dfrac{1}{2}$;   **(d)** $-\dfrac{1}{\sqrt{2}}$

EXERCISES 7.2 (PAGE 382)

1. (a) 0.3335;   **(b)** 0.9428;   **(c)** 0.2875;   **(d)** -0.9578 **3. (a)** -0.9121;   **(b)** -0.4099;   **(c)** -0.9779;   **(d)** 0.2092

5. (a) 0.9898;   **(b)** 0.9238;   **(c)** 0.3420;   **(d)** 0.8090   **7. (a)** $\dfrac{1}{\sqrt{2}}$;   **(b)** $\dfrac{1}{\sqrt{2}}$;   **(c)** $-\dfrac{\sqrt{3}}{2}$;   **(d)** $-\dfrac{1}{2}$

9. (a) $-\dfrac{1}{2}$;   **(b)** $\dfrac{\sqrt{3}}{2}$;   **(c)** $\dfrac{1}{\sqrt{2}}$;   **(d)** $-\dfrac{1}{\sqrt{2}}$ **11. (a)** $-\dfrac{\sqrt{3}}{2}$;   **(b)** $\dfrac{1}{2}$;   **(c)** 0;   **(d)** 1   **13. (a)** 1;   **(b)** 0;   **(c)** 1;   **(d)** 0

15. (a) $\frac{2}{3}\pi$;   **(b)** $\frac{1}{3}\pi$;   **(c)** 8π;   **(d)** 10π **17. (a)** $\frac{2}{3}\pi$;   **(b)** $\frac{7}{7}\pi$;   **(c)** $\frac{1}{2}$;   **(d)** 6   **19. (a)** 10;   **(b)** 2   **21.** 0.7317

23. -0.8994 **25.** -0.7243 **27. (a)** $\frac{1}{25}$;   **(b)** -2 volts;   **(c)** 0;   **(d)** 2 volts;   **(e)** -2 volts

29. (a) $\frac{1}{1400}\pi$;   **(b)** 3.35 amperes;   **(c)** 8.55 amperes;   **(d)** 9.91 amperes;   **(e)** 2.71 amperes

31. (a) $\frac{2}{3}\pi$;   **(b)** at the central position;   **(c)** 0.28 cm above the central position;   **(d)** 0.56 cm below the central position;   **(e)** 1.30 cm above the central position   **33. (a)** $\frac{1}{750}$;   **(b)** 0.0173 dynes/cm²;   **(c)** -0.02 dynes/cm²;   **(d)** 0.0156 dynes/cm²;   **(e)** 0

EXERCISES 7.3 **(PAGE 398)**

1. **(a)** 0.519;   **(b)** −0.916   **3.** **(a)** −0.958;   **(b)** −0.0423   **5.** **(a)** 1.35, 1.79;   **(b)** 1.06, 5.22   **7.** **(a)** 3.84, 5.59;
(b) 2.10, 4.18

9. (a)

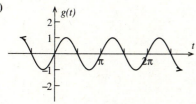

11. (a)

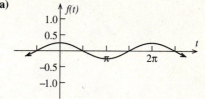

(b)

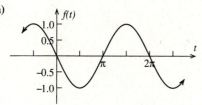

(b)

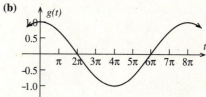

13. (a)

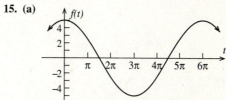

15. (a)

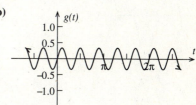

(b)

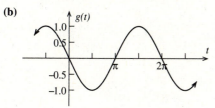

(b)

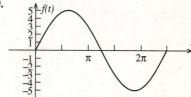

17.

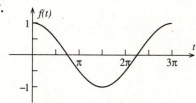

19.

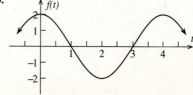

21.

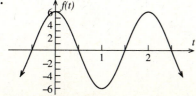

23.

25.

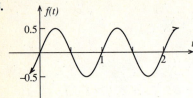

27.

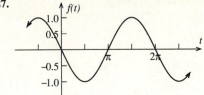

29.

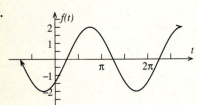

31.

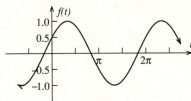

33.

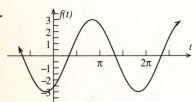

35.

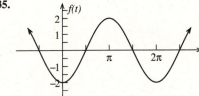

37.

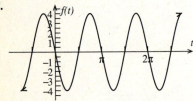

39.

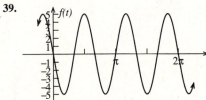

41.

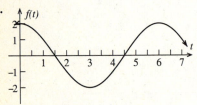

43.

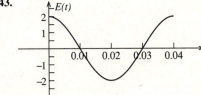

45.

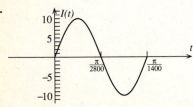

47. (a) $T(t) = -5 \cos \frac{1}{12} \pi t$;   (b) $-2.5°$ Celsius;
(c) $4.33°$ Celsius;   (d) $4.33°$ Celsius;   (e) $-2.5°$ Celsius
49. (a) 0.998, 0.998;   (b) 0.999, 0.999;
(c) 0.9999, 0.9999;   (d) 0.99998, 0.99998;
(e) 0.9999998, 0.9999998;   (f) 1

EXERCISES 7.4 (PAGE 406)

1. (a) 3; **(b)** 12; **(c)** $\frac{1}{12}$ of a vibration per second; **(d)** $f(0) = +3, f(2) = +\frac{3}{2}, f(4) = -\frac{3}{2}, f(6) = -3$

3. (a) 5; **(b)** π; **(c)** $\frac{1}{\pi}$ of a vibration per second; **(d)** $f(0) = 0, f(\frac{1}{4}\pi) = +5, f(\frac{1}{2}\pi) = 0, f(\frac{3}{4}\pi) = -5$

5. (a) 8; **(b)** 1; **(c)** 1 vibration per second; **(d)** $f(0) = +4, f(\frac{1}{6}) = +8, f(\frac{1}{3}) = +4, f(\frac{1}{2}) = -4$

7. The table below describes the weight's directed distance from the central position, its direction of motion, and whether its speed is increasing from or decreasing toward 0, for the first 12 seconds of motion. At 12 seconds, the weight has returned to its starting position, and the cycle repeats indefinitely.

sec.	cm from central position	direction	speed is . . .
0	+3	at rest	0
0-3		down	increasing
3	0	down	constant
3-6		down	decreasing
6	−3	at rest	0
6-9		up	increasing
9	0	up	constant
9-12		up	decreasing

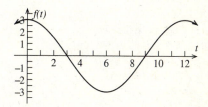

9. The table below describes the weight's directed distance from the central position, its direction of motion, and whether its speed is increasing from or decreasing toward 0, for the first 1 second of motion. At 1 second, the weight has returned to its starting position, and the cycle repeats indefinitely.

sec.	cm from central position	direction	speed is . . .
0-$\frac{1}{6}$		up	decreasing
$\frac{1}{6}$	+8	at rest	0
$\frac{1}{6}$ - $\frac{5}{12}$		down	increasing
$\frac{5}{12}$	0	down	constant
$\frac{5}{12}$ - $\frac{2}{3}$		down	decreasing
$\frac{2}{3}$	−8	at rest	0
$\frac{2}{3}$ - $\frac{11}{12}$		up	increasing
$\frac{11}{12}$	0	up	constant
$\frac{11}{12}$ - 1		up	decreasing

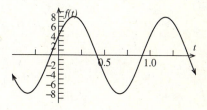

11. (b) 5 amperes; **(c)** 15 cycles per second; **(d)** at 0.05 sec, −5 amperes; at 0.1 sec, 0; at 0.15 sec, 5 amperes; at 0.2 sec, 0

13. (b) 0.6 amperes; **(c)** $\frac{200}{\pi}$ cycles per second; **(d)** at 1 sec, −0.5106 amperes; at 3.5 sec, −0.5477 amperes; at 8 sec, 0.5753 amperes; at 10 sec, −0.4101 amperes

15. (b) 220 volts; **(c)** 60 cycles per second; **(d)** at 0.01 sec, −129.3 volts; at 0.05 sec, 0; at 0.06 sec, −129.3 volts; at 0.1 sec, 0

17. (b) 8 volts; **(c)** $\frac{166}{\pi}$ cycles per second; **(d)** at 0.05 sec, −6.227 volts; at 0.15 sec, −3.591 volts; at 0.25 sec, 7.747 volts; at 0.35 sec, 0.3113 volts

19. (b) 0.005 dynes/cm²; **(c)** 440 cycles per second; **(d)** at 0.0025 sec, 0.002939 dynes/cm²; at 0.005 sec, 0.004755 dynes/cm²; at 0.0075 sec, 0.004755 dynes/cm²; at 0.01 sec, 0.002939 dynes/cm²

21. (b) 0.02 dynes/cm²; **(c)** $\frac{300}{\pi}$ cycles per second; **(d)** at 0.002 sec, 0.01864 dynes/cm²; at 0.006 sec, −0.008850 dynes/cm²; at 0.018 sec, −0.01962 dynes/cm²; at 0.054 sec, 0.01666 dynes/cm²

23. (a) $f(t) = 2 \sin 4\pi(t + \frac{1}{8})$ or, equivalently,
$f(t) = 2 \cos 4\pi t$;

(b)

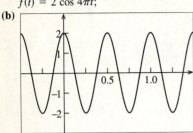

(c) 1 cm above the central position;
(d) 2 cm below the central position

25. $f(t) = 9 \cos \frac{5}{6}\pi(t - \frac{1}{5})$

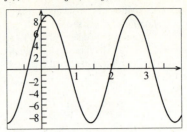

27. $I(t) = 20 \sin 120\pi(t - \frac{1}{720})$

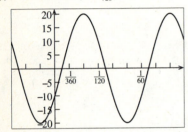

29. The current lags the electromotive force by $\frac{1}{240}$ sec

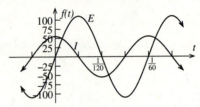

31. (a)

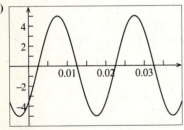

(b) 5 dynes/cm²;   (c) 50 cycles per second;

(d) $-\dfrac{5}{\sqrt{2}}$ dynes/cm² ≈ -3.536 dynes/cm²;

(e) $\dfrac{5}{\sqrt{2}}$ dynes/cm² ≈ 3.536 dynes/cm²;

(f) $-\dfrac{5}{\sqrt{2}}$ dynes/cm² ≈ -3.536 dynes/cm²

33. (a) $-\sqrt{3}$ cm from the origin with a velocity of
$\frac{1}{3}\pi$ cm/sec and an acceleration of $\dfrac{\sqrt{3}}{9}\pi^2$ cm/sec²;
(b) at the origin with a velocity of $\frac{2}{3}\pi$ cm/sec
and an acceleration of 0;
(c) $\sqrt{3}$ cm from the origin with a velocity of
$\frac{1}{3}\pi$ cm/sec and an acceleration of $-\dfrac{\sqrt{3}}{9}\pi^2$ cm/sec²;
(d) $\sqrt{3}$ cm from the origin with a velocity of
$-\frac{1}{3}\pi$ cm/sec and an acceleration of $-\dfrac{\sqrt{3}}{9}\pi^2$ cm/sec²;
(e) at the origin with a velocity of $-\frac{2}{3}\pi$ cm/sec
and an acceleration of 0

EXERCISES 7.5 (PAGE 412)

In Exercises 1-15, there is a point of intersection for every integer value of k.

1. (b)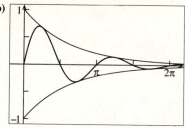

(c) $t = \frac{1}{4}(2k + 1)\pi; \frac{1}{4}\pi, \frac{3}{4}\pi, \frac{5}{4}\pi, \frac{7}{4}\pi;$
(d) $t = \frac{1}{2}k\pi; 0, \frac{1}{2}\pi, \pi, \frac{3}{2}\pi, 2\pi$

3. (b)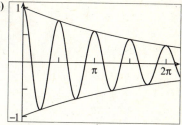

(c) $t = \frac{1}{4}k\pi; 0, \frac{1}{4}\pi, \frac{1}{2}\pi, \frac{3}{4}\pi, \pi, \frac{5}{4}\pi, \frac{3}{2}\pi, \frac{7}{4}\pi, 2\pi;$
(d) $t = \frac{1}{8}(2k + 1)\pi; \frac{1}{8}\pi, \frac{3}{8}\pi, \frac{5}{8}\pi, \frac{7}{8}\pi, \frac{9}{8}\pi, \frac{11}{8}\pi, \frac{13}{8}\pi, \frac{15}{8}\pi$

5. (b)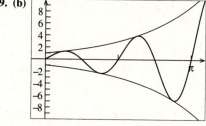

(c) $t = \frac{1}{6}(2k + 1)\pi; \frac{1}{6}\pi, \frac{1}{2}\pi, \frac{5}{6}\pi, \frac{7}{6}\pi, \frac{3}{2}\pi, \frac{11}{6}\pi;$
(d) $t = \frac{1}{3}k\pi; 0, \frac{1}{3}\pi, \frac{2}{3}\pi, \pi, \frac{4}{3}\pi, \frac{5}{3}\pi, 2\pi$

7. (b)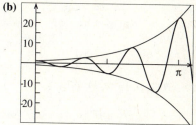

(c) $t = \frac{1}{8}k\pi; 0, \frac{1}{8}\pi, \frac{1}{4}\pi, \frac{3}{8}\pi, \ldots, \frac{7}{4}\pi, \frac{15}{8}\pi, 2\pi;$
(d) $t = \frac{1}{16}(2k + 1)\pi; \frac{1}{16}\pi, \frac{3}{16}\pi, \frac{5}{16}\pi, \ldots, \frac{29}{16}\pi, \frac{31}{16}\pi$

9. (b)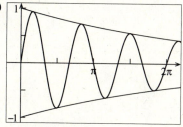

(c) $t = \frac{1}{8}(2k + 1)\pi; \frac{1}{8}\pi, \frac{3}{8}\pi, \frac{5}{8}\pi, \frac{7}{8}\pi;$
(d) $t = \frac{1}{4}k\pi; 0, \frac{1}{4}\pi, \frac{1}{2}\pi, \frac{3}{4}\pi, \pi$

11. (b)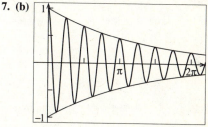

(c) $t = \frac{1}{6}k\pi; 0, \frac{1}{6}\pi, \frac{1}{3}\pi, \frac{1}{2}\pi, \frac{2}{3}\pi, \frac{5}{6}\pi, \pi;$
(d) $t = \frac{1}{12}(2k + 1)\pi; \frac{1}{12}\pi, \frac{1}{4}\pi, \frac{5}{12}\pi, \frac{7}{12}\pi, \frac{3}{4}\pi$

13. (b)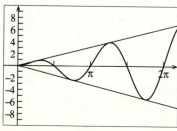

(c) $t = \frac{1}{4}(2k + 1)\pi; \frac{1}{4}\pi, \frac{3}{4}\pi, \frac{5}{4}\pi, \frac{7}{4}\pi;$
(d) $t = \frac{1}{2}k\pi; 0, \frac{1}{2}\pi, \pi, \frac{3}{2}\pi, 2\pi$

15. (b)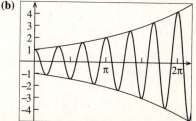

(c) $t = \frac{1}{6}k\pi; 0, \frac{1}{6}\pi, \frac{1}{3}\pi, \frac{1}{2}\pi, \ldots, \frac{5}{3}\pi, \frac{11}{6}\pi, 2\pi;$
(d) $t = \frac{1}{12}(2k + 1)\pi; \frac{1}{12}\pi, \frac{1}{4}\pi, \frac{5}{12}\pi, \frac{7}{12}\pi, \frac{3}{4}\pi,$
$\frac{11}{12}\pi, \frac{13}{12}\pi, \frac{5}{4}\pi, \frac{17}{12}\pi, \frac{19}{12}\pi, \frac{7}{4}\pi, \frac{23}{12}\pi$

17.

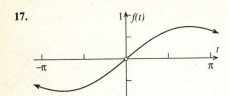

19. (b) $\frac{1}{2}$; **(c)** 0.4996, 0.4996;
(d) 0.499996, 0.499996;
(e) 0.49999996, 0.49999996;
(f) yes
21. (b) 2; **(c)** 2.0067, 2.0067;
(d) 2.000067, 2.000067;
(e) 2.00000067, 2.00000067;
(f) yes

23. (a-b)

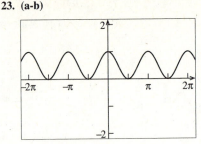

(c) They are identical

EXERCISES 7.6 (PAGE 423)

1. (a) $\sqrt{3}$; **(b)** $\dfrac{1}{\sqrt{3}}$; **(c)** 2; **(d)** $\dfrac{2}{\sqrt{3}}$ **3. (a)** $-\dfrac{2}{\sqrt{5}}$; **(b)** $-\dfrac{\sqrt{5}}{2}$; **(c)** $-\dfrac{3}{\sqrt{5}}$; **(d)** $\dfrac{3}{2}$ **5. (a)** $\frac{12}{5}$; **(b)** $\frac{5}{12}$; **(c)** $-\frac{13}{5}$;

(d) $-\frac{13}{12}$ **7. (a)** $-\dfrac{3}{5}$; **(b)** $-\dfrac{5}{3}$; **(c)** $\dfrac{\sqrt{34}}{5}$; **(d)** $-\dfrac{\sqrt{34}}{3}$ **9. (a)** undefined; **(b)** 0; **(c)** undefined; **(d)** -1

11. $\sin \frac{7}{6}\pi = -\frac{1}{2}$; $\cos \frac{7}{6}\pi = -\dfrac{\sqrt{3}}{2}$; $\tan \frac{7}{6}\pi = \dfrac{1}{\sqrt{3}}$; $\cot \frac{7}{6}\pi = \sqrt{3}$; $\sec \frac{7}{6}\pi = -\dfrac{2}{\sqrt{3}}$; $\csc \frac{7}{6}\pi = -2$

13. $\sin(-\frac{1}{4}\pi) = -\dfrac{1}{\sqrt{2}}$; $\cos(-\frac{1}{4}\pi) = \dfrac{1}{\sqrt{2}}$; $\tan(-\frac{1}{4}\pi) = -1$; $\cot(-\frac{1}{4}\pi) = -1$; $\sec(-\frac{1}{4}\pi) = \sqrt{2}$; $\csc(-\frac{1}{4}\pi) = -\sqrt{2}$

15. $\sin \frac{2}{3}\pi = \dfrac{\sqrt{3}}{2}$; $\cos \frac{2}{3}\pi = -\frac{1}{2}$; $\tan \frac{2}{3}\pi = -\sqrt{3}$; $\cot \frac{2}{3}\pi = -\dfrac{1}{\sqrt{3}}$; $\sec \frac{2}{3}\pi = -2$, $\csc \frac{2}{3}\pi = \dfrac{2}{\sqrt{3}}$ **17. (a)** -1; **(b)** 1

19. (a) $-\dfrac{2}{\sqrt{3}}$; **(b)** $-\dfrac{2}{\sqrt{3}}$ **21. (a)** $\dfrac{1}{\sqrt{3}}$; **(b)** 2 **23. (a)** -1; **(b)** 0 **25. (a)** 3.113; **(b)** 0.3212; **(c)** 1.060; **(d)** 2.998

27. (a) 1.743; **(b)** 0.5736; **(c)** 1.022; **(d)** 4.797 **29. (a)** -3.624; **(b)** 1.037; **(c)** 1.928; **(d)** -1.126

31. (a) 6.955; **(b)** 1.082; **(c)** -2.613 **33. (a)** -1.376; **(b)** -4.494; **(c)** -1.236

35. (a) fourth quadrant; **(b)** third quadrant **37. (a)** second quadrant; **(b)** third quadrant

39. $\sin t = \frac{3}{5}$; $\tan t = -\frac{3}{4}$; $\cot t = -\frac{4}{3}$; $\sec t = -\frac{5}{4}$; $\csc t = \frac{5}{3}$ **41.** $\sin t = -\frac{15}{17}$; $\cos t = -\frac{8}{17}$; $\cot t = \frac{8}{15}$; $\sec t = -\frac{17}{8}$; $\csc t = -\frac{17}{15}$

43. $\sin t = -\frac{12}{13}$; $\cos t = \frac{5}{13}$; $\tan t = -\frac{12}{5}$; $\sec t = \frac{13}{5}$; $\csc t = -\frac{13}{12}$ **45. (a)** $\dfrac{1 - \sin^2 t}{\sin^3 t}$; **(b)** $\dfrac{1}{\cos t}$

47. (a) $\sin t - \sin^3 t$; **(b)** $-\cos^2 t$ **49. (a)** $\dfrac{1}{\sin t}$; **(b)** $\dfrac{\pm 1}{(1 - \cos^2 t)^{3/2}}$ **51.** $\sin^2 t + 1 - \dfrac{1}{\sin^2 t}$

53. k represents any integer; $\sin t \csc t = 1$, $t \neq k\pi$; $\cos t \sec t = 1$, $t \neq (k + \frac{1}{2})\pi$; $\tan t = \dfrac{\sin t}{\cos t}$, $t \neq (k + \frac{1}{2})\pi$; $\cot t = \dfrac{\cos t}{\sin t}$, $t \neq k\pi$;

$\sin^2 t + \cos^2 t = 1$, all t; $1 + \tan^2 t = \sec^2 t$, $t \neq (k + \frac{1}{2})\pi$; $1 + \cot^2 t = \csc^2 t$, $t \neq k\pi$

EXERCISES 7.7 (PAGE 435)

1. (a) -1; **(b)** -1; **(c)** -1; **(d)** -1 **3. (a)** $\dfrac{1}{\sqrt{3}}$; **(b)** $\sqrt{3}$; **(c)** $\dfrac{1}{\sqrt{3}}$; **(d)** $\sqrt{3}$ **5. (a)** $-\sqrt{3}$; **(b)** $-\dfrac{1}{\sqrt{3}}$;

(c) $-\sqrt{3}$; **(d)** $-\dfrac{1}{\sqrt{3}}$ **7. (a)** 0; **(b)** undefined; **(c)** undefined; **(d)** 0 **9. (a)** $\sqrt{2}$; **(b)** $\sqrt{2}$; **(c)** -2; **(d)** $-\dfrac{2}{\sqrt{3}}$

11. (a) $\dfrac{2}{\sqrt{3}}$; **(b)** -2; **(c)** $-\sqrt{2}$; **(d)** $\sqrt{2}$ **13. (a)** -1; **(b)** undefined; **(c)** undefined; **(d)** -1 **15. (a)** $\frac{1}{3}\pi$;

(b) $\frac{1}{6}\pi$; **(c)** 5π; **(d)** 4π **17. (a)** $\frac{2}{7}\pi$; **(b)** $\frac{7}{2}\pi$; **(c)** $\frac{1}{7}\pi$; **(d)** $\frac{4}{7}\pi$ **19. (a)** $\frac{2}{3}$; **(b)** 4; **(c)** $\frac{1}{3}$; **(d)** $\frac{4}{3}$

13. (a) -1; **(b)** undefined; **(c)** undefined; **(d)** -1 **15. (a)** $\frac{1}{3}\pi$; **(b)** $\frac{1}{6}\pi$; **(c)** 5π; **(d)** 4π

17. (a) $\frac{2}{7}\pi$; **(b)** $\frac{7}{2}\pi$; **(c)** $\frac{1}{7}\pi$; **(d)** $\frac{4}{7}\pi$ **19. (a)** $\frac{2}{3}$; **(b)** 4; **(c)** $\frac{1}{3}$; **(d)** $\frac{4}{3}$

21. (a) 1.24; **(b)** -1.42; **(c)** -0.703; **(d)** 0.433; **(e)** -3.75; **(f)** 1.00

23. (a) 0.659, 3.80; **(b)** 2.23, 5.37; **(c)** 1.84, 4.98; **(d)** 0.240, 3.38; **(e)** 1.65, 4.63; **(f)** 0.366, 2.77

25. (a)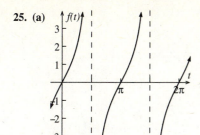

(b) $t = \frac{1}{2}\pi$, $t = \frac{3}{2}\pi$

27. (a)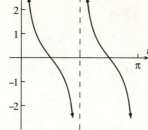

(b) $t = \frac{1}{2}\pi$

29. (a)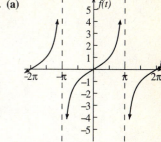

(b) $t = -\pi$, $t = \pi$

31. (a)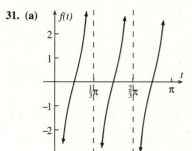

(b) $t = \frac{1}{3}\pi$, $t = \frac{2}{3}\pi$

33. (a)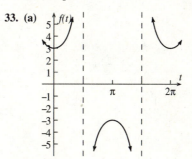

(b) $t = \frac{1}{2}\pi$, $t = \frac{3}{2}\pi$

35. (a)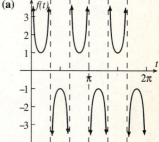

(b) $t = \frac{1}{3}\pi$, $t = \frac{2}{3}\pi$, $t = \pi$, $t = \frac{4}{3}\pi$, $t = \frac{5}{3}\pi$

37. (a)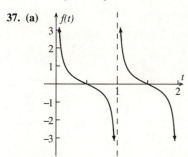

(b) $t = 1$

39. (a)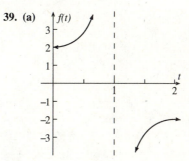

(b) $t = 1$

41.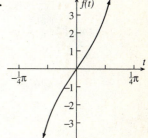

43.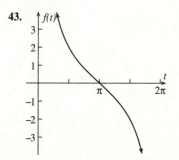

45.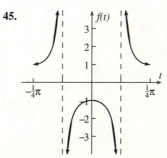

47.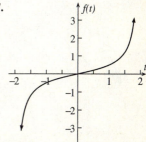

49.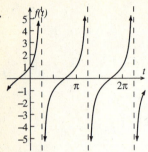

51.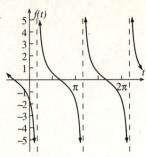

53.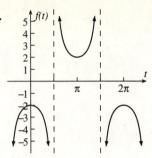

55.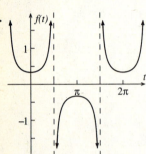

57.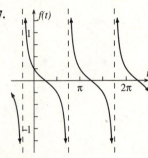

59. By stretching vertically by a factor of a; no amplitude.

REVIEW EXERCISES FOR CHAPTER 7 (PAGE 437)

1. (a) 1; (b) 0; (c) -1; (d) $\frac{1}{2}$; (e) $-\frac{1}{2}$; (f) -1 **3.** (a) $\dfrac{1}{\sqrt{2}}$; (b) $-\dfrac{1}{\sqrt{2}}$; (c) $\sqrt{3}$; (d) $\sqrt{3}$; (e) 2; (f) -2

5. $\cos t = \frac{12}{13}$, $\tan t = \frac{5}{12}$, $\cot t = \frac{12}{5}$, $\sec t = \frac{13}{12}$, $\csc t = \frac{13}{5}$ **7.** $\sin t = \frac{15}{17}$, $\cos t = -\frac{8}{17}$, $\cot t = -\frac{8}{15}$, $\sec t = -\frac{17}{8}$, $\csc t = \frac{17}{15}$

9. $\sin t = -\frac{3}{5}$, $\cos t = -\frac{4}{5}$, $\tan t = \frac{3}{4}$, $\cot t = \frac{4}{3}$, $\csc t = -\frac{5}{3}$ **11.** $\sin t = -\frac{5}{13}$, $\tan t = -\frac{5}{12}$, $\cot t = -\frac{12}{5}$, $\sec t = \frac{13}{12}$, $\csc t = -\frac{13}{5}$

13. $\sin t = -\frac{8}{17}$, $\cos t = -\frac{15}{17}$, $\tan t = \frac{8}{15}$, $\cot t = \frac{15}{8}$, $\sec t = -\frac{17}{15}$ **15.** (a) $\pm\dfrac{1}{\sqrt{1 - \sin^2 t}}$; (b) $\dfrac{\cos t}{1 - \cos^2 t}$

17. (a) $\dfrac{\sin^2 t(1 - 2\sin^2 t)}{(1 - \sin^2 t)^2}$; (b) $-\dfrac{(1 - \cos^2 t)^2}{\cos^2 t}$ **19.** (a) $1 + \sin t$; (b) $\dfrac{2}{\cos^2 t}$ **21.** (a) 0.6606; (b) -0.7508; (c) 0.4567;

(d) -1.099; (e) -0.7884; (f) -1.273 **23.** (a) 0.8579; (b) -0.7499; (c) -0.6113 **25.** (a) -0.9399; (b) -1.064

(c) 3.078; (d) 0.5774   **27.** (a) $\frac{2}{3}\pi$; (b) 10π; (c) $\frac{4}{3}\pi$; (d) $\frac{8}{3}\pi$ **29.** (a) $\frac{1}{3}$; (b) $\frac{10}{3}$; (c) $\frac{1}{2}$; (d) 3

31. (a) $-\dfrac{1}{\sqrt{2}}$; (b) $-\dfrac{1}{2}$; (c) $\dfrac{2}{\sqrt{3}}$; (d) 0; (e) 0; (f) 1   **33.** (a) 1; (b) $-\dfrac{1}{\sqrt{3}}$; (c) undefined; (d) $-\sqrt{3}$; (e) $\sqrt{3}$;

(f) undefined   **35.** (a) 0.955; (b) 0.762   **37.** (a) -12.6; (b) 0.688   **39.** (a) 0.573, 2.57; (b) 1.86, 4.42

41. (a) 1.46, 4.60; (b) 4.47, 4.95

43. (a) $\frac{2}{3}\pi$;

(b)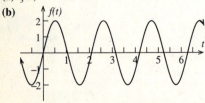

(c) at the central position;

(d) 0.28 cm above the central position;

(e) 0.56 cm below the central position;

(f) 1.50 cm below the central position

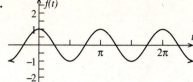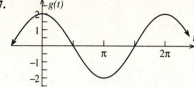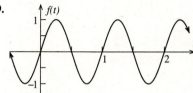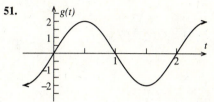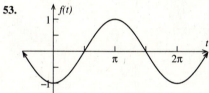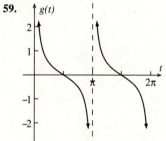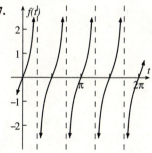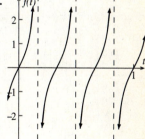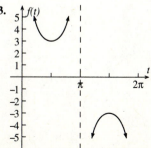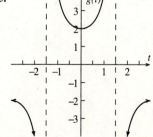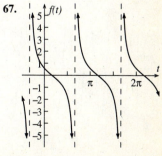

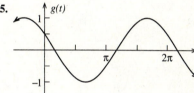

45.

47.

49.

51.

53.

55.

57.

59.

61.

63.

65.

67.

In Exercises 69-71, there is a point of intersection for every integer value of k.

69. (b)

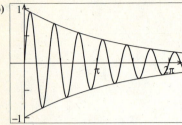

(c) $t = \frac{1}{12}(2k + 1)\pi$; $\frac{1}{12}\pi$, $\frac{1}{4}\pi$, $\frac{5}{12}\pi$, $\frac{7}{12}\pi$, $\frac{3}{4}\pi$, $\frac{11}{12}\pi$, $\frac{13}{12}\pi$, $\frac{5}{4}\pi$, $\frac{17}{12}\pi$, $\frac{19}{12}\pi$, $\frac{7}{4}\pi$, $\frac{23}{12}\pi$;

(d) $t = \frac{1}{6}k\pi$; 0, $\frac{1}{6}\pi$, $\frac{1}{3}\pi$, $\frac{1}{2}\pi$, $\frac{2}{3}\pi$, $\frac{5}{6}\pi$, π, $\frac{7}{6}\pi$, $\frac{4}{3}\pi$, $\frac{3}{2}\pi$, $\frac{5}{3}\pi$, $\frac{11}{6}\pi$, 2π

71. (b)

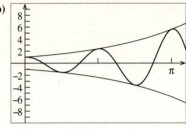

(c) $t = \frac{1}{4}k\pi$; 0, $\frac{1}{4}\pi$, $\frac{1}{2}\pi$, $\frac{3}{4}\pi$, π;

(d) $t = \frac{1}{8}(2k + 1)\pi$; $\frac{1}{8}\pi$, $\frac{3}{8}\pi$, $\frac{5}{8}\pi$, $\frac{7}{8}\pi$

73. (a) $f(t) = -6 \cos \frac{10}{9}\pi t$;

(b)

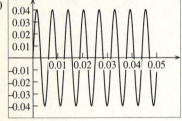

(c) 5.64 cm above the central position;

(d) 1.04 cm below the central position

75. (a)

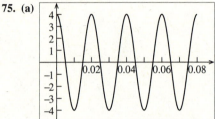

(b) 4 volts;   **(c)** $\frac{1}{50}$;   **(d)** 50 cycles per sec;

(e) −4 volts;   **(f)** 4 volts;   **(g)** −4 volts;   **(h)** 4 volts

77. (a)

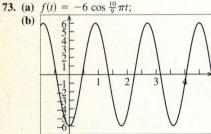

(b) 0.04 dynes/cm²;

(c) $\frac{500}{\pi}$ cycles per sec;

(d) 0.03366 dynes/cm²;

(e) −0.02176 dynes/cm²;

(f) 0.03652 dynes/cm²;

(g) −0.01049 dynes/cm²

79. $I(t) = 40 \sin 60\pi(t + \frac{1}{360})$

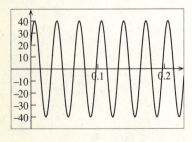

81. (a) 0.4706;   **(b)** −0.8720

83. (a)

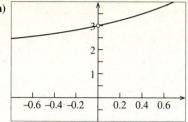

(b) 3;
(c) 3.105, 2.905;
(d) 3.010, 2.990;
(e) 3.001, 2.999;
(f) Yes

85. (a)

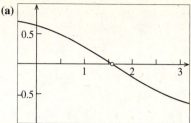

(b) 0;
(c) 0.005, −0.0046;
(d) 0.000898, −0.001;
(e) 0.000048, −0.00005;
(f) Yes

EXERCISES 8.1 (PAGE 450)

1. (a) in first quadrant; **(b)** in second quadrant; **(c)** quadrantal angle; **(d)** in third quadrant; **(e)** in fourth quadrant
3. (a) in fourth quadrant; **(b)** quadrantal angle; **(c)** in third quadrant; **(d)** in second quadrant; **(e)** in first quadrant
5. (a) in first quadrant; **(b)** in second quadrant; **(c)** in third quadrant; **(d)** in fourth quadrant; **(e)** in second quadrant
7. (a) $\frac{5}{4}\pi$; **(b)** $\frac{7}{6}\pi$; **(c)** $\frac{3}{2}\pi$; **(d)** $\frac{5}{3}\pi$ **9. (a)** 1.00; **(b)** 2.72; **(c)** 2.03; **(d)** 1.56
11. (a) $\frac{1}{3}\pi$; **(b)** $\frac{3}{4}\pi$; **(c)** $\frac{7}{6}\pi$; **(d)** $-\frac{5}{6}\pi$ **13. (a)** $\frac{1}{9}\pi$; **(b)** $\frac{5}{2}\pi$; **(c)** $-\frac{5}{12}\pi$; **(d)** $\frac{5}{9}\pi$
15. (a) 45°; **(b)** 120°; **(c)** 330°; **(d)** −90°   **17. (a)** 28.65°; **(b)** −114.59°; **(c)** 273.87°; **(d)** 13.18°
19. (a) 35.37°, 0.62; **(b)** 102.52°, 1.79   **21. (a)** 315°; **(b)** 150°; **(c)** 180°; **(d)** 240°
23. (a) 22.56°; **(b)** 241.76°; **(c)** 106.15°; **(d)** 302.36°   **25. (a)** 6π in.; **(b)** 27π in.²   **27. (a)** $\frac{9}{2}\pi$ cm; **(b)** $\frac{27}{2}\pi$ cm²
29. (a) 1.82 in.; **(b)** 4.30 in.²   **31.** 21.70° N   **33.** 1009 mi   **35.** 560   **37.** 2.79 in.   **39. (a)** 611.56; **(b)** 2342.76

EXERCISES 8.2 (PAGE 466)

1. (a) $\frac{1}{2}\sqrt{3}$; **(b)** $\frac{1}{2}\sqrt{3}$; **(c)** 1   **3. (a)** $-\frac{1}{\sqrt{2}}$; **(b)** $-\frac{1}{2}$; **(c)** $-\sqrt{3}$ **5. (a)** $-\frac{1}{2}$; **(b)** $-\frac{1}{\sqrt{2}}$; **(c)** $\sqrt{3}$

7. $\sin 0° = 0$; $\cos 0° = 1$; $\tan 0° = 0$; $\sec 0° = 1$; $\cot 0°$ and $\csc 0°$ are not defined
9. $\sin(-90°) = -1$; $\cos(-90°) = 0$; $\cot(-90°) = 0$; $\csc(-90°) = -1$; $\tan(-90°)$ and $\sec(-90°)$ are not defined
11. $\sin \theta = \frac{4}{5}$; $\cos \theta = \frac{3}{5}$; $\tan \theta = \frac{4}{3}$; $\cot \theta = \frac{3}{4}$; $\sec \theta = \frac{5}{3}$; $\csc \theta = \frac{5}{4}$
13. $\sin \theta = \frac{8}{17}$; $\cos \theta = -\frac{15}{17}$; $\tan \theta = -\frac{8}{15}$; $\cot \theta = -\frac{15}{8}$; $\sec \theta = -\frac{17}{15}$; $\csc \theta = \frac{17}{8}$
15. $\sin \theta = -\frac{1}{\sqrt{5}}$; $\cos \theta = \frac{2}{\sqrt{5}}$; $\tan \theta = -\frac{1}{2}$; $\cot \theta = -2$; $\sec \theta = \frac{\sqrt{5}}{2}$; $\csc \theta = -\sqrt{5}$
17. $\sin \theta = -\frac{4}{\sqrt{17}}$; $\cos \theta = -\frac{1}{\sqrt{17}}$; $\tan \theta = 4$; $\cot \theta = \frac{1}{4}$; $\sec \theta = -\sqrt{17}$; $\csc \theta = -\frac{\sqrt{17}}{4}$
19. $\sin \theta = \frac{4}{\sqrt{65}}$; $\cos \theta = \frac{7}{\sqrt{65}}$; $\tan \theta = \frac{4}{7}$; $\cot \theta = \frac{7}{4}$; $\sec \theta = \frac{\sqrt{65}}{7}$; $\csc \theta = \frac{\sqrt{65}}{4}$
21. $\sin \theta = \frac{2}{3}$; $\cos \theta = \frac{\sqrt{5}}{3}$; $\tan \theta = \frac{2}{\sqrt{5}}$; $\cot \theta = \frac{\sqrt{5}}{2}$; $\sec \theta = \frac{3}{\sqrt{5}}$; $\csc \theta = \frac{3}{2}$
23. (a) 0.5358; **(b)** 0.5358; **(c)** 3.4197; **(d)** 3.4197; **(e)** 1.4020; **(f)** 1.4020
25. (a) 0.9677; **(b)** 0.9252; **(c)** 0.6950; **(d)** 0.0332; **(e)** 1.6464; **(f)** 21.2285
27. (a) 0.9239; **(b)** 0.8974; **(c)** 3.1566; **(d)** 0.3249

29. (a) $\sin 45° = \frac{1}{\sqrt{2}}$; **(b)** $-\cos 30° = -\frac{\sqrt{3}}{2}$; **(c)** $-\tan 60° = -\sqrt{3}$; **(d)** $-\cot 30° = -\sqrt{3}$;

(e) $-\sec 0° = -1$; **(f)** $-\csc 60° = -\frac{2}{\sqrt{3}}$

31. (a) $-\sin \frac{1}{3}\pi = -\frac{\sqrt{3}}{2}$; **(b)** $\cos \frac{1}{4}\pi = \frac{1}{\sqrt{2}}$; **(c)** $\tan \frac{1}{4}\pi = 1$; **(d)** $\cot \frac{1}{2}\pi = 0$; **(e)** $-\sec \frac{1}{6}\pi = -\frac{2}{\sqrt{3}}$; **(f)** $-\csc \frac{1}{6}\pi = -2$

33. (a) $\sin 55°42' \approx 0.8261$; **(b)** $-\cos 63°36' \approx -0.4446$; **(c)** $-\tan 15°6' \approx -0.2698$
35. (a) $-\cos 7.6° \approx -0.9912$; **(b)** $\sin 83.8° \approx 0.9942$; **(c)** $\tan 20.2° \approx 0.3679$
37. (a) $\cot 10°30' \approx 5.3955$; **(b)** $\sec 67.4° \approx 2.6022$; **(c)** $\csc 4.5° \approx 12.746$ **39. (a)** $30°, 150°$; **(b)** $120°, 240°$; **(c)** $45°, 225°$
41. (a) $\frac{5}{4}\pi, \frac{7}{4}\pi$; **(b)** $\frac{1}{6}\pi, \frac{11}{6}\pi$; **(c)** $0, \pi$ **43. (a)** $31.1°, 148.9°$; **(b)** $80.3°, 279.7°$; **(c)** $67.1°, 247.1°$; **(d)** $43.5°, 223.5°$
45. (a) $0.4342, 2.707$; **(b)** $1.938, 4.346$; **(c)** $1.142, 4.284$; **(d)** $2.917, 6.058$

EXERCISES 8.3 (PAGE 475)

1. (a) $\cos 30°$; **(b)** $\sin 5.7°$; **(c)** $\cot 40.2°$; **(d)** $\tan 37.9°$; **(e)** $\sec 22.5°$
3. (a) $\cos 42°42'$; **(b)** $\sin 18°18'$; **(c)** $\cot 44°$; **(d)** $\tan 34°54'$; **(e)** $\csc 25°30'$ **5.** $\beta = 30°$; $a = 2.3$; $b = 1.4$
7. $\alpha = 45°$; $b = 56$; $c = 79$ **9.** $\beta = 66°$; $b = 36$; $c = 39$ **11.** $\alpha = 19°$; $a = 14$; $b = 42$ **13.** $\alpha = 47°$; $\beta = 43°$; $a = 28$
15. $\alpha = 52.6°$; $b = 3.20$; $c = 5.26$ **17.** $\beta = 73.1°$; $b = 448$; $c = 468$ **19.** $\alpha = 47.6°$; $\beta = 42.4°$; $c = 86.1$
21. $\beta = 31.57°$; $a = 532.8$; $b = 327.4$ **23.** $\alpha = 49.08°$; $a = 42.36$; $c = 56.06$ **25.** $\alpha = 22.74°$; $\beta = 67.26°$; $b = 746.1$
27. $\alpha = 24°42'$; $a = 16.4$; $b = 35.6$ **29.** $\beta = 60°24'$; $a = 163$; $c = 330$ **31.** $(3.0)10^2$ **33.** $(1.167)10^5$ **35.** 333 m
37. $13.32°$ **39. (a)** 50 nautical miles; **(b)** $162°$; **(c)** $342°$ **41.** 108 m **43.** 367 m

EXERCISES 8.4 (PAGE 488)

1. $b = 5.6$ **3.** $a = 51$ **5.** $\gamma = 75°$; $b = 41$; $c = 41$ **7.** $\alpha = 70.0°$; $a = 56.9$; $b = 45.5$ **9.** $\beta = 83.0°$; $a = 2.22$;
$c = 0.935$ **11.** $\gamma = 51°42'$; $a = 335$; $c = 331$ **13.** one triangle; $\beta = 29°$; $\gamma = 109°$; $c = 9.0$
15. one triangle; $\alpha = 60°$; $\gamma = 90°$; $a = 29$ **17.** two triangles; $\alpha_1 = 67.3°$; $\beta_1 = 77.5°$; $b_1 = 72.0$; $\alpha_2 = 112.7°$; $\beta_2 = 32.1°$;
$b_2 = 39.2$ **19.** one triangle; $\alpha = 27.2°$; $\gamma = 36.4°$; $c = 560$ **21.** no triangle **23.** two triangles; $\alpha_1 = 64.48°$; $\gamma_1 = 65.29°$;
$a_1 = 4.182$; $\alpha_2 = 15.06°$; $\gamma_2 = 114.71°$; $a_2 = 1.204$ **25.** $(4.7)10^2$ **27.** 1.03 **29.** 14 **31.** $(1.08)10^5$ **33.** 61.8 m
35. 296 m **37.** 9.80 ft

EXERCISES 8.5 (PAGE 495)

1. $c = 5.6$ **3.** $b = 34$ **5.** $\beta = 55°$; $\gamma = 43°$; $a = 4.1$ **7.** $\alpha = 26.4°$; $\beta = 37.2°$; $c = 22.6$
9. $\alpha = 18.51°$; $\gamma = 150.97°$; $b = 1176$ **11.** $\alpha = 26°23'$; $\beta = 18°25'$; $c = 8.34$ **13.** 12.3 **15.** 76.7 **17.** $(5.836)10^5$
19. $\alpha = 44°$; $\beta = 108°$; $\gamma = 28°$ **21.** $\alpha = 57.5°$; $\beta = 24.6°$; $\gamma = 97.9°$ **23.** $\alpha = 98.3°$; $\beta = 38.4°$; $\gamma = 43.3°$
25. $\alpha = 144.36°$; $\beta = 27.74°$; $\gamma = 7.90°$ **27. (a)** $33°$; **(b)** $(3.9)10^2$ **29. (a)** 30.4 cm; **(b)** 194 cm^2
31. (a) 17.3 nautical miles; **(b)** $188.1°$ **33.** 58.7 m **35.** $31.3°$ **37.** $56°$ **39. (a)** 39.2 mi; **(b)** $212.5°$ **45.** $(3.83)10^4$
47. 20

REVIEW EXERCISES FOR CHAPTER 8 (PAGE 498)

1. (a) $\frac{1}{6}\pi$; **(b)** $\frac{5}{4}\pi$ **3. (a)** $-\frac{2}{3}\pi$; **(b)** 1.75 **5. (a)** $60°$; **(b)** $315°$ **7. (a)** $-19.1°$; **(b)** $135.2°$
9. (a) 10π cm; **(b)** 60π cm^2 **11.** $\sin\theta = \frac{15}{17}$; $\cos\theta = -\frac{8}{17}$; $\tan\theta = -\frac{15}{8}$; $\cot\theta = -\frac{8}{15}$; $\sec\theta = -\frac{17}{8}$; $\csc\theta = \frac{17}{15}$
13. $\sin\theta = \frac{12}{13}$; $\cos\theta = -\frac{5}{13}$; $\tan\theta = -\frac{12}{5}$; $\cot\theta = -\frac{5}{12}$; $\sec\theta = -\frac{13}{5}$; $\csc\theta = \frac{13}{12}$
15. $\sin 45° = \dfrac{1}{\sqrt{2}}$; $\cos 45° = \dfrac{1}{\sqrt{2}}$; $\tan 45° = 1$; $\cot 45° = 1$; $\sec 45° = \sqrt{2}$; $\csc 45° = \sqrt{2}$
17. $\sin\frac{2}{3}\pi = \dfrac{\sqrt{3}}{2}$; $\cos\frac{2}{3}\pi = -\dfrac{1}{2}$; $\tan\frac{2}{3}\pi = -\sqrt{3}$; $\cot\frac{2}{3}\pi = -\dfrac{1}{\sqrt{3}}$; $\sec\frac{2}{3}\pi = -2$; $\csc\frac{2}{3}\pi = \dfrac{2}{\sqrt{3}}$
19. $\sin(-150°) = -\dfrac{1}{2}$; $\cos(-150°) = -\dfrac{\sqrt{3}}{2}$; $\tan(-150°) = \dfrac{1}{\sqrt{3}}$; $\cot(-150°) = \sqrt{3}$; $\sec(-150°) = -\dfrac{2}{\sqrt{3}}$; $\csc(-150°) = -2$
21. $\sin(-\frac{9}{4}\pi) = -\dfrac{1}{\sqrt{2}}$; $\cos(-\frac{9}{4}\pi) = \dfrac{1}{\sqrt{2}}$; $\tan(-\frac{9}{4}\pi) = -1$; $\cot(-\frac{9}{4}\pi) = -1$; $\sec(-\frac{9}{4}\pi) = \sqrt{2}$; $\csc(-\frac{9}{4}\pi) = -\sqrt{2}$
23. $\sin 270° = -1$; $\cos 270° = 0$; $\cot 270° = 0$; $\csc 270° = -1$; $\tan 270°$ and $\sec 270°$ are not defined
25. $\sin(-\pi) = 0$; $\cos(-\pi) = -1$; $\tan(-\pi) = 0$; $\sec(-\pi) = -1$; $\cot(-\pi)$ and $\csc(-\pi)$ are not defined
27. (a) $45°$; **(b)** $38°$; **(c)** $\frac{1}{6}\pi$; **(d)** 1.13; **(e)** $\frac{4}{9}\pi$; **(f)** 1.33 **29. (a)** $\sin 30° = \dfrac{1}{2}$; **(b)** $-\cos 45° = -\dfrac{1}{\sqrt{2}}$;
(c) $-\tan 60° = -\sqrt{3}$; **(d)** $-\cot\frac{1}{4}\pi = -1$; **(e)** $-\sec\frac{1}{3}\pi = -2$; **(f)** $-\csc\frac{1}{6}\pi = -2$

31. (a) sin 43.6° ≈ 0.6896; **(b)** cos 10.8° ≈ 0.9823; **(c)** −tan 32.9° ≈ −0.6469; **(d)** −cot 25.7° ≈ −2.0778;
(e) −sec 14.8° ≈ −1.0343; **(f)** −csc 81.4° ≈ −1.0114   **33. (a)** 60°, 300°; **(b)** 135°, 315°   **35. (a)** $\frac{4}{3}\pi, \frac{5}{3}\pi$; **(b)** π
37. (a) 18.8°, 161.2°; **(b)** 61.3°, 298.7°; **(c)** 71.3°, 251.3°; **(d)** 168.0°, 348.0°   **39. (a)** 3.924, 5.501; **(b)** 1.351, 4.933;
(c) 2.206; 5.348; **(d)** 0.8685, 4.010   **41.** 32   **43.** 3.7   **45.** 70   **47.** $\alpha = 56°$; $\beta = 34°$; $c = 5.8$
49. $\gamma = 75.4°$; $a = 20.9$; $c = 29.5$ **51.** $\alpha = 28.7°$; $\beta = 46.0°$; $c = 43.4$ **53.** $\alpha = 54.46°$; $\beta = 43.67°$; $\gamma = 81.87°$
55. 271   **57.** $(1.128)10^5$ **59.** two triangles; $\beta_1 = 72.0°$; $\gamma_1 = 53.6°$; $c_1 = 111$; $\beta_2 = 108.0°$; $\gamma_2 = 17.6°$; $c_2 = 41.6$
61. one triangle; $\alpha = 107.7°$; $\gamma = 32.6°$; $a = 19.1$ **63.** 336   **65. (a)** 111 ft; **(b)** 256 ft   **67.** 60°
69. (a) 241 mi; **(b)** 73.0°   **71.** 40.6 ft   **73.** 27.5   **75.** 78.8

EXERCISES 9.1 (PAGE 510)

1. (a) sin x; **(b)** cos² x   **3. (a)** $\frac{1}{\sin x}$; **(b)** cos² x   **5. (a)** $\frac{1}{\sin x \cos x}$; **(b)** $\frac{1}{\sin x}$ **43.** let $t = \frac{1}{4}\pi$ **45.** let $y = \frac{1}{4}\pi$
47. not an identity; let $\theta = \frac{1}{3}\pi$ **49.** identity   **51.** not an identity; let $x = \frac{3}{4}\pi$ **53.** identity

EXERCISES 9.2 (PAGE 522)

1. (a) $\frac{\sqrt{3}-1}{2\sqrt{2}}$; **(b)** $\frac{-\sqrt{3}-1}{2\sqrt{2}}$ **3. (a)** $\frac{-\sqrt{3}-1}{2\sqrt{2}}$; **(b)** $\frac{\sqrt{3}+1}{2\sqrt{2}}$ **5. (a)** $-2-\sqrt{3}$; **(b)** $2-\sqrt{3}$ **7. (a)** 1; **(b)** −1
9. (a) $\frac{1}{2}$; **(b)** $-\frac{1}{\sqrt{2}}$ **11. (a)** 1; **(b)** $-\sqrt{3}$ **13. (a)** $\frac{323}{325}$; **(b)** $\frac{36}{325}$; **(c)** $\frac{253}{325}$; **(d)** $\frac{204}{325}$; **(e)** first; **(f)** first
15. (a) $\frac{16}{65}$; **(b)** $-\frac{63}{65}$; **(c)** $\frac{56}{65}$; **(d)** $-\frac{33}{65}$; **(e)** second; **(f)** second   **17. (a)** $-\frac{37}{9}$; **(b)** $\frac{19}{33}$; **(c)** second; **(d)** first
19. (a) $-\frac{38}{55}$; **(b)** $\frac{62}{25}$; **(c)** fourth; **(d)** first   **31. (a)** cos 7x; **(b)** sin 7x   **33. (a)** −cos 10x; **(b)** −sin x
35. (a) tan 10x; **(b)** −tan 2x
43. (a) $f(t) = 2\sin(t + \frac{1}{6}\pi)$;
  (b) amplitude is 2, period is 2π, phase shift is $\frac{1}{6}\pi$;
  (c)
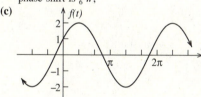

45. (a) $f(t) = 3\sqrt{2}\sin(t - \frac{1}{4}\pi)$;
  (b) amplitude is $3\sqrt{2}$, period is 2π; phase shift is $-\frac{1}{4}\pi$;
  (c)
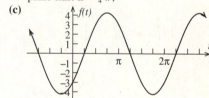

47. (a) $f(t) = 2.94\sin(4t + 0.45)$;
  (b) amplitude is 2.94; period is $\frac{1}{2}\pi$, phase shift is 0.11;
  (c)
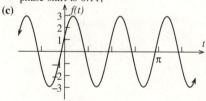

49. (a) $f(t) = 5\sin(6t + 0.93)$;
  (b) amplitude is 5, period is $\frac{1}{3}\pi$, frequency is $\frac{3}{\pi}$;
  (c)
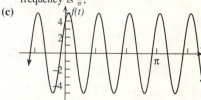

51. $f(t) = 0.043\sin(2\pi nt - 0.72)$
55. $10\sqrt{5}$ feet   **57. (b)** cos x
53. (a) $s = \frac{1+\sqrt{3}}{\sqrt{2}}\sin(4t + \frac{1}{4}\pi)$; **(b)** amplitude is $\frac{1+\sqrt{3}}{\sqrt{2}}$, frequency is $\frac{2}{\pi}$

EXERCISES 9.3 (PAGE 534)

1. (a) $\frac{24}{25}$; (b) $-\frac{7}{25}$; (c) $-\frac{24}{7}$ 3. (a) $-\frac{120}{169}$; (b) $-\frac{119}{169}$; (c) $\frac{120}{119}$ 5. (a) $\frac{240}{289}$; (b) $\frac{161}{289}$; (c) $\frac{240}{161}$

7. (a) $-\frac{336}{625}$; (b) $\frac{527}{625}$; (c) $-\frac{336}{527}$ 9. $\cos 3x = 4\cos^3 x - 3\cos x$ 11. $\sin 4x = \pm 4\sin x\sqrt{1 - \sin^2 x}\,(1 - 2\sin^2 x)$

13. $\tan 3x = \dfrac{3\tan x - \tan^3 x}{1 - 3\tan^2 x}$ 15. $\sec 2x = \dfrac{\sec^2 x}{2 - \sec^2 x}$ 17. $\frac{3}{8} + \frac{1}{2}\cos 2t + \frac{1}{8}\cos 4t$ 19. $\frac{1}{8} - \frac{1}{8}\cos 12t$

23. (a) $\dfrac{\sqrt{2 - \sqrt{3}}}{2}$ 25. $\frac{1}{2}\sqrt{2 + \sqrt{2}}$ 27. $-1 - \sqrt{2}$ 29. $\dfrac{1}{\sqrt{3}}$ 31. $\frac{3}{5}$ 33. 5

35. (a) $\cos 2x$; (b) $\cos 4x$; (c) $\sin 3x$ 37. (a) $\sin 2x$; (b) $2\sin 8x$; (c) $\sin x$

39. (a) $\tan 2x$; (b) $-2\tan 4x$; (c) $3\tan 3x$ 41. (a) $\tan t$; (b) $\tan 2t$; (c) $-\tan 2t$

43. (a) $\sin 2t$; (b) $\sin 4t$; (c) $2\sin 4t$ 55. $\dfrac{2 - 2z}{1 + z^2}$ 57. $\dfrac{4z(1 - z^2)}{(1 + z^2)^2} + \dfrac{2z}{1 - z^2}$ 59. $\dfrac{z^4 + 6z^2 + 1}{(1 - z^2)^2}$

61. (a) $s = 6\sin(8t + \frac{1}{2}\pi)$; (b) amplitude is 6, frequency is $\frac{4}{\pi}$

EXERCISES 9.4 (PAGE 541)

1. (a) $\frac{1}{2}(\sin 5x + \sin 3x)$; (b) $\frac{1}{2}(\cos 3x - \cos 5x)$; (c) $\frac{1}{2}(\cos 5x + \cos 3x)$

3. (a) $\frac{1}{2}(\sin 8x - \sin 4x)$; (b) $\frac{1}{2}(\cos 4x - \cos 8x)$; (c) $\frac{1}{2}(\cos 8x + \cos 4x)$ 5. (a) $\frac{1}{4}(\sqrt{3} + \sqrt{2})$; (b) $\dfrac{-2 - \sqrt{2}}{4}$

7. (a) $\dfrac{-2 - \sqrt{2}}{4}$; (b) $\dfrac{\sqrt{2} - \sqrt{3}}{4}$ 9. (a) $2\sin 4t\cos 2t$; (b) $2\cos 4t\sin 2t$; (c) $2\cos 4t\cos 2t$; (d) $-2\sin 4t\sin 2t$

11. (a) $2\sin 5\theta\cos 2\theta$; (b) $-2\cos 5\theta\sin 2\theta$; (c) $2\cos 5\theta\cos 2\theta$; (d) $2\sin 5\theta\sin 2\theta$

13. (a) $\dfrac{\sqrt{6}}{2}$; (b) $-\dfrac{1}{\sqrt{2}}$ 15. (a) $-\dfrac{1}{\sqrt{2}}$; (b) $\dfrac{\sqrt{6}}{2}$

35. $\sin 21\pi t + \sin 19\pi t = 2\sin 20\pi t\cos \pi t$

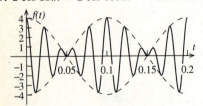

37. $\cos 9\pi t - \cos 11\pi t = 2\sin 10\pi t\sin \pi t$

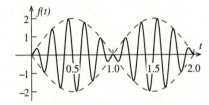

39. $2\cos 80\pi t + 2\cos 100\pi t = 4\cos 90\pi t\cos 10\pi t$

41. $3\sin 95t - 3\sin 105t = -6\cos 100t\sin 5t$

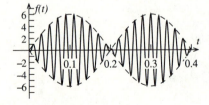

47. $f(t) = 0.04\sin 500\pi t\cos 4\pi t$

EXERCISES 9.5 (PAGE 554)

1. (a) $\frac{1}{6}\pi$; (b) $-\frac{1}{6}\pi$; (c) $\frac{1}{3}\pi$; (d) $\frac{2}{3}\pi$ 3. (a) $\frac{1}{6}\pi$; (b) $-\frac{1}{3}\pi$ (c) $\frac{1}{6}\pi$; (d) $\frac{7}{6}\pi$

5. (a) $\frac{1}{2}\pi$; (b) $-\frac{1}{2}\pi$; (c) $\frac{1}{2}\pi$; (d) $-\frac{1}{2}\pi$; (e) 0   7. (a) 0.51; (b) -0.51; (c) 1.06; (d) 2.08

9. (a) 0.41; (b) -0.41; (c) 1.16; (d) 1.98   11. (a) 1.07; (b) 4.21; (c) 0.50; (d) -2.64

13. (a) $\dfrac{2\sqrt{2}}{3}$; (b) $\dfrac{1}{2\sqrt{2}}$; (c) $2\sqrt{2}$; (d) $\dfrac{3}{2\sqrt{2}}$; (e) 3   **15.** (a) $\dfrac{2\sqrt{2}}{3}$; (b) $-\dfrac{1}{2\sqrt{2}}$; (c) $-2\sqrt{2}$; (d) $\dfrac{3}{2\sqrt{2}}$; (e) -3

17. (a) $-\dfrac{2}{\sqrt{5}}$; (b) $\dfrac{1}{\sqrt{5}}$; (c) $-\dfrac{1}{2}$; (d) $\sqrt{5}$; (e) $-\dfrac{\sqrt{5}}{2}$ **19.** (a) $-\dfrac{2}{3}$; (b) $-\dfrac{\sqrt{5}}{3}$; (c) $\dfrac{2}{\sqrt{5}}$; (d) $\dfrac{\sqrt{5}}{2}$; (e) $-\dfrac{3}{\sqrt{5}}$

21. (a) $\frac{1}{6}\pi$; (b) $-\frac{1}{6}\pi$; (c) $\frac{1}{6}\pi$; (d) $-\frac{1}{6}\pi$ **23.** (a) $\frac{1}{3}\pi$; (b) $\frac{1}{3}\pi$; (c) $\frac{2}{3}\pi$; (d) $\frac{2}{3}\pi$
25. (a) $\frac{1}{6}\pi$; (b) $-\frac{1}{3}\pi$; (c) $\frac{1}{6}\pi$; (d) $-\frac{1}{3}\pi$ **27.** (a) $\frac{1}{6}\pi$; (b) $\frac{2}{3}\pi$; (c) $\frac{1}{6}\pi$; (d) $\frac{2}{3}\pi$

29. (a) $\frac{1}{3}\pi$; (b) $\frac{1}{3}\pi$; (c) $\frac{4}{3}\pi$; (d) $\frac{4}{3}\pi$ **31.** (a) $\frac{1}{6}\pi$; (b) $-\frac{5}{6}\pi$; (c) $\frac{1}{6}\pi$; (d) $-\frac{5}{6}\pi$ **33.** (a) $\sqrt{3}$; (b) $\dfrac{\sqrt{3}}{\sqrt{7}}$

35. (a) $\dfrac{\sqrt{3}}{2}$; (b) $\dfrac{\sqrt{3}}{2}$ **37.** $\frac{119}{169}$ **39.** $\dfrac{2+2\sqrt{10}}{9}$ **41.** $\dfrac{7\sqrt{5}+8\sqrt{2}}{27}$ **43.** $\dfrac{3\sqrt{3}-4}{4\sqrt{3}+3}=\dfrac{48-25\sqrt{3}}{39}$ **45.** $\dfrac{4\sqrt{2}+1}{3\sqrt{5}}$

49.
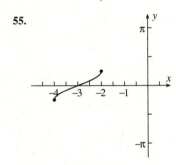

51.
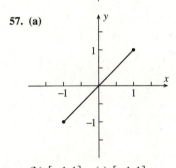

53.
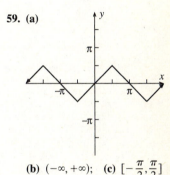

55.

57. (a)

(b) $[-1, 1]$; (c) $[-1, 1]$

59. (a)

(b) $(-\infty, +\infty)$; (c) $\left[-\dfrac{\pi}{2}, \dfrac{\pi}{2}\right]$

61. (a) $\theta = \tan^{-1}\dfrac{5}{x} - \tan^{-1}\dfrac{2}{x}$; (b) $\theta \approx 0.4049$; (c) $\theta \approx 0.4424$; (d) $\theta \approx 0.4324$; (e) $\sqrt{10} \approx 3.16$ ft

EXERCISES 9.6 (PAGE 564)
1. (a) $\{\frac{1}{2}\pi\}$; (b) $\{\frac{1}{3}\pi\}$ **3.** (a) $\{\frac{1}{4}\pi\}$; (b) $\{\frac{1}{3}\pi\}$ **5.** (a) $\{0\}$; (b) $\{\frac{1}{4}\pi\}$ **7.** (a) $\{0, \frac{1}{2}\pi\}$; (b) $\{\frac{1}{2}\pi\}$
9. (a) $\{0, \frac{1}{3}\pi\}$; (b) $\{0, \frac{1}{4}\pi\}$ **11.** (a) $\{\frac{1}{6}\pi, \frac{5}{6}\pi, \frac{7}{6}\pi, \frac{11}{6}\pi\}$; (b) $\{\frac{1}{4}\pi, \frac{3}{4}\pi, \frac{5}{4}\pi, \frac{7}{4}\pi\}$ **13.** (a) $\{\frac{2}{3}\pi, \pi, \frac{4}{3}\pi\}$; (b) $\{\frac{7}{6}\pi, \frac{11}{6}\pi\}$
15. (a) $\{\frac{1}{4}\pi, \frac{5}{4}\pi\}$ (b) $\{\frac{1}{3}\pi, \pi, \frac{5}{3}\pi\}$ **17.** (a) $\{\frac{3}{4}\pi, \frac{7}{4}\pi\}$; (b) $\{\frac{3}{4}\pi, \frac{7}{4}\pi\}$ **19.** $\{2.14, 4.14\}$ **21.** $\{0.87, 2.44, 4.01, 5.58\}$
23. $\{48.2°, 311.8°\}$ **25.** $\{0°, 41.4°, 180°, 318.6°\}$ **27.** $\{71.6°, 135°, 251.6°, 315°\}$
29. (a) $\{t \mid t = \frac{1}{6}\pi + k\pi\} \cup \{t \mid t = \frac{5}{6}\pi + k\pi\}, k \in J$; (b) $\{t \mid t = \frac{1}{4}\pi + k \cdot \frac{1}{2}\pi\}, k \in J$
31. (a) $\{t \mid t = \frac{1}{4}\pi + k\pi\}, k \in J$; (b) $\{t \mid t = \frac{1}{3}\pi + k \cdot \frac{2}{3}\pi\}, k \in J$ **33.** $\{t \mid t = 2.14 + 2k\pi\} \cup \{t \mid t = 4.14 + 2k\pi\}, k \in J$
35. $\{\theta \mid \theta = 48.2° + k \cdot 360°\} \cup \{\theta \mid \theta = 311.8° + k \cdot 360°\}, k \in J$
37. $\{\theta \mid \theta = k \cdot 180°\} \cup \{\theta \mid \theta = 41.4° + k \cdot 360°\} \cup \{\theta \mid \theta = 318.6° + k \cdot 360°\}, k \in J$
39. $\{\theta \mid \theta = 71.6° + k \cdot 180°\} \cup \{\theta \mid \theta = 135° + k \cdot 180°\}, k \in J$
41. (a) $\{\frac{1}{4}\pi, \frac{7}{12}\pi, \frac{11}{12}\pi, \frac{5}{4}\pi, \frac{19}{12}\pi, \frac{23}{12}\pi\}$; (b) $\{\frac{1}{18}\pi, \frac{5}{18}\pi, \frac{13}{18}\pi, \frac{17}{18}\pi, \frac{25}{18}\pi, \frac{29}{18}\pi\}$
43. (a) $\{\frac{5}{24}\pi, \frac{11}{24}\pi, \frac{17}{24}\pi, \frac{23}{24}\pi, \frac{29}{24}\pi, \frac{35}{24}\pi, \frac{41}{24}\pi, \frac{47}{24}\pi\}$; (b) $\{\frac{1}{15}\pi, \frac{1}{3}\pi, \frac{7}{15}\pi, \frac{11}{15}\pi, \frac{13}{15}\pi, \frac{17}{15}\pi, \frac{19}{15}\pi, \frac{23}{15}\pi, \frac{5}{3}\pi, \frac{29}{15}\pi\}$
45. (a) $\{\frac{1}{4}\pi, \frac{3}{4}\pi, \frac{5}{4}\pi, \frac{7}{4}\pi\}$; (b) $\{\frac{4}{3}\pi\}$ **47.** (a) $\{\frac{2}{3}\pi, \frac{4}{3}\pi\}$; (b) $\{\frac{1}{2}\pi\}$

49. (a) $\{x \mid x = \frac{1}{4}\pi + k \cdot \frac{1}{3}\pi\}, k \in J$; **(b)** $\{x \mid x = \frac{1}{18}\pi + k \cdot \frac{2}{3}\pi\} \cup \{x \mid x = \frac{5}{18}\pi + k \cdot \frac{2}{3}\pi\}, k \in J$

51. (a) $\{x \mid x = \frac{1}{4}\pi + k \cdot \frac{1}{2}\pi\}, k \in J$; **(b)** $\{x \mid x = \frac{4}{3}\pi + k \cdot 2\pi\}, k \in J$

53. $\{0°, 20°, 90°, 100°, 140°, 180°, 220°, 260°, 270°, 340°\}$ **55.** $\{22.5°, 73.2°, 112.5°, 163.2°, 202.5°, 253.2°, 292.5°, 343.2°\}$

57. (a) identity **59. (a)** not an identity; **(b)** $\{\frac{1}{4}\pi, \frac{1}{6}\pi, \frac{3}{4}\pi, \frac{5}{6}\pi, \frac{5}{4}\pi, \frac{7}{4}\pi\}$

61. (a) not an identity; **(b)** $\{0, \frac{1}{6}\pi, \frac{1}{3}\pi, \frac{1}{2}\pi, \frac{2}{3}\pi, \frac{5}{6}\pi, \pi, \frac{7}{6}\pi, \frac{4}{3}\pi, \frac{3}{2}\pi, \frac{5}{3}\pi, \frac{11}{6}\pi\}$ **63. (a)** $\frac{1}{25}$; **(b)** $\frac{1}{50}$ **65. (a)** $\frac{5}{9}$; **(b)** 0.61

67. (a) $\left\{t \mid t = \frac{1}{4\pi}\sin^{-1}\frac{y}{2} - \frac{1}{8} + k \cdot \frac{1}{2}\right\} \cup \left\{t \mid t = \frac{1}{8} - \frac{1}{4\pi}\sin^{-1}\frac{y}{2} + k \cdot \frac{1}{2}\right\}, k \in J$; **(b)** $\frac{1}{12}, \frac{5}{12}, \frac{7}{12}$

69. For $[-\pi, \pi]$ the solution set of $\sin^2 x - \frac{1}{4}\pi = 0$ is $\{-\frac{5}{6}\pi, -\frac{1}{6}\pi, \frac{1}{6}\pi, \frac{5}{6}\pi\}$; that of $x^2 - (\sin^{-1}\frac{1}{2})^2 = 0$ is $\{-\frac{1}{6}\pi, \frac{1}{6}\pi\}$.

REVIEW EXERCISES FOR CHAPTER 9 (PAGE 566)

29. (a) $\dfrac{\sqrt{6} - \sqrt{2}}{4}$; **(b)** $\dfrac{\sqrt{2 - \sqrt{3}}}{2}$ **31. (a)** $\dfrac{1 + \sqrt{3}}{1 - \sqrt{3}}$; **(b)** $-2 - \sqrt{3}$ **33. (a)** $\frac{253}{355}$; **(b)** $\frac{204}{325}$; **(c)** $-\frac{323}{325}$; **(d)** $-\frac{36}{325}$;

(e) first; **(f)** third **35.** -2 **37.** $\frac{1}{16} - \frac{1}{32}\cos 4t - \frac{1}{16}\cos 8t + \frac{1}{32}\cos 12t$ **39.** $\dfrac{6 - 4z}{1 + z^2}$

41.

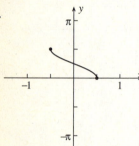

43.

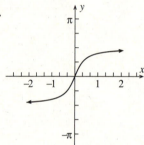

45. (a) $\{\frac{1}{3}\pi, \frac{5}{3}\pi\}$; **(b)** $\{\frac{1}{4}\pi, \frac{3}{4}\pi, \frac{5}{4}\pi, \frac{7}{4}\pi\}$

47. $\{\frac{1}{6}\pi, \frac{5}{6}\pi\}$

49. $\{1.25, \frac{3}{4}\pi, 4.39, \frac{7}{4}\pi\}$

51. $\{\frac{1}{18}\pi, \frac{5}{18}\pi, \frac{7}{18}\pi, \frac{11}{18}\pi, \frac{13}{18}\pi, \frac{17}{18}\pi, \frac{19}{18}\pi, \frac{23}{18}\pi, \frac{25}{18}\pi, \frac{29}{18}\pi, \frac{31}{18}\pi, \frac{35}{18}\pi\}$ **53.** $\{0.23, 1.34, 1.80, 2.91, 3.37, 4.48, 4.94, 6.05\}$

55. (a) $\{t \mid t = \frac{1}{3}\pi + k \cdot 2\pi\} \cup \{t \mid t = \frac{5}{3}\pi + k \cdot 2\pi\}, k \in J$; **(b)** $\{t \mid t = \frac{1}{4}\pi + k \cdot \frac{1}{2}\pi\}, k \in J$

57. $\{t \mid t = 1.25 + k\pi\} \cup \{t \mid t = \frac{3}{4}\pi + k\pi\}, k \in J$ **59.** $\{45°, 135°, 225°, 315°\}$ **61.** $\{90°, 270°\}$ **63.** $\{120°, 180°, 240°\}$

65. $\{\theta \mid \theta = 45° + k \cdot 90°\} \cup \{\theta \mid \theta = \frac{1}{4}\pi + k \cdot \frac{1}{2}\pi\}, k \in J$ **67.** $\{\theta \mid \theta = 90° + k \cdot 180°\} \cup \{\theta \mid \theta = \frac{1}{2}\pi + k\pi\}, k \in J$

69. (a) identity **71. (a)** not an identity; **(b)** $\{\frac{1}{4}\pi, \frac{5}{4}\pi\}$

73. (a) 0.65; **(b)** -0.65; **(c)** 0.92; **(d)** 2.22; **(e)** 0.54; **(f)** -0.54 **75. (a)** $\frac{4}{5}$,; **(b)** $\frac{3}{5}$; **(c)** $\frac{3}{4}$; **(d)** $-\frac{3}{4}$

77. (a) $\frac{1}{4}\pi$; **(b)** $\frac{1}{4}\pi$, **(c)** $\frac{3}{4}\pi$; **(d)** $-\frac{1}{4}\pi$; **(e)** $\frac{1}{4}\pi$ **79. (a)** $-\frac{1}{6}\pi$; **(b)** $-\frac{1}{3}\pi$; **(c)** $\frac{1}{6}\pi$; **(d)** $\frac{1}{3}\pi$ **81. (a)** $-\frac{120}{169}$; **(b)** $\frac{3}{4}$

83. (a) $f(t) = 3.03 \sin(2t + 1.06)$;

(b) amplitude is 3.030, period is π, phase shift is 0.53;

(c)

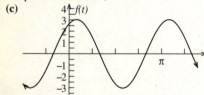

85.

87. (a) $f(t) = 2 \sin 40\pi t \cos \pi t$;

(b)

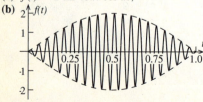

89. $f(t) = 0.061 \sin(200\pi t + 0.158)$

91. (a) $s = \sqrt{2} \sin(6t - 0.26)$;

(b) amplitude is $\sqrt{2}$, frequency is $\dfrac{3}{\pi}$

93. (a) $\left\{ t \mid t = \dfrac{1}{2\pi} \sin^{-1} \dfrac{y}{4} - \dfrac{1}{6} + k \right\} \cup \left\{ t \mid t = \dfrac{1}{3} - \dfrac{1}{2\pi} \sin^{-1} \dfrac{y}{4} + k \right\}, k \in J;$ **(b)** 0.20, 0.97

95. (a) $\theta = \tan^{-1}\!\left(\dfrac{5x}{x^2 + 36} \right);$ **(b)** 0.3890;   **(c)** 0.3948;   **(d)** 0.3906;   **(e)** 6.0 ft

97. (a) $f(t) = 2 \cos(2t + \tfrac{1}{6}\pi);$ **(c)**
(b) amplitude is 2, period is π, phase shift is $\tfrac{1}{12}\pi;$

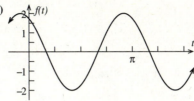

EXERCISES 10.1 (PAGE 581)

1. $|\mathbf{A}| = 5;$ **(b)** $|\mathbf{A}| = 2$ **3. (a)** $\langle 2, -3 \rangle;$ **(b)** $\langle -2, -7 \rangle$ **5. (a)** $(-4, 3);$ **(b)** $(12, -5)$ **7. (a)** $\langle -1, 9 \rangle;$ **(b)** $\langle 1, -5 \rangle$
9. (a) $\langle 7, 3 \rangle;$ **(b)** $\langle -9, -4 \rangle$ **11. (a)** $\langle 6, 1 \rangle;$ **(b)** $\sqrt{74};$ **(c)** $\sqrt{1061}$
13. (a) $10\mathbf{i} + 15\mathbf{j};$ **(b)** $-24\mathbf{i} + 6\mathbf{j};$ **(c)** $6\mathbf{i} + 2\mathbf{j};$ **(d)** $2\sqrt{10}$
15. (a) $\sqrt{13} + \sqrt{17};$ **(b)** $-14\mathbf{i} + 21\mathbf{j};$ **(c)** $7\sqrt{13};$ **(d)** $5\sqrt{13} - 6\sqrt{17}$ **17.** $\langle 14.6, 10.6 \rangle$ **19.** $\langle -12.0, -32.9 \rangle$
21. (a) $5(\tfrac{3}{5}\mathbf{i} - \tfrac{4}{5}\mathbf{j});$ **(b)** $2\sqrt{2}(\cos \tfrac{1}{4}\pi \mathbf{i} + \sin \tfrac{1}{4}\pi \mathbf{j})$ **23. (a)** $8(\cos \tfrac{2}{3}\pi \mathbf{i} + \sin \tfrac{2}{3}\pi \mathbf{j});$ **(b)** $16(\cos \pi \mathbf{i} + \sin \pi \mathbf{j})$
25. (a) 135 lb; **(b)** 17°   **27. (a)** 135 lb; **(b)** 17°   **29.** 29.0°   **31. (a)** 28.1°; **(b)** 1.7 mi/hr; **(c)** 0.53 mi
33. (a) 346.1°; **(b)** 243 mi/hr   **35.** 15.3 knots, 191.3°

EXERCISES 10.2 (PAGE 590)

1. $(-\infty, 0) \cup (0, 4]$ **3.** $(0, +\infty)$ **5.** $\{ t \mid t \neq k \cdot \tfrac{1}{2}\pi \}, k \in J$ **7.** $[-1, 1]$ **9.** $(-\infty, -4] \cup [3, +\infty)$

11.

$x^2 + y^2 = 1$

13.

$x^2 + y^2 = 16,\ y \geq 0$

15.

$y = \tfrac{2}{9}x^2$

17. (a)

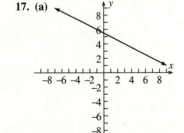

(b) $x + 2y = 11$

19. (a)

(b) $y = (4x)^{2/3}$

21. (a)

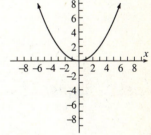

(b) $\dfrac{x^2}{81} + \dfrac{y^2}{16} = 1$

23. (a) 5 sec; **(b)** 300 ft; **(c)** 100 ft;
(d) $y = \frac{4}{3}x - \frac{1}{225}x^2$;
(e)

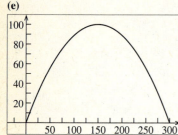

25. If t is the number of seconds after the shot, then
(a) $1250\sqrt{2}t\,\mathbf{i} + (1250\sqrt{2}t - 16t^2)\mathbf{j}$;
(b) $x = 1250\sqrt{2}t$, $y = 1250\sqrt{2}t - 16t^2$;
(c) 110.5 sec;
(d) 195,312.5 ft $\approx$ 37 mi;
(e) 48,828.125 ft $\approx$ 9.25 mi;
(f) $y = x - \dfrac{2x^2}{390,625}$;
(g)

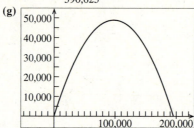

27. (a) 270 ft; **(b)** 108 ft;
(c)

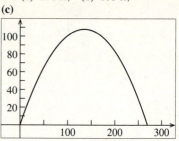

(d) $y = 1.60x - 0.00593x^2$

29. (a) $x = 4(t - \sin t)$, $y = 4(1 - \cos t)$;
(b)

31.

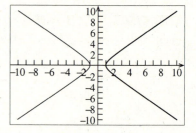

33. The graph of part (b) is shown lighter than that of part (a).

EXERCISES 10.3 (PAGE 597)

5. (a) $(-4, \frac{5}{4}\pi)$; **(b)** $(4, -\frac{7}{4}\pi)$; **(c)** $(-4, -\frac{3}{4}\pi)$ **7. (a)** $(-2, \frac{3}{2}\pi)$; **(b)** $(2, -\frac{3}{2}\pi)$; **(c)** $(-2, -\frac{1}{2}\pi)$
9. (a) $(-\sqrt{2}, \frac{3}{4}\pi)$; **(b)** $(\sqrt{2}, -\frac{1}{4}\pi)$; **(c)** $(-\sqrt{2}, \frac{5}{4}\pi)$ **11. (a)** $(-2, \frac{3}{4}\pi)$; **(b)** $(-2, -\frac{5}{4}\pi)$; **(c)** $(2, \frac{15}{4}\pi)$
13. $(3, \frac{4}{3}\pi)$; $(-3, \frac{1}{3}\pi)$ **15.** $(-4, -\frac{7}{6}\pi)$; $(4, -\frac{1}{6}\pi)$ **17.** $(-2, \frac{3}{4}\pi)$; $(2, \frac{7}{4}\pi)$ **19.** $(2, 2\pi + 6)$; $(-2, 6 - \pi)$
21. (a) $(-3, 0)$; **(b)** $(-1, -1)$; **(c)** $(2, -2\sqrt{3})$; **(d)** $(\frac{1}{2}\sqrt{3}, -\frac{1}{2})$ **23. (a)** $(\sqrt{2}, \frac{7}{4}\pi)$; **(b)** $(2, \frac{5}{6}\pi)$; **(c)** $(2\sqrt{2}, \frac{1}{4}\pi)$; **(d)** $(5, \pi)$
25. $r = |a|$ **27.** $r = \dfrac{2}{1 - \cos\theta}$ **29.** $r = 6\sin\theta$ **31.** $r^2 = 4\cos 2\theta$ **33.** $r = \dfrac{3a\sin 2\theta}{2(\sin^3\theta + \cos^3\theta)}$
35. $(x^2 + y^2)^2 = 4xy$ **37.** $(x^2 + y^2)^3 = x^2$ **39.** $y = x\tan(x^2 + y^2)$ **41.** $x = -1$ **43.** $4x^2 - 5y^2 - 36y - 36 = 0$

EXERCISES 10.4 (PAGE 608)

1. (a) line through the pole with slope $\sqrt{3}$; **(b)** circle with center at the pole and radius $\frac{1}{3}\pi$
3. (a) line through the pole with slope $\tan^{-1}2$; **(b)** circle with center at the pole and radius 2
5. (a) line parallel to the $\frac{1}{2}\pi$ axis and 4 units to the right of it;
  (b) circle tangent to the $\frac{1}{2}\pi$ axis with center on the polar axis and radius 2.
7. (a) line parallel to the polar axis and 4 units below it;
  (b) circle tangent to the polar axis with center on the extension of the $\frac{1}{2}\pi$ axis and radius 2

9. cardioid; symmetric with respect to polar axis; points to left;

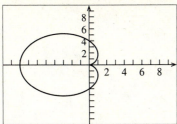

11. cardioid; symmetric with respect to $\frac{1}{2}\pi$ axis; points upward;

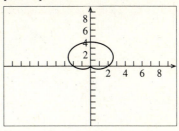

13. limaçon with a loop; symmetric with respect to $\frac{1}{2}\pi$ axis; points downward;

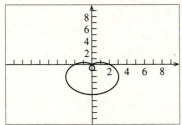

15. limaçon with a dent; symmetric with respect to polar axis; points to left;

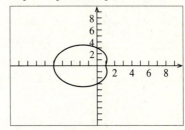

17. convex limaçon; symmetric with respect to $\frac{1}{2}\pi$ axis; points upward;

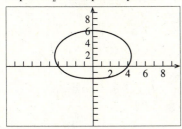

19. 3-leafed rose

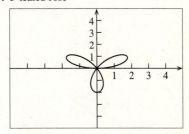

21. 8-leafed rose

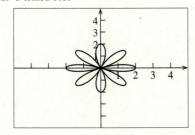

23. 4-leafed rose

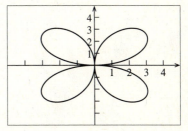

25. logarithmic spiral, some of whose points (r, θ) are given in the following table

r	1	$e^{\pi/2} \approx 5$	$e^{\pi} \approx 23$	$e^{3\pi/2} \approx 111$	$e^{2\pi} \approx 535$	$e^{5\pi/2} \approx 2576$	$e^{3\pi} \approx 12{,}392$
θ	0	$\frac{1}{2}\pi$	π	$\frac{3}{2}\pi$	2π	$\frac{5}{2}\pi$	3π

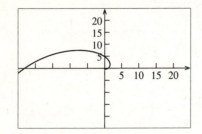

27. reciprocal spiral, some of whose points (r, θ) are given in the following table

r	$\frac{6}{\pi} \approx 1.9$	$\frac{3}{\pi} \approx 0.95$	$\frac{2}{\pi} \approx 0.63$	$\frac{1}{\pi} \approx 0.32$	$\frac{1}{2\pi} \approx 0.16$	$\frac{1}{3\pi} \approx 0.12$	$\frac{1}{4\pi} \approx 0.08$	$\frac{1}{6\pi} \approx 0.05$
θ	$\frac{1}{6}\pi$	$\frac{1}{3}\pi$	$\frac{1}{2}\pi$	π	2π	3π	4π	6π

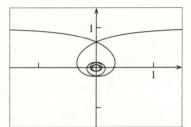

29.

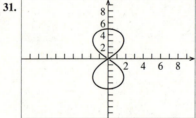

31.

33.

35.

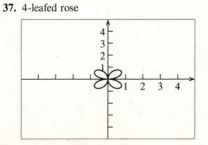

37. 4-leafed rose

39.

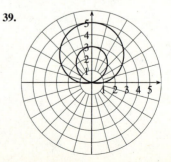

41.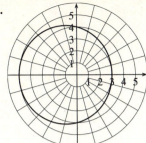

43.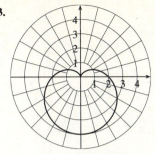

45.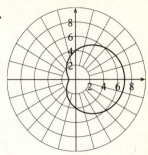

47.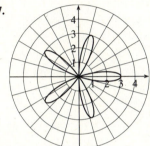

49.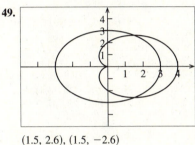

(1.5, 2.6), (1.5, −2.6)

51.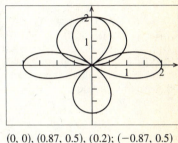

(0, 0), (0.87, 0.5), (0.2); (−0.87, 0.5)

53. $(3, \frac{1}{3}\pi), (3, -\frac{1}{3}\pi)$

55. pole, $(1, \frac{1}{6}\pi), (2, \frac{1}{2}\pi), (1, \frac{5}{6}\pi)$

EXERCISES 10.5 (PAGE 616)

9. (a) $\sqrt{29}$; **(b)** $\sqrt{5}$ **11. (a)** 3; **(b)** 3   **13. (a)** 2; **(b)** 2   **15. (a)** 10; **(b)** 10
17. (a) $\frac{3}{2} + \frac{3}{2}\sqrt{3}i$; **(b)** $-\frac{3}{2} - \frac{3}{2}\sqrt{3}i$ **19. (a)** $-3\sqrt{3} + 3i$; **(b)** $3\sqrt{3} - 3i$ **21. (a)** $2i$; **(b)** $-\frac{2}{3}$
23. (a) $4\sqrt{2}(\cos\frac{7}{4}\pi + i\sin\frac{7}{4}\pi)$; **(b)** $6(\cos 0 + i\sin 0)$; **(c)** $\cos\frac{1}{2}\pi + i\sin\frac{1}{2}\pi$
25. (a) $6(\cos\frac{5}{6}\pi + i\sin\frac{5}{6}\pi)$; **(b)** $7(\cos\pi + i\sin\pi)$; **(c)** $7(\cos\frac{3}{2}\pi + i\sin\frac{3}{2}\pi)$
27. (a) $5[\cos(\text{arc tan}\frac{4}{3}) + i\sin(\text{arc tan}\frac{4}{3})]$; **(b)** $5(\cos 0.93 + i\sin 0.93)$; **(c)** $5(\cos 53.1° + i\sin 53.1°)$
29. (a) $\sqrt{5}[\cos(\text{arc tan}(-2) + 2\pi) + i\sin(\text{arc tan}(-2) + 2\pi)]$; **(b)** $\sqrt{5}(\cos 5.18 + i\sin 5.18)$; **(c)** $\sqrt{5}(\cos 296.6° + i\sin 296.6°)$
31. $0 + 10i$ **33.** $1 + \sqrt{3}i$ **35.** $-2\sqrt{2} + 2\sqrt{2}i$ **37.** $2\sqrt{3} - 2i$ **39.** $\sqrt{3} + i$ **41.** $-2\sqrt{2} + 2\sqrt{2}i$ **43.** $\frac{5}{2} - \frac{5}{2}\sqrt{3}i$
45. $0 - \frac{2}{3}i$

EXERCISES 10.6 (PAGE 629)

1. $2 + 2\sqrt{3}i$ **3.** 64   **5.** -1 **7.** $-16 - 16\sqrt{3}i$ **9.** $-2 + 2i$ **11.** 64   **13.** $4096 - 4096i$ **15.** -1 **17.** 1
19. $-6^{30} \approx (-2.2107)10^{23}$ **21. (a)** $1, -\frac{1}{2} \pm \frac{1}{2}\sqrt{3}i$; **(b)** $3, -\frac{3}{2} \pm \frac{3}{2}\sqrt{3}i$
23. (a) $1, \cos\frac{2}{7}\pi + i\sin\frac{2}{7}\pi, \cos\frac{4}{7}\pi + i\sin\frac{4}{7}\pi, \cos\frac{6}{7}\pi + i\sin\frac{6}{7}\pi, \cos\frac{8}{7}\pi + i\sin\frac{8}{7}\pi, \cos\frac{10}{7}\pi + i\sin\frac{10}{7}\pi, \cos\frac{12}{7}\pi + i\sin\frac{12}{7}\pi$;
(b) $2, 2(\cos\frac{2}{7}\pi + i\sin\frac{2}{7}\pi), 2(\cos\frac{4}{7}\pi + i\sin\frac{4}{7}\pi), 2(\cos\frac{6}{7}\pi + i\sin\frac{6}{7}\pi), 2(\cos\frac{8}{7}\pi + i\sin\frac{8}{7}\pi), 2(\cos\frac{10}{7}\pi + i\sin\frac{10}{7}\pi),$
$2(\cos\frac{12}{7}\pi + i\sin\frac{12}{7}\pi)$ **25.** $\pm 2, \pm 2i$ **27.** $i, \frac{1}{2}\sqrt{3} - \frac{1}{2}i, -\frac{1}{2}\sqrt{3} - \frac{1}{2}i$ **29.** $-\sqrt{2} + \sqrt{2}i, \sqrt{2} - \sqrt{2}i$
31. $-i, \frac{1}{2}\sqrt{3} + \frac{1}{2}i, -\frac{1}{2}\sqrt{3} + \frac{1}{2}i$ **33.** $-0.35 + 1.97i, -1.53 - 1.29i, 1.88 - 0.68i$
35. $\frac{1}{2}\sqrt{3} + \frac{1}{2}i \approx 0.87 + 0.5i; -\frac{1}{2} + \frac{1}{2}\sqrt{3}i \approx -0.5 + 0.87i; -\frac{1}{2}\sqrt{3} - \frac{1}{2}i \approx -0.87 - 0.5i; \frac{1}{2} - \frac{1}{2}\sqrt{3}i \approx 0.5 - 0.87i$
37. $1.21 + 0.66i, -0.25 + 1.36i, -1.37 + 0.18i, -0.59 - 1.25i, 1.00 - 0.95i$
41. $\cos 2\theta = \cos^2\theta - \sin^2\theta; \sin 2\theta = 2\sin\theta\cos\theta$ **43.** $\sin 3\theta = 3\sin\theta - 4\sin^3\theta$ **45.** $\cos 4\theta = 8\cos^4\theta - 8\cos^2\theta + 1$

REVIEW EXERCISES FOR CHAPTER 10 (PAGE 631)

1. (a) $\langle 5, 3\rangle$; **(b)** $\langle -1, 3\rangle$ **3. (a)** $\langle 6, 7\rangle$; **(b)** $\langle -6, 13\rangle$ **5. (a)** $-25\mathbf{i} + 63\mathbf{j}$; **(b)** $\sqrt{4594}$; **(c)** $15\sqrt{2} - 14\sqrt{13}$
7. (a) $\langle 6\sqrt{3}, 6\rangle$; **(b)** $\langle -13.5, 33.4\rangle$ **9. (a)** $6\sqrt{2}(\cos\frac{1}{4}\pi\mathbf{i} + \sin\frac{1}{4}\pi\mathbf{j})$; **(b)** $6(\cos 131.8°\mathbf{i} + \sin 131.8°\mathbf{j})$ **11.** $(0, 2) \cup (2, +\infty)$
13. $\{t \mid \neq (k + \frac{1}{2})\pi\}, k \in J$

15.

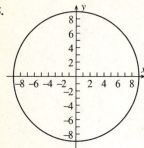

$x^2 + y^2 = 81$

17. (a)

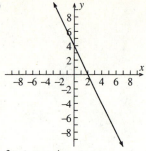

(b) $2x + y = 4$

19. (a)

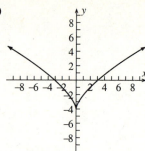

(b) $y = 3(\frac{1}{2}x)^{2/3} - 4$

21. (a) $(2, \frac{11}{4}\pi), (-2, \frac{7}{4}\pi)$; **(b)** $(-3, -\frac{5}{6}\pi), (3, \frac{1}{6}\pi)$ **23. (a)** $(0, 1)$; **(b)** $(1, -\sqrt{3})$; **(c)** $(-2\sqrt{2}, -2\sqrt{2}), (-\frac{3}{2}\sqrt{3}, -\frac{3}{2})$

25. (a) $(4\sqrt{2}, \frac{3}{4}\pi)$; **(b)** $(2, \frac{5}{3}\pi)$; **(c)** $(6, \frac{1}{2}\pi)$; **(d)** $(4, \frac{7}{6}\pi)$ **27.** $r^2(4\cos^2\theta - 9\sin^2\theta) = 36$ **29.** $r = 9\cos\theta - 8\sin\theta$

31. $xy = 2$ **33.** $x^4 + 2x^2y^2 + y^4 - y^2 = 0$

35. (a) line through the pole with slope 1; **(b)** circle with center at the pole and radius 4

37. (a) line perpendicular to polar axis through the point $(3, 0)$; **(b)** circle with center at $(\frac{3}{2}, 0)$ and tangent to the $\frac{1}{2}\pi$ axis at the pole

39. limaçon with a dent; symmetric with respect to polar axis; points to right;

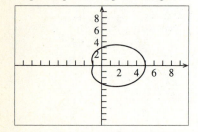

41. cardioid, symmetric with respect to polar axis, points to left;

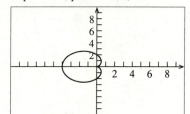

43. limaçon with a loop; symmetric with respect to $\frac{1}{2}\pi$ axis; points downward;

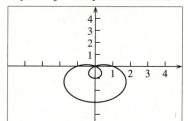

45. 4-leafed rose

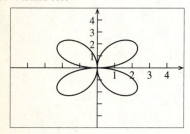

47.

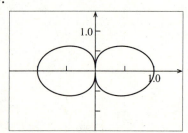

49. lemniscate

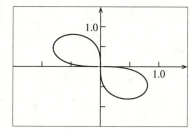

51. (a)

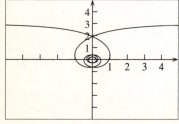

(b)

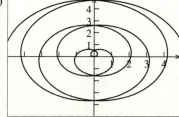

55. (a) 5; **(b)** $2\sqrt{10}$

57. (a) $2\sqrt{3}(\cos\frac{5}{6}\pi + i\sin\frac{5}{6}\pi)$; **(b)** $4(\cos\frac{1}{6}\pi + i\sin\frac{1}{6}\pi)$; **(c)** $4(\cos\frac{3}{2}\pi + i\sin\frac{3}{2}\pi)$; **(d)** $\sqrt{2}(\cos\frac{5}{4}\pi + i\sin\frac{5}{4}\pi)$

59. $\sqrt{3} + i$ **61.** $-2\sqrt{2} + 2\sqrt{2}i$ **63.** $-\frac{3}{2} + \frac{3}{2}\sqrt{3}i$ **65.** $-\sqrt{2} - \sqrt{2}i$ **67.** $0 + 8i$ **69.** $-\frac{1}{512}i$ **71.** $2, -1 \pm \sqrt{3}i$

73. $1.21 + 0.14i, -0.14 + 1.21i, -1.21 - 0.14i, 0.14 - 1.21i$

75. $0.73 + 1.19i, -0.91 + 1.07i, -1.29 - 0.53i, 0.11 - 1.40i, 1.36 - 0.33i$

77. (a) 104.4 lb; **(b)** 35.5°   **79. (a)** 245 mi/hr; **(b)** 105°

81. (a) 2.5 sec; **(b)** 75 ft; **(c)** 25 ft;
(d) $y = \frac{4}{3}x - \frac{4}{225}x^2$;
(e)

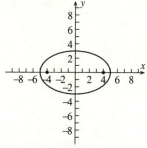

83. (a) 603 ft; **(b)** 5.62 sec; **(c)** 126 ft;
(d)

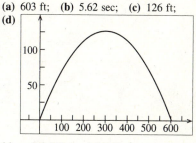

(e) $y = 0.839x - 0.00139x^2$

EXERCISES 11.1 (PAGE 646)

1. (a) $(0, 0)$; **(b)** x axis;
(c) $(-5, 0), (5, 0)$;
(d) $(0, -3), (0, 3)$;
(e) $(-4, 0), (4, 0)$;
(f)

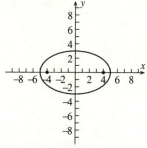

3. (a) $(0, 0)$; **(b)** y axis;
(c) $(0, -4), (0, 4)$;
(d) $(-2, 0), (2, 0)$;
(e) $(0, -2\sqrt{3}), (0, 2\sqrt{3})$;
(f)

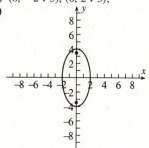

5. (a) $(0, 0)$; **(b)** x axis;
(c) $(-10, 0), (10, 0)$;
(d) $(0, -6), (0, 6)$;
(e) $(-8, 0), (8, 0)$;
(f)

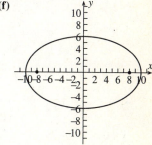

7. (a) $(0, 0)$; **(b)** y axis;
(c) $(0, -3), (0, 3)$;
(d) $(-1, 0), (1, 0)$;
(e) $(0, -2\sqrt{2}), (0, 2\sqrt{2})$;
(f)

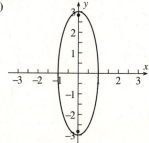

9. (a) $(2, 1)$; **(b)** $y = 1$;
(c) $(-1, 1), (5, 1)$;
(d) $(2, -1), (2, 3)$;
(e) $(2 - \sqrt{5}, 1), (2 + \sqrt{5}, 1)$;
(f)

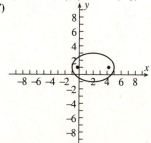

11. (a) $(-1, 2)$; **(b)** $x = -1$;
(c) $(-1, -8), (-1, 12)$;
(d) $(-6, 2), (4, 2)$;
(e) $(-1, 2 - 5\sqrt{3}), (-1, 2 + 5\sqrt{3})$;
(f)

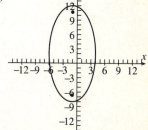

13. point $(-\frac{5}{2}, 4)$

15. (a) $(2, 0)$; (b) x axis;
(c) $(2 - 3\sqrt{3}, 0), (2 + 3\sqrt{3}, 0)$;
(d) $(2, -3), (2, 3)$;
(e) $(2 - 3\sqrt{2}, 0), (2 + 3\sqrt{2}, 0)$;
(f)

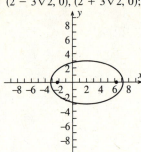

17. point $(1, -1)$

19. $16x^2 + 25y^2 = 100$

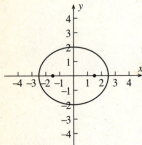

21. $3x^2 + 2y^2 = 54$

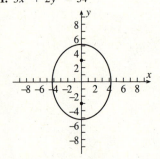

23. $x^2 + 4y^2 = 4$

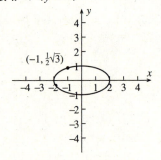

25. $\dfrac{(x - 4)^2}{25} + \dfrac{(y + 2)^2}{9} = 1$

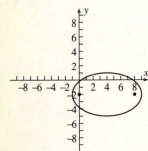

27. $\dfrac{(x + 1)^2}{48} + \dfrac{(y - 3)^2}{64} = 1$

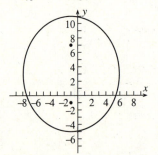

29. 8.4 m

31. $9x^2 + 25y^2 = 562,500$, where the unit is 1 million miles

EXERCISES 11.2 (PAGE 659)

1. (a) $(0, 0)$;   (b) x axis;
(c) $(-8, 0)$, $(8, 0)$;
(d) $(-10, 0)$, $(10, 0)$;
(e)

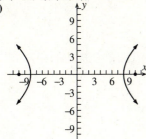

3. (a) $(0, 0)$;   (b) y axis;
(c) $(0, -5)$, $(0, 5)$;
(d) $(0, -13)$, $(0, 13)$;
(e)

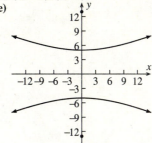

5. (a) $(0, 0)$;   (b) x axis;
(c) $(-2, 0)$, $(2, 0)$;
(d) $(-\sqrt{13}, 0)$, $(\sqrt{13}, 0)$;
(e)

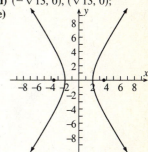

7. (a) $(0, 0)$;   (b) x axis;
(c) $(-5, 0)$, $(5, 0)$;
(d)

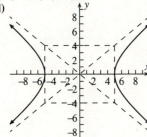

9. (a) $(0, 0)$;   (b) y axis;
(c) $(0, -2)$, $(0, 2)$;
(d)

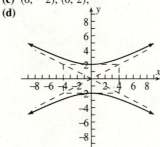

11. (a) $(0, 0)$;   (b) y axis;
(c) $(0, -6)$, $(0, 6)$;
(d)

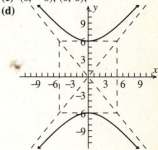

13. (a) $(-3, -2)$;   (b) $y = -2$;
(c) $(-6, -2)$, $(0, -2)$;
(d)

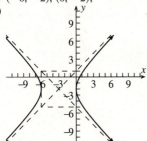

15. (a) $(-3, -1)$;   (b) $y = -1$;
(c) $(-7, -1)$, $(1, -1)$;
(d)

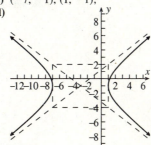

17. (a) $(-1, 4)$,   (b) $x = -1$;
(c) $(-1, 4 - 2\sqrt{7})$, $(-1, 4 + 2\sqrt{7})$;
(d)

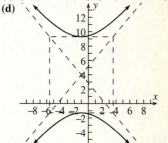

19. (a) $(1, -2)$; **(b)** $x = 1$;
(c) $(1, -5), (1, 1)$;
(d)

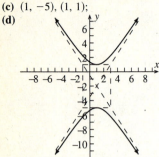

21. $y = -\frac{4}{5}x, y = \frac{4}{5}x$
23. $y = -x - 5, y = x + 1$
25. $y = -\frac{3}{2}x - \frac{1}{2}, y = \frac{3}{2}x - \frac{7}{2}$

27. $\dfrac{x^2}{4} - \dfrac{y^2}{9} = 1$

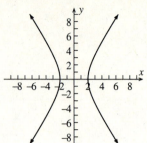

29. (a) $32y^2 - 33x^2 = 380$

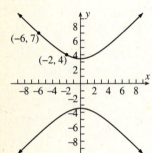

31. $\dfrac{x^2}{576} - \dfrac{y^2}{100} = 1$

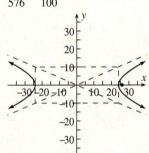

33. $\dfrac{(y + 1)^2}{144} - \dfrac{(x + 2)^2}{81} = 1$

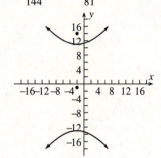

35. $72(x - 3)^2 - 9(y - 4)^2 = 128$

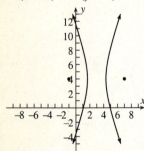

37. (a) $\dfrac{(x + 2)^2}{1} - \dfrac{(y + 1)^2}{4} = 1$;
   (b) $y = 2x + 3, y = -2x - 5$

39. $\dfrac{x^2}{4} - \dfrac{y^2}{7} = 1$

41. the right branch of the hyperbola $\dfrac{x^2}{900} - \dfrac{y^2}{1600} = 1$

45. $x = \dfrac{8}{\cos t}, y = 6 \tan t$; or, $x = 8 \cosh t, y = 6 \sinh t$ and $x = -8 \cosh t, y = 6 \sinh t$

47. $x = 12 \tan t, y = \dfrac{5}{\cos t}$; or, $x = 12 \sinh t, y = 5 \cosh t$ and $x = 12 \sinh t, y = -5 \cosh t$

49. $x = 3 + \dfrac{3}{\cos t}, y = -2 + 3 \tan t$; or, $x = 3 + 3 \cosh t, y = -2 + 3 \sinh t$ and $x = 3 - 3 \cosh t, y = -2 + 3 \sinh t$

51. $x = -1 + \sqrt{21} \tan t, y = 4 + \dfrac{2\sqrt{7}}{\cos t}$; or, $x = -1 + \sqrt{21} \sinh t, y = 4 + 2\sqrt{7} \cosh t$ and $x = -1 + \sqrt{21} \sinh t,$
   $y = 4 - 2\sqrt{7} \cosh t$

EXERCISES 11.3 (PAGE 670)

1. (a) hyperbola; **(b)** ellipse; **(c)** parabola; **(d)** two intersecting lines

3. (a) point; **(b)** hyperbola; **(c)** parabola; **(d)** empty set

In part (a) of Exercises 5-25, the conic may be degenerate.

5. (a) hyperbola;
(b) $\bar{x}^2 - \bar{y}^2 = 16$;
(c)

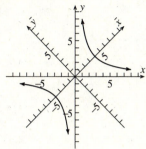

7. (a) hyperbola;
(b) $\bar{x}\,\bar{y} = -4$;
(c)

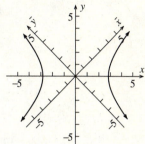

In Exercises 9-25, the $\bar{x}$ and $\bar{y}$ axes have been rotated through angle α.

9. (a) hyperbola;
(b) $\alpha \approx 36.9°$,
$16\bar{y}^2 - 9\bar{x}^2 = 36$;
(c)

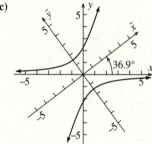

11. (a) parabola;
(b) $\alpha = \frac{1}{4}\pi$,
$\bar{x}^2 + 4\sqrt{2}\,\bar{y} = 0$;
(c)

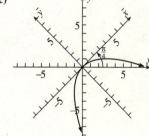

13. (a) hyperbola;
(b) $\alpha = \frac{1}{4}\pi$,
$\bar{y}^2 - \bar{x}^2 = 32$;
(c)

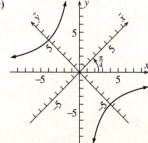

15. (a) ellipse;
(b) $\alpha = \frac{1}{6}\pi$,
$\dfrac{\bar{x}^2}{4} + \dfrac{\bar{y}^2}{9} = 1$
(c)

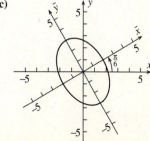

In Exercises 17-25, the x' and y' axes have also been translated such that $x' = \bar{x} - h$, $y' = \bar{y} - k$.

17. (a) ellipse; **(b)** $\alpha = \frac{1}{4}\pi$,
$h = \frac{1}{2}\sqrt{2}$, $k = \frac{3}{2}\sqrt{2}$, $\dfrac{x'^2}{6} + \dfrac{y'^2}{18} = 1$;

(c)

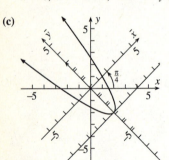

19. (a) ellipse; **(b)** $\alpha \approx 63.4°$,
$h = \dfrac{2}{\sqrt{5}}$, $k = -\dfrac{4}{\sqrt{5}}$, $\dfrac{x'^2}{16} + \dfrac{y'^2}{4} = 1$;

(c)

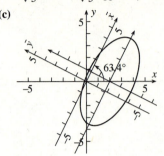

21. (a) parabola; **(b)** $\alpha = \frac{1}{4}\pi$,
$h = 0$, $k = -2\sqrt{2}$, $\sqrt{2}x'^2 = y'$;

(c)

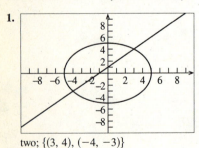

23. (a) hyperbola; **(b)** $\alpha \approx 53.1°$;
$h = 5$, $k = 0$, $\dfrac{x'^2}{16} - \dfrac{y'^2}{4} = 1$;

(c)

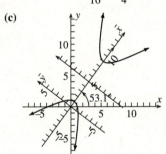

25. (a) parabola; **(b)** $\alpha \approx 26.6°$,
$\sqrt{5}\bar{x}^2 + 6\sqrt{5}\bar{y} = 0$;

(c)

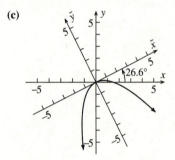

EXERCISES 11.4 (PAGE 679)

1.

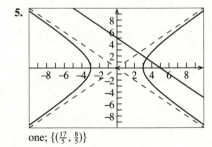

two; $\{(3, 4), (-4, -3)\}$

3.

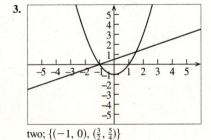

two; $\{(-1, 0), (\frac{3}{2}, \frac{5}{4})\}$

5.

one; $\{(\frac{17}{5}, \frac{8}{5})\}$

7.

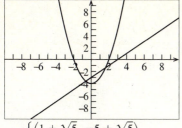

two; $\left\{ \left(\dfrac{1 + \sqrt{5}}{2}, \dfrac{-5 + \sqrt{5}}{2} \right), \left(\dfrac{1 - \sqrt{5}}{2}, \dfrac{-5 - \sqrt{5}}{2} \right) \right\}$

9.

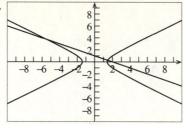

two; $\{(-2 + 2\sqrt{3}, 2 - \sqrt{3}), (-2 - 2\sqrt{3}, 2 + \sqrt{3})\}$

11.

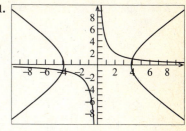

two; $\{(-4, -1), (4, 1), (i, -4i), (-i, 4i)\}$

13.

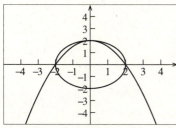

three; $\{(-2, 0), (2, 0), (0, 2)\}$

15.

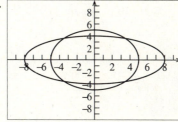

four; $\{(-2\sqrt{3}, -\sqrt{13}), (2\sqrt{3}, -\sqrt{13}), (-2\sqrt{3}, \sqrt{13}), (2\sqrt{3}, \sqrt{13})\}$

17.

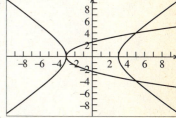

three; $\{(-3, 0), (5, -4), (5, 4)\}$

19.

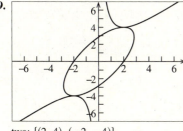

two; $\{(2, 4), (-2, -4)\}$

21.

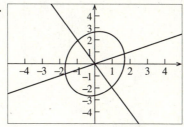

four; $\{(1, -2), (-1, 2), (\tfrac{2}{3}\sqrt{6}, \tfrac{1}{3}\sqrt{6}), (-\tfrac{2}{3}\sqrt{6}, -\tfrac{1}{3}\sqrt{6})\}$

23. 6, 10
25. 12 cm, 35 cm
27. 8 m, 12 m
29. 25, $6.00
31. 9.9 cm, 49.1 cm
33. 4 ft by 4 ft by 7 ft

35. $\{(-1, -2), (-1, 2), (1, -2), (1, 2)\}$
37. **(a)**

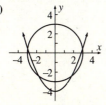

(b)

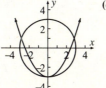

(c)

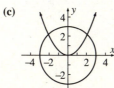

(d)

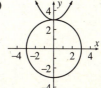

(e)

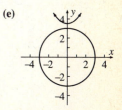

EXERCISES 11.5 (PAGE 689)

1. (a) $e = \frac{1}{3}\sqrt{5}$;
foci: $(\pm\sqrt{5}, 0)$;
directrices: $x = \pm\frac{9}{5}\sqrt{5}$;
(b)

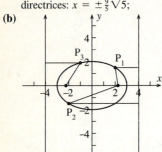

3. (a) $e = \frac{1}{5}\sqrt{21}$;
foci: $(0, \pm\sqrt{21})$;
directrices: $y = \pm\frac{25}{21}\sqrt{21}$;
(b)

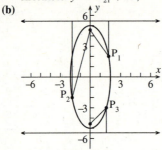

5. (a) $e = \frac{1}{5}\sqrt{29}$;
foci: $(\pm\sqrt{29}, 0)$;
directrices: $x = \pm\frac{25}{29}\sqrt{29}$;
(b)

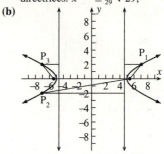

7. (a) $e = \frac{5}{3}$;
foci: $(\pm 5, 0)$;
directrices: $x = \pm\frac{9}{5}$;
(b)

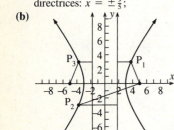

9. (a) parabola; **(b)** hyperbola; **(c)** ellipse; **(d)** circle

11. (a) 1; **(b)** parabola;
(c) $r\cos\theta = -2$;
(d)

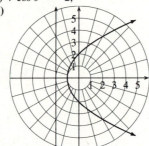

13. (a) $\frac{1}{2}$; **(b)** ellipse;
(c) $r\sin\theta = 5$;
(d)

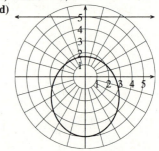

15. (a) $\frac{2}{3}$; **(b)** ellipse;
(c) $r\cos\theta = -3$;
(d)

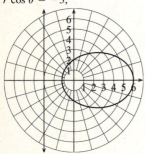

17. (a) $\frac{6}{5}$; **(b)** hyperbola;
(c) $r\sin\theta = -\frac{3}{2}$;
(d)

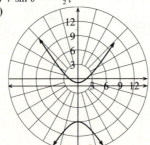

19. (a) $\frac{2}{7}$; **(b)** ellipse;
(c) $r\sin\theta = -5$;
(d)

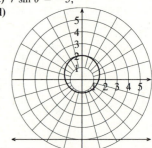

21. (a) $\frac{5}{4}$;   **(b)** hyperbola;
(c) $r\cos\theta = 2$;
(d)

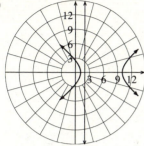

23. $r = \dfrac{8}{1 - \sin\theta}$

25. $r = \dfrac{36}{3 + 4\cos\theta}$

27. $r = \dfrac{3}{2 - \cos\theta}$

29. (a) $r = \dfrac{5}{1 - 3\cos\theta}$;   **(b)** $r\cos\theta = -\frac{5}{3}$

35. (a) $r = \dfrac{40{,}000{,}000}{1 - \cos\theta}$;   **(b)** 20,000,000 miles

REVIEW EXERCISES FOR CHAPTER 11 (PAGE 691)

1.

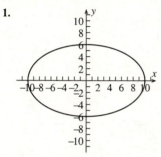

(a) $(0, 0)$;   **(b)** $y = 0$;
(c) $(-10, 0), (10, 0)$;
(d) $(0, -6), (0, 6)$;
(e) $(-8, 0), (8, 0)$

3.

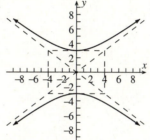

(a) $(0, 0)$;   **(b)** $x = 0$;
(c) $(0, -3), (0, 3)$

5.

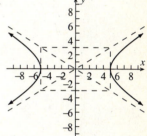

(a) $(0, 0)$;   **(b)** $y = 0$;
(c) $(-5, 0), (5, 0)$

7.

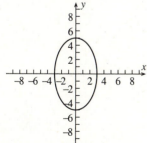

(a) $(0, 0)$;   **(b)** $x = 0$;
(c) $(0, -5), (0, 5)$;
(d) $(-3, 0), (3, 0)$;
(e) $(0, -4), (0, 4)$

9.

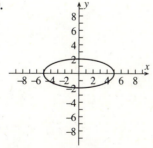

(a) $(0, 0)$;   **(b)** $y = 0$;
(c) $(-5, 0), (5, 0)$
(d) $(0, -2), (0, 2)$;
(e) $(-\sqrt{21}, 0), (\sqrt{21}, 0)$

11.

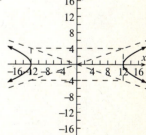

(a) $(0, 0)$;   **(b)** $y = 0$;
(c) $(-12, 0), (12, 0)$

13.

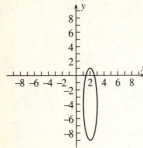

(a) $(2, -4)$;   (b) $x = 2$;
(c) $(2, -9), (2, 1)$;
(d) $(1, -4), (3, -4)$;
(e) $(2, -4 - 2\sqrt{6}), (2, -4 + 2\sqrt{6})$

15.

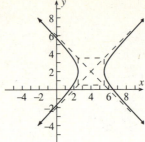

(a) $(4, 2)$;   (b) $y = 2$;
(c) $(\frac{5}{2}, 2), (\frac{11}{2}, 2)$

17.

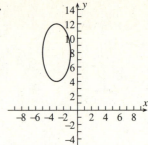

(a) $(-3, 8)$;   (b) $x = -3$;
(c) $(-3, 4), (-3, 12)$;
(d) $(-5, 8), (-1, 8)$;
(e) $(-3, 8 - 2\sqrt{3}), (-3, 8 + 2\sqrt{3})$

19.

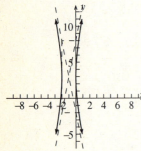

(a) $(-1, 3)$;   (b) $y = 3$;
(c) $(-2, 3), (0, 3)$

21. $y = \frac{3}{4}x, y = -\frac{3}{4}x$

23. $y = -x + 6, y = x - 2$

25. $x = 10 \cos t, y = 6 \sin t$

27. $x = 4 \tan t, y = \dfrac{3}{\cos t}$

29. $x = 2 + \cos t, y = -4 + 4 \sin t$

31. $x = -1 + \dfrac{1}{\cos t}, y = 3 + 5 \tan t$

33. $\dfrac{x^2}{25} + \dfrac{y^2}{9} = 1$

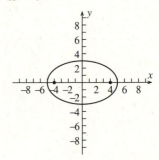

35. $\dfrac{x^2}{20} + \dfrac{(y - 2)^2}{36} = 1$

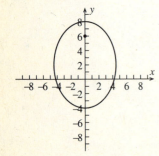

37. $\dfrac{y^2}{8} - \dfrac{x^2}{4} = 1$

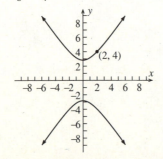

$(2, 4)$

39. $\dfrac{(x + 2)^2}{4} - \dfrac{(y - 1)^2}{5} = 1$

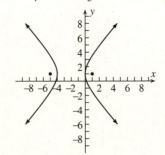

In Exercises 41-47, the $\bar{x}$ and $\bar{y}$ axes have been rotated through angle α.

41. (a) ellipse or degenerate;

(b) $\alpha = \frac{1}{4}\pi$, $\dfrac{\bar{x}^2}{4} + \bar{y}^2 = 1$;

(c)

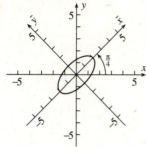

43. (a) parabola or degenerate;

(b) $\alpha = \frac{1}{4}\pi$, $\bar{x}^2 = 4\sqrt{2}\,\bar{y}$;

(c)

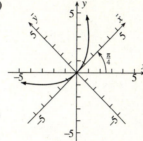

In Exercises 45-47, the x' and y' axes have also been translated such that $x' = \bar{x} - h$, $y' = \bar{y} - k$.

45. (a) hyperbola or two intersecting lines;

(b) $\alpha \approx 71.6°$, $h = \dfrac{-6}{\sqrt{10}}$, $k = \dfrac{6}{7\sqrt{10}}$,

$\frac{7}{24}x'^2 - \frac{49}{72}y'^2 = 1$;

(c)

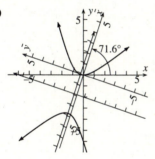

47. (a) hyperbola or two intersecting lines;

(b) $\alpha \approx 26.6°$, $h = -\dfrac{3}{\sqrt{5}}$, $k = -\dfrac{16}{\sqrt{5}}$,

$\dfrac{y'^2}{63} - \dfrac{4x'^2}{63} = 1$;

(c)

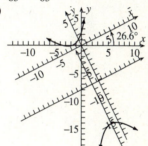

53.

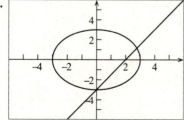

two; $\{(0, -3), (\frac{36}{13}, \frac{15}{13})\}$

55.

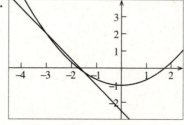

two; $\{(-3, 2), (-\frac{3}{2}, -\frac{1}{4})\}$

57.

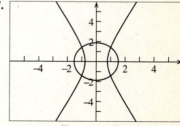

four; $\{(1, \sqrt{2}), (-1, \sqrt{2}),$
$(1, -\sqrt{2}), (-1, -\sqrt{2})\}$

59.

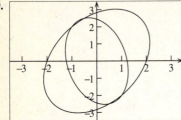

four; $\{(-\frac{1}{2}, \frac{5}{2}), (\frac{1}{2}, -\frac{5}{2}), (-1, 2), (1, -2)\}$

61. (a) $\frac{1}{2}$;   **(b)** ellipse;
(c) $r \sin \theta = -2$;
(d)

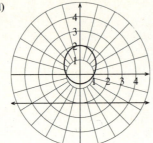

63. (a) $\frac{3}{2}$;   **(b)** hyperbola;
(c) $r \cos \theta = \frac{4}{3}$;
(d)

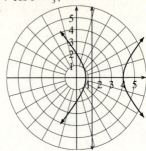

65. $r = \dfrac{8}{1 - 3 \cos \theta}$

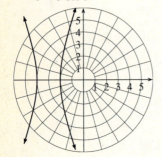

67. $r = \dfrac{6}{1 - \sin \theta}$

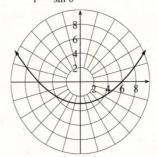

69. (a) $e = \frac{1}{3}\sqrt{34}$;
foci: $(\pm\sqrt{34}, 0)$;
directrices: $x = \pm\frac{9}{34}\sqrt{34}$;
(b)

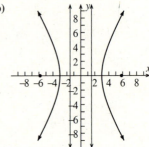

73. 18.9 m   **75.** 12.0 in. by 10.0 in.   **77.** $r = \dfrac{1}{1 - \cos \theta}$

81. first quadrant parabola

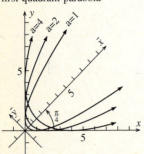

EXERCISES 12.1 (PAGE 709)

1. $\{(3, 1, 0)\}$ **3.** $\{(1, 2, 1)\}$ **5.** $\emptyset$; inconsistent   **7.** $\{(3, 1, -2)\}$ **9.** $\{(\frac{1}{4}, \frac{1}{2}, 1)\}$ **11.** $\{(3, 1, 0)\}$ **13.** $\{(3, -1, 2)\}$

15. $\{(2, -3, 8)\}$ **17.** $\{(-10 - 7t, 8 + 6t, t)\}$; dependent   **19.** $\{(-2, 2, -3, 1)\}$ **21.** $\emptyset$; inconsistent

23. solution set is $\{(\frac{20}{3} + \frac{7}{3}t, \frac{35}{3} + \frac{22}{3}t, 9 + 4t, t)\}$ **25.** $\{(t - 3, 2t - 4, t)\}$ **27.** $\emptyset$ **29.** inconsistent

31. consistent; $\{(-3, -2)\}$ **33.** consistent; $\{(3, 2, -4)\}$ **35.** inconsistent   **37.** $x^2 + y^2 + 4x - 6y - 12 = 0$

39. $y = 4x^2 - 8x + 3$ **41.** \$10,000 at 10%, \$5,000 at 12%, \$10,000 at 6%

43. 20 oz. of first alloy, 40 oz. of second alloy, 10 oz. of third alloy

45. 2 large, 7 medium, and 1 small; 3 large, 5 medium, and 2 small; 4 large, 3 medium, and 3 small; or 5 large, 1 medium, and 4 small

47. (b) $0 \le x \le \frac{5}{2}; \frac{5}{3} \le y \le \frac{5}{2}; 0 \le z \le \frac{10}{3}$

EXERCISES 12.2 (PAGE 725)

1. (a) 17;   **(b)** 16   **3.** -1 **5. (a)** 2×4;   **(b)** 3×1;   **(c)** -2;   **(d)** $\begin{bmatrix} 9 & -6 & 21 & 0 \\ 12 & 15 & -3 & 24 \end{bmatrix}$;   **(e)** $\begin{bmatrix} 12 \\ -4 \\ -10 \end{bmatrix}$

7. (a) $\begin{bmatrix} -4 & 8 \\ 12 & 20 \\ -24 & 4 \end{bmatrix}$;   **(b)** $\begin{bmatrix} 35 & 0 \\ 20 & -30 \end{bmatrix}$ **9. (a)** $\begin{bmatrix} 4 & 7 \\ -2 & 16 \end{bmatrix}$;   **(b)** $\begin{bmatrix} 2 & 2 \\ 7 & 3 \end{bmatrix}$ **11. (a)** $[7]$;   **(b)** $\begin{bmatrix} 2 & 4 & -6 \\ 1 & 2 & -3 \\ -1 & -2 & 3 \end{bmatrix}$

13. $\begin{bmatrix} -8 & 15 \\ 0 & -6 \\ -10 & 12 \end{bmatrix}$ **15.** $\begin{bmatrix} 5 & -26 & 2 \\ 8 & -1 & 13 \end{bmatrix}$ **23.** $H^{-1} = \begin{bmatrix} \frac{1}{3} & -\frac{2}{3} \\ -\frac{2}{3} & \frac{7}{3} \end{bmatrix}$ **25.** invertible   **27.** not invertible   **29.** $x = 1, y = 4$

31. $x = -3, y = 1, z = 4$ **33.** $\{(-3, 4)\}$ **35.** $\{(\frac{58}{7}, -\frac{18}{7}, -8)\}$ **37.** $\begin{bmatrix} 3 & 4 \\ 2 & 3 \end{bmatrix}$

EXERCISES 12.3 (PAGE 734)

1. $\dfrac{3}{x - 2} - \dfrac{3}{x + 2}$ **3.** $-\dfrac{1}{x} + \dfrac{2}{x + 1}$ **5.** $\dfrac{4}{x - 3} - \dfrac{3}{x - 1}$ **7.** $\dfrac{-5}{3x + 1} + \dfrac{2}{x - 2}$ **9.** $\dfrac{2}{x} - \dfrac{3}{3x - 2} - \dfrac{1}{2x + 3}$

11. $2x + \dfrac{5}{x - 2} + \dfrac{3}{x + 2}$ **13.** $2x - 5 + \dfrac{3}{2x - 3} + \dfrac{2}{x + 2}$ **15.** $\dfrac{5}{x^2} - \dfrac{4}{x} + \dfrac{7}{x - 2}$ **17.** $\dfrac{2}{x + 2} - \dfrac{3}{(x - 2)^2} - \dfrac{1}{x - 2}$

19. $\dfrac{6}{(2x + 1)^2} + \dfrac{\frac{14}{3}}{2x + 1} + \dfrac{2}{(x - 1)^2} - \dfrac{\frac{7}{3}}{x - 1}$ **21.** $1 + \dfrac{16}{(x - 2)^3} + \dfrac{18}{(x - 2)^2} + \dfrac{6}{x - 2}$ **23.** $\dfrac{4}{x} - \dfrac{x + 5}{x^2 + x + 1}$

25. $\dfrac{1}{x - 2} + \dfrac{2x + 4}{x^2 + 2x + 4}$ **27.** $-\dfrac{2}{x - 1} + \dfrac{4x - 3}{x^2 + 1}$ **29.** $\dfrac{4}{x + 4} + \dfrac{3x - 1}{2x^2 + 3}$ **31.** $\dfrac{x + 2}{x^2 + 1} - \dfrac{x - 2}{x^2 - x - 1}$

33. $\dfrac{2}{x^2} - \dfrac{1}{x} + \dfrac{x}{x^2 + 2x + 3}$ **35.** $\dfrac{x - 1}{x^2 + 1} - \dfrac{x - 1}{(x^2 + 1)^2}$ **37.** $\dfrac{x - 15}{(x^2 + 2)^2} + \dfrac{3x + 4}{x^2 + 2} - \dfrac{2}{x - 1}$

EXERCISES 12.4 (PAGE 742)

1. 5, 7, 9, 11, 13, 15, 17, 19   **3.** 2, $\frac{5}{2}, \frac{10}{3}, \frac{17}{4}, \frac{26}{5}, \frac{37}{6}, \frac{50}{7}, \frac{65}{8}$ **5.** 2, $-1, \frac{4}{5}, -\frac{5}{7}, \frac{6}{9}, -\frac{7}{11}, \frac{8}{13}, -\frac{9}{15}$

7. $-\frac{2}{3}, \frac{4}{5}, -\frac{8}{9}, \frac{16}{17}, -\frac{32}{33}, \frac{64}{65}, -\frac{128}{129}, \frac{256}{257}$ **9.** $\frac{1}{3}x, -\frac{1}{4}x^2, \frac{1}{5}x^3, -\frac{1}{6}x^4, \frac{1}{7}x^5, -\frac{1}{8}x^6, \frac{1}{9}x^7, -\frac{1}{10}x^8$ **11.** 0, 4, 0, 8, 0, 12, 0, 16

13. 1, 2, $\frac{1}{2}$, 2, $\frac{1}{3}$, 2, $\frac{1}{4}$, 2   **15.** 1, 1, 2, 2, 3, 3, 4, 4   **17.** $1 + 5 + 9 + 13 + 17 = 45$ **19.** $2 + \frac{3}{2} + \frac{4}{3} + \frac{5}{4} + \frac{6}{5} = \frac{437}{60}$

21. $5 + 5 + 5 + \ldots + 5$ (100 terms); 500   **23.** $1 + \frac{1}{2} + \frac{1}{4} + \frac{1}{8} + \frac{1}{16} + \frac{1}{32} = \frac{63}{32}$

25. $x - x^3 + x^5 - x^7 + x^9 - x^{11} + x^{13} - x^{15}$ **27.** $a_0 - 2a_1 + 3a_2 - 4a_3 + 5a_4 - 6a_5 + 7a_6 - 8a_7$

29. $f(x_0) + f(x_1) + f(x_2) + \ldots + f(x_{n-1})$ **31.** $\sum_{i=0}^{5}(2i + 1)$ **33.** $\sum_{i=1}^{5}(-1)^{i-1}(3i + 1)$ **35.** $\sum_{i=1}^{6}\dfrac{2i - 1}{i^2}$

37. $\sum_{i=0}^{4}\dfrac{(-x^2)^i}{2i + 2}$ **39.** $a_n = 2n - 1$: 1, 3, 5, 7, 9, 11; $a_n = 2n - 1 + (n - 1)(n - 2)(n - 3)$: 1, 3, 5, 13, 33, 71;
$a_n = 2n - 1 - (n - 1)(n - 2)(n - 3)$: 1, 3, 5, 1, -15, -49

EXERCISES 12.6 (PAGE 767)

1. (a) 5, 8, 11, 14, 17; **(b)** 10, 6, 2, -2, -6 **3. (a)** -5, -12, -19, -26, -33; **(b)** x, $x + 2y$, $x + 4y$, $x + 6y$, $x + 8y$

5. (a) 5, 15, 45, 135, 405; **(b)** 8, -4, 2, -1, $\frac{1}{2}$ **7. (a)** $-\frac{9}{16}$, $\frac{3}{8}$, $-\frac{1}{4}$, $\frac{1}{6}$, $-\frac{1}{9}$; **(b)** $\frac{x}{y}$, -1, $\frac{y}{x}$, $-\frac{y^2}{x^2}$, $\frac{y^3}{x^3}$

9. (a) an arithmetic sequence; -9, -13; **(b)** a geometric sequence; 27, 81; **(c)** a geometric sequence; $3\sqrt{6}$, $9\sqrt{2}$
    (d) neither an arithmetic nor a geometric sequence

11. (a) an arithmetic sequence; 0, -1.11; **(b)** an arithmetic sequence, $\frac{1}{12}$, 0; **(c)** a geometric sequence; $\frac{2}{9}$, $-\frac{2}{27}$;
    (d) an arithmetic sequence, $4x + 3y$, $5x + 4y$

13. 35 **15.** 16 **17.** thirtieth **19.** $\frac{729}{4}$ **21.** -3 **23.** eleventh **25. (a)** $\frac{26}{5}$, $\frac{27}{5}$, $\frac{28}{5}$, $\frac{29}{5}$; **(b)** 2, 4, 8, 16, 32

27. (a) -2; **(b)** 21 **29. (a)** 100; **(b)** 510 **31. (a)** -66; **(b)** $\frac{5465}{729}$ **33. (a)** 13.412; **(b)** 1030 **35.** 71.5 **37.** 2430

39. $280, $260, $240, $220 **41.** 120 **43.** $\frac{117,649}{262,144}$ **45.** 18 **47.** $33,065.95 **49.** 23, 27, 31

EXERCISES 12.7 (PAGE 780)

1. (a) $\frac{1}{504}$; **(b)** 863,040 **3. (a)** $\frac{33}{2}$; **(b)** $\frac{1}{168}$ **5. (a)** 35; **(b)** 1 **7. (a)** 220; **(b)** 1225

9. $a^8 + 8a^7b + 28a^6b^2 + 56a^5b^3 + 70a^4b^4 + 56a^3b^5 + 28a^2b^6 + 8ab^7 + b^8$

11. $x^9 - 9x^8y + 36x^7y^2 - 84x^6y^3 + 126x^5y^4 - 126x^4y^5 + 84x^3y^6 - 36x^2y^7 + 9xy^8 - y^9$

13. $x^5 + 15x^4y + 90x^3y^2 + 270x^2y^3 + 405xy^4 + 243y^5$

15. $4096 - 6144ab + 3840a^2b^2 - 1280a^3b^3 + 240a^4b^4 - 24a^5b^5 + a^6b^6$

17. $243e^{5x} - 810e^{3x} + 1080e^x - 720e^{-x} + 240e^{-3x} - 32e^{-5x}$

19. $a^4 + 8a^{7/2}b^{1/2} + 28a^3b + 56a^{5/2}b^{3/2} + 70a^2b^2 + 56a^{3/2}b^{5/2} + 28ab^3 + 8a^{1/2}b^{7/2} + b^4$

21. $4096x^{24} + 24{,}576x^{22}y^2 + 67{,}584x^{20}y^4 + 112{,}640x^{18}y^6$ **23.** $u^{-11} - 33u^{-10}v^2 + 495u^{-9}v^4 - 4455u^{-8}v^6$

25. $a^3 - 9a^{8/3}b^{1/3} + 36a^{7/3}b^{2/3} - 84a^2b$ **27.** $3003a^8b^6$ **29.** $-489{,}888x^4$ **31.** $\dfrac{924x^{18}}{y^{12}}$ **33.** $\dfrac{231}{16}a^6$ **35.** -8064

37. $a^9 + 9a^8b + 36a^7b^2 + 84a^6b^3 + 126a^5b^4 + 126a^4b^5 + 84a^3b^6 + 36a^2b^7 + 9ab^8 + b^9$

39. $r^{12} - 12r^{11}t + 66r^{10}t^2 - 220r^9t^3 + 495r^8t^4 - 792r^7t^5 + 924r^6t^6 - 792r^5t^7 + 495r^4t^8 - 220r^3t^9 + 66r^2t^{10} - 12rt^{11} + t^{12}$

41. 28 **43. (a)** 120; **(b)** 120 **47.** $32x^5 - 240x^4 + 720x^3 - 1080x^2 + 810x - 243$

EXERCISES 12.8 (PAGE 789)

1. $\frac{64}{3}$ **3.** $\frac{600}{11}$ **5.** $\frac{4}{5}$ **7.** $\frac{16}{21}$ **9.** $\dfrac{9 + 3\sqrt{3}}{2}$ **11.** $\frac{2}{3}$ **13.** $\frac{9}{11}$ **15.** 3 **17.** $\frac{137}{111}$ **19.** $\frac{47}{101}$ **21.** $\dfrac{29{,}029}{9{,}000}$ **23.** $\frac{244}{99}$

25. $1 - x^2 + x^4 - x^6 + \ldots$ **27.** $1 - x - \frac{1}{2}x^2 - \frac{1}{2}x^3 - \ldots$ **29.** $2 + \frac{1}{12}x - \frac{1}{288}x^2 + \frac{5}{20,736}x^3 - \ldots$ **31.** 1.020

33. 3.979 **35.** 4.990 **37.** 2.7183 **39.** 0.9211 **41.** 0.3746 **43.** 84 m **45.** 400 cm **47.** 500

REVIEW EXERCISES FOR CHAPTER 12 (PAGE 791)

1. $\{(3, 1, -2)\}$ **3.** $\{(3t + 1, 2t - 1, t)\}$; dependent **5.** $\{(-4, -2, 3)\}$ **7.** $\{(1, -1, 2, \frac{1}{3})\}$ **9.** $\{(-4t - 5, -2t - 3, t)\}$

11. $\{(5t + 7, 39t + 43, -26t - 29, t)\}$ **13.** consistent; $\{(4, \frac{1}{2}, -\frac{2}{3}\}$ **15. (a)** 1; **(b)** -35

17. (a) $\begin{bmatrix} -8 & 0 \\ 2 & -6 \\ 16 & -4 \end{bmatrix}$ **(b)** $\begin{bmatrix} -8 & 4 & 0 \\ -12 & -4 & 8 \end{bmatrix}$ **19. (a)** $\begin{bmatrix} 0 & 2 \\ -1 & 1 \end{bmatrix}$ **(b)** $\begin{bmatrix} 2 & -1 & 3 \\ -2 & -4 & -3 \end{bmatrix}$ **21.** $H^{-1} = \begin{bmatrix} \frac{4}{11} & \frac{1}{11} \\ -\frac{3}{11} & \frac{2}{11} \end{bmatrix}$

23. invertible; $H^{-1} = \begin{bmatrix} -\frac{5}{151} & \frac{30}{151} & -\frac{12}{151} \\ \frac{26}{151} & -\frac{5}{151} & \frac{2}{151} \\ -\frac{3}{151} & \frac{18}{151} & \frac{23}{151} \end{bmatrix}$ **25.** $\{(-1, 2, 1)\}$ **27.** $\{(\frac{3}{4}t - \frac{1}{2}, 1 - \frac{3}{4}t, t)\}$, dependent

29. $-\dfrac{3}{x - 1} + \dfrac{6}{x + 4} - \dfrac{1}{x + 2}$ **31.** $\dfrac{6}{7(x - 2)} + \dfrac{x + 10}{7(x^2 + x + 1)}$ **33.** $1 + \dfrac{x - 2}{x^2 + 1} - \dfrac{3}{(x^2 + 1)^2}$

35. $\frac{1}{3}$, $-\frac{1}{3}$, $\frac{5}{27}$, $-\frac{7}{81}$, $\frac{1}{27}$, $-\frac{11}{729}$ **37.** $3 + 2 + \frac{5}{3} + \frac{3}{2} + \frac{7}{5} + \frac{4}{3} = \frac{109}{10}$ **39.** $\sum_{i=1}^{4} (-1)^{i-1} \dfrac{x^{2i}}{2^i}$ **41.** $\frac{3069}{1024}$ **43.** 130

45. (a) 55; **(b)** $\frac{8}{7}$ **47. (a)** 84; **(b)** 91,390 **49. (a)** arithmetic sequence, 24, 28; **(b)** geometric sequence, -1, $\frac{1}{3}$

51. (a) geometric sequence, $\dfrac{1}{\sqrt[4]{2}}, \dfrac{1}{\sqrt{2}}$;   **(b)** neither an arithmetic nor a geometric sequence   **53.** $\frac{7}{16}$ **55.** $\frac{1}{9}$ or $-\frac{1}{9}$

57. twenty-first   **59.** $\frac{13}{24}, \frac{7}{12}, \frac{5}{8}$ **61.** 96, 48, 24, 12, 6   **63. (a)** 20;   **(b)** -16 **65.** $\frac{87}{5}$ **67.(a)** 50;   **(b)** 15,000

69. $\frac{8}{19}$ **71.** $\frac{8}{11}$ **77.** $128x^7 - 448x^6y + 672x^5y^2 - 560x^4y^3 + 280x^3y^4 - 84x^2y^5 + 14xy^6 - y^7$

79. $x^{20} + 40x^{18} + 760x^{16} + 9120x^{14}$ **81.** $-5005t^{-3/2}$ **83.** $-\frac{231}{16}z^{12}$ **85.** 2.924   **87.** 0.5646   **89.** 1.6487

91. $y = 2x^2 - 3x + 1$ **93.** \$4000 in 6% bonds, \$8000 in 5% savings, \$3000 in business   **95.** 455

97. (a) 1,049,000;   **(b)** 2,097,000   **99.** 120   **101.** 107 km   **103.** 20, 10, 5 or 5, 10, 20

EXERCISES A.1 (PAGE A-6)

1. commutative law for multiplication (Axiom 2)   **3.** existence of additive identity element (Axiom 5)

5. commutative law for addition (Axiom 2)   **7.** associative law for multiplication (Axiom 3)

9. existence of additive inverse (Axiom 6)   **11.** commutative law for addition (Axiom 2)

13. existence of multiplicative inverse (Axiom 7)   **15.** commutative law for multiplication (Axiom 2)

17. existence of additive identity element (Axiom 5)   **19.** existence of multiplicative identity element (Axiom 5)

21. distributive law (Axiom 4)   **23.** associative law for addition (Axiom 3)   **25.** closed; $2 + 3 = 5$ **27.** closed; $3 \cdot 5 = 15$

29. closed; $0 + 0 = 0$ **31.** closed; $1 \cdot 1 = 1$ **33.** not closed; $1 + 2 = 3$ **35.** closed; $0 \cdot 1 = 0, 0 \cdot 0 = 0, 1 \cdot 1 = 1$

Index